Handbook on Networked Multipoint Multimedia Conferencing and Multistream Immersive Telepresence using SIP

Handbook on Networked Multipoint Multimedia Conferencing and Multistream Immersive Telepresence using SIP

Scalable Distributed Applications and Media Control over Internet

Radhika Ranjan Roy

CRC Press
Taylor & Francis Group
Boca Raton London New York

CRC Press is an imprint of the
Taylor & Francis Group, an **informa** business

First edition published 2021

by CRC Press
6000 Broken Sound Parkway NW, Suite 300, Boca Raton, FL 33487-2742
and by CRC Press
2 Park Square, Milton Park, Abingdon, Oxon, OX14 4RN

CRC Press is an imprint of Taylor & Francis Group, LLC

ISBN: 9780367565800 (hbk)
ISBN: 9781003098454 (ebk)

Typeset in Times
by Deanta Global Publishing Services, Chennai, India

The publication of this book is coming out at an opportune moment, when our beloved son Ajanta has just married Nageen, where our gracious daughter Elora and son-in-law Nick played a memorable leading role, binding all of the friends and relatives together and making the marriage ceremony the most joyous occasion ever in our life. We have been blessed to have a daughter-in-law like Nageen, whose grace and love have won the hearts of all of us.

To our dearest son, Debasri Roy, Medicinae Doctoris (MD) (January 20, 1988 – October 31, 2014), who had a brilliant career (Summa Cum Laude – undergraduate), had so much more to contribute to this country and the world as a whole in the future, had been so eager to see this book published, saying, "Daddy, you are my hero," pictured with his fiancée, "Love is Forever," wrote his life-long wishes, "I am ready for my life, I like to see the whole world, I will read spiritual scriptures to find the mystery of life, and I want to make a difference in the world," and loved all of us to the deepest of his kindest heart, including his life-long fiancée, who was also an MD and his classmate and to whom he was engaged and wanted to be married on October 30, 2015, relatives, friends, colleagues, and neighbors, until his last breath, and spiritualized us in so many ways to love God, along with his prophetic words, "Mom, I am extremely happy that you are with me and I want to be happy in life." May God let his soul live in peace in His abode.

To my GrandMa for her unconditional love, my parents Rakesh Chandra Roy and Sneholota Roy, whose spiritual inspiration remains vividly alive within all of us, my late sisters GitaSree Roy, Anjali Roy, and Aparna Roy and their spouses, and my brother Raghunath Roy and his wife Nupur for their inspiration, my daughter Elora and my son-in-law Nick, my son Ajanta, and finally, my beloved wife Jharna for their love.

And our heartfelt thanks to my son Ajanta for inspiring me with his wonderful creativity in his first major thoughtful invention of a multimedia telephony model, using cupboard paper along with a vivid description, written on a piece of paper, of how the phone would work when he was only in fifth grade, while I was the technical lead of the "AT&T Vision 2001 Multimedia Architecture" project at AT&T Bell Laboratories in the year 1993. Immediately, I took the model to AT&T Bell Labs, and all my colleagues were surprised to see his wonderful imagination about multimedia telephony. I wish I could patent his wonderful idea. Since then, I have been so inspired that I have been working in VoIP/Multimedia telephony ever since, and I still am today. Today, he himself is an energetic electrical engineer, progressing towards a bright future in his own right and keeping his colleagues and all of us amazed. This book is the culmination of the vivid inspiration that he has embedded inside me.

Contents

Preface

The worldwide catastrophic impact of the coronavirus pandemic has made the prospect of the touchless global economy a reality. Real-time multipoint multimedia conferencing along with high-quality immersive telepresence is poised to play a vital role in advancing the creation of the touchless global economy, offering economies-of-scale for doing business, replacing costly physical face-to-face meetings and eliminating travel time. This is the first book that provides a full treatment of building products and services over the Internet in multivendor environments for real-time multipoint conferencing using interoperable technical standards that have been created over two decades in standards fora like the Internet Engineering Task Force (IETF).

In the 1980s and 1990s, the advent of the networked real-time multipoint multimedia conferencing application, which deals with audio, video, and/or application sharing, where signaling and media are sent between the source and the destination using two different paths and where media (audio, video, and/or application sharing) bridging is needed, has posed serious technical challenges for the legacy circuit switching-based public switched telephone network (PSTN)/integrated services digital network (ISDN). Although constant bit rate (CBR) audio and video are quite efficient for circuit-switching networks, application-sharing traffic that may consist of text, graphics, and/or animations will be highly bursty, and its bandwidth, which may vary from a few kilobits to multi-gigabits, is scalable and efficient only for the packet-switching network [1]. Moreover, local area networks (LANs) that use variable packet sizes for data or frames and have multi-megabits/multi-gigabits bandwidth available locally can also support variable bit rate (VBR) video and audio and/or highly bursty application sharing–based multimedia conferencing services conveniently. Unlike many other applications, the unique requirement for providing media bridging dynamically in real time for multipoint multimedia conferencing has made this application hugely complex.

Moreover, a key technical finding has been that a network that is suitable for multipoint multimedia conferencing can also support all other applications, because the technical requirements of all other applications are only a subset of the multimedia conferencing application. Asynchronous transfer mode (ATM) [2], which uses asynchronous time-division multiplexing to encode data into small, fixed-sized cells, has been proposed to support both circuit-switched (that is, time-division multiplexing [TDM] trunks/TDM hierarchies emulations) and packet/cell-switched applications. ATM has been standardized by both the ATM Forum and the International Telecommunications Union-Telecommunication (ITU-T).

In the 1990s, the public Internet, which uses variable packet sizes for packet-switching, has grown worldwide for data applications such as emails, file transfer, and electronic data interchange (EDI) and has started experimenting with offering real-time multipoint multimedia conferencing services directly, connecting the premises using LANs. Although performances for multipoint conferencing over the Internet were inferior, especially for real-time CBR-based audio and video, to those of the ATM network, the ease of usage and cost advantages of multipoint multimedia conference services over the Internet have been very popular. Consequently, the Internet has established itself as the dominant and most popular worldwide communications network, including multipoint communications.

The IETF has eventually standardized both point-to-point, for example, session initiation protocol (SIP), and multipoint multimedia conferencing. I have been working on networked multimedia communications all along, including at my present position at United States Command, Control, Communications, Computers, Combat Systems, Intelligence, Surveillance, and Reconnaissance (C5ISR)-Space and Terrestrial Communications Directorate (S&TCD) Laboratories for large-scale global SIP-based Voice-over-Internet Protocol (VoIP)/Multimedia networks since 1993, before which I was at AT&T Bell Laboratories. I was the editor of the Multimedia Communications Forum

(MMCF) when it was created by many participating companies throughout the world, including AT&T, for promoting the technical standards for networked multimedia communications to fill an important gap of that time when no standard bodies came forward to do so. Later on, I had the opportunity to participate in ITU-T on behalf of AT&T for standardization of H.323. H.323 was the first successful technical standard for VoIP/multimedia telephony. However, SIP, which was standardized in the IETF much later than H.323 and has emulated the simplicity of the protocol architecture of hypertext transfer protocol (HTTP), has started to be popularized for VoIP over the Internet because of its friendliness to being meshed with Web services. In this connection, I have published two books: *Handbook on Session Initiation Protocol: Networked Multimedia Communications for IP Telephony*, CRC Press, 2016 and *Handbook on SDP for Multimedia Session Negotiations: SIP and WebRTC IP Telephony, CRC Press*, 2018.

Session description protocol (SDP) is being used for negotiations of media capabilities to be supported among the endpoints for multimedia conferencing. However, the protocol architecture of SDP is such that it is not possible to negotiate media capabilities dynamically evolving from a point-to-point call established using SIP to a multipoint multimedia conferencing call where, unlike a point-to-point call, media (audio, video, and/or application sharing) bridging is needed. Furthermore, multipoint multimedia conferencing needs many more protocols, such as multipoint multimedia conference control protocol, media bridging protocols, and a dynamic session floor control protocol needed among multipoint conference participants in real time by selecting a chairperson who will establish centralized control for a given duration of time, notifications of messages among the participants, side conferencing among a subset of participants when needed, and other requirements. Different market studies show that centralized multipoint multimedia conferencing is the most popular, used by customers worldwide, as opposed to distributed multipoint conferencing.

The IETF created the XCON working group (WG) for standardization only of the centralized multipoint multimedia conferencing where the location of the centralized controller, termed a focus, is known a priori by the conference participants setting up the multipoint conferencing, which creates a logical star-like traffic flow around the centralized controller. By the time the XCON WG had created standards for multipoint conferencing, it was too late, because many renowned commercial vendors have developed their proprietary multipoint multimedia conference control protocols offering solutions of their own to meet the market demand, although many of them use SIP/SDP for the call setup and media negotiations in a point-to-point fashion by each participant with the centralized controller. The use of the interoperable standard-based multipoint multimedia conferencing signaling protocols will open the door, offering complex conferencing services even by small vendors with ease without developing any costly proprietary conferencing protocols, just by choosing and picking standardized products from the open commercial market.

Although XCON has developed the centralized multipoint multimedia conferencing standards, the scalable distributive audio-video bridging standards are the key hallmark. This has ushered in a new era for offering interoperable multipoint multimedia conferencing services. Vendors will now be able to build interoperable media bridging server products using the standards developed by the IETF, and service providers will now be able to offer scalable multipoint multimedia conference services over the global worldwide network with economics-of-scale.

However, no book has yet been published that provides a full treatment of the IETF XCON standards for centralized multipoint multimedia conferencing. As a result, many in industry, universities, product development, research, and other institutions do not know that a complete IETF XCON standard-based centralized multipoint multimedia conferencing can be offered. Recently, a more exciting development has taken place in the IETF for standardization of the High-Quality Multistream Immersive Telepresence Systems by the ControLling mUltiple streams for tElepresence (CLUE) WG. The CLUE WG has standardized the use cases for Immersive Telepresence for multiple streams. The IETF drafts for the schema for the CLUE Data Model, CLUE Protocol, CLUE Protocol Data Channel, Mapping between RTP streams to CLUE Media Captures, and Call between two CLUE-capable Endpoints, as well as between a CLUE-capable and a non-CLUE Endpoint, which were proposed some time ago, are on the verge of creation of Request For Comments (RFCs).

This is the first book ever published on centralized multipoint multimedia conferencing and the high-quality multistream immersive telepresence systems based on the standards developed by the IETF XCON and CLUE WG. After so many years of working experience on SIP for building large-scale VoIP networks, I have found that it is an urgent requirement to have a complete book that integrates all RFCs/draft standards developed by XCON and CLUE WG in a systematic way that network designers, software developers, product manufacturers, implementors, interoperability testers, professionals, professors, and researchers can use for multipoint conferencing, providing interoperability and offering scalability with economics-of-scale operating in multivendor and multi-product environments for large-scale global networks.

Each RFC or draft in the IETF is reviewed by all members of a given WG throughout the world, and a rough consensus is arrived at on which parts of texts of the RFC/draft need to be "mandatory" and "optional," including whether an RFC or draft needs to be "standards track," "informational," or "experimental." The key point is that when one tries to put together all SIP/SDP, Multipoint Multimedia Conference Control Protocol, CLUE Protocol, and others related to RFCs or drafts, making a textbook has serious challenges in how to put all texts together, because it is not simply a case of putting one RFC/draft after another chronologically. Texts of each RFC/draft need to be put together for each of these particular functionalities, capabilities, and features, maintaining integrity. Since this book is planned as if were a single super-RFC/draft for the IETF XCON/CLUE WG standard-based centralized multipoint multimedia conferencing/high-quality multi-streams immersive telepresence systems, primarily using SIP (complemented by other protocols as well), I have very limited freedom to change any texts of RFCs/drafts other than some editorial text to make it look like a book. I have used texts, figures, tables, and references from RFCs/drafts as much as necessary, so that everyone can use each of those as they are found in RFCs/drafts. All RFCs/drafts, along with their authors, are provided in references, and all credits of this book go primarily to those authors of these RFCs and the many IETF WG members who shaped final RFCs/drafts with their invaluable comments and input.

In this connection, I also extend my sincere thanks to Dr. Stephan Wenger, IETF Trustee, for his kind consent to reproduce texts, figures, and tables with IETF copyright notification. My only credit, as I mentioned earlier, is to put all those RFCs together in a way that looks like one complete RFC for building products and offering multipoint multimedia conference services, including high-quality immersive telepresence, based on technical standards ushering interoperability in multivendor environments and thereby, economies-of-scale.

The book is organized into 16 sections (or chapters) based on their major functionalities, features, and capabilities, which exemplifies multipoint multimedia conferencing, including scalable media bridging and high-quality multistream immersive telepresence, as follows:

- Chapter 1: Scalable Networked Multipoint Multimedia Conferencing and Telepresence
- Chapter 2: Centralized Conferencing Architecture
- Chapter 3: Media Server Control Architecture
- Chapter 4: Conferencing Information Data Model for Centralized Conferencing
- Chapter 5: Centralized Conference Manipulation Protocol
- Chapter 6: Binary Floor Control Protocol
- Chapter 7: XCON Notification Service
- Chapter 8: Media Channel Control Framework
- Chapter 9: Mixer Package for Media Control Channel Framework
- Chapter 10: Media Session Recording
- Chapter 11: Media Resource Brokering
- Chapter 12: Media Control Channel Framework for Interactive Voice Response

- Chapter 13: SIP Interface to VoiceXML Media Services
- Chapter 14: Media Resource Control Protocol Version 2
- Chapter 15: Media Control Channel Framework (CFW) Call Flow Examples
- Chapter 16: Multistream Immersive Telepresence Conferencing Systems

Chapter 1 provides the legacy centralized non-scalable proprietary centralized controller-based multipoint conferencing before the actual development of the standard-based multipoint multimedia conference systems that are described in Chapters 2 through 16. I am looking forward to seeing whether readers will validate my judgment. In addition, I am providing a general statement for the IETF copyright information as follows:

"IETF RFCs have texts that are 'mandatory' and 'optional' including the use of words like 'shall,' 'must,' 'may,' 'should,' and 'recommended.' These texts are critical for providing interoperability for implementation using products from different vendors as well as for inter-carrier communications. The main objective of this book is, as explained earlier, to create, as it were a single book that contains IETF XCON/CLUE WG standards–based integrated multipoint multimedia conference and high-quality multistream immersive telepresence systems. The texts have been reproduced from the IETF RFCs/drafts for providing interoperability with permission from IETF in Chapters 1 through 16. The copyright © for the texts that are being reproduced (with permission) in different sections and sub-sections of this part belongs to the IETF. It is recommended that readers should consult original RFCs/drafts posted on the IETF website."

I am greatly indebted to many researchers, professionals, software and product developers, network designers, professors, intellectuals, and individual authors and contributors of technical standard documents, drafts, and RFCs/drafts throughout the world for learning from their high-quality technical papers and discussions in group meetings, conferences, and emails in WGs for more than two decades. In addition, I had the privilege to meet many of those great souls in person during MMCF, ITU-T, IETF, and other technical standards conferences held in different countries of the world. Their unforgettable personal touch has enriched my heart very deeply as well.

I admire Ms Gabriella Williams, Editor, Information Security, Networking, Communication and Emerging Technologies, Taylor & Francis/CRC Press, for her appreciative approach to publishing this book. Finally, I provide my heartfelt thanks to the editorial team members for helping me in numerous ways with the publication of my book with Taylor & Francis/CRC Press, including presentation material for outlining the publication proposal.

REFERENCES

1. Roy, R. R. et al. (1994). An Analysis of Universal Multimedia Switching Architectures, vol. 73. No. 6, AT&T Technical Journal.
2. Roy, R. R. (1994). Networking Constraints in Multimedia Conferencing and the Role of ATM Networks, vol. 73, no. 4, AT&T Technical Journal.

Author

Radhika Ranjan Roy has been an electronics engineer, United States Army Research, United States Command, Control, Communications, Computers, Combat Systems, Intelligence, Surveillance, and Reconnaissance (C5ISR)-S&TCD Laboratories (previously known as CERDEC), Aberdeen Proving Ground (APG), Maryland, United States since 2009. Dr. Roy is leading research and development efforts in the development of scalable large-scale SIP-based VoIP/Multimedia networks and services, cybersecurity, artificial intelligence and machine/deep learning (AI/ML/DL) architecture, mobile ad hoc networks (MANETs), Peer-to-Peer (P2P) networks, cybersecurity detecting application software and network vulnerability, jamming detection, supporting an array of the Army/ Department of Defense's Nationwide and Worldwide Warfighter Networking Architectures and participating in technical standards development in Multimedia/Real-Time Services Collaboration, IPv6, Radio Communications, Enterprise Services Management, and Information Transfer of Department of Defense (DoD) Technical Working Groups (TWGs). He received his PhD in Electrical Engineering with Major in Computer Communications from the City University of New York (CUNY), NY, United States in 1984 and his MS in Electrical Engineering from the Northeastern University, Boston, MA, United States in 1978. He received his BS in Electrical Engineering from the Bangladesh University of Engineering & Technology (BUET), Dhaka, Bangladesh in 1967. He was born in the renowned country town of Derai, Bangladesh.

Prior to joining CERDEC, Dr. Roy worked as the lead system engineer at CACI, Eatontown, NJ from 2007 to 2009 and developed Army Technical Resource Model (TRM), Army Enterprise Architecture (AEA), DoD Architecture Framework (DoDAF), and Army LandWarNet (LWN) Capability Sets, and technical standards for Joint Tactical Radio System (JTRS), Mobile IPv6, MANET, and SIP, supporting Army Chief Information Officer (CIO)/G-6. Dr. Roy worked as senior system engineer, SAIC, Abingdon, MD from 2004 to 2007, supporting Modeling, Simulations, Architectures, and System Engineering of many Army projects: WIN-T, FCS, and JNN.

During his career, Dr. Roy worked in AT&T/Bell Laboratories, Middletown, NJ as senior consultant from 1990 to 2004 and led a team of engineers in designing AT&T's Worldwide SIP-based VoIP/Multimedia Communications Network Architecture, consisting of wired and wireless, from the preparation of Request for Information (RFI) to the evaluation of vendor RFI responses and interactions with all selected major vendors related to their products. He participated in and contributed to the development of VoIP/H.323/SIP multimedia standards in ITU-T, IETF, ATM, and Frame Relay standard organizations.

Dr. Roy worked as senior principal engineer in CSC, Falls Church, VA from 1984 to 1990 and worked in the design and performance analysis of the US Treasury nationwide X.25 packet-switching network. In addition, he designed many network architectures of many proposed U.S. Government and Commercial Worldwide and Nationwide Networks: Department of State Telecommunications Network (DOSTN), U.S. Secret Service Satellite Network, Veteran Communications Network, and Ford Company's Dealership Network. Prior to CSC, he worked from 1967 to 1977 as deputy director, Design, in PDP, Dhaka, Bangladesh.

Dr. Roy's research interests include the areas of artificial intelligence, machine/deep learning, blockchain cloud and fog computing, mobile ad hoc networks, multimedia communications, peer-to-peer networking, and quality-of-service. He has published over 60 technical papers and is holding or pending over 35 patents. He also participates in many IETF working groups. Dr. Roy authored three books *Handbook on SDP for Multimedia Session Negotiations: SIP and WebRTC Telephony* (CRC Press/Taylor & Francis, 2018), *Handbook on Session Initiation Protocol: Networked Multimedia Communications for IP Telephony* (CRC Press/Taylor & Francis, 2016), and *Handbook of Mobile Ad Hoc Networks for Mobility Models*, Springer, 2010. He lives in the historical district of Howell Township, New Jersy, with his wife Jharna.

1 Scalable Networked Multipoint Multimedia Conferencing and Telepresence

1.1 NETWORKED MULTIPOINT MULTIMEDIA CONFERENCING

Multipoint conferencing where three or more parties communicate adds another new complex capability that requires media bridging (or mixing) of the multiple parties, and then, the bridged media needs to be sent to the conference participants. It also implies a lot of other things for management of a conference consisting of multiple parties, such as conference creation and termination, adding and removal of parties, dial-in and dial-out conferencing, privacy of attendees, side conferencing, application sharing, conference recording, floor control, media manipulation, whispering, announcements, and many other functionalities that are not usually relevant to the two-party conferencing, although the two-party conferencing system will be able to use all these multipoint conferencing features if applicable. In this book, we are describing the multipoint conferencing using the session initiation protocol (SIP) [1] and session description protocol (SDP) [2]. In this book, we are assuming that the readers already have some basic knowledge about both SIP and SDP.

A basic multipoint conferencing system can have centralized or distributed architecture considering the flows of signaling and media. In a centralized system, all signaling and media flow in a star configuration, where both signaling and media are controlled by a centralized entity, known as the conference application server, and all participants communicate with the centralized entity in a point-to-point fashion. In a purely distributive conferencing system, there is no centralized entity in setting up the multipoint conferencing, and it is a little difficult to build such a conferencing system. There are a couple of technical limitations in the SIP/SDP protocol architecture.

Although SDP serves quite well for point-to-point multimedia conferencing, its protocol architecture is such that it cannot be extended for distributed multipoint multimedia conferencing as of today. The only multipoint conferencing is the centralized star-like topology with the a priori known address of the centralized controller. In this conference architecture, again, capability negotiations are done using SDP in point-to-point fashion only. When we work with SDP, we have to know its limitations regarding how and where we can use SDP for setting up the multimedia conferencing dynamically. As a result, the Internet Engineering Task Force (IETF) multipoint working group known as XCON has developed the SIP/SDP-based multipoint conference standards as the centralized ones, where the centralized conference control is a functional entity named "focus/foci," whose address is known a priori. It is assumed that foci control all conferencing resources, such as application servers, media servers, and multipoint bridging (mixing). Before we go into detail on multipoint conferencing, we provide a brief description of the multimedia conferencing protocol architecture.

A multimedia conversational communications session can be very feature-rich, with a lot of very complex functionalities. The multimedia session control is an application layer control protocol and is designed to provide services to multimedia applications, as shown in Figure 1.1. The establishment of a real-time multimedia communications session, especially with humans, needs a lot of intelligence.

Even an automaton can be a conference participant. Multiple users located in different geographical locations may participate in the same session. Each participant may play a different role

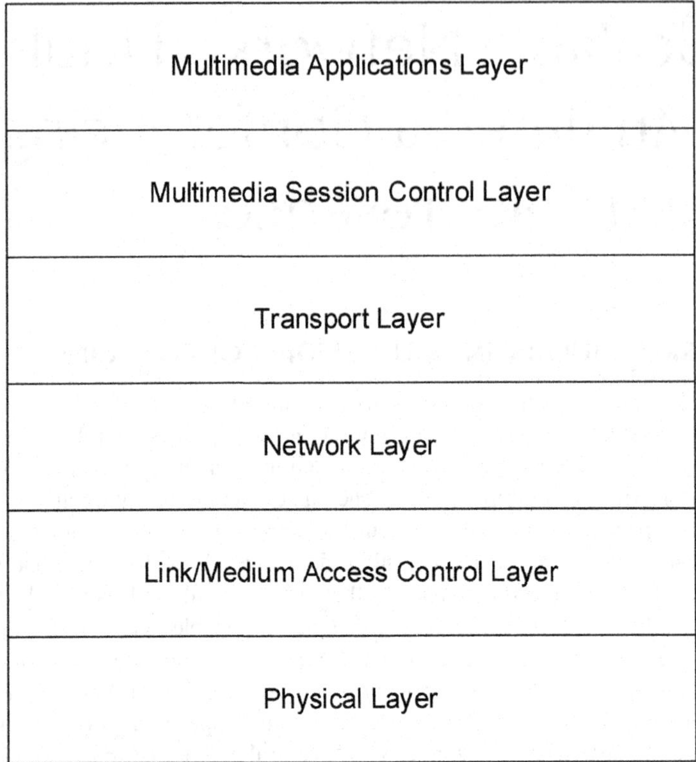

FIGURE 1.1 Relationship of Multimedia Session Control Layer to Other Layers.

in the conference based on the conference policy. The multimedia communications may consist of different kinds of media, and each media needs to be negotiated with each participant before establishment of the call. Different codecs may be used for audio or video by each participant, and negotiations may require agreeing on a common codec, or transcoding services may need to be offered for dissimilar codecs. Application sharing may require a variety of controls among the call participants. Even multimedia files may be shared or created through collaboration among conference participants. In addition, each media has its own quality-of-service (QOS) requirements, and QOS for each media needs to be guaranteed during the call setup, if that is what the conference participants expect per service level agreement (SLA). The multimedia session security, which includes authentication, integrity, confidentiality, non-repudiation, and authorization, is paramount for conference participants in relationship to both signaling and media (audio, video, and data/application sharing).

Early media (e.g., audio and video) is another feature that is used to indicate the progress of the multimedia session before the call is accepted by the called party. It may be unidirectional or bidirectional and can be generated by the calling party, the called party, or both. For example, early media generated by the caller can be a ringing tone and announcements (e.g., queuing status). Early media typically may consist of voice commands or dual-tone multi-frequency (DTMF) tones to drive interactive voice response (IVR) systems. Early media cannot be declined, modified, or identified. Consequently, it becomes very problematic to accommodate all the complex functionalities of the early media in the call setup.

If more than two participants join in the conference call, media bridging is required. Although audio bridging is straightforward, the video bridging can range from simple functionality, like video switching of the loudest speaker, to the composition of a very complex composite video of the conference participants, maintaining audio and video intermedia synchronization so that lip synchronization can be maintained. It may happen that the multipoint multimedia conference can be set up dynamically. For example, two participants are on a point-to-point conference call, and then a third or more conference participants need to be added to the same conference call. In this situation, the call needs to be diverted to a conference bridge for bridging of audio, video, and/or data dynamically without tearing down the two-party conference call.

Much complex functionality, like virtual meeting rooms, may be introduced to make the multimedia conferencing more resourceful and powerful. For example, there may be opportunities to join in a virtual meeting room for conferencing, while there may be many virtual meeting rooms available, where many other conferences may be going on simultaneously, each of which is separate and independent. A participant may even have the option to be a part of multiple conferences simultaneously while coming in and out of each conference at different times. Even a conference participant joining late may also dial-in a conference recording server to listen to and can see what has already happened while they were not in the conference. Even high-quality video with eye-contact features during the meeting may be needed in case human personal interactions need to be known among the conference participants as if they were in a face-to-face meeting.

The multimedia session establishment and teardown also needs to support both user and terminal mobility. In the case of user mobility, the session establishment mechanisms will have to deal with the recent address where a user has moved, and it is the user who will have to work proactively to update his or her recent address. The conferencing system and the session establishment mechanisms should have in-built schemes to deal with this. In the case of terminal mobility, the terminal itself may break and reestablish the network point of attachment as it moves from one place to another while the session is going on. However, a session object resides in the application layer, in Open Standard International (OSI) terminology, and may not be aware of the lower network layer's change in point of attachment. It is expected that the lower-layer protocols, such as the transport layer, network layer, and medium access control layer, and physical layer protocols will work transparently mitigating any changes in the network point of attachment. However, multimedia service portability, which demands the end user's ability to obtain subscribed services in a transparent manner regardless of the end user's point of attachment to the network, needs transparent support in the application layer session establishment.

A host of new service features related to the same call may also need to be satisfied even after the session establishment. Sub-conferencing, side bars, call transfer, call consultation transfer, call conference out-of-consultation, call diversion, call hold, call parking, call completion services for unsuccessful calls, pre-paid call services, invoking of new applications, and many others are examples of this category of services. The signaling protocol for the session establishment needs to have all the intelligence to satisfy the whole variety of requirements of the conference participants for how each media/application will be sent, received, and shared. In addition, the multimedia signaling protocol needs to have the ability to offer many other services to the conference participants within the same session even long after the establishment of the call. Because of the complexity of the real-time multimedia conversational services, the session establishment signaling messages have to be separated from the media (audio, video, and data/application sharing). The paths, through which signaling messages, audio, video, and data/application sharing traffic is transferred, can be completely independent.

The emergence of a new kind of feature-rich web-based communications for application visualization and sharing has enriched users' experiences over the Internet. It implies that the internet protocol (IP) has not only emerged as the chosen universal communications protocol over which audio, video, and data will be transferred using a single network. Users also want to keep the same experiences of communications over the Internet even for audio and video conferencing services for application sharing. These criteria have made web services applications an integral part of application sharing for audio and video conferencing. A secondary consequence of this has been the popularity among developers of using text encoding for multimedia conferencing services, because text encoding has been used for web services because of its simplicity to debug, modify, and integrate with many existing applications.

SIP has been standardized in the IETF as the call control protocol for establishment of the multimedia conversational session between the conference participants. SIP has embraced the simplicity of web-based communications protocol architecture as well as the text encoding. SIP is a very attractive protocol for multimedia session establishment for the time- and mission-critical point-to-point conference call because of its human-understandable text encoding of signaling messages and use of the hypertext transfer protocol (HTTP)/hypertext markup language (HTML)-based web-services like protocol architecture separating the signaling messages into two parts: Header and body. This inherent in-built capability of SIP has been used to create an enormous amount of new applications services not only for time- and mission-critical conversational audio and video services but also for integration of the non-time-critical web services defined within the framework of service oriented architecture (SOA), primarily as a part of the application sharing under the same audio and video conferencing session.

The building of multipoint multimedia conferencing services evolving from simple point-to-point call dynamically using SIP will be very difficult, if not impossible, unless the present SIP protocol architecture is drastically changed. One of the observations is that the SIP protocol architecture is very weak in its conference negotiation capabilities in multipoint multimedia communications environments, which require much more imbedded intelligence in the signaling protocol architecture. For example, initially set up as a point-to-point two-party conference call, the SIP protocol cannot be used to construct a multipoint conference call dynamically when third or more users join in the same conference call, which needs media bridging. As a result, knowing the address of the conference server a priori, a centralized multipoint conferencing with star-like connectivity architecture is set up, whereby the conference is established between each user of the multipoint conference participants and the conference server in a point-to-point fashion. In this respect, SIP is an application layer protocol with the capability to establish and tear down point-to-point and multipoint-to-multipoint sessions using unicast and/or multicast communications environments. SIP, being the application layer protocol, can support both user and terminal mobility; because the application layer mobility does not require any changes to any of the application layer operating systems of any of the participants, and thereby can be deployed widely, other lower-layer protocols, including mobile IP, can take care of mobility such as changes in network point of attachments transparently.

1.2 BASIC CONFERENCING PROTOCOL ARCHITECTURE

SIP, as described earlier, is an application layer multimedia signaling protocol that supports the establishment, management, and teardown of multimedia sessions between the conference participants accompanying SDP and other protocols but does not provide services. SIP only performs these specific functions and relies heavily on other protocols to describe the media sessions, transport the media, and provide the QOS. Figure 1.2 shows the relationship between SIP/SDP and other protocols.

An SIP message consists of two parts: Header and message body. The header is primarily used to route the signaling messages from the caller to the called party and contains a request line composed of the request type, the SIP Uniform Resource Identifier of the destination or next hop, and the version of SIP being used. The message body is optional, depending on the type of message and where it falls within the establishment process. A blank line is used to separate the header and the message body part. If SIP invitations used to create sessions carry session descriptions that allow participants to agree on a set of compatible media types and compatible codecs, the message body part will include all this information as described in the SDP.

It should be noted that SIP is the application layer protocol that provides services in setting up the sessions between conference participants as directed by multimedia applications like teleconferencing (TC), video teleconferencing (VTC), videoconferencing (VC), application sharing, and web conferencing. Only audio is used in TC, both audio and video are used in VTC, while audio, video,

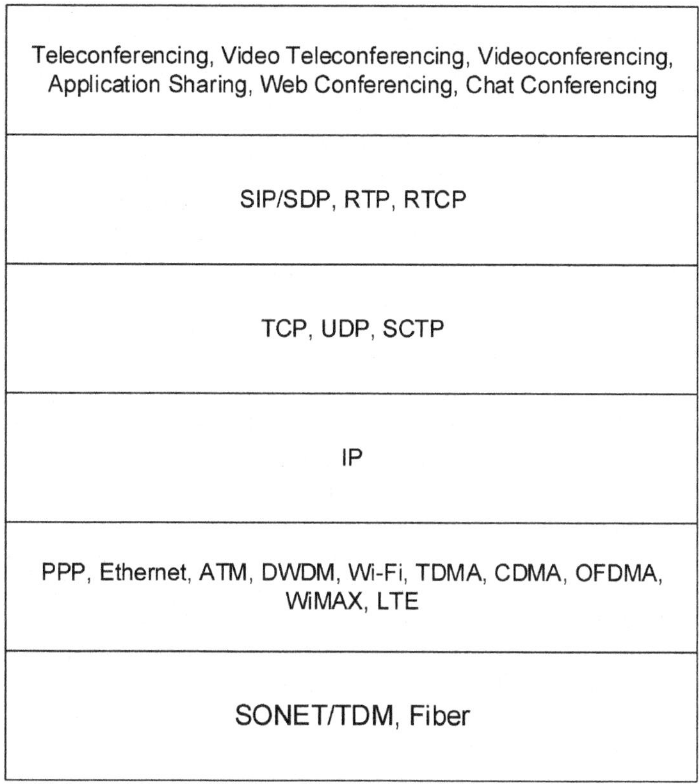

FIGURE 1.2 SIP/SDP and Its Relationship to Other Protocols.

and application(s) sharing are used in VC. Sometimes, application sharing may be used as a part of the web services integrated (or decoupling) with audio/video and can be termed web conferencing. Chat conferencing deals with real-time text messaging among two or multiple parties.

SIP signaling messages can be sent using any transport protocol, such as transmission control protocol (TCP), user datagram protocol (UDP), or stream control transmission protocol (SCTP). Of course, IP can run over point-to-point protocol (PPP), Ethernet, asynchronous transfer mode (ATM), dense wavelength division multiplexing (DWDM), Wi-Fi, time division multiple access (TDMA), code division multiple access (CDMA), orthogonal frequency division multiple access (OFDMA), worldwide interoperability for microwave access (WiMAX), or long term evolution (LTE) wireless networking protocol. It may be worthwhile to mention that ATM can run over a synchronous optical network (SONET)/time division multiplexing (TDM) network, which runs over fiber, and DWDM running over fiber increases the bandwidth by combining and transmitting multiple signals simultaneously at different wavelengths on the same fiber.

Different audio (e.g., International Telegraph Union – Telecommunication [ITU-T] G-series) and video (e.g., Moving Picture Expert Group [MPEG], Joint Photographic Expert Group [JPEG], and ITU-T H-series) codecs are used in multimedia sessions by conference participants. The bit streams of each codec are transferred over the real-time transport protocol (RTP) for transferring over UDP/IP. However, the common codec type either for audio or for video is negotiated by the SIP signaling messages that contain the information for each codec type that is proposed by the conference participants. SIP does not mandate any audio codec for any media that should be used by each participant if no common codec is supported among the conferees. In this situation, the conferees may use the transcoding services for preventing failures of the session establishment. The real-time transport control protocol (RTCP) is based on the periodic transmission of control packets to all participants in the session and provides feedback on the quality of the data (e.g., RTP packets of audio/video) distribution. This is an integral part of the RTP's role as a transport protocol and is related to the flow and congestion control functions of other transport protocols.

The domain name system (DNS) and dynamic host configuration protocol (DHCP) are the integral tools for IP address resolution and allocation, respectively, for routing of the SIP messages between the conference participants over the IP network. For example, a host can discover and contact a DHCP server to provide it with an IP address as well as the addresses of the DNS server and default router that can be used to route SIP messages over the IP network.

1.2.1 Performance Characteristics

Audio and video of multimedia applications can be continuous, while data consisting of text, still images, and/or graphics can be discrete. However, animation that is also considered as a part of data is continuous, consisting of audio, video, still images, graphics, and/or texts, and needs inter-/intramedia synchronization. Audio, video, and still images are usually captured from the real world, while text, graphics, and animation are synthesized by computers.

Based on the performance characteristics of communications, multimedia applications can be categorized as follows: Real-Time (RT), Near-Real-Time (Near-RT), and Non-Real-Time (Non-RT). The RT applications will have strict bounds on packet loss, packet delay, and delay jitter, while the Near-RT applications will have less strict bounds on those performance parameters than those of the RT applications. For example, TC and VTC/VC are considered RT services because of real-time two-way, point-to-point/multipoint conversations between users, and the audio and video performance requirements can be stated as follows [3]:

- One-way end-to-end delay (including propagation, network, and equipment) for audio or video should be between 100 and 150 milliseconds.
- Mean-opinion-score (MOS) level for audio should be between 4.0 and 5.0.
- MOS level for video should be between 3.5 and 5.0.
- End-to-end delay jitter should be very low, less than 250 microseconds in some cases.
- Bit-error-rate (BER) should be very low for good-quality audio or video, although some BER can be tolerated.
- Intermedia and intramedia synchronization needs to be maintained using suitable algorithms.
- Differential delay between audio and video transmission should be between no more than –20 milliseconds and +40 milliseconds for maintaining proper intermedia synchronization. This is a very important requirement, especially for lip-synchronization.

One-way video-on-demand (VOD) [4], which is considered Near-RT communication, can have much less stringent performances than those of the TC or VTC. The text or graphics are a Non-RT application, and the one-way delay requirement can be of the order of few seconds, although unlike audio or video, it cannot tolerate any BER.

The synchronization requirements between different media of multimedia applications impose a heavy burden on the multimedia transport networks, especially for packet networks such as IP. The RT applications are also considered live multimedia applications with the generation of live audio, video, and/or data from live sources of microphones, video cameras, and/or application sharing by human/machine, while the Near-RT applications are usually retrieved from databases and can be considered as retrieval multimedia applications. Consequently, the synchronization requirements between RT and Near-RT applications are also significantly different. The transmission side of the RT applications does not require much control, while Near-RT applications must have some defined relationships between media and require some scheduling mechanisms for guaranteed synchronization between the retrieved and transmitted media. The end-to-end delay requirements for RT applications are more stringent than for Near-RT applications. RT applications require synchronization between media generated by different live sources, although this may not be so common. However, Near-RT applications are commonly retrieved from different multimedia servers and are presented to users synchronously.

In general, multimedia synchronization can deal with many aspects: Temporal, spatial, or even logical aspect relationships between objects, data entities, or media streams [5]. In the context of multimedia computing and communications, the synchronization accuracy is critical for high-quality multimedia applications and can be measured by the following performance parameters: Delay, delay jitter, intermedia skew, and tolerable error rate. In RT applications, delay is measured in end-to-end delay from the live source to the destination, while in Near-RT applications, delay is measured in retrieval time; that is, delay from the time a request is made to the time the application is retrieved from the server and reaches its destination. It implies that excessive buffering should not be used in RT applications. Delay jitter measures the deviation of presentation time of the continuous media samples from their fixed or desired presentation time. Intermedia skew measures the time shift between related media from the desired temporal relationship. The acceptable value for intermedia skew is determined by the media types concerned. Table 1.1 shows some examples of intermedia skew tolerance [3–9].

However, the implementation of multimedia synchronization services can be done using a variety of techniques depending on different modes of application. Each synchronization implementation model can be quite complex [10]. As mentioned earlier, continuous media can tolerate some errors, error-free transmission is not essential to achieve acceptable quality, and the tolerable error rate measures the allowable BER and packet error rate (PER) for a particular media in a specific application. The multimedia traffic can be very bursty, and the burstiness can vary from 0.1 to 1. If constant bit rate (CBR) audio or video codec is used, there will be no variation in bit rates of the

TABLE 1.1
Intermedia Time Skew/Jitter Tolerance

Intermedia		Media Communication Mode in Different Applications	Tolerable Time Skew/Jitter (Note: Some variations of the following figures are also reported [1–7])
Audio	Audio	Tightly coupled (e.g., stereo)	±11 microseconds
		Loosely coupled (e.g., teleconferencing)	±80 to ±150 milliseconds
		Loosely coupled (e.g., background music)	±500 milliseconds
	Animation	Event correlation (e.g., dancing)	±80 milliseconds
	Image	Tightly coupled (e.g., music with notes)	±5 milliseconds
		Loosely coupled (e.g. Slide Show)	±500 milliseconds
	Text	Text annotation	±240 milliseconds
	Pointer	Audio relates to shown item	–500 to +750 milliseconds
Video	Audio	Lip synchronization (e.g., videoconferencing, video streaming)	±80 to ±150 milliseconds (Note: out-of-sync region spans somewhat at –160 milliseconds and +160 milliseconds)
	Animation	Correlated	±120 milliseconds
	Image	Overlay	±2400 milliseconds
		Non-overlay	±500 milliseconds
	Text	Overlay	±240 milliseconds
		Non-overlay	±500 milliseconds

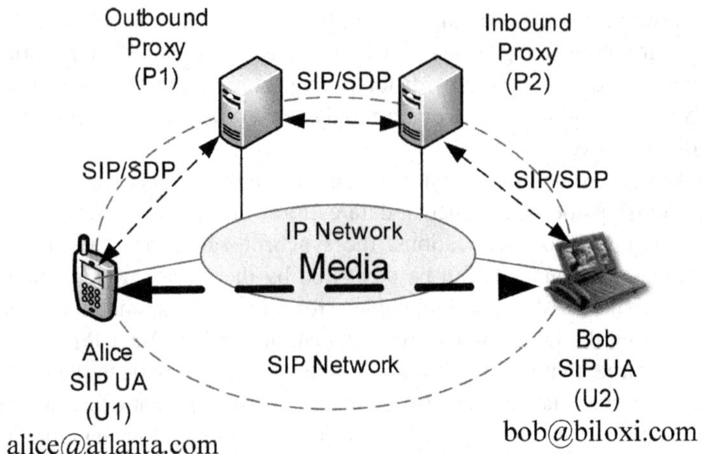

FIGURE 1.3 SIP Trapezoidal Network.

codec, and the burstiness will be 1. The multimedia call duration can vary from a few seconds to few hours.

1.2.2 Point-to-Point Conferencing

The SIP [1]/SDP [2] protocol architecture has originally been defined for point-to-point and point-to-multipoint multimedia conferencing, as shown in Figure 1.3, where outgoing and incoming proxies are used to route the call between the two conference participants.

This particular call flow is known as the SIP trapezoidal network call flow. The key is that once the call is set up, the media (audio, video, and/or text/graphics) goes directly between the two conference participants, and no proxy resides in the media path. Because there are only two participants, no mixing of any media (audio, video, or text/graphics) is needed. Consequently, no multipoint control unit (MCU) is needed for media mixing.

1.2.3 Multipoint Conferencing

We have described earlier that multipoint conferencing consists of three or more conference participants, where media (audio, video, and text/graphics) needs to be bridged. The functional entity that does media bridging is known as the MCU. Figure 1.4 shows a simple three-party multipoint conference where a centralized controller is controlling all conference resources, including media bridging functions.

The call flow architecture is a star topology centralizing the controller, which acts as the SIP back-to-back user agent (B2BUA). The controller not only controls the call setup and teardown but also functions as the MCU for providing media mixing. After each type of media is bridged, the bridged media is also sent back to each conference participant. It should be noted that the audio, video, or text bridging is hugely complex. For simplicity, we have not shown any of those functionalities, assuming a simple black box. Further, the controller also controls all application servers used for application sharing and other functions. However, this media bridging function, application sharing, and controlling of the conference are the heart of the multipoint conferencing. We will describe each of these functions in a step-by-step manner in the subsequent sections to deal with these immensely complex functions of multipoint conferencing.

1.2.3.1 Audio

Audio with a full-duplex interoperability feature, that is, the ability of participants to speak at any time, is the most important medium for multimedia conferencing [3]. General human factor

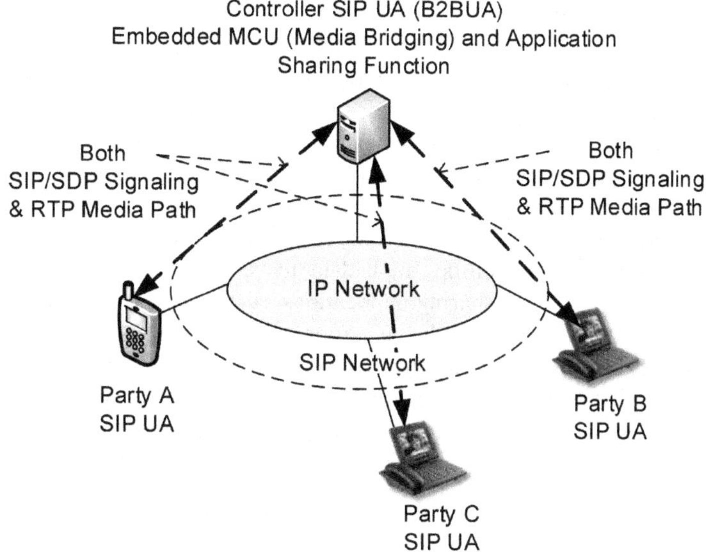

FIGURE 1.4 Multipoint Conferencing with Three Conference Participants.

studies also indicate that among all the media of multimedia conferencing, audio is the most critical. Depending on the application, audio bandwidth can vary significantly. ITU-T and other standard bodies have created a variety of audio codec standards for both constant bit rate (CBR) and variable bit rate (VBR) codecs. However, it is appropriate to maintain an MOS of 4.0 or above for multimedia conferencing [3]. The one-way, encoding-decoding delay for compact disk (CD) or frequency modulation (FM) stereo audio codec coders using proprietary algorithms may be about 85 milliseconds. The codec delay for pulse-coded modulation (PCM), or adaptive delays PCM (ADPCM), is negligible, while the delay for low delay-code excited linear predictive (LD-CELP) coding may be about 2 milliseconds. Just note that the media server dealing with audio, for example, audio bridging as a part of the MCU function, will also be under the control of the controller, as shown in Figure 1.4.

1.2.3.2 Video

Video is the most bandwidth-intensive application for multimedia conferencing. The actual video bit rates will depend on the following factors [3]:

- The activity in the captured scene
- The number bits per pixel – for a given number of samples per line and a total number of lines per frame
- The number of video frames transmitted per second
- The chosen level of video quality
- The types of compression algorithms

ISO, ITU-T, and other standards organizations have created many commercial video codec standards. An MOS of 3.5 or above will provide video quality between fair and good; that is, the perceptible impairment may or may not be slightly annoying. Thus, it would be appropriate to maintain an MOS level of 3.5 or above for multimedia conferencing [3]. Like audio, video codec can also be CBR or VBR depending on the application. The back-to-back encoding-decoding delay for CBR video codec may exceed 150 milliseconds, upper bound, before a user perceives the delay. However, VBR video codec can reduce the encoding-decoding delay, as well as maintaining superior video quality, but at much higher peak bit rates. It should be noted that like audio, the media server dealing with video, for example, video bridging as a part of the MCU function, will also be under the control of the controller, as shown in Figure 1.4.

1.2.3.3 Application Sharing

The application sharing in multimedia conferencing is huge, as there are numerous applications available that need to be shared with others in both the commercial and the private world interactively. The traffic generated during the conference through sharing of text files, still images, and graphics can be from a few bits to multi-gigabits per second (uncompressed) [3]. Although this traffic can be bursty with a high peak-to-average ratio, the network needs to be designed for the peak traffic. In addition, multimedia files can contain audio, video, and/or text/graphics and even timed texts for sharing among the conference participants, and the bandwidth required for this application can be significantly higher than that of the simple data (texts, still images, graphics) sharing file. In addition, there are many other data sharing applications, such as presence and instant messaging (IM), emails, conference management, web applications, short message service (SMS), multimedia messaging service (MMS), calendar and scheduling, and a host of others. Finally, like audio and video, in many cases, shared applications may also be bridged. This shared application bridging will also belong to the MCU function, as shown in Figure 1.4. Whether or not a shared application is bridged, the control of each media server dealing with application sharing will reside with the controller.

1.2.3.4 Timed Text

Timed text is real-time text with timing information embedded, which is usually synchronized with captioning and subtitling functions of application sharing (for example, foreign language movies on web, devices lacking audio). That is, textual information is intrinsically or extrinsically associated with timing information. There are a few timed text standards available: MIME type, specified in IETF RFC 3839, MPEG-4 Part 17 Timed Text, W3C Timed Text Markup Language (TTM), WebVTT, SMPTE-TT, and others. Like audio and video, the timed text bridging function will also belong to the MCU function, as shown in Figure 1.4.

1.2.3.5 Media Mixing and Transcoding

We see that audio, video, application sharing, and timed text, as discussed in the preceding subsections, needs media mixing or media bridging, which belong to the MCU. However, the media mixing requirement for each kind of media is fundamentally different. For example, when audio is bridged or mixed from multiple sources for sending the endpoints, the destination endpoint to which the audio is sent should not be bridged or mixed. However, video bridging can be done in a variety of ways depending on the users' needs and conference policy. The most easy and popular way is the video-switching that is performed in sending the video of the speaker. In another example, a composite video of all participants or a sub-set of the participants is formed, and then the composite video is sent to all or some of the participants, depending on the policy and capability of the endpoints, while in the case of data (text, timed text, graphics, and still images) bridging, all data received from all endpoints are combined together based on some policy. For example, a document can be jointly written in real time combining all data of all endpoints.

Sometimes, dissimilar audio and video codecs are used by the endpoints. In those situations, the MCU needs to do the transcoding of audio and video as appropriate. Note that non-compatible audio/video of each endpoint will be sent to the MCU for transcoding. In turn, the MCU transcodes audio/video and sends it to all other endpoints, possibly with versions consistent with those endpoints. In this situation, the MCU will be burdened with the huge computational load for transcoding operations.

1.2.3.6 Media Server

We have discussed audio and video applications, why these media need to be bridged for multipoint (three or more) conferencing, and their relationship to the centralized controller, MCU, and the end-users. In the real world, endpoints will have audio and video codecs, while MCU will provide the media bridging functions under the control of the controller (Figure 1.4). In addition to the media bridging function, there are many other functions, such as DTMF, announcement, prompt and collect (voice dialog using VoiceXML script), early media, conference recoding, text-to-speech (TTS), speech-to-text (STT), transcoding, and many other services. In general, all these services are offered by one or more media servers under the control of the centralized controller for multipoint conferencing.

1.2.3.7 Application Server

The application server is a broad term that usually refers to all kinds of servers that provide services to clients. However, we have differentiated here, in that a server that is related to audio and video is termed the media server, while all other servers are termed application servers, because media servers have some special service characteristics that deal with RT and near-RT services, while the application servers deal with non-RT. As described earlier, IM, emails, conference management, floor control, web applications, SMS, MMS, calendar and scheduling, and a host of other services are offered by the application servers. Figure 1.5 depicts how the controller, media server (MS), application server (AS), and MCU are inter-related in the context of SIP-based multipoint (three participants/endpoints) conferencing at a very high level.

In contrast to Figure 1.4, it is interesting to note that the conferencing architecture shown in Figure 1.5 has expanded the black box controller with three more functional entities, MS, AS, and MCU, based on their respective functions. In addition to the controller, other functional entities like MS, AS, and MCU may also interact directly with the endpoints/users/participants if the controller authorizes them.

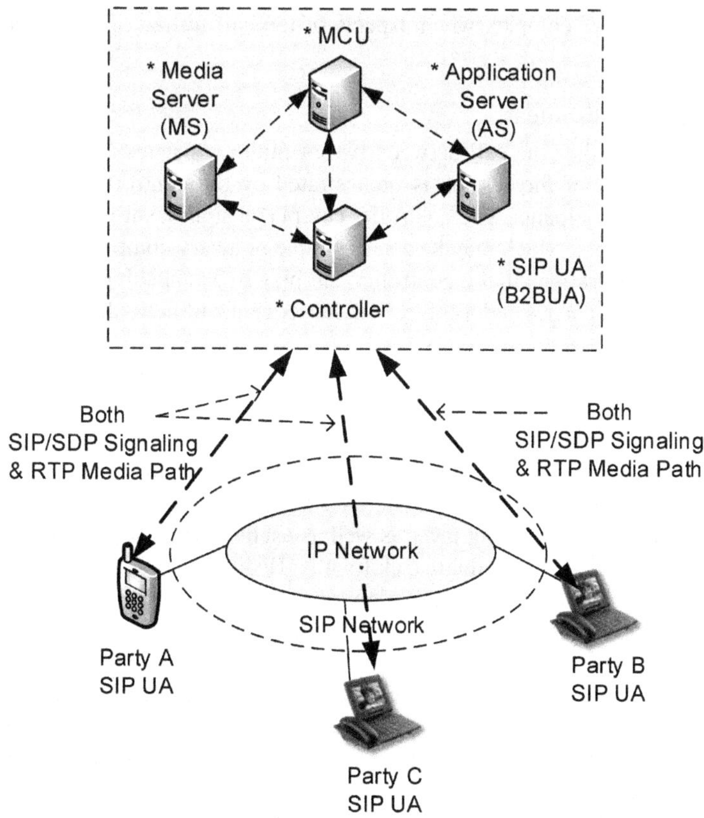

FIGURE 1.5 High-Level SIP-based Multipoint (Three-party) Conference Architecture with Controller, MS, AS, and MCU.

1.3 THIRD-PARTY MULTIPARTY CONFERENCING

1.3.1 Multiparty Conferencing

Third-party call control (3pcc) provides mechanisms for the creation of multiparty conference calls in a star-like point-to-point fashion from a centralized controller. In fact, 3pcc refers to the general ability to manipulate calls between other parties. It is assumed that the controller knows the addresses of all parties and when to dial-out the participants, or the address of the controller is priori known to all conference participants for dial-in. The call control, including modifications of signaling messages acting as the B2BUA, is done by the controller. It is also possible that the media bridge can be used to dial-out or dial-in as another party by the controller. A controller is an SIP user agent that wishes to create a session between two other user agents.

3pcc is often used for operator services (where an operator creates a call that connects two or more participants together) and multimedia conferencing. Many SIP services are possible through 3pcc. These include not only the traditional ones on the public switched telephone network (PSTN) but also new ones such as click-to-dial, mid-call announcements, media transcoding, call transfer, and others. For example, click-to-dial allows a user to click on a web page when they wish to speak to a customer service representative. The web server then creates a call between the user and a customer service representative. The call can be between two phones, a phone and an IP host, or two IP hosts.

3pcc is possible using only the mechanisms specified within RFC 3261. In addition, many other services are possible using other methods and mechanisms defined in other SIP-related RFCs. Indeed, many possible approaches can be used for the 3pcc, but we have described here some examples of the call setup following RFC 3725, each with different benefits and drawbacks. The usage of 3pcc also becomes more complex when aspects of the call utilize SIP extensions or optional features of SIP.

1.3.1.1 3pcc Call Establishment

The controller that establishes the call between two or multiple parties is termed the third party, because the establishment of the session is orchestrated by the controller, which is not a party among the conference participants. The controller can play a significant role in setting up the sessions among the participants, which may be a very simple or a very complex one. Figure 1.6 shows different primitives of operations that a controller can play in establishing the calls in an SIP network that consists of three parties and a controller. Subsequent call flows described here are similar to RFC 3725.

It should be noted that we have used the connection address of 0.0.0.0, which though recommended by RFC 3264, has numerous drawbacks. It is anticipated that a future specification will recommend usage of a domain within the .invalid DNS top-level domain instead of the 0.0.0.0 IP address. As a result, implementors are encouraged to track such developments once they arise. In addition, RFC 6157 describes how the IPv4 SIP user agents can communicate with IPv6 SIP user agents (and vice versa) at the signaling layer as well as exchange media once the session has been successfully set up. Both single- and dual-stack (that is, IPv4-only and IPv4/IPv6) user agents are also considered. We have not discussed these scenarios in the 3pcc call flows here.

1.3.1.2 Simplest Flows

The simplest 3pcc is depicted in Figure 1.6b. The controller first sends a SIP message INVITE (F1) with no session description protocol (SDP) to Party A. That is, this INVITE message has no session description. Party A's phone rings, and Party A answers. This results in a 200 OK (F2) message that contains an offer as defined in RFC 3264. The controller needs to send its answer in the ACK message, as mandated by RFC 3261 [1]. To obtain the answer, it sends the offer it got from Party A (offerl) in an INVITE message (F3) to Party B. Party B's phone rings.

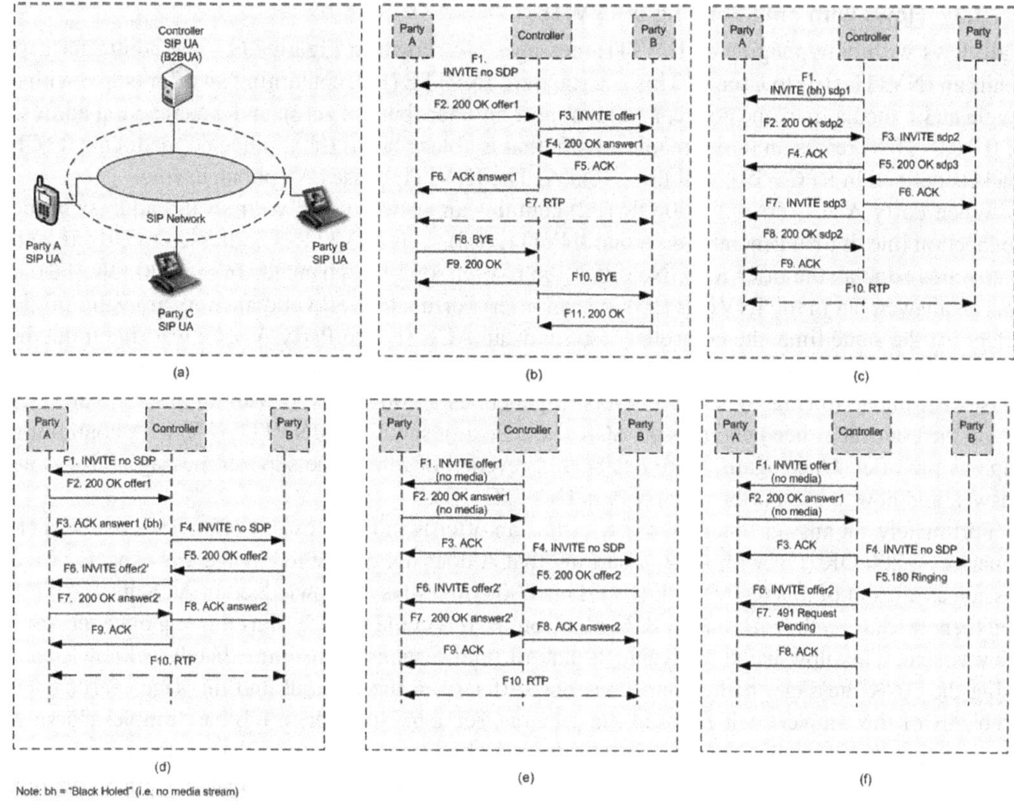

FIGURE 1.6 Third-Party Call Control Call Flows in SIP Network. (a) SIP Network, (b) Simplest Flows, (c) Flows with Ping-Ponging of INVITEs, (d) Flows with Unknown Entities, (e) Efficient Flows with Unknown Entities, and (f) Flows with Error Handling. (Copyright: IETF.)

When Party B answers, the 200 OK message (F4) contains the answer to this offer, answer1. The controller sends an ACK message (F5) to Party B and then passes answer1 to A in an ACK message (F6) sent to it. Because the offer was generated by A, and the answer was generated by B, the actual media session is between A and B. Therefore, RTP (F7) media flows between Parties A and B.

This flow is simple, requires no manipulation of the SDP by the controller, and works for any media types supported by both endpoints. However, it has a serious timeout problem. User B may not answer the call immediately. The result is that the controller cannot send the ACK to A right away. This causes A to retransmit the 200 OK response periodically. As specified in Section 13.3.1.4 of RFC 3261 [1], the 200 OK will be retransmitted for 64*T1 seconds. If an ACK does not arrive by then, the call is considered to have failed. This limits the applicability of this flow to scenarios where the controller knows that B will answer the INVITE immediately.

Once the calls are established, both participants believe they are in a single point-to-point call. However, they are exchanging media directly with each other rather than with the controller. The controller is involved in two dialogs yet sees no media. Since the controller is still a central point for signaling, it now has complete control over the call. If it receives a BYE from one of the participants, it can create a new BYE and hang up with the other participant. This is shown in Figure 1.6b as well in messages F8 through F11. It should be noted that this will be the general behavior of the controller in dealing with BYE messages sent by one of the participants. Later, we will also describe how the continued call processing can be done by the controller through use of the re-INVITE message.

1.3.1.3 Flows with Ping-Ponging of INVITEs

Call flows with ping-ponging of INVITE messages are shown in Figure 1.6c. The controller first sends an INVITE (F1) to User A. This is a standard INVITE (F1), containing an offer (sdp1) with a single audio media line, one codec, a random port number (but not zero), and a connection address of 0.0.0.0. This creates an initial media stream that is "black holed (bh)," since no media (or RTCP packets defined in RFC 3550) will flow from A. The INVITE causes A's phone to ring.

When Party A answers, the 200 OK (F2) contains an answer, sdp2, with a valid address in the connection line. It then generates a second INVITE (F3). This INVITE is addressed to User B, and it contains sdp2 as the offer to B. Note that the role of sdp2 has changed. In the 200 OK (F2), it was an answer, but in the INVITE (F3), it is an offer. Fortunately, all valid answers are valid initial offers. At the same time, the controller also sends an ACK (F4) to Party A, as it was due immediately in response to A's 200 OK (F2) for stopping retransmissions or timeout. The INVITE (F3) causes B's phone to ring. When it answers, it generates a 200 OK (F5) with an answer, sdp3. As usual, the controller then generates an ACK (F6). Next, it sends a re-INVITE (F7) to A containing sdp3 as the offer. Once again, there has been a reversal of roles. The sdp3 of message F5 was an answer, and now, it is an offer to Party A in INVITE (F7).

Fortunately, an answer to an answer recast as an offer is in turn, a valid offer. This re-INVITE generates a 200 OK (F8) with sdp2, assuming that A does not decide to change any aspects of the session as a result of this re-INVITE (F7). This 200 OK (F8) is acknowledged in ACK message F9, and then, media can flow from A to B. Media from B to A could already start flowing once message F5 was sent. This flow has the advantage that all final responses are immediately acknowledged using the ACK message. It therefore does not suffer from the timeout and message inefficiency problems of the simplest call flows in the previous section. However, it too has troubles. First, it requires that the controller know the media types to be used for the call (since it must generate a "black holed (bh)" SDP, which requires media lines). Second, the first INVITE (F1) to A contains media with a 0.0.0.0 connection address. The controller expects that the response contains a valid, non-zero connection address for A.

However, experience has shown that many user agents (UAs) respond to an offer of a 0.0.0.0 connection address with an answer containing a 0.0.0.0 connection address. The offer-answer specification of RFC 3264 explicitly tells implementors not to do this, but at the time of publication of this RFC 3725, many implementations still did. If A should respond with a 0.0.0.0 connection address in sdp2, the flow will not work.

The most serious flaw in this flow is the assumption that the 200 OK (F8) to the re-INVITE (F7) contains the same SDP as in message 2. This may not be the case. If it is not, the controller needs to re-INVITE party B with that SDP (say, sdp4), which may result in getting a different SDP, say sdp5, in the 200 OK from party B. Then, the controller needs to re-INVITE Party A again, and so on. The result is an infinite loop of re-INVITEs. It is possible to break this cycle by having very smart UAs that can return the same SDP whenever possible or really smart controllers that can analyze the SDP to determine whether a re-INVITE is really needed. However, RFC 3725 recommends keeping this mechanism simple and avoids SDP awareness in the controller. As a result, this flow is not really workable. This 3pcc call flow is therefore not recommended by RFC 3725.

1.3.1.4 Flows with Unknown Entities

The call flows with parties whose media compositions for the sessions are not known to the controller are shown in Figure 1.6d. First, the controller sends an INVITE (F1) to User A without any SDP, because the controller does not need to assume anything about the media composition of the session. Party A's phone rings. When Party A answers, a 200 OK (F2) is generated containing its offer, offer1. The controller generates an immediate ACK (F3) containing an answer (F3). This answer (F3) is a "black holed (bh)" SDP with its connection address equal to 0.0.0.0.

The controller then sends an INVITE (F4) to B without SDP. This causes Party B's phone to ring. When Party B answers, a 200 OK (F5) is sent, containing their offer, offer2. This SDP is used to create a re-INVITE (F6) back to Party A. That re-INVITE (F6) is based on offer2, but may need to be reorganized to match up media lines or to trim media lines. For example, if offerl contained an audio and a video line, in that order, but offer2 contained just an audio line, the controller would need to add a video line to the offer (setting its port to zero) to create offer2′. Since this is a re-INVITE (F6), it should complete quickly in the general case. That is good, since User B is retransmitting their 200 OK (F5), waiting for an ACK. The SDP in the 200 OK (F7) from A, answer2′, may also need to be reorganized or trimmed before sending it as an ACK (F8) to B as answer2. Finally, an ACK (F9) is sent to A, and then, media can flow. This flow has many benefits. First, it will usually operate without any spurious retransmissions or timeouts (although this may still happen if a re-INVITE is not responded to quickly). Second, it does not require the controller to guess the media that will be used by the participants.

There are some drawbacks. The controller does need to perform SDP manipulations. Specifically, it must take some SDP and generate another SDP that has the same media composition but has connection addresses equal to 0.0.0.0. This is needed for message F3. Second, it may need to reorder and trim SDP Y so that its media lines match up with those in some other SDP Z. Third, the offer from B (offer2 – message F5) may have no codecs or media streams in common with the offer from A (offerl – message F2). The controller will need to detect this condition and terminate the call. Finally, the flow is far more complicated than the simplest and elegant call flows shown in Figure 1.6b.

1.3.1.5 Efficient Flows with Unknown Entities

The call flows with reduced complexities shown in Figure 1.6e are a variation of the call flows shown in Figure 1.6d with the parties whose media compositions for the sessions are not known to the controller. The actual message flow is identical, but the SDP placement and construction differ. The initial INVITE (Fl) contains SDP with no media at all, meaning that there are no m lines. This is valid and implies that the media makeup of the session will be established later through a re-INVITE described in RFC 3264. Once the INVITE is received, User A is alerted. When Party A answers the call, the 200 OK (F2) has an answer with no media either. This is acknowledged by the controller sending an ACK message (F3).

The flow from this point onwards is identical to flows shown in Figure 1.6d. However, the manipulations required to convert offer2 to offer2′ and answer2′ to answer2 are much simpler. Indeed, no media manipulations are needed at all. The only change that is needed is to modify the origin lines so that the origin line in offer2′ is valid based on the value in offerl (validity requires that the version increments by one and that the other parameters remain unchanged).

There are some limitations associated with this flow. First, User A will be alerted without any media having been established yet. This means that User A will not be able to reject or accept the call based on its media composition. Second, both A and B will end up answering the call (that is, generating a 200 OK) before it is known whether there is compatible media. If there is no media in common, the call can be terminated later with a BYE. However, the users will have already been alerted, resulting in user annoyance and possibly resulting in billing events.

1.3.2 Recommendations for 3pcc Call setups

The call flows shown in Figure 1.6b represent the simplest and the most efficient flows. This flow should be used by a controller if it knows with certainty that user B is actually an automaton that will answer the call immediately. This is the case for devices such as media servers, conferencing servers, and messaging servers, for example. Since we expect a great deal of third-party call control to be to automata, special casing in this scenario is reasonable.

For calls to unknown entities, or to entities known to represent people, it is recommended that the flows shown in Figure 1.6e be used for 3pcc. the call flows shown in Figure 1.6d may be used instead but provide no additional benefits over the flows of Figure 1.6e. However, the flows shown in Figure 1.6c should not be used because of the potential for infinite ping-ponging of re-INVITEs.

Several of these flows use a "black holed (bh)" connection address of 0.0.0.0. This is an IPv4 address with the property that packets sent to it will never leave the host that sent them; they are just discarded. Those flows are therefore specific to IPv4. For other network or address types, an address with an equivalent property should be used. In most cases, including the recommended flows, User A will hear silence while the call to B completes. This may not always be ideal. It can be remedied by connecting the caller to a music-on-hold source while the call to B occurs. In addition, RFC 6157 can be used for IPv4 and IPv6 network scenarios, which are not shown here.

1.3.3 Error Handling

There are numerous error cases that merit discussion. With all the call flows described earlier, one call is established to Party A, and then the controller attempts to establish a call to Party B. However, this call attempt may fail for any number of reasons. User B may be busy (resulting in a 486 Busy Here response to the INVITE), there may not be any media in common, the requests may time out, and so on. If the call attempt to B should fail, it is recommended that the controller send a BYE to A. This BYE should include a Reason header, defined in RFC 3326, which carries the status code from the error response. This will inform User A of the precise reason for the failure. The information is important from a user interface perspective. For example, if A was calling from a black phone, and B generated a 486 Busy Here, the BYE will contain a Reason code of 486 Busy Here, and this could be used to generate a local busy signal so that A knows that B is busy.

Another error condition worth discussing is shown in Figure 1.6f. After the controller establishes the dialog with Party A (messages F1–F3), it attempts to contact B (message F4). Contacting Party B may take some time. During that interval, Party A could possibly attempt a re-INVITE, providing an updated offer. However, the controller cannot pass this offer on to B, since it has an INVITE transaction pending with it. As a result, the controller needs to reject the request. It is recommended that a 491 Request Pending response be used. The situation here is similar to the glare condition described in RFC 3261 [1], and thus, the same error handling is sensible. However, User A is likely to retry its request (as a result of the 491 Request Pending), and this may occur before the exchange with B is completed. In that case, the controller would respond with another 491 Request Pending.

1.3.4 Continued Processing in 3PCC

In continuation of the call flows of Figure 1.6e after setting the RTP (F10), if the controller receives a re-INVITE from one of the participants, it can forward it to the other participant. Depending on which flow was used, this may require some manipulation on the SDP before passing it on. However, the controller need not "proxy" the SIP messages received from one of the parties. Since it is a B2BUA, it can invoke any signaling mechanism on each dialog as it sees fit. For example, if the controller receives a BYE from A, it can generate a new INVITE to a third party, C, and connect B to that participant instead. A call flow for this is shown in Figure 1.7a, assuming the case where C represents an end-user, not an automaton.

From here, new parties can be added, removed, transferred, and so on as the controller sees fit. In many cases, the controller will be required to modify the SDP exchanged between the participants in order to effect these changes. In particular, the version number in the SDP will need to be changed by the controller in certain cases. Should the controller issue an SDP offer on its own (for example, to place a call on hold), it will need to increment the version number in the SDP offer. The other participant in the call will not know that the controller has done this, and any subsequent offer it generates will have the wrong version number as far as its peer is concerned. As a result, the controller will be required to modify the version number in SDP messages to match what the recipient is expecting. It is important to point out that the call need not have been established by the controller in order for the processing of this section to be used. Rather, the controller could have acted as a B2BUA during a call established by A towards B (or vice versa).

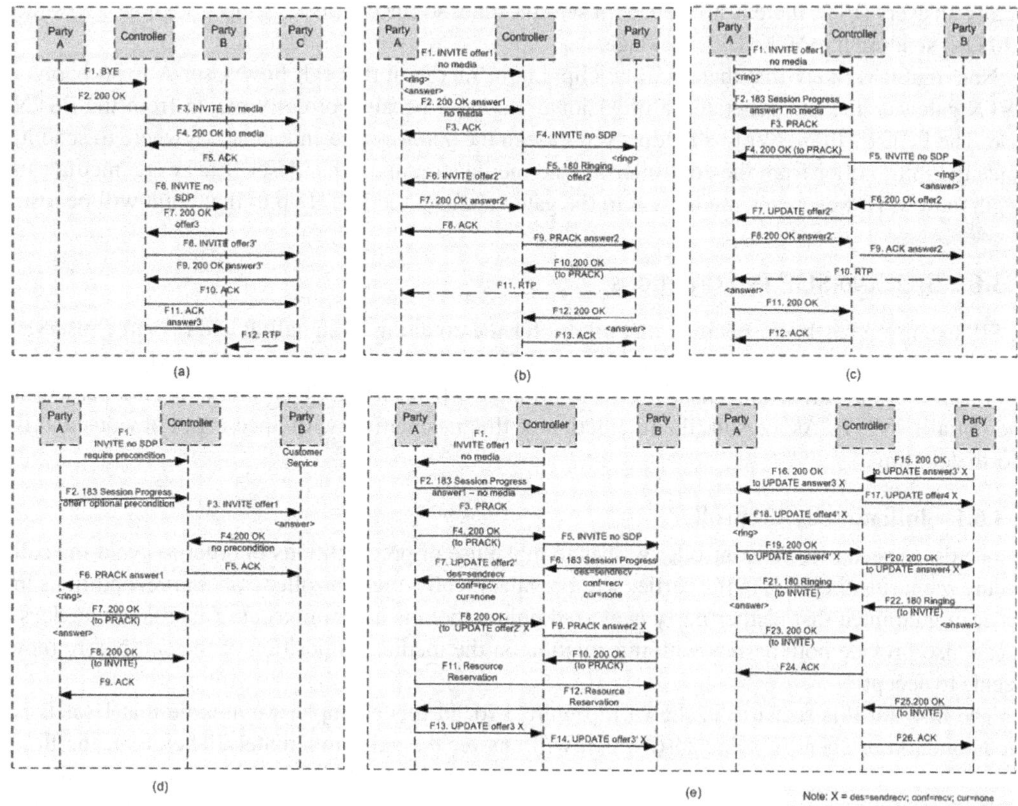

FIGURE 1.7 Third-Party Call Control: (a) Continued Call Processing, (b) Simple Early Media, (c) Complex Early Media, (d) Controller Initiated SDP Preconditions, and (e) Party Initiated SDP Preconditions. (Copyright: IETF.)

1.3.5 3PCC and Early Media

Early media represents the condition where the session is established (as a result of the completion of an offer/answer exchange), yet the call itself has not been accepted. This is usually used to convey tones or announcements regarding the progress of the call. Handling of early media in a third-party call is straightforward. Figure 1.7b shows the case where User B generates early media before answering the call. The flow is almost identical to the flows of Figure 1.7e. The only difference is that User B generates a reliable provisional response (F5) defined in RFC 3262 instead of a final response, and answer2 is carried in a PRACK (F9) instead of an ACK. When Party B finally does accept the call (F12), there is no change in the session state, and therefore, no signaling needs to be done with User A. The controller simply acknowledges the 200 OK (F13) to confirm the dialog.

The case where User A generates early media is more complicated and is shown in Figure 1.7c. The flow is based on call flows of Figure 1.6e. The controller sends an INVITE (F1) to User A with an offer containing no media streams. User A generates a reliable provisional response (F2) containing an answer with no media streams. The controller acknowledges this provisional response (F2), sending PRACK (F3). Now, the controller sends an INVITE (F5) without SDP to User B. User B's phone rings, and they answer, resulting in a 200 OK (F6) with an offer, offer2. The controller now needs to update the session parameters with User A. However, since the call has not been answered, it cannot use a re-INVITE. Rather, it uses a SIP UPDATE request (F7) defined in RFC 3311, passing the offer (after modifying it to get the origin field correct). User A generates its answer in the 200 OK (F8) to the UPDATE. This answer is passed to User B in the ACK (F9) message. When User A finally answers (F11), there is no change in session state, so the controller simply acknowledges the 200 OK, sending the ACK (F12) message.

Note that it is likely that there will be clipping of media in this call flow. User A is probably a PSTN gateway and has generated a provisional response because of early media from the PSTN side. The PSTN will deliver this media even though the gateway does not have anywhere to send it, since the initial offer from the controller had no media streams. When User B answers, media can begin to flow. However, any media sent to the gateway from the PSTN up to that point will be lost.

1.3.6 3PCC AND SDP PRECONDITIONS

A SIP extension has been specified that allows for the coupling of signaling and resource reservation described in RFC 3312. This specification relies on exchanges of session descriptions before completion of the call setup. These flows are initiated when certain SDP parameters are passed in the initial INVITE. As a result, the interaction of this mechanism with 3pcc is not obvious and is worth detailing.

1.3.6.1 Initiation by Controller

In one usage scenario, the controller wishes to make use of preconditions in order to avoid the call failure scenarios specified in the earlier section. Specifically, the controller can use preconditions in order to guarantee that neither party is alerted unless there is a common set of media and codecs. It can also provide both parties with information on the media composition of the call before they decide to accept it.

The flow for this scenario is shown in Figure 1.7d. In this example, we assume that User B is an automaton or agent of some sort, which will answer the call immediately. Therefore, the flow is based on call flows of Figure 1.6b. The controller sends an INVITE to User A containing no SDP but with a Require header indicating that preconditions are required. This specific scenario (an INVITE without an offer but with a Require header indicating preconditions) is not described in RFC 3312. It is recommended that the UAS respond with an offer in a 1xx including the media streams it wishes to use for the call, and for each, list all preconditions it supports as optional. Of course, the user is not alerted at this time. The controller takes this offer and passes it to User B (F3). User B does not support preconditions, or if it does, it is not interested in them. Therefore, when it answers the call, the 200 OK contains an answer without any preconditions listed (F4). This answer is passed to User A in the PRACK (F6). At this point, User A knows that there are no preconditions actually in use for the call, and therefore, it can alert the user. When the call is answered, User A sends a 200 OK (F8) to the controller, and the call is complete.

In the event that the offer generated by User A was not acceptable to User B (because of non-overlapping codecs or media, for example), User B would immediately reject the INVITE (F3). The controller would then CANCEL the request to User A. In this situation, neither User A nor User B would have been alerted, achieving the desired effect. It is interesting to note that this property is achieved using preconditions, even though it does not matter what specific types of preconditions are supported by User A. It is also entirely possible that User B does actually desire preconditions. In that case, it might generate a 1xx of its own with an answer containing preconditions. That answer would still be passed to User A, and both parties would proceed with whatever measures are necessary to meet the preconditions. Neither user would be alerted until the preconditions were met.

1.3.6.2 Initiation by Party A

In the previous section, the controller requested the use of preconditions to achieve a specific goal. It is also possible that the controller does not care (or perhaps does not even know) about preconditions, but one of the participants in the call does care. A call flow for this case is shown in Figure 1.7e. The controller follows the call flows of Figure 1.6e; it has no specific requirements for support of the preconditions specification of RFC 3312. Therefore, it sends an INVITE (Fl) with SDP that contains no media lines. User A is interested in supporting preconditions and does not want to ring its phone until resources are reserved. Since there are no media streams in the INVITE, it cannot reserve resources for media streams, and therefore, it cannot ring the phone until they are conveyed in a subsequent offer and then reserved. Therefore, it generates a 183 Session Progress (F2) message with the answer and does not alert the user. The controller acknowledges this 183 Session Progress provisional response, sending the PRACK message (F3), and A responds to the PRACK (F3) message with 200 OK (F4). At this point, the controller attempts to bring B into the call. It sends B an INVITE without SDP (F5). B is interested in having preconditions for this call. Therefore, it generates its offer in a 183 Session Progress (F6) message that contains the appropriate SDP attributes. The controller passes this offer to A in an UPDATE request (F7).

The controller uses UPDATE because the call has not been answered yet, and therefore, it cannot use a re-INVITE. User A sees that its peer is capable of supporting preconditions. Since it desires preconditions for the call, it generates an answer in the 200 OK (F8) to the UPDATE. This answer, in turn, is passed to B in the PRACK (F9) for the provisional response. Now, both sides perform resource reservation. User A succeeds first and passes an updated session description in an UPDATE request (F13). The controller simply passes this to A after the manipulation of the origin field, as required in Call Flows of Figure 1.7e, in an UPDATE (F14), and the answer (F15) is passed back to A (F16). The same flow happens, but from B to A, when B's reservation succeeds (F17–F20). Since the preconditions have been met, both sides ring (F21 and F22), and then both answer (F23 and F25), completing the call.

What is important about this flow is that the controller does not know anything about preconditions. It merely passes the SDP back and forth as needed. The trick is the usage of UPDATE and PRACK to pass the SDP when needed. That determination is made entirely based on the offer/answer rules described in RFCs 3311 and 3262 and is independent of preconditions.

1.3.7 3pcc Services Example

We have considered two applications for offering using the 3pcc mechanisms: click-to-dial and mid-call announcement. Figure 1.8 shows the call flows for both these services.

For the click-to-dial service, Figure 1.8a shows the SIP network that contains User A's phone and browser, controller, and customer service, while Figure 1.8b shows the call flows. Note that both phone and browser capability can also be implemented in a single device. Figure 1.8c depicts the SIP network with User A's phone, Called party B's phone, controller, and media server, while Figure 1.8d describes the call flows for the Mid-Call Announcement services.

1.3.7.1 Click-to-Dial

In the click-to-dial service, a user is browsing the web page of an e-commerce site and would like to speak to a customer service representative. The user clicks on a link, and a call is placed to a customer service representative. When the representative picks up, the phone on the user's desk rings. When the user picks up, the customer service representative is there, ready to talk to the user.

The call flow for this service is given in Figure 1.8b. It is identical to that of Figure 1.6e, with the exception that the service is triggered through an HTTP POST request when the user clicks on the link. Normally, this POST request would contain neither the number of the user nor that of the customer service representative. The user's number would typically be obtained by the web application from back-end databases, since the user would have presumably logged into the site, giving the server the needed context. The customer service number would typically be obtained through provisioning. Thus, the HTTP POST is actually providing the server nothing more than an indication that a call is desired.

We note that this service can be provided through other mechanisms, namely PSTN/Internet Interworking (PINT), described in RFC 2848. However, there are numerous differences between

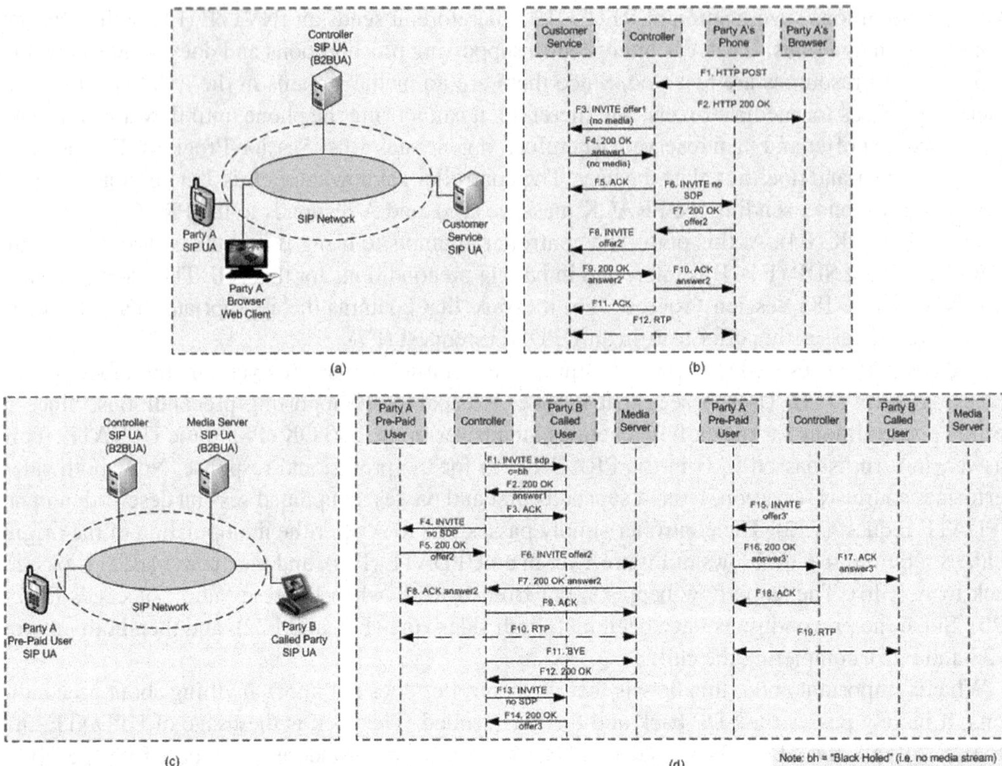

FIGURE 1.8 3pcc Services Example: (a) Click-to-Dial SIP Network, (b) Click-to-Dial Call Flows, (c) Mid-Call Announcement SIP Network, and (d) Mid-Call Announcement Call Flows. (Copyright: IETF.)

the way in which the service is provided by PINT and the way in which it is provided here, as described in RFC 3725:

- The PINT solution enables calls only between two PSTN endpoints. The solution described here allows calls between PSTN phones (through SIP enabled gateways) and native IP phones.
- When used for calls between two PSTN phones, the solution here may result in a portion of the call being routed over the Internet. In PINT, the call is always routed only over the PSTN. This may result in better-quality calls with the PINT solution, depending on the codec in use and the QOS capabilities of the network routing the Internet portion of the call.
- The PINT solution requires extensions to SIP (PINT is an extension to SIP), whereas the solution described here is done with baseline SIP.
- The PINT solution allows the controller (acting as a PINT client) to "step out" once the call is established. The solution described here requires the controller to maintain call state for the entire duration of the call.

1.3.7.2 Mid-Call Announcement Capability

The 3pcc mechanism described here can also be used to enable mid-call announcements. Consider a service for pre-paid calling cards. Once the pre-paid call is established, the system needs to set a timer to fire when they run out of minutes. We would like the user to hear an announcement that tells them to enter a credit card to continue when this timer fires. Once they enter the credit card info, more money is added to the pre-paid card, and the user is reconnected to the destination party.

We consider here the usage of 3pcc just for playing the mid-call dialog to collect the credit card information.

We assume the call is set up so that the controller is in the call as a B2BUA. We wish to connect the caller to a media server when the timer fires. The flow for this is shown in Figure 1.8d. When the timer expires, the controller places the called party with a connection address of 0.0.0.0 (F1). This effectively "disconnects" the called party. The controller then sends an INVITE without SDP to the pre-paid caller (F4). The offer returned from the caller (F5) is used in an INVITE to the media server, which will be collecting digits (F6). This is an instantiation of the call flows of Figure 1.8b. This flow can only be used here because the media server is an automaton and will answer the INVITE immediately. If the controller were connecting the pre-paid user with another end user, the call flows of Figure 1.8d would need to be used. The media server returns an immediate 200 OK (F7) with an answer, which is passed to the caller in an ACK (F8). The result is that the media server and the pre-paid caller have their media streams connected.

The media server plays an announcement and prompts the user to enter a credit card number. After collection, the card number is validated. The media server then passes the card number to the controller (using some means outside the scope of this specification) and then hangs up the call (F11). After hanging up with the media server, the controller reconnects the user to the original called party. To do this, the controller sends an INVITE without SDP to the called party (F13). The 200 OK (F14) contains an offer, offer3. The controller modifies the SDP as is done in Figure 1.8d and passes the offer in an INVITE to the pre-paid user (F15). The pre-paid user generates an answer in a 200 OK (F16), which the controller passes to user B in the ACK (F17). At this point, the caller and the called party are reconnected.

1.3.8 3pcc Implementation Recommendations

Most of the work involved in supporting 3pcc is within the controller. A standard SIP UA should be controllable using the mechanisms described here. However, 3pcc relies on a few features that might not be implemented. As such, we recommend that implementors of UA servers support the following, as described in RFC 3725:

- Offers and answers that contain a connection line with an address of 0.0.0.0.
- Re-INVITE requests that change the port to which media should be sent
- Re-INVITEs that change the connection address
- Re-INVITEs that add a media stream
- Re-INVITEs that remove a media stream (setting its port to zero)
- Re-INVITEs that add a codec among the set in a media stream
- SDP Connection address of zero
- Initial INVITE requests with a connection address of zero
- Initial INVITE requests with no SDP
- Initial INVITE requests with SDP but no media lines
- Re-INVITEs with no SDP
- The UPDATE method described in RFC 3311
- Reliability of provisional responses in RFC 3262
- Integration of resource management and SIP, described in RFC 3312

1.3.9 Security Considerations

1.3.9.1 Authorization and Authentication

In most uses of SIP INVITE, whether or not a call is accepted is based on a decision made by a human when presented with information about the call, such as the identity of the caller. In other cases, automata answer the calls, and whether or not they do so may depend on the particular application to which SIP is applied. For example, if a caller makes a SIP call to a voice portal service, the call may be rejected unless the caller has previously signed up (perhaps via a website). In other cases, call handling policies are based on automated scripts, such as those described by the Call Processing Language (CPL) in RFC 3880. Frequently, those decisions are also made based on the identity of the caller.

These authorization mechanisms would be applied to normal first-party calls and third-party calls, as these two are indistinguishable. As a result, it is important for these authorization policies to continue to operate correctly for third-party calls. Of course, third-party calls introduce a new party – the one initiating the third-party call. Do the authorization policies apply based on the identity of that third party, or do they apply based on the participants in the call? Ideally, the participants would be able to know the identities of both other parties, and authorization policies would be based on those, as appropriate. However, this is not possible using existing mechanisms. As a result, the next best thing is for the INVITE requests to contain the identity of the third party. Ultimately, this is the user who is requesting communication, and it makes sense for call authorization policies to be based on that identity.

This requires, in turn, that the controller authenticate itself as that third party. This can be challenging, and the appropriate mechanism depends on the specific application scenario. In one common scenario, the controller is acting on behalf of one of the participants in the call. A typical example is click-to-dial, where the controller and the customer service representative are run by the same administrative domain. Indeed, for the purposes of identification, the controller can legitimately claim to be the customer service representative. In this scenario, it would be appropriate for the INVITE to the end-user to contain a From field identifying the customer service representative and authenticate the request using S/MIME (see section 23 of RFC 3261 [1]) signed by the key of the customer service representative (which is held by the controller). This requires the controller to actually have credentials with which it can authenticate itself as the customer support representative.

In many other cases, the controller is representing one of the participants but does not possess their credentials. Unfortunately, there are currently no standardized mechanisms that allow a user to delegate credentials to the controller in a way that limits their usage to specific 3pcc operations. In the absence of such a mechanisms, the best that can be done is to use the display name in the From field to indicate the identity of the user on whose behalf the call is being made. It is recommended that the display name be set to "[controller] on behalf of [user]," where user and controller are textual identities of the user and controller, respectively. In this case, the uniform resource identifier (URI) in the From field would identify the controller.

In other situations, there is no real relationship between the controller and the participants in the call. In these situations, ideally, the controller would have a means to assert that the call is from a particular identity (which could be one of the participants, or even a third party, depending on the application) and to validate that assertion with a signature using the key of the controller. The security features described in RFCs 4474 and 4916 can be used for authenticated identity management and connected identity in 3pcc scenarios, respectively.

1.3.9.2 End-to-End Encryption and Integrity

With 3pcc, the controller is actually one of the participants as far as the SIP dialog is concerned. Therefore, encryption and integrity of the SIP messages, as provided by S/MIME, will occur between participants and the controller rather than directly between participants. However, the integrity, authenticity, and confidentiality of the media sessions can be provided through a controller. End-to-end media security is based on the exchange of keying material within SDP [2], described in RFC 4568.

The proper operation of these mechanisms with 3pcc depends on the controller behaving properly. As long as it is not attempting to explicitly disable these mechanisms, the protocols will properly operate between the participants, resulting in a secure media session that even the controller cannot eavesdrop or modify. Since 3pcc is based on a model of trust between the users and the controller, it is reasonable to assume it is operating in a well-behaved manner. However, there is no cryptographic means that can prevent the controller from interfering with the initial exchanges of keying materials. As a result, it is trivially possible for the controller to insert itself as an intermediary on the media exchange if it should so desire.

1.4 MULTISTREAM IMMERSIVE TELEPRESENCE

Telepresence is conceived of to let the conference participants have the sensation of being in a shared environment, so that they all perceive that they are immersed in a face-to-face meeting in the same place instead of their true locations (RFC 7205). Telepresence requires a set of special technologies allowing real-time communications by all remote and local participants to be immersed in a properly equipped conference room providing them with multi-sensorial stimuli replicating in space and time. The equipment within the specially designed conference room of all sites must be able to capture and send all the information needed to allow a high-fidelity reproduction in all sites. The conference participants' movements, voice, appearance, eye contacts, three-dimensional (3-D) audio and video, life-size images with multiple video and audio streams using multiple cameras and microphones/loudspeakers, background sounds, and other features need to be reproduced in remote locations.

Therefore, telepresence goes far beyond what multipoint conferencing does for creating an immersive experience among the participants. Indeed, telepresence rooms are equipped with several capture devices wisely distributed to best capture the local scene from multiple viewpoints. Each of these streams should be sent to the remote site in order to make it able to choose the set of streams that best fits its rendering capabilities (that is, the displays and loudspeakers installed at the remote location). In this context, a new kind of session description feature needs to be accommodated in addition to what SDP protocol does. However, SIP can still be used as the call control protocol for session setup and termination. We have just introduced the telepresence system in this section, while details are provided in Chapter 16.

1.4.1 CONFERENCE ROOM

The specially designed conference room is the heart of the telepresence system. The room is equipped with several capture devices distributed within the room in such a way as to best capture the local scene within the room from multiple viewpoints. The audio, video, and/or texts within a capture, along with the spatial distribution of streams, are then converted into electrical signals, usually fed into an media encoder. Each of these streams should be sent to the remote site in order to make it able to choose the set of streams that best fits its rendering capabilities, that is, in accordance with capabilities of the displays and loudspeakers installed in the remote conference room site. First, the remote sites are informed about the contents of all received media streams. Second, the remote sites will then select the most suitable combination they wish to be provided with, including the original spatial distribution of the streams. Finally, only the selected streams including their spatial distribution need to be transferred to the remote site to better duplicate the source environment.

1.4.2 MEDIA CAPTURE

Media capture is the fundamental representation of a source media stream transmitted from one or more capture devices, where media can be any data, including audio, video, and/or timed text, conveyed over the RTP output of either a physical source (for example, a camera or microphone) or a synthetic source (for example, a laptop, computer, or DVD player), or also concepts such as "the loudest speaker stream" or constructed from other media streams.

1.4.3 Encoding Group

A set of encoding parameters represents a total media encoding capability to be sub-divided across potentially multiple individual encodings. Groups of encodings will share a certain maximum bandwidth. Each media capture is indeed associated with an encoding group, meaning that it can be encoded with the encodings included in the group. Media captures sharing the same group must respect the maximum bandwidth constraints when encoded.

1.4.4 Capture Scene

This is a structure representing a spatial region captured by one or more capture devices, each capturing media representing a portion of the region. The spatial region represented by a capture scene may correspond to a real region in physical space, such as a room. A capture scene includes attributes and one or more capture scene views, with each view including one or more media captures.

1.4.5 Simultaneous Transmission Sets

A simultaneous transmission set is a set of media captures that can be transmitted simultaneously from a source transmitter.

The telepresence system that creates the immersive sense, as if all remote and local conference participants were in a face-to-face meeting, is primarily based on the good design of the conference room equipped with 3-D audio and video capture devices, far beyond the traditional multipoint conferencing as discussed in Section 1.1. It should be noted that the basic multipoint conferencing architecture described in Section 1.1 is still applicable to telepresence systems, but the telepresence architecture is built on top of the multipoint conference architecture that uses SIP/SDP. That is why IETF's ControLling mUltiple streams for tElepresence (CLUE) working group has defined a new session description protocol, termed a CLUE protocol, on top of SDP, because the present SDP and its extensions are not capable of capturing the capabilities of CLUE. In Chapter 16, we describe the CLUE immersive telepresence protocol using SIP/(SDP + CLUE) protocol architecture.

The key criterion for scalability of multipoint conferencing is the distribution in each functional entity of different kinds of MSs, ASs, and MCU, and finally, the controller itself. However, it is still called centralized multipoint conferencing because a central controller controls the multipoint conferencing. We can term this architecture "logically" centralized but physically distributive. In the following sections, we start with 3pcc, where a single centralized controller controls the multipoint conference, acting as the black box for providing all services: call control, different application services, media services, and media bridging services. The 3pcc architecture is not scalable for the large-scale multipoint conference architecture. We then expand step by step how the scalable multipoint conference architecture can be built, solving many of those problems, if not all, using an IETF XCON WG multipoint conference.

1.5 SCALABLE CENTRALIZED MULTIPOINT MULTIMEDIA CONFERENCING SYSTEMS AND PROTOCOLS FRAMEWORK

1.5.1 Scalability

The multipoint multimedia conferencing architecture is hugely complex because it is dealing with a large set of RT, near-RT, and non-RT feature-rich applications. A huge variety of conferencing service features are needed to satisfy the conference participants' needs. In fact, the discussion of all conferencing features and the development of scalable architecture is at the heart of the subsequent sections. For example, Figure 1.5 (see Section 1.2.3.7) has shown three additional functional entities, MS, AS, and MCU, which are distinctly separate physical servers due to their completely different types of capabilities. This separation of the functional entities are only a first step for the achievement of scalability for the multipoint conferencing architecture. Interestingly, in earlier days in 3pcc centralized conferencing systems (see Section 1.3), those functional entities are not addressed at all, as if all of them were put into a proprietary black box.

If we consider the large-scale global multipoint conferencing architecture, the architecture described in Section 1.2 or Section 1.3 is not good enough to provide scalability unless each of those MS, AS, and MCU servers is further distributed. That is, there need to be multiple MSs, multiple ASs and multiple MCUs to make the large-scale conference architecture scalable. Furthermore, as described earlier, there will be many different types of ASs, MSs, and MCUs. Again, there have to be multiple physical servers of each kind of AS, MS, and MCU server, and each of those needs to be distributed across the network. More importantly, standard protocols need to be used between different kinds of MS, AS, and MCU servers for interoperability.

1.5.2 Conferencing Systems and Protocols Framework

A centralized conference is an association of endpoints, called conference participants, with a central endpoint, called a conference focus. The focus has direct peer relationships with the participants by maintaining a separate call signaling interface with each. Consequently, in this centralized conferencing model, the call signaling graph is always a star. The most basic conference supported in this model would be an ad hoc, unmanaged conference, which would not necessarily require any of the functionality defined within this framework. For example, it could be supported using basic SIP signaling functionality with a participant serving as the focus; the SIP Conferencing Framework (RFC 4353) together with the SIP Call Control Conferencing for User Agents (RFC 4579) specifications address these types of scenarios.

In addition to the basic features, however, a conferencing system supporting the centralized conferencing model proposed in this framework specification can offer richer functionality by including dedicated conferencing applications with explicitly defined capabilities and reserved recurring conferences, along with providing the standard protocols for managing and controlling the different attributes of these conferences. The core requirements for centralized conferencing are outlined in RFC 4245. These requirements are applicable for conferencing systems using various call signaling protocols, including SIP. Additional conferencing requirements are provided in RFC 4376 and RFC 4597.

The centralized conferencing system framework described here is built around a fundamental concept of a conference object (RFC 5239 – see Chapter 2). A conference object provides the data representation of a conference during each of the various stages of a conference, such as creation, reservation, active, completed, and others. A conference object is accessed via the logical functional elements, with which a conferencing client interfaces, using the various protocols identified in Figure 1.9. In fact, Figure 1.9 provides a high-level representation of the detail standardized functional entities and protocols that has been conceived in the XCON conferencing framework (RFC 5239 – see Section 2).

The functional elements defined for a conferencing system consist of a conference control server, a floor control server, any number of foci, and a notification service. The centralized conference manipulation protocol (CCMP) defined in RFC 6503 (see Chapter 5) provides the interface between a conference and a media control client and the conference control server. The floor control protocol, for example, binary floor control protocol (BFCP) described in RFC 4582 (see Chapter 6), provides the interface between a floor control client and the floor control server. A call signaling protocol such as SIP provides the interface between a call signaling client and a focus. The notification protocol, for example, SIP Subscribe/Notify, defined in RFCs 3265, 4575 and 6502 (see Chapter 7), provides the interface between the conferencing client and the notification service.

The conferencing system can support a subset of the conferencing functions depicted in the conferencing system logical decomposition in Figure 1.9 and described in subsequent sections. However, there are some essential components that would typically be used by most other advanced

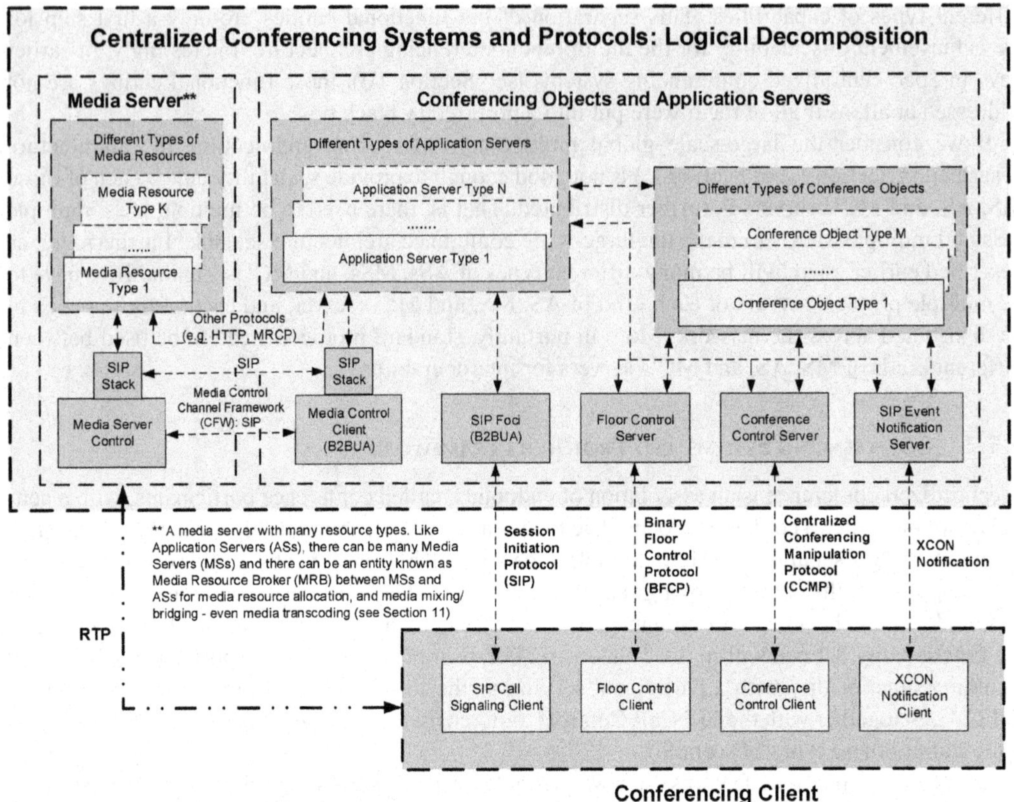

FIGURE 1.9 Centralized Conferencing System Logical Functional Entities (Systems) and Protocols.

functions, such as the notification service. For example, the notification service is used to correlate information, such as the list of participants with their media streams, between the various other components.

Multipoint multimedia conferencing can be a highly complex composite application that may consist of many sophisticated diverse audio, video, and/or data applications during a given session. Each of these audio, video, or data applications may use its own application protocol(s) within the framework of a given conference. Even many sub-conferences, known as sidebars, may take place under a given conference. In addition, audio, video, and/or data media of many different conference participants may also be bridged and then need to be sent to multiple parties. Figure 1.9 depicts the logical functional entities for the centralized conferencing system and protocols described in RFCs 5239 (see Chapter 2), 5567 (see Chapter 3), and 6230 (see Chapter 8).

The media graph that represents the signaling messaging and audio/video/data flows of a conference can be centralized, decentralized, or any combination of both and potentially can differ per media type. In the centralized case, the media sessions are established between a media mixer (or bridge) controlled by the focus and each one of the participants. In the decentralized (that is, distributed) case, the media graph is a multicast or multi-unicast mesh among the participants. Consequently, the media processing (e.g., mixing) can be controlled either by the focus alone or by the participants using a media channel control framework using SIP as specified in RFC 6230 (see Sections 8), and we term this CFW. In Figure 1.9, the media control client (MCC) that acts as the SIP B2BUA is used for communications between the SIP foci and the media server control (MSC), and the implementation of the CFW can be very flexible, either in the conference application server containing the SIP foci or in the media server. However, we are describing a conferencing architecture that clearly maps to a centralized media model.

Figure 1.9 depicts only the logical view of the conferencing object and application servers as well the media servers. However, in practice, there can be many application and media servers controlling many different application and media resources distributed across different geographical areas for a large-scale conferencing system. Although all these servers are geographically distributed, it is assumed that the respective application and media servers are logically centralized, serving the same conferencing system. In the same vein, there can be many foci located in the same or many different geographical locations acting as a centralized logical entity controlling the same or different conferences. These kinds of multipoint conferencing architectures that create decentralized signaling and media cases offer opportunities for building more scalable large-scale conferencing systems. For example, we can consider that each of the foci is geographically distributed but logically centralized. Issues will include: What should the communications protocol be between the foci? What would the communications architecture be: peer-to-peer (P2P) or client-server (C/S) protocol? Which of the foci will be controlling the conference call: source focus or destination focus? A lot of possibilities are available in designing this kind of large-scale multipoint conference call. For example, we can choose a given P2P protocol, like P2P-SIP, for the inter-foci protocol or develop new ones. We have not addressed this in this book, as no standardization efforts have yet emerged.

It should be noted that we have discussed the multistream immersive telepresence conference system. The detail of the architecture of the telepresence system is described in Chapter 16. The telepresence architecture is a super set of the traditional multipoint conference architecture as depicted in Figure 1.9. For example, the specially designed telepresence conference room will have high-quality multiple 3-D video cameras, multiple loudspeakers and microphones, special effects of background images, spatial relationships between multiple video and audio streams, and many other complexities. The basic rules and principles of conferencing architecture that are shown in Figure 1.9 are still applicable but with many more complexities that are beyond the traditional scope of multipoint conferencing.

1.5.3 Organization of the Book

In fact, the overall unified view of the architectural framework of centralized conferencing systems and protocols provided in Figure 1.9 is described in different chapters of this book as follows:

- Chapter 2 Centralized Conferencing System Architecture
- Chapter 3 Media Server Control Architecture
- Chapter 4 Conference Information Data Model for Centralized Conferencing
- Chapter 5 Centralized Conference Manipulation Protocol
- Chapter 6 Binary Floor Control Protocol
- Chapter 7 XCON Notification Service
- Chapter 8 Media Channel Control Framework
- Chapter 9 Mixer Package for Media Control Channel Framework
- Chapter 10 Media Session Recording
- Chapter 11 Media Resource Brokering
- Chapter 12 Media Control Channel Framework for Interactive Voice Response
- Chapter 13 SIP Interface to VoiceXML Media Services
- Chapter 14 Media Resource Control Protocol Version 2
- Chapter 15 Media Control Channel Framework (CFW) Call Flow Examples
- Chapter 16 Multistream Immersive Telepresence Conferencing Systems

This is the first book that has ever been published to capture a unified view of the centralized multipoint multimedia conferencing architecture as described earlier, relating each of the standardized systems (or functional entities) and protocols. The key is that call control signal protocols for multimedia conferencing can use this unified architecture, although SIP has been used as an example in the IETF. Over more than two decades, many researchers, educationists, scientists, product vendors/manufacturers, and others have published numerous papers on multimedia conferencing. Numerous proposals had been submitted to the IETF for standardization of multipoint multimedia conferencing, the ultimate mother of the hugely complex application. It turns out that SIP/SDP has the protocol architectural limitations that cannot be used for multipoint multimedia conferencing dynamically evolving even from point-to-point conference call or not to speak about for creation of distributed multimedia conferencing.

Re-architecting the SIP/SDP protocol, including media bridging for creation of distributed multipoint conferencing, is not worthwhile considering the entrenched investment of the present SIP/SDP in the industry and lack of market demands for distributed multipoint conferencing. At this moment, centralized multipoint conferencing is good enough to meet the market demands.

Therefore, the creation of standards for multipoint multimedia conferencing, even for centralized conferencing with a known focus (i.e., location of the conferencing point), took a long time in the IETF. In the meantime, different product vendors have built their own proprietary conferencing systems and have created silos of non-interoperable conferencing networks. The multipoint conferencing architecture described in this book will enable industry and research to implement the conferencing architecture using systems and protocols based on standards for fostering interoperability in multi-vendor conferencing networking environments and thereby offering economies of scale. Now, we describe each of those standardized systems and protocols for multipoint conferencing in the subsequent chapters.

1.6 TERMINOLOGY

The terminology and functional entities of centralized conferencing have been defined by RFCs 4353, 5239, 5567, and 6230. Table 1.2 shows the definitions of all entities/terminology.

1.7 SUMMARY

We have provided a brief overview of networked multipoint multimedia conferencing and immersive telepresence. The basic functional and performance characteristics of networked multipoint multimedia conferencing are defined. The roles of the SIP for call control and SDP for session negotiations are described. The relationship between the application layer SIP/SDP protocol and other protocol layers (transport, network, link/medium access control, and physical) along with other application layer protocols is explained. The basic difference between point-to-point and multipoint conferencing is described. We have described at great length how the media (audio, video, text/timed text/still images/graphics) bridging and centralized controller functions make multipoint conferencing fundamentally different from point-to-point conferencing.

In addition, media transcoding is described for providing interoperability between dissimilar codecs. In this context, the different variety of media services provided by media servers and application sharing services provided by application servers makes multipoint multimedia conferencing hugely complex. We have briefly described the scalability of the highly complicated multipoint multimedia conferencing, distributing each kind of physical servers across the large-scale network while still keeping the conferencing architecture logically centralized.

The multiparty 3pcc conference architecture is described using SIP/SDP per RFC 3725, where the controller sets up the multipart call in a star-like point-to-point centralized conferencing topology under many assumptions. Even then, the multiparty conferencing call establishment may fail under many circumstances. We have explained each of those error cases, which need to be taken care of. The simple SDP media description architecture, which has been very attractive for the initial, less media-rich point-to-point call, has posed a huge challenge for multiparty multimedia conferencing.

TABLE 1.2
Centralized Conferencing Terminologies

Terminology	Definition
aai	Used to specify a "JSON value" (RFC 4627) that is mapped to the session.connection.aai VoiceXML session variable – see Section 13.2.4. (RFC 5552: Standards Track)
Active Conference	The term "active conference" refers to a conference object that has been created and activated via the allocation of its identifiers (e.g., conference object identifier and conference identifier) and the associated focus. An active conference is created based on either a system default conference blueprint or a specific conference reservation. (RFC 5239: Standards Track)
Advertisement	Advertisement is a CLUE message a media provider sends to a media consumer describing specific aspects of the content of the media and any restrictions it has in terms of being able to provide certain Streams simultaneously. (See Section 16.3 – IETF Draft: Telepresence)
Application Server (AS)	An SIP (RFC 3261– also see Section 1.2) application server (AS) hosts and executes services such as interactive media and conferencing in an operator's network. An AS influences and impacts the SIP session, in particular by terminating SIP sessions on a media server, which is under its control. (RFC 6231: Standards Track)An entity that requests media processing and manipulation from a media server; typical examples are back-to-back user agents (B2BUAs) and endpoints requesting manipulation of a third party's media stream. (RFC 7058: Informational)A functional entity that hosts one or more instances of a communication application. The application server may include the conference policy server, the focus, and the conference notification server, as defined in RFC 4353. Also, it may include communication applications that use IVR or announcement services. (RFC 5567 – see Section 3: Informational)
Attributes	Attributes represent the elements listed in each of the classes. The attributes of a class are listed in the second compartment below the class name. Each instance of a class conveys values for the attributes of that class. These values are added to the recording's metadata. (RFC 7865: Standards Track)
Audio Capture	Media Capture for audio. Denoted as ACn in the examples in this specification. (See Section 16.3 – IETF Draft: Telepresence)
Back-to-Back User Agent (B2BUA)	A B2BUA is a back-to-back SIP user agent (RFC 3261 – also see Section 1.2). (RFC 6230: Standards Track)
Call Signaling Protocol	The call signaling protocol is used between a participant and a focus. In this context, the term "call" means a channel or session used for media streams. (RFC 5239 – see Section 2: Standards Track)
Capture	Same as Media Capture. (See Section 16.3 – IETF Draft: Telepresence)
Capture Device	A device that converts physical input, such as audio, video, or text, into an electrical signal, in most cases to be fed into a media encoder. (See Section 16.3 – IETF Draft: Telepresence)
Capture Encoding	A specific encoding of a media capture, to be sent by a media provider to a media consumer via RTP. (See Section 16.3 – IETF Draft: Telepresence.)Capture encoding is a specific encoding of a media capture, to be sent via RTP (RFC 3550). CLUE participant (CP): An entity able to use the CLUE protocol within a telepresence session. It can be an endpoint or an MCU able to use the CLUE protocol. (See Section 16.5 – IETF Draft: Telepresence)

(Continued)

TABLE 1.2 (CONTINUED)
Centralized Conferencing Terminologies

Terminology	Definition
Capture Scene	Capture Scene is a structure representing a spatial region captured by one or more Capture Devices, each capturing media representing a portion of the region. The spatial region represented by a Capture Scene may correspond to a real region in physical space, such as a room. A Capture Scene includes attributes and one or more Capture Scene Views, with each view including one or more Media Captures. (See Section 16.3 – IETF Draft: Telepresence)A structure representing a spatial region captured by one or more Capture Devices, each capturing media representing a portion of the region. The spatial region represented by a Capture Scene MAY correspond to a real region in physical space, such as a room. A Capture Scene includes attributes and one or more Capture Scene Views, with each view including one or more Media Captures. (See Section 16.4 – IETF Draft: Telepresence)
Capture Scene View (CSV)	Capture Scene View (CSV) is a list of Media Captures of the same media type that together form one way to represent the entire Capture Scene. (See Section 16.3 – IETF Draft: Telepresence)A list of Media Captures of the same media type that together form one way to represent the entire Capture Scene. CLUE Participant: This term is imported from the CLUE protocol (see Section 16.5). (See Section 16.4 – IETF Draft: Telepresence)
ccxml	Used to specify a "JSON value" (RFC 4627) that is mapped to the session.connection. ccxml VoiceXML session variable – see Section 13.2.4. (RFC 5552: Standards Track)
Centralized Conference Manipulation Protocol (CCMP)	A centralized conference manipulation protocol provides the interface for data manipulation and state retrieval for the centralized conferencing data, represented by the conference object. (RFC 5239 – see Section 2: Standards Track)
Client	A floor participant or a floor chair that communicates with a floor control server using BFCP. (RFC 4582: Standards Track)
CLUE-capable device	A device that supports the CLUE data channel (see Section 16.6), the CLUE protocol (see Section 16.5), and the principles of CLUE negotiation and seeks CLUE-enabled calls. (See Section 16.3 – IETF Draft: Telepresence)A CLUE-capable device is a device that supports the CLUE data channel (see Section 16.6), the CLUE protocol, and the principles of CLUE negotiation and seeks CLUE-enabled calls. (see Section 16.5 – IETF Draft: Telepresence)
CLUE-controlled media	A media "m=" line that is under CLUE control; the Capture Source that provides the media on this "m=" line is negotiated in CLUE. See Section 16.8.4 for details of how this control is signaled in SDP. There is a corresponding "non-CLUE-controlled" media term. (See Section 16.8 – IETF Draft: Telepresence)
CLUE-enabled call	A call in which two CLUE-capable devices have successfully negotiated support for a CLUE data channel in SDP (RFC 4566). A CLUE-enabled call is not necessarily immediately able to send CLUE-controlled media; negotiation of the data channel and of the CLUE protocol must complete first. Calls between two CLUE-capable devices that have not yet successfully completed negotiation of support for the CLUE data channel in SDP are not considered CLUE-enabled. (See Section 16.3 – IETF Draft: Telepresence)
Communication Session (CS)	A session created between two or more SIP User Agents (UAs) that is the subject of recording. (RFC 6341: Informational)
Conference	It is used as defined in "A Framework for Conferencing within the Session Initiation Protocol (SIP)" (RFC 4353). (RFC 7262 – Informational)Conference is used as defined in RFC 4353, A Framework for Conferencing within the Session Initiation Protocol (SIP). (See Section 16.3 – IETF Draft: Telepresence)

(Continued)

TABLE 1.2 (CONTINUED)
Centralized Conferencing Terminologies

Terminology	Definition
Conference and Media Control Client (CMCC)	The CMCC is logically equivalent to the use of a User Agent Client (UAC) as the client notation in the media control call flows. A CMCC differs from a generic media client in being an XCON-aware entity and thus, also being able to issue CCMP requests. (RFC 5239 – see Section 2: Standards Track)
Conference Blueprint	A conference blueprint is a static conference object within a conferencing system, which describes a typical conference setting supported by the system. A conference blueprint is the basis for creation of dynamic conference objects. A system may maintain multiple blueprints. Each blueprint is comprised of the initial values and ranges for the elements in the object, conformant to the data schemas for the conference information. (RFC 5239 – see Section 2: Standards Track)
Conference Factory	A conference factory is a logical entity that generates unique URI(s) to identify and represent a conference focus.
Conference Identifier (ID)	A conference identifier is a call signaling protocol–specific URI that identifies a conference focus and its associated conference instance. (RFC 5239 – see Section 2: Standards Track)
Conference Information	The conference information includes definitions for basic conference features, such as conference identifiers, membership, signaling, capabilities, and media types applicable to a wide range of conferencing applications. The conference information also includes the media and application-specific data for enhanced conferencing features or capabilities, such as media mixers. The conference information is the data type (i.e., the XML schema) for a conference object. (RFC 5239 – see Section 2: Standards Track)
Conference Instance	A conference instance refers to an internal implementation of a specific conference, represented as a set of logical conference objects and associated identifiers. (RFC 5239 – see Section 2: Standards Track)
Conference Object	A conference object represents a conference at a certain stage (e.g., description upon conference creation, reservation, activation, etc.), which a conferencing system maintains in order to describe the system capabilities and to provide access to the services available for each object independently. The conference object schema is based on the conference information. (RFC 5239 – see Section 2: Standards Track)
Conference Object Identifier (ID)	A conference object identifier is a URI that uniquely identifies a conference object and is used by a conference control protocol to access and modify the conference information. (RFC 5239 – see Section 2: Standards Track)
Conference Policies	Conference policies collectively refer to a set of rights, permissions, and limitations pertaining to operations being performed on a certain conference object. (RFC 5239 – see Section 2: Standards Track)
Conference Reservation	A conference reservation is a conference object, which is created from either a system default or a client-selected blueprint. (RFC 5239 – see Section 2: Standards Track)
Conference Server (ConfS)	The term "conference server" is used interchangeably with the term "Application Server (AS)" as used in the media control architectural framework in RFC 5567. A conference server is intended to be able to act as a conference control server as defined in the XCON framework; that is, it is able to handle CCMP requests and issue CCMP responses. (RFC 5239 – see Section 2: Standards Track)
Conference State	The conference state reflects the state of a conference instance and is represented using a specific, well-defined schema. (RFC 5239 – see Section 2: Standards Track)
Conference User Identifier (ID)	A unique identifier for a user within the scope of a conferencing system. A user may have multiple conference user identifiers within a conferencing system (e.g., to represent different roles).

(*Continued*)

TABLE 1.2 (CONTINUED)
Centralized Conferencing Terminologies

Terminology	Definition
Conferencing System	Conferencing system refers to a conferencing solution based on the data model discussed in this framework specification and built using the protocol specifications referenced in this framework specification. (RFC 5239 – see Section 2: Standards Track)
Configure Message	A CLUE message a Media Consumer sends to a Media Provider specifying which content and Media Streams it wants to receive, based on the information in a corresponding Advertisement message. (See Section 16.3 – IETF Draft: Telepresence)
Connection Oriented Media (COMEDIA)	Connection-oriented media (i.e., TCP and Transport Layer Security [TLS]). Also used to signify the support in SDP for connection-oriented media and the RFCs that define that support (RFC 4145 and RFC 4572). (RFC 7058: Informational)
Connection-oriented media (COMEDIA)	COMEDIA – Connection-oriented media (i.e., TCP and Transport Layer Security [TLS]). Also used to signify the support in SDP for connection-oriented media and the RFCs that define that support (RFC 4145 and RFC 4572).
Consumer	Consumer is short for Media Consumer. (See Section 16.3 – IETF Draft: Telepresence)
Control Channel	A Control Channel is a reliable connection between a Client and a Server that is used to exchange Framework messages. The term "Connection" is used synonymously within this specification. (RFC 6230 – see Section 8: Standards Track)A reliable connection between an Application Server and a Media Server that is used to exchange framework messages. (RFC 7058: Informational)
Control Client	A Control Client is an entity that requests processing from a Control Server. Note that the Control Client might not have any processing capabilities whatsoever. For example, the Control Client may be an application server (B2BUA) or other endpoint requesting manipulation of a third party's media stream that terminates on a media server acting in the role of a Control Server. In this specification, we often refer to the Control Client simply as "the Client." (RFC 6230 – see Section 8: Standards Track)
Control Command	A Control Command is an application-level request from a Client to a Server. Control Commands are carried in the body of CONTROL messages. Control Commands are defined in separate specifications known as "Control Packages." (RFC 6230 – see Section 8: Standards Track)
Control Server	A Control Server is an entity that performs a service, such as media processing, on behalf of a Control Client. For example, a media server offers mixing, announcement, tone detection and generation, and playing and recording of services. The Control Server has a direct Real-Time Transport Protocol (RTP), specified in RFC 3550, relationship with the source or sink of the media flow. In this specification, we often refer to the Control Server simply as "the Server." (RFC 6230 – see Section 8: Standards Track)
Dialog	A dialog performs media interaction with a user following the concept of an IVR (Interactive Voice Response) dialog (this sense of "dialog" is completely unrelated to a SIP dialog). A dialog is specified as inline XML or via a URI reference to an external dialog specification. Traditional IVR dialogs typically feature capabilities such as playing audio prompts, collecting DTMF input, and recording audio input from the user. More inclusive definitions include support for other media types, runtime controls, synthesized speech, recording and playback of video, recognition of spoken input, and mixed initiative conversations. (RFC 6231: Standards Track)
DTMF	Dual-Tone Multi-Frequency; a method of transmitting key presses in-band, either as actual tones or as named tone events (RFC 4733). (RFC 6787: Standards Track)
Encoding	Encoding is short for Individual Encoding. (See Section 16.3 – IETF Draft: Telepresence)

(*Continued*)

TABLE 1.2 (CONTINUED)
Centralized Conferencing Terminologies

Terminology	Definition
Encoding Group	A set of encoding parameters representing a total media encoding capability to be sub-divided across potentially multiple Individual Encodings. (See Section 16.3 – IETF Draft: Telepresence)A set of encoding parameters representing a total media encoding capability to be sub-divided across potentially multiple Individual Encodings. Endpoint: A CLUE-capable device which is the logical point of final termination through receiving, decoding and rendering, and/or initiation through capturing, encoding, and sending of media streams. An endpoint consists of one or more physical devices that source and sink media streams, and exactly one RFC 4553 Participant (which in turn, includes exactly one SIP User Agent). Endpoints can be anything from multiscreen/multicamera rooms to handheld devices. (See Section 16.4 – IETF Draft: Telepresence)
Encoding or Individual Encoding	A set of parameters representing a way to encode a Media Capture to become a Capture Encoding.
Endpoint	The logical point of final termination through receiving, decoding and rendering, and/or initiation through capturing, encoding, and sending of media streams. An endpoint consists of one or more physical devices that source and sink media streams, and exactly one participant (RFC 4353) (which in turn, includes exactly one SIP user agent). In contrast to an endpoint, an MCU may also send and receive media streams, but it is not the initiator or the final terminator in the sense that media is captured or rendered. Endpoints can be anything from multiscreen/multicamera rooms to handheld devices. Endpoint characteristics include placement of capture and rendering devices, capture/render angle, resolution of cameras and screens, spatial location, and mixing parameters of microphones. Endpoint characteristics are not specific to individual media streams sent by the endpoint. (RFC 7262 – Informational)
Endpoint	A CLUE-capable device that is the logical point of final termination through receiving, decoding and rendering, and/or initiation through capturing, encoding, and sending of media streams. An endpoint consists of one or more physical devices that source and sink media streams, and exactly one (RFC 4353) Participant (which in turn, includes exactly one SIP User Agent). Endpoints can be anything from multiscreen/multicamera rooms to handheld devices. (See Section 16.3 – IETF Draft: Telepresence)An endpoint is a CLUE-capable device which is the logical point of final termination through receiving, decoding and rendering, and/or initiation through capturing, encoding, and sending of media streams. An endpoint consists of one or more physical devices that source and sink media streams, and exactly one (RFC 4353) Participant (which in turn, includes exactly one SIP User Agent). Endpoints can be anything from multiscreen/multicamera rooms to handheld devices. (See Section 16.5 – IETF Draft: Telepresence)
Endpointing	The process of automatically detecting the beginning and end of speech in an audio stream. This is critical both for speech recognition and for automated recording such as one would find in voice mail systems. (RFC 6787: Standards Track)
Event Package	An event package is an additional specification that defines a set of state information to be reported by a notifier to a subscriber. Event packages also define further syntax and semantics that are based on the framework defined by this specification and are required to convey such state information. (RFC 4582 – see Section 6: Standards Track)
Event Template-Package	An event template-package is a special kind of event package that defines a set of states that may be applied to all possible event packages, including itself. (RFC 4582 – see Section 6: Standards Track)
First-Party Request	A request issued by the client to manipulate its own conferencing data. (RFC 6501 see Section 4: Standards Track)

(*Continued*)

TABLE 1.2 (CONTINUED)
Centralized Conferencing Terminologies

Terminology	Definition
Floor	Floor refers to a set of data or resources associated with a conference instance, for which a conference participant, or group of participants, is granted temporary access. (RFC 5239 – see Section 2: Standards Track)A temporary permission to access or manipulate a specific shared resource or set of resources. (RFC 4582 – see Section 6: Standards Track)
Floor Chair	A floor chair is a floor control protocol–compliant client, either a human participant or an automated entity, who is authorized to manage access to one floor and can grant, deny, or revoke access. The floor chair does not have to be a participant in the conference instance. (RFC 5239: Standards Track)A logical entity that manages one floor (grants, denies, or revokes a floor). An entity that assumes the logical role of a floor chair for a given transaction may assume a different role (e.g., floor participant) for a different transaction. The roles of floor chair and floor participant are defined on a transaction-by-transaction basis. BFCP transactions are defined in Section 8. Floor Control: A mechanism that enables applications or users to gain safe and mutually exclusive or non-exclusive input access to the shared object or resource. (RFC 4582: Standards Track)
Floor Control	A mechanism that enables applications or users to gain safe and mutually exclusive or non-exclusive input access to the shared object or resource. (RFC 4582: Standards Track)
Floor Control Server	A logical entity that maintains the state of the floor(s), including which floors exist, who the floor chairs are, who holds a floor, etc. Requests to manipulate a floor are directed at the floor control server. The floor control server of a conference may perform other logical roles (e.g., floor participant) in another conference. (RFC 4582: Standards Track)
Floor Participant	A logical entity that requests floors, and possibly information about them, from a floor control server. An entity that assumes the logical role of a floor participant for a given transaction may assume a different role (e.g., a floor chair) for a different transaction. The roles of floor participant and floor chair are defined on a transaction-by-transaction basis. BFCP transactions are defined in Chapter 8. In floor-controlled conferences, a given floor participant is typically co-located with a media participant, but it does not need to be. Third-party floor requests consist of having a floor participant request a floor for a media participant when they are not co-located. (RFC 4582: Standards Track)
Focus	A focus is a logical entity that maintains the call signaling interface with each participating client and the conference object representing the active state. As such, the focus acts as an endpoint for each of the supported signaling protocols and is responsible for all primary conference membership operations (e.g., join, leave, or update the conference instance) and for media negotiation/maintenance between a conference participant and the focus. (RFC 5239: Standards Track)
Framework Message	A Framework Message is a message on a Control Channel that has a type corresponding to one of the Methods defined in this specification. A Framework Message is often referred to by its method, such as a "CONTROL message."(RFC 6230: Standards Track)
Framework Transaction	A Framework Transaction is defined as a sequence composed of a Control Framework message originated by either a Control Client or a Control Server and responded to with a Control Framework response code message. Note that the Control Framework has no "provisional" responses. A Control Framework transaction is referenced throughout the specification as a "Transaction-Timeout." (RFC 6230: Standards Track)
Global View	A set of references to one or more Capture Scene Views of the same media type that are defined within Scenes of the same advertisement. A Global View is a suggestion from the Provider to the Consumer for one set of CSVs that provide a useful representation of all the scenes in the advertisement. (See Section 16.3 – IETF Draft: Telepresence)

(Continued)

TABLE 1.2 (CONTINUED)
Centralized Conferencing Terminologies

Terminology	Definition
Global View List	A list of Global Views included in an Advertisement. A Global View List may include Global Views of different media types. (see Section 16.3 – IETF Draft: Telepresence)
Hotword Mode	A mode of speech recognition where a stream of utterances is evaluated for match against a small set of command words. This is generally employed either to trigger some action or to control the subsequent grammar to be used for further recognition. (RFC 6787: Standards Track)
Individual Encoding	Individual Encoding is a set of parameters representing a way to encode a Media Capture to become a Capture Encoding. (See Section 16.3 – IETF Draft: Telepresence)
In-line MRB (Media Resource Broker)	An instantiation of an MRB that directly receives requests on the signaling path. There is no separate query. (RFC 6917: Standards Track)
Layout	How rendered media streams are spatially arranged with respect to each other on a telepresence endpoint with a single screen and a single loudspeaker, and how rendered media streams are arranged with respect to each other on a telepresence endpoint with multiple screens or loudspeakers. Note that audio as well as video are encompassed by the term layout – in other words, included is the placement of audio streams on loudspeakers as well as video streams on video screens. (RFC 7262 – Informational)
Linkages	Linkages represent the relationship between the classes in the model. Each linkage represents a logical connection between classes (or objects) in class diagrams (or object diagrams). The linkages used in the metadata model of this specification are associations. This specification also refers to the terminology defined in RFC 6341 (see Section 10.1). (RFC 7865: Standards Track)
Local	Sender and/or receiver physically co-located ("local") in the context of the discussion. (RFC 7262 – Informational)
maxstale	Used to set the max-stale value of the Cache-Control header in conjunction with VoiceXML specifications fetched using HTTP, as per RFC 2616. If omitted, the VoiceXML Media Server will use a default value. (RFC 5552: Standards Track)
Media	Any data that, after suitable encoding, can be conveyed over RTP, including audio, video, or timed text. (RFC 7262 – Informational)
Media Capture (MC)	A source of Media, such as from one or more Capture Devices or constructed from other Media streams. (See Section 16.4 – IETF Draft: Telepresence)Media Capture is a source of Media, such as from one or more Capture Devices or constructed from other Media streams. (See Section 16.5 – IETF Draft: Telepresence)
Media Consumer	Media Consumer is a CLUE-capable device that intends to receive Capture Encodings. (See Section 16.3 – IETF Draft: Telepresence)A CLUE-capable device that intends to receive Capture Encodings. (See Section 16.4 – IETF Draft: Telepresence)
Media Control Channel Framework (CFW)	Media Control Channel Framework, as specified in RFC 6230 (see Chapter 8). (RFC 6917: Standards Track)
Media Control Unit (MCU)	MCU is a CLUE-capable device that connects two or more endpoints together into one single multimedia conference (RFC 5117). An MCU includes an RFC 4353-like Mixer, without the RFC 4353 requirement to send media to each participant (See Section 16.5 – IETF Draft: Telepresence)
Media Functions	Functions available on a Media Server that are used to supply media services to the AS. Some examples are Dual-Tone Multi-Frequency (DTMF) detection, mixing, transcoding, playing announcement, recording, etc. (RFC 5567: Informational)
Media Graph	The media graph is the logical representation of the flow of media for a conference. (RFC 5239: Standards Track)

(Continued)

TABLE 1.2 (CONTINUED)
Centralized Conferencing Terminologies

Terminology	Definition
Media Mixer	A media mixer is the logical entity with the capability to combine media inputs of the same type, transcode the media, and distribute the result(s) to a single or multiple outputs. In this context, the term "media" means any type of data being delivered over the network using appropriate transport means, such as RTP/RTP Control Protocol (RTCP), defined in RFC 3550, or Message Session Relay Protocol, defined in RFC 4975. (RFC 5239: Standards Track)
Media Participant	An entity that has access to the media resources of a conference (e.g., it can receive a media stream). In floor-controlled conferences, a given media participant is typically co-located with a floor participant, but it does not need to be. Third-party floor requests consist of having a floor participant request a floor for a media participant when they are not co-located. The protocol between a floor participant and a media participant (that are not co-located) is outside the scope of this specification. (RFC 4582: Standards Track)
Media Provider	A CLUE-capable device that intends to send Capture Encodings. (See Section 16.4 – IETF Draft: Telepresence)
Media Provider (MP)	Media Provider is a CLUE-capable device that intends to send Capture Encodings. (See Section 16.3 – IETF Draft: Telepresence)Media Provider (MP) is a CLUE Participant (i.e., an Endpoint or an MCU) able to send Capture Encodings. (See Section 16.5 – IETF Draft: Telepresence)
Media Resource	An entity on the speech processing server that can be controlled through MRCPv2. (RFC 6787: Standards Track)
Media Resource Broker (MRB)	A logical entity that is responsible for both the collection of appropriate published Media Server (MS) information and supplying of appropriate MS information to consuming entities. The MRB is an optional entity and will be discussed in a separate specification. (RFC 5567: Informational)
Media Server	The Media Server includes the mixer as defined in RFC 4353. The Media Server plays announcements; it processes media streams for functions like DTMF detection and transcoding. The media server may also record media streams for supporting IVR functions like announcing conference participants. In the architecture for the 3GPP IP Multimedia Subsystem (IMS), a Media Server is referred to as a Media Resource Function (MRF). (RFC 5567: Informational)A Media Server (MS) processes media streams on behalf of an AS by offering functionality such as interactive media, conferencing, and transcoding to the end user. Interactive media functionality is realized by way of dialogs that are initiated by the application server. (RFC 6231: Standards Track)An entity that performs a service, such as media processing, on behalf of an Application Server; typical provided functions are mixing, announcement, tone detection and generation, and play and record services. (RFC 7058: Informational)
Media Services	Application service requiring media functions such as Interactive Voice Response (IVR) or media conferencing. (RFC 5567: Informational)
Media Session	From the Session Description Protocol (SDP) specification of RFC 4566: "A multimedia session is a set of multimedia senders and receivers and the data streams flowing from senders to receivers. A multimedia conference is an example of a multimedia session." (RFC 5567: Informational)
Media Stream	A media stream on a Media Server represents a media flow either between a connection and a conference, between two connections, or between two conferences. Streams can be audio or video and can be bidirectional or unidirectional. (RFC 6505: Standards Track)

(Continued)

TABLE 1.2 (CONTINUED)
Centralized Conferencing Terminologies

Terminology	Definition
Metadata Classes	Each block in the model represents a class. A class is a construct that is used as a blueprint to create instances (called "objects") of itself. The description of each class also has a representation of its attributes in a second compartment below the class name. (RFC 7865: Standards Track)
Metadata Model	A metadata model is an abstract representation of metadata using a Unified Modeling Language (UML) [1] class diagram. (RFC 7865: Standards Track)
Method	A Method is the type of a Framework message. Four Methods are defined in this specification: SYNC, CONTROL, REPORT, and K-ALIVE. (RFC 6230: Standards Track)"method" is used to set the HTTP method applied in the fetch of the initial VoiceXML specification. Allowed values are "get" or "post" (case-insensitive). Default is "get." (RFC 5552: Standards Track)
Model	Model: A set of assumptions a telepresence system of a given vendor adheres to and expects the remote telepresence system(s) to also adhere to. (RFC 7262 – Informational)
MRCP Client	An entity controlling one or more Media Resources through MRCPv2 ("Client" for short). (RFC 6787: Standards Track)
MRCP Server	Aggregate of one or more "Media Resource" entities on a server, exposed through MRCPv2. Often, "server" in this specification refers to an MRCP server. (RFC 6787: Standards Track)
MS (Media Server) Conference	An MS Conference provides the media-related mixing resources and services for conferences. In this specification, an MS Conference is often referred to simply as a conference. (RFC 6505: Standards Track)
MS (Media Server) Connection	An MS connection represents the termination on a Media Server of one or more RTP (RFC 3550) sessions that are associated to a single SIP dialog. A Media Server receives media from the output(s) of a connection, and it transmits media on the input(s) of a connection. (RFC 6505: Standards Track)
MS (Media Server) Control Channel	A reliable transport connection between the AS and the MS used to exchange MS Control PDUs. Implementations must support the Transport Control Protocol (TCP) of RFC 0793 and may support the Stream Control Transmission Protocol (SCTP) of RFC 4960. Implementations must support TLS of RFC 5246 as a transport-level security mechanism, although its use in deployments is optional. (RFC 5567: Informational)
MS (Media Server) Control Dialog	A SIP dialog that is used for establishing a control channel between the User Agent (UA) and the MS. (RFC 5567: Informational)
MS (Media Server) Control Protocol	The protocol used by an AS to control an MS. The MS Control Protocol assumes a reliable underlying transport protocol for the MS Control Channel. (RFC 5567: Informational)
MS (Media Server) Media Dialog	A SIP dialog between the AS and MS that is used for establishing media sessions between a user device such as a SIP phone and the MS. The preceding definitions for AS, MS, and MRB are taken from RFC 5167. (RFC 5567: Informational)
Multiple Content Capture	A Capture that mixes and/or switches other Captures of a single type (e.g., all audio or all video). Particular Media Captures may or may not be present in the resultant Capture Encoding depending on time or space. Denoted as MCCn in the example cases in this specification. (See Section 16.3 – IETF Draft: Telepresence)
Multipoint Control Unit (MCU)	An MCU is a device that connects two or more endpoints together into one single multimedia conference (RFC 5117). An MCU may include a mixer (RFC 4353). (RFC 7262 – Informational).A CLUE-capable device that connects two or more endpoints together into one single multimedia conference (RFC 5117). An MCU includes an RFC 4353-like mixer without the RFC 4353 requirement to send media to each participant. (See Section 16.3 – IETF Draft: Telepresence)

(Continued)

TABLE 1.2 (CONTINUED)
Centralized Conferencing Terminologies

Terminology	Definition
Non-CLUE device	A device that supports standard SIP and SDP but either does not support CLUE or does support it but does not currently wish to invoke CLUE capabilities. (See Section 16.8 – IETF Draft: Telepresence)
Notification	Notification is the act of a notifier sending a NOTIFY request to a subscriber to inform the subscriber of the state of a resource. (RFC 4582: Standards Track)
Notifier	A notifier is a user agent that generates NOTIFY requests for the purpose of notifying subscribers of the state of a resource. Notifiers typically also accept SUBSCRIBE requests to create subscriptions. (RFC 4582: Standards Track)
Participant	An entity that acts as a floor participant, a media participant, or as both. (RFC 4582: Standards Track)
Plane of Interest	The spatial plane within a scene containing the most relevant subject matter. (See Section 16.3 – IETF Draft: Telepresence)
postbody	Used to set the application/x-www-form-urlencoded HTML 4.01 encoded HTTP body for "post" requests (or is otherwise ignored). (RFC 5552: Standards Track)
Provider	Same as Media Provider. (See Section 16.4 – IETF Draft: Telepresence)
Query MRB	An instantiation of an MRB (see previous definition) that provides an interface for an Application Server to retrieve the address of an appropriate Media Server. The result returned to the Application Server can be influenced by information contained in the query request. (RFC 6917: Standards Track)
Recording Metadata	The metadata describing the CS that is required by the SRS. This will include, for example, the identities of users that participate in the CS and dialog state. Typically, this metadata is archived with the Replicated Media at the SRS. The recording metadata is delivered in real time to the SRS. (RFC 7245: Informational)
Recording Session (RS)	The SIP session created between an SRC and SRS for the purpose of recording a Communication Session. (RFC 6341: Informational)
Recording-aware User Agent (UA)	An SIP User Agent that is aware of SIP extensions associated with the CS. Such extensions may be used to notify the recording-aware UA that a session is being recorded, or by a recording-aware UA to express preferences as to whether a recording should be started, paused, resumed, or stopped. (RFC 7245: Informational)
Recording-unaware User Agent (UA)	An SIP User Agent that is unaware of SIP extensions associated with the CS. Such a recording-unaware UA will be notified that a session is being recorded or will express preferences as to whether a recording should be started, paused, resumed, or stopped via some other means that is out of scope for the SIP media recording architecture. (RFC 7245: Informational)
Remote	Sender and/or receiver on the other side of the communication channel (depending on context); i.e., not local. A remote can be an endpoint or an MCU. (RFC 7262 – Informational)
Render	Render is the process of generating a representation from a media, such as displayed motion video or sound emitted from loudspeakers. (RFC 7262 – Informational)The process of reproducing the received streams, like for instance, displaying of the remote video on the Media Consumer's screens or playing of the remote audio through loudspeakers. (See Section 16.4 – IETF Draft: Telepresence)
Replicated Media	A copy of the media that is associated with the CS, was created by the SRC, and was sent to the SRS. It may contain all the media associated with the CS (e.g., audio and video) or just a subset (e.g., audio). Replicated Media is part of the Recording Session. (RFC 7245: Informational)

(Continued)

TABLE 1.2 (CONTINUED)
Centralized Conferencing Terminologies

Terminology	Definition
Role	A role provides the context for the set of conference operations that a participant can perform. A default role (e.g., standard conference participant) will always exist, providing a user with a set of basic conference operations. Based on system-specific authentication and authorization, a user may take on alternate roles, such as conference moderator, allowing access to a wider set of conference operations. (RFC 5239: Standards Track)
Scene	Same as Capture Scene. (See Section 16.3 – IETF Draft: Telepresence)
Session Recording Client (SRC)	A Session Recording Client (SRC) is an SIP User Agent (UA) that acts as the source of the recorded media, sending it to the SRS. An SRC is a logical function. Its capabilities may be implemented across one or more physical devices. In practice, an SRC could be a personal device (such as an SIP phone), an SIP Media Gateway (MG), a Session Border Controller (SBC), or an SIP Media Server (MS) integrated with an Application Server (AS). This specification defines the term "SRC" such that all such SIP entities can be generically addressed under one definition. The SRC provides metadata to the SRS. (RFC 6341: Informational)
Session Recording Server (SRS)	A Session Recording Server (SRS) is an SIP User Agent (UA) that is a specialized media server or collector that acts as the sink of the recorded media. An SRS is typically implemented as a multi-port device that is capable of receiving media from multiple sources simultaneously. An SRS is the sink of the recorded session metadata. (RFC 6341: Informational)
Sidebar	A sidebar is a separate conference instance that only exists within the context of a parent conference instance. The objective of a sidebar is to be able to provide additional or alternate media only to specific participants. (RFC 5239: Standards Track)
Simultaneous Transmission Set	A set of Media Captures that can be transmitted simultaneously from a Media Provider. (See Section 16.3 – IETF Draft: Telepresence)
Single Media Capture	A capture that contains media from a single source capture device, e.g., an audio capture from a single microphone or a video capture from a single camera. (See Section 16.3 – IETF Draft: Telepresence)
Spatial Relation	The arrangement in space of two objects, in contrast to relation in time or other relationships. (See Section 16.3 – IETF Draft: Telepresence)
Stream	Stream is a Capture Encoding sent from a Media Provider to a Media Consumer via RTP (RFC 3550). (See Section 16.3 – IETF Draft: Telepresence)
Stream	A Capture Encoding sent from a Media Provider to a Media Consumer via RTP (RFC 3550). (See Section 16.4 – IETF Draft: Telepresence)
Stream Characteristics	The media stream attributes commonly used in non-CLUE SIP/SDP environments (such as media codec, bit rate, resolution, profile/level, etc.) as well as CLUE-specific attributes, such as the Capture ID or a spatial location. (See Section 16.3 – IETF Draft: Telepresence)The union of the features used to describe a stream in the CLUE environment and in the SIP-SDP environment.
Subscriber	A subscriber is a user agent that receives NOTIFY requests from notifiers; these NOTIFY requests contain information about the state of a resource in which the subscriber is interested. Subscribers typically also generate SUBSCRIBE requests and send them to notifiers to create subscriptions. (RFC 4582: Standards Track)
Subscription Migration	Subscription migration is the act of moving a subscription from one notifier to another notifier. (RFC 4582: Standards Track)

(*Continued*)

TABLE 1.2 (CONTINUED)
Centralized Conferencing Terminologies

Terminology	Definition
Telepresence	Telepresence is an environment that gives non-co-located users or user groups a feeling of (co-located) presence – the feeling that a local user is in the same room with other local users and the remote parties. The inclusion of remote parties is achieved through multimedia communication including at least audio and video signals of high fidelity. (RFC 7262 – Informational)
Third-Party Request	A request issued by a client to manipulate the conference data of another client. (RFC 6501: Standards Track)
Transaction-Timeout	The maximum allowed time between a Control Client or Server issuing a Framework message and its arrival at the destination. The value for "Transaction-Timeout" is 10 seconds. (RFC 6230: Standards Track)
User Agent Client (UAC)	As specified in RFC 3261 (also see Section 1.2). (RFC 6230: Standards Track)
User Agent Server (UAS)	As specified in RFC 3261 (also see Section 1.2). (RFC 6230: Standards Track)
VCR Controls	Runtime control of aspects of an audio playback, like speed and volume, via dual-tone multi-frequency (DTMF) signals sent by the user, in a manner that resembles the functions of a VCR (video cassette recorder) controller. (RFC 7058: Informational)
Video Capture	Media Capture for video. Denoted as VCn in the example cases in this specification. (See Section 16.3 – IETF Draft: Telepresence)
Video Composite	A single image that is formed, normally by an RTP mixer inside an MCU, by combining visual elements from separate sources. (See Section 16.3 – IETF Draft: Telepresence)
voicexml	URI of the initial VoiceXML specification to fetch. This will typically contain an HTTP URI but may use other URI schemes, for example, to refer to local, static VoiceXML specifications. If the "voicexml" parameter is omitted, the VoiceXML Media Server may select the initial VoiceXML specification by other means, such as by applying a default, or may reject the request. (RFC 5552: Standards Track)
Whisper	A whisper involves a one-time media input to (a) specific participant(s) within a specific conference instance, accomplished using a sidebar. An example of a whisper would be an announcement injected only to the conference chair or to a new participant joining a conference. (RFC 5239: Standards Track)
XCON-aware client	An XCON conferencing system client that is able to issue CCMP requests. (XCON = Centralized Conferencing.) (RFCs 6501 and 6503: Standards Track)

The multistream immersive telepresence system, which goes far beyond the traditional multipoint multimedia conferencing system described so far, is briefly addressed. We have described briefly how the specially designed telepresence conference rooms, life-size images, media capture, encoding groups, capture scenes from different angles, and simultaneous sets provide an immersive experience, as if all remote and local participants were in a face-to-face meeting in the same space.

Finally, we have defined the scalable multipoint multimedia conferencing and telepresence systems and protocols architecture framework, which will be designed based on standards. Accordingly, we have articulated each of the functional entities of all conferencing systems and each of the protocols that will be using these systems with each other. We have explained that telepresence also fits into the same conferencing architectural framework with the addition of special effects in the conference rooms. In fact, the subsequent sections of this book describe each of these conferencing systems and protocols. We have briefly discussed how the worldwide geographically distributed but logically centralized multipoint conferencing architecture can be constructed with multiple foci. However, some additional standards need to be chosen or developed for interoperability.

1.8 PROBLEMS

1. Describe differences between SIP/SDP and multimedia conferencing in terms of the OSI protocol layer point of view. What are the implications of physical, link/message authentication code (MAC), network, and transport layer when SIP/SDP is used for multimedia conferencing?
2. What are the functional and performance characteristics of multimedia conferencing? What are the implications for network design in offering conferencing services?
3. What are the differences between teleconferencing, videoconferencing, application sharing, web conferencing, and chat conferencing? Describe in detail.
4. What are the salient differences between point-to-point and multipoint multimedia conferencing? How does centralized multipoint conferencing differ from distributed multipoint conferencing? Explain in detail using illustrations.
5. Explain briefly the key characteristics of each application: audio, video, application sharing, timed text, media mixing and transcoding, media server, and application server.
6. Describe in detail, using illustrations, the roles of audio, video, application sharing, timed text, media mixing, media transcoding, media server, and application server in offering multipoint conferencing services.
7. How can each of these applications be made logically centralized but physically distributed in view of the centralized multipoint conferencing service: media mixing, media transcoding, media server, and application server?
8. What are the networking features that need to be taken care of in offering scalable multipoint multimedia conferencing?
9. What are the fundamental functional differences between two-party point-to-point and multiparty conference calls?
10. What are the limitations of single or multiple media negotiations with the media description architecture of multiparty conferencing?
11. Describe the problems of developing the multiparty call control architecture, which needs media bridging, evolving from the initial two-party point-to-point call using SIP/SDP.
12. What assumptions are made with the third-party centralized controller to set up the multiparty conference call using SIP/SDP, which is described in RFC 3725?
13. Describe the 3pcc call flows using SIP/SDP for the following: (a) simplest flows, (b) flows with ping-ponging of INVITEs, (c) flows with unknown entities, and (d) efficient flows with unknown entities. Provide detailed descriptions of every header and message body field of each SIP/SDP message of each call establishment flow of (a) through (d).
14. Describe all limitations of each call establishment flow of (a) through (d) of Problem 13.
15. What are the recommendations of 3pcc call setups per RFC 3725?
16. How can 3pcc call establishment errors be handled in general?
17. Describe the 3pcc call flows using SIP/SDP for continued call processing along with each header and message body field of each message.
18. Describe the 3pcc call flows with early media using SIP/SDP for continued call processing along with each header and message body field of each message.
19. Describe the 3pcc call flows with SDP preconditions using SIP/SDP for the following along with their limitations, if any: (a) Initial by controller and (b) initiation by the first party (say, A). Provide detailed descriptions of every header and message body field of each SIP/SDP message of each call establishment flow of (a) and (b).
20. Describe the 3pcc call flows using SIP/SDP for the following services along with their limitations, if any: (a) Click-to-dial and (b) mid-call announcement. Provide detailed descriptions of every header and message body field of each SIP/SDP message of each call establishment flow of (a) and (b).
21. What are the implementation recommendations of 3pcc call setups per RFC 3725?

22. Describe the security recommendations, using some examples, for the 3pcc call setups and services per RFC 3725: Authentication, authorization, end-to-end encryption, and integrity.
23. Describe the limitations and suggest probable solutions for the worldwide large-scale 3pcc multiparty audio, video, and application sharing (including media bridging) conferencing services architecture recommended in RFC 3725 for providing scalability and interoperability in multivendor environments.
24. What is multistream immersive telepresence? How do the conference room, media capture, encoding schemes, capture scenes, and simultaneous transmission features differ from those of conventional multipoint multimedia conferencing?
25. Describe the scalable multipoint multimedia conference and telepresence systems and protocols framework. Draw its architecture. Articulate each of the systems and protocols. How does it different from 3pcc?
26. Describe briefly how a large-scale worldwide geographically distributed conferencing system with multiple foci can be designed. How could it be geographically distributed but logically centralized?

REFERENCES

1. Roy, R. R. (2016) *Handbook on Session Initiation Protocol: Networked Multimedia Communications for IP Telephony*, CRC Press – Taylor & Francis Group.
2. Roy, R. R. (2018) *Handbook on SDP for Multimedia Session Negotiations: SIP and WebRTC IP Telephony*, CRC Press – Taylor & Francis Group.
3. Roy, R. R. (1994) Networking Constraints in Multimedia Conferencing and the Role of ATM Networks. *AT&T Technical Journal*, 73(4).
4. Roy, R. R. et al. (1994) An Analysis of Universal Multimedia Switching Architectures. *AT&T Technical Journal*, 73(6).
5. Georganas, N. D. et al. Editors. (1996) Synchronization Issues in Multimedia Communications. *IEEE SAC*, 14(1).
6. Fluckinger, F. (1995) *Understanding Networked Multimeida – Applications and Technology*, Prentice Hall.
7. Steinmetz, R., T. Meyer (1992) Multimedia Synchronization Techniques: Experience-Based on Different System Structures. *ACM SIGCOMM Computer Communication Review*, 22(1).
8. Hehmann, D. et al. (1990) Transport Services for Multimedia Applications on Broadband Networks. *Computer Communications*, 13(4).
9. Ghinea, G., O. A. Ademoye (2010) Perceived Synchronization of Olfactory Multimedia. *IEEE Transactions on Systems, Man and Cybernetics, – Part A: Systems and Humans.*
10. Blakowski, R. Steinmetz (1996) A Media Synchronization Survey: Reference Model, Specification, and Case Studies. *IEEE SAC*, 14(1).

2 Centralized Conferencing System Architecture

2.1 INTRODUCTION

In this chapter, we cover the centralized multipoint multimedia conferencing systems architecture described in RFC 5239. The centralized multipoint multimedia systems and protocols framework described earlier in Section 1.5 assumes that the centralized controller is a black box. However, this chapter defines the logical conferencing systems and protocols between the entities that are described in subsequent chapters based on specific standards defining all the standardized logical functional entities of the controller itself. Specifically, this chapter is structured as follows: First, more details on the conference (data) objects; second, the constructs and identifiers that MUST be implemented to manage the conference objects, instances, and users associated with a conferencing system; third, how a conferencing system is logically built using the defined high-level data model and how the conference objects are maintained; fourth, definitions of the fundamental conferencing mechanisms and high-level overview of the protocols; fifth, realizations of various conferencing scenarios, detailing the manipulation of the conference objects using the defined protocols; sixth, the relationship between this centralized conferencing framework and the session initiation protocol (SIP) conferencing framework; and finally, the security of the centralized conferencing system architecture.

The centralized multipoint multimedia conferencing systems architecture provides the fundamental basis for designing the scalable large-scale centralized multipoint multimedia conferencing framework. The centralized conference system described earlier in Section 1.5 is based on a centralized controller that happens to use SIP to the outside world without disclosing the inner working of its internal functionalities. As a result, this type of proprietary centralized controller-based conferencing system is not scalable for the large-scale conferencing system. In this section, we define the standardized logical functional entities, including the controller of the conferencing systems, and the corresponding protocols between them. We describe the conferencing architecture, which consists of conference data objects, constructs, and identities of the objects for managing the conference objects, instances, and users associated with a conferencing system; building of the data model and maintenance of conference objects, conference mechanisms, and protocols; and security. SIP is used as the call control protocol for the centralized multipoint conferencing per RFC 5239 (although other call control protocols such as H.323, Jabber, Q.931, or Integrated Signaling Digital Network User Part [ISUP] can also use this framework). In addition, Section 2.10 describes the relationship between the SIP conferencing framework and the centralized multipoint conferencing framework (described in RFC 5239). The subsequent chapters of this book describe each of these logical systems and corresponding protocols based on technical standards developed in the Internet Engineering Task Force (IETF) and other standards fora.

2.2 CONVENTION

As described in RFC 2119.

2.3 TERMINOLOGY

See Table 1.2, Section 1.6.

2.4 OVERVIEW

A centralized conference is an association of endpoints, called conference participants, with a central endpoint, called a conference focus. The focus has direct peer relationships with the participants by maintaining a separate call signaling interface with each. Consequently, in this centralized conferencing model, the call signaling graph is always a star. The most basic conference supported in this model would be an ad hoc, unmanaged conference, which would not necessarily require any of the functionality defined within this framework. For example, it could be supported using basic SIP signaling functionality with a participant serving as the focus; the SIP Conferencing Framework (RFC 4353) together with the SIP Call Control Conferencing for User Agents (RFC 4579) specifications addresses these types of scenarios.

In addition to the basic features, however, a conferencing system supporting the centralized conferencing model proposed in this framework specification can offer richer functionality by including dedicated conferencing applications with explicitly defined capabilities and reserved recurring conferences, along with providing the standard protocols for managing and controlling the different attributes of these conferences. The core requirements for centralized conferencing are outlined in RFC 4245. These requirements are applicable for conferencing systems using various call signaling protocols, including SIP. Additional conferencing requirements are provided in RF C4376 and RFC 4597.

The centralized conferencing system proposed by this framework is built around the fundamental concept of a conference object. A conference object provides the data representation of a conference during each of the various stages of a conference (e.g., creation, reservation, active, completed, etc.). A conference object is accessed via the logical functional elements with which a conferencing client interfaces, using the various protocols identified in Figure 2.1. The functional elements defined for a conferencing system described by the framework are a conference control server, a floor control server, any number of foci, and a notification service. A conference control protocol (CCP) provides the interface between a conference and media control client and the conference control server. A floor control protocol (e.g., binary floor control protocol [BFCP] described in Chapter 6) provides the interface between a floor control client and the floor control server. A call signaling protocol (e.g., SIP, H.323, Jabber, Q.931, ISUP, etc.) provides the interface between a call signaling client and a focus. A notification protocol (RFC 6502 – see Chapter 7) provides the interface between the conferencing client and the notification service.

A conferencing system can support a subset of the conferencing functions depicted in the conferencing system logical decomposition in Figure 2.1 and described in this specification. However, there are some essential components that would typically be used by most other advanced functions, such as the notification service. For example, the notification service is used to correlate information, such as the list of participants with their media streams, between the various other components.

The media graph of a conference can be centralized, decentralized, or any combination of both and can potentially differ per media type. In the centralized case, the media sessions are established between a media mixer controlled by the focus and each one of the participants.

In the decentralized (i.e., distributed) case, the media graph is a multicast or multi-unicast mesh among the participants. Consequently, the media processing (e.g., mixing) can be controlled either by the focus alone or by the participants. The concepts in this framework specification clearly map to a centralized media model.

The concepts can also apply to the decentralized media case; however, the details of such are left for future study. Section 2.5 of this chapter provides more details on the conference object. Section 2.6 defines the constructs and identifiers that MUST be implemented to manage the conference objects, instances, and users associated with a conferencing system. Section 2.7 describes how a conferencing system is logically built using the defined high-level data model and how the conference objects are maintained. Section 2.8 describes the fundamental conferencing mechanisms and provides a high-level overview of the protocols. Section 2.9 then provides realizations of various

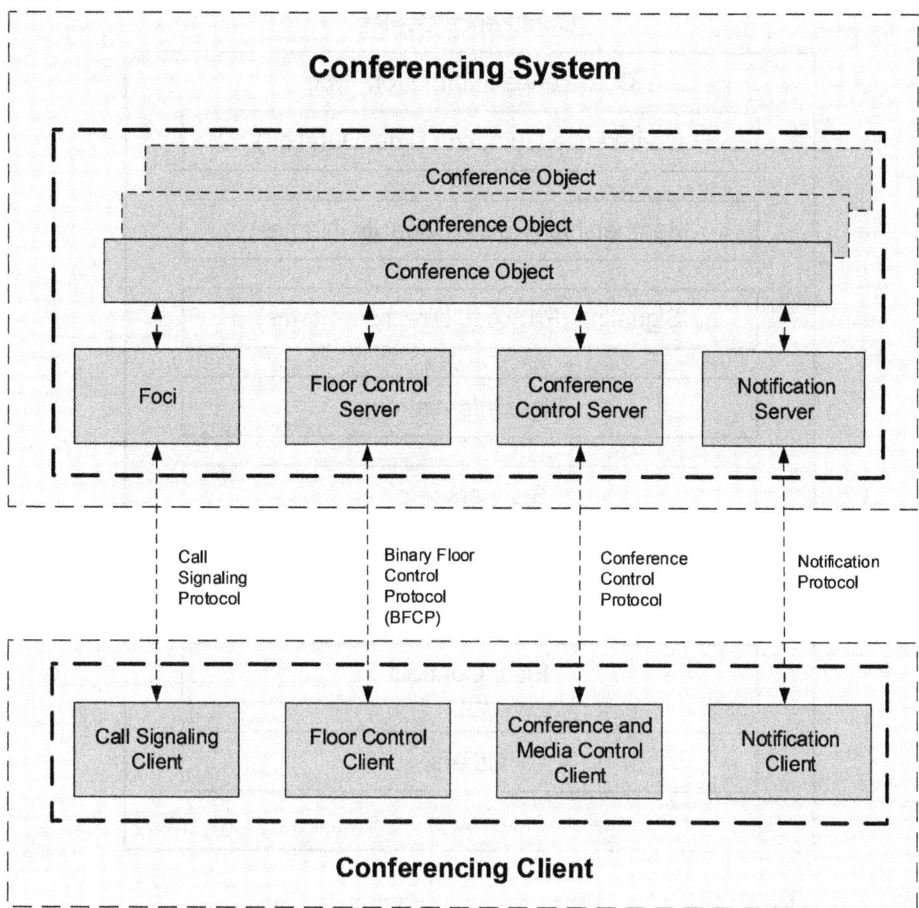

FIGURE 2.1 Conferencing System Logical Decomposition.

conferencing scenarios, detailing the manipulation of the conference objects using the defined protocols. Section 2.10 summarizes the relationship between this centralized conferencing framework and the SIP conferencing framework.

2.5 CENTRALIZED CONFERENCING DATA

The centralized conference data is logically represented by the conference object. A conference object is of "Conference information type," as illustrated in Figure 2.2. The conference information type is extensible.

In a system based on this conferencing framework, the same conference object type is used for representation of a conference during different stages of a conference, such as expressing conferencing system capabilities, reserving conferencing resources, or reflecting the state of ongoing conferences. Section 2.7 describes the usage semantics of the conference objects. The exact XML schema of the conference object, including the organization of the conference information, is detailed in RFC 6501 (see Chapter 4).

Along with the basic data model, as defined in RFC 6501 (see Chapter 4), the realization of this framework requires a policy infrastructure. The policies required by this framework to manage and control access to the data include local, system-level boundaries associated with specific data elements, such as the membership, and the ranges and limitations of other data elements. Additional policy considerations for a system realization based on this data model are discussed in Section 2.5.2.

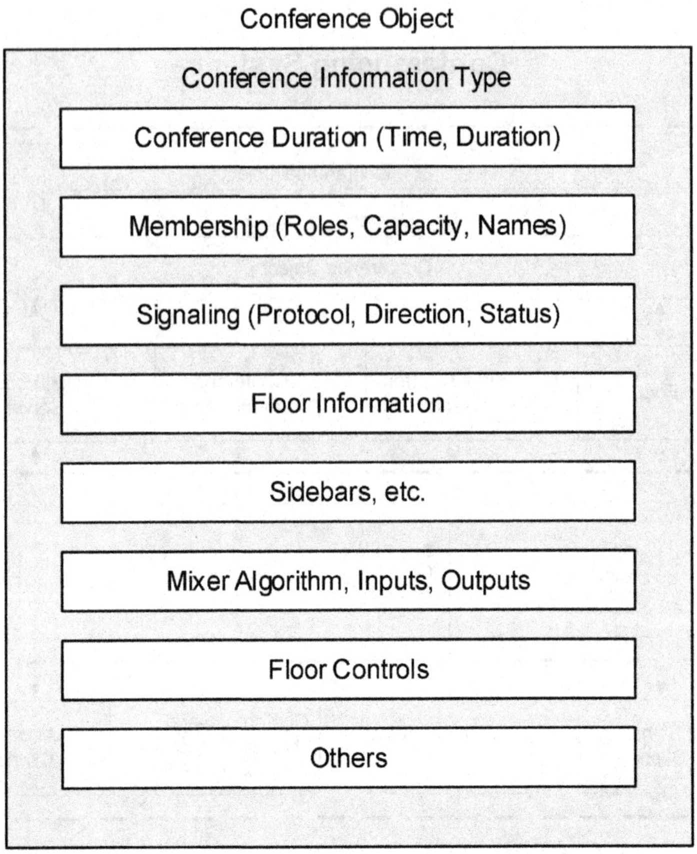

Conference Object

Conference Information Type
Conference Duration (Time, Duration)
Membership (Roles, Capacity, Names)
Signaling (Protocol, Direction, Status)
Floor Information
Sidebars, etc.
Mixer Algorithm, Inputs, Outputs
Floor Controls
Others

FIGURE 2.2 Conference Object Type Decomposition. (Copyright: IETF.)

2.5.1 Conference Information

A core set of data in the conference information is utilized in any conference, which is independent of the specific conference media nature (e.g., the mixing algorithms performed, the advanced floor control applied, etc.). This core set of data in the conference information contains the definitions representing the conference object capabilities, membership, roles, call signaling, and media status relevant to different stages of the conference lifecycle. This core set of conference information may be represented using the conference-type, as defined in the SIP conference event package (RFC 4575). Typically, participants with read-only access to the conference information would be interested in this core set of conference information only.

In order to support more complex media manipulations and enhanced conferencing features, the conference information, as defined in the data model (see Chapter 4), contains additional data beyond that defined in the SIP conference event package (see Chapter 4). The information defined in the data model (see Chapter 4) provides specific media mixing details, available floor controls, and other data necessary to support enhanced conferencing features. This information allows authorized clients to manipulate the mixer's behavior via the focus, with the resultant distribution of the media to all or individual participants. By doing so, a client can change its own state and/or the state of other participants in the conference. New centralized conferencing specifications can extend the basic conference-type, as defined in the data model (see Chapter 4), and introduce additional data elements to be used within the conference information type.

2.5.2 Conference Policies

Conference policies collectively refer to a set of rights, permissions, and limitations pertaining to operations being performed on a certain conference object. The set of rights describes the read/write access privileges for the conference object as a whole. This access would usually be granted and defined in terms of giving the read-only or read/write access to clients with certain roles in the conference. Managing this access would require a conferencing system to have access to basic policy information to make the decisions but does not necessarily require an explicit representation in the policy model.

The permissions and limits require explicit policy mechanisms and are outside the scope of the data model (see Chapter 4) and this framework specification. However, there are some important policy considerations for a conferencing system. A conferencing system associates specific policies in the form of permissions and limitations with each user in a conferencing system. The permissions may vary depending upon the role associated with a specific conference user identifier. A conferencing system may provide a default user role that only allows participation in a conference through the default signaling means.

The conference object identifier provides access to the data associated with a specific conference. It is important to ensure that elements in the data have individual policy controls to provide flexibility in defining the various roles and specific data elements that may be manipulated by users with specific roles. In addition, the conference notification interface allows specific data elements to be sent to users that register for such notifications. It is important that the appropriate access control is provided so that only users that are authorized to view specific data elements receive the data in the notifications.

2.6 CENTRALIZED CONFERENCING CONSTRUCTS AND IDENTIFIERS

We describe details of the identifiers associated with the centralized conferencing framework constructs and the identifiers that are required to address and manage the clients associated with a conferencing system. An overview of the allocation, characteristics, and functional role of the identifiers is provided.

2.6.1 Conference Identifier

The conference identifier (conference ID) is a call signaling protocol-specific uniform resource identifier (URI) that identifies a specific conference focus and its associated conference instance. A conference factory is one method for generating a unique conference ID to identify and address a conference focus using a call signaling interface. Details on the use of a conference factory and "isfocus" feature tag for SIP signaling are described in RFC 4579. The conference identifier can also be obtained using the conference control protocol or other, including proprietary, out-of-band mechanisms. To realize the centralized conferencing described here, a conferencing system is required to support SIP as the default call signaling protocol. Other call signaling protocols (e.g., ISUP) are optional.

2.6.2 Conference Object

A conference object provides the logical representation of a conference instance in a certain stage, such as a conference blueprint representing a conferencing system's capabilities, the data representing a conference reservation, and the conference state during an active conference. Each conference object is independently addressable through the conference control protocol interface (see Section 2.8.3). A conferencing system MUST provide a default blueprint representing the basic capabilities provided by that specific conferencing system. Figure 2.3 illustrates the relationships between the conference identifier, the focus, and the conference object ID within the context of a logical conference instance, with the conference object corresponding to an active conference.

A conference object representing a conference in the active state can have multiple call signaling conference identifiers; for example, one for each call signaling protocol supported. There is a one-to-one mapping between an active conference object and a conference focus. The focus is addressed by explicitly associating unique conference IDs for each signaling protocol supported by the active conference object.

2.6.2.1 Conference Object Identifier

In order to make each conference object externally accessible, the conferencing system must allocate a unique URI per distinct conference object in the system. The conference object identifier is defined in RFC 6501 (see Chapter 4). A conferencing system allocates a conferencing object identifier for every conference blueprint, for every conference reservation, and for every active conference. The distribution of the conference object identifier depends upon the specific use case and includes a variety of mechanisms, such as through the conference control protocol mechanism, the data model and conference package, or out-of-band mechanisms such as email.

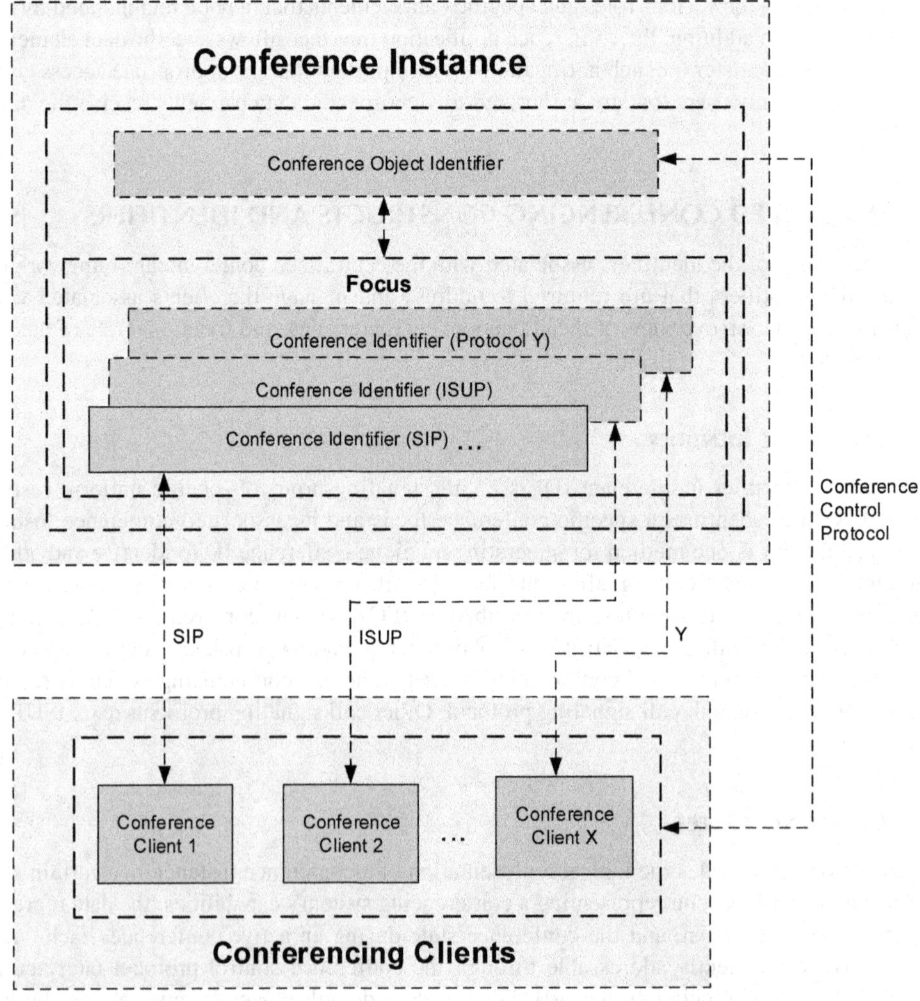

FIGURE 2.3 Identifier Relationship for an Active Conference. (Copyright: IETF.)

When a user wishes to create or join a conference, and the user does not have the conference object identifier for the specific conference, more general signaling mechanisms apply. A user may have a pre-configured conference object identifier to access the conferencing system, or other signaling protocols may be used and the conferencing system maps those to a specific conference object identifier. Once a conference is established, a conference object identifier is required for the user to manipulate any of the conferencing data or take advantage of any of the advanced conferencing features.

The same notion applies to users joining a conference using other signaling protocols. They are able to initially join a conference using any of the other signaling protocols supported by the specific conferencing system, but the conference object identifier must be used to manipulate any of the conferencing data or take advantage of any of the advanced conferencing features. As mentioned previously, the mechanism by which the user learns of the conference object identifier varies and could be via the conference control protocol, using the data model and conference package, or entirely out-of-band mechanisms such as email or a web interface.

The conference object identifier logically maps to other protocol-specific identifiers associated with the conference instance, such as the BFCP "confid" (see Chapter 6). The mapping of the conference object identifier can be viewed to contain sensitive information in many conferencing systems. The conferencing system ensures that the data is protected, that only authorized users can manipulate that information via the conferencing control protocol, and that only the appropriate users receive the information through the notification protocol. In general, this information would not be expected to be distributed to the average conference participant.

2.6.3 Conference User Identifier

Each user within a conferencing system is allocated a unique conference user identifier, which is described in detail in Chapter 4. The conference user identifier is used in association with the conference object identifier to uniquely identify a user within the scope of the conferencing system. The mechanism is also in place for identifying conferencing system users who may not be participating in a conference instance. Examples of these users would be a non-participating "floor control chair" or "media policy controller." The conference user identifier is required, in conference control protocol requests, to uniquely determine who is issuing commands, so that appropriate policies can be applied to the requested command.

A typical mode for distributing the user identifier is out of band during conferencing client configuration; thus, the mechanism is outside the scope of the centralized conferencing system and protocols. However, the conferencing system is capable of allocating and distributing a user identifier during the first signaling interaction with the conferencing system, such as an initial request for blueprints or adding a new user to an existing conference using the conference control protocol. When a user joins a conference using a signaling-specific protocol, such as SIP for a dial-in conference, a conference user identifier is assigned if one is not already associated with that user. While this conference user identifier is not required for the participant to join the conference, it is allocated and assigned by the conferencing system such that it is available for use for any subsequent conference control protocol operations and/or notifications associated with that conference. For example, the conference user identifier would be sent in any notifications that may be sent to existing participants, such as the moderator, when this user joins.

The conference user identifier is logically associated with the other user identifiers assigned to the conferencing client for other protocol interfaces, such as an authenticated SIP user. The mapping of the conference user identifier to signaling-specific user identifiers requires that methods for protecting and securing a user's identity are considered. In addition, the conferencing system ensures the appropriate access control around any internal data structure that maintains this persistent data. This information would typically only be available to a conferencing system administrator.

2.7 CONFERENCING SYSTEM REALIZATION

Implementations based on this centralized conferencing framework can range from systems supporting ad hoc conferences, with default behavior only, to sophisticated systems with the ability to schedule recurring conferences, each with distinct characteristics, being integrated with external resource reservation tools, and providing snapshots of the conference information at any of the stages of the conference lifecycle.

A conference object is the logical representation of a conference instance at a certain stage, such as description of capabilities upon conference creation, reservation, activation, etc., which a conferencing system maintains in order to describe the system capabilities and to provide access to the available services provided by the conferencing system. Consequently, it is not mandatory for the actual usage of the conference object for the centralized conferencing; rather, we are defining the general cloning tree concept and the mechanisms required for its realization, as described in detail in Section 2.7.1.

Ad hoc and advanced conferencing examples are provided in Sections 2.7.2 and 2.7.3, with the latter providing additional description of the conference object in terms of the stages of a conference to support scheduled and other advanced conference capabilities. The scheduling of a conference based on these concepts and mechanisms is then detailed in Section 2.7.4. The overall policy, as described in Section 2.5.2, in terms of permissions and limitations has been addressed. The policies applicable to the conference object as a whole in terms of read/write access would require a conferencing system have access to basic policy information to make the decisions. In the examples in this section, the policies are shown logically associated with the conference objects to emphasize the general requirement for policy functionality necessary for the realization of this framework.

2.7.1 Cloning Tree

The concept defined here is a logical representation only, as it is reflected through the centralized conferencing mechanisms: the URIs and the protocols. Of course, the actual system realization can differ from the presented model. The intent is to illustrate the role of the logical elements in providing an interface to the data, based on conferencing system and conferencing client actions, and describe the resultant protocol implications.

Any conference object in a conferencing system is created by either being explicitly cloned from an existing parent object or being implicitly cloned from a default system conference blueprint. A conference blueprint is a static conference object used to describe a typical conference setting supported by the system. Each system can maintain multiple blueprints, typically each describing a different conferencing type using the conference information format. This specification uses the "cloning" metaphor instead of the "inheritance" metaphor, because it more closely fits the idea of object replication rather than a data type re-usage and extension concept.

The cloning operation specifies whether or not the link between the parent and child needs to be maintained in the system. If no link between the parent and child exists, the objects become independent, and the child is not impacted by any operations on the parent object or subject to any limitations of the parent object. Once the new object is created, it can be addressed by a unique conference object URI assigned by the system, as described in Section 2.6.2.1. By default, the newly created object contains all the data existing in the parent object. The newly created object can expand the data it contains within the schema types supported by the parent. It can also restrict the read/write access to its objects.

However, unless the object is independent, it cannot modify the access restrictions imposed by the parent object. Any piece of data in the child object can be independently accessed and, by default, can be independently modified without affecting the parent data. Unless the object is independent, the parent object can enforce a different policy by marking certain data elements as "parent enforceable." The values of these data elements cannot be changed by directly accessing the

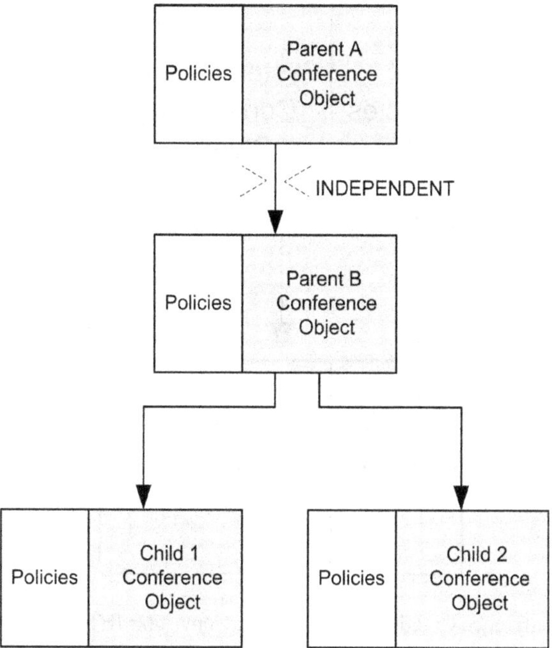

FIGURE 2.4 Cloning Tree of an Object. (Copyright: IETF.)

child object, nor can they be expanded in the child object alone. Figure 2.4 illustrates an example of a conference (Parent B) that is created independent of its parent (Parent A). Parent B creates two child objects, Child 1 and Child 2.

Any of the data elements of Parent B can be modified (i.e., there are no "parent enforceable" data elements), and depending upon the element, the changes will be reflected in Child 1 and Child 2, whereas changes to Parent A will not impact the data elements of Parent B. Any "parent enforceable" data elements, as defined by Parent B, cannot be modified in the child objects. Using the defined cloning model and its tools, the following sections show examples of how different systems based on this framework can be realized.

2.7.2 Ad Hoc Example

Figure 2.5 illustrates how an ad hoc conference can be created and managed in a conferencing system. A client can create a conference by establishing a call signaling channel with a conference factory, as specified in Section 2.6.1.

The conference factory can internally select one of the system-supported conference blueprints based on the requesting client privileges and the media lines included in the session description protocol (SDP) body. The selected blueprint with its default values is copied by the server into a newly created conference object, referred to as an "active conference." At this point, the conference object becomes independent of its blueprint. A new conference object identifier, a new conference identifier, and a new focus are allocated by the server. During the conference lifetime, an authorized client can manipulate the conference object by performing operations such as adding participants using the conference control manipulating protocol (see Chapter 5).

2.7.3 Advanced Example

Figure 2.6 illustrates how a recurring conference can be specified according to system capabilities, scheduled, reserved, and managed in a conferencing system. A client would first query a conferencing system for its capabilities.

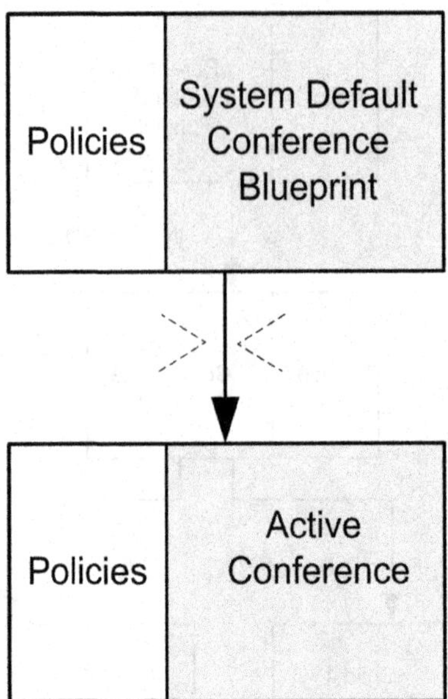

FIGURE 2.5 Ad Hoc Conference Creation and Lifetime. (Copyright: IETF.)

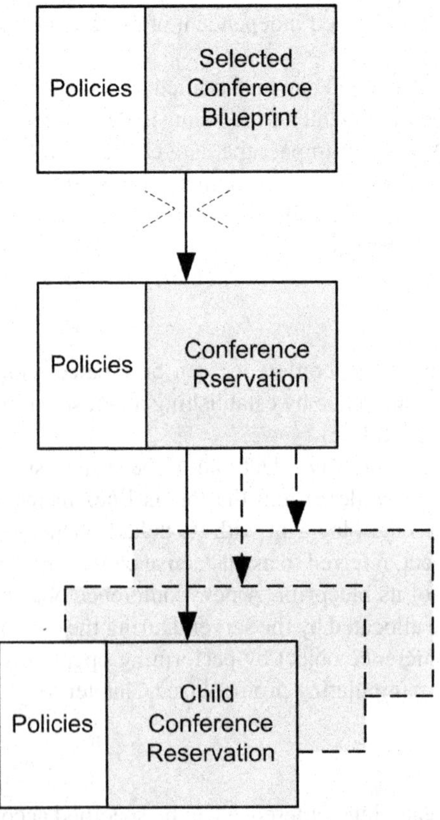

FIGURE 2.6 Advance Conference Definition, Creation, and Lifecycle. (Copyright: IETF.)

This can be done by requesting a list of the conference blueprints the system supports. Each blueprint contains a specific combination of capabilities and limitations of the conference server in terms of supported media types (e.g., audio, video, text, or combinations of these), participant roles, maximum number of participants of each role, availability of floor control, controls available for participants, availability and type of sidebars, the definitions and names of media streams, etc.

The selected blueprint with its default values is cloned by the client into a newly created conference object, referred to as a conference reservation, that specifies the resources needed from the system for this conference instance. At this point, the conference reservation becomes independent of its blueprint. The client can also change the default values, within the system ranges, and add additional information, such as the list of participants and the conference "start" time, to the conference reservation.

At this point, the client can ask the conference server to create new conference reservations by attaching the conference reservation to the request. As a result, the server can allocate the needed resources, create the additional conference objects for the child conference reservations, and allocate the conference object identifiers for the original conference reservation and for each child conference reservation.

From this point on, any authorized client is able to access and modify each of the conference objects independently. By default, changes to an individual child conference reservation will affect neither the parent conference reservation, from which it was created, nor its siblings.

On the other hand, some of the conference sub-objects, such as the maximum number of participants and the participants list, can be defined by the system as parent enforceable. As a result, these objects can be modified by accessing the parent conference reservation only. The changes to these objects can be applied automatically to each of the child reservations, subject to local policy.

When the time comes to schedule the conference reservation, either via the system determination that the "start" time has been reached or via client invocation, an active conference is cloned based on the conference reservation. As in the ad hoc example, the active conference is independent of the parent, and changes to the conference reservation will not impact the active conference. Any desired changes must be targeted towards the active conference. An example of this interaction is shown in Section 2.9.1.

2.7.4 Scheduling a Conference

The capability to schedule conferences forms an important part of the conferencing system solution. An individual conference reservation typically has a specified "start" and "end" time, with the times being specified relative to a single specified "fixed" time (e.g., "start" = 09.00 GMT, "end" = "start"+2), subject to system considerations. In most advanced conferencing solutions, it is possible to not only schedule an individual occurrence of a conference reservation but also schedule a series of related conferences (e.g., a weekly meeting that starts on Thursday at 09.00 GMT).

To be able to achieve such functionality, a conferencing system needs to be able to appropriately schedule and maintain conference reservations that form part of a recurring conference. The mechanism proposed in this specification makes use of the "Internet Calendaring and Scheduling Core Object" specification defined in RFC 2445 in combination with the concepts introduced in Section 2.5 for the purpose of achieving advanced conference scheduling capability.

Figure 2.7 illustrates a simplified view of a client interacting with a conferencing system. The client is using the conference control protocol to add a new conference reservation to the conferencing system by interfacing with the conference control server. The CCP (Conference Control Manipulation Protocol (CCMP) – see Chapter 5) request contains a valid conference reservation and reference by value to an "iCal" object that contains scheduling information about the conference (e.g., "start" time, "end" time).

The CCP (i.e., CCMP – see Chapter 5) request to create a new conference reservation is validated, including the associated iCal object, and the resultant conference reservation is created.

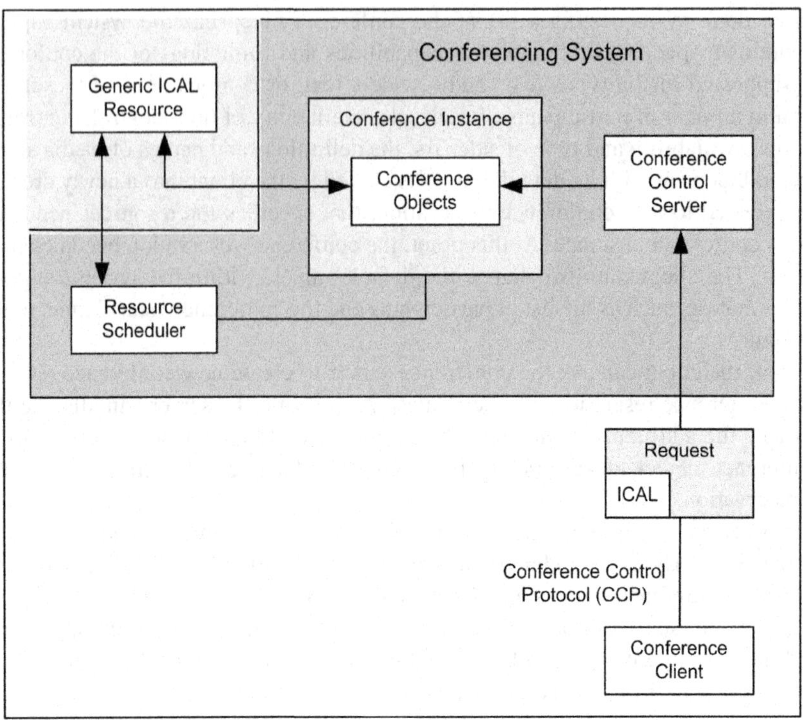

FIGURE 2.7 Resource Scheduling. (Copyright: IETF.)

The conference reservation is uniquely represented within the conferencing system by a conference object identifier (e.g., xcon:hd87928374), as introduced in Section 2.6.2.1 and defined in Chapter 4 (Conf Object and Data Model). This unique URI is returned to the client and can be used to reference the conference reservation if any future manipulations are required (e.g., alter "start" time) using the CCP (i.e., CCMP – see Chapter 5) request.

The previous example explains how a client creates a basic conference reservation using an iCal reference in association with a conference control protocol. Figure 1.9 (see Chapter 1) can also be applied when explaining how a series of conferences are scheduled in the system. The description is almost identical with the exception that the iCal definition that is included in the CCP (i.e., CCMP – see Chapter 5) request represents a series of recurring conference instances (e.g., conference "start" time, "end" time, occur weekly). The conferencing system will treat this request in the same way as the first example. The CCP (i.e., CCMP – see Chapter 5) request will be validated, along with the associated iCal object, and the conference reservation is created. The conference reservation and its conference object ID, created for this example, represent the entire series of recurring conference instances rather than a single conference. If the client uses the conference object ID provided and a CCP (i.e., CCMP – see Chapter 5) request to adjust the conference reservation, every conference instance in the series will be altered. This includes all future occurrences, such as a conference scheduled as an infinite series, subject to the limitations of the available calendaring interface.

A conferencing system that supports the scheduling of a series of conference instances should also be able to support manipulation within a specific range of the series. A good example is a conference reservation that has been scheduled to occur every Monday at 09.00 GMT. For the next 3 weeks only, the meeting has been altered to occur at 10.00 GMT in an alternative venue. With Figure 2.7 in mind, the client will construct a CCP request whose purpose is to modify the existing conference reservation for the recurring conference instance. The client will include the conference object ID provided by the conferencing system to explicitly reference the conference reservation within the conferencing system. A CCP request will contain all the required changes to the conference reservation (e.g., change of venue).

The conferencing system matches the incoming CCP request to the existing conference reservation but identifies that the associated iCal object only refers to a range of the existing series. The conferencing system creates a child, by cloning the original conference reservation, to represent the altered conference instances within the series. The cloned child object is not independent of the original parent object, thus preventing any potential conflicts in scheduling (e.g., a change to the whole series "'start' time"). The cloned conference reservation, representing the altered series of conference instances, has its own associated conference object ID, which is returned to the client using the CCP response. This conference object ID is then used by the client to make any future alterations on the newly defined sub-series. This process can be repeated any number of times, as the newly returned conference object ID, representing an altered (cloned) series of conference instances, can itself be manipulated using the CCP request for the newly created conference object ID. This provides a flexible approach to the scheduling of recurring conference instances.

2.8 CONFERENCING MECHANISMS

2.8.1 Call Signaling

The focus is the central component of the conference. Participants interface with the focus using an appropriate call signaling protocol (CSP) such as SIP. Participants request to establish or join a conference using the SIP. After checking the applicable policies, a focus then either accepts the request, sends a progress indication related to the status of the request (e.g., for a parked call while awaiting moderator approval to join), or rejects that request using the call signaling interface.

During an active conference, a conference control protocol can be used to affect the conference state. For example, CCP (i.e., CCMP – see Chapter 5) requests to add and delete participants are communicated to the focus and checked against the conference policies. If approved, the participants are added or deleted using the call signaling to/from the focus.

2.8.2 Notifications

A conferencing system is responsible for implementing a conference notification service. The conference notification service provides updates about the conference instance state to authorized parties, including participants. A model for notifications using SIP is defined in RFC 3265 (also see Chapter 7) with the specifics to support conferencing defined in RFC 4575 (also see Chapter 7). The conference user identifier and associated role are used by the conferencing system to filter the notifications such that they contain only information that is allowed to be sent to that user.

2.8.3 Conference Control Protocol

The conference control protocol provides for data manipulation and state retrieval for the centralized conferencing data, represented by the conference object. The details of the conference control protocol are provided in separate specifications.

2.8.4 Floor Control

A floor control protocol allows an authorized client to manage access to a specific floor and to grant, deny, or revoke access of other conference users to that floor. Floor control is not a mandatory mechanism for a conferencing system implementation, but it provides advanced media input control features for conference users. The BFCP defined in RFC 4582 (see Chapter 6) is used for floor control within a conferencing system.

Within this framework, a client-supporting floor control needs to obtain information for connecting to a floor control server to enable it to issue floor requests. This connection information can be retrieved using information provided by mechanisms such as negotiation using the SDP specified in RFC 4566 along with exchange of offer/answer defined in RFC 3264 on the signaling interface with the focus. Section 2.11.3 provides a discussion of client authentication of a floor control server.

As well as the client to the floor control server connection information, a client wishing to interact with a floor control server requires access to additional information. This information associates floor control interactions with the appropriate floor instance. Once a connection has been established and authenticated (RFC 4582 – see Chapter 6), a specific floor control message requires detailed information to uniquely identify a conference, a user, and a floor.

The conference is uniquely identified by the conference object ID per Section 2.6.2.1. This conference object ID must be included in all floor control messages. When the SDP model is used as described in RFC 4583, this identifier maps to the "confid" SDP attribute. Each authorized user associated with a conference object is uniquely represented by a conference user ID per Section 2.6.3. This conference user ID must be included in all floor control messages. When using SDP offer/answer exchange to negotiate a floor control connection with the focus using the call signaling protocol, the unique conference user identifier is contained in the "userid" SDP attribute, as defined in RFC 4583.

A media session within a conferencing system can have any number of floors (0 or more) that are represented by the conference identifier. When using SDP offer/answer exchange to negotiate a floor control connection with the focus using the call signaling interface, the unique conference identifier is contained in the "floorid" SDP attribute, as defined in RFC 4583; for example, a=floorid:1 m-stream:10. Each "floorid" attribute, representing a unique floor, has an "m-stream" tag containing one or more identifiers. The identifiers represent individual SDP media sessions (as defined using "m=" from SDP) using the SDP "Label" attribute, as defined in RFC 4574.

2.9 CONFERENCE SCENARIO REALIZATIONS

This section addresses how advanced conferencing scenarios, many of which have been described in RFC 4597, are realized using this centralized conferencing framework. The objective of this section is to further illustrate the model, mechanisms, and protocols presented in the previous sections and also serves to validate that the model, mechanisms, and protocols are sufficient to support advanced conferencing scenarios.

The scenarios provide a high-level primitive view of the necessary operations and general logic flow. The details shown in the scenarios are for illustrative purposes only and do not necessarily reflect the actual structure of the conference control protocol messages or the detailed data, including states, which are defined in separate specifications. It should be noted that not all entities impacted by the request are shown in the diagram (e.g., focus); rather, the emphasis is on the new entities introduced by this centralized conferencing framework.

2.9.1 Conference Creation

There are different ways to create a conference. A participant can create a conference using call signaling means only, such as SIP, detailed in RFC 4579. For a conferencing client to have more flexibility in defining the characteristics and capabilities of a conference, a conferencing client would implement a conference control protocol client. By using a conference control protocol, the client can determine the capabilities of a conferencing system and its various resources.

Figure 2.8 provides an example of one client, "Alice," determining the conference blueprints available for a particular conferencing system and creating a conference based on the desired blueprint.

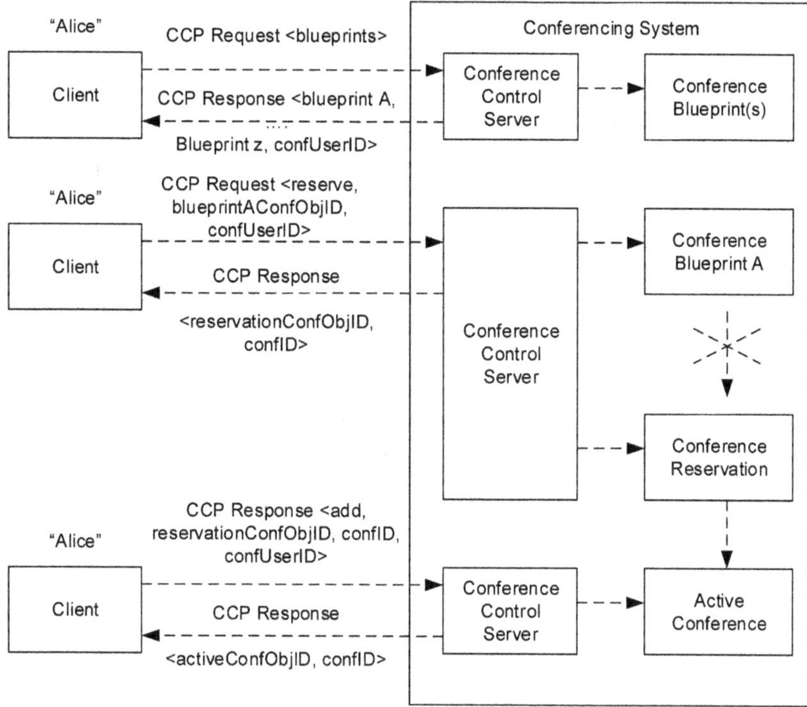

FIGURE 2.8 Client Creation of Conference Using Blueprints. (Copyright: IETF.)

Upon receipt of the conference control protocol request for blueprints, the conferencing system would first authenticate "Alice" (and allocate a conference user identifier, if necessary) and then ensure that "Alice" has the appropriate authority based on system policies to receive any blueprints supported by that system. Any blueprints that "Alice" is authorized to use are returned in a response, along with the conference user ID.

Upon receipt of the conference control protocol response containing the blueprints, "Alice" determines which blueprint to use for the conference to be created. "Alice" creates a conference object based on the blueprint (i.e., clones) and modifies applicable fields, such as membership list and "start" time. "Alice" then sends a request to the conferencing system to create a conference reservation based upon the updated blueprint.

Upon receipt of the conference control protocol request to "reserve" a conference based upon the blueprint in the request, the conferencing system ensures that the blueprint received is a valid blueprint (i.e., the values of the various field are within range). The conferencing system determines the appropriate read/write access of any users to be added to a conference based on this blueprint (using membership, roles, etc.). The conferencing system uses the received blueprint to clone a conference reservation. The conferencing system also reserves or allocates a conference ID to be used for any subsequent protocol requests from any of the members of the conference. The conferencing system maintains the mapping between this conference ID and the conference object ID associated with the reservation through the conference instance.

Upon receipt of the conference control protocol response to reserve the conference, "Alice" can now create an active conference using that reservation or create additional reservations based upon the existing reservations. In this example, "Alice" has reserved a meetme conference bridge. Thus, "Alice" provides the conference information, including the necessary conference ID, to desired participants. When the first participant, including "Alice," requests to be added to the conference, an active conference and focus are created. The focus is associated with the conference ID received in the request. Any participants that have the authority to manipulate the conference would receive the conference object identifier of the active conference object in the response.

2.9.2 Participant Manipulations

There are different ways to affect a participant state in a conference. A participant can join and leave the conference using call signaling means only, such as SIP. This kind of operation is called "first party signaling" and does not affect the state of other participants in the conference.

Limited operations for controlling other conference participants (a so-called "third party control") through the focus, using call signaling only, may also be available for some signaling protocols. For example, "Conferencing for SIP User Agents" (RFC 4579) shows how SIP with REFER can be used to achieve this functionality. In order to perform richer conference control, a user client needs to implement a conference control protocol client. By using a conference control protocol, the client can affect its own state, the state of other participants, and the state of various resources (such as media mixers) that may indirectly affect the state of any of the conference participants.

Figure 2.9 provides an example of one client, "Alice," impacting the state of another client, "Bob." This example assumes an established conference. In this example, "Alice" wants to add "Bob" to the conference.

Upon receipt of the conference control protocol request to "add" a party ("Bob") in the specific conference as identified by the conference object ID, the conferencing system ensures that "Alice" has the appropriate authority based on the policies associated with that specific conference object to perform the operation. The conferencing system must also determine whether "Bob" is already a user of this conferencing system or whether he is a new user.

If "Bob" is a new user for this conferencing system, a conference user identifier is created for "Bob." Based upon the addressing information provided for "Bob" by "Alice," the call signaling to add "Bob" to the conference is instigated through the focus. Once the call signaling indicates that "Bob" has been successfully added to the specific conference, per updates to the state, and depending upon the policies, other participants (including "Bob") may be notified of the addition of "Bob" to the conference via the conference notification service.

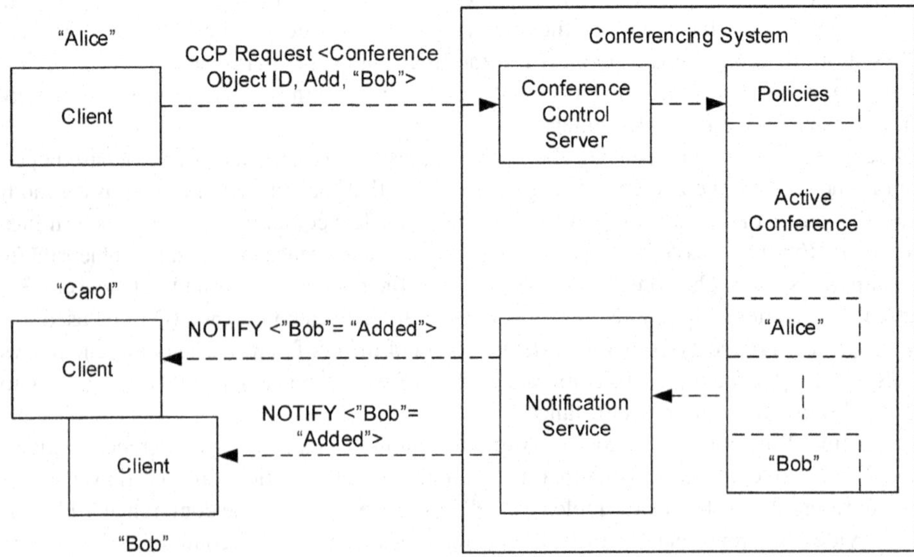

FIGURE 2.9 Client Manipulation of Conference – Add a Party. (Copyright: IETF.)

2.9.3 Media Manipulations

There are different ways to manipulate the media in a conference. A participant can change its own media streams by, for example, sending re-INVITE with new SDP content using SIP only. This kind of operation is called "first party signaling" and does not affect the state of other participants in the conference. In order to perform richer conference control, a user client needs to implement a conference control protocol client. By using a conference control protocol, the client can manipulate the state of various resources, such as media mixers, which may indirectly affect the state of any of the conference participants. Figure 2.10 provides an example of one client, "Alice," impacting the media state of another client, "Bob." This example assumes an established conference. In this example, the client "Alice," whose role is "moderator" of the conference, wants to mute "Bob" on a medium-size multiparty conference, as his device is not muted (and he is obviously not listening to the call), and background noise in his office environment is disruptive to the conference.

Upon receipt of the conference control protocol request to "mute" a party ("Bob") in the specific conference as identified by the conference object ID, the conference server ensures that "Alice" has the appropriate authority based on the policies associated with that specific conference object to perform the operation. "Bob"'s status is marked as "recvonly," and the conference object is updated to reflect that "Bob"'s media is not to be "mixed" with the conference media. Depending upon the policies, other participants (including "Bob") may be notified of this change via the conference notification service.

2.9.4 Sidebar Manipulations

A sidebar can be viewed as a separate conference instance that only exists within the context of a parent conference instance. Although viewed as an independent conference instance, it cannot exist without a parent. A sidebar is created using the same mechanisms employed for a standard conference, as described in Section 2.7.1. A conference object representing a sidebar is created by cloning the parent associated with the existing conference and updating any information specific to the sidebar. A sidebar conference object is implicitly linked to the parent conference object (i.e., it is not an independent object) and is associated with the parent conference object identifier, as shown in Figure 2.11. A conferencing system manages and enforces the parent and appropriate localized

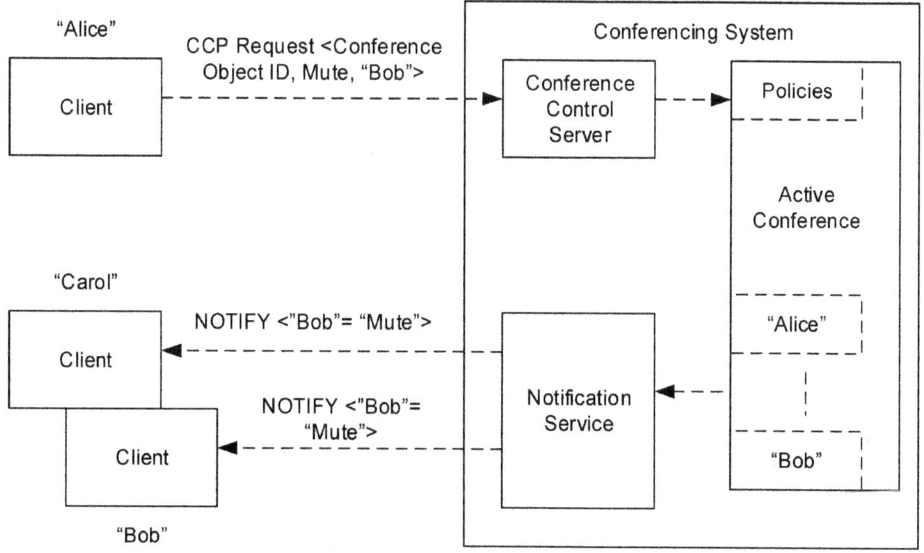

FIGURE 2.10 Client Manipulation of Conference – Mute a Party. (Copyright: IETF.)

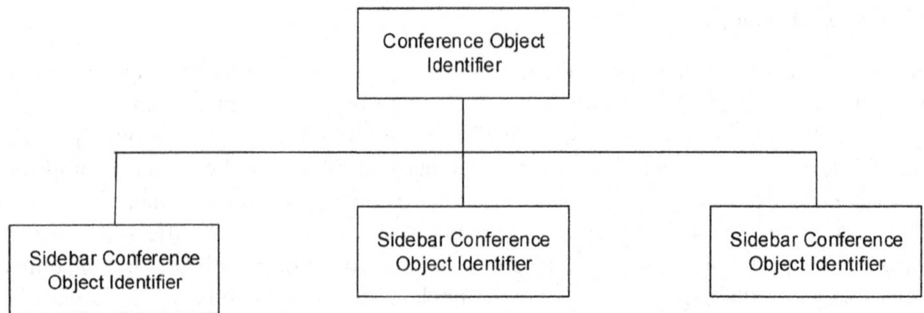

FIGURE 2.11 Conference Object Mapping.

restrictions on the sidebar conference object (e.g., no members from outside the parent conference instance can join, sidebar conference cannot exist if parent conference is terminated, etc.).

Figure 2.11 illustrates the relationship between a conference object and associated sidebar conference objects within a conferencing system. Each sidebar conference object has a unique conference object identifier, as described in Section 2.6.2.1. The main conference object identifier acts as a top-level identifier for associated sidebars.

A sidebar conference object identifier follows many of the concepts outlined in the cloning tree model described in Section 2.7.1. A sidebar conference object contains a subset of members from the original conference object. Properties of the sidebar conference object can be manipulated by a conference control protocol using the unique conference object identifier for the sidebar. It is also possible for the top-level conference object to enforce policy on the sidebar object (similar to parent enforceable, as discussed in Section 2.7.1).

2.9.4.1 Internal Sidebar

Figure 2.12 provides an example of one client, "Alice," involved in active conference with "Bob" and "Carol." "Alice" wants to create a sidebar to have a side discussion with "Bob" while still viewing the video associated with the main conference. Alternatively, the audio from the main conference could be maintained at a reduced volume. "Alice" initiates the sidebar by sending a request to the conferencing system to create a conference reservation based upon the active conference object. "Alice" and "Bob" would remain on the roster of the main conference, such that other participants could be aware of their participation in the main conference, while an internal-sidebar conference is occurring.

Upon receipt of the conference control protocol request to "reserve" a new sidebar conference, based upon the active conference received in the request, the conferencing system uses the received active conference to clone a conference reservation for the sidebar. As discussed previously, the sidebar reservation is NOT independent of the active conference (i.e., parent). The conferencing system also reserves or allocates a conference ID to be used for any subsequent protocol requests from any of the members of the conference. The conferencing system maintains the mapping between this conference ID and the conference object ID associated with the sidebar reservation through the conference instance.

Upon receipt of the conference control protocol response to reserve the conference, "Alice" can now create an active conference using that reservation or create additional reservations based upon the existing reservations. In this example, "Alice" wants only "Bob" to be involved in the sidebar; thus, she manipulates the membership. "Alice" also only wants the video from the original conference and wants the audio to be restricted to the participants in the sidebar.

Alternatively, "Alice" could manipulate the media values to receive the audio from the main conference at a reduced volume. "Alice" sends a conference control protocol request to update the information in the reservation and to create an active conference. Upon receipt of the conference control

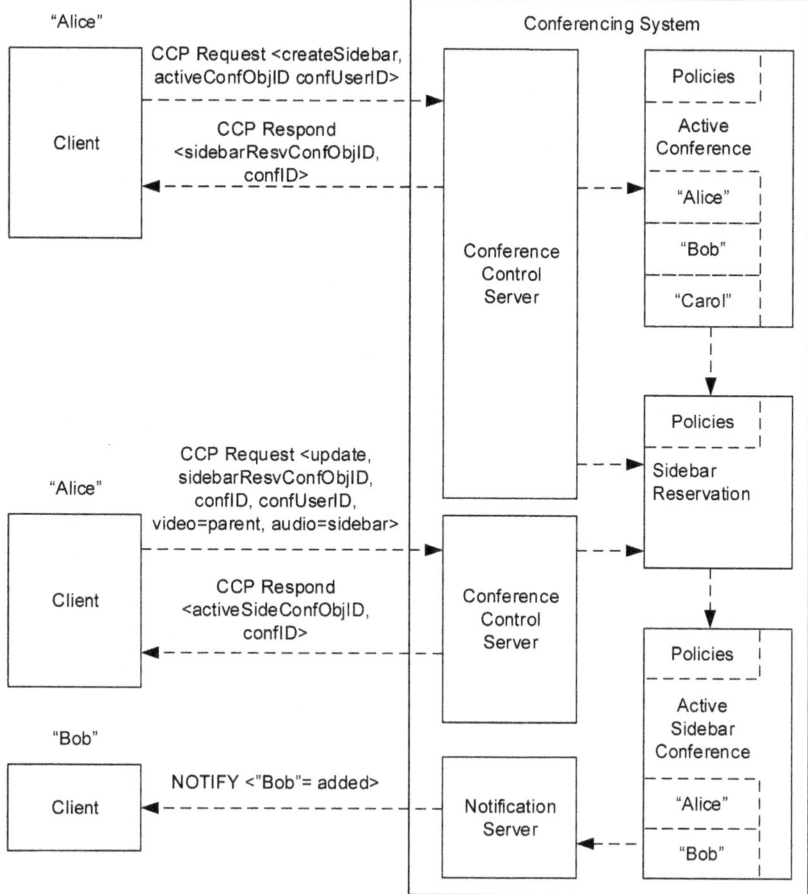

FIGURE 2.12 Client Creation of a Sidebar Conference. (Copyright: IETF.)

protocol request to update the reservation and to create an active conference for the sidebar, as identified by the conference object ID, the conferencing system ensures that "Alice" has the appropriate authority based on the policies associated with that specific conference object to perform the operation. The conferencing system must also validate the updated information in the reservation, ensuring that a member like "Bob" is already a user of this conferencing system. Depending upon the policies, the initiator of the request (i.e., "Alice") and the participants in the sidebar (i.e., "Bob") may be notified of his addition to the sidebar via the conference notification service.

2.9.4.2 External Sidebar

Figure 2.13 provides an example of one client, "Alice," involved in an active conference with "Bob," "Carol," "David," and "Ethel." "Alice" gets an important text message via a whisper from "Bob" that a critical customer needs to talk to "Alice," "Bob," and "Ethel." "Alice" creates a sidebar to have a side discussion with the customer "Fred," including the participants in the current conference with the exception of "Carol" and "David," who remain in the active conference. "Alice" initiates the sidebar by sending a request to the conferencing system to create a conference reservation based upon the active conference object. "Alice," "Bob," and "Ethel" would remain on the roster of the main conference in a hold state. Whether or not the hold state of these participants is visible to other participants depends upon the individual and local policy.

Upon receipt of the conference control protocol request to "reserve" a new sidebar conference, based upon the active conference received in the request, the conferencing system uses the received

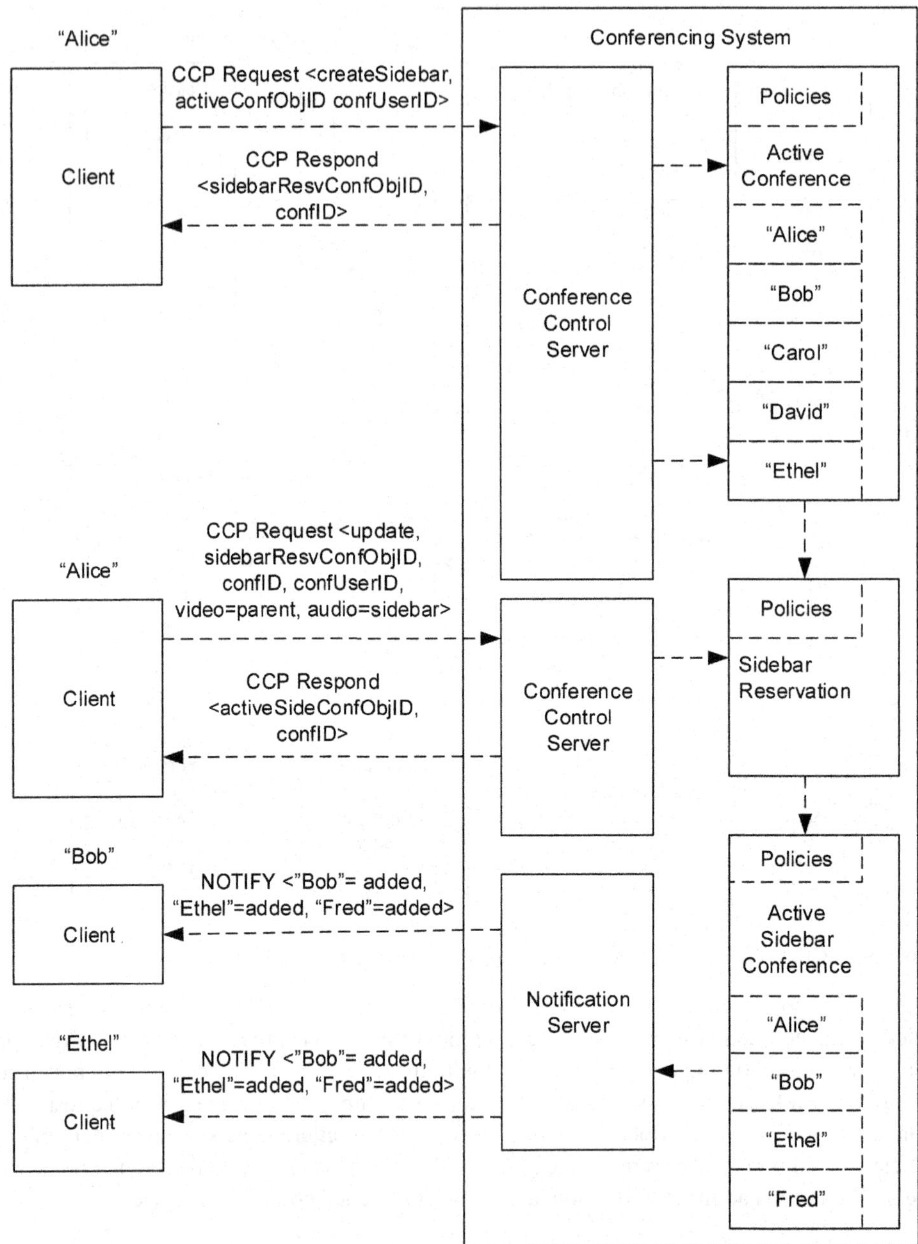

FIGURE 2.13 Client Creation of an External Sidebar. (Copyright: IETF.)

active conference to clone a conference reservation for the sidebar. As discussed previously, the sidebar reservation is NOT independent of the active conference (i.e., parent). The conferencing system also reserves or allocates a conference ID to be used for any subsequent protocol requests from any of the members of the conference. The conferencing system maintains the mapping between this conference ID and the conference object ID associated with the sidebar reservation through the conference instance.

Upon receipt of the conference control protocol response to reserve the conference, "Alice" can now create an active conference using that reservation or create additional reservations based upon the existing reservations. In this example, "Alice" wants only "Bob" and "Ethel," along with the new participant "Fred," to be involved in the sidebar; thus, she manipulates the membership. "Alice" sets the media such that the participants in the sidebar do not receive any media from the main conference. "Alice" sends a conference control protocol request to update the information in the reservation and to create an active conference.

Upon receipt of the conference control protocol request to update the reservation and to create an active conference for the sidebar, as identified by the conference object ID, the conferencing system ensures that "Alice" has the appropriate authority based on the policies associated with that specific conference object to perform the operation. The conferencing system must also validate the updated information in the reservation, ensuring whether the member "Fred" is already a user of this conferencing system or whether he is a new user. Since "Fred" is a new user for this conferencing system, a conference user identifier is created for "Fred." Based upon the addressing information provided for "Fred" by "Alice," the call signaling to add "Fred" to the conference is instigated through the focus. Depending upon the policies, the initiator of the request (i.e., "Alice") and the participants in the sidebar (i.e., "Bob" and "Ethel") may be notified of his addition to the sidebar via the conference notification service.

2.9.5 Floor Control Using Sidebars

Floor control with sidebars can be used to realize conferencing scenarios such as an analyst briefing. In this scenario, the conference call has a panel of speakers who are allowed to talk in the main conference. The other participants are the analysts, who are not allowed to speak unless they have the floor. To request access to the floor, they have to join a new sidebar with the moderator and ask their question. The moderator can also whisper to each analyst what their status/position is in the floor control queue, similarly to the example in Figure 2.15.

Figure 2.14 provides an example of the configuration involved for this type of conference. As in the previous sidebar examples, there is the main conference along with a sidebar. "Alice" and "Bob" are the main participants in the conference, with "A1," "A2," and "A3" representing the analysts. The sidebar remains active throughout the conference, with the moderator, "Carol," serving as the chair. As discussed previously, the sidebar conference is NOT independent of the active conference (i.e., parent). The analysts are provided with the conference object ID associated with the active sidebar when they join the main conference. The conferencing system also allocates a conference ID to be used for any subsequent manipulations of the sidebar conference. The conferencing system maintains the mapping between this conference ID and the conference object ID associated with the active sidebar conference through the conference instance.

The analysts are permanently muted while in the main conference. The analysts are moved to the sidebar when they wish to speak. Only one analyst is given the floor at a given time. All participants in the main conference receive audio from the sidebar conference as well as audio provided by the panelists in the main conference.

When "A1" wishes to ask a question, he sends a Floor Request message to the floor control server. Upon receipt of the request, the floor control server notifies the moderator, "Carol," of the active sidebar conference, who is serving as the floor chair. Note that this signaling flow is not shown in the diagram. Since no other analysts have yet requested the floor, "Carol" indicates to the floor control server that "A1" may be granted the floor.

2.9.6 Whispering or Private Messages

The case of private messages can be handled as a sidebar with just two participants, similarly to the example in Section 2.9.4.1, but rather than using audio within the sidebar, "Alice" could add an additional text-based media stream to the sidebar. The other context, referred to as whisper, in this specification refers to situations involving one-time media targeted to specific user(s). An example

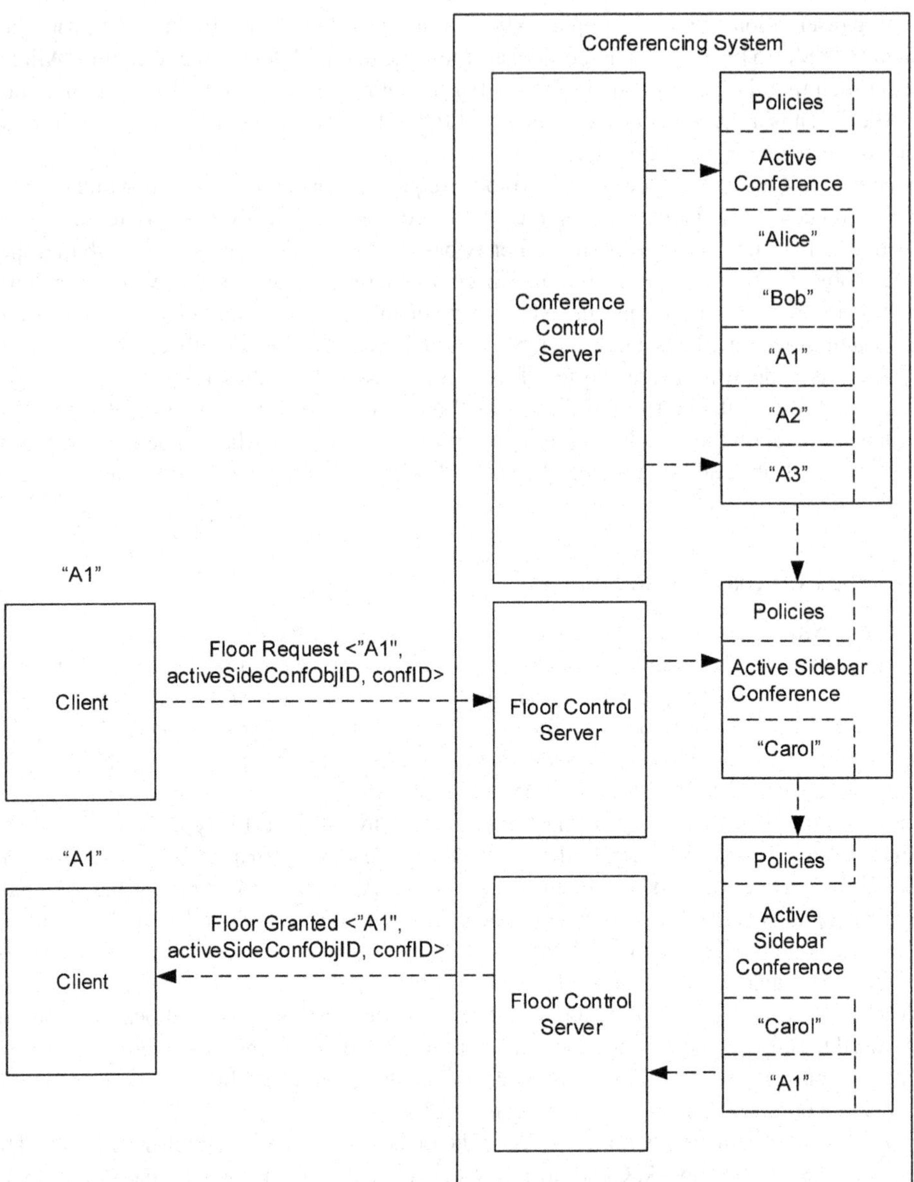

FIGURE 2.14 Floor Control with Sidebars. (Copyright: IETF.)

of a whisper would be an announcement injected only to the conference chair or to a new participant joining a conference.

Figure 2.15 provides an example of one user, "Alice," who is chairing a fixed-length conference with "Bob" and "Carol." The configuration is such that only the chair is provided with a warning when there are only 10 minutes left in the conference. At that time, "Alice" is moved into a sidebar created by the conferencing system, and only "Alice" receives the announcement.

When the conferencing system determines that there are only 10 minutes left in the conference which "Alice" is chairing, rather than creating a reservation as was done for the sidebar in Section 2.9.4.1, the conferencing system directly creates an active sidebar conference, based on the active conference associated with "Alice." As discussed previously, the sidebar conference is NOT

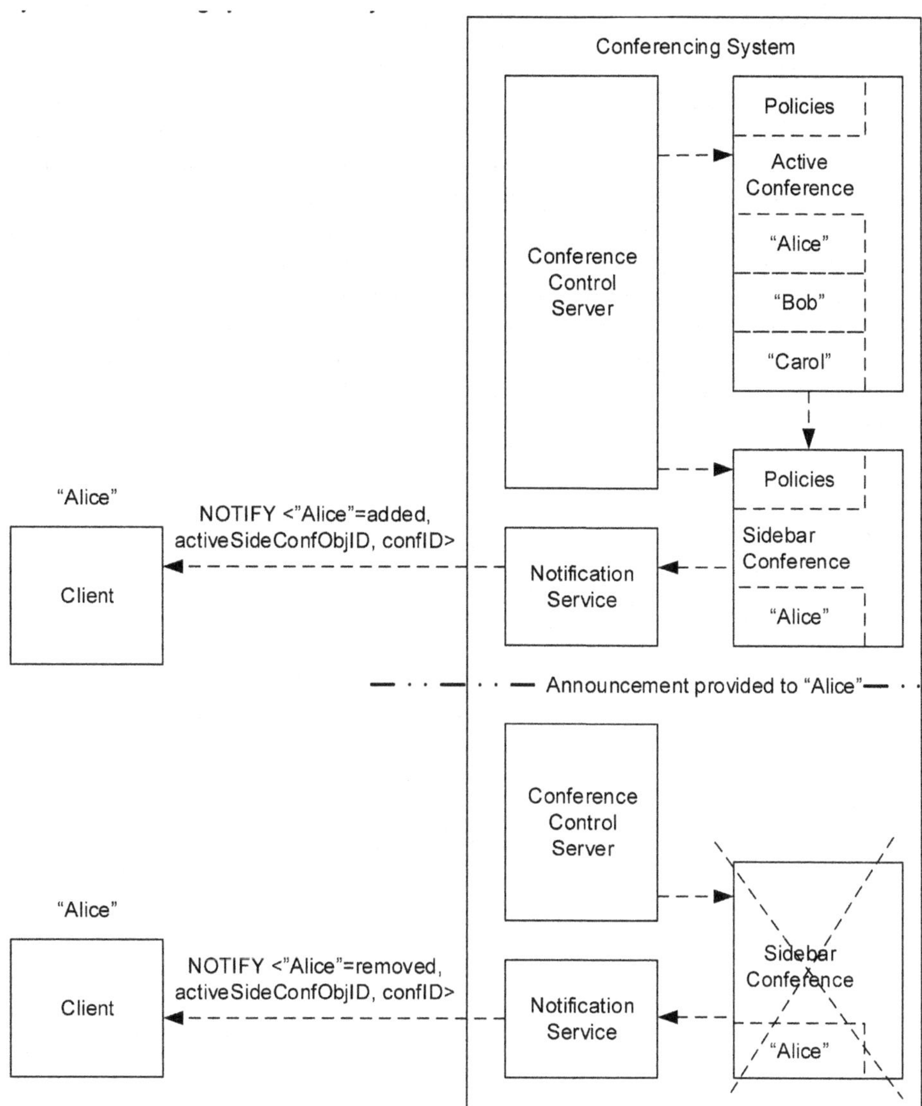

FIGURE 2.15 Whisper. (Copyright: IETF.)

independent of the active conference (i.e., parent). The conferencing system also allocates a conference ID to be used for any subsequent manipulations of the sidebar conference. The conferencing system maintains the mapping between this conference ID and the conference object ID associated with the active sidebar conference through the conference instance.

Immediately upon creation of the active sidebar conference, the announcement media is provided to "Alice." Depending upon the policies, "Alice" may be notified of her addition to the sidebar via the conference notification service. "Alice" continues to receive the media from the main conference. Upon completion of the announcement, "Alice" is removed from the sidebar, and the sidebar conference is deleted. Depending upon the policies, "Alice" may be notified of her removal from the sidebar via the conference notification service.

2.9.7 Conference Announcements and Recordings

Each participant can require a different type of announcement and/or recording service from the system. For example, "Alice," the conference chair, could be listening to a roll call, while "Bob" may be using a telephony user interface to create a sidebar. Some announcements would apply to all the participants, such as "This conference will end in 10 minutes." Recording is often required to capture the names of participants as they join a conference, typically after the participant has entered an access code, as discussed in Section 2.9.8. These recorded names are then announced to all the participants as the new participant is added to the active conference.

An example of a conferencing recording and announcement, along with collecting the dual-tone multi-frequency (DTMF), within the context of this framework, is shown in Figure 2.16.

Upon receipt of the conference control protocol request from "Alice" to join "Bob"'s conference, the conferencing system maps the identifier received in the request to the conference object representing "Bob"'s active conference. The conferencing system determines that a password is

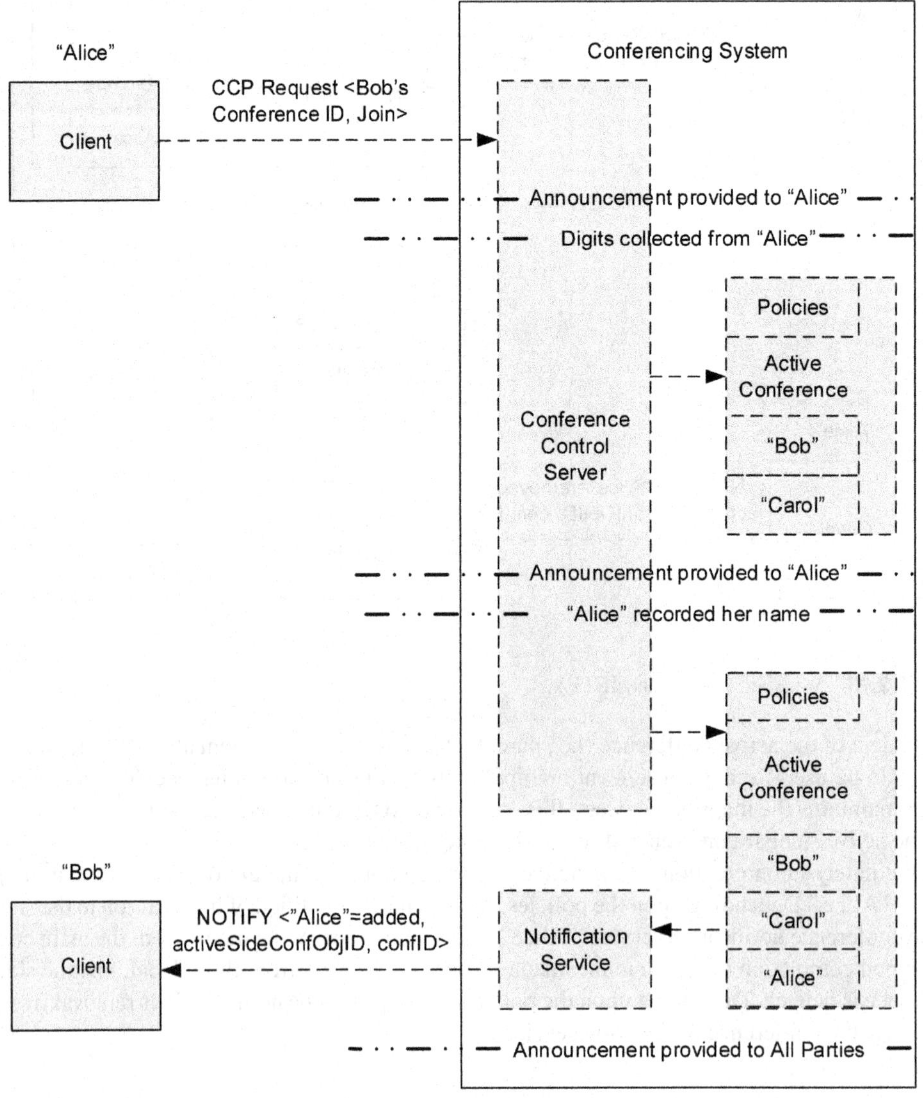

FIGURE 2.16 Recording and Announcements. (Copyright: IETF.)

required for this specific conference; thus, an announcement asking "Alice" to enter the password is provided to "Alice." Once "Alice" enters the password, it is validated against the policies associated with "Bob"'s active conference. The conferencing system then connects to a server that prompts and records "Alice"'s name. The conferencing system must also determine whether "Alice" is already a user of this conferencing system or whether she is a new user.

If "Alice" is a new user for this conferencing system, a conference user identifier is created for "Alice." Based upon the addressing information provided by "Alice," the call signaling to add "Alice" to the conference is instigated through the focus. Once the call signaling indicates that "Alice" has been successfully added to the specific conference, per updates to the state, and depending upon the policies, other participants (e.g., "Bob") are notified of the addition of "Alice" to the conference via the conference notification service, and an announcement is provided to all the participants indicating that "Alice" has joined the conference.

2.9.8 Monitoring for DTMF

The conferencing system also needs the capability to monitor for DTMF from each individual participant. This would typically be used to enter the identifier and/or access code for joining a specific conference. An example of DTMF monitoring, within the context of the framework elements, is shown in Figure 2.16.

2.9.9 Observing and coaching

The capability to observe a conference allows a participant with the appropriate authority to listen to the conference, typically without being an active participant and often as a hidden participant. When such a capability is available on a conferencing system, there is often an announcement provided to each participant as they join the conference indicating that the call may be monitored. This capability is useful in the context of conferences that might be experiencing technical difficulties, thus allowing a technician to listen in to evaluate the type of problem.

This capability could also apply to call center applications, as it provides a mechanism for a supervisor to observe how the agent is handling a particular call with a customer. This scenario can be handled by a supervisor adding themselves to the existing active conference with a listen-only audio media path. Whether the agent is aware of when the supervisor joins the call should be configurable.

Taking the supervisor capability one step further introduces a scenario whereby the agent can hear the supervisor as well as the customer. The customer can still only hear the agent. This scenario would involve the creation of a sidebar involving the agent and the supervisor. Both the agent and supervisor receive the audio from the main conference. When the agent speaks, it is heard by the customer in the main conference. When the supervisor speaks, it is heard only by the agent in the sidebar conference. An example of observing and coaching is shown in Figure 2.17. In this example, call center agent "Bob" is involved in a conference with customer "Carol." Since "Bob" is a new agent and "Alice" sees that he has been on the call with "Carol" for longer than normal, she decides to observe the call and coach "Bob" as necessary.

Upon receipt of the conference control protocol request from "Alice" to "reserve" a new sidebar conference, based upon the active conference received in the request, the conferencing system uses the received active conference to clone a conference reservation for the sidebar. The conferencing system also reserves or allocates a conference ID to be used for any subsequent protocol requests from any of the members of the conference. The conferencing system maintains the mapping between this conference ID and the conference object ID associated with the sidebar reservation through the conference instance.

Upon receipt of the conference control protocol response to reserve the conference, "Alice" can now create an active conference using that reservation or create additional reservations based upon

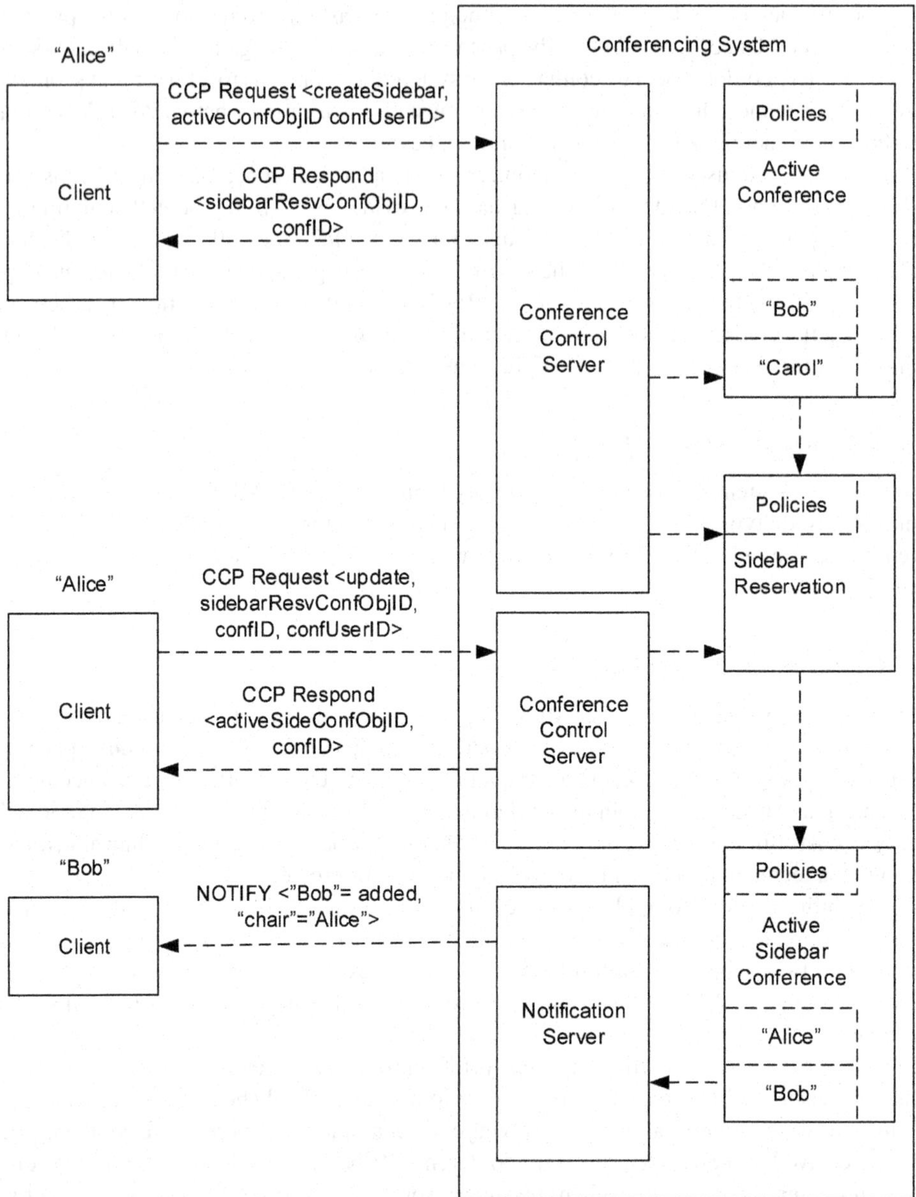

FIGURE 2.17 Supervisor Creating a Sidebar for Observing/Coaching. (Copyright: IETF.)

the existing reservations. In this example, "Alice" wants only "Bob" to be involved in the sidebar; thus, she manipulates the membership. "Alice" also wants the audio to be received by herself and "Bob" from the original conference, but she wants any outgoing audio from herself to be restricted to the participants in the sidebar, whereas "Bob"'s outgoing audio should go to the main conference, so that both "Alice" and the customer "Carol" hear the same audio from "Bob." "Alice" sends a conference control protocol request to update the information in the reservation and to create an active conference.

Upon receipt of the conference control protocol request to update the reservation and to create an active conference for the sidebar, as identified by the conference object ID, the conferencing system ensures that "Alice" has the appropriate authority based on the policies associated with that specific conference object to perform the operation. Based upon the addressing information provided for "Bob" by "Alice," the call signaling to add "Bob" to the sidebar with the appropriate media characteristics is instigated through the focus. "Bob" is notified of his addition to the sidebar via the conference notification service; thus, he is aware that "Alice," the supervisor, is available for coaching him through this call.

2.10 RELATIONSHIPS BETWEEN SIP AND CENTRALIZED CONFERENCING SYSTEM FRAMEWORKS

The SIP Conferencing Framework (RFC 4353) provides an overview of a wide range of centralized conferencing solutions known today in the conferencing industry. The specification introduces a terminology and logical entities in order to systemize the overview and to show the common core of many of these systems. The logical entities and the listed scenarios in the SIP Conferencing Framework are used to illustrate how SIP (RFC 3261) can be used as a signaling means in these conferencing systems. The SIP Conferencing Framework does not define new conference control protocols to be used by the general conferencing system. It uses only basic SIP (RFC 3261), the SIP Conferencing for User Agents (RFC 4579), and the SIP Conference Package (RFC 4575) for basic SIP conferencing realization.

This centralized conferencing framework specification defines a particular centralized conferencing system and the logical entities implementing it. It also defines a particular data model and refers to the set of protocols (beyond call signaling means) to be used among the logical entities for implementing advanced conferencing features. The purpose of the XCON Working Group and this framework is to achieve interoperability between the logical entities from different vendors for controlling different aspects of advanced conferencing applications.

The logical entities defined in the two frameworks are not intended to be mapped one-to-one. The two frameworks differ in the interpretation of the internal conferencing system decomposition and the corresponding operations. Nevertheless, the basic SIP (RFC 3261), the SIP Conferencing for User Agents (RFC 4579), and the SIP Conference Package (RFC 4575) are fully compatible with both framework specifications. The basis for compatibility is provided by including the basic data elements defined in RFC 4575 in the Conference Information Data Model for Centralized Conferencing (XCON) (RFC 6501 – see Chapter 4). User agents that only support RFC 4579 and do not support the conferencing control protocol are still provided with basic SIP conferencing but cannot take advantage of any of the advanced features.

2.11 SECURITY CONSIDERATIONS

There are a wide variety of potential attacks related to conferencing due to the natural involvement of multiple endpoints and the many, often user-invoked, capabilities provided by the conferencing system. Examples of attacks include the following: an endpoint attempting to listen to conferences in which it is not authorized to participate, an endpoint attempting to disconnect or mute other users, and theft of service by an endpoint in attempting to create conferences it is not allowed to create.

There are several issues surrounding the security of this conferencing framework. One set of issues involves securing the actual protocols and the associated authorization mechanisms. This first set of issues should be addressed in the specifications specific to the protocols described in Section 2.8 and policy control. The protocols used for manipulation and retrieval of confidential information need to support a confidentiality and integrity mechanism.

Similar requirements apply for the floor control protocols. Section 2.11.3 discusses an approach for client authentication of a floor control server. It is recommended that all the protocols that interface with the conferencing system implement transport layer security (TLS). There are also security issues associated with the authorization to perform actions on the conferencing system to invoke specific capabilities. Section 2.5.2 discusses the policies associated with the conference object to ensure that only authorized entities are able to manipulate the data to access the capabilities. Another set of issues involves the privacy and security of the identity of a user in the conference, which is discussed in Section 2.11.2.

A final issue is related to denial of service (DoS) attacks on the conferencing system itself. In order to minimize the potential for DoS attacks, it is recommended that conferencing systems require user authentication and authorization for any client participating in a conference. It is recommended that the specific signaling and media protocols include mechanisms to minimize the potential for DoS.

2.11.1 User Authentication and Authorization

Many policy authorization decisions are based on the identity of the user or the role that a user may have. Conferencing systems typically require authentication of users to validate their identity. There are several ways that a user might authenticate its identity to the system. For users joining a conference using one of the call signaling protocols, the user authentication mechanisms for the specific protocol often suffice. For the case of users joining the conference via SIP signaling or using the conference control protocol, TLS is recommended.

The conferencing system may also know (by e.g., out-of-band mechanisms) about specific users and assign passwords to allow these users to be authorized. In some cases, for example, public switched telephone network (PSTN) users, additional authorization may be required to allow the user to participate in the conference. This may be in the form of an interactive voice response (IVR) system or other means. The users may also be authorized by knowing a particular conference ID and a personal identification (PIN) for it. Sometimes, a PIN is not required, and the conference ID is used as a shared secret.

In the cases where a user is authorized via multiple mechanisms, it is up to the conferencing system to correlate (if desired) the authorization of the call signaling interface with other authorization mechanisms. A conferencing system can avoid the problem with multiple mechanisms by restricting the methods by which a conference can be joined. For example, many conferencing systems that provide a web interface for conferences correlate the PSTN call signaling by forcing a dial-out mode for joining the conference. Thus, there is only the need for a single PIN or password to join the conference.

When a conferencing system presents the identity of authorized users, it may choose to provide information about the way the identity was proven or verified by the system. A user may also come as a completely unauthenticated user into the system – this fact needs also to be communicated to interested parties.

When guest users interact with the system, it is often in the context of a particular conference. In this case, the user may provide a PIN or a password that is specific to the conferences and authorizes the user to take on a certain role in that conference. The guest user can then perform actions that are allowed to any user with that role. The term "password" refers to the usual reasonably sized and hard-to-predict shared secret. Today, users often have passwords containing up to 30 bits (8–16 characters) of entropy.

A PIN is a special password case – a shared secret that is only numeric and often contains a fairly small number of bits (often as few as 10 bits or three digits). When conferencing systems are used for audio on the PSTN, there is often a need to authenticate using a PIN. Typically, if the user fails to provide the correct PIN a few times in a row, the PSTN call is disconnected. The rate of making the calls and getting to the point to enter a PIN makes it fairly hard to do an exhaustive search of the PIN space even for four-digit PINs. When using a high-speed interface to connect to a conferencing system, it is often possible to make thousands of attempts per second, and the PIN space could quickly be searched. Because of this, it is not appropriate to use PINs for authorization on any of the interfaces that provide fast queries or many simultaneous queries.

Once a user is authenticated and authorized through the various mechanisms available on the conferencing system, a conference user identifier is associated with any signaling-specific user identifiers that may have been used for authentication and authorization. This conference user identifier may be provided to a specific user through the conference notification interface and will be provided to users that interact with the conferencing system using the conference control protocol. This conference user identifier is required for any subsequent operations on the conference object.

2.11.2 Security and Privacy of Identity

This conferencing system has an idea of the identity of a user, but this does not mean it can reveal this identity to other users, due to privacy considerations. Users can select various options for revealing their identity to other users. A user can be "hidden," such that other users cannot see they are a participant in the conference, "anonymous," such that users can see that another user is there but not see the identity of the user, or "public," where other users can see their identity. If there are multiple "anonymous" users, other parties will be able to see them as independent "anonymous" parties and will be able to tell how many "anonymous" parties are in the conference. Note that the visibility to other participants is dependent on their roles. For example, users' identity (including "anonymous" and "hidden") may be displayed to the moderator or administrator, subject to a conferencing system's local policies. "Hidden" status is often used by automated or machine participants of a conference (e.g., call recording) and is also used in many call center situations.

Since a centralized conferencing system based on the isfocus and conference factory allocates a unique conference user identifier for each user of the conferencing system, it is not necessary to distribute any signaling-specific user identifier to other users or participants. Access to any signaling-specific user identifiers can be controlled by applying the appropriate access control to the signaling-specific user identifiers in the data schema.

2.11.3 Floor Control Server Authentication

The floor control protocol contains mechanisms that clients can use to authenticate servers and that servers can use to authenticate clients, as described in RFC 4582 (see Chapter 6). The precise mechanisms used for such authentication can vary depending on the call control protocol used. Clients using call control protocols that employ an SDP offer/answer model, such as SIP, use the mechanism described in RFC 4583. Clients using other call control protocols make use of the mechanisms described in the BFCP Connection Establishment specification of RFC 5018 (see Section 6.15).

2.12 SUMMARY

We have described the architectural framework for the centralized multipoint multimedia conferencing that is applicable for all call signaling protocols. In fact, this architectural framework has been realized with more specific detail standardization of each of these functional entities and protocols described in the following chapters of this book (as outlined in Section 1.5 and specifically indicated in Figure 1.9). Any call signaling protocol such as SIP, H.323, Jabber, Q.931 or ISUP that exchanges media in a centralized unicast conference can use this framework. The global architectural framework shows a road map of how the conference object and information data model, CCP/CCMP, BFCP, SIP event notification, Media Control Channel Framework (CFW) (see Chapter 8), conference creation using CCP/CCMP, participant manipulations using CCP/CCMP, media manipulations using CFW, sidebar manipulations using CCP/CCMP, floor control using BFCP, whispering, conference recording and announcements, monitoring DTMF, observing and coaching, and multistream immersive telepresence that are described in subsequent chapters of this book are seamlessly integrated in the context of multipoint multimedia conferencing using standardized protocols and logical functional entities fostering interoperability. Note that we have not described the specificity of each of these protocols and functional entities.

Rather, in this framework, we outline a set of conferencing protocols, which are complementary to the call signaling protocols, for building advanced conferencing applications. The framework binds all the defined components together for the benefit of builders of conferencing systems. In this section, we have specifically described the centralized conferencing system architecture that includes conferencing constructs and identifiers, conferencing systems realization, conferencing mechanisms, conferencing scenarios realization, conferencing announcements and recordings, usage of the SIP call control signaling protocol within the context of a centralized conferencing framework, and security mechanisms.

2.13 PROBLEMS

1. What are the key objectives of the centralized conferencing architectural framework? Show and describe how all the functional entities and protocols are inter-related and complement each other.
2. What are the logical functional entities that are contained in the centralized conferencing system architecture?
3. What is a conference object? What is the conference information type of object? What is the conference policy infrastructure? Explain each of these features in detail along with conference information and conference policies.
4. What are conference constructs and conference identifiers of object, object identifiers, and user identifiers? Describe in detail.
5. How do you realize the conference system of cloning tree, ad hoc example, advanced example, and scheduling? Describe in detail.
6. What are the conference mechanisms of multipoint conferencing systems? Describe the call signaling, notification, conference control protocol, and conference floor control in detail.
7. How did you implement the following conference features: conference creation, participant manipulations, internal and external sidebar manipulations, floor control using sidebars, whispering or private messages, conference announcements and recordings, monitoring of DTMF, and observing and caching?
8. Is the relationship between SIP and centralized conferencing system frameworks complementary? Justify your findings.
9. Describe the security features for the following centralized multipoint conferencing systems: user authentication and authorization, and security and privacy of identity.

3 Media Server Control Architecture

3.1 INTRODUCTION

Application servers (ASs) host one or more instances of a communications application. Media Servers (MSs) provide real-time media processing functions. This Chapter (RFC 5567) describes media control, media mixing, and interactive voice response requirements. RFC 5167 (see Section 8.2) defines the requirements for a media server control protocol that enables an AS to control an MS. It will address the aspects of announcements, interactive voice response (IVR), and conferencing media services. The requirements are for the protocol and do not address the AS or MS functionality discussed in the media control framework. The requirements for MS control are defined in RFC 5167 (see Section 8.2) as follows:

Media Control Requirements:

1. The MS Control Protocol shall enable one or more application servers to request media services from one or more media servers.
2. The MS Control Protocol shall use a reliable transport protocol.
3. The applications supported by the protocol shall include conferencing and Interactive Voice Response media services. Note that, though the protocol enables these services, the functionality is invoked through other mechanisms.
4. Media types supported in the context of the applications shall include audio, tones, text, and video. Tones media include in-band audio or RFC 4733 payload.
5. The MS control protocol should allow, but must not require, a media resource broker (MRB) or intermediate proxy to exist with the application server and media server.
6. On the MS control channel, there shall be requests to the MS, responses from the MS, and notifications to the AS.
7. SIP/SDP (Session Initiation Protocol / Session Description Protocol) shall be used to establish and modify media connections to a media server.
8. It should be possible to support a single conference spanning multiple media servers. Note: It is probable that spanning multiple MSs can be accomplished by the AS and does not require anything in the protocol for the scenarios we have in mind. However, the concern is that if this requirement is treated too lightly, one may end up with a protocol that precludes its support.
9. It must be possible to split call legs individually, or in groups, away from a main conference on a given media server, without performing re-establishment of the call legs to the MS (e.g., for purposes such as performing IVR with a single call leg or creating sub-conferences, not for creating entirely new conferences).
10. The MS control protocol should be extendable, facilitating forward and backward compatibility.
11. The MS control protocol shall include an authentication component to ensure that only an authorized AS can communicate with the MS, and vice versa.
12. The MS control protocol shall use some form of transport protection to ensure the confidentiality and integrity of the data between the AS and MS.
13. Different application servers may have different privileges for using an MS. The protocol should prevent the AS from doing unauthorized operations on a MS.

14. The MS control protocol requires mechanisms to protect the MS resources used by one AS from another AS since the solution needs to support multiple ASs controlling one MS.
15. During session establishment, there shall be a capability to negotiate parameters that are associated with media streams. This requirement should also enable an AS managing conference to specify the media streams allowed in the conference.
16. The AS shall be able to instruct the MS to perform stream operations like mute and gain control.
17. The AS shall be able to instruct the MS to play a specific announcement.
18. The AS shall be able to request the MS to create, delete, and manipulate a mixing, IVR, or announcement session.
19. The AS shall be able to instruct the MS to play announcements to a single user or to a conference mix.
20. The MS control protocol should enable the AS to ask the MS for a session summary report. The report may include resource usage and quality metrics.
21. The MS shall be able to notify the AS of events received in the media stream if requested by the AS. (Examples – STUN request, Flow Control, etc.)

Media Mixing Requirements:

22. The AS shall be able to define a conference mix; the MS may offer different mixing topologies. The conference mix may be defined on a conference or user level.
23. The AS may be able to define a custom video layout built of rectangular sub-windows.
24. For video, the AS shall be able to map a stream to a specific sub-window or to define to the MS how to decide which stream will go to each sub-window.
25. The MS shall be able to notify the ASs of who the active sources of the media are; for example, who the active speaker is or who is being viewed in a conference. The speaker and the video source may be different, for example, a person describing a video stream from a remote camera managed by a different user.
26. The MS shall be able to inform the AS which layouts it supports.
27. The MS control protocol should enable the AS to instruct the MS to record a specific conference mix.

IVR Requirements:

28. The AS shall be able to instruct the MS to perform one or more IVR scripts and receive the results. The script may be in a server or contained in the control message.
29. The AS shall be able to manage the IVR session by sending requests to play announcements to the MS and receiving the response (e.g., DTMF). The IVR session flow, in this case, is handled by the AS by starting a next phase based on the response it receives from the MS on the current phase.
30. The AS should be able to instruct the MS to record a short participant stream and play it back. This is not a recording requirement.

Operational Requirements:

31. The MS control protocol must allow the AS to audit the MS state during an active session.
32. The MS shall be able to inform the AS about its status during an active session.

Note: These requirements may be applicable to the MRB, but they can be used by an AS if it has a one-to-one connection to the MS.

This section (RFC 5567) presents the core architectural framework to allow ASs to control MSs. An overview of the architecture describing the core logical entities and their interactions is presented in Section 3.3. The SIP (RFC 3261 – also see Section 1.2) is used as the session establishment protocol within this architecture. ASs use it both to terminate media streams on MSs and to create and manage control channels for MS control between themselves and MSs. The detailed model for MS control together with a description of SIP usage is presented in Section 3.4. Several services are described using the framework defined in this specification. Use cases for IVR services are described in Section 3.5, and conferencing use cases are described in Section 3.6.

3.2 TERMINOLOGY

See Section 1.6, Table 1.2.

3.3 ARCHITECTURE OVERVIEW

An MS is a network device that processes media streams. Examples of media processing functionality may include:

- Control of the real-time protocol (RTP) (RFC 3550) streams using the Extended RTP Profile for Real-time Transport Control Protocol (RTCP)-Based Feedback (RTP/AVPF) (RFC 4585)
- Mixing of incoming media streams
- Media stream source (for multimedia announcements)
- Media stream processing (e.g., transcoding, dual-tone multi-frequency [DTMF] detection)
- Media stream sink (for multimedia recordings)

An MS supplies one or more media processing functionalities, which may include others than those illustrated in the list, to an AS. An AS is able to send a particular call to a suitable MS, either through discovery of the capabilities that a specific MS provides or through the use of an MRB. The type of processing that a MS performs on media streams is specified and controlled by an AS. ASs are logical entities that are capable of running one or more instances of a communications application. Examples of ASs that may interact with a MS are an AS acting as a conference "focus" (as defined in RFC 4353) or an IVR application using a MS to play announcements and detect DTMF key presses.

Application servers use SIP to establish control channels between themselves and MSs. An MS control channel implements a reliable transport protocol that is used to carry the MS control protocol. A SIP dialog used to establish a control channel is referred to as an MS control dialog. ASs terminate SIP (RFC 3261 – also see Section 1.1) signaling from SIP user agents and may terminate other signaling outside the scope of this specification. They use SIP third-party call control (RFC 3725 – see Section 1.3) (3pcc) to establish, maintain, and tear down media streams from those SIP UAs to an MS.

A SIP dialog used by an AS to establish a media session on an MS is referred to as an MS media dialog. Media streams go directly between SIP UAs and MSs. MSs support multiple types of media. Common supported RTP media types include audio and video, but others such as text and the binary floor control protocol (BFCP) (RFC 4583 – see Chapter 6) are also possible. This basic architecture, showing session establishment signaling between a single AS and MS, is shown in Figure 3.1.

The architecture must support a many-to-many relationship between ASs and MSs. In real-world deployments, an AS may interact with multiple MSs, and/or an MS may be controlled by more than one AS. ASs can use the SIP uniform resource identifier (URI) as described in RFC 4240 to request basic functions from MSs. Basic functions are characterized as requiring no mid-call interactions

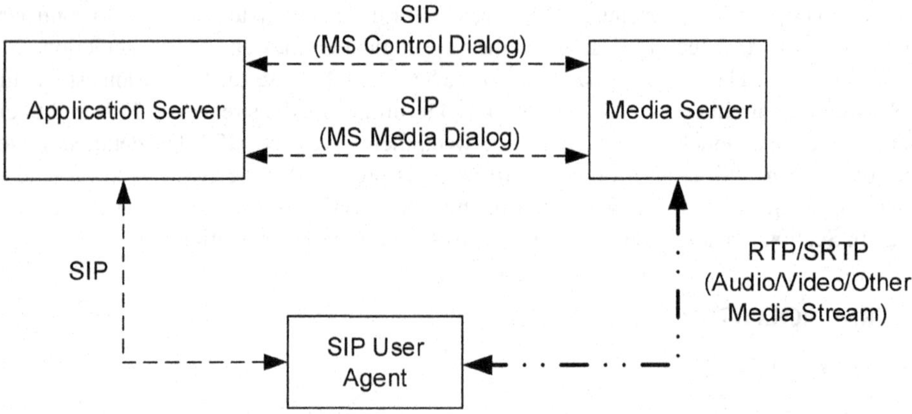

FIGURE 3.1 Basic Signaling Architecture. (Copyright: IETF.)

between the AS and the MS. Examples of these functions are simple announcement-playing or basic conference-mixing, where the AS does not need to explicitly control the mixing.

Most services, however, have interactions between the AS and the MS during a call or conference. The type of interactions can be generalized as follows:

- Commands from an AS to an MS to request the application or configuration of a function. The request may apply to a single media stream, multiple media streams associated with multiple SIP dialogs, or properties of a conference mix.
- Responses from an MS to an AS reporting on the status of particular commands.
- Notifications from an MS to an AS that report results from commands or notify changes to subscribed status. Commands, responses, and notifications are transported using one or more dedicated control channels between the AS and the MS. Dedicated control channels provide reliable, sequenced, peer-to-peer transport for MS control interactions. Implementations must support the transport control protocol (TCP) (RFC 0793) and may support the stream control transmission protocol (SCTP) (RFC 4960). Because MS control requires sequenced reliable delivery of messages, unreliable protocols such as the user datagram protocol (UDP) are not suitable. Implementations must support TLS (RFC 5246) as a transport-level security mechanism, although its use in deployments is optional. A dedicated control channel is shown in Figure 3.2.

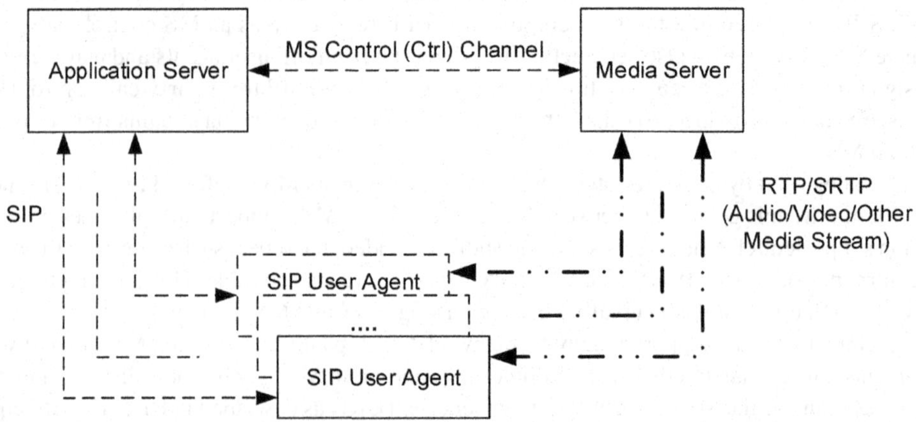

FIGURE 3.2 Media Server Control Architecture. (Copyright: IETF.)

Both ASs and MSs may interact with other servers for specific purposes beyond the scope of this specification. For example, ASs will often communicate with other infrastructure components, which are usually based on deployment requirements with links to back-office data stores and applications. MSs will often retrieve announcements from external file servers. Also, many MSs support IVR dialog services using Voice Extensible Markup Language (VoiceXML) (Version 2.0, W3C Recommendation 16 March 2004).

In this case, the MS interacts with other servers using hypertext transfer protocol (HTTP) during standard VoiceXML processing. VoiceXML MSs may also interact with speech engines (for example, using the media resource control protocol version 2 [MRCPv2] – see Chapter 14) for speech recognition and generation purposes. Some specific types of interactions between ASs and MSs are also out of scope for this specification. MS resource reservation is one such interaction. Any interactions between ASs, or between MSs, are also out of scope.

3.4 NSIP USAGE

The SIP (RFC 3261 – also see Section 1.1) was developed by the Internet Engineering Task Force (IETF) for the purposes of initiating, managing, and terminating multimedia sessions. The popularity of SIP has grown dramatically since its inception, and it is now the primary Voice over IP (VoIP) protocol. This includes being selected as the basis for architectures such as the IP Multimedia Subsystem (IMS) in the 3rd Generation Partnership Project (3GPP) and included in many of the early live deployments of VoIP-related systems. Media servers are not a new concept in Internet protocol (IP) telephony networks, and there have been numerous signaling protocols and techniques proposed for their control. The most popular techniques to date have used a combination of SIP and various markup languages to convey media service requests and responses.

As discussed in Chapter 3 and illustrated in Figure 3.1, the logical architecture described by this specification involves interactions between an AS and a MS. The SIP interactions can be broken into "MS media dialogs," which are used between an AS and an MS to establish media sessions between an endpoint and an MS, and "MS control dialogs," which are used to establish and maintain MS control channels.

SIP is the primary signaling protocol for session signaling and is used for all media sessions directed towards an MS as described in this specification. MSs may support other signaling protocols, but this type of interaction is not considered here. ASs may terminate non-SIP signaling protocols but must gateway those requests to SIP when interacting with an MS.

SIP will also be used for the creation, management, and termination of the dedicated MS control channel(s). Control channel(s) provide reliable sequenced delivery of MS Control Protocol messages. The ASs and MSs use the session description protocol (SDP) attributes defined in RFC 4145 to allow SIP negotiation of the control channel. A control channel is closed when SIP terminates the corresponding MS control dialog. Further details and example flows are provided in the SIP Control Framework (RFC 6230 – see Chapter 8). The SIP Control Framework also includes basic control message semantics corresponding to the types of interactions identified in Section 3.3. It uses the concept of "packages" to allow domain-specific protocols to be defined using the Extensible Markup Language (XML) (Fourth Edition), World Wide Web Consortium Recommendation RECxml-20060816, August 2006) format. The MS Control Protocol is made up of one or more packages for the SIP Control Framework. Using SIP for both media and control dialogs provides a number of inherent benefits over other potential techniques. These include:

1. The use of SIP location and rendezvous capabilities, as defined in RFC 3263. This provides core mechanisms for routing a SIP request based on techniques such as Domain Name System service record (DNS SRV) and Name Authority Pointer (NAPTR) records. The SIP infrastructure makes heavy use of such techniques.

2. The security and identity properties of SIP; for example, using transport layer security (TLS) for reliably and securely connecting to another SIP-based entity. The SIP protocol has a number of identity mechanisms that can be used. RFC 3261 (also see Section 1.2) provides an intradomain digest-based mechanism, and RFC 4474 defines a certificate-based interdomain identity mechanism. SIP with S/MIME provides the ability to secure payloads using encrypted and signed certificate techniques.
3. SIP has extremely powerful and dynamic media-negotiation properties as defined in RFC 3261 (also see Section 1.2) and RFC 3264.
4. The ability to select an appropriate SIP entity based on capability sets as discussed in RFC 3840. This provides a powerful function that allows MSs to convey a specific capability set. An AS is then free to select an appropriate MS based on its requirements.
5. Using SIP also provides consistency with IETF protocols and usages. SIP was intended to be used for the creation and management of media sessions, and this provides a correct usage of the protocol.

As mentioned previously in this section, media services using SIP are fairly well understood. Some previous proposals suggested using the SIP INFO (RFC 2976) method as the transport vehicle between the AS and the MS. Using SIP INFO in this way is not advised for a number of reasons, which include:

- INFO is an opaque request with no specific semantics. A SIP endpoint that receives an INFO request does not know what to do with it based on SIP signaling.
- SIP INFO was not created to carry generic session control information along the signaling path, and it should only really be used for optional application information, e.g., carrying mid-call public switched telephone network (PSTN) signaling messages between PSTN gateways.
- SIP INFO traverses the signaling path, which is an inefficient use for control messages that can be routed directly between the AS and the MS.
- RFC 3261 (also see Section 1.1) contains rules when using an unreliable protocol such as UDP. When a packet reaches a size close to the maximum transmission unit (MTU), the protocol should be changed to TCP. This type of operation is not ideal when constantly dealing with large payloads such as XML-formatted MS control messages.

3.5 MEDIA CONTROL FOR IVR SERVICES

One of the functions of an MS is to assist an AS that is implementing IVR services by performing media processing functions on media streams. Although "IVR" is somewhat generic terminology, the scope of media functions provided by an MS addresses the needs for user interaction dialogs. These functions include media transcoding, basic announcements, user input detection (via DTMF or speech), and media recording.

A particular IVR or user dialog application typically requires the use of several specific media functions, as described earlier. The range and complexity of IVR dialogs can vary significantly, from a simple single announcement play-back to complex voice mail applications. As previously discussed, an AS uses SIP (RFC 3261 – also see Section 1.1) and SDP (RFC 4566) to establish and configure media sessions to a MS. An AS uses the MS control channel, established using SIP, to invoke IVR requests and to receive responses and notifications. This topology is shown in Figure 3.3.

The variety in complexity of AS IVR services requires support for different levels of media functions from the MS, as described in the following sub-sections.

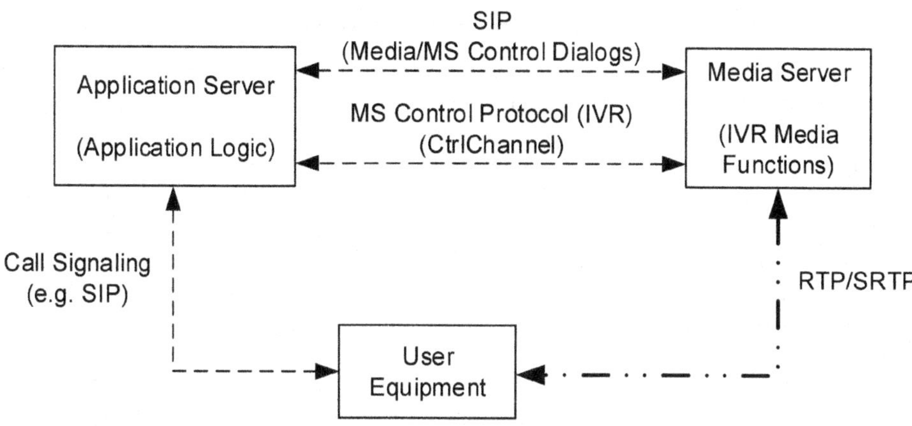

FIGURE 3.3 IVR Topology. (Copyright: IETF.)

3.5.1 Basic IVR Services

For simple basic announcement requests, the MS control channel, as depicted in Figure 3.3, is not required. Simple announcement requests may be invoked on the MS using the SIP URI mechanism defined in RFC 4240. This interface allows no digit detection or collection of user input and no mid-call dialog control. However, many applications only require basic media services, and the processing burden on the MS to support more complex interactions with the AS would not be needed in that case.

3.5.2 IVR Services with Mid-Call Controls

For more complex IVR dialogs, which require mid-call interaction and control between the AS and the MS, the MS control channel (as shown in Figure 3.3) is used to invoke specific media functions on the MS. These functions include, but are not limited to, complex announcements with barge-in facility, user input detection and reporting (e.g., DTMF) to an AS, DTMF and voice-activity controlled recordings, etc.

Composite services, such as play-collect and play-record, are also addressed by this model. Mid-call control also allows ASs to subscribe to IVR-related events and the MS to notify the AS when these events occur. Examples of such events are announcement completion events, record completion events, and reporting of collected DTMF digits.

3.5.3 Advanced IVR Services

Although IVR services with mid-call control, as described earlier, provide a comprehensive set of media functions expected from a MS, the advanced IVR services model allows a higher level of abstraction describing application logic, as provided by VoiceXML, to be executed on the MS. Invocation of VoiceXML IVR dialogs may be via the "Prompt and Collect" mechanism of RFC 4240.

Additionally, the IVR control protocol can be extended to allow VoiceXML requests to also be invoked over the MS control channel. VoiceXML IVR services invoked on the MS may require an HTTP interface (not shown in Figure 3.3) between the MS and one or more back-end servers that host or generate VoiceXML specifications. The back-end server(s) may or may not be physically separate from the AS.

3.6 MEDIA CONTROL FOR CONFERENCING SERVICES

RFC 4353 describes the overall architecture and protocol components needed for multipoint conferencing using SIP. The framework for centralized conferencing (RFC 5239 – see Chapter 2) extends the framework to include a protocol between the user and the conferencing server. RFC 4353 describes the conferencing server decomposition but leaves the specifics open.

This section describes the decomposition and discusses the functionality of the decomposed functional units. The conferencing factory and the conference focus are part of the AS described in this specification. An AS uses SIP 3pcc (RFC 3725 – see Section 1.3) to establish media sessions from SIP user agents to an MS. The same mechanism is used by the AS, as described in this section, to add/remove participants to/from a conference as well as to handle the involved media streams set up on a per-user basis.

Since the XCON framework has been conceived as protocol-agnostic when talking about the call signaling protocol used by users to join a conference, an XCON-compliant AS will have to take care of gatewaying non-SIP signaling negotiations. This is in order to set up and make available valid SIP media sessions between itself and the MS while still keeping the non-SIP interaction with the user in a transparent way.

To complement the functionality provided by 3pcc and by the XCON control protocol, the AS makes use of a dedicated MS control channel in order to set up and manage media conferences on the MS. Figure 3.4 shows the signaling and media paths for a two-participant conference. The three SIP dialogs between the AS and MS establish one control session (1c) and two media sessions (2m) from the participants (one originally signaled using H.323 and then gatewayed into SIP and one signaled directly in SIP).

As a conference focus, the AS is responsible for setting up and managing a media conference on the MSs in order to make sure that all the media streams provided in a conference are available to its participants. This is achieved by using the services of one or more mixer entities (as described in RFC 4353), whose role as part of the MS is described in this section. Services required by the AS include, but are not limited to, means to set up, handle, and destroy a new media conference, adding and removing participants from a conference, managing media streams in a conference, controlling the layout and the mixing configuration for each involved media, allowing per-user custom media profiles, and so on.

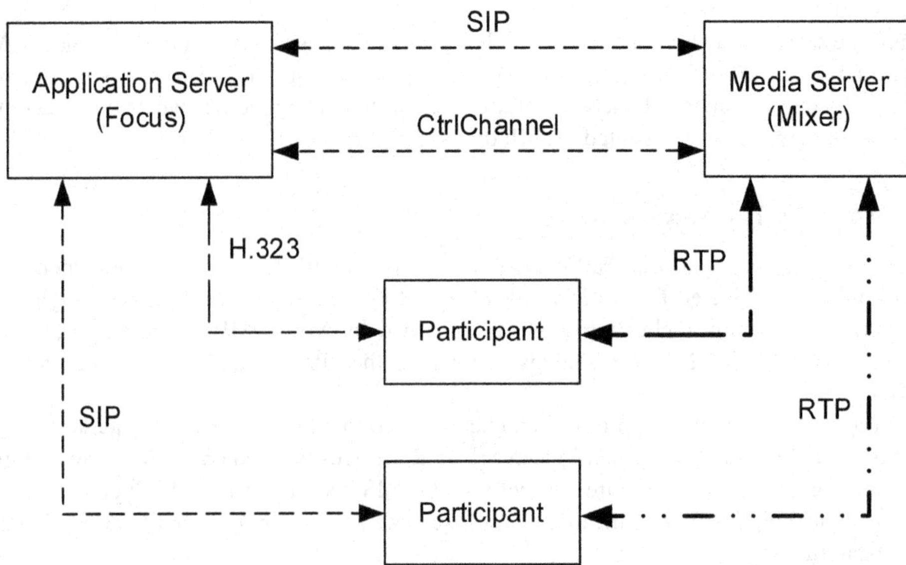

FIGURE 3.4 Conference Topology. (Copyright: IETF.)

As a mixer entity, in such a multimedia conferencing scenario, the MS receives a set of media streams of the same type (after transcoding if needed) and then takes care of combining the received media in a type-specific manner, redistributing the result to each authorized participant. The way each media stream is combined, as well as the media-related policies, is properly configured and handled by the AS by means of a dedicated MS control channel. To summarize, the AS needs to be able to manage MSs at both conference and participant level.

3.6.1 Creating a New Conference

When a new conference is created, as a result of a previous conference scheduling or of the first participant dialing in to a specified URI, the AS must take care of appropriately creating a media conference on the MS. It does so by sending an explicit request to the MS. This can be by means of an MS control channel message. This request may contain detailed information upon the desired settings and policies for the conference (e.g., the media to involve, the mixing configuration for them, the relevant identifiers, etc.). The MS validates such a request and takes care of allocating the needed resources to set up the media conference.

ASs may use mechanisms other than sending requests over the control channel to establish conferences on an MS and then subsequently use the control channel to control the conference. Examples of other mechanisms to create a conference include using the Request-URI mechanism of RFC 4240 or the procedures defined in RFC 4579. Once done, the MS informs the AS about the result of the request. Each conference will be referred to by a specific identifier, which both the AS and the MS will include in subsequent transactions related to the same conference (e.g., to modify the settings of an extant conference).

3.6.2 Adding a Participant to a Conference

As stated before, an AS uses SIP 3pcc to establish media sessions from SIP user agents to an MS. The URI that the AS uses in the INVITE to the MS may be one associated with the conference on the MS. More likely, however, the media sessions are first established to the MS using a URI for the MS and then subsequently joined to the conference using the MS control protocol. This allows IVR dialogs to be performed prior to joining the conference. The AS as a 3pcc correlates the media session negotiation between the UA and the MS in order to appropriately establish all the needed media streams based on the conference policies.

3.6.3 Media Controls

The XCON Common Data Model (RFC 6501 – see Chapter 4) currently defines some basic media-related controls, which conference-aware participants can take advantage of in several ways, e.g., by means of an XCON conference control protocol or IVR dialogs. These controls include the possibility to modify the participants' own volume for audio in the conference, configure the desired layout for incoming video streams, mute/unmute oneself, and pause/unpause one's own video stream. Such controls are exploited by conference-aware participants through the use of dedicated conference control protocol requests to the AS. The AS takes care of validating such requests and translates them into the MS control protocol before forwarding them over the MS control channel to the MS. According to the directives provided by the AS, the MS manipulates the involved media streams (Figure 3.5).

The MS may need to inform the AS of events like in-band DTMF tones during the conference.

3.6.4 Floor Control

The XCON framework introduces "floor control" functionality as an enhancement upon RFC 4575. Floor control is a means to manage joint or exclusive access to shared resources in a (multiparty)

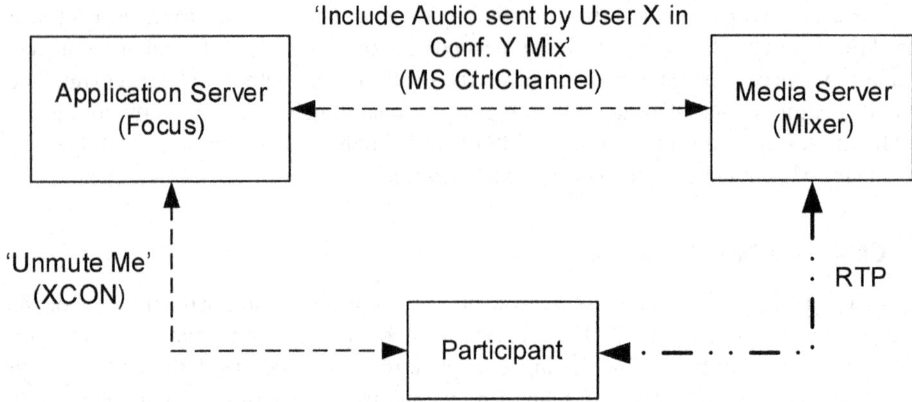

FIGURE 3.5 Conferencing Example: Unmuting a Participant. (Copyright: IETF.)

conferencing environment. Floor control is not a mandatory mechanism for a conferencing system implementation, but it provides advanced media input control features for conference-aware participants. Such a mechanism allows coordinated and moderated access to any set of resources provided by the conferencing system. To do so, a so-called floor is associated to a set of resources, thus representing for participants the right to access and manipulate the related resources themselves. In order to take advantage of the floor control functionality, a specific protocol, the BFCP, has been specified (RFC 4582 – Section 6). RFC 4583 provides a way for SIP UAs to set up a BFCP connection towards the Floor Control Server (FCS) and exploit floor control by means of a Connection-Oriented Media (COMEDIA) (RFC 4145) negotiation.

In the context of the AS–MS interaction, floor control constitutes a further means to control participants' media streams. A typical example is a floor associated with the right to access the shared audio channel in a conference. A participant who is granted such a floor is granted by the conferencing system the right to talk, which means that its audio frames are included by the MS in the overall audio conference mix. Similarly, when the floor is revoked, the participant is muted in the conference, and its audio is excluded from the final mix.

The BFCP defines an FCS and the floor chair. It is clear that the floor chair making decisions about floor requests is part of the application logic. This implies that when the role of floor chair in a conference is automated, it will normally be part of the AS. The example makes it clear that there can be a direct or indirect interaction between the FCS and the MS in order to correctly bind each floor to its related set of media resources.

A similar interaction is needed between the FCS and the AS as well, since the latter must be aware of all the associations between floors and resources in order to opportunely orchestrate the related bindings with the element responsible for such resources (e.g., the MS when talking about audio and/or video streams) and the operations upon them (e.g., mute/unmute a participant in a conference). For this reason, the FCS can be co-located with either the MS or the AS as long as both elements are allowed to interact with the FCS by means of some kind of protocol.

In the following text, both the approaches will be described in order to better explain the interactions between the involved components in both the topologies. When the AS and the FCS are co-located, the scenario is quite straightforward. In fact, it can be considered as a variation of the case depicted in Figure 3.6. The only relevant difference is that in this case, the action the AS commands on the control channel is triggered by a change in the floor control status instead of a specific control requested by a participant himself. The sequence diagram in Figure 3.6 describes the interaction between the involved parties in a typical scenario. It assumes that a BFCP connection between the UA and the FCS (which we assume is co-located with the AS) has already been negotiated and established, and that the UA has been made aware of all the relevant identifiers and floors–resources

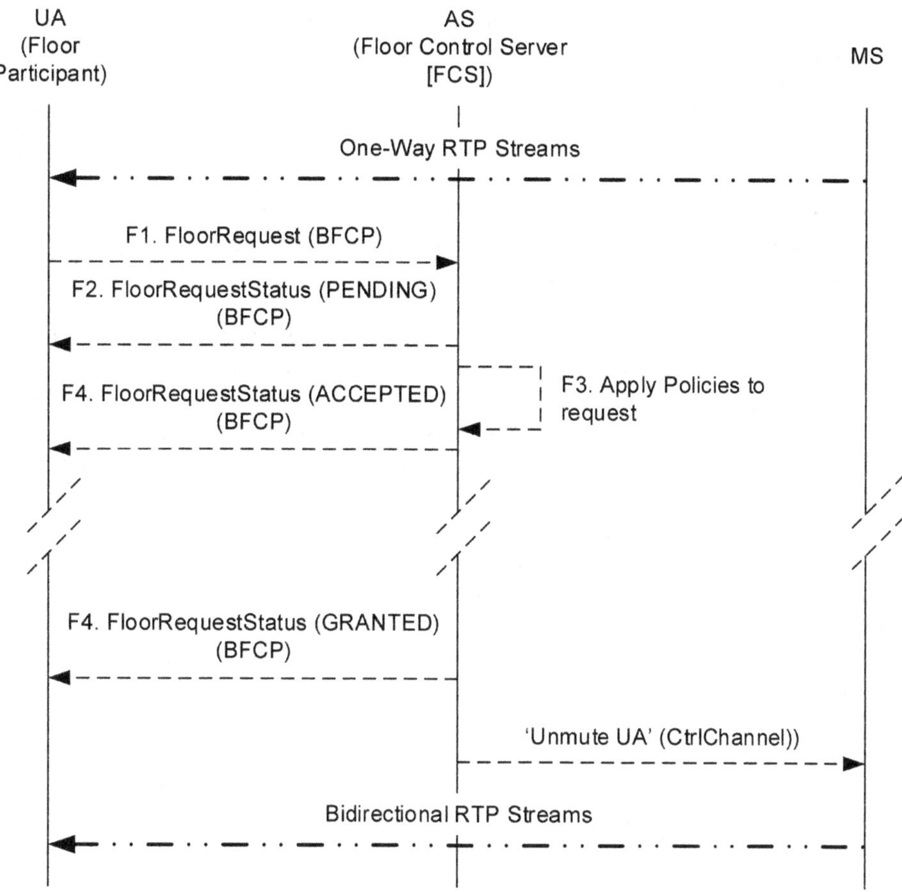

FIGURE 3.6 Conferencing Example: Floor Control Call Flow. (Copyright: IETF.)

associations (e.g., by means of RFC 4583). It also assumes that the AS has previously configured the media mixing on the MS using the MS control channel. Every frame the UA might be sending on the related media stream is currently being dropped by the MS, since the UA still is not authorized to use the resource. For an SIP UA, this state could be consequent to a "sendonly" field associated to the media stream in a re-INVITE originated by the MS. It is worth pointing out that the AS has to make sure that no user media control mechanisms, such as mentioned in the previous sub-section, can override the floor control.

A UA, which also acts as a floor participant, sends a "FloorRequest" to the FCS (which is co-located with the AS), stating his wish to be granted the floor associated with the audio stream in the conference. The AS answers the UA with a "FloorRequestStatus" message with a PENDING status, meaning that a decision on the request has not been made yet. The AS, according to the BFCP policies for this conference, makes a decision on the request, i.e., accepting it. Note that this decision might be relayed to another participant if he has previously been assigned as chair of the floor.

Assuming the request has been accepted, the AS notifies the UA about the decision with a new "FloorRequestStatus," this time with an ACCEPTED status in it. The ACCEPTED status, of course, only means that the request has been accepted, which does not mean that the floor has been granted yet. Once the queue management in the FCS, according to the specified algorithms for scheduling, states that the floor request previously made by the UA can be granted, the AS sends a new "FloorRequestStatus" to the UA with a GRANTED status and takes care of unmuting the participant in the conference by sending a directive to the MS through the control channel.

Once the UA receives the notification stating that his request has been granted, he can start sending his media, aware of the fact that now, his media stream will not be dropped by the MS. If the session has been previously updated with a "sendonly" associated to the media stream, the MS must originate a further re-INVITE stating that the media stream flow is now bidirectional ("sendrecv").

As mentioned before, this scenario envisages an automated floor chair role, where it is the AS, according to some policies, that makes decisions on floor requests. The case of a chair role performed by a real person is exactly the same, with the difference that the incoming request is not directly handled by the AS according to its policies but is instead forwarded to the floor control participant that the chair UA is exploiting. The decision on the request is then communicated by the chair UA to the AS–FCS by means of a "ChairAction" message.

The rest of this section will instead explore the other scenario, which assumes that the interaction between AS and FCS happens through the MS control channel. This scenario is compliant with the H.248.19 specification related to conferencing in 3GPP. The following sequence diagram describes the interaction between the involved parties in the same use-case scenario that has been explored for the previous topology: consequently, the diagram makes exactly the same assumptions that have been made for the previously described scenario.

This means that the scenario again assumes that a BFCP connection between the UA and the FCS has already been negotiated and established, and that the UA has been made aware of all the relevant identifiers and floors–resources associations. It also assumes that the AS has previously configured the media mixing on the MS using the MS control channel. This time, it includes identifying the BFCP-moderated resources, establishing basic policies and instructions about chair identifiers for each resource, and subscribing to events of interest, because the FCS is not co-located with the AS any more.

Additionally, a BFCP session (Figure 3.7) has been established between the AS (which in this scenario acts as a floor chair) and the FCS (MS). Every frame the UA might be sending on the related media stream is currently being dropped by the MS, since the UA still is not authorized to use the resource. For an SIP UA, this state could be consequent to a "sendonly" field associated to the media stream in a re-INVITE originated by the MS. Again, it is worth pointing out that the AS has to make sure that no user media control mechanisms, such as mentioned in the previous subsection, can override the floor control.

A UA, which also acts as a floor participant, sends a "FloorRequest" to the FCS (which is co-located with the MS), stating his wish to be granted the floor associated with the audio stream in the conference. The MS answers the UA with a "FloorRequestStatus" message with a PENDING status, meaning that a decision on the request has not been made yet. It then notifies the AS, which in this example handles the floor chair role, about the new request by forwarding the received request. The AS, according to the BFCP policies for this conference, makes a decision on the request, i.e., accepting it. It informs the MS about its decision through a BFCP "ChairAction" message.

The MS acknowledges the "ChairAction" message and then notifies the UA about the decision with a new "FloorRequestStatus," this time with an ACCEPTED status in it. The ACCEPTED status, of course, only means that the request has been accepted, which does not mean the floor has been granted yet. Once the queue management in the MS, according to the specified algorithms for scheduling, states that the floor request previously made by the UA can be granted, the MS sends a new "FloorRequestStatus" to the UA with a GRANTED status and takes care of unmuting the participant in the conference.

Once the UA receives the notification stating that his request has been granted, he can start sending his media, aware of the fact that now, his media stream will not be dropped by the MS. If the session has been previously updated with a "sendonly" associated to the media stream, the MS must originate a further re-INVITE stating that the media stream flow is now bidirectional ("sendrecv"). This scenario envisages an automated floor chair role, where it is the AS, according to some policies, that makes decisions on floor requests. Again, the case of a chair role performed by a real person is exactly the same, with the difference that the incoming request is not forwarded to the

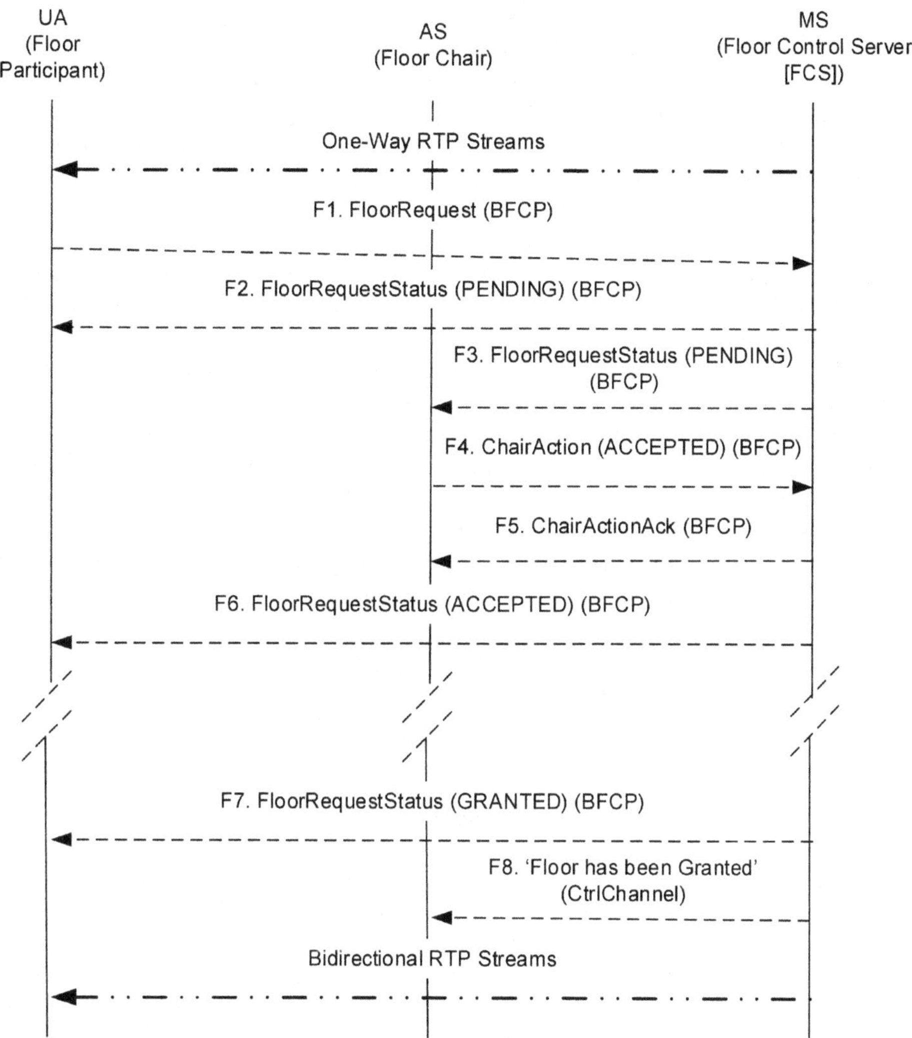

FIGURE 3.7 Conferencing Example: Floor Control Call Flow. (Copyright: IETF.)

AS but to the floor control participant that the chair UA is exploiting. The decision on the request is communicated by means of a "ChairAction" message in the same way.

Another typical scenario is a BFCP-moderated conference with no chair to manage floor requests. In such a scenario, the MS has to take care of incoming requests according to some predefined policies, e.g., always accepting new requests. In this case, no decisions are required by external entities, since all are instantly decided by means of policies in the MS. As stated before, the case of the FCS co-located with the AS is much simpler to understand and exploit. When the AS has full control of the FCS, including its queue management, the AS directly instructs the MS according to the floor status changes, e.g., by instructing the MS through the control channel to unmute a participant who has been granted the floor associated to the audio media stream.

3.7 SECURITY CONSIDERATIONS

This section describes the architectural framework to be used for MS control. Its focus is the interactions between ASs and MSs. User agents interact with ASs by means of signaling protocols such as SIP. These interactions are beyond the scope of this section. ASs are responsible for utilizing the security mechanisms of their signaling protocols, combined with application-specific policy, to ensure that they grant service only to authorized users.

Media interactions between user agents and MSs are also outside the scope of this specification. Those interactions are at the behest of ASs, which must ensure that appropriate security mechanisms are used. For example, if the MS is acting as the FCS, then the BFCP connection between the user agent and the MS is established to the MS by the AS using SIP and the SDP mechanisms described in RFC 4583. BFCP (RFC 4582 – see Chapter 6) strongly imposes the use of TLS for BFCP.

MSs are valuable network resources and need to be protected against unauthorized access. ASs use SIP and related standards both to establish control channels to MSs and to establish media sessions, including BFCP sessions, between an MS and end-users. MSs use the security mechanisms of SIP to authenticate requests from ASs and to ensure the integrity of those requests. Leveraging the security mechanisms of SIP ensures that only authorized ASs are allowed to establish sessions to an MS and to access MS resources through those sessions.

Control channels between an AS and MS carry the MS control protocol, which affects both the service seen by end-users and the resources used on an MS. TLS (RFC 5246) must be implemented as the transport-level security mechanism for control channels to guarantee the integrity of MS control interactions. The resources of an MS can be shared by more than one AS. MSs must prevent one AS from accessing and manipulating the resources that have been assigned to another AS. This may be achieved by an MS associating ownership of a resource to the AS that originally allocates it and then ensuring that future requests involving that resource correlate to the AS that owns and is responsible for it.

3.8 SUMMARY

We have specified the MS signaling architecture and MS control for control services using SIP (RFC 3261 – also see Section 1.1) call control protocol. Note that the functionalities of the MS architecture that is explained in this section are part of the multipoint conferencing framework described in Section 1.5 and Chapter 2. We have described the architectural framework for MS control, defining the logical entities that exist within the context of MS control based on MS control requirements specified in RFC 5167. We have defined all the functional entities, such as MS, AS, MRB, and IVR, that are needed for media control of XCON conferencing systems, along with some specific functions such as media functions, media services, media session, MS control channel, MS control dialog, and MS media dialog. The basic signaling architecture with interactions between MS, AS, and UA, and the media server control architecture with establishment of MS control channel between MS and AS, are defined in detail. These media control architectures specifically describe how a media server processes the media streams, such as RTP streams, along with RTCP feedback signaling, mixing of incoming media streams, media stream source for multimedia announcements, media stream for transcoding and DTMF detection, and media stream sink for multimedia recordings. We have specified how SIP signaling is used for interactions between MS and AS. In addition, we have discussed the IVR services architecture framework or basic services, mid-call controls, and advanced services.

More importantly, we have specified the media control architecture for conferencing services, showing that an MS is a separate physical functional entity that is not a black box within an AS, ensuring scalability and creating distributed media architecture. We have then compared this distributed conference architecture with that of the non-scalable 3pcc (see Section 1.3). Furthermore, creating a new conference, adding a participant to a conference, media controls, and floor control are described, which are the precursors for the development of actual XCON conference systems architecture specifying a standardized MS control protocol. However, we have developed two detailed call flows specifically for controlling the floor control of XCON conference systems using already standardized BFCP protocol (specified in Chapter 6): One for connection between the UA and the

FCS (MS) co-located with the AS and another for the AS itself playing the automated floor chair role. We have articulated the differences between these conference floor architectures, discussing a variety of different networking configurations that can be developed by placing the MS in different locations once the MS control protocol is specified. Finally, we have described the implications of interactions between the MS and the AS.

3.9 PROBLEMS

1. What is the basic purpose of the MS control architecture? What are the control, IVR, media mixing, and operational requirements of the MS control architecture? Describe each of the requirements in detail.
2. How does the centralized media server control architecture belong to the centralized multipoint multimedia conferencing framework described in Section 1.5 and Chapter 2? Justify your answer in detail.
3. Explain the role of the centralized MS architecture in delaying the control of RTP audio and/or video streams along with RTP/Audio-Visual Profile with Feedback (AVPF).
4. Articulate in detail how the SIP call control protocol is used for controlling of the media of the conference participants/devices by the MS under the direction of the AS (where the application logic resides), taking IVR services as an example.
5. Describe in detail the basic signaling schemes for the MS control architecture, showing AS and UA.
6. Draw the MS control server with a dedicated control channel between the MS and the AS. Explain all the services that can be offered by MS to UA, including VoiceXML.
7. Explain in detail how SIP/SDP signaling mechanisms can be used for offering services to the UA using the MS by ASs.
8. Describe in detail how IVR services can be offered by controlling the MS by the AS: Basic IVR services, IVR services with mid-call controls, and advanced IVR services.
9. Draw the basic topology of media control for conferencing services showing AS that also works as focus, MS acting as media mixture, and multiple UAs. Describe the following functions of conference services balancing with SIP/SDP, control channel (CtrlChannel), and RTP: Creating a new conference, adding a participant to a conference, and media controls.
10. What is the conference floor control service? Describe all the functionalities of the floor control in detail. Articulate the role of FCS, floor chair, and floor participant. Describe floor control call flows using BFCP protocol for a conferencing example where AS acts as FCS, UA as floor participant, and MS for sending RTP media streams.
11. Explain floor control call flows in detail where a BFCP session is established between the AS acting as a floor chair (automated) and the MS playing the role of FCS (i.e. FCS is co-located with MS) along with RTP media functions, and UAs as floor participants. How do SIP/SDP signaling schemes play a role in this conference scenario? What happens to these call flows when a real person plays the role of the chair?
12. What are the security implications that impact all functional entities that are involved in the MS control architecture? How can security attacks be prevented?

4 Conferencing Information Data Model for Centralized Conferencing

4.1 INTRODUCTION

A core data set of conference information is utilized in any conference independently of the specific conference media. This core data set, called the "conference information data model," (RFC 6501) is defined in this chapter using an XML-based format. The conference information data model defined in this chapter is logically represented by the conference object. Conference objects are a fundamental concept in centralized conferencing, as described in the centralized conferencing framework (RFC 5239 – see Chapter 2). The conference object represents a particular instantiation of a conference information data model. Consequently, conference objects use the XML format defined in this specification.

The session initiation protocol (SIP) event package for conference state, specified in RFC 4575, already defines a data format for conferences. However, that model is SIP specific and lacks elements related to some of the functionality defined by the centralized conferencing framework in RFC 5239 (see Chapter 2) (for example, floor control). The data model defined in this chapter constitutes a superset of the data format defined in RFC 4575. The result is a data format that supports more call signaling protocols (CSPs) besides SIP and that covers all the functionality defined in the centralized conferencing framework (RFC 5239 – see Chapter 2).

The data model specified in this chapter is the result of extending the data format defined in RFC 4575 with new elements. Examples of such extensions include scheduling elements, media control elements, floor control elements, non-SIP uniform resource identifiers (URIs), and the addition of localization extensions to text elements. This data model can be used by conference servers providing different types of basic conferences. It is expected that this data model can be further extended with new elements in the future in order to implement additional advanced features.

4.2 TERMINOLOGY

See Section 1.6.

4.3 OVERVIEW

The conference information data model provides a core data set of conference information expressed using an XML-based format that is utilized in any conference, independently of the specific conference media. The conference information data model (RFC 6501) described here is logically represented by the conference object. Note that RFC 6501 extends the data format defined in RFC 4575, and the data model described in the subsequent sections is based on RFC 6501 unless otherwise mentioned. The conference object, as described earlier, represents a particular instantiation of a conference information data model. This data model can be used by conference servers providing different types of basic conferences. It is expected that this data model can be further extended with new elements in the future in order to implement additional advanced features.

4.3.1 Data Model Format

A conference object specification is an XML [1] specification. Conference object specifications must be based on XML 1.0 and must be encoded using UTF-8. The normative description of the syntax of the conference object specification, for use by implementers of parsers and generators, is found in the RELAX NG schema provided in Section 4.5. Compliant messages must meet the requirements of that schema.

4.3.2 Data Model Namespace

A new namespace specification is specified for identifying the elements defined in the data model. This namespace is as follows:

urn:ietf:params:xml:ns:xcon-conference-info

4.3.3 Conference Object Identifier

The conference object identifier specified as XCON-URI can be viewed as a key to accessing a specific conference object. It can be used, for instance, by the conference control protocol to access, manipulate, and delete a conference object. A conference object identifier is provided to the conferencing client by the conference notification service or through out-of-band mechanisms (e.g., email). A conferencing system may maintain a relationship between the conference object identifiers and the identifiers associated with each of the complementary centralized conferencing protocols (e.g., call signaling protocol such as SIP, BFCP, and others). To facilitate the maintenance of these relationships, the conference object identifier acts as a top-level identifier within the conferencing system for the purpose of identifying the interfaces for these other protocols. This implicit binding provides a structured mapping of the various protocols with the associated conference object identifier. Figure 4.1 illustrates the relationship between the identifiers used for the protocols and the general conference object identifier (XCON-URI).

In Figure 4.1, the conference object identifier acts as the top-level key in the identification process. The CSPs have an associated conference user identifier, often represented in the form of a

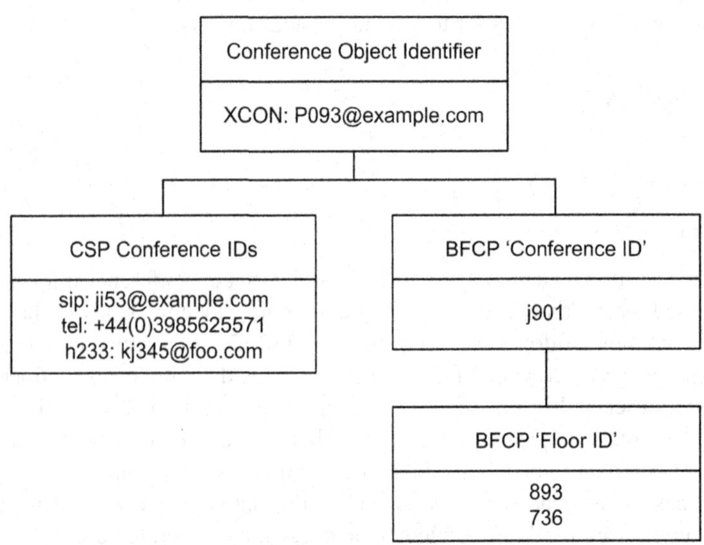

FIGURE 4.1 Conferencing Object Model. (Copyright: IETF.)

URI. The BFCP, as defined in RFC 4582 (see Chapter 6), defines the "conference ID" identifier, which represents a conference instance within floor control. When created within the conferencing system, the "conference ID" has a 1:1 mapping to the unique conference object identifier (XCON-URI). Operations associated with the conference control protocols are directly associated with the conference object; thus, the primary identifier associated with these protocols is the conference object identifier (XCON-URI). The mappings between additional protocols/interfaces are not strictly 1:1 and do allow for multiple occurrences. For example, multiple CSPs will each have a representation that is implicitly linked to the top-level conference object identifier, e.g., SIP, and tel URIs that represent a conference instance. It should be noted that a conferencing system is free to structure such relationships as required, and this information is just included as a guideline that can be used. Further elements can be added to the tree representation in Figure 4.1 to enable a complete representation of a conference instance within a conferencing system.

4.3.3.1 Conference Object URI Definition

The syntax is defined by the following augmented Backus–Naur form (ABNF) specified in RFC 5234 rules as follows:

```
XCON-URI = "xcon" ":" [conf-object-id "@"] host
conf-object-id = 1*( unreserved / "+" / "=" / "/" )
```

Note: host and unreserved are defined in RFC 3986.

An XCON-URI is not designed to be resolved, and an application must not attempt to perform a standard domain name system (DNS) lookup on the host portion of such a URI in an attempt to discover an Internet protocol (IP) address or port at which to connect.

4.3.3.2 Normalization and Conference Object URI Comparison

In order to facilitate the comparison of the XCON-URI identifiers, all the components of the identifiers must be converted to lowercase. After normalizing the URI strings, the URI comparison must be applied on a character-by-character basis as prescribed by Section 6.2.1 of RFC 3986. The host construction, as defined in RFC 3986, can take the form of an IP address, which is not conventionally compared on a character-by-character basis. The host part of an XCON-URI serves only as an identifier; that is, it is never used as an address. The character-by-character comparison still applies.

4.3.4 Data Model Structure

The information in this data model is structured in the following manner. All the information related to a conference is contained in a <conference-info> element. The <conference-info> element contains the following child elements:

- The <conference-description> element describes the conference as a whole. It has, for instance, information about the URI of the conference, maximum number of users allowed in the conference, media available in the conference, or the time the conference will start.
- The <host-info> element contains information about the entity hosting the conference (e.g., its URI).
- The <conference-state> element informs the subscribers about the changes in the overall conference information.
- The <floor-information> element contains information about the status of the different floors in the conference.
- The <users> element describes the membership information as a whole. The <users> element contains a set of <user> child elements, each describing a single participant in the conference.

- If a participant in the main conference joins a sidebar, a new element is created in the conference referenced from the <sidebars-by-ref> element or under one of the <sidebars-by-val> elements.

Note that some of the elements described, such as <conference-info>, <conference-description>, <sidebars-by-ref>, or <sidebars-by-val>, are not defined in the data model in this specification but are defined in the data format of RFC 4575. We describe them here because they are part of the basic structure of the data model.

4.4 DATA MODEL DEFINITION

The non-normative diagram in Figure 4.2 shows the structure of conference object specifications. The symbol "!" preceding an element indicates that the element is REQUIRED in the data model. The symbol "*" following an element indicates that the element is introduced and defined here (RFC 6501). That is, elements without a "*" have already been defined in RFC 4575.

The following sub-sections describe these elements in detail. The full RELAX NG schema is provided in Section 4.5.

4.4.1 <CONFERENCE-INFO>

A conference object specification begins with the root element <conference-info>, which is defined in RFC 4575. The "state" and "version" attributes of the <conference-info> element are defined in RFC 4575 and are not used in the context of the XCON Conference Information Model, since they apply only to notification mechanisms. In addition, RFC 4575 defines an "entity" attribute that contains the SIP URI identifier. This XCON specification (RFC 6501) extends the meaning of the "entity" attribute to the conference object identifier (XCON-URI) explained in Section 4.3.3. This specification adds to the <conference-info> element the child elements of the <floor-information> element.

4.4.2 <CONFERENCE-DESCRIPTION>

The <conference-description> element, which is defined in RFC 4575, describes the conference as a whole. It should have an attribute "lang" to specify the language used in the contents of this element. It is comprised of <language>, <display-text>, <subject>, <freetext>, <keywords>, <allow-sidebars>, <cloning-parent>, <sidebar-parent>, <conference-time>, <conf-uris>, <service-uris>, <maximumuser-count>, and <available-media>. The <display-text>, <subject>, <freetext>, <keywords>, <service-uris>, and <maximum-user-count> elements are described in Section 5.3 of RFC 4575. The following sub-sections describe these elements in more detail. Other child elements may be defined in the future to extend the <conference-description> element.

4.4.2.1 <language>

The <language> element indicates the predominant language that is expected to be employed within a conference. This element contains only one language. The possible values of this element are the values of the "Subtag" column of the "Language Subtag Registry" at [2] originally defined in RFC 5646. This element does not enforce the language of the conference: it only informs the participants about the desirable language that they should use in the conference. Participants are free to switch to other languages if they like.

4.4.2.2 <allow-sidebars>

The <allow-sidebars> element represents a Boolean value. If it is set to true or "1," the conference is allowed to create sidebar conferences. If it is absent, or set to "false" or "0," the conference cannot create sidebar conferences.

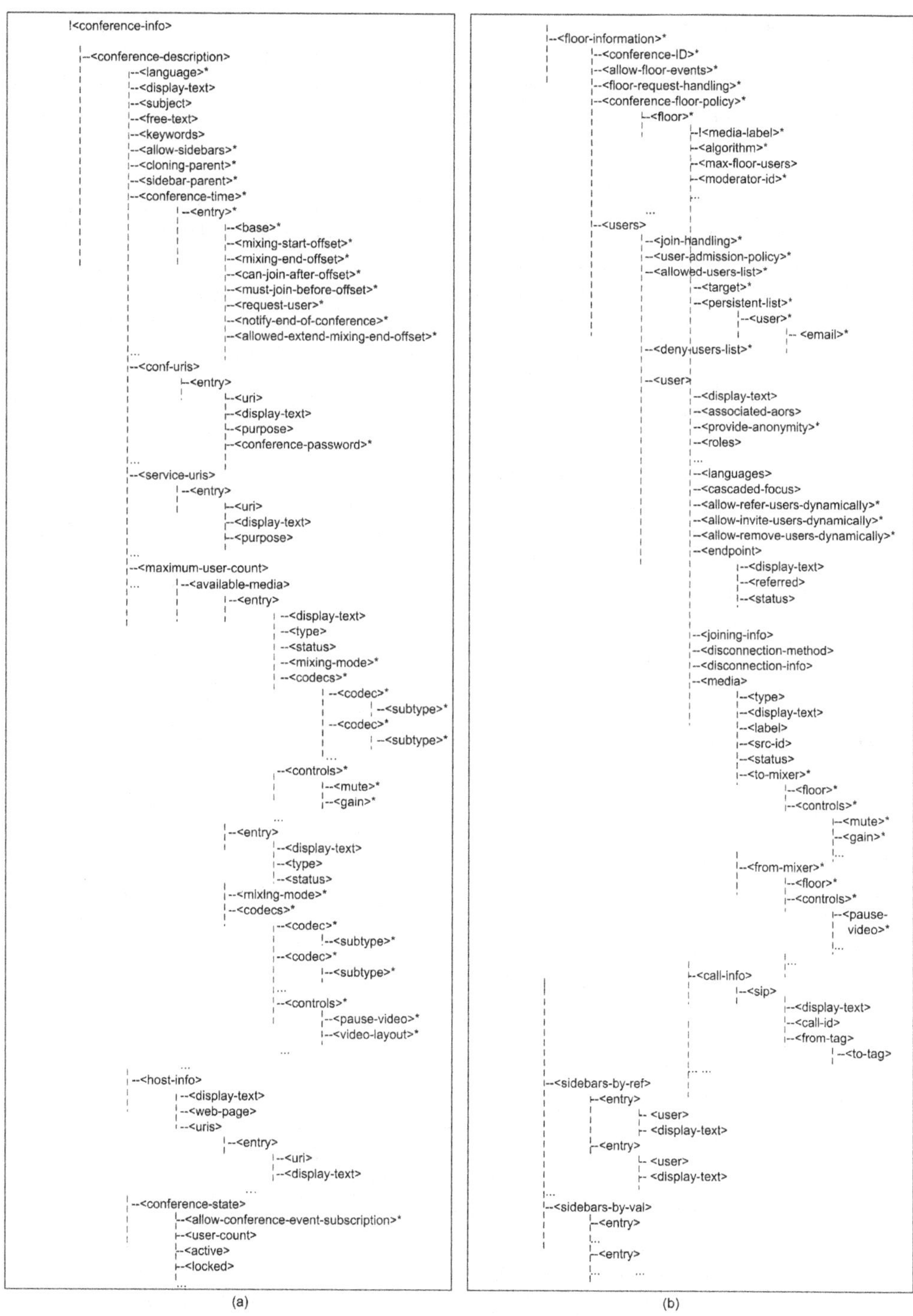

FIGURE 4.2 Structure of Conference Object Specification. (Copyright: IETF.)

4.4.2.3 <cloning-parent>

When the <cloning-parent> is present, it indicates that the conference object is a child of a parent conference. The <cloning-parent> element contains the conference object identifier (XCON-URI) (different from the main XCON-URI) of the parent.

4.4.2.4 <sidebar-parent>

When the <sidebar-parent> is present, it indicates that the conference object represents a sidebar of another conference. The <sidebar-parent> element contains the conference object identifier (XCON-URI) (different from the main XCON-URI) of the parent.

4.4.2.5 <conference-time>

The <conference-time> element contains the information related to time, policy, and duration of a conference. The <conference-time> element contains one or more <entry> elements, each defining the time and policy information specifying a single conference occurrence. The <conference-time> element differs from the iCalendar objects defined in RFC 5545 in that it has the ability to define different policies (<can-join-after-offset>, <must-join-before-offset>) for the same conference at different times. Every <entry> element contains the following child elements:

- <base>: The <base> child element specifies the iCalendar object of the conference. The iCalendar object components are defined in RFC 5545.
- <mixing-start-offset>: The <mixing-start-offset> child element specifies when the conference media mixing starts before the conference starts. The <mixing-start-offset> element specifies an absolute value rather than an offset value. If the <mixing-startoffset> element is not present, it indicates that the conference media mixing starts immediately. The <mixing-start-offset> must include the "required-participant" attribute. This attribute contains one of the following values: "none," "administrator," "moderator," "user," "observer," and "participant." The roles' semantic definitions are out of the scope of this specification and are subject to future policy specifications. More values can be specified in the future. The "required-participant" attribute allows a privileged user to define when media mixing starts based on the later of the mixing start time and the time the first participant arrives. If the value is set to "none," mixing starts according to the mixing start time.
- <mixing-end-offset>: The <mixing-end-offset> child element specifies the time conference media mixing stops after the conference stops. If the <mixing-end-offset> element is not present, it indicates that the conference occurrence is not bounded. The <mixing-end-offset> element must include the "required-participant" attribute. This attribute contains one of the following values: "none," "administrator," "moderator," "user," "observer," and "participant." More values can be specified in the future. The "required-participant" attribute allows a privileged user to define when media mixing ends based on the earlier of the <mixing-end-offset> and the time the last participant leaves. If the value is set to "none," mixing stops according to the <mixing-end-offset>. If the conference policy was modified so that the last privileged user is now a normal conference participant, and the conference requires a privileged user to continue, that conference must terminate.
- <can-join-after-offset>: An administrator can indicate the time when users can join a conference by populating the <can-join-after-offset> element.
- <must-join-before-offset>: An administrator can define the time after which new users are not allowed to join the conference any more.
- <request-user>: This element defines the time when users or resources on the <allowed-users-list> are requested to join the conference by using the <request-users> element.
- <notify-end-of-conference>: The <notify-end-of-conference> element defines in seconds when the system must send a notification that the end of the conference is approaching. If the <notify-end-of-conference> element is not present, this indicates that the system does not notify the users when the end of the conference is approaching.
- <allowed-extend-mixing-end-offset>: The <allowed-extend-mixing-end-offset> element indicates whether the conference is allowed to be extended. It has a Boolean value.

4.4.2.6 <conf-uris>

The <conf-uris> contains a set of <entry> child elements – each containing a new element: <conference-password>. This element contains the password(s) of the conference; for instance, a public switched telephone network (PSTN) conference will store the "PIN code" in this element. All the other <conf-uris> child elements are described in Section 5.3.1 of RFC 4575.

The RELAX NG schema (extension) in Section 4.5 allows <conference-password> to appear anywhere uris-type is expanded. This specification only provides meaning for <conference-password> appearing as a descendant of the <conf-uris> element. Future standardization may give meaning to <conference-password> appearing in other elements of type "uris-type." In the absence of such standardization, <conference-password> must not appear in elements of type "uris-type" other than <conf-uris>.

4.4.2.7 <available-media>

The <available-media> element consists of a sequence of <entry> child elements. Each <entry> element MAY contain the following child elements:

- The <display-text>, <type>, and <status> elements are described in Section 5.3.4 of RFC 4575.
- The child element <mixing-mode> describes a default scheduling policy by which the mixer will build the outgoing stream from the incoming streams. Note that this policy is different from the policy describing the floors for each media. The <mixing-mode> child element MUST contain one and only one of the "moderator-controlled," "FCFS," and "automatic" values, indicating the default algorithm to use with every media stream. The "moderator-controlled" value indicates that the moderator of the conference controls the media stream policy. The "FCFS" value indicates a "first-come-first-served" policy. The "automatic" value means that the mixer must choose the best scheduling policy for the conference.
- The <codecs> element specifies the allowed codecs in the conference. It has an attribute "decision" that specifies whether the focus decides the common codec automatically or needs the approval of the moderator of the conference ("automatic" or "moderator-controlled"). The <codecs> element contains <codec> elements. A <codec> element can have the attributes "name" and "policy." The "name" attribute is a codec identifier assigned by the conferencing server. The "policy" attribute contains the policy for that codec (allowed or disallowed). The <codec> element has the child element <subtype>, which stores the codec's name. The possible values of this element are the values of the "subtype" column of the "RTP Payload Format media types" registry [3] originally defined in RFC 4855. It is expected that future conferencing specifications will define corresponding schema extensions, as appropriate.
- The <controls> element contains the basic audio and video global control elements for a conference. These controls are sufficient for the majority of basic conferences. If the conference server wants to support more-advanced controls, then it is recommended that an extension to the data model be used. In the <controls> element, the schema is extensible; hence, new control types can be added in the future. So, moderator controls that affect all media output would go under the <available-media> element. The following child elements are defined for <controls>:
  - The <mute> element is used in conjunction with an audio stream to cease transmission of any audio from the associated stream. That means that for the entire duration where mute is applicable, all current and future participants of the conference are muted and will not send any audio. It has a Boolean value. If this control is not specified, access to the control is not available to the client.

- The <pause-video> element is used in conjunction with a video stream to cease transmission of associated media. It has a Boolean value. If this control is not specified, the access to the control is not available to the client.
- The <gain> element is used in conjunction with a media output stream to indicate the amount of amplification of an audio stream. The value is an integer number that ranges from −127 to +127. If this control is not specified, access to the control is not available to the client.
- The <video-layout> element is used in conjunction with a video stream to specify how the video streams (of participants) are viewed by each participant. Only one layout type can be specified for each output stream. If there are fewer participants than panels in the specified layout, then blanking (black screen) may be mixed into the stream on the behalf of the missing input streams. If unspecified, the <video-layout> default type should be "single-view." The <video-layout> types are as follows, although any number of custom layouts may be specified in future extensions:
  - single-view: Only one stream is presented by the focus to all participants in one panel.
  - dual-view: This dual-view option will present the video side by side in two panels and not alter the aspect ratio of the streams. This will require the focus to introduce blanking on parts of the overall image as viewed by the participants.
  - dual-view-crop: This side-by-side layout option instructs the focus to alter the aspect ratio of the streams (alter-aspect-ratio=true) so that blanking is not necessary. The focus handles the cropping of the streams.
  - + dual-view-2×1: This layout option instructs the focus to place one stream above the other, in essence, with two rows and one column. In this option, the aspect ratio is not altered, and blanking is introduced.
  - dual-view-2×1-crop: This layout option also instructs the focus to place one stream above the other, in essence, with two rows and one column. In this option, the aspect ratio is altered, and the video streams are cropped.
  - quad-view: Four equal-sized panels in a 2×2 layout are presented by the focus to all participants. Typically, the aspect ratio of the streams is maintained (alter-aspect-ratio=FALSE).
  - + multiple-3×3: Nine equal-sized panels in a 3×3 layout are presented by the focus to all participants. Typically, the aspect ratio of the streams is preserved.
  - multiple-4×4: 16 equal-sized panels in a 4×4 layout are presented by the focus to all participants. Typically, the aspect ratio of the streams is preserved.
  - multiple-5×1: This option refers to a 5×1 layout where one panel will occupy 4/9 of the mixed video stream, while the others will each occupy 1/9 of the stream. Typically, the aspect ratio of the streams is preserved.
  - automatic: This option allows the focus to add panels as streams are added.

4.4.3 <HOST-INFO>

The <host-info> element and its child elements are described in Section 5.4 of RFC 4575.

4.4.4 <CONFERENCE-STATE>

The <conference-state> is introduced in RFC 4575. The <conference-state> element contains the <allow-conference-event-subscription>, <user-count>, <active>, and <locked> child elements. The <user-count>, <active>, and <locked> child elements are defined in Section 5.5 of RFC 4575.

4.4.4.1 <allow-conference-event-subscription>

The <allow-conference-event-subscription> element represents a Boolean action. If it is set to true, the focus is instructed to allow the subscription to conference state events, such as "SIP event package for conference state" of RFC 4575. If it is set to FALSE, the subscription to conference state events must be rejected. If this element is undefined, it has a default value of true, causing the subscription to conference state events to be accepted.

4.4.5 <FLOOR-INFORMATION>

The <floor-information> element contains the <conference-ID>, <allow-floor-events>, <floor-request-handling>, and <conference-floor-policy> child elements. The absence of this element from an XML specification indicates that the conference does not have a floor.

4.4.5.1 <conference-ID>

The <conference-ID> represents a conference instance within floor control. When BFCP serves as the floor control protocol, the <conference-ID> is a 32-bit BFCP conference identifier defined in RFC 4582 (see Sections 6 and 6.5.1). Note that when it is created within the conferencing system, there is a 1:1 mapping between this <conference-ID> and the unique conference object identifier (XCON-URI).

4.4.5.2 <allow-floor-events>

The <allow-floor-events> element represents a Boolean action. If it is set to true, the focus is instructed to accept the subscription to floor control events. If it is set to FALSE, the focus is instructed to reject the subscription. If this element is undefined, it has a default value of FALSE, causing the subscription to floor control events to be rejected.

A conference participant can subscribe himself to a floor control event in two different ways: one method is by using an offer/answer exchange mechanism of RFC 3264 using SIP INVITE and BFCP parameters in the SDP specified in RFC 4583, and the other is a general authorization mechanism described in RFC 4582 (see Sections 6 and 6.9) and in RFC 5018. Future specifications may define additional connection mechanisms.

4.4.5.3 <floor-request-handling>

The <floor-request-handling> element defines the actions used by the conference focus to control floor requests. This element defines the action that the focus is to take when processing a particular request to a floor within a conference. This element defines values of the following:

- "block:" This action instructs the focus to deny the floor request. This action is the default action taken in the absence of any other actions.
- "confirm:" This action instructs the focus to allow the request. The focus then uses the defined floor algorithm to further allow or deny the floor. The algorithms used are outside the scope of this specification.

Note that this section discusses floor control information; therefore, the value "block" in a <floor-request-handling> element is not related to the "block" value in the <join-handling> element (see Section 4.4.6.1).

4.4.5.4 <conference-floor-policy>

The <conference-floor-policy> element has one or more <floor> child elements. Every <floor> child element has an attribute "id," which uniquely identifies a floor within a conference. In the case of BFCP specified in RFC 4582 (see Chapter 6), the "id" attribute corresponds to the floor-id identifier defined in RFC 4582 (see Section 6.5.2.2).

- <media-label>: Every floor is identified for one or more mandatory <media-label> elements. If the <available-media> information is included in the conference specification, the value of this element must be equal to the "label" value of the corresponding media stream <entry> in the <available-media> container. The number of those elements indicates how many floors the conference can have. A floor can be used for one or more media types:
- <algorithm>: A floor can be controlled using many algorithms; the mandatory <algorithm> element must be set to any of the "moderator-controlled," "FCFS," or "random" values indicating the algorithm. The "moderator-controlled" value indicates that the moderator of the conference controls the floor. The "FCFS" value indicates a "first-come-first-served" policy.
- <max-floor-users>: The <max-floor-users> child element in the <floor> element is optional and if present, dictates the maximum number of users who can have the floor at one time.
- <moderator-id>: The optional <moderator-id> indicates the "User ID" of the moderator(s). It must be set if the element <algorithm> is set to the "moderator-controlled" value. When the floor is created within the conferencing system, the XCON-USERID may be used as the <moderator-id>. In the case where the BFCP (see Chapter 6) is the floor control protocol, RFC 4582 refers to the moderator role as a "floor chair."

4.4.6 <USERS>

The <users> element is described in RFC 4575 and contains the <join-handling>, <user-admission-policy>, <allowed-users-list>, and <deny-users- list> defined in this specification and <user> child elements defined in RFC 4575. When the <users> element is used in the context of the XCON Conference Information Model, the "state" and "version" attributes defined in RFC 4575 are not used, since they apply only to notification mechanisms. The following sections describe these elements in more detail. Other child elements and attributes can be used to extend <users> in the future.

4.4.6.1 <join-handling>

The <join-handling> element defines the actions used by the conference focus to control conference participation. This element defines the action that the focus is to take when processing a particular request to join a conference. This element defines values of:

- "block:" This action instructs the focus to deny access to the conference. This action is the default action taken in the absence of any other actions.
- "confirm:" This action instructs the focus to place the participant on a pending list (e.g., by parking the call on a music-on-hold server), awaiting moderator input for further actions.
- "allow:" This action instructs the focus to accept the conference join request and grant access to the conference within the instructions specified in the transformations of this rule.
- "authenticate:" This action instructs the focus that the user has to provide a combination of username/password.
- "directed-operator:" This action instructs the focus to direct the user to an operator.

4.4.6.2 <user-admission-policy>

The <user-admission-policy> is an element that lets an organizer (or a participant with appropriate rights) choose a policy for the conference that controls how users are authenticated into the conference, using a mechanism of the conference's choosing. Since a variety of signaling protocols are possible, a variety of authentication mechanisms – determined by every individual conference server – may need to be mapped from the different protocols. The specific types of authentication mechanisms are beyond the scope of describing here. The lists of possible values are as follows:

- "closedAuthenticated:" A "closedAuthenticated" policy must have each conference participant in the allowed users list (listed under the <allowed-users-list> element), with each participant being sufficiently (up to local policy) authenticated. Conference join requests for users not in the allowed users list or participants not authenticated should be rejected unless a <join-handling> action of "confirm" is selected, in which case, the user is placed on a pending list as indicated earlier. A "closedAuthenticated" policy must not include a <deny-users-list>. If <deny-users-list> appears in the data model, it must be ignored.
- "openAuthenticated:" An "openAuthenticated" policy requires each conferencing participant to be sufficiently authenticated. Typically, this implies that anyone capable of authenticating with the conferencing system may join the conference. The "openAuthenticated" policy permits the specification of "banned" conferencing participants. Such banned users are prevented from re-joining the conference until they have been un-banned. An "openAuthenticated" policy should have a deny users list (listed under the <deny-users-list> XML element) to support the banning of conferencing participants from a conference. An "openAuthenticated" policy must not include an <allowed-users-list>. If <allowed-users-list> appears in the data model, it must be ignored.
- "anonymous:" An "anonymous" policy grants any join requests and is the least restrictive policy. An "anonymous" policy must not include either an <allowed-users-list> or a <deny-users-list>. If either of these lists appears in the data model, it MUST be ignored. In all other cases, the appearance of an <allowed-users-list> and <deny-users-list> must be ignored except as otherwise described in a future specification. Future specifications describing the use of these lists must provide clear guidance on how to process the lists when they occur concurrently, especially when both lists contain the same user. For example, such a specification could disallow both lists from appearing at the same time similar to <user-admission-policy> values defined in this specification.

4.4.6.3 <allowed-users-list>

The <allowed-users-list> child element contains a list of user URIs (e.g., XCON-USERID, as defined in Section 4.4.6.5), roles (defined in Section 4.4.6.5.2), or domains (e.g., *@example.com) that the focus uses to determine who can join the conference, who can be invited to join a conference, or who the focus needs to "refer to" the conference. The <allowed-users-list> element includes zero or more <target> child elements. This child element includes the mandatory "uri" attribute and the mandatory "method" attribute. The same "uri" attribute with different method values can appear in the list more than once. The "method" attribute is a list with the following values:

- "dial-in:" The value "dial-in" is used by the focus to determine who can join the conference.
- "dial-out:" The value "dial-out" contains a list of resources with which the focus will initiate a session.
- "refer:" The value "refer" is used by the focus to determine the resources that the focus needs to "refer to" the conference. In SIP, this is achieved by the focus sending a REFER request to those potential participants. In a different paradigm, this could also mean that the focus sends a short message service (SMS) or an email to the referred user. This list can be updated during the conference lifetime, so it can be used for mid-conference refers as well.

The "refer" value differs from "dial-out" in that the resources on the "refer" value are expected to initiate the session establishment toward the focus themselves. It is also envisioned that different users will have different access rights to those lists, and therefore, a separation between the two is needed.

The <allowed-users-list> element has a <persistent-list> child element as well. Some chat room systems allow – and some require – registration of detailed information about a user before they are allowed to join a chat room. The <persistent-list> child element stores persistent information about users who are not actively part of an ongoing chat room session. The <persistent-list> element stores the following information:

- user: The <user> element stores the name, nickname, conference user identifier (XCON-USERID), and email address of a user. It has three attributes, "name," "nickname," and "id," and an <email> element. Future extensions to this schema may define new elements for the <user> element.

Future extensions to this schema may define new elements for the <target> element.

4.4.6.4 <deny-users-list>

The <deny-users-list> child element contains a list of user URIs (e.g., SIP URI, XCON-USERID defined in Section 4.4.6.5), roles (defined in Section 4.4.6.5.2), or domains (e.g., *@example.com) that the focus uses to determine who has been "banned" from the conference. Such banned users are prevented from re-joining the chat room until the ban has been lifted.

4.4.6.5 <user> and Its <user> Sub-Elements

The element <user> is described in RFC 4575 and describes a single participant in the conference. The <user> element has an attribute "entity." However, when the <user> element is used in the context of the XCON Conference Information Model, the "state" and "version" attributes defined in RFC 4575 are not used, since they only apply to notification mechanisms. The attribute "entity" contains a unique conference user identifier (XCON-USERID) within the scope of the conference. The URI format of this identifier is as follows (using ABNF of RFC 5234):

```
XCON-USERID = "xcon-userid" ":" conf-user-id
conf-user-id = 1*unreserved
Note: unreserved is defined in RFC 3986.
```

In order to facilitate the comparison of the XCON-USERID identifiers, all the components of the identifiers must be converted to lowercase. After normalizing the URI strings, the URIs comparison must be applied codepoint-by-codepoint after conversion to a common character encoding, as prescribed in Section 6.2.1 of RFC 3986.

Other user identifiers can be associated with this conference user identifier and enable the conferencing system to correlate and map these multiple authenticated user identities to a single global user identifier. Figure 4.3 illustrates an example using the conference user identifier in association

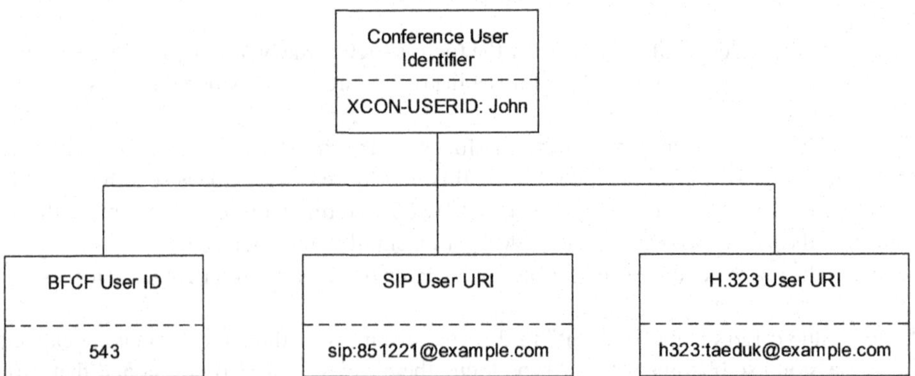

FIGURE 4.3 Conference User Mapping.

with the user identity defined for BFCP, SIP, and H323 user identity. It should be noted that a conferencing system is free to structure such relationships as required, and this information is just included as a guideline.

The element <user> element contains the <display-text>, <associated-aors>, <provide-anonymity>, <roles>, <languages>, <cascaded-focus>, <allow-refer-users-dynamically>, <allow-invite-users-dynamically>, <allow-remove-users-dynamically>, and <endpoint>. The following sub-sections describe these elements in more detail. The <display-text>, <associated-aors>, <languages>, and <cascaded-focus> are defined in Section 5.6 of RFC 4575.

4.4.6.5.1 <provide-anonymity>

The <provide-anonymity> element specifies what level of anonymity the server should provide to the user. In this case, the focus provides the rest of the participants with an anonymous identity for that user, for example, anonymousX, or it does not provide any information for that user, such that other users cannot see that he is a participant in the conference. This element only affects the way the user information is provided to the other participants. The real user information is stored in the data model but should not be provided to the other participants of the conference. This can be achieved by using the <provide-anonymity> element. This element has three values: "private," "semi-private," and "hidden." The "private" value specifies that this user is completely anonymous in the conference. The "semi-private" value specifies that this user is anonymous to all users who have not been granted permission to see him. The "hidden" value specifies that other users cannot see this participant in the conference.

4.4.6.5.2 <roles>

A <role> provides the context for the set of conference operations that a participant can perform. This element can contain one or more of the following values: "administrator," "moderator," "user," "participant," "observer," and "none". A role of "none" indicates that any role is assigned. The <roles> semantic definition is out of the scope of this specification and is subject to future policy specifications. This element can be extended with new roles in future specifications.

4.4.6.5.3 <allow-refer-users-dynamically>

The <allow-refer-users-dynamically> element represents a Boolean value. If it is set to true, a participant is allowed to instruct the focus to refer a user to the conference without modifying the <allowed-users-list> (in SIP terms, a participant is allowed to send a REFER request specified in RFC 3515 [updated by RFCs 7647 and 8217] to the focus, which results in the focus sending a REFER request to the user the referrer wishes to join the conference). If it is set to FALSE, the REFER request is rejected. If this element is undefined, it has a value of FALSE, causing the REFER request to be rejected.

4.4.6.5.4 <allow-invite-users-dynamically>

The <allow-invite-users-dynamically> element represents a Boolean action. If it is set to true, a participant is allowed to instruct the focus to invite a user to the conference without modifying the <allowed-users-list> list (in SIP terms, a participant is allowed to send a REFER request specified in RFC 3515 [updated by RFCs 7647 and 8217] to the focus, which results in the focus sending an INVITE request to the user the referrer wishes to join the conference). If it is set to FALSE, the REFER request is rejected. If this element is undefined, it has a value of FALSE, causing the REFER request to be rejected.

4.4.6.5.5 <allow-remove-users-dynamically>

The <allow-remove-users-dynamically> element represents a Boolean action. If it is set to true, a participant is allowed to instruct the focus to remove a user from the conference without modifying the ruleset (in SIP terms, a participant is allowed to send a REFER request [RFC 3515 – updated by

RFCs 7647 and 8217] to the focus, which results in the focus sending a BYE request to the user the referrer wishes to leave the conference). If it is set to FALSE, the REFER request is rejected. If this element is undefined, it has a value of FALSE, causing the REFER request to be rejected.

4.4.6.5.6 *<endpoint>*

The <endpoint> child element is identical to the element with the same name in RFC 4575 except that the "state" attribute is not included. When the <endpoint> element is used in the context of the XCON Conference Information Model, the "state" and "version" attributes defined in RFC 4575 are not used, since they apply only to notification mechanisms. The <endpoint> element can provide the desired level of detail about the user's devices and their signaling sessions taking part in the conference.

The <endpoint> element has the following child elements: <display-text>, <referred>, <status>, <joining-method>, <joining-info>, <disconnection-method>, <disconnection-info>, <media>, and <callinfo>. All the <endpoint> child elements are defined in of RFC 4575 with the exception of the <to-mixer> element and the <from-mixer> element. The <endpoint>/<media> element has two other child elements defined in this specification: the <to-mixer> and the <from-mixer>:

- <from-mixer>, <to-mixer>: These are controls that apply to a user's media stream being sent from the mixer to the participant's endpoint or to the mixer from the participant's endpoint. The <to-mixer> element details properties associated with the incoming streams to the mixer (streams sent to the mixer from the participant). The <from-mixer> element details properties associated with the outgoing streams from the mixer (sent from the mixer to the participant). Both of these elements have the attribute "name." The "name" attribute has the values "VideoIn," "VideoOut," "AudioOut," and "AudioIn." The "VideoOut" and "AudioOut" media streams detail properties associated with the outgoing video and audio from the mixer. The "VideoIn" and "AudioIn" media stream details properties associated with the incoming video and audio to the mixer. Both these elements can have the <floor> child element defined:
  - The <floor> element refers to the floor assigned to a certain participant in the conference. If a participant, for instance, needs to talk in the conference, it first needs to get the floor from the chair of the conference. The <floor> element has an attribute "id," which uniquely identifies a floor within a conference. The "id" attribute corresponds to the floor-id identifier defined in RFC 4582 (see Sections 6 and 6.5.2.2). The <floor> element has a Boolean value. A value of FALSE indicates that this user does not hold the floor in this moment. If this control is not specified, this user SHOULD NOT specify the floor option.
  - The <to-mixer> and <from-mixer> elements can have the <controls> child element:
- Controls that apply to a specific user would appear under the <controls> element.
- More values can be defined in the future.

4.4.7 <SIDEBARS-BY-REF>

The <sidebars-by-ref> element contains a set of <entry> child elements. This element is described in Section 5.9.1 of RFC 4575. When the <sidebars-by-ref> element is used in the context of the XCON Conference Information Model, the "state" and "version" attributes defined in RFC 4575 are not used, since they apply only to notification mechanisms.

4.4.8 <SIDEBARS-BY-VAL>

The <sidebars-by-val> element contains a set of <entry> child elements, each containing information about a single sidebar. This element is described in Section 5.9.2 of RFC 4575. When the

<sidebars-by-val> element is used in the context of the XCON Conference Information Model, the "state" and "version" attributes defined in RFC 4575 are not used, since they apply only to notification mechanisms.

4.5 RELAX NG SCHEMA

In accordance with the centralized conferencing framework specification of RFC 5239 (see Chapter 2), the conference object is a logical representation of a conference instance. The conference information schema contains core information that is utilized in any conference. It also contains the variable information part of the conference object. The normative schema is backwards compatible with RFC 5239; in other words, valid RFC 5239 instance specifications are also valid according to this RELAX NG schema shown in Figure 4.4 [4]. In addition to approximately similar RELAX NG [2] definitions of RFC 5239, this schema contains extension elements in the "urn:ietf:params:xml :ns:xcon-conference-info" namespace.

```
default namespace = "urn:ietf:params:xml:ns:conference-info"
namespace xcon = "urn:ietf:params:xml:ns:xcon-conference-info"
```

4.6 XML SCHEMA EXTENSIBILITY

The conference information data model defined here (Section 4.7) is meant to be extensible. Extensions are accomplished by defining elements or attributes qualified by namespaces other than "urn:ietf:params:xml:ns:conference-info" and "urn:ietf:params:xml:ns:xcon-conference-info" for use wherever the schema allows such extensions (i.e., where the RELAX NG definition specifies "anyAttribute" or "anyElement"). Elements or attributes from unknown namespaces must be ignored.

4.7 XML EXAMPLE

The following is an example (Figure 4.5) of a conference information specification. The conference starts on October 17, 2007, at 10:30 a.m. in New York City, finishes the same day at 12:30 p.m. every week, and repeats every week. In this example, there are currently three participants in the conference: one administrator, one moderator, and one participant. Sidebars are allowed in this conference, and consequently, there is one sidebar in the conference. In addition, Alice and Carol are using a floor in the main conference to manage the audio and video resources. At the moment, Alice is assigned to use the floor.

4.8 NON-NORMATIVE RELAX NG SCHEMA IN XML SYNTAX (APPENDIX A OF RFC 6501)

Figure 4.6 shows the non-normative RELAX NG schema in XML syntax. Note that is defined in Appendix A of RFC 6501.

4.9 RELAX SCHEMA IN XML SYNTAX (APPENDIX B OF RFC 6501)

The non-normative W3C XML schema defines extension elements in the "urn:ietf:params:xml:ns:x con-conference-info" namespace. Note that <xs:any> extensions in this schema are stricter than in the normative RELAX NG schema [4], and the normative RELAX NG schema [4] allows unordered child elements, unlike this schema (and the RFC 4575 schema). Also, note that this schema allows otherwise valid extension elements to appear in the non-allowed positions. Likewise, the cardinalities of these extension elements cannot be constrained with this schema (Figure 4.7).

```
start = element conference-info { conference-type }
# CONFERENCE TYPE
conference-type =
    attribute entity { text }
    & anyAttribute
    & conference-description-type?
    & element host-info { host-type }?
    & element conference-state { conference-state-type }?
    & element users { users-type }?
    & element sidebars-by-ref { uris-type }?
    & element sidebars-by-val { sidebars-by-val-type }?
    & element xcon:floor-information { floor-information-type }?
    & anyElement*
# CONFERENCE DESCRIPTION TYPE
conference-description-type =
  element conference-description {
    attribute xml:lang { xsd:language }?
    & anyAttribute
    & element display-text { text }?
    & element subject { text }?
    & element free-text { text }?
    & element keywords {
          list { xsd:string* }
       }?
    & element conf-uris { uris-type }?
    & element service-uris { uris-type }?
    & element maximum-user-count { xsd:int }?
    & element available-media { conference-media-type }?
    & element xcon:language { xsd:language }?
    & element xcon:allow-sidebars { xsd:boolean }?
    & element xcon:cloning-parent { xsd:anyURI }?
    & element xcon:sidebar-parent { xsd:anyURI }?
    & element xcon:conference-time { conferencetime-type }?
    & anyElement*
  }
# HOST TYPE
host-type =
    element display-text { text }?
    & element web-page { xsd:anyURI }?
    & element uris { uris-type }?
    & anyElement*
    & anyAttribute
# CONFERENCE STATE TYPE
conference-state-type =
    anyAttribute
    & element user-count { xsd:unsignedInt }?
    & element active { xsd:boolean }?
    & element locked { xsd:boolean }?
    & element xcon:allow-conference-event-subscription { xsd:boolean }?
    & anyElement*
# CONFERENCE MEDIA TYPE
conference-media-type =
    anyAttribute
    & element entry { conference-medium-type }*
    & anyElement*
# CONFERENCE MEDIUM TYPE
conference-medium-type =
    attribute label { text }
    & anyAttribute
    & element display-text { text }?
    & element type { text }?
    & element status { media-status-type }?
    & element xcon:mixing-mode { mixing-mode-type }?
    & element xcon:codecs { codecs-type }?
    & element xcon:controls { control-type }?
    & anyElement*
# URIs TYPE
uris-type =
    anyAttribute
    & element entry { uri-type }*
    & anyElement*
# URI TYPE
uri-type =
    element uri { xsd:anyURI }
    & element display-text { text }?
    & element purpose { text }?
    & element modified { execution-type }?
    & element xcon:conference-password { text }*
    & anyElement*
    & anyAttribute
# USERS TYPE
users-type =
    anyAttribute
    & element user { user-type }*
    & element xcon:join-handling { join-handling-type }?
    & element xcon:user-admission-policy { user-admission-policy-type }?
    & element xcon:allowed-users-list { allowed-users-list-type }?
    & element xcon:deny-users-list { deny-user-list-type }?
```

(a)

```
    & anyElement*
# USER TYPE
    user-type =
    attribute entity { xsd:anyURI }
    & anyAttribute
    & element display-text { text }?
    & element associated-aors { uris-type }?
    & element roles {
          element entry { single-role-type }+
      }?
    & element languages {
          list { xsd:language }
      }?
    & element cascaded-focus { xsd:anyURI }?
    & element endpoint { endpoint-type }*
    & element xcon:provide-anonymity { provide-anonymity-type }?
    & element xcon:allow-refer-users-dynamically { xsd:boolean }?
    & element xcon:allow-invite-users-dynamically { xsd:boolean }?
    & element xcon:allow-remove-users-dynamically { xsd:boolean }?
    & anyElement*
# ENDPOINT TYPE
endpoint-type =
    attribute entity { text }
    & anyAttribute
    & element display-text { text }?
    & element referred { execution-type }?
    & element status { endpoint-status-type }?
    & element joining-method { joining-type }?
    & element joining-info { execution-type }?
    & element disconnection-method { disconnection-type }?
    & element disconnection-info { execution-type }?
    & element media { media-type }*
    & element call-info { call-type }?
    & anyElement*
# ENDPOINT STATUS TYPE
endpoint-status-type =
    "pending"
    | "dialing-out"
    | "dialing-in"
    | "alerting"
    | "on-hold"
    | "connected"
    | "muted-via-focus"
    | "disconnecting"
    | "disconnected"
    | free-text-extension
# JOINING TYPE
joining-type =
    "dialed-in" | "dialed-out" | "focus-owner" | free-text-extension
# DISCONNECTION TYPE
disconnection-type =
    "departed" | "booted" | "failed" | "busy" | free-text-extension
# EXECUTION TYPE
execution-type =
    element when { xsd:dateTime }?
    & element reason { text }?
    & element by { xsd:anyURI }?
    & anyAttribute
# CALL TYPE
call-type =
    element sip { sip-dialog-id-type }
    & anyElement*
    & anyAttribute
# SIP DIALOG ID TYPE
sip-dialog-id-type =
    element display-text { text }?
    & element call-id { text }
    & element from-tag { text }
    & element to-tag { text }
    & anyElement*
    & anyAttribute
# MEDIA TYPE
media-type =
    attribute id { xsd:int }
    & anyAttribute
    & element display-text { text }?
    & element type { text }?
    & element label { text }?
    & element src-id { text }?
    & element status { media-status-type }?
    & element xcon:to-mixer { mixer-type }?
    & element xcon:from-mixer { mixer-type }?
    & anyElement*
# MEDIA STATUS TYPE
media-status-type =
    "recvonly"
    | "sendonly"
    | "sendrecv"
    | "inactive"
```

(b)

FIGURE 4.4 Relax NG Schema. (Copyright: IETF.)

4.10 SECURITY CONSIDERATIONS

There are numerous security considerations for this specification. Overall, the security considerations for authentication and the Security and Privacy of Identity described in RFC 5239 (see Sections 2, 2.11, and 2.11.3) apply to this specification. This specification defines a data model for conference objects. Different conferencing systems may use different protocols to provide access to

```
| free-text-extension
# SIDEBARS-BY-VAL TYPE
sidebars-by-val-type =
anyAttribute
& element entry { conference-type }*
& anyElement*
# CONFERENCE TIME
conferencetime-type =
    anyAttribute
    & element xcon:entry {
        element xcon:base { text },
        element xcon:mixing-start-offset {
        time-type,
        attribute required-participant { single-role-type },
        anyAttribute
    }?,
    element xcon:mixing-end-offset {
        time-type,
        attribute required-participant { single-role-type },
        anyAttribute
    }?,
    element xcon:can-join-after-offset { time-type }?,
    element xcon:must-join-before-offset { time-type }?,
    element xcon:request-user { time-type }?,
    element xcon:notify-end-of-conference { xsd:nonNegativeInteger }?,
    element xcon:allowed-extend-mixing-end-offset { xsd:boolean }?,
    anyElement*
  }*
# TIME TYPE
time-type = xsd:dateTime { pattern = ".+T.+Z.*" }
# SINGLE ROLE TYPE
single-role-type =
    xsd:string "none"
    | xsd:string "administrator"
    | xsd:string "moderator"
    | xsd:string "user"
    | xsd:string "observer"
    | sd:string "participant"
    | free-text-extension
# MIXING MODE TYPE
mixing-mode-type =
    xsd:string "moderator-controlled"
    | xsd:string "FCFS"
    | xsd:string "automatic"
    | free-text-extension
# CODECS TYPE
codecs-type =
    attribute decision { decision-type }
    & anyAttribute
    & element xcon:codec { codec-type }*
    & anyElement*
# CODEC TYPE
codec-type =
    attribute name { text }
    & attribute policy { policy-type }
    & anyAttribute
    & element xcon:subtype { text }?
    & anyElement*
# DECISION TYPE
decision-type =
    xsd:string "automatic"
    | xsd:string "moderator-controlled"
    | free-text-extension
# POLICY TYPE
policy-type =
    xsd:string "allowed" | xsd:string "disallowed" | free-text-extension
# CONTROL TYPE
control-type =
    anyAttribute
    & element xcon:mute { xsd:boolean }?
    & element xcon:pause-video { xsd:boolean }?
    & element xcon:gain { gain-type }?
    & element xcon:video-layout { video-layout-type }?
    & anyElement*
# GAIN TYPE
gain-type = xsd:int { minInclusive = "-127" maxInclusive = "127" }
# VIDEO LAYOUT TYPE
video-layout-type =
    xsd:string "single-view"
    | xsd:string "dual-view"
    | xsd:string "dual-view-crop"
    | xsd:string "dual-view-2x1"
    | xsd:string "dual-view-2x1-crop"
    | xsd:string "quad-view"
    | xsd:string "multiple-3x3"
    | xsd:string "multiple-4x4"
    | xsd:string "multiple-5x1"
    | xsd:string "automatic"
    | free-text-extension
```

(c)

```
# FLOOR INFORMATION TYPE
floor-information-type =
    anyAttribute
    & element xcon:conference-ID { xsd:unsignedLong }?
    & element xcon:allow-floor-events { xsd:boolean }?
    & element xcon:floor-request-handling { floor-request-type }?
    & element xcon:conference-floor-policy { conference-floor-policy }?
    & anyElement*
# FLOOR REQUEST TYPE
floor-request-type =
    xsd:string "block" | xsd:string "confirm" | free-text-extension
# CONFERENCE FLOOR POLICY
conference-floor-policy =
    anyAttribute
    & element xcon:floor {
    attribute id { text }
    & anyAttribute
    & element xcon:media-label { xsd:nonNegativeInteger }+
    & element xcon:algorithm { algorithm-type }?
    & element xcon:max-floor-users { xsd:nonNegativeInteger }?
    & element xcon:moderator-id { xsd:nonNegativeInteger }?
    & anyElement*
    }*
# ALGORITHM POLICY
algorithm-type =
    xsd:string "moderator-controlled"
    | xsd:string "FCFS"
    | xsd:string "random"
    | free-text-extension
# USERS ADMISSION POLICY
user-admission-policy-type =
    xsd:string "closedAuthenticated"
    | xsd:string "openAuthenticated"
    | xsd:string "anonymous"
    | free-text-extension
# JOIN HANDLING TYPE
join-handling-type =
    xsd:string "block"
    | xsd:string "confirm"
    | xsd:string "allow"
    | xsd:string "authenticate"
    | xsd:string "directed-operator"
    | free-text-extension
# DENY USERLIST
deny-user-list-type =
    anyAttribute
    & element xcon:target {
        attribute uri { xsd:anyURI },
        anyAttribute
      }*
    & anyElement*
# ALLOWED USERS LIST TYPE
allowed-users-list-type =
    anyAttribute
    & element xcon:target { target-type }*
    & element xcon:persistent-list { persistent-list-type }?
    & anyElement*
# PERSISTENT LIST TYPE
persistent-list-type =
    element xcon:user {
        attribute name { text }
        & attribute nickname { text }
        & attribute id { text }
        & anyAttribute
        & element xcon:e-mail { text }*
        & anyElement*

      }*
    & anyElement*
# TARGET TYPE
target-type =
    attribute uri { xsd:anyURI },
    attribute method { method-type },
    anyAttribute
# METHOD TYPE
method-type =
    xsd:string "dial-in"
    | xsd:string "dial-out"
    | xsd:string "refer"
    | free-text-extension
# ANONYMITY TYPE
provide-anonymity-type =
    "private" | "semi-private" | "hidden" | free-text-extension
# MIXER TYPE
mixer-type =
    attribute name { mixer-name-type }
    & anyAttribute
    & element xcon:controls { control-type }*
    & element xcon:floor {
```

(d)

FIGURE 4.4 Continued.

these conference objects. This section contains general security considerations for the conference objects and for the protocols. The specification of each particular protocol needs to discuss how the specific protocol meets the security requirements provided in this section.

A given conferencing system usually supports different protocols in order to implement different functions (e.g., SIP for session control and BFCP for floor control). Each of these protocols may use its own authentication mechanism. In cases where a user is authenticated using multiple

```
          attribute id { text },
          anyAttribute,
          xsd:boolean
        }*
      & anyElement*
# MIXER NAME TYPE
mixer-name-type =
      "VideoIn" | "VideoOut" | "AudioOut" | "AudioIn" | free-text-extension
# FREE TEXT EXTENSION
#
free-text-extension = text
# *********************************
# EXTENSIBILITY OF THE SCHEMA
# *********************************
# EXTENSIBILITY ELEMENTS
#
anyElement =
    element * - (conference-description
                 | host-info
                 | conference-state
                 | users
                 | sidebars-by-ref
                 | sidebars-by-val
                 | display-text
                 | subject
                 | free-text
                 | keywords
                 | conf-uris
                 | service-uris
                 | maximum-user-count
                 | available-media
                 | web-page
                 | uris
                 | uri
                 | user-count
                 | active
                 | locked
                 | entry
                 | type
                 | status
                 | purpose
                 | modified
                 | user
                 | associated-aors
                 | roles
                 | languages
                 | cascaded-focus
                 | endpoint
                 | referred
                 | joining-method
                 | joining-info
                 | disconnection-method
                 | disconnection-info
                 | media
                 | call-info
                 | when
                 | reason
                 | by
                 | sip
                 | call-id
                 | from-tag
                 | to-tag
                 | label
                 | src-id
                 | xcon:conference-password
                 | xcon:mixing-mode
                 | xcon:codecs
                 | xcon:controls
```

(e)

```
                 | xcon:language
                 | xcon:allow-sidebars
                 | xcon:cloning-parent
                 | xcon:sidebar-parent
                 | xcon:allow-conference-event-subscription
                 | xcon:to-mixer
                 | xcon:provide-anonymity
                 | xcon:allow-refer-users-dynamically
                 | xcon:allow-invite-users-dynamically
                 | xcon:allow-remove-users-dynamically
                 | xcon:from-mixer
                 | xcon:join-handling
                 | xcon:user-admission-policy
                 | xcon:allowed-users-list
                 | xcon:deny-users-list
                 | xcon:floor-information
                 | xcon:conference-time
                 | xcon:provide-anonymity
                 | xcon:floor
                 | xcon:entry
                 | xcon:mixing-start-offset

                 | xcon:mixing-end-offset
                 | xcon:can-join-after-offset
                 | xcon:must-join-before-offset
                 | xcon:request-user
                 | xcon:notify-end-of-conference
                 | xcon:allowed-extend-mixing-end-offset
                 | xcon:codec
                 | xcon:subtype
                 | xcon:mute
                 | xcon:pause-video
                 | xcon:gain
                 | xcon:video-layout
                 | xcon:conference-ID
                 | xcon:allow-floor-events
                 | xcon:floor-request-handling
                 | xcon:conference-floor-policy
                 | xcon:media-label
                 | xcon:algorithm
                 | xcon:max-floor-users
                 | xcon:moderator-id
                 | xcon:target
                 | xcon:persistent-list
                 | xcon:e-mail
                 | xcon:user) { anyExtension }
anyExtension =
    (attribute * { text }
     | any)*
    any =
        element * {
           (attribute * { text }
            | text
            | any)*
        }
# EXTENSIBILITY ATTRIBUTES
#

anyAttribute =
    attribute * - (xml:lang
                   | entity
                   | required-participant
                   | label
                   | decision
                   | name
                   | policy
                   | uri
                   | method
                   | id
                   | nickname) { text }*
```

(f)

FIGURE 4.4 Continued.

authentication mechanisms, it is up to the conferencing system to map all the different authentications to the same user. Discussing the specifics of different authentication mechanisms is beyond the scope of this chapter.

Furthermore, users may use different identifiers to access a conference, as explained in Section 4.4.6.5. These different namespaces can be associated with a unique conference user identifier (XCON-USERID). A mapping database is used to map all these authenticated user namespaces to the XCON-USERID. There are several threats against this database. In order to minimize these threats, the administrator of the conferencing system must ensure that only authorized users can connect to this database (e.g., by using access control rules). In particular, the integrity of the database MUST be protected against unauthorized modifications. In addition, the XCON-USERID or XCON-URI should be hard to guess. It is critical that the URI remain difficult to "guess" via brute force methods. Generic security considerations for usage of URIs are discussed in RFC 3986.

It is recommended that the database uses encryption mechanisms if the information is stored in long-term storage (e.g., disk). If the database contains sensitive elements such as passwords, the confidentiality of the database must be protected from unauthorized users. If no sensitive elements are

```
<?xml version="1.0" encoding="UTF-8"?>
<conference-info
        xmlns="urn:ietf:params:xml:ns:conference-info"
        xmlns:xcon="urn:ietf:params:xml:ns:xcon-conference-info"
        entity="conference123@example.com">
        <!--
                CONFERENCE DESCRIPTION
        -->
        <conference-description xml:lang="en-us">
                <display-text>Discussion of Formula-1 racing</display-text>
                <subject>Sports:Formula-1</subject>
                <free-text>This is a conference example</free-text>
                <keywords>Formula-1 cars</keywords>
                <!--
                CONFERENCE UNIQUE IDENTIFIERS
        -->
        <conf-uris>
                <entry>
                        <uri>tel:+3585671234</uri>
                        <display-text>Conference Bridge</display-text>
                        <purpose>participation</purpose>
                        <xcon:conference-password
                                >5678</xcon:conference-password>
                </entry>
                <entry>
                        <uri>http://www.example.com/live.ram</uri>
                        <purpose>streaming</purpose>
                </entry>
        </conf-uris>
        <!--
                SERVICE URIS
        -->
        <service-uris>
                <entry>
                        <uri>mailto:bob@example.com</uri>
                        <display-text>email</display-text>
                </entry>
        </service-uris>
        <!--
                MAXIMUM USER COUNT
        -->
        <maximum-user-count>50</maximum-user-count>
        <!--
                AVAILABLE MEDIA
        -->
        <available-media>
                <entry label="10234">
                        <display-text>main audio</display-text>
                        <type>audio</type>
                        <status>sendrecv</status>
                        <xcon:mixing-mode>automatic</xcon:mixing-mode>
                                <xcon:codecs decision="automatic">
                                        <xcon:codec name="122" policy="allowed">
                                                <xcon:subtype>PCMU</
xcon:subtype>
                                        </xcon:codec>
                                        </xcon:codecs>
                                        <xcon:controls>
                                                <xcon:mute>true</xcon:mute>
                                                <xcon:gain>50</xcon:gain>
                                        </xcon:controls>
                </entry>
                <entry label="10235">
                        <display-text>main video</display-text>
                        <type>video</type>
                        <status>sendrecv</status>
                        <xcon:mixing-mode>automatic</xcon:mixing-mode>
                        <xcon:codecs decision="automatic">
                                <xcon:codec name="123" policy="allowed">
```

(a)

```
                                        <xcon:subtype>H.263</xcon:subtype>
                                </xcon:codec>
                        </xcon:codecs>
                        <xcon:controls>
                                <xcon:video-layout
                                        >single-view</xcon:video-layout>
                        </xcon:controls>
                </entry>
        </available-media>
<xcon:language>En-us</xcon:language>

<xcon:allow-sidebars>true</xcon:allow-sidebars>

<!--
CONFERENCE TIME
-->
 <xcon:conference-time>
        <xcon:entry>
                <xcon:base>BEGIN:VCALENDAR
PRODID:-//LlamaSpinner Inc.//NONSGML CamelCall//EN
VERSION:2.0
BEGIN:VEVENT
DTSTAMP:20071003T140728Z
UID:20071003T140728Z-345FDA-carol@example.com
ORGANIZER:MAILTO:carol@example.com
DTSTART:20071017T143000Z
RRULE:FREQ=WEEKLY
DTEND:20071217T163000Z
END:VEVENT
END:VCALENDAR</xcon:base>
                <xcon:mixing-start-offset
                        required-participant="moderator"
                      >2007-10-17T14:29:00Z</xcon:mixing-start-offset>
                <xcon:mixing-end-offset
                        required-participant="participant"
                      >2007-10-17T16:31:00Z</xcon:mixing-end-offset>
                <xcon:must-join-before-offset
                      >2007-10-17T15:30:00Z
                </xcon:must-join-before-offset>
        </xcon:entry>
 </xcon:conference-time>

</conference-description>
<!--
HOST INFO
-->
<host-info>
        <display-text>Formula1</display-text>
        <web-page>http://www.example.com/formula1/</web-page>
        <uris>
                <entry>
                        <uri>sip:alice@example.com</uri>
                </entry>
                <entry>
                        <uri>sip:carol@example.com</uri>
                </entry>
        </uris>
</host-info>
<!--
CONFERENCE STATE
-->
<conference-state>
        <user-count>3</user-count>
        <active>true</active>
        <locked>false</locked>
        <xcon:allow-conference-event-subscription
                >true</xcon:allow-conference-event-subscription>
</conference-state>
```

(b)

FIGURE 4.5 Conference Information Specification. (Copyright: IETF.) Note: Due to RFC formatting conventions, this specification splits lines whose content would exceed 72 characters.

present, then confidentiality is not needed. In addition to implementing access control, as discussed earlier, it is recommended that administrators of conferencing systems only provide access to the database over encrypted channels (e.g., using transport layer security [TLS] encryption) in order to avoid eavesdroppers. Administrators of conferencing systems should also avoid disclosing information to unauthorized parties when a conference is being cloned or when a sidebar is being created. For example, an external sidebar as defined in RFC 5239 (see Chapter 2 and Section 2.9.4.2) may include participants who were not authorized for the parent conference.

The security considerations for authentication described in RFC 5239 (see Sections 2 and 2.11.1) of the centralized conferencing framework specification specified in RFC 5239 also apply to this specification. Similarly, the security considerations for authorization described in Section 5.2 of the SIP REFER Method specified in RFC 3515 (updated by RFCs 7647 and 8217) apply to this specification as well. Note that the specification of the privacy policy is outside the scope of this specification. Saying that, a privacy policy will be needed in the real implementation of the data model and therefore, is subject to future policy specifications.

```
<!--
USERS
-->
<users>
        <!--
        USER BOB
        -->
        <user entity="xcon-userid:bob534">
                <display-text>Bob Hoskins</display-text>
                <associated-aors>
                        <entry>
                                <uri>mailto:bob@example.com</uri>
                                <display-text>email</display-text>
                        </entry>
                </associated-aors>
                <roles>
                        <entry>participant</entry>
                </roles>
                <languages>en-us</languages>
<!--
        ENDPOINTS
-->
<endpoint entity="sip:bob@example.com">
        <display-text>Bob's Laptop</display-text>
        <referred>
                <when>2007-10-17T14:00:00Z</when>
                <reason>expert required</reason>
                <by>sip:alice@example.com</by>
        </referred>
        <status>connected</status>
        <joining-method>dialed-out</joining-method>
        <joining-info>
                <when>2007-10-17T14:00:00Z</when>
                <reason>invitation</reason>
                <by>sip:alice@example.com</by>
        </joining-info>
<!--
        MEDIA
-->
<media id="1">
        <type>video</type>
        <label>10235</label>
        <src-id>432424</src-id>
        <status>sendrecv</status>
        <xcon:to-mixer name="VideoIn">
                <xcon:controls>
                <xcon:video-layout
                        >single-view</xcon:video-layout>
                </xcon:controls>
        </xcon:to-mixer>
</media>
<!--
        CALL INFO
-->
  <call-info>
        <sip>
                <display-text>full info</display-text>
                <call-id>hsjh8980vhsb78</call-id>
                <from-tag>vav738dvbs</from-tag>
                <to-tag>8954jgjg8432</to-tag>
        </sip>
  </call-info>
</endpoint>
<xcon:provide-anonymity
        >semi-private</xcon:provide-anonymity>
        <xcon:allow-refer-users-dynamically
                >false</xcon:allow-refer-users-dynamically>
        <xcon:allow-invite-users-dynamically
                >false</xcon:allow-invite-users-dynamically>
        <xcon:allow-remove-users-dynamically
                >false</xcon:allow-remove-users-dynamically>
</user>

<!--
        USER ALICE
-->
<user entity="xcon-userid:alice334">
        <display-text>Alice Kay</display-text>
        <associated-aors>
                <entry>
                        <uri>mailto:alice@example.com</uri>
                        <display-text>email</display-text>
                </entry>
        </associated-aors>
        <roles>
                <entry>moderator</entry>
        </roles>
        <languages>en-us</languages>
```

(c)

```
<!--
        ENDPOINTS
-->
<endpoint entity="sip:alice@example.com">
        <display-text>Alice's Desktop</display-text>
        <status>connected</status>
        <joining-method>dialed-in</joining-method>
        <joining-info>
                <when>2007-10-17T13:35:08Z</when>
                        <reason>invitation</reason>
                <by>sip:conference@example.com</by>
        </joining-info>
<!--
        MEDIA
-->
<media id="1">
        <type>video</type>
        <label>10235</label>
        <src-id>432424</src-id>
        <status>sendrecv</status>
        <xcon:to-mixer name="VideoIn">
                <xcon:controls>
                        <xcon:video-layout
                                >single-view</xcon:video-layout>
                </xcon:controls>
        </xcon:to-mixer>
</media>
<media id="2">
        <type>audio</type>
        <label>10234</label>
        <src-id>532535</src-id>
        <status>sendrecv</status>
        <xcon:to-mixer name="AudioIn">
                <xcon:controls>
                        <xcon:gain>50</xcon:gain>
                </xcon:controls>
        </xcon:to-mixer>
        <xcon:from-mixer name="AudioOut">
                <xcon:controls>
                        <xcon:gain>50</xcon:gain>
                </xcon:controls>
        </xcon:from-mixer>
</media>
<!--
        CALL INFO
-->
  <call-info>
        <sip>
                <display-text>full info</display-text
                <by>sip:conference@example.com</by>
        </joining-info>
<!--
        MEDIA
-->
<media id="1">
        <type>video</type>
        <label>10235</label>
        <src-id>432424</src-id>
        <status>sendrecv</status>
        <xcon:to-mixer name="VideoIn">
                <xcon:controls>
                        <xcon:video-layout
                                >single-view</xcon:video-layout>
                </xcon:controls>
        </xcon:to-mixer>
</media>
<media id="2">
        <type>audio</type>
        <label>10234</label>
        <src-id>532535</src-id>
        <status>sendrecv</status>
        <xcon:to-mixer name="AudioIn">
                <xcon:controls>
                        <xcon:gain>50</xcon:gain>
                </xcon:controls>
        </xcon:to-mixer>
        <xcon:from-mixer name="AudioOut">
                <xcon:controls>
                        <xcon:gain>50</xcon:gain>
                </xcon:controls>
        </xcon:from-mixer>
</media>
<!--
        CALL INFO
-->
  <call-info>
        <sip>
                <display-text>full info</display-text
```

(d)

FIGURE 4.5 Continued.

4.11 SUMMARY

We have described the XML-based conference information data model to be used for conference objects that are described in RFC 5239 (see Chapter 2). First, we have defined the conference information data model, that consists of data model format and namespace, conference object identifier (object URI and normalization and URI comparison), and data model structure. Second, we have

```
<!--
    FLOOR INFORMATION
-->
<xcon:floor-information>
    <xcon:conference-ID>567</xcon:conference-ID>
    <xcon:allow-floor-events>true</xcon:allow-floor-events>
    <xcon:floor-request-handling
        >confirm</xcon:floor-request-handling>

<xcon:conference-floor-policy>
    <xcon:floor id="345">
        <xcon:media-label>10234</xcon:media-label>
        <xcon:media-label>10235</xcon:media-label>
        <xcon:algorithm
            >moderator-controlled</xcon:algorithm>
        <xcon:max-floor-users>1</xcon:max-floor-users>
        <xcon:moderator-id>234</xcon:moderator-id>
    </xcon:floor>
  </xcon:conference-floor-policy>
 </xcon:floor-information>

</conference-info>
```

(e)

```
        </xcon:deny-users-list>
</users>

 <!--
     SIDEBARS BY REFERENCE
 -->
 <sidebars-by-ref>
     <entry>
          <uri>xcon:conf223</uri>
          <display-text>private with Bob</display-text>
     </entry>
 </sidebars-by-ref>

 <!--
     SIDEBARS BY VALUE
 -->
 <sidebars-by-val>
     <entry entity="conf223">
          <users>
               <user entity="xcon-userid:bob534"/>
               <user entity="xcon-userid:carol233"/>
          </users>
     </entry>
 </sidebars-by-val>
```

(f)

```
<! ---
   FLOOR INFORMATION
--->

<xcon:floor-information>
   <xcon:conference-ID>567</xcon:conference-ID>
   <xcon:allow-floor-events>true</xcon:allow-floor-events>
   <xcon:floor-request-handling
    >confirm</xcon:floor-request-handling>

   <xcon:conference-floor-policy>
      <xcon:floor id="345">
         <xcon:media-label>10234</xcon:media-label>
         <xcon:media-label>10235</xcon:media-label>
         <xcon:algorithm
            >moderator-controlled</xcon:algorithm>
         <xcon:max-floor-users>1</xcon:max-floor-users>
         <xcon:moderator-id>234</xcon:moderator-id>
         </xcon:floor>
   </xcon:conference-floor-policy>
</xcon:floor-information>

</conference-info>
```

(g)

FIGURE 4.5 Continued.

specified the XML-based data model for each conference object, extending the data format specified in the SIP event package for conference state. The normative RELAX NG schema, which is backwards compatible with RFC 5239 (see Chapter 2), is specified in this section, adding more resourceful features of the conference objects. The XML schema defined here is extensible. The non-normative RELAX NG Schema in XML Syntax and non-normative W3C XML Schema for conference objects are also described. Finally, security schemes for the conference data model schema are described considering when conferencing objects defined in the schema are transferred over protocols.

4.12 PROBLEMS

1. What are the conference objects? Explain in detail using centralized multimedia conferencing information data architectural model. Why is an XML-based format used for the conference information data model? Is this data model independent of specific conference media (e.g. audio, video, and data)? If not, explain why not.

```
<?xml version="1.0" encoding="UTF-8" ?>
 <grammar
   ns="urn:ietf:params:xml:ns:conference-info"
   xmlns="http://relaxng.org/ns/structure/1.0"
   xmlns:xcon="urn:ietf:params:xml:ns:xcon-conference-info"
   datatypeLibrary="http://www.w3.org/2001/XMLSchema-datatypes">
   <start>
    <element name="conference-info">
     <ref name="conference-type"/>
    </element>
   </start>
   <!--
     CONFERENCE TYPE

   -->
   <define name="conference-type">
    <interleave>
    <attribute name="entity">
     <text/>
    </attribute>
    <ref name="anyAttribute"/>
    <optional>
     <ref name="conference-description-type"/>
    </optional>
    <optional>
     <element name="host-info">
      <ref name="host-type"/>
     </element>
    </optional>
    <optional>
     <element name="conference-state">
      <ref name="conference-state-type"/>
     </element>
    </optional>
    <optional>
     <element name="users">
      <ref name="users-type"/>
     </element>
    </optional>
    <optional>
     <element name="sidebars-by-ref">
      <ref name="uris-type"/>
     </element>
    </optional>
    <optional>
     <element name="sidebars-by-val">
     <ref name="sidebars-by-val-type"/>
```

(a)

```
    </element>
   </optional>
   <optional>
    <element name="xcon:floor-information">
     <ref name="floor-information-type"/>
    </element>
   </optional>
   <zeroOrMore>
    <ref name="anyElement"/>
   </zeroOrMore>
   </interleave>
  </define>
  <!--
    CONFERENCE DESCRIPTION TYPE
  -->
  <define name="conference-description-type">
   <element name="conference-description">
    <interleave>
    <optional>
     <attribute name="xml:lang">
      <data type="language"/>
     </attribute>
    </optional>
    <ref name="anyAttribute"/>
    <optional>
     <element name="display-text">
      <text/>
     </element>
    </optional>
    <optional>
     <element name="subject">
      <text/>
     </element>
    </optional>
    <optional>
     <element name="free-text">
      <text/>
     </element>
    </optional>
    <optional>
     <element name="keywords">
      <list>
       <zeroOrMore>
        <data type="string"/>
       </zeroOrMore>
      </list>
     </element>
```

(b)

FIGURE 4.6 Non-Normative RELAX NG Schema in XML Syntax. (Copyright: IETF.)

2. What are the salient differences between the data model defined in the SIP event package (RFC 4575) and the call signaling independent centralized conferencing framework (RFC 5239)?
3. Describe the conference information data model, including namespace, object identifier, URI and normalization and conference object URI comparison, and data model structure.
4. Explain in detail the important characteristics of all conference objects defined in the data model.
5. Describe in detail each of the following conference objects: <conference-info>, <conference-description>, <host-info>, <conference-state>, <floor-information>, <floor-information>, <users>, <sidebars-by-ref>, and <sidebars-by-val>.
6. What is the RELAX NG Schema? How does it relate to the conference information schema (RFC 5239)?
7. What are the benefits offered through extensibility of the XML schema for conference objects?
8. Explain in detail the centralized conferencing scenarios for the XML example described in Section 4.7.
9. Explain in detail the conferencing object features for the non-normative RELAX NG W3C XML schema described in Sections 4.8 and 4.9, respectively.

```
</optional>
<optional>
<element name="conf-uris">
<ref name="uris-type"/>
</element>
</optional>
<optional>
<element name="service-uris">
<ref name="uris-type"/>
</element>
</optional>
<optional>
<element name="maximum-user-count">
<data type="int"/>
</element>
</optional>
<optional>
<element name="available-media">
<ref name="conference-media-type"/>
</element>
</optional>
<optional>
<element name="xcon:language">
<data type="language"/>
</element>
</optional>
<optional>
<element name="xcon:allow-sidebars">
<data type="boolean"/>
</element>
</optional>
<optional>
<element name="xcon:cloning-parent">
<data type="anyURI"/>
</element>
</optional>
<optional>
<element name="xcon:sidebar-parent">
<data type="anyURI"/>
</element>
</optional>
<optional>
<element name="xcon:conference-time">
<ref name="conferencetime-type"/>
</element>
</optional>
<zeroOrMore>
<ref name="anyElement"/>
</zeroOrMore>
</interleave>
</element>
```

(c)

```
</define>
<!--
HOST TYPE
-->
<define name="host-type">
<interleave>
<optional>
<element name="display-text">
<text/>
</element>
</optional>
<optional>
<element name="web-page">
<data type="anyURI"/>
</element>
</optional>
<optional>
<element name="uris">
<ref name="uris-type"/>
</element>
</optional>
<zeroOrMore>
<ref name="anyElement"/>
</zeroOrMore>
<ref name="anyAttribute"/>
</interleave>
</define>
<!--
CONFERENCE STATE TYPE
-->
<define name="conference-state-type">
<interleave>
<ref name="anyAttribute"/>
<optional>
<element name="user-count">
<data type="unsignedInt"/>
</element>
</optional>
<optional>
<element name="active">
<data type="boolean"/>
</element>
</optional>
<optional>
<element name="locked">
```

(d)

FIGURE 4.6 Continued.

10. What are the security capabilities that have been taken care of for the centralized information data model? Explain each of those security capabilities. Can you describe any more security weaknesses that need to be addressed for the centralized conference object data model?

REFERENCES

1. Bray, T. et al. (2008, November) Extensible Markup Language (XML) 1.0 (Fifth Edition). *World Wide Web Consortium Recommendation RECxml-20081126.* <http://www.w3.org/TR/2008/REC-xml-20081126>.
2. IANA. Language Subtag Registry. <http://www.iana.org/assignments/ language-subtag-registry>.
3. IANA. RTP Payload Types. <http://www.iana.org/assignments/rtp-parameters>.
4. RELAX NG Home Page. ISO/IEC 19757-2:2008.

```
<data type="boolean"/>
 </element>
 </optional>
 <optional>
 <element name="xcon:allow-conference-event-subscription">
  <data type="boolean"/>
 </element>
 </optional>
 <zeroOrMore>
 <ref name="anyElement"/>
 </zeroOrMore>
 </interleave>
</define>
<!--
  CONFERENCE MEDIA TYPE
-->
<define name="conference-media-type">
 <interleave>
 <ref name="anyAttribute"/>
 <zeroOrMore>
  <element name="entry">
  <ref name="conference-medium-type"/>
  </element>
 </zeroOrMore>
 <zeroOrMore>
  <ref name="anyElement"/>
 </zeroOrMore>
 </interleave>
</define>
<!--
  CONFERENCE MEDIUM TYPE
-->
<define name="conference-medium-type">
 <interleave>
 <attribute name="label">
  <text/>
 </attribute>
 <ref name="anyAttribute"/>
 <optional>
  <element name="display-text">
  <text/>
  </element>
 </optional>
 <optional>
  <element name="type">
  <text/>
  </element>
 </optional>
 <optional>
   <element name="status">
   <ref name="media-status-type"/>
```

(e)

```
  </element>
 </optional>
 <optional>
  <element name="xcon:mixing-mode">
  <ref name="mixing-mode-type"/>
  </element>
 </optional>
 <optional>
  <element name="xcon:codecs">
  <ref name="codecs-type"/>
  </element>
 </optional>
 <optional>
  <element name="xcon:controls">
  <ref name="control-type"/>
  </element>
 </optional>
 <zeroOrMore>
  <ref name="anyElement"/>
 </zeroOrMore>
 </interleave>
</define>
<!--
  URIs TYPE
-->
<define name="uris-type">
 <interleave>
 <ref name="anyAttribute"/>
 <zeroOrMore>
  <element name="entry">
  <ref name="uri-type"/>
  </element>
 </zeroOrMore>
 <zeroOrMore>
  <ref name="anyElement"/>
 </zeroOrMore>
 </interleave>
</define>
<!--
  URI TYPE
-->
<define name="uri-type">
 <interleave>
 <element name="uri">
  <data type="anyURI"/>
```

(f)

FIGURE 4.6 Continued.

```
</element>
<optional>
 <element name="display-text">
  <text/>
 </element>
</optional>
<optional>
 <element name="purpose">
  <text/>
 </element>
</optional>
<optional>
 <element name="modified">
  <ref name="execution-type"/>
 </element>
</optional>
<zeroOrMore>
 <element name="xcon:conference-password">
  <text/>
 </element>
</zeroOrMore>
<zeroOrMore>
 <ref name="anyElement"/>
</zeroOrMore>
<ref name="anyAttribute"/>
</interleave>

</define>
<!--
  USERS TYPE
-->
<define name="users-type">
 <interleave>
 <ref name="anyAttribute"/>
 <zeroOrMore>
  <element name="user">
   <ref name="user-type"/>
  </element>
 </zeroOrMore>
 <optional>
  <element name="xcon:join-handling">
   <ref name="join-handling-type"/>
  </element>
 </optional>
 <optional>
  <element name="xcon:user-admission-policy">
   <ref name="user-admission-policy-type"/>
  </element>
  </optional>
  <optional>
   <element name="xcon:allowed-users-list">
  <ref name="allowed-users-list-type"/>
```

(g)

```
  </element>
 </optional>
 <optional>
  <element name="xcon:deny-users-list">
   <ref name="deny-user-list-type"/>
  </element>
 </optional>
 <zeroOrMore>
  <ref name="anyElement"/>
 </zeroOrMore>
 </interleave>
</define>
<!--
  USER TYPE
-->
<define name="user-type">
 <interleave>
 <attribute name="entity">
  <data type="anyURI"/>
 </attribute>
 <ref name="anyAttribute"/>
 <optional>
  <element name="display-text">
   <text/>
  </element>
 </optional>
 <optional>
  <element name="associated-aors">
   <ref name="uris-type"/>
  </element>
 </optional>
 <optional>
  <element name="roles">
   <oneOrMore>
    <element name="entry">
     <ref name="single-role-type"/>
    </element>
   </oneOrMore>
  </element>
 </optional>
 <optional>
  <element name="languages">
   <list>
    <data type="language"/>
```

(h)

FIGURE 4.6 Continued.

```
  </list>
  </element>
 </optional>
 <optional>
  <element name="cascaded-focus">
   <data type="anyURI"/>
  </element>
 </optional>
 <zeroOrMore>
  <element name="endpoint">
   <ref name="endpoint-type"/>
  </element>
 </zeroOrMore>
 <optional>
  <element name="xcon:provide-anonymity">
   <ref name="provide-anonymity-type"/>
  </element>
 </optional>
<optional>
  <element name="xcon:allow-refer-users-dynamically">
   <data type="boolean"/>
  </element>
 </optional>
 <optional>
  <element name="xcon:allow-invite-users-dynamically">
   <data type="boolean"/>
  </element>
 </optional>
 <optional>
  <element name="xcon:allow-remove-users-dynamically">
   <data type="boolean"/>
  </element>
 </optional>
  <zeroOrMore>
  <ref name="anyElement"/>
 </zeroOrMore>
 </interleave>
</define>
<!--
  ENDPOINT TYPE
-->
<define name="endpoint-type">
 <interleave>
 <attribute name="entity">
  <text/>
 </attribute>
 <ref name="anyAttribute"/>
 <optional>
```

(i)

```
<element name="display-text">
  <text/>
 </element>
</optional>
<optional>
 <element name="referred">
  <ref name="execution-type"/>
 </element>
</optional>
<optional>
 <element name="status">
  <ref name="endpoint-status-type"/>
 </element>
</optional>
<optional>
 <element name="joining-method">
  <ref name="joining-type"/>
 </element>
</optional>
<optional>
 <element name="joining-info">
  <ref name="execution-type"/>
 </element>
</optional>
<optional>
 <element name="disconnection-method">
  <ref name="disconnection-type"/>
 </element>
</optional>
<optional>
 <element name="disconnection-info">
  <ref name="execution-type"/>
 </element>
</optional>
<zeroOrMore>
 <element name="media">
  <ref name="media-type"/>
 </element>
</zeroOrMore>
<optional>
 <element name="call-info">
  <ref name="call-type"/>
 </element>
</optional>
<zeroOrMore>
 <ref name="anyElement"/>
</zeroOrMore>
</interleave>
```

(j)

FIGURE 4.6 Continued.

```
</define>
<!--
  ENDPOINT STATUS TYPE
-->
<define name="endpoint-status-type">
 <choice>
  <value>pending</value>
  <value>dialing-out</value>
  <value>dialing-in</value>
  <value>alerting</value>
  <value>on-hold</value>
  <value>connected</value>
  <value>muted-via-focus</value>
  <value>disconnecting</value>
  <value>disconnected</value>
  <ref name="free-text-extension"/>
 </choice>
</define>
<!--
  JOINING TYPE
-->
<define name="joining-type">
 <choice>
  <value>dialed-in</value>
  <value>dialed-out</value>
  <value>focus-owner</value>
  <ref name="free-text-extension"/>
 </choice>
</define>
<!--
  DISCONNECTION TYPE
-->
<define name="disconnection-type">
 <choice>
  <value>departed</value>
  <value>booted</value>
  <value>failed</value>
  <value>busy</value>
  <ref name="free-text-extension"/>
 </choice>
</define>
<!--
  EXECUTION TYPE
-->
<define name="execution-type">
 <interleave>
 <optional>
  <element name="when">
```

(k)

```
<data type="dateTime"/>
  </element>
 </optional>
 <optional>
  <element name="reason">
   <text/>
  </element>
 </optional>
 <optional>
  <element name="by">
   <data type="anyURI"/>
  </element>
 </optional>
 <ref name="anyAttribute"/>
 </interleave>
</define>
<!--
  CALL TYPE
-->
<define name="call-type">
 <interleave>
  <element name="sip">
   <ref name="sip-dialog-id-type"/>
  </element>
  <zeroOrMore>
   <ref name="anyElement"/>
  </zeroOrMore>
  <ref name="anyAttribute"/>
 </interleave>
</define>
<!--
  SIP DIALOG ID TYPE
-->
<define name="sip-dialog-id-type">


 <interleave>
 <optional>
 <element name="display-text">
  <text/>
 </element>
 </optional>
 <element name="call-id">
  <text/>
 </element>
 <element name="from-tag">
  <text/>
 </element>
```

(l)

FIGURE 4.6 Continued.

```
<element name="to-tag">
 <text/>
</element>
<zeroOrMore>
 <ref name="anyElement"/>
</zeroOrMore>
<ref name="anyAttribute"/>
</interleave>
</define>
<!--
  MEDIA TYPE
-->
<define name="media-type">
 <interleave>
 <attribute name="id">
  <data type="int"/>
 </attribute>
 <ref name="anyAttribute"/>
 <optional>
  <element name="display-text">
   <text/>
  </element>
 </optional>
 <optional>
  <element name="type">
   <text/>
  </element>
 </optional>
 <optional>
  <element name="label">
   <text/>
  </element>
 </optional>
 <optional>
  <element name="src-id">
   <text/>
  </element>
 </optional>
 <optional>
  <element name="status">
   <ref name="media-status-type"/>
  </element>
 </optional>
 <optional>
  <element name="xcon:to-mixer">
   <ref name="mixer-type"/>
  </element>
 </optional>
```

(m)

```
 <optional>
  <element name="xcon:from-mixer">
   <ref name="mixer-type"/>
  </element>
 </optional>
 <zeroOrMore>
  <ref name="anyElement"/>
 </zeroOrMore>
 </interleave>
</define>
<!--
  MEDIA STATUS TYPE
-->
<define name="media-status-type">
 <choice>
  <value>recvonly</value>
  <value>sendonly</value>
  <value>sendrecv</value>
  <value>inactive</value>
  <ref name="free-text-extension"/>
 </choice>
</define>
<!--
  SIDEBARS-BY-VAL TYPE
-->
<define name="sidebars-by-val-type">
 <interleave>
 <ref name="anyAttribute"/>
 <zeroOrMore>
  <element name="entry">
   <ref name="conference-type"/>
  </element>
 </zeroOrMore>
 <zeroOrMore>
  <ref name="anyElement"/>
 </zeroOrMore>
 </interleave>
</define>
<!--
  CONFERENCE TIME
-->
<define name="conferencetime-type">
 <interleave>
 <ref name="anyAttribute"/>
 <zeroOrMore>
  <element name="xcon:entry">
   <element name="xcon:base">
    <text/>
```

(n)

FIGURE 4.6 Continued.

```
</element>
  <optional>
   <element name="xcon:mixing-start-offset">
    <ref name="time-type"/>
    <attribute name="required-participant">
     <ref name="single-role-type"/>
    </attribute>
    <ref name="anyAttribute"/>
   </element>
  </optional>
  <optional>
   <element name="xcon:mixing-end-offset">
    <ref name="time-type"/>
    <attribute name="required-participant">
     <ref name="single-role-type"/>
    </attribute>
    <ref name="anyAttribute"/>
   </element>
  </optional>
  <optional>
   <element name="xcon:can-join-after-offset">
    <ref name="time-type"/>
   </element>
  </optional>
  <optional>
   <element name="xcon:must-join-before-offset">
    <ref name="time-type"/>
   </element>
  </optional>
  <optional>
   <element name="xcon:request-user">
    <ref name="time-type"/>
   </element>
  </optional>
  <optional>
   <element name="xcon:notify-end-of-conference">
    <data type="nonNegativeInteger"/>
   </element>
  </optional>
  <optional>
   <element name="xcon:allowed-extend-mixing-end-offset">
    <data type="boolean"/>
   </element>
  </optional>
  <zeroOrMore>
   <ref name="anyElement"/>
  </zeroOrMore>
 </element>
```

(o)

```
  </zeroOrMore>
  </interleave>
 </define>
 <!--
   TIME TYPE
 -->
 <define name="time-type">
  <data type="dateTime">
   <param name="pattern">.+T.+Z.*</param>
  </data>
 </define>
 <!--
   SINGLE ROLE TYPE
 -->
 <define name="single-role-type">
  <choice>
   <value type="string">none</value>
   <value type="string">administrator</value>
   <value type="string">moderator</value>
   <value type="string">user</value>
   <value type="string">observer</value>
   <value type="string">participant</value>
   <ref name="free-text-extension"/>
  </choice>
 </define>
 <!--
   MIXING MODE TYPE
 -->
 <define name="mixing-mode-type">
  <choice>
   <value type="string">moderator-controlled</value>
   <value type="string">FCFS</value>
   <value type="string">automatic</value>
   <ref name="free-text-extension"/>
  </choice>
 </define>
 <!--
   CODECS TYPE
 -->
 <define name="codecs-type">
  <interleave>
  <attribute name="decision">
   <ref name="decision-type"/>
  </attribute>
  <ref name="anyAttribute"/>
  <zeroOrMore>
   <element name="xcon:codec">
    <ref name="codec-type"/>
```

(p)

FIGURE 4.6 Continued.

```
</element>
</zeroOrMore>
<zeroOrMore>
 <ref name="anyElement"/>
</zeroOrMore>
</interleave>
</define>
<!--
  CODEC TYPE
-->
<define name="codec-type">
 <interleave>
 <attribute name="name">
  <text/>
 </attribute>
 <attribute name="policy">
  <ref name="policy-type"/>
 </attribute>
 <ref name="anyAttribute"/>
 <optional>
  <element name="xcon:subtype">
   <text/>
  </element>
 </optional>
 <zeroOrMore>
  <ref name="anyElement"/>
 </zeroOrMore>
 </interleave>
</define>
<!--
  DECISION TYPE
-->
<define name="decision-type">
 <choice>
  <value type="string">automatic</value>
  <value type="string">moderator-controlled</value>
  <ref name="free-text-extension"/>
 </choice>
</define>
<!--
  POLICY TYPE
-->
<define name="policy-type">
 <choice>
  <value type="string">allowed</value>
  <value type="string">disallowed</value>
  <ref name="free-text-extension"/>
 </choice>
```

(q)

```
</define>
<!--
  CONTROL TYPE
-->
<define name="control-type">
 <interleave>
 <ref name="anyAttribute"/>
 <optional>
  <element name="xcon:mute">
   <data type="boolean"/>
  </element>
 </optional>
 <optional>
  <element name="xcon:pause-video">
   <data type="boolean"/>
  </element>
 </optional>
 <optional>
  <element name="xcon:gain">
   <ref name="gain-type"/>
  </element>
 </optional>
 <optional>
  <element name="xcon:video-layout">
   <ref name="video-layout-type"/>
  </element>
 </optional>

 <zeroOrMore>
  <ref name="anyElement"/>
 </zeroOrMore>
 </interleave>
</define>
<!--
  GAIN TYPE
-->
<define name="gain-type">
 <data type="int">
   <param name="minInclusive">-127</param>
   <param name="maxInclusive">127</param>
 </data>
</define>
<!--
  VIDEO LAYOUT TYPE
-->
```

(r)

FIGURE 4.6 Continued.

```
<define name="video-layout-type">
  <choice>
   <value type="string">single-view</value>
   <value type="string">dual-view</value>
   <value type="string">dual-view-crop</value>
   <value type="string">dual-view-2x1</value>
   <value type="string">dual-view-2x1-crop</value>
   <value type="string">quad-view</value>
   <value type="string">multiple-3x3</value>
   <value type="string">multiple-4x4</value>
   <value type="string">multiple-5x1</value>
   <value type="string">automatic</value>
   <ref name="free-text-extension"/>
  </choice>
</define>
<!--
   FLOOR INFORMATION TYPE
-->
<define name="floor-information-type">
 <interleave>
 <ref name="anyAttribute"/>
 <optional>
  <element name="xcon:conference-ID">
   <data type="unsignedLong"/>
  </element>
 </optional>
 <optional>
  <element name="xcon:allow-floor-events">
   <data type="boolean"/>
  </element>
 </optional>
 <optional>
  <element name="xcon:floor-request-handling">
   <ref name="floor-request-type"/>
  </element>
 </optional>
 <optional>
  <element name="xcon:conference-floor-policy">
   <ref name="conference-floor-policy"/>
  </element>
 </optional>
 <zeroOrMore>
  <ref name="anyElement"/>
 </zeroOrMore>
 </interleave>
</define>
<!--
   FLOOR REQUEST TYPE
```

(s)

```
-->
<define name="floor-request-type">
 <choice>
  <value type="string">block</value>
  <value type="string">confirm</value>
  <ref name="free-text-extension"/>
 </choice>
</define>
<!--
   CONFERENCE FLOOR POLICY
-->
<define name="conference-floor-policy">
 <interleave>
 <ref name="anyAttribute"/>
 <oneOrMore>
  <element name="xcon:floor">
   <interleave>
   <attribute name="id">
    <text/>
   </attribute>
   <ref name="anyAttribute"/>
   <oneOrMore>
    <element name="xcon:media-label">
     <data type="nonNegativeInteger"/>
    </element>
   </oneOrMore>
   <optional>
    <element name="xcon:algorithm">
     <ref name="algorithm-type"/>
    </element>
   </optional>
   <optional>
    <element name="xcon:max-floor-users">
     <data type="nonNegativeInteger"/>
    </element>
   </optional>
   <optional>
    <element name="xcon:moderator-id">
     <data type="nonNegativeInteger"/>
    </element>
   </optional>
   <zeroOrMore>
    <ref name="anyElement"/>
   </zeroOrMore>
   </interleave>
  </element>
 </oneOrMore>
 </interleave>
```

(t)

FIGURE 4.6 Continued.

```
</define>
 <!--
   ALGORITHM POLICY
 -->
 <define name="algorithm-type">
  <choice>
   <value type="string">moderator-controlled</value>
   <value type="string">FCFS</value>
   <value type="string">random</value>
   <ref name="free-text-extension"/>
  </choice>
 </define>
 <!--
   USERS ADMISSION POLICY
 -->
 <define name="user-admission-policy-type">
  <choice>
   <value type="string">closedAuthenticated</value>
   <value type="string">openAuthenticated</value>
   <value type="string">anonymous</value>
   <ref name="free-text-extension"/>
  </choice>

 </define>
 <!--
   JOIN HANDLING TYPE
 -->
 <define name="join-handling-type">
  <choice>
   <value type="string">block</value>
   <value type="string">confirm</value>
   <value type="string">allow</value>
   <value type="string">authenticate</value>
   <value type="string">directed-operator</value>
   <ref name="free-text-extension"/>
  </choice>
 </define>
 <!--
   DENY USERLIST
 -->
 <define name="deny-user-list-type">
  <interleave>
  <ref name="anyAttribute"/>
  <zeroOrMore>
   <element name="xcon:target">
    <attribute name="uri">
     <data type="anyURI"/>
    </attribute>
```

(u)

```
    <ref name="anyAttribute"/>
    </element>
   </zeroOrMore>
   <zeroOrMore>
    <ref name="anyElement"/>
   </zeroOrMore>
   </interleave>
  </define>
  <!--
    ALLOWED USERS LIST TYPE
  -->
  <define name="allowed-users-list-type">
   <interleave>
   <ref name="anyAttribute"/>
   <zeroOrMore>
    <element name="xcon:target">
     <ref name="target-type"/>
    </element>
   </zeroOrMore>
   <optional>
    <element name="xcon:persistent-list">
     <ref name="persistent-list-type"/>
    </element>
   </optional>
   <zeroOrMore>
    <ref name="anyElement"/>
   </zeroOrMore>
   </interleave>
  </define>
  <!--
    PERSISTENT LIST TYPE
  -->
  <define name="persistent-list-type">
   <interleave>
    <zeroOrMore>
     <element name="xcon:user">
      <interleave>
      <attribute name="name">
       <text/>
      </attribute>
      <attribute name="nickname">
       <text/>
      </attribute>
      <attribute name="id">
       <text/>
      </attribute>
      <ref name="anyAttribute"/>
      <zeroOrMore>
```

(v)

FIGURE 4.6 Continued.

```
      <element name="xcon:e-mail">
        <text/>
       </element>
      </zeroOrMore>
      <zeroOrMore>
       <ref name="anyElement"/>
      </zeroOrMore>
      </interleave>
     </element>
    </zeroOrMore>
    <zeroOrMore>
     <ref name="anyElement"/>
    </zeroOrMore>
   </interleave>
  </define>
  <!--
    TARGET TYPE
  -->
  <define name="target-type">
   <attribute name="uri">
    <data type="anyURI"/>
   </attribute>
   <attribute name="method">
    <ref name="method-type"/>
   </attribute>
   <ref name="anyAttribute"/>
  </define>
  <!--
    METHOD TYPE
  -->
  <define name="method-type">
   <choice>
    <value type="string">dial-in</value>
    <value type="string">dial-out</value>
    <value type="string">refer</value>
    <ref name="free-text-extension"/>
   </choice>
  </define>
  <!--
    ANONYMITY TYPE
  -->
  <define name="provide-anonymity-type">
    <choice>
     <value>private</value>
     <value>semi-private</value>
     <value>hidden</value>
     <ref name="free-text-extension"/>
    </choice>
```

(s)

```
</define>
<!--
  MIXER TYPE
-->
<define name="mixer-type">
 <interleave>
 <attribute name="name">
  <ref name="mixer-name-type"/>
 </attribute>

 <ref name="anyAttribute"/>
 <zeroOrMore>
  <element name="xcon:floor">
   <attribute name="id">
    <text/>
   </attribute>
   <ref name="anyAttribute"/>
   <data type="boolean"/>
  </element>
 </zeroOrMore>
 <zeroOrMore>
  <element name="xcon:controls">
   <ref name="control-type"/>
  </element>
 </zeroOrMore>
 <zeroOrMore>
  <ref name="anyElement"/>
 </zeroOrMore>
 </interleave>
</define>
<!--
  MIXER NAME TYPE
-->
<define name="mixer-name-type">
  <choice>
   <value>VideoIn</value>
   <value>VideoOut</value>
   <value>AudioOut</value>
   <value>AudioIn</value>
   <ref name="free-text-extension"/>
  </choice>
</define>
<!--
  FREE TEXT EXTENSION
-->
<define name="free-text-extension">
 <text/>
</define>
```

(t)

FIGURE 4.6 Continued.

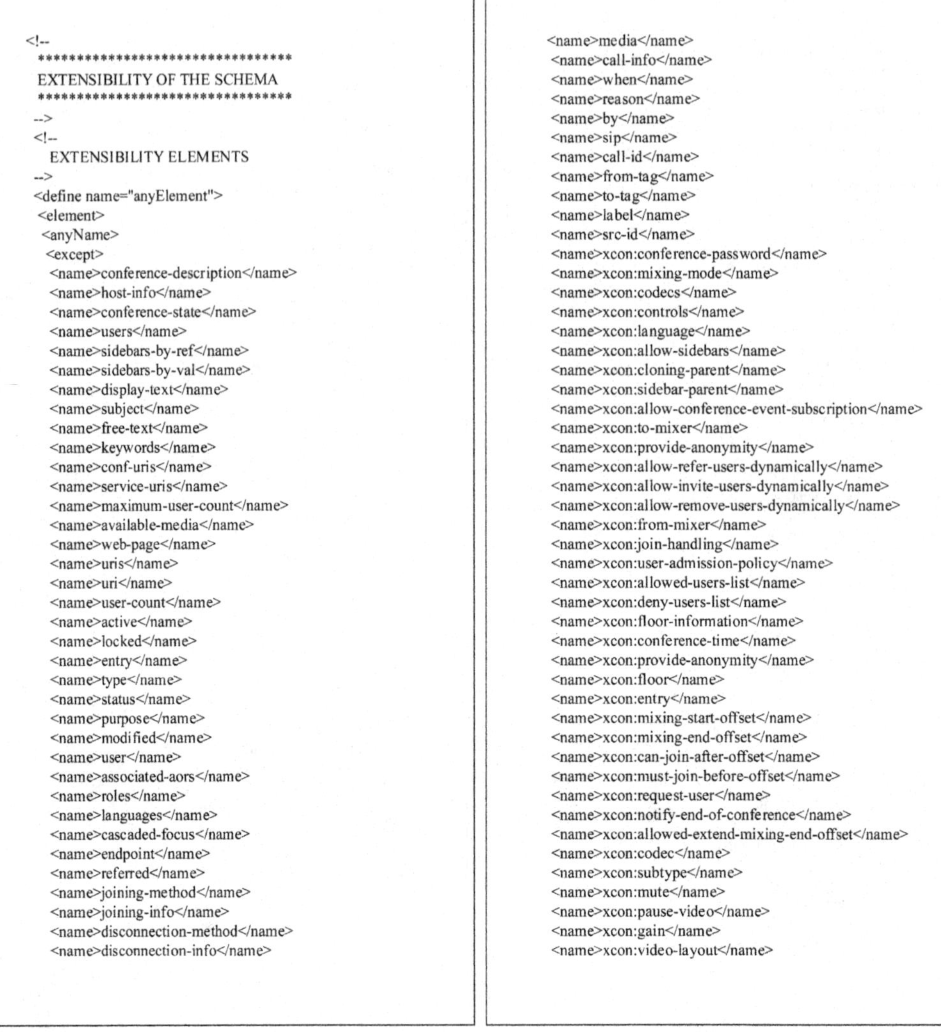

```
<!--
 ********************************
 EXTENSIBILITY OF THE SCHEMA
 ********************************
-->
<!--
  EXTENSIBILITY ELEMENTS
-->
<define name="anyElement">
 <element>
  <anyName>
   <except>
    <name>conference-description</name>
    <name>host-info</name>
    <name>conference-state</name>
    <name>users</name>
    <name>sidebars-by-ref</name>
    <name>sidebars-by-val</name>
    <name>display-text</name>
    <name>subject</name>
    <name>free-text</name>
    <name>keywords</name>
    <name>conf-uris</name>
    <name>service-uris</name>
    <name>maximum-user-count</name>
    <name>available-media</name>
    <name>web-page</name>
    <name>uris</name>
    <name>uri</name>
    <name>user-count</name>
    <name>active</name>
    <name>locked</name>
    <name>entry</name>
    <name>type</name>
    <name>status</name>
    <name>purpose</name>
    <name>modified</name>
    <name>user</name>
    <name>associated-aors</name>
    <name>roles</name>
    <name>languages</name>
    <name>cascaded-focus</name>
    <name>endpoint</name>
    <name>referred</name>
    <name>joining-method</name>
    <name>joining-info</name>
    <name>disconnection-method</name>
    <name>disconnection-info</name>
```

(u)

```
   <name>media</name>
    <name>call-info</name>
    <name>when</name>
    <name>reason</name>
    <name>by</name>
    <name>sip</name>
    <name>call-id</name>
    <name>from-tag</name>
    <name>to-tag</name>
    <name>label</name>
    <name>src-id</name>
    <name>xcon:conference-password</name>
    <name>xcon:mixing-mode</name>
    <name>xcon:codecs</name>
    <name>xcon:controls</name>
    <name>xcon:language</name>
    <name>xcon:allow-sidebars</name>
    <name>xcon:cloning-parent</name>
    <name>xcon:sidebar-parent</name>
    <name>xcon:allow-conference-event-subscription</name>
    <name>xcon:to-mixer</name>
    <name>xcon:provide-anonymity</name>
    <name>xcon:allow-refer-users-dynamically</name>
    <name>xcon:allow-invite-users-dynamically</name>
    <name>xcon:allow-remove-users-dynamically</name>
    <name>xcon:from-mixer</name>
    <name>xcon:join-handling</name>
    <name>xcon:user-admission-policy</name>
    <name>xcon:allowed-users-list</name>
    <name>xcon:deny-users-list</name>
    <name>xcon:floor-information</name>
    <name>xcon:conference-time</name>
    <name>xcon:provide-anonymity</name>
    <name>xcon:floor</name>
    <name>xcon:entry</name>
    <name>xcon:mixing-start-offset</name>
    <name>xcon:mixing-end-offset</name>
    <name>xcon:can-join-after-offset</name>
    <name>xcon:must-join-before-offset</name>
    <name>xcon:request-user</name>
    <name>xcon:notify-end-of-conference</name>
    <name>xcon:allowed-extend-mixing-end-offset</name>
    <name>xcon:codec</name>
    <name>xcon:subtype</name>
    <name>xcon:mute</name>
    <name>xcon:pause-video</name>
    <name>xcon:gain</name>
    <name>xcon:video-layout</name>
```

(v)

FIGURE 4.6 Continued.

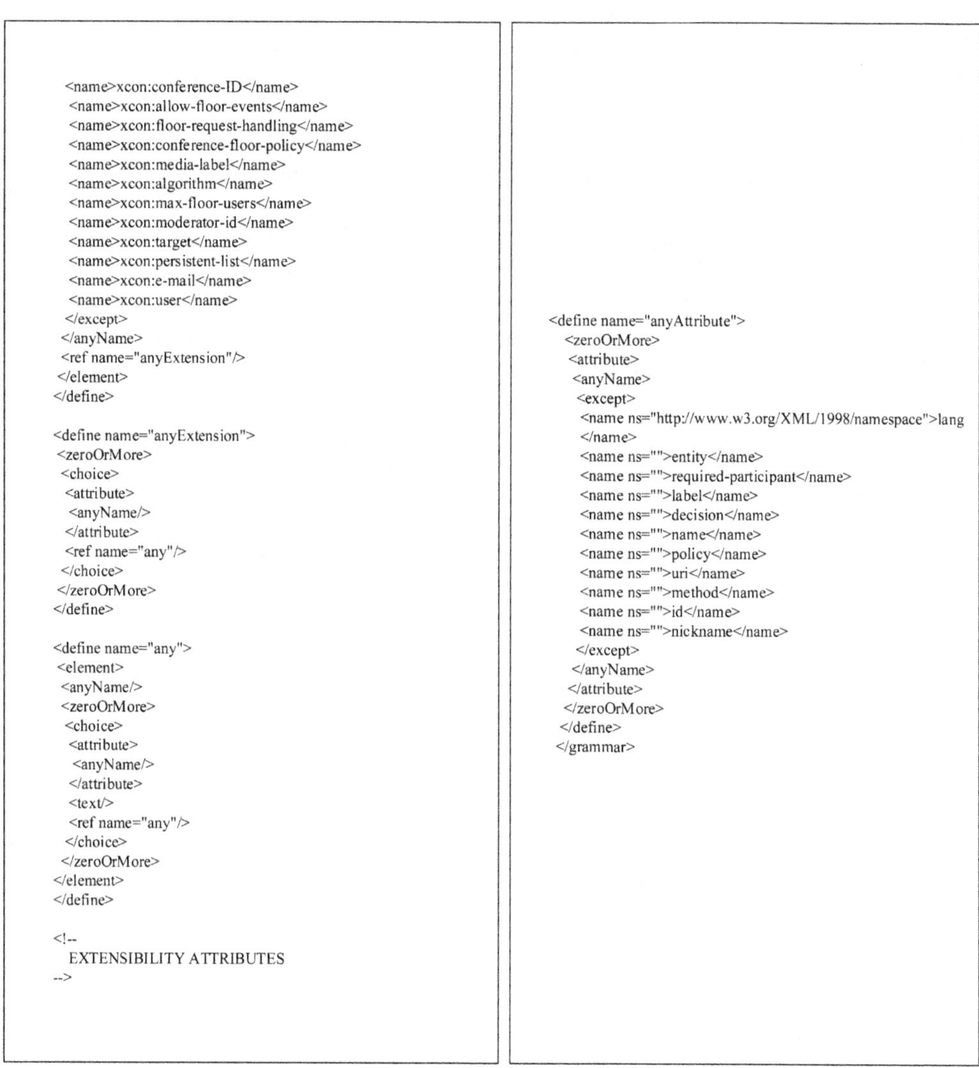

```
  <name>xcon:conference-ID</name>
  <name>xcon:allow-floor-events</name>
  <name>xcon:floor-request-handling</name>
  <name>xcon:conference-floor-policy</name>
  <name>xcon:media-label</name>
  <name>xcon:algorithm</name>
  <name>xcon:max-floor-users</name>
  <name>xcon:moderator-id</name>
  <name>xcon:target</name>
  <name>xcon:persistent-list</name>
  <name>xcon:e-mail</name>
  <name>xcon:user</name>
 </except>
 </anyName>
 <ref name="anyExtension"/>
 </element>
</define>

<define name="anyExtension">
 <zeroOrMore>
  <choice>
   <attribute>
    <anyName/>
   </attribute>
   <ref name="any"/>
  </choice>
 </zeroOrMore>
</define>

<define name="any">
 <element>
  <anyName/>
  <zeroOrMore>
   <choice>
    <attribute>
     <anyName/>
    </attribute>
    <text/>
    <ref name="any"/>
   </choice>
  </zeroOrMore>
 </element>
</define>

<!--
  EXTENSIBILITY ATTRIBUTES
-->
```

(w)

```
<define name="anyAttribute">
  <zeroOrMore>
   <attribute>
    <anyName>
     <except>
      <name ns="http://www.w3.org/XML/1998/namespace">lang
      </name>
      <name ns="">entity</name>
      <name ns="">required-participant</name>
      <name ns="">label</name>
      <name ns="">decision</name>
      <name ns="">name</name>
      <name ns="">policy</name>
      <name ns="">uri</name>
      <name ns="">method</name>
      <name ns="">id</name>
      <name ns="">nickname</name>
     </except>
    </anyName>
   </attribute>
  </zeroOrMore>
 </define>
</grammar>
```

(x)

FIGURE 4.6 Continued.

```
<?xml version="1.0" encoding="UTF-8"?>
 <xs:schema
   xmlns="urn:ietf:params:xml:ns:xcon-conference-info"
   xmlns:info="urn:ietf:params:xml:ns:conference-info"
   xmlns:xs="http://www.w3.org/2001/XMLSchema"
   attributeFormDefault="unqualified" elementFormDefault="qualified"
   targetNamespace="urn:ietf:params:xml:ns:xcon-conference-info">

  <xs:import namespace="urn:ietf:params:xml:ns:conference-info"
    schemaLocation="rfc4575.xsd"/>

  <xs:import namespace="http://www.w3.org/XML/1998/namespace"
    schemaLocation="http://www.w3.org/2001/03/xml.xsd"/>

  <xs:element name="mixing-mode" type="mixing-mode-type"/>
  <xs:element name="codecs" type="codecs-type"/>
  <xs:element name="conference-password" type="xs:string"/>
  <xs:element name="controls" type="controls-type"/>
  <xs:element name="language" type="xs:language"/>
  <xs:element name="allow-sidebars" type="xs:boolean"/>
  <xs:element name="cloning-parent" type="xs:anyURI"/>
  <xs:element name="sidebar-parent" type="xs:anyURI"/>
  <xs:element name="conference-time" type="conference-time-type"/>
  <xs:element name="allow-conference-event-subscription"
    type="xs:boolean"/>
  <xs:element name="to-mixer" type="mixer-type"/>
  <xs:element name="provide-anonymity"
    type="provide-anonymity-type"/>
  <xs:element name="allow-refer-users-dynamically"
    type="xs:boolean"/>
  <xs:element name="allow-invite-users-dynamically"
    type="xs:boolean"/>
  <xs:element name="allow-remove-users-dynamically"
    type="xs:boolean"/>
  <xs:element name="from-mixer" type="mixer-type"/>
  <xs:element name="join-handling" type="join-handling-type"/>
```

(a)

```
  <xs:element name="user-admission-policy"
    type="user-admission-policy-type"/>
  <xs:element name="allowed-users-list"
    type="allowed-users-list-type"/>
  <xs:element name="deny-users-list" type="deny-users-list-type"/>
  <xs:element name="floor-information" type="floor-information-
type"/>

  <!-- CONFERENCE TIME -->

  <xs:complexType name="conference-time-type">
   <xs:sequence>
    <xs:element name="entry"
     minOccurs="0" maxOccurs="unbounded">
     <xs:complexType>
      <xs:sequence>
       <xs:element name="base"
        type="xs:string" minOccurs="1"/>
       <xs:element name="mixing-start-offset" minOccurs="0">
        <xs:complexType>
         <xs:simpleContent>
          <xs:extension base="time-type">
           <xs:attribute name="required-participant"
            type="role-type" use="required"/>
           <xs:anyAttribute namespace="##any"
            processContents="lax"/>
          </xs:extension>
         </xs:simpleContent>
        </xs:complexType>
       </xs:element>
       <xs:element name="mixing-end-offset" minOccurs="0">
        <xs:complexType>
         <xs:simpleContent>
          <xs:extension base="time-type">
           <xs:attribute name="required-participant"
            type="role-type" use="required"/>
           <xs:anyAttribute namespace="##any"
            processContents="lax"/>
          </xs:extension>
         </xs:simpleContent>
        </xs:complexType>
       </xs:element>
       <xs:element name="can-join-after-offset" type="time-type"
        minOccurs="0"/>
       <xs:element name="must-join-before-offset"
        type="time-type" minOccurs="0"/>
       <xs:element name="request-user" type="time-type"
        minOccurs="0"/>
       <xs:element name="notify-end-of-conference"
```

(b)

FIGURE 4.7 Non-Normative W3C RELAX NG Schema in XML Syntax. (Copyright: IETF.)

```
      type="xs:nonNegativeInteger" minOccurs="0"/>
       <xs:element name="allowed-extend-mixing-end-offset"
        type="xs:boolean" minOccurs="0"/>
       <xs:any namespace="##other" processContents="lax"
        minOccurs="0" maxOccurs="unbounded"/>
      </xs:sequence>
     </xs:complexType>
    </xs:element>
    <xs:any namespace="##other" processContents="lax"
     minOccurs="0" maxOccurs="unbounded"/>
   </xs:sequence>
   <xs:anyAttribute namespace="##any" processContents="lax"/>
  </xs:complexType>

  <!-- TIME TYPE -->

  <xs:simpleType name="time-type">
   <xs:restriction base="xs:dateTime">
    <xs:pattern value=".+T.+Z.*"/>
   </xs:restriction>
  </xs:simpleType>

  <!-- ROLE-TYPE -->

  <xs:simpleType name="role-type">
   <xs:restriction base="xs:string">
    <xs:pattern value="none"/>
    <xs:pattern value="administrator"/>
    <xs:pattern value="moderator"/>
    <xs:pattern value="user"/>
    <xs:pattern value="observer"/>
    <xs:pattern value="participant"/>
    <xs:pattern value=".+"/>
   </xs:restriction>
  </xs:simpleType>

  <!-- MIXING MODE TYPE -->

  <xs:simpleType name="mixing-mode-type">
   <xs:restriction base="xs:string">
    <xs:pattern value="moderator-controlled"/>
    <xs:pattern value="FCFS"/>
    <xs:pattern value="automatic"/>
    <xs:pattern value=".+"/>
   </xs:restriction>
  </xs:simpleType>

  <!-- CODECS TYPE -->
```

(c)

```
  <xs:complexType name="codecs-type">
   <xs:sequence>
    <xs:element name="codec" type="codec-type"/>
    <xs:any namespace="##other" processContents="lax"
     minOccurs="0" maxOccurs="unbounded"/>
   </xs:sequence>
   <xs:attribute name="decision"
    type="decision-type" use="required"/>
   <xs:anyAttribute namespace="##any" processContents="lax"/>
  </xs:complexType>

  <!-- CODEC TYPE -->

  <xs:complexType name="codec-type">
   <xs:sequence>
    <xs:element name="subtype" type="xs:string" minOccurs="0"/>
    <xs:any namespace="##other" processContents="lax"
     minOccurs="0" maxOccurs="unbounded"/>
   </xs:sequence>
   <xs:attribute name="name"
    type="xs:string" use="required"/>
   <xs:attribute name="policy"
    type="policy-type" use="required"/>
   <xs:anyAttribute namespace="##any" processContents="lax"/>
  </xs:complexType>

  <!-- DECISION TYPE -->

  <xs:simpleType name="decision-type">
   <xs:restriction base="xs:string">
    <xs:pattern value="automatic"/>
    <xs:pattern value="moderator-controlled"/>
    <xs:pattern value=".+"/>
   </xs:restriction>
  </xs:simpleType>

  <!-- POLICY TYPE -->

  <xs:simpleType name="policy-type">
   <xs:restriction base="xs:string">
    <xs:pattern value="allowed"/>
    <xs:pattern value="disallowed"/>
    <xs:pattern value=".+"/>
   </xs:restriction>
  </xs:simpleType>

  <!-- CONTROL TYPE -->
```

(d)

FIGURE 4.7 Continued.

```
<xs:complexType name="controls-type">
  <xs:sequence>
    <xs:element name="mute"
      type="xs:boolean" minOccurs="0"/>
    <xs:element name="pause-video"
      type="xs:boolean" minOccurs="0"/>
    <xs:element name="gain"
      type="gain-type" minOccurs="0"/>
    <xs:element name="video-layout"
      type="video-layout-type" default="single-view" minOccurs="0"/>


    <xs:any namespace="##other" processContents="lax"
      minOccurs="0" maxOccurs="unbounded"/>
  </xs:sequence>
  <xs:anyAttribute namespace="##any" processContents="lax"/>
</xs:complexType>

<!-- GAIN TYPE -->

<xs:simpleType name="gain-type">
  <xs:restriction base="xs:integer">
    <xs:minInclusive value="-127"/>
    <xs:maxInclusive value="127"/>
  </xs:restriction>
</xs:simpleType>


<!-- VIDEO LAYOUT TYPE -->

<xs:simpleType name="video-layout-type">
  <xs:restriction base="xs:string">
    <xs:pattern value="single-view"/>
    <xs:pattern value="dual-view"/>
    <xs:pattern value="dual-view-crop"/>
    <xs:pattern value="dual-view-2x1"/>
    <xs:pattern value="dual-view-2x1-crop"/>
    <xs:pattern value="quad-view"/>
    <xs:pattern value="multiple-3x3"/>
    <xs:pattern value="multiple-4x4"/>

    <xs:pattern value="multiple-5x1"/>
    <xs:pattern value="automatic"/>
    <xs:pattern value=".+"/>
  </xs:restriction>
</xs:simpleType>

<!-- FLOOR INFORMATION TYPE -->
```

(e)

```
<xs:complexType name="floor-information-type">
  <xs:sequence>
    <xs:element name="conference-ID"
      type="xs:unsignedLong" minOccurs="0"/>
    <xs:element name="allow-floor-events"
      type="xs:boolean" default="false" minOccurs="0"/>
    <xs:element name="floor-request-handling"
      type="floor-request-handling-type" minOccurs="0"/>
    <xs:element name="conference-floor-policy"
      type="conference-floor-policy" minOccurs="0"/>
    <xs:any namespace="##other" processContents="lax"
      minOccurs="0" maxOccurs="unbounded"/>
  </xs:sequence>
  <xs:anyAttribute namespace="##any" processContents="lax"/>
</xs:complexType>

<!-- FLOOR REQUEST TYPE -->

<xs:simpleType name="floor-request-handling-type">
  <xs:restriction base="xs:string">
    <xs:pattern value="block"/>
    <xs:pattern value="confirm"/>
    <xs:pattern value=".+"/>
  </xs:restriction>
</xs:simpleType>

<!-- CONFERENCE FLOOR POLICY -->

<xs:complexType name="conference-floor-policy">
  <xs:sequence>
    <xs:element name="floor" maxOccurs="unbounded">
      <xs:complexType>
        <xs:sequence>
          <xs:element name="media-label"
            type="xs:string" minOccurs="1" maxOccurs="unbounded"/>
          <xs:element name="algorithm"
            type="algorithm-type" minOccurs="0"/>
          <xs:element name="max-floor-users"
            type="xs:nonNegativeInteger" minOccurs="0"/>
          <xs:element name="moderator-id"
            type="xs:nonNegativeInteger" minOccurs="0"/>
            <xs:any namespace="##other" processContents="lax"
              minOccurs="0" maxOccurs="unbounded"/>
        </xs:sequence>
        <xs:attribute name="id" type="xs:string" use="required"/>
        <xs:anyAttribute namespace="##any" processContents="lax"/>
      </xs:complexType>
    </xs:element>
```

(f)

FIGURE 4.7 Continued.

```
 </xs:sequence>
  <xs:anyAttribute namespace="##any" processContents="lax"/>
</xs:complexType>

<!-- ALGORITHM TYPE -->

<xs:simpleType name="algorithm-type">
  <xs:restriction base="xs:string">
    <xs:pattern value="moderator-controlled"/>
    <xs:pattern value="FCFS"/>
    <xs:pattern value="random"/>
    <xs:pattern value=".+"/>
  </xs:restriction>
</xs:simpleType>

<!-- USER ADMISSION POLICY TYPE -->

<xs:simpleType name="user-admission-policy-type">
  <xs:restriction base="xs:string">
    <xs:pattern value="closedAuthenticated"/>
    <xs:pattern value="openAuthenticated"/>
    <xs:pattern value="anonymous"/>
    <xs:pattern value=".+"/>
  </xs:restriction>
</xs:simpleType>

<!-- JOIN HANDLING TYPE -->

<xs:simpleType name="join-handling-type">
  <xs:restriction base="xs:string">
    <xs:pattern value="block"/>
    <xs:pattern value="confirm"/>
    <xs:pattern value="allow"/>
    <xs:pattern value="authenticate"/>
    <xs:pattern value="directed-operator"/>
    <xs:pattern value=".+"/>
  </xs:restriction>
</xs:simpleType>

<!-- DENY USER LIST TYPE -->

<xs:complexType name="deny-users-list-type">
  <xs:sequence>
    <xs:element name="target" minOccurs="0" maxOccurs="unbounded">
      <xs:complexType>
        <xs:attribute name="uri" use="required" type="xs:anyURI"/>
        <xs:anyAttribute namespace="##any" processContents="lax"/>
      </xs:complexType>
```

(g)

```
    </xs:element>
    <xs:any namespace="##other" processContents="lax"
      minOccurs="0" maxOccurs="unbounded"/>
  </xs:sequence>
  <xs:anyAttribute namespace="##any" processContents="lax"/>
</xs:complexType>

<!-- ALLOWED USERS LIST TYPE -->

<xs:complexType name="allowed-users-list-type">
  <xs:sequence>
    <xs:element name="target" type="target-type"
      minOccurs="0" maxOccurs="unbounded">
    </xs:element>
    <xs:element name="persistent-list"
      type="persistent-list-type"
      minOccurs="0"/>
    <xs:any namespace="##other" processContents="lax"
      minOccurs="0" maxOccurs="unbounded"/>
  </xs:sequence>
  <xs:anyAttribute namespace="##any" processContents="lax"/>
</xs:complexType>

<!-- PERSISTENT LIST TYPE -->

<xs:complexType name="persistent-list-type">
  <xs:sequence>
    <xs:element name="user" minOccurs="0"
maxOccurs="unbounded">
      <xs:complexType>
        <xs:sequence>
          <xs:element name="email" type="xs:string"
            minOccurs="0" maxOccurs="unbounded"/>
          <xs:any namespace="##other" processContents="lax"
            minOccurs="0" maxOccurs="unbounded"/>
        </xs:sequence>
        <xs:attribute name="name"
          use="required" type="xs:anyURI"/>
        <xs:attribute name="nickname"
          use="required" type="xs:string"/>
        <xs:attribute name="id"
          use="required" type="xs:string"/>
        <xs:anyAttribute namespace="##any" processContents="lax"/>
      </xs:complexType>
    </xs:element>
    <xs:any namespace="##other" processContents="lax"
      minOccurs="0" maxOccurs="unbounded"/>
  </xs:sequence>
  <xs:anyAttribute namespace="##any" processContents="lax"/>
```

(h)

FIGURE 4.7 Continued.

```
</xs:complexType>

<!-- TARGET TYPE -->

<xs:complexType name="target-type">
 <xs:attribute name="uri" use="required"
  type="xs:anyURI"/>
 <xs:attribute name="method" use="required"
  type="method-type"/>
 <xs:anyAttribute namespace="##any" processContents="lax"/>
</xs:complexType>

<!-- METHOD TYPE -->

<xs:simpleType name="method-type">
 <xs:restriction base="xs:string">
  <xs:pattern value="dial-in"/>
  <xs:pattern value="dial-out"/>
  <xs:pattern value="refer"/>
  <xs:pattern value=".+"/>
 </xs:restriction>
</xs:simpleType>

<!-- ANONYMITY TYPE -->

<xs:simpleType name="provide-anonymity-type">
 <xs:restriction base="xs:string">
  <xs:pattern value="private"/>
  <xs:pattern value="semi-private"/>
  <xs:pattern value="hidden"/>
  <xs:pattern value=".+"/>
 </xs:restriction>
</xs:simpleType>

<!-- MIXER TYPE -->

<xs:complexType name="mixer-type">
 <xs:sequence>
  <xs:element name="floor">
   <xs:complexType>
    <xs:simpleContent>
     <xs:extension base="xs:boolean">
      <xs:attribute name="id" type="xs:string"
       use="required"/>
      <xs:anyAttribute namespace="##any"
       processContents="lax"/>
     </xs:extension>
    </xs:simpleContent>
```

(i)

```
    </xs:complexType>
   </xs:element>
   <xs:element name="controls" type="controls-type"
    minOccurs="0" maxOccurs="unbounded"/>
   <xs:any namespace="##other" processContents="lax"
    minOccurs="0" maxOccurs="unbounded"/>
  </xs:sequence>
  <xs:attribute name="name" type="mixer-name-type"
   use="required"/>
  <xs:anyAttribute namespace="##any" processContents="lax"/>
 </xs:complexType>

 <!-- MIXER NAME TYPE -->

 <xs:simpleType name="mixer-name-type">
  <xs:restriction base="xs:string">
   <xs:pattern value="VideoIn"/>
   <xs:pattern value="VideoOut"/>
   <xs:pattern value="AudioOut"/>
   <xs:pattern value="AudioIn"/>
   <xs:pattern value=".+"/>
  </xs:restriction>
 </xs:simpleType>

</xs:schema>
```

(j)

FIGURE 4.7 Continued.

5 Centralized Conferencing Manipulation Protocol

5.1 INTRODUCTION

The centralized conferencing manipulation protocol (CCMP) (RFC 6503) described here allows authenticated and authorized users to create, manipulate, and delete conference objects. Operations on conferences include adding and removing participants and changing their roles, as well as adding and removing media streams and associated endpoints. CCMP implements the client-server model within the centralized conferencing (XCON) framework defined in RFC 5239 (see Chapter 2), where the conferencing client and conference server act as client and server, respectively. CCMP uses the hypertext transfer (HTTP) protocol defined in RFC 2616 to transfer requests and responses, which contain the domain-specific XML-encoded data objects described in RFC 6501 (see Chapter 4).

5.2 CONVENTIONS AND TERMINOLOGY

For conventions, see RFC 2119.

For terminology, see Table 1.1, Section 1.6.

- XCON-aware client: An XCON conferencing system client that is able to issue CCMP requests.
- First-Party Request: A request issued by the client to manipulate its own conferencing data.
- Third-Party Request: A request issued by a client to manipulate the conference data of another client.

5.3 XCON CONFERENCE CONTROL SYSTEM ARCHITECTURE

The logical view of the CCMP that supports the XCON framework is depicted in Figure 5.1 as a subset of the conferencing system logical decomposition architecture described earlier. It illustrates the role that CCMP plays within the framework of the overall centralized architecture.

The CCMP allows the conference control client (conferencing client) to interface with the conference object maintained by the conferencing system. We are describing the CCMP protocol that interacts with the conferencing client, the conferencing control server (conferencing server associated with a conferencing system).

5.3.1 Conference Objects

Conference objects feature a simple dynamic inheritance-and-override mechanism. Conference objects are linked into a tree known as a "cloning tree" described in RFC 5239 (see Sections 2 and 2.7.1). Each cloning tree node inherits attributes from its parent node. The roots of these inheritance trees are conference templates, also known as "blueprints." Nodes in the inheritance tree can be active conferences or simply descriptions that do not currently have any resources associated with

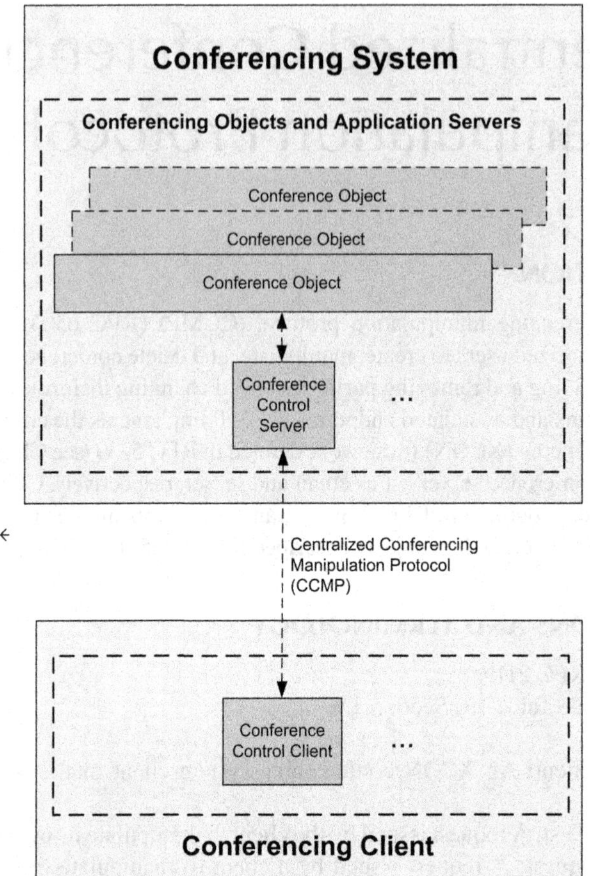

FIGURE 5.1 Conference Client Information. (Copyright: IETF.)

them (i.e., conference reservations). An object can mark certain of its properties as unalterable, so that they cannot be overridden. A client may specify a parent object (a conference or blueprint) from which to inherit values when a conference is created using the conference control protocol.

Conference objects are uniquely identified by the XCON-URI within the scope of the conferencing system. The XCON-URI is introduced in the XCON framework and defined in the XCON common data model. Conference objects are comprehensively represented through XML specifications compliant with the XML schema defined in the XCON data model (RFC 6501 – see Chapter 4). The root element of such specifications, called <conference-info>, is of type "conference-type." It encompasses other XML elements describing different conference features and users as well. Using CCMP, conferencing clients can use these XML structures to express their preferences in creating or updating a conference. A conference server can convey conference information back to the clients using the XML elements.

5.3.2 Conference Users

Each conference can have zero or more users. All conference participants are users, but some users may have only administrative functions and do not contribute or receive media. Users are added one user at a time to simplify error reporting. When a conference is cloned from a parent object, users are inherited as well, so that it is easy to set up a conference that has the same set of participants or a common administrator. The conference server creates individual users, assigning them a unique conference user identifier (XCON-USERID). The XCON-USERID as identifier of each conferencing system client is introduced in the XCON framework and defined in the XCON common data model. Each CCMP request, with an exception pointed out in Section 5.5.3.6 representing the case of a user at his first entrance in the system as a conference participant, must carry the XCON-USERID of the requestor in the proper <confUserID> parameter.

The XCON-USERID acts as a pointer to the user's profile as a conference actor, e.g., her signaling uniform resource identifier (URI) and other XCON protocol URIs in general, her role (moderator, participant, observer, etc.), her display text, her joining information, and so on. A variety of elements defined in the common <conference-info> element as specified in the XCON data model are used to describe the users related to a conference, including the <users> element as well as each <user> element included within it. For example, it is possible to determine how a specific user expects and is allowed to join a conference by looking at the <allowed-users-list> in <users>: each <target> element involved in such a list represents a user and shows a "method" attribute defining how the user is expected to join the conference, i.e., "dial-in" for users that are allowed to dial and "dial-out" for users that the conference focus will be trying to reach (with "dial-in" being the default mode). If the conference is currently active, dial-out users are contacted immediately; otherwise, they are contacted at the start of the conference. CCMP, acting as the conference control protocol, provides a means to manipulate these and other kinds of user-related features.

As a consequence of an explicit user registration to a specific XCON conferencing system, conferencing clients are usually provided (besides the XCON-USERID) with log-in credentials (i.e., username and password). Such credentials can be used to authenticate the XCON-aware client issuing CCMP requests. Thus, both username and password should be carried in a CCMP request as part of the "subject" parameter whenever a registered conferencing client wishes to contact a CCMP server. CCMP does not maintain a user's subscriptions at the conference server; hence, it does not provide any specific mechanism allowing clients to register their conferencing accounts. The "subject" parameter is just used for carrying authentication data associated with pre-registered clients, with the specific registration modality outside the scope of this specification.

5.4 PROTOCOL OVERVIEW

CCMP is a client-server, XML-based protocol for user creation, retrieval, modification, and deletion of conference objects. CCMP is a stateless protocol, such that implementations can safely handle transactions independently of each other. CCMP messages are XML specifications or XML specification fragments compliant with the XCON data model representation (RFC 6501 – see Chapter 4).

Section 5.4.1 specifies the basic operations that can create, retrieve, modify, and delete conference-related information in a centralized conference. The core set of objects manipulated by CCMP includes conference blueprints, the conference object, users, and sidebars. Each operation in the protocol model, as summarized in Section 5.4.1, is atomic and either succeeds or fails as a whole. The conference server MUST ensure that the operations are atomic in that the operation invoked by a specific conferencing client completes prior to another client's operation on the same conference object. While the details for this data locking functionality are out of scope of the CCMP specification and are implementation specific for a conference server, some core functionality for ensuring the integrity of the data is provided by CCMP, as described in Section 5.4.2. While the XML specifications that are carried in CCMP need to comply with the XCON data model, there are situations in which the values for mandatory elements are unknown to the client. The mechanism for ensuring compliance with the data model in these cases is described in Section 5.4.3.

CCMP is completely independent of underlying protocols, which means that there can be different ways to carry CCMP messages from a conferencing client to a conference server. The specification describes the use of HTTP as a transport solution, including CCMP requests in HTTP POST messages and CCMP responses in HTTP 200 OK replies. This implementation approach is further described in Section 5.4.4.

5.4.1 Protocol Operations

The main operations provided by CCMP belong in four general categories:

- **Create**: for the creation of a conference object, a conference user, a sidebar, or a blueprint.
- **Retrieve**: to get information about the current state of either a conference object (be it an actual conference, a blueprint, or a sidebar) or a conference user. A retrieve operation can also be used to obtain the XCON-URIs of the current conferences (active or registered) handled by the conferencing server and/or the available blueprints.
- **Update**: to modify the current features of a specified conference or conference user.
- **Delete**: to remove from the system a conference object or a conference user.

Thus, the main targets of CCMP operations are as follows:

- Conference objects associated with either active or registered conferences
- Conference objects associated with blueprints
- Conference objects associated with sidebars, both embedded in the main conference (i.e., <entry> elements in <sidebars-by-value>) and external to it (i.e., whose XCON-URIs are included in the <entry> elements of <sidebars-by-ref>)
- <user> Elements associated with conference users
- The list of XCON-URIs related to conferences and blueprints available at the server, for which only retrieval operations are allowed

5.4.2 Data Management

The XCON framework defines a model whereby the conference server centralizes and maintains the conference information. Since multiple clients can modify the same conference objects, a conferencing client might not have the latest version of a specific conference object when it initiates operations. To determine whether the client has the most up-to-date conference information, CCMP defines a versioning approach. Each conference object is associated with a version number. All CCMP response messages containing a conference specification (or a fragment thereof) MUST contain a <version> parameter. When a client sends an update message to the server, which includes modifications to a conference object, if the modifications are all successfully applied, the server MUST return a response, with a <response-code> of "200," containing the version number of the modified object. With this approach, a client working on version "X" of a conference object that receives a response, with a <response-code> of "200," with a version number that is "X+1" can be certain that the version it manipulated was the most up to date. However, if the response contains a version that is at least "X+2," the client knows that the object modified by the server was more up to date than the object the client was manipulating. In order to ensure that the client always has the latest version of the modified object, the client can send a request to the conference server to retrieve the conference object. The client can then update the relevant data elements in the conference object prior to invoking a specific operation. Note that a client subscribed to the XCON event package (RFC 6502 – see Chapter 7) notifications about conference object modifications will receive the most up-to-date version of that object upon receipt of a notification.

The "version" parameter is OPTIONAL for requests, since it is not needed by the server: as long as the required modifications can be applied to the target conference object without conflicts, the server does not care whether the client has stored an up-to-date view of the information. In addition, to ensure the integrity of the data, the conference server first checks all the parameters before making any changes to the internal representation of the conference object. For example, it would be undesirable to change the <subject> of the conference but then detect an invalid URI in one of the <serviceuris> and abort the remaining updates.

5.4.3 Data Model Compliance

The XCON data model (RFC 6501 – see Chapter 4) identifies some elements and attributes as mandatory. Since the XML specifications carried in the body of the CCMP requests and responses need to be compliant with the XCON data model, there can be a problem in cases of client-initiated operations, such as the initial creation of conference objects and cases whereby a client updates a conference object, adding new elements, such as a new user. In such cases, not all the mandatory data can be known in advance by the client issuing a CCMP request. As an example, a client cannot know, at the time it issues a conference creation request, the XCON-URI that the server will assign to the yet-to-be-created conference; hence, it is not able to populate the mandatory "entity" attribute of the conference specification contained in the request with the correct value. To solve this issue, the CCMP client fills all mandatory data model fields, for which no value is available at the time the request is constructed, with placeholder values in the form of a wildcard string, AUTO_GENERATE_X (all uppercase), with X being a unique numeric index for each data model field for which the value is unknown. This form of wildcard string is chosen, rather than the use of random unique strings (e.g., FOO_BAR_LA) or non-numeric values for X, to simplify processing at the server. The values of AUTO_GENERATE_X are only unique within the context of the specific request. The placeholder AUTO_GENERATE_X values MUST be within the value part of an attribute or element (e.g., <userinfo entity="xconuserid:AUTO_GENERATE_1@example.com">).

When the server receives requests containing values in the form of AUTO_GENERATE_X, the server does the following:

(a) Generates the proper identifier for each instance of AUTO_GENERATE_X in the specification. If an instance of AUTO_GENERATE_X is not within the value part of the attribute/element, the server MUST send a <response-code> of "400 Bad Request." In cases where AUTO_GENERATE_X appears only in the user part of a URI (i.e., in the case of XCON-USERIDs or XCON-URIs), the server needs to ensure that the domain name is one that is within the server's domain of responsibility. If the domain name is not within the server's domain of responsibility, then the server MUST send a <response-code> of "427 Invalid Domain Name." The server MUST replace each instance of a specific wildcard field (e.g., AUTO_GENERATE_1) with the same identifier. The identifiers MUST be unique for each instance of AUTO_GENERATE_X within the same XML specification received in the request; for example, the value that replaces AUTO_GENERATE_1 MUST NOT be the same as the value that replaces AUTO_GENERATE_2. Note that the values that replace the instances of AUTO_GENERATE_X are not the same across all conference objects; for example, different values can be used to replace AUTO_GENERATE_1 in two different specifications.

(b) Sends a response in which all values of AUTO_GENERATE_X received in the request have been replaced by the newly created one(s). With this approach, compatibility with the data model requirements is maintained while allowing for client-initiated manipulation of conference objects at the server's side. Note that the use of this mechanism could be avoided in some cases by using multiple operations, such as creating a new user and then adding the new user to an existing conference. However, the AUTO_GENERATE_X mechanism allows a single operation to be used to effect the same change on the conference object.

5.4.4 Implementation Approach

CCMP is implemented using HTTP, placing the CCMP request messages into the body of an HTTP POST operation and placing the CCMP responses into the body of the HTTP response messages. A nonexhaustive summary of the other approaches that were considered and the perceived advantages of the HTTP solution described in this specification are provided in Section 5.14 (Appendix A of RFC 6503).

Most CCMP commands can pend indefinitely, thus increasing the potential that pending requests can continue to increase when a server is receiving more requests than it can process within a specific time period. In this case, a server SHOULD return a <response-code> of “510” to the pending requests. In addition, to mitigate the situation, clients MUST NOT wait indefinitely for a response and MUST implement a timer such that when it expires, the client MUST close the connection. Thirty seconds is RECOMMENDED as the default value for this timer. Sixty seconds is considered a reasonable upper range. Note that there may be cases where a response message is lost and a request has been successful (e.g., user added to a conference); yet, the client will be unaware and close the connection. However, as described in Section 5.4.2, there is a versioning mechanism for the conference objects; thus, there is a mechanism for the conference object stored by the client to be brought up to date.

CCMP messages have a MIME type of “application/ccmp+xml,” which appears inside the Content-Type and Accept header fields of HTTP requests and responses. The XML specifications in the CCMP messages MUST be encoded in UTF-8. This specification follows the recommendations and conventions described in RFC 3023, including the naming convention of the type (“+xml” suffix) and the usage of the “charset” parameter. The “charset” parameter MUST be included with the XML specification. Section 5.9 provides the complete requirements for an HTTP implementation to support CCMP.

5.5 CCMP MESSAGES

CCMP messages are either requests or responses. The general CCMP request message is defined in Section 5.5.1. The general CCMP response message is defined in Section 5.5.2. The details of the specific message type that is carried in the CCMP request and response messages are described in Section 5.5.3. CCMP response codes are listed in Section 5.5.4.

5.5.1 CCMP REQUEST MESSAGE TYPE

A CCMP request message is comprised of the following parameters:

- **subject**: An OPTIONAL parameter containing the username and password of the client registered at the conferencing system (Figure 5.2). Each user who subscribes to the conferencing system is assumed to be equipped with those credentials and SHOULD enclose

```
<!-- Definition of CCMP Request -->

<xs:element name="ccmpRequest" type="ccmp-request-type" />

<!-- Definition of ccmp-request-type-->

<xs:complexType name="ccmp-request-type">
    <xs:sequence>
        <xs:element name="ccmpRequest"
            type="ccmp-request-message-type" />
    </xs:sequence>
</xs:complexType>
```

```
<!-- Definition of ccmp-request-message-type -->

<xs:complexType abstract="true"
        name="ccmp-request-message-type">
    <xs:sequence>
        <xs:element name="subject" type="subject-type"
            minOccurs="0" maxOccurs="1" />
        <xs:element name="confUserID" type="xs:string"
            minOccurs="0" maxOccurs="1" />
        <xs:element name="confObjID" type="xs:string"
            minOccurs="0" maxOccurs="1" />
        <xs:element name="operation" type="operationType"
            minOccurs="0" maxOccurs="1" />
        <xs:element name="conference-password" type="xs:string"
            minOccurs="0" maxOccurs="1" />
        <xs:any namespace="##other" processContents="lax"
            minOccurs="0" maxOccurs="unbounded"/>
    </xs:sequence>
    <xs:anyAttribute namespace="##any" processContents="lax"/>
</xs:complexType>
```

FIGURE 5.2 Structure of CCMP Request Messages. (Copyright: IETF.)

them in each CCMP request she issues. These fields can be used to control that the user sending the CCMP request has the authority to perform the requested operation. The same fields can also be used for other authorization and authentication procedures.

- **confUserID**: An OPTIONAL parameter containing the XCON-USERID of the client. The XCON-USERID is used to identify any conferencing client within the context of the conferencing system, and it is assigned by the conference server for each conferencing client who interacts with it. The <confUserID> parameter is REQUIRED in the CCMP request and response messages, with the exception of the case of a user who has no XCON-USERID and who wants to enter, via CCMP, a conference whose identifier is known. In such case, a side effect of the request is that the user is provided with an appropriate XCON-USERID. An example of the aforementioned case will be provided in Section 5.5.3.6.
- **confObjID**: An OPTIONAL parameter containing the XCON-URI of the target conference object.
- **operation**: An OPTIONAL parameter refining the type of specialized request message. The <operation> parameter is REQUIRED in all requests except for the blueprintsRequest and confsRequest specialized messages.
- **conference-password**: The parameter is OPTIONAL except that it MUST be inserted in all requests whose target conference object is password protected, i.e., contains the <conference-password> element in RFC 6501 (see Chapter 4). A CCMP <response-code> of "423" MUST be returned if a conference-password is not included in the request when required.
- **specialized request message:** This is a specialization of the generic request message (e.g., blueprintsRequest), containing parameters that are dependent on the specific request sent to the server. A specialized request message MUST be included in the CCMP request message. The details for the specialized messages and associated parameters are provided in Section 5.5.3 (Figure 5.3).

5.5.2 CCMP Response Message Type

A CCMP response message is comprised of the following parameters:

- **confUserID**: A REQUIRED parameter in CCMP response messages containing the XCON-USERID of the conferencing client that issued the CCMP request message.

```
<!-- Definition of CCMP Response -->

<xs:element name="ccmpResponse" type="ccmp-response-type" />

<!-- Definition of ccmp-response-type -->

<xs:complexType name="ccmp-response-type">
        <xs:sequence>
                <xs:element name="ccmpResponse"
                        type="ccmp-response-message-type" />
        </xs:sequence>
</xs:complexType>
```

```
<!-- Definition of ccmp-response-message-type -->

<xs:complexType abstract="true"
                name="ccmp-response-message-type">
        <xs:sequence>
                <xs:element name="confUserID" type="xs:string"
                        minOccurs="1" maxOccurs="1" />
                <xs:element name="confObjID" type="xs:string"
                        minOccurs="0" maxOccurs="1" />
                <xs:element name="operation" minOccurs="0"
                        maxOccurs="1" />
                <xs:element name="response-code"
                        type="response-codeType"
                        minOccurs="1" maxOccurs="1" />
                <xs:element name="response-string" type="xs:string"
                        minOccurs="0" maxOccurs="1" />
                <xs:element name="version" type="xs:positiveInteger"
                        minOccurs="0" maxOccurs="1" />
                <xs:any namespace="##other" processContents="lax"
                        minOccurs="0" maxOccurs="unbounded"/>
        </xs:sequence>
        <xs:anyAttribute namespace="##any" processContents="lax"/>
</xs:complexType>
```

FIGURE 5.3 Structure of CCMP Response Message. (Copyright: IETF.)

- **confObjID**: An OPTIONAL parameter containing the XCON-URI of the target conference object.
- **operation**: An OPTIONAL parameter for CCMP response messages. This parameter is REQUIRED in all responses except for the "blueprintsResponse" and "confsResponse" specialized messages.
- **response-code**: A REQUIRED parameter containing the response code associated with the request. The response code MUST be chosen from the codes listed in Section 5.5.4.
- **response-string**: An OPTIONAL reason string associated with the response. In case of an error, in particular, this string can be used to provide the client with detailed information about the error itself.
- **version**: An OPTIONAL parameter reflecting the current version number of the conference object referred by the confObjID. This number is contained in the "version" attribute of the <conference-info> element related to that conference. This parameter is REQUIRED in CCMP response messages and SHOULD NOT be included in CCMP request messages.
- **specialized response message**: This is a specialization of the generic response message, containing parameters that are dependent on the specific request sent to the server (e.g., "blueprintsResponse"). A specialized response message SHOULD be included in the CCMP response message, except in an error situation where the CCMP request message did not contain a valid specialized message. In this case, the conference server MUST return a <response-code> of "400." The details for the specialized messages and associated parameters are provided in Section 5.5.3.

5.5.3 Detailed Messages

Based on the request and response message structures described in Sections 5.5.1 and 5.5.2, the following summarizes the specialized CCMP request and response types described in this specification:

1. blueprintsRequest/blueprintsResponse
2. confsRequest/confsResponse
3. blueprintRequest/blueprintResponse
4. confRequest/confResponse
5. usersRequest/usersResponse
6. userRequest/userResponse
7. sidebarsByValRequest/sidebarsByValResponse
8. sidebarsByRefRequest/sidebarsByRefResponse
9. sidebarByValRequest/sidebarByValResponse
10. sidebarByRefRequest/sidebarByRefResponse
11. extendedRequest/extendedResponse
12. optionsRequest/optionsResponse

These CCMP request/response pairs use the fundamental CCMP operations as defined in Section 5.4.1 to manipulate the conference data. These request/response pairs are included in an Internet Assigned Numbers Authority (IANA) registry as defined in Section 5.12.5. Table 5.1 summarizes the remaining CCMP operations and corresponding actions that are valid for a specific CCMP request type, noting that neither blueprintsRequest/blueprintsResponse nor confsRequest/confsResponse requires an <operation> parameter. An entity MUST support the response message for each of the request messages that is supported. The corresponding response message MUST contain the same <operation> parameter. Note that some entries are labeled "N/A," indicating that the operation is invalid for that request type. In the case of an "N/A*" label, the operation MAY be allowed

TABLE 5.1
Request Type Operation-Specific Processing

	Operation			
Request Type	**Retrieve**	**Create**	**Update**	**Delete**
blueprintsRequest	Get list of blueprints	N/A	N/A	N/A
blueprintRequest	Get blueprint	N/A	N/A	N/A
confsRequest	Get list of confs	N/A	N/A	N/A
confRequest	Get conference object	Create conference object	Change conference object	Delete conference object
usersRequest	Get <users>	N/A (**)	Change <users>	N/A (**)
userRequest	Get specified <user>	Add a <user> to a conf (***)	Change specified <user>	Delete specified <user>
sidebarsByValRequest	Get <sidebars-by-val>	N/A	N/A	N/A
sidebarsByRefRequest	Get <sidebars-by-ref>	N/A	N/A	N/A
sidebarByValRequest	Get sidebar-by-val	Create sidebar-by-val	Change sidebar-by-val	Delete sidebar-by-val
sidebarByRefRequest	Get sidebar-by-ref	Create sidebar-by-ref	Change sidebar-by-ref	Delete sidebar-by-ref

(**): These operations are not allowed for a usersRequest message, since the <users> section, which is the target element of such a request, is created and removed in conjunction with the creation and deletion, respectively, of the associated conference specification. Thus, "update" and "retrieve" are the only semantically correct operations for such a message.

(***): This operation can involve the creation of an XCON-USERID if the sender does not add it in the <confUserID> parameter and/or if the entity field of the <userInfo> parameter is void.

for specific privileged users or system administrators but is not part of the functionality included in this specification.

Additional parameters included in the specialized CCMP request and response messages are detailed in the subsequent sections. If a required parameter is not included in a request, the conference server MUST return a <response-code> of "400" per Section 5.5.4.

5.5.3.1 blueprintsRequest and blueprintsResponse

A blueprintsRequest (Figure 5.4) message is sent to request the list of XCON-URIs associated with the available blueprints from the conference server. These XCON-URIs can be subsequently used by the client to access detailed information about a specified blueprint with a specific blueprintRequest message per Section 5.5.3.3.

The <confUserID> parameter MUST be included in every blueprintsRequest/Response message and reflect the XCON-USERID of the conferencing client issuing the request. Since a blueprintsRequest message is not targeted to a specific conference instance and is a "retrieve-only" request, the <confObjID> and <operation> parameters MUST NOT be included in the blueprintsRequest/Response messages. In order to obtain a specific subset of the available blueprints, a client may specify a selection filter providing an appropriate xpath query in the OPTIONAL "xpathFilter" parameter of the request. The information in the blueprints typically represents general capabilities and characteristics. For example, to select blueprints having both audio and video stream support, a possible xpathFilter value could be: "/conference-info[conference-description/available-media/entry/type="audio" and conference-description/available-media/entry/type="vide

```
<!-- blueprintsRequest -->

<xs:complexType name="ccmp-blueprints-request-message-type">
        <xs:complexContent>
                <xs:extension base="tns:ccmp-request-message-type">
                        <xs:sequence>
                                <xs:element ref="blueprintsRequest" />
                        </xs:sequence>
                </xs:extension>
        </xs:complexContent>
</xs:complexType>

<!-- blueprintsRequestType -->

<xs:element name="blueprintsRequest" type="blueprintsRequestType"/>

<xs:complexType name="blueprintsRequestType">
        <xs:sequence>
                <xs:element name="xpathFilter" type="xs:string"
minOccurs="0"/>
                <xs:any namespace="##other" processContents="lax"
                        minOccurs="0" maxOccurs="unbounded"/>
        </xs:sequence>
        <xs:anyAttribute namespace="##any" processContents="lax"/>
</xs:complexType>
```

```
<!-- blueprintsResponse -->

<xs:complexType name="ccmp-blueprints-response-message-type">
        <xs:complexContent>
                <xs:extension base="tns:ccmp-response-message-type">
                        <xs:sequence>
                                <xs:element ref="blueprintsResponse" />
                        </xs:sequence>
                </xs:extension>
        </xs:complexContent>
</xs:complexType>

<!-- blueprintsResponseType -->

<xs:element name="blueprintsResponse" type="blueprintsResponseType"/>

<xs:complexType name="blueprintsResponseType">
        <xs:sequence>
                <xs:element name="blueprintsInfo"
                        type="info:uris-type" minOccurs="0"/>
                <xs:any namespace="##other" processContents="lax"
                        minOccurs="0" maxOccurs="unbounded"/>
        </xs:sequence>
        <xs:anyAttribute namespace="##any" processContents="lax"/>
</xs:complexType>
```

FIGURE 5.4 Structure of the blueprintsRequest and blueprintsResponse Messages. (Copyright: IETF.)

```
<!-- confsRequest -->

<xs:complexType name="ccmp-confs-request-message-type">
        <xs:complexContent>
                <xs:extension base="tns:ccmp-request-message-type">
                        <xs:sequence>
                                <xs:element ref="confsRequest" />
                        </xs:sequence>
                </xs:extension>
        </xs:complexContent>
</xs:complexType>

<!-- confsRequestType -->

<xs:element name="confsRequest" type="confsRequestType" />

<xs:complexType name="confsRequestType">
        <xs:sequence>
                <xs:element name="xpathFilter" type="xs:string"
minOccurs="0"/>
                <xs:any namespace="##other" processContents="lax"
                        minOccurs="0" maxOccurs="unbounded"/>
        </xs:sequence>
        <xs:anyAttribute namespace="##any" processContents="lax"/>
</xs:complexType>
```

```
<!-- confsResponse -->

<xs:complexType name="ccmp-confs-response-message-type">
        <xs:complexContent>
                <xs:extension base="tns:ccmp-response-message-type">
                        <xs:sequence>
                                <xs:element ref="confsResponse" />
                        </xs:sequence>
                </xs:extension>
        </xs:complexContent>
</xs:complexType>

<!-- confsResponseType -->

<xs:element name="confsResponse" type="confsResponseType"/>

<xs:complexType name="confsResponseType">
        <xs:sequence>
                <xs:element name="confsInfo" type="info:uris-type"
                        minOccurs="0"/>
                <xs:any namespace="##other" processContents="lax"
                        minOccurs="0" maxOccurs="unbounded"/>
        </xs:sequence>
        <xs:anyAttribute namespace="##any" processContents="lax"/>
</xs:complexType>
```

FIGURE 5.5 Structure of the confsRequest and confsResponse Messages. (Copyright: IETF.)

o"]." A conference server SHOULD NOT provide any sensitive information (e.g., passwords) in the blueprints. The associated blueprintsResponse message SHOULD contain, as shown in Figure 5.4, a "blueprintsInfo" parameter containing the above-mentioned XCON-URI list.

5.5.3.2 confsRequest and confsResponse

A confsRequest message (Figure 5.5) is used to retrieve, from the server, the list of XCON-URIs associated with active and registered conferences currently handled by the conferencing system. The <confUserID> parameter MUST be included in every confsRequest/Response message and reflect the XCON-USERID of the conferencing client issuing the request. The <confObjID> parameter MUST NOT be included in the confsRequest message. The confsRequest message is of a retrieve-only type, since the sole purpose is to collect information available at the conference server.

Thus, an <operation> parameter MUST NOT be included in a confsRequest message. In order to retrieve a specific subset of the available conferences, a client may specify a selection filter providing an appropriate xpath query in the OPTIONAL "xpathFilter" parameter of the request. For example, to select only the registered conferences, a possible xpathFilter value could be "/conferen ce-info[conference-description/conference-state/active="false"]." The associated confsResponse message SHOULD contain the list of XCON-URIs in the "confsInfo" parameter. A user, upon receipt of the response message, can interact with the available conference objects through further CCMP messages.

5.5.3.3 blueprintRequest and blueprintResponse

Through a blueprintRequest, a client can manipulate the conference object associated with a specified blueprint. Along with the <confUserID> parameter, the request MUST include the <confObjID> and the <operation> parameters. Again, the <confUserID> parameter MUST be included in every blueprintRequest/Response message and reflect the XCON-USERID of the conferencing client issuing the request. The <confObjID> parameter MUST contain the XCON-URI of the blueprint, which might have been previously retrieved through a blueprintsRequest message.

The blueprintRequest message SHOULD NOT contain an <operation> parameter with a value other than "retrieve." An <operation> parameter with a value of "create," "update," or "delete" SHOULD NOT be included in a blueprintRequest message except in the case of privileged users (e.g., the conference server administration staff), who might authenticate themselves by the mean of the "subject" request parameter.

A blueprintRequest/retrieve carrying a <confObjID> parameter whose value is not associated with one of the available system's blueprints will generate, on the server's side, a blueprintResponse message containing a <response-code> of "404." This also holds for the case in which the mentioned <confObjID> parameter value is related to an existing conference specification stored at the server but associated with an actual conference (be it active or registered) or with a sidebar rather than a blueprint.

For a <response-code> of "200" in a "retrieve" operation, the <blueprintInfo> parameter MUST be included in the blueprintResponse message. The <blueprintInfo> parameter contains the conference specification associated with the blueprint as identified by the <confObjID> parameter specified in the blueprintRequest (Figure 5.6).

5.5.3.4 confRequest and confResponse

With a confRequest message, CCMP clients can manipulate conference objects associated with either active or registered conferences. The <confUserID> parameter MUST be included in every confRequest/Response message and reflect the XCON-USERID of the conferencing client issuing

```
<!-- blueprintRequest -->

<xs:complexType name="ccmp-blueprint-request-message-type">
      <xs:complexContent>
            <xs:extension base="tns:ccmp-request-message-type">
                  <xs:sequence>
                        <xs:element ref="blueprintRequest" />
                  </xs:sequence>
            </xs:extension>
      </xs:complexContent>
</xs:complexType>

<!-- blueprintRequestType -->

<xs:element name="blueprintRequest" type="blueprintRequestType" />

<xs:complexType name="blueprintRequestType">
      <xs:sequence>
            <xs:element name="blueprintInfo"
                  type="info:conference-type" minOccurs="0"/>
            <xs:any namespace="##other" processContents="lax"
                  minOccurs="0" maxOccurs="unbounded"/>
      </xs:sequence>
      <xs:anyAttribute namespace="##any" processContents="lax"/>
</xs:complexType>
```

```
<!-- blueprintResponse -->

<xs:complexType name="ccmp-blueprint-response-message-type">
      <xs:complexContent>
            <xs:extension base="tns:ccmp-response-message-type">
                  <xs:sequence>
                        <xs:element ref="blueprintResponse" />
                  </xs:sequence>
            </xs:extension>
            </xs:complexContent>
            </xs:complexType>

<!-- blueprintResponseType -->

<xs:element name="blueprintResponse" type="blueprintResponseType"/>

<xs:complexType name="blueprintResponseType">
      <xs:sequence>
            <xs:element name="blueprintInfo" type="info:conference-type"
                  minOccurs="0"/>
            <xs:any namespace="##other" processContents="lax"
                  minOccurs="0" maxOccurs="unbounded">
      </xs:sequence>
      <xs:anyAttribute namespace="##any" processContents="lax"/>
</xs:complexType>
```

FIGURE 5.6 Structure of the blueprintRequest and blueprintResponse Messages. (Copyright: IETF.)

the request. confRequest and confResponse messages MUST also include an <operation> parameter. ConfResponse messages MUST return to the requestor a <response-code> and MAY contain a <response-string> explaining it. Depending upon the type of operation, a <confObjID> and <confInfo> parameter MAY be included in the confRequest and response. For each type of operation, the following text describes whether the <confObjID> and <confInfo> parameters need to be included in the confRequest and confResponse messages.

The creation case deserves care. To create a new conference through a confRequest message, two approaches can be considered:

1. Creation through explicit cloning: The <confObjID> parameter MUST contain the XCON-URI of the blueprint or of the conference to be cloned, while the <confInfo> parameter MUST NOT be included in the confRequest. Note that cloning of an active conference is only done in the case of a sidebar operation per the XCON framework and as described in Section 5.5.3.8.
2. Creation through implicit cloning (also known as "direct creation"): The <confObjID> parameter MUST NOT be included in the request, and the CCMP client can describe the desired conference to be created using the <confInfo> parameter. If no <confInfo> parameter is provided in the request, the new conference will be created as a clone of the system default blueprint.

In both creation cases, the confResponse, for a successful completion of a "create" operation, contains a <response-code> of "200" and MUST contain the XCON-URI of the newly created conference in the <confObjID> parameter in order to allow the conferencing client to manipulate that conference through following CCMP requests. In addition, the <confInfo> parameter containing the conference specification created MAY be included, at the discretion of the conferencing system implementation, along with the REQUIRED <version> parameter initialized at "1," since, at creation time, the conference object is at its first version.

In the case of a confRequest with an <operation> parameter of "retrieve," the <confObjID> parameter representing the XCON-URI of the target conference MUST be included and the <confInfo> parameter MUST NOT be included in the request. The conference server MUST ignore any <confInfo> parameter that is received in a confRequest "retrieve" operation. If the confResponse for the retrieve operation contains a <response-code> of "200," the <confInfo> parameter MUST be included in the response. The <confInfo> parameter MUST contain the entire conference specification describing the target conference object in its current state. The current state of the retrieved conference object MUST also be reported in the proper "version" response parameter.

In the case of a confRequest with an <operation> parameter of "update," the <confInfo> and <confObjID> parameters MUST be included in the request. The <confInfo> represents an object of type "conference-type" containing all the changes to be applied to the conference whose identifier has the same value as the <confObjID> parameter. Note that in such a case, though the <confInfo> parameter indeed has to follow the rules indicated in the XCON data model, it does not represent the entire updated version of the target conference, since it conveys just the modifications to apply to that conference. For example, in order to change the conference title, the <confInfo> parameter will be of the form shown in Figure 5.7).

Similarly, to remove the title of an existing conference, a confRequest/update carrying the following <confInfo> parameter would do the job (Figure 5.8).

In the case of a confResponse/update with a <response-code> of "200," no additional information is REQUIRED in the response message, which means that the return of a <confInfo> parameter is OPTIONAL. A subsequent confRequest/retrieve transaction might provide the CCMP client with the current status of the conference after the modification, or the notification protocol might address that task as well. A <response-code> of "200" indicates that the conference object has been changed according to the request by the conference server. The <version> parameter MUST be enclosed in the

```
<confInfo entity="xcon:8977777@example.com">
<conference-description>
<display-text> *** NEW CONFERENCE TITLE *** </display-text>
</conference-description>
</confInfo>
```

FIGURE 5.7 Updating a Conference Object: Modifying the Title of a Conference.

```
<confInfo entity="xcon:8977777@example.com">
<conference-description>
<display-text/>
</conference-description>
</confInfo>
```

FIGURE 5.8 Updating a Conference Object: Removing the Title of a Conference.

confResponse/update message, in order to let the client understand what the current conferenceobject version is, upon the applied modifications. A <response-code> of "409" indicates that the changes reflected in the <confInfo> parameter of the request are not feasible. This could be due to policies, requestor roles, specific privileges, unacceptable values, etc., with the reason specific to a conferencing system and its configuration. Together with a <response-code> of "409," the <version> parameter MUST be attached in the confResponse/update, allowing the client to later retrieve the current version of the target conference if the one she attempted to modify was not the most up to date.

In the case of a confRequest with an <operation> parameter of "delete," the <confObjID> parameter representing the XCON-URI of the target conference MUST be included, while the <confInfo> parameter MUST NOT be included in the request. The conference server MUST ignore any <confInfo> parameter that is received within such a request. The confResponse MUST contain the same value for the <confObjID> parameter that was included in the confRequest. If the confResponse/delete operation contains a <response-code> of "200," the conference indicated in the <confObjID> parameter has been successfully deleted. A confResponse/delete with a <response-code> of "200" MUST NOT contain the <confInfo> parameter. The <version> parameter SHOULD NOT be returned in any confResponse/delete. If the conference server cannot delete the conference referenced by the <confObjID> parameter received in the confRequest because it is the parent of another conference object that is in use, the conference server MUST return a <response-code> of "425."

A confRequest with an <operation> parameter of "retrieve," "update," or "delete" carrying a <confObjID> parameter that is not associated with one of the conferences (active or registered) that the system is holding will generate, on the server's side, a confResponse message containing a <response-code> of "404." This also holds for the case in which the mentioned <confObjID> parameter is related to an existing conference object stored at the server but associated with a blueprint or with a sidebar rather than an actual conference. The schema for the confRequest/confResponse pair is shown in Figure 5.9.

5.5.3.5 usersRequest and usersResponse

The usersRequest message allows a client to manipulate the <users> element of the conference object represented by the <confObjID> parameter. The <users> element contains the list of <user> elements associated with conference participants, the list of the users to which access to the conference is allowed/denied, conference participation policies, etc. The <confObjID> parameter MUST be included in a usersRequest message.

```
<!-- confRequest -->

<xs:complexType name="ccmp-conf-request-message-type">
    <xs:complexContent>
        <xs:extension base="tns:ccmp-request-message-type">
            <xs:sequence>
                <xs:element ref="confRequest" />
            </xs:sequence>
        </xs:extension>
    </xs:complexContent>
</xs:complexType>

<!-- confRequestType -->

<xs:element name="confRequest" type="confRequestType" />

<xs:complexType name="confRequestType">
    <xs:sequence>
        <xs:element name="confInfo" type="info:conference-type"
            minOccurs="0"/>
        <xs:any namespace="##other" processContents="lax"
            minOccurs="0" maxOccurs="unbounded"/>
    </xs:sequence>
    <xs:anyAttribute namespace="##any" processContents="lax"/>
</xs:complexType>
```

```
<!-- confResponse -->

<xs:complexType name="ccmp-conf-response-message-type">
    <xs:complexContent>
        <xs:extension base="tns:ccmp-response-message-type">
            <xs:sequence>
                <xs:element ref="confResponse" />
            </xs:sequence>
        </xs:extension>
    </xs:complexContent>
</xs:complexType>

<!-- confResponseType -->

<xs:element name="confResponse" type="confResponseType" />

<xs:complexType name="confResponseType">
    <xs:sequence>
        <xs:element name="confInfo" type="info:conference-type"
            minOccurs="0"/>
        <xs:any namespace="##other" processContents="lax"
            minOccurs="0" maxOccurs="unbounded"/>
    </xs:sequence>
    <xs:anyAttribute namespace="##any" processContents="lax"/>
</xs:complexType>
```

FIGURE 5.9 Structure of the confRequest and confResponse Messages. (Copyright: IETF.)

A <usersInfo> parameter MAY be included in a usersRequest message depending upon the operation. If the <usersInfo> parameter is included in the usersRequest message, the parameter MUST be compliant with the <users> field of the XCON data model. Two operations are allowed for a usersRequest message:

1. "retrieve:" In this case, the request MUST NOT include a <usersInfo> parameter, while the successful response MUST contain the desired <users> element in the <usersInfo> parameter. The conference server MUST ignore a <usersInfo> parameter if it is received in a request with an <operation> parameter of "retrieve."
2. "update:" In this case, the <usersInfo> parameter MUST contain the modifications to be applied to the <users> element indicated. If the <response-code> is "200," then the <usersInfo> parameter SHOULD NOT be returned. Any <usersInfo> parameter that is returned SHOULD be ignored. A <response-code> of "426" indicates that the conferencing client is not allowed to make the changes reflected in the <usersInfo> contained in the usersRequest message. This could be due to policies, roles, specific privileges, etc., with the reason being specific to a conferencing system and its configuration.

Operations of "create" and "delete" are not applicable to a usersRequest message and MUST NOT be considered by the server, which means that a <response-code> of "403" MUST be included in the usersResponse message (Figure 5.10).

5.5.3.6 userRequest and userResponse

A userRequest message is used to manipulate <user> elements inside a conference specification associated with a conference identified by the <confObjID> parameter. Besides retrieving information about a specific conference user, the message is used to request that the conference server either create, modify, or delete information about a user. A userRequest message MUST include the <confObjID> and the <operation> parameters, and it MAY include a <userInfo> parameter containing the detailed user's information, depending upon the operation and whether the <userInfo> has already been populated for a specific user. Note that a user may not necessarily be a conferencing control client (i.e., some participants in a conference are not "XCON aware").

An XCON-USERID SHOULD be assigned to each and every user subscribed to the system. In such a way, a user who is not a conference participant can make requests (provided she has

```
<!-- usersRequest -->

<xs:complexType name="ccmp-users-request-message-type">
      <xs:complexContent>
            <xs:extension base="tns:ccmp-request-message-type">
                  <xs:sequence>
                        <xs:element ref="usersRequest" />
                  </xs:sequence>
            </xs:extension>
      </xs:complexContent>
</xs:complexType>

<!-- usersRequestType -->

<xs:element name="usersRequest" type="usersRequestType" />

<xs:complexType name="usersRequestType">
      <xs:sequence>
            <xs:element name="usersInfo"
                  type="info:users-type" minOccurs="0" />
            <xs:any namespace="##other" processContents="lax"
                  minOccurs="0" maxOccurs="unbounded"/>
      </xs:sequence>
      <xs:anyAttribute namespace="##any" processContents="lax"/>
</xs:complexType>
```

```
<!-- usersResponse -->

<xs:complexType name="ccmp-users-response-message-type">
      <xs:complexContent>
            <xs:extension base="tns:ccmp-response-message-type">
                  <xs:sequence>
                        <xs:element ref="usersResponse" />
                  </xs:sequence>
            </xs:extension>
      </xs:complexContent>
</xs:complexType>

<!-- usersResponseType -->

<xs:element name="usersResponse" type="usersResponseType" />

<xs:complexType name="usersResponseType">
      <xs:sequence>
            <xs:element name="usersInfo" type="info:users-type"
                  minOccurs="0"/>
            <xs:any namespace="##other" processContents="lax"
                  minOccurs="0" maxOccurs="unbounded"/>
      </xs:sequence>
      <xs:anyAttribute namespace="##any" processContents="lax"/>
</xs:complexType>
```

FIGURE 5.10 Structure of the usersRequest and usersResponse Messages. (Copyright: IETF.)

successfully passed authorization and authentication checks), like creating a conference or retrieving conference information.

Conference users can be created in a number of different ways. In each of these cases, the <operation> parameter MUST be set to "create" in the userRequest message. Each of the userResponse messages for these cases MUST include the <confObjID>, <confUserID>, <operation>, and <response-code> parameters. In the case of a <response-code> of "200," the userResponse message MAY include the <userInfo> parameter, depending upon the manner in which the user was created:

- A conferencing client with an XCON-USERID adds itself to the conference: In this case, the <userInfo> parameter MAY be included in the userRequest. The <userInfo> parameter MUST contain a <user> element (compliant with the XCON data model), and the "entity" attribute MUST be set to a value that represents the XCON-USERID of the user initiating the request. No additional parameters beyond those previously described are required in the userResponse message in the case of a <response-code> of "200."
- A conferencing client acts on behalf of another user whose XCON-USERID is known: In this case, the <userInfo> parameter MUST be included in the userRequest. The <userInfo> parameter MUST contain a <user> element, and the "entity" attribute value MUST be set to the XCON-USERID of the other user in question. No additional parameters beyond those previously described are required in the userResponse message in the case of a <response-code> of "200."
- A conferencing client who has no XCON-USERID and who wants to enter, via CCMP, a conference whose identifier is known: In this case, a side effect of the request is that the user is provided with a new XCON-USERID (created by the server) carried inside the <confUserID> parameter of the response. This is the only case in which a CCMP request can be valid though carrying a void <confUserID> parameter. A <userInfo> parameter MUST be enclosed in the request, providing at least a contact URI of the joining client, in order to let the focus initiate the signaling phase needed to add her to the conference. The mandatory "entity" attribute of the <userInfo> parameter in the request MUST be filled with a placeholder value with the user part of the XCON-USERID containing a value of AUTO_GENERATE_X as described in Section 4.3 to conform to the rules contained in the XCON data model XML schema. The messages (userRequest and userResponse) in this case should look like Figure 5.11).

```
confUserID=null;
confObjID=confXYZ;
operation=create;
userInfo=
<userInfo entity="xcon-userid:AUTO_GENERATE_1@example.com">
        <endpoint entity="sip:GHIL345@example.com">
        ...

Response fields (in case of success):

confUserID=user345;
confObjID=confXYZ;
operation=create;
response-code=200;
userInfo=null; //or the entire userInfo object
```

FIGURE 5.11 userRequest and userResponse in the Absence of an xcon-userid. (Copyright: IETF.)

Request fields:

- A conferencing client is unaware of the XCON-USERID of a third user: In this case, the XCON-USERID in the request, <confUserID>, is the sender's, and the "entity" attribute of the attached <userInfo> parameter is filled with the placeholder value "xcon-userid:AUTO_GENERATE_1@example.com." The XCON-USERID for the third user MUST be returned to the client issuing the request in the "entity" attribute of the response <userInfo> parameter if the <response-code> is "200." This scenario is intended to support both the case where a brand new conferencing system user is added to a conference by a third party (i.e., a user who has not yet been provided with an XCON-USERID) and the case where the CCMP client issuing the request does not know the to-be-added user's XCON-USERID (which means that such an identifier could already exist on the server's side for that user). In this latter case, the conference server is in charge of avoiding XCON-URI duplicates for the same conferencing client, looking at key fields in the request-provided <userInfo> parameter, such as the signaling URI. If the joining user is brand new, then the generation of a new XCON-USERID is needed; otherwise, if that user exists already, the server must recover the corresponding XCON-USERID.

In the case of a userRequest with an <operation> parameter of "retrieve," the <confObjID> parameter representing the XCON-URI of the target conference MUST be included. The <confUserID>, containing the CCMP client's XCON-USERID, MUST also be included in the userRequest message. If the client wants to retrieve information about her profile in the specified conference, no <userInfo> parameter is needed in the retrieve request. On the other hand, if the client wants to obtain someone else's info within the given conference, she MUST include in the userRequest/retrieve a <userInfo> parameter whose "entity" attribute conveys the desired user's XCON-USERID. If the userResponse for the retrieve operation contains a <response-code> of "200," the <userInfo> parameter MUST be included in the response.

In the case of a userRequest with an <operation> parameter of "update," the <confObjID>, <confUserID>, and <userInfo> parameters MUST be included in the request. The <userInfo> parameter is of type "usertype" and contains all the changes to be applied to a specific <user> element in the conference object identified by the <confObjID> parameter in the userRequest message. The user to be modified is identified through the "entity" attribute of the <userInfo> parameter included in the request. In the case of a userResponse with a <response-code> of "200," no additional information is required in the userResponse message. A <response-code> of "200" indicates that the referenced <user> element has been updated by the conference server. A <response-code> of "426" indicates that the conferencing client is not allowed to make the changes reflected in the <userInfo> in the initial request. This could be due to policies, roles, specific privileges, etc., with the reason specific to a conferencing system and its configuration.

In the case of a userRequest with an <operation> parameter of "delete," the <confObjID> representing the XCON-URI of the target conference MUST be included. The <confUserID> parameter, containing the CCMP client's XCON-USERID, MUST be included in the userRequest message. If the client wants to exit the specified conference, no <userInfo> parameter is needed in the delete request. On the other hand, if the client wants to remove another participant from the given conference, she MUST include in the userRequest/delete a <userInfo> parameter whose "entity" attribute conveys the XCON-USERID of that participant. The userResponse MUST contain the same value for the <confObjID> parameter that was included in the <confObjID> parameter in the userRequest. The userResponse MUST contain a <response-code> of "200" if the target <user> element has been successfully deleted. If the userResponse for the delete operation contains a <response-code> of "200," the userResponse MUST NOT contain the <userInfo> parameter (Figure 5.12).

5.5.3.7 sidebarsByValRequest and sidebarsByValResponse

A sidebarsByValRequest message is used to execute a retrieve-only operation on the <sidebars-by-val> field of the conference object represented by the <confObjID>. The sidebarsByValRequest message is of a retrieve-only type, so an <operation> parameter MUST NOT be included in a sidebarsByValRequest message. As with blueprints and conferences, CCMP allows for the use of xpath filters whenever a selected subset of the sidebars available at the server's side has to be retrieved by the client. This applies to both sidebars by reference and sidebars by value. A sidebarsByValResponse message with a <response-code> of "200" MUST contain a <sidebarsByValInfo> parameter containing the desired <sidebars-by-val> element. A sidebarsByValResponse message MUST return to the client a <version> element related to the current version of the main conference object (i.e., the one whose identifier is contained in the <confObjID> field of the request) with which the sidebars in question are associated. The <sidebarsByValInfo> parameter contains the list of the conference objects associated with the sidebars by value derived from the main conference. The retrieved sidebars can then be updated or deleted using the sidebarByValRequest message (Figure 5.13), which is described in Section 5.5.3.8.

```
<!-- userRequest -->

<xs:complexType name="ccmp-user-request-message-type">
    <xs:complexContent>
        <xs:extension base="tns:ccmp-request-message-type">
            <xs:sequence>
                <xs:element ref="userRequest" />
            </xs:sequence>
        </xs:extension>
    </xs:complexContent>
</xs:complexType>

<!-- userRequestType -->

<xs:element name="userRequest" type="userRequestType" />

<xs:complexType name="userRequestType">
    <xs:sequence>
        <xs:element name="userInfo"
            type="info:user-type" minOccurs="0" />
        <xs:any namespace="##other" processContents="lax"
            minOccurs="0" maxOccurs="unbounded"/>
    </xs:sequence>
    <xs:anyAttribute namespace="##any" processContents="lax"/>
</xs:complexType>
```

```
<!-- userResponse -->

<xs:complexType name="ccmp-user-response-message-type">
    <xs:complexContent>
        <xs:extension base="tns:ccmp-response-message-type">
            <xs:sequence>
                <xs:element ref="userResponse" />
            </xs:sequence>
        </xs:extension>
    </xs:complexContent>
</xs:complexType>

<!-- userResponseType -->

<xs:element name="userResponse" type="userResponseType" />
    <xs:complexType name="userResponseType">
        <xs:sequence>
            <xs:element name="userInfo" type="info:user-type"
                minOccurs="0"/>
            <xs:any namespace="##other" processContents="lax"
                minOccurs="0" maxOccurs="unbounded"/>
        </xs:sequence>
        <xs:anyAttribute namespace="##any" processContents="lax"/>
</xs:complexType>
```

FIGURE 5.12 Structure of the userRequest and userResponse Messages. (Copyright: IETF.)

```
<!-- sidebarsByValRequest -->

<xs:complexType name="ccmp-sidebarsByVal-request-message-type">
    <xs:complexContent>
        <xs:extension base="tns:ccmp-request-message-type">
            <xs:sequence>
                <xs:element ref="sidebarsByValRequest"/>
            </xs:sequence>
        </xs:extension>
    </xs:complexContent>
</xs:complexType>

<!-- sidebarsByValRequestType -->

<xs:element name="sidebarsByValRequest"
        type="sidebarsByValRequestType" />

<xs:complexType name="sidebarsByValRequestType">
    <xs:sequence>
        <xs:element name="xpathFilter" type="xs:string"
                minOccurs="0"/>
        <xs:any namespace="##other" processContents="lax"
                minOccurs="0" maxOccurs="unbounded"/>
    </xs:sequence>
    <xs:anyAttribute namespace="##any" processContents="lax"/>
</xs:complexType>
```

```
<!-- sidebarsByValResponse -->

<xs:complexType name="ccmp-sidebarsByVal-response-message-type">
    <xs:complexContent>
        <xs:extension base="tns:ccmp-response-message-type">
            <xs:sequence>
                <xs:element ref="sidebarsByValResponse"/>
            </xs:sequence>
        </xs:extension>
    </xs:complexContent>
</xs:complexType>

<!-- sidebarsByValResponseType -->

<xs:element name="sidebarsByValResponse"
        type="sidebarsByValResponseType" />

<xs:complexType name="sidebarsByValResponseType">
    <xs:sequence>
        <xs:element name="sidebarsByValInfo"
            type="info:sidebars-by-val-type" minOccurs="0"/>
        <xs:any namespace="##other" processContents="lax"
            minOccurs="0" maxOccurs="unbounded"/>
    </xs:sequence>
    <xs:anyAttribute namespace="##any" processContents="lax"/>
</xs:complexType>
```

FIGURE 5.13 Structure of the sidebarsByValRequest and sidebarsByValResponse Messages. (Copyright: IETF.)

5.5.3.8 sidebarByValRequest and sidebarByValResponse

A sidebarByValRequest message MUST contain the <operation> parameter, which distinguishes among retrieval, creation, modification, and deletion of a specific sidebar. The other required parameters depend upon the type of operation. In the case of a "create" operation, the <confObjID> parameter MUST be included in the sidebyValRequest message. In this case, the <confObjID> parameter contains the XCON-URI of the main conference in which the sidebar has to be created. If no "sidebarByValInfo" parameter is included, the sidebar is created by cloning the main conference, as envisioned in the XCON framework (RFC 5239 – see Chapter 2) following the implementation-specific cloning rules. Otherwise, similarly to the case of direct creation, the sidebar conference object is built on the basis of the "sidebarByValInfo" parameter provided by the requestor. As a consequence of a sidebar-by-val creation, the conference server MUST update the main conference object reflected by the <confObjID> parameter in the sidebarbyValRequest/create message introducing the new sidebar object as a new <entry> in the proper section <sidebars-by-val>. The newly created sidebar conference object MAY be included in the sidebarByValResponse in the <sidebarByValInfo> parameter if the <response-code> is "200." The XCON-URI of the newly created sidebar MUST appear in the <confObjID> parameter of the response. The conference server can notify any conferencing clients that have subscribed to the conference event package and that are authorized to receive the notification of the addition of the sidebar to the conference.

In the case of a sidebarByValRequest message with an <operation> parameter of "retrieve," the URI for the conference object created for the sidebar (received in response to a create operation or in a sidebarsByValResponse message) MUST be included in the <confObjID> parameter in the request. This retrieve operation is handled by the conference server in the same manner as in the case of an <operation> parameter of "retrieve" included in a confRequest message, as described in Section 5.5.3.4.

In the case of a sidebarByValRequest message with an <operation> parameter of "update", the <sidebarByValInfo> MUST also be included in the request. The <confObjID> parameter contained in the request message identifies the specific sidebar instance to be updated. An update operation on the specific sidebar instance contained in the <sidebarByValInfo> parameter is handled by the conference server in the same manner as an update operation on the conference instance as reflected by the <confInfo> parameter included in a confRequest message, as detailed in Section 5.3.4. A sidebarByValResponse message MUST return to the client a <version> element related to the current version of the sidebar whose identifier is contained in the <confObjID> field of the request. If an <operation> parameter of "delete" is included in the sidebarByVal request, the <sidebarByValInfo> parameter MUST NOT be included in the request. Any <sidebarByValInfo> included in the request MUST be ignored by the conference server. The URI for the conference object associated with the sidebar MUST be included in the <confObjID> parameter in the request. If the specific conferencing user, as reflected by the <confUserID> parameter, in the request is authorized to delete the conference, the conference server deletes the conference object reflected by the <confObjID> parameter and updates the data in the conference object from which the sidebar was cloned. The conference server can notify any conferencing clients that have subscribed to the conference event package and that are authorized to receive the notification of the deletion of the sidebar from the conference.

If a sidebarByValRequest with an <operation> parameter of "retrieve," "update," or "delete" carries a <confObjID> parameter that is not associated with any existing sidebar-by-val, a confResponse message containing a <response-code> of "404" will be generated on the server's side. This also holds for the case in which the mentioned <confObjID> parameter is related to an existing conference object stored at the server but associated with a blueprint or with an actual conference or with a sidebar-by-ref rather than a sidebar-by-val (Figure 5.14).

5.5.3.9 sidebarsByRefRequest and sidebarsByRefResponse

Similarly to the sidebarsByValRequest, a sidebarsByRefRequest can be invoked to retrieve the <sidebars-by-ref> element of the conference object identified by the <confObjID> parameter. The sidebarsByRefRequest message is of a retrieve-only type, so an <operation> parameter MUST NOT be included in a sidebarsByRefRequest message. In the case of a <response-code> of "200," the <sidebarsByRefInfo> parameter, containing the <sidebars-by-ref> element of the conference object, MUST be included in the response.

The <sidebars-by-ref> element represents the set of URIs of the sidebars associated with the main conference, whose description (in the form of a standard XCON conference specification) is external to the main conference itself. Through the retrieved URIs, it is then possible to access single sidebars using the sidebarByRefRequest message, described in Section 5.5.3.10. A sidebarsByRefResponse message MUST carry back to the client a <version> element related to the current

```
<!-- sidebarByValRequest -->

<xs:complexType name="ccmp-sidebarByVal-request-message-type">
        <xs:complexContent>
                <xs:extension base="tns:ccmp-request-message-type">
                        <xs:sequence>
                                <xs:element ref="sidebarByValRequest"/>
                        </xs:sequence>
                </xs:extension>
        </xs:complexContent>
</xs:complexType>

<!-- sidebarByValRequestType -->

<xs:element name="sidebarByValRequest"
                type="sidebarByValRequestType" />

<xs:complexType name="sidebarByValRequestType">
        <xs:sequence>
                <xs:element name="sidebarByValInfo"
                                type="info:conference-type" minOccurs="0"/>
                <xs:any namespace="##other" processContents="lax"
                                minOccurs="0" maxOccurs="unbounded"/>
        </xs:sequence>
        <xs:anyAttribute namespace="##any" processContents="lax"/>
</xs:complexType>
```

```
<!-- sidebarByValResponse -->

<xs:complexType name="ccmp-sidebarByVal-response-message-type">
        <xs:complexContent>
                <xs:extension base="tns:ccmp-response-message-type">
                        <xs:sequence>
                                <xs:element ref="sidebarByValResponse"/>
                        </xs:sequence>
                </xs:extension>
        </xs:complexContent>
</xs:complexType>

<!-- sidebarByValResponseType -->

<xs:element name="sidebarByValResponse"
                type="sidebarByValResponseType" />

<xs:complexType name="sidebarByValResponseType">
        <xs:sequence>
                <xs:element name="sidebarByValInfo"
                                type="info:conference-type" minOccurs="0"/>
                <xs:any namespace="##other" processContents="lax"
                                minOccurs="0" maxOccurs="unbounded"/>
        </xs:sequence>
        <xs:anyAttribute namespace="##any" processContents="lax"/>
</xs:complexType>
```

FIGURE 5.14 Structure of the sidebarByValRequest and sidebarByValResponse Messages. (Copyright: IETF.)

```
<!-- sidebarsByRefRequest -->

<xs:complexType name="ccmp-sidebarsByRef-request-message-type">
    <xs:complexContent>
        <xs:extension base="tns:ccmp-request-message-type">
            <xs:sequence>
                <xs:element ref="sidebarsByRefRequest"/>
            </xs:sequence>
        </xs:extension>
    </xs:complexContent>
</xs:complexType>

<!-- sidebarsByRefRequestType -->

<xs:element name="sidebarsByRefRequest"
type="sidebarsByRefRequestType" />

<xs:complexType name="sidebarsByRefRequestType">
    <xs:sequence>
        <xs:element name="xpathFilter" type="xs:string"
minOccurs="0"/>
            <xs:any namespace="##other" processContents="lax"
                minOccurs="0" maxOccurs="unbounded"/>
    </xs:sequence>
    <xs:anyAttribute namespace="##any" processContents="lax"/>
</xs:complexType>
```

```
<!-- sidebarsByRefResponse -->

<xs:complexType name="ccmp-sidebarsByref-response-message-type">
    <xs:complexContent>
        <xs:extension base="tns:ccmp-response-message-type">
            <xs:sequence>
                <xs:element ref="sidebarsByRefResponse"/>
            </xs:sequence>
        </xs:extension>
    </xs:complexContent>
</xs:complexType>

<!-- sidebarsByRefResponseType -->

<xs:element name="sidebarsByRefResponse"
type="sidebarsByRefResponseType" />

<xs:complexType name="sidebarsByRefResponseType">
    <xs:sequence>
        <xs:element name="sidebarsByRefInfo"
                type="info:uris-type" minOccurs="0"/>
        <xs:any namespace="##other" processContents="lax"
                minOccurs="0" maxOccurs="unbounded"/>
    </xs:sequence>
    <xs:anyAttribute namespace="##any" processContents="lax"/>
</xs:complexType>
```

FIGURE 5.15 Structure of the sidebarsByRefRequest and sidebarsByRefResponse Messages. (Copyright: IETF.)

version of the main conference object (i.e., the one whose identifier is contained in the <confObjId> field of the request) with which the sidebars in question are associated (Figure 5.15).

5.5.3.10 sidebarByRefRequest and sidebarByRefResponse

A sidebarByValResponse message MUST return to the client a <version> element related to the current version of the sidebar whose identifier is contained in the <confObjID> field of the request. In the case of a create operation, the <confObjID> parameter MUST be included in the sidebyRefRequest message. In this case, the <confObjID> parameter contains the XCON-URI of the main conference in which the sidebar has to be created. If no <sidebarByRefInfo> parameter is included, following the XCON framework (RFC 5239 – see Chapter 2), the sidebar is created by cloning the main conference, observing the implementation-specific cloning rules. Otherwise, similarly to the case of direct creation, the sidebar conference object is built on the basis of the <sidebarByRefInfo> parameter provided by the requestor. If the creation of the sidebar is successful, the conference server MUST update the <sidebars-by-ref> element in the conference object from which the sidebar was created (i.e., as identified by the <confObjID> in the original sidebarByRefRequest) with the URI of the newly created sidebar. The newly created conference object MAY be included in the response in the <sidebarByRefInfo> parameter with a <response-code> of "200." The URI for the conference object associated with the newly created sidebar object MUST appear in the <confObjID> parameter of the response. The conference server can notify any conferencing clients that have subscribed to the conference event package and that are authorized to receive the notification of the addition of the sidebar to the conference.

In the case of a sidebarByRefRequest message with an <operation> parameter of "retrieve," the URI for the conference object created for the sidebar MUST be included in the <confObjID> parameter in the request. A retrieve operation on the <sidebarByRefInfo> is handled by the conference server in the same manner as a retrieve operation on the confInfo included in a confRequest message as detailed in Section 5.3.4.

In the case of a sidebarByRefRequest message with an <operation> parameter of "update," the URI for the conference object created for the sidebar MUST be included in the <confObjID> parameter in the request. The <sidebarByRefInfo> MUST also be included in the request in the case of an "update" operation. An update operation on the <sidebarByRefInfo> is handled by the conference server in the same manner as an update operation on the <confInfo> included in a confRequest message, as detailed in Section 5.5.3.4. A sidebarByRefResponse message MUST carry back to the client a <version> element related to the current version of the sidebar whose identifier is contained in the <confObjID> field of the request.

If an <operation> parameter of "delete" is included in the sidebarByRefRequest, the <sidebar-ByRefInfo> parameter MUST NOT be included in the request. Any <sidebarByRefInfo> included in the request MUST be ignored by the conference server. The URI for the conference object for the sidebar MUST be included in the <confObjID> parameter in the request. If the specific conferencing user as reflected by the <confUserID> parameter in the request is authorized to delete the conference, the conference server SHOULD delete the conference object reflected by the <confObjID> parameter and SHOULD update the <sidebars-by-ref> element in the conference object from which the sidebar was originally cloned. The conference server can notify any conferencing clients that have subscribed to the conference event package and that are authorized to receive the notification of the deletion of the sidebar.

If a sidebarByRefRequest with an <operation> parameter of "retrieve," "update," or "delete" carries a <confObjID> parameter that is not associated with any existing sidebar-by-ref, a confResponse message containing a <response-code> of "404" will be generated on the server's side. This also holds for the case in which the value of the mentioned <confObjID> parameter is related to an existing conference object stored at the server but associated with a blueprint or with an actual conference or with a sidebar-by-val rather than a sidebar-by-ref (Figure 5.16).

5.5.3.11 extendedRequest and extendedResponse

In order to allow specifying new request and response pairs for conference control, CCMP defines the extendedRequest and extendedResponse messages. Such messages constitute a CCMP skeleton in which implementers can transport the information needed to realize conference control mechanisms not explicitly envisioned in the CCMP specification; these mechanisms are called, in this context, "extensions." Each extension is assumed to be characterized by an appropriate name that MUST be carried in the extendedRequest/ extendedResponse pair in the provided <extensionName> field. Extension-specific information can be transported in the form of schema-defined XML elements inside the <any> element present in both extendedRequest and extendedResponse.

The conferencing client SHOULD be able to determine the extensions supported by a CCMP server and to recover the XML schema defining the related specific elements by means of an

```
<!-- sidebarByRefRequest -->

<xs:complexType name="ccmp-sidebarByRef-request-message-type">
	<xs:complexContent>
		<xs:extension base="tns:ccmp-request-message-type">
			<xs:sequence>
				<xs:element ref="sidebarByRefRequest"/>
			</xs:sequence>
		</xs:extension>
	</xs:complexContent>
</xs:complexType>

<!-- sidebarByRefRequestType -->

<xs:element name="sidebarByRefRequest"
		type="sidebarByRefRequestType" />

<xs:complexType name="sidebarByRefRequestType">
	<xs:sequence>
		<xs:element name="sidebarByRefInfo"
				type="info:conference-type" minOccurs="0"/>
		<xs:any namespace="##other" processContents="lax"
				minOccurs="0" maxOccurs="unbounded"/>
	</xs:sequence>
	<xs:anyAttribute namespace="##any" processContents="lax"/>
</xs:complexType>
```

```
<!-- sidebarByRefResponse -->

<xs:complexType name="ccmp-sidebarByRef-response-message-type">
	<xs:complexContent>
		<xs:extension base="tns:ccmp-response-message-type">
			<xs:sequence>
				<xs:element ref="sidebarByRefResponse"/>
			</xs:sequence>
		</xs:extension>
	</xs:complexContent>
</xs:complexType>

<!-- sidebarByRefResponseType -->

<xs:element name="sidebarByRefResponse"
type="sidebarByRefResponseType" />

<xs:complexType name="sidebarByRefResponseType">
	<xs:sequence>
		<xs:element name="sidebarByRefInfo"
				type="info:conference-type" minOccurs="0"/>
		<xs:any namespace="##other" processContents="lax"
				minOccurs="0" maxOccurs="unbounded"/>
	</xs:sequence>
	<xs:anyAttribute namespace="##any" processContents="lax"/>
</xs:complexType
```

FIGURE 5.16 Structure of the sidebarByRefRequest and sidebarByRefResponse Messages. (Copyright: IETF.)

```
<!-- extendedRequest -->

<xs:complexType name="ccmp-extended-request-message-type">
<xs:complexContent>
    <xs:extension base="tns:ccmp-request-message-type">
        <xs:sequence>
            <xs:element ref="extendedRequest"/>
        </xs:sequence>
    </xs:extension>
</xs:complexContent>
</xs:complexType>

<!-- extendedRequestType -->

<xs:element name="extendedRequest" type="extendedRequestType"/>
<xs:complexType name="extendedRequestType">
    <xs:sequence>
        <xs:element name="extensionName" type="xs:string" minOccurs="1"/>
        <xs:any namespace="##other" processContents="lax"
                minOccurs="0" maxOccurs="unbounded" />
    </xs:sequence>
</xs:complexType>
```

(a)

```
<!-- extendedResponse -->

<xs:complexType name="ccmp-extended-response-message-type">
    <xs:complexContent>
        <xs:extension base="tns:ccmp-response-message-type">
            <xs:sequence>
                <xs:element ref="extendedResponse"/>
            </xs:sequence>
        </xs:extension>
    </xs:complexContent>
</xs:complexType>

<!-- extendedResponseType -->

<xs:element name="extendedResponse" type="extendedResponseType"/>
<xs:complexType name="extendedResponseType">
    <xs:sequence>
        <xs:element name="extensionName" type="xs:string"
                                          minOccurs="1"/>
        <xs:any namespace="##other" processContents="lax"
                          minOccurs="0" maxOccurs="unbounded" />
    </xs:sequence>
</xs:complexType>
```

(b)

FIGURE 5.17 Structure of the extendedRequest and extendedResponse Messages. (Copyright: IETF.)

optionsRequest/optionsResponse CCMP transaction (see Section 5.5.3.12). The meaning of the common CCMP parameters inherited by the extendedRequest and extendedResponse from the basic CCMP request and response messages SHOULD be preserved and exploited appropriately while defining an extension (Figure 5.17).

5.5.3.12 optionsRequest and optionsResponse

The optionsRequest message (Figure 5.18) retrieves general information about conference server capabilities. These capabilities include the standard CCMP messages (request/response pairs) and potential extension messages supported by the conference server. As such, it is a basic CCMP message rather than a specialization of the general CCMP request. The optionsResponse returns, in the appropriate <options> field, a list of the supported CCMP message pairs as defined in this specification. These messages are in the form of a list: <standardmessage-list> including each of the supported messages as reflected by <standard-message> elements. The optionsResponse message also allows for an <extended-message-list>, which is a list of additional message types in the form of <extended-message-list> elements that are currently undefined to allow for future extensibility. The following information is provided for both types of messages:

- **<name>** (REQUIRED): In the case of standard messages, it can be one of the 10 standard message names defined in this specification (i.e., "blueprintsRequest," "confsRequest," etc.). In the case of extensions, this element MUST carry the same value of the <extension-name> inserted in the corresponding extendedRequest/extendedResponse message pair.
- **<operations>** (OPTIONAL): This field is a list of <operation> entries, each representing the Create, Read, Update, Delete (CRUD) operation supported by the server for the message. If this element is absent, the client SHOULD assume that the server is able to handle the entire set of CRUD operations or in the case of standard messages, all the operations envisioned for that message in this specification.
- **<schema-ref>** (OPTIONAL): Since all CCMP messages can potentially contain XML elements not envisioned in the CCMP schema (due to the presence of <any> elements and attributes), a reference to a proper schema definition specifying such new elements/attributes can also be sent back to the clients by means of such a field. If this element is absent, no new elements are introduced in the messages other than those explicitly defined in the CCMP specification.
- **<description>** (OPTIONAL): Human-readable information about the related message.

```
<!-- optionsRequest -->

<xs:complexType name="ccmp-options-request-message-type">
    <xs:complexContent>
        <xs:extension base="tns:ccmp-request-message-type"/>
    </xs:complexContent>
</xs:complexType>

<!-- optionsResponse -->

<xs:complexType name="ccmp-options-response-message-type">
    <xs:complexContent>
        <xs:extension base="tns:ccmp-response-message-type">
            <xs:sequence>
                <xs:element ref="optionsResponse"/>
            </xs:sequence>
        </xs:extension>
    </xs:complexContent>
</xs:complexType>

<!-- optionsResponseType -->

<xs:element name="optionsResponse" type="optionsResponseType" />
<xs:complexType name="optionsResponseType">
    <xs:sequence>
        <xs:element name="options" type="options-type" minOccurs="0"/>
        <xs:any namespace="##other" processContents="lax" minOccurs="0"
                                                maxOccurs="unbounded"/>
    </xs:sequence>
    <xs:anyAttribute namespace="##any" processContents="lax"/>
</xs:complexType>

<!-- options-type -->

<xs:complexType name="options-type">
    <xs:sequence>
        <xs:element name="standard-message-list"
                type="standard-message-list-type" minOccurs="1"/>
        <xs:element name="extended-message-list"
                type="extended-message-list-type" minOccurs="0"/>
        <xs:any namespace="##other" processContents="lax"
                minOccurs="0" maxOccurs="unbounded"/>
    </xs:sequence>
    <xs:anyAttribute namespace="##any" processContents="lax"/>
</xs:complexType>

<!-- standard-message-list-type -->

<xs:complexType name="standard-message-list-type">
<xs:sequence>
```

(a)

```
        <xs:element name="standard-message"
            type="standard-message-type" minOccurs="1" maxOccurs="10"/>
        <xs:any namespace="##other" processContents="lax"
            minOccurs="0" maxOccurs="unbounded"/>
    </xs:sequence>
    <xs:anyAttribute namespace="##any" processContents="lax"/>
</xs:complexType>

<!-- standard-message-type -->

<xs:complexType name="standard-message-type">
    <xs:sequence>
        <xs:element name="name" type="standard-message-name-type"
                minOccurs="1"/>
        <xs:element name="operations" type="operations-type"
                minOccurs="0"/>
        <xs:element name="schema-def" type="xs:string" minOccurs="0"/>
        <xs:element name="description" type="xs:string" minOccurs="0"/>
        <xs:any namespace="##other" processContents="lax"
                minOccurs="0" maxOccurs="unbounded"/>
    </xs:sequence>
    <xs:anyAttribute namespace="##any" processContents="lax"/>
</xs:complexType>

<!-- standard-message-name-type -->

<xs:simpleType name="standard-message-name-type">
    <xs:restriction base="xs:token">
        <xs:enumeration value="confsRequest"/>
        <xs:enumeration value="confRequest"/>
        <xs:enumeration value="blueprintsRequest"/>
        <xs:enumeration value="blueprintRequest"/>
        <xs:enumeration value="usersRequest"/>
        <xs:enumeration value="userRequest"/>
        <xs:enumeration value="sidebarsByValRequest"/>
        <xs:enumeration value="sidebarByValRequest"/>
        <xs:enumeration value="sidebarsByRefRequest"/>
        <xs:enumeration value="sidebarByRefRequest"/>
    </xs:restriction>
</xs:simpleType>

<!-- operations-type -->

<xs:complexType name="operations-type">
    <xs:sequence>
        <xs:element name="operation" type="operationType"
                minOccurs="1" maxOccurs="4"/>
    </xs:sequence>
    <xs:anyAttribute namespace="##any" processContents="lax"/>
    </xs:complexType>
```

(b)

FIGURE 5.18 Structure of the optionsRequest and optionsResponse Messages. (Copyright: IETF.)

The only parameter needed in the optionsRequest is the sender confUserID, which is mirrored in the same parameter of the corresponding optionsResponse. The CCMP server MUST include the <standard-message-list> containing at least one <operation> element in the optionsResponse, since a CCMP server is REQUIRED to be able to handle both the request and response messages for at least one of the operations (Figure 5.18).

5.5.4 CCMP Response Codes

All CCMP response messages MUST include a <response-code>. This specification defines an IANA registry for the CCMP response codes, as described in Section 5.12.5.2. The following summarizes the CCMP response codes:

- 200 Success: Successful completion of the requested operation.
- 400 Bad Request: Syntactically malformed request.
- 401 Unauthorized: User not allowed to perform the required operation.
- 403 Forbidden: Operation not allowed (e.g., cancellation of a blueprint).
- 404 Object Not Found: The target conference object does not exist at the server (the object in the error message refers to the <confObjID> parameter in the generic request message).

- 409 Conflict: A generic error associated with all those situations in which a requested client operation cannot be successfully completed by the server. An example of such a situation is when the modification of an object cannot be applied due to conflicts arising at the server's side, e.g., because the client version of the object is an obsolete one and the requested modifications collide with the up-to-date state of the object stored at the server. Such code would also be used if a client attempts to create an object (conference or user) with an entity that already exists.
- 420 User Not Found: Target user missing at the server (it is related to the XCON-USERID in the "entity" attribute of the <userInfo> parameter when it is included in userRequests).
- 421 Invalid confUserID: User does not exist at the server (this code is returned for requests where the <confUserID> parameter is invalid).
- 422 Invalid Conference Password: The password for the target conference object contained in the request is wrong.
- 423 Conference Password Required: "conference-password" missing in a request to access a password-protected conference object.
- 424 Authentication Required: User's authentication information is missing or invalid.
- 425 Forbidden Delete Parent: Cancel operation failed since the target object is a parent of child objects that depend on it, or because it affects, based on the "parent-enforceable" mechanism, the corresponding element in a child object.
- 426 Forbidden Change Protected: Update refused by the server because the target element cannot be modified due to its implicit dependence on the value of a parent object ("parent-enforceable" mechanism).
- 427 Invalid Domain Name: The domain name in an AUTO_GENERATE_X instance in the conference object is not within the CCMP server's domain of responsibility.
- 500 Server Internal Error: The server cannot complete the required service due to a system internal error.
- 501 Not Implemented: Operation defined by the protocol but not implemented by this server.
- 510 Request Timeout: The time required to serve the request has exceeded the configured service threshold.
- 511 Resources Not Available: This code is used when the CCMP server cannot execute a command because of resource issues, e.g., it cannot create a sub-conference because the system has reached its limits on the number of sub-conferences, or if a request for adding a new user fails because the maximum number of users has been reached for the conference or the maximum number of users has been reached for the conferencing system.

The handling of a <response-code> of "404," "409," "420," "421," "425," "426," or "427" is only applicable to specific operations for specialized message responses, and the details are provided in Section 5.5.3. Table 5.2 summarizes these response codes and the specialized message and operation to which they are applicable:

In the case of a <response-code> of "510," a conferencing client MAY re-attempt the request within a period of time that would be specific to a conferencing client or conference server.

A <response-code> of "400" indicates that the conferencing client sent a malformed request, which is indicative of an error in the conferencing client or in the conference server. The handling is specific to the conferencing client implementation (e.g., generate a log; display an error message, etc.). It is NOT RECOMMENDED that the client re-attempt the request in this case.

A <response-code> of "401" or "403" indicates that the client does not have the appropriate permissions or there is an error in the permissions: re-attempting the request would likely not succeed, and thus, it is NOT RECOMMENDED.

Any unexpected or unknown <response-code> SHOULD be treated by the client in the same manner as a <response-code> of "500," the handling of which is specific to the conferencing client implementation.

TABLE 5.2
Response Codes and Associated Operations

Response	Create	Retrieve	Update	Delete
404	userRequest sidebarByValRequest, sidebarByRefRequest	All retrieve requests EXCEPT: blueprintsRequest, confsRequest	All update requests	All delete requests
409	N/A	N/A	All update requests	N/A
420	userRequest (third-party invite with third-user entity) (*)	userRequest	userRequest	userRequest
421	All create requests EXCEPT: userRequest with no confUserID (**)	All update requests	All update requests	All delete requests
425	N/A	N/A	N/A	All delete requests
426	N/A	N/A	All update requests	N/A
427	ConfRequest UserRequest	N/A	All update requests	N/A

(*) "420" in answer to a "userRequest/create" operation: In the case of a third-party invite, this code can be returned if the <confUserID> (contained in the "entity" attribute of the <userInfo> parameter) of the user to be added is unknown. In this case, if instead it is the <confUserID> parameter of the sender of the request that is invalid, a <response-code> of "421" is returned to the client.

(**) "421" is not sent in answer to userRequest/create messages having a "null" confUserID, since this case is associated with a user who is unaware of his own XCON-USERID but wants to enter a known conference.

5.6 EXAMPLE OF CCMP IN ACTION

In this section a typical, non-normative scenario in which CCMP comes into play is described by showing the actual composition of the various CCMP messages. In the call flows of the example, the conferencing client is a CCMP-enabled client, and the conference server is a CCMP-enabled server. The XCON-USERID of the client, Alice, is "xcon-userid:alice@example.com," and it appears in the <confUserID> parameter in all requests. The sequence of operations is as follows:

1. Alice retrieves the list of available blueprints from the server (Section 5.6.1).
2. Alice asks for detailed information about a specific blueprint (Section 5.6.2).
3. Alice decides to create a new conference by cloning the retrieved blueprint (Section 5.6.3).
4. Alice modifies information (e.g., XCON-URI, name, and description) associated with the newly created blueprint (Section 5.6.4).
5. Alice specifies a list of users to be contacted when the conference is activated (Section 5.6.5).
6. Alice joins the conference (Section 5.6.6).
7. Alice lets a new user, Ciccio (whose XCON-USERID is "xcon-userid:Ciccio@example.com"), join the conference (Section 5.6.7).
8. Alice asks for the CCMP server capabilities (Section 5.6.8).
9. Alice exploits an extension of the CCMP server (Section 5.6.9).

Note that the examples do not include any details beyond the basic operation.

In the following sections, we deal with each of the aforementioned actions separately.

5.6.1 Alice Retrieves the Available Blueprints

This section illustrates the transaction associated with retrieval of the blueprints together with a dump of the two messages exchanged (blueprintsRequest and blueprintsResponse). As shown in Figure 5.19, the blueprintsResponse message contains, in the <blueprintsInfo> parameter, information about the available blueprints, in the form of the standard XCON-URI of the blueprint, plus additional (and optional) information, like its display text and purpose. After receiving the reply, Alice retrieves from the server the list of available blueprints.

5.6.2 Alice Gets Detailed Information about a Specific Blueprint

This section illustrates the second transaction in the overall flow. In this case, Alice, who now knows the XCON-URIs of the blueprints available at the server, makes a drill-down query, in the form of a CCMP blueprintRequest message (Figure 5.20), to get detailed information about one of

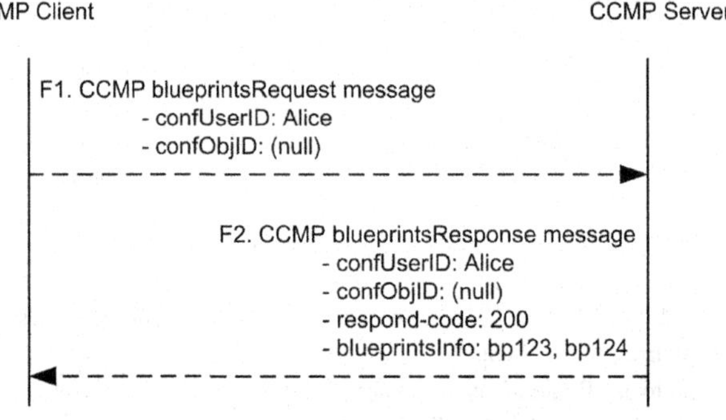

F1. blueprintsRequest message:

```
<?xml version="1.0" encoding="UTF-8" standalone="yes"?>
<ccmp:ccmpRequest
        xmlns:info="urn:ietf:params:xml:ns:conference-info"
        xmlns:ccmp="urn:ietf:params:xml:ns:xcon-ccmp"
        xmlns:xcon="urn:ietf:params:xml:ns:xcon-conference-info">
    <ccmpRequest xmlns:xsi="http://www.w3.org/2001/XMLSchema-instance"
            xsi:type="ccmp:ccmp-blueprints-request-message-type">
        <confUserID>xcon-userid:alice@example.com</confUserID>
        <ccmp:blueprintsRequest/>
    </ccmpRequest>
</ccmp:ccmpRequest>
```

F2. blueprintsResponse message from the server:

```
<?xml version="1.0" encoding="UTF-8" standalone="yes"?>
<ccmp:ccmpResponse
        xmlns:xcon="urn:ietf:params:xml:ns:xcon-conference-info"
        xmlns:info="urn:ietf:params:xml:ns:conference-info"
        xmlns:ccmp="urn:ietf:params:xml:ns:xcon-ccmp">
    <ccmpResponse
            xmlns:xsi="http://www.w3.org/2001/XMLSchema-instance"
            xsi:type="ccmp:ccmp-blueprints-response-message-type">
        <confUserID>xcon-userid:alice@example.com</confUserID>
        <response-code>200</response-code>
        <ccmp:blueprintsResponse>
            <blueprintsInfo>
                <info:entry>
                    <info:uri>xcon:AudioRoom@example.com</info:uri>
                    <info:display-text>AudioRoom</info:display-text>
                    <info:purpose>Simple Room: conference room with
                        public access, where only audio is available, more
                        users can talk at the same time and the requests for
                        the AudioFloor are automatically accepted.
                    </info:purpose>
                </info:entry>
                <info:entry>
                    <info:uri>xcon:VideoRoom@example.com</info:uri>
                    <info:display-text>VideoRoom</info:display-text>
                    <info:purpose>Video Room: conference room with public
                        access, where both audio and video are available,
                        8 users can talk and be seen at the same time, and the
                        floor requests are automatically accepted.
                    </info:purpose>
                </info:entry>
                <info:entry>
                    <info:uri>xcon:AudioConference1@example.com</info:uri>
                    <info:display-text>AudioConference1</info:display-text>
                    <info:purpose>Public Audio Conference: conference
                        with public access, where only audio is available, only
                        one user can talk at the same time, and the requests for
                        the AudioFloor must be accepted by a Chair.
                    </info:purpose>
                </info:entry>
                <info:entry>
                    <info:uri>xcon:VideoConference1@example.com</info:uri>
                    <info:display-text>VideoConference1</info:display-text>
                    <info:purpose>Public Video Conference: conference
                            where both audio and video are available, only
                            one user can talk.
                    </info:purpose>
                </info:entry>
                <info:entry>
                    <info:uri>xcon:AudioConference2@example.com</info:uri>
                    <info:display-text>AudioConference2</info:display-text>
                    <info:purpose>Basic Audio Conference: conference
                            with private access, where only audio is available,
                            only one user can talk at the same time, and the
                            requests for the AudioFloor must be accepted by a
                            Chair.
                    </info:purpose>
                </info:entry>
            </blueprintsInfo>
        </ccmp:blueprintsResponse>
    </ccmpResponse>
</ccmp:ccmpResponse>
```


FIGURE 5.19 Getting Blueprints from the Server. (Copyright: IETF.)

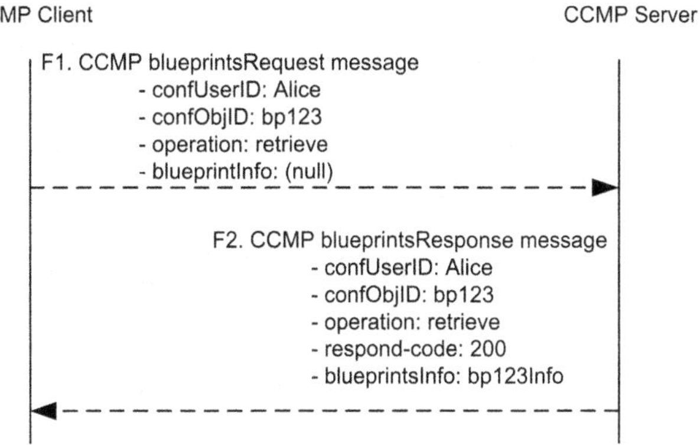


```
F1. blueprintRequest message:

<?xml version="1.0" encoding="UTF-8" standalone="yes"?>

<ccmp:ccmpRequest
        xmlns:info="urn:ietf:params:xml:ns:conference-info"
        xmlns:ccmp="urn:ietf:params:xml:ns:xcon-ccmp"
        xmlns:xcon="urn:ietf:params:xml:ns:xcon-conference-info">
    <ccmpRequest xmlns:xsi="http://www.w3.org/2001/XMLSchema-instance"
            xsi:type="ccmp:ccmp-blueprint-request-message-type">
        <confUserID>xcon-userid:alice@example.com</confUserID>
        <confObjID>xcon:AudioRoom@example.com</confObjID>
        <operation>retrieve</operation>
        <ccmp:blueprintRequest/>
    </ccmpRequest>
</ccmp:ccmpRequest>

F2. blueprintResponse message from the server:

<?xml version="1.0" encoding="UTF-8" standalone="yes"?>

<ccmp:ccmpResponse
          xmlns:xcon="urn:ietf:params:xml:ns:xcon-conference-info"
          xmlns:info="urn:ietf:params:xml:ns:conference-info"
          xmlns:ccmp="urn:ietf:params:xml:ns:xcon-ccmp">
    <ccmpResponse xmlns:xsi="http://www.w3.org/2001/
          XMLSchema-instancexsi:
          type="ccmp:ccmp-blueprint-response-message-type">
        <confUserID>xcon-userid:alice@example.com</confUserID>
        <confObjID>xcon:AudioRoom@example.com</confObjID>
        <operation>retrieve</operation>
        <response-code>200</response-code>
        <response-string>Success</response-string>
        <ccmp:blueprintResponse>
            <blueprintInfo entity="xcon:AudioRoom@example.com">
                <info:conference-description>
                    <info:display-text>AudioRoom</info:display-text>
                    <info:available-media>
                        <info:entry label="audioLabel">
                            <info:display-text>audio stream</info:display-text>
                            <info:type>audio</info:type>
                        </info:entry>
                    </info:available-media>
                </info:conference-description>
                <info:users>
                    <xcon:join-handling>allow</xcon:join-handling>
                </info:users>
                <xcon:floor-information>
                    <xcon:floor-request-handling>confirm</xcon:floor-request-
                                                                 handling>
                    <xcon:conference-floor-policy>
                        <xcon:floor id="audioFloor">
                            <xcon:media-label>audioLabel</xcon:media-label>
                        </xcon:floor>
                    </xcon:conference-floor-policy>
                </xcon:floor-information>
            </blueprintInfo>
        </ccmp:blueprintResponse>
    </ccmpResponse>
</ccmp:ccmpResponse>
```

FIGURE 5.20 Getting Information about a Specific Blueprint. (Copyright: IETF.)

them (the one called with XCON-URI "xcon:AudioRoom@example.com"). The picture shows such a transaction. Notice that the response contains, in the <blueprintInfo> parameter, a specification compliant with the standard XCON data model.

Alice retrieves detailed information about a specified blueprint:

5.6.3 Alice Creates a New Conference through a Cloning Operation

This section illustrates the third transaction in the overall flow. Alice decides to create a new conference by cloning the blueprint having XCON-URI "xcon:AudioRoom@example.com," for which she just retrieved detailed information through the blueprintRequest message.

This is achieved by sending a confRequest/create message having the blueprint's URI in the <confObjID> parameter. The picture shows such a transaction. Notice that the response contains, in the <confInfo> parameter, the specification associated with the newly created conference, which is compliant with the standard XCON data model. The <confObjID> parameter in the response is set to the XCON-URI of the new conference (in this case, "xcon:8977794@example.com"). We also notice that this value is equal to the value of the "entity" attribute of the <conference-info> element of the specification representing the newly created conference object.

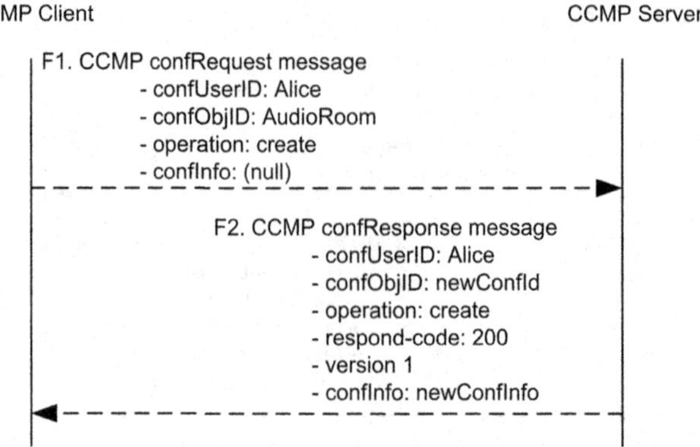


```
F1. confRequest message:

<?xml version="1.0" encoding="UTF-8" standalone="yes"?>

<ccmp:ccmpRequest
        xmlns:info="urn:ietf:params:xml:ns:conference-info"
        xmlns:ccmp="urn:ietf:params:xml:ns:xcon-ccmp"
        xmlns:xcon="urn:ietf:params:xml:ns:xcon-conference-info">
    <ccmpRequest
        xmlns:xsi="http://www.w3.org/2001/XMLSchema-instance"
        xsi:type="ccmp:ccmp-conf-request-message-type">
        <confUserID>xcon-userid:alice@example.com</confUserID>
        <confObjID>xcon:AudioRoom@example.com</confObjID>
        <operation>create</operation>
        <ccmp:confRequest/>
    </ccmpRequest>
</ccmp:ccmpRequest>

F2. confResponse message from the server:

<?xml version="1.0" encoding="UTF-8" standalone="yes"?>

<ccmp:ccmpResponse
        xmlns:xcon="urn:ietf:params:xml:ns:xcon-conference-info"
        xmlns:info="urn:ietf:params:xml:ns:conference-info"
        xmlns:ccmp="urn:ietf:params:xml:ns:xcon-ccmp">
    <ccmpResponse xmlns:xsi="http://www.w3.org/
        2001/XMLSchema-instance"
        xsi:type="ccmp:ccmp-conf-response-message-type">
        <confUserID>xcon-userid:alice@example.com</confUserID>
        <confObjID>xcon:8977794@example.com</confObjID>
        <operation>create</operation>
```

```
        <response-code>200</response-code>
        <response-string>Success</response-string>
        <version>1</version>
        <ccmp:confResponse>
            <confInfo entity="xcon:8977794@example.com">
                <info:conference-description>
                    <info:display-text>
                        New conference by Alice cloned from AudioRoom
                    </info:display-text>
                    <info:available-media>
                        <info:entry label="333">
                            <info:display-text>audio stream</info:display-text>
                            <info:type>audio</info:type>
                        </info:entry>
                    </info:available-media>
                </info:conference-description>
                <info:users>
                    <xcon:join-handling>allow</xcon:join-handling>
                </info:users>
                <xcon:floor-information>
                    <xcon:floor-request-handling>confirm</
                                        xcon:floor-request-handling>
                    <xcon:conference-floor-policy>
                        <xcon:floor id="11">
                            <xcon:media-label>333</xcon:media-label>
                        </xcon:floor>
                    </xcon:conference-floor-policy>
                </xcon:floor-information>
            </confInfo>
        </ccmp:confResponse>
    </ccmpResponse>
</ccmp:ccmpResponse>
```

FIGURE 5.21 Creating a New Conference by Cloning a Blueprint. (Copyright: IETF.)

Alice creates a new conference by cloning the "xcon:AudioRoom@example.com" blueprint (Figure 5.21).

5.6.4 Alice Updates Conference Information

This section illustrates the fourth transaction in the overall flow. Alice decides to modify some of the details associated with the conference she just created. More precisely, she changes the <display-text> element under the <conference-description> element of the specification representing the conference. This is achieved through a confRequest/update message carrying the fragment of the conference specification to which the required changes have to be applied. As shown in the picture, the response contains a code of "200," which acknowledges the modifications requested by the client, while also updating the conference version number from 1 to 2, as reflected in the "version" parameter.

Alice updates information about the conference she just created (Figure 5.22).

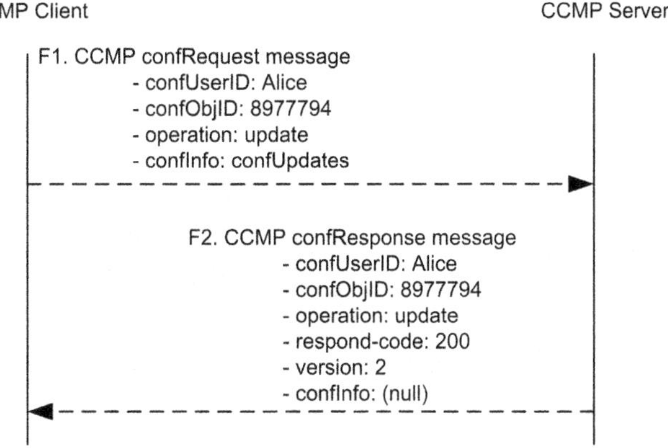


```
F1. confRequest message:

<?xml version="1.0" encoding="UTF-8" standalone="yes"?>

<ccmp:ccmpRequest
          xmlns:info="urn:ietf:params:xml:ns:conference-info"
          xmlns:ccmp="urn:ietf:params:xml:ns:xcon-ccmp"
          xmlns:xcon="urn:ietf:params:xml:ns:xcon-conference-info">
   <ccmpRequest
          xmlns:xsi="http://www.w3.org/2001/XMLSchema-instance"
          xsi:type="ccmp:ccmp-conf-request-message-type">
       <confUserID>xcon-userid:alice@example.com</confUserID>
       <confObjID>xcon:8977794@example.com</confObjID>
       <operation>update</operation>
       <ccmp:confRequest>
          <confInfo entity="xcon:8977794@example.com">
             <info:conference-description>
                <info:display-text>
                   Alice's conference
                </info:display-text>
             </info:conference-description>
          </confInfo>
       </ccmp:confRequest>
   </ccmpRequest>
</ccmp:ccmpRequest>
```

```
F2. confResponse message from the server:

<?xml version="1.0" encoding="UTF-8" standalone="yes"?>

<ccmp:ccmpResponse
          xmlns:xcon="urn:ietf:params:xml:ns:xcon-conference-info"
          xmlns:info="urn:ietf:params:xml:ns:conference-info"
          xmlns:ccmp="urn:ietf:params:xml:ns:xcon-ccmp">
   <ccmpResponse xmlns:xsi="http://www.w3.org/2001/
          XMLSchema-instance"
          xsi:type="ccmp:ccmp-conf-response-message-type">
       <confUserID>xcon-userid:alice@example.com</confUserID>
       <confObjID>xcon:8977794@example.com</confObjID>
       <operation>update</operation>
       <response-code>200</response-code>
       <response-string>Success</response-string>
       <version>2</version>
       <ccmp:confResponse/>
   </ccmpResponse>
</ccmp:ccmpResponse>
```

FIGURE 5.22 Updating Conference Information. (Copyright: IETF.)

5.6.5 Alice Inserts a List of Users into the Conference Object

This section illustrates the fifth transaction in the overall flow. Alice modifies the <allowed-users-list> under the <users> element in the specification associated with the conference she created. To achieve this, she makes use of the usersRequest message provided by CCMP.

Alice updates information about the list of users to whom access to the conference is permitted (Figure 5.23).

5.6.6 Alice Joins the Conference

This section illustrates the sixth transaction in the overall flow. Alice uses CCMP to add herself to the newly created conference. This is achieved through a userRequest/create message containing, in the <userInfo> parameter, a <user> element compliant with the XCON data model representation. Notice that such an element includes information about the user's Addresses of Record as well as her current endpoint. Figure 5.24 shows the transaction. Notice how the <confUserID> parameter is equal to the "entity" attribute of the <userInfo> element, which indicates that the request issued by the client is a first-party one.

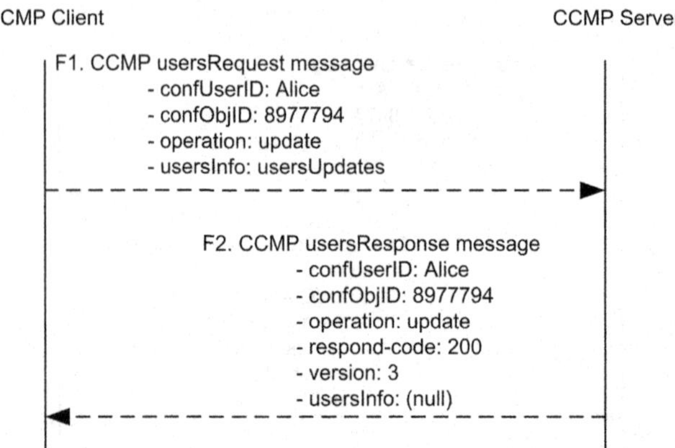


```
F1. usersRequest message:

<?xml version="1.0" encoding="UTF-8" standalone="yes"?>

<ccmp:ccmpRequest
        xmlns:xcon="urn:ietf:params:xml:ns:xcon-conference-info"
        xmlns:info="urn:ietf:params:xml:ns:conference-info"
        xmlns:ccmp="urn:ietf:params:xml:ns:xcon-ccmp">
    <ccmpRequest xmlns:xsi="http://www.w3.org/2001/XMLSchema-instance"
        xsi:type="ccmp:ccmp-users-request-message-type">
      <confUserID>xcon-userid:alice@example.com</confUserID>
      <confObjID>xcon:8977794@example.com</confObjID>
      <operation>update</operation>
      <ccmp:usersRequest>
        <usersInfo>
          <xcon:allowed-users-list>
            <xcon:target method="dial out"
                uri="xmpp:cicciolo@pippozzo.com"/>
            <xcon:target method="refer"
                uri="tel:+1-972-555-1234"/>
            <xcon:target method="refer"
                uri="sip:Carol@example.com"/>
          </xcon:allowed-users-list>
        </usersInfo>
      </ccmp:usersRequest>
    </ccmpRequest>
</ccmp:ccmpRequest>
```

```
F2. usersResponse message from the server:

<?xml version="1.0" encoding="UTF-8" standalone="yes"?>

<ccmp:ccmpResponse
        xmlns:xcon="urn:ietf:params:xml:ns:xcon-conference-info"
        xmlns:info="urn:ietf:params:xml:ns:conference-info"
        xmlns:ccmp="urn:ietf:params:xml:ns:xcon-ccmp">
    <ccmpResponse xmlns:xsi="http://www.w3.org/2001/XMLSchema-
            instance" xsi:type="ccmp:ccmp-users-response-message-type">
      <confUserID>xcon-userid:alice@example.com</confUserID>
      <confObjID>xcon:8977794@example.com</confObjID>
      <operation>retrieve</operation>
      <response-code>200</response-code>
      <response-string>Success</response-string>
      <version>3</version>
      <ccmp:usersResponse/>
    </ccmpResponse>
</ccmp:ccmpResponse>
```

FIGURE 5.23 Updating the List of Allowed Users for the Conference "xcon:8977794@example.com." (Copyright: IETF.)

Alice joins the conference by issuing a userRequest/create message with her own ID to the server (Figure 5.24).

5.6.7 Alice Adds a New User to the Conference

This section illustrates the seventh and last transaction in the overall flow. Alice uses CCMP to add a new conferencing system user, Ciccio, to the conference. This "third-party" request is realized through a userRequest/create message containing, in the <userInfo> parameter, a <user> element compliant with the XCON data model representation. Notice that such an element includes information about Ciccio's Addresses of Record, as well as his current endpoint, but has a placeholder "entity" attribute, "AUTO_GENERATE_1@example.com" as discussed in Section 5.4.3, since the XCON-USERID is initially unknown to Alice. Thus, the conference server is in charge of generating a new XCON-USERID for the user Alice indicates (i.e., Ciccio) and returning it in the "entity" attribute of the <userInfo> parameter carried in the response as well as adding the user to the conference. Figure 5.25 shows the transaction.

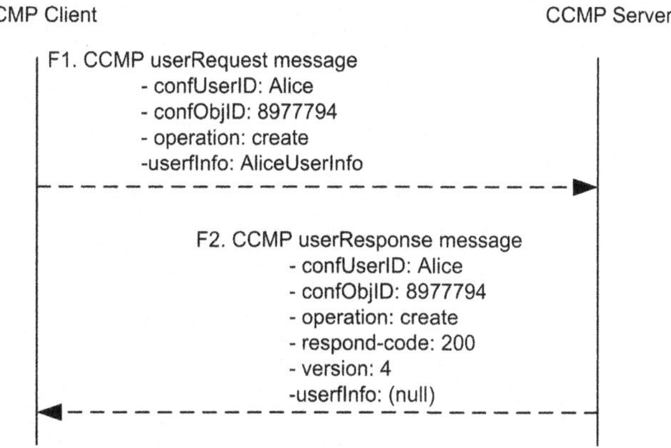


```
F1. userRequest message:

<?xml version="1.0" encoding="UTF-8" standalone="yes"?>

<ccmp:ccmpRequest
        xmlns:info="urn:ietf:params:xml:ns:conference-info"
        xmlns:ccmp="urn:ietf:params:xml:ns:xcon-ccmp"
        xmlns:xcon="urn:ietf:params:xml:ns:xcon-conference-info">
   <ccmpRequest xmlns:xsi="http://www.w3.org/2001/XMLSchema-instance"
            xsi:type="ccmp:ccmp-user-request-message-type">
      <confUserID>xcon-userid:alice@example.com</confUserID>
      <confObjID>xcon:8977794@example.com</confObjID>
      <operation>create</operation>
      <ccmp:userRequest>
         <userInfo entity="xcon-userid:alice@example.com">
            <info:associated-aors>
               <info:entry>
                  <info:uri>
                     mailto:Alice83@example.com
                  </info:uri>
                  <info:display-text>email</info:display-text>
               </info:entry>
            </info:associated-aors>
            <info:endpoint entity="sip:alice_789@example.com"/>
         </userInfo>
      </ccmp:userRequest>
   </ccmpRequest>
</ccmp:ccmpRequest>
```

```
F2. userResponse message from the server:

<?xml version="1.0" encoding="UTF-8" standalone="yes"?>

<ccmp:ccmpResponse
        xmlns:xcon="urn:ietf:params:xml:ns:xcon-conference-info"
        xmlns:info="urn:ietf:params:xml:ns:conference-info"
        xmlns:ccmp="urn:ietf:params:xml:ns:xcon-ccmp">
   <ccmpResponse xmlns:xsi="http://www.w3.org/2001/XMLSchema-
        instance" xsi:type="ccmp:ccmp-user-response-message-type">
      <confUserID>xcon-userid:alice@example.com</confUserID>
      <confObjID>xcon:8977794@example.com</confObjID>
      <operation>create</operation>
      <response-code>200</response-code>
      <response-string>Success</response-string>
      <version>4</version>
      <ccmp:userResponse/>
   </ccmpResponse>
</ccmp:ccmpResponse>
```

FIGURE 5.24 Alice Joins the Conference through CCMP. (Copyright: IETF.)

Alice adds user "Ciccio" to the conference by issuing a third-party userRequest/create message to the server (Figure 5.25).

5.6.8 Alice Asks for the CCMP Server Capabilities

This section illustrates how Alice can discover which standard CCMP messages and what extensions are supported by the CCMP server with which she interacts through an optionsRequest/optionsResponse transaction. To prepare the optionsRequest, Alice just puts her XCON-USERID in the <confUserID> parameter. Looking at the <options> element in the received optionsResponse, Alice infers the following server capabilities as regards standard CCMP messages:

- The server does not support sidebarsByValRequest or the sidebarByValRequest messages, since they do not appear in the <standard-message-list>.
- The only implemented operation for the blueprintRequest message is "retrieve," since no other <operation> entries are included in the related <operations> field.

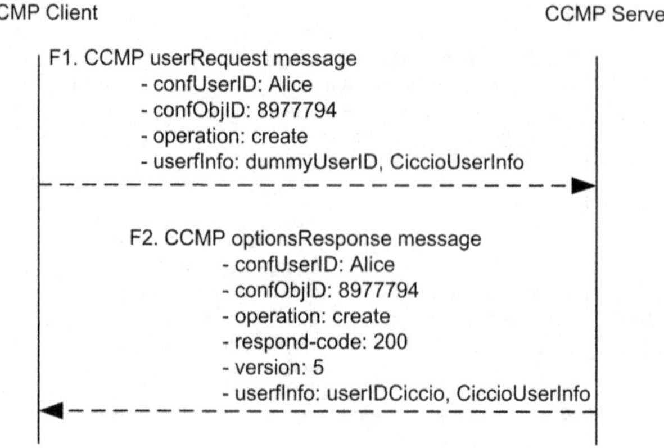


```
F1. "third-party" userRequest message from Alice:

<?xml version="1.0" encoding="UTF-8" standalone="yes"?>

<ccmp:ccmpRequest
        xmlns:info="urn:ietf:params:xml:ns:conference-info"
        xmlns:ccmp="urn:ietf:params:xml:ns:xcon-ccmp"
        xmlns:xcon="urn:ietf:params:xml:ns:xcon-conference-info">
  <ccmpRequest xmlns:xsi="http://www.w3.org/2001/XMLSchema-instance"
        xsi:type="ccmp:ccmp-user-request-message-type">
    <confUserID>xcon-userid:alice@example.com</confUserID>
    <confObjID>xcon:8977794@example.com</confObjID>
    <operation>create</operation>
    <ccmp:userRequest>
      <userInfo entity="xcon-userid:
              AUTO_GENERATE_1@example.com">
        <info:associated-aors>
          <info:entry>
            <info:uri>
              mailto:Ciccio@example.com
            </info:uri>
            <info:display-text>email</info:display-text>
          </info:entry>
        </info:associated-aors>
        <info:endpoint entity="sip:Ciccio@example.com"/>
      </userInfo>
    </ccmp:userRequest>
  </ccmpRequest>
</ccmp:ccmpRequest>
```

```
F2. "third-party" userResponse message from the server:

<?xml version="1.0" encoding="UTF-8" standalone="yes"?>

<ccmp:ccmpResponse
        xmlns:info="urn:ietf:params:xml:ns:conference-info"
        xmlns:ccmp="urn:ietf:params:xml:ns:xcon-ccmp"
        xmlns:xcon="urn:ietf:params:xml:ns:xcon-conference-info">
  <ccmpResponse xmlns:xsi="http://www.w3.org/2001/XMLSchema-
        instance" xsi:type="ccmp:ccmp-user-response-message-type">
    <confUserID>xcon-userid:alice@example.com</confUserID>
    <confObjID>xcon:8977794@example.com</confObjID>
    <operation>create</operation>
    <response-code>200</response-code>
    <version>5</version>
    <ccmp:userResponse>
      <userInfo entity="xcon-userid:Ciccio@example.com">
        <info:associated-aors>
          <info:entry>
            <info:uri>
              mailto:Ciccio@example.com
            </info:uri>
            <info:display-text>email</info:display-text>
          </info:entry>
        </info:associated-aors>
        <info:endpoint entity="sip:Ciccio@example.com"/>
      </userInfo>
    </ccmp:userResponse>
  </ccmpResponse>
</ccmp:ccmpResponse>
```

FIGURE 5.25 Alice Adds a New User to the Conference through CCMP. (Copyright: IETF.)

By analyzing the <extended-message-list>, Alice discovers that the server implements a bluePrint extension, referred to as "confSummaryRequest" in this example. This extension allows Alice to recover via CCMP a brief description of a specific conference; the XML elements involved in this extended conference control transaction are available at the URL indicated in the <schema-ref> element, and the only operation provided by this extension is "retrieve." To better understand how Alice can exploit the "confSummaryRequest" extension via CCMP, see Section 5.6.9.

Figure 5.26 shows the optionsRequest/optionsResponse message exchange between Alice and the CCMP server.

5.6.9 Alice Makes Use of a CCMP Server Extension

In this section, a very simple example of CCMP extension support is provided. Alice can recover information about this and other server-supported extensions by issuing an optionsRequest (see Section 5.6.8). The extension in question is named "confSummaryRequest" and allows a CCMP client to obtain from the CCMP server synthetic information about a specific conference. The conference summary is in the form of an XML element (Figure 5.27).

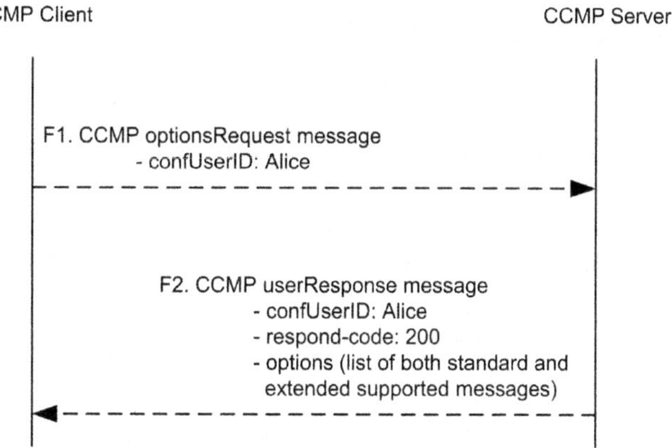


```
F1. optionsRequest (Alice asks for CCMP server capabilities)

<?xml version="1.0" encoding="UTF-8" standalone="yes"?>

<ccmp:ccmpRequest
        xmlns:info="urn:ietf:params:xml:ns:conference-info"
        xmlns:ccmp="urn:ietf:params:xml:ns:xcon-ccmp"
        xmlns:xcon="urn:ietf:params:xml:ns:xcon-conference-info">
    <ccmpRequest xmlns:xsi="http://www.w3.org/2001/XMLSchema-instance"
            xsi:type="ccmp:ccmp-options-request-message-type">
        <confUserID>xcon-userid:alice@example.com</confUserID>
    </ccmpRequest>
</ccmp:ccmpRequest>

F2. optionsResponse (the server returns the list of its conference control
    capabilities)

<?xml version="1.0" encoding="UTF-8" standalone="yes"?>

<ccmp:ccmpResponse
    xmlns:info="urn:ietf:params:xml:ns:conference-info"
    xmlns:ccmp="urn:ietf:params:xml:ns:xcon-ccmp"
    xmlns:xcon="urn:ietf:params:xml:ns:xcon-conference-info">
    <ccmpResponse xmlns:xsi="http://www.w3.org/2001/XMLSchema-
            instance" xsi:type="ccmp:ccmp-options-response-message-
            type">
        <confUserID>xcon-userid:alice@example.com</confUserID>
        <response-code>200</response-code>
        <response-string>success</response-string>
        <ccmp:optionsResponse>
            <options>
                <standard-message-list>
                    <standard-message>
                    <name>blueprintsRequest</name>
                    </standard-message>
                    <standard-message>
                        <name>blueprintRequest</name>
                        <operations>
                            <operation>retrieve</operation>
                        </operations>
                    </standard-message>
```

```
                    <standard-message>
                        <name>confsRequest</name>
                    </standard-message>
                    <standard-message>
                        <name>confRequest</name>
                    </standard-message>
                    <standard-message>
                        <name>usersRequest</name>
                    </standard-message>
                    <standard-message>
                        <name>userRequest</name>
                    </standard-message>
                    <standard-message>
                        <name>sidebarsByRefRequest</name>
                    </standard-message>
                    <standard-message>
                        <name>sidebarByRefRequest</name>
                    </standard-message>
                </standard-message-list>
                <extended-message-list>
                    <extended-message>
                        <name>confSummaryRequest</name>
                        <operations>
                            <operation>retrieve</operation>
                        </operations>
                        <schema-def>
                            http://example.com/ccmp-extension-schema.xsd
                        </schema-def>
                        <description>
                            confSummaryRequest is intended to allow the
                            requestor to retrieve a brief description of the
                            conference indicated in the confObjID request
                            parameter
                        </description>
                    </extended-message>
                </extended-message-list>
            </options>
        </ccmp:optionsResponse>
    </ccmpResponse>
</ccmp:ccmpResponse>
```

FIGURE 5.26 Alice Asks for the Server Control Capabilities. (Copyright: IETF.)

As can be inferred from the schema file, the <confSummary> element contains conference information related to the following:

- Title
- Status (active or registered)
- Participation modality (if everyone is allowed to participate, the Boolean <public> element is set to "true")
- Involved media

```
<?xml version="1.0" encoding="UTF-8"?>

<xs:schema xmlns:xs="http://www.w3.org/2001/XMLSchema"
      targetNamespace="http://example.com/ccmp-extension"
      xmlns="http://example.com/ccmp-extension">

   <xs:element name="confSummary" type="conf-summary-type"/>

   <xs:complexType name="conf-summary-type">
      <xs:sequence>
         <xs:element name="title" type="xs:string"/>
         <xs:element name="status" type="xs:string"/>
         <xs:element name="public" type="xs:boolean"/>
         <xs:element name="media" type="xs:string"/>
      </xs:sequence>
   </xs:complexType>

</xs:schema>
```

FIGURE 5.27 Example of XML Schema Defining an Extension Parameter (ccmp-extension-schema.xsd). (Copyright: IETF.)

In order to retrieve a conference summary related to the conference she participates in, Alice sends to the CCMP server an extendedRequest with a "confSummaryRequest" <extensionName>, specifying the conference XCON-URI in the confObjID request parameter, as depicted in Figure 5.28).

5.7 LOCATING A CONFERENCE SERVER

If a conferencing client is not pre-configured to use a specific conference server for the requests, the client MUST first discover the conference server before it can send any requests. The result of the discovery process is the address of the server supporting conferencing. In this specification, the result is an http: or https: URI, which identifies a conference server. It is RECOMMENDED to use domain name system (DNS) to locate a conference server if the client is not pre-configured to use a specific conference server. URI-Enabled Name Authority Pointer (NAPTR) (U-NAPTR) resolution for conferencing takes a domain name as input and produces a URI that identifies the conference server. This process also requires an Application Service tag and an Application Protocol tag, which differentiate conferencing-related NAPTR records from other records for that domain.

Section 5.12.4.1 defines an Application Service tag of "XCON," which is used to identify the centralized conferencing (XCON) server for a particular domain. The Application Protocol tag "CCMP," defined in Section 5.12.4.2, is used to identify an XCON server that understands CCMP.

The NAPTR records in the following example (Figure 5.29) demonstrate the use of the Application Service and Application Protocol tags. Iterative NAPTR resolution is used to delegate responsibility for the conferencing service from "zonea.example.com." and "zoneb.example.com." to "outsource.example.com."

Details for the "XCON" Application Service tag and the "CCMP" Application Protocol tag are included in Section 5.12.4.

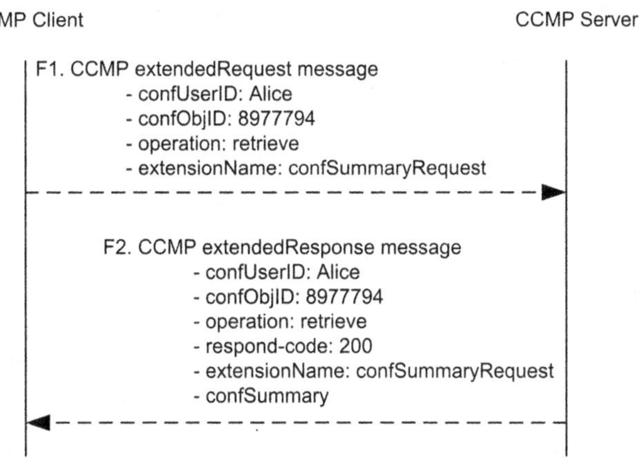

F1. extendedRequest (Alice makes use of the "confSummaryRequest")

```
<?xml version="1.0" encoding="UTF-8" standalone="yes"?>

<ccmp:ccmpRequest xmlns:info="urn:ietf:params:xml:ns:conference-info"
        xmlns:ccmp="urn:ietf:params:xml:ns:xcon-ccmp"
        xmlns:xcon="urn:ietf:params:xml:ns:xcon-conference-info"
        xmlns:example="http://example.com/ccmp-extension">

<ccmpRequest xmlns:xsi="http://www.w3.org/2001/XMLSchema-instance"
        xsi:type="ccmp:ccmp-extended-request-message-type">
    <confUserID>xcon-userid:alice@example.com</confUserID>
    <confObjID>xcon:8977794@example.com</confObjID>
    <operation>retrieve</operation>
    <ccmp:extendedRequest>
        <extensionName>confRequestSummary</extensionName>
    </ccmp:extendedRequest>
</ccmpRequest>

</ccmp:ccmpRequest>
```

F2. extendedResponse (the server provides Alice with a brief description of the desired conference)

```
<?xml version="1.0" encoding="UTF-8" standalone="yes"?>

<ccmp:ccmpResponse xmlns:info="urn:ietf:params:xml:ns:conference-info"
        xmlns:ccmp="urn:ietf:params:xml:ns:xcon-ccmp"
        xmlns:xcon="urn:ietf:params:xml:ns:xcon-conference-info"
        xmlns:example="http://example.com/ccmp-extension">

    <ccmpResponse xmlns:xsi="http://www.w3.org/2001/XMLSchema-
            instance" xsi:type="ccmp:ccmp-extended-response-message-type">
        <confUserID>xcon-userid:alice@example.com</confUserID>
        <confObjID>xcon:8977794@example.com</confObjID>
        <operation>retrieve</operation>
        <response-code>200</response-code>
        <response-string>success</response-string>
        <ccmp:extendedResponse>
            <extensionName>confSummaryRequest</extensionName>
            <example:confSummary>
                <title> Alice's conference </title>
                <status> active </status>
                <public> true </public>
                <media> audio </media>
            </example:confSummary>
        </ccmp:extendedResponse>
    </ccmpResponse>

</ccmp:ccmpResponse>
```

FIGURE 5.28 Alice Exploits the "confSummaryRequest" Extension. (Copyright: IETF.)

5.8 MANAGING NOTIFICATIONS

As per RFC 5239 (see Chapter 2), CCMP is one of the following four protocols, which have been formally identified within the XCON framework:

- **Conference Control Manipulation Protocol (CCMP):** Mediates between conference and media control client (conferencing client) and conference server. This section describes such a protocol (RFC 6503).
- **Binary Floor Control Protocol (BFCP):** Operates between the floor control client and the floor control server. An example of such a protocol is the BFCP, specified in RFC 4582 (see Chapter 6).
- **Call Signaling Protocol (CSP):** Operates between the Call Signaling Client and the focus. Examples of CSPs include SIP, H.323, and IAX. Such protocols are capable of negotiating a conferencing session.

```
zonea.example.com.
;;          order  pref   flags
IN NAPTR    100    10     "" "XCON-CCMP"  (    ; service
""                                              ; regex
outsource.example.com.                          ; replacement
)
zoneb.example.com.
;;          order  pref   flags
IN NAPTR    100    10     "" "XCON-CCMP"  (    ; service
""                                              ; regex
outsource.example.com.                          ; replacement
)
outsource.example.com.
;;          order  pref   flags
IN NAPTR    100    10     "u" "XCON-CCMP" (    ; service
"!*.!https://confs.example.com/!"               ; regex
.                                               ; replacement
)
```

FIGURE 5.29 Sample XCON-CCMP Service NAPTR Records. (Copyright: IETF.)

- **Notification Protocol:** Operates between the Notification Client and the XCON Notification Service. This specification does not define a new notification protocol. For clients that use SIP as the call signaling protocol, the XCON event package (RFC 6502) MUST be used by the client for notifications of changes in the conference data as described later.

The protocol specified in this specification is a proactive one and is used by a conferencing client to send requests to a conference server in order to retrieve information about the conference objects stored by the server and possibly to manipulate them. However, a complete conferencing solution is not prohibited from providing clients with a means for receiving asynchronous updates about the status of the objects available at the server. The notification protocol, while conceptually independent of all the mentioned companion protocols, can nonetheless be chosen in a way that is consistent with the overall protocol architecture characterizing a specific deployment, as discussed in the following.

When the conferencing control client uses SIP (RFC 3261 – also see Section 1.2) as the signaling protocol to participate in the conference, SIP event notification can be used. In such a case, the conferencing control client MUST implement the conference event package for XCON (RFC 6502 – see Chapter 7). This is the default mechanism for conferencing clients, as is SIP for signaling per the XCON framework (RFC 5239 – see Chapter 2).

In the case where the interface to the conference server is entirely web based, there is a common mechanism for web-based systems that could be used – a "call back." With this mechanism, the conferencing client provides the conference server with an HTTP URL that is invoked when a change occurs. This is a common implementation mechanism for e-commerce. This works well in the scenarios where the conferencing client is a web server that provides the graphical HTML user interface and uses CCMP as the back-end interface to the conference server. This model can coexist with the SIP event notification model. PC-based clients behind network address translations (NATs) could provide a SIP event URI, whereas web-based clients using CCMP in the back-end would probably find the HTTP call back approach much easier. The details of this approach are out of scope for CCMP; thus, we expect that a future specification will specify this solution.

5.9 HTTP TRANSPORT

This section describes the use of HTTP (RFC 2616) and HTTP over transport layer security (TLS) (RFC 2818) as transport mechanisms for CCMP, which a conforming conference server and conferencing client MUST support. Although CCMP uses HTTP as a transport, it uses a strict subset of HTTP features, and due to the restrictions of some features, a conferencing server might not be a fully compliant HTTP server. It is intended that a conference server can easily be built using an HTTP server with extensibility mechanisms and that a conferencing client can trivially use existing HTTP libraries. This subset of requirements helps implementers avoid ambiguity with the many options the full HTTP protocol offers.

Support of HTTP authentication (RFC 2617) and cookies (RFC 6265) is OPTIONAL for a conferencing client that conforms to this specification. These mechanisms are unnecessary because CCMP requests carry their own authentication information (in the "subject" field; see Section 5.5.1). A conferencing client SHOULD include support for HTTP proxy authentication.

A CCMP request is carried in the body of an HTTP POST request. The conferencing client MUST include a Host header in the request. The MIME type of CCMP request and response bodies is "application/ ccmp+xml." The conference server and conferencing client MUST provide this value in the HTTP Content-Type and Accept header fields. If the conference server does not receive the appropriate Content-Type and Accept header fields, the conference server SHOULD fail the request, returning a 406 (Not Acceptable) response. CCMP responses SHOULD include a Content-Length header.

Conferencing clients MUST NOT use the Expect header or the Range header in CCMP requests. The conference server MAY return 501 (Not Implemented) errors if either of these HTTP features is used. In the case that the conference server receives a request from the conferencing client containing an If-* (conditional) header, the conference server SHOULD return a 412 (precondition failed) response.

The POST method is the only method REQUIRED for CCMP. If a conference server chooses to support GET or HEAD, it SHOULD consider the kind of application doing the GET. Since a conferencing client only uses a POST method, the GET or HEAD MUST be a URL that was found outside its normal context (e.g., somebody found a URL in protocol traces or log files and fed it into their browser), or somebody is testing or debugging a system. The conference server could provide information in the CCMP response indicating that the URL corresponds to a conference server and only responds to CCMP POST requests, or the conference server could instead try to avoid any leak of information by returning a very generic HTTP error message such as 405 (Method Not Allowed).

The conference server populates the HTTP headers of responses so that they are consistent with the contents of the message. In particular, the CacheControl header SHOULD be set to disable caching of any conference information by HTTP intermediaries. Otherwise, there is the risk of stale information and/or the unauthorized disclosure of the information. The HTTP status code MUST indicate a 2xx series response for all CCMP Response and Error messages.

The conference server MAY redirect a CCMP request. A conference server MUST NOT include CCMP responses in a 3xx response. A conferencing client MUST handle redirects by using the Location header provided by the server in a 3xx response. When redirecting, the conferencing client MUST observe the delay indicated by the Retry-After header. The conferencing client MUST authenticate the server that returns the redirect response before following the redirect. A conferencing client SHOULD authenticate the conference server indicated in a redirect.

The conference server SHOULD support persistent connections and request pipelining. If pipelining is not supported, the conference server MUST NOT allow persistent connections. The conference server MUST support termination of a response by the closing of a connection. Implementations of CCMP that implement HTTP transport MUST implement transport over TLS (RFC 2818). TLS provides message integrity and confidentiality between the conferencing client and the conference server. The conferencing client MUST implement the server authentication method described in HTTPS (RFC 2818). The device uses the URI obtained during conference server discovery to authenticate the server. The details of this authentication method are provided in Section 3.1 of HTTPS RFC 2818. When TLS is used, the conferencing client SHOULD fail a request if server authentication fails.

5.10 SECURITY CONSIDERATIONS

As identified in the XCON framework (RFC 5239 – see Chapter 2), there are a wide variety of potential attacks related to conferencing due to the natural involvement of multiple endpoints and the capability to manipulate the data on the conference server using CCMP. Examples of attacks include the following: an endpoint attempting to listen to conferences in which it is not authorized to participate, an endpoint attempting to disconnect or mute other users, and an endpoint theft of service in attempting to create conferences it is not allowed to create.

The following summarizes the security considerations for CCMP:

1. The client MUST determine the proper conference server. The conference server discovery is described in Section 5.7.
2. The client MUST connect to the proper conference server. The mechanisms for addressing this security consideration are described in Section 5.10.1.
3. The protocol MUST support a confidentiality and integrity mechanism. As described in Section 5.9, implementations of CCMP MUST implement the HTTP transport over TLS (RFC 2818).
4. There are security issues associated with the authorization to perform actions on the conferencing system to invoke specific capabilities. A conference server SHOULD ensure that only authorized entities can manipulate the conference data. The mechanisms for addressing this security consideration are described in Section 5.10.2.
5. The privacy and security of the identity of a user in the conference MUST be assured. The mechanisms to ensure the security and privacy of identity are discussed in Section 5.10.3.
6. A final issue is related to denial-of-service (DoS) attacks on the conference server itself. The recommendations to minimize the potential and impact of DoS attacks are discussed in Section 5.10.4.

Of the considerations listed here, items 1 and 3 are addressed within the referenced sections earlier in this specification. The remaining security considerations are addressed in detail in the following sections.

5.10.1 Ensuring That the Proper Conference Server Has Been Contacted

Section 5.7 describes a mechanism using DNS by which a conferencing client discovers a conference server. A primary concern is spoofed DNS replies; thus, the use of DNS Security (DNSSEC) is RECOMMENDED to ensure that the client receives a valid response from the DNS server in cases where this is a concern. When the CCMP transaction is conducted using TLS (RFC 5246), the conference server can authenticate its identity, either as a domain name or as an IP address, to the conferencing client by presenting a certificate containing that identifier as a subjectAltName (i.e., as an iPAddress or dNSName, respectively). Any implementation of CCMP MUST be capable of being transacted over TLS so that the client can request the above authentication. Note that in order for the presented certificate to be valid at the client, the client MUST be able to validate the certificate following the procedures in RFC 2818 in the case of HTTP as a transport. In particular, the validation path of the certificate must end in one of the client's trust anchors, even if that trust anchor is the conference server certificate itself. If the client has external information as to the expected identity or credentials of the proper conference server, the authentication checks described above MAY be omitted.

5.10.2 User Authentication and Authorization

Many policy authorization decisions are based on the identity of the user or the role that a user may have. The conference server MUST implement mechanisms for authentication of users to validate their identity. There are several ways that a user might authenticate its identity to the system. For users joining a conference using one of the call signaling protocols, the user authentication mechanisms for the specific protocol can be used. For example, in the case of a user joining the conference using SIP signaling, the user authentication as defined in RFC 3261 (also see Section 1.2) MUST be used. For the case of users joining the conference using CCMP, the CCMP Request messages provide a subject field that contains a username and password, which can be used for authentication. Since the CCMP messages are RECOMMENDED to be carried over TLS, this information can be sent securely.

The XCON framework (RFC 5239 – see Chapter 2) provides an overview of other authorization mechanisms. In the cases where a user is authorized via multiple mechanisms, it is RECOMMENDED that the conference server associate the authorization of the CCMP interface with other authorization mechanisms; for example, public switched telephone network (PSTN) users that join with a PIN and control the conference using CCMP. When a conference server presents the identity of authorized users, it MAY provide information about the way the identity was proven or verified by the system. A conference server can also allow a completely unauthenticated user into the system – this information SHOULD also be communicated to interested parties.

Once a user is authenticated and authorized through the various mechanisms available on the conference server, the conference server MUST allocate a conference user identifier (XCON-USERID) and SHOULD associate the XCON-USERID with any signaling-specific user identifiers that were used for authentication and authorization. This XCON-USERID can be provided to a specific user through the conference notification interface and MUST be provided to users that interact with the conferencing system using CCMP (i.e., in the appropriate CCMP response messages). The XCON-USERIDs for each user/participant in the conference are contained in the "entity" attribute in the <user> element in the conference object. The XCON-USERID is REQUIRED for any subsequent operations by the user on the conference object and is carried in the confUserID parameter in the CCMP requests and responses.

Note that the policy management of an XCON-compliant conferencing system is out of the scope of this specification as well as of the XCON working group (WG). However, the specification of a policy management framework is realizable with the overall XCON architecture, in particular with regard to a role-based access control (RBAC) approach. In RBAC, the following elements are identified: (i) Users; (ii) Roles; (iii) Objects; (iv) Operations; (v) Permissions. For all these elements, a direct mapping exists onto the main XCON entities. As an example, RBAC objects map onto XCON data model objects, and RBAC operations map onto CCMP operations.

Future specifications can define an RBAC framework for XCON by first focusing on the definition of roles and then specifying the needed permission policy sets and role policy sets (used to associate policy permission sets with specific roles). With these policies in place, access to a conference object compliant with the XCON data model can be appropriately controlled. As far as assigning users to roles, the Users in the RBAC model relate directly to the <users> element in the conference object. The <users> element is comprised of <user> elements representing a specific user in the conferencing system. Each <user> element contains an "entity" attribute with the XCON-USERID and a <role> element. Thus, each authorized user (as represented by an XCON-USERID) can be associated with a <role> element.

5.10.3 Security and Privacy of Identity

An overview of the required privacy and anonymity for users of a conferencing system is provided in the XCON framework (RFC 5239 – see Chapter 2). The security of the identity in the form of the XCON-USERID is provided in CCMP through the use of TLS. The conference server SHOULD support the mechanism to ensure the privacy of the XCON-USERID. The conferencing client indicates the desired level of privacy by manipulation of the <provide-anonymity> element defined in the XCON data model (RFC 6501 – see Chapter 4). The <provide-anonymity> element controls the degree to which a user reveals their identity.

The following summarizes the values for the <provide-anonymity> element that the client includes in their requests:

- "hidden:" Ensures that other participants are not aware that there is an additional participant (i.e., the user issuing the request) in the conference. This could be used in cases of users that are authorized with a special role in a conference (e.g., a supervisor in a call center environment).
- "anonymous:" Ensures that other participants are aware that there is another participant (i.e., the user issuing the request); however, the other participants are not provided with information as to the identity of the user.
- "semi-private:" Ensures that the user's identity is only to be revealed to other participants or users that have a higher-level authorization (e.g., a conferencing system can be configured such that a human administrator can see all users).

If the client desires privacy, the conferencing client SHOULD include the <provide-anonymity> element in the <confInfo> parameter in a CCMP confRequest message with an <operation> parameter of "update" or "create" or in the <userInfo> parameter in a CCMP userRequest message with an <operation> parameter of "update" or "create." If the <provide-anonymity> element is not included in the conference object, then other users can see the participant's identity. Participants are made aware of other participants that are "anonymous" or "semi-private" when they perform subsequent operations on the conference object or retrieve the conference object or when they receive subsequent notifications.

Note that independently of the level of anonymity requested by the user, the identity of the user is always known by the conferencing system, as that is required to perform the necessary authorization as described in Section 5.10.2. The degree to which human administrators can see the information can be controlled using policies (e.g., some information in the data model can be hidden from human administrators).

5.10.4 Mitigating DoS Attacks

RFC 4732 provides an overview of possible DoS attacks. In order to minimize the potential for DoS attacks, it is RECOMMENDED that conferencing systems require user authentication and authorization for any client participating in a conference. This can be accomplished through the use of the mechanisms described in Section 5.10.2 as well as by using the security mechanisms associated with the specific signaling (e.g., session initiation protocol secure [SIPS]) and media protocols (e.g., secure realtime transport protocol [SRTP]). In addition, Section 5.4.4 describes the use of a timer mechanism to alleviate the situation whereby CCMP messages pend indefinitely, thus increasing the potential for pending requests to continue to increase when a server is receiving more requests than it can process.

5.11 XML Schema

This section gives the XML schema definition [1–2] of the "application/ccmp+xml" format. This is presented as a formal definition of the "application/ccmp+xml" format. A new XML namespace, a new XML schema (Figure 5.30), and the MIME type for this schema are registered with IANA (for detail see RFC 6503 – not included here). Note that this XML schema definition is not intended to be used with on-the-fly validation of the presence XML specification. Whitespaces are included in the schema to conform to the line length restrictions of the RFC format without having a negative

```
<?xml version="1.0" encoding="utf-8"?>

  <xs:schema
    targetNamespace="urn:ietf:params:xml:ns:xcon-ccmp"
    xmlns:xsi="http://www.w3.org/2001/XMLSchema-instance"
    xmlns="urn:ietf:params:xml:ns:xcon-ccmp"
    xmlns:tns="urn:ietf:params:xml:ns:xcon-ccmp"
    xmlns:dm="urn:ietf:params:xml:ns:xcon-conference-info"
    xmlns:info="urn:ietf:params:xml:ns:conference-info"
    xmlns:xs="http://www.w3.org/2001/XMLSchema">

  <xs:import namespace="urn:ietf:params:xml:ns:xcon-conference-info"
      schemaLocation="DataModel.xsd"/>
  <xs:import namespace="urn:ietf:params:xml:ns:conference-info"
      schemaLocation="rfc4575.xsd"/>

  <xs:element name="ccmpRequest" type="ccmp-request-type" />
  <xs:element name="ccmpResponse" type="ccmp-response-type" />

  <!-- CCMP request definition -->

  <xs:complexType name="ccmp-request-type">
    <xs:sequence>
      <xs:element name="ccmpRequest"
          type="ccmp-request-message-type" />
    </xs:sequence>
  </xs:complexType>

<!-- ccmp-request-message-type -->

  <xs:complexType abstract="true"
      name="ccmp-request-message-type">
    <xs:sequence>
      <xs:element name="subject" type="subject-type"
          minOccurs="0" maxOccurs="1" />
      <xs:element name="confUserID" type="xs:string"
          minOccurs="0" maxOccurs="1" />
      <xs:element name="confObjID" type="xs:string"
          minOccurs="0" maxOccurs="1" />
      <xs:element name="operation" type="operationType"
          minOccurs="0" maxOccurs="1" />
      <xs:element name="conference-password" type="xs:string"
          minOccurs="0" maxOccurs="1" />
      <xs:any namespace="##other" processContents="lax"
          minOccurs="0" maxOccurs="unbounded"/>
    </xs:sequence>
    <xs:anyAttribute namespace="##any" processContents="lax"/>
  </xs:complexType>

<!-- CCMP response definition -->

  <xs:complexType name="ccmp-response-type">
    <xs:sequence>
      <xs:element name="ccmpResponse"
          type="ccmp-response-message-type" />
    </xs:sequence>
  </xs:complexType>

<!-- ccmp-response-message-type -->

  <xs:complexType abstract="true" name="ccmp-response-message-type">
    <xs:sequence>
      <xs:element name="confUserID" type="xs:string"
          minOccurs="1" maxOccurs="1" />
      <xs:element name="confObjID" type="xs:string"
          minOccurs="0" maxOccurs="1" />
      <xs:element name="operation" type="operationType"
          minOccurs="0" maxOccurs="1" />
```

(a)

```
      <xs:element name="response-code"
          type="response-codeType"
          minOccurs="1" maxOccurs="1" />
      <xs:element name="response-string" type="xs:string"
      minOccurs="0" maxOccurs="1" />
      <xs:element name="version" type="xs:positiveInteger"
          minOccurs="0" maxOccurs="1" />
      <xs:any namespace="##other" processContents="lax"
          minOccurs="0" maxOccurs="unbounded"/>
    </xs:sequence>
    <xs:anyAttribute namespace="##any" processContents="lax"/>
  </xs:complexType>

<!-- CCMP REQUESTS -->

  <!-- blueprintsRequest -->

  <xs:complexType name="ccmp-blueprints-request-message-type">
    <xs:complexContent>
      <xs:extension base="tns:ccmp-request-message-type">
        <xs:sequence>
          <xs:element ref="blueprintsRequest" />
        </xs:sequence>
      </xs:extension>
    </xs:complexContent>
  </xs:complexType>

  <!-- blueprintsRequestType -->

  <xs:element name="blueprintsRequest" type="blueprintsRequestType"/>
  <xs:complexType name="blueprintsRequestType">
    <xs:sequence>
      <xs:element name="xpathFilter" type="xs:string"
          minOccurs="0"/>
      <xs:any namespace="##other" processContents="lax"
          minOccurs="0" maxOccurs="unbounded"/>
    </xs:sequence>
    <xs:anyAttribute namespace="##any" processContents="lax"/>
  </xs:complexType>

  <!-- blueprintRequest -->

  <xs:complexType name="ccmp-blueprint-request-message-type">
    <xs:complexContent>
      <xs:extension base="tns:ccmp-request-message-type">
        <xs:sequence>
          <xs:element ref="blueprintRequest" />
        </xs:sequence>
      </xs:extension>
    </xs:complexContent>
  </xs:complexType>

  <!-- blueprintRequestType -->

  <xs:element name="blueprintRequest" type="blueprintRequestType" />
  <xs:complexType name="blueprintRequestType">
    <xs:sequence>
      <xs:element name="blueprintInfo"
          type="info:conference-type" minOccurs="0"/>
      <xs:any namespace="##other" processContents="lax"
          minOccurs="0" maxOccurs="unbounded"/>
    </xs:sequence>
    <xs:anyAttribute namespace="##any" processContents="lax"/>
  </xs:complexType>
```

(b)

FIGURE 5.30 CCMP XML Schema. (Copyright: IETF.)

impact on the readability of the specification. Any conforming processor should remove leading and trailing white spaces.

5.12 IANA CONSIDERATIONS

This specification registers a new XML namespace, a new XML schema, and the MIME type for the schema. This specification also registers the "XCON" Application Service tag and the "CCMP" Application Protocol tag and defines registries for the CCMP operation types and response codes. The detail can be seen in RFC 6503.

```
<!-- confsRequest -->

<xs:complexType name="ccmp-confs-request-message-type">
    <xs:complexContent>
        <xs:extension base="tns:ccmp-request-message-type">
            <xs:sequence>
                <xs:element ref="confsRequest" />
            </xs:sequence>
        </xs:extension>
    </xs:complexContent>
</xs:complexType>

<!-- confsRequestType -->

<xs:element name="confsRequest" type="confsRequestType" />
<xs:complexType name="confsRequestType">
    <xs:sequence>
        <xs:element name="xpathFilter" type="xs:string"
                minOccurs="0"/>
        <xs:any namespace="##other" processContents="lax"
                minOccurs="0" maxOccurs="unbounded"/>
    </xs:sequence>
    <xs:anyAttribute namespace="##any" processContents="lax"/>
</xs:complexType>

<!-- confRequest -->

<xs:complexType name="ccmp-conf-request-message-type">
    <xs:complexContent>
        <xs:extension base="tns:ccmp-request-message-type">
            <xs:sequence>
                <xs:element ref="confRequest" />
            </xs:sequence>
        </xs:extension>
    </xs:complexContent>
</xs:complexType>

<!-- confRequestType -->

<xs:element name="confRequest" type="confRequestType" />
<xs:complexType name="confRequestType">
    <xs:sequence>
        <xs:element name="confInfo" type="info:conference-type"
                minOccurs="0"/>
        <xs:any namespace="##other" processContents="lax"
                minOccurs="0" maxOccurs="unbounded"/>
    </xs:sequence>
    <xs:anyAttribute namespace="##any" processContents="lax"/>
</xs:complexType>

<!-- usersRequest -->

<xs:complexType name="ccmp-users-request-message-type">
    <xs:complexContent>
        <xs:extension base="tns:ccmp-request-message-type">
            <xs:sequence>
                <xs:element ref="usersRequest" />
            </xs:sequence>
        </xs:extension>
    </xs:complexContent>
</xs:complexType>

<!-- usersRequestType -->

<xs:element name="usersRequest" type="usersRequestType" />
<xs:complexType name="usersRequestType">
    <xs:sequence>
        <xs:element name="usersInfo" type="info:users-type"
                minOccurs="0" />
```

(c)

```
                <xs:any namespace="##other" processContents="lax"
                        minOccurs="0" maxOccurs="unbounded"/>
    </xs:sequence>
    <xs:anyAttribute namespace="##any" processContents="lax"/>
</xs:complexType>

<!-- userRequest -->

<xs:complexType name="ccmp-user-request-message-type">
    <xs:complexContent>
        <xs:extension base="tns:ccmp-request-message-type">
            <xs:sequence>
                <xs:element ref="userRequest" />
            </xs:sequence>
        </xs:extension>
    </xs:complexContent>
</xs:complexType>

<!-- userRequestType -->

<xs:element name="userRequest" type="userRequestType" />
<xs:complexType name="userRequestType">
    <xs:sequence>
        <xs:element name="userInfo" type="info:user-type"
                minOccurs="0" />
        <xs:any namespace="##other" processContents="lax"
                minOccurs="0" maxOccurs="unbounded"/>
    </xs:sequence>
    <xs:anyAttribute namespace="##any" processContents="lax"/>
</xs:complexType>

<!-- sidebarsByValRequest -->

<xs:complexType name="ccmp-sidebarsByVal-request-message-type">
    <xs:complexContent>
        <xs:extension base="tns:ccmp-request-message-type">
            <xs:sequence>
                <xs:element ref="sidebarsByValRequest" />
            </xs:sequence>
        </xs:extension>
    </xs:complexContent>
</xs:complexType>

<!-- sidebarsByValRequestType -->

<xs:element name="sidebarsByValRequest"
    type="sidebarsByValRequestType" />
<xs:complexType name="sidebarsByValRequestType">
    <xs:sequence>
        <xs:element name="xpathFilter"
            type="xs:string" minOccurs="0"/>
        <xs:any namespace="##other" processContents="lax"
                minOccurs="0" maxOccurs="unbounded"/>
    </xs:sequence>
    <xs:anyAttribute namespace="##any" processContents="lax"/>
</xs:complexType>

<!-- sidebarsByRefRequest -->

<xs:complexType name="ccmp-sidebarsByRef-request-message-type">
    <xs:complexContent>
        <xs:extension base="tns:ccmp-request-message-type">
            <xs:sequence>
                <xs:element ref="sidebarsByRefRequest" />
            </xs:sequence>
        </xs:extension>
    </xs:complexContent>
</xs:complexType>
```

(d)

FIGURE 5.30 Continued.

5.12.1 URN Sub-Namespace Registration

This section registers a new XML namespace, "urn:ietf:params:xml:ns:xcon-ccmp".
URI: urn:ietf:params:xml:ns:xcon-ccmp

- Registrant Contact: IETF XCON working group (xcon@ietf.org), Mary Barnes (mary.ietf.barnes@gmail.com).

For more detailed information on URN Sub-Namespace Registration, see RFC 6503.

```
<!-- sidebarsByRefRequestType -->

<xs:element name="sidebarsByRefRequest"
        type="sidebarsByRefRequestType" />
<xs:complexType name="sidebarsByRefRequestType">
    <xs:sequence>
        <xs:element name="xpathFilter" type="xs:string"
                minOccurs="0"/>
        <xs:any namespace="##other" processContents="lax"
                minOccurs="0" maxOccurs="unbounded"/>
    </xs:sequence>
    <xs:anyAttribute namespace="##any" processContents="lax"/>
</xs:complexType>

<!-- sidebarByValRequest -->

<xs:complexType name="ccmp-sidebarByVal-request-message-type">
    <xs:complexContent>
        <xs:extension base="tns:ccmp-request-message-type">
            <xs:sequence>
                <xs:element ref="sidebarByValRequest" />
            </xs:sequence>
        </xs:extension>
    </xs:complexContent>
</xs:complexType>

<!-- sidebarByValRequestType -->

<xs:element name="sidebarByValRequest"
        type="sidebarByValRequestType"/>
<xs:complexType name="sidebarByValRequestType">
    <xs:sequence>
        <xs:element name="sidebarByValInfo"
                type="info:conference-type" minOccurs="0"/>
        <xs:any namespace="##other" processContents="lax"
                minOccurs="0" maxOccurs="unbounded"/>
    </xs:sequence>
    <xs:anyAttribute namespace="##any" processContents="lax"/>
</xs:complexType>

<!-- sidebarByRefRequest -->

<xs:complexType name="ccmp-sidebarByRef-request-message-type">
    <xs:complexContent>
        <xs:extension base="tns:ccmp-request-message-type">
            <xs:sequence>
                <xs:element ref="sidebarByRefRequest" />
            </xs:sequence>
    </xs:extension>
    </xs:complexContent>
</xs:complexType>

<!-- sidebarByRefRequestType -->

<xs:element name="sidebarByRefRequest"
        type="sidebarByRefRequestType" />
<xs:complexType name="ccmp-extended-request-message-type">
    <xs:complexContent>
        <xs:extension base="tns:ccmp-request-message-type">
            <xs:sequence>
                <xs:element ref="extendedRequest"/>
            </xs:sequence>
        </xs:extension>
    </xs:complexContent>
</xs:complexType>

<!-- extendedRequest -->

<xs:complexType name="ccmp-extended-request-message-type">
    <xs:complexContent>
```

(e)

```
        <xs:extension base="tns:ccmp-request-message-type">
            <xs:sequence>
                <xs:element ref="extendedRequest"/>
            </xs:sequence>
        </xs:extension>
    </xs:complexContent>
</xs:complexType>

<!-- extendedRequestType -->

<xs:element name="extendedRequest" type="extendedRequestType"/>
<xs:complexType name="extendedRequestType">
    <xs:sequence>
        <xs:element name="extensionName"
                type="xs:string" minOccurs="1"/>
        <xs:any namespace="##other" processContents="lax"
                minOccurs="0"
                maxOccurs="unbounded" />
    </xs:sequence>
</xs:complexType>

<!-- optionsRequest -->

<xs:complexType name="ccmp-options-request-message-type">
    <xs:complexContent>
        <xs:extension base="tns:ccmp-request-message-type">
                            </xs:extension>
    </xs:complexContent>
</xs:complexType>

<!-- CCMP RESPONSES -->

<!-- blueprintsResponse -->

<xs:complexType name="ccmp-blueprints-response-message-type">
    <xs:complexContent>
        <xs:extension base="tns:ccmp-response-message-type">
            <xs:sequence>
                <xs:element ref="blueprintsResponse" />
            </xs:sequence>
        </xs:extension>
    </xs:complexContent>
</xs:complexType>

<!-- blueprintsResponseType -->

<xs:element name="blueprintsResponse" type="blueprintsResponseType"/>
<xs:complexType name="blueprintsResponseType">
    <xs:sequence>
        <xs:element name="blueprintsInfo" type="info:uris-type"
                minOccurs="0"/>
        <xs:any namespace="##other" processContents="lax"
                minOccurs="0" maxOccurs="unbounded"/>
    </xs:sequence>
    <xs:anyAttribute namespace="##any" processContents="lax"/>
</xs:complexType>

<!-- blueprintResponse -->

<xs:complexType name="ccmp-blueprint-response-message-type">
    <xs:complexContent>
        <xs:extension base="tns:ccmp-response-message-type">
            <xs:sequence>
                <xs:element ref="blueprintResponse" />
            </xs:sequence>
        </xs:extension>
    </xs:complexContent>
</xs:complexType>
```

(f)

FIGURE 5.30 Continued.

5.12.2 XML Schema Registration

This section registers an XML schema per the guidelines in RFC 3688.

- URI: urn:ietf:params:xml:schema:xcon-ccmp

For more detailed information on XML Schema Registration, see RFC 6503.

```
<!-- blueprintResponseType -->

<xs:element name="blueprintResponse" type="blueprintResponseType"/>
<xs:complexType name="blueprintResponseType">
    <xs:sequence>
        <xs:element name="blueprintInfo" type="info:conference-type"
                minOccurs="0"/>
        <xs:any namespace="##other" processContents="lax"
                minOccurs="0" maxOccurs="unbounded"/>
    </xs:sequence>
    <xs:anyAttribute namespace="##any" processContents="lax"/>
</xs:complexType>

<!-- confsResponse -->

<xs:complexType name="ccmp-confs-response-message-type">
    <xs:complexContent>
        <xs:extension base="tns:ccmp-response-message-type">
            <xs:sequence>
                <xs:element ref="confsResponse" />
            </xs:sequence>
        </xs:extension>
    </xs:complexContent>
</xs:complexType>

<!-- confsResponseType -->

<xs:element name="confsResponse" type="confsResponseType" />
<xs:complexType name="confsResponseType">
    <xs:sequence>
        <xs:element name="confsInfo" type="info:uris-type"
                minOccurs="0"/>
        <xs:any namespace="##other" processContents="lax"
                minOccurs="0" maxOccurs="unbounded"/>
    </xs:sequence>
    <xs:anyAttribute namespace="##any" processContents="lax"/>
</xs:complexType>

<!-- confResponse -->

<xs:complexType name="ccmp-conf-response-message-type">
    <xs:complexContent>
        <xs:extension base="tns:ccmp-response-message-type">
            <xs:sequence>
                <xs:element ref="confResponse"/>
            </xs:sequence>
        </xs:extension>
    </xs:complexContent>
</xs:complexType>

<!-- confResponseType -->

<xs:element name="confResponse" type="confResponseType" />
<xs:complexType name="confResponseType">
    <xs:sequence>
        <xs:element name="confInfo" type="info:conference-type"
                minOccurs="0"/>
        <xs:any namespace="##other" processContents="lax"
                minOccurs="0" maxOccurs="unbounded"/>
    </xs:sequence>
    <xs:anyAttribute namespace="##any" processContents="lax"/>
</xs:complexType>

<!-- usersResponse -->

<xs:complexType name="ccmp-users-response-message-type">
    <xs:complexContent>
        <xs:extension base="tns:ccmp-response-message-type">
            <xs:sequence>
```

(g)

```
                <xs:element ref="usersResponse" />
            </xs:sequence>
        </xs:extension>
    </xs:complexContent>
</xs:complexType>

<!-- usersResponseType -->

<xs:element name="usersResponse" type="usersResponseType" />
<xs:complexType name="usersResponseType">
    <xs:sequence>
        <xs:element name="usersInfo" type="info:users-type"
                minOccurs="0"/>
        <xs:any namespace="##other" processContents="lax"
                minOccurs="0" maxOccurs="unbounded"/>
    </xs:sequence>
    <xs:anyAttribute namespace="##any" processContents="lax"/>
</xs:complexType>

<!-- userResponse -->

<xs:complexType name="ccmp-user-response-message-type">
    <xs:complexContent>
        <xs:extension base="tns:ccmp-response-message-type">
            <xs:sequence>
                <xs:element ref="userResponse" />
            </xs:sequence>
        </xs:extension>
    </xs:complexContent>
</xs:complexType>

<!-- userResponseType -->

<xs:element name="userResponse" type="userResponseType" />
<xs:complexType name="userResponseType">
    <xs:sequence>
        <xs:element name="userInfo" type="info:user-type"
                minOccurs="0"/>
        <xs:any namespace="##other" processContents="lax"
                minOccurs="0" maxOccurs="unbounded"/>
    </xs:sequence>
    <xs:anyAttribute namespace="##any" processContents="lax"/>
</xs:complexType>

<!-- sidebarsByValResponse -->

<xs:complexType name="ccmp-sidebarsByVal-response-message-type">
    <xs:complexContent>
        <xs:extension base="tns:ccmp-response-message-type">
            <xs:sequence>
                <xs:element ref="sidebarsByValResponse" />
            </xs:sequence>
        </xs:extension>
    </xs:complexContent>
</xs:complexType>

<!-- sidebarsByValResponseType -->

<xs:element name="sidebarsByValResponse"
        type="sidebarsByValResponseType" />
<xs:complexType name="sidebarsByValResponseType">
    <xs:sequence>
    <xs:element name="sidebarsByValInfo"
            type="info:sidebars-by-val-type" minOccurs="0"/>
    <xs:any namespace="##other" processContents="lax"
            minOccurs="0" maxOccurs="unbounded"/>
    </xs:sequence>
    <xs:anyAttribute namespace="##any" processContents="lax"/>
</xs:complexType>
```

(h)

FIGURE 5.30 Continued.

5.12.3 MIME Media Type Registration for "application/ccmp+xml"

This section registers the "application/ccmp+xml" MIME type.

- To: ietf-types@iana.org
- Subject: Registration of MIME media type application/ccmp+xml
- MIME media type name: application MIME subtype name: ccmp+xml
- Required parameters: (none)

```
<!-- sidebarsByRefResponse -->

<xs:complexType name="ccmp-sidebarsByRef-response-message-type">
   <xs:complexContent>
      <xs:extension base="tns:ccmp-response-message-type">
         <xs:sequence>
            <xs:element ref="sidebarsByRefResponse" />
         </xs:sequence>
      </xs:extension>
   </xs:complexContent>
</xs:complexType>

<!-- sidebarsByRefResponseType -->

<xs:element name="sidebarsByRefResponse"
      type="sidebarsByRefResponseType" />
<xs:complexType name="sidebarsByRefResponseType">
   <xs:sequence>
      <xs:element name="sidebarsByRefInfo" type="info:uris-type"
            minOccurs="0"/>
      <xs:any namespace="##other" processContents="lax"
            minOccurs="0" maxOccurs="unbounded"/>
   </xs:sequence>
   <xs:anyAttribute namespace="##any" processContents="lax"/>
</xs:complexType>

<!-- sidebarByValResponse -->

<xs:complexType name="ccmp-sidebarByVal-response-message-type">
   <xs:complexContent>
      <xs:extension base="tns:ccmp-response-message-type">
         <xs:sequence>
            <xs:element ref="sidebarByValResponse" />
         </xs:sequence>
      </xs:extension>
   </xs:complexContent>
</xs:complexType>

<!-- sidebarByValResponseType -->

<xs:element name="sidebarByValResponse"
      type="sidebarByValResponseType" />
<xs:complexType name="sidebarByValResponseType">
   <xs:sequence>
      <xs:element name="sidebarByValInfo"
            type="info:conference-type" minOccurs="0"/>
      <xs:any namespace="##other" processContents="lax"
            minOccurs="0" maxOccurs="unbounded"/>
   </xs:sequence>
   <xs:anyAttribute namespace="##any" processContents="lax"/>
</xs:complexType>

<!-- sidebarByRefResponse -->

<xs:complexType name="ccmp-sidebarByRef-response-message-type">
<xs:complexContent>
<xs:extension base="tns:ccmp-response-message-type">
<xs:sequence>
<xs:element ref="sidebarByRefResponse" />
</xs:sequence>
</xs:extension>
</xs:complexContent>
</xs:complexType>

<!-- sidebarByRefResponseType -->

<xs:element name="sidebarByRefResponse"
      type="sidebarByRefResponseType" />
<xs:complexType name="sidebarByRefResponseType">
   <xs:sequence>
```

(i)

```
      <xs:element name="sidebarByRefInfo"
      type="info:conference-type" minOccurs="0"/>
      <xs:any namespace="##other" processContents="lax"
      minOccurs="0" maxOccurs="unbounded"/>
   </xs:sequence>
   <xs:anyAttribute namespace="##any" processContents="lax"/>
</xs:complexType>
<xs:complexType name="ccmp-sidebarByRef-response-message-type">
   <xs:complexContent>
      <xs:extension base="tns:ccmp-response-message-type">
         <xs:sequence>
            <xs:element ref="sidebarByRefResponse" />
         </xs:sequence>
      </xs:extension>
   </xs:complexContent>
</xs:complexType>

<!-- sidebarByRefResponseType -->

<xs:element name="sidebarByRefResponse"
      type="sidebarByRefResponseType" />
<xs:complexType name="sidebarByRefResponseType">
   <xs:sequence>
      <xs:element name="sidebarByRefInfo"
            type="info:conference-type" minOccurs="0"/>
      <xs:any namespace="##other" processContents="lax"
            minOccurs="0" maxOccurs="unbounded"/>
   </xs:sequence>
   <xs:anyAttribute namespace="##any" processContents="lax"/>
</xs:complexType>

<!-- extendedResponse -->

<xs:complexType name="ccmp-extended-response-message-type">
   <xs:complexContent>
      <xs:extension base="tns:ccmp-response-message-type">
         <xs:sequence>
            <xs:element ref="extendedResponse"/>
         </xs:sequence>
      </xs:extension>
   </xs:complexContent>
</xs:complexType>

<!-- extendedResponseType -->

<xs:element name="extendedResponse" type="extendedResponseType"/>
<xs:complexType name="extendedResponseType">
   <xs:sequence>
      <xs:element name="extensionName"
            type="xs:string" minOccurs="1"/>
      <xs:any namespace="##other" processContents="lax"
            minOccurs="0"
            maxOccurs="unbounded" />
   </xs:sequence>
</xs:complexType>

<!-- optionsResponse -->

<xs:complexType name="ccmp-options-response-message-type">
   <xs:complexContent>
      <xs:extension base="tns:ccmp-response-message-type">
         <xs:sequence>
            <xs:element ref="optionsResponse"/>
         </xs:sequence>
      </xs:extension>
   </xs:complexContent>
</xs:complexType>
```

(j)

FIGURE 5.30 Continued.

For more detailed information on MIME Media Type Registration, see RFC 6503.

5.12.4 DNS Registrations

Section 5.12.4.1 defines an Application Service tag of "XCON," which is used to identify the centralized conferencing (XCON) server for a particular domain. The Application Protocol tag "CCMP," defined in Section 5.12.4.2, is used to identify an XCON server that understands CCMP.

```
    <!-- optionsResponseType -->

    <xs:element name="optionsResponse"
          type="optionsResponseType" />
    <xs:complexType name="optionsResponseType">
        <xs:sequence>
            <xs:element name="options"
                  type="options-type" minOccurs="0"/>
            <xs:any namespace="##other" processContents="lax"
                  minOccurs="0" maxOccurs="unbounded"/>
        </xs:sequence>
        <xs:anyAttribute namespace="##any" processContents="lax"/>
    </xs:complexType>

<!-- CCMP ELEMENT TYPES -->

    <!-- response-codeType-->

    <xs:simpleType name="response-codeType">
        <xs:restriction base="xs:positiveInteger">
            <xs:pattern value="[0-9][0-9][0-9]" />
        </xs:restriction>
    </xs:simpleType>

    <!-- operationType -->

    <xs:simpleType name="operationType">
        <xs:restriction base="xs:token">
            <xs:enumeration value="retrieve"/>
            <xs:enumeration value="create"/>
            <xs:enumeration value="update"/>
            <xs:enumeration value="delete"/>
        </xs:restriction>
    </xs:simpleType>

    <!-- subject-type -->

    <xs:complexType name="subject-type">
        <xs:sequence>
            <xs:element name="username" type="xs:string"
                  minOccurs="0" maxOccurs="1" />
            <xs:element name="password" type="xs:string"
                  minOccurs="0" maxOccurs="1" />
            <xs:any namespace="##other" processContents="lax"
                  minOccurs="0" maxOccurs="unbounded"/>
        </xs:sequence>
        <xs:anyAttribute namespace="##any" processContents="lax"/>
    </xs:complexType>

    <!-- options-type -->

    <xs:complexType name="options-type">
        <xs:sequence>
            <xs:element name="standard-message-list"
                  type="standard-message-list-type"
                  minOccurs="1"/>
            <xs:element name="extended-message-list"
                  type="extended-message-list-type"
                  minOccurs="0"/>
            <xs:any namespace="##other" processContents="lax"
                  minOccurs="0" maxOccurs="unbounded"/>
        </xs:sequence>
        <xs:anyAttribute namespace="##any" processContents="lax"/>
    </xs:complexType>

    <!-- standard-message-list-type -->

    <xs:complexType name="standard-message-list-type">
        <xs:sequence>
```

(k)

```
            <xs:element name="standard-message"
                  type="standard-message-type"
                  minOccurs="1" maxOccurs="10"/>
            <xs:element name="extended-message-list"
                  type="extended-message-list-type"
                  minOccurs="0"/>
            <xs:any namespace="##other" processContents="lax"
                  minOccurs="0" maxOccurs="unbounded"/>
        </xs:sequence>
        <xs:anyAttribute namespace="##any" processContents="lax"/>
    </xs:complexType>

    <!-- standard-message-list-type -->

    <xs:complexType name="standard-message-list-type">
        <xs:sequence>
            <xs:element name="standard-message"
                  type="standard-message-type"
                  minOccurs="1" maxOccurs="10"/>
            <xs:any namespace="##other" processContents="lax"
                  minOccurs="0" maxOccurs="unbounded"/>
        </xs:sequence>
        <xs:anyAttribute namespace="##any" processContents="lax"/>
    </xs:complexType>

    <!-- standard-message-type -->

    <xs:complexType name="standard-message-type">
        <xs:sequence>
            <xs:element name="name"
                  type="standard-message-name-type"
                  minOccurs="1"/>
            <xs:element name="operations"
                  type="operations-type"
                  minOccurs="0"/>
            <xs:element name="schema-def" type="xs:string" minOccurs="0"/>
            <xs:element name="description" type="xs:string" minOccurs="0"/>
            <xs:any namespace="##other" processContents="lax"
                  minOccurs="0" maxOccurs="unbounded"/>
        </xs:sequence>
        <xs:anyAttribute namespace="##any" processContents="lax"/>
    </xs:complexType>

    <!-- standard-message-name-type -->

    <xs:simpleType name="standard-message-name-type">
        <xs:restriction base="xs:token">
            <xs:enumeration value="confsRequest"/>
            <xs:enumeration value="confRequest"/>
            <xs:enumeration value="blueprintsRequest"/>
            <xs:enumeration value="blueprintRequest"/>
            <xs:enumeration value="usersRequest"/>
            <xs:enumeration value="userRequest"/>
            <xs:enumeration value="sidebarsByValRequest"/>
            <xs:enumeration value="sidebarByValRequest"/>
            <xs:enumeration value="sidebarsByRefRequest"/>
            <xs:enumeration value="sidebarByRefRequest"/>
        </xs:restriction>
    </xs:simpleType>

    <!-- operations-type -->

    <xs:complexType name="operations-type">
        <xs:sequence>
            <xs:element name="operation" type="operationType"
                  minOccurs="1" maxOccurs="4"/>
        </xs:sequence>
        <xs:anyAttribute namespace="##any" processContents="lax"/>
    </xs:complexType>
```

(l)

FIGURE 5.30 Continued.

5.12.4.1 Registration of a Conference Server Application Service Tag

This section registers a new S-NAPTR/U-NAPTR Application Service tag for XCON, as mandated by RFC 3958.

- Application Service Tag: XCON
- Intended usage: Identifies a server that supports centralized conferencing
- Defining publication: RFC 6503

```
<!-- operations-type -->

<xs:complexType name="operations-type">
    <xs:sequence>
        <xs:element name="operation" type="operationType"
                minOccurs="1" maxOccurs="4"/>
    </xs:sequence>
    <xs:anyAttribute namespace="##any" processContents="lax"/>
</xs:complexType>

<!-- extended-message-list-type -->

<xs:complexType name="extended-message-list-type">
    <xs:sequence>
        <xs:element name="extended-message"
                type="extended-message-type"
                minOccurs="0"/>
        <xs:any namespace="##other" processContents="lax"
                minOccurs="0" maxOccurs="unbounded"/>
    </xs:sequence>
    <xs:anyAttribute namespace="##any" processContents="lax"/>
</xs:complexType>
```

(m)

```
<!-- extended-message-type -->

<xs:complexType name="extended-message-type">
    <xs:sequence>
        <xs:element name="name" type="xs:string" />
        <xs:element name="operations"
                type="operations-type"
                minOccurs="0"/>
        <xs:element name="schema-def" type="xs:string" />
        <xs:element name="description"
                type="xs:string"
                minOccurs="0"/>
        <xs:any namespace="##other" processContents="lax"
                minOccurs="0" maxOccurs="unbounded"/>
    </xs:sequence>
    <xs:anyAttribute namespace="##any" processContents="lax"/>
</xs:complexType>
</xs:schema>
```

(n)

FIGURE 5.30 Continued.

For more information on Registration of a Conference Server Application Service Tag, see RFC 6503.

5.12.4.2 Registration of a Conference Server Application Protocol Tag for CCMP

This section registers a new S-NAPTR/U-NAPTR Application Protocol tag for CCMP, as mandated by RFC 3958.

- Application Service Tag: CCMP
- Intended Usage: Identifies the Centralized Conferencing (XCON) Manipulation Protocol
- Applicable Service Tag(s): XCON
- Terminal NAPTR Record Type(s): U
- Defining Publication: RFC 6503

For more information on Registration of a Conference Server Application Protocol Tag for CCMP, see RFC 6503.

5.12.5 CCMP Protocol Registry

The IANA has created a new registry for CCMP: http://www.iana.org/assignments/ccmp-paramete rs. The document creates initial sub-registries for CCMP operation types and response codes.

5.12.5.1 CCMP Message Types

The following summarizes the registry for CCMP messages:

- Related Registry: CCMP Message Types Registry
- Defining RFC: RFC 6503
- Registration/Assignment Procedures: Following the policies outlined in RFC 5226, the IANA policy for assigning new values for the CCMP message types for CCMP is Specification Required

For more information on CCMP Protocol Registry, see RFC 6503.

5.12.5.2 CCMP Response Codes

The following summarizes the requested registry for CCMP response codes:

- Related Registry: CCMP Response Code Registry
- Defining RFC: RFC 6503
- Registration/Assignment Procedures: Following the policies outlined in RFC 5226, the IANA policy for assigning new values for the Response codes for CCMP shall be Specification Required
- This specification establishes the Response-code sub-registry under http://www.iana.org/assignments/ccmp-parameters. The initial Response-code table is populated using the Response codes defined in Section 5.4

For more detailed information on CCMP Response Codes sub-registry, see RFC 6503.

5.13 SUMMARY

The CCMP specified in this chapter offers the fundamental capabilities for manipulations of the entire centralized conferencing system in a variety of ways that conference participants want in real time. The XCON system client can use the CCMP protocol to create, retrieve, change, and delete objects that describe a centralized conference for controlling both basic and advanced conference features, such as conference state and capabilities, participants, relative roles, and details. Moreover, CCMP is a stateless, XML-based, client-server protocol that carries, in its request and response messages, conference information in the form of XML specifications and fragments conforming to the centralized conferencing data model schema.

We have described the CCMP control system architecture, detailing conference objects and participants. The CCMP protocol description is explained showing protocol operations, data management, and data model compliance along with implementation approach. The CCMP request and response message types are defined. In addition,. each of these CCMP protocol messages is specified in complete detail. The CCMP response codes that are used to indicate the result of operations of the protocol between different conference functional entities are also specified. A complete example of how CCMP operates in practice in a centralized conference system is provided for better understanding the insights of the protocol.

A key need of the multipoint centralized conference system is to locate a conference server across the global network. The detailed procedures for locating the sever are provided here. The notifications that are required to be used in a conference system by different entities are explained using notification messages of the conference event package defined in RFC 6502. We have described in great detail how the HTTP transport protocol is used to transfer requests and responses, which contain the domain-specific XML-encoded data objects described in RFC 6501, by the CCMP messages. Finally, we have described the security mechanisms of a conferencing system that uses CCMP protocol. Note that RFC 6503 specification registers a new XML namespace, a new XML schema, and the MIME type for the schema in IANA. It also registers the "XCON" Application Service tag and the "CCMP" Application Protocol tag and defines registries for the CCMP operation types and response codes. Interested readers need to see RFC 6503 for the detail, as it is not included here for the sake of brevity.

5.14 EVALUATION OF OTHER PROTOCOL MODELS AND TRANSPORTS CONSIDERED FOR CCMP (APPENDIX A OF RFC 6503)

This section provides some background on the selection of HTTP as the transport for the CCMP requests/responses. In addition to HTTP, the operations on the objects can be implemented in at least two different ways, namely as remote procedure calls – using SOAP, as described in Section 5.14.1 (Appendix A.1 of RFC 6503) – and by defining resources following a RESTful architecture (Section 5.14.2) (Appendix A.2 of RFC 6503). In both the SOAP and RESTFUL approaches, servers will have to recreate their internal state representation of the object with each update request, checking parameters and triggering function invocations. In the SOAP approach, it would be possible to describe a separate operation for each atomic element, but that would greatly increase the complexity of the protocol. A coarser-grained approach to CCMP does require that the server process XML elements in updates that have not changed and that there can be multiple changes in one update.

For CCMP, the resource (REST) model might appear more attractive, since the conference operations fit the CRUD approach. However, neither of these approaches was considered ideal. SOAP was considered to bring additional overhead. It is quite awkward to apply a RESTful approach, since CCMP requires a more complex request/ response protocol in order to maintain the data both in the server and at the client. This does not map very elegantly to the basic request/response model, whereby a response typically indicates whether the request was successful or not, rather than providing additional data to maintain the synchronization between the client and server data. In addition, the CCMP clients may also receive the data in notifications.

While the notification method or protocol used by some conferencing clients can be independent of CCMP, the same data in the server is used for both CCMP and notifications – this requires a server application above the transport layer (e.g., HTTP) for maintaining the data, which in the CCMP model is transparent to the transport protocol.

Thus, the solution for CCMP defined in this specification is viewed as a good compromise among the most notable past candidates and is referred to as "HTTP single-verb transport plus CCMP body." With this approach, CCMP is able to take advantage of existing HTTP functionality. As with SOAP, CCMP uses a "single HTTP verb" for transport (i.e., a single transaction type for each request/response pair); this allows CCMP messages to be decoupled from HTTP messages.

Similarly, as with any RESTful approach, CCMP messages are inserted directly in the body of HTTP messages, thus avoiding any unnecessary processing and communication burden associated with further intermediaries. With this approach, no modification to the CCMP messages/operations is required to use a different transport protocol.

5.14.1 Using SOAP for CCMP (Appendix A.1 of RFC 6503)

A remote procedure call (RPC) mechanism for CCMP could use SOAP (Simple Object Access Protocol [1, 2]), where conferences and the other objects are modeled as services with associated operations. Conferences and other objects are selected by their own local identifiers, such as email-like names for users. This approach has the advantage that it can easily define atomic operations that have well-defined error conditions. All SOAP operations would use a single HTTP verb. While the RESTful approach requires the use of a URI for each object, SOAP can use any token.

5.14.2 A RESTful Approach for CCMP (Appendix A.2 of RFC 6503)

Conference objects can also be modeled as resources identified by URIs, with the basic CRUD operations mapped to the HTTP methods POST/ PUT for creating objects, GET for reading objects, PATCH/POST/PUT for changing objects, and DELETE for deleting them. Many of the objects, such as conferences, already have natural URIs. CCMP can be mapped into the CRUD design pattern. The basic CRUD operations are used to manipulate conference objects, which are XML specifications containing the information characterizing a specified conference instance, be it an active conference or a conference blueprint used by the conference server to create new conference instances through a simple clone operation.

Following the CRUD approach, CCMP could use a general-purpose protocol such as HTTP (RFC 2616) to transfer domain-specific XML-encoded data objects defined in the "Conference Information Data Model for Centralized Conferencing" (RFC 6501 – see Chapter 4). Following on from the CRUD approach, CCMP could follow the well-known REST (REpresentational State Transfer) architectural style [3]. CCMP could map onto the REST philosophy by specifying resource URIs, resource formats, methods supported at each URI, and status codes that have to be returned when a certain method is invoked on a specific URI. A REST-style approach must ensure that all operations can be mapped to HTTP operations.

The following summarizes the specific HTTP method that could be used for each of the CCMP Requests:

Retrieve: HTTP GET could be used on XCON-URIs, so that clients can obtain data about conference objects in the form of XML data model specifications.

Create: HTTP PUT could be used to create a new object as identified by the XCON-URI or XCON-USERID.

Change: Either HTTP PATCH or HTTP POST could be used to change the conference object identified by the XCON-URI.

Delete: HTTP DELETE could be used to delete conference objects and parameters within conference objects identified by the XCON-URI.

5.15 PROBLEMS

1. What are the operations that users (conference participants) can perform on conferences using the CCMP protocol that allows users to create, manipulate, and delete conference objects?
2. Describe the interaction of the conferencing client (user) with conference objects via the conference control server in detail using the CCMP protocol.
3. Why is the client-server CCMP protocol designed as a stateless protocol instead of a stateful one? What is the atomic operation model of a protocol? What are the core set of conference objects that CCMP can manipulate? Does CCMP operate in atomic mode? If so, explain in detail. Why do the XML specifications that are carried by CCMP protocol need to comply with XCON data model?
4. What are four operations of CCMP? Explain the data management and XCON data model compliance in operations of CCMP protocol. How does you implement the CCMP for operations of a centralized multipoint conference? Use an example to explain this.
5. Explain the functions of CCMP request and response messages. Show the use of each of these message types, including response codes, with detailed call flows for typical XCON conferences.
6. Explain the probable security features used in CCMP protocol operations that can be associated with each step of the call flows described in the conference example of Section 5.6.
7. Describe in detail how an XCON conference server can be located in the network.
8. Explain in detail, using a XCON conference architecture along with call flows, how notifications can be managed using a conference event package (RFC 6502).
9. Articulate in detail how CCMP protocol messages are carried over the HTTP/HTTPS transport protocol, including security for both CCMP application and HTTPS transport.
10. What are the potential attacks that can occur related to the XCON conference? How can the CCMP server and the conference clients be protected against all those attacks? Explain in detail.
11. Explain the salient features of the new CCMP XML schema that has been registered with IANA in relation to the XCON conference.

REFERENCES

1. Nielsen, H. et al. (2007, April) SOAP Version 1.2 Part 1: Messaging Framework (Second Edition). *World Wide Web Consortium Recommendation REC-soap12-part1-20070427.* <http://www.w3.org/TR/2007/REC-soap12-part1-20070427>.
2. Moreau, J. et al. (2007, April) SOAP Version 1.2 Part 2: Adjuncts (Second Edition). *World Wide Web Consortium Recommendation REC-soap12-part2-20070427.* <http://www.w3.org/TR/2007/REC-soap12-part2-20070427>.
3. REST Fielding. (2000) *Architectural Styles and the Design of Networkbased Software Architectures.* University of California.

6 Binary Floor Control Protocol

6.1 INTRODUCTION

Within a conference, some applications need to manage the access to a set of shared resources, such as the right to send media to a particular media session. Floor control enables such applications to provide users with coordinated (shared or exclusive) access to these resources. The Requirements for Floor Control Protocol (RFC 4376) lists a set of requirements that need to be met by floor control protocols. The binary floor control protocol (BFCP) (RFC 4582), which is specified in this chapter, meets these requirements.

In addition, BFCP has been designed so that it can be used in low-bandwidth environments. The binary encoding used by BFCP achieves a small message size (when message signatures are not used) that keeps the time it takes to transmit delay-sensitive BFCP messages to a minimum. Delay-sensitive BFCP messages include FloorRequest, FloorRelease, FloorRequestStatus, and ChairAction. It is expected that future extensions to these messages will not increase the size of these messages in a significant way.

The remainder of this chapter is organized as follows: Section 6.2 defines the terminology used throughout this specification, Section 6.3 discusses the scope of BFCP (i.e., which tasks fall within the scope of BFCP and which ones are performed using different mechanisms), Section 6.4 provides a non-normative overview of BFCP operation, and subsequent sections provide the normative specification of BFCP.

6.2 TERMINOLOGY

See Table 1.2, Section 1.6.

6.3 SCOPE

As stated earlier, BFCP is a protocol to coordinate access to shared resources in a conference following the requirements defined in RFC 4376. Floor control complements other functions defined in the XCON conferencing framework (RFC 5239 – see Chapter 2). The floor control protocol BFCP defined in this specification only specifies a means to arbitrate access to floors. The rules and constraints for floor arbitration and the results of floor assignments are outside the scope of this specification and are defined by other protocols (RFC 5239 – see Chapter 2). Figure 6.1 shows the tasks that BFCP can perform.

BFCP provides a means:

- for floor participants to send floor requests to floor control servers
- for floor control servers to grant or deny requests to access a given resource from floor participants
- for floor chairs to send floor control servers decisions regarding floor requests
- for floor control servers to keep floor participants and floor chairs informed about the status of a given floor or a given floor request

Even though tasks that do not belong to the previous list are outside the scope of BFCP, some of these out-of-scope tasks relate to floor control and are essential for creating floors and establishing BFCP connections between different entities. In the following sub-sections, we discuss some of these tasks and mechanisms to perform them.

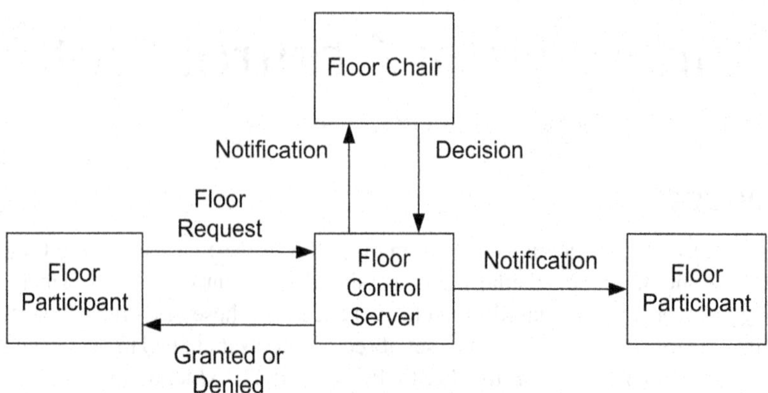

FIGURE 6.1 Functionality Provided by BFCP.

6.3.1 Floor Creation

The association of a given floor with a resource or a set of resources (e.g., media streams) is out of the scope of BFCP as described in RFC 5239 (see Chapter 2). Floor creation and termination are also outside the scope of BFCP; these aspects are handled using the conference control protocol for manipulating the conference object. Consequently, the floor control server needs to stay up to date on changes to the conference object (e.g., when a new floor is created).

6.3.2 Obtaining Information to Contact a Floor Control Server

A client needs a set of data in order to establish a BFCP connection to a floor control server. These data include the transport address of the server, the conference identifier, and a user identifier. Clients can obtain this information in different ways. One is to use a session description protocol (SDP) offer/answer (RFC 3264) exchange, which is described in RFC 4583. Other mechanisms are described in the XCON framework (RFC 5239 – see Chapter 2) (and other related specifications).

6.3.3 Obtaining Floor–Resource Associations

Floors are associated with resources. For example, a floor that controls who talks at a given time has a particular audio session as its associated resource. Associations between floors and resources are part of the conference object. Floor participants and floor chairs need to know which resources are associated with which floors. They can obtain this information by using different mechanisms, such as an SDP offer/answer (RFC 3264) [8] exchange. How to use an SDP offer/answer exchange to obtain these associations is described in RFC 4583.

Note that floor participants perform SDP offer/answer exchanges with the conference focus of the conference. So, the conference focus needs to obtain information about associations between floors and resources in order to be able to provide this information to a floor participant in an SDP offer/answer exchange. Other mechanisms for obtaining this information, including discussion of how the information is made available to a session initiation protocol (SIP) focus, are described in the XCON framework (RFC 5239 – see Chapter 2) [10] (and other related specifications).

6.3.4 Privileges of Floor Control

A participant whose floor request is granted has the right to use (in a certain way) the resource or resources associated with the floor that was requested. For example, the participant may have the right to send media over a particular audio stream. Nevertheless, holding a floor does not imply that others will not be able to use its associated resources at the same time, even if they do not have the right to do so. Determination of which media participants can actually use the resources in the conference is discussed in the XCON Framework (RFC 5239 – see Chapter 2) [10].

6.4 OVERVIEW OF OPERATION

This section provides a non-normative description of BFCP operations. Section 6.4.1 describes the interface between floor participants and floor control servers, and Section 6.4.2 describes the interface between floor chairs and floor control servers. BFCP messages, which use a TLV (type-length-value) binary encoding, consist of a common header followed by a set of attributes. The common header contains, among other information, a 32-bit conference identifier. Floor participants, media participants, and floor chairs are identified by 16-bit user identifiers. BFCP supports nested attributes (i.e., attributes that contain attributes). These are referred to as grouped attributes.

There are two types of transactions in BFCP: Client-initiated transactions and server-initiated transactions. Client-initiated transactions consist of a message from a client to the floor control server and a response from the floor control server to the client. Both messages can be related because they carry the same Transaction ID value in their common headers. Server-initiated transactions consist of a single message, whose Transaction ID is 0, from the floor control server to a client.

6.4.1 Floor Participant to Floor Control Server Interface

Floor participants request a floor by sending a FloorRequest message to the floor control server. BFCP supports third-party floor requests. That is, the floor participant sending the floor request need not be co-located with the media participant that will get the floor once the floor request is granted. FloorRequest messages carry the identity of the requester in the User ID field of the common header and the identity of the beneficiary of the floor (in third-party floor requests) in a BENEFICIARY-ID attribute.

Third-party floor requests can be sent, for example, by floor participants that have a BFCP connection to the floor control server but that are not media participants (i.e., they do not handle any media). FloorRequest messages identify the floor or floors being requested by carrying their 16-bit floor identifiers in FLOOR-ID attributes. If a FloorRequest message carries more than one floor identifier, the floor control server treats all the floor requests as an atomic package. That is, the floor control server either grants or denies all the floors in the FloorRequest message.

Floor control servers respond to FloorRequest messages with FloorRequestStatus messages, which provide information about the status of the floor request. The first FloorRequestStatus message is the response to the FloorRequest message from the client and therefore has the same Transaction ID as the FloorRequest.

Additionally, the first FloorRequestStatus message carries the Floor Request ID in a FLOOR-REQUEST-INFORMATION attribute. Subsequent FloorRequestStatus messages related to the same floor request will carry the same Floor Request ID. This way, the floor participant can associate them with the appropriate floor request. Messages from the floor participant related to a particular floor request also use the same Floor Request ID as the first FloorRequestStatus Message from the floor control server. Figure 6.2 shows how a floor participant requests a floor, obtains it, and at a later time, releases it. This figure illustrates the use, among other things, of the Transaction ID and the FLOOR-REQUEST-ID attribute.

Figure 6.3 shows how a floor participant requests to be informed on the status of a floor. The first FloorStatus message from the floor control server is the response to the FloorQuery message and as such, has the same Transaction ID as the FloorQuery message. Subsequent FloorStatus messages consist of server-initiated transactions, and therefore their Transaction ID is 0. FloorStatus message (2) indicates that there are currently two floor requests for the floor whose Floor ID is 543. FloorStatus message (3) indicates that the floor requests with Floor Request ID 764 have been granted, and the floor request with Floor Request ID 635 is the first in the queue. FloorStatus message (4) indicates that the floor request with Floor Request ID 635 has been granted.

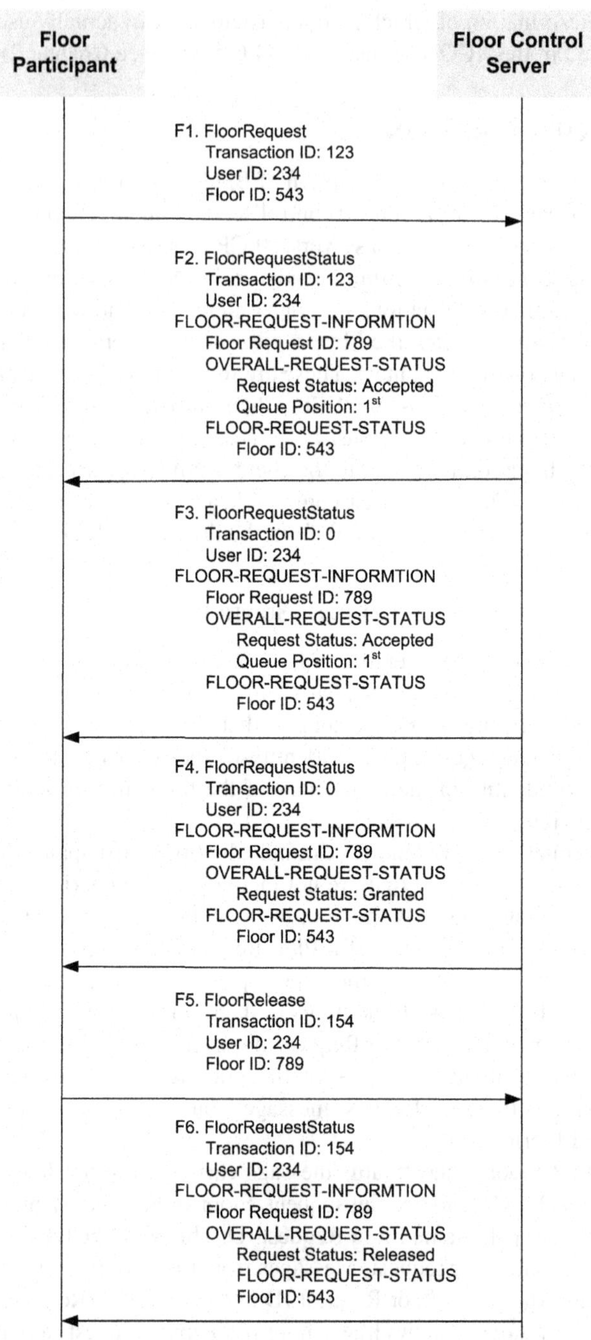

FIGURE 6.2 Requesting and Releasing a Floor. (Copyright: IETF.)

FloorStatus messages contain information about the floor requests they carry. For example, FloorStatus message (4) indicates that the floor request with Floor Request ID 635 has as the beneficiary (i.e., the participant that holds the floor when a particular floor request is granted) the participant whose User ID is 154. The floor request applies only to the floor whose Floor ID is 543. That is, this is not a multi-floor floor request.

A multi-floor floor request applies to more than one floor (e.g., a participant wants to be able to speak and write on the whiteboard at the same time). The floor control server treats a multi-floor

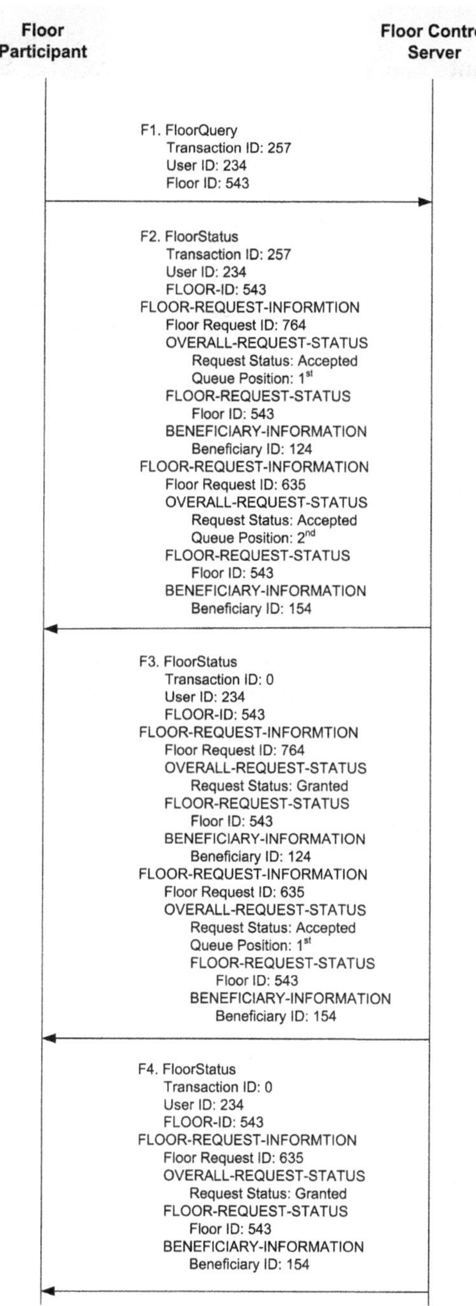

FIGURE 6.3 Obtaining Status Information about a Floor. (Copyright: IETF.)

floor request as an atomic package. That is, the floor control server either grants the request for all floors or denies the request for all floors.

6.4.2 Floor Chair to Floor Control Server Interface

Figure 6.4 shows a floor chair instructing a floor control server to grant a floor. Note, however, that although the floor control server needs to take into consideration the instructions received in ChairAction messages (e.g., granting a floor), it does not necessarily need to perform them exactly

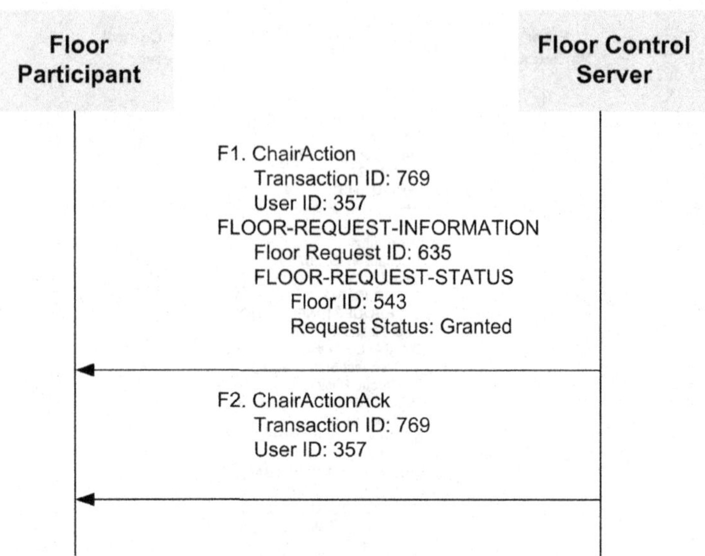

FIGURE 6.4 Chair Instructing the Floor Control Server. (Copyright: IETF.)

as requested by the floor chair. The operation that the floor control server performs depends on the ChairAction message and on the internal state of the floor control server.

For example, a floor chair may send a ChairAction message granting a floor that was requested as part of an atomic floor request operation that involved several floors. Even if the chair responsible for one of the floors instructs the floor control server to grant the floor, the floor control server will not grant it until the chairs responsible for the other floors agree to grant them as well. In another example, a floor chair may instruct the floor control server to grant a floor to a participant. The floor control server needs to revoke the floor from its current holder before granting it to the new participant. So, the floor control server is ultimately responsible for keeping a coherent floor state using instructions from floor chairs as input to this state.

6.5 PACKET FORMAT

BFCP packets consist of a 12-octet common header followed by attributes. All the protocol values MUST be sent in network byte order.

6.5.1 COMMON-HEADER FORMAT

The following is the format of the common header (Figure 6.5):

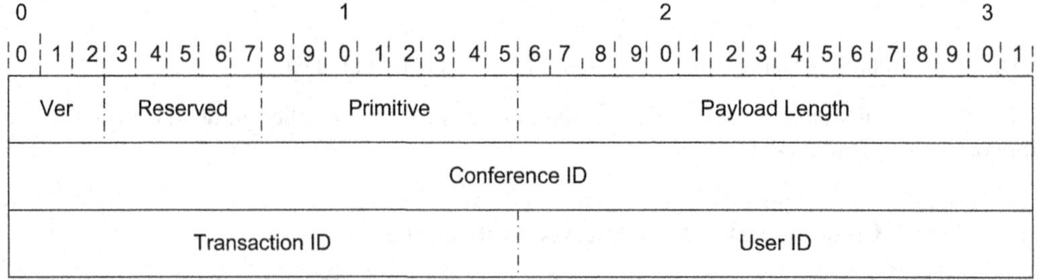

FIGURE 6.5 COMMON-HEADER Format.

TABLE 6.1
Requesting and Releasing a Floor

Value	Primitive	Direction	
1	FloorRequest	P -> S	
2	FloorRelease	P -> S	
3	FloorRequestQuery	P -> S;	Ch -> S
4	FloorRequestStatus	P <- S ;	Ch <-S
5	UserQuery	P -> S;	Ch -> S
6	UserStatus	P <- S ;	Ch <- S
7	FloorQuery	P -> S;	Ch -> S
8	FloorStatus	P<- S ;	Ch <- S
9	ChairAction	Ch -> S	
10	ChairActionAct	Ch <-S	
11	Hello	P -> S;	Ch -> S
12	HelloAck	P <- S ;	Ch <- S
13	Error	P <- S ;	Ch <- S

S: FloorControl Server; P:FloorParicipant; Ch: FloorChair

- **Ver:** The 3-bit version field MUST be set to 1 to indicate this version of BFCP.
- **Reserved:** At this point, the 5 bits in the reserved field SHOULD be set to zero by the sender of the message and MUST be ignored by the receiver.
- **Primitive:** This 8-bit field identifies the main purpose of the message. The following primitive values are defined (Tables 6.1 and 6.2):
- **Payload Length:** This 16-bit field contains the length of the message in 4-octet units, excluding the common header.
- **Conference ID:** This 32-bit field identifies the conference the message belongs to.
- **Transaction ID:** This field contains a 16-bit value that allows users to match a given message with its response. The value of the Transaction ID in server-initiated transactions is 0 (see Chapter 8).
- **User ID**: This field contains a 16-bit value that uniquely identifies a participant within a conference.

The identity used by a participant in BFCP, which is carried in the User ID field, is generally mapped to the identity used by the same participant in the session establishment protocol (e.g., in SIP). The way this mapping is performed is outside the scope of this specification.

6.5.2 Attribute Format

BFCP attributes are encoded in TLV format. Attributes are 32-bit aligned (Figure 6.6).

- **Type:** This 7-bit field contains the type of the attribute. Each attribute, identified by its type, has a particular format. The attribute formats defined are:
- **Unsigned16:** The contents of the attribute consist of a 16-bit unsigned integer.
- **OctetString16**: The contents of the attribute consist of 16 bits of arbitrary data.
- **OctetString:** The contents of the attribute consist of arbitrary data of variable length.
- **Grouped:** The contents of the attribute consist of a sequence of attributes.

Note that extension attributes defined in the future may define new attribute formats.

TABLE 6.2
Obtaining Status Information about a Floor

Value	Primitive	Format
1	BENEFICIARY-ID	Unsigned16
2	FLOOR-ID	Unsigned16
3	FLOOR-REQUEST-ID	Unsigned16
4	PRIORITY	OctetString16
5	REQUEST-STATUS	OctetString16
6	ERROR-CODE	OctetString
7	ERROR-INFO	OctetString
8	PARTICIPANT-PROVIDED-INFO	OctetString
9	STATUS-INFO	OctetString
10	SUPPORTED-ATTRIBUTES	OctetString
11	SUPPORTED-PRIMITIVES	OctetString
12	USER-DISPLAY-NAME	OctetString
13	USER-URI	OctetString
14	BENEFICIARY-INFORMATION	Grouped
15	FLOOR-REQUEST-INFORMATION	Grouped
16	REQUESTED-BY-INFORMATION	Grouped
17	FLOOR-REQUEST-STATUS	Grouped
18	OVERALL-REQUEST-STATUS	Grouped

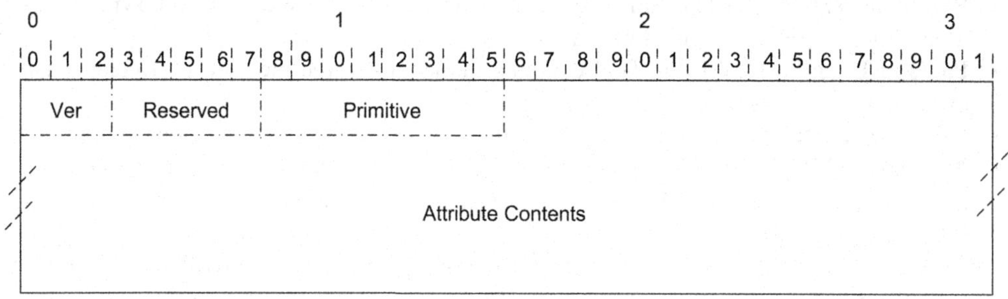

FIGURE 6.6 Attribute Format.

The BFCP attribute types are defined in Table 6.2.

- **M**: The "M" bit, known as the Mandatory bit, indicates whether support of the attribute is required. If an unrecognized attribute with the "M" bit set is received, the message is rejected. The "M" bit is significant for extension attributes defined in other specifications only. All attributes specified in this specification MUST be understood by the receiver, so that the setting of the "M" bit is irrelevant for these. In all other cases, the unrecognized attribute is ignored, but the message is processed.
- **Length:** This 8-bit field contains the length of the attribute in octets, excluding any padding defined for specific attributes. The length of attributes that are not grouped includes the Type, "M" bit, and Length fields. The Length in grouped attributes is the length of the grouped attribute itself (including Type, "M" bit, and Length fields) plus the total length (including padding) of all the included attributes.
- **Attribute Contents:** The contents of the different attributes are defined in the following sections.

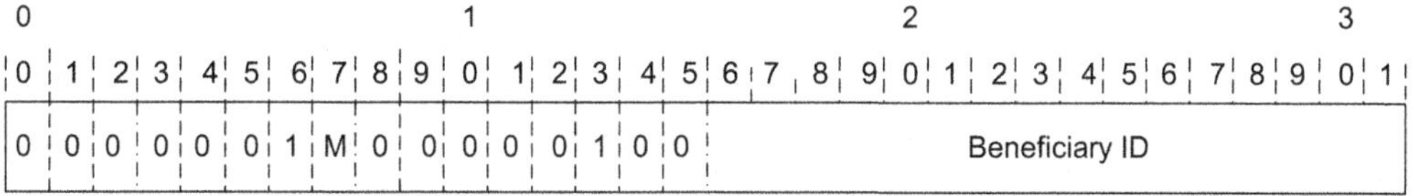

FIGURE 6.7 BENEFICIARY-ID Format.

6.5.2.1 BENEFICIARY-ID

The following is the format of the BENEFICIARY-ID attribute (Figure 6.7):

- **Beneficiary ID:** This field contains a 16-bit value that uniquely identifies a user within a conference.

Note that although the formats of the Beneficiary ID and of the User ID field in the common header are similar, their semantics are different. The Beneficiary ID is used in third-party floor requests and to request information about a particular participant.

6.5.2.2 FLOOR-ID

The following is the format of the FLOOR-ID attribute (Figure 6.8):

- **Floor ID:** This field contains a 16-bit value that uniquely identifies a floor within a conference.

6.5.2.3 FLOOR-REQUEST-ID

The following is the format of the Floor-Request-ID attribute (Figure 6.9):

- **Floor Request ID:** This field contains a 16-bit value that identifies a floor request at the floor control server.

6.5.2.4 PRIORITY

The following is the format of the PRIORITY attribute (Figure 6.10):

- **Prio:** This field contains a 3-bit priority value, as shown in Table 6.3. Senders SHOULD NOT use values higher than 4 in this field. Receivers MUST treat values higher than 4 as if the value received were 4 (Highest). The default priority value when the PRIORITY attribute is missing is 2 (Normal).

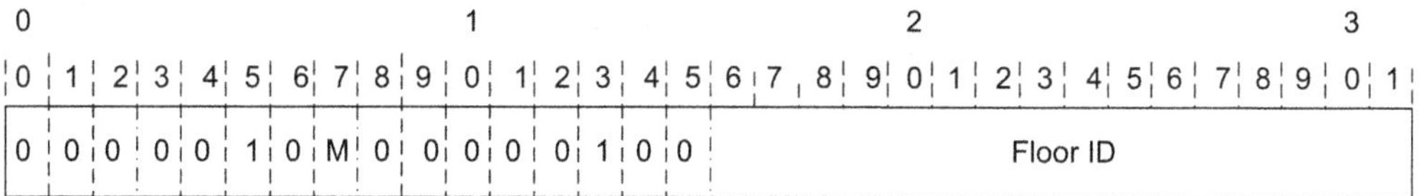

FIGURE 6.8 FLOOR-ID Format.

0 1 2 3
0 1 2 3 4 5 6 7 8 9 0 1 2 3 4 5 6 7 8 9 0 1 2 3 4 5 6 7 8 9 0 1
0 0 0 0 0 1 1 M 0 0 0 0 0 1 0 0
Floor Request ID

FIGURE 6.9 FLOOR-REQUEST-ID Format.

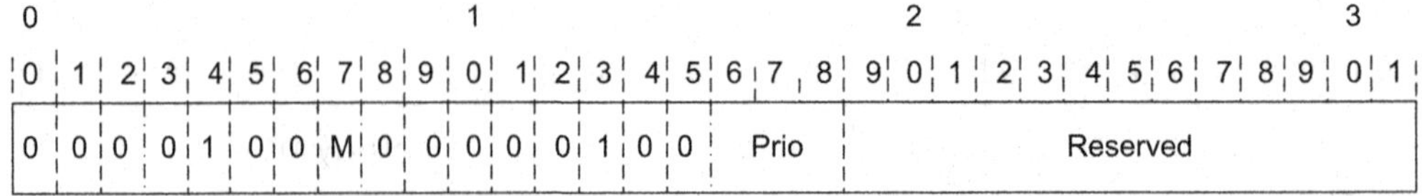

FIGURE 6.10 PRIORITY Format.

TABLE 6.3
Priority Values

Value	Priority
0	Lowest
1	Low
2	Normal
3	High
4	Highest

- **Reserved:** At this point, the 13 bits in the reserved field SHOULD be set to zero by the sender of the message and MUST be ignored by the receiver.

6.5.2.5 REQUEST-STATUS

The following is the format of the REQUEST-STATUS attribute (Figure 6.11):

- **Request Status:** This 8-bit field contains the status of the request, as described in Table 6.4.
- **Queue Position:** This 8-bit field contains, when applicable, the position of the floor request in the floor request queue at the server. If the Request Status value is different from Accepted, if the floor control server does not implement a floor request queue, or if the floor control server does not want to provide the client with this information, all the bits of this field SHOULD be set to zero.

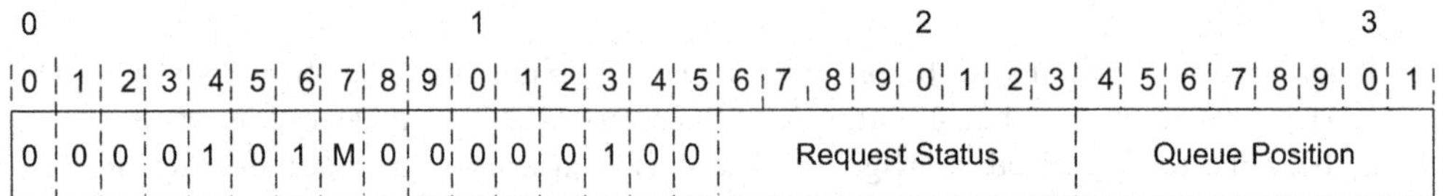

FIGURE 6.11 REQUEST-STATUS Format.

TABLE 6.4
Request Status Values

Value	Status
1	Pending
2	Accepted
3	Granted
4	Denied
5	Cancelled
6	Released
7	Revoked

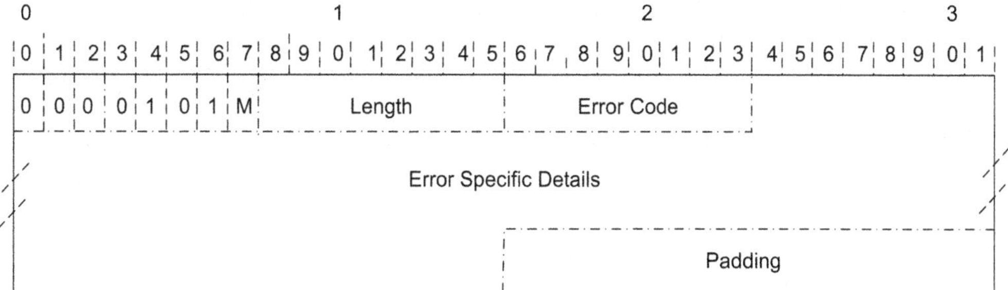

FIGURE 6.12 ERROR-CODE Format.

TABLE 6.5
Error Code Meaning

Value	Status
1	Conference Does Not Exist
2	Conference Does Not Exist
3	Unknown Primitive
4	Unknown Mandatory Attribute
5	Unauthorized Operation
6	Invalid Floor ID
7	Floor Request ID Does Not Exist
8	You Have Already Reached the Maximum Number of Ongoing Floor Requests for This Floor
9	Use TLS

A floor request is in Pending state if the floor control server needs to contact a floor chair in order to accept the floor request but has not done it yet. Once the floor control chair accepts the floor request, the floor request is moved to the Accepted state.

6.5.2.6 ERROR-CODE

The following is the format of the ERROR-CODE attribute (Figure 6.12):

- **Error Code:** This 8-bit field contains an error code from Table 6.5. If an error code is not recognized by the receiver, then the receiver MUST assume that an error exists and therefore that the message is processed, but the nature of the error is unclear.
- **Error-Specific Details:** Present only for certain Error Codes. In this specification, only for Error Code 4 (Unknown Mandatory Attribute). See Section 5.2.6.1 for its definition.
- **Padding:** One, two, or three octets of padding added so that the contents of the ERROR-CODE attribute are 32-bit aligned. If the attribute is already 32-bit aligned, no padding is needed. The Padding bits SHOULD be set to zero by the sender and MUST be ignored by the receiver.

6.5.2.6.1 Error-Specific Details for Error Code 4

The following is the format of the Error-Specific Details field for Error Code 4 (Figure 6.13):

- **Unknown Type:** These 7-bit fields contain the Types of the attributes (which were present in the message that triggered the Error message) that were unknown to the receiver.

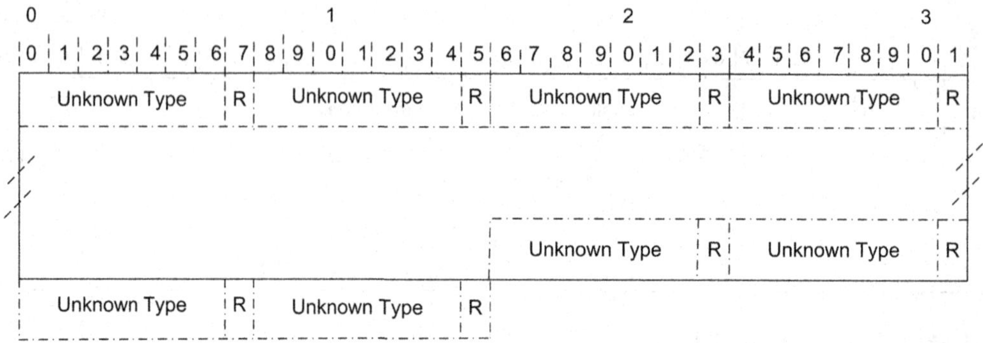

FIGURE 6.13 Unknown Attributes Format.

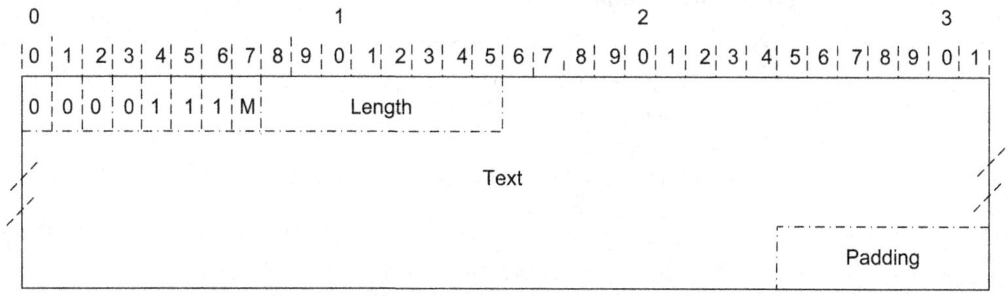

FIGURE 6.14 ERROR-INFO Format.

- **R:** At this point, this bit is reserved. It SHOULD be set to zero by the sender of the message and MUST be ignored by the receiver.

6.5.2.7 ERROR-INFO

The following is the format of the ERROR-INFO attribute (Figure 6.14):

- **Text:** This field contains UTF-8 (RFC 3629) [6] encoded text.
  In some situations, the contents of the Text field may be generated by an automaton. If this automaton has information about the preferred language of the receiver of a particular ERROR-INFO attribute, it MAY use this language to generate the Text field.
- **Padding:** One, two, or three octets of padding added so that the contents of the ERROR-INFO attribute is 32-bit aligned. The Padding bits SHOULD be set to zero by the sender and MUST be ignored by the receiver. If the attribute is already 32-bit aligned, no padding is needed.

6.5.2.8 PARTICIPANT-PROVIDED-INFO

The following is the format of the PARTICIPANT-PROVIDED-INFO attribute (Figure 6.15):

- **Text:** This field contains UTF-8 (RFC 3629) [6] encoded text.
- **Padding:** One, two, or three octets of padding added so that the contents of the PARTICIPANT-PROVIDED-INFO attribute is 32-bit aligned. The Padding bits SHOULD be set to zero by the sender and MUST be ignored by the receiver. If the attribute is already 32-bit aligned, no padding is needed.

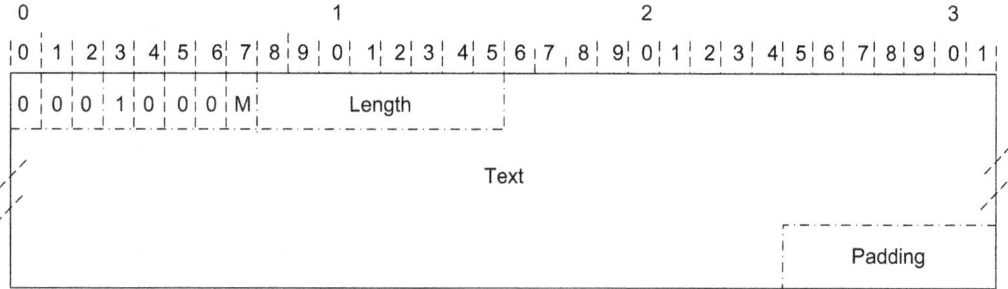

FIGURE 6.15 PARTICIPANT-PROVIDED-INFO Format.

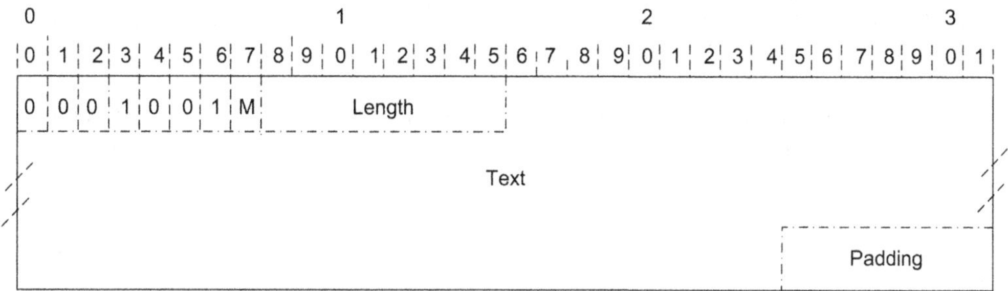

FIGURE 6.16 STATUS-INFO Format.

6.5.2.9 STATUS-INFO

The following is the format of the STATUS-INFO attribute (Figure 6.16).

- **Text**: This field contains UTF-8 (RFC 3629) [6] encoded text.
  In some situations, the contents of the Text field may be generated by an automaton. If this automaton has information about the preferred language of the receiver of a particular STATUS-INFO attribute, it MAY use this language to generate the Text field.
- **Padding**: One, two, or three octets of padding added so that the contents of the STATUS-INFO attribute is 32-bit aligned. The Padding bits SHOULD be set to zero by the sender and MUST be ignored by the receiver. If the attribute is already 32-bit aligned, no padding is needed.

6.5.2.10 SUPPORTED-ATTRIBUTES

The following is the format of the SUPPORTED-ATTRIBUTES attribute (Figure 6.17).

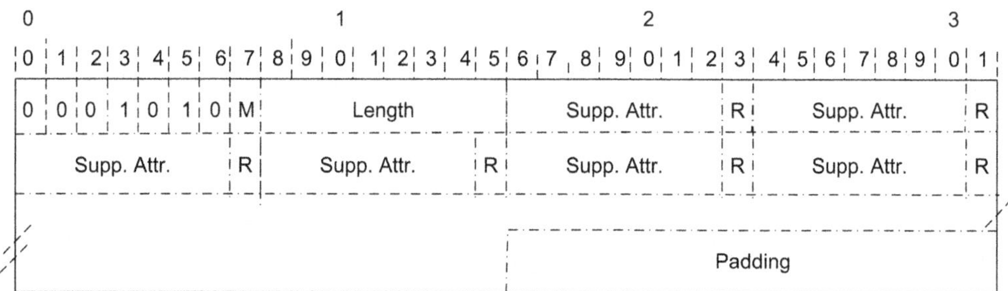

FIGURE 6.17 SUPPORTED-ATTRIBUTES Format.

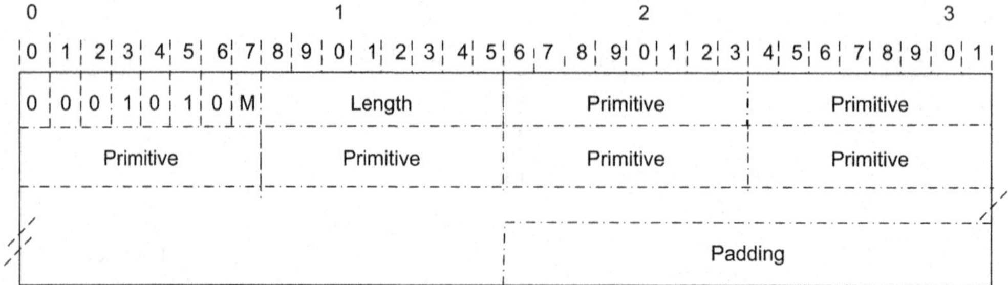

FIGURE 6.18 SUPPORTED-PRIMITIVES Format.

- **Supp. Attr.**: These fields contain the Types of the attributes that are supported by the floor control server in the following format:
- **R**: Reserved: This bit MUST be set to zero upon transmission and MUST be ignored upon reception.
- **Padding:** Two octets of padding added so that the contents of the SUPPORTED-ATTRIBUTES attribute is 32-bit aligned. If the attribute is already 32-bit aligned, no padding is needed.

The Padding bits SHOULD be set to zero by the sender and MUST be ignored by the receiver.

6.5.2.11 SUPPORTED-PRIMITIVES

The following is the format of the SUPPORTED-PRIMITIVES attribute (Figure 6.18).

- **Primitive:** These fields contain the types of the BFCP messages that are supported by the floor control server. See Table 6.1 for the list of BFCP primitives.
- **Padding:** One, two, or three octets of padding added so that the contents of the SUPPORTED-PRIMITIVES attribute is 32-bit aligned. If the attribute is already 32-bit aligned, no padding is needed. The Padding bits SHOULD be set to zero by the sender and MUST be ignored by the receiver.

6.5.2.12 USER-DISPLAY-NAME

The following is the format of the USER-DISPLAY-NAME attribute (Figure 6.19).

- **Text:** This field contains the UTF-8 encoded name of the user.
- **Padding:** One, two, or three octets of padding added so that the contents of the USER-DISPLAY-NAME attribute is 32-bit aligned. The Padding bits SHOULD be set to zero by the sender and MUST be ignored by the receiver. If the attribute is already 32-bit aligned, no padding is needed.

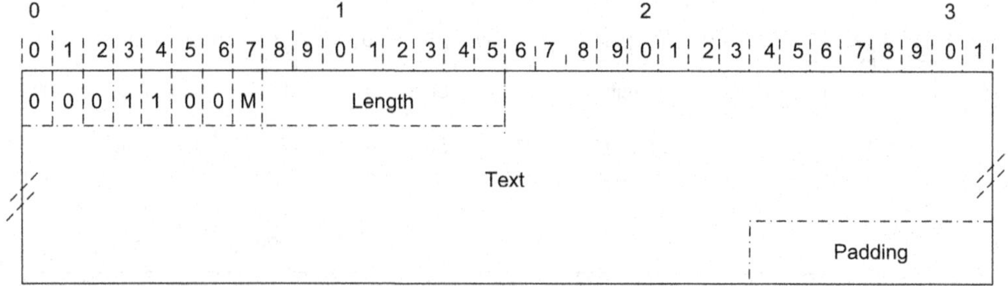

FIGURE 6.19 USER-DISPLAY-NAME Format.

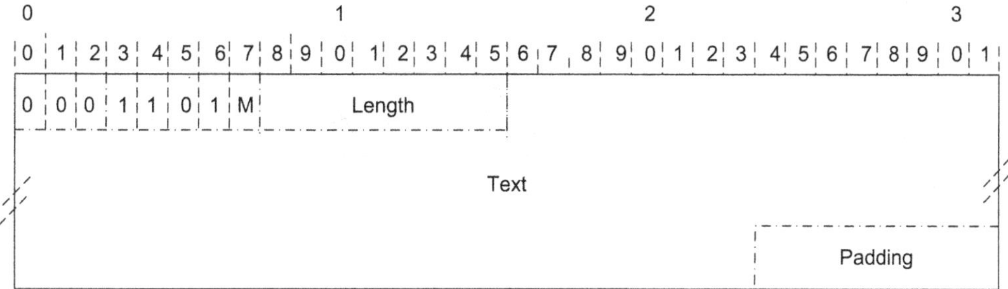

FIGURE 6.20 USER-URI Format.

6.5.2.13 USER-URI

The following is the format of the USER-URI attribute (Figure 6.20).

- **Text:** This field contains the UTF-8 encoded user's contact URI, that is, the URI used by the user to set up the resources (e.g., media streams) that are controlled by BFCP. For example, in the context of a conference set up by SIP, the USER-URI attribute would carry the SIP URI of the user.

  Messages containing a user's URI in a USER-URI attribute also contain the user's User ID. In this way, a client receiving such a message can correlate the user's URI (e.g., the SIP URI the user used to join a conference) with the user's User ID.
- **Padding:** One, two, or three octets of padding added so that the contents of the USER-URI attribute is 32-bit aligned. The Padding bits SHOULD be set to zero by the sender and MUST be ignored by the receiver. If the attribute is already 32-bit aligned, no padding is needed.

6.5.2.14 BENEFICIARY-INFORMATION

The BENEFICIARY-INFORMATION attribute is a grouped attribute that consists of a header, which is referred to as BENEFICIARYINFORMATION-HEADER, followed by a sequence of attributes. The following is the format of the BENEFICIARY-INFORMATION-HEADER (Figure 6.21):

- **Beneficiary ID:** This field contains a 16-bit value that uniquely identifies a user within a conference. The following is the ABNF (Augmented Backus–Naur Form) [2] of the BENEFICIARY-INFORMATION grouped attribute. (EXTENSION-ATTRIBUTE refers to extension attributes that may be defined in the future.) (Figure 6.22).

6.5.2.15 FLOOR-REQUEST-INFORMATION

The FLOOR-REQUEST-INFORMATION attribute is a grouped attribute that consists of a header, which is referred to as FLOOR-REQUESTINFORMATION- HEADER, followed by a sequence of attributes. The following is the format of the FLOOR-REQUEST-INFORMATION-HEADER (Figure 6.23):

- **Floor Request ID:** This field contains a 16-bit value that identifies a floor request at the floor control server. The following is the ABNF of the FLOOR-REQUEST-INFORMATION

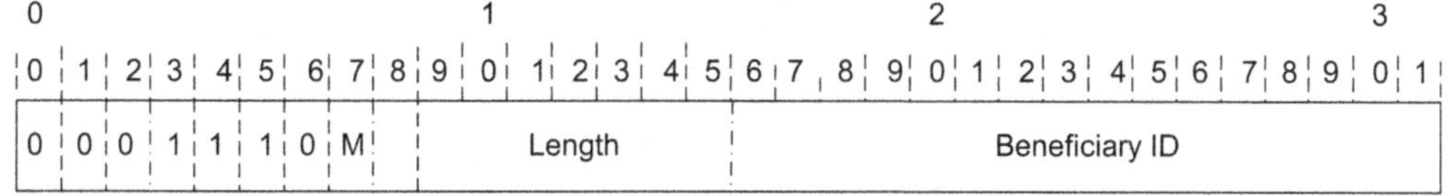

FIGURE 6.21 BENEFICIARY-INFORMATION-HEADER Format.

```
BENEFICIARY-INFORMATION = (BENEFICIARY-INFORMATION-HEADER)
                          [USER-DISPLAY-NAME]
                          [USER-URI]
                          *[EXTENSION-ATTRIBUTE]
```

FIGURE 6.22 BENEFICIARY-INFORMATION Format.

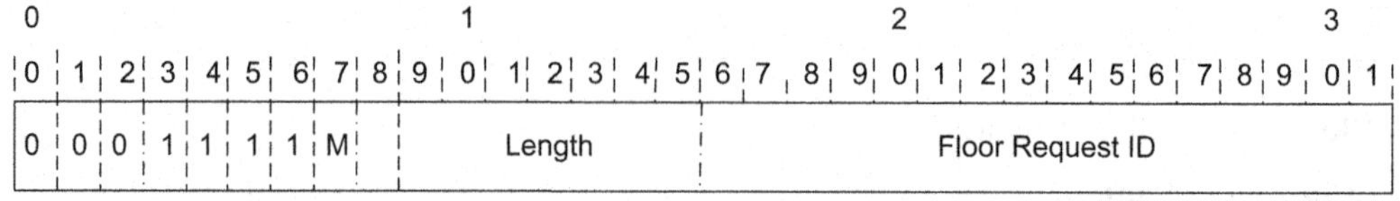

```
 0                   1                   2                   3
 0 1 2 3 4 5 6 7 8 9 0 1 2 3 4 5 6 7 8 9 0 1 2 3 4 5 6 7 8 9 0 1
+-+-+-+-+-+-+-+-+-+-+-+-+-+-+-+-+-+-+-+-+-+-+-+-+-+-+-+-+-+-+-+-+
|0 0 0 1 1 1 1 M|    Length     |       Floor Request ID        |
+-+-+-+-+-+-+-+-+-+-+-+-+-+-+-+-+-+-+-+-+-+-+-+-+-+-+-+-+-+-+-+-+
```

FIGURE 6.23 FLOOR-REQUEST-INFORMATION-HEADER Format.

```
FLOOR-REQUEST-INFORMATION =   (FLOOR-REQUEST-INFORMATION-
                              HEADER)
                              [OVERALL-REQUEST-STATUS]
                              1*(FLOOR-REQUEST-STATUS)
                              [BENEFICIARY-INFORMATION]
                              [REQUESTED-BY-INFORMATION]
                              [PRIORITY]
                              [PARTICIPANT-PROVIDED-INFO]
                              *[EXTENSION-ATTRIBUTE]
```

FIGURE 6.24 FLOOR-REQUEST-INFORMATION Format.

grouped attribute. (EXTENSION-ATTRIBUTE refers to extension attributes that may be defined in the future.) (Figure 6.24).

6.5.2.16 REQUESTED-BY-INFORMATION

The REQUESTED-BY-INFORMATION attribute is a grouped attribute that consists of a header, which is referred to as REQUESTED-BYINFORMATION-HEADER, followed by a sequence of attributes. The following is the format of the REQUESTED-BY-INFORMATION-HEADER (Figure 6.25):

- **Requested-by ID:** This field contains a 16-bit value that uniquely identifies a user within a conference. The following is the ABNF of the REQUESTED-BY-INFORMATION grouped attribute. (EXTENSION-ATTRIBUTE refers to extension attributes that may be defined in the future.) (Figure 6.26).

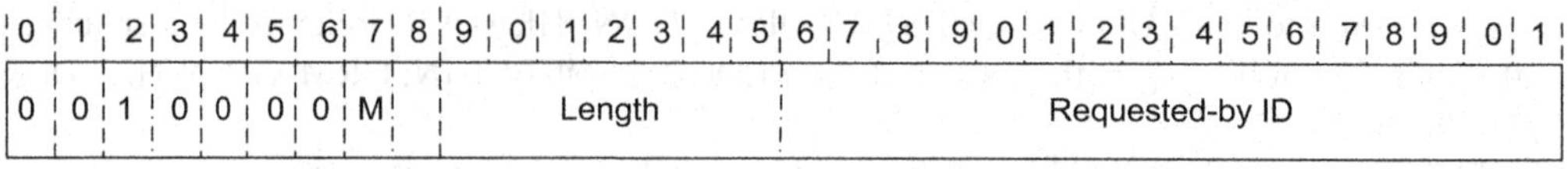

```
 0 1 2 3 4 5 6 7 8 9 0 1 2 3 4 5 6 7 8 9 0 1 2 3 4 5 6 7 8 9 0 1
+-+-+-+-+-+-+-+-+-+-+-+-+-+-+-+-+-+-+-+-+-+-+-+-+-+-+-+-+-+-+-+-+
|0 0 1 0 0 0 0 M|    Length     |        Requested-by ID        |
+-+-+-+-+-+-+-+-+-+-+-+-+-+-+-+-+-+-+-+-+-+-+-+-+-+-+-+-+-+-+-+-+
```

FIGURE 6.25 REQUESTED-BY-INFORMATION-HEADER Format.

```
REQUESTED-BY-INFORMATION = (REQUESTED-BY-INFORMATION-HEADER)
                           [USER-DISPLAY-NAME]
                           [USER-URI]
                           *[EXTENSION-ATTRIBUTE]
```

FIGURE 6.26 REQUESTED-BY-INFORMATION Format.

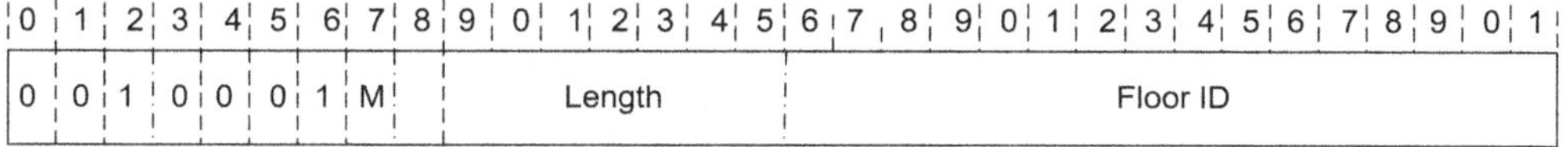

FIGURE 6.27 FLOOR-REQUEST-STATUS-HEADER Format.

```
FLOOR-REQUEST-STATUS =     (FLOOR-REQUEST-STATUS-HEADER)
                           [REQUEST-STATUS]
                           [STATUS-INFO]
                           *[EXTENSION-ATTRIBUTE]
```

FIGURE 6.28 FLOOR-REQUEST-STATUS Format.

6.5.2.17 FLOOR-REQUEST-STATUS

The FLOOR-REQUEST-STATUS attribute is a grouped attribute that consists of a header, which is referred to as FLOOR-REQUEST-STATUS-HEADER, followed by a sequence of attributes. The following is the format of the FLOOR-REQUEST-STATUS-HEADER (Figure 6.27):

- **Floor ID:** this field contains a 16-bit value that uniquely identifies a floor within a conference. The following is the ABNF of the FLOOR-REQUEST-STATUS grouped attribute. (EXTENSION-ATTRIBUTE refers to extension attributes that may be defined in the future.) (Figure 6.28).

6.5.2.18 OVERALL-REQUEST-STATUS

The OVERALL-REQUEST-STATUS attribute is a grouped attribute that consists of a header, which is referred to as OVERALL-REQUEST-STATUS-HEADER, followed by a sequence of attributes. The following is the format of the OVERALL-REQUEST-STATUS-HEADER (Figure 6.29):

- **Floor Request ID:** this field contains a 16-bit value that identifies a floor request at the floor control server. The following is the ABNF of the OVERALL-REQUEST-STATUS grouped attribute. (EXTENSION-ATTRIBUTE refers to extension attributes that may be defined in the future.) (Figure 6.30).

6.5.3 Message Format

This section contains the normative ABNF [2] of the BFCP messages. Extension attributes that may be defined in the future are referred to as EXTENSION-ATTRIBUTE in the ABNF.

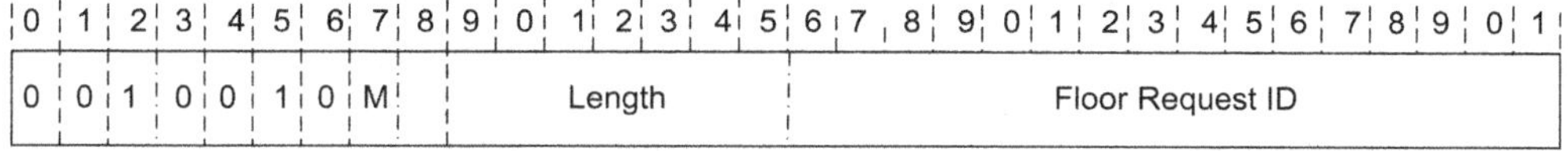

FIGURE 6.29 OVERALL-REQUEST-STATUS-HEADER Format.

```
OVERALL-REQUEST-STATUS =  (OVERALL-REQUEST-STATUS-HEADER)
                          [REQUEST-STATUS]
                          [STATUS-INFO]
                          *[EXTENSION-ATTRIBUTE]
```

FIGURE 6.30 OVERALL-REQUEST-STATUS Format.

```
FloorRequest =      (COMMON-HEADER)
                    1*(FLOOR-ID)
                    [BENEFICIARY-ID]
                    [PARTICIPANT-PROVIDED-INFO]
                    [PRIORITY]
                    *[EXTENSION-ATTRIBUTE]
```

FIGURE 6.31 FloorRequest Format.

```
FloorRelease =      (COMMON-HEADER)
                    (FLOOR-REQUEST-ID)
                    *[EXTENSION-ATTRIBUTE]
```

FIGURE 6.32 FloorRelease Format.

6.5.3.1 FloorRequest

Floor participants request a floor by sending a FloorRequest message to the floor control server. The following is the format of the FloorRequest message (Figure 6.31):

6.5.3.2 FloorRelease

Floor participants release a floor by sending a FloorRelease message to the floor control server. Floor participants also use the FloorRelease message to cancel pending floor requests. The following is the format of the FloorRelease message (Figure 6.32):

6.5.3.3 FloorRequestQuery

Floor participants and floor chairs request information about a floor request by sending a FloorRequestQuery message to the floor control server. The following is the format of the FloorRequestQuery message (Figure 6.33):

6.5.3.4 FloorRequestStatus

The floor control server informs floor participants and floor chairs about the status of their floor requests by sending them FloorRequestStatus messages. The following is the format of the FloorRequestStatus message (Figure 6.34):

6.5.3.5 UserQuery

Floor participants and floor chairs request information about a participant and the floor requests related to this participant by sending a UserQuery message to the floor control server. The following is the format of the UserQuery message (Figure 6.35):

```
FloorRequestQuery =     (COMMON-HEADER)
                        (FLOOR-REQUEST-ID)
                        *[EXTENSION-ATTRIBUTE]
```

FIGURE 6.33 FloorRequestQuery Format.

```
FloorRequestStatus =    (COMMON-HEADER)
                        (FLOOR-REQUEST-INFORMATION)
                        *[EXTENSION-ATTRIBUTE]
```

FIGURE 6.34 FloorRequestStatus Format.

```
UserQuery =        (COMMON-HEADER)
                   [BENEFICIARY-ID]
                   *[EXTENSION-ATTRIBUTE]
```

FIGURE 6.35 UserQuery Format.

```
UserStatus =       (COMMON-HEADER)
                   [BENEFICIARY-INFORMATION]
                   *(FLOOR-REQUEST-INFORMATION)
                   *[EXTENSION-ATTRIBUTE]
```

FIGURE 6.36 UserStatus Format.

6.5.3.6 UserStatus

The floor control server provides information about participants and their related floor requests to floor participants and floor chairs by sending them UserStatus messages. The following is the format of the UserStatus message (Figure 6.36):

6.5.3.7 FloorQuery

Floor participants and floor chairs request information about a floor or floors by sending a FloorQuery message to the floor control server. The following is the format of the FloorRequest message (Figure 6.37):

6.5.3.8 FloorStatus

The floor control server informs floor participants and floor chairs about the status (e.g., the current holder) of a floor by sending them FloorStatus messages. The following is the format of the FloorStatus message (Figure 6.38):

6.5.3.9 ChairAction

Floor chairs send instructions to floor control servers by sending ChairAction messages. The following is the format of the ChairAction message (Figure 6.39):

6.5.3.10 ChairActionAck

Floor control servers confirm that they have accepted a ChairAction message by sending a ChairActionAck message. The following is the format of the ChairActionAck message (Figure 6.40):

```
FloorQuery =       (COMMON-HEADER)
                   *(FLOOR-ID)
                   *[EXTENSION-ATTRIBUTE]
```

FIGURE 6.37 FloorQuery Format.

```
FloorStatus =      (COMMON-HEADER)
                   *1(FLOOR-ID)
                   *[FLOOR-REQUEST-INFORMATION]
                   *[EXTENSION-ATTRIBUTE]
```

FIGURE 6.38 FloorStatus Format.

```
ChairAction =       (COMMON-HEADER)
                    (FLOOR-REQUEST-INFORMATION)
                    *[EXTENSION-ATTRIBUTE]
```

FIGURE 6.39 ChairAction Format.

```
ChairActionAck =    (COMMON-HEADER)
                    *[EXTENSION-ATTRIBUTE]
```

FIGURE 6.40 ChairActionAck Format.

```
Hello =             (COMMON-HEADER)
                    *[EXTENSION-ATTRIBUTE]
```

FIGURE 6.41 Hello Format.

6.5.3.11 Hello

Floor participants and floor chairs check the liveliness of floor control servers by sending a Hello message. The following is the format of the Hello message (Figure 6.41):

6.5.3.12 HelloAck

Floor control servers confirm that they are alive on reception of a Hello message by sending a HelloAck message. The following is the format of the HelloAck message (Figure 6.42):

6.5.3.13 Error

Floor control servers inform floor participants and floor chairs about errors in processing requests by sending them Error messages. The following is the format of the Error message (Figure 6.43):

6.6 TRANSPORT

BFCP entities exchange BFCP messages using transmission control protocol (TCP) connections. TCP provides an in-order reliable delivery of a stream of bytes. Consequently, message framing

```
HelloAck =          (COMMON-HEADER)
                    (SUPPORTED-PRIMITIVES)
                    (SUPPORTED-ATTRIBUTES)
                    *[EXTENSION-ATTRIBUTE]
```

FIGURE 6.42 HelloAck Format.

```
Error =             (COMMON-HEADER)
                    (ERROR-CODE)
                    [ERROR-INFO]
                    *[EXTENSION-ATTRIBUTE]
```

FIGURE 6.43 Error Format.

is implemented in the application layer. BFCP implements application-layer framing using TLV-encoded attributes. A client MUST NOT use more than one TCP connection to communicate with a given floor control server within a conference. Nevertheless, if the same physical box handles different clients (e.g., a floor chair and a floor participant), which are identified by different User IDs, a separate connection per client is allowed.

If a BFCP entity (a client or a floor control server) receives data from TCP that cannot be parsed, the entity MUST close the TCP connection, and the connection SHOULD be reestablished. Similarly, if a TCP connection cannot deliver a BFCP message and times out, the TCP connection SHOULD be reestablished.

The way connection reestablishment is handled depends on how the client obtains information to contact the floor control server (e.g., using an SDP offer/answer exchange (RFC 4583) [7]). Once the TCP connection is reestablished, the client MAY resend those messages for which it did not get a response from the floor control server. If a floor control server detects that the TCP connection towards one of the floor participants is lost, it is up to the local policy of the floor control server what to do with the pending floor requests of the floor participant. In any case, it is RECOMMENDED that the floor control server keep the floor requests (i.e., that it does not cancel them) while the TCP connection is reestablished.

If a client wishes to end its BFCP connection with a floor control server, the client closes (i.e., a graceful close) the TCP connection towards the floor control server. If a floor control server wishes to end its BFCP connection with a client (e.g., the focus of the conference informs the floor control server that the client has been kicked out of the conference), the floor control server closes (i.e., a graceful close) the TCP connection towards the client.

6.7 LOWER-LAYER SECURITY

BFCP relies on lower-layer security mechanisms to provide replay and integrity protection and confidentiality. BFCP floor control servers and clients (which include both floor participants and floor chairs) MUST support transport layer security (TLS) (RFC 4346) [3]. Any BFCP entity MAY support other security mechanisms.

BFCP entities MUST support, at a minimum, the TLS TLS_RSA_WITH_AES_128_CBC_SHA ciphersuite (RFC 3268) [5]. Which party, the client or the floor control server, acts as the TLS server depends on how the underlying TCP connection is established. For example, when the TCP connection is established using an SDP offer/answer exchange (RFC 4583) [7], the answerer (which may be the client or the floor control server) always acts as the TLS server.

6.8 PROTOCOL TRANSACTIONS

In BFCP, there are two types of transactions: client-initiated transactions and server-initiated transactions (notifications). Client-initiated transactions consist of a request from a client to a floor control server and a response from the floor control server to the client. The request carries a Transaction ID in its common header, which the floor control server copies into the response. Clients use Transaction ID values to match responses with previously issued requests. Server-initiated transactions consist of a single message from a floor control server to a client. Since they do not trigger any response, their Transaction ID is set to 0.

6.8.1 CLIENT BEHAVIOR

A client starting a client-initiated transaction MUST set the Conference ID in the common header of the message to the Conference ID for the conference that the client obtained previously. The client MUST set the Transaction ID value in the common header to a number that is different from 0 and that MUST NOT be reused in another message from the client until a response from the server is received for the transaction. The client uses the Transaction ID value to match this message with the response from the floor control server.

6.8.2 Server Behavior

A floor control server sending a response within a client-initiated transaction MUST copy the Conference ID, the Transaction ID, and the User ID from the request received from the client into the response. Server-initiated transactions MUST contain a Transaction ID equal to 0.

6.9 AUTHENTICATION AND AUTHORIZATION

BFCP clients SHOULD authenticate the floor control server before sending any BFCP message to it or accepting any BFCP message from it. Similarly, floor control servers SHOULD authenticate a client before accepting any BFCP message from it or sending any BFCP message to it. BFCP supports TLS-based mutual authentication between clients and floor control servers, as specified in Section 6.9.1. This is the RECOMMENDED authentication mechanism in BFCP.

Note that future extensions may define additional authentication mechanisms. In addition to authenticating BFCP messages, floor control servers need to authorize them. On receiving an authenticated BFCP message, the floor control server checks whether the client sending the message is authorized. If the client is not authorized to perform the operation being requested, the floor control server generates an Error message, as described in Section 13.8, with an Error code with a value of 5 (Unauthorized Operation). Messages from a client that cannot be authorized MUST NOT be processed further.

6.9.1 TLS-Based Mutual Authentication

BFCP supports TLS-based mutual authentication between clients and floor control servers. BFCP assumes that there is an integrity-protected channel between the client and the floor control server that can be used to exchange their self-signed certificates or more commonly, the fingerprints of these certificates. These certificates are used at TLS establishment time.

The implementation of such an integrity-protected channel using SIP and the SDP offer/answer model is described in RFC 4583 [7]. BFCP messages received over an authenticated TLS connection are considered authenticated. A floor control server that receives a BFCP message over TCP (no TLS) can request the use of TLS by generating an Error message, as described in Section 6.13.8, with an Error code with a value of 9 (Use TLS). Clients SHOULD simply ignore unauthenticated messages.

Note that future extensions may define additional authentication mechanisms that may not require an initial integrity-protected channel (e.g., authentication based on certificates signed by a certificate authority). As described in Section 6.9, floor control servers need to perform authorization before processing any message. In particular, the floor control server SHOULD check that messages arriving over a given authenticated TLS connection use an authorized User ID (i.e., a User ID that the user that established the authenticated TLS connection is allowed to use).

6.10 FLOOR PARTICIPANT OPERATIONS

This section specifies how floor participants can perform different operations, such as requesting a floor, using the protocol elements described in earlier sections. Section 6.11 specifies operations that are specific to floor chairs, such as instructing the floor control server to grant or revoke a floor, and Section 6.12 specifies operations that can be performed by any client (i.e., both floor participants and floor chairs).

6.10.1 Requesting a Floor

A floor participant that wishes to request one or more floors does so by sending a FloorRequest message to the floor control server.

6.10.1.1 Sending a FloorRequest Message

The ABNF in Section 6.5.3.1 describes the attributes that a FloorRequest message can contain. In addition, the ABNF specifies normatively which of these attributes are mandatory and which ones are optional. The floor participant sets the Conference ID and the Transaction ID in the common header following the rules given in Section 6.8.1.

The floor participant sets the User ID in the common header to the floor participant's identifier. This User ID will be used by the floor control server to authenticate and authorize the request. If the sender of the FloorRequest message (identified by the User ID) is not the participant that would eventually get the floor (i.e., a third-party floor request), the sender SHOULD add a BENEFICIARY-ID attribute to the message identifying the beneficiary of the floor.

Note that the name space for both the User ID and the Beneficiary ID is the same. That is, a given participant is identified by a single 16-bit value that can be used in the User ID in the common header and in several attributes: BENEFICIARY-ID, BENEFICIARYINFORMATION, and REQUESTED-BY-INFORMATION. The floor participant must insert at least one FLOOR-ID attribute in the FloorRequest message. If the client inserts more than one FLOOR-ID attribute, the floor control server will treat all the floor requests as an atomic package. That is, the floor control server will either grant or deny all the floors in the FloorRequest message.

The floor participant may use a PARTICIPANT-PROVIDED-INFO attribute to state the reason why the floor or floors are being requested. The Text field in the PARTICIPANT-PROVIDED-INFO attribute is intended for human consumption. The floor participant may request that the server handle the floor request with a certain priority using a PRIORITY attribute.

6.10.1.2 Receiving a Response

A message from the floor control server is considered a response to the FloorRequest message if the message from the floor control server has the same Conference ID, Transaction ID, and User ID as the FloorRequest message, as described in Section 6.8.1. On receiving such a response, the floor participant follows the rules in Section 6.9 that relate to floor control server authentication.

The successful processing of a FloorRequest message at the floor control server involves generating one or several FloorRequestStatus messages. The floor participant obtains a Floor Request ID in the Floor Request ID field of a FLOOR-REQUEST-INFORMATION attribute in the first FloorRequestStatus message from the floor control server.

Subsequent FloorRequestStatus messages from the floor control server regarding the same floor request will carry the same Floor Request ID in a FLOOR-REQUEST-INFORMATION attribute as the initial FloorRequestStatus message. In this way, the floor participant can associate subsequent incoming FloorRequestStatus messages with the ongoing floor request.

The floor participant obtains information about the status of the floor request in the FLOOR-REQUEST-INFORMATION attribute of each of the FloorRequestStatus messages received from the floor control server. This attribute is a grouped attribute, and as such it includes a number of attributes that provide information about the floor request.

The OVERALL-REQUEST-STATUS attribute provides information about the overall status of the floor request. If the Request Status value is Granted, all the floors that were requested in the FloorRequest message have been granted. If the Request Status value is Denied, all the floors that were requested in the FloorRequest message have been denied. A floor request is considered to be ongoing while it is in the Pending, Accepted, or Granted state. If the floor request value is unknown, then the response is still processed. However, no meaningful value can be reported to the user.

The STATUS-INFO attribute, if present, provides extra information that the floor participant MAY display to the user. The FLOOR-REQUEST-STATUS attributes provide information about the status of the floor request as it relates to a particular floor. The STATUS-INFO attribute, if present, provides extra information that the floor participant MAY display to the user.

The BENEFICIARY-INFORMATION attribute identifies the beneficiary of the floor request in third-party floor requests. The REQUESTED-BY-INFORMATION attribute need not be present in FloorRequestStatus messages received by the floor participant that requested the floor, as this floor participant is already identified by the User ID in the common header.

The PRIORITY attribute, when present, contains the priority that was requested by the generator of the FloorRequest message. If the response is an Error message, the floor control server could not process the FloorRequest message for some reason, which is described in the Error message.

6.10.2 Cancelling a Floor Request and Releasing a Floor

A floor participant that wishes to cancel an ongoing floor request does so by sending a FloorRelease message to the floor control server. The FloorRelease message is also used by floor participants that hold a floor and would like to release it.

6.10.2.1 Sending a FloorRelease Message

The ABNF in Section 5.3.2 describes the attributes that a FloorRelease message can contain. In addition, the ABNF specifies normatively which of these attributes are mandatory and which ones are optional. The floor participant sets the Conference ID and the Transaction ID in the common header following the rules given in Section 6.8.1. The floor participant sets the User ID in the common header to the floor participant's identifier. This User ID will be used by the floor control server to authenticate and authorize the request.

Note that the FloorRelease message is used to release a floor or floors that were granted and to cancel ongoing floor requests (from the protocol perspective, both are ongoing floor requests). Using the same message in both situations helps resolve the race condition that occurs when the FloorRelease message and the FloorGrant message cross each other on the wire. The floor participant uses the FLOOR-REQUEST-ID that was received in the response to the FloorRequest message that the FloorRelease message is cancelling. Note that if the floor participant requested several floors as an atomic operation (i.e., in a single FloorRequest message), all the floors are released as an atomic operation as well (i.e., all are released at the same time).

6.10.2.2 Receiving a Response

A message from the floor control server is considered a response to the FloorRelease message if the message from the floor control server has the same Conference ID, Transaction ID, and User ID as the FloorRequest message, as described in Section 6.8.1. On receiving such a response, the floor participant follows the rules in Section 6.9 that relate to floor control server authentication.

If the response is a FloorRequestStatus message, the Request Status value in the OVERALL-REQUEST-STATUS attribute (within the FLOORREQUEST-INFORMATION grouped attribute) will be Cancelled or Released. If the response is an Error message, the floor control server could not process the FloorRequest message for some reason, which is described in the Error message.

It is possible that the FloorRelease message crosses on the wire with a FloorRequestStatus message from the server with a Request Status different from Cancelled or Released. In any case, such a FloorRequestStatus message will not be a response to the FloorRelease message, as its Transaction ID will not match that of the FloorRelease.

6.11 CHAIR OPERATIONS

This section specifies how floor chairs can instruct the floor control server to grant or revoke a floor using the protocol elements described in earlier sections. Floor chairs that wish to send instructions to a floor control server do so by sending a ChairAction message.

6.11.1 SENDING A CHAIRACTION MESSAGE

The ABNF in Section 6.5.3.9 describes the attributes that a ChairAction message can contain. In addition, the ABNF specifies normatively which of these attributes are mandatory and which ones are optional. The floor chair sets the Conference ID and the Transaction ID in the common header following the rules given in Section 6.8.1. The floor chair sets the User ID in the common header to the floor participant's identifier. This User ID will be used by the floor control server to authenticate and authorize the request.

The ChairAction message contains instructions that apply to one or more floors within a particular floor request. The floor or floors are identified by the FLOOR-REQUEST-STATUS attributes, and the floor request is identified by the FLOOR-REQUEST-INFORMATION-HEADER, which are carried in the ChairAction message.

For example, if a floor request consists of two floors that depend on different floor chairs, each floor chair will grant its floor within the floor request. Once both chairs have granted their floor, the floor control server will grant the floor request as a whole. On the other hand, if one of the floor chairs denies its floor, the floor control server will deny the floor request as a whole, regardless of the other floor chair's decision.

The floor chair provides the new status of the floor request as it relates to a particular floor using a FLOOR-REQUEST-STATUS attribute. If the new status of the floor request is Accepted, the floor chair MAY use the Queue Position field to provide a queue position for the floor request. If the floor chair does not wish to provide a queue position, all the bits of the Queue Position field SHOULD be set to zero. The floor chair SHOULD use the Status Revoked to revoke a floor that was granted (i.e., Granted status) and SHOULD use the Status Denied to reject floor requests in any other status (e.g., Pending and Accepted).

The floor chair MAY add an OVERALL-REQUEST-STATUS attribute to the ChairAction message to provide a new overall status for the floor request. If the new overall status of the floor request is Accepted, the floor chair MAY use the Queue Position field to provide a queue position for the floor request. Note that a particular floor control server may implement a different queue for each floor containing all the floor requests that relate to that particular floor, a general queue for all floor requests, or both. Also, note that a floor request may involve several floors and that a ChairAction message may only deal with a subset of these floors (e.g., if a single floor chair is not authorized to manage all the floors). In this case, the floor control server will combine the instructions received from the different floor chairs in FLOOR-REQUEST-STATUS attributes to come up with the overall status of the floor request. Note that while the action of a floor chair may communicate information in the OVERALL-REQUEST-STATUS attribute, the floor control server may override, modify, or ignore this field's content. The floor chair may use STATUS-INFO attributes to state the reason why the floor or floors are being accepted, granted, or revoked. The Text in the STATUS-INFO attribute is intended for human consumption.

6.11.2 RECEIVING A RESPONSE

A message from the floor control server is considered a response to the ChairAction message if the message from the server has the same Conference ID, Transaction ID, and User ID as the ChairAction message, as described in Section 6.8.1. On receiving such a response, the floor chair follows the rules in Section 6.9 that relate to floor control server authentication.

A ChairActionAck message from the floor control server confirms that the floor control server has accepted the ChairAction message. An Error message indicates that the floor control server could not process the ChairAction message for some reason, which is described in the Error message.

6.12 GENERAL CLIENT OPERATIONS

This section specifies operations that can be performed by any client. That is, they are not specific to floor participants or floor chairs. They can be performed by both.

6.12.1 Requesting Information about Floors

A client can obtain information about the status of a floor or floors in different ways, which include using BFCP and using out-of-band mechanisms. Clients using BFCP to obtain such information use the procedures described in this section. Clients request information about the status of one or several floors by sending a FloorQuery message to the floor control server.

6.12.1.1 Sending a FloorQuery Message

The ABNF in Section 6.5.3.7 describes the attributes that a FloorQuery message can contain. In addition, the ABNF specifies normatively which of these attributes are mandatory and which ones are optional. The client sets the Conference ID and the Transaction ID in the common header following the rules given in Section 6.8.1. The client sets the User ID in the common header to the client's identifier. This User ID will be used by the floor control server to authenticate and authorize the request.

The client inserts in the message all the Floor IDs it wants to receive information about. The floor control server will send periodic information about all these floors. If the client does not want to receive information about a particular floor any longer, it sends a new FloorQuery message removing the FLOOR-ID of this floor. If the client does not want to receive information about any floor any longer, it sends a FloorQuery message with no FLOOR-ID attribute.

6.12.1.2 Receiving a Response

A message from the floor control server is considered a response to the FloorQuery message if the message from the floor control server has the same Conference ID, Transaction ID, and User ID as the FloorRequest message, as described in Section 6.8.1. On receiving such a response, the client follows the rules in Section 6.9 that relate to floor control server authentication.

On reception of the FloorQuery message, the floor control server will respond with a FloorStatus message or with an Error message. If the response is a FloorStatus message, it will contain information about one of the floors the client requested information about. If the client did not include any FLOOR-ID attribute in its FloorQuery message (i.e., the client does not want to receive information about any floor any longer), the FloorStatus message from the floor control server will not include any FLOOR-ID attribute either. FloorStatus messages that carry information about a floor contain a FLOOR-ID attribute that identifies the floor.

After this attribute, FloorStatus messages contain information about existing (one or more) floor requests that relate to that floor. The information about each particular floor request is encoded in a FLOOR-REQUEST-INFORMATION attribute. This grouped attribute carries a Floor Request ID that identifies the floor request, followed by a set of attributes that provide information about the floor request. After the first FloorStatus, the floor control server will continue sending FloorStatus messages, periodically informing the client about changes on the floors the client requested information about.

6.12.2 Requesting Information about Floor Requests

A client can obtain information about the status of one or several floor requests in different ways, which include using BFCP and using out-of-band mechanisms. Clients using BFCP to obtain such information use the procedures described in this section. Clients request information about the current status of a floor request by sending a FloorRequestQuery message to the floor control server.

Requesting information about a particular floor request is useful in a number of situations. For example, on reception of a FloorRequest message, a floor control server may choose to return FloorRequestStatus messages only when the floor request changes its state (e.g., from Accepted to Granted) but not when the floor request advances in its queue. In this situation, if the user requests it, the floor participant can use a FloorRequestQuery message to poll the floor control server for the status of the floor request.

6.12.2.1 Sending a FloorRequestQuery Message

The ABNF in Section 6.5.3.3 describes the attributes that a FloorRequestQuery message can contain. In addition, the ABNF specifies normatively which of these attributes are mandatory and which ones are optional. The client sets the Conference ID and the Transaction ID in the common header following the rules given in Section 6.8.1.

The client sets the User ID in the common header to the client's identifier. This User ID will be used by the floor control server to authenticate and authorize the request. The client must insert a FLOOR-REQUEST-ID attribute that identifies the floor request at the floor control server.

6.12.2.2 Receiving a Response

A message from the floor control server is considered a response to the FloorRequestQuery message if the message from the floor control server has the same Conference ID, Transaction ID, and User ID as the FloorRequestQuery message, as described in Section 6.8.1. On receiving such a response, the client follows the rules in Section 6.9 that relate to floor control server authentication.

If the response is a FloorRequestStatus message, the client obtains information about the status of the FloorRequest the client requested information about in a FLOOR-REQUEST-INFORMATION attribute. If the response is an Error message, the floor control server could not process the FloorRequestQuery message for some reason, which is described in the Error message.

6.12.3 Requesting Information about a User

A client can obtain information about a participant and the floor requests related to this participant in different ways, which include using BFCP and using out-of-band mechanisms. Clients using BFCP to obtain such information use the procedures described in this section. Clients request information about a participant and the floor requests related to this participant by sending a UserQuery message to the floor control server.

This functionality may be useful for floor chairs or floor participants interested in the display name and the URI of a particular floor participant. In addition, a floor participant may find it useful to request information about itself. For example, a floor participant, after experiencing connectivity problems (e.g., its TCP connection with the floor control server was down for a while and eventually was reestablished), may need to request information about all the floor requests associated to itself that still exist.

6.12.3.1 Sending a UserQuery Message

The ABNF in Section 6.5.3.5 describes the attributes that a UserQuery message can contain. In addition, the ABNF specifies normatively which of these attributes are mandatory and which ones are optional. The client sets the Conference ID and the Transaction ID in the common header following the rules given in Section 6.8.1. The client sets the User ID in the common header to the client's identifier.

This User ID will be used by the floor control server to authenticate and authorize the request. If the floor participant the client is requesting information about is not the client issuing the UserQuery message (which is identified by the User ID in the common header of the message), the client MUST insert a BENEFICIARY-ID attribute.

6.12.3.2 Receiving a Response

A message from the floor control server is considered a response to the UserQuery message if the message from the floor control server has the same Conference ID, Transaction ID, and User ID as the UserQuery message, as described in Section 6.8.1. On receiving such a response, the client follows the rules in Section 6.9 that relate to floor control server authentication.

If the response is a UserStatus message, the client obtains information about the floor participant in a BENEFICIARY-INFORMATION grouped attribute and about the status of the floor requests associated with the floor participant in FLOOR-REQUEST-INFORMATION attributes. If the response is an Error message, the floor control server could not process the UserQuery message for some reason, which is described in the Error message.

6.12.4 Obtaining the Capabilities of a Floor Control Server

A client that wishes to obtain the capabilities of a floor control server does so by sending a Hello message to the floor control server.

6.12.4.1 Sending a Hello Message

The ABNF in Section 6.5.3.11 describes the attributes that a Hello message can contain. In addition, the ABNF specifies normatively which of these attributes are mandatory and which ones are optional. The client sets the Conference ID and the Transaction ID in the common header following the rules given in Section 6.8.1. The client sets the User ID in the common header to the client's identifier. This User ID will be used by the floor control server to authenticate and authorize the request.

6.12.4.2 Receiving Responses

A message from the floor control server is considered a response to the Hello message by the client if the message from the floor control server has the same Conference ID, Transaction ID, and User ID as the Hello message, as described in Section 6.8.1. On receiving such a response, the client follows the rules in Section 6.9 that relate to floor control server authentication.

If the response is a HelloAck message, the floor control server could process the Hello message successfully. The SUPPORTED-PRIMITIVES and SUPPORTED-ATTRIBUTES attributes indicate which primitives and attributes, respectively, are supported by the server. If the response is an Error message, the floor control server could not process the Hello message for some reason, which is described in the Error message.

6.13 FLOOR CONTROL SERVER OPERATIONS

This section specifies how floor control servers can perform different operations, such as granting a floor, using the protocol elements described in earlier sections. On reception of a message from a client, the floor control server MUST check whether the value of the Primitive is supported. If it does not, the floor control server SHOULD send an Error message, as described in Section 6.13.8, with Error code 3 (Unknown Primitive).

On reception of a message from a client, the floor control server MUST check whether the value of the Conference ID matched an existing conference. If it does not, the floor control server SHOULD send an Error message, as described in Section 6.13.8, with Error code 1 (Conference does not Exist).

On reception of a message from a client, the floor control server follows the rules in Section 6.9 that relate to the authentication of the message. On reception of a message from a client, the floor control server MUST check whether it understands all the mandatory ("M" bit set) attributes in the message. If the floor control server does not understand all of them, the floor control server

SHOULD send an Error message, as described in Section 6.13.8, with Error code 2 (Authentication Failed). The Error message SHOULD list the attributes that were not understood.

6.13.1 Reception of a FloorRequest Message

On reception of a FloorRequest message, the floor control server follows the rules in Section 6.9 that relate to client authentication and authorization. If while processing the FloorRequest message, the floor control server encounters an error, it SHOULD generate an Error response following the procedures described in Section 6.13.8. BFCP allows floor participants to have several ongoing floor requests for the same floor (e.g., the same floor participant can occupy more than one position in a queue at the same time). A floor control server that only supports a certain number of ongoing floor requests per floor participant (e.g., one) can use Error Code 8 (You have Already Reached the Maximum Number of Ongoing Floor Requests for this Floor) to inform the floor participant.

6.13.1.1 Generating the First FloorRequestStatus Message

The successful processing of a FloorRequest message by a floor control server involves generating one or several FloorRequestStatus messages, the first of which SHOULD be generated as soon as possible. If the floor control server cannot accept, grant, or deny the floor request right away (e.g., a decision from a chair is needed), it SHOULD use a Request Status value of Pending in the OVERALL-REQUESTSTATUS attribute (within the FLOOR-REQUEST-INFORMATION grouped attribute) of the first FloorRequestStatus message it generates. The policy that a floor control server follows to grant or deny floors is outside the scope of this specification. A given floor control server may perform these decisions automatically, while another may contact a human acting as a chair every time a decision needs to be made.

The floor control server MUST copy the Conference ID, the Transaction ID, and the User ID from the FloorRequest into the FloorRequestStatus, as described in Section 6.8.2. Additionally, the floor control server MUST add a FLOOR-REQUEST-INFORMATION grouped attribute to the FloorRequestStatus. The attributes contained in this grouped attribute carry information about the floor request.

The floor control server MUST assign an identifier that is unique within the conference to this floor request, and MUST insert it in the Floor Request ID field of the FLOOR-REQUEST-INFORMATION attribute. This identifier will be used by the floor participant (or by a chair or chairs) to refer to this specific floor request in the future.

The floor control server MUST copy the Floor IDs in the FLOOR-ID attributes of the FloorRequest into the FLOOR-REQUEST-STATUS attributes in the FLOOR-REQUEST-INFORMATION grouped attribute. These Floor IDs identify the floors being requested (i.e., the floors associated with this particular floor request).

The floor control server SHOULD copy (if present) the contents of the BENEFICIARY-ID attribute from the FloorRequest into a BENEFICIARY-INFORMATION attribute inside the FLOOR-REQUEST-INFORMATION grouped attribute. Additionally, the floor control server MAY provide the display name and the URI of the beneficiary in this BENEFICIARY-INFORMATION attribute.

The floor control server MAY provide information about the requester of the floor in a REQUESTED-BY-INFORMATION attribute inside the FLOOR-REQUEST-INFORMATION grouped attribute. The floor control server MAY copy (if present) the PARTICIPANTPROVIDED-INFO attribute from the FloorRequest into the FLOORREQUEST-INFORMATION grouped attribute.

Note that this attribute carries the priority requested by the participant. The priority that the floor control server assigns to the floor request depends on the priority requested by the participant and the rights the participant has according to the policy of the conference. For example, a participant that is only allowed to use the Normal priority may request Highest priority for a floor request. In that case, the floor control server would ignore the priority requested by the participant. The floor control server MAY copy (if present) the PARTICIPANT-PROVIDED-INFO attribute from the FloorRequest into the FLOOR-REQUEST-INFORMATION grouped attribute.

6.13.1.2 Generation of Subsequent FloorRequestStatus Messages

A floor request is considered to be ongoing as long as it is not in the Cancelled, Released, or Revoked state. If the OVERALL-REQUESTSTATUS attribute (inside the FLOOR-REQUEST-INFORMATION grouped attribute) of the first FloorRequestStatus message generated by the floor control server did not indicate any of these states, the floor control server will need to send subsequent FloorRequestStatus messages.

When the status of the floor request changes, the floor control server SHOULD send new FloorRequestStatus messages with the appropriate Request Status. The floor control server MUST add a FLOOR-REQUEST-INFORMATION attribute with a Floor Request ID equal to the one sent in the first FloorRequestStatus message to any new FloorRequestStatus related to the same floor request. (The Floor Request ID identifies the floor request to which the FloorRequestStatus applies.)

The floor control server MUST set the Transaction ID of subsequent FloorRequestStatus messages to 0. The rate at which the floor control server sends FloorRequestStatus messages is a matter of local policy. A floor control server may choose to send a new FloorRequestStatus message every time the floor request moves in the floor request queue, while another may choose only to send a new FloorRequestStatus message when the floor request is Granted or Denied.

The floor control server may add a STATUS-INFO attribute to any of the FloorRequestStatus messages it generates to provide extra information about its decisions regarding the floor request (e.g., why it was denied). Floor participants and floor chairs may request to be informed about the status of a floor following the procedures in Section 6.12.1.

If the processing of a floor request changes the status of a floor (e.g., the floor request is granted and consequently, the floor has a new holder), the floor control server needs to follow the procedures in Section 6.13.5 to inform the clients that have requested that information. The common header and the rest of the attributes are the same as in the first FloorRequestStatus message. The floor control server can discard the state information about a particular floor request when this reaches a status of Cancelled, Released, or Revoked.

6.13.2 Reception of a FloorRequestQuery Message

On reception of a FloorRequestQuery message, the floor control server follows the rules in Section 6.9 that relate to client authentication and authorization. If while processing the FloorRequestQuery message, the floor control server encounters an error, it SHOULD generate an Error response following the procedures described in Section 6.13.8. The successful processing of a FloorRequestQuery message by a floor control server involves generating a FloorRequestStatus message, which SHOULD be generated as soon as possible. The floor control server MUST copy the Conference ID, the Transaction ID, and the User ID from the FloorRequestQuery message into the FloorRequestStatus message, as described in Section 6.8.2. Additionally, the floor control server MUST include information about the floor request in the FLOOR-REQUEST-INFORMATION grouped attribute to the FloorRequestStatus.

The floor control server MUST copy the contents of the FLOOR-REQUEST-ID attribute from the FloorRequestQuery message into the Floor Request ID field of the FLOOR-REQUEST-INFORMATION attribute. The floor control server MUST add FLOOR-REQUEST-STATUS attributes to the FLOOR-REQUEST-INFORMATION grouped attribute identifying the floors being requested (i.e., the floors associated with the floor request identified by the FLOOR-REQUEST-ID attribute).

The floor control server SHOULD add a BENEFICIARY-ID attribute to the FLOOR-REQUEST-INFORMATION grouped attribute identifying the beneficiary of the floor request. Additionally, the floor control server MAY provide the display name and the URI of the beneficiary in this BENEFICIARY-INFORMATION attribute.

The floor control server MAY provide information about the requester of the floor in a REQUESTED-BY-INFORMATION attribute inside the FLOOR-REQUEST-INFORMATION grouped attribute. The floor control server MAY provide the reason why the floor participant requested the floor in a PARTICIPANT-PROVIDED-INFO. The floor control server MAY also add to the FLOOR-REQUEST-INFORMATION grouped attribute a PRIORITY attribute with the Priority value requested for the floor request and a STATUS-INFO attribute with extra information about the floor request. The floor control server MUST add an OVERALL-REQUEST-STATUS attribute to the FLOOR-REQUEST-INFORMATION grouped attribute with the current status of the floor request. The floor control server MAY provide information about the status of the floor request as it relates to each of the floors being requested in the FLOOR-REQUEST-STATUS attributes.

6.13.3 Reception of a UserQuery Message

On reception of a UserQuery message, the floor control server follows the rules in Section 6.9 that relate to client authentication and authorization. If while processing the UserQuery message, the floor control server encounters an error, it SHOULD generate an Error response following the procedures described in Section 6.13.8.

The successful processing of a UserQuery message by a floor control server involves generating a UserStatus message, which SHOULD be generated as soon as possible. The floor control server MUST copy the Conference ID, the Transaction ID, and the User ID from the UserQuery message into the UserStatus message, as described in Section 6.8.2. The sender of the UserQuery message is requesting information about all the floor requests associated with a given participant (i.e., the floor requests where the participant is either the beneficiary or the requester). This participant is identified by a BENEFICIARY-ID attribute or in the absence of a BENEFICIARY-ID attribute, by the User ID in the common header of the UserQuery message.

The floor control server MUST copy, if present, the contents of the BENEFICIARY-ID attribute from the UserQuery message into a BENEFICIARY-INFORMATION attribute in the UserStatus message. Additionally, the floor control server MAY provide the display name and the URI of the participant about which the UserStatus message provides information in this BENEFICIARY-INFORMATION attribute.

The floor control server SHOULD add to the UserStatus message a FLOOR-REQUEST-INFORMATION grouped attribute for each floor request related to the participant about which the message provides information (i.e., the floor requests where the participant is either the beneficiary or the requester). For each FLOOR-REQUEST-INFORMATION attribute, the floor control server follows the following steps.

The floor control server MUST identify the floor request the FLOOR-REQUEST-INFORMATION attribute applies to by filling the Floor Request ID field of the FLOOR-REQUEST-INFORMATION attribute. The floor control server MUST add FLOOR-REQUEST-STATUS attributes to the FLOOR-REQUEST-INFORMATION grouped attribute identifying the floors being requested (i.e., the floors associated with the floor request identified by the FLOOR-REQUEST-ID attribute).

The floor control server SHOULD add a BENEFICIARY-ID attribute to the FLOOR-REQUEST-INFORMATION grouped attribute identifying the beneficiary of the floor request. Additionally, the floor control server MAY provide the display name and the URI of the beneficiary in this BENEFICIARY-INFORMATION attribute. The floor control server MAY provide information about the requester of the floor in a REQUESTED-BY-INFORMATION attribute inside the FLOOR-REQUEST-INFORMATION grouped attribute.

The floor control server MAY provide the reason why the floor participant requested the floor in a PARTICIPANT-PROVIDED-INFO. The floor control server MAY also add to the FLOOR-REQUESTINFORMATION grouped attribute a PRIORITY attribute with the Priority value requested for the floor request. The floor control server MUST include the current status of the floor request in an OVERALL-REQUEST-STATUS attribute to the FLOOR-REQUESTINFORMATION grouped attribute. The floor control server MAY add a STATUS-INFO attribute with extra information about the floor request. The floor control server MAY provide information about the status of the floor request as it relates to each of the floors being requested in the FLOOR-REQUEST-STATUS attributes.

6.13.4 Reception of a FloorRelease Message

On reception of a FloorRelease message, the floor control server follows the rules in Section 6.9 that relate to client authentication and authorization. If while processing the FloorRelease message, the floor control server encounters an error, it SHOULD generate an Error response following the procedures described in Section 6.13.8. The successful processing of a FloorRelease message by a floor control server involves generating a FloorRequestStatus message, which SHOULD be generated as soon as possible. The floor control server MUST copy the Conference ID, the Transaction ID, and the User ID from the FloorRelease message into the FloorRequestStatus message, as described in Section 6.8.2.

The floor control server MUST add a FLOOR-REQUEST-INFORMATION grouped attribute to the FloorRequestStatus. The attributes contained in this grouped attribute carry information about the floor request. The FloorRelease message identifies the floor request it applies to using a FLOOR-REQUEST-ID. The floor control server MUST copy the contents of the FLOOR-REQUEST-ID attribute from the FloorRelease message into the Floor Request ID field of the FLOOR-REQUEST-INFORMATION attribute. The floor control server MUST identify the floors being requested (i.e., the floors associated with the floor request identified by the FLOOR-REQUEST-ID attribute) in FLOOR-REQUEST-STATUS attributes to the FLOOR-REQUEST-INFORMATION grouped attribute. The floor control server MUST add an OVERALL-REQUEST-STATUS attribute to the FLOOR-REQUEST-INFORMATION grouped attribute. The Request Status value SHOULD be Released, if the floor (or floors) had been previously granted, or Cancelled, if the floor (or floors) had not been previously granted. The floor control server MAY add a STATUSINFO attribute with extra information about the floor request.

6.13.5 Reception of a FloorQuery Message

On reception of a FloorQuery message, the floor control server follows the rules in Section 6.9 that relate to client authentication. If while processing the FloorRelease message, the floor control server encounters an error, it SHOULD generate an Error response following the procedures described in Section 6.13.8.

A floor control server receiving a FloorQuery message from a client SHOULD keep this client informed about the status of the floors identified by FLOOR-ID attributes in the FloorQuery message. Floor Control Servers keep clients informed by using FloorStatus messages. An individual FloorStatus message carries information about a single floor. So when a FloorQuery message requests information about more than one floor, the floor control server needs to send separate FloorStatus messages for different floors. The information FloorQuery messages carry may depend on the user requesting the information. For example, a chair may be able to receive information about pending requests, while a regular user may not be authorized to do so.

6.13.5.1 Generation of the First FloorStatus Message

The successful processing of a FloorQuery message by a floor control server involves generating one or several FloorStatus messages, the first of which SHOULD be generated as soon as possible.

The floor control server MUST copy the Conference ID, the Transaction ID, and the User ID from the FloorQuery message into the FloorStatus message, as described in Section 6.8.2.

If the FloorQuery message did not contain any FLOOR-ID attribute, the floor control server sends the FloorStatus message without adding any additional attribute and does not send any subsequent FloorStatus message to the floor participant. If the FloorQuery message contained one or more FLOOR-ID attributes, the floor control server chooses one from among them and adds this FLOOR-ID attribute to the FloorStatus message. The floor control server SHOULD add a FLOOR-REQUEST-INFORMATION grouped attribute for each floor request associated to the floor. Each FLOOR-REQUEST-INFORMATION grouped attribute contains a number of attributes that provide information about the floor request. For each FLOOR-REQUEST-INFORMATION attribute, the floor control server follows the following steps.

The floor control server MUST identify the floor request the FLOOR-REQUEST-INFORMATION attribute applies to by filling the Floor Request ID field of the FLOOR-REQUEST-INFORMATION attribute. The floor control server MUST add FLOOR-REQUEST-STATUS attributes to the FLOOR-REQUEST-INFORMATION grouped attribute identifying the floors being requested (i.e., the floors associated with the floor request identified by the FLOOR-REQUEST-ID attribute).

The floor control server SHOULD add a BENEFICIARY-ID attribute to the FLOOR-REQUEST-INFORMATION grouped attribute identifying the beneficiary of the floor request. Additionally, the floor control server MAY provide the display name and the URI of the beneficiary in this BENEFICIARY-INFORMATION attribute. The floor control server MAY provide information about the requester of the floor in a REQUESTED-BY-INFORMATION attribute inside the FLOOR-REQUEST-INFORMATION grouped attribute. The floor control server MAY provide the reason why the floor participant requested the floor in a PARTICIPANT-PROVIDED-INFO. The floor control server MAY also add to the FLOOR-REQUEST-INFORMATION grouped attribute a PRIORITY attribute with the Priority value requested for the floor request.

The floor control server MUST add an OVERALL-REQUEST-STATUS attribute to the FLOOR-REQUEST-INFORMATION grouped attribute with the current status of the floor request. The floor control server MAY add a STATUS-INFO attribute with extra information about the floor request. The floor control server MAY provide information about the status of the floor request as it relates to each of the floors being requested in the FLOOR-REQUEST-STATUS attributes.

6.13.5.2 Generation of Subsequent FloorStatus Messages

If the FloorQuery message carried more than one FLOOR-ID attribute, the floor control server SHOULD generate a FloorStatus message for each of them (except for the FLOOR-ID attribute chosen for the first FloorStatus message) as soon as possible. These FloorStatus messages are generated following the same rules as those for the first FloorStatus message (see Section 6.13.5.1), but their Transaction ID is 0.

After generating these messages, the floor control server sends FloorStatus messages, periodically keeping the client informed about all the floors for which the client requested information. The Transaction ID of these messages MUST be 0. The rate at which the floor control server sends FloorStatus messages is a matter of local policy. A floor control server may choose to send a new FloorStatus message every time a new floor request arrives, while another may choose to only send a new FloorStatus message when a new floor request is Granted.

6.13.6 Reception of a ChairAction Message

On reception of a ChairAction message, the floor control server follows the rules in Section 6.9 that relate to client authentication and authorization. If while processing the ChairAction message, the floor control server encounters an error, it SHOULD generate an Error response following the procedures described in Section 6.13.8. The successful processing of a ChairAction message by a floor control server involves generating a ChairActionAck message, which SHOULD be generated as soon as possible.

The floor control server MUST copy the Conference ID, the Transaction ID, and the User ID from the ChairAction message into the ChairActionAck message, as described in Section 6.8.2. The floor control server needs to take into consideration the operation requested in the ChairAction message (e.g., granting a floor) but does not necessarily need to perform it as requested by the floor chair. The operation that the floor control server performs depends on the ChairAction message and on the internal state of the floor control server.

For example, a floor chair may send a ChairAction message granting a floor that was requested as part of an atomic floor request operation that involved several floors. Even if the chair responsible for one of the floors instructs the floor control server to grant the floor, the floor control server will not grant it until the chairs responsible for the other floors agree to grant them as well. So, the floor control server is ultimately responsible for keeping a coherent floor state using instructions from floor chairs as input to this state. If the new Status in the ChairAction message is Accepted, and all the bits of the Queue Position field are zero, the floor chair is requesting that the floor control server assign a queue position (e.g., the last in the queue) to the floor request based on the local policy of the floor control server. (Of course, such a request only applies if the floor control server implements a queue.)

6.13.7 Reception of a Hello Message

On reception of a Hello message, the floor control server follows the rules in Section 6.9 that relate to client authentication. If while processing the Hello message, the floor control server encounters an error, it SHOULD generate an Error response following the procedures described in Section 6.13.8. The successful processing of a Hello message by a floor control server involves generating a HelloAck message, which SHOULD be generated as soon as possible. The floor control server MUST copy the Conference ID, the Transaction ID, and the User ID from the Hello into the HelloAck, as described in Section 6.8.2.

The floor control server MUST add a SUPPORTED-PRIMITIVES attribute to the HelloAck message listing all the primitives (i.e., BFCP messages) supported by the floor control server. The floor control server MUST add a SUPPORTED-ATTRIBUTES attribute to the HelloAck message listing all the attributes supported by the floor control server.

6.13.8 Error Message Generation

Error messages are always sent in response to a previous message from the client as part of a client-initiated transaction. The ABNF in Section 6.5.3.13 describes the attributes that an Error message can contain. In addition, the ABNF specifies normatively which of these attributes are mandatory and which ones are optional. The floor control server MUST copy the Conference ID, the Transaction ID, and the User ID from the message from the client into the Error message, as described in Section 6.8.2. The floor control server MUST add an ERROR-CODE attribute to the Error message. The ERROR-CODE attribute contains an Error Code from Table 6.5. Additionally, the floor control server may add an ERROR-INFO attribute with extra information about the error.

6.14 SECURITY CONSIDERATIONS

BFCP uses TLS to provide mutual authentication between clients and servers. TLS also provides replay and integrity protection and confidentiality. It is RECOMMENDED that TLS with non-null encryption always be used. BFCP entities MAY use other security mechanisms as long as they provide similar security properties. The remainder of this section analyzes some of the threats against BFCP and how they are addressed. An attacker may attempt to impersonate a client (a floor participant or a floor chair) in order to generate forged floor requests or to grant or deny existing floor requests. Client impersonation is avoided by having servers only accept BFCP messages over authenticated TLS connections. The floor control server assumes that attackers cannot hijack the TLS connection and therefore, that messages over the TLS connection come from the client that was initially authenticated.

An attacker may attempt to impersonate a floor control server. A successful attacker would be able to make clients think that they hold a particular floor so that they would try to access a resource (e.g., sending media) without having legitimate rights to access it. Floor control server impersonation is avoided by having servers only accept BFCP messages over authenticated TLS connections. Attackers may attempt to modify messages exchanged by a client and a floor control server. The integrity protection provided by TLS connections prevents this attack.

An attacker may attempt to fetch a valid message sent by a client to a floor control server and replay it over a connection between the attacker and the floor control server. This attack is prevented by having floor control servers check that messages arriving over a given authenticated TLS connection use an authorized user ID (i.e., a user ID that the user that established the authenticated TLS connection is allowed to use). Attackers may attempt to pick messages from the network to get access to confidential information between the floor control server and a client (e.g., why a floor request was denied). TLS confidentiality prevents this attack. Therefore, it is RECOMMENDED that TLS be used with a non-null encryption algorithm.

6.15 BFCP CONNECTION ESTABLISHMENT

RFC 5018 specifies how a BFCP client establishes a connection to a BFCP floor control server outside the context of an offer/answer exchange. Client and server authentication are based on TLS. As discussed in the BFCP specification earlier, a given BFCP client needs a set of data in order to establish a BFCP connection to a floor control server. These data include the transport address of the server, the conference identifier, and the user identifier. Once a client obtains this information, it needs to establish a BFCP connection to the floor control server. The way this connection is established depends on the context of the client and the floor control server. How to establish such a connection in the context of an SDP (RFC 4566) offer/answer (RFC 3264) exchange between a client and a floor control server is specified in RFC 4583. This specification specifies how a client establishes a connection to a floor control server outside the context of an SDP offer/answer exchange.

BFCP entities establishing a connection outside an SDP offer/answer exchange need different authentication mechanisms than entities using offer/answer exchanges. This is because offer/answer exchanges provide parties with an initial integrity-protected channel that clients and floor control servers can use to exchange the fingerprints of their self-signed certificates. Outside the offer/answer model, such a channel is not typically available. This specification specifies how to authenticate clients using PSK (Pre-Shared Key)-TLS (RFC 4279) and how to authenticate servers using server certificates.

6.15.1 TERMINOLOGY

The key word conventions used in this specification are to be interpreted as described in RFC 2119.

6.15.2 TCP CONNECTION ESTABLISHMENT

As stated earlier, a given BFCP client needs a set of data in order to establish a BFCP connection to a floor control server. These data include the transport address of the server, the conference identifier, and the user identifier. It is outside the scope of this specification to specify how a client obtains this information. This specification assumes that the client obtains this information using an out-of-band method. Once the client has the transport address (i.e., Internet protocol [IP] address and port) of the floor control server, it initiates a TCP connection towards it. That is, the client performs an active TCP open.

If the client is provided with the floor control server's host name instead of with its IP address, the client MUST perform a DNS lookup in order to resolve the host name into an IP address. Clients eventually perform an A or AAAA DNS lookup (or both) on the host name.

In order to translate the target to the corresponding set of IP addresses, IPv6-only or dual-stack clients MUST use name resolution functions that implement the Source and Destination Address Selection algorithms specified in RFC 3484. (On many hosts that support IPv6, APIs like getaddrinfo() provide this functionality and subsume existing APIs like gethostbyname().) The advantage of the additional complexity is that this technique will output an ordered list of IPv6/IPv4 destination addresses based on the relative merits of the corresponding source/destination pairs. This will result in the selection of a preferred destination address.

However, the Source and Destination Selection algorithms of RFC 3484 are dependent on broad operating system support and uniform implementation of the application programming interfaces that implement this behavior. Developers should carefully consider the issues described by RFC 4943 with respect to address resolution delays and address selection rules. For example, implementations of getaddrinfo() may return address lists containing IPv6 global addresses at the top of the list and IPv4 addresses at the bottom, even when the host is only configured with an IPv6 local scope (e.g., link-local) and an IPv4 address. This will, of course, introduce a delay in completing the connection.

The BFCP specification describes a number of situations when the TCP connection between a client and the floor control server needs to be reestablished. However, that specification does not describe the reestablishment process, because this process depends on how the connection was established in the first place.

When the existing TCP connection is closed following the rules in RFC 4582, the client SHOULD reestablish the connection towards the floor control server. If a TCP connection cannot deliver a BFCP message from the client to the floor control server and times out, the client SHOULD reestablish the TCP connection.

6.15.3 TLS Usage

This specification (RFC 4582) requires that all BFCP entities implement TLS (RFC 4346) and recommends that they use it in all their connections. TLS provides integrity and replay protection and optional confidentiality. The floor control server MUST always act as the TLS server. A floor control server that receives a BFCP message over TCP (no TLS) SHOULD request the use of TLS by generating an Error message with an Error code with a value of 9 (Use TLS).

6.15.4 Authentication

BFCP supports client authentication based on pre-shared secrets and server authentication based on server certificates.

6.15.4.1 Certificate-Based Server Authentication

At TLS connection establishment, the floor control server MUST present its certificate to the client. The certificate provided at the TLS level MUST either be directly signed by one of the other party's trust anchors or be validated using a certification path that terminates at one of the other party's trust anchors (RFC 3280).

A client establishing a connection to a server knows the server's host name or IP address. If the client knows the server's host name, the client MUST check it against the server's identity as presented in the server's Certificate message in order to prevent man-in-the-middle attacks.

If a subjectAltName extension of type dNSName is present, that MUST be used as the identity. Otherwise, the (most specific) Common Name field in the Subject field of the certificate MUST be used. Although the use of the Common Name is existing practice, it is deprecated, and Certification Authorities are encouraged to use the subjectAltName instead.

Matching is performed using the matching rules specified by RFC 3280. If more than one identity of a given type is present in the certificate (e.g., more than one dNSName name), a match in any one of the set is considered acceptable. Names in Common Name fields may contain the wildcard character *, which is considered to match any single domain name component or component fragment (e.g., *.a.com matches foo.a.com but not bar.foo.a.com; f*.com matches foo.com but not bar.com).

If the client does not know the server's host name and contacts the server directly using the server's IP address, the iPAddress, subjectAltName must be present in the certificate and must exactly match the IP address known to the client.

If the host name or IP address known to the client does not match the identity in the certificate, user-oriented clients MUST either notify the user (clients MAY give the user the opportunity to continue with the connection in any case) or terminate the connection with a bad certificate error. Automated clients MUST log the error to an appropriate audit log (if available) and SHOULD terminate the connection (with a bad certificate error). Automated clients MAY provide a configuration setting that disables this check but MUST provide a setting that enables it.

6.15.4.2 Client Authentication Based on a Pre-Shared Secret

Client authentication is based on a pre-shared secret between client and server. Authentication is performed using PSK-TLS (RFC 4279). The BFCP specification mandates support for the TLS_RSA_WITH_AES_128_CBC_SHA ciphersuite. Additionally, clients and servers supporting this specification MUST support the TLS_RSA_PSK_WITH_AES_128_CBC_SHA ciphersuite as well.

6.15.5 Security Considerations

Client and server authentication as specified in this specification are based on the use of TLS. Therefore, it is strongly RECOMMENDED that TLS with non-null encryption is always used. Clients and floor control servers MAY use other security mechanisms as long as they provide similar security properties (i.e., replay and integrity protection, confidentiality, and client and server authentication).

TLS PSK simply relies on a pre-shared key without specifying the nature of the key. In practice, such keys have two sources: text passwords and randomly generated binary keys. When keys are derived from passwords, TLS PSK mode is subject to offline dictionary attacks. In DHE (Diffie-Hellman Exchange) and RSA modes, an attacker who can mount a single man-in-the-middle attack on a client/server pair can then mount a dictionary attack on the password. In modes without DHE or RSA, an attacker who can record communications between a client/server pair can mount a dictionary attack on the password. Accordingly, it is RECOMMENDED that where possible, clients use certificate-based server authentication ciphersuites with password derived PSKs in order to defend against dictionary attacks.

In addition, passwords SHOULD be chosen with enough entropy to provide some protection against dictionary attacks. Because the entropy of text varies dramatically and is generally far lower than that of an equivalent random bitstring, no hard and fast rules about password length are possible. However, in general, passwords SHOULD be chosen to be at least 8 characters and selected from a pool containing both upper and lower case, numbers, and special keyboard characters (note that an 8-character ASCII password has a maximum entropy of 56 bits and in general, far lower). FIPS PUB 112 (National Institute of Standards and Technology [NIST], "Password Usage," FIPS PUB 112, May 1985) provides some guidance on the relevant issues. If possible, passphrases are preferable to passwords. It is RECOMMENDED that implementations support, at minimum, 16-character passwords or passphrases. In addition, a cooperating client and server pair MAY choose to derive the TLS PSK shared key from the passphrase via a password-based key derivation function such as PBKDF2 (RFC 2898).

Because such key derivation functions may incorporate iteration functions for key strengthening, they provide some additional protection against dictionary attacks by increasing the amount of work that the attacker must perform. When the keys are randomly generated and of sufficient length, dictionary attacks are not effective, because such keys are highly unlikely to be in the attacker's dictionary. Where possible, keys SHOULD be generated using a strong random number generator as specified in RFC 4086. A minimum key length of 80 bits SHOULD be used.

The remainder of this section analyzes some of the threats against BFCP and how they are addressed. An attacker may attempt to impersonate a client (a floor participant or a floor chair) in order to generate forged floor requests or to grant or deny existing floor requests. Client impersonation is avoided by using TLS. The floor control server assumes that attackers cannot hijack TLS connections from authenticated clients.

An attacker may attempt to impersonate a floor control server. A successful attacker would be able to make clients think that they hold a particular floor so that they would try to access a resource (e.g., sending media) without having legitimate rights to access it. Floor control server impersonation is avoided by having floor control servers present their server certificates at TLS connection establishment time.

Attackers may attempt to modify messages exchanged by a client and a floor control server. The integrity protection provided by TLS connections prevents this attack. Attackers may attempt to pick messages from the network to get access to confidential information between the floor control server and a client (e.g., why a floor request was denied). TLS confidentiality prevents this attack. Therefore, it is RECOMMENDED that TLS is used with a non-null encryption algorithm.

6.16 IANA CONSIDERATIONS

The IANA has created a new registry for BFCP parameters called "Binary Floor Control Protocol (BFCP) Parameters (RFC 4582." This new registry has a number of subregistries, which are described in RFC 4582.

6.16.1 BFCP Connection Establishment (RFC 5018)

RFC 5018, which is described in this section, specifies how a BFCP client establishes a connection to a BFCP floor control server outside the context of an offer/answer exchange. Client and server authentication are based on TLS. As discussed in the BFCP specification (RFC 4582 – see Sections 6 through 6.14), a given BFCP client needs a set of data in order to establish a BFCP connection to a floor control server. These data include the transport address of the server, the conference identifier, and the user identifier. Once a client obtains this information, it needs to establish a BFCP connection to the floor control server. The way this connection is established depends on the context of the client and the floor control server. How to establish such a connection in the context of an SDP (RFC 4566) using offer/answer (RFC 3264) exchange between a client and a floor control server is specified in RFC 4583.

This specification specifies how a client establishes a connection to a floor control server outside the context of an SDP offer/answer exchange. BFCP entities establishing a connection outside an SDP offer/answer exchange need different authentication mechanisms than entities using offer/answer exchanges. This is because offer/answer exchanges provide parties with an initial integrity-protected channel that clients and floor control servers can use to exchange the fingerprints of their self-signed certificates. Outside the offer/ answer model, such a channel is not typically available.

This specification specifies how to authenticate clients using PSK-TLS (RFC 4279) and how to authenticate servers using server certificates.

6.16.2 TCP Connection Establishment

As stated earlier in Section 6.16, a given BFCP client needs a set of data in order to establish a BFCP connection to a floor control server. These data include the transport address of the server, the conference identifier, and the user identifier. It is outside the scope of this specification to specify how a client obtains this information. This specification assumes that the client obtains this information using an out-of-band method.

Once the client has the transport address (i.e., IP address and port) of the floor control server, it initiates a TCP connection towards it. That is, the client performs an active TCP open. If the client is provided with the floor control server's host name instead of with its IP address, the client MUST perform a DNS lookup in order to resolve the host name into an IP address. Clients eventually perform an A or AAAA DNS lookup (or both) on the host name.

In order to translate the target to the corresponding set of IP addresses, IPv6-only or dual-stack clients MUST use name resolution functions that implement the Source and Destination Address Selection algorithms specified in RFC 3484. (On many hosts that support IPv6, APIs like getaddrinfo() provide this functionality and subsume existing APIs like gethostbyname().)

The advantage of the additional complexity is that this technique will output an ordered list of IPv6/IPv4 destination addresses based on the relative merits of the corresponding source/destination pairs. This will result in the selection of a preferred destination address. However, the Source and Destination Selection algorithms of RFC 3484 are dependent on broad operating system support and uniform implementation of the application programming interfaces that implement this behavior.

Developers should carefully consider the issues described by Roy et al. in RFC 4943 with respect to address resolution delays and address selection rules. For example, implementations of getaddrinfo() may return address lists containing IPv6 global addresses at the top of the list and IPv4 addresses at the bottom even when the host is only configured with an IPv6 local scope (e.g., link-local) and an IPv4 address. This will, of course, introduce a delay in completing the connection.

The BFCP specification (RFC 4582 see Sections 6 through 6.16) describes a number of situations when the TCP connection between a client and the floor control server needs to be reestablished. However, that specification does not describe the reestablishment process, because this process depends on how the connection was established in the first place. When the existing TCP connection is closed following the rules in RFC 4582 (see Sections 6 through 6.16), the client SHOULD reestablish the connection towards the floor control server. If a TCP connection cannot deliver a BFCP message from the client to the floor control server and times out, the client SHOULD reestablish the TCP connection.

6.16.3 TLS Usage

This specification (RFC 4582) requires that all BFCP entities implement TLS (RFC 4346) and recommends that they use it in all their connections. TLS provides integrity and replay protection and optional confidentiality. The floor control server MUST always act as the TLS server. A floor control server that receives a BFCP message over TCP (no TLS) SHOULD request the use of TLS by generating an Error message with an Error code with a value of 9 (Use TLS).

6.16.4 Authentication

BFCP supports client authentication based on pre-shared secrets and server authentication based on server certificates.

6.16.4.1 Certificate-Based Server Authentication

At TLS connection establishment, the floor control server MUST present its certificate to the client. The certificate provided at the TLS level MUST either be directly signed by one of the other party's trust anchors or be validated using a certification path that terminates at one of the other party's trust anchors (RFC 3280). A client establishing a connection to a server knows the server's host name or IP address. If the client knows the server's host name, the client MUST check it against the server's identity as presented in the server's Certificate message in order to prevent man-in-the middle attacks.

If a subjectAltName extension of type dNSName is present, that MUST be used as the identity. Otherwise, the (most specific) Common Name field in the Subject field of the certificate MUST be used. Although the use of the Common Name is existing practice, it is deprecated, and Certification Authorities are encouraged to use the subjectAltName instead.

Matching is performed using the matching rules specified by RFC 3280. If more than one identity of a given type is present in the certificate (e.g., more than one dNSName name), a match in any one of the set is considered acceptable. Names in Common Name fields may contain the wildcard character *, which is considered to match any single domain name component or component fragment (e.g., *.a.com matches foo.a.com but not bar.foo.a.com; f*.com matches foo.com but not bar.com).

If the client does not know the server's host name and contacts the server directly using the server's IP address, the iPAddress subjectAltName must be present in the certificate and must exactly match the IP address known to the client. If the host name or IP address known to the client does not match the identity in the certificate, user-oriented clients MUST either notify the user (clients MAY give the user the opportunity to continue with the connection in any case) or terminate the connection with a bad certificate error. Automated clients MUST log the error to an appropriate audit log (if available) and SHOULD terminate the connection (with a bad certificate error). Automated clients MAY provide a configuration setting that disables this check but MUST provide a setting that enables it.

6.16.4.2 Client Authentication Based on a Pre-Shared Secret

Client authentication is based on a pre-shared secret between client and server. Authentication is performed using PSK-TLS (RFC 4279). The BFCP specification mandates support for the TLS_RSA_WITH_AES_128_CBC_SHA ciphersuite. Additionally, clients and servers supporting this specification MUST support the TLS_RSA_PSK_WITH_AES_128_CBC_SHA ciphersuite as well.

6.16.5 Security Considerations

Client and server authentication as specified in this specification are based on the use of TLS. Therefore, it is strongly RECOMMENDED that TLS with non-null encryption is always used. Clients and floor control servers MAY use other security mechanisms as long as they provide similar security properties (i.e., replay and integrity protection, confidentiality, and client and server authentication). TLS PSK simply relies on a pre-shared key without specifying the nature of the key. In practice, such keys have two sources: text passwords and randomly generated binary keys. When keys are derived from passwords, TLS PSK mode is subject to offline dictionary attacks. In DHE and RSA modes, an attacker who can mount a single man-in-the-middle attack on a client/server pair can then mount a dictionary attack on the password. In modes without DHE or RSA, an attacker who can record communications between a client/server pair can mount a dictionary attack on the password. Accordingly, it is RECOMMENDED that where possible, clients use certificate-based server authentication ciphersuites with password-derived PSKs in order to defend against dictionary attacks.

In addition, passwords SHOULD be chosen with enough entropy to provide some protection against dictionary attacks. Because the entropy of text varies dramatically and is generally far lower than that of an equivalent random bitstring, no hard and fast rules about password length are possible. However, in general, passwords SHOULD be chosen to be at least 8 characters and selected from a pool containing both upper and lower case, numbers, and special keyboard characters (note that an 8-character ASCII password has a maximum entropy of 56 bits and in general, far lower). FIPS PUB 112 (National Institute of Standards and Technology [NIST], "Password Usage," FIPS PUB 112, May 1985) provides some guidance on the relevant issues. If possible, passphrases are preferable to passwords. It is RECOMMENDED that implementations support, at minimum, 16-character passwords or passphrases. In addition, a cooperating client and server pair MAY choose to derive the TLS PSK shared key from the passphrase via a password-based key derivation function such as PBKDF2 (RFC 2898).

Because such key derivation functions may incorporate iteration functions for key strengthening, they provide some additional protection against dictionary attacks by increasing the amount of work that the attacker must perform. When the keys are randomly generated and of sufficient length, dictionary attacks are not effective, because such keys are highly unlikely to be in the attacker's dictionary. Where possible, keys SHOULD be generated using a strong random number generator as specified in RFC 4086. A minimum key length of 80 bits SHOULD be used.

The remainder of this section analyzes some of the threats against BFCP and how they are addressed. An attacker may attempt to impersonate a client (a floor participant or a floor chair) in order to generate forged floor requests or to grant or deny existing floor requests. Client impersonation is avoided by using TLS. The floor control server assumes that attackers cannot hijack TLS connections from authenticated clients. An attacker may attempt to impersonate a floor control server. A successful attacker would be able to make clients think that they hold a particular floor so that they would try to access a resource (e.g., sending media) without having legitimate rights to access it. Floor control server impersonation is avoided by having floor control servers present their server certificates at TLS connection establishment time.

Attackers may attempt to modify messages exchanged by a client and a floor control server. The integrity protection provided by TLS connections prevents this attack. Attackers may attempt to pick messages from the network to get access to confidential information between the floor control server and a client (e.g., why a floor request was denied). TLS confidentiality prevents this attack. Therefore, it is RECOMMENDED that TLS is used with a non-null encryption algorithm.

6.17 SUMMARY

We have described the BFCP that manages joint or exclusive access to shared resources in an XCON (multiparty) conferencing environment. BFCP is a unique application that is used between floor participants and floor control servers, and between floor chairs (that is, moderators) and floor control servers, specifically for coordinated sharing conference resources among the multiple conference participants simultaneously in real time. BFCP protocol meets the requirements specified in RFC 4376 in a way that complements other functions, such as conference and media session setup, conference policy manipulation, and media control, that are realized by other protocols. Binary encoding of BFCP keeps its messages smaller in size, achieving low-bandwidth capacity and enabling low transit-time delay, which are prime performance requirements for signaling messages of a real-time XCON conferencing system.

Specifically, we have explained the functionalities of the BFCP protocol: Floor Creation, Obtaining Information to Contact a Floor Control Server, Obtaining Floor–Resource Associations, and Privileges of Floor Control, as well as BFCP packet format: Common-Header Format, Attribute Format, and Message Format. In addition, the following are described in great detail: Floor Participant Operations in BFCP: Requesting a Floor and Cancelling a Floor Request and Releasing a Floor, and Chair Operations in Sending a ChairAction Message and Receiving a Response using the BFCP protocol.

Details of sending and receiving messages of the BFCP protocol for general client operations are provided for the following cases: a. Requesting Information about Floors, b. Requesting Information about Floor Requests, c. Requesting Information about a User, and d. Obtaining the Capabilities of a Floor Control Server. Floor Control Server Operations are also explained in detail using call flows using the BFCP protocol: a. Reception of a FloorRequest Message, b. Reception of a FloorRequestQuery Message, c. Reception of a UserQuery Message, d. Reception of a FloorRelease Message, e. Reception of a FloorQuery Message, f. Reception of a ChairAction Message, g. Reception of a Hello Message, and h. Error Message Generation.

A key section discusses how the connection-oriented TCP transport protocol is used for the BFCP for controlling XCON multiparty conference system where lower-layer security protocols like TLS are used for protecting the BFCP, noting that TLS needs a reliable transport. Finally, we have articulated the authentication schemes between the floor control client and server using TLS for BFCP protocol messages with detail call flows: Certificate-Based Server Authentication and Client Authentication Based on a Pre-Shared Secret. The protection against various attacks under different scenarios for the BFCP control operations is explained. IANA Considerations note that this RFC 4582 requires IANA registration for Attribute Subregistry, Primitive Subregistry, Request Status Subregistry, and Error Code Subregistry, and interested readers need to see the RFC 4582 itself. We have also described the BFCP connection establishment and its security in great detail.

6.18 PROBLEMS

1. What is the floor control that is used for the XCON multipoint conference system? What are the requirements that a floor control protocol needs to satisfy for a multiparty conference system such as XCON?
2. Explain in detail the functionalities of the BFCP protocol: Floor Creation, Obtaining Information to Contact a Floor Control Server, Obtaining Floor–Resource Associations, and Privileges of Floor Control.
3. Describe the BFCP protocol operations using the interfaces specifically for Floor Participant to Floor Control Server and Floor Chair to Floor Control Server.
4. Articulate the functionalities of BFCP packet format in detail: Common-Header Format, Attribute Format, and Message Format.
5. Why does the BFCP use a connection-oriented transport protocol like TCP? Explain in detail how the TCP transport protocol is used by the BFCP for controlling an XCON multiparty conference system. Can connectionless UDP transport protocol be used for BFCP?
6. Does BFCP protocol have any security features for protecting itself? If not, how is a lower-layer security protocol like TLS used for protecting the BFCP?
7. How are transactions initiated using the BFCP? Describe both client and server behavior in detail using call flows.
8. How are TLS-based mutual authentication and authorization supported for the BFCP? Describe in detail using call flows.
9. Describe in detail with call flows for Floor Participant Operations in BFCP: Requesting a Floor and Cancelling a Floor Request and Releasing a Floor.
10. How do the Chair Operations occur in Sending a ChairAction Message and Receiving a Response using the BFCP protocol? Describe in detail using call flows.
11. Describe the call flows in sending and receiving messages of the BFCP protocol for general client operations for the following cases: a. Requesting Information about Floors, b. Requesting Information about Floor Requests, c. Requesting Information about a User, and d. Obtaining the Capabilities of a Floor Control Server.

12. Articulate the Floor Control Server Operations in detail using call flows for the BFCP protocol: a. Reception of a FloorRequest Message, b. Reception of a FloorRequestQuery Message, c. Reception of a UserQuery Message, d. Reception of a FloorRelease Message, e. Reception of a FloorQuery Message, f. Reception of a ChairAction Message, g. Reception of a Hello Message, and h. Error Message Generation.
13. How does the BFCP connection establishment happen over the TCP connection and usage of TLS for security? Describe in detail using call flows.
14. Describe authentication schemes between the floor control client and server using TLS for BFCP protocol messages with detailed call flows: Certificate-Based Server Authentication and Client Authentication Based on a Pre-Shared Secret.
15. Describe in detail how security is provided against various attacks under different scenarios for the BFCP control operations.
16. Describe the BFCP connection establishment and the security implications using detailed call flows.

7 XCON Notification Service

7.1 INTRODUCTION

The XCON (Centralized Conferencing) framework (RFC 5239 – see Chapter 2) defines a notification service that provides updates about a conference instance's state to authorized parties using a notification protocol, as shown in Figure 7.1. This section specifies how to use the SIP (session initiation protocol – specified in RFC 3261) event package for conference state defined in RFC 4575 as a notification protocol between a client and a conference's notification server.

In addition to specifying the SIP event package for conference state, RFC 4575 specifies a data format to be used with the event package. The XCON data model defined in RFC 6501 (see Chapter 4) extends that format with new elements and attributes so that the extended format supports more functionality (e.g., floor control – see Chapter 6). The notification protocol specified in this specification supports all the data defined in the XCON data model (i.e., the data originally defined in RFC 4575 plus all the extensions defined in RFC6501 (see Chapter 4) plus a partial notification mechanism based on XML patch operations specified in RFC 5261.

7.2 TERMINOLOGY

See Table 1.2, Section 1.6.

7.3 NOTIFICATION FORMATS

In order to obtain notifications from a conference server's notification service, a client subscribes to the "conference" event package at the server as specified in RFC 4575. Per RFC 4575, NOTIFY requests within this event package can carry an XML specification in the "application/conference-info+xml" format. Additionally, per this specification, NOTIFY requests can also carry XML specifications in the "application/xcon-conference-info+xml" and the "application/xcon-confere nce-info-diff+xml" formats. A specification in the "application/xcon-conference-info+xml" format provides the user agent with the whole state of a conference instance. A specification in the "appl ication/xcon-conference-info-diff+xml" format provides the user agent with the changes the state of the conference instance has experienced since the last notification sent to the user agent.

7.4 FULL NOTIFICATIONS

Subscribers signal support for full notifications by including the "application/xcon-conference-info+xml" format in the Accept header field of the SUBSCRIBE requests they generate. If a client subscribing to the "conference" event package generates an Accept header field that includes the MIME type "application/xcon-conference-info+xml," the server has the option of returning specifications that follow the XML format specified in RFC6501 (see Chapter 4) and are carried in "application/xcon-conference-info+xml" message bodies.

7.4.1 Backwards Compatibility

Conference servers that implement the SIP event package for conference state and support the "application/event eveeevnteventxcon-conference-info+xml" MIME type MUST also support the "application/conference-info+xml" MIME type. In this way, legacy clients, which only support

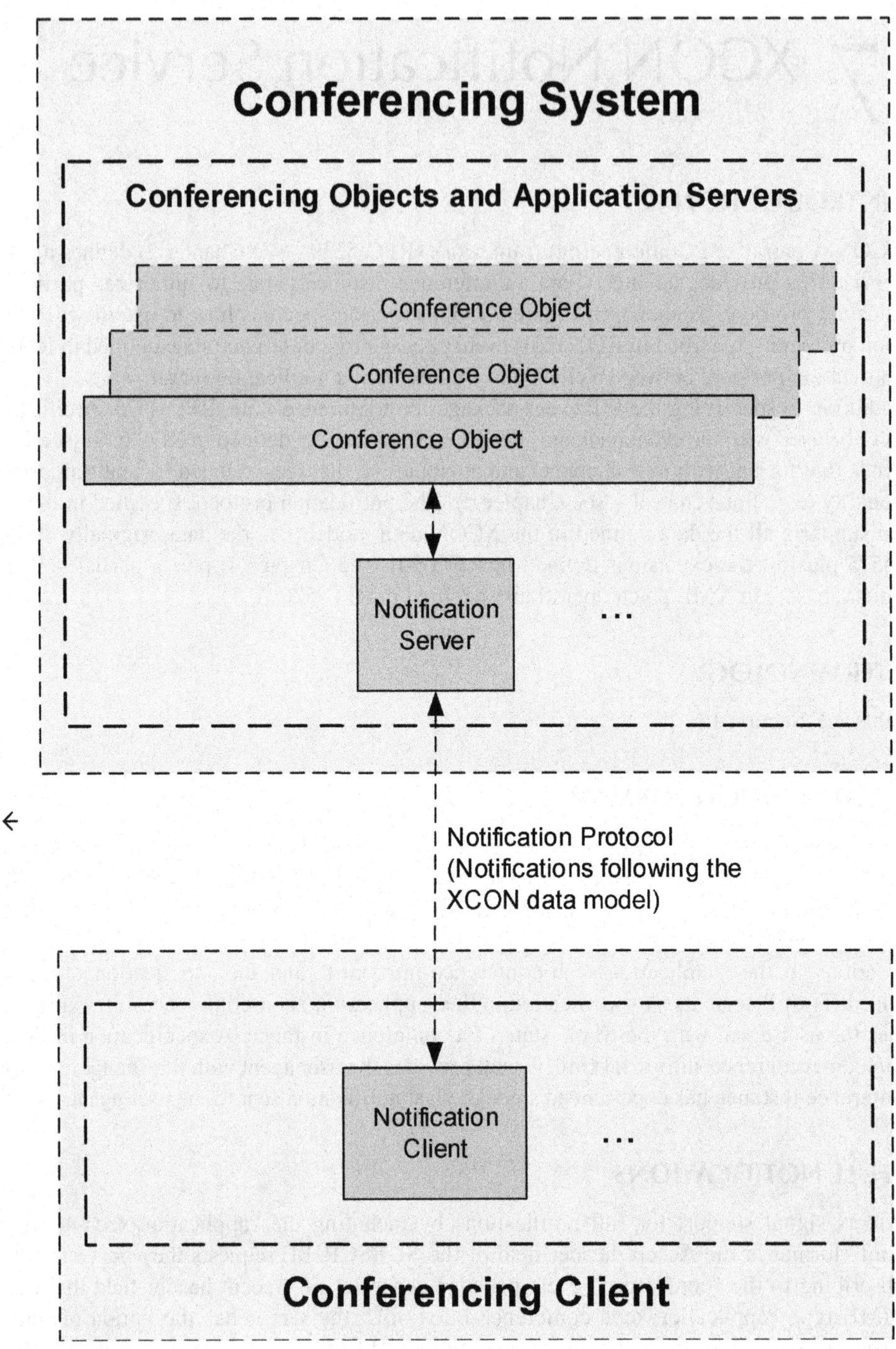

FIGURE 7.1 Notification service and protocol in the XCON architecture.(Copyright: IETF.)

"application/conference-info+xml," are able to receive notifications in a format they understand. Clients that implement the SIP event package for conference state and support the "application/xcon-conference-info+xml" MIME type SHOULD also support the "application/conference-info+xml" MIME type. In this way, these clients are able to receive notifications from legacy servers, which only support "application/conference-info+xml," in a format they understand.

7.5 PARTIAL NOTIFICATIONS

The conference state reported by this event package may contain many elements. When the "xcon conference-info+xml" format is used and there is a change in the state of an element, the server generates a notification with the whole conference state. Generating large notifications to report small changes does not meet the efficiency requirements of some bandwidth-constrained environments. The partial notifications mechanism specified in this section is a more efficient way to report changes in the conference state. The SIP event package for conference state defined a partial notification mechanism based on <state> elements. Servers compliant with this specification MUST NOT use that partial notification mechanism. Instead, they MUST use the mechanism specified in this section.

Subscribers signal support for partial notifications by including the "application/xcon-confere nce-info-diff+xml" format in the Accept header field of the SUBSCRIBE requests they generate. If a client subscribing to the "conference" event package generates an Accept header field that includes the MIME type "application/xcon-conference-info-diff+xml," the server has the option of returning specifications that follow the XML format specified in Section 7.5.4 and are carried in "application/ xcon-conference-diff-info+xml" message bodies.

7.5.1 Generation of Partial Notifications

Once a subscription is accepted and installed, the server MUST deliver full state in its first notification. To report full state, the server MUST set the Content-Type header field to the value "application/xcon-conference-info+xml." In order to deliver a partial notification, the server MUST set the Content-Type header field to the value "application/xcon-conference-info-diff+xml." When the server generates a partial notification, the server SHOULD only include the information that has changed compared with the previous notification. It is up to the server's local policy to determine what is considered as a change to the previous state. The server MUST construct partial notifications according to the following logic: All the information that has been added to the specification is listed inside <add> elements. All information that has been removed from the specification is listed inside <remove> elements, and all information that has been changed is listed under <replace> elements.

The server MUST NOT send a new NOTIFY request with a partial notification until it has received a final response from the subscriber for the previous one or the previous NOTIFY request has timed out. When the server receives a SUBSCRIBE request (refresh or termination) within the associated subscription, it SHOULD send a NOTIFY request containing the full specification using the "application/xcon-conference-info+xml" content type. If the server has used a content type other than "application/ xcon-conference-info+xml" in notifications within the existing subscription and changes to deliver partial notifications, the server MUST deliver full state using the "application/xcon-conference-info+xml" content type before generating its first partial notification.

7.5.2 Processing of Partial Notifications

When a subscriber receives the first notification containing full state in an "application/xcon-conference-info+xml" MIME body, the subscriber MUST store the received full specification as its local copy. When the subscriber receives a subsequent notification, the subscriber MUST modify its locally stored information according to the following logic:

- If the notification carries an "application/xcon-conference-info+xml" specification, the subscriber MUST replace its local copy of the specification with the specification received in the notification.

- If the notification carries an "application/xcon-conference-info-diff+xml" specification, the subscriber MUST apply the changes indicated in the received "application/ xcon-conference-info-diff+xml" specification to its local copy of the full specification.

If the subscriber encounters a processing error while processing an "application/xcon-conference-info-diff+xml" encoded specification, the subscriber SHOULD renew its subscription. A subscriber can fall back to normal operations by not including the "application/xcon-conference-info-diff+xml" format in a new SUBSCRIBE request. If the server changes the content type used in notifications within the existing subscription, the subscriber MUST discard all the previously received information and process the new content as specified for that content type.

7.5.3 Partial Notification Format

An xcon-conference-info-diff specification is an XML [1] specification that MUST be well-formed and SHOULD be valid. The namespace URI for the <conference-info-diff> root specification element is defined in RFC 6501 (see Chapter 4): urn:ietf:params:xml:ns:xcon-conference-info. The root specification element <conference-info-diff> has a single mandatory attribute, "entity." The value of this attribute is the conference object identifier (XCON-URI) that identifies the conference being described in the specification. The content of the <conference-info-diff> element is an unordered sequence of <add>, <replace>, and <remove> elements followed by elements from other namespaces for the purposes of extensibility. Any such unknown elements MUST be ignored by the client. The <add>, <replace>, and <remove> elements can contain other extension attributes than those that are defined in the corresponding base types of RFC 5261 (also see Section 1.2).

7.5.4 XML Schema for Partial Notifications

This is the XML schema for the "application/xcon-conference-info-diff+xml" data format. The "urn:ietf:params:xml:schema:xml-patch-ops" schema is defined in RFC 5261.

```
<?xml version="1.0" encoding="UTF-8"?>
<xs:schema
   targetNamespace="urn:ietf:params:xml:ns:xcon-conference-info"
   xmlns="urn:ietf:params:xml:ns:xcon-conference-info"
   xmlns:xs="http://www.w3.org/2001/XMLSchema"
   elementFormDefault="qualified">
   <!-- include patch-ops type definitions -->
   <xs:include
   schemaLocation="urn:ietf:params:xml:schema:patch-ops"/>
   <!-- partial updates -->
   <xs:element name="conference-info-diff">
      <xs:complexType>
         <xs:sequence minOccurs="0" maxOccurs="unbounded">
            <xs:choice>
               <!-- add some content -->
               <xs:element name="add">
                  <xs:complexType mixed="true">
                     <xs:complexContent>
                        <xs:extension base="add">
                           <xs:anyAttribute processContents="lax"/>
                        </xs:extension>
                     </xs:complexContent>
                  </xs:complexType>
               </xs:element>
```

```
                <!-- remove some content -->
                <xs:element name="remove">
                    <xs:complexType>
                        <xs:complexContent>
                            <xs:extension base="remove">
                                <xs:anyAttribute processContents="lax"/>
                            </xs:extension>
                        </xs:complexContent>
                    </xs:complexType>
                </xs:element>
                <!-- replace some content -->
                <xs:element name="replace">
                    <xs:complexType mixed="true">
                        <xs:complexContent>
                            <xs:extension base="replace">
                                <xs:anyAttribute processContents="lax"/>
                            </xs:extension>
                        </xs:complexContent>
                    </xs:complexType>
                </xs:element>
                <!-- allow extension elements from other namespaces -->
                <xs:any namespace="##other" processContents="lax"/>
            </xs:choice>
        </xs:sequence>
        <xs:attribute name="entity" type="xs:anyURI" use="required"/>
        <xs:anyAttribute processContents="lax"/>
      </xs:complexType>
    </xs:element>
</xs:schema>
```

7.5.5 Examples

The following is an "application/xcon-conference-info-diff+xml" partial update specification:

```
<?xml version="1.0" encoding="UTF-8"?>
<conference-info-diff
          xmlns="urn:ietf:params:xml:ns:xcon-conference-info"
          entity="conference123@example.com">
          <add
                   sel="*/users/allowed-users-list"> <target
                   uri="sip:john@example.com" method="refer"/>
          </add>
          <replace sel="*/conference-state/user-count/text()">5</replace>
</conference-info-diff>
```

7.6 IANA Considerations

Note that this specification needs registration of MIME type, URN Sub-Namespace, and XML to the IANA. Interested readers need to see RFC 6502 for registration procedures.

7.7 Security Considerations

This specification (RFC 6502) specifies how to deliver notifications using the SIP event package for conference state in two new formats. The fact that notifications are encoded in a different format does not have security implications All security considerations that are described in Section 8 of

RFC 4575 are also applicable to this XCON event package. Subscriptions to conference state information can reveal very sensitive information. For this reason, it is RECOMMENDED that a focus use a strong means for authentication and conference information protection and that it apply comprehensive authorization rules when using the conference notification mechanism defined in this specification. The following sections discuss each of these aspects in more detail.

7.7.1 Connection Security

It is RECOMMENDED that a focus authenticate a conference package subscriber using the normal SIP authentication mechanisms, such as Digest as defined in Section 22 of RFC 3261 (also see Section 1.2). The mechanism used for conveying the conference information MUST ensure integrity and SHOULD ensure confidentiality of the information. In order to achieve these, an end-to-end SIP encryption mechanism, such as S/MIME, described in Section 26.2.4 of RFC 3261 (also see Section 1.2), SHOULD be used. If a strong end-to-end security means (such as this) is not available, it is RECOMMENDED that a focus use mutual hop-by-hop transport layer security (TLS) authentication and encryption mechanisms, described in Section 26.2.2 "SIPS URI Scheme" and Section 26.3.2.2 "Inter-domain Requests" of RFC 3261.

7.7.2 Authorization Considerations

Generally speaking, conference applications are very concerned about authorization decisions. Mechanisms for establishing and enforcing such authorization rules are a central concept throughout the SIP Conferencing Framework (RFC 4353). Because most of the information about a conference can be presented using the conference package, many of the authorization rules directly apply to this specification. As a result, a notification server MUST be capable of generating distinct conference information views to different subscribers, subject to a subscriber's role in a conference, personal access rights, etc. – all subject to local authorization policies and rules. Since a focus provides participant identity information using this event package, participant privacy needs to be taken into account. A focus MUST support requests by participants for privacy. Privacy can be indicated by the conference policy – for every participant or selected participants. It can also be indicated in the session signaling. In SIP, this can be done using the Privacy header field described in RFC 3323. For a participant requesting privacy, no identity information SHOULD be revealed by the focus in any included uniform resource identifier (URI) (e.g., the Address of Record, Contact, or globally routable user agent URI [GRUU]). For these cases, the anonymous URI generation method outlined in Section 5.6 of RFC 4575 MUST be followed.

7.8 SUMMARY

We have described the XCON event notification service that uses SIP protocol. Note that the basic SIP event notification service is defined in RFC 3265 (superseded by 4566), which extends SIP with two event notification–specific methods: SUBSCRIBE and NOTIFY. Again, RFC 4575 defines a conference event package for tightly coupled conferences using the SIP events framework along with a data format used in notifications for this package specifically based on RFC 3265 (superseded by 4566) before the finalization of XCON specifications. However, the RFC 4575 event conference package is not sufficient to meet all the requirements of XCON. Hence, the RFC 6502 XCON conference package is defined as a superset of RFC 4575 providing interoperability.

The XCON notification format is defined here along with full and partial notification formats. The backwards compatibility of different notification formats with those of XCON formats is explained. In an XCON conference, it may happen that only a subset of the entire event elements changes. In this situation, only partial notifications need to be generated for saving bandwidths. We have described in detail how the generation and processing of partial notifications need to be performed in view of the partial notification formats. The XML schema for partial notification is also defined for XCON in addition to explanation using an example. Finally, we have described in detail the connection security and authorization for XCON event notifications, because subscriptions to conference state information can reveal very sensitive information.

7.9 PROBLEMS

1. What are the capabilities of XCON notification services? Describe in detail using the XCON notification architecture with possible call flows.
2. Why is the backwards compatibility in XCON full formats needed? Describe the XCON full notification formats.
3. What is the use of partial notification as opposed to full notification? Describe the partial notification formats. How are the partial notifications generated? How are the partial notifications generated?
4. Are there any differences between XML schemas of partial and full notifications? Justify your answer. Describe in detail the functions of XML schema of partial notifications.
5. Why is security important for XCON notifications? How can a focus offer strong authentication and conference information protection? Describe in detail using call flows for XCON conferencing.

REFERENCE

1. Bray, T. et al. (2008, November) Extensible Markup Language (XML) 1.0 (Fifth Edition). *World Wide Web Consortium Recommendation RECxml-20081126.* <http://www.w3.org/TR/2008/REC-xml-20081126>.

8 Media Channel Control Framework

8.1 INTRODUCTION

Real-time applications such as multimedia conferencing need to be developed using an architecture where the application logic and media processing activities are distributed for scalability of services. The usual case is that application logic runs on application servers while the processing runs on external servers, such as Media Servers (MSs). The Media Channel Control Framework defined in RFC 6230 is a framework and protocol for communications between the Application Server (AS) and the external processing server using session initiation protocol (SIP), for example RFCs 3261 (also see Section 1.2) and others, along with some extensions of new methods. It is due to the fact that SIP provides the ideal rendezvous mechanism for establishing and maintaining control connections to external server components.

8.2 CFW CAPABILITIES

The media server channel control framework CFW has considered the control of the audio/video/data MS that is used for media manipulation, including media bridging in the context of multimedia conferencing, as the primary use case, but it has been developed for much broader usages for a variety of control scenarios, including conferencing. The control connections can then be used to exchange explicit command/response interactions that allow media control and associated command response results. The primary capabilities that the CFW needs to have developed are listed in RFC 5167 and are summarized as follows:

- **Media Control Capabilities (MCC)**
  1. Enables media services from one or more media servers for conferencing
  2. Enables interactive voice response (IVR), media resource brokering (MRB), or intermediate proxy existing with the AS and the MS (e.g. audio, tones in-band audio (RFC 4733) payload, text, and video) using reliable transport protocol
  3. Uses the session initiation protocol/session description protocol (SIP/SDP) to establish and modify media connections to a MS
  4. Enables reporting of media sources, supports a single conference spanning multiple MSs
  5. Enables splitting of call legs individually, or in groups, away from a main conference on a given MS without performing reestablishment of the call legs to the MS (e.g., for purposes such as performing IVR with a single call leg or creating sub-conferences, not for creating entirely new conferences)
  6. Allows extensibility with forwards and backwards compatibility of the protocol, authentication component to ensure that only an authorized AS can communicate with the MS, and vice versa
- **Security Capabilities (SC)**
  7. Uses transport layer security (TLS) transport protection to ensure the confidentiality and integrity of the data between the AS and the MS

8. Prevents the AS from doing unauthorized operations on a MS. while different ASs will have different privileges with the MS
9. Provides mechanisms to protect the MS resources used by one AS from another AS, since the solution needs to support multiple ASs controlling one MS

- **Capability Negotiations (CN)**

10. Can negotiate parameters that are associated with media streams during the call establishment, enabling an AS managing conference to specify the media streams allowed in the conference with the use of SIP/SDP
11. Has the ability to renegotiate a connection, ensure it is active, and so forth

- **AS Capabilities (ASC)**

12. Can instruct the MS to perform stream operations like mute and gain control and playing specific announcements, request the MS to create, delete, and manipulate a mixing, IVR, or announcement session, and instruct the MS to play announcements to a single user or to a conference mix
13. Can ask the MS for a session summary report including resource usage and quality metrics
14. Can instruct the MS to notify the AS itself of events received in the media stream if requested by it (e.g., STUN request, Flow Control, etc.)
15. Can make necessary extensions, as SIP is generic and application agnostic

- **Media Mixing Capabilities (MMC)**

16. Enables the AS to define a conference mix; the MS may offer different mixing topologies. The conference mix may be defined on a conference or user level.
17. Enables the AS to define a custom video layout built of rectangular sub-windows.
18. Enables the AS to map a stream to a specific sub-window or to define to the MS how to decide which stream will go to each sub-window in the case of video.
19. Enables the MS to notify the ASs of who the active sources of the media are; for example, who is the active speaker or who is being viewed in a conference. The speaker and the video source may be different; for example, a person describing a video stream from a remote camera managed by a different user.
20. Enables the MS to inform the AS which layouts it supports.
21. Enables the AS to instruct the MS to record a specific conference mix.

- **IVR Capabilities (IVRC)**

22. Enables the AS to instruct the MS to perform one or more IVR scripts and receive the results. The script may be in a server or contained in the control message.
23. Enables the AS to manage the IVR session by sending requests to play announcements to the MS and receiving the response (e.g., dual-tone multi-frequency [DTMF]). The IVR session flow, in this case, is handled by the AS starting a next phase based on the response it receives from the MS on the current phase.
24. Enables the AS to instruct the MS to record a short participant stream and play it back. This is not a recording requirement.

- **Operational Capabilities (OC)**

25. Allows the MRB/AS to audit the MS state during an active session
26. Enables the MS to inform the MRB/AS about its status during an active session

- **Location Service (LS)**

27. Uses SIP proxies and back-to-back user agents (B2BUA) for locating Control Servers applicable both at a routing level, where SIP RFC 3263 provides the physical location of devices, and at the service level, using Caller Preferences specified in RFC 3840 and Callee Capabilities specified in RFC 3841; for example, selection of voice, video, instant messaging (IM) server considering a distributed, clustered architecture containing varying services

The CFW defines additional terms as follows (also see Section 1.3, Table 1.2:

- User Agent Client (UAC): As specified in RFC 3261 (also see Section 1.2).
- User Agent Server (UAS): As specified in RFC 3261 (also see Section 1.2).
- B2BUA: A B2BUA is a back-to-back SIP UA (RFC 3261 – also see Section 1.2).
- Control Server: A Control Server is an entity that performs a service, such as media processing, on behalf of a Control Client. For example, an MS offers mixing, announcement, tone detection and generation, and playing and recording of services. The Control Server has a direct real-time transport protocol (RTP), specified in RFC 3550, relationship with the source or sink of the media flow. In this specification, we often refer to the Control Server simply as "the Server."
- Control Client: A Control Client is an entity that requests processing from a Control Server. Note that the Control Client might not have any processing capabilities whatsoever. For example, the Control Client may be an application server (B2BUA) or other endpoint requesting manipulation of a third party's media stream that terminates on an MS acting in the role of a Control Server. In this specification, we often refer to the Control Client simply as "the Client."
- Control Channel: A Control Channel is a reliable connection between a Client and a Server that is used to exchange Framework messages. The term "Connection" is used synonymously within this specification.
- Framework Message: A Framework message is a message on a Control Channel that has a type corresponding to one of the Methods defined in this specification. A Framework message is often referred to by its method, such as a "CONTROL message."
- Method: A Method is the type of a Framework message. Four Methods are defined in this specification: SYNC, CONTROL, REPORT, and K-ALIVE.
- Control Command: A Control Command is an application-level request from a Client to a Server. Control Commands are carried in the body of CONTROL messages. Control Commands are defined in separate specifications known as "Control Packages."
- Framework Transaction: A Framework Transaction is defined as a sequence composed of a Control Framework message originated by either a Control Client or Control Server and responded to with a Control Framework response code message. Note that the Control Framework has no "provisional" responses. A Control Framework transaction is referenced throughout the specification as a "Transaction-Timeout."
- Transaction-Timeout: The maximum allowed time between a Control Client or Server issuing a Framework message and it arriving at the destination. The value for "Transaction-Timeout" is 10 seconds.

8.3 OVERVIEW

In fact, CFW details mechanisms for establishing, using, and terminating a reliable transport connection channel using SIP and the SDP offer/answer exchange specified in RFC 3264. The established connection is then used for controlling an external server. Control Channels are negotiated using standard SIP mechanisms that would be used in a similar manner to creating a SIP multimedia session. Figure 8.1 illustrates a simplified view of the mechanism. It highlights a separation of the SIP signaling traffic and the associated Control Channel that is established as a result of the SIP interactions.

The example from Figure 8.1 conveys a 1:1 connection between the Control Client and the Control Server. It is possible, if required, for the client to request multiple Control Channels using separate SIP INVITE dialogs between the Control Client and the Control Server entities. Any of the connections created between the two entities can then be used for Server control interactions.

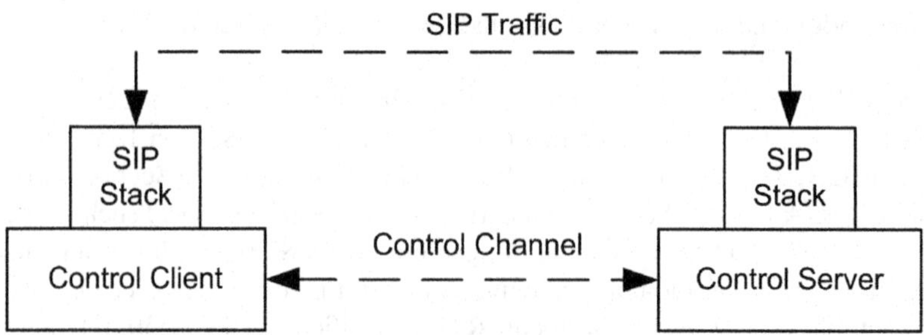

FIGURE 8.1 Basic Architecture.

The control connections are orthogonal to any given media session. Specific media session information is incorporated in control interaction commands, which themselves are defined in external packages, using the XML schema defined in Section 8.14 (Appendix A of RFC 6230). The ability to have multiple Control Channels allows stronger redundancy and the ability to manage high volumes of traffic in busy systems. It should be noted that there can be many media connections under a given conferencing session.

Consider the following simple example for session establishment between a Client and a Server. In this example, some lines are removed for clarity and brevity. The roles discussed here are logical and can change during a session if the Control Package allows. The Client constructs and sends a standard SIP INVITE request, as defined in RFC 3261 (also see Section 1.2), to the external Server. The SDP payload includes the required information for Control Channel negotiation and is the primary mechanism for conveying support for this specification. The application/cfw MIME type is defined in this specification to convey the appropriate SDP format for compliance to this specification. The Connection-Oriented Media (COMEDIA) in the RFC 4145 specification for setting up and maintaining reliable connections is used as part of the negotiation mechanism (more detail available in later sections). The Client also includes the "cfw-id" SDP attribute, as defined in this specification, which is a unique identifier used to correlate the underlying Media Control Channel with the offer/answer exchange.

Client Sends to External Server:

```
INVITE sip:External-Server@example.com SIP/2.0
To: <sip:External-Server@example.com>
From: <sip:Client@example.com>;tag=64823746
Via: SIP/2.0/UDP client.example.com;branch=z9hG4bK72d
Call-ID: 7823987HJHG6
Max-Forwards: 70
CSeq: 1 INVITE
Contact: <sip:Client@clientmachine.example.com>
Content-Type: application/sdp
Content-Length: [..]
v=0
o=originator 2890844526 2890842808 IN IP4 controller.example.com
s=-
c=IN IP4 controller.example.com
m=application 49153 TCP cfw
a=setup:active
a=connection:new
a=cfw-id:H839quwhjdhegvdga
```

On receiving the INVITE request, an external Server supporting this mechanism generates a 200 OK response containing appropriate SDP and formatted using the application/cfw MIME type specified in this specification. The Server inserts its own unique "cfw-id" SDP attribute, which differs from the one received in the INVITE (offer).

External Server Sends to Client:

```
SIP/2.0 200 OK
To: <sip:External-Server@example.com>;tag=28943879
From: <sip:Client@example.com>;tag=64823746
Via: SIP/2.0/UDP client.example.com;branch=z9hG4bK72d;received=192.0.2.4
Call-ID: 7823987HJHG6
CSeq: 1 INVITE
Contact: <sip:External-Server@servermachine.example.com>
Content-Type: application/sdp
Content-Length: [..]
v=0
o=responder 2890844526 2890842808 IN IP4 server.example.com
s=-
c=IN IP4 mserver.example.com
m=application 7563 TCP cfw
a=setup:passive
a=connection:new
a=cfw-id:U8dh7UHDushsdu32uha
```

The Control Client receives the SIP 200 OK response and extracts the relevant information (also sending a SIP ACK). It creates an outgoing (as specified by the SDP "setup" attribute of "active") transmission control protocol (TCP) connection to the Control Server. The connection address (taken from "c=") and port (taken from "m=") are used to identify the remote port in the new connection. Once established, the newly created connection can be used to exchange requests and responses as defined in this specification. If required, after the Control Channel has been set up, media sessions can be established using standard SIP third-party call control (3pcc) of RFC 3725 (see Section 1.3) described earlier. Figure 8.2 provides a simplified example where the framework is used to control a UA's RTP session as follows:

- The link (1) represents the SIP INVITE dialog usage and dedicated Control Channel previously described in this overview section.
- The link (2) from Figure 8.2 represents the UA SIP INVITE dialog usage interactions and associated media flow.

A UA creates a SIP INVITE dialog usage with the Control Client entity. The Control Client entity then creates a SIP INVITE dialog usage to the Control Server, using B2BUA-type functionality. Using the interaction illustrated by (2), the Control Client negotiates media capabilities with the

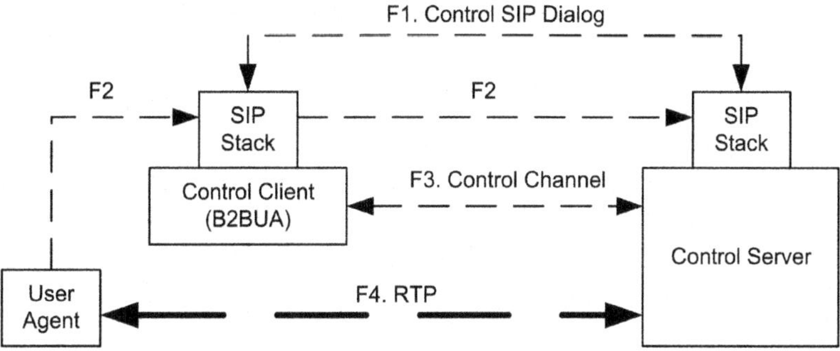

FIGURE 8.2 Participant Architecture.

Control Server, on behalf of the UA, using SIP 3PCC in RFC 3725 (see Section 1.3) described earlier.

8.4 CONTROL CHANNEL SETUP

This section describes the setup, using SIP, of the dedicated Control Channel. Once the Control Channel has been established, commands can be exchanged that are described in the subsequent section.

8.4.1 Control Client SIP UAC Behavior

When a UAC wishes to establish a Control Channel, it must construct and transmit a new SIP INVITE request for Control Channel setup. The UAC must construct the INVITE request as defined in RFC 3261 (also see Section 1.2) described earlier. If a reliable response is received as defined in RFCs 3261 and 3262, the mechanisms defined in this specification are applicable to the newly created SIP INVITE dialog usage. The UAC should include a valid session description (an "offer" as well as "response" defined in RFC 3264) in an INVITE request using the SDP defined in RFC 4566 but may choose an offer-less INVITE as per RFC 3261 (also see Section 1.2). The SDP should be formatted in accordance with the steps described later and using the MIME type application/cfw, which is registered in Section 8.13. The following information defines the composition of specific elements of the SDP payload the offerer must adhere to when used in an SIP-based offer/answer exchange using SDP and the application/cfw MIME type. The SDP being constructed must contain only a single occurrence of a Control Channel definition outlined in this specification but can contain other media lines if required. The Connection Data line in the SDP payload is constructed as specified in RFC 4566: c=<nettype> <addrtype> <connection-address>. The first sub-field, <nettype>, must equal the value "IN." The second sub-field, <addrtype>, must equal either "IP4" or "IP6." The third sub-field for Connection Data is <connection-address>.

This supplies a representation of the SDP originator's address, for example, domain name system (DNS)/Internet protocol (IP) representation. The address is the address used for connections.

Example:

```
c=IN IP4 controller.example.com
```

The SDP must contain a corresponding Media Description entry:

```
m=<media> <port> <proto> <fmt>
```

The first "sub-field," <media>, must equal the value "application." The second sub-field, <port>, must represent a port on which the constructing client can receive an incoming connection if required. The port is used in combination with the address specified in the Connection Data line defined previously to supply connection details. If the entity constructing the SDP cannot receive incoming connections, it must still enter a valid port entry. The use of the port value "0" has the same meaning as defined in an SIP offer/answer exchange described in RFC 3264. The Control Framework has a default port defined in Section 8.13.5. This value is default, although a client is free to choose explicit port numbers. However, SDP should use the default port number unless local policy prohibits its use. Using the default port number allows network administrators to manage firewall policy for Control Framework interactions. The third sub-field, <proto>, compliant to this specification, must support the values "TCP" and "TCP/TLS." Implementations must support transport layer security (TLS) as a transport-level security mechanism for the Control Channel, although use of TLS in specific deployments is optional. Control Framework implementations must support TCP as a transport protocol. When an entity identifies a transport value but is not willing to establish the session, it must respond using the appropriate SIP mechanism. The <fmt> sub-field must contain the value "cfw."

The SDP must also contain a number of SDP media attributes (a=) that are specifically defined in the COMEDIA of RFC 4145 specification. The attributes provide connection negotiation and maintenance parameters. It is RECOMMENDED that a Controlling UAC initiate a connection to an external Server but that an external Server may negotiate and initiate a connection using COMEDIA if network topology prohibits initiating connections in a certain direction. An example of the COMEDIA attributes is:

```
a=setup:active
a=connection:new
```

This example demonstrates a new connection that will be initiated from the owner of the SDP payload. The connection details are contained in the SDP answer received from the UAS. A full example of an SDP payload compliant to this specification can be viewed in Section 8.3. Once the SDP has been constructed along with the remainder of the SIP INVITE request as defined in RFC 3261 (also see Section 1.2), it can be sent to the appropriate location. The SIP INVITE dialog usage and appropriate control connection are then established. An SIP UAC constructing an offer must include the "cfw-id" SDP attribute as defined in Section 8.9.2. The "cfw-id" attribute indicates an identifier that can be used within the Control Channel to correlate the Control Channel with this SIP INVITE dialog usage.

The "cfw-id" attribute must be unique in the context of the interaction between the UAC and the UAS and must not clash with instances of the "cfw-id" used in other SIP offer/answer exchanges. The value chosen for the "cfw-id" attribute must be used for the entire duration of the associated SIP INVITE dialog usage and not be changed during updates to the offer/answer exchange. This applies specifically to the "connection" attribute as defined in RFC 4145. If an SIP UAC wants to change some other parts of the SDP but reuse the already established connection, it uses the value of "existing" in the "connection" attribute (for example, a=connection:existing). If it has noted that a connection has failed and wants to reestablish the connection, it uses the value of "new" in the "connection" attribute (for example, a=connection:new). Throughout this, the connection identifier specified in the "cfw-id" SDP parameter must not change. One is simply negotiating the underlying TCP connection between endpoints but always using the same Control Framework session, which is 1:1 for the lifetime of the SIP INVITE dialog usage.

A non-2xx-class final SIP response (3xx, 4xx, 5xx, and 6xx) received for the INVITE request indicates that no SIP INVITE dialog usage has been created and is treated as specified by SIP in RFC 3261 (also see Section 1.2). Specifically, support of this specification is negotiated through the presence of the media type defined in this specification. The receipt of an SIP error response such as "488" indicates that the offer contained in a request is not acceptable. The inclusion of the media line associated with this specification in such a rejected offer indicates to the client generating the offer that this could be due to the receiving client not supporting this specification. The client generating the offer must act as it would normally on receiving this response, as per RFC 3261 (also see Section 1.2). Media streams can also be rejected by setting the port to "0" in the "m=" line of the session description, as defined in RFC 3264. A client using this specification must be prepared to receive an answer where the "m=" line it inserted for using the Control Framework has been set to "0." In this situation, the client will act as it would for any other media type with a port set to "0."

8.4.2 Control Server SIP UAS Behavior

On receiving a SIP INVITE request, an external server (SIP UAS) inspects the message for indications of support for the mechanisms defined in this specification. This is achieved through inspection of the session description of the offer message and identifying support for the application/cfw

MIME type in the SDP. If the SIP UAS wishes to construct a reliable response that conveys support for the extension, it must follow the mechanisms defined in RFC 3261 (also see Section 1.2). If support is conveyed in a reliable SIP provisional response, the mechanisms in RFC 3262 must also be used. It should be noted that the SDP offer is not restricted to the initial INVITE request and may appear in any series of messages that are compliant to RFCs 3261 (also see Section 1.2), 3262, 3311, and 3264.

When constructing an answer, the SDP payload must be constructed using the semantic (connection, media, and attribute) defined in Section 8.4.1 using valid local settings and also with full compliance to the COMEDIA in the RFC 4145 specification. For example, the SDP attributes included in the answer constructed for the example offer provided in Section 8.4.1 would look as follows:

```
a=setup:passive
a=connection:new
```

A client constructing an answer must include the "cfw-id" SDP attribute as defined in Section 8.9.2. This attribute must be unique in the context of the interaction between the UAC and UAS and must not clash with instances of the "cfw-id" used in other SIP offer/answer exchanges. The "cfw-id" must be different from the "cfw-id" value received in the offer as it is used to uniquely identify and distinguish between multiple endpoints that generate SDP answers. The value chosen for the "cfw-id" attribute must be used for the entire duration of the associated SIP INVITE dialog usage and not be changed during updates to the offer/answer exchange.

Once the SDP answer has been constructed, it is sent using standard SIP mechanisms. Depending on the contents of the SDP payloads that were negotiated using the offer/answer exchange, a reliable connection will be established between the Controlling UAC and External Server UAS entities. The newly established connection is now available to exchange Control Command primitives. The state of the SIP INVITE dialog usage and the associated Control Channel are now implicitly linked. If either party wishes to terminate a Control Channel, it simply issues an SIP termination request, for example, a SIP BYE request or appropriate response in an early SIP INVITE dialog usage. The Control Channel therefore lives for the duration of the SIP INVITE dialog usage. A UAS receiving an SIP OPTIONS request must respond appropriately as defined in RFC 3261 (also see Section 1.2). The UAS must include the media types supported in the SIP 200 OK response in an SIP "Accept" header to indicate the valid media types.

8.5 ESTABLISHING MEDIA STREAMS – CONTROL CLIENT SIP UAC BEHAVIOR

It is intended that the Control Framework will be used within a variety of architectures for a wide range of functions. One of the primary functions will be the use of the Control Channel to apply multiple specific Control Package commands to media sessions established by SIP INVITE dialogs (media dialogs) with a given remote server. For example, the Control Server might send a command to generate audio media (such as an announcement) on an RTP stream between a UA and an MS. SIP INVITE dialogs used to establish media sessions (see Figure 8.2) on behalf of UAs may contain more than one Media Description (as defined by "m=" in the SDP). The Control Client must include a media label attribute, as defined in RFC 4574, for each "m=" definition received that is to be directed to an entity using the Control Framework. This allows the Control Client to later explicitly direct commands on the Control Channel at a specific media line (m=).

This framework identifies the referencing of such associated media dialogs as extremely important. A connection reference attribute has been specified that can optionally be imported into any Control Package. It is intended that this will reduce the repetitive specifying of dialog reference language. The schema can be found in Section 8.14.1 (Appendix A.1 of RFC 6230). Similarly, the ability to identify and apply commands to a group of associated media dialogs (multiparty) is also identified as a common structure that could be defined and reused; for example, playing a prompt to all participants in a conference. The schema for such operations can also be found in Section 8.14 (Appendix A.1 of RFC 6230). Support for both the common attributes described here is specified as part of each Control Package definition, as detailed in Section 9.8.

8.6 CONTROL FRAMEWORK INTERACTIONS

In this specification, the use of the COMEDIA specification allows a Control Channel to be set up in either direction as a result of an SIP INVITE transaction. SIP provides a flexible negotiation mechanism to establish the Control Channel, but there needs to be a mechanism within the Control Channel to correlate it with the SIP INVITE dialog usage implemented for its establishment. A Control Client receiving an incoming connection (whether it be acting in the role of UAC or UAS) has no way of identifying the associated SIP INVITE dialog usage, as it could be simply listening for all incoming connections on a specific port. The following steps, which implementations must support, allow a connecting UA (that is, the UA with the active role in COMEDIA) to identify the associated SIP INVITE dialog usage that triggered the connection. Unless there is an alternative dialog association mechanism used, the UAs must carry out these steps before any other signaling on the newly created Control Channel.

- Once the connection has been established, the UA acting in the active role (active UA) to initiate the connection must send a Control Framework SYNC request. The SYNC request must be constructed as defined in Section 8.9.1 and must contain the "Dialog-ID" message header.
- The "Dialog-ID" message header is populated with the value of the local "cfw-id" media-level attribute that was inserted by the same client in the SDP offer/answer exchange to establish the Control Channel. This allows a correlation between the Control Channel and its associated SIP INVITE dialog usage.
- On creating the SYNC request, the active UA must follow the procedures outlined in Section 8.6.3.3. This provides details of connection keep-alive messages.
- On creating the SYNC request, the active UA must also follow the procedures outlined in Section 8.6.3.4.2. This provides details of the negotiation mechanism used to determine the protocol data units (PDUs) that can be exchanged on the established Control Channel connection.
- The UA in the active role for the connection creation must then send the SYNC request. If the UA in the active role for the connection creation is an SIP UAS and has generated its SDP response in a 2xx-class SIP response, it must wait for an incoming SIP ACK message before issuing the SYNC. If the UA in the active role for the connection creation is an SIP UAS and has generated its SDP response in a reliable 1XX class SIP response, it must wait for an incoming SIP PRACK message before issuing the SYNC. If the UA in the active role for the connection creation is an SIP UAC, it must send the SYNC message immediately on establishment of the Control Channel. It must then wait for a period of at least 2*"Transaction-Timeout" to receive a response. It may choose a longer time to wait, but it must not be shorter than "Transaction-Timeout." In general, a Control Framework transaction must complete within 20 (2*"Transaction-Timeout") seconds and is referenced throughout the specification as "Transaction-Timeout."
- If no response is received for the SYNC message, a timeout occurs, and the Control Channel is terminated along with the associated SIP INVITE dialog usage. The active UA must issue a BYE request to terminate the SIP INVITE dialog usage.
- If the active UA receives a 481 response from the passive UA, this means the SYNC request was received, but the associated SIP INVITE dialog usage specified in the SYNC message does not exist. The active client must terminate the Control Channel. The active UA must issue a SIP BYE request to terminate the SIP INVITE dialog usage.

- All other error responses received for the SYNC request are treated as detailed in this specification and also result in the termination of the Control Channel and the associated SIP INVITE dialog usage. The active UA must issue a BYE request to terminate the SIP INVITE dialog usage.
- The receipt of a 200 response to a SYNC message implies that the SIP INVITE dialog usage and control connection have been successfully correlated. The Control Channel can now be used for further interactions.

SYNC messages can be sent at any point while the Control Channel is open from either side once the initial exchange is complete. If present, the contents of the "Keep-Alive" and "Dialog-ID" headers must not change. New values of the "Keep-Alive" and "Dialog-ID" headers have no relevance, as they are negotiated for the lifetime of the Media Control Channel Framework session.

Once a successful Control Channel has been established, as defined in Sections 8.4.1 and 8.4.2, and the connection has been correlated, as described in previous paragraphs, the two entities are now in a position to exchange Control Framework messages. The following sub-sections specify the general behavior for constructing Control Framework requests and responses. Section 8.6.3 specifies the core Control Framework methods and their transaction processing.

8.6.1 General Behavior for Constructing Requests

An entity acting as a Control Client that constructs and sends requests on a Control Channel must adhere to the syntax defined in Section 8.9. Note that either entity can act as a Control Client depending on individual package requirements. Control Commands must also adhere to the syntax defined by the Control Packages negotiated in Sections 8.4.1 and 8.4.2 of this specification. A Control Client must create a unique transaction and associated identifier for insertion in the request. The transaction identifier is then included in the first line of a Control Framework message along with the method type, as defined in the Augmented Backus–Naur Form (ABNF) in Section 8.9. The first line starts with the "CFW" token for the purpose of easily extracting the transaction identifier. The transaction identifier must be unique in the context of the interaction between the Control Client and the Control Server.

This unique property helps avoid clashes when multiple client entities could be creating transactions to be carried out on a single receiving server. All required, mandatory, and optional Control Framework headers are then inserted into the request with appropriate values (see relevant individual header information for explicit detail). A "Control-Package" header must also be inserted with the value indicating the Control Package to which this specific request applies. Multiple packages can be negotiated per Control Channel using the SYNC message discussed in Section 8.6.3.4.2.

Any Framework message that contains an associated payload must also include the "Content-Type" and "Content-Length" message headers, which indicate the MIME type of the payload specified by the individual Control Framework packages and the size of the message body represented as a whole decimal number of octets, respectively. If no associated payload is to be added to the message, the "Content- Length" header must have a value of "0." A Server receiving a Framework message request must respond with an appropriate response (as defined in Section 8.6.2). Control Clients must wait for a minimum of 2*"Transaction-Timeout" for a response before considering the transaction a failure and tidying state appropriately depending on the extension package being used.

8.6.2 General Behavior for Constructing Responses

An entity acting as a Control Server, on receiving a request, must generate a response within the "Transaction-Timeout" as measured from the Control Client. The response must conform to the ABNF defined in Section 8.9. The first line of the response must contain the transaction identifier used in the first line of the request, as defined in Section 6.1. Responses must not include the "Status" or "Timeout" message headers, and these must be ignored if received by a Client in a response.

A Control Server must include a status code in the first line of the response. If there is no error, the Server responds with a 200 Control Framework status code, as defined in Section 8.7.1. The 200 response may include message bodies. If the response contains a payload, the message must include the "Content-Length" and "Content-Type" headers. When the Control Client receives a 2xx-class response, the Control Command transaction is complete. If the Control Server receives a request, like CONTROL, that the Server understands, but the Server knows processing the command will exceed the "Transaction-Timeout," then the Server must respond with a 202 status code in the first line of the response. Following the initial response, the server will send one or more REPORT messages as described in Section 8.6.3.2. A Control Package must explicitly define the circumstances under which the server sends 200 and 202 messages.

If a Control Server encounters problems with a Control Framework request (like REPORT or CONTROL), an appropriate error code must be used in the response, as listed in Section 8.7. The generation of a non-2xx-class response code to a Control Framework request (like CONTROL or REPORT) will indicate failure of the transaction, and all associated transaction state and resources must be terminated. The response code may provide an explicit indication of why the transaction failed, which might result in a re-submission of the request, depending on the extension package being used.

8.6.3 Transaction Processing

The Control Framework defines four types of requests (methods): CONTROL, REPORT, K-ALIVE, and SYNC. Implementations must support sending and receiving these four methods. The following sub-sections specify each Control Framework method and its associated transaction processing.

8.6.3.1 CONTROL Transactions

A CONTROL message is used by the Control Client to pass control-related information to a Control Server. It is also used as the event-reporting mechanism in the Control Framework. Reporting events is simply another usage of the CONTROL message, which is permitted to be sent in either direction between two participants in a session, carrying the appropriate payload for an event. The message is constructed in the same way as any standard Control Framework message, as discussed in Section 8.6.1 and defined in Section 8.9. A CONTROL message may contain a message body. The explicit Control Command(s) of the message payload contained in a CONTROL message are specified in separate Control Package specifications. Separate Control Package specifications must conform to the format defined in Section 8.8.4. A CONTROL message containing a payload must include a "Content-Type" header. The payload must be one of the payload types defined by the Control Package. Individual packages may allow a CONTROL message that does not contain a payload. This could, in fact, be a valid message exchange within a specific package; if it is not, an appropriate package-level error message must be generated.

8.6.3.2 REPORT Transactions

A "REPORT" message is used by a Control Server when processing of a CONTROL command extends beyond the "Transaction-Timeout," as measured from the Client. In this case, the Server returns a 202 response. The Server returns status updates and the final results of the command in subsequent REPORT messages. All REPORT messages must contain the same transaction ID in the request start line that was present in the original CONTROL transaction. This correlates extended transactions with the original CONTROL transaction. A REPORT message containing a payload must include the "Content-Type" and "Content-Length" headers indicating the payload MIME type of RFC 2045 defined by the Control Package and the length of the payload, respectively.

8.6.3.2.1 Reporting the Status of Extended Transactions

On receiving a CONTROL message, a Control Server must respond within "Transaction-Timeout" with a status code for the request, as specified in Section 8.6.2. If the processing of the command completes within that time, a 200 response code must be sent. If the command does not complete within that time, the response code 202 must be sent, indicating that the requested command is still being processed and the CONTROL transaction is being extended. The REPORT method is then used to update and terminate the status of the extended transaction. The Control Server should not wait until the last possible opportunity to make the decision of issuing a 202 response code and should ensure that it has plenty of time for the response to arrive at the Control Client. If it does not have time, transactions will be terminated (timed out) at the Control Client before completion.

A Control Server issuing a 202 response must ensure that the message contains a "Timeout" message header. This header must have a value in seconds that is the amount of time the recipient of the 202 message must wait before assuming that there has been a problem and terminating the extended transaction and associated state. The initial REPORT message must contain a "Seq" (Sequence) message header with a value equal to "1." Note: the "Seq" numbers at both Control Client and Control Server for Framework messages are independent.

All REPORT messages for an extended CONTROL transaction must contain a "Timeout" message header. This header will contain a value in seconds that is the amount of time the recipient of the REPORT message must wait before assuming that there has been a problem and terminating the extended transaction and associated state. On receiving a REPORT message with a "Status" header of "update," the Control Client must reset the timer for the associated extended CONTROL transaction to the indicated timeout period. If the timeout period approaches and no intended REPORT messages have been generated, the entity acting as a Control Framework UAS for the interaction must generate a REPORT message containing, as defined in this paragraph, a "Status" header of "update" with no associated payload. Such a message acts as a timeout refresh and in no way impacts the extended transaction, because no message body or semantics are permitted. It is RECOMMENDED that a minimum value of 10 and a maximum value of 15 seconds be used for the value of the "Timeout" message header. It is also RECOMMENDED that a Control Server refresh the timeout period of the CONTROL transaction at an interval that is not too close to the expiry time. A value of 80% of the timeout period could be used. For example, if the timeout period is 10 seconds, the Server would refresh the transaction after 8 seconds.

Subsequent REPORT messages that provide additional information relating to the extended CONTROL transaction must also include and increment by 1 the "Seq" header value. A REPORT message received that has not been incremented by 1 must be responded to with a 406 response, and the extended transaction must be considered terminated. On receiving a 406 response, the extended transaction must be terminated. REPORT messages must also include a "Status" header with a value of "update." These REPORT messages sent to update the extended CONTROL transaction status may contain a message body, as defined by individual Control Packages and specified in Section 8.8.5.

A REPORT message sent updating the extended transaction also acts as a timeout refresh, as described earlier in this section. This will result in a transaction timeout period at the initiator of the original CONTROL request being reset to the interval contained in the "Timeout" message header. When all processing for an extended CONTROL transaction has taken place, the entity acting as a Control Server must send a terminating REPORT message. The terminating REPORT message must increment the value in the "Seq" message header by the value of "1" from the previous REPORT message. It must also include a "Status" header with a value of "terminate" and may contain a message body. It must also contain a "Timeout" message header with a valid value. The inclusion of the "Timeout" header is for consistency, and its value is ignored. A Control Framework UAC can then clean up any pending state associated with the original CONTROL transaction.

8.6.3.3 K-ALIVE Transactions

The protocol defined in this specification may be used in various network architectures. These include a wide range of deployments where the clients could be co-located in a secured, private domain or spread across disparate domains that require traversal of devices such as network address translators (NATs) and firewalls. A keep-alive mechanism enables the Control Channel to be kept active during times of inactivity. This is because many firewalls have a timeout period after which connections are closed. This mechanism also provides the ability for application-level failure detection. It should be noted that the following procedures apply only to the Control Channel being created. For details relating to the SIP keep-alive mechanism, implementers should seek guidance from SIP Outbound, described in RFC 5626.

The following keep-alive procedures must be implemented. Specific deployments may choose not to use the keep-alive mechanism if both entities are in a co-located domain. Note that choosing not to use the keep-alive mechanism defined in this section, even when in a co-located architecture, will reduce the ability to detect application-level errors, especially during long periods of inactivity.

Once the SIP INVITE dialog usage has been established and the underlying Control Channel has been set up, including the initial correlation handshake using SYNC as discussed in Section 8.6, both entities acting in the active and passive roles, as defined in COMEDIA, described in RFC 4145, must start a keep-alive timer equal to the value negotiated during the Control Channel SYNC request/response exchange. This is the value from the "Keep-Alive" header in seconds.

8.6.3.3.1 *Behavior for an Entity in an Active Role*

When in an active role, a K-ALIVE message must be generated before the local keep-alive timer fires. An active entity is free to send the K-ALIVE message whenever it chooses. It is RECOMMENDED for the entity to issue a K-ALIVE message after 80% of the local keep-alive timer. On receiving a 200 OK Control Framework message for the K-ALIVE request, the active entity must reset the local keep-alive timer. If no 200 OK response is received to the K-ALIVE message, or a transport-level problem is detected by some other means, before the local keep-alive timer fires, the active entity may use COMEDIA renegotiation procedures to recover the connection. Otherwise, the active entity must tear down the SIP INVITE dialog and recover the associated Control Channel resources.

8.6.3.3.2 *Behavior for an Entity in a Passive Role*

When acting as a passive entity, a K-ALIVE message must be received before the local keep-alive timer fires. When a K-ALIVE request is received, the passive entity must generate a 200 OK Control Framework response and reset the local keep-alive timer. No other Control Framework response is valid. If no K-ALIVE message is received (or a transport-level problem is detected by some other means) before the local keep-alive timer fires, the passive entity must tear down the SIP INVITE dialog and recover the associated Control Channel resources.

8.6.3.4 SYNC Transactions

The initial SYNC request on a Control Channel is used to negotiate the timeout period for the Control Channel keep-alive mechanism and to allow clients and servers to learn the Control Packages that each supports. Subsequent SYNC requests may be used to change the set of Control Packages that can be used on the Control Channel.

8.6.3.4.1 *Timeout Negotiation for the Initial SYNC Transaction*

The initial SYNC request allows the timeout period for the Control Channel keep-alive mechanism to be negotiated. The following rules must be followed for the initial SYNC request:

- If the Client initiating the SDP offer has a COMEDIA "setup" attribute equal to active, the "Keep-Alive" header must be included in the SYNC message generated by the offerer. The value of the "Keep-Alive" header should be in the range of 95 to 120 seconds (this is consistent with SIP Outbound, described in RFC 5626). The value of the "Keep-Alive" header must not exceed 600 seconds. The client that generated the SDP "Answer" (the passive client) must copy the "Keep-Alive" header into the 200 response to the SYNC message with the same value.
- If the Client initiating the SDP offer has a COMEDIA "setup" attribute equal to passive, the "Keep-Alive"' header parameter must be included in the SYNC message generated by

the answerer. The value of the "Keep-Alive" header should be in the range of 95 to 120 seconds. The client that generated the SDP offer (the passive client) must copy the "Keep-Alive" header into the 200 response to the SYNC message with the same value.
- If the Client initiating the SDP offer has a COMEDIA "setup" attribute equal to "actpass," the "Keep-Alive" header parameter must be included in the SYNC message of the entity who is the active participant in the SDP session. If the client generating the subsequent SDP answer places a value of "active" in the COMEDIA SDP "setup" attribute, it will generate the SYNC request and include the "Keep-Alive" header. The value should be in the range of 95 to 120 seconds. If the client generating the subsequent SDP answer places a value of "passive" in the COMEDIA "setup" attribute, the original UA making the SDP will generate the SYNC request and include the "Keep-Alive" header. The value should be in the range of 95 to 120 seconds.
- If the initial negotiated offer/answer results in a COMEDIA "setup" attribute equal to "holdconn," the initial SYNC mechanism will occur when the offer/answer exchange is updated and the active/passive roles are resolved using COMEDIA.

The previous steps ensure that the entity initiating the Control Channel connection is always the one specifying the keep-alive timeout period. It will always be the initiator of the connection who generates the K-ALIVE messages.

Once negotiated, the keep-alive timeout applies for the remainder of the Control Framework session. Any subsequent SYNC messages generated in the Control Channel do not impact the negotiated keep-alive property of the session. The "Keep-Alive" header must not be included in subsequent SYNC messages, and if it is received, it must be ignored.

8.6.3.4.2 Package Negotiation

As part of the SYNC message exchange, a client generating the request must include a "Packages" header, as defined in Section 8.9. The "Packages" header contains a list of all Control Framework packages that can be supported within this control session from the perspective of the client creating the SYNC message. All Channel Framework package names must be tokens that adhere to the rules set out in Section 8.8. The "Packages" header of the initial SYNC message must contain at least one value. A server receiving the initial SYNC request must examine the contents of the "Packages" header. If the server supports at least one of the packages listed in the request, it must respond with a 200 response code. The response must contain a "Packages" header that lists the supported packages that are in common with those from the "Packages" header of the request (either all or a subset). This list forms a common set of Control Packages that are supported by both parties.

Any Control Packages supported by the server that are not listed in the "Packages" header of the SYNC request may be placed in the "Supported" header of the response. This provides a hint to the client that generated the SYNC request about additional packages supported by the server. If no common packages are supported by the server receiving the SYNC message, it must respond with a 422 error response code. The error response must contain a "Supported" header indicating the packages that are supported. The initiating client can then choose to either re-submit a new SYNC message based on the 422 response or consider the interaction a failure. This would lead to termination of the associated SIP INVITE dialog by sending a SIP BYE request, as per RFC 3261 (also see Section 1.2).

Once the initial SYNC transaction is completed, either client may choose to send a subsequent new SYNC message to renegotiate the packages that are supported within the Control Channel. A new SYNC message whose "Packages" header has different values from the previous SYNC message can effectively add and delete the packages used in the Control Channel. If a client receiving a subsequent SYNC message does not wish to change the set of packages, it must respond with a 421 Control Framework response code. Subsequent SYNC messages must not change the value of the "Dialog-ID" and "Keep-Alive" Control Framework headers that appeared in the original SYNC negotiation. An entity may honor Control Framework commands relating to a Control Package it no longer supports after package renegotiation. When the entity does not wish to honor such commands, it must respond to the request with a 420 response.

8.7 RESPONSE CODE DESCRIPTIONS

The following response codes are defined for transaction responses to methods defined in Section 8.6.1. All response codes in this section must be supported and can be used in response to both CONTROL and REPORT messages except that a 202 must not be generated in response to a REPORT message. Note that these response codes apply to Framework Transactions only. Success or error indications for Control Commands must be treated as the result of a Control Command and returned in either a 200 response or a REPORT message.

- **200 Response Code:** The framework protocol transaction completed successfully.
- **202 Response Code:** The framework protocol transaction completed successfully, and additional information will be provided at a later time through the REPORT mechanism defined in Section 8.6.3.2.
- **400 Response Code:** The request was syntactically incorrect.
- **403 Response Code:** The server understood the request but is refusing to fulfill it. The client should not rcpcat the request.
- **405 Response Code:** Method not allowed. The primitive is not supported.
- **406 Response Code:** Message out of sequence.
- **420 Response Code:** Intended target of the request is for a Control Package that is not valid for the current session.
- **421 Response Code:** Recipient does not wish to renegotiate Control Packages at this moment in time.
- **422 Response Code:** Recipient does not support any Control Packages listed in the SYNC message.
- **423 Response Code:** Recipient has an existing transaction with the same transaction ID.
- **481 Response Code:** The transaction of the request does not exist. In response to a SYNC request, the 481 response code indicates that the corresponding SIP INVITE dialog usage does not exist.
- **500 Response Code:** The recipient does not understand the request.

8.8 CONTROL PACKAGES

Control Packages specify behavior that extends the capability defined in this specification. Control Packages must not weaken statements of "must" and "should" strength in this specification. A Control Package may strengthen "should," "RECOMMENDED," and "may" to "must" if justified by the specific usage of the framework. In addition to the usual sections expected in Standards-Track RFCs and SIP extension specifications, authors of Control Packages need to address each of the issues detailed in the following sub-sections.

The following sections must be used as a template and included appropriately in all Control-Package specifications. To reiterate, the following sections do not solely form the basis of all Control-Package specifications but are included as a minimum to provide essential package-level information. A Control-Package specification can take any valid form it wishes as long as it includes at least the following information listed in this section.

8.8.1 Control Package Name

This section must be present in all extensions to this specification and provides a token name for the Control Package. The section must include information that appears in the Internet Assigned

Numbers Authority (IANA) registration of the token. Information on registering Control Package tokens is contained in Section 8.13.

8.8.2 Framework Message Usage

The Control Framework defines a number of message primitives that can be used to exchange commands and information. There are no limitations restricting the directionality of messages passed down a Control Channel. This section of a Control Package specification must explicitly detail the types of Framework messages (Methods) that can be used as well as provide an indication of directionality between entities. This will include which role type is allowed to initiate a request type.

8.8.3 Common XML Support

This optional section is only included in a Control Package if the attributes for media dialog or conference reference are required, as defined and discussed in Appendix A.1. The Control Package will make strong statements using language from RFC 2119 if the XML schema defined in Appendix A.1 is to be supported. If only part of the schema is required (for example, just "connectionid" or "conferenceid"), the Control Package will make equally strong statements using language from RFC 2119.

8.8.4 CONTROL Message Bodies

This mandatory section of a Control Package defines the control body that can be contained within a CONTROL command request, as defined in Chapter 6, or that no Control Package body is required. This section must indicate the location of detailed syntax definitions and semantics for the appropriate MIME specified in RFC 2045 body type that apply to a CONTROL command request and optionally, the associated 200 response. For Control Packages that do not have a Control Package body, making such a statement satisfies the "must" strength of this section in the Control Package specification.

8.8.5 REPORT Message Bodies

This mandatory section of a Control Package defines the REPORT body that can be contained within a REPORT command request, as defined in Chapter 6, or that no report package body is required. This section must indicate the location of detailed syntax definitions and semantics for the appropriate MIME specified in RFC 2045 body type. It should be noted that the Control Framework specification does allow payloads to exist in 200 responses to CONTROL messages (as defined in this specification). An entity that is prepared to receive a payload type in a REPORT message must also be prepared to receive the same payload in a 200 response to a CONTROL message. For Control Packages that do not have a Control Package body, stating such satisfies the "must" strength of this section in the Control Package specification.

8.8.6 Audit

Auditing of various Control Package properties such as capabilities and resources (package-level meta-information) is extremely useful. Such meta-data usually has no direct impact on Control Framework interactions but allows contextual information to be learnt. Control Packages are encouraged to make use of Control Framework interactions to provide relevant package audit information.

This section should include the following information:

- Whether an auditing capability is available in this package
- How auditing information is triggered (for example, using a Control Framework CONTROL message) and delivered (for example, in a Control Framework 200 response)

- The location of the audit query and response format for the payload (for example, it could be a separate XML schema OR part of a larger XML schema)

8.8.7 Examples

It is strongly RECOMMENDED that Control Packages provide a range of message flows that represent common flows using the package and this framework specification.

8.9 FORMAL SYNTAX

8.9.1 Control Framework Formal Syntax

The Control Framework interactions use the UTF-8 transformation format as defined in RFC 3629. The syntax in this section uses the ABNF as defined in RFC 5234 including types "DIGIT," "CRLF," and "ALPHA." Unless otherwise stated in the definition of a particular header field, field values, parameter names, and parameter values are not case-sensitive.

```
control-req-or-resp   =   control-request / control-response
control-request       =   control-req-start *headers CRLF
                          [control-content]
control-response      =   control-resp-start *headers CRLF
                          [control-content]
control-req-start     =   pCFW SP trans-id SP method CRLF
control-resp-start    =   pCFW SP trans-id SP status-code CRLF
pCFW                  =   %x43.46.57; CFW in caps
trans-id              =   alpha-num-token
method                =   mCONTROL / mREPORT / mSYNC / mK-ALIVE /
                          other-method
mCONTROL              =   %x43.4F.4E.54.52.4F.4C ; CONTROL in caps
mREPORT               =   %x52.45.50.4F.52.54 ; REPORT in caps
mSYNC                 =   %x53.59.4E.43 ; SYNC in caps
mK-ALIVE              =   %x4B.2D.41.4C.49.56.45 ; K-ALIVE in caps
other-method          =   1*UPALPHA
status-code           =   3*DIGIT ; any code defined in this and other
                          specifications
headers               =   header-name CRLF
header-name           =   (Content-Length
                          /Content-Type
                          /Control-Package
                          /Status
                          /Seq
                          /Timeout
                          /Dialog-ID
                          /Packages
                          /Supported
                          /Keep-alive
                          /ext-header)
Content-Length        =   "Content-Length:" SP 1*DIGIT
Control-Package       =   "Control-Package:" SP 1*alpha-num-token
Status                =   "Status:" SP ("update" / "terminate" )
Timeout               =   "Timeout:" SP 1*DIGIT
Seq                   =   "Seq:" SP 1*DIGIT
Dialog-ID             =   "Dialog-ID:" SP dialog-id-string
Packages              =   "Packages:" SP package-name *(COMMA package-name)
Supported             =   "Supported:" SP supprtd-alphanum *(COMMA
                          supprtd-alphanum)
```

```
Keep-alive              =   "Keep-Alive:" SP kalive-seconds
dialog-id-string        =   alpha-num-token
package-name            =   alpha-num-token
supprtd-alphanum        =   alpha-num-token
kalive-seconds          =   1*DIGIT
alpha-num-token         =   ALPHANUM 3*31alpha-num-tokent-char
alpha-num-tokent-char =     ALPHANUM / "." / "-" / "+" / "%" / "=" / "/"
control-content         =   *OCTET
Content-Type            =   "Content-Type:" SP media-type
media-type              =   type "/" subtype *(SP ";" gen-param )
type                    =   token ; Section 4.2 of RFC 4288
subtype                 =   token ; Section 4.2 of RFC 4288
gen-param               =   pname [ "=" pval ]
pname                   =   token
pval                    =   token / quoted-string
token                   =   1*(%x21 / %x23-27 / %x2A-2B / %x2D-2E
                            / %x30-39 / %x41-5A / %x5E-7E)
quoted-string           =   DQUOTE *(qdtext / qd-esc) DQUOTE
qdtext                  =   SP / HTAB / %x21 / %x23-5B / %x5D-7E
                            / UTF8-NONASCII
qd-esc                  =   (BACKSLASH BACKSLASH) /
                            (BACKSLASH DQUOTE)
BACKSLASH               =   "\"
UPALPHA                 =   %x41-5A
ALPHANUM                =   ALPHA / DIGIT
ext-header              =   hname ":" SP hval CRLF
hname                   =   ALPHA *token
hval                    =   utf8text
utf8text                =   *(HTAB / %x20-7E / UTF8-NONASCII)
UTF8-NONASCII           =   UTF8-2 / UTF8-3 / UTF8-4 ; From RFC 3629
```

Table 8.1 details a summary of the headers that can be contained in Control Framework interactions.

The notation used in Table 8.1 is as follows:

- R: Header field may only appear in requests.
- r: Header field may only appear in responses.
- 2xx, 4xx, etc.: Response codes with which the header field can be used.

TABLE 8.1
Summary of Headers in Control Framework Interactions

Header field Where CONTROL REPORT SYNC K-ALIVE

Content-Length o - -

Control-Package R m - - -

Seq - m - -

Status R - m - -

Timeout R - m - -

Timeout 202 - m - -

Dialog-ID R - - m -

Packages - - m -

Supported r - - o -

Keep-Alive R - - o -

Content-Type o - -

- [blank]: Header field may appear in either requests or responses.
- m: Header field is mandatory.
- o: Header field is optional.
- -: Header field is not applicable (ignored if present).

8.9.2 Control Framework Dialog Identifier SDP Attribute

This specification defines a new media-level value attribute: "cfw-id." Its formatting in SDP is described by the following ABNF described in RFC 5234.

```
cfw-dialog-id = "a=cfw-id:" 1*(SP cfw-id-name) CRLF
cfw-id-name = token
token = 1*(token-char)
token-char = %x21 / %x23-27 / %x2A-2B / %x2D-2E / %x30-39
/ %x41-5A / %x5E-7E
```

The token-char and token elements are defined in RFC 4566 but included here to provide support for the implementer of this SDP feature.

8.10 EXAMPLES

The following examples provide an abstracted flow of Control Channel establishment and Control Framework message exchange. The SIP signaling is prefixed with the token "SIP." All other messages are Control Framework interactions defined in this specification. In this example, the Control Client establishes a Control Channel, SYNCs with the Control Server, and issues a CONTROL request that cannot be completed within the "Transaction-Timeout," so the Control Server returns a 202 response code to extend the transaction. The Control Server then follows with REPORTs until the requested action has been completed. The SIP INVITE dialog is then terminated.

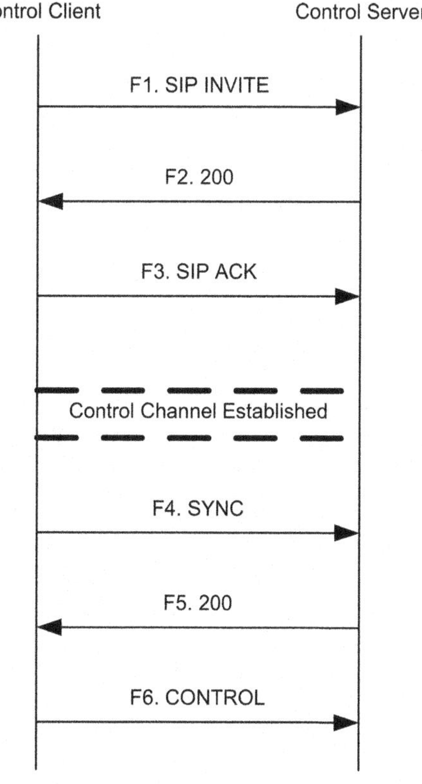

F1. Control Client-->Control Server (SIP): INVITE

```
sip:control-server@example.com
INVITE sip:control-server@example.com SIP/2.0
To: <sip:control-server@example.com>
From: <sip:control-client@example.com>;tag=8937498
Via: SIP/2.0/UDP client.example.com;branch=z9hG4bK123
CSeq: 1 INVITE
Max-Forwards: 70
Call-ID: 893jhoeihjr8392@example.com
Contact: <sip:control-client@pc1.example.com>
Content-Type: application/sdp
Content-Length: 206
v=0
o=originator 2890844526 2890842808 IN IP4 controller.example.com
s=-
c=IN IP4 control-client.example.com
m=application 49153 TCP cfw
a=setup:active
a=connection:new
a=cfw-id:fndskuhHKsd783hjdla
```

F2. Control Server-->Control Client (SIP): 200 OK

```
SIP/2.0 200 OK
To: <sip:control-server@example.com>;tag=023983774
From: <sip:control-client@example.com>;tag=8937498
Via: SIP/2.0/UDP client.example.com;branch=z9hG4bK123;received=192.0.2.5
CSeq: 1 INVITE
Call-ID: 893jhoeihjr8392@example.com
Contact: <sip:control-server@pc2.example.com>
Content-Type: application/sdp
Content-Length: 203
v=0
o=responder 2890844600 2890842900 IN IP4 controller.example.com
s=-
c=IN IP4 control-server.example.com
m=application 49153 TCP cfw
a=setup:passive
a=connection:new
a=cfw-id:7JeDi23i7eiysi32
```

F3. Control Client-->Control Server (SIP): ACK

F4. Control Client opens a TCP connection to the Control Server. The connection can now be used to exchange Control Framework messages. Control Client-->Control Server (Control Framework message): SYNC.

```
CFW 8djae7khauj SYNC
Dialog-ID: fndskuhHKsd783hjdla
Keep-Alive: 100
Packages: msc-ivr-basic/1.0
```

F5. Control Server-->Control Client (Control Framework message): 200.

```
CFW 8djae7khauj 200
Keep-Alive: 100
```

```
Packages: msc-ivr-basic/1.0
Supported: msc-ivr-vxml/1.0,msc-conf-audio/1.0
```

F6. Once the SYNC process has completed, the connection can now be used to exchange Control Framework messages. Control Client-->Control Server (Control Framework message): CONTROL.

```
CFW i387yeiqyiq CONTROL
Control-Package: <package-name>
Content-Type: example_content/example_content

Content-Length: 11
<XML BLOB/>
```

F7. Control Server-->Control Client (Control Framework message): 202.

```
CFW i387yeiqyiq 202
Timeout: 10
```

F8. Control Server-->Control Client (Control Framework message): REPORT.

```
CFW i387yeiqyiq REPORT
Seq: 1
Status: update
Timeout: 10
```

F9. Control Client-->Control Server (Control Framework message): 200.

```
CFW i387yeiqyiq 200
Seq: 1
```

F10. Control Server-->Control Client (Control Framework message): REPORT.

```
CFW i387yeiqyiq REPORT
Seq: 2
Status: update
Timeout: 10
Content-Type: example_content/example_content
Content-Length: 11
<XML BLOB/>
```

F11. Control Client-->Control Server (Control Framework message): 200.

```
CFW i387yeiqyiq 200
Seq: 2
```

F12. Control Server-->Control Client (Control Framework message): REPORT.

```
CFW i387yeiqyiq REPORT
Seq: 3
Status: terminate
Timeout: 10
Content-Type: example_content/example_content
Content-Length: 11
<XML BLOB/>
```

F13. Control Client-->Control Server (Control Framework message):200.

```
CFW i387yeiqyiq 200
Seq: 3
```

```
F14. Control Client-->Control Server (SIP): BYE

    BYE sip:control-server@pc2.example.com SIP/2.0
    To: <sip:control-server@example.com>;tag=023983774
    From: <sip:client@example.com>;tag=8937498
    Via: SIP/2.0/UDP client.example.com;branch=z9hG4bK234
    CSeq: 2 BYE
    Max-Forwards: 70
    Call-ID: 893jhoeihjr8392@example.com
    Contact: <sip:control-client@pc1.example.com>
    Content-Length: 0

F15. Control Server-->Control Client (SIP): 200 OK

    SIP/2.0 200 OK
    To: <sip:control-server@example.com>;tag=023983774
    From: <sip:client@example.com>;tag=8937498
    Via: SIP/2.0/UDP client.example.com;branch=z9hG4bK234;received=192.0.2.5
    CSeq: 2 BYE
    Call-ID: 893jhoeihjr8392@example.com
    Contact: <sip:control-server@pc1.example.com>
    Content-Length: 0
```

8.11 EXTENSIBILITY

The Media Control Channel Framework was designed to be only minimally extensible. New methods, header fields, and status codes can be defined in Standards-Track RFCs. The Media Control Channel Framework does not contain a version number or any negotiation mechanism to require or discover new features. If an extension is specified in the future that requires negotiation, the specification will need to describe how the extension is to be negotiated in the encapsulating signaling protocol. If a non-interoperable update or extension occurs in the future, it will be treated as a new protocol, and it must describe how its use will be signaled.

In order to allow extension header fields without breaking interoperability, if a Media Control Channel device receives a request or response containing a header field that it does not understand, it must ignore the header field and process the request or response as if the header field were not present. If a Media Control Channel device receives a request with an unknown method, it must return a 500 response.

8.12 SECURITY CONSIDERATIONS

The Channel Framework provides confidentiality and integrity for the messages it transfers. It also provides assurances that the connected host is the host that it meant to connect to and that the connection has not been hijacked, as discussed in the remainder of this section. In design, the Channel Framework complies with the security-related requirements specification in "Media Server Control Protocol Requirements" specified in RFC 5167 (also see Section 8.2) – more specifically, REQ-MCP-11, REQ-MCP-12, REQ-MCP-13, and REQ-MCP-14. Specific security measures employed by the Channel Framework are summarized in the following sub-sections.

8.12.1 Session Establishment

Channel Framework sessions are established as media sessions described by SDP within the context of a SIP INVITE dialog. In order to ensure secure rendezvous between Control Framework clients and servers, the Media Channel Control Framework should make full use of mechanisms provided by SIP. The use of the "cfw-id" SDP attribute results in important session information being carried across the SIP network. For this reason, SIP clients using this specification must use appropriate security mechanisms, such as TLS, specified in RFC 5246, and SMIME, described in RFC 5751, when deployed in open networks.

8.12.2 Transport-Level Protection

When using only TCP connections, the Channel Framework security is weak. Although the Channel Framework requires the ability to protect this exchange, there is no guarantee that the protection will be used all the time. If such protection is not used, anyone can see data exchanges. Sensitive data, such as private and financial data, is carried over the Control Framework channel. Clients and servers must be properly authenticated/authorized, and the Control Channel must permit the use of confidentiality, replay protection, and integrity protection for the data. To ensure Control Channel protection, Control Framework clients and servers must support TLS and should use it by default unless alternative Control Channel protection is used or a protected environment is guaranteed by the administrator of the network.

Alternative Control Channel protection may be used if desired, for example, IPsec, defined in RFC 8446 (obsoleted RFC 5246). TLS is used to authenticate devices and to provide integrity, replay protection, and confidentiality for the header fields being transported on the Control Channel. Channel Framework elements must implement TLS and must also implement the TLS ClientExtendedHello extended hello information for server name indication as described in RFC 8446 (obsoleted RFC 5246). A TLS ciphersuite of TLS_RSA_WITH_AES_128_CBC_SHA of RFC 3261 (also see Section 1.2) must be supported. Other ciphersuites may also be supported.

When a TLS client establishes a connection with a server, it is presented with the server's X.509 certificate. Authentication proceeds as described in RFC 8446 (obsoleted RFC 5246) and in Section 7.3 ("Client Behavior") of RFC 5922. A TLS server conformant to this specification must ask for a client certificate; if the client possesses a certificate, it will be presented to the server for mutual authentication, and authentication proceeds as described in Section 7.4 ("Server Behavior") of RFC 5922.

8.12.3 Control Channel Policy Management

This specification permits the establishment of a dedicated Control Channel using SIP. It is also permitted for entities to create multiple channels for the purpose of failover and redundancy. As a general solution, the ability for multiple entities to create connections and have access to resources could be the cause of potential conflict in shared environments. It should be noted that this specification does not carry any specific mechanism to overcome such conflicts but will provide a summary of how to do so.

It can be determined that access to resources and use of Control Channels relate to policy. It can be considered that implementation and deployment detail dictates the level of policy that is adopted. The authorization and associated policy of a Control Channel can be linked to the authentication mechanisms described in this section. For example, strictly authenticating a Control Channel using TLS authentication allows entities to protect resources and ensure the required level of granularity. Such policy can be applied at the package level or even as low as a structure like a conference instance (Control Channel X is not permitted to issue commands for Control Package Y OR Control Channel A is not permitted to issue commands for conference instance B). Systems should ensure that if required, an appropriate policy framework is adopted to satisfy the requirements for implemented packages. The most robust form of policy can be achieved using a strong authentication mechanism such as mutual TLS authentication on the Control Channel. This specification provides a Control Channel response code (403) to indicate to the issuer of a command that it is not permitted. The 403 response must be issued to Control Framework requests that are not permitted under the implemented policy. If a 403 response is received, a Control Framework client may choose to re-submit the request with differing requirements or to abandon the request. The 403 response does not provide any additional information on the policy failure due to the generic nature of this specification. Individual Control Packages can supply additional information if required. The mechanism for providing such additional information is not mandated in this specification. It should be noted that additional policy requirements to those covered in this section might be defined and applied in individual packages that specify a finer granularity for access to resources, etc.

8.13 IANA CONSIDERATIONS

Note that Control Packages Registration Information, Control Framework Header Fields, Control Framework Status Codes, Control Framework Header Fields, Control Framework Port, Media Type, "cfw-id" SDP Attribute, URN Sub-Namespace, and XML Schema of this specification need to be registered with the IANA. Interested readers are requested to see RFC 6230.

8.14 COMMON PACKAGE COMPONENTS (APPENDIX A OF RFC 6230)

During the creation of the Control Framework, it has become clear that there are a number of components that are common across multiple packages. It has become apparent that it would be useful to collect such reusable components in a central location. In the short term, this appendix provides the placeholder for the utilities, and it is the intention that this section will eventually form the basis of an initial "Utilities Specification" that can be used by Control Packages.

8.14.1 Common Dialog/Multiparty Reference Schema (Appendix A1 of RFC 6230)

The following schema provides some common attributes for allowing Control Packages to apply specific commands to a particular SIP media dialog (also referred to as "Connection") or conference. If used within a Control Package, the Connection and multiparty attributes will be imported and used appropriately to specifically identify either a SIP dialog or a conference instance. If used within a package, the value contained in the "connectionid" attribute MUST be constructed by concatenating the "Local" and "Remote" SIP dialog identifier tags as defined in RFC 3261 (also see Section 1.2). They MUST then be separated using the ":" character. So, the format would be: "Local Dialog tag" + ":" + "Remote Dialog tag."

As an example, for an entity that has a SIP Local dialog identifier of "7HDY839" and a Remote dialog identifier of "HJKSkyHS," the "connectionid" attribute for a Control Framework command would be 7HDY839:HJKSkyHS. It should be noted that Control Framework requests initiated in conjunction with a SIP dialog will produce a different "connectionid" value depending on the directionality of the request; for example, Local and Remote tags are locally identifiable.

As with the Connection attribute previously defined, it is useful to have the ability to apply specific Control Framework commands to a number of related dialogs, such as a multiparty call. This typically consists of a number of media dialogs that are logically bound by a single identifier. The following schema allows Control Framework commands to explicitly reference such a grouping through a "conferenceid" XML container. If used by a Control Package, any control XML referenced by the attribute applies to all related media dialogs. Unlike the dialog attribute, the "conferenceid" attribute does not need to be constructed based on the overlying SIP dialog.

The "conferenceid" attribute value is system specific and should be selected with relevant context and uniqueness. It should be noted that the values contained in both the "connectionid" and "conferenceid" identifiers MUST be compared in a case-sensitive manner.

The full schema follows:

```
<?xml version="1.0" encoding="UTF-8"?>
<xsd:schema
   targetNamespace="urn:ietf:params:xml:ns:control:framework-attributes"
   xmlns:xsd="http://www.w3.org/2001/XMLSchema"
   xmlns="urn:ietf:params:xml:ns::control:framework-attributes"
   elementFormDefault="qualified" attributeFormDefault="unqualified">
   <xsd:attributeGroup name="framework-attributes">
      <xsd:annotation>
         <xsd:specificationation>
              SIP Connection and Conf Identifiers
         </xsd:specificationation>
      </xsd:annotation>
      <xsd:attribute name="connectionid" type="xsd:string"/>
      <xsd:attribute name="conferenceid" type="xsd:string"/>
   </xsd:attributeGroup>
</xsd:schema>
```

8.15 SUMMARY

We have described a media control channel protocol that allows the multimedia conferencing functional entities to be distributed physically without being co-located even though the conferencing system is logically centralized. This is a very important milestone for providing scalability for audio and video media servers for XCON conferencing systems that need to serve conference participants across the world, because audio and especially, video are very bandwidth-intensive. Moreover, audio and video codecs are primarily built with constant bit rate (CBR) with no room for saving bandwidths through statistical multiplexing, and the end-to-end bandwidths need to be available in real time within only hundreds of milliseconds constantly for the entire duration of the transmission of their activities. More importantly, the media control capabilities, such as Security Capabilities, Capability Negotiations, Capability Negotiations, Application Server Capabilities, Media Mixing Capabilities, IVR Capabilities, Operational Capabilities, and Location Service, that can be controlled by the CFW are explained in detail.

In addition, the Control Server, Control Client, Framework Message, Method, Control Command, and Transaction-Timeout that are relevant to CFW are defined. The Control Channel setup, Control Server and client behavior, establishing media streams, control framework interactions, general behavior of constructing requests and responses, detailed transaction processing, and response codes are specified for the CFW.

Control Packages can be defined for extensions of CFW (RFC 6230) capabilities in the future. The guidelines for Control Package Name, Framework Message Usage, Common XML Support, CONTROL Message Bodies, REPORT Message Bodies, and Audit are described. In this context, an example of Control Package is provided. The formal syntax for CFW is defined. We have also discussed the schema that may be common across multiple packages for media mixing. Finally, security for Session Establishment, Transport-Level Protection, and Control Channel Policy Management of CFW is articulated.

8.16 PROBLEMS

1. Explain why we need a media channel control protocol. Explain in detail how an XCOM conferencing system can be logically centralized while the conference functional entities (e.g. audio and video media servers) can be distributed physically across different geographical locations (that is, without being co-located).

2. How does the CFW protocol provide scalability in designing the large-scale XCON conferencing systems serving worldwide conference participants using audio and video media? Draw the conferencing architecture and explain in detail to justify.
3. What are the CFW capabilities? Articulate in detail with examples of the XCON conferencing systems, including architecture and call flows.
4. Describe in detail how the CFW sets up the Control Channel. Describe Control Client and server behavior.
5. How does the CFW control framework interact? Describe general behavior for constructing requests and responses.
6. What is the CFW transaction processing? Describe in detail for different transaction methods: CONTROL, REPORT, K-ALIVE, and SYNC.
7. Describe each of the CFW response codes with call flows using XCON conference systems in detail.
8. What are the objectives of CFW packages? Develop a complete CFW package for a capability that you think will need to be used in the future.
9. Describe in detail the security for CFW that uses SIP protocol for the following: Session Establishment Transport-Level Protection and Control Channel Policy Management.

9 Mixer Package for Media Control Channel Framework

9.1 INTRODUCTION

The Media Control Channel Framework (RFC 6230 – see Chapter 8) provides a generic approach for establishment and reporting capabilities of remotely initiated commands. The Control Framework (CFW) – an equivalent term for the Media Control Channel Framework – utilizes many functions provided by the session initiation protocol (SIP) (RFC 3261 – also see Section 1.2) for the rendezvous and establishment of a reliable channel for control interactions. The Control Framework also introduces the concept of a Control Package. A Control Package is an explicit usage of the Control Framework for a particular interaction set. This specification defines a package for media conference mixers and media connection mixers (RFC 6505).

This package defines mixer management elements for creating, modifying, and deleting conference mixers, elements for joining, modifying, and unjoining media streams between connections and conferences (including mixers between connections), and associated responses and notifications. The package also defines elements for auditing package capabilities and mixers. This package has been designed to satisfy media-mixing requirements specified in the Media Server Control Protocol Requirements specification (RFC 5167 – also see Section 8.2); more specifically, REQ-22, REQ-23, REQ-24, REQ-25, REQ-26, and REQ-27 (see Section 8.2). The package provides the major conferencing functionality of SIP media server languages such as Media Server Control Markup Language (MSCML) (RFC 5022: Informational) and (Media Server Markup Language) MSML (RFC 5707: Informational). A key differentiator is that the package (RFC 6505: Standards Track) described in this section provides such functionality using the CFW.

Out of scope for this mixer package are more advanced functions, including personalized video mixes for conference participants, support for floor control protocols, and support for video overlays and text insertion. Such functionality can be addressed by extensions to this package (through addition of foreign elements or attributes from another namespace) or use of other Control Packages that could build upon this package. The functionality of this package is defined by messages, containing XML [1] elements and transported using the CFW. The XML elements can be divided into two types: mixer management elements and audit elements (for auditing package capabilities and mixers managed by the package).

The specification is organized as follows. Section 9.3 describes how this Control Package fulfills the requirements for a Media Control Channel Framework Control Package. Section 9.4 describes the syntax and semantics of defined elements, including mixer management (Section 9.4.2) and audit elements (Section 9.4.3). Section 9.5 describes an XML schema for these elements and provides extensibility by allowing attributes and elements from other namespaces. Section 9.6 provides examples of package usage. Section 9.7 describes important security considerations for use of this Control Package. Section 9.8 provides information on Internet Assigned Numbers Authority (IANA) registration of this Control Package, including its name, XML namespace, and MIME media type.

9.2 CONVENTIONS AND TERMINOLOGY

The following additional terms are defined for use in this specification:

- **Application Server**: An SIP (RFC 3261 – also see Section 1.2) application server (AS) is a Control Client that hosts and executes services such as interactive media and conferencing in an operator's network. An AS controls the media server (MS), influencing and impacting the SIP sessions terminating on an MS, which the AS can have established, for example, using SIP third-party call control. Media Server: An MS processes media streams on behalf of an AS by offering functionality such as interactive media, conferencing, and transcoding to the end user. Interactive media functionality is realized by way of dialogs, which are identified by a uniform resource identifier (URI) and initiated by the AS.
- **MS Conference:** An MS Conference provides the media-related mixing resources and services for conferences. In this specification, an MS Conference is often referred to simply as a conference.
- **MS Connection:** An MS connection represents the termination on a media server of one or more RTP (RFC 3550) sessions that are associated to a single SIP dialog. An MS receives media from the output(s) of a connection, and it transmits media on the input(s) of a connection.
- **Media Stream:** A media stream on an MS represents a media flow between a connection and a conference, between two connections, or between two conferences. Streams can be audio or video and can be bidirectional or unidirectional.

9.3 CONTROL PACKAGE DEFINITION

This section fulfills the mandatory requirement for information that MUST be specified during the definition of a Control Framework Package, as detailed in Section 8.8 (Section 8 of RFC 6230).

9.3.1 Control Package Name

The Control Framework requires a Control Package definition to specify and register a unique name. The name and version of this Control Package is "msc-mixer/1.0" (Media Server Control – Mixer – version 1.0). Its IANA registration is specified in Section 9.8.1. Since this is the initial ("1.0") version of the Control Package, there are no backwards compatibility issues to address.

9.3.2 Framework Message Usage

The Control Framework requires a Control Package to explicitly detail the control messages that can be used as well as provide an indication of directionality between entities. This will include which role type is allowed to initiate a request type. This package specifies CONTROL and response messages in terms of XML elements defined in Section 9.4, where the message bodies have the MIME media type defined in Section 9.8.4. These elements describe requests, responses, and notifications, and all are contained within a root <mscmixer> element (Section 9.4.1).

In this package, the MS operates as a Control Server in receiving requests from, and sending responses to, the AS (operating as a Control Client). Mixer management requests and responses are defined in Section 9.4.2. Audit requests and responses are defined in Section 9.4.3. Mixer management and audit responses are carried in a framework 200 response or REPORT message bodies. This package's response codes are defined in Section 9.4.6.

Note that package responses are different from framework response codes. Framework error response codes provided in Section 8.7 (Section 7 of RFC 6230) are used when the request or event notification is invalid, for example, a request is invalid XML (400) or not understood (500).

The MS also operates as a Control Client in sending event notification to the AS (Control Server). Event notifications (Section 9.4.2.4) are carried in CONTROL message bodies. The AS MUST respond with a Control Framework 200 response.

9.3.3 Common XML Support

The Control Framework requires a Control Package definition to specify whether the attributes for media dialog or conference references are required. This package requires that the XML schema in Section 8.14.1 (Appendix A.1 of RFC 6230) MUST be supported for media dialogs and conferences. The package uses "connectionid" and "conferenceid" attributes for various element definitions (Section 9.4). The XML schema (Section 9.5) imports the definitions of these attributes from the framework schema.

9.3.4 CONTROL Message Body

The Control Framework requires a Control Package to define the control body that can be contained within a CONTROL command request and to indicate the location of detailed syntax definitions and semantics for the appropriate body types. When operating as a Control Server, the MS receives CONTROL messages with the MIME media type defined in Section 9.8.4 and a body containing a <mscmixer> element (Section 9.4.1) with either a mixer management or audit request child element. The following mixer management request elements are carried in CONTROL message bodies to MS: <createconference> (Section 9.4.2.1.1), <modifyconference> (Section 9.4.2.1.2), <destroyconference> (Section 9.4.2.1.3), <join> (Section 9.4.2.2.2), <modifyjoin> (Section 9.4.2.2.3), and <unjoin> (Section 9.4.2.2.4) elements. The <audit> request element (Section 9.4.3.1) is also carried in CONTROL message bodies. When operating as a Control Client, the MS sends CONTROL messages with the MIME media type defined in Section 9.8.4 and a body containing a <mscmixer> element (Section 9.4.1) with a notification <event> child element (Section 9.4.2.4).

9.3.5 REPORT Message Body

The Control Framework requires a Control Package definition to define the REPORT body that can be contained within a REPORT command request or to indicate that no report package body is required. This section indicates the location of detailed syntax definitions and semantics for the appropriate body types. When operating as a Control Server, the MS sends REPORT bodies with the MIME media type defined in Section 9.8.4 and a <mscmixer> element with a response child element. The response element for mixer management requests is a <response> element (Section 9.4.2.3). The response element for an audit request is an <auditresponse> element (Section 9.4.3.2).

9.3.6 Audit

The Control Framework encourages Control Packages to specify whether auditing is available, how it is triggered, and the query/response formats. This Control Package supports auditing of package capabilities and mixers on the MS. An audit request is carried in a CONTROL message and an audit response in a REPORT message (or a 200 response to the CONTROL if it can execute the audit in time). The syntax and semantics of audit request and response elements are defined in Section 9.4.3.

9.3.7 Examples

The Control Framework recommends Control Packages to provide a range of message flows that represent common flows using the package and this framework specification. This Control Package provides examples of such message flows in Section 9.6.

9.4 ELEMENT DEFINITIONS

This section defines the XML elements for this package. The elements are defined in the XML namespace specified in Section 9.8.2. The root element is <mscmixer> (Section 9.4.1). All other XML elements (requests, responses, and notification elements) are contained within it. Child elements describe mixer management (Section 9.4.2) and audit (Section 9.4.3) functionality. Response status codes are defined in Section 9.4.6 and type definitions in Section 9.4.7. Implementation of this Control Package MUST address the security considerations described in Section 9.7. Implementation of this Control Package MUST adhere to the syntax and semantics of XML elements defined in this section and the schema (Section 9.5). The XML schema supports extensibility by allowing attributes and elements from other namespaces. Implementations MAY support attributes and elements from other (foreign) namespaces. If an MS implementation receives a <mscmixer> element containing attributes or elements from another namespace, which it does not support, the MS sends a 428 response (Section 9.4.6).

Extensible attributes and elements are not described in this section. In all other cases where there is a difference in constraints between the XML schema and the textual description of elements in this section, the textual definition takes priority. Some elements in this Control Package contain attributes whose value is descriptive text primarily for diagnostic use. The implementation can indicate the language used in the descriptive text by means of a "desclang" attribute (RFC 2277). The "desclang" attribute can appear on the root element as well as selected subordinate elements (see Section 9.4.1). The "desclang" attribute value on the root element applies to all "desclang" attributes in subordinate elements unless the subordinate element has an explicit "desclang" attribute that overrides it. Usage examples are provided in Section 9.6.

9.4.1 <MSCMIXER>

The <mscmixer> element has the following attributes (in addition to standard XML namespace attributes such as "xmlns"): version: a string specifying the mscmixer package version. The value is fixed as "1.0" for this version of the package. The attribute is mandatory. desclang: specifies the language used in descriptive text attributes of subordinate elements (unless the subordinate element provides a "desclang" attribute that overrides the value for its descriptive text attributes). The descriptive text attributes on subordinate elements include: the "reason" attribute on <response> (Section 9.4.2.3), <unjoin-notify> (Section 9.4.2.4.2), <conferenceexit> (Section 9.4.2.4.3), and <auditresponse> (Section 9.4.3.2). A valid value is a language identifier (Section 9.4.7.7). The attribute is optional. The default value is "i-default" (BCP 47 – RFC 5646). The <mscmixer> element has the following defined child elements, only one of which can occur:

1. Mixer management elements defined in Section 9.4.2:
   - <createconference>: Create and configure a new conference mixer. See Section 9.4.2.1.1. <modifyconference>: modify the configuration of an existing conference mixer. See Section 9.4.2.1.2 <destroyconference>: destroy an existing conference mixer. See Section 9.4.2.1.3
   - <join>: Create and configure media streams between connections and/or conferences (for example, add a participant to a conference). See Section 9.4.2.2.2
   - <modifyjoin>: modify the configuration of joined media streams. See Section 9.4.2.2.3
   - <unjoin>: Delete a media stream (for example, remove a participant from a conference). See Section 9.4.2.2.4 <response>: response to a mixer request. See Section 9.4.2.3
   - <event>: Mixer or subscription notification. See Section 9.4.2.4
2. Audit elements defined in Section 9.4.3:
   - <audit>: Audit package capabilities and managed mixers. See Section 9.4.3.1

- <auditresponse>: Response to an audit request. See Section 9.4.3.2. For example, a request to the MS to create a conference mixer is as follows:
  - <mscmixer version="1.0" xmlns="urn:ietf:params:xml:ns:msc-mixer">
  - <createconference/>
  - </mscmixer>

And a response from the MS that the conference was successfully created is as follows:

```
<mscmixer version="1.0" xmlns="urn:ietf:params:xml:ns:msc-mixer"
      desclang="en">
   <response status="200" conferenceid="conference1"
      reason="conference created"/>
</mscmixer>
```

9.4.2 Mixer Elements

This section defines the mixer management XML elements for this Control Package. These elements are divided into requests, responses, and notifications. Request elements are sent to the MS to request a specific mixer operation to be executed. The following request elements are defined:

- <createconference>: Create and configure a new a conference mixer. See Section 9.4.2.1.1.
- <modifyconference>: Modify the configuration of an existing conference mixer. See Section 9.4.2.1.2.
- <destroyconference>: Destroy an existing conference mixer. See Section 9.4.2.1.3.
- <join>: Create and configure media streams between connections and/or conferences (for example, add a participant to a conference). See Section 9.4.2.2.2.
- <modifyjoin>: Modify the configuration of joined media streams. See Section 9.4.2.2.3.
- <unjoin>: Delete a media stream (for example, remove a participant from a conference). See Section 9.4.2.2.4.

Responses from the MS describe the status of the requested operation. Responses are specified in a <response> element (Section 9.4.2.3) that includes a mandatory attribute describing the status in terms of a numeric code. Response status codes are defined in Section 9.4.6. The MS MUST respond to a request message with a response message. If the MS is not able to process the request and carry out the mixer operation (in whole or in part), then the request has failed: The MS MUST ensure that no part of the requested mixer operation is carried out, and the MS MUST indicate the class of failure using an appropriate 4xx response code. Unless an error response code is specified for a class of error within this section, implementations follow Section 9.4.6 in determining the appropriate status code for the response. Notifications are sent from the MS to provide updates on the status of a mixer operation or subscription. Notifications are specified in an <event> element (Section 9.4.2.4).

9.4.2.1 Conference Elements

9.4.2.1.1 <createconference>

The <createconference> element is sent to the MS to request creation of a new conference (multiparty) mixer. The <createconference> element has the following attributes: conferenceid: string indicating a unique name for the new conference. If this attribute is not specified, the MS MUST create a unique name for the conference. The value is used in subsequent references to the conference (e.g., as conferenceid in a <response>). The attribute is optional. There is no default value. reserved-talkers: indicates the requested number of guaranteed speaker slots to be reserved for the conference. A valid value is a non-negative integer (see Section 9.4.7.2). The attribute is optional.

The default value is 0. reserved-listeners: indicates the requested number of guaranteed listener slots to be reserved for the conference. A valid value is a non-negative integer (see Section 9.4.7.2). The attribute is optional. The default value is 0.

The <createconference> element has the following sequence of child elements:

- <codecs>: An element to configure the codecs supported by the conference (see Section 9.4.4). If codecs are specified, then they impose limitations on media capability when the MS attempts to join the conference to other entities (see Sections 9.4.2.2.2 and 9.4.2.2.3). The element is optional.
- <audio-mixing>: An element to configure the audio mixing characteristics of a conference (see Section 9.4.2.1.4.1). The element is optional.
- <video-layouts>: An element to configure the video layouts of a conference (see Section 9.4.2.1.4.2). The element is optional.
- <video-switch>: An element to configure the video switch policy for the layout of a conference (see Section 9.4.2.1.4.3). The element is optional. <subscribe>: an element to request subscription to conference events. (see Section 9.4.2.1.4.4). The element is optional.

If the "conferenceid" attribute specifies a value that is already used by an existing conference, the MS reports an error (405), MUST NOT create a new conference, and MUST NOT affect the existing conference. If the MS is unable to configure the conference according to specified "reserved-talkers" or "reserved-listeners" attributes, the MS reports an error (420) and MUST NOT create the conference. If the MS is unable to configure the conference according to a specified <audio-mixing> element, the MS reports an error (421) and MUST NOT create the conference.

If the MS is unable to configure the conference according to a specified <video-layouts> element, the MS reports an error (423) and MUST NOT create the conference. If the MS is unable to configure the conference according to a specified <video-switch> element, the MS reports an error (424) and MUST NOT create the conference. If the MS is unable to configure the conference according to a specified <codecs> element, the MS reports an error (425) and MUST NOT create the conference.

When a MS has finished processing a <createconference> request, it MUST reply with an appropriate <response> element (Section 9.4.2.3). For example, a request to create an audio-video conference mixer with specified codecs, video layout, video switch, and subscription is as follows:

```
<mscmixer version="1.0" xmlns="urn:ietf:params:xml:ns:msc-mixer">
<createconference conferenceid="conference1"
      reserved-talkers="1" reserved-listeners="10">
   <codecs>
   <codec name="video">
      <subtype>H264</subtype>
   </codec>
   <codec name="audio">
      <subtype>PCMA</subtype>
   </codec>
   </codecs>
   <audio-mixing type="nbest"/>
   <video-layouts>
      <video-layout min-participants="1"><single-view/</video-layout>
      <video-layout min-participants="2"><dual-view/></video-layout>
      <video-layout min-participants="3"><quad-view/></video-layout>
   </video-layouts>
   <video-switch interval="5"><vas/></video-switch>
   <subscribe>
      <active-talkers-sub interval="4"/>
```

```
        </subscribe>
</createconference>
</mscmixer>
```

A response from the MS if the conference was successfully created is as follows:

```
<mscmixer version="1.0" xmlns="urn:ietf:params:xml:ns:msc-mixer">
      <response status="200" conferenceid="conference1"/>
</mscmixer>
```

Alternatively, a response if the MS could not create the conference due to a lack of support for the H264 codec is as follows:

```
<mscmixer version="1.0" xmlns="urn:ietf:params:xml:ns:msc-mixer">
   <response status="425" conferenceid="conference1"
      reason="H264 codec not supported"/>
</mscmixer>
```

9.4.2.1.2 <modifyconference>

The <modifyconference> element is sent to the MS to request modification of an existing conference. The <modifyconference> element has the following attribute:

conferenceid: String indicating the name of the conference to modify. This attribute is mandatory. The <modifyconference> element has the following sequence of child elements (one or more):

- <codecs>: An element to configure the codecs supported by the conference (see Section 9.4.4). If codecs are specified, then they impose limitations in media capability when the MS attempts to join the conference to other entities (see Sections 9.4.2.2.2 and 9.4.2.2.3). Existing conference participants are unaffected by any policy change. The element is optional.
- <audio-mixing>: An element to configure the audio mixing characteristics of a conference (see Section 9.4.2.1.4.1). The element is optional.
- <video-layouts>: An element to configure the video layouts of a conference (see Section 9.4.2.1.4.2). The element is optional.
- <video-switch>: An element to configure the video switch policy for the layout of a conference (see Section 9.4.2.1.4.3). The element is optional.
- <subscribe>: An element to request subscription to conference events (see Section 9.4.2.1.4.4). The element is optional. If the "conferenceid" attribute specifies the name of a conference that does not exist, the MS reports an error (406). If the MS is unable to configure the conference according to a specified <audio-mixing> element, the MS reports an error (421) and MUST NOT modify the conference in any way. If the MS is unable to configure the conference according to a specified <video-layouts> element, the MS reports an error (423) and MUST NOT modify the conference in any way.

If the MS is unable to configure the conference according to a specified <video-switch> element, the MS reports an error (424) and MUST NOT modify the conference in any way. If the MS is unable to configure the conference according to a specified <codecs> element, the MS reports an error (425) and MUST NOT modify the conference. When a MS has finished processing a <modifyconference> request, it MUST reply with an appropriate <response> element (Section 9.4.2.3).

9.4.2.1.3 <destroyconference>

The <destroyconference> element is sent to the MS to request destruction of an existing conference. The <destroyconference> element has the following attribute:

conferenceid: String indicating the name of the conference to destroy. This attribute is mandatory. The <destroyconference> element does not specify any child elements. If the "conferenceid" attribute specifies the name of a conference that does not exist, the MS reports an error (406). When a MS has finished processing a <destroyconference> request, it MUST reply with an appropriate <response> element (Section 9.4.2.3). Successfully destroying the conference (status code 200) will result in all connection or conference participants being removed from the conference mixer, <unjoin-notify> notification events (Section 9.4.2.4.2) being sent for each conference participant, and a <conferenceexit> notification event (Section 9.4.2.4.3) indicating that conference has exited. A <response> with any other status code indicates that the conference mixer still exists and participants are still joined to the mixer.

9.4.2.1.4 Conference Configuration

The elements in this section are used to establish and modify the configuration of conferences.

9.4.2.1.4.1 <audio-mixing> The <audio-mixing> element defines the configuration of the conference audio mix. The <audio-mixing> element has the following attributes:

- type: Is a string indicating the audio stream mixing policy. Defined values are "nbest" (where the N best [loudest] participant signals are mixed) and "controller" (where the contributing participant[s] is/are selected by the controlling AS via an external floor control protocol). The attribute is optional. The default value is "nbest."
- n: Indicates the number of eligible participants included in the conference audio mix. An eligible participant is a participant who contributes audio to the conference. Inclusion is based on having the greatest audio energy. A valid value is a non-negative integer (see Section 9.4.7.2). A value of 0 indicates that all participants contributing audio to the conference are included in the audio mix. The default value is 0. The element is optional. If the "type" attribute does not have the value "nbest," the MS ignores the "n" attribute.

The <audio-mixing> element has no child elements. For example, a fragment where the audio-mixing policy is set to "nbest" with three participants to be included is as follows:

```
<audio-mixing type="nbest" n="3"/>
```

If the conference had 200 participants, of whom 30 contributed audio, then there would be 30 eligible participants for the audio mix. Of these, the three loudest participants would have their audio included in the conference.

9.4.2.1.4.2 <video-layouts> The <video-layouts> element describes the video presentation layout configuration for participants providing a video input stream to the conference. This element allows multiple video layouts to be specified so that the MS automatically changes layout depending on the number of video-enabled participants. The <video-layouts> element has no attributes. The <video-layouts> element has the following sequence of child elements (one or more):

```
<video-layout>: Element describing a video layout (Section 9.4.2.1.4.2.1).
```

If the MS does not support video conferencing at all, or does not support multiple video layouts, or does not support a specific video layout, the MS reports an 423 error in the response to the request element containing the <video-layouts> element. An MS MAY support more than one <video-layout> element, although only one layout can be active at a time. A <video-layout> is active if the number of participants in the conference is equal to or greater than the value of its "min-participants" attribute but lower than the value of the "min-participants" attribute for any other <video-layout> element. An MS reports an error (400) if more than one <video-layout> has the same value for the "min-participants" attribute. When the number of regions within the active layout is greater than the number of participants in the conference, the display of unassigned regions is implementation specific. The assignment of participant video streams to regions within the layout is according to the video switch policy specified by the <video-switch> element (Section 9.4.2.1.4.3). For example, a fragment describing a single layout is as follows:

```
<video-layouts>
    <video-layout><single-view/></video-layout>
</video-layouts>
```

A fragment describing a sequence of layouts is as follows:

```
<video-layouts>
    <video-layout min-participants="1"><single-view/></video-layout>
    <video-layout min-participants="2"><dual-view/></video-layout>
    <video-layout min-participants="3"><quad-view/></video-layout>
    <video-layout min-participants="5"><multiple-3x3/></video-layout>
</video-layouts>
```

When the conference has one participant providing a video input stream to the conference, then the single-view format is used. When the conference has two such participants, the dual-view layout is used. When the conference has three or four participants, the quad-view layout is used. When the conference has five or more participants, the multiple-3×3 layout is used.

9.4.2.1.4.2.1 <video-layout> The <video-layout> element describes a video layout containing one or more regions in which participant video input streams are displayed. The <video-layout> element has the following attribute:

- min-participants: The minimum number of conference participants needed to allow this layout to be active. A valid value is a positive integer (see Section 9.4.7.3). The attribute is optional. The default value is 1.

The <video-layout> element has one child element specifying the video layout. An MS MAY support the predefined video layouts defined in the conference information data model for centralized conferencing (XCON) (RFC 6501 – see Chapter 4): <single-view>, <dual-view>, <dual-view-crop>, <dual-view-2x1>, <dual-view-2x1-crop>, <quad-view>, <multiple-3x3>, <multiple-4x4>, and <multiple-5x1>. The MS MAY support other video layouts. Non-XCON layouts MUST be specified using an element from a namespace other than the one used in this specification, for example:

```
<video-layout>
    <mylayout xmlns='http://example.com/foo'>
        my-single-view</mylayout>
</video-layout>
```

If the MS does not support the specified video layout configuration, then the MS reports a 423 error (Section 9.4.6) in the response to the request element containing the <video-layout> element. Each video layout has associated with it one or more regions. The XCON layouts are associated with the following named regions:

<single-view/>: Layout with one stream in a single region as shown in Figure 9.1.

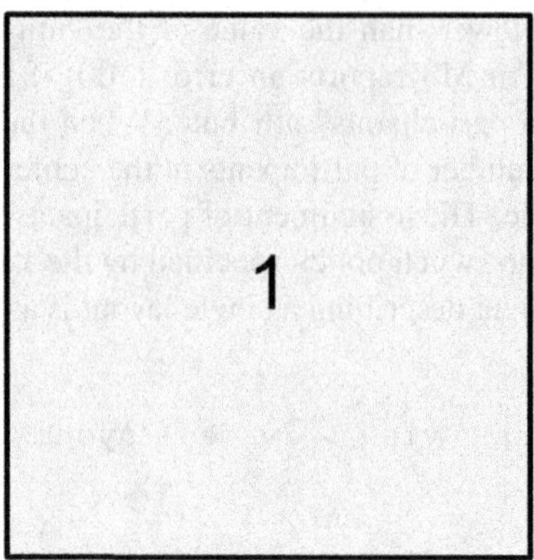

FIGURE 9.1 Single-View Video Layout

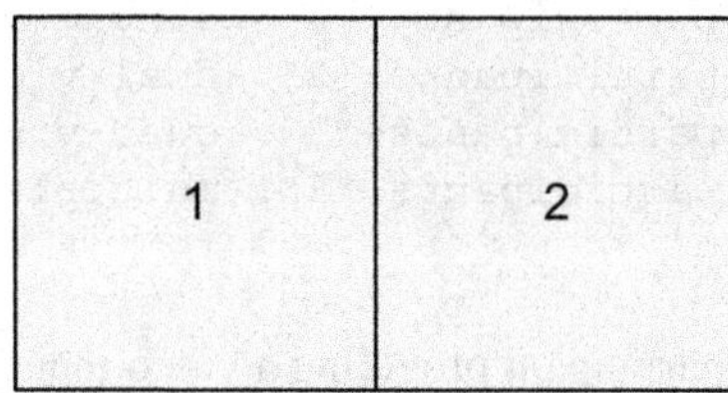

FIGURE 9.2 Dual-View Video Layout

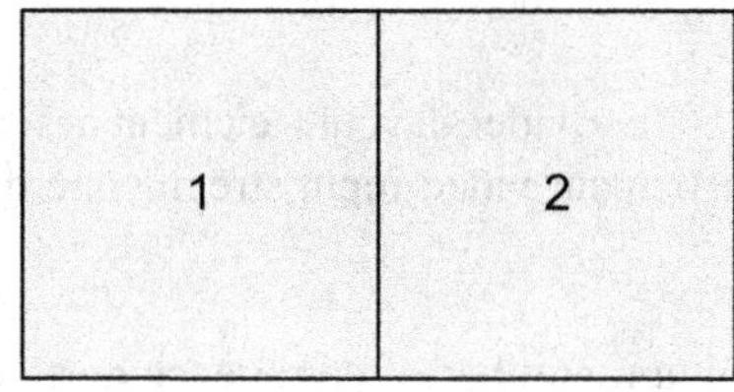

FIGURE 9.3 Dual-View-Crop Video Layout

<dual-view/>: Layout presenting two streams side-by-side in two regions as shown in Figure 9.2. The MS MUST NOT alter the aspect ratio of each stream to fit the region, and hence, the MS might need to blank out part of each region.

<dual-view-crop/>: Layout presenting two streams side-by-side in two regions as shown in Figure 9.3. The MS MUST alter the aspect ratio of each stream to fit its region so that no blanking is required.

<dual-view-2x1/>: Layout presenting two streams, one above the other, in two regions as shown in Figure 9.4. The MS MUST NOT alter the aspect ratio of each stream to fit its region, and hence, the MS might need to blank out part of each region.

<dual-view-2x1-crop/>: Layout presenting two streams one above the other in two regions as shown in Figure 9.5. The MS MUST alter the aspect ratio of each stream to fit its region so that no blanking is required.

<quad-view/>: Layout presenting four equal-sized regions in a 2×2 layout as shown in Figure 9.6. Typically, the aspect ratio of the streams is preserved, so blanking is required.

<multiple-3x3/>: Layout presenting nine equal-sized regions in a 3×3 layout as shown in Figure 9.7. Typically, the aspect ratio of the streams is preserved, so blanking is required.

<multiple-4x4/>: Layout presenting 16 equal-sized regions in a 4×4 layout as shown in Figure 9.8. Typically, the aspect ratio of the streams is preserved, so blanking is required.

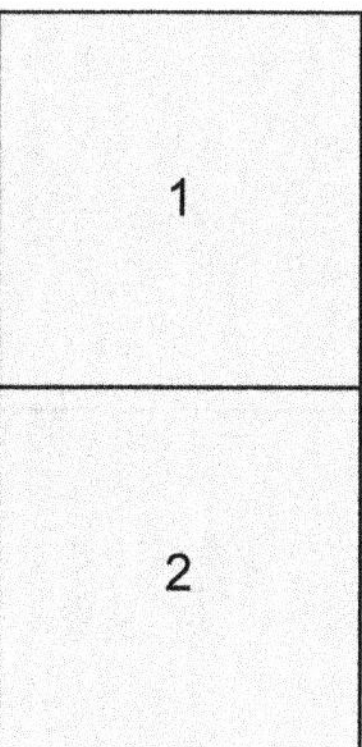

FIGURE 9.4 Dual-View-2×1 Video Layout

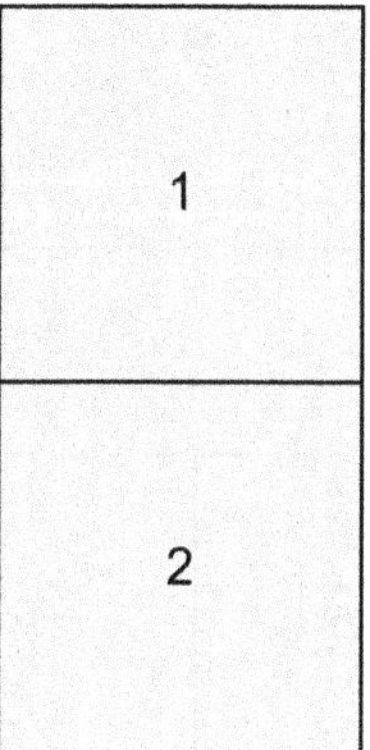

FIGURE 9.5 Dual-View-2×1-Crop Video Layout

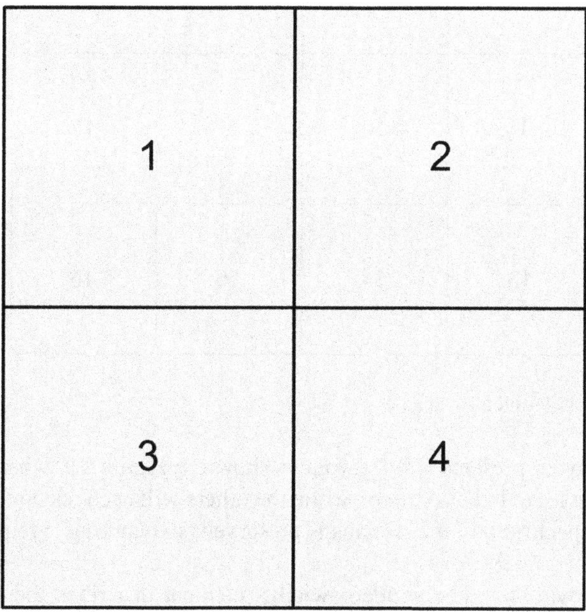

FIGURE 9.6 Quad-View Video Layout

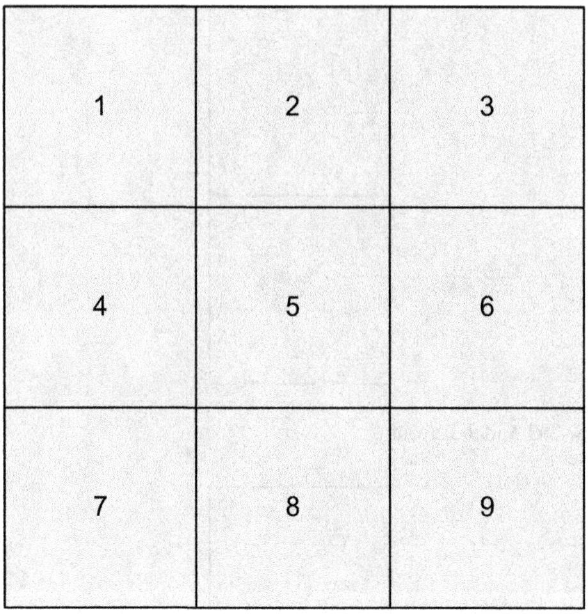

1	2	3
4	5	6
7	8	9

FIGURE 9.7 Multiple-3×3 Video Layout

1	2	3	4
5	6	7	8
9	10	11	12
13	14	15	16

FIGURE 9.8 Multiple-4×4 Video Layout

<multiple-5x1/>: layout presents a 5×1 layout as shown in Figure 9.9, where one region will occupy 4/9 of the mixed video stream, while the others will each occupy 1/9 of the stream. Typically, the aspect ratio of the streams is preserved, so blanking is required.

9.4.2.1.4.3 <video-switch> The <video-switch> element describes the configuration of the conference policy for how participants' input video streams are assigned to regions within the

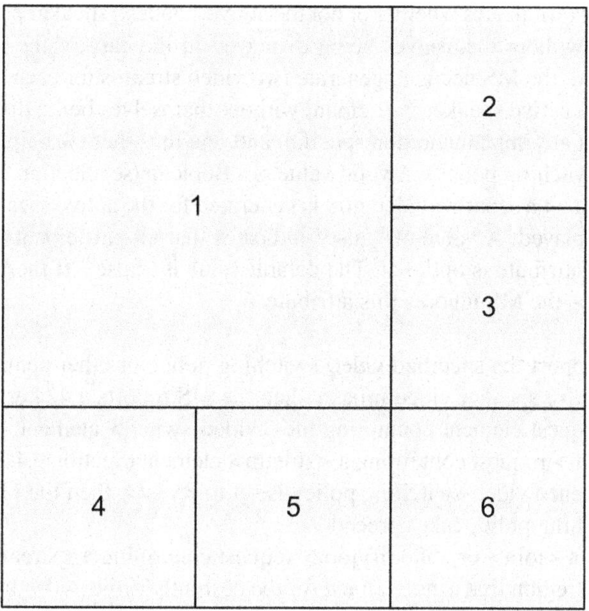

FIGURE 9.9 Multiple-5×1 Video Layout

active video layout. The <video-switch> element has the following child elements defined (one child occurrence only) to indicate the video-switching policy of the conference:

<vas/>: (Voice-Activated Switching) enables automatic display of the loudest speaker participant that is contributing both audio and video to the conference mix. Participants who do not provide an audio stream are not considered for automatic display. If a participant provides more than one audio stream, then the policy for inclusion of such a participant in the VAS is implementation specific; an MS could select one stream, sum audio streams, or ignore the participant for VAS consideration. If there is only one region in the layout, then the loudest speaker is displayed there. If more than one region is available, then the loudest speaker is displayed in the largest region (if any), and then in the first region from the top-left corner of the layout. The MS assigns the remaining regions based on the priority mechanism described in Section 9.4.2.1.4.3.1. <controller/>: Enables manual control over video switching. The controller AS determines how the regions are assigned based on an external floor control policy. The MS receives <join>, <modifyjoin>, and <unjoin> commands with a <stream> element (Section 9.4.2.2.5) indicating the region where the stream is displayed. If no explicit region is specified, the MS assigns the region based on the priority mechanism described in Section 9.4.2.1.4.3.1. An MS MAY support other video-switching policies. Other policies MUST be specified using an element from a namespace other than the one used in this specification. For example:

```
<video-switch>
   <mypolicy xmlns='http://example.com/foo'/>
</video-switch>
```

The <video-switch> element has the following attributes:

- interval: Specifies the period between video switches as a number of seconds. In the case of <vas/> policy, a speaker needs to be the loudest speaker for the interval before the switch takes place. A valid value is a non-negative integer (see Section 9.4.7.2). A value of 0 indicates that switching is applied immediately. The attribute is optional. The default value is 3 (seconds).

- activespeakermix: Indicates whether or not the active (loudest) speaker participant receives a video stream without themselves being displayed in the case of the <vas/> switching policy. If enabled, the MS needs to generate two video streams for each conference.
- mix: One for the active speaker participant without themselves being displayed (details of this video layout are implementation specific) and one for other participants (as described in the <vas/> switching policy). A valid value is a Boolean (see Section 9.4.7.1). A value of "true" indicates that a separate video mix is generated for the active speaker without themselves being displayed. A value of "false" indicates that all participants receive the same video mix. The attribute is optional. The default value is "false." If the "type" attribute is not set to <vas/>, the MS ignores this attribute.

If the MS does not support the specified video-switching policy or other configuration parameters (including separate active speaker video mixes), then the MS reports a 424 error (Section 9.4.6) in the response to the request element containing the <video-switch> element. If the MS receives a <join> or <modifyjoin> request containing a <stream> element (Section 9.4.2.2.5) that specifies a region, and the conference video-switching policy is set to <vas/>, then the MS ignores the region (i.e., conference-switching policy takes precedence).

If the MS receives a <join> or <modifyjoin> request containing a <stream> element (Section 9.4.2.2.5) specifying a region that is not defined for the currently active video layout, the MS MUST NOT report an error. Even though the participant is not currently visible, the MS displays the participant if the layout changes to one that defines the specified region. For example, a fragment specifying a <vas/> video-switching policy with an interval of 2 s:

```
<video-switch interval="2"><vas/></video-switch>
```

For example, a fragment specifying a <controller/> video-switching policy where video switching takes place immediately is as follows:

```
<video-switch interval="0"><controller/></video-switch>
```

9.4.2.1.4.3.1 Priority Assignment In cases where the video-switching policy does not explicitly determine the region to which a participant is assigned, the following priority assignment mechanism applies:

1. Each participant has a (positive integer) priority value: The lower the value, the higher the priority. The priority value is determined by the <priority> child element (Section 9.4.2.2.5.4) of <stream>. If not explicitly specified, the default priority value is 100.
2. The MS uses priority values to assign participants to regions in the video layout that remain unfilled after application of the video-switching policy. The MS MUST dedicate larger and/or more prominent portions of the layout to participants with higher priority values first (e.g., first, all participants with priority 1, then those with 2, 3, etc.).
3. The policy for displaying participants with the same priority is implementation specific. The MS applies this priority policy each time the video layout is changed or updated. It is RECOMMENDED that the MS does not move a participant from one region to another unless this is required by the video-switching policy when an active video layout is updated. This model allows the MS to apply default video layouts after applying the video-switching policy. For example, region 2 is statically assigned to Bob, so the priority mechanism only applies to regions 1, 3, 4, etc.

9.4.2.1.4.4 <subscribe> The <subscribe> element is a container for specifying conference notification events to which a controlling entity subscribes. Notifications of conference events are delivered using the <event> element (see Section 9.4.2.4). The <subscribe> element has no attributes but has the following child element:

- <active-talkers-sub>: Subscription to active talker events (Section 9.4.2.1.4.4.1). The element is optional.

The MS MUST support a <active-talkers-sub> subscription. It MAY support other event subscriptions (specified using attributes and child elements from a foreign namespace). If the MS does not support a subscription specified in a foreign namespace, it sends a <response> with a 428 status code (see Section 9.4.6).

9.4.2.1.4.4.1 <active-talkers-sub> The <active-talkers-sub> element has the following attribute: interval: The minimum amount of time (in seconds) that elapses before further active talker events can be generated. A valid value is a non-negative integer (see Section 9.4.7.2). A value of 0 suppresses further notifications. The attribute is optional. The default value is 3 (seconds). The <active-talkers-sub> element has no child elements. Active talker notifications are delivered in the <active-talkersnotify> element (Section 9.4.2.4.1).

9.4.2.2 Joining Elements

This section contains definitions of the joining model (Section 9.4.2.2.1) as well as the <join> (Section 9.4.2.2.2), <modifyjoin> (Section 9.4.2.2.3), <unjoin> (Section 9.4.2.2.4), and <stream> (Section 9.4.2.2.5) elements.

9.4.2.2.1 Joining Model

The <join> operation creates a media stream between a connection and a conference, between connections, or between conferences. This section describes the model of conferences and connections and specifies the behavior for join requests to targets that already have an associated media stream. Conferences support multiple inputs and have resources to mix them together. An MS conference in essence is a mixer that combines media streams. A simple audio mixer simply sums its input audio signals to create a single common output. Conferences, however, use a more complex algorithm so that participants do not hear themselves as part of the mix. That algorithm, sometimes called an "n-minus mix," subtracts each participant's input signal from the summed input signals, creating a unique output for each contributing participant. Each <join> operation to a conference uses one of the conference's available inputs and/or outputs to the maximum number of supported participants.

A connection is the termination of one or more real-time transport protocol (RTP) sessions on an MS. It has a single input and output for each media session established by its SIP dialog. The output of a connection can feed several different inputs, such as both a conference mixer and a recording of that participant's audio. Joining two connections that are not joined to anything else simply creates a media stream from the outputs(s) of one connection to the corresponding inputs(s) of the other connection. It is not necessary to combine media from multiple sources in this case. There are, however, several common scenarios where combining media from several sources to create a single input to a connection is needed. In the first case, a connection can be receiving media from one source (for example, a conference), and it is necessary to play an announcement to the connection so that both the conference audio and the announcement can be heard by the conference participant. This is sometimes referred to as a "whisper announcement." An alternative to a whisper announcement is to have the announcement preempt the conference media.

Another common case is the call-center coaching scenario, where a supervisor can listen to the conversation between an agent and a customer and provide hints to the agent that are not heard by the customer. Both these cases can be solved by having the controlling AS create one or more conferences for audio mixing and then join and unjoin the media streams as required. A better solution is to have the MS automatically mix media streams that are requested to be joined to a common input when only the simple summing of audio signals, as described earlier, is required. This is the case for both the use cases presented here. Automatically mixing streams has several benefits. Conceptually, it is straightforward and simple, requiring no indirect requests on the part of the controlling AS. This increases transport efficiency and reduces the coordination complexity and the latency of the overall operation. Therefore, it is RECOMMENDED that an MS be able to automatically mix at least two audio streams where only the simple summing of signals is required.

When an MS receives a <join> request, it MUST automatically mix all the media streams included in the request with any streams already joined to one of the entities identified in the request, or it MUST fail the request and MUST NOT join any of the streams (and MUST NOT change existing streams of the entities). A controlling AS uses the <createconference> request for generic conferences where the complex mixing algorithm is required. Specifications that extend this package to handle additional media types such as text MUST define the semantics of the join operation when multiple streams are requested to be joined to a single input, such as that for a connection with a single RTP session per media type.

9.4.2.2.2 <join>

The <join> element is sent to the MS to request creation of one or more media streams between a connection and a conference, between connections, or between conferences. The two entities to join are specified by the attributes of <join>. Streams can be of any media type and can be bidirectional or unidirectional. A bidirectional stream is implicitly composed of two unidirectional streams that can be manipulated independently. The streams to be established are specified by child <stream> elements (see Section 9.4.2.2.5). The <join> element has the following attributes:

- id1: An identifier for either a connection or a conference. The identifier MUST conform to the syntax defined in Appendix A.1 of RFC 6230 (see Section 8.14.1). The attribute is mandatory.
- id2: An identifier for either a connection or a conference. The identifier MUST conform to the syntax defined in Section 8.14.1 (Appendix A.1 of RFC 6230). The attribute is mandatory. Note: Section 8.14.1 (Appendix A.1 of RFC 6230) defines the semantics for a conference identifier but not its syntax.

MS implementations need to distinguish between conferences and connections based upon the values of the "id1" and "id2" attributes. If id1 or id2 specifies a conference identifier, and the conference does not exist on the MS, the MS reports an error (406). If id1 or id2 specifies a connection identifier, and the connection does not exist on the MS, the MS reports an error (412). The <join> element has the following child element (zero or more):

- <stream>: An element that both identifies the media streams to join and defines the way that they are to be joined (see Section 9.4.2.2.5). The element is optional. If no <stream> elements are specified, then the default is to join all streams between the entities according to the media configuration of the connection or conference.

One or more <stream> elements can be specified so that individual media streams can be controlled independently. For example, if a connection supports both audio and video streams, a <stream> element could be used to indicate that only the audio stream is used in receive mode. In cases where there are multiple media streams of the same type for a connection or conference, the configuration MUST be explicitly specified using <stream> elements.

Multiple <stream> elements can be specified for precise control over the media flow in different directions within the same media stream. One <stream> element can be specified for the receiving media flow and another element for the sending media flow, where each independently controls features such as volume (see child element of <stream> in Section 9.4.2.2.5). If there is only one <stream> element for a given media specifying a "sendonly" or "recvonly" direction, then the media flow in the opposite direction is inactive (established but there is no actual flow of media) unless this leads to a stream conflict.

If the MS is unable to execute the join as specified in <stream> because a <stream> element is in conflict with (a) another <stream> element, (b) specified connection or conference media capabilities (including supported or available codec information), or (c) a session description protocol (SDP) label value as part of the connection-id defined in Section 8.14.1 (Appendix A.1 of RFC 6230), then the MS reports an error (407), MUST NOT join the entities, and MUST NOT change existing streams of the entities. If the MS is unable to execute the join as specified in <stream> elements because the MS does not support the media stream configuration, the MS reports an error (422), MUST NOT join the entities, and MUST NOT change existing streams of the entities.

- If the MS is unable to join an entity to a conference because it is full, then the MS reports an error (410).
- If the specified entities are already joined, then the MS reports an error (408).
- If the MS does not support joining two specified connections together, the MS reports an error (426).
- If the MS does not support joining two specified conferences together, the MS reports an error (427).
- If the MS is unable to join the specified entities for any other reason, the MS reports an error (411).

When the MS has finished processing a <join> request, it MUST reply with a <response> element (Section 9.4.2.3). For example, a request to join two connections together is as follows:

```
<mscmixer version="1.0" xmlns="urn:ietf:params:xml:ns:msc-mixer">
   <join id1="1536067209:913cd14c" id2="1536067209:913cd14c"/>
</mscmixer>
```

The response if the MS does not support joining media streams between connections is as follows:

```
<mscmixer version="1.0" xmlns="urn:ietf:params:xml:ns:msc-mixer">
   <response status="426" reason="mixing connections not supported"/>
</mscmixer>
```

9.4.2.2.3 <modifyjoin>

The <modifyjoin> element is sent to the MS to request changes in the configuration of media stream(s) that were previously established between a connection and a conference, between two connections, or between two conferences. The <modifyjoin> element has the following attributes:

- id1: An identifier for either a connection or a conference. The identifier MUST conform to the syntax defined in Section 8.14.1 (Appendix A.1 of RFC 6230). The attribute is mandatory.
- id2: An identifier for either a connection or a conference. The identifier MUST conform to the syntax defined in Section 8.14.1 (Appendix A.1 of RFC 6230). The attribute is mandatory. The <modifyjoin> element has the following child elements (one or more):

- <stream>: An element that both identifies the media streams to modify and defines the way that each stream is to be configured from this point forward (see Section 9.4.2.2.5).

The MS MUST support <modifyjoin> for any stream that was established using <join>. The MS MUST configure the streams that are included within <modifyjoin> to that stated by the child elements. If the MS is unable to modify the join as specified in <stream> elements because a <stream> element is in conflict with (a) another <stream> element, (b) specified connection or conference media capabilities (including supported or available codec information), or (c) a SDP label value as part of the connection-id defined in Section 8.14.1 (Appendix A.1 of RFC 6230), then the MS reports an error (407), MUST NOT modify the join between the entities, and MUST NOT change existing streams of the entities.

- If the MS is unable to modify the join as specified in <stream> elements because the MS does not support the media stream configuration, the MS reports an error (422), MUST NOT modify the join between the entities, and MUST NOT change existing streams of the entities.
- If the specified entities are not already joined, then the MS reports an error (409).
- If the MS is unable to modify the join between the specified entities for any other reason, the MS reports an error (411).

When an MS has finished processing a <modifyjoin> request, it MUST reply with an appropriate <response> element (Section 9.4.2.3). In cases where stream characteristics are controlled independently for each direction, then a <modifyjoin> request needs to specify a child element for each direction in order to retain the original stream directionality. For example, if a <join> request establishes independent control for each direction of an audio stream (see Section 9.4.2.2.5):

```
<mscmixer version="1.0" xmlns="urn:ietf:params:xml:ns:msc-mixer">
   <join id1="1536067209:913cd14c" id2="conference1">
      <stream media="audio" direction="sendonly">
         <volume controltype="setgain" value="-3"/>
      </stream>
      <stream media="audio" direction="recvonly">
         <volume controltype="setgain" value="+3"/>
      </stream>
   </join>
</mscmixer>
```

then the following <modifyjoin> request

```
<mscmixer version="1.0" xmlns="urn:ietf:params:xml:ns:msc-mixer">
   <modifyjoin id1="1536067209:913cd14c" id2="conference1">
      <stream media="audio" direction="sendonly">
         <volume controltype="setgain" value="0"/>
      </stream>
   </modifyjoin>
</mscmixer>
```

would cause, in addition to the modification of the sendonly volume, the overall stream directionality to change from sendrecv to sendonly, since there is no <stream> element in this <modifyjoin> request for the recvonly direction. The following would change the sendonly volume and retain the recvonly stream together with its original characteristics, such as volume:

```
<mscmixer version="1.0" xmlns="urn:ietf:params:xml:ns:msc-mixer">
   <modifyjoin id1="1536067209:913cd14c" id2="conference1">
      <stream media="audio" direction="sendonly">
         <volume controltype="setgain" value="0"/>
      </stream>
      <stream media="audio" direction="recvonly"/>
   </modifyjoin>
</mscmixer>
```

9.4.2.2.4 <unjoin>

The <unjoin> element is sent to the MS to request removal of previously established media stream(s) from between a connection and a conference, between two connections, or between two conferences. The <unjoin> element has the following attributes:

- id1: An identifier for either a connection or a conference. The identifier MUST conform to the syntax defined in Section 8.14.1 (Appendix A.1 of RFC 6230). The attribute is mandatory.
- id2: An identifier for either a connection or a conference. The identifier MUST conform to the syntax defined in Section 8.14.1 (Appendix A.1 of RFC 6230). The attribute is mandatory.

The <unjoin> element has the following child element (zero or more occurrences):

- <stream>: An element that identifies the media stream(s) to remove (see Section 9.4.2.2.5). The element is optional. When it is not present, all currently established streams between "id1" and "id2" are removed.

The MS MUST support <unjoin> for any stream that was established using <join> and that has not already been removed by a previous <unjoin> on the same stream.

- If the MS is unable to terminate the join as specified in <stream> elements because a <stream> element is in conflict with (a) another <stream> element, (b) specified connection or conference media capabilities, or (c) a SDP label value as part of the connection-id defined in Section 8.14.1 (Appendix A.1 of RFC 6230), then the MS reports an error (407), MUST NOT terminate the join between the entities, and MUST NOT change existing streams of the entities.
- If the MS is unable to terminate the join as specified in <stream> elements because the MS does not support the media stream configuration, the MS reports an error (422), MUST NOT terminate the join between the entities, and MUST NOT change existing streams of the entities.
- If the specified entities are not already joined, then the MS reports an error (409).
- If the MS is unable to terminate the join between the specified entities for any other reason, the MS reports an error (411).
- When an MS has successfully processed a <unjoin> request, it MUST reply with a successful <response> element (Section 9.4.2.3).

9.4.2.2.5 <stream>

<join>, <modifyjoin>, and <unjoin> require the identification and manipulation of media streams. Media streams represent the flow of media between a participant connection and a conference, between two connections, or between two conferences. The <stream> element is used (as a child to

<join>, <modifyjoin>, and <unjoin>) to identify the media stream(s) for the request and to specify the configuration of the media stream. The <stream> element has the following attributes:

- media: A string indicating the type of media associated with the stream. A valid value is a MIME type name as defined in Section 4.2 of RFC 4288. The following values MUST be used for common types of media: "audio" for audio media and "video" for video media. See IANA [2] for registered MIME type names. The attribute is mandatory.
- label: A string indicating the SDP label associated with a media stream RFC 4574. The attribute is optional.
- direction: A string indicating the allowed media flow of the stream relative to the value of the "id1" attribute of the parent element. Defined values are: "sendrecv" (media can be sent and received), "sendonly" (media can only be sent), "recvonly" (media can only be received), and "inactive" (stream established but no media flow). The default value is "sendrecv". The attribute is optional.

The <stream> element has the following sequence of child elements:

- <volume>: An element (Section 9.4.2.2.5.1) to configure the volume or gain of the media stream. The element is optional.
- <clamp>: An element (Section 9.4.2.2.5.2) to configure filtering and removal of tones from the media stream. The element is optional.
- <region>: An element (Section 9.4.2.2.5.3) to configure a region within a video layout where the media stream is displayed. The element is optional.
- <priority>: An element (Section 9.4.2.2.5.4) to configure priority associated with the stream in the media mix. The element is optional.

In each child element, the media stream affected is indicated by the value of the "direction" attribute of the parent element.

- If the "media" attribute does not have the value of "audio," then the MS ignores and <clamp> elements.
- If the "media" attribute does not have the value of "video," then the MS ignores a <region> element.

For example, a request to join a connection to conference in both directions with volume control is as follows:

```
<mscmixer version="1.0" xmlns="urn:ietf:params:xml:ns:msc-mixer">
   <join id1="1536067209:913cd14c" id2="conference1">
      <stream media="audio" direction="sendrecv">
         <volume controltype="setgain" value="-3"/>
      </stream>
   </join>
</mscmixer>
```

where audio flow from the connection (id1) to the conference (id2) has the volume lowered by 3 dB, and likewise, the volume of the audio flow from the conference to the connection is lowered by 3 dB. In this example, the volume is independently controlled for each direction.

```
<mscmixer version="1.0" xmlns="urn:ietf:params:xml:ns:msc-mixer">
   <join id1="1536067209:913cd14c" id2="conference1">
      <stream media="audio" direction="sendonly">
```

```
            <volume controltype="setgain" value="-3"/>
         </stream>
         <stream media="audio" direction="recvonly">
            <volume controltype="setgain" value="+3"/>
         </stream>
      </join>
   </mscmixer>
```

where audio flow from the connection (id1) to the conference (id2) has the volume lowered by 3 dB, but the volume of the audio flow from the conference to the connection is raised by 3 dB.

9.4.2.2.5.1 <volume> The element is used to configure the volume of an audio media stream. It can be set to a specific gain amount, to automatically adjust the gain to a desired target level, or to mute the volume. The element has no child elements but has the following attributes:

- controltype: A string indicating the type of volume control to use for the stream. Defined values are: "automatic" (the volume will be adjusted automatically to the level specified by the "value" attribute), "setgain" (use the value of the "value" attribute as a specific gain measured in decibels to apply), and "setstate" (set the state of the stream to "mute" or "unmute" as specified by the value of the "value" attribute). The attribute is mandatory.
- value: A string specifying the amount or state for the volume control defined by the value of the "controltype" attribute. The attribute is optional. There is no default value.

If the audio media stream is in a muted state, then the MS also automatically changes the state to unmuted with an "automatic" or "setgain" volume control. For example, assume an audio stream has been muted with <volume controltype="setstate" value="mute"/>. If the gain on the same stream is changed with <volume controltype="setgain" value="+3"/>, then the volume is increased, and stream state is also changed to unmuted.

9.4.2.2.5.2 <clamp> The <clamp> element is used to configure whether tones are filtered and removed from a media stream. The <clamp> element has no child elements but has the following attribute:

- tones: A space-separated list of the tones to remove. The attribute is optional. The default value is "1 2 3 4 5 6 7 8 9 0 * # A B C D" (i.e., all DTMF [dual-tone multi-frequency] tones are removed).

9.4.2.2.5.3 <region> As described in Section 9.4.2.1.4.2.1, each <video-layout> is composed of one or more named regions (or areas) in which video media can be presented. For example, the XCON layout <dual-view> has two regions named "1" and "2." The <region> element is used to explicitly specify the name of the area within a video layout where a video media stream is displayed. The <region> element has no attributes, and its content model specifies the name of the region.

9.4.2.2.5.4 <priority> The <priority> element is used to explicitly specify the priority of a participant. The MS uses this priority to determine where the media stream is displayed within a video layout (Section 9.4.2.1.4.3.1). The <priority> element has no attributes, and its content model specifies a positive integer (see Section 9.4.7.3). The lower the value, the higher the priority.

9.4.2.3 <response>

Responses to requests are indicated by a <response> element. The <response> element has the following attributes:

- status: Numeric code indicating the response status. Valid values are defined in Section 9.4.6. The attribute is mandatory. reason: String specifying a reason for the response status. The attribute is optional.
- desclang: Specifies the language used in the value of the "reason" attribute. A valid value is a language identifier (Section 9.4.7.7). The attribute is optional. If it is not specified, the value of the "desclang" attribute on <mscmixer> (Section 9.4.1) applies.
- conferenceid: String identifying the conference (see Section 8.14.1 of this book — Appendix A.1 of RFC 6230). The attribute is optional.
- connectionid: String identifying the SIP dialog connection defined in Section 8.14.1 (Appendix A.1 of RFC 6230). The attribute is optional.

For example, a response when a conference was created successfully is as follows:

```
<response code="200"/>
```

If conference creation failed due to the requested conference ID already existing, the response is:

```
<response code="405" reason="Conference already exists"/>
```

9.4.2.4 <event>

When a mixer generates a notification event, the MS sends the event using an <event> element. The <event> element has no attributes but has the following sequence of child elements (zero or more instances of each child):

- <active-talkers-notify>: Specifies an active talkers notification (Section 9.4.2.4.1).
- <unjoin-notify>: Notifies that a connection or conference has been completely unjoined (Section 9.4.2.4.2).
- <conferenceexit>: Notifies that a conference has exited (Section 9.4.2.4.3).

9.4.2.4.1 <active-talkers-notify>

The <active-talkers-notify> element describes zero or more speakers that have been active in a conference during the specified interval (see Section 9.4.2.1.4.4.1).

The <active-talkers-notify> element has the following attribute:

- conferenceid: String indicating the name of the conference from which the event originated. This attribute is mandatory.

The <active-talkers-notify> element has the following sequence of child elements (zero or more occurrences): <active-talker>: element describing an active talker (Section 9.4.2.4.1.1).

9.4.2.4.1.1 <active-talker> The <active-talker> element describes an active talker, associated with either a connection or conference participant in a conference. The <active-talker> element has the following attributes:

- connectionid: String indicating the connectionid of the active talker. This attribute is optional. There is no default value.
- conferenceid: String indicating the conferenceid of the active talker. This attribute is optional. There is no default value.

Note that the element does not describe an active talker if both the "connectionid" and "conferenceid" attributes are specified, or if neither attribute is specified. The <active-talker> element has no child elements.

9.4.2.4.2 <unjoin-notify>

The <unjoin-notify> element describes a notification event where a connection and/or conference have been completely unjoined. The <unjoin-notify> element has the following attributes:

- status: A status code indicating why the unjoin occurred. A valid value is a non-negative integer (see Section 9.4.7.2).

The MS MUST support the following values:

- 0 indicates the join has been terminated by a <unjoin> request.
- 1 indicates that the join terminated due to an execution error.
- 2 indicates that the join terminated because a connection or conference has terminated.

All other valid but undefined values are reserved for future use, where new status codes are assigned using the Standards Action process defined in RFC 5226. The AS MUST treat any status code it does not recognize as being equivalent to 1 (join execution error). The attribute is mandatory.

- reason: A textual description providing a reason for the status code, e.g., details about an error. A valid value is a string (see Section 9.4.7.4). The attribute is optional. There is no default value.
- desclang: Specifies the language used in the value of the "reason" attribute. A valid value is a language identifier (Section 9.4.7.7). The attribute is optional. If it is not specified, the value of the "desclang" attribute on <mscmixer> (Section 9.4.1) applies.
- id1: An identifier for either a connection or a conference. The identifier MUST conform to the syntax defined in Section 8.14.1 (Appendix A.1 of RFC 6230). The attribute is mandatory.
- id2: An identifier for either a connection or a conference. The identifier MUST conform to the syntax defined in Section 8.14.1 (Appendix A.1 of RFC 6230). The attribute is mandatory. The <unjoin-notify> element has no child elements.

9.4.2.4.3 <conferenceexit>

The <conferenceexit> element indicates that a conference has exited because it has been terminated or because an error occurred (for example, a hardware error in the conference mixing unit). This event MUST be sent by the MS whenever a successfully created conference exits. The <conferenceexit> element has the following attributes:

- conferenceid: String indicating the name of the conference. This attribute is mandatory.
- status: A status code indicating why the conference exited. A valid value is a non-negative integer (see Section 9.4.7.2).

The MS MUST support the following values:

- 0 indicates the conference has been terminated by a <destroyconference> request.
- 1 indicates the conference terminated due to an execution error.
- 2 indicates the conference terminated due to exceeding the maximum duration for a conference.

All other valid but undefined values are reserved for future use, where new status codes are assigned using the Standards Action process defined in RFC 5226. The AS MUST treat any status code it does not recognize as being equivalent to 1 (conference execution error). The attribute is mandatory.

reason: A textual description providing a reason for the status code, e.g., details about an error. A valid value is a string (see Section 9.4.7.4). The attribute is optional. There is no default value.

desclang: Specifies the language used in the value of the "reason" attribute. A valid value is a language identifier (Section 9.4.7.7). The attribute is optional. If it is not specified, the value of the "desclang" attribute on <mscmixer> (Section 9.4.1) applies.

The <conferenceexit> element has no child elements. When an MS sends a <conferenceexit> event, the identifier for the conference ("conferenceid" attribute) is no longer valid on the MS and can be reused for another conference. For example, the following notification event would be sent from the MS when the conference with identifier "conference99" exits due to a successful <destroyconference/>:

```
<mscmixer version="1.0" xmlns="urn:ietf:params:xml:ns:msc-mixer">
   <event>
      <conferenceexit conferenceid="conference99"
         status="0"/>
   </event>
</mscmixer>
```

9.4.3 Audit Elements

The audit elements defined in this section allow the MS to be audited for package capabilities as well as mixers managed by the package. Auditing is particularly important for two use cases. First, it enables discovery of package capabilities supported on an MS before an AS creates a conference mixer or joins connections and conferences. The AS can then use this information to create request elements using supported capabilities and in the case of codecs, to negotiate an appropriate SDP for a user agent's connection. Second, auditing enables discovery of the existence and status of mixers currently managed by the package on the MS. This could be used when one AS takes over management of mixers if the AS that created the mixers fails or is no longer available (see the security considerations in Section 9.7).

9.4.3.1 <audit>

The <audit> request element is sent to the MS to request information about the capabilities of, and mixers currently managed with, this Control Package. Capabilities include supported conference codecs and video layouts. Mixer information includes the status of managed mixers as well as codecs. The <audit> element has the following attributes:

- capabilities: Indicates whether package capabilities are to be audited. A valid value is a Boolean (see Section 9.4.7.1). A value of "true" indicates that capability information is to be reported. A value of "false" indicates that capability information is not to be reported. The attribute is optional. The default value is "true."
- mixers: Indicates whether mixers currently managed by the package are to be audited. A valid value is a Boolean (see Section 9.4.7.1). A value of "true" indicates that mixer information is to be reported. A value of "false" indicates that mixer information is not to be reported. The attribute is optional. The default value is "true."
- conferenceid: String identifying a specific conference mixer to audit. It is an error (406) if the "conferenceid" attribute is specified and the conference identifier is not valid. The attribute is optional. There is no default value.

If the "mixers" attribute has the value "true," and "conferenceid" attribute is specified, then only audit information about the specified conference mixer is reported. If the "mixers" attribute has the value "false," then no mixer audit information is reported even if a "conferenceid" attribute is specified. The <audit> element has no child elements. When the MS receives an <audit> request, it MUST reply with a <auditresponse> element (Section 9.4.3.2) that includes a mandatory attribute describing the status in terms of a numeric code. Response status codes are defined in Section 9.4.6. If the request is successful, the <auditresponse> contains (depending on attribute values) a <capabilities> element (Section 9.4.3.2.1) reporting package capabilities and a <mixers> element (Section 9.4.3.2.2) reporting managed mixer information. If the MS is not able to process the request and carry out the audit operation, the audit request has failed, and the MS MUST indicate the class of failure using an appropriate 4xx response code. Unless an error response code is specified for a class of error within this section, implementations follow Section 4.6 in determining the appropriate status code for the response. For example, a request to audit capabilities and mixers managed by the package is as follows:

```
<mscmixer version="1.0" xmlns="urn:ietf:params:xml:ns:msc-mixer">
   <audit/>
</mscmixer>
```

In this example, only capabilities are to be audited:

```
<mscmixer version="1.0" xmlns="urn:ietf:params:xml:ns:msc-mixer">
   <audit mixers="false"/>
</mscmixer>
```

With this example, only a specific conference mixer is to be audited:

```
<mscmixer version="1.0" xmlns="urn:ietf:params:xml:ns:msc-mixer">
   <audit capabilities="false" conferenceid="conf4"/>
</mscmixer>
```

9.4.3.2 <auditresponse>

The <auditresponse> element describes a response to an <audit> request. The <auditresponse> element has the following attributes:

- status: Numeric code indicating the audit response status. The attribute is mandatory. Valid values are defined in Section 9.4.6. reason: String specifying a reason for the status. The attribute is optional. desclang: Specifies the language used in the value of the "reason" attribute. A valid value is a language identifier (Section 9.4.7.7). The attribute is optional. If it is not specified, the value of the "desclang" attribute on <mscmixer> (Section 9.4.1) applies.

The <auditresponse> element has the following sequence of child elements:

- <capabilities>: Element describing capabilities of the package (see Section 9.4.3.2.1). The element is optional. <mixers>: Element describing information about managed mixers (see Section 9.4.3.2.2). The element is optional.

For example, a successful response to an <audit> request for capabilities and mixer information is as follows:

```
<mscmixer version="1.0" xmlns="urn:ietf:params:xml:ns:msc-mixer">
   <auditresponse status="200">
      <capabilities>
         <codecs>
            <codec name="video">
```

```
                <subtype>H263</subtype>
            </codec>
            <codec name="video">
                <subtype>H264</subtype>
            </codec>
            <codec name="audio">
                <subtype>PCMU</subtype>
            </codec>
            <codec name="audio">
                <subtype>PCMA</subtype>
            </codec>
        </codecs>
    </capabilities>
    <mixers>
    <conferenceaudit conferenceid="conf1">
        <codecs>
            <codec name="audio">
                <subtype>PCMA</subtype>
            </codec>
        </codecs>
        <participants>
            <participant id="1536067209:913cd14c"/>
            </participants>
    </conferenceaudit>
    <joinaudit id1="1536067209:913cd14c" id2="conf1"/>
    <joinaudit id1="1636067209:113cd14c"
        id2="1836067209:313cd14c"/>
    <joinaudit id1="1736067209:213cd14c"
        id2="1936067209:413cd14c"/>
    </mixers>
  </auditresponse>
</mscmixer>
```

9.4.3.2.1 <capabilities>

The <capabilities> element provides audit information about package capabilities. The <capabilities> element has no attributes. The <capabilities> element has the following sequence of child elements:

- <codecs>: Element (Section 9.4.4) describing codecs available to the package. The element is mandatory.

For example, a fragment describing capabilities is as follows:

```
<capabilities>
    <codecs>
        <codec name="video">
            <subtype>H263</subtype>
        </codec>
        <codec name="video">
            <subtype>H264</subtype>
        </codec>
        <codec name="audio">
            <subtype>PCMU</subtype>
        </codec>
```

```
        <codec name="audio">
            <subtype>PCMA</subtype>
        </codec>
    </codecs>
</capabilities>
```

9.4.3.2.2 <mixers>

The <mixers> element provides audit information about mixers. The <mixers> element has no attributes. The <mixers> element has the following sequence of child elements (zero or more occurrences, any order):

- <conferenceaudit>: Audit information for a conference mixer (Section 9.4.3.2.2.1). The element is optional.
- <joinaudit>: Audit information for a join mixer (Section 9.4.3.2.2.2). The element is optional.

9.4.3.2.2.1 <conferenceaudit> The <conferenceaudit> element has the following attribute: conferenceid: String identifying the conference defined in Appendix A.1 of RFC 6230 (see Section 8.14.1). The attribute is mandatory. The <conferenceaudit> element has the following sequence of child elements:

<codecs> element describing codecs used in the conference. See Section 9.4.4. The element is optional. <participants> element listing connections or conferences joined to the conference. See Section 9.4.3.2.2.1.1. The element is optional. <video-layout> element describing the active video layout for the conference. See Section 9.4.2.1.4.2.1. The element is optional.

For example, a fragment describing a conference that has been created but has no participants is as follows:

```
<conferenceaudit conferenceid="conference1"/>
      A fragment when the same conference has three participants (two
      connections and another conference) joined to it is as follows:
<conferenceaudit conferenceid="conference1">
   <codecs>
   <codec name="audio">
      <subtype>PCMU</subtype>
   </codec>
   </codecs>
   <participants>
      <participant id="connection1"/>
      <participant id="connection2"/>
      <participant id="conference2"/>
   </participants>
</conferenceaudit>
```

9.4.3.2.2.1.1 <participants> The <participants> element is a container for <participant> elements (Section 4.3.2.2.1.1.1).\ The <participants> element has no attributes, but the following child elements are defined (zero or more):

- <participant>: specifies a participant (Section 9.4.3.2.2.1.1.1).

9.4.3.2.2.1.1.1 <participant> The <participant> element describes a participant. The <participant> element has the following attribute:

- id: An identifier for either a connection or a conference. The identifier MUST conform to the syntax defined in Section 8.14.1 (Appendix A.1 of RFC 6230). The attribute is mandatory. The <participant> element has no children.

9.4.3.2.2.3 <joinaudit> The <joinaudit> element has the following attributes:

- id1: An identifier for either a connection or a conference. The identifier MUST conform to the syntax defined in Section 8.14.1 (Appendix A.1 of RFC 6230). The attribute is mandatory.
- id2: An identifier for either a connection or a conference. The identifier MUST conform to the syntax defined in Section 8.14.1 (Appendix A.1 of RFC 6230). The attribute is mandatory.

The <joinaudit> element has no children. For example, a fragment describing an audit of two join mixers, one between connections and the second between conferences, is as follows:

```
<mixers>
    <joinaudit id1="1536067209:913cd14"id2
          ="1636067209:413cd14"/>
    <joinaudit id1="conference1" id2="conference2"/>
</mixers>
```

9.4.4 <CODECS>

The <codecs> element is a container for one or more codec definitions. Codec definitions are used by an AS to specify the codecs allowed for a conference (e.g., when used as a child of <createconference> or <modifyconference). Codec definitions are used by an MS to provide audit information about the codecs supported by an MS and used in specific conferences. The <codecs> element has no attributes. The <codecs> element has the following sequence of child elements (zero or more occurrences):

- <codec>: Defines a codec and optionally, its policy (Section 9.4.4.1). The element is optional.

For example, a fragment describing two codecs is as follows:

```
<codecs>
    <codec name="audio">
        <subtype>PCMA</subtype>
    </codec>
    <codec name="video">
        <subtype>H263</subtype>
    </codec>
</codecs>
```

9.4.4.1 <codec>

The <codec> element describes a codec. The element is modeled on the <codec> element in the XCON conference information data model (RFC 6501 – see Chapter 4) and allows additional information (e.g., rate, speed, etc.) to be specified. The <codec> element has the following attribute:

- name: Indicates the type name of the codec's media format as defined in IANA [2]. A valid value is a "type-name" as defined in Section 4.2 of RFC 4288. The attribute is mandatory.

The <codec> element has the following sequence of child elements:

- <subtype>: Element whose content model describes the subtype of the codec's media format as defined in IANA [2]. A valid value is a "subtype-name" as defined in Section 4.2 of RFC 4288. The element is mandatory.
- <params>: Element (Section 9.4.5) describing additional information about the codec. This package is agnostic to the names and values of the codec parameters supported by an implementation. The element is optional.

For example, a fragment with a <codec> element describing the H263 codec is as follows:

```
<codec name="video">
   <subtype>H263</subtype>
</codec>
```

A fragment where the <codec> element describes the H264 video codec with additional information about the profile level and packetization mode is as follows:

```
<codec name="video">
   <subtype>H264</subtype>
   <params>
      <param name="profile-level-id">42A01E</param>
      <param name="packetization-mode">0</param>
   </params>
</codec>
```

9.4.5 <PARAMS>

The <params> element is a container for <param> elements (Section 9.4.5.1). The <params> element has no attributes, but the following child elements are defined (zero or more):

- <param>: specifies a parameter name and value (Section 9.4.5.1).

9.4.5.1 <param>

The <param> element describes a parameter name and value. The <param> element has the following attributes:

- name: A string indicating the name of the parameter. The attribute is mandatory.
- type: Specifies a type indicating how the in-line value of the parameter is to be interpreted. A valid value is a MIME media type (see Section 9.4.7.6). The attribute is optional. The default value is "text/plain."
- encoding: Specifies a content-transfer-encoding schema applied to the in-line value of the parameter on top of the MIME media type specified with the "type" attribute. A valid value is a content transfer-encoding schema as defined by the "mechanism" token in Section 6.1 of RFC 2045. The attribute is optional. There is no default value. The <param> element content model is the value of the parameter.

Note that a value that contains XML characters (e.g., "<") needs to be escaped following standard XML conventions.

9.4.6 Response Status Codes

This section describes the response codes in Table 9.1 for the "status" attribute of mixer management <response> (Section 9.4.2.3) and <auditresponse> (Section 9.4.3.2). The MS MUST support the status response codes defined here. All other valid but undefined values are reserved for future use, where new status codes are assigned using the Standards Action process defined in RFC 5226. The AS MUST treat any responses it does not recognize as being equivalent to the x00 response code for all classes. For example, if an AS receives an unrecognized response code of 499, it can safely assume that there was something wrong with its request and treat the response as if it had received a 400 (Syntax error) response code. 4xx responses are definite failure responses from a particular MS. The "reason" attribute in the response SHOULD identify the failure in more detail, for example, "Mandatory attribute missing: id2 join element" for a 400 (Syntax error) response code.

The AS SHOULD NOT retry the same request without modification (for example, correcting a syntax error or changing the conferenceid to use one available on the MS). However, the same request to a different MS might be successful, for example, if another MS supports a capability required in the request. 4xx failure responses can be grouped into three classes: failure due to a syntax error in the request (400); failure due to an error executing the request on the MS (405–419); and failure due to the request requiring a capability not supported by the MS (420–435).

In cases where more than one request code could be reported for a failure, the MS SHOULD use the most specific error code of the failure class for the detected error. For example, if the MS detects that the conference identifier in the request is invalid, then it uses a 406 status code. However, if the MS merely detects that an execution error occurred, then 419 is used.

9.4.7 Type Definitions

This section defines types referenced in attribute definitions.

9.4.7.1 Boolean

The value space of Boolean is the set {true, false, 1, 0} as defined in Section 3.2.2 of Biron and Malhotra [3]. In accordance with this definition, the concept of false can be lexically represented by the strings "0" and "false" and the concept of true by the strings "1" and "true;" implementations MUST support both styles of lexical representation.

9.4.7.2 Non-Negative Integer

The value space of non-negative integer is the infinite set {0,1,2,...} as defined in Section 3.3.20 of Biron and Malhotra [3].

9.4.7.3 Positive Integer

The value space of positive integer is the infinite set {1,2,...} as defined in Section 3.3.25 of Biron and Malhotra [3].

9.4.7.4 String

A string in the character encoding associated with the XML element as defined in Section 3.2.1 of Biron and Malhotra [3].

9.4.7.5 Time Designation

A time designation consists of a non-negative real number followed by a time unit identifier. The time unit identifiers are: "ms" (milliseconds) and "s" (seconds). Examples include: "3s," "850ms," "0.7s," "0.5s," and "+1.5s."

TABLE 9.1
Status Codes

Code	Summary	Description	Informational: AS Possible Recovery Action
200	OK	Request has succeeded.	
400	Syntax Error	Request is syntactically invalid; it is not valid with respect to the XML schema specified in Section 9.5 or it violates a co-occurrence constraint for a request element defined in Section 9.4.	Change the request so that it is syntactically valid.
405	Conference already exists	Request uses an identifier to create a new conference (Section 9.4.2.1.1) that is already used by another conference on the MS.	Send an <audit> request (Section 9.4.3.1) requesting the list of conference mixer identifiers already used by the MS and then use a conference identifier that is not listed.
406	Conference does not exist	Request uses an identifier for a conference that does not exist on the MS.	Send an <audit> request (Section 9.4.3.1) requesting the list of conference mixer identifiers used by the MS and then use a conference identifier that is listed.
407	Incomplete stream configuration	Request specifies a media stream configuration that is in conflict with itself, the connection, or conference capabilities (see Section 9.4.2.2.2).	Change the media stream confirmation to match the capabilities of the connection or conference.
408	Joining the entities already joined	Request attempts to create a join mixer (Section 9.4.2.2.2) where the entities are already joined.	Send an <audit> request (Section 9.4.3.1) requesting the list of join mixers on the MS and then use entities that are not listed.
409	Joining entities not joined	Request attempts to manipulate a join mixer where the entities are not joined.	Send an <audit> request (Section 9.4.3.1) requesting the list of join mixers on the MS and then use entities that are not listed.
410	Unable to join – conference full	Request attempts to join a participant to a conference (Section 9.4.2.2.2) but the conference are already full.	
411	Unable to perform join mixer operation	Request attempts to create, modify, or delete a join between entities but fails.	
412	Connection does not exist	Request uses as identifier for a connection that does not exist on the MS.	
419	Other execution error	Requested operation cannot be executed by the MS.	
420	Conference reservation failed	Request to create a new conference (see Section 9.4.2.1.1) failed due to unsupported reservation of talkers or listeners.	
421	Unable to configure audio mix	Request to create or modify a conference failed due to unsupported audio mix.	
422	Unsupported media stream configuration	Request contains one or more <stream> elements (Section 9.4.2.2.5) whose configuration is not supported by the MS.	
423	Unable to configure video layouts	Request to create or modify a conference failed due to unsupported video layout configuration.	

(Continued)

TABLE 9.1 (CONTINUED)
Status Codes

Code	Summary	Description	Informational: AS Possible Recovery Action
424	Unable to configure video switch	Request to create or modify a conference failed due to unsupported video switch configuration.	
425	Unable to configure codecs	Request to create or modify a conference failed due to unsupported codec.	
426	Unable to join – mixing connections not supported	Request to join connection entities (Section 9.4.2.2.2) failed due to lack of support for mixing connections.	
427	Unable to join – mixing conferences not supported	Request to join conference entities (Section 9.4.2.2.2) failed due to lack of support for mixing connections.	
428	Unsupported foreign namespace attribute or element	The request contains attributes or elements from another namespace that the MS does not support.	
435	Other unsupported capability	Request requires another capability not supported by the MS.	

9.4.7.6 MIME Media Type

A string formatted as an IANA MIME media type [4]. The augmented Backus–Naur form (ABNF) (RFC 5234) production for the string is: media-type = type-name "/" subtype-name *(";" parameter) parameter = parameter-name "=" value, where "type-name" and "subtype-name" are defined in section 4.2 of RFC 4288, "parameter-name" is defined in section 4.3 of RFC 4288, and "value" is defined in Section 5.1 of RFC 2045.

9.4.7.7 Language Identifier

A language identifier labels information content as being of a particular human language variant. Following the XML specification for language identification [1], a legal language identifier is identified by a RFC 5646 code and matched according to RFC 4647.

9.5 FORMAL SYNTAX

This section defines the XML schema for the Mixer Control Package. The schema is normative. The schema defines datatypes, attributes, and mixer elements in the urn:ietf:params:xml:ns:msc-mixer namespace. In most elements, the order of child elements is significant. The schema is extensible: elements allow attributes and child elements from other namespaces. Elements from outside this package's namespace can occur after elements defined in this package. The schema is dependent upon the schema (framework.xsd) defined in Section 8.14.1 (Appendix A.1 of the Control Framework RFC 6230) (Figure 9.10).

```
<?xml version="1.0" encoding="UTF-8"?>
<xsd:schema targetNamespace="urn:ietf:params:xml:ns:msc-mixer"
 xmlns:fw="urn:ietf:params:xml:ns:control:framework-attributes"
 elementFormDefault="qualified"
 xmlns:xs="http://www.w3.org/2001/XMLSchema"
 xmlns="urn:ietf:params:xml:ns:msc-mixer"
 xmlns:xsd="http://www.w3.org/2001/XMLSchema">

 <xsd:annotation>
  <xsd:documentation>
   IETF MediaCtrl Mixer 1.0 (20110104)

   This is the schema of the Mixer Control Package.  It
   defines request, response, and notification elements for
   mixing.

   The schema namespace is urn:ietf:params:xml:ns:msc-mixer

  </xsd:documentation>
 </xsd:annotation>

 <!--
  ##############################################################

  SCHEMA IMPORTS

  ##############################################################
 -->


 <xsd:import
  namespace="urn:ietf:params:xml:ns:control:framework-attributes"
  schemaLocation="framework.xsd">
  <xsd:annotation>
   <xsd:documentation>
    This import brings in the framework attributes for
    conferenceid and connectionid.
   </xsd:documentation>
  </xsd:annotation>
 </xsd:import>

 <!--
```

FIGURE 9.10 Mixer Package XML Schema

9.6 EXAMPLES

This section provides examples of the Mixer Control Package.

9.6.1 AS-MS Framework Interaction Examples

The following example assumes a Control Channel has been established and synced as described in the Media Control Channel Framework (RFC 6230 – see Chapter 8). The XML messages are in angled brackets (with the root <mscmixer> and other details omitted for clarity); the REPORT status is in parentheses. Other aspects of the protocol are omitted for readability.

```
 ####################################################

 Extensible core type

 ####################################################
-->


<xsd:complexType name="Tcore">
 <xsd:annotation>
  <xsd:documentation>
   This type is extended by other (non-mixed) component types to
   allow attributes from other namespaces.
  </xsd:documentation>
 </xsd:annotation>
 <xsd:sequence/>
 <xsd:anyAttribute namespace="##other" processContents="lax" />
</xsd:complexType>

<!--
 ####################################################

 TOP-LEVEL ELEMENT: mscmixer

 ####################################################
-->

<xsd:complexType name="mscmixerType">
 <xsd:complexContent>
  <xsd:extension base="Tcore">
   <xsd:sequence>
    <xsd:choice>
     <xsd:element ref="createconference" />
     <xsd:element ref="modifyconference" />
     <xsd:element ref="destroyconference" />
     <xsd:element ref="join" />
     <xsd:element ref="unjoin" />
     <xsd:element ref="modifyjoin" />
     <xsd:element ref="response" />
     <xsd:element ref="event" />
     <xsd:element ref="audit" />
     <xsd:element ref="auditresponse" />
     <xsd:any namespace="##other" minOccurs="0"
      maxOccurs="unbounded" processContents="lax" />
    </xsd:choice>
   </xsd:sequence>
   <xsd:attribute name="version" type="version.datatype"
    use="required" />
```

FIGURE 9.10 Continued

```
<xsd:attribute name="desclang" type="xsd:language"
        default="i-default" />
      </xsd:extension>
     </xsd:complexContent>
    </xsd:complexType>

    <xsd:element name="mscmixer" type="mscmixerType" />

    <!--
     #####################################################

     CONFERENCE MANAGEMENT TYPES

     #####################################################
    -->

    <!--  createconference -->

    <xsd:complexType name="createconferenceType">
     <xsd:complexContent>
      <xsd:extension base="Tcore">
       <xsd:sequence>
        <xsd:element ref="codecs" minOccurs="0"
         maxOccurs="1" />
        <xsd:element ref="audio-mixing" minOccurs="0"
         maxOccurs="1" />
        <xsd:element ref="video-layouts" minOccurs="0"
         maxOccurs="1" />
        <xsd:element ref="video-switch" minOccurs="0"
         maxOccurs="1" />
        <xsd:element ref="subscribe" minOccurs="0"
         maxOccurs="1" />
        <xsd:any namespace="##other"
         processContents="lax" minOccurs="0" maxOccurs="unbounded" />
       </xsd:sequence>
       <xsd:attribute name="conferenceid" type="xsd:string" />
       <xsd:attribute name="reserved-talkers"
        type="xsd:nonNegativeInteger" default="0" />
       <xsd:attribute name="reserved-listeners"
        type="xsd:nonNegativeInteger" default="0" />
      </xsd:extension>
     </xsd:complexContent>
    </xsd:complexType>


    <xsd:element name="createconference" type="createconferenceType" />

  <!-- modifyconference -->
```

FIGURE 9.10 Continued

```
<xsd:complexType name="modifyconferenceType">
 <xsd:complexContent>
  <xsd:extension base="Tcore">
   <xsd:sequence>
    <xsd:element ref="codecs" minOccurs="0"
     maxOccurs="1" />
    <xsd:element ref="audio-mixing" minOccurs="0"
     maxOccurs="1" />
    <xsd:element ref="video-layouts" minOccurs="0"
     maxOccurs="1" />
    <xsd:element ref="video-switch" minOccurs="0"
     maxOccurs="1" />
    <xsd:element ref="subscribe" />
    <xsd:any namespace="##other"
     processContents="lax" minOccurs="0" maxOccurs="unbounded" />
   </xsd:sequence>
   <xsd:attribute name="conferenceid" type="xsd:string"
    use="required" />
  </xsd:extension>
 </xsd:complexContent>
</xsd:complexType>

<xsd:element name="modifyconference" type="modifyconferenceType" />


<!--  destroyconference -->

<xsd:complexType name="destroyconferenceType">
<xsd:complexContent>
  <xsd:extension base="Tcore">
  <xsd:sequence>
   <xsd:any namespace="##other" minOccurs="0"
     maxOccurs="unbounded" processContents="lax" />
   </xsd:sequence>
  <xsd:attribute name="conferenceid" type="xsd:string"
  use="required" />
  </xsd:extension>
  </xsd:complexContent>
</xsd:complexType>

<xsd:element name="destroyconference"
 type="destroyconferenceType" />

<!--
 #####################################################

 JOIN TYPES
```

FIGURE 9.10 Continued

```
 #####################################################
-->

<xsd:complexType name="joinType">
 <xsd:complexContent>
  <xsd:extension base="Tcore">
   <xsd:sequence>
    <xsd:element ref="stream" minOccurs="0"
     maxOccurs="unbounded" />
    <xsd:any namespace="##other"
     processContents="lax" minOccurs="0" maxOccurs="unbounded" />
   </xsd:sequence>
   <xsd:attribute name="id1" type="xsd:string"
    use="required" />
   <xsd:attribute name="id2" type="xsd:string"
    use="required" />
  </xsd:extension>
 </xsd:complexContent>
</xsd:complexType>

<xsd:element name="join" type="joinType" />

<xsd:complexType name="modifyjoinType">
 <xsd:complexContent>
  <xsd:extension base="Tcore">
   <xsd:sequence>
    <xsd:element ref="stream" minOccurs="0"
     maxOccurs="unbounded" />
    <xsd:any namespace="##other"
     processContents="lax" minOccurs="0" maxOccurs="unbounded" />
   </xsd:sequence>
   <xsd:attribute name="id1" type="xsd:string"
    use="required" />
   <xsd:attribute name="id2" type="xsd:string"
    use="required" />
  </xsd:extension>
 </xsd:complexContent>
</xsd:complexType>

<xsd:element name="modifyjoin" type="modifyjoinType" />


<xsd:complexType name="unjoinType">
 <xsd:complexContent>
  <xsd:extension base="Tcore">
   <xsd:sequence>
    <xsd:element ref="stream" minOccurs="0"
     maxOccurs="unbounded" />
```

FIGURE 9.10 Continued

```
    <xsd:any namespace="##other"
     processContents="lax" minOccurs="0" maxOccurs="unbounded" />
   </xsd:sequence>
   <xsd:attribute name="id1" type="xsd:string"
    use="required" />
   <xsd:attribute name="id2" type="xsd:string"
    use="required" />
  </xsd:extension>
 </xsd:complexContent>
</xsd:complexType>

<xsd:element name="unjoin" type="unjoinType" />

<!--
 ####################################################

 OTHER TYPES

 ####################################################
-->

<xsd:complexType name="eventType">
 <xsd:complexContent>
  <xsd:extension base="Tcore">
   <xsd:sequence>
    <xsd:choice>
     <xsd:element ref="active-talkers-notify"
      minOccurs="0" maxOccurs="1" />
     <xsd:element ref="unjoin-notify"
      minOccurs="0" maxOccurs="1" />
     <xsd:element ref="conferenceexit"
      minOccurs="0" maxOccurs="1" />
     <xsd:any namespace="##other" minOccurs="0"
      maxOccurs="unbounded" processContents="lax" />
    </xsd:choice>
   </xsd:sequence>
  </xsd:extension>
 </xsd:complexContent>
</xsd:complexType>

<xsd:element name="event" type="eventType" />

<xsd:complexType name="activetalkersnotifyType">
 <xsd:complexContent>
  <xsd:extension base="Tcore">
   <xsd:sequence>
    <xsd:element ref="active-talker" minOccurs="0"
     maxOccurs="unbounded" />
```

FIGURE 9.10 Continued

```
<xsd:any namespace="##other" minOccurs="0"
        maxOccurs="unbounded" processContents="lax" />
      </xsd:sequence>
      <xsd:attribute name="conferenceid" type="xsd:string"
       use="required" />
     </xsd:extension>
    </xsd:complexContent>
   </xsd:complexType>

   <xsd:element name="active-talkers-notify"
    type="activetalkersnotifyType" />


   <xsd:complexType name="activetalkerType">
    <xsd:complexContent>
     <xsd:extension base="Tcore">
      <xsd:sequence>
       <xsd:any namespace="##other" minOccurs="0"
        maxOccurs="unbounded" processContents="lax" />
      </xsd:sequence>
      <xsd:attributeGroup ref="fw:framework-attributes" />
     </xsd:extension>
    </xsd:complexContent>
   </xsd:complexType>

   <xsd:element name="active-talker" type="activetalkerType" />


   <xsd:complexType name="unjoinnotifyType">
    <xsd:complexContent>
     <xsd:extension base="Tcore">
      <xsd:sequence>
       <xsd:any namespace="##other" minOccurs="0"
        maxOccurs="unbounded" processContents="lax" />
      </xsd:sequence>
      <xsd:attribute name="status" type="xsd:nonNegativeInteger"
        use="required" />
      <xsd:attribute name="reason" type="xsd:string" />
        <xsd:attribute name="desclang" type="xsd:language"/>
      <xsd:attribute name="id1" type="xsd:string"
       use="required" />
      <xsd:attribute name="id2" type="xsd:string"
       use="required" />
     </xsd:extension>
    </xsd:complexContent>
   </xsd:complexType>

   <xsd:element name="unjoin-notify" type="unjoinnotifyType" />
```

FIGURE 9.10 Continued

```
<!--  conferenceexit-->

<xsd:complexType name="conferenceexitType">
 <xsd:complexContent>
  <xsd:extension base="Tcore">
   <xsd:sequence>
    <xsd:any namespace="##other" minOccurs="0"
     maxOccurs="unbounded" processContents="lax" />
   </xsd:sequence>
   <xsd:attribute name="conferenceid" type="xsd:string"
    use="required" />
   <xsd:attribute name="status"
    type="xsd:nonNegativeInteger" use="required" />
   <xsd:attribute name="reason" type="xsd:string" />
     <xsd:attribute name="desclang" type="xsd:language"/>
  </xsd:extension>
 </xsd:complexContent>
</xsd:complexType>

<xsd:element name="conferenceexit" type="conferenceexitType" />


<xsd:complexType name="responseType">
 <xsd:complexContent>
  <xsd:extension base="Tcore">
  <xsd:sequence>
   <xsd:any namespace="##other" minOccurs="0"
     maxOccurs="unbounded" processContents="lax" />
   </xsd:sequence>
   <xsd:attribute name="status" type="status.datatype"
    use="required" />
   <xsd:attribute name="reason" type="xsd:string" />
     <xsd:attribute name="desclang" type="xsd:language"/>
   <xsd:attributeGroup ref="fw:framework-attributes" />
  </xsd:extension>
 </xsd:complexContent>
</xsd:complexType>


<xsd:element name="response" type="responseType" />

<xsd:complexType name="subscribeType">
 <xsd:complexContent>
  <xsd:extension base="Tcore">
   <xsd:sequence>
    <xsd:element ref="active-talkers-sub"
     minOccurs="0" maxOccurs="1" />
    <xsd:any namespace="##other" minOccurs="0"
```

FIGURE 9.10 Continued

```
<xsd:extension base="Tcore">
       <xsd:sequence>
        <xsd:element ref="video-layout" minOccurs="0"
         maxOccurs="unbounded" />
        <xsd:any namespace="##other" minOccurs="0"
         maxOccurs="unbounded" processContents="lax" />
       </xsd:sequence>
      </xsd:extension>
     </xsd:complexContent>
    </xsd:complexType>

    <xsd:element name="video-layouts" type="videolayoutsType" />


    <!-- video-layout -->
    <xsd:complexType name="videolayoutType">
    <xsd:complexContent>
      <xsd:extension base="Tcore">
       <xsd:sequence>
        <xsd:choice>
          <xsd:element name="single-view" type="Tcore"/>
          <xsd:element name="dual-view" type="Tcore"/>
          <xsd:element name="dual-view-crop" type="Tcore"/>
          <xsd:element name="dual-view-2x1" type="Tcore"/>
          <xsd:element name="dual-view-2x1-crop" type="Tcore"/>
          <xsd:element name="quad-view" type="Tcore"/>
          <xsd:element name="multiple-3x3" type="Tcore"/>
          <xsd:element name="multiple-4x4" type="Tcore"/>
          <xsd:element name="multiple-5x1" type="Tcore"/>
          <xsd:any namespace="##other" processContents="lax" />
        </xsd:choice>
       </xsd:sequence>
      <xsd:attribute name="min-participants"
        type="xsd:positiveInteger" default="1" />
      </xsd:extension>
     </xsd:complexContent>
    </xsd:complexType>

    <xsd:element name="video-layout" type="videolayoutType" />

    <xsd:complexType name="auditType">
     <xsd:complexContent>
      <xsd:extension base="Tcore">
      <xsd:sequence>
      <xsd:any namespace="##other" minOccurs="0"
       maxOccurs="unbounded" processContents="lax" />
      </xsd:sequence>
       <xsd:attribute name="capabilities"
```

FIGURE 9.10 Continued

```
</xsd:complexType>

<xsd:element name="stream" type="streamType" />


<xsd:complexType name="volumeType">
 <xsd:complexContent>
  <xsd:extension base="Tcore">
   <xsd:sequence>
    <xsd:any namespace="##other" minOccurs="0"
     maxOccurs="unbounded" processContents="lax" />
   </xsd:sequence>
   <xsd:attribute name="controltype"
    type="volumecontroltype.datatype" use="required" />
   <xsd:attribute name="value" type="xsd:string" />
  </xsd:extension>
 </xsd:complexContent>
</xsd:complexType>

<xsd:element name="volume" type="volumeType" />


<xsd:complexType name="clampType">
 <xsd:complexContent>
  <xsd:extension base="Tcore">
   <xsd:sequence>
    <xsd:any namespace="##other" minOccurs="0"
     maxOccurs="unbounded" processContents="lax" />
   </xsd:sequence>
   <xsd:attribute name="tones" type="xsd:string"
    default="1 2 3 4 5 6 7 8 9 0 * # A B C D"/>
  </xsd:extension>
 </xsd:complexContent>
</xsd:complexType>

<xsd:element name="clamp" type="clampType" />


<!--  region  -->
<xsd:simpleType name="regionType">
 <xsd:restriction base="xsd:NMTOKEN" />
</xsd:simpleType>

<xsd:element name="region" type="regionType" />


<!--  priority  -->
<xsd:simpleType name="priorityType">
```

FIGURE 9.10 Continued

```
<xsd:restriction base="xsd:positiveInteger" />
    </xsd:simpleType>

    <xsd:element name="priority" type="priorityType" />

    <xsd:complexType name="audiomixingType">
     <xsd:complexContent>
      <xsd:extension base="Tcore">
      <xsd:sequence>
        <xsd:any namespace="##other" minOccurs="0"
         maxOccurs="unbounded" processContents="lax" />
       </xsd:sequence>
       <xsd:attribute name="type" type="audiomix.datatype"
        default="nbest" />
       <xsd:attribute name="n" type="xsd:nonNegativeInteger"
        default="0" />
      </xsd:extension>
     </xsd:complexContent>
    </xsd:complexType>

    <xsd:element name="audio-mixing" type="audiomixingType" />

    <!-- video-switch -->

    <xsd:complexType name="videoswitchType">
     <xsd:complexContent>
      <xsd:extension base="Tcore">
       <xsd:sequence>
         <xsd:choice>
          <xsd:element name="vas" type="Tcore"/>
          <xsd:element name="controller" type="Tcore"/>
          <xsd:any namespace="##other" processContents="lax" />
        </xsd:choice>
       </xsd:sequence>
       <xsd:attribute name="interval"
        type="xsd:nonNegativeInteger" default="3" />
       <xsd:attribute name="activespeakermix"
        type="xsd:boolean" default="false" />
      </xsd:extension>
     </xsd:complexContent>
    </xsd:complexType>

    <xsd:element name="video-switch" type="videoswitchType" />

    <!-- video-layouts -->

    <xsd:complexType name="videolayoutsType">
  <xsd:complexContent>
```

FIGURE 9.10 Continued

```
<xsd:extension base="Tcore">
      <xsd:sequence>
       <xsd:element ref="video-layout" minOccurs="0"
        maxOccurs="unbounded" />
       <xsd:any namespace="##other" minOccurs="0"
        maxOccurs="unbounded" processContents="lax" />
      </xsd:sequence>
     </xsd:extension>
    </xsd:complexContent>
   </xsd:complexType>

   <xsd:element name="video-layouts" type="videolayoutsType" />


   <!-- video-layout -->
   <xsd:complexType name="videolayoutType">
   <xsd:complexContent>
     <xsd:extension base="Tcore">
      <xsd:sequence>
       <xsd:choice>
         <xsd:element name="single-view" type="Tcore"/>
         <xsd:element name="dual-view" type="Tcore"/>
         <xsd:element name="dual-view-crop" type="Tcore"/>
         <xsd:element name="dual-view-2x1" type="Tcore"/>
         <xsd:element name="dual-view-2x1-crop" type="Tcore"/>
         <xsd:element name="quad-view" type="Tcore"/>
         <xsd:element name="multiple-3x3" type="Tcore"/>
         <xsd:element name="multiple-4x4" type="Tcore"/>
         <xsd:element name="multiple-5x1" type="Tcore"/>
         <xsd:any namespace="##other" processContents="lax" />
       </xsd:choice>
      </xsd:sequence>
     <xsd:attribute name="min-participants"
       type="xsd:positiveInteger" default="1" />
     </xsd:extension>
    </xsd:complexContent>
   </xsd:complexType>

   <xsd:element name="video-layout" type="videolayoutType" />

   <xsd:complexType name="auditType">
    <xsd:complexContent>
     <xsd:extension base="Tcore">
     <xsd:sequence>
     <xsd:any namespace="##other" minOccurs="0"
      maxOccurs="unbounded" processContents="lax" />
     </xsd:sequence>
      <xsd:attribute name="capabilities"
```

FIGURE 9.10 Continued

```
type="xsd:boolean" default="true" />
      <xsd:attribute name="mixers" type="xsd:boolean"
       default="true" />
      <xsd:attribute name="conferenceid" type="xsd:string" />
     </xsd:extension>
    </xsd:complexContent>
   </xsd:complexType>

   <xsd:element name="audit" type="auditType" />


   <xsd:complexType name="auditresponseType">
    <xsd:complexContent>
     <xsd:extension base="Tcore">
      <xsd:sequence>
       <xsd:element ref="capabilities" minOccurs="0"
        maxOccurs="1" />
       <xsd:element ref="mixers" minOccurs="0"
        maxOccurs="1" />
       <xsd:any namespace="##other" minOccurs="0"
        maxOccurs="unbounded" processContents="lax" />
      </xsd:sequence>
      <xsd:attribute name="status" type="status.datatype"
       use="required" />
      <xsd:attribute name="reason" type="xsd:string" />
        <xsd:attribute name="desclang" type="xsd:language"/>
     </xsd:extension>
    </xsd:complexContent>
   </xsd:complexType>

   <xsd:element name="auditresponse" type="auditresponseType" />

   <!-- mixers -->

   <xsd:complexType name="mixersType">
    <xsd:complexContent>
     <xsd:extension base="Tcore">
      <xsd:sequence>
       <xsd:element ref="conferenceaudit" minOccurs="0"
        maxOccurs="unbounded" />
       <xsd:element ref="joinaudit" minOccurs="0"
        maxOccurs="unbounded" />
       <xsd:any namespace="##other" minOccurs="0"
        maxOccurs="unbounded" processContents="lax" />
      </xsd:sequence>
     </xsd:extension>
    </xsd:complexContent>
   </xsd:complexType>
```

FIGURE 9.10 Continued

```
<xsd:element name="mixers" type="mixersType" />

<!--  joinaudit -->

<xsd:complexType name="joinauditType">
 <xsd:complexContent>
  <xsd:extension base="Tcore">
   <xsd:sequence>
    <xsd:any namespace="##other"
     processContents="lax" minOccurs="0" maxOccurs="unbounded" />
   </xsd:sequence>
   <xsd:attribute name="id1" type="xsd:string"
    use="required" />
   <xsd:attribute name="id2" type="xsd:string"
    use="required" />
  </xsd:extension>
 </xsd:complexContent>
</xsd:complexType>

<xsd:element name="joinaudit" type="joinauditType" />

<!-- conferenceaudit -->

<xsd:complexType name="conferenceauditType">
 <xsd:complexContent>
  <xsd:extension base="Tcore">
   <xsd:sequence>
    <xsd:element ref="codecs" minOccurs="0"
     maxOccurs="1" />
    <xsd:element ref="participants" minOccurs="0"
     maxOccurs="1" />
    <xsd:element ref="video-layout" minOccurs="0"
     maxOccurs="1" />
    <xsd:any namespace="##other" minOccurs="0"
     maxOccurs="unbounded" processContents="lax" />
   </xsd:sequence>
   <xsd:attribute name="conferenceid" type="xsd:string"
    use="required" />
  </xsd:extension>
 </xsd:complexContent>
</xsd:complexType>

<xsd:element name="conferenceaudit" type="conferenceauditType" />


<!-- participants -->

<xsd:complexType name="participantsType">
```

FIGURE 9.10 Continued

```
  <xsd:complexContent>
   <xsd:extension base="Tcore">
    <xsd:sequence>
     <xsd:element ref="participant" minOccurs="0"
      maxOccurs="unbounded" />
     <xsd:any namespace="##other" minOccurs="0"
      maxOccurs="unbounded" processContents="lax" />
    </xsd:sequence>
   </xsd:extension>
  </xsd:complexContent>
 </xsd:complexType>

 <xsd:element name="participants" type="participantsType" />

 <!-- participant -->

 <xsd:complexType name="participantType">
  <xsd:complexContent>
   <xsd:extension base="Tcore">
    <xsd:sequence>
     <xsd:any namespace="##other" minOccurs="0"
      maxOccurs="unbounded" processContents="lax" />
    </xsd:sequence>
    <xsd:attribute name="id" type="xsd:string"
     use="required" />
   </xsd:extension>
  </xsd:complexContent>
 </xsd:complexType>

 <xsd:element name="participant" type="participantType" />

 <!-- capabilities -->

 <xsd:complexType name="capabilitiesType">
  <xsd:complexContent>
   <xsd:extension base="Tcore">
    <xsd:sequence>
     <xsd:element ref="codecs" minOccurs="1"
      maxOccurs="1" />
     <xsd:any namespace="##other" minOccurs="0"
      maxOccurs="unbounded" processContents="lax" />
    </xsd:sequence>
   </xsd:extension>
  </xsd:complexContent>
 </xsd:complexType>

 <xsd:element name="capabilities" type="capabilitiesType" />
```

FIGURE 9.10 Continued

```
<!-- codecs -->

<xsd:complexType name="codecsType">
 <xsd:complexContent>
  <xsd:extension base="Tcore">
   <xsd:sequence>
    <xsd:element ref="codec" minOccurs="0"
     maxOccurs="unbounded" />
    <xsd:any namespace="##other" minOccurs="0"
     maxOccurs="unbounded" processContents="lax" />
   </xsd:sequence>
  </xsd:extension>
 </xsd:complexContent>
</xsd:complexType>

<xsd:element name="codecs" type="codecsType" />

<!-- codec -->

<xsd:complexType name="codecType">
 <xsd:complexContent>
  <xsd:extension base="Tcore">
   <xsd:sequence>
    <xsd:element ref="subtype" minOccurs="1"
     maxOccurs="1" />
    <xsd:element ref="params" minOccurs="0"
     maxOccurs="1" />
    <xsd:any namespace="##other" minOccurs="0"
     maxOccurs="unbounded" processContents="lax" />
   </xsd:sequence>
    <xsd:attribute name="name" type="xsd:string"
    use="required" />
  </xsd:extension>
 </xsd:complexContent>
</xsd:complexType>

<xsd:element name="codec" type="codecType" />

<!-- subtype -->

<xsd:simpleType name="subtypeType">
 <xsd:restriction base="xsd:string" />
</xsd:simpleType>

<xsd:element name="subtype" type="subtypeType" />


<!-- params -->
```

FIGURE 9.10 Continued

```
<xsd:complexType name="paramsType">
     <xsd:complexContent>
      <xsd:extension base="Tcore">
       <xsd:sequence>
        <xsd:element ref="param" minOccurs="0"
         maxOccurs="unbounded" />
        <xsd:any namespace="##other" minOccurs="0"
         maxOccurs="unbounded" processContents="lax" />
       </xsd:sequence>
      </xsd:extension>
     </xsd:complexContent>
    </xsd:complexType>

    <xsd:element name="params" type="paramsType" />


    <!--  param -->
       <!--  doesn't extend tCore since its content model is mixed -->
    <xsd:complexType name="paramType" mixed="true">
     <xsd:sequence/>
     <xsd:attribute name="name" type="xsd:string" use="required" />
     <xsd:attribute name="type" type="mime.datatype"
     default="text/plain" />
        <xsd:attribute name="encoding" type="xsd:string"/>
    </xsd:complexType>

    <xsd:element name="param" type="paramType" />


    <!--
     #####################################################

     DATATYPES

     #####################################################
    -->

   <xsd:simpleType name="version.datatype">
     <xsd:restriction base="xsd:NMTOKEN">
      <xsd:enumeration value="1.0" />
     </xsd:restriction>
    </xsd:simpleType>

    <xsd:simpleType name="eventname.datatype">
     <xsd:restriction base="xsd:NMTOKEN">
      <xsd:pattern value="[a-zA-Z0-9\.]+" />
     </xsd:restriction>
    </xsd:simpleType>
```

FIGURE 9.10 Continued

```
<xsd:simpleType name="audiomix.datatype">
 <xsd:restriction base="xsd:NMTOKEN">
  <xsd:enumeration value="nbest" />
  <xsd:enumeration value="controller" />
 </xsd:restriction>
</xsd:simpleType>

<xsd:simpleType name="media.datatype">
 <xsd:restriction base="xsd:string" />
</xsd:simpleType>

<xsd:simpleType name="label.datatype">
 <xsd:restriction base="xsd:string" />
</xsd:simpleType>

<xsd:simpleType name="status.datatype">
 <xsd:restriction base="xsd:positiveInteger">
  <xsd:pattern value="[0-9][0-9][0-9]" />
 </xsd:restriction>
</xsd:simpleType>

<xsd:simpleType name="direction.datatype">
 <xsd:restriction base="xsd:NMTOKEN">
  <xsd:enumeration value="sendonly" />
  <xsd:enumeration value="recvonly" />
  <xsd:enumeration value="sendrecv" />
  <xsd:enumeration value="inactive" />
 </xsd:restriction>
</xsd:simpleType>


<xsd:simpleType name="mime.datatype">
 <xsd:restriction base="xsd:string" />
</xsd:simpleType>


<xsd:simpleType name="volumecontroltype.datatype">
 <xsd:restriction base="xsd:NMTOKEN">
  <xsd:enumeration value="automatic" />
  <xsd:enumeration value="setgain" />
  <xsd:enumeration value="setstate" />
 </xsd:restriction>
</xsd:simpleType>


</xsd:schema>
```

FIGURE 9.10 Continued

9.6.1.1 Creating a Conference Mixer and Joining a Participant

A conference mixer is created successfully and a participant is joined. Application Server

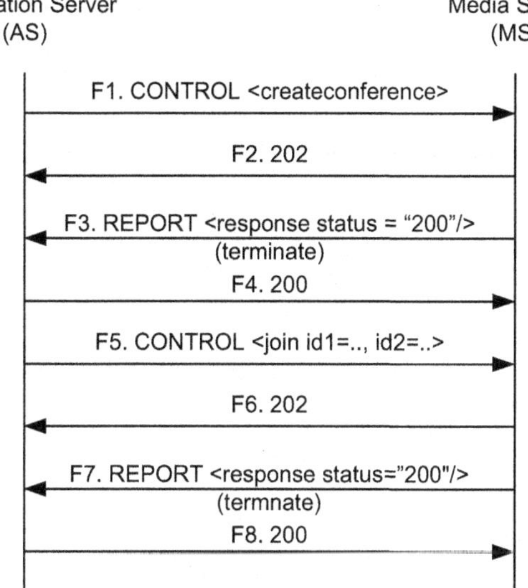

9.6.1.2 Receiving Active Talker Notifications

An active talker notification event is sent by the MS.

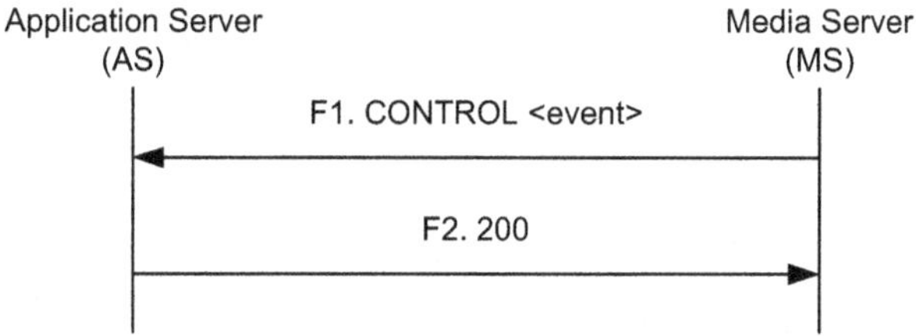

9.6.1.3 Conference Termination

The MS receives a request to terminate the conference, resulting in conference exit and participant unjoined notifications.

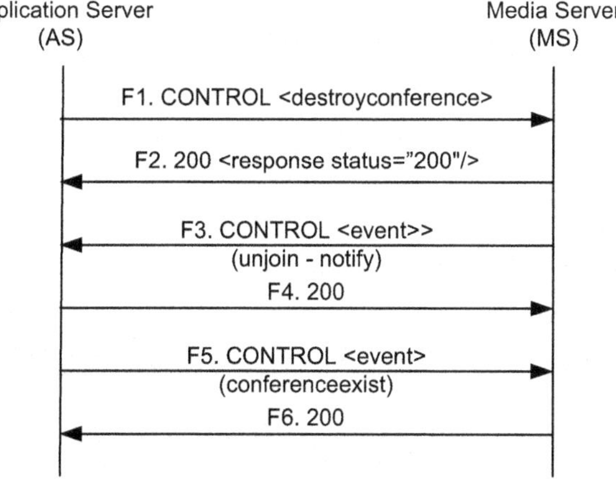

9.6.2 Mixing Examples

The following examples show how the mixing package can be used to create audio conferences, bridge connections, and video conferences. The examples do not specify all messages between the AS and the MS.

9.6.2.1 Audio Conferencing

The AS sends a request to create a conference mixer:

```
<mscmixer version="1.0" xmlns="urn:ietf:params:xml:ns:msc-mixer">
   <createconference conferenceid="conf1"
         reserved-talkers="2" reserved-listeners="3">
      <audio-mixing type="nbest"/>
      <subscribe>
         <active-talkers-sub interval="5"/>
      </subscribe>
   </createconference>
</mscmixer>
```

The request specifies that the conference is assigned the conference id "conf1" and is configured with two reserved talkers, three reserved listener slots, audio-mixing policy set to nbest, and active talkers notifications set to 5 seconds.

If the MS is able to create this conference mixer, it sends a 200 response:

```
<mscmixer version="1.0" xmlns="urn:ietf:params:xml:ns:msc-mixer">
   <response status="200" reason="conference created"
         conferenceid="conf1"/>
</mscmixer>
```

The AS is now able to join connections to the conference as participants. A participant able to contribute to the audio mix would be joined as follows:

```
<mscmixer version="1.0" xmlns="urn:ietf:params:xml:ns:msc-mixer">
   <join id1="1536067209:913cd14c" id2="conf1">
      <stream media="audio" direction="sendrecv"/>
   </join>
</mscmixer>
```

If the MS can join the participant 1536067209:913cd14c to the conference conf1 with audio in both directions, then it sends a successful response:

```
<mscmixer version="1.0" xmlns="urn:ietf:params:xml:ns:msc-mixer">
   <response status="200" reason="join successful"/>
</mscmixer>
```

The AS could also join listener-only participants to the conference by setting the stream direction to receive only:

```
<mscmixer version="1.0" xmlns="urn:ietf:params:xml:ns:msc-mixer">
   <join id1="9936067209:914cd14c" id2="conf1">
      <stream media="audio" direction="recvonly"/>
   </join>
</mscmixer>
```

If the MS can join the participant 9936067209:914cd14c to the conference conf1, then it will send a successful response (not shown). As the active talker changes, the MS sends an active talker notification to the AS:

```
<mscmixer version="1.0" xmlns="urn:ietf:params:xml:ns:msc-mixer">
   <event>
      <active-talkers-notify conferenceid="conf1">
         <active-talker connectionid="1536067209:913cd14c"/>
      </active-talkers-notify>
   </event>
</mscmixer>
```

The AS could decide to change the status of a talker connection so that they can only listen:

```
<mscmixer version="1.0" xmlns="urn:ietf:params:xml:ns:msc-mixer">
   <modifyjoin id1="1536067209:913cd14c" id2="conf1">
      <stream media="audio" direction="recvonly"/>
   </modifyjoin>
</mscmixer>
```

where the participant 1536067209:913cd14c is no longer able to contribute to the audio mix on the conference. If the MS is able to execute this request, it will send a 200 response. The AS could decide to remove this participant from the conference:

```
<mscmixer version="1.0" xmlns="urn:ietf:params:xml:ns:msc-mixer">
   <unjoin id1="1536067209:913cd14c" id2="conf1"/>
</mscmixer>
```

Again, if the MS can execute this request, a 200 response will be sent. Finally, the AS terminates the conference:

```
<mscmixer version="1.0" xmlns="urn:ietf:params:xml:ns:msc-mixer">
   <destroyconference conferenceid="conf1"/>
</mscmixer>
```

If the MS is able to destroy the conference conf1, it sends a 200 response:

```
<mscmixer version="1.0" xmlns="urn:ietf:params:xml:ns:msc-mixer">
   <response status="200" conferenceid="conf1"/>
</mscmixer>
```

For each participant attached to the conference when it is destroyed, the MS sends an unjoin notification event:

```
<mscmixer version="1.0" xmlns="urn:ietf:params:xml:ns:msc-mixer">
   <event>
      <unjoin-notify status="2" id1="9936067209:914cd14c"
            id2="conf1"/>
   </event>
</mscmixer>
```

And the MS sends a conferenceexit notification event when the conference finally exits:

```
<mscmixer version="1.0" xmlns="urn:ietf:params:xml:ns:msc-mixer">
   <event>
      <conferenceexit status="0" conferenceid="conf1"/>
   </event>
</mscmixer>
```

9.6.2.2 Bridging Connections

The mixer package can be used to join connections to one another. In a call-center scenario, for example, this package can be used to set up and modify connections between a caller, agent, and supervisor. A caller is joined to an agent with bidirectional audio:

```
<mscmixer version="1.0" xmlns="urn:ietf:params:xml:ns:msc-mixer">
   <join id1="caller:001" id2="agent:002">
      <stream media="audio" direction="sendrecv"/>
   </join>
</mscmixer>
```

If the MS is ablc to cstablish this connection, then it will send a 200 response:

```
<mscmixer version="1.0" xmlns="urn:ietf:params:xml:ns:msc-mixer">
   <response status="200"/>
</mscmixer>
```

Now, assume that the AS wants a supervisor to listen into the agent conversation with the caller and provide whispered guidance to the agent. First, the AS would send a request to join the supervisor and the caller connections:

```
<mscmixer version="1.0" xmlns="urn:ietf:params:xml:ns:msc-mixer">
   <join id1="supervisor:003" id2="caller:001">
      <stream media="audio" direction="recvonly"/>
   </join>
</mscmixer>
```

If this request was successful, audio output from the caller connection would now be sent to both the agent and the supervisor. Second, the AS would send a request to join the supervisor and the agent connections:

```
<mscmixer version="1.0" xmlns="urn:ietf:params:xml:ns:msc-mixer">
   <join id1="supervisor:001" id2="agent:002">
      <stream media="audio" direction="sendrecv"/>
   </join>
</mscmixer>
```

If this request was successful, the audio mixing would occur on both the agent and supervisor connections: the agent would hear the caller and the supervisor, and the supervisor would hear the agent and the caller. The caller would only hear the agent. If the MS is unable to join and mix connections in this way, it will send a 426 response.

9.6.2.3 Video Conferencing

In this example, an audio-video conference is created, where the loudest participant has the most prominent region in the video layout. The AS sends a request to create an audio-video conference:

```
<mscmixer version="1.0" xmlns="urn:ietf:params:xml:ns:msc-mixer">
   <createconference conferenceid="conf2">
      <audio-mixing type="nbest"/>
      <video-layouts>
         <video-layout min-participants="1">
            <single-view/>
         </video-layout>
         <video-layout min-participants="2">
            <dual-view/>
         </video-layout>
         <video-layout min-participants="3">
            <quad-view/>
         </video-layout>
         <video-layout min-participants="5">
            <multiple-5x1/>
         </video-layout>
      </video-layouts>
      <video-switch><vas/></video-switch>
   </createconference>
</mscmixer>
```

In this configuration, the conference uses an nbest audio mixing policy and a <vas/> video-switching policy, so that the loudest speaker receives the most prominent region in the layout. Multiple video layouts are specified, and the active one depends on the number of participants. Assume that four participants are already joined to the conference. In that case, the video layout will be quad-view (Figure 9.6), with the most active speaker displayed in region 1. When a fifth participant joins, the video layout automatically switches to a multiple-5×1 layout (Figure 9.9), again with the most active speaker in region 1. The AS can manipulate which participants are displayed in the remaining regions. For example, it could force an existing conference participant to be displayed in region 2:

```
<mscmixer version="1.0" xmlns="urn:ietf:params:xml:ns:msc-mixer">
   <modifyjoin id1="1536067209:913cd14c" id2="conf2">
      <stream media="video">
         <region>2</region>
      </stream>
   </modifyjoin>
</mscmixer>
```

9.7 SECURITY CONSIDERATIONS

As this Control Package processes XML markup, implementations MUST address the security considerations of RFC 3023. As a Control Package of the Media Control Channel Framework, security, confidentiality, and integrity of messages transported over the Control Channel MUST be addressed as described in Section 8.12 (Section 12 of the Media Control Channel Framework – RFC 6230), including transport level protection, Control Channel policy management, and session establishment. In addition, implementations MUST address the security, confidentiality, and integrity of user agent sessions with the MS in terms of both SIP signaling and the associated RTP media flow; RFC 6230 (see Chapter 8) provides further details on this topic.

Adequate transport protection and authentication are critical, especially when the implementation is deployed in open networks. If the implementation fails to correctly address these issues, it risks exposure to malicious attacks, including (but not limited to):

Denial of Service: An attacker could insert a request message into the transport stream causing specific conferences or join mixers on the MS to be destroyed: for example, <destroyconference conferenceid="XXXX">, where the value of "XXXX" could be guessed or discovered by auditing active mixers on the MS using an <audit> request. Likewise, an attacker could impersonate the MS and insert error responses into the transport stream, thereby denying the AS access to package capabilities.

Resource Exhaustion: An attacker could insert into the Control Channel new request messages (or modify existing ones) with, for instance, <createconference> elements causing large numbers of conference mixer resources to be allocated. At some point, this will exhaust the number of conference mixers that the MS is able to allocate. The Media Control Channel Framework permits additional policy management (beyond that specified for the Media Control Channel Framework), including resource access and Control Channel usage, to be specified at the Control Package level as defined in Section 8.12.3 (Section 12.3 of RFC 6230).

Since the creation of conference and join mixers is associated with media-mixing resources on the MS, the security policy for this Control Package needs to address how such mixers are securely managed across more than one Control Channel. Such a security policy is only useful for secure, confidential, and integrity-protected channels. The identity of Control Channels is determined by the channel identifier, i.e., the value of the "cfw-id" attribute in the SDP and Dialog-ID header in the channel protocol (RFC 6230 – see Chapter 8). Channels are the same if they have the same identifier; otherwise, they are different. This Control Package imposes the following additional security policies:

- **Responses:** The MS MUST only send a response to a mixer management or audit request using the same Control Channel as the one used to send the request.
- **Notifications:** The MS MUST only send notification events for conference and join mixers using the same Control Channel by which it received the request creating the mixer.
- **Auditing:** The MS MUST only provide audit information about conference and join mixers that have been created on the same Control Channel as the one upon which the <audit> request is sent. For example, if a join between two connections has been created on one channel, then a request on another channel to audit all mixers – <audit mixers="true"/> – would not report on this join mixer.
- **Rejection:** The MS SHOULD reject requests to audit or manipulate an existing conference or join mixer on the MS if the channel is not the same as the one used when the mixer was created. The MS rejects a request by sending a Control Framework 403 response in Sections 8.7.4 and 8.12.3 (Sections 7.4 and 12.3 of RFC 6230), respectively. For example, if a channel with identifier "cfw1234" has been used to send a request to create a particular conference, and the MS receives on channel "cfw98969" a request to audit or destroy this particular conference, then the MS sends a Control Framework 403 response. There can be valid reasons why an implementation does not reject an audit or mixer manipulation request on a different channel from the one that created the mixer. For example, a system administrator might require a separate channel to audit mixer resources created by system users and to terminate mixers consuming excessive system resources. Alternatively, a system monitor or resource broker might require a separate channel to audit mixers managed by this package on a MS. However, the full implications need to be understood by the implementation and carefully weighed before accepting these reasons as valid. If the reasons are not valid in their particular circumstances, the MS rejects such requests.

There can also be valid reasons for "channel handover," including high availability support or when one AS needs to take over management of mixers after the AS that created them has failed. This could be achieved by the Control Channels using the same channel identifier, one after another.

For example, assume a channel is created with the identifier "cfw1234," and the channel is used to create mixers on the MS. This channel (and associated SIP dialog) then terminates due to a failure on the AS. As permitted by the Control Framework, the channel identifier "cfw1234" could then be reused so that another channel is created with the same identifier "cfw1234," allowing it to "take over" management of the mixers on the MS. Again, the implementation needs to understand the full implications and carefully weigh them before accepting these reasons as valid. If the reasons are not valid for their particular circumstances, the MS uses the appropriate SIP mechanisms to prevent session establishment when the same channel identifier is used in setting up another Control Channel, described in Section 8.4 (Section 4 of RFC 6230).

9.8 IANA CONSIDERATIONS

Per this specification, IANA has registered a new Media Control Channel Framework Package, a new XML namespace, a new XML schema, and a new MIME type. IANA has further created a new registry for the response codes for the MEDIACTRL Mixer Control Package, RFC 6505. Interested readers are requested to see RFC 6505 for details.

9.9 SUMMARY

In this section, we have described the media mixture control package for the CFW (RFC 6230 – see Chapter 8) that is using SIP protocol. We have also explained how this package meets the CFW requirements. The control package manages mixers for media conferences and connections, where a media conference is defined as the media-related mixing resources and services for conferences provided by an MS acting as the Control Server. The elements of the media mixture package for the XCON conference are defined. In addition, the syntax and semantics of defined elements, including mixer management and audit elements, are specified in complete detail. An XML schema for these elements is defined, including their extensibility by allowing attributes and elements from other namespaces. In addition, examples of the media mixture package usage are provided: (a) AS-MS Framework Interaction that includes Creation of a Conference Mixer and Joining, a Participant Receiving Active Talker Notifications and Conference Termination and (b) Media Mixing for Audio Conferencing, Bridging Connections, and Video Conferencing. Finally, we have explained important security considerations such as Denial of Service (DoS), Resource Exhaustion, and Security Policies for use of this media mixture control package.

9.10 PROBLEMS

1. What is the media mixture control package? Why does the XCON conference system need it? Explain in detail.
2. What are the specific CFW requirements that the media mixture control package need to satisfy? Explain in detail.
3. How does the MS act as the Control Server for the media-related mixing resources and services for the conference?
4. What are the specific elements of the media mixture control package for the XCON conference defined? Narrate each element in detail.
5. Describe in detail the syntax and semantics of each element of the media mixture control package.
6. What are the mixer management and audit elements of the media mixture control package? Describe their syntax and semantics in detail.
7. What are the specific features of the example XML schema of the elements of the media mixture control package? Explain how their extensibility has been defined by allowing attributes and elements from other namespaces.

8. Explain the media mixture package usage for the following scenarios: (a) AS-MS Framework Interaction that includes Creation of a Conference Mixer and Joining, a Participant Receiving Active Talker Notifications and Conference Termination and (b) Media Mixing for Audio Conferencing, Bridging Connections, and Video Conferencing.
9. How does the media mixture control package take care of the DoS and Resource Exhaustion security attacks? What are the security policies that need to be used for prevention of these attacks? Explain in detail.

REFERENCES

1. Bray, T. et al. (2008, November). Extensible Markup Language (XML) 1.0 (Fifth Edition). *World Wide Web Consortium Recommendation REC-xml-20081126.* <http://www.w3.org/TR/2008/REC-xml-20081126>.
2. IANA. RTP Payload Types. <http://www.iana.org/assignments/rtp-parameters>.
3. Biron, P. and A. Malhotra. (2004 October). XML Schema Part 2: Datatypes Second Edition. *W3C Recommendation.*
4. IANA. MIME Media Types. <http://www.iana.org/assignments/media-types>.

10 Media Session Recording

10.1 USE CASES AND REQUIREMENTS FOR SIP-BASED MEDIA RECORDING

10.1.1 Introduction

Session recording (RFCs 6341 – see Section 10.1, 7245 – see Section 10.2, 7865 – see Section 10.3, and 7866 – see Section 10.4) is a critical operational requirement in many businesses, especially where voice is used as a medium for commerce and customer support. A prime example where voice is used for trade is the financial industry. The call recording requirements in this industry are quite stringent. The recorded calls are used for dispute resolution and compliance. Other businesses, such as customer support call centers, typically employ call recording for quality control or business analytics, with different requirements.

Depending on the country and its regulatory requirements, financial trading floors typically must record all calls. In contrast, call centers typically only record a subset of the calls, and calls must not fail, regardless of the availability of the recording device. Respecting the privacy rights and wishes of users engaged in a call is of paramount importance. In many jurisdictions, participants have a right to know that the session is being recorded or might be recorded, and they have a right to opt out, either by terminating the call or by demanding that the call not be recorded.

Therefore, this specification contains requirements for being able to notify users that a call is being recorded and for users to be able to request that a call not be recorded. Use cases where users participating in a call are not informed that the call is or might be recorded are outside the scope of this specification. In particular, lawful intercept is outside the scope of this specification.

Furthermore, a one-size-fits-all model will not fit all markets, where the scale and cost burdens vary widely and where needs differ for such solution capabilities as media injection, transcoding, and security. If a standardized solution supports all the requirements from every recording market, but doing so would be expensive for markets with lesser needs, then proprietary solutions for those markets will continue to propagate. Care must be taken, therefore, to make a standards-based solution support optionality and flexibility.

This specification specifies requirements for using session initiation protocol (SIP) (RFC 3261 – also see Section 1.2) between a Session Recording Client (SRC) and a Session Recording Server (SRS) to control the recording of media that has been transmitted in the context of a Communication Session (CS). This is termed SIP-based Media Recording (SIPREC). A CS is the "call" between participants. The SRC is the source of the recorded media. The SRS is the sink of recorded media. It should be noted that the requirements for the protocol between an SRS and an SRS have very similar requirements (such as codec and transport negotiation, encryption key interchange, and firewall traversal) as compared with regular SIP media sessions. The choice of SIP for session recording provides reuse of an existing protocol. The recorded sessions can be any real-time transport protocol (RTP) media sessions, including voice, dual-tone multifrequency (DTMF) (as defined by RFC 4733), video, and text (as defined by RFC 4103).

An archived session recording is typically comprised of the CS media content and the CS Metadata. The CS Metadata allows recording archives to be searched and filtered at a later time and allows a session to be played back in a meaningful way, e.g., with correct synchronization between the media. The CS Metadata needs to be conveyed from the SRC to the SRS.

This specification only considers active recording, where the SRC purposefully streams media to an SRS. Passive recording, where a recording device detects media directly from the network, is outside the scope of this specification.

10.1.2 Requirements Notation

See RFC 2119.

10.1.3 Definitions

(Also see Section 1.6, Table 1.2.)

- Session Recording Server (SRS): An SRS is an SIP user agent (UA) that is a specialized Media Server (MS) or collector that acts as the sink of the recorded media. An SRS is typically implemented as a multi-port device that is capable of receiving media from multiple sources simultaneously. An SRS is the sink of the recorded session metadata.
- Session Recording Client (SRC): An SRC is an SIP UA that acts as the source of the recorded media, sending it to the SRS. An SRC is a logical function. Its capabilities may be implemented across one or more physical devices. In practice, an SRC could be a personal device (such as an SIP phone), an SIP Media Gateway (MG), a Session Border Controller (SBC), or an SIP MS integrated with an Application Server (AS). This specification defines the term "SRC" such that all such SIP entities can be generically addressed under one definition. The SRC provides metadata to the SRS.
- Communication Session (CS): A session created between two or more SIP UAs that is the subject of recording.
- Recording Session (RS): The SIP session created between an SRC and an SRS for the purpose of recording a CS.

Figure 10.1 pictorially represents the relationship between an RS and a CS.

Metadata: Information that describes recorded media and the CS to which they relate.

Pause and Resume during a CS:

- Pause: The action of temporarily discontinuing the transmission and collection of RS media.
- Resume: The action of recommencing the transmission and collection of RS media.

Most security-related terms in this specification are to be understood in the sense defined in RFC 4949; such terms include, but are not limited to, "authentication," "confidentiality," "encryption," "identity," and "integrity."

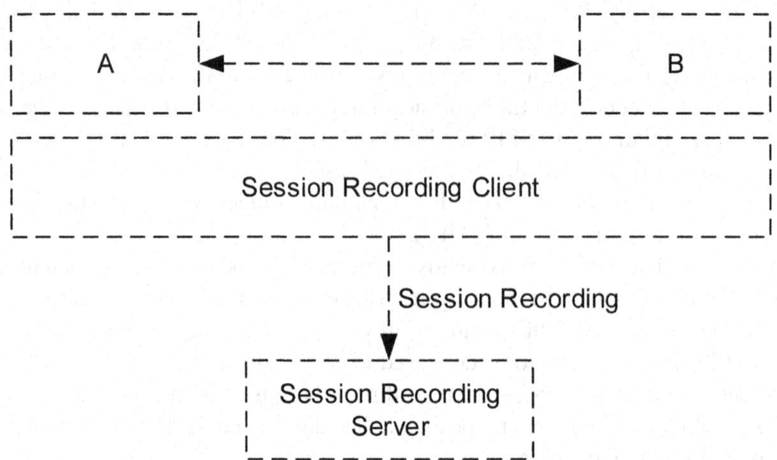

FIGURE 10.1 Media Session Recording.

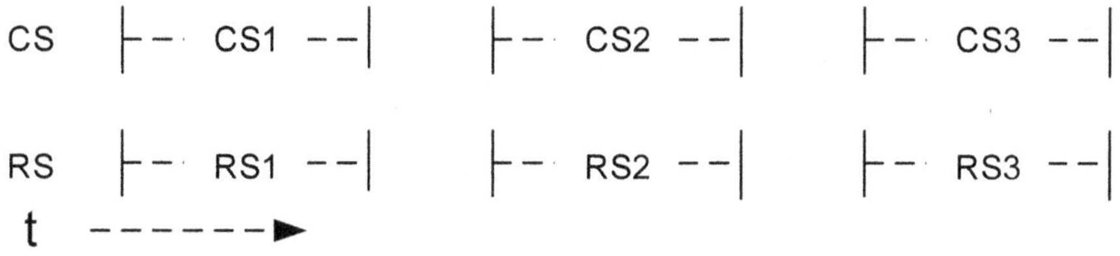

```
CS   |-- CS1 --|    |-- CS2 --|    |-- CS3 --|

RS   |-- RS1 --|    |-- RS2 --|    |-- RS3 --|
t    ------►
```

FIGURE 10.2 Lifecycle of the CSs and the Relationship to the RSs.

10.1.4 Use Cases

Use Case 1: Full-time Recording: One Recording Session for each Communication Session.

For example, Figure 10.2 shows the life cycle of CSs and their relationship to the RSs.

Record every CS for each specific extension/person. The need to record all calls is typically due to business process purposes (such as transaction confirmation or dispute resolution) or to ensure compliance with governmental regulations. Applications include enterprise, contact center, and financial trading floors. This is also commonly known as Total Recording.

Use Case 2: Selective Recording: Start a Recording Session when a Communication Session to be recorded is established.

In this example (Figure 10.3), CSs 1 and 3 are recorded, but CS 2 is not.

Use Case 3: Start/Stop a Recording Session during a Communication Session.

The RS (Figure 10.4) starts during a CS, either manually via a user-controlled mechanism (e.g., a button on a user's phone) or automatically via an application (e.g., a contact center customer service application) or business event. An RS ends either during the CS or when the CS ends. One or more RSs may record each CS.

Use Case 4: Persistent Recording: A single Recording Session captures one or more Communication Sessions.

An RS records continuously without interruption (Figure 10.5). Periods when there is no CS in progress must be reproduced upon playback (e.g., by recording silence during such periods, or by not recording such periods but marking them by means of metadata for utilization on playback, etc.). Applications include financial trading desks and emergency (first-responder) service bureaus. The length of a Persistent RS is independent of the length

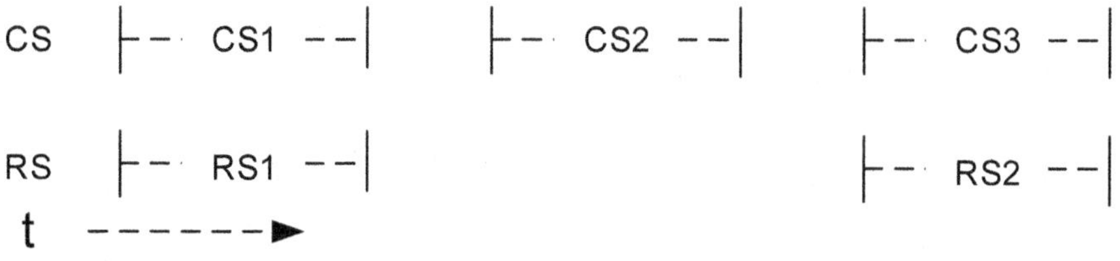

```
CS   |-- CS1 --|    |-- CS2 --|    |-- CS3 --|

RS   |-- RS1 --|                   |-- RS2 --|
t    ------►
```

FIGURE 10.3 CSs 1 and 3 Recorded But Not CS2.

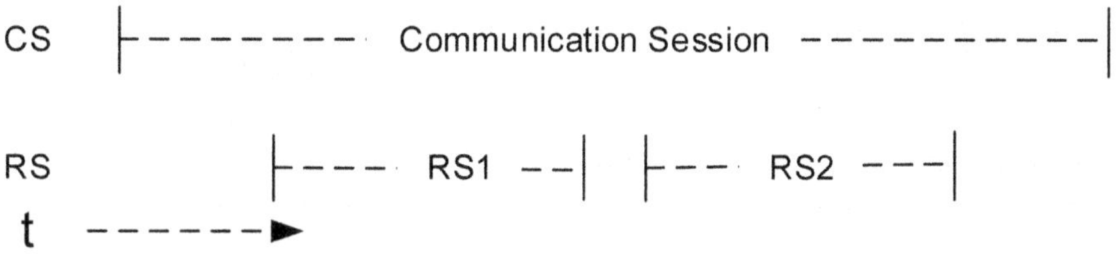

```
CS   |--------  Communication Session  ----------|

RS            |--- RS1 --|  |--- RS2 ---|
t    ------►
```

FIGURE 10.4 Start/Stop of an RS during a CS.

CS CS1 CS2 CS3

RS Recording Session

t

FIGURE 10.5 Persistent Recording Capturing Multiple CSs.

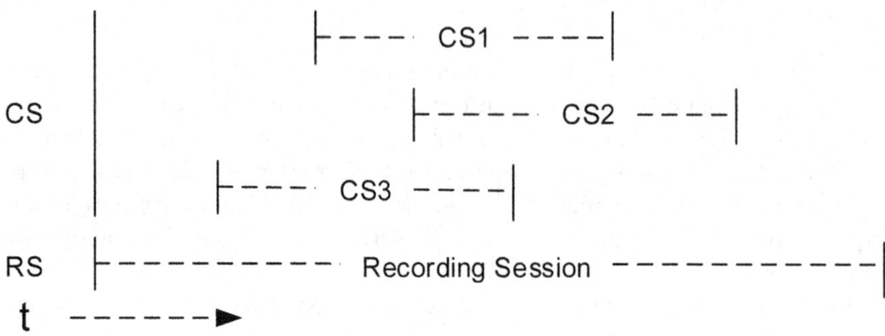

FIGURE 10.6 Mixing of Multiple Concurrent Sessions into One Recording Session.

of the actual CSs. Persistent RSs avoid issues such as media clipping that can occur due to delays in RS establishment.

The connection and attributes of media in the RS are not dynamically signaled for each CS before it can be recorded; however, codec renegotiation is possible.

In some cases, more than one concurrent CS (Figure 10.6) (on a single end-user apparatus, e.g., trading-floor turret) is mixed into one RS.

Use Case 5: Real-time Recording Controls.

For an active RS, privacy or security reasons may demand not capturing a specific portion of a conversation. An example is for PCI (payment card industry) compliance, where credit card information must be protected. One solution is not to record a caller speaking their credit card information.

An example of a real-time control is Pause/Resume.

Use Case 6: Interactive Voice Response (IVR)/Voice Portal Recording.

Self-service IVR applications may need to be recorded for application performance tuning or to meet compliance requirements.

Metadata about an IVR session recording must include session information and may include application context information (e.g., VoiceXML session variables, dialog names, etc.).

Use Case 7: Enterprise Mobility Recording.

Many agents and enterprise workers whose calls are to be recorded are not located on company premises.

Examples:

- Home-based agents or enterprise workers.
- Mobile phones of knowledge workers (e.g., insurance agents, brokers, or physicians) when they conduct work-related (and legally required recording) calls.

Use Case 8: Geographically distributed or centralized recording.

Enterprises such as banks, insurance agencies, and retail stores may have many locations, possibly up to thousands of small sites. Frequently, only phones and network infrastructure are installed in branches without local recording services. In cases where calls inside or between branches must be recorded, a centralized recording system in data centers together with telephony infrastructure (e.g., Private Branch Exchange (PBX)) may be deployed.

Use Case 9: Record complex call scenarios.

The following is an example of a scenario where one call that is recorded must be associated with a related call that also must be recorded.

- A Customer is in a conversation with a Customer Service Agent.
- The Agent puts the Customer on hold in order to consult with a Supervisor.
- The Agent enters into a conversation with the Supervisor.
- The Agent disconnects from the Supervisor, then reconnects with the Customer.
- The Supervisor call must be associated with the original Customer call.

Use Case 10: High availability and continuous recording.

Specific deployment scenarios present different requirements for system availability, error handling, etc., including the following:

- An SRS must always be available at call setup time.
- No loss of media recording can occur, including during failure of an SRS.
- The CS must be terminated (or suitable notification given to parties) in the event of a recording failure.

Use Case 11: Record multi-channel, multimedia session.

Some applications require the recording of more than one media stream, possibly of different types. Media are synchronized either at storage or at playback. Speech analytics technologies (e.g., word spotting, emotion detection, and speaker identification) may require speaker-separated recordings for optimum performance. Multi-modal contact centers may include audio, video, instant messaging (IM), or other interaction modalities.

In trading-floor environments, in order to minimize storage and recording system resources, it may be preferable to mix multiple concurrent calls (Communication Sessions) on different handsets/speakers on the same turret into a single recording session.

Use Case 12: Real-time media processing.

It must be possible for an SRS to support real-time media processing, such as speech analytics of trading-floor interactions. Real-time analytics may be employed for automatic intervention (stopping interaction or alerting) if, for example, a trader is not following regulations. Speaker separation is required in order to reliably detect who is saying specific phrases.

10.1.5 Requirements

The following are requirements for SIP-based Media Recording:

- REQ-001: The mechanism MUST provide a means for using the SIP protocol for establishing, maintaining, and terminating RSs between an SRC and an SRS.
- REQ-002: The mechanism MUST support the ability to record all CSs in their entirety.
- REQ-003: The mechanism MUST support the ability to record selected CSs in their entirety, according to policy.
- REQ-004: The mechanism MUST support the ability to record selected parts of selected CSs.
- REQ-005: The mechanism MUST support the ability to record a CS without loss of media of RS (for example, clipping media at the beginning of the CS) due to RS recording preparation and also without impacting the quality or timing of the CS (for example, delaying the start of the CS while preparing for an RS). See Use Case 4 in Section 10.1.4 for more details.
- REQ-006: The mechanism MUST support the recording of IVR sessions.

- REQ-007: The mechanism MUST support the recording of the following RTP media types: voice, DTMF (as defined by RFC 4733), video, and text (as defined by RFC 4103).
- REQ-008: The mechanism MUST support the ability for an SRC to deliver mixed audio streams from multiple CSs to an SRS. Note: A mixed audio stream is where several related CRs are carried in a single RS. A mixed-media stream is typically produced by a mixer function. The RS MAY be informed about the composition of the mixed streams through session metadata.
- REQ-009: The mechanism MUST support the ability for an SRC to deliver mixed audio streams from different parties of a given CS to an SRS.
- REQ-010: The mechanism MUST support the ability to deliver to the SRS multiple media streams for a given CS.
- REQ-011: The mechanism MUST support the ability to pause and resume the transmission and collection of RS media.
- REQ-012: The mechanism MUST include a means for providing the SRS with metadata describing CSs that are being recorded, including the media being used and the identifiers of parties involved.
- REQ-013: The mechanism MUST include a means for the SRS to be able to correlate RS media with CS participant media.
- REQ-014: Metadata format must be agnostic of the transport protocol.
- REQ-015: The mechanism MUST support a means to stop the recording.
- REQ-016: The mechanism MUST support a means for a recording-aware UA involved in a CS to request at session establishment time that the CS should be recorded or should not be recorded, the honoring of such a request being dependent on policy.
- REQ-017: The mechanism MUST support a means for a recording-aware UA involved in a CS to request during a session that the recording of the CS should be started, paused, resumed, or stopped, the honoring of such a request being dependent on policy. Such recording-aware UAs MUST be notified about the outcome of such requests.
- REQ-018: The mechanism MUST NOT prevent the application of tones or announcements during recording or at the start of a CS to support notification to participants that the call is being recorded or may be recorded.
- REQ-019: The mechanism MUST provide a means of indicating to recording-aware UAs whether recording is taking place for appropriate rendering at the user interface.
- REQ-020: The mechanism MUST provide a way for metadata to be conveyed to the SRS incrementally during the CS.
- REQ-021: The mechanism MUST NOT prevent high-availability deployments.
- REQ-022: The mechanism MUST provide means for facilitating synchronization of the recorded media streams and metadata.
- REQ-023: The mechanism MUST provide means for facilitating synchronization among the recorded media streams.
- REQ-024: The mechanism MUST provide means to relate recording and recording controls, such as start/stop/pause/resume, to the wall clock time.
- REQ-025: The mechanism MUST provide means for an SRS to authenticate the SRC on RS initiation.
- REQ-026: The mechanism MUST provide means for an SRC to authenticate the SRS on RS initiation.
- REQ-027: The mechanism MUST include a means for ensuring that the integrity of the metadata sent from the SRC to the SRS is an accurate representation of the original CS metadata.
- REQ-028: The mechanism MUST include a means for ensuring that the integrity of the media sent from the SRC to the SRS is an accurate representation of the original CS media.

- REQ-029: The mechanism MUST include a means for ensuring the confidentiality of the metadata sent from the SRC to the SRS.
- REQ-030: The mechanism MUST provide a means to support RS confidentiality.
- REQ-031: The mechanism MUST support the ability to deliver to the SRS multiple media streams of the same media type (e.g., audio, video). One example is the case of delivering unmixed audio for each participant in the CS.

10.1.6 Privacy Considerations

Respecting the privacy rights and wishes of users engaged in a call is of paramount importance. In many jurisdictions, participants have a right to know that the session is being recorded or might be recorded, and they have a right to opt out, either by terminating the call or by demanding that the call not be recorded. Therefore, this specification contains requirements for being able to notify users that a call is being recorded and for users to be able to request that a call not be recorded. Use cases where users participating in a call are not informed that the call is or might be recorded are outside the scope of this specification. In particular, lawful intercept is outside the scope of this specification.

Requirements for participant notification of recording vary widely by jurisdiction. In a given deployment, not all users will be authorized to stop the recording of a CS (although any user can terminate its participation in a CS). Typically, users within the domain that is carrying out the recording will be subject to policies of that domain concerning whether CSs are recorded. For example, in a call center, agents will be subject to policies of the call center and may or may not have the right to prevent the recording of a CS or part of a CS. Users calling into the call center, on the other hand, will typically have to ask the agent not to record the CS. If the agent is unable to prevent recording, or if the caller does not trust the agent, the only option generally is to terminate the CS.

Privacy considerations also extend to what happens to a recording once it has been created. Typical issues are who can access the recording (e.g., receive a copy of the recording, view the metadata, play back the media, etc.), for what purpose the recording can be used (e.g., for training purposes, for quality control purposes, etc.), and for how long the recording is to be retained before deletion. These are typically policies of the domain that makes the recording rather than policies of individual users involved in a recorded CS, whether those users be in the same domain or in a different domain. Taking the call center example again, agents might be made aware of call center policy regarding retention and use of recordings as part of their employment contract, and callers from outside the call center might be given some information about policy when notified that a CS will be recorded (e.g., through an announcement that says that calls may be recorded for quality purposes).

This specification does not specify any requirements for a user engaged in a CS to be able to dictate policy for what happens to a recording, or for such information to be conveyed from an SRC to an SRS. It is assumed that the SRS has access to policy applicable to its environment and can ensure that recordings are stored and used in accordance with that policy.

10.1.7 Security Considerations

Session recording has substantial security implications for the SIP UAs being recorded, the SRC, and the SRS. For the SIP UAs involved in the CS, the requirements in this specification enable the UA to identify that a CS is being recorded and to request that a given CS not be subject to recording. Since humans do not typically look at or know about protocol signaling such as SIP, and indeed, the SIP session might have originated through a public switched telephone network (PSTN) gateway without any ability to pass on in-signaling indications of recording, users can be notified of recording in the media itself through voice announcements, a visual indicator on the endpoint, or other means. With regard to security implications of the protocol(s), clearly, there is a need for authentication, authorization, and eavesdropping protection for the solution. The SRC needs to know the SRS it is communicating with is legitimate, and vice versa, even if they are in different domains. Both the signaling and the media for the RS need the ability to be authenticated and protected from eavesdropping. Requirements are detailed in Section 10.1.5.

CSs and RSs can require different security levels for both signaling and media, depending on deployment configurations. For some environments, e.g., the SRS and SRC will be collocated in a secure network region, and therefore, the RS will not require the same protection level as a CS that extends over a public network, for example. For other environments, the SRS can be located in a public cloud, for example, and the RS will require a higher protection level than the CS. For these reasons, there is not a direct relationship between the security level of CSs and the security level of RSs.

A malicious or corrupt SRC can tamper with media and metadata relating to a CS before sending the data to an SRS. Also, CS media and signaling can be tampered with in the network prior to reaching an SRC, unless proper means are provided to ensure integrity protection during transmission on the CS. Means for ensuring the correctness of media and metadata emitted by an SRC are outside the scope of this work. Other organizational and technical controls will need to be used to prevent tampering.

10.2 MEDIA RECORDING ARCHITECTURE

10.2.1 Introduction

Session recording is a critical requirement in many communications environments, such as call centers and financial trading. In some of these environments, all calls must be recorded for regulatory, compliance, and consumer protection reasons. Recording of a session is typically performed by sending a copy of a media stream to a recording device. This specification describes architectures for deploying session recording solutions as defined in "Use Cases and Requirements for SIP-Based Media Recording (SIPREC)" (RFC 6341 – see Section 10.1).

However, this section (RFC 7245) describes architectures for deploying session recording solutions based on the use cases and requirements for SIPREC specified in RFC 6341 (see Section 10.1). This specification focuses on how sessions are established between an SRC and the SRS for the purpose of conveying the Replicated Media and Recording Metadata (e.g., identity of the parties involved) relating to the CS.

Once the Replicated Media and Recording Metadata have been received by the SRS, they will typically be archived for retrieval at a later time. The procedures relating to the archiving and retrieval of this information are outside the scope of this specification. This specification only considers active recording, where the SRC purposefully streams media to an SRS. Passive recording, where a recording device detects media directly from the network (e.g., using port-mirroring techniques), is outside the scope of this specification.

In addition, lawful intercept is outside the scope of this specification, which takes account of the Internet Engineering Task Force (IETF) policy on wiretapping (RFC 2804). The RS that is established between the SRC and the SRS uses the normal procedures for establishing INVITE-initiated dialogs as specified in RFC 3261 (also Section 1.2) and uses the session description protocol (SDP) for describing the media to be used during the session, as specified in RFC 4566. However, it is intended that some extensions to SIP (e.g., Headers, Option Tags, etc.) will be defined to support the requirements for media recording. The Replicated Media is required to be sent in real time to the SRS and is not buffered by the SRC to allow for real-time analysis of the media by the SRS.

10.2.2 Definitions

The following terms are defined by RFC 7245 for the SIP-based media recording architecture (also see Section 1.6, Table 1.2):

- Recording-aware User Agent (UA): An SIP UA that is aware of SIP extensions associated with the CS. Such extensions may be used to notify the recording-aware UA that a session is being recorded, or by a recording-aware UA to express preferences as to whether a recording should be started, paused, resumed, or stopped.
- Recording-unaware User Agent (UA): An SIP UA that is unaware of SIP extensions associated with the CS. Such a recording-unaware UA will be notified that a session is being recorded or will express preferences as to whether a recording should be started, paused, resumed, or stopped via some other means that is out of scope for the SIP media recording architecture.
- Recording Metadata: The metadata describing the CS that is required by the SRS. This will include, for example, the identities of users that participate in the CS and dialog state. Typically, this metadata is archived with the Replicated Media at the SRS. The recording metadata is delivered in real time to the SRS.

Replicated Media: A copy of the media that is associated with the CS, was created by the SRC, and was sent to the SRS. It may contain all the media associated with the CS (e.g., audio and video) or just a subset (e.g., audio). Replicated Media is part of the Recording Session.

10.2.3 Session Recording Architecture

10.2.3.1 Location of the SRC

This section contains some example session recording architectures showing how the SRC is a logical function that can be located in or split between various physical components.

10.2.3.1.1 Back-to-Back User Agent (B2BUA) Acts as a SRC

An SIP B2BUA that has access to the media to be recorded may act as an SRC (Figure 10.7). The B2BUA may already be aware that a session needs to be recorded before the initial establishment of the CS, or the decision to record the session may occur after the session has been established. If the SRC makes the decision to initiate the RS, then it will do so by sending an SIP INVITE request to

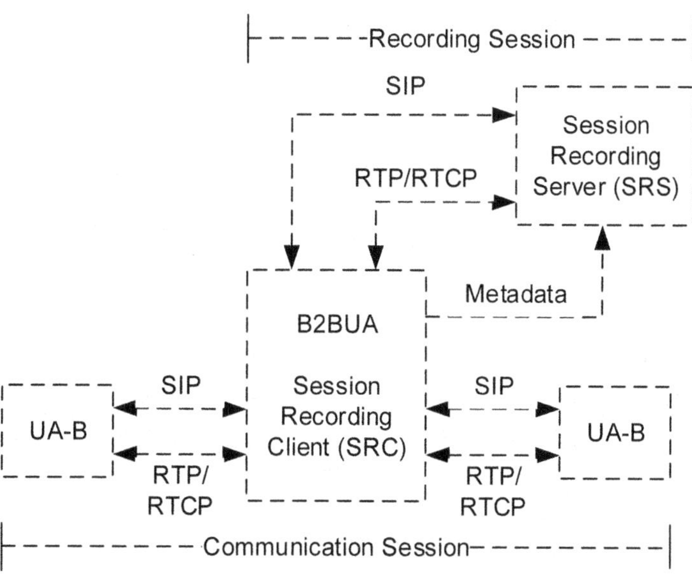

FIGURE 10.7 B2BUA Acts as the Session Recording Client. (Copyright: IETF.)

the SRS. If the SRS makes the decision to initiate the RS, then it will initiate the establishment of a SIP RS by sending an INVITE to the SRC.

The RS INVITE contains information that identifies the session as being established for the purposes of recording and prevents the session from being accidentally rerouted to a UA that is not an SRS if the RS was initiated by the SRC, or vice versa. The B2BUA/SRC is responsible for notifying the UAs involved in the CS that the session is being recorded.

The B2BUA/SRC is responsible for complying with requests from recording aware UAs or through some configured policies indicating that the CS should not be recorded.

10.2.3.1.2 Endpoint Acts as SRC

A SIP endpoint / UA may act as a SRC (Figure 10.8). In that case, the endpoint sends the Replicated Media to the SRS. If the endpoint makes the decision to initiate the RS, then it will initiate the establishment of an SIP Session by sending an INVITE to the SRS. If the SRS makes the decision to initiate the RS, then it will initiate the establishment of an SIP Session by sending an INVITE to the endpoint. The actual decision mechanism is out of scope for the SIP media recording architecture.

10.2.3.1.3 An SIP Proxy Cannot Be an SRC

An SIP Proxy is unable to act as an SRC because it does not have access to the media and therefore has no way of enabling the delivery of the Replicated Media to the SRS.

10.2.3.1.4 Interaction with MEDIACTRL

The MEDIACTRL architecture (RFC 5567 – see Chapter 3) describes an architecture in which an AS controls an MS, which may be used for purposes such as conferencing and recording media streams. In the architecture described in RFC 5567 (see Chapter 3), the AS typically uses SIP third-party call control (3pcc) to instruct the SIP UAs to direct their media to the MS. The SRC or the SRS described in this specification may be architected according to RFC 5567 (see Chapter 3); therefore, when further decomposed, they may be made up of an AS that uses a MEDIACTRL interface to control an MS. As shown in Figure 10.9, when the SRS is architected according to RFC 5567 (see Chapter 3), the MS acts as a sink of the recording media, and the AS acts as a sink of the metadata and the termination point for RS SIP signaling. As shown in Figure 10.10, when the SRC

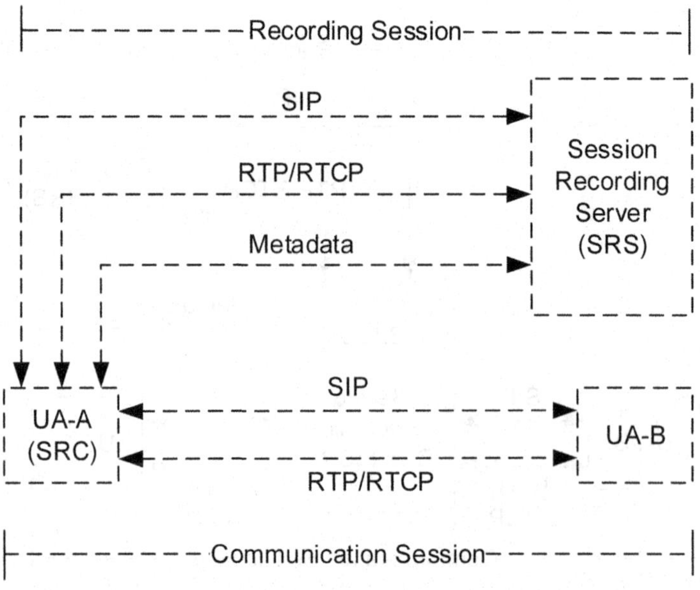

FIGURE 10.8 SIP Endpoint Acts as the Session Recording Client. (Copyright: IETF.)

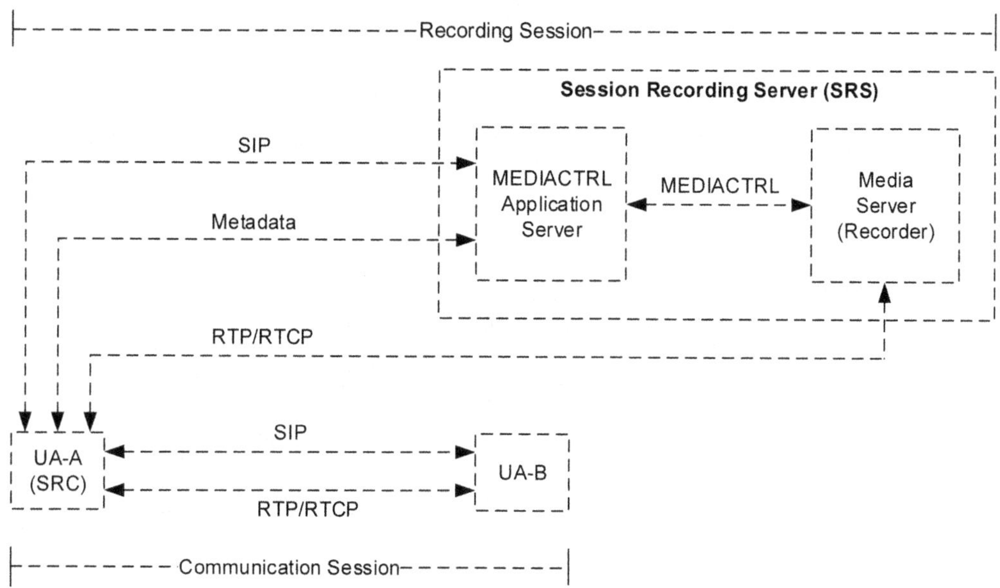

FIGURE 10.9 Example of Session Recording Server Using MEDIACTRL. (Copyright: IETF.)

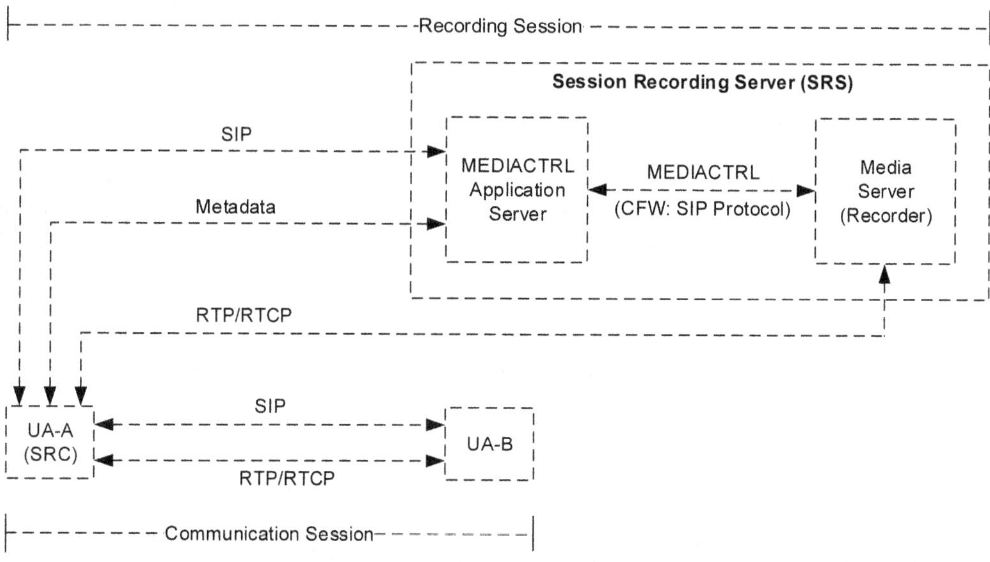

FIGURE 10.10 Example of Session Recording Client Decomposition. (Copyright: IETF.)

is architected according to RFC 5567 (see Chapter 3), the MS acts as a source of recording media, and the AS acts as a source of the metadata and the termination point for RS SIP signaling.

10.2.3.1.5 Interaction with Conference Focus

In the case of a centralized conference, a combination of the conference focus and mixer (RFC 5567 – see Chapter 3) may act as a SRC and therefore provide the SRS with the Replicated Media and associated recording metadata. In this arrangement, the SRC is able to provide media and metadata relating to each of the participants, including, for example, any side conversations where the media passes through the mixer. The conference focus can either provide mixed Replicated Media or separate streams per conference participant (as depicted in Figure 10.11).

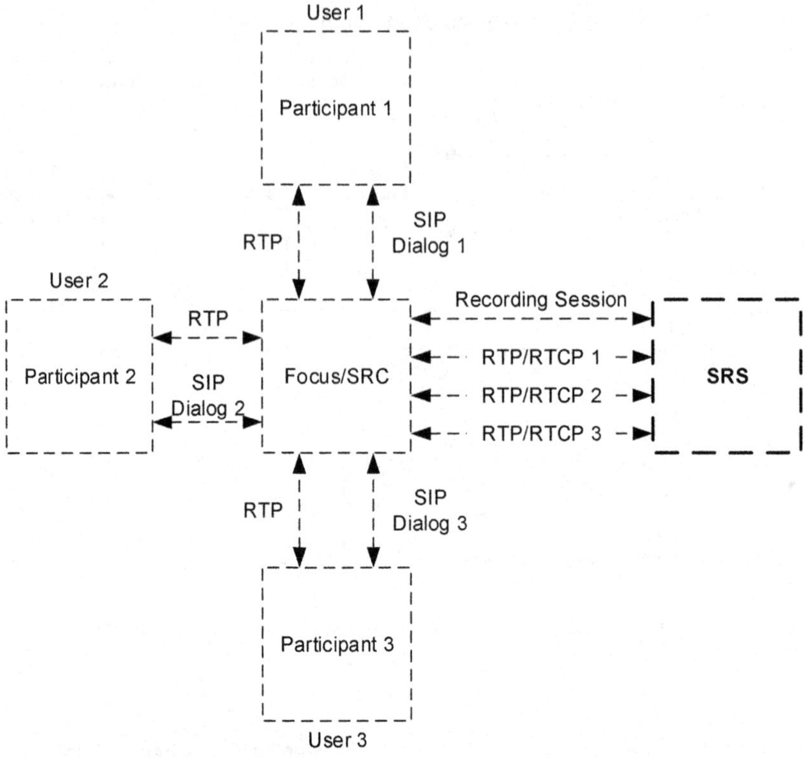

FIGURE 10.11 Conference Focus Acting as an SRC. (Copyright: IETF.)

The conference focus may also act as a recording-aware UA in the case when one of the participants acts as a SRC. In an alternative arrangement, an SIP endpoint that is a conference participant can act as an SRC. The SRC will in this case have access to the media and metadata relating to that particular participant and may be able to obtain additional metadata from the conference focus. The SRC may, for example, use the conference event package as described in RFC 5567 (see Chapter 3) to obtain information about other participants that it provides to the SRS within the recording metadata. The SRC may be involved in the conference from the very beginning or may join at some later point of time.

10.2.3.2 Establishing the Recording Session

The SRC or the SRS may initiate the RS. It should be noted that the RS is independent of the CS that is being recorded at both the SIP dialog level and the session level. Concerning media negotiation, regular SIP/SDP capabilities should be used, and existing transcoding capabilities and media encryption should not be precluded.

10.2.3.2.1 SRC-Initiated Recording

When the SRC initiates the RS for the purpose of conveying media to the SRS, it performs the following actions:

- Is provisioned with a unified resource identifier (URI) for the SRS; the URI is resolved through normal RFC 3263 procedures.
- Initiates the dialog by sending an INVITE request to the SRS. The dialog is established according to the normal procedures for establishing an INVITE-initiated dialog as specified in RFC 3261 (also see Section 1.2).
- Includes in the INVITE an indication that the session is established for the purpose of recording the associated media.

- Includes an SDP attribute of "a=sendonly" for each media line if the Replicated Media is to be started immediately, or includes "a=inactive" if it is not ready to transmit the media.
- Replicates the media streams that are to be recorded and transmits the media to the SRS. The RS may replicate all media associated with the CS or only a subset.

10.2.3.2.2 SRS-Initiated Recording

When the SRS initiates the media RS with the SRC, it performs the following actions:

- Is provisioned with a URI for the SRC; the URI is resolved through normal (RFC 3263) procedures.
- Sends an INVITE request to the SRC.
- Includes in the INVITE an indication that the session is established for the purpose of recording the associated media.
- Identifies the sessions that are to be recorded. The actual mechanism of the identification depends on SRC policy.
- Includes an SDP attribute of "a=recvonly" for each media line if the RS is to be started immediately, or includes "a=inactive" if it is not ready to receive the media. If the SRS does not have prior knowledge of what media streams are available to be recorded, it can make use of an offerless INVITE, which allows the SRC to make the initial SDP offer.

10.2.3.2.3 Pause/Resume Recording Session

The SRS or the SRC may pause the recording by changing the SDP direction attribute to "inactive" and resume the recording by changing the direction back to "recvonly" or "sendonly."

10.2.3.2.4 Media Stream Mixing

In a basic session involving only audio, there are typically two audio/RTP streams between the two UAs involved in transporting media in each direction. When recording this media, the two streams may be mixed or not mixed at the SRC before being transmitted to the SRS. In the case when they are not mixed, two separate streams are sent to the SRS, and the SDP offer sent to the SRS must describe two separate media streams. In the mixed case, a single mixed media stream is sent to the SRS.

10.2.3.2.5 Media Transcoding

The CS and the RS are negotiated separately using the standard SDP offer/answer exchange, which may result in the SRC having to perform media transcoding between the two sessions. If the SRC is not capable of performing media transcoding, it may limit the media formats in the offer to the SRS depending on what media is negotiated on the CS or may limit what it includes in the offer on the CS if it has prior knowledge of the media formats supported by the SRS. However, typically, the SRS will be a more capable device, which can provide a wide range of media format options to the SRC and may also be able to make use of a media transcoder as detailed in RFC 5369.

10.2.3.2.6 Lossless Recording

Session recording may be a regulatory requirement in certain communication environments. Such environments may impose a requirement generally known as "lossless recording." An overall solution for lossless recording may involve multiple layers of solutions. Individual aspects of the solutions may range from administering networks for appropriate quality of service (QOS), reliable transmission of recorded media, and perhaps certain SIPREC protocol-level capabilities in SRC and SRS.

10.2.3.3 Recording Metadata

10.2.3.3.1 Contents of Recording Metadata

The metadata model is defined in RFC 7865 (see Section 10.3).

10.2.3.3.2 Mechanisms for Delivery of Metadata to SRS

The SRS obtains session recording metadata from the SRC. The metadata is transported via SIP-based mechanisms as specified in RFC 7866 (see Section 10.4). It is also possible that metadata is transported via non-SIP-based mechanisms, but these are considered out of scope. It is also possible to have an RS session without the metadata; in that case, the SRS will be receiving the metadata by some other means or not at all.

10.2.3.4 Notifications to the Recorded User Agents

Typically, a user that is involved in a session that is to be recorded is notified by an announcement at the beginning of the session or may receive some warning tones within the media. However, SIPREC enables an indication that the call is being recorded to be included in the SIP requests and responses associated with that CS. The SRC provides the notification to all SIP UAs for which it is replicating received media for the purpose of recording. If the SRC is acting as a SIP endpoint, as described in Section 10.2.3.1.2, then it also provides a notification to the local user.

10.2.3.5 Preventing the Recording of a SIP Session

During the initial session establishment or during an established session, a recording-aware UA may provide an indication of its preference with regard to recording the media in the CS. The mechanisms for this are specified in RFC 7866 (see Section 10.4).

10.2.4 IANA Considerations

This specification has no actions for IANA. This specification mentions SIP/SDP extensions. The associated IANA considerations are addressed in RFC 7866 (see Section 10.4), which defines them.

10.2.5 Security Considerations

The RS is fundamentally a standard SIP dialog and media session and therefore makes use of existing SIP security mechanisms for securing the RS and Recording Metadata. The intended use of this architecture is only for the case where the users are aware that they are being recorded, and the architecture provides the means for the SRC to notify users that they are being recorded. This architectural solution is not intended to support lawful intercept, which in contrast, requires that users are not informed.

It is the responsibility of the SRS to protect the Replicated Media and Recording Metadata once it has been received and archived. The stored content must be protected using a cipher at least as strong as (or stronger than) the original content; however, the mechanism for protecting the storage and retrieval from the SRS is out of scope of this work. The keys used to store the data must also be securely maintained by the SRS and should only be released, securely, to authorized parties. How to secure these keys, properly authorize a receiving party, or securely distribute the keying material is also out of scope of this work.

Protection of the RS should not be weaker than protection of the CS and may need to be stronger because the media is retransmitted (allowing more possibility for interception). This applies to both the signaling and media paths.

It is essential that the SRC will authenticate the SRS, because the client must be certain that it is recording on the right recording system. It is less important that the SRS authenticate the SRC, but implementations must have the ability to perform mutual authentication. In some environments, it is desirable not to decrypt and re-encrypt the media. This means that the same media encryption key is negotiated and used within the CS and RS. If for any reason the media are decrypted on the CS and are re-encrypted on the RS, a new key must be used.

The retrieval mechanism for media recorded by this protocol is out of scope. Implementations of retrieval mechanisms should consider the security implications carefully, as the retriever is not usually a party to the call that was recorded. Retrievers should be authenticated carefully. The cryptosuites on the retrieval should be no less strong than those used on the RS and may need to be stronger.

10.3 MEDIA RECORDING METADATA

10.3.1 Introduction

Session recording is a critical requirement in many communications environments, such as call centers and financial trading organizations. In some of these environments, all calls must be recorded for regulatory, compliance, and consumer protection reasons. The recording of a session is typically performed by sending a copy of a media stream to a recording device. This specification focuses on the recording metadata, which describes the CS. The specification describes a metadata model as viewed by the SRS and the recording metadata format, the requirements for which are described in RFC 6341 (see Section 10.1) and the architecture for which is described in RFC 7245 (see Section 10.2).

10.3.2 Terminology

See Section 1.6.

10.3.3 Definitions

- Metadata model: A metadata model is an abstract representation of metadata using a Unified Modeling Language (UML) [1] class diagram.
- Metadata classes: Each block in the model represents a class. A class is a construct that is used as a blueprint to create instances (called "objects") of itself. The description of each class also has a representation of its attributes in a second compartment below the class name.
- Attributes: Attributes represent the elements listed in each of the classes. The attributes of a class are listed in the second compartment below the class name. Each instance of a class conveys values for the attributes of that class. These values are added to the recording's metadata.
- Linkages: Linkages represent the relationship between the classes in the model. Each linkage represents a logical connection between classes (or objects) in class diagrams (or object diagrams). The linkages used in the metadata model of this specification are associations. This specification also refers to the terminology defined in RFC 6341 (see Section 10.1).

10.3.4 Metadata Model

Metadata is the information that describes recorded media and the CS to which they relate. Figure 10.12 shows a model for metadata as viewed by an SRS.

The metadata model is a class diagram in UML. The model describes the structure of metadata in general by showing the classes, their attributes, and the relationships among the classes. Each block in the model represents a class. The linkages between the classes represent the relationships, which can be associations or compositions. The metadata is conveyed from the SRC to the SRS.

The model allows metadata describing CSs to be communicated to the SRS as a series of snapshots, each representing the state as seen by a single SRC at a particular instant in time. Metadata changes from one snapshot to another reflect changes in what is being recorded. For example, if a participant joins a conference, then the SRC sends the SRS a snapshot of metadata having that participant information (with attributes like (Name, AoR) tuple and associate-time). (Note: "AoR" means "Address-of-Record.")

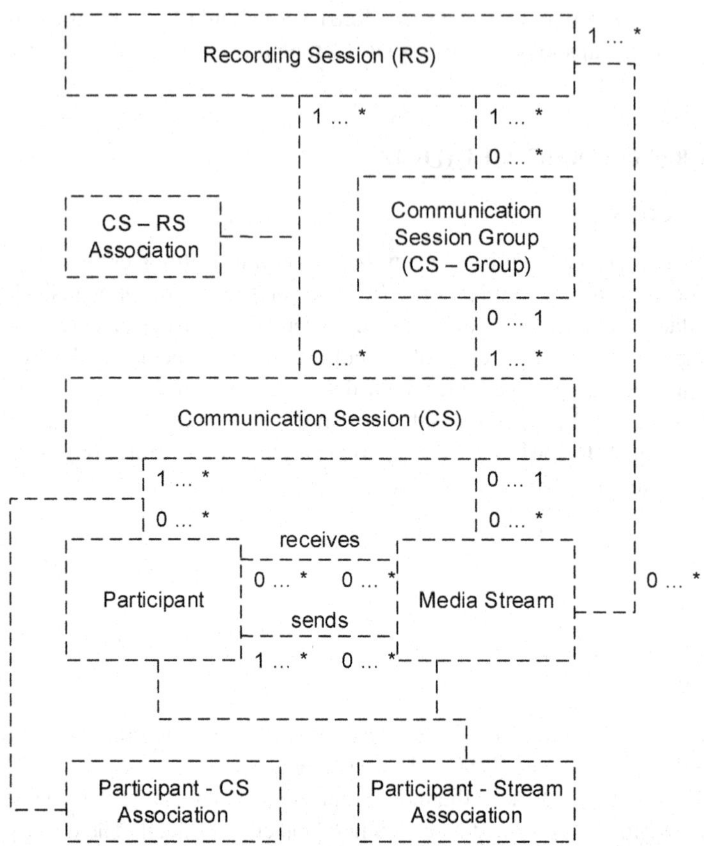

FIGURE 10.12 Model for Metadata as Viewed by SRS. (Copyright: IETF.)

Some of the metadata is not required to be conveyed explicitly from the SRC to the SRS if it can be obtained contextually by the SRS (e.g., from SIP or SDP signaling). For example, the "label" attribute within the "stream" XML element references an SDP "a=label" attribute that identifies an m-line within the RS SDP. The SRS would learn the media properties from the media line.

10.3.5 Recording Metadata Format from SRC to SRS

This section gives an overview of the recording metadata format. Some data from the metadata model is assumed to be made available to the SRS through SDP (RFC 4566), and therefore, this data is not represented in the XML specification format specified in this specification. SDP attributes describe different media formats like audio and video. The other metadata attributes, such as participant details, are represented in a new recording-specific XML specification of type "application/rs-metadata+xml." The SDP "label" attribute (RFC 4574) provides an identifier by which a metadata XML specification can refer to a specific media description in the SDP sent from the SRC to the SRS. The XML specification format can be used to represent either the complete metadata or a partial update to the metadata. The latter includes only elements that have changed compared with the previously reported metadata.

10.3.5.1 XML Data Format

Every recording metadata XML specification sent from the SRC to the SRS contains a "recording" element. The "recording" element acts as a container for all other elements in this XML specification. A "recording" object is an XML specification. It has the XML declaration and contains an encoding declaration in the XML declaration, e.g., "<?xml version="1.0" encoding="UTF-8"?>". If the charset parameter of the MIME content type declaration is present, and it is different from the encoding declaration, the charset parameter takes precedence. Every application conforming to this specification MUST accept the UTF-8 character encoding to ensure minimal interoperability. Syntax and semantic errors in an XML specification should be reported to the originator using application-specific mechanisms.

10.3.5.1.1 Namespace

With the following uniform resource name (URN), this specification defines a new namespace URI for elements defined herein:

urn:ietf:params:xml:ns:recording:1

10.3.5.1.2 "recording" Element

The "recording" element MUST contain an xmlns namespace attribute with a value of urn:ietf:params:xml:ns:recording:1. Exactly one "recording" element MUST be present in every recording metadata XML specification. A "recording" element MAY contain a "dataMode" element indicating whether the XML specification is a complete specification or a partial update. If no "dataMode" element is present, then the default value is "complete."

10.3.6 Recording Metadata Classes

This section describes each class of the metadata model and the attributes of each class. This section also describes how different classes are linked and the XML element for each of them.

10.3.6.1 Recording Session

Each instance of an RS class (Figure 10.13), namely the RS object, represents a SIP session created between an SRC and SRS for the purpose of recording a CS.

The RS object is represented in the XML schema using the "recording" element, which in turn relies on the SIP/SDP session with which the XML specification is associated to provide the attributes of the RS element.

10.3.6.1.1 Attributes

An RS class has the following attributes:

- start-time – Represents the start time of an RS object.
- end-time – Represents the end time of an RS object.

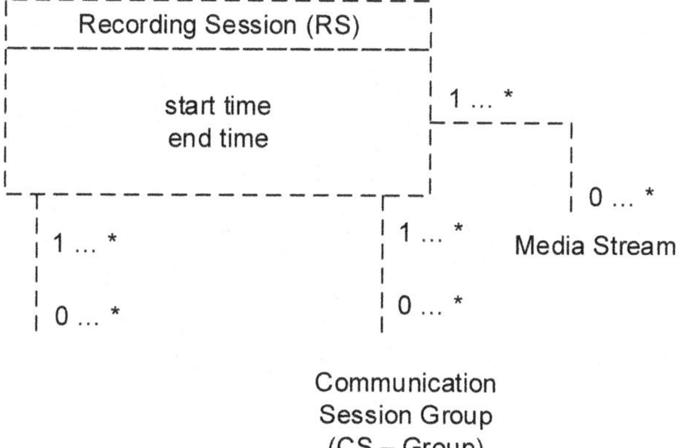

FIGURE 10.13 Recording a Communication Session.

"start-time" and "end-time" attribute values are derivable from the Date header (if present in the SIP message) in the RS. In cases where the Date header is not present, "start-time" is derivable from the time at which the SRS receives the notification of the SIP message to set up the RS, and "end-time" is derivable from the time at which the SRS receives a disconnect on the RS SIP dialog.

10.3.6.1.2 Linkages

Each instance of an RS has:

- Zero or more instances of CS-Groups
- Zero or more instances of CS objects
- Zero or more instances of MediaStream objects

Zero instances of CSs and CS-Groups in a "recording" element are allowed to accommodate persistent recording scenarios. A persistent RS is an SIP dialog that is set up between the SRC and the SRS even before any CS is set up. The metadata sent from the SRC to the SRS when the persistent RS SIP dialog is set up may not have any CS (and the related CS-Group) elements in the XML, as there may not be a session that is associated to the RS yet. For example, a phone acting as an SRC can set up an RS with the SRS, possibly even before the phone is part of a CS. Once the phone joins a CS, the same RS would be used to convey the CS metadata.

10.3.6.2 Communication Session Group

One instance of a CS-Group class (Figure 10.14), namely the CS-Group object, provides association or grouping of all related CSs. For example, in a contact center flow, a call can get transferred to multiple agents.

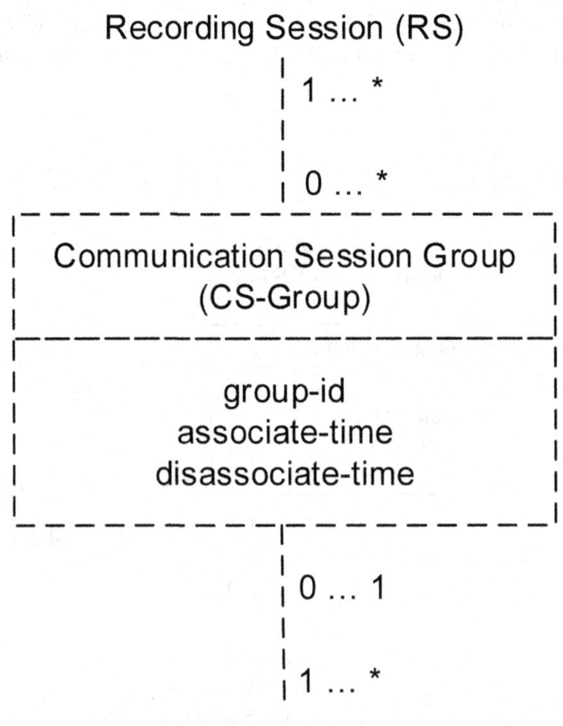

FIGURE 10.14 Recording Session for Communication Session Group (CS-Group). (Copyright: IETF.)

Each of these can trigger the setup of a new CS. In cases where the SRC knows the related CSs, it can group them using the CS-Group element. The CS-Group object is represented in the XML schema using the "group" element.

10.3.6.2.1 Attributes

A CS-Group has the following attributes:

- **group_id** – This attribute groups different CSs that are related. The SRC (or the SRS) is responsible for ensuring the uniqueness of "group_id" in cases where multiple SRCs interact with the same SRS. The mechanism by which the SRC groups the CS is outside the scope of this specification.
- **associate-time** – This is the time when a grouping is formed. The rules that determine how a grouping of different CS objects is done by the SRC are outside the scope of this specification.
- **disassociate-time** – "disassociate-time" for the CS-Group is calculated by the SRC as the time when the grouping ends.

10.3.6.2.2 Linkages

The linkages between a CS-Group class and other classes are associations. A CS-Group is associated with the RS and CS in the following manner:

- There are one or more RS objects per CS-Group.
- Each CS-Group object has to be associated with one or more RSs. Here, each RS can be set up by the potentially different SRCs.
- There are one or more CSs per CS-Group (for example, in cases where the call is transferred). A CS cannot be associated with more than one CS-Group.

10.3.6.3 Communication Session

A CS class (Figure 10.15) and its object in the metadata model represent the CS and its properties as seen by the SRC.

The CS object is represented in the XML schema using the "session" element.

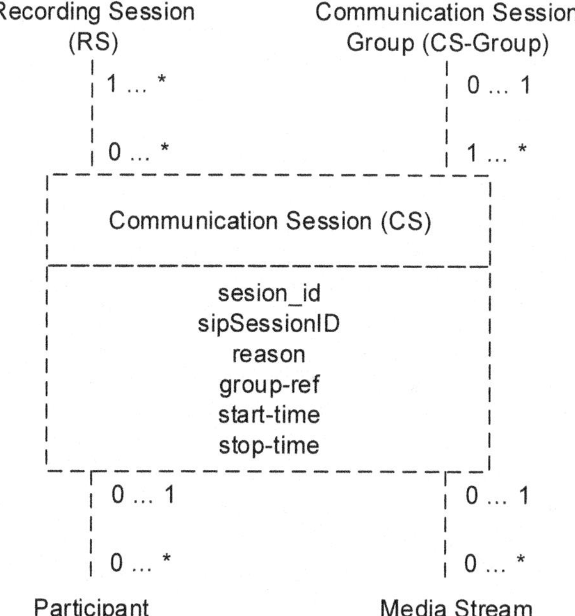

FIGURE 10.15 CS Class and Metadata Model.

10.3.6.3.1 Attributes

A CS class has the following attributes:

- **session_id** – This attribute is used to uniquely identify an instance of a CS object, namely the "session" XML element within the metadata XML specification. "session_id" is generated using the rules mentioned in Section 10.3.6.10.
- **reason** – This represents the reason why a CS was terminated. The value for this attribute is derived from the SIP Reason header RFC 3326 of the CS. There MAY be multiple instances of the "reason" XML element inside a "session" element. The "reason" XML element has "protocol" as an attribute, which indicates the protocol from which the reason string is derived. The default value for the "protocol" attribute is "SIP." The "reason" element can be derived from a SIP Reason header in the CS.
- **sipSessionID** – This attribute carries a SIP Session-ID as defined in RFC 7989 (obsolete RFC 7329). Each CS object can have zero or more "sipSessionID" elements. More than one "sipSessionID" attribute may be present in a CS. For example, if three participants – A, B, and C – are in a conference that has a focus acting as an SRC, the metadata sent from the SRC to the SRS will likely have three "sipSessionID" elements that correspond to the SIP dialogs that the focus has with each of the three participants.
- **group-ref** – A "group-ref" attribute MAY be present to indicate the group (identified by "group_id") to which the enclosing session belongs.
- **start-time** – This optional attribute represents the start time of the CS as seen by the SRC.
- **stop-time** – This optional attribute represents the stop time of the CS as seen by the SRC.

This specification does not specify attributes relating to what should happen to a recording of a CS after it has been delivered to the SRS (e.g., how long to retain the recording or what access controls to apply). The SRS is assumed to behave in accordance with its local policy. The ability of the SRC to influence this policy is outside the scope of this specification. However, if there are implementations where the SRC desires to specify its own policy preferences, this information could be sent as extension data attached to the CS.

10.3.6.3.2 Linkages

A CS is linked to the CS-Group, participant, MediaStream, and RS classes by using the association relationship. The association between the CS and the participant allows the following:

- A CS will have zero or more participants.
- A participant is associated with zero or more CSs. This includes participants who are not directly part of any CS. An example of such a case is participants in a pre-mixed media stream. The SRC may have knowledge of such participants but not have any signaling relationship with them. This might arise if one participant in a CS is a conference focus. To summarize, even if the SRC does not have direct signaling relationships with all participants in a CS, it should nevertheless create a participant object for each participant that it knows about.
- The model also allows participants in a CS that are not participants in the media. An example is the identity of a 3pcc that has initiated a CS to two or more participants in the CS.

Another example is the identity of a conference focus. Of course, a focus is probably in the media, but since it may only be there as a mixer, it may not report itself as a participant in any of the media streams. The association between the CS and the media stream allows the following:

- A CS will have zero or more streams.
- A stream can be associated with at most one CS. A stream in a persistent RS is not required to be associated with any CS before the CS is created, and hence, the zero association is allowed.

The association between the CS and the RS allows the following:

- Each instance of an RS has zero or more instances of CS objects.
- Each CS has to be associated with one or more RSs. Each RS can be potentially set up by different SRCs.

10.3.6.4 CS-RS Association

The CS-RS Association class (Figure 10.16) describes the association of a CS to an RS for a period of time. A single CS may be associated with different RSs (perhaps by different SRCs) and may be associated and dissociated several times.

The CS-RS Association class is represented in XML using the "sessionrecordingassoc" XML element.

10.3.6.4.1 Attributes

The CS-RS Association class has the following attributes:

- associate-time – Associate-time is calculated by the SRC as the time it sees a CS associated to an RS.
- disassociate-time – Disassociate-time is calculated by the SRC as the time it sees a CS disassociate from an RS.
- session_id – Each instance of this class MUST have a "session_id" attribute that identifies the CS to which this association belongs.

10.3.6.4.2 Linkages

The CS-RS Association class is linked to the CS and RS classes.

10.3.6.5 Participant

A participant class (Figure 10.17) and its objects have information about a device that is part of a CS and/or contributes/consumes media stream(s) belonging to a CS.

The participant object is represented in the XML schema using the "participant" element.

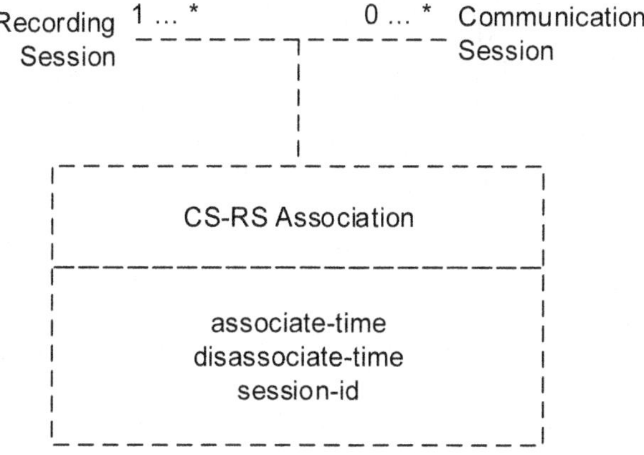

FIGURE 10.16 CS-RS Association Class.

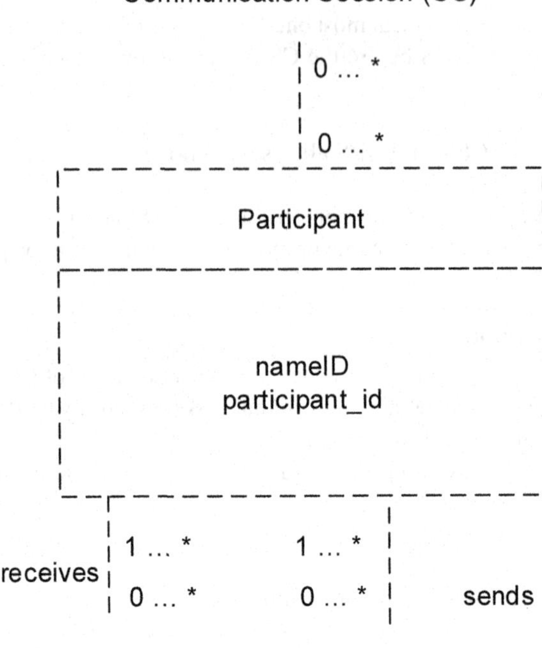

FIGURE 10.17 Participant Class and Its Objects.

10.3.6.5.1 Attributes

A participant class has two attributes:

- **nameID** – This attribute is a list of (Name, AoR) tuples. An AoR of Section 6 of RFC 3261 (also see Section 1.2) can be either a SIP/SIPS/tel URI ("SIPS" means "SIP Secure;" the tel URI is discussed in RFC 3966), a Fully Qualified Domain Name (FQDN), or an Internet protocol (IP) address.
  - For example, the AoR may be drawn from the From header field or the P-Asserted-Identity header (RFC 3325) field. The SRC's local policy is used to decide where to draw the AoR from. The Name parameter represents the participant name (SIP display name) or dialed number (when known). Multiple tuples are allowed for cases where a participant has more than one AoR. For example, a P-Asserted-Identity header can have both SIP and tel URIs.
- **participant_id** – This attribute is used to identify the "participant" XML element within the XML specification. It is generated using the rules mentioned in Section 10.3.6.10. This attribute MUST be used for all references to a participant within a CS-Group and MAY be used to reference the same participant more globally.

This specification does not specify other attributes relating to participants (e.g., participant role or participant type). An SRC that has information regarding these attributes can provide this information as part of extension data to the "participant" XML element from the SRC to the SRS.

10.3.6.5.2 Linkages

The participant class is linked to the MS and CS classes by using an association relationship. The association between the participant and the MS allows the following:

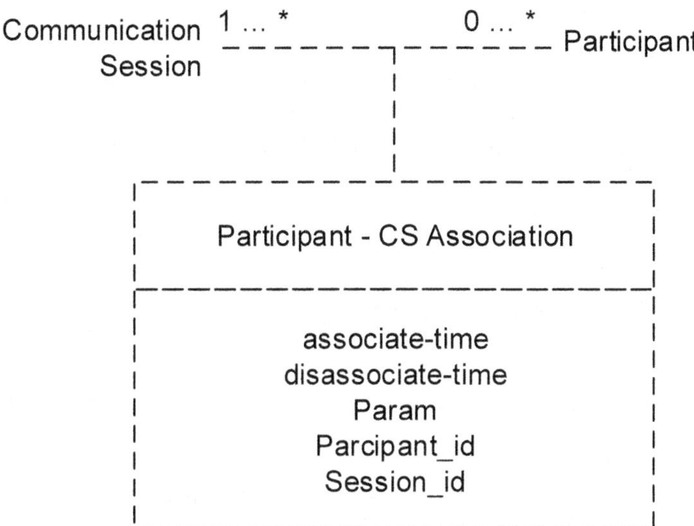

FIGURE 10.18 Participation – CS Association Class.

- A participant will receive zero or more media streams.
- A participant will send zero or more media streams. (The same participant provides multiple streams, e.g., audio and video.)
- A media stream will be received by zero or more participants. It is possible, though perhaps unlikely, that a stream is generated but sent only to the SRC and SRS, not to any participant – for example, in conferencing where all participants are on hold and the SRC is co-located with the focus. Also, a media stream may be received by multiple participants (e.g., "whisper" calls, side conversations).
- A media stream will be sent by one or more participants (pre-mixed streams).

An example of a case where a participant receives zero or more streams is where a supervisor may have a side conversation with an agent while the agent converses with a customer.

10.3.6.6 Participant-CS Association

The Participant-CS Association class (Figure 10.18) describes the association of a participant to a CS for a period of time. A participant may be associated to and dissociated from a CS several times (for example, connecting to a conference, then disconnecting, then connecting again).

The Participant-CS Association object is represented in the XML schema using the "participant-sessionassoc" element.

10.3.6.6.1 Attributes

The Participant-CS Association class has the following attributes:

- **associate-time** – associate-time is calculated by the SRC as the time it sees a participant associated to a CS.
- **disassociate-time** – disassociate-time is calculated by the SRC as the time it sees a participant disassociate from a CS. It is possible that a given participant can have multiple associate times/disassociate times within a given communication session.
- **param** – The capabilities here are those that are indicated in the Contact header as defined in Section 9 of RFC 3840. For example, in a CS (which can be a conference), you can have participants who are playing the role of "focus." These participants do not contribute to media in the CS; however, they switch the media received from one participant to every

other participant in the CS. Indicating the capabilities of the participants (here, "focus") would be useful for the recorder to learn about these kinds of participants. The capabilities are represented using the "param" XML element in the metadata. The "param" XML element encoding defined in RFC 4235 is used to represent the capability attributes in metadata. Each participant may have zero or more capabilities. A participant may use different capabilities, depending on the role it plays at a particular instance – for example, if a participant moves across different CSs (e.g., due to transfer) or is simultaneously present in different CSs with different roles.

- **participant_id** – This attribute identifies the participant to which this association belongs.
- **session_id** – This attribute identifies the session to which this association belongs.

10.3.6.6.2 Linkages

The Participant-CS Association class is linked to the participant and CS classes.

10.3.6.7 Media Stream

A MS class (and its objects) (Figure 10.19) has the properties of media as seen by the SRC and sent to the SRS. Different snapshots of MS objects may be sent whenever there is a change in media (e.g., a direction change, like pause/resume, codec change, and/or participant change).

The MS object is represented in the XML schema using the "stream" element.

10.3.6.7.1 Attributes

An MS class has the following attributes:

- **label** – The "label" attribute within the "stream" XML element references an SDP "a=label" attribute that identifies an m-line within the RS SDP. That m-line carries the media stream from the SRC to the SRS.
- **content-type** – The content of a MS element will be described in terms of the "a=content" attribute defined in Section 5 of RFC 4796. If the SRC wishes to convey the content-type to the SRS, it does so by including an "a=content" attribute with the m-line in the RS SDP.
- **stream_id** – Each "stream" element has a unique "stream_id" attribute that helps to uniquely identify the stream. This identifier is generated using the rules mentioned in Section 10.3.6.10.
- **session_id** – This attribute associates the stream with a specific 'session' element.

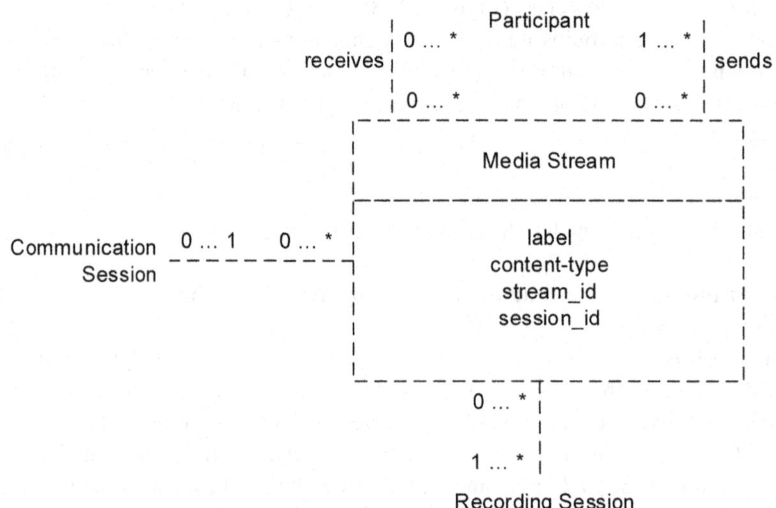

FIGURE 10.19 Media Session Class and Its Objects.

The metadata model can include media streams that are not being delivered to the SRS. For example, an SRC offers audio and video towards an SRS that accepts only audio in response. The metadata snapshots sent from the SRC to the SRS can continue to indicate the changes to the video stream as well.

10.3.6.7.2 Linkages

An MS class is linked to the participant and CS classes by using the association relationship. Details regarding associations with the participant are described in Section 10.3.6.5. Details regarding associations with the CS are mentioned in Section 10.3.6.3.

10.3.6.8 Participant-Stream Association

A Participant-Stream Association class (Figure 10.20) describes the association of a participant to a MS for a period of time, as a sender or as a receiver, or both.

This class is represented in XML using the "participantstreamassoc" element.

10.3.6.8.1 Attributes

A Participant-Stream Association class has the following attributes:

- **associate-time** – This attribute indicates the time a participant started contributing to a MS.
- **disassociate-time** – This attribute indicates the time a participant stopped contributing to a MS.
- **send** – This attribute indicates whether a participant is contributing to a stream or not. This attribute has a value that points to a stream represented by its unique_id. The presence of this attribute indicates that a participant is contributing to a stream. If a participant stops contributing to a stream due to changes in a CS, a snapshot MUST be sent from the SRC to the SRS with no "send" element for that stream.
- **recv** – This attribute indicates whether a participant is receiving a media stream or not. This attribute has a value that points to a stream represented by its unique_id. The presence of this attribute indicates that a participant is receiving a stream. If the participant stops receiving a stream due to changes in a CS (like hold), a snapshot MUST be sent from the SRC to the SRS with no "recv" element for that stream.
- **participant_id** – This attribute points to the participant with which a "stream" element is associated.

The "participantstreamassoc" XML element is used to represent a participant association with a stream. The "send" and "recv" XML elements MUST be used to indicate whether a participant is contributing to a stream or receiving a stream. There MAY be multiple instances of the "send" and "recv" XML elements inside a "participantstreamassoc" element. If a metadata snapshot is

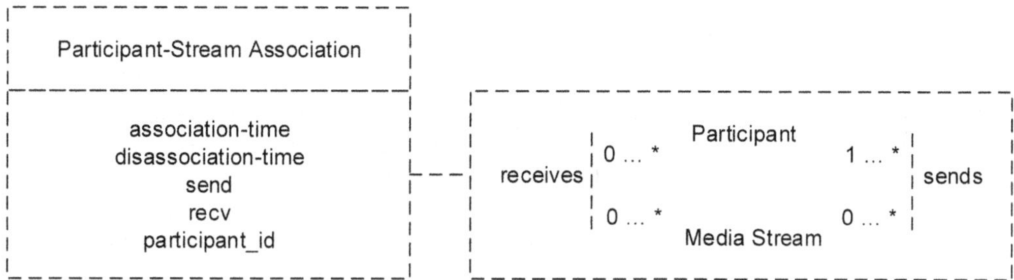

FIGURE 10.20 Participant-Stream Association and Its Class.

sent with a "participantstreamassoc" element that does not have any "send" and "recv" elements, it means that the participant is neither contributing to any streams nor receiving any streams.

10.3.6.8.2 Linkages

The Participant-Stream Association class is linked to the participant and MS classes.

10.3.6.9 Syntax of XML Elements for Date and Time

The XML elements "associate-time," "disassociate-time," "start-time," and "stop-time" contain strings representing the date and time. The value of these elements MUST follow the Instant Messaging and Presence Protocol (IMPP) date-time format (RFC 3339). Timestamps that contain "T" or "Z" MUST use the capitalized forms. As a security measure, the "timestamp" element MUST be included in all tuples unless the exact time of the status change cannot be determined.

10.3.6.10 Format of Unique ID

A **unique_id** is generated in two steps:

- The Universally Unique Identifier (UUID) is created using any of the procedures mentioned in Sections 4,3, 4,4, and 4.5 of RFC 4122. The algorithm MUST ensure that it does not use any potentially personally identifying information to generate the UUIDs. If implementations are using a Name-Based UUID as defined in Section 4.3 of RFC 4122, a namespace ID generated using the guidance in Section 4.2 or 4.3 of RFC 4122 might be a good choice.
- The UUID is encoded using base64 as defined in RFC 4648.

The above-mentioned unique_id mechanism SHOULD be used for each metadata element. Multiple SRCs can refer to the same element/UUID (how each SRC learns the UUID here is beyond the scope of this specification). If two SRCs use the same UUID, they MUST retain the UUID/element mapping. If the SRS detects that a UUID is mapped to more than one element at any point in time, it MUST treat this as an error. For example, the SRS may choose to reject or ignore the portions of metadata where it detects that the same UUID is mapped to an element that is different than the expected element (the SRS learns the mapped UUID when it sees an element for the first time in a metadata instance).

10.3.6.11 Metadata Version Indicator

The Metadata version is defined to help the SRC and SRS know the version of metadata XML schema used. SRCs and SRSs that support this specification MUST use version 1 in the namespace (urn:ietf:params:xml:ns:recording:1) in all the XML specifications. Implementations may not interoperate if the version implemented by the sender is not known by the receiver. No negotiation of versions is provided. The version number has no significance, although specifications that update or obsolete this specification (possibly including drafts of such specifications) should include a higher version number if the metadata XML schema changes.

10.3.7 Recording Metadata Snapshot Request Format

The SRS can explicitly request a metadata snapshot from the SRC. To request a metadata snapshot, the SRS MUST send an SIP request message with an XML specification having the namespace urn :ietf:params:xml:ns:recording:1. The XML specification has the following elements:

- A "requestsnapshot" XML element MUST be present as the top-level element in the XML specification.
- A "requestreason" XML element that indicates the reason (as a string) for requesting the snapshot MAY be present as a child XML element of "requestsnapshot."

The following example shows a metadata snapshot request from the SRS to SRC:

```
<?xml version="1.0" encoding="UTF-8"?>
   <requestsnapshot xmlns='urn:ietf:params:xml:ns:recording:1'>
   <requestreason xml:lang="it">SRS internal error</requestreason>
</requestsnapshot>
```

10.3.8 SIP Recording Metadata Examples

10.3.8.1 Complete SIP Recording Metadata Example

The following example provides all the tuples involved in the recording metadata XML body.

```
<?xml version="1.0" encoding="UTF-8"?>
<recording xmlns='urn:ietf:params:xml:ns:recording:1'>
   <datamode>complete</datamode>
   <group group_id="7+OTCyoxTmqmqyA/1weDAg==">
      <associate-time>2010-12-16T23:41:07Z</associate-time>
      <!-- Standardized extension -->
      <call-center xmlns='urn:ietf:params:xml:ns:callcenter'>
         <supervisor>sip:alice@atlanta.com</supervisor>
      </call-center>
      <mydata xmlns='http://example.com/my'>
         <structure>FOO!</structure>
         <whatever>bar</whatever>
      </mydata>
   </group>
   <session session_id="hVpd7YQgRW2nD22h7q60JQ==">
      <sipSessionID>ab30317f1a784dc48ff824d0d3715d86;
            remote=47755a9de7794ba387653f2099600ef2</sipSessionID>
   <group-ref>7+OTCyoxTmqmqyA/1weDAg==</group-ref>
   <!-- Standardized extension -->
   <mydata xmlns='http://example.com/my'>
      <structure>FOO!</structure>
      <whatever>bar</whatever>
   </mydata>
   </session>
   <participant participant_id="srfBElmCRp2QB23b7Mpk0w==">
      <nameID aor="sip:bob@biloxi.com">
         <name xml:lang="it">Bob</name>
      </nameID>
      <!-- Standardized extension -->
      <mydata xmlns='http://example.com/my'>
         <structure>FOO!</structure>
         <whatever>bar</whatever>
      </mydata>
   </participant>
   <participant participant_id="zSfPoSvdSDCmU3A3TRDxAw==">
      <nameID aor="sip:Paul@biloxi.com">
         <name xml:lang="it">Paul</name>
      </nameID>
      <!-- Standardized extension -->
      <mydata xmlns='http://example.com/my'>
         <structure>FOO!</structure>
         <whatever>bar</whatever>
      </mydata>
   </participant>
```

```
<stream stream_id="UAAMm5GRQKSCMVvLyl4rFw=="
   session_id="hVpd7YQgRW2nD22h7q60JQ==">
   <label>96</label>
</stream>
<stream stream_id="i1Pz3to5hGk8fuXl+PbwCw=="
   session_id="hVpd7YQgRW2nD22h7q60JQ==">
   <label>97</label>
</stream>
<stream stream_id="8zc6e0lYTlWIINA6GR+3ag=="
   session_id="hVpd7YQgRW2nD22h7q60JQ==">
   <label>98</label>
</stream>
<stream stream_id="EiXGlc+4TruqqoDaNE76ag=="
   session_id="hVpd7YQgRW2nD22h7q60JQ==">
   <label>99</label>
</stream>
<sessionrecordingassoc session_id="hVpd7YQgRW2nD22h7q60JQ==">
   <associate-time>2010-12-16T23:41:07Z</associate-time>
</sessionrecordingassoc>
<participantsessionassoc
   participant_id="srfBElmCRp2QB23b7Mpk0w=="
   session_id="hVpd7YQgRW2nD22h7q60JQ==">
   <associate-time>2010-12-16T23:41:07Z</associate-time>
</participantsessionassoc>
<participantsessionassoc
   participant_id="zSfPoSvdSDCmU3A3TRDxAw=="
   session_id="hVpd7YQgRW2nD22h7q60JQ==">
   <associate-time>2010-12-16T23:41:07Z</associate-time>
</participantsessionassoc>
<participantstreamassoc
   participant_id="srfBElmCRp2QB23b7Mpk0w==">
   <send>i1Pz3to5hGk8fuXl+PbwCw==</send>
   <send>UAAMm5GRQKSCMVvLyl4rFw==</send>
   <recv>8zc6e0lYTlWIINA6GR+3ag==</recv>
   <recv>EiXGlc+4TruqqoDaNE76ag==</recv>
</participantstreamassoc>
<participantstreamassoc
   participant_id="zSfPoSvdSDCmU3A3TRDxAw==">
   <send>8zc6e0lYTlWIINA6GR+3ag==</send>
   <send>EiXGlc+4TruqqoDaNE76ag==</send>
   <recv>UAAMm5GRQKSCMVvLyl4rFw==</recv>
   <recv>i1Pz3to5hGk8fuXl+PbwCw==</recv>
</participantstreamassoc>
</recording>
```

Example Metadata Snapshot from SRC to SRS

10.3.8.2 Partial Update of Recording Metadata XML Body

The following example provides a partial update in the recording metadata XML body for the previous example. The example has a snapshot that carries the disassociate-time for a participant from a session.

```
<?xml version="1.0" encoding="UTF-8"?>
<recording xmlns='urn:ietf:params:xml:ns:recording:1'>
   <datamode>partial</datamode>
   <participant participant_id="srfBElmCRp2QB23b7Mpk0w==">
```

```
        <nameID aor="sip:bob@biloxi.com">
            <name xml:lang="it">Bob</name>
        </nameID>
    </participant>
    <participantsessionassoc
        participant_id="srfBElmCRp2QB23b7Mpk0w=="
        session_id="hVpd7YQgRW2nD22h7q60JQ==">
        <disassociate-time>2010-12-16T23:41:07Z</disassociate-time>
    </participantsessionassoc>
</recording>
```

10.3.9 XML Schema Definition for Recording Metadata

This section defines the XML schema for the recording metadata specification.

```
<?xml version="1.0" encoding="UTF-8"?>
<xs:schema targetNamespace="urn:ietf:params:xml:ns:recording:1"
    xmlns:xs="http://www.w3.org/2001/XMLSchema"
    xmlns:tns="urn:ietf:params:xml:ns:recording:1"
    elementFormDefault="qualified"
    attributeFormDefault="unqualified">
    <!-- This import brings in the XML language attribute xml:lang -->
    <xs:import namespace="http://www.w3.org/XML/1998/namespace"
        schemaLocation="https://www.w3.org/2001/xml.xsd"/>
    <xs:element name="recording" type="tns:recording"/>
    <xs:complexType name="recording">
        <xs:sequence>
            <xs:element name="datamode" type="tns:dataMode"
                minOccurs="0"/>
            <xs:element name="group" type="tns:group"
                minOccurs="0" maxOccurs="unbounded"/>
            <xs:element name="session" type="tns:session"
                minOccurs="0" maxOccurs="unbounded"/>
            <xs:element name="participant" type="tns:participant"
                minOccurs="0" maxOccurs="unbounded"/>
            <xs:element name="stream" type="tns:stream"
                minOccurs="0" maxOccurs="unbounded"/>
            <xs:element name="sessionrecordingassoc"
                type="tns:sessionrecordingassoc"
                minOccurs="0" maxOccurs="unbounded"/>
            <xs:element name="participantsessionassoc"
                type="tns:participantsessionassoc"
                minOccurs="0" maxOccurs="unbounded"/>
            <xs:element name="participantstreamassoc"
                type="tns:participantstreamassoc"
                minOccurs="0" maxOccurs="unbounded"/>
            <xs:any namespace='##other'
                minOccurs='0'
                maxOccurs='unbounded'
                processContents='lax'/>
        </xs:sequence>
    </xs:complexType>
    <xs:complexType name="group">
        <xs:sequence>
            <xs:element name="associate-time" type="xs:dateTime"
                minOccurs="0"/>
```

```
        <xs:element name="disassociate-time" type="xs:dateTime"
           minOccurs="0"/>
        <xs:any namespace='##other'
           minOccurs='0'
           maxOccurs='unbounded'
           processContents='lax'/>
     </xs:sequence>
     <xs:attribute name="group_id" type="xs:base64Binary"
        use="required"/>
  </xs:complexType>
  <xs:complexType name="session">
     <xs:sequence>
        <xs:element name="sipSessionID" type="xs:string"
           minOccurs="0" maxOccurs="unbounded"/>
        <xs:element name="reason" type="tns:reason"
           minOccurs="0" maxOccurs="unbounded"/>
        <xs:element name="group-ref" type="xs:base64Binary"
           minOccurs="0" maxOccurs="1"/>
        <xs:element name="start-time" type="xs:dateTime"
           minOccurs="0" maxOccurs="1"/>
        <xs:element name="stop-time" type="xs:dateTime"
           minOccurs="0" maxOccurs="1"/>
        <xs:any namespace='##other'
           minOccurs='0'
           maxOccurs='unbounded'
           processContents='lax'/>
     </xs:sequence>
     <xs:attribute name="session_id" type="xs:base64Binary"
        use="required"/>
  </xs:complexType>
  <xs:complexType name="sessionrecordingassoc">
     <xs:sequence>
        <xs:element name="associate-time" type="xs:dateTime"
           minOccurs="0"/>
        <xs:element name="disassociate-time" type="xs:dateTime"
           minOccurs="0"/>
        <xs:any namespace='##other'
           minOccurs='0'
           maxOccurs='unbounded'
           processContents='lax'/>
     </xs:sequence>
     <xs:attribute name="session_id" type="xs:base64Binary"
        use="required"/>
  </xs:complexType>
  <xs:complexType name="participant">
     <xs:sequence>
        <xs:element name="nameID" type="tns:nameID"
           maxOccurs='unbounded'/>
        <xs:any namespace='##other'
           minOccurs='0'
           maxOccurs='unbounded'
           processContents='lax'/>
     </xs:sequence>
     <xs:attribute name="participant_id" type="xs:base64Binary"
        use="required"/>
  </xs:complexType>
```

```
<xs:complexType name="participantsessionassoc">
    <xs:sequence>
        <xs:element name="associate-time" type="xs:dateTime"
            minOccurs="0"/>
        <xs:element name="disassociate-time" type="xs:dateTime"
            minOccurs="0"/>
        <xs:element name="param" minOccurs="0" maxOccurs="unbounded">
            <xs:complexType>
                <xs:attribute name="pname" type="xs:string"
                    use="required"/>
                <xs:attribute name="pval" type="xs:string"
                    use="required"/>
            </xs:complexType>
        </xs:element>
        <xs:any namespace='##other'
            minOccurs='0'
            maxOccurs='unbounded'
            processContents='lax'/>
    </xs:sequence>
    <xs:attribute name="participant_id" type="xs:base64Binary"
        use="required"/>
    <xs:attribute name="session_id" type="xs:base64Binary"
        use="required"/>
</xs:complexType>
<xs:complexType name="participantstreamassoc">
    <xs:sequence>
        <xs:element name="send" type="xs:base64Binary"
            minOccurs="0" maxOccurs="unbounded"/>
        <xs:element name="recv" type="xs:base64Binary"
            minOccurs="0" maxOccurs="unbounded"/>
        <xs:element name="associate-time" type="xs:dateTime"
            minOccurs="0"/>
        <xs:element name="disassociate-time" type="xs:dateTime"
            minOccurs="0"/>
        <xs:any namespace='##other'
            minOccurs='0'
            maxOccurs='unbounded'
            processContents='lax'/>
    </xs:sequence>
    <xs:attribute name="participant_id" type="xs:base64Binary"
        use="required"/>
</xs:complexType>
<xs:complexType name="stream">
    <xs:sequence>
        <xs:element name="label" type="xs:string"
            minOccurs="0" maxOccurs="1"/>
        <xs:any namespace='##other'
            minOccurs='0'
            maxOccurs='unbounded'
            processContents='lax'/>
    </xs:sequence>
    <xs:attribute name="stream_id" type="xs:base64Binary"
        use="required"/>
    <xs:attribute name="session_id" type="xs:base64Binary"/>
</xs:complexType>
```

```
    <xs:simpleType name="dataMode">
      <xs:restriction base="xs:string">
        <xs:enumeration value="complete"/>
        <xs:enumeration value="partial"/>
      </xs:restriction>
    </xs:simpleType>
    <xs:complexType name="nameID">
      <xs:sequence>
        <xs:element name="name" type ="tns:name" minOccurs="0"
          maxOccurs="1"/>
      </xs:sequence>
      <xs:attribute name="aor" type="xs:anyURI" use="required"/>
    </xs:complexType>
    <xs:complexType name="name">
      <xs:simpleContent>
        <xs:extension base="xs:string">
          <xs:attribute ref="xml:lang" use="optional"/>
        </xs:extension>
      </xs:simpleContent>
    </xs:complexType>
    <xs:complexType name="reason">
      <xs:simpleContent>
        <xs:extension base="xs:string">
          <xs:attribute type="xs:short" name="cause"
            use="required"/>
          <xs:attribute type="xs:string" name="protocol"
            default="SIP"/>
        </xs:extension>
      </xs:simpleContent>
    </xs:complexType>
    <xs:element name="requestsnapshot" type="tns:requestsnapshot"/>
    <xs:complexType name="requestsnapshot">
      <xs:sequence>
        <xs:element name="requestreason" type="tns:name"
          minOccurs="0"/>
        <xs:any namespace='##other'
          minOccurs='0'
          maxOccurs='unbounded'
          processContents='lax'/>
      </xs:sequence>
    </xs:complexType>
  </xs:schema>
```

10.3.10 Security Considerations

This specification describes an extensive set of metadata that may be recorded by the SRS. Most of the metadata could be considered private data. The procedures mentioned in the Security Considerations section of RFC 7866 (see Section 10.4) MUST be followed by the SRC and the SRS for mutual authentication and to protect the content of the metadata in the RS. An SRC MAY, by policy, choose to limit the parts of the metadata sent to the SRS for recording. Also, the policy of the SRS might not require recording all the metadata it receives. For the sake of data minimization, the SRS MUST NOT record additional metadata that is not explicitly required by local policy. Metadata in storage needs to be provided with a level of security that is comparable to that of the RS.

10.3.11 IANA Considerations

This specification registers a new XML namespace and a new XML schema. For details, readers are requested to see RFC 7865.

10.4 MEDIA SESSION RECODING PROTOCOL

10.4.1 Introduction

This specification specifies the mechanism to record a CS by delivering real-time media and metadata from the CS to a recording device. In accordance with the architecture (RFC 7245 – see Section 10.2), the Session Recording Protocol specifies the use of SIP, the SDP, and RTP to establish an RS between the SRC, which is on the path of the CS, and an SRS at the recording device. SIP is also used to deliver metadata to the recording device, as specified in RFC 7865 (see Section 10.3). Metadata is information that describes recorded media and the CS to which they relate. The Session Recording Protocol is intended to satisfy the SIPREC) requirements listed in RFC 6341 (see Section 10.1). In addition to the Session Recording Protocol, this specification specifies extensions for UAs that are participants in a CS to receive recording indications and to provide preferences for recording. This specification considers only active recording, where the SRC purposefully streams media to an SRS, and all participating UAs are notified of the recording. Passive recording, where a recording device detects media directly from the network (e.g., using port-mirroring techniques), is outside the scope of this specification. In addition, lawful intercept is outside the scope of this specification, in accordance with RFC 2804.

10.4.2 Terminology

See Section 1.6.

10.4.3 Definitions

This specification refers to the core definitions provided in the architecture specification (RFC 7245 – see Section 10.2). Section 10.4.8 uses the definitions provided in "RTP: A Transport Protocol for Real-Time Applications" (RFC 3550).

10.4.4 Scope

The scope of the Session Recording Protocol includes the establishment of the RSs and the reporting of the metadata. The scope also includes extensions supported by UAs participating in the CS, such as an indication of recording. The UAs need not be recording aware in order to participate in a CS being recorded. The items in the following list, which is not exhaustive, do not represent the protocol itself and are considered out of scope for the Session Recording Protocol:

- Delivering recorded media in real time as the CS media
- Specifications of criteria to select a specific CS to be recorded or triggers to record a certain CS in the future
- Recording policies that determine whether the CS should be recorded and whether parts of the CS are to be recorded
- Retention policies that determine how long a recording is stored
- Searching and accessing the recorded media and metadata
- Policies governing how CS users are made aware of recording
- Delivering additional RS metadata through a non-SIP mechanism

10.4.5 Overview of Operations

This section is informative and provides a description of recording operations. Section 10.4.6 describes the SIP communication in an RS between an SRC and an SRS as well as the procedures for recording-aware UAs participating in a CS. Section 10.4.7 describes SDP handling in an RS and the procedures for recording indications and recording preferences. Section 10.4.8 describes RTP handling in an RS. Section 10.4.9 describes the mechanism to deliver recording metadata from the SRC to the SRS. As mentioned in the architecture specification (RFC 7245 – see Section 10.2), there are a number of types of call flows based on the location of the SRC. The sample call flows discussed in Section 10.4.5 provide a quick overview of the operations between the SRC and the SRS.

10.4.5.1 Delivering Recorded Media

When an SIP B2BUA with SRC functionality routes a call from UA A to UA B, the SRC has access to the media path between the UAs. When the SRC is aware that it should be recording the conversation, the SRC can cause the B2BUA to relay the media between UA A and UA B. The SRC then establishes the RS with the SRS and sends replicated media towards the SRS. An endpoint may also have SRC functionality, whereby the endpoint itself establishes the RS to the SRS. Since the endpoint has access to the media in the CS, the endpoint can send replicated media towards the SRS.

The example call flows in Figures 10.21 and 10.22 show an SRC establishing an RS towards an SRS. Figure 10.21 illustrates UA A acting as the SRC. Figure 10.22 illustrates a B2BUA acting as the SRC. Note that the SRC can choose when to establish the RS independently of the CS, even though the example call flows suggest that the SRC is establishing the RS (message 5 in Figure 10.21) after the CS is established.

The call flow shown in Figure 10.22 can also apply to the case of a centralized conference with a mixer. For clarity, ACKs to INVITEs and 200 OKs to BYEs are not shown. The conference focus

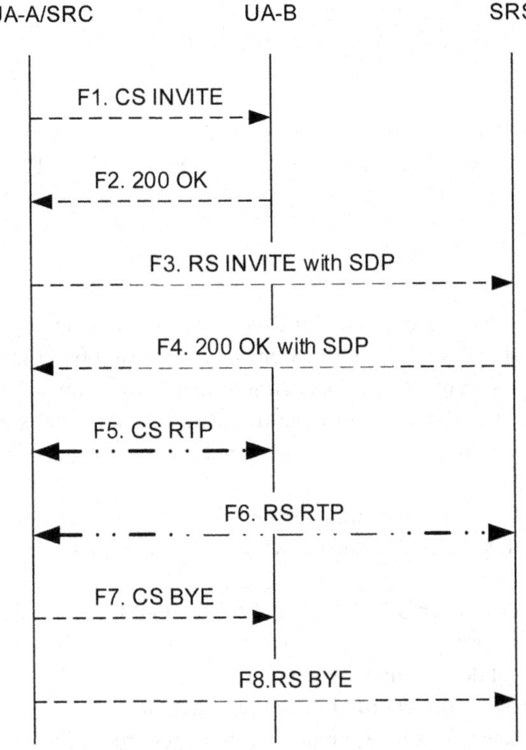

FIGURE 10.21 Basic Recording Call Flow with UA as SRC.

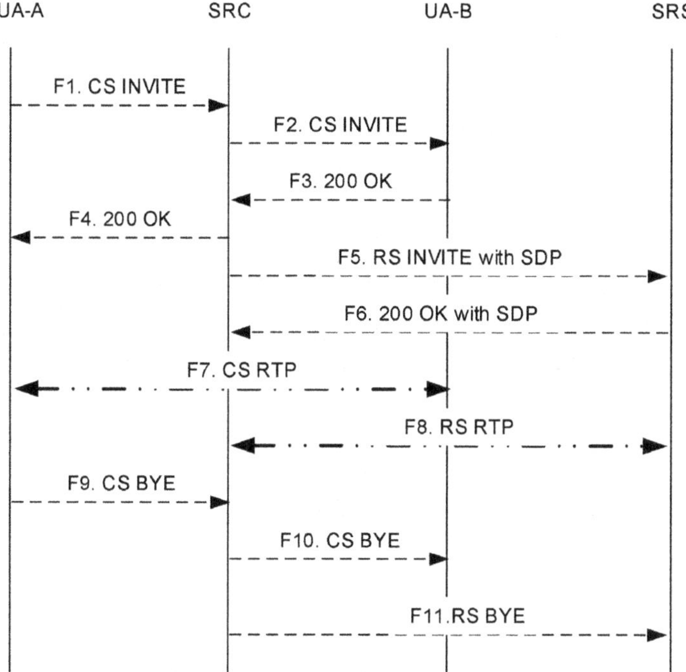

FIGURE 10.22 Basic Recording Call Flow with B2BUA as SRC.

can provide the SRC functionality, since the conference focus has access to all the media from each conference participant. When a recording is requested, the SRC delivers the metadata and the media streams to the SRS. Since the conference focus has access to a mixer, the SRC may choose to mix the media streams from all participants as a single mixed media stream towards the SRS.

An SRC can use a single RS to record multiple CSs. Every time the SRC wants to record a new call, the SRC updates the RS with a new SDP offer to add new recorded streams to the RS and to correspondingly also update the metadata for the new call. An SRS can also establish an RS to an SRC, although it is beyond the scope of this specification to define how an SRS would specify which calls to record.

10.4.5.2 Delivering Recording Metadata

The SRC is responsible for the delivery of metadata to the SRS (Figure 10.23). The SRC may provide an initial metadata snapshot about recorded media streams in the initial INVITE content in the RS. Subsequent metadata updates can be represented as a stream of events in UPDATE (RFC 3311) or re-INVITE requests sent by the SRC. These metadata updates are normally incremental updates to the initial metadata snapshot to optimize the size of updates. However, the SRC may also decide to send a new metadata snapshot at any time. Metadata is transported in the body of INVITE or UPDATE messages.

Certain metadata, such as the attributes of the recorded media stream, is located in the SDP of the RS. The SRS has the ability to send a request to the SRC to ask for a new metadata snapshot update from the SRC. This can happen when the SRS fails to understand the current stream of incremental updates for whatever reason – for example, when the SRS loses the current state due to internal failure. The SRS may optionally attach a reason along with the snapshot request. This request allows both the SRC and the SRS to synchronize the states with a new metadata snapshot so that further incremental metadata updates will be based on the latest metadata snapshot. Similarly to the metadata content, the metadata snapshot request is transported as content in UPDATE or INVITE messages sent by the SRS in the RS.

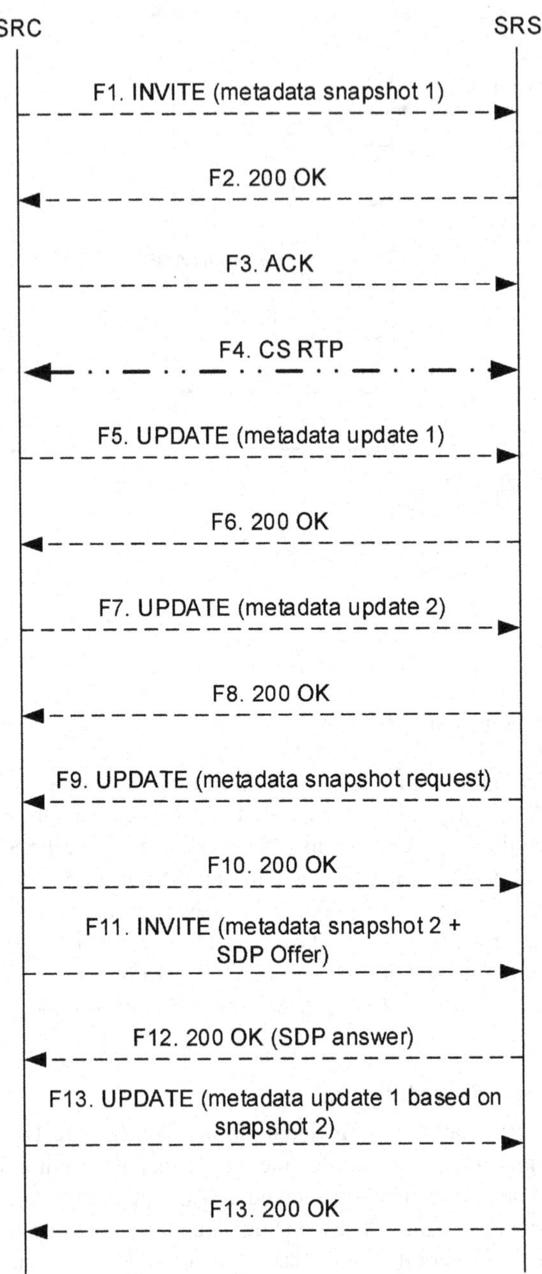

FIGURE 10.23 Delivering Metadata via SIP UPDATE. (Copyright: IETF.)

10.4.5.3 Receiving Recording Indications and Providing Recording Preferences

The SRC is responsible for providing recording indications to the participants in the CS. A recording-aware UA supports receiving recording indications via the SDP "a=record" attribute, and it can specify a recording preference in the CS by including the SDP "a=recordpref" attribute. The recording attribute is a declaration by the SRC in the CS to indicate whether recording is taking place. The recording preference attribute is a declaration by the recording-aware UA in the CS to indicate its recording preference. A UA that does not want to be recorded may still be notified that recording is occurring for a number of reasons (e.g., it was not capable of indicating its preference

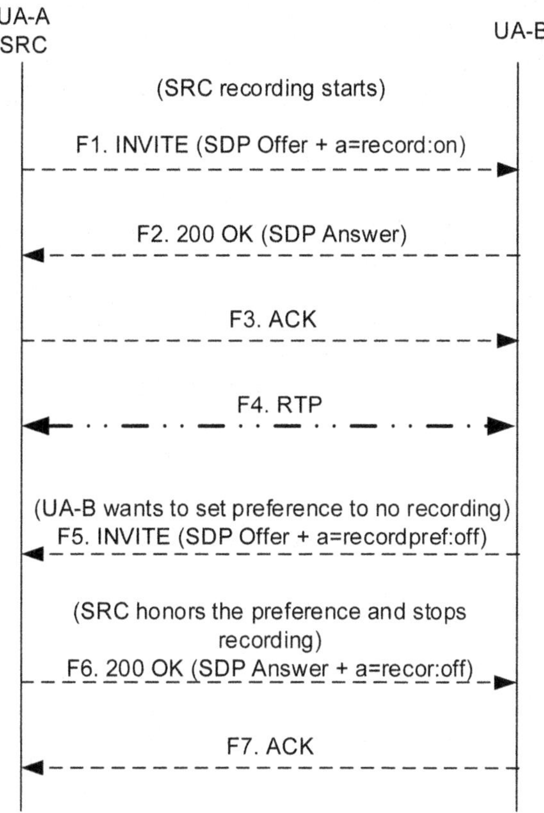

FIGURE 10.24 Recording Indication and Recording Preference. (Copyright: IETF.)

or its preference was ignored). If this occurs, the UA's only mechanism to avoid being recorded is to terminate its participation in the session.

To illustrate how the attributes are used, if UA A is initiating a call to UA B, and UA A is also an SRC that is performing the recording (Figure 10.24), then UA A provides the recording indication in the SDP offer with a=record:on. Since UA A is the SRC, UA A receives the recording indication from the SRC directly. When UA B receives the SDP offer, UA B will see that recording is happening on the other endpoint of this session. Since UA B is not an SRC and does not provide any recording preference, the SDP answer does not contain a=record or a=recordpref.

After the call is established and recording is in progress, UA B later decides to change the recording preference to no recording and sends a re-INVITE with the "a=recordpref" attribute. It is up to the SRC to honor the preference, and in this case, the SRC decides to stop the recording and updates the recording indication in the SDP answer. Note that UA B could have explicitly indicated a recording preference in message 2, the 200 OK (Figure 10.24) for the original INVITE. Indicating a preference of no recording in an initial INVITE or an initial response to an INVITE may reduce the chance of a user being recorded in the first place.

10.4.6 SIP Handling

10.4.6.1 Procedures at the SRC

10.4.6.1.1 Initiating a Recording Session

An RS is a SIP session with specific extensions applied, and these extensions are listed in the following procedures for the SRC and the SRS. When an SRC or an SRS receives a SIP session that is not an RS, it is up to the SRC or the SRS to determine what to do with the SIP session. The SRC can initiate an RS by sending a SIP INVITE request to the SRS. The SRC and the SRS are identified in the From and To headers, respectively. The SRC MUST include the "+sip.src" feature tag in the Contact URI, defined in this specification as an extension to RFC 3840, for all RSs. An SRS uses the presence of the "+sip.src" feature tag in dialog creating and modifying requests and responses to confirm that the dialog being created is for the purpose of an RS.

In addition, when an SRC sends a REGISTER request to a registrar, the SRC MAY include the "+sip.src" feature tag to indicate that it is an SRC. Since SIP Caller Preferences extensions are optional to implement for routing proxies, there is no guarantee that an RS will be routed to an SRC or SRS. A new option tag, "siprec," is introduced. As per RFC 3261 (also see Section 1.2), only an SRC or an SRS can accept this option tag in an RS. An SRC MUST include the "siprec" option tag in the Require header when initiating an RS so that UAs that do not support the Session Recording Protocol extensions will simply reject the INVITE request with a 420 (Bad Extension) response. When an SRC receives a new INVITE, the SRC MUST only consider the SIP session as an RS when both the "+sip.srs" feature tag and the "siprec" option tag are included in the INVITE request.

10.4.6.1.2 SIP Extensions for Recording Indications and Preferences

For the CS, the SRC MUST provide recording indications to all participants in the CS. A participant UA in a CS can indicate that it is recording aware by providing the "record-aware" option tag, and the SRC MUST provide recording indications in the new SDP "a=record" attribute described in Section 10.4.7. In the absence of the "record-aware" option tag – meaning that the participant UA is not recording aware – an SRC MUST provide recording indications through other means, such as playing a tone in-band or having a signed participant contract in place. An SRC in the CS may also indicate itself as a session recording client by including the "+sip.src" feature tag. A recording-aware participant can learn that an SRC is in the CS and can set the recording preference for the CS with the new SDP "a=recordpref" attribute described in Section 10.4.7.

10.4.6.2 Procedures at the SRS

When an SRS receives a new INVITE, the SRS MUST only consider the SIP session as an RS when both the "+sip.src" feature tag and the "siprec" option tag are included in the INVITE request. The SRS can initiate an RS by sending a SIP INVITE request to the SRC. The SRS and the SRC are identified in the From and To headers, respectively. The SRS MUST include the "+sip.srs" feature tag in the Contact URI, as per RFC 3840, for all RSs. An SRC uses the presence of this feature tag in dialog creation and modification requests and responses to confirm that the dialog being created is for the purpose of an RS (REQ-030 in RFC 6341 – see Section 10.1). In addition, when an SRS sends a REGISTER request to a registrar, the SRS SHOULD include the "+sip.srs" feature tag to indicate that it is an SRS. An SRS MUST include the "siprec" option tag in the Require header as per RFC 3261 (also see Section 1.2) when initiating an RS so that UAs that do not support the Session Recording Protocol extensions will simply reject the INVITE request with a 420 (Bad Extension) response.

10.4.6.3 Procedures for Recording-Aware User Agents

A recording-aware UA is a participant in the CS that supports the SIP and SDP extensions for receiving recording indications and for requesting recording preferences for the call. A recording-aware UA MUST indicate that it can accept the reporting of recording indications provided by the SRC with a new "record-aware" option tag when initiating or establishing a CS; this means including the "record-aware" option tag in the Supported header in the initial INVITE request or response.

A recording-aware UA MUST provide a recording indication to the end-user through an appropriate user interface, indicating whether recording is on, off, or paused for each medium. Appropriate user interfaces may include real-time notification or previously established agreements that use of the device is subject to recording. Some UAs that are automata (e.g., IVR, MS, or PSTN gateway) may not have a user interface to render a recording indication. When such a UA indicates recording awareness, the UA SHOULD render the recording indication through other means, such as passing an in-band tone on the PSTN gateway, putting the recording indication in a log file, or raising an application event in a VoiceXML dialog. These UAs MAY also choose not to indicate recording awareness, thereby relying on whatever mechanism an SRC chooses to indicate recording, such as playing a tone in-band.

10.4.7 SDP Handling

10.4.7.1 Procedures at the SRC

The SRC and SRS follow the SDP offer/answer model described in RFC 3264. The procedures for the SRC and SRS describe the conventions used in an RS.

10.4.7.1.1 SDP Handling in the RS

Since the SRC does not expect to receive media from the SRS, the SRC typically sets each media stream of the SDP offer to only send media by qualifying them with the “a=sendonly” attribute according to the procedures in RFC 3264. The SRC sends recorded streams of participants to the SRS, and the SRC MUST provide a “label” attribute (“a=label”), as per RFC 4574, on each media stream in order to identify the recorded stream with the rest of the metadata. The “a=label” attribute identifies each recorded media stream, and the label name is mapped to the Media Stream Reference in the metadata as per RFC 7865 (see Section 10.3). The scope of the “a=label” attribute only applies to the SDP and metadata conveyed in the bodies of the SIP request or response that the label appeared in. Note that a recorded stream is distinct from a CS stream; the metadata provides a list of participants that contribute to each recorded stream.

Figure 10.25 shows an example SDP offer from an SRC with both audio and video recorded streams. Note that this example contains unfolded lines longer than 72 characters; these lines are captured between <allOneLine> tags.

10.4.7.1.1.1 Handling Media Stream Updates Over the lifetime of an RS, the SRC can add and remove recorded streams to and from the RS for various reasons – for example, when a CS stream is added to or removed from the CS, or when a CS is created or terminated if an RS handles multiple CSs. To remove a recorded stream from the RS, the SRC sends a new SDP offer, where the port of the media stream to be removed is set to zero according to the procedures in RFC 3264. To add a recorded stream to the RS, the SRC sends a new SDP offer by adding a new media stream description or by reusing an old media stream that had been previously disabled, according to the procedures in RFC 3264.

The SRC can temporarily discontinue streaming and collection of recorded media from the SRC to the SRS for reasons such as masking the recording. In this case, the SRC sends a new SDP offer and sets the media stream to inactive (a=inactive) for each recorded stream to be paused, as per the procedures in RFC 3264. To resume streaming and collection of recorded media, the SRC sends a new SDP offer and sets the media stream to sendonly (a=sendonly). Note that a CS may itself change the media stream direction by updating the SDP – for example, by setting a=inactive for SDP hold. Media stream direction changes in the CS are conveyed in the metadata by the SRC. When a CS media stream is changed to or from inactive, the effect on the corresponding RS media stream is governed by SRC policy. The SRC MAY have a local policy to pause an RS media stream when the corresponding CS media stream is inactive, or it MAY leave the RS media stream as sendonly.

10.4.7.1.2 Recording Indication in the CS

While there are existing mechanisms for providing an indication that a CS is being recorded, these mechanisms are usually delivered on the CS media streams, such as playing an in-band tone or an announcement to the participants. A new “record” SDP attribute is introduced to allow the SRC to indicate recording state to a recording-aware UA in a CS. The “record” SDP attribute appears at the

```
v=0
o=SRC 2890844526 2890844526 IN IP4 198.51.100.1
s=-
c=IN IP4 198.51.100.1
t=0 0
m=audio 12240 RTP/AVP 0 4 8
a=sendonly
a=label:1
m=video 22456 RTP/AVP 98
a=rtpmap:98 H264/90000
<allOneLine>
a=fmtp:98 profile-level-id=42A01E;
sprop-parameter-sets=Z0IACpZTBYmI,aMljiA==
</allOneLine>
a=sendonly
a=label:2
m=audio 12242 RTP/AVP 0 4 8
a=sendonly
a=label:3
m=video 22458 RTP/AVP 98
a=rtpmap:98 H264/90000
<allOneLine>
a=fmtp:98 profile-level-id=42A01E;
sprop-parameter-sets=Z0IACpZTBYmI,aMljiA==
</allOneLine>
a=sendonly
a=label:4
```

FIGURE 10.25 Sample SDP Offer from SRC with Audio and Video Streams.

media level or session level in either an SDP offer or answer. When the attribute is applied at the session level, the indication applies to all media streams in the SDP. When the attribute is applied at the media level, the indication applies to that one media stream only, and that overrides the indication if also set at the session level.

Whenever the recording indication needs to change, such as termination of recording, the SRC MUST initiate a re-INVITE or UPDATE to update the SDP "a=record" attribute. The following is the ABNF (RFC 5234) of the "record" attribute:

```
attribute     =  / record-attr
                 ; attribute defined in RFC 4566
record-attr   =  "record:" indication
indication    =  "on" / "off" / "paused"

on: Recording is in progress.
off: No recording is in progress.
paused: Recording is in progress but media is paused.
```

10.4.7.1.3 Recording Preference in the CS

When the SRC receives the "a=recordpref" SDP in an SDP offer or answer, the SRC chooses to honor the preference to record based on local policy at the SRC. If the SRC makes a change in recording state, the SRC MUST report the new recording state in the "a=record" attribute in the SDP answer or in a subsequent SDP offer.

10.4.7.2 Procedures at the SRS

Typically, the SRS only receives RTP streams from the SRC; therefore, the SDP offer/answer from the SRS normally sets each media stream to receive media, by setting them with the "a=recvonly" attribute, according to the procedures of RFC 3264. When the SRS is not ready to receive a recorded stream, the SRS sets the media stream as inactive in the SDP offer or answer by setting it with an "a=inactive" attribute according to the procedures of RFC 3264. When the SRS is ready to receive recorded streams, the SRS sends a new SDP offer and sets the media streams with an "a=recvonly" attribute.

Figure 10.26 shows an example of an SDP answer from the SRS for the SDP offer from Figure 10.25. Note that this example contains unfolded lines longer than 72 characters; these lines are captured between <allOneLine> tags.

```
v=0
o=SRS 0 0 IN IP4 198.51.100.20
s=-
c=IN IP4 198.51.100.20
t=0 0
m=audio 10000 RTP/AVP 0
a=recvonly
a=label:1
m=video 10002 RTP/AVP 98
a=rtpmap:98 H264/90000
<allOneLine>
a=fmtp:98 profile-level-id=42A01E;
sprop-parameter-sets=Z0IACpZTBYmI,aMljiA==
</allOneLine>
a=recvonly
a=label:2
m=audio 10004 RTP/AVP 0
a=recvonly
a=label:3
m=video 10006 RTP/AVP 98
a=rtpmap:98 H264/90000
<allOneLine>
a=fmtp:98 profile-level-id=42A01E;
sprop-parameter-sets=Z0IACpZTBYmI,aMljiA==
</allOneLine>
a=recvonly
a=label:4
```

FIGURE 10.26 Sample SDP Answer from SRS with Audio and Video Streams.

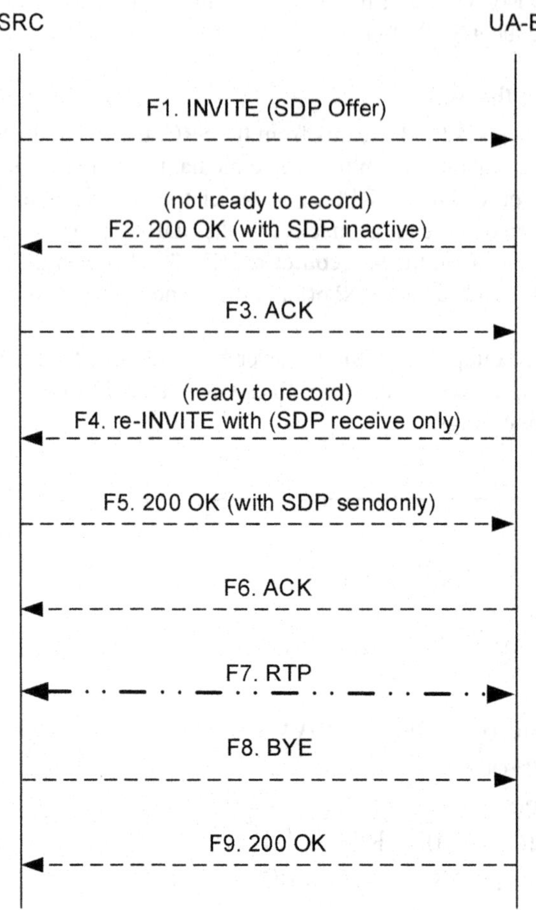

FIGURE 10.27 SRS Responding to Offer with a=inactive. (Copyright: IETF.)

Over the lifetime of an RS, the SRS can remove recorded streams from the RS for various reasons. To remove a recorded stream from the RS, the SRS sends a new SDP offer where the port of the media stream to be removed is set to zero, according to the procedures in RFC 3264. The SRS MUST NOT add recorded streams in the RS when the SRS sends a new SDP offer. Similarly, when the SRS starts an RS, the SRS MUST initiate the INVITE without an SDP offer to let the SRC generate the SDP offer with the streams to be recorded.

The sequence diagram in Figure 10.27 shows an example where the SRS is initially not ready to receive recorded streams and later updates the RS when the SRS is ready to record.

10.4.7.3 Procedures for Recording-Aware User Agents

10.4.7.3.1 Recording Indication

When a recording-aware UA receives an SDP offer or answer that includes the “a=record” attribute, the UA provides to the end user an indication as to whether the recording is on, off, or paused for each medium based on the most recently received “a=record” SDP attribute for that medium. When a CS is traversed through multiple UAs, such as a B2BUA or a conference focus, each UA involved in the CS that is aware that the CS is being recorded MUST provide the recording indication through the “a=record” attribute to all other parties in the CS.

It is possible that more than one SRC is in the call path of the same CS, but the recording indication attribute does not provide any hint as to which SRC or how many SRCs are recording. An endpoint knows only that the call is being recorded. Furthermore, this attribute is not used as a request for a specific SRC to start or stop recording.

10.4.7.3.2 Recording Preference

A participant in a CS MAY set the recording preference in the CS to be recorded or not recorded at session establishment or during the session. A new "recordpref" SDP attribute is introduced, and the participant in the CS may set this recording preference attribute in any SDP offer/answer at session establishment time or during the session. The SRC is not required to honor the recording preference from a participant, based on local policies at the SRC, and the participant can learn the recording indication through the "a=record" SDP attribute as described in Section 10.4.7.3.1.

The SDP "a=recordpref" attribute can appear at the media level or the session level and can appear in an SDP offer or answer. When the attribute is applied at the session level, the recording preference applies to all media streams in the SDP. When the attribute is applied at the media level, the recording preference applies to that one media stream only, and that overrides the recording preference if also set at the session level. The UA can change the recording preference by changing the "a=recordpref" attribute in a subsequent SDP offer or answer. The absence of the "a=recordpref" attribute in the SDP indicates that the UA has no recording preference.

The following is the ABNF of the "recordpref" attribute:

```
attribute =/ recordpref-attr
                          ; attribute defined in RFC 4566
recordpref-attr = "a=recordpref:" pref
pref = "on" / "off" / "pause" / "nopreference"
```

- **on:** Sets the preference to record if it has not already been started. If the recording is currently paused, the preference is to resume recording.
- **off:** Sets the preference for no recording. If recording has already been started, then the preference is to stop the recording.
- **pause:** If the recording is currently in progress, sets the preference to pause the recording.
- **nopreference**: Indicates that the UA has no preference regarding recording.

10.4.8 RTP Handling

This section provides recommendations and guidelines for RTP and the real-time transport control protocol (RTCP) in the context of SIPREC (RFC 6341 – see Section 10.1). In order to communicate most effectively, the SRC, the SRS, and any recording-aware UAs should utilize the mechanisms provided by RTP in a well-defined and predictable manner. It is the goal of this specification to make the reader aware of these mechanisms and to provide recommendations and guidelines.

10.4.8.1 RTP Mechanisms

This section briefly describes important RTP/RTCP constructs and mechanisms that are particularly useful within the context of SIPREC.

10.4.8.1.1 RTCP

The RTP data transport is augmented by a control protocol (RTCP) to allow monitoring of the data delivery. RTCP, as defined in RFC 3550, is based on the periodic transmission of control packets to all participants in the RTP session using the same distribution mechanism as the data packets. Support for RTCP is REQUIRED, per RFC 3550, and it provides, among other things, the following important functionality in relation to SIPREC:

1. Feedback on the quality of the data distribution. This feedback from the receivers may be used to diagnose faults in the distribution. As such, RTCP is a well-defined and efficient mechanism for the SRS to inform the SRC, and for the SRC to inform recording-aware UAs, of issues that arise with respect to the reception of media that is to be recorded. Including a persistent transport-level identifier – the CNAME, or canonical name – for an RTP source.

The synchronization source (SSRC) (RFC 3550) identifier may change if a conflict is discovered or a program is restarted, in which case receivers can use the CNAME to keep track of each participant. Receivers may also use the CNAME to associate multiple data streams from a given participant in a set of related RTP sessions – for example, to synchronize audio and video. Synchronization of media streams is also facilitated by the network time protocol (NTP) and RTP timestamps included in RTCP packets by data senders.

10.4.8.1.2 RTP Profile

The RECOMMENDED RTP profiles for the SRC, SRS, and recording-aware UAs are "Extended Secure RTP Profile for Real-time Transport Control Protocol (RTCP)-Based Feedback (RTP/SAVPF)" (RFC 5124) when using encrypted RTP streams and "Extended RTP Profile for Real-time Transport Control Protocol (RTCP)-Based Feedback (RTP/AVPF)" (RFC 4585) when using non-encrypted media streams. However, as these are not requirements, some implementations may use "The Secure Real-time Transport Protocol (SRTP)" (RFC 3371) and "RTP Profile for Audio and Video Conferences with Minimal Control" (RFC 3551).

Therefore, it is RECOMMENDED that the SRC, SRS, and recording-aware UAs not rely entirely on RTP/SAVPF or RTP/AVPF for core functionality that may be at least partially achievable using RTP/SAVP and RTP/AVP. AVPF and SAVPF provide an improved RTCP timer model that allows more flexible transmission of RTCP packets in response to events rather than strictly according to bandwidth. AVPF-based codec control messages provide efficient mechanisms for an SRC, an SRS, and recording-aware UAs to handle events such as scene changes, error recovery, and dynamic bandwidth adjustments. These messages are discussed in more detail later in this specification. SAVP and SAVPF provide media encryption, integrity protection, replay protection, and a limited form of source authentication. They do not contain or require a specific keying mechanism.

10.4.8.1.3 SSRC

The SSRC, as defined in RFC 3550, is carried in the RTP header and in various fields of RTCP packets. It is a random 32-bit number that is required to be globally unique within an RTP session. It is crucial that the number be chosen with care in order that participants on the same network or starting at the same time are not likely to choose the same number. Guidelines regarding SSRC value selection and conflict resolution are provided in RFC 3550. The SSRC may also be used to separate different sources of media within a single RTP session. For this reason, as well as for conflict resolution, it is important that the SRC, SRS, and recording-aware UAs handle changes in SSRC values and properly identify the reason for the change. The CNAME values carried in RTCP facilitate this identification.

10.4.8.1.4 CSRC

The contributing source (CSRC), as defined in RFC 3550, identifies the source of a stream of RTP packets that has contributed to the combined stream produced by an RTP mixer. The mixer inserts a list of the SSRC identifiers of the sources that contributed to the generation of a particular packet into the RTP header of that packet. This list is called the CSRC list. It is RECOMMENDED that an SRC or recording-aware UA, when acting as a mixer, set the CSRC list accordingly and that the SRC and SRS interpret the CSRC list per RFC 3550 when received.

10.4.8.1.5 SDES

The source description (SDES), as defined in RFC 3550, contains an SSRC/CSRC identifier followed by a list of zero or more items that carry information about the SSRC/CSRC. End systems send one SDES packet containing their own source identifier (the same as the SSRC in the fixed RTP header). A mixer sends one SDES packet containing a chunk for each CSRC from which it is receiving SDES information, or multiple complete SDES packets if there are more than 31 such sources. The ability to identify individual CSRCs is important in the context of SIPREC. Metadata (RFC 7865 – see Section 10.3) provides a mechanism to achieve this at the signaling level. SDES provides a mechanism at the RTP level.

10.4.8.1.5.1 CNAME The Canonical End-Point Identifier (CNAME), as defined in RFC 3550, provides the binding from the SSRC identifier to an identifier for the source (sender or receiver) that remains constant. It is important that the SRC and recording-aware UAs generate CNAMEs appropriately and that the SRC and SRS interpret and use them for this purpose. Guidelines for generating CNAME values are provided in "Guidelines for Choosing RTP Control Protocol (RTCP) Canonical Names (CNAMEs)" (RFC 7022).

10.4.8.1.6 Keepalive

It is anticipated that media streams in SIPREC may exist in an inactive state for extended periods of time for any of a number of valid reasons. In order for the bindings and any pinholes in network address translators (NATs)/firewalls to remain active during such intervals, it is RECOMMENDED that the SRC, SRS, and recording-aware UAs follow the keepalive procedure recommended in "Application Mechanism for Keeping Alive the NAT Mappings Associated with RTP / RTP Control Protocol (RTCP) Flows" (RFC 6263) for all RTP media streams.

10.4.8.1.7 RTCP Feedback Messages

"Codec Control Messages in the RTP Audio-Visual Profile with Feedback (AVPF)" (RFC 5104) specifies extensions to the messages defined in AVPF (RFC 4585). Support for and proper usage of these messages are important to SRC, SRS, and recording-aware UA implementations. Note that these messages are applicable only when using the AVPF or SAVPF RTP profiles.

10.4.8.1.7.1 Full Intra Request A Full Intra Request (FIR) command, when received by the designated media sender, requires that the media sender send a decoder refresh point at the earliest opportunity. Using a decoder refresh point implies refraining from using any picture sent prior to that point as a reference for the encoding process of any subsequent picture sent in the stream.

Decoder refresh points, especially Intra or Instantaneous Decoding Refresh (IDR) pictures for H.264 video codecs, are in general several times larger in size than predicted pictures. Thus, in scenarios in which the available bit rate is small, the use of a decoder refresh point implies a delay that is significantly longer than the typical picture duration.

10.4.8.1.7.1.1 Deprecated Usage of SIP INFO Instead of FIR "XML Schema for Media Control" (RFC 5168) defines an Extensible Markup Language (XML) Schema for video fast update. Implementations are discouraged from using the method described in RFC 5168 except for purposes of backwards compatibility. Implementations SHOULD use FIR messages instead. To make sure that a common mechanism exists between the SRC and the SRS, the SRS MUST support both mechanisms (FIR and SIP INFO), using FIR messages when negotiated successfully with the SRC and using SIP INFO otherwise.

10.4.8.1.7.2 Picture Loss Indication Picture Loss Indication (PLI), as defined in RFC 4585, informs the encoder of the loss of an undefined amount of coded video data belonging to one or more pictures. RFC 4585 recommends using PLI instead of FIR messages to recover from errors. FIR is appropriate only in situations where not sending a decoder refresh point would render the video unusable for the users. Examples of where sending FIR messages is appropriate include a multipoint conference when a new user joins the conference and no regular decoder refresh point interval is established, and a video-switching multipoint control unit (MCU) that changes streams. Appropriate use of PLI and FIR is important to ensure, with minimum overhead, that the recorded video is usable (e.g., the necessary reference frames exist for a player to render the recorded video).

10.4.8.1.7.3 Temporary Maximum Media Stream Bit Rate Request A receiver, translator, or mixer uses the Temporary Maximum Media Stream Bit Rate Request (TMMBR) (RFC 5104) to request a sender to limit the maximum bit rate for a media stream to the provided value. Appropriate use of TMMBR facilitates rapid adaptation to changes in available bandwidth.

10.4.8.1.7.3.1 Renegotiation of SDP Bandwidth Attribute If it is likely that the new value indicated by TMMBR will be valid for the remainder of the session, the TMMBR sender is expected to perform a renegotiation of the session upper limit using the session signaling protocol. Therefore, for SIPREC, implementations are RECOMMENDED to use TMMBR for temporary changes and renegotiation of bandwidth via SDP offer/answer for more permanent changes.

10.4.8.1.8 Symmetric RTP/RTCP for Sending and Receiving

Within an SDP offer/answer exchange, RTP entities choose the RTP and RTCP transport addresses (i.e., IP addresses and port numbers) on which to receive packets. When sending packets, the RTP entities may use the same source port or a different source port than those signaled for receiving packets. When the transport address used to send and receive RTP is the same, it is termed "symmetric RTP" (RFC 4961). Likewise, when the transport address used to send and receive RTCP is the same, it is termed "symmetric RTCP" (RFC 4961).

When sending RTP, the use of symmetric RTP is REQUIRED. When sending RTCP, the use of symmetric RTCP is REQUIRED. Although an SRS will not normally send RTP, it will send RTCP as well as receive RTP and RTCP. Likewise, although an SRC will not normally receive RTP from the SRS, it will receive RTCP as well as send RTP and RTCP.

Note: Symmetric RTP and symmetric RTCP are different from RTP/RTCP multiplexing (RFC 5761).

10.4.8.2 Roles

An SRC has the task of gathering media from the various UAs in one or more CSs and forwarding the information to the SRS within the context of a corresponding RS (Figure 10.28). There are numerous ways in which an SRC may do this, including, but not limited to, appearing as a UA within a CS or as a B2BUA between UAs within a CS.

The following subsections define a set of roles an SRC may choose to play, based on its position with respect to a UA within a CS and an SRS within an RS. A CS and a corresponding RS are independent sessions; therefore, an SRC may play a different role within a CS than it does within the corresponding RS.

10.4.8.2.1 SRC Acting as an RTP Translator

The SRC may act as a translator as defined in RFC 3550. A defining characteristic of a translator is that it forwards RTP packets with their SSRC identifier intact. There are two types of translator: one that simply forwards and another that performs transcoding (e.g., from one codec to another) in addition to forwarding.

10.4.8.2.1.1 Forwarding Translator When acting as a forwarding translator, RTP received as separate streams from different sources (e.g., from different UAs with different SSRCs) cannot be

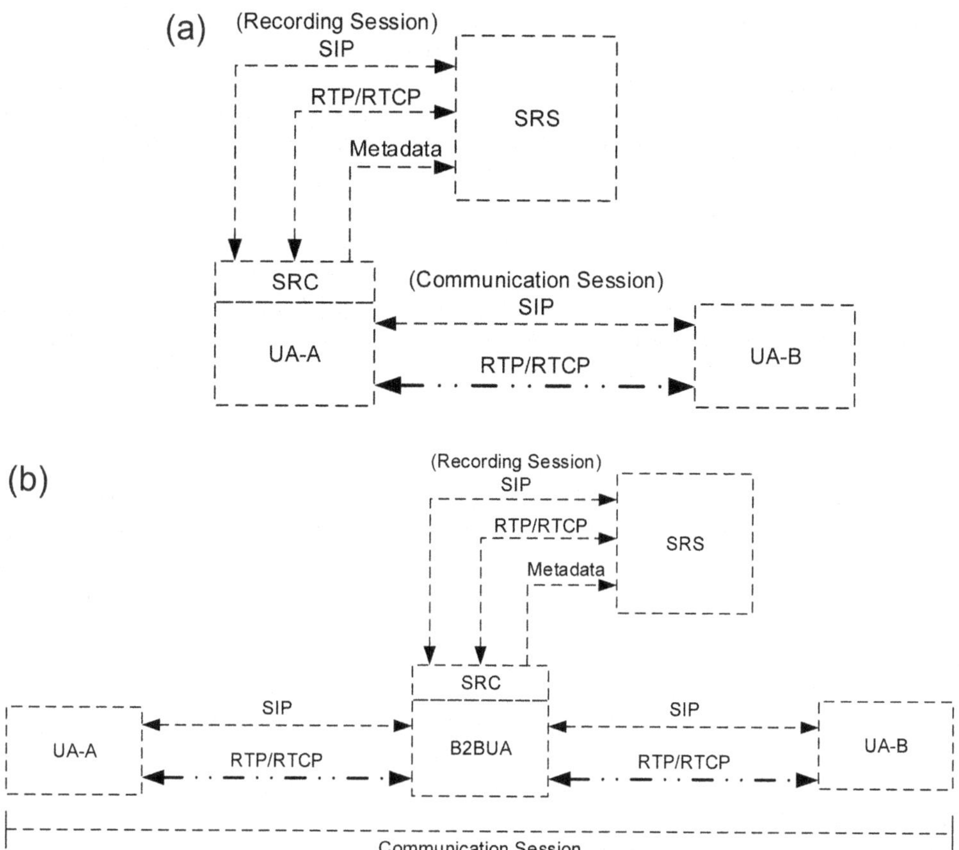

FIGURE 10.28 (a) UA as SRC. (b) B2BUA as SRC. (Copyright: IETF.)

mixed by the SRC and MUST be sent separately to the SRS. All RTCP reports MUST be passed by the SRC between the UAs and the SRS, such that the UAs and SRS are able to detect any SSRC collisions.

RTCP Sender Reports generated by a UA sending a stream MUST be forwarded to the SRS. RTCP Receiver Reports generated by the SRS MUST be forwarded to the relevant UA. UAs may receive multiple sets of RTCP Receiver Reports – one or more from other UAs participating in the CS and one from the SRS participating in the RS. A UA SHOULD process the RTCP Receiver Reports from the SRS if it is recording aware. If SRTP is used on both the CS and the RS, decryption and/or re-encryption may occur. For example, if different keys are used, it will occur. If the same keys are used, it need not occur. Section 10.4.12 provides additional information on secure real-time transport protocol (SRTP) and keying mechanisms. If packet loss occurs, either from the UA to the SRC or from the SRC to the SRS, the SRS SHOULD detect and attempt to recover from the loss. The SRC does not play a role in this other than forwarding the associated RTP and RTCP packets.

10.4.8.2.1.2 Transcoding Translator When acting as a transcoding translator, an SRC MAY perform transcoding (e.g., from one codec to another), and this may result in a different rate of packets between what the SRC receives on the CS and what the SRC sends on the RS. As when acting as a forwarding translator, RTP received as separate streams from different sources (e.g., from different UAs with different SSRCs) cannot be mixed by the SRC and MUST be sent separately to the SRS. All RTCP reports MUST be passed by the SRC between the UAs and the SRS, such that the UAs and SRS are able to detect any SSRC collisions.

RTCP Sender Reports generated by a UA sending a stream MUST be forwarded to the SRS. RTCP Receiver Reports generated by the SRS MUST be forwarded to the relevant UA. The SRC may need to manipulate the RTCP Receiver Reports to take into account any transcoding that has taken place. UAs may receive multiple sets of RTCP Receiver Reports – one or more from other UAs participating in the CS and one from the SRS participating in the RS. A recording-aware UA SHOULD be prepared to process the RTCP Receiver Reports from the SRS, whereas a recording-unaware UA may discard such RTCP packets as irrelevant.

If SRTP is used on both the CS and the RS, decryption and/or re-encryption may occur. For example, if different keys are used, it will occur. If the same keys are used, it need not occur. Section 10.4.12 provides additional information on SRTP and keying mechanisms. If packet loss occurs, either from the UA to the SRC or from the SRC to the SRS, the SRS SHOULD detect and attempt to recover from the loss. The SRC does not play a role in this other than forwarding the associated RTP and RTCP packets.

10.4.8.2.2 SRC Acting as an RTP Mixer

In the case of the SRC acting as an RTP mixer, as defined in RFC 3550, the SRC combines RTP streams from different UAs and sends them towards the SRS using its own SSRC. The SSRCs from the contributing UA SHOULD be conveyed as CSRC identifiers within this stream. The SRC may make timing adjustments among the received streams and generate its own timing on the stream sent to the SRS. Optionally, an SRC acting as a mixer can perform transcoding and can even cope with different codings received from different UAs. RTCP Sender Reports and Receiver Reports are not forwarded by an SRC acting as a mixer, but there are requirements for forwarding RTCP SDES packets. The SRC generates its own RTCP Sender Reports and Receiver Reports towards the associated UAs and SRS.

The use of SRTP between the SRC and the SRS for the RS is independent of the use of SRTP between the UAs and the SRC for the CS. Section 10.4.12 provides additional information on SRTP and keying mechanisms. If packet loss occurs from the UA to the SRC, the SRC SHOULD detect and attempt to recover from the loss. If packet loss occurs from the SRC to the SRS, the SRS SHOULD detect and attempt to recover from the loss.

10.4.8.2.3 SRC Acting as an RTP Endpoint

The case of the SRC acting as an RTP endpoint, as defined in RFC 3550, is similar to the mixer case except that the RTP session between the SRC and the SRS is considered completely independent from the RTP session that is part of the CS. The SRC can, but need not, mix RTP streams from different participants prior to sending to the SRS. RTCP between the SRC and the SRS is completely independent of RTCP on the CS.

The use of SRTP between the SRC and the SRS for the RS is independent of the use of SRTP between the UAs and SRC for the CS. Section 10.4.12 provides additional information on SRTP and keying mechanisms. If packet loss occurs from the UA to the SRC, the SRC SHOULD detect and attempt to recover from the loss. If packet loss occurs from the SRC to the SRS, the SRS SHOULD detect and attempt to recover from the loss.

10.4.8.3 RTP Session Usage by SRC

There are multiple ways that an SRC may choose to deliver recorded media to an SRS. In some cases, it may use a single RTP session for all media within the RS, whereas in others, it may use multiple RTP sessions. The following sub-sections provide examples of basic RTP session usage by the SRC, including a discussion of how the RTP constructs and mechanisms covered previously are used. An SRC may choose to use one or more of the RTP session usages within a single RS. For the purpose of base interoperability between SRC and SRS, an SRC MUST support separate m-lines in SDP, one per CS media direction. The set of RTP session usages described is not meant to be exhaustive.

10.4.8.3.1 SRC Using Multiple m-lines

When using multiple m-lines, an SRC includes each m-line in an SDP offer to the SRS. The SDP answer from the SRS MUST include all m-lines, with any rejected m-lines indicated with a zero port per RFC 3264. Having received the answer, the SRC starts sending media to the SRS as indicated in the answer. Alternatively, if the SRC deems the level of support indicated in the answer to be unacceptable, it may initiate another SDP offer/answer exchange in which an alternative RTP session usage is negotiated.

In order to preserve the mapping of media to participant within the CSs in the RS, the SRC SHOULD map each unique CNAME within the CSs to a unique CNAME within the RS. Additionally, the SRC SHOULD map each unique combination of CNAME/SSRC within the CSs to a unique CNAME/SSRC within the RS. In doing so, the SRC may act as an RTP translator or as an RTP endpoint.

Figure 10.29 illustrates a case in which each UA represents a participant contributing two RTP sessions (e.g., one for audio and one for video), each with a single SSRC. The SRC acts as an RTP translator and delivers the media to the SRS using four RTP sessions, each with a single SSRC. The CNAME and SSRC values used by the UAs within their media streams are preserved in the media streams from the SRC to the SRS.

10.4.8.3.2 SRC Using Mixing

When using mixing, the SRC combines RTP streams from different participants and sends them towards the SRS using its own SSRC. The SSRCs from the contributing participants SHOULD be conveyed as CSRC identifiers. The SRC includes one m-line for each RTP session in an SDP offer to the SRS. The SDP answer from the SRS MUST include all m-lines, with any rejected m-lines indicated with a zero port per RFC 3264. Having received the answer, the SRC starts sending media to the SRS as indicated in the answer. In order to preserve the mapping of media to participant within the CSs in the RS, the SRC SHOULD map each unique CNAME within the CSs to a unique CNAME within the RS. Additionally, the SRC SHOULD map each unique combination of CNAME/SSRC within the CSs to a unique CNAME/SSRC within the RS. The SRC MUST avoid SSRC collisions, rewriting SSRCs if necessary when used as CSRCs in the RS. In doing so, the SRC acts as an RTP mixer. In the event that the SRS does not support this usage of CSRC values, it relies entirely on the SIPREC metadata to determine the participants included within each mixed stream. Figure 10.30 illustrates a case in which each UA represents a participant contributing two RTP sessions (e.g., one for audio and one for video), each with a single SSRC. The SRC acts as an RTP mixer and delivers the media to the SRS using two RTP

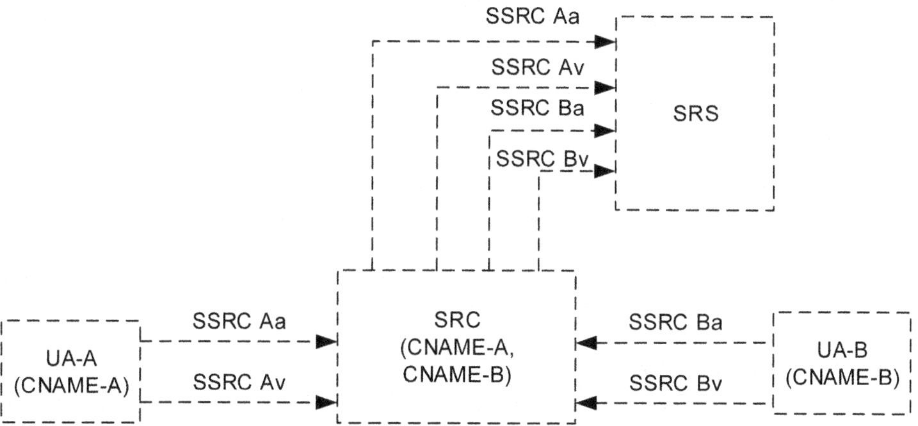

FIGURE 10.29 SRC Using Multiple m-Lines. (Copyright: IETF.)

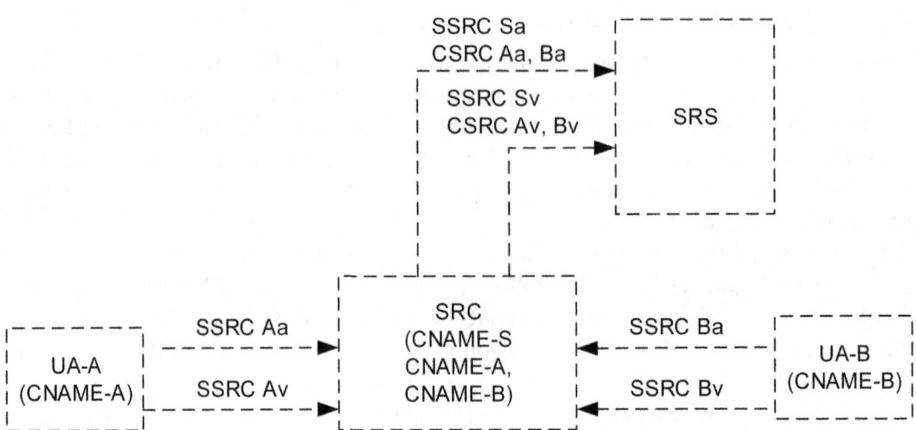

FIGURE 10.30 SRC Using Mixing. (Copyright: IETF.)

sessions, mixing media from each participant into a single RTP session containing a single SSRC and two CSRCs.

10.4.8.4 RTP Session Usage by SRS

An SRS that supports recording an audio CS MUST support SRC usage of separate audio m-lines in SDP, one per CS media direction. An SRS that supports recording a video CS MUST support SRC usage of separate video m-lines in SDP, one per CS media direction. Therefore, for an SRS supporting a typical audio call, the SRS has to support receiving at least two audio m-lines. For an SRS supporting a typical audio and video call, the SRS has to support receiving at least four total m-lines in the SDP – two audio m-lines and two video m-lines.

These requirements allow an SRS to be implemented that supports video only without requiring support for audio recording. They also allow an SRS to be implemented that supports recording only one direction of one stream in a CS – for example, an SRS designed to record security monitoring cameras that only send (not receive) video without any audio. These requirements were not written to prevent other modes from being implemented and used, such as using a single m-line and mixing the separate audio streams together. Rather, the requirements were written to provide a common base mode to implement for the sake of interoperability. It is important to note that an SRS implementation supporting the common base mode may not record all media streams in a CS if a participant supports more than one m-line in a video call, such as one for camera and one for presentation. SRS implementations may support other modes as well, but they have to at least support the modes discussed earlier, such that they interoperate in the common base mode for basic interoperability.

10.4.9 Metadata

Some metadata attributes are contained in SDP, and others are contained in a new content type called "application/rs-metadata." The format of the metadata is described as part of the mechanism in RFC 7865 (see Section 10.3). A new "disposition-type" of Content-Disposition is defined for the purpose of carrying metadata. The value is "recording-session," which indicates that the "application/rs-metadata" content contains metadata to be handled by the SRS.

10.4.9.1 Procedures at the SRC

The SRC MUST send metadata to the SRS in an RS. The SRC SHOULD send metadata as soon as it becomes available and whenever it changes. Cases in which an SRC may be justified in waiting temporarily before sending metadata include:

- waiting for a previous metadata exchange to complete (i.e., the SRC cannot send another SDP offer until the previous offer/answer completes and may also prefer not to send an UPDATE during this time)
- constraining the signaling rate on the RS
- sending metadata when key events occur rather than for every event that has any impact on metadata

The SRC may also be configured to suppress certain metadata out of concern for privacy or perceived lack of need for it to be included in the recording. Metadata sent by the SRC is categorized as either a full metadata snapshot or a partial update. A full metadata snapshot describes all metadata associated with the RS. The SRC MAY send a full metadata snapshot at any time. The SRC MAY send a partial update only if a full metadata snapshot has been sent previously. The SRC MAY send metadata (either a full metadata snapshot or a partial update) in an INVITE request, an UPDATE request (RFC 3311), or a 200 response to an offerless INVITE from the SRS. If the metadata contains a reference to any SDP labels, the request containing the metadata MUST also contain an SDP offer that defines those labels.

When a SIP message contains both an SDP offer and metadata, the request body MUST have content type "multipart/mixed," with one subordinate body part containing the SDP offer and another containing the metadata. When a SIP message contains only an SDP offer or metadata, the "multipart/mixed" container is optional. The SRC SHOULD include a full metadata snapshot in the initial INVITE request establishing the RS. If metadata is not yet available (e.g., an RS established in the absence of a CS), the SRC SHOULD send a full metadata snapshot as soon as metadata becomes available. If the SRC receives a snapshot request from the SRS, it MUST immediately send a full metadata snapshot.

Figure 10.31 illustrates an example of a full metadata snapshot sent by the SRC in the initial INVITE request.

10.4.9.2 Procedures at the SRS

The SRS receives metadata updates from the SRC in INVITE and UPDATE requests. Since the SRC can send partial updates based on the previous update, the SRS needs to keep track of the sequence of updates from the SRC. In the case of an internal failure at the SRS, the SRS may fail to recognize a partial update from the SRC. The SRS may be able to recover from the internal failure by requesting a full metadata snapshot from the SRC. Certain errors, such as syntax errors or semantic errors in the metadata information, are likely caused by an error on the SRC side, and it is likely that the same error will occur again even when a full metadata snapshot is requested. In order to avoid repeating the same error, the SRS can simply terminate the RS when a syntax error or semantic error is detected in the metadata.

The SRS MAY explicitly request a full metadata snapshot by sending an UPDATE request. This request MUST contain a body with Content-Disposition type "recording-session" and MUST NOT contain an SDP body. The SRS MUST NOT request a full metadata snapshot in an UPDATE response or in any other SIP transaction. The format of the content is "application/rs-metadata," and the body is an XML specification, the format of which is defined in RFC 7865 (see Section 10.3). Figure 10.32 shows an example.

Note that UPDATE was chosen for the SRS to request a metadata snapshot, because it can be sent regardless of the state of the dialog. This was seen as better than requiring support for both UPDATE and re-INVITE messages for this operation. When the SRC receives a request for a metadata snapshot, it MUST immediately provide a full metadata snapshot in a separate INVITE or UPDATE transaction. Any subsequent partial updates will not be dependent on any metadata sent prior to this full metadata snapshot.

The metadata received by the SRS can contain ID elements used to cross-reference one element to another. An element containing the definition of an ID and an element containing

```
INVITE sip:recorder@example.com SIP/2.0
Via: SIP/2.0/TCP src.example.com;branch=z9hG4bKdf6b622b648d9
From: <sip:2000@example.com>;tag=35e195d2-947d-4585-946f-09839247
To: <sip:recorder@example.com>
Call-ID: d253c800-b0d1ea39-4a7dd-3f0e20a
CSeq: 101 INVITE
Max-Forwards: 70
Require: siprec
Accept: application/sdp, application/rs-metadata
Contact: <sip:2000@src.example.com>;+sip.src
Content-Type: multipart/mixed;boundary=foobar
Content-Length: [length]
--foobar
Content-Type: application/sdp
v=0
o=SRS 2890844526 2890844526 IN IP4 198.51.100.1
s=-
c=IN IP4 198.51.100.1
t=0 0
m=audio 12240 RTP/AVP 0 4 8
a=sendonly
a=label:1
--foobar
Content-Type: application/rs-metadata
Content-Disposition: recording-session
[metadata content]
```

FIGURE 10.31 Sample INVITE Request for the Recording Session.

a reference to that ID will often be received from the same SRC. It is also valid for those elements to be received from different SRCs – for example, when each endpoint in the same CS acts as an SRC to record the call, and a common ID refers to the same CS. The SRS MUST NOT consider this an error.

10.4.10 Persistent Recording

Persistent recording is a specific use case addressing REQ-005 in RFC 6341 (see Section 10.1), where an RS can be established in the absence of a CS. The SRC continuously records media in an RS to the SRS even in the absence of a CS for all UAs that are part of persistent recording. By allocating recorded streams and continuously sending recorded media to the SRS, the SRC does not have to prepare new recorded streams with a new SDP offer when a new CS is created and also does not impact the timing of the CS. The SRC only needs to update the metadata when new CSs are created.

When there is no CS running on the devices with persistent recording, there is no recorded media to stream from the SRC to the SRS. In certain environments where an NAT is used, a minimum amount of flow activity is typically required to maintain the NAT binding for each port opened. Agents that support Interactive Connectivity Establishment (ICE) solve this problem. For non-ICE

```
UPDATE sip:2000@src.example.com SIP/2.0
Via: SIP/2.0/UDP srs.example.com;branch=z9hG4bKdf6b622b648d9
To: <sip:2000@example.com>;tag=35e195d2-947d-4585-946f-098392474
From: <sip:recorder@example.com>;tag=1234567890
Call-ID: d253c800-b0d1ea39-4a7dd-3f0e20a
CSeq: 1 UPDATE
Max-Forwards: 70
Require: siprec
Contact: <sip:recorder@srs.example.com>;+sip.srs
Accept: application/sdp, application/rs-metadata
Content-Disposition: recording-session
Content-Type: application/rs-metadata
Content-Length: [length]
<?xml version="1.0" encoding="UTF-8"?>
<requestsnapshot xmlns='urn:ietf:params:xml:ns:recording:1'>
        <requestreason xml:lang="it">SRS internal error</requestreason>
</requestsnapshot>
```

FIGURE 10.32 Metadata Request.

agents, in order not to lose the NAT bindings for the RTP/RTCP ports opened for the recorded streams, the SRC and SRS SHOULD follow the recommendations provided in RFC 6263 to maintain the NAT bindings.

10.4.11 IANA Considerations

This specification registers some tags and attributes in IANA. The detail can be seen in RFC 7866.

10.4.12 Security Considerations

The RS is fundamentally a standard SIP dialog (RFC 3261 – also see Section 1.2); therefore, the RS can reuse any of the existing SIP security mechanisms available for securing the session signaling, the recorded media, and the metadata. The use cases and requirements specification (RFC 6341 – see Section 10.1) outlines the general security considerations, and this specification describes specific security recommendations. The SRC and SRS MUST support SIP with transport layer security (TLS) version 1.2, SHOULD follow the best practices when using TLS as per RFC 7525, and MAY use session initiation protocol secure (SIPS) with TLS as per RFC 5630. The RS MUST be at least as secure as the CS; this means using at least the same strength of ciphersuite as the CS if the CS is secured. For example, if the CS uses SIPS for signaling and RTP/SAVP for media, then the RS may not use SIP or plain RTP unless other equivalent security measures are in effect, since doing so would mean an effective security downgrade. Examples of other potentially equivalent security mechanisms include mutually authenticated TLS for the RS signaling channel or an appropriately protected network path for the RS media component.

10.4.12.1 Authentication and Authorization

At the transport level, the RS uses TLS authentication to validate the authenticity of the SRC and SRS. The SRC and SRS MUST implement TLS mutual authentication for establishing the RS.

Whether the SRC/SRS chooses to use TLS mutual authentication is a deployment decision. In deployments where a UA acts as its own SRC, this requires that the UA have its own certificate as needed for TLS mutual authentication. In deployments where the SRC and the SRS are in the same administrative domain and have some other means of ensuring authenticity, the SRC and SRS may choose not to authenticate each other or to have the SRC authenticate the SRS only. In deployments where the SRS can be hosted on a different administrative domain, it is important to perform mutual authentication to ensure the authenticity of both the SRC and the SRS before transmitting any recorded media. The risk of not authenticating the SRS is that the recording may be sent to an entity other than the intended SRS, allowing a sensitive call recording to be received by an attacker. On the other hand, the risk of not authenticating the SRC is that an SRS will accept calls from an unknown SRC and allow potential forgery of call recordings.

There may be scenarios in which the signaling between the SRC and SRS is not direct; e.g., an SIP proxy exists between the SRC and the SRS. In such scenarios, each hop is subject to the TLS mutual authentication constraint, and transitive trust at each hop is utilized. Additionally, an SRC or SRS may use other existing SIP mechanisms available, including, but not limited to, Digest authentication (RFC 3261), asserted identity (RFC 3325), and connected identity (RFC 4916). The SRS may have its own set of recording policies to authorize recording requests from the SRC. The use of recording policies is outside the scope of the Session Recording Protocol.

10.4.12.2 RTP Handling

In many scenarios, it will be critical for the media transported between the SRC and the SRS to be protected. Media encryption is an important element in the overall SIPREC solution; therefore, the SRC and the SRS MUST support RTP/SAVP (RFC 3711) and RTP/SAVPF (RFC 5124). RTP/SAVP and RTP/SAVPF provide media encryption, integrity protection, replay protection, and a limited form of source authentication. They do not contain or require a specific keying mechanism. At a minimum, the SRC and SRS MUST support the SDP security descriptions key negotiation mechanism (RFC 4568). For cases in which Datagram Transport Layer Security for Secure RTP (DTLS-SRTP) is used to encrypt a CS media stream, an SRC may use SRTP Encrypted Key Transport (EKT) [2] in order to use SRTP-SDES in the RS without needing to re-encrypt the media.

Note: When using EKT in this manner, it is possible for participants in the CS to send traffic that appears to be from other participants and have this forwarded by the SRC to the SRS within the RS. If this is a concern (e.g., if the RS is intended for audit or compliance purposes), EKT is not an appropriate choice. When RTP/SAVP or RTP/SAVPF is used, an SRC can choose to use the same keys or different keys in the RS than those used in the CS. Some SRCs are designed to simply replicate RTP packets from a CS media stream to the SRS, in which case the SRC will use the same key in the RS as the key used in the CS.

In this case, the SRC MUST secure the SDP containing the keying material in the RS with at least the same level of security as in the CS. The risk of lowering the level of security in the RS is that it will effectively become a downgrade attack on the CS, since the same key is used for both the CS and the RS. SRCs that decrypt an encrypted CS media stream and re-encrypt it when sending it to the SRS MUST use a different key than the one used for the CS media stream to ensure that it is not possible for someone who has the key for the CS media stream to access recorded data they are not authorized to access. In order to maintain a comparable level of security, the key used in the RS SHOULD be of equivalent strength to, or greater strength than, that used in the CS.

10.4.12.3 Metadata

Metadata contains sensitive information, such as the AoR of the participants and other extension data placed by the SRC. It is essential to protect the content of the metadata in the RS. Since metadata is a content type transmitted in SIP signaling, metadata SHOULD be protected at the transport level by SIPS/TLS.

10.4.12.4 Storage and Playback

While storage and playback of the call recording are beyond the scope of this specification, it is worthwhile to mention here that it is also important for the recording storage and playback to provide a level of security that is comparable to the CS. It would defeat the purpose of securing both the CS and the RS, mentioned in the previous sections, if the recording can be easily played back with a simple, unsecured HTTP interface without any form of authentication or authorization.

10.5 SUMMARY

We have described the use case scenarios for the media session recording or simply "media recording" to exemplify the critical requirements in XCOM conference communications environments such as business teleconferences and videoconferences, call centers, and financial trading floors for business purposes, quality control, and analytics, including recording of calls for regulatory and compliance reasons. We have explained how recording is performed by sending a copy of the session media to the recording devices, and the SIP with extensions chosen for communications between the SRS and the devices termed as SIP UAs. The detail requirements for SIP-based media recording are defined, including privacy and security.

The media recording architecture using SP, SDP, and RTP is articulated, satisfying the SIP-based media recording requirements. The SIPREC architectures are developed, showing how the SRC is a logical function that can be located in or split between various physical components, how an SIP endpoint or SIP UA can act as a SRC, how an SIP UA is architected as the SRS where the MS acts as a sink of the recording media and the AS acts as a sink of the metadata and the termination point for RS SIP signaling, how an SIP UA is architected as the SRS where the MS acts as a source of recording media and the AS acts as a source of the metadata and the termination point for RS SIP signaling, and interaction of the conference focus where a combination of the focus and MS can act as the SRS. The architectures for establishing the recorded session are described in various scenarios: SRC-Initiated Recording, SRS-Initiated Recording, Pause/Resume Recording Session, Media Stream Mixing, Media Transcoding, and Lossless Recording. For Recording Metadata, two kinds of architectures are discussed: Contents of Recording Metadata and Mechanisms for Delivery of Metadata to SRS. The Notifications to the Recorded User Agents, either through an announcement or providing some warning tones, are explained. We also discuss media recording architecture for (a) locating a SRC that can be a B2BUA or an endpoint including interaction with media control channel conference focus, (b) establishing the recording session that includes SRC- or SRS-initiated recording, pause/resume of the recording session, media stream mixing, media transcoding, and lossless recording, (c) recording metadata that includes metadata recording and mechanisms for delivering of metadata to SRS, (d) notifications to recording user agents, (e) indication by a recording-aware UA's preference with regard to recording the media in the CS, whether media recording could be performed or not, and (f) the security for these media session architectures is explained using SIP-based media recording signaling session.

We have described the metadata information that describes the recorded media and the CS to which they relate. The metadata model, which is a class diagram in unified modeling language, is described, explaining the structure of metadata in general by showing the classes, their attributes, the relationships among the classes, and linkages between the classes representing the relationships while the metadata is conveyed from the SRC to the SRS. The Recording Metadata Classes with recording session, communication session, CS-RS association, participant, media stream, participant-stream association, syntax of XML elements, format of unique ID, and metadata version indicator are described here.

The media session recording protocol that uses SIP with extensions is specified, describing how SIP, SDP, and RTP are used for delivering real-time media and metadata from a CS to a recording device in good detail. Specifically, protocol operations, SIP handling, SDP handling, RTP handling, metadata, and persistent recording of the media session recording protocol are explained in great length. The security for RTP handling, metadata, storage, and playback for the media session recording protocol are described along with authentication and authorization.

10.6 PROBLEMS

1. Why do we need media session recording in XCON conference systems? How is the media recording performed? Explain in detail.
2. Describe the following use case scenarios for media recording along with call flows: Full-time Recording, Selective Recording, Start/Stop a Recording Session during a Communication Session, Persistent Recording, Real-time Recording Controls, IVR/Voice Portal Recording, Enterprise Mobility Recording, Geographically distributed or centralized recording, Record complex call scenarios, High availability and continuous recording, Record multi-channel, multimedia session, and Real-time media processing.
3. What are the reasons why SIP has been chosen for media recording? Justify your answer with detailed technical explanations, including drawing architectures and call flows.
4. Describe the requirements for SIP-based media recording, pointing to the capabilities needed for the use case scenarios that are described in Problem 2.
5. What are the privacy and security concerns for media recording? Explain in detail, considering the fact of the use case scenarios that are listed in Problem 2.
6. Develop a media recording architecture using SP, SDP, and RTP for a use case scenario. Justify how the architecture meets the media recording requirements described in Problem 4.
7. Describe in detail, drawing the media recording architectures for the following scenarios: (a) SRC, a logical function, which can be located in or split between various physical components; an SIP endpoint (i.e., SIP UA) can act as a SRC; (b) an SIP UA is architected as the SRS where the MS acts as a sink of the recording media, and the AS acts as a sink of the metadata and the termination point for RS SIP signaling; (c) an SIP UA architected as the SRS where the MS acts as a source of recording media and the AS acts as a source of the metadata and the termination point for RS SIP signaling; and (d) interaction of the conference focus where a combination of the focus and the MS can act as the SRS.
8. Explain in detail, drawing the media recording architectures for establishing the recorded session for the following scenarios: (a) SRC-Initiated Recording, (b) SRS-Initiated Recording, (c) Pause/Resume Recording Session, (d) Media Stream Mixing, (e) Media Transcoding, and (f) Lossless Recording.
9. Draw architectures and explain in detail Recording Metadata for the following: (a) Contents of Recording Metadata and (b) Mechanisms for Delivery of Metadata to SRS.
10. Explain in detail Notifications to the Recorded User Agents, either through an announcement or by providing some warning tones, drawing the architecture.
11. Draw media recording architectures and explain in detail for the following scenario: A recording-aware UA providing an indication of its preference with regard to recording the media in the CS, whether or not media recording could be performed during the initial session establishment as well as during an established session.
12. Describe in detail the security for all these media session architectures of the SIP-based media recording signaling session: Problems 7 through 11.
13. What is metadata? What are the metadata model and metadata format? What is the utility of the metadata model and metadata format for using in media recording? Justify your answer, showing any media recording architecture.
14. Draw UML class diagrams showing the classes, their attributes, the relationships among the classes, and linkages between the classes and explain in detail for the following as applicable: (a) Recording Metadata Classes with recording session, (b) Communication

session, (c) CS-RS Association, (d) Participant, (e) Media Stream, (f) Participant-Stream Association, (g) Syntax of XML Elements, (h) Format of Unique ID, (i) Metadata Version Indicator, (j) Recording Metadata Snapshot Request Format, (k) Recording Metadata Snapshot Request Format, and (l) XML Schema Definition for Recording Metadata.

15. What are the extensions that are made in SIP in order to satisfy the SIPREC requirements? Explain, using an SIP-based media recording architecture with the use of SIP, SDP, and RTP, for the following, drawing a media recording architecture: Delivering real-time media and metadata from a CS to a recording device.
16. Describe in detail the media recording architectures with call flows for the SIP-SDP-RTP-based media recording architecture for the following: (a) Protocol Operations, (b) SIP Handling, (c) SDP Handling, (d) RTP Handling, (e) Metadata, and (f) Persistent Recording of the Media Session.
17. What are the security concerns for the SIP-based media recording protocol for RTP handling, metadata, storage, and playback for the media session? Explain in detail.

REFERENCES

1. [UML] Object Management Group. (2011). OMG Unified Modeling Language (UML). <http://www.omg.org/spec/UML/2.4/>.
2. EKT-SRTP et al. (2019, July). Encrypted Key Transport for DTLS and Secure RTP. Work in Progress, draft-ietf-perc-srtp-ekt-diet-10.

REFERENCES

11 Media Resource Brokering

11.1 INTRODUCTION

As Internet protocol (IP)-based multimedia infrastructures mature, the complexity of and the demands from deployments increase. Such complexity will result in a wide variety of capabilities from a range of vendors, which should all be interoperable using the architecture and protocols produced by the MediaCtrl working group. It should be possible for a controlling entity to be assisted in Media Server (MS) selection so that the most appropriate resource is selected for a particular operation. The importance increases when one introduces a flexible level of deployment scenarios, as specified in RFC 5167 (see Section 8.2) and RFC 5567 (see Chapter 3). These specifications make statements like "it should be possible to have a many-to-many relationship between Application Servers and Media Servers that use this protocol." This leads to the following deployment architectures being possible when considering media resources to provide what can be effectively described as media resource brokering (RFC 6917). The simplest deployment view is illustrated in Figure 11.1.

This simply involves a single Application Server (AS) and MS. Expanding on this view, it is also possible for an AS to control multiple (more than 1) MS instances at any one time. This deployment view is illustrated in Figure 11.2. Typically, such architectures are associated with application logic that requires high-demand media services. It is more than possible that each MS possesses a different media capability set. MSs may offer different media services, as specified in the MediaCtrl architecture specification RFC 5567 (see Section 3). An MS may have similar media functionality but may have different capacity or media codec support.

Figure 11.3 conveys the opposite view to that in Figure 11.2. In this model, there are a number of (more than 1) ASs, possibly supporting dissimilar applications, controlling a single MS. Typically, such architectures are associated with application logic that requires low-demand media services.

The final deployment view is the most complex (Figure 11.4). In this model (M:N), there exist any number of ASs and any number of MSs. It is again possible in this model that MSs might not be homogeneous; they might have different capability sets and capacities.

The remaining sections in this specification will focus on a new entity called a Media Resource Broker (MRB), which can be utilized in the deployment architectures described previously in this section. The MRB entity provides the ability to obtain media resource information and appropriately allocate (broker) on behalf of client applications. The high-level deployment options discussed in this section rely on network architecture and policy to prohibit inappropriate use. Such policies are out of scope for this specification. This specification will take a look at the specific problem areas related to such deployment architectures. It is recognized that the solutions proposed in this specification should be equally adaptable to all the previously described deployment models. It is also recognized that the solution is far more relevant to some of the previously discussed deployment models and can almost be viewed as redundant for others.

11.2 CONVENTIONS AND TERMINOLOGY

This specification inherits terminology proposed in RFC 5567 (see Chapter 3) and in "Media Control Channel Framework" (RFC 6230 – see Chapter 8). In addition, the following terms (also see Section 1.6, Table 1.2) are defined for use in this specification and for use in the context of the MediaCtrl working group in the Internet Engineering Task Force (IETF):

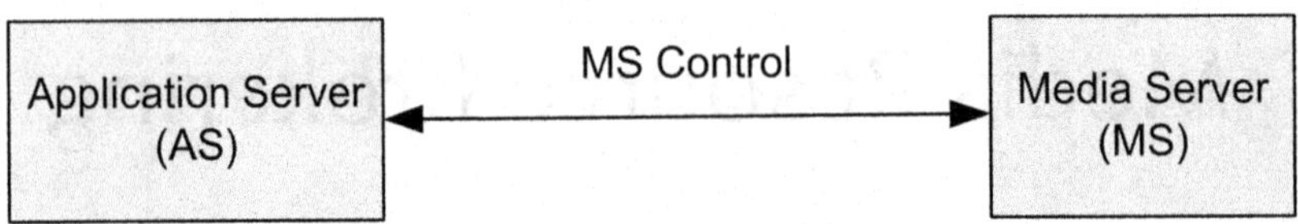

FIGURE 11.1 Basic Architecture.

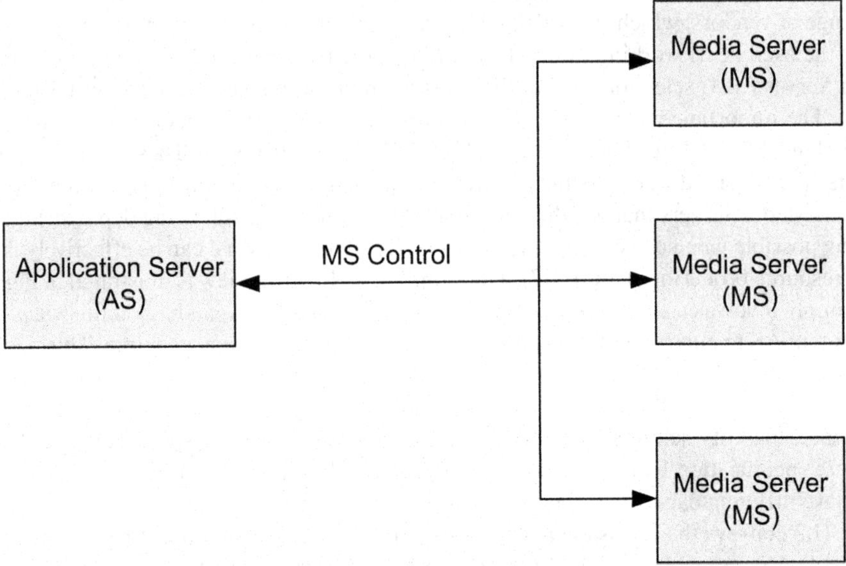

FIGURE 11.2 Multiple Media Servers.

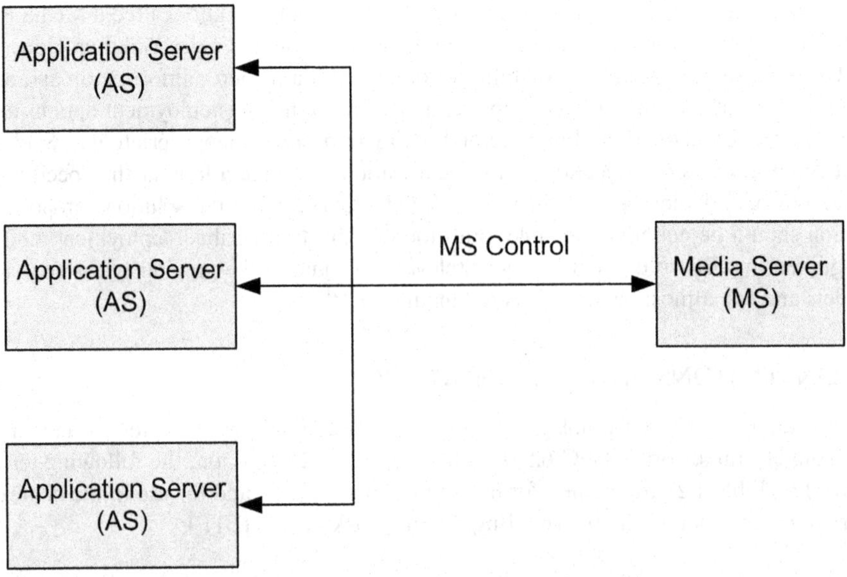

FIGURE 11.3 Multiple Application Servers.

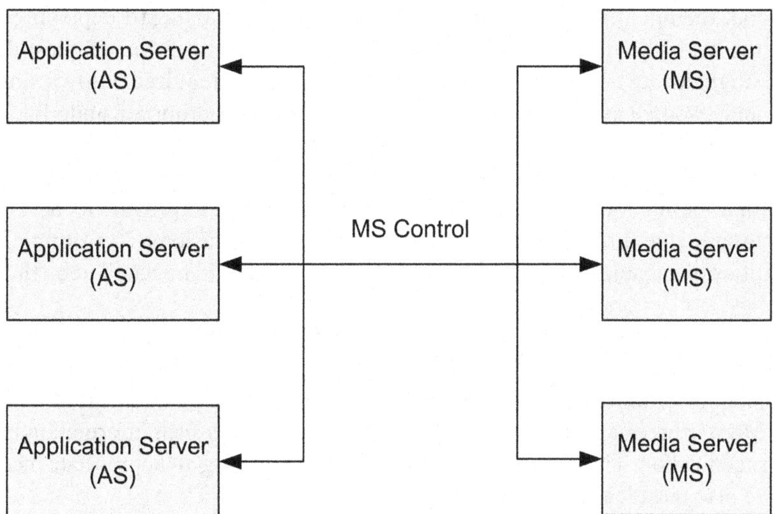

FIGURE 11.4 Many-to-Many Architecture.

- Media Resource Broker (MRB): A logical entity that is responsible for both collection of appropriate published MS information and selecting appropriate MS resources on behalf of consuming entities.
- Query MRB: An instantiation of an MRB (see previous definition) that provides an interface for an AS to retrieve the address of an appropriate MS. The result returned to the AS can be influenced by information contained in the query request.
- In-line MRB: An instantiation of an MRB (see previous definition) that directly receives requests on the signaling path. There is no separate query.
- CFW: Media Control Channel Framework, as specified in RFC 6230 (see Chapter 8).

Within the context of in-line MRBs, additional terms are defined:

- In-line Aware MRB Mode (IAMM): Defined in Section 11.5.2.2.1.
- In-line Unaware MRB Mode (IUMM): Defined in Section 11.5.3.

The specification will often specify when a specific identifier in a protocol message needs to be unique. Unless stated otherwise, such uniqueness will always be within the scope of the MSs controlled by the same MRB. The interaction between different MRB instances, e.g., the partitioning of a logical MRB, is out of scope for this specification.

11.3 PROBLEM DISCUSSION

As discussed in Chapter 1, a goal of the MediaCtrl working group is to produce a solution that will service a wide variety of deployment architectures. Such architectures range from the simplest 1:1 relationship between MSs and ASs to potentially linearly scaling 1:M, M:1, and M:N deployments. Managing such deployments is itself non-trivial for the proposed solution until an additional number of factors that increase complexity are included in the equation. As MSs evolve, it must be taken into consideration that where many can exist in a deployment, they may not have been produced by the same vendor and may not have the same capability set. It should be possible for an AS that exists in a deployment to select a media service based on a common, appropriate capability set. In conjunction with capabilities, it is also important to take available resources into consideration. The ability to select an appropriate media service function is an extremely useful feature but becomes even more powerful when considered with available resources for servicing a request.

In conclusion, the intention is to create a toolkit that allows MediaCtrl deployments to effectively utilize the available media resources. It should be noted that in the simplest deployments where only a single MS exists, an MRB function is probably not required. Only a single capability set exists, and resource availability can be handled using the appropriate underlying signaling, e.g., session initiation protocol (SIP) response. This specification does not prohibit such uses of an MRB; it simply provides the tools for various entities to interact where appropriate. It is also worth noting that the functions specified in this specification aim to provide a "best effort" view of media resources at the time of request for initial MS routing decisions. Any dramatic change in media capabilities or capacity after a request has taken place should be handled by the underlying protocol.

It should be noted that there may be additional information that is desirable for the MRB to have for the purposes of selecting a MS resource, such as resource allocation rules across different applications, planned or unplanned downtime of MS resources, the planned addition of future MS resources, or MS resource capacity models. How the MRB acquires such information is outside the scope of this specification. The specific techniques used for selecting an appropriate media resource by an MRB are also outside the scope of this specification.

11.4 DEPLOYMENT SCENARIO OPTIONS

Research into media resource brokering concluded that a couple of high-level models provided an appropriate level of flexibility. The general principles of "in-line" and "query" MRB concepts are discussed in the rest of this section. It should be noted that while the interfaces are different, they both use common underlying mechanisms defined in this specification.

11.4.1 Query MRB

The "Query" model for MRB interactions provides the ability for a client of media services (for example, an AS) to "ask" an MRB for an appropriate MS, as illustrated in Figure 11.5.

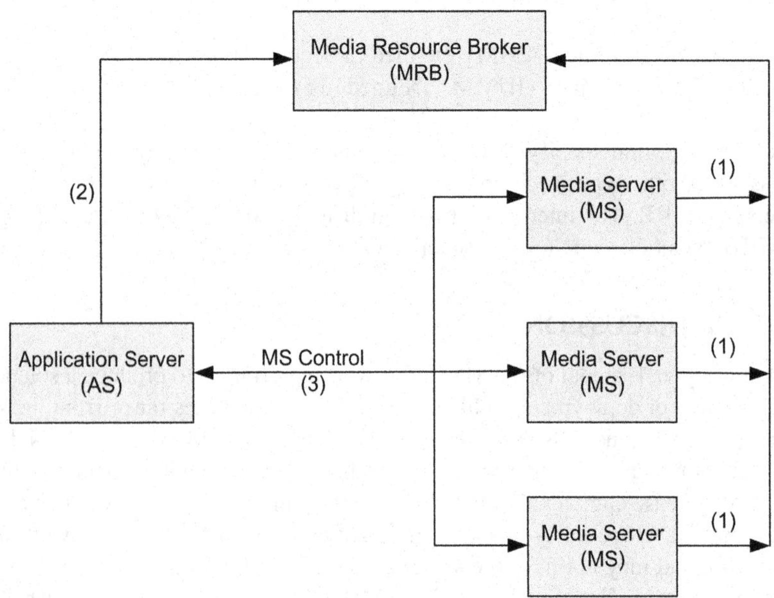

FIGURE 11.5 Query MRB.

In this deployment, the MSs use the MS Resource Publish interface, as discussed in Section 11.5.1, to convey capability sets as well as resource information. This is depicted by (1) in Figure 11.5. It is then the MRB's responsibility to accumulate all appropriate information relating to media services in the logical deployment cluster. The AS (or other media services client) is then able to query the MRB for an appropriate resource, as identified by (2) in Figure 11.5. Such a query would carry specific information related to the media service required and enable the MRB to provide increased accuracy in its response. This particular interface is discussed in "Media Service Resource Consumer Interface" (Section 11.5.2). The AS is then able to direct control commands (for example, create a conference) and media dialogs to the appropriate MS, as shown by (3) in Figure 11.5. Additionally, with Query mode, the MRB is not directly in the signaling path between the AS and the selected MS resource.

11.4.1.1 Hybrid Query MRB

As mentioned previously, it is the intention that a toolkit is provided for MRB functionality within a MediaCtrl architecture. It is expected that in specific deployment scenarios, the role of the MRB might be co-hosted as a hybrid logical entity with an AS, as shown in Figure 11.6.

This diagram is identical to that in Figure 11.5 with the exception that the MRB is now hosted on the AS. The MS Publish interface is still being used to accumulate resource information at the MRB, but as it is co-hosted on the AS, the MS Consumer interface has collapsed. It might still exist within the AS/MRB interaction, but this is an implementation issue. This type of deployment suits a single-AS environment, but it should be noted that an MS Consumer interface could then be offered from the hybrid if required. In a similar manner, the MS could also act as a hybrid for the deployment cluster, as illustrated in Figure 11.7.

In this example, the MRB has collapsed and is co-hosted by the MS. The MS Consumer interface is still available to the ASs (1) to query MS resources. The MS Publish interface has collapsed onto the MS. It might still exist within the MS/MRB interaction, but this is an implementation issue. This type of deployment suits a single-MS environment, but it should be noted that an MS Publish interface could then be offered from the hybrid if required. A typical use case scenario for such a topology would be a single MS representing a pool of MSs in a cluster. In this case, the MRB would actually be handling a cluster of MSs rather than one.

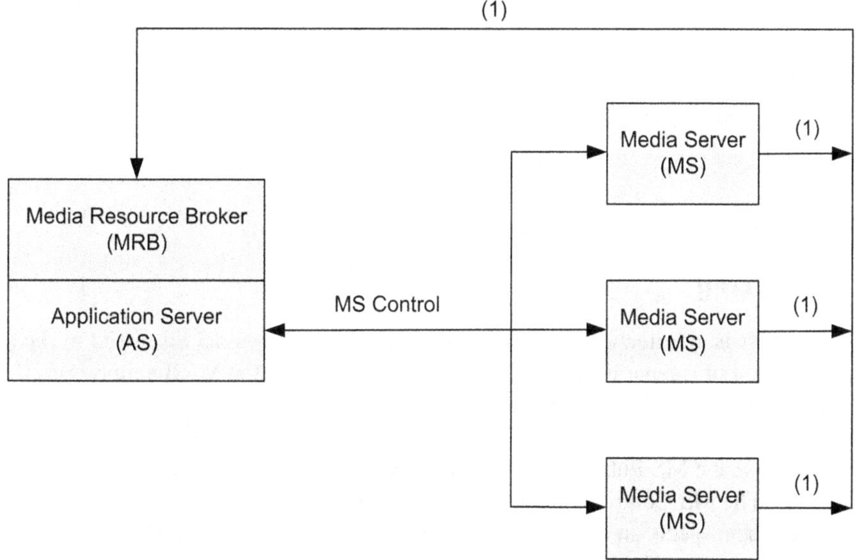

FIGURE 11.6 Hybrid Query MRB – Application Server Hosted.

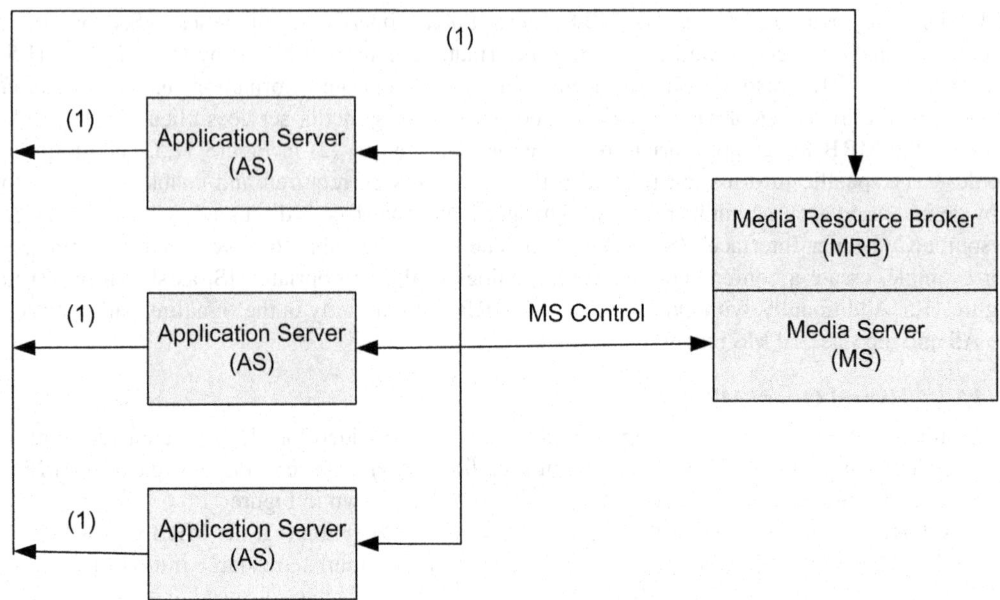

FIGURE 11.7 Hybrid Query MRB – Media Server Hosted.

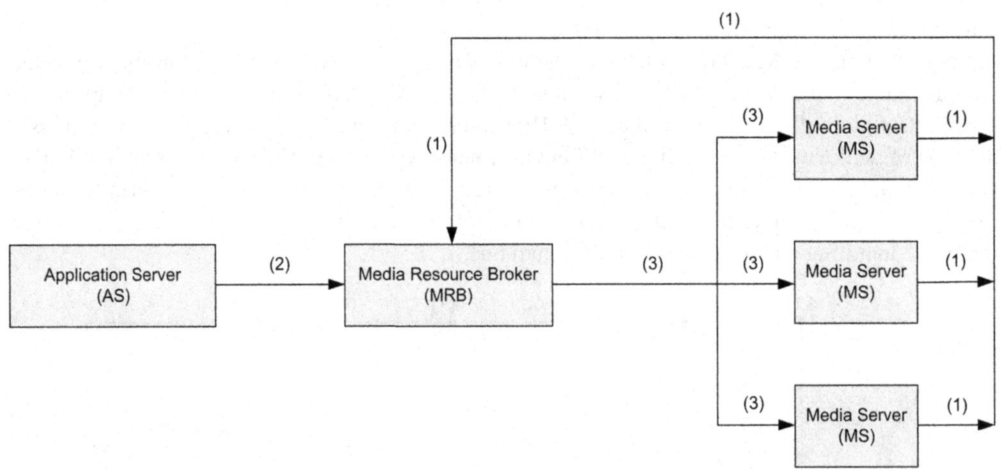

FIGURE 11.8 In-line MRB.

11.4.2 In-Line MRB

The "In-line" MRB is architecturally different from the "Query" model discussed in the previous section. The concept of a separate query disappears. The client of the MRB simply uses the media resource control and media dialog signaling to involve the MRB. This type of deployment is illustrated in Figure 11.8.

The MSs still use the MS Publish interface to convey capabilities and resources to the MRB, as illustrated by (1). The MS Control Channels (and media dialogs as well, if required) are sent to the MRB (2), which then selects an appropriate MS (3) and remains in the signaling path between the AS and the MS resources. The in-line MRB can be split into two distinct logical roles that can be applied on a per-request basis. They are:

- In-line Unaware MRB Mode (IUMM): Allows an MRB to act on behalf of clients requiring media services who are not aware of an MRB or its operation. In this case, the AS does not provide explicit information on the kind of MS resource it needs (as in Section 11.5.2), and the MRB is left to deduce it by potentially inspecting other information in the request from the AS (for example, session description protocol [SDP] content, or address of the requesting AS, or additional Request-URI parameters as per RFC 4240).
- In-line Aware MRB Mode (IAMM): Allows an MRB to act on behalf of clients requiring media services who are aware of an MRB and its operation. In particular, it allows the AS to explicitly convey matching characteristics to those provided by MSs, as does the Query MRB mode (as in Section 11.5.2).

In either of the previously described roles, signaling as specified by the Media Control Channel Framework (RFC 6230 – see Chapter 8) would be involved, and the MRB would deduce that the selected MS resources are no longer needed when the AS or MS terminates the corresponding SIP dialog. The two modes are discussed in more detail in Section 11.5.3.

11.5 MRB INTERFACE DEFINITIONS

The intention of this specification is to provide a toolkit for a variety of deployment architectures where media resource brokering can take place. Two main interfaces are required to support the differing requirements. The two interfaces are described in the remainder of this section and have been named the MS Resource Publish and MS Resource Consumer interfaces. It is beyond the scope of this specification to define exactly how to construct an MRB using the interfaces described. It is, however, important that the two interfaces are complementary so that the development of appropriate MRB functionality is supported.

11.5.1 Media Server Resource Publish Interface

The MS Resource Publish interface is responsible for providing an MRB with appropriate MS resource information. As such, this interface is assumed to provide both general and specific details related to MS resources. This information needs to be conveyed using an industry standard mechanism to provide increased levels of adoption and interoperability. A Control Package for the CFW will be specified to fulfill this interface requirement. It provides an establishment and monitoring mechanism to enable an MS to report appropriate statistics to an MRB. The Publish interface is used with both the Query mode and the In-line mode of MRB operation.

As already discussed in Section 11.1, the MRB view of MS resource availability will in reality be approximate — i.e., partial and imperfect. The MRB Publish interface does not provide an exhaustive view of current MS resource consumption; the MS may in some cases provide a best-effort computed view of resource consumption parameters conveyed in the Publish interface (e.g., digital signal processors [DSPs] with a fixed number of streams versus graphics processing units [GPUs] with central processing unit [CPU] availability). Media resource information may only be reported periodically over the Publish interface to an MRB.

It is also worth noting that while the scope of the MRB is in providing interested ASs with the available resources, the MRB also allows the retrieval of information about consumed resources. While this is, of course, a relevant piece of information (e.g., for monitoring purposes), such functionality inevitably raises security considerations, and implementations should take this into account. See Section 11.12 for more details. The MRB Publish interface uses the CFW (RFC 6230 – see Chapter 8) as the basis for interaction between a MS and an MRB. The CFW uses an extension mechanism to allow specific usages that are known as Control Packages. Section 11.5.1.1 defines the Control Package that MUST be implemented by any MS wanting to interact with an MRB entity.

11.5.1.1 Control Package Definition

This section fulfills the requirement for information that must be specified during the definition of a Control Framework package, as detailed in Section 8.8 (Section 8 of RFC 6230).

11.5.1.1.1 Control Package Name

The CFW requires a Control Package definition to specify and register a unique name and version. The name and version of this Control Package is "mrb-publish/1.0."

11.5.1.1.2 Framework Message Usage

The MRB Publish interface allows an MS to convey available capabilities and resources to an MRB entity. This package defines XML elements in Section 11.5.1.2 and provides an XML schema in Section 11.10. The XML elements in this package are split into requests, responses, and event notifications. Requests are carried in CONTROL message bodies; the <mrbrequest> element is defined as a package request. This request can be used for creating new subscriptions and updating/removing existing subscriptions. Event notifications are also carried in CONTROL message bodies; the <mrbnotification> element is defined for package event notifications. Responses are carried either in REPORT message or in Control Framework 200 response bodies; the <mrbresponse> element is defined as a package-level response.

Note that package responses are different from framework response codes. Framework error response codes (see Section 8.7 [Section 7 of RFC 6230]) are used when the request or event notification is invalid; for example, a request has invalid XML (400) or is not understood (500). Package-level responses are carried in framework 200 response or REPORT message bodies. This package's response codes are defined in Section 11.5.1.4.

11.5.1.1.3 Common XML Support

The CFW (RFC 6230 – see Chapter 8) requires a Control Package definition to specify whether the attributes for media dialog or conference references are required. The Publish interface defined in Chapter 10 imports and makes use of the common XML schema defined in the CFW. The Consumer interface defined in Section 11.11 imports and makes use of the common XML schema defined in the CFW.

11.5.1.1.4 CONTROL Message Body

A valid CONTROL message body MUST conform to the schema defined in Section 11.10 and described in Section 11.5.1.2. XML messages appearing in CONTROL messages MUST contain either an <mrbrequest> or an <mrbnotification> element.

11.5.1.1.5 REPORT Message Body

A valid REPORT message body MUST conform to the schema defined in Section 11.10 and described in Section 11.5.1.2. XML messages appearing in REPORT messages MUST contain an <mrbresponse> element.

11.5.1.1.6 Audit

The "mrb-publish/1.0" CFW package does not require any additional auditing capability.

11.5.1.2 Element Definitions

This section defines the XML elements for the Publish interface Media Control Channel package defined in Section 11.5.1. The formal XML schema definition for the Publish interface can be found in Section 11.10. The root element is <mrbpublish>. All other XML elements (requests, responses, and notifications) are contained within it. The MRB Publish interface request element is detailed in Section 11.5.1.3. The MRB Publish interface notification element is detailed in Section 11.5.1.5. The MRB Publish interface response element is detailed in Section 11.5.1.4.

The <mrbpublish> element has the following attributes: version: A token specifying the mrbpublish package version. The value is fixed as "1.0" for this version of the package. The attribute MUST be present. The <mrbpublish> element has the following child elements, and there MUST NOT be more than one such child element in any <mrbpublish> message:

- <mrbrequest> for sending an MRB request. See Section 11.5.1.3.
- <mrbresponse> for sending an MRB response. See Section 11.5.1.4.
- <mrbnotification> for sending an MRB notification. See Section 11.5.1.5.

11.5.1.3 <mrbrequest>

This section defines the <mrbrequest> element used to initiate requests from an MRB to an MS. The element describes information relevant for the interrogation of an MS. The <mrbrequest> element has no defined attributes. The <mrbrequest> element has the following child element: <subscription> for initiating a subscription to a MS from an MRB. See Section 11.5.1.3.1.

11.5.1.3.1 <subscription>

The <subscription> element is included in a request from an MRB to an MS to provide the details relating to the configuration of updates (known as a subscription session). This element can be used either to request a new subscription or to update an existing one (e.g., to change the frequency of the updates) and to remove ongoing subscriptions as well (e.g., to stop an indefinite update). The MRB will inform the MS regarding how long it wishes to receive updates and the frequency at which updates should be sent. Updates related to the subscription are sent using the <mrbnotification> element. The <subscription> element has the following attributes:

- id: Indicates a unique token representing the subscription session between the MRB and the MS. The attribute MUST be present.
- seqnumber: Indicates a sequence number to be used in conjunction with the subscription session ID to identify a specific subscription command. The first subscription MUST contain a non-zero number "seqnumber," and subsequent subscriptions MUST contain a higher number than the previous "seqnumber" value. If a subsequent "seqnumber" is not higher, a 405 response code is generated as per Section 11.5.1.4. The attribute MUST be present.
  - action: Provides the operation that should be carried out on the subscription:
  - The value of "create" instructs the MS to attempt to set up a new subscription.
  - The value of "update" instructs the MS to attempt to update an existing subscription.
  - The value of "remove" instructs the MS to attempt to remove an existing subscription and consequently stop any ongoing related notification. The attribute MUST be present.

The <subscription> element has zero or more of the following child elements:

- <expires>: Provides the amount of time in seconds that a subscription should be installed for notifications at the MS. Once the amount of time has passed, the subscription expires, and the MRB has to subscribe again if it is still interested in receiving notifications from the MS. The element MAY be present.
- <minfrequency>: Provides the minimum frequency in seconds that the MRB wishes to receive notifications from the MS. The element MAY be present.
- <maxfrequency>: Provides the maximum frequency in seconds that the MRB wishes to receive notifications from the MS. The element MAY be present.

Please note that these three optional pieces of information provided by the MRB only act as a suggestion: The MS MAY change the proposed values if it considers the suggestions unacceptable (e.g., if the MRB has requested a notification frequency that is too high). In such a case, the request would not fail, but the updated, acceptable values would be reported in the <mrbresponse> accordingly.

11.5.1.4 <mrbresponse>

Responses to requests are indicated by an <mrbresponse> element. The <mrbresponse> element has the following attributes:

- status: Numeric code indicating the response status. The attribute MUST be present.
- reason: String specifying a reason for the response status. The attribute MAY be present.

The <mrbresponse> element has a single child element:

- <subscription> for providing details related to a subscription requested by an MS (see later in this section).

The following status codes are defined for "status" in Table 11.1:

If a new subscription request made by an MRB (action="create") has been accepted, the MS MUST reply with an <mrbresponse> with status code 200. The same rule applies whenever a request to update (action="update") or remove (action="remove") an existing transaction can be fulfilled by the MS. A subscription request, nevertheless, may fail for several reasons. In such a case, the status codes defined in Table 11.1 must be used instead. Specifically, if the MS fails to handle a request due to a syntax error in the request itself (e.g., incorrect XML, violation of the schema constraints, or invalid values in any of the attributes/elements), the MS MUST reply with an <mrbresponse> with status code 400. If a syntactically correct request fails because the request also includes any attribute/element the MS does not understand, the MS MUST reply with an <mrbresponse> with status code 420.

If a syntactically correct request fails because the MRB wants to create a new subscription, but the provided unique "id" for the subscription already exists, the MS MUST reply with an <mrbresponse> with status code 406. If a syntactically correct request fails because the MRB wants to update/remove a subscription that does not exist, the MS MUST reply with an <mrbresponse> with status code 404. If the MS is unable to accept a request for any other reason (e.g., the MRB has no more resources to fulfill the request), the MS MUST reply with an <mrbresponse> with status code 401/402/403, depending on the action the MRB provided in its request:

- action='create' → 401
- action='update' → 402
- action='remove' → 403

TABLE 11.1
<mrbresponse> Status Codes

Code	Description
200	OK
400	Syntax error
401	Unable to create Subscription
402	Unable to update Subscription
403	Unable to remove Subscription
404	Subscription does not exist
405	Wrong sequence number
406	Subscription already exists
420	Unsupported attribute or element

A response to a subscription request that has a status code of 200 indicates that the request is successful. The response MAY also contain a <subscription> child that describes the subscription. The <subscription> child MAY contain "expires," "minfrequency," and "maxfrequency" values even if they were not contained in the request.

The MS can choose to change the suggested "expires," "minfrequency," and "maxfrequency" values provided by the MRB in its <mrbrequest> if it considers them unacceptable (e.g., the requested frequency range is too high). In such a case, the response MUST contain a <subscription> element describing the subscription as the MS accepted it, and the MS MUST include in the <subscription> element all those values that it modified relative to the request, to inform the MRB about the change.

11.5.1.5 <mrbnotification>

The <mrbnotification> element is included in a request from an MS to an MRB to provide the details relating to current status. The MS will inform the MRB of its current status as defined by the information in the <subscription> element. Updates are sent using the <mrbnotification> element. The <mrbnotification> element has the following attributes:

id: Indicates a unique token representing the session between the MRB and the MS and is the same as the one appearing in the <subscription> element. The attribute MUST be present. seqnumber: Indicates a sequence number to be used in conjunction with the subscription session ID to identify a specific notification update. The first notification update MUST contain a non-zero number "seqnumber," and subsequent notification updates MUST contain a higher number than the previous "seqnumber" value. If a subsequent "seqnumber" is not higher, the situation should be considered an error by the entity receiving the notification update. How the receiving entity deals with this situation is implementation specific. The attribute MUST be present.

It is important to point out that the "seqnumber" that appears in an <mrbnotification> is not related to the "seqnumber" appearing in a <subscription>. In fact, the latter is associated with subscriptions and would increase at every command issued by the MRB, while the former is associated with the asynchronous notifications the MS would trigger according to the subscription and as such, would increase at every notification message to enable the MRB to keep track of them. The following sub-sections provide details of the child elements that make up the contents of the <mrbnotification> element.

11.5.1.5.1 <media-server-id>

The <media-server-id> element provides a unique system-wide identifier for a MS instance. The element MUST be present and MUST be chosen such that it is extremely unlikely that two different MSs would present the same id to a given MRB.

11.5.1.5.2 <supported-packages>

The <supported-packages> element provides the list of Media Control Channel packages supported by the MS. The element MAY be present. The <supported-packages> element has no attributes. The <supported-packages> element has a single child element: <package>: Gives the name of a package supported by the MS. The <package> element has a single attribute, "name," which provides the name of the supported CFW package, compliant with Section 8.13.1.1 (Section 13.1.1 of RFC 6230).

11.5.1.5.3 <active-rtp-sessions>

The <active-rtp-sessions> element provides information detailing the current active real-time transport protocol (RTP) sessions. The element MAY be present. The <active-rtp-sessions> element has no attributes. The <active-rtp-sessions> element has a single child element:

- <rtp-codec>: Describes a supported codec and the number of active sessions using that codec. The <rtp-codec> element has one attribute. The value of the attribute, "name," is a media type (which can include parameters per RFC 6381). The <rtp-codec> element has two child elements. The child element <decoding> has as content the decimal number of RTP sessions being decoded using the specified codec, and the child element <encoding> has as content the decimal number of RTP sessions being encoded using the specified codec.

11.5.1.5.4 <active-mixer-sessions>

The <active-mixer-sessions> element provides information detailing the current active mixed RTP sessions. The element MAY be present. The <active-mixer-sessions> element has no attributes. The <active-mixer-sessions> element has a single child element:

- <active-mix>: Describes a mixed active RTP session. The <active-mix> element has one attribute. The value of the attribute, "conferenceid," is the name of the mix. The <active-mix> element has one child element. The child element, <rtp-codec>, contains the same information relating to RTP sessions as that defined in Section 5.1.5.3. The element MAY be present.

11.5.1.5.5 <non-active-rtp-sessions>

The <non-active-rtp-sessions> element provides information detailing the currently available inactive RTP sessions, that is, how many more RTP streams this MS can support. The element MAY be present. The <non-active-rtp-sessions> element has no attributes. The <non-active-rtp-sessions> element has a single child element:

- <rtp-codec>: Describes a supported codec and the number of non-active sessions for that codec. The <rtp-codec> element has one attribute. The value of the attribute, "name," is a media type (which can include parameters per RFC 6381). The <rtp-codec> element has two child elements. The child element <decoding> has as content the decimal number of RTP sessions available for decoding using the specified codec, and the child element <encoding> has as content the decimal number of RTP sessions available for encoding using the specified codec.

11.5.1.5.6 <non-active-mixer-sessions>

The <non-active-mixer-sessions> element provides information detailing the current inactive mixed RTP sessions, that is, how many more mixing sessions this MS can support. The element MAY be present. The <non-active-mixer-sessions> element has no attributes. The <non-active-mixer-sessions> element has a single child element:

- <non-active-mix>: Describes available mixed RTP sessions. The <non-active-mix> element has one attribute. The value of the attribute, "available," is the number of mixes that could be used with that profile. The <non-active-mix> element has one child element. The child element, <rtp-codec>, contains the same information relating to RTP sessions as that defined in Section 11.5.1.5.5. The element MAY be present.

11.5.1.5.7 <media-server-status>

The <media-server-status> element provides information detailing the current status of the MS. The element MUST be present. It can return one of the following values:

- active: Indicates that the MS is available for service. deactivated: Indicates that the MS has been withdrawn from service, and as such, requests should not be sent to it before it becomes "active" again.
- unavailable: Indicates that the MS continues to process past requests but cannot accept new requests, and as such, should not be contacted before it becomes "active" again.

The <media-server-status> element has no attributes.

The <media-server-status> element has no child elements.

11.5.1.5.8 <supported-codecs>

The <supported-codecs> element provides information detailing the current codecs supported by an MS and associated actions. The element MAY be present. The <supported-codecs> element has no attributes. The <supported-codecs> element has a single child element:

- <supported-codec>: Has a single attribute, "name," which provides the name of the codec about which this element provides information. A valid value is a media type that, depending on its definition, can include additional parameters (e.g., RFC 6381).

The <supported-codec> element then has a further child element, <supported-codec-package>. The <supported-codec-package> element has a single attribute, "name," which provides the name of the CFW package, compliant with Section 8.13.1.1 (Section 13.1.1 of RFC 6230), for which the codec support applies.

The <supported-codec-package> element has zero or more <supported-action> children, each of which describes an action that a MS can apply to this codec:

- "decoding," meaning that a decoder for this codec is available
- "encoding," meaning that an encoder for this codec is available
- "passthrough," meaning that the MS is able to pass a stream encoded using that codec through without re-encoding

11.5.1.5.9 <application-data>

The <application-data> element provides an arbitrary string of characters as application-level data. This data is meant to only have meaning at the application-level logic and as such, is not otherwise restricted by this specification. The set of allowed characters is the same as those in XML (tab, carriage return, line feed, and the legal characters of Unicode and ISO/IEC 10646) [1] (see also Section 2.2 of < www.w3.org/TR/xml/>). The element MAY be present.

The <application-data> element has no attributes.

The <application-data> element has no child elements.

11.5.1.5.10 <file-formats>

The <file-formats> element provides a list of file formats supported for the purpose of playing media. The element MAY be present. The <file-formats> element has no attributes. The <file-formats> element has zero or more of the following child elements:

- <supported-format>: Has a single attribute, "name," which provides the type of file format that is supported. A valid value is a media type that, depending on its definition, can include additional parameters (e.g., RFC 6381).

The <supported-format> element then has a further child element, <supported-file-package>. The <supported-file-package> element provides the name of the CFW package, compliant with Section 8.13.1.1 (Section 13.1.1 of RFC 6230), for which the file format support applies.

11.5.1.5.11 <max-prepared-duration>

The <max-prepared-duration> element provides the maximum amount of time a media dialog will be kept in the prepared state before timing out per Section 12.4.4.2.2.6 (Section 4.4.2.2.6 of RFC 6231). The element MAY be present. The <max-prepared-duration> element has no attributes.

The <max-prepared-duration> element has a single child element:

- <max-time>: Has a single attribute, "max-time-seconds," which provides the amount of time in seconds that a media dialog can be in the prepared state. The <max-time> element then has a further child element, <max-time-package>. The <max-time-package> element provides the name of the CFW package, compliant with Section 8.13.1.1 (Section 13.1.1 of RFC 6230), for which the time period applies.

11.5.1.5.12 <dtmf-support>

The <dtmf-support> element specifies the supported methods to detect dual-tone multi-frequency (DTMF) tones and to generate them. The element MAY be present. The <dtmf-support> element has no attributes.

The <dtmf-support> element has zero or more of the following child elements:

- <detect>: Indicates the support for DTMF detection. The <detect> element has no attributes. The <detect> element then has a further child element, <dtmf-type>. The <dtmf-type> element has two attributes: "name" and "package." The "name" attribute provides the type of DTMF being used, and it can only be a case-insensitive string containing either "RFC 4733" or "Media" (detecting tones as signals from the audio stream). The "package" attribute provides the name of the CFW package, compliant with Section 8.13.1.1 (Section 13.1.1 of RFC 6230), for which the DTMF type applies.
- <generate>: Indicates the support for DTMF generation. The <generate> element has no attributes. The <generate> element then has a further child element, <dtmf-type>. The <dtmf-type> element has two attributes: "name" and "package." The "name" attribute provides the type of DTMF being used, and it can only be a case-insensitive string containing either "RFC 4733" or "Media" (generating tones as signals in the audio stream). The "package" attribute provides the name of the CFW package, compliant with Section 8.13.1.1 (Section 13.1.1 of RFC 6230), for which the DTMF type applies.
- <passthrough>: Indicates the support for passing DTMF through without re-encoding. The <passthrough> element has no attributes. The <passthrough> element then has a further child element, <dtmf-type>. The <dtmf-type> element has two attributes: "name" and "package." The "name" attribute provides the type of DTMF being used, and it can only be a case-insensitive string containing either "RFC 4733" or "Media" (passing tones as signals through the audio stream). The "package" attribute provides the name of the CFW package, compliant with Section 8.13.1.1 (Section 13.1.1 of RFC 6230), for which the DTMF type applies.

11.5.1.5.13 <mixing-modes>

The <mixing-modes> element provides information about the support for audio and video mixing of a MS, specifically a list of supported algorithms to mix audio and a list of supported video presentation layouts. The element MAY be present. The <mixing-modes> element has no attributes. The <mixing-modes> element has zero or more of the following child elements:

- <audio-mixing-modes>: Describes the available algorithms for audio mixing. The <audio-mixing-modes> element has no attributes. The <audio-mixing-modes> element has one child element. The child element, <audio-mixing-mode>, contains a specific available algorithm. Valid values for the <audio-mixing-mode> element are algorithm names, e.g., "nbest" and "controller" as defined in RFC 6505 (see Chapter 9). The element has a single attribute, "package." The attribute "package" provides the name of the CFW package, compliant with Section 8.13.1.1 (Section 13.1.1 of RFC 6230), for which the algorithm support applies.

- <video-mixing-modes>: Describes the available video presentation layouts and the supported functionality related to video mixing. The <video-mixing-modes> element has two attributes: "vas" and "activespeakermix." The "vas" attribute is of type Boolean with a value of "true" indicating that the MS supports automatic Voice Activated Switching. The "activespeakermix" is of type Boolean with a value of "true" indicating that the MS is able to prepare an additional video stream for the loudest speaker participant without its contribution. The <video-mixing-modes> element has one child element. The child element, <video-mixing-mode>, contains the name of a specific video presentation layout. The name may refer to one of the predefined video layouts defined in the XCON conference information data model (RFC 6501 – see Chapter 4), or to non-XCON layouts as well, as long as they are properly prefixed according to the schema they belong to. The <video-mixing-mode> element has a single attribute, "package." The attribute "package" provides the name of the CFW package, compliant with Section 8.13.1.1 (Section 13.1.1 of RFC 6230), for which the algorithm support applies.

11.5.1.5.14 <supported-tones>

The <supported-tones> element provides information about which tones an MS is able to play and recognize. In particular, the support is reported by referring to both support for country codes (ISO 3166-1) [2] and supported functionality (ITU-T Recommendation Q.1950 [3]). The element MAY be present. The <supported-tones> element has no attributes. The <supported-tones> element has zero or more of the following child elements:

- <supported-country-codes>: Describes the supported country codes with respect to tones. The <supported-country-codes> element has no attributes. The <supported-country-codes> element has one child element. The child element, <country-code>, reports support for a specific country code, compliant with the ISO 3166-1 [2] specification. The <country-code> element has a single attribute, "package." The attribute "package" provides the name of the CFW package, compliant with Section 8.13.1.1 (Section 13.1.1 of RFC 6230), in which the tones from the specified country code are supported.
- <supported-h248-codes>: Describes the supported H.248 codes with respect to tones. The <supported-h248-codes> element has no attributes. The <supported-h248-codes> element has one child element. The child element, <h248-code>, reports support for a specific H.248 code, compliant with the ITU-T Recommendation Q.1950 [3] specification. The codes can be either specific (e.g., cg/dt to only report the Dial Tone from the Call Progress Tones package) or generic (e.g., cg/* to report all the tones from the Call Progress Tones package) using wildcards. The <h248-code> element has a single attribute, "package." The attribute "package" provides the name of the CFW package, compliant with Section 8.13.1.1 (Section 13.1.1 of RFC 6230), in which the specified codes are supported.

11.5.1.5.15 <file-transfer-modes>

The <file-transfer-modes> element allows the MS to specify which scheme names are supported for transferring files to an MS for each CFW package type; for example, whether the MS supports fetching resources via HTTP, HTTPS, NFS, etc. The element MAY be present. The <file-transfer-modes> element has no attributes. The <file-transfer-modes> element has a single child element:

- <file-transfer-mode>: Has two attributes: "name" and "package." The "name" attribute provides the scheme name of the protocol that can be used for file transfer (e.g., HTTP, HTTPS, NFS, etc.); the value of the attribute is case insensitive. The "package" attribute provides the name of the CFW package, compliant with the specification in the related Internet Assigned Numbers Authority (IANA) registry (e.g., "msc-ivr/1.0"), for which the scheme name applies.

It is important to point out that this element provides no information about whether or not the MS supports any flavor of live streaming: for instance, a value of "HTTP" for the IVR (interactive voice response) package would only mean that the "http" scheme makes sense to the MS within the context of that package. Whether or not the MS can make use of HTTP to only fetch resources, or also to attach an HTTP live stream to a call, is to be considered implementation specific to the MS and irrelevant to the AS and/or MRB. Besides, the MS supporting a scheme does not imply that it also supports the related secure versions: for instance, if the MS supports both HTTP and HTTPS, both the schemes will appear in the element. A lack of the "HTTPS" value would need to be interpreted as a lack of support for the "https" scheme.

11.5.1.5.16 <asr-tts-support>

The <asr-tts-support> element provides information about the support for automatic speech recognition (ASR) and text-to-speech (TTS) functionality in an MS. The functionality is reported by referring to the supported languages (using ISO 639-1 [4] codes) regarding both ASR and TTS. The element MAY be present. The <asr-tts-support> element has no attributes. The <asr-tts-support> element has zero or more of the following child elements:

- <asr-support>: Describes the available languages for ASR. The <asr-support> element has no attributes. The <asr-support> element has one child element. The child element, <language>, reports that the MS supports ASR for a specific language. The <language> element has a single attribute, "xml:lang." The attribute "xml:lang" contains the ISO 639-1 [2] code of the supported language.
- <tts-support>: Describes the available languages for TTS. The <tts-support> element has no attributes. The <tts-support> element has one child element. The child element, <language>, reports that the MS supports TTS for a specific language. The <language> element has a single attribute, "xml:lang." The attribute "xml:lang" contains the ISO 639-1 [2] code of the supported language.

11.5.1.5.17 <vxml-support>

The <vxml-support> element specifies when the MS supports VoiceXML (VXML) and if it does, through which protocols the support is exposed (e.g., via the control framework, RFC 4240, or RFC 5552 [see Chapter 13]). The element MAY be present. The <vxml-support> element has no attributes. The <vxml-support> element has a single child element:

- <vxml-mode>: Has two attributes: "package" and "support." The "package" attribute provides the name of the CFW package, compliant with Section 8.13.1.1 (Section 13.1.1 of RFC 6230), for which the VXML support applies. The "support" attribute provides the type of VXML support provided by the MS (e.g., RFC 5552 – see Chapter 13), RFC 4240, or the IVR Package (RFC 6231 – also see Section 1.2), and valid values are case-insensitive RFC references (e.g., "rfc 6231" to specify that the MS supports VoiceXML as provided by the IVR Package (RFC 6231 – see Chapter 12).

The presence of at least one <vxml-mode> child element would indicate that the MS does support VXML as specified by the child element itself. An empty <vxml> element would otherwise indicate that the MS does not support VXML at all.

11.5.1.5.18 <media-server-location>

The <media-server-location> element provides information about the civic location of an MS. Its description makes use of the Civic Address Schema standardized in RFC 5139. The element MAY be present. More precisely, this section is entirely optional, and it is implementation specific to fill it with just the details each implementer deems necessary for any optimization that may be needed.

The <media-server-location> element has no attributes. The <media-server-location> element has a single child element:

- <civicAddress>: Describes the civic address location of the MS, whose representation refers to Section 4 of RFC 5139.

11.5.1.5.19 <label>

The <label> element allows an MS to declare a piece of information that will be understood by the MRB. For example, the MS can declare whether it is a blue or green one. It is a string to allow arbitrary values to be returned in order to allow arbitrary classification. The element MAY be present.

The <label> element has no attributes.

The <label> element has no child elements.

11.5.1.5.20 <media-server-address>

The <media-server-address> element allows an MS to provide a direct SIP uniform resource identifier (URI) where it can be reached (e.g., the URI that the AS would call in order to set up a Control Channel and relay SIP media dialogs). The element MAY be present.

The <media-server-address> element has no attributes.

The <media-server-address> element has no child elements.

11.5.1.6 <encryption>

The <encryption> element allows an MS to declare support for encrypting RTP media streams using RFC 3711. The element MAY be present. If the element is present, then the MS supports DTLS-SRTP (a secure real-time transport protocol [SRTP] extension for datagram transport layer security [DTLS]) (RFC 5763).

The <encryption> element has no attributes.

The <encryption> element has no child elements.

11.5.2 Media Service Resource Consumer Interface

The MS Consumer interface provides the ability for clients of an MRB, such as ASs, to request an appropriate MS to satisfy specific criteria. This interface allows a client to pass detailed meta-information to the MRB to help select an appropriate MS. The MRB is then able to make an informed decision and provide the client with an appropriate MS resource. The MRB Consumer interface includes both 1) the IAMM, which uses the SIP, and 2) the Query mode, which uses the hypertext transfer protocol (HTTP) (RFC 2616). The MRB Consumer interface does not include the IUMM, which is further explained in Section 11.5.3. The following sub-sections provide guidance on using the Consumer interface, which is represented by the "application/mrb-consumer+xml" media type in Section 11.11, with HTTP and SIP.

11.5.2.1 Query Mode/HTTP Consumer Interface Usage

An appropriate interface for such a "query"-style interface is in fact an HTTP usage. Using HTTP and XML combined reduces complexity and encourages the use of common tools that are widely available in the industry today. The following information explains the primary operations required to request and then receive information from an MRB, by making use of HTTP (RFC 2616) and HTTPS (RFC 2818) as transport for a query for a media resource, and the appropriate response.

The media resource query, as defined by the <mediaResourceRequest> element from Section 11.11, MUST be carried in the body of an HTTP/HTTPS POST request. The media type contained in the HTTP/HTTPS request/ response MUST be "application/mrb-consumer+xml." This value MUST be reflected in the appropriate HTTP headers, such as "Content-Type" and "Accept." The body of the HTTP/HTTPS POST request MUST only contain an <mrbconsumer> root element with only one child <mediaResourceRequest> element as defined in Section 11.11.

The media resource response to a query, as defined by the <mediaResourceResponse> element from Section 11.11, MUST be carried in the body of an HTTP/HTTPS 200 response to the original HTTP/HTTPS POST request. The media type contained in the HTTP/HTTPS request/response MUST be "application/mrb-consumer+xml." This value MUST be reflected in the appropriate HTTP headers, such as "Content-Type" and "Accept." The body of the HTTP/HTTPS 200 response MUST only contain an <mrbconsumer> root element with only one child <mediaResourceResponse> element as defined in Section 11.11. When an AS wants to release previously awarded media resources granted through a prior request/response exchange with an MRB, it will send a new request with an <action> element with value "remove," as described in Section 11.5.2.3 ("Consumer Interface Lease Mechanism").

11.5.2.2 In-Line Aware Mode/SIP Consumer Interface Usage

This specification provides a complete toolkit for MRB deployment that includes the ability to interact with an MRB using SIP for the Consumer interface. The following information explains the primary operations required to request and then receive information from an MRB, by making use of SIP (RFC 3261 – also see Section 1.2) as transport for a request for media resources, and the appropriate response when using IAMM as the mode of operation (as discussed in Section 11.5.2.2.1).

The use of IAMM, besides having the MRB select appropriate media resources on behalf of a client application, includes setting up either a Control Framework Control Channel between an AS and one of the MSs (Section 11.5.2.2.1) or a media dialog session between an AS and one of the MSs (Section 11.5.2.2.2). Note that in either case, the SIP URIs of the selected MSs are made known to the requesting AS in the SIP 200 OK response by means of one or more <media-server-address> child elements in the <response-session-info> element (Section 11.5.2.6).

11.5.2.2.1 IAMM and Setting Up a Control Framework Control Channel

The media resource request information, as defined by the <mediaResourceRequest> element from Section 11.11, is carried in an SIP INVITE request. The INVITE request will be constructed as it would have been to connect to a MS, as defined by the CFW [RFC6230]. It should be noted that this specification does not exclude the use of an offerless INVITE as defined in RFC 3261 (also see Section 1.2) Using offerless INVITE messages to an MRB can potentially cause confusion when applying resource selection algorithms, and an MRB, like any other SIP device, can choose to reject with a 4xx response. For an offerless INVITE to be treated appropriately, additional contextual information would need to be provided with the request; this is out of scope for this specification. The following additional steps MUST be followed when using the Consumer interface:

- The Consumer client will include a payload in the SIP INVITE request of type "multipart/mixed" (RFC 2046). One of the parts to be included in the "multipart/mixed" payload MUST be the "application/sdp" format, which is constructed as specified in the CFW (RFC 6230 – see Chapter 8).
- Another part of the "multipart/mixed" payload MUST be of type "application/mrb-consumer+xml," as specified in this specification and defined in Section 11.11. The body part MUST be an XML specification without prolog and whose root element is <mediaResourceRequest>.
- The INVITE request will then be dispatched to the MRB, as defined by RFC 6230 (see Chapter 8). On receiving an SIP INVITE request containing the multipart/mixed payload as specified previously, the MRB will complete a number of steps to fulfill the request. It will:

- Extract the multipart MIME payload from the SIP INVITE request. It will then use the contextual information provided by the client in the "application/mrb-consumer+xml" part to determine which MS (or MSs, if more than one is deemed to be needed) should be selected to service the request.
- Extract the "application/sdp" part from the payload and use it as the body of a new SIP INVITE request for connecting the client to one of the selected MSs, as defined in the CFW (RFC 6230 – see Chapter 8).

The policy the MRB follows to pick a specific MS out of the MSs it selects is implementation specific and out of scope for this specification. It is important to configure the SIP elements between the MRB and the MS in such a way that the INVITE will not fork. In the case of a failure in reaching the chosen MS, the MRB SHOULD proceed to the next one, if available. If none of the available MSs can be reached, the MRB MUST reply with an SIP 503 error message that includes a Retry-After header with a non-zero value. The AS MUST NOT attempt to set up a new session before the time that the MRB asked it to wait has passed.

If at least one MS is reachable, the MRB acts as a back-to-back user agent (B2BUA) that extracts the "application/ mrb-consumer+xml" information from the SIP INVITE request and then sends a corresponding SIP INVITE request to the MS it has selected to negotiate a Control Channel as defined in the CFW (RFC 6230 – see Chapter 8). In the case of a failure in negotiating the Control Channel with the MS, the MRB SHOULD proceed to the next one, if available, as explained earlier. If none of the available MSs can be reached, or the negotiations of the Control Channel with all of them fail, the MRB MUST reply with an SIP 503 error message that includes a Retry-After header with a non-zero value. The AS MUST NOT attempt to set up a new session before the time that the MRB asked it to wait has expired.

Once the MRB receives the SIP response from the selected media resource (i.e., MS), it will in turn respond to the requesting client (i.e., AS). The media resource response generated by an MRB to a request, as defined by the <mediaResourceResponse> element from Section 11.11, MUST be carried in the payload of an SIP 200 OK response to the original SIP INVITE request. The SIP 200 OK response will be constructed as it would have been to connect from an MS, as defined by the CFW (RFC 6230 – see Chapter 8). The following additional steps MUST be followed when using the Consumer interface:

- Include a payload in the SIP 200 response of type "multipart/ mixed" as per RFC 2046. One of the parts to be included in the "multipart/mixed" payload MUST be the "application/sdp" format, which is constructed as specified in the CFW (RFC 6230 – see Chapter 8) and based on the incoming response from the selected media resource.
- Another part of the "multipart/mixed" payload MUST be of type "application/mrb-consumer+xml," as specified in this specification and defined in Section 11.11. Only the <mediaResourceResponse> and its child elements can be included in the payload.
- The SIP 200 response will then be dispatched from the MRB.
- An SIP ACK to the 200 response will then be sent back to the MRB. Considering that the use of SIP as a transport for Consumer transactions may result in failure, the IAMM relies on a successful INVITE transaction to address the previously discussed sequence (using the "seq" XML element) increment mechanism. This means that if the INVITE is unsuccessful for any reason, the AS MUST use the same "seq" value as previously used for the next Consumer request that it may want to send to the MRB for the same session. An MRB implementation may be programmed to conclude that the requested resources are no longer needed when it receives an SIP BYE from the Application Server or MS that concludes the SIP dialog that initiated the request, or when the lease (Section 11.5.2.3) interval expires.

11.5.2.2.2 IAMM and Setting Up a Media Dialog

This scenario is identical to the description in the previous section for setting up a CFW Control Channel, with the exception that the application/sdp payload conveys content appropriate for setting up the media dialog to the media resource, as per RFC 3261 (also see Section 1.2), instead of setting up a Control Channel.

11.5.2.3 Consumer Interface Lease Mechanism

The Consumer interface defined in Sections 11.5.2 and 11.11 allows a client to request an appropriate media resource based on information included in the request (either an HTTP POST or an SIP INVITE message). In the case of success, the response that is returned to the client MUST contain a <response-session-info> element in either the SIP 200 or the HTTP 200 response. The success response contains the description of certain resources that have been reserved to a specific Consumer client in a (new or revised) "resource session," which is identified in the <response-session-info>. The resource session is a "lease", in that the reservation is scheduled to expire at a particular time in the future, releasing the resources to be assigned for other uses. The lease may be extended or terminated earlier by future Consumer client requests that identify and reference a specific resource session.

Before delving into the details of such a lease mechanism, it is worth clarifying its role within the context of the Consumer interface. As explained in Section 11.5.1, the knowledge the MRB has of the resources of all the MSs it is provisioned to manage is not real time. How an MRB actually manages such resources is implementation specific – for example, an implementation may choose to have the MRB keeping track and state of the allocated resources or simply rely on the MSs themselves to provide the information using the Publish interface. Further information may also be inferred by the signaling in the case where an MRB is in the path of media dialogs.

The <mediaResourceResponse> element returned from the MRB contains a <response-session-info> element if the request is successful. The <response-session-info> element has zero or more of the following child elements, which provide the appropriate resource session information:

- <session-id> is a unique identifier that enables a Consumer client and an MRB to correlate future media resource requests related to an initial media resource request. The <session-id> MUST be included in all future related requests (see the <session-id> paragraph later in this section, where constructing a subsequent request is discussed).
- <seq> is a numeric value returned to the Consumer client. On issuing any future requests related to the media resource session (as determined by the <session-id> element), the Consumer client MUST increment the value returned in the <seq> element and include it in the request (see the <seq> paragraph later in this section, where constructing a subsequent request is discussed). Its value is a non-negative integer that MUST be limited within the 0..2^31-1 range.
- <expires> provides a value indicating the number of seconds for which the request for media resources is deemed alive. The Consumer client should issue a refresh of the request, as discussed later in this section, if the expiry is due to fire and the media resources are still required.
- <media-server-address> provides information representing an assigned MS. More instances of this element may appear should the MRB assign more MSs to a Consumer request. The <mediaResourceRequest> element is used in subsequent Consumer interface requests if the client wishes to manipulate the session. The Consumer client MUST include the <session-info> element, which enables the receiving MRB to determine an existing media resource allocation session. The <session-info> element has the following child elements, which provide the appropriate resource session information to the MRB:
- <session-id> is a unique identifier that allows a Consumer client to indicate the appropriate existing media resource session to be manipulated by the MRB for this request. The

value was provided by the MRB in the initial request for media resources, as discussed earlier in this section (<session-id> element included as part of the <session-info> element in the initial <mediaResourceResponse>).

- <seq> is a numeric value returned to the Consumer client in the initial request for media resources, as discussed earlier in this section (<seq> element included as part of the <session-info> element in the initial <mediaResourceResponse>). On issuing any future requests related to the specific media resource session (as determined by the <session-id> element), the Consumer client MUST increment the value returned in the <seq> element from the initial response (contained in the <mediaResourceResponse>) for every new request. The value of the <seq> element in requests acts as a counter and when used in conjunction with the unique <session-id>, allows unique identification of a request. As anticipated before, the <seq> value is limited to the 0..2^31-1 range: In the unlikely case that the counter increases to reach the highest allowed value, the <seq> value MUST be set to 0. The first numeric value for the <seq> element is not meant to be "1" but SHOULD be generated randomly by the MRB: This is to reduce the chances of a malicious MRB disrupting the session created by this MRB, as explained in Section 11.12.
- <action> provides the operation to be carried out by the MRB on receiving the request:
  - The value of "update" is a request by the Consumer client to update the existing session on the MRB with alternate media resource requirements. If the requested resource information is identical to the existing MRB session, the MRB will attempt a session refresh. If the information has changed, the MRB will attempt to update the existing session with the new information. If the operation is successful, the 200 status code in the response is returned in the status attribute of the <mediaResourceResponseType> element. If the operation is not successful, a 409 status code in the response is returned in the status attribute of the <mediaResourceResponseType> element.
  - The value of "remove" is a request by the Consumer client to remove the session on the MRB. This provides a mechanism for Consumer clients to release unwanted resources before they expire. If the operation is successful, a 200 status code in the response is returned in the status attribute of the <mediaResourceResponseType> element. If the operation is not successful, a 410 status code in the response is returned in the status attribute of the <mediaResourceResponseType> element. Omitting the "action" attribute means requesting a new set of resources.

When used with HTTP, the <session-info> element MUST be included in an HTTP POST message (as defined in RFC 2616). When used with SIP, the <session-info> element MUST instead be included in either an SIP INVITE or an SIP re-INVITE (as defined in RFC 3261 – also see Section 1.6), or in an SIP UPDATE (as defined in RFC 3311) request: in fact, any SIP dialog, be it a new or an existing one, can be exploited to carry leasing information, and as such, new SIP INVITE messages can update other leases as well as request a new one. With IAMM, the AS or MS will eventually send an SIP BYE to end the SIP session, whether it was for a Control Channel or a media dialog. That BYE contains no Consumer interface lease information.

11.5.2.4 <mrbconsumer>

This section defines the XML elements for the Consumer interface. The formal XML schema definition for the Consumer interface can be found in Section 11.11. The root element is <mrbconsumer>. All other XML elements (requests, responses) are contained within it. The MRB Consumer interface request element is detailed in Section 11.5.2.5.1. The MRB Consumer interface response element is detailed in Section 11.5.2.6.1. The <mrbconsumer> element has the following attributes:

- version: A token specifying the mrb-consumer package version. The value is fixed as "1.0" for this version of the package. The attribute MUST be present.

The <mrbconsumer> element may have zero or more children of one of the following child element types: <mediaResourceRequest> for sending a Consumer request. See Section 11.5.2.5.1; <mediaResourceResponse> for sending a Consumer response. See Section 11.5.2.6.1.

11.5.2.5 Media Service Resource Request

This section provides the element definitions for use in Consumer interface requests. The requests are carried in the <mediaResourceRequest> element.

11.5.2.5.1 <mediaResourceRequest>

The <mediaResourceRequest> element provides information for clients wishing to query an external MRB entity. The <mediaResourceRequest> element has a single mandatory attribute, "id:" this attribute contains a random identifier, generated by the client, that will be included in the response in order to map it to a specific request. The <mediaResourceRequest> element has <generalInfo>, <ivrInfo>, and <mixerInfo> as child elements. These three elements are used to describe the requirements of a client requesting a MS and are covered in Sections 11.5.2.5.1.1, 11.5.2.5.1.2, and 11.5.2.5.1.3, respectively.

11.5.2.5.1.1 <generalInfo> The <generalInfo> element provides general Consumer request information that is neither IVR specific nor mixer specific. This includes session information that can be used for subsequent requests as part of the leasing mechanism described in Section 11.5.2.3. The following sub-sections describe the <session-info> and <packages> elements, as used by the <generalInfo> element.

11.5.2.5.1.1.1 <session-info> The <session-info> element is included in Consumer requests when an update is being made to an existing media resource session. The ability to change and remove an existing media resource session is described in more detail in Section 11.5.2.3. The element MAY be present.

The <session-info> element has no attributes.

The <session-info> element has zero or more of the following child elements:

- <session-id>: A unique identifier that explicitly references an existing media resource session on the MRB. The identifier is included to update the existing session and is described in more detail in Section 11.5.2.3.
- <seq>: Used in association with the <session-id> element in a subsequent request to update an existing media resource session on an MRB. The <seq> number is incremented from its original value returned in response to the initial request for media resources. Its value is a non-negative integer that MUST be limited within the 0..2^31-1 range. In the unlikely case that the counter increases to reach the highest allowed value, the <seq> value MUST be set to 0. More information about its use is provided in Section 11.5.2.3.
- <action>: Provides the operation that should be carried out on an existing media resource session on an MRB:
    - The value of "update" instructs the MRB to attempt to update the existing media resource session with the information contained in the <ivrInfo> and <mixerInfo> elements.
    - The value of "remove" instructs the MRB to attempt to remove the existing media resource session. More information on its use is provided in Section 11.5.2.3.

11.5.2.5.1.1.2 <packages> The <packages> element provides a list of CFW-compliant packages that are required by the Consumer client. The element MAY be present.

The <packages> element has no attributes.

The <packages> element has a single child element:

- <package>: Contains a string representing the CFW package required by the Consumer client. The <package> element can appear multiple times. A valid value is a Control Package name compliant with Section 8.13.1.1 (Section 13.1.1 of RFC 6230).

11.5.2.5.1.2 <ivrInfo> The <ivrInfo> element provides information for general Consumer request information that is IVR specific. The following sub-sections describe the elements of the <ivrInfo> element: <ivr-sessions>, <file-formats>, <dtmf>, <tones>, <asr-tts>, <vxml>, <location>, <encryption>, <application-data>, <max-prepared-duration>, and <file-transfer-modes>.

11.5.2.5.1.2.1 <ivr-sessions> The <ivr-sessions> element indicates the number of IVR sessions that a Consumer client requires from a media resource. The element MAY be present.

The <ivr-sessions> element has no attributes.

The <ivr-sessions> element has a single child element:

- <rtp-codec>: Describes a required codec and the number of sessions using that codec. The <rtp-codec> element has one attribute. The value of the attribute, "name," is a media type (which can include parameters per RFC 6381).

The <rtp-codec> element has two child elements. The child element <decoding> contains the number of RTP sessions required for decoding using the specified codec, and the child element <encoding> contains the number of RTP sessions required for encoding using the specified codec.

11.5.2.5.1.2.2 <file-formats> The <file-formats> element provides a list of file formats required for the purpose of playing media. It should be noted that this element describes media types and might better have been named "media-formats," but due to existing implementations, the name "file-formats" is used. The element MAY be present.

The <file-formats> element has no attributes.

The <file-formats> element has a single child element:

- <required-format>: Has a single attribute, "name," which provides the type of file format that is required. A valid value is a media type that depending on its definition, can include additional parameters (e.g., RFC 6381). The <required-format> element then has a further child element, <required-file-package>.

The <required-file-package> element has a single attribute, "required-file-package-name," which contains the name of the CFW package, compliant with Section 8.13.1.1 (Section 13.1.1 of RFC 6230), for which the file format support applies.

11.5.2.5.1.2.3 <dtmf> The <dtmf> element specifies the required methods to detect DTMF tones and to generate them. The element MAY be present.

The <dtmf> element has no attributes.

The <dtmf> element has zero or more of the following child elements:

- <detect>: Indicates the required support for DTMF detection. The <detect> element has no attributes. The <detect> element has a further child element, <dtmf-type>. The <dtmf-type> element has two attributes: "name" and "package." The "name" attribute provides the type of DTMF required and is a case-insensitive string containing either "RFC 4733" or "Media" (detecting tones as signals from the audio stream). The "package" attribute provides the name of the CFW package, compliant with Section 8.13.1.1 (Section 13.1.1 of RFC 6230), for which the DTMF type applies.

- <generate>: Indicates the required support for DTMF generation. The <generate> element has no attributes. The <generate> element has a single child element, <dtmf-type>. The <dtmf-type> element has two attributes: "name" and "package." The "name" attribute provides the type of DTMF required and is a case-insensitive string containing either "RFC 4733" or "Media" (generating tones as signals in the audio stream). The "package" attribute provides the name of the CFW package, compliant with Section 8.13.1.1 (Section 13.1.1 of RFC 6230), for which the DTMF type applies.
- <passthrough>: Indicates the required support for passing DTMF through without re-encoding. The <passthrough> element has no attributes. The <passthrough> element then has a further child element, <dtmf-type>. The <dtmf-type> element has two attributes: "name" and "package." The "name" attribute provides the type of DTMF required and is a case-insensitive string containing either "RFC 4733" or "Media" (passing tones as signals through the audio stream). The "package" attribute provides the name of the CFW package, compliant with Section 8.13.1.1 (Section 13.1.1 of RFC 6230), for which the DTMF type applies.

11.5.2.5.1.2.4 <tones> The <tones> element provides requested tones that a MS must support for IVR. In particular, the request refers to both support for country codes (ISO 3166-1 [2]) and requested functionality (ITU-T Recommendation Q.1950 [3]). The element MAY be present.

The <tones> element has no attributes.

The <tones> element has zero or more of the following child elements:

- <country-codes>: Describes the requested country codes in relation to tones. The <country-codes> element has no attributes. The <country-codes> element has one child element. The child element, <country-code>, requests a specific country code, compliant with the ISO 3166-1 [2] specification. The <country-code> element has a single attribute, "package." The attribute "package" provides the name of the Media Control Channel Framework package, compliant with Section 8.13.1.1 (Section 13.1.1 of RFC 6230), in which the tones from the specified country code are requested.
- <h248-codes>: Describes the requested H.248 codes in relation to tones. The <h248-codes> element has no attributes. The <h248-codes> element has one child element. The child element, <h248-code>, requests a specific H.248 code, compliant with the ITU-T Recommendation Q.1950 [3] specification. The codes can be either specific (e.g., cg/dt to only report the Dial Tone from the Call Progress Tones package) or generic (e.g., cg/* to report all the tones from the Call Progress Tones package) using wildcards. The <h248-code> element has a single attribute, "package." The attribute "package" provides the name of the CFW package, compliant with Section 8.13.1.1 (Section 13.1.1 of RFC 6230), in which the specified codes are requested.

11.5.2.5.1.2.5 <asr-tts> The <asr-tts> element requests information about the support for ASR and TTS functionality in a MS. The functionality is requested by referring to the supported languages (using ISO 639-1 [4] codes) in relation to both ASR and TTS. The <asr-tts> element has no attributes. The <asr-tts> element has zero or more of the following child elements:

- <asr-support>: Describes the available languages for ASR. The <asr-support> element has no attributes. The <asr-support> element has one child element. The child element, <language>, requests that the MS supports ASR for a specific language. The <language> element has a single attribute, "xml:lang." The attribute "xml:lang" contains the ISO 639-1 [4] code of the supported language.

- <tts-support>: Describes the available languages for TTS. The <tts-support> element has no attributes. The <tts-support> element has one child element. The child element, <language>, requests that the MS supports TTS for a specific language. The <language> element has a single attribute, "xml:lang." The attribute "xml:lang" contains the ISO 639-1 [4] code of the supported language.

11.5.2.5.1.2.6 <vxml> The <vxml> element specifies whether the Consumer client requires VoiceXML and if so, which protocols are supported (e.g., via the control framework, RFC 4240, or RFC 5552 [see Chapter 13]). The element MAY be present. The <vxml> element has a single child element:

- <vxml-mode>: Has two attributes: "package" and "require." The "package" attribute provides the name of the CFW package, compliant with Section 8.13.1.1 (Section 13.1.1 of RFC 6230), for which the VXML support applies. The "require" attribute specifies the type of VXML support required by the Consumer client (e.g., RFC 5552 [see Chapter 13]), RFC 4240, or IVR Package (RFC 6231 – see Chapter 12), and valid values are case-insensitive RFC references (e.g., "rfc 6231" to specify that the client requests support for VoiceXML as provided by the IVR Package [RFC 6231 – see Chapter 13]).

The presence of at least one <vxml> child element would indicate that the Consumer client requires VXML support as specified by the child element itself. An empty <vxml> element would otherwise indicate that the Consumer client does not require VXML support.

11.5.2.5.1.2.7 <location> The <location> element requests a civic location for an IVR MS. The request makes use of the Civic Address Schema standardized in RFC 5139. The element MAY be present. More precisely, this section is entirely optional and is implementation specific in its level of population.

The <location> element has no attributes.

The <location> element has a single child element:

- <civicAddress>: Describes the civic address location of the requested MS, whose representation refers to Section 4 of RFC 5139.

11.5.2.5.1.2.8 <encryption> The <encryption> element allows a Consumer client to request support for encrypting RTP media streams using RFC 3711. The element MAY be present. If the element is present, then the MS supports DTLS-SRTP (RFC 5763).

The <encryption> element has no attributes.

The <encryption> element has no child elements.

11.5.2.5.1.2.9 <application-data> The <application-data> element provides an arbitrary string of characters as IVR application-level data. This data is meant to have meaning only at the application-level logic and as such, is not otherwise restricted by this specification. The set of allowed characters is the same as those in XML (tab, carriage return, line feed, and the legal characters of Unicode and ISO/IEC 10646 [1] [see also Section 2.2 of www.w3.org/TR/xml/]). The element MAY be present.

The <application-data> element has no attributes.

The <application-data> element has no child elements.

11.5.2.5.1.2.10 <max-prepared-duration> The <max-prepared-duration> element indicates the amount of time required by the Consumer client representing media dialog preparation in the system before it is executed. The element MAY be present. The <max-prepared-duration> element has no attributes. The <max-prepared-duration> element has a single child element: <max-time>. It has a single attribute, "max-time-seconds," which provides the amount of time in seconds that a media dialog can be in the prepared state. The <max-time> element then has a further child element, <max-time-package>. The <max-time-package> element provides the name of the CFW package, compliant with Section 8.13.1.1 (Section 13.1.1 of RFC 6230), for which the time period applies.

11.5.2.5.1.2.11 <file-transfer-modes> The <file-transfer-modes> element allows the Consumer client to specify which scheme names are required for file transfer to a MS for each CFW package type. For example, does the MS support fetching media resources via HTTP, HTTPS, NFS, etc.? The element MAY be present.

The <file-transfer-modes> element has no attributes.

The <file-transfer-modes> element has a single child element:

<file-transfer-mode>: Has two attributes: "name" and "package." The "name" attribute provides the scheme name of the protocol required for fetching resources: valid values are case-insensitive scheme names (e.g., HTTP, HTTPS, NFS, etc.). The "package" attribute provides the name of the CFW package, compliant with Section 8.13.1.1 (Section 13.1.1 of RFC 6230), for which the scheme name applies. The same considerations relating to file transfer and live streaming are explained further in Section 11.5.1.5.15 and apply here as well.

11.5.2.5.1.3 <mixerInfo> The <mixerInfo> element provides information for general Consumer request information that is mixer specific. The following sub-sections describe the elements of the <mixerInfo> element: <mixers>, <file-formats>, <dtmf>, <tones>, <mixing-modes>, <application-data>, <location>, and <encryption>.

11.5.2.5.1.3.1 <mixers> The <mixers> element provides information detailing the required mixed RTP sessions. The element MAY be present. The <mixers> element has no attributes. The <mixers> element has a single child element:

- <mix>: Describes the required mixed RTP sessions. The <mix> element has one attribute. The value of the attribute, "users," is the number of participants required in the mix.

The <mix> element has one child element. The child element, <rtp-codec>, contains the same information relating to RTP sessions as that defined in Section 11.5.1.5.3. The element MAY be present.

11.5.2.5.1.3.2 <file-formats> The <file-formats> element provides a list of file formats required by the Consumer client for the purpose of playing media to a mix. The element MAY be present. The <file-formats> element has no attributes. The <file-formats> element has a single child element:

- <required-format>: Has a single attribute, "name," which provides the type of file format supported. A valid value is a media type that, depending on its definition, can include additional parameters (e.g., RFC 6381).

The <required-format> element has a child element, <required-file-package>. The <required-file-package> element contains a single attribute, "required-file-package-name," which contains the name of the CFW package, compliant with Section 8.13.1.1 (Section 13.1.1 of RFC 6230), for which the file format support applies.

11.5.2.5.1.3.3 <dtmf> The <dtmf> element specifies the required methods to detect DTMF tones and to generate them in a mix. The element MAY be present. The <dtmf> element has no attributes. The <dtmf> element has zero or more of the following child elements:

- <detect>: Indicates the required support for DTMF detection. The <detect> element has no attributes.

The <detect> element then has a further child element, <dtmf-type>. The <dtmf-type> element has two attributes: "name" and "package." The "name" attribute provides the type of DTMF being used and is a case-insensitive string containing either "RFC 4733" or "Media" (detecting tones as signals from the audio stream). The "package" attribute provides the name of the CFW package, compliant with Section 8.13.1.1 (Section 13.1.1 of RFC 6230), for which the DTMF type applies.

- <generate>: Indicates the required support for DTMF generation. The <generate> element has no attributes.

The <generate> element has a single child element, <dtmf-type>. The <dtmf-type> element has two attributes: "name" and "package." The "name" attribute provides the type of DTMF being used and is a case-insensitive string containing either "RFC 4733" or "Media" (generating tones as signals in the audio stream). The "package" attribute provides the name of the CFW package, compliant with Section 8.13.1.1 (Section 13.1.1 of RFC 6230), for which the DTMF type applies.

- <passthrough>: Indicates the required support for passing DTMF through without re-encoding.

The <passthrough> element has no attributes. The <passthrough> element has a single child element, <dtmf-type>. The <dtmf-type> element has two attributes: "name" and "package." The "name" attribute provides the type of DTMF being used and is a case-insensitive string containing either "RFC 4733" or "Media" (passing tones as signals through the audio stream). The "package" attribute provides the name of the CFW package, compliant with Section 8.13.1.1 (Section 13.1.1 of RFC 6230), for which the DTMF type applies.

11.5.2.5.1.3.4 <tones> The <tones> element provides requested tones that a MS must support for a mix. In particular, the request refers to both support for country codes (ISO 3166-1 [2]) and requested functionality (ITU-T Recommendation Q.1950 [3]). The element MAY be present. The <tones> element has no attributes. The <tones> element has zero or more of the following child elements:

- <country-codes>: Describes the requested country codes in relation to tones. The <country-codes> element has no attributes.

The <country-codes> element has a single child element. The child element, <country-code>, requests a specific country code, compliant with the ISO 3166-1 [2] specification. The <country-code> element has a single attribute, "package." The attribute "package" provides the name of the CFW package, compliant with the specification in the related IANA registry (e.g., "msc-ivr/1.0"), in which the tones from the specified country code are requested.

- <h248-codes>: Describes the requested H.248 codes with respect to tones. The <h248-codes> element has no attributes.

The <h248-codes> element has a single child element. The child element, <h248-code>, requests a specific H.248 code, compliant with the ITU-T Recommendation Q.1950 [3] specification. The codes can be either specific (e.g., cg/dt to only report the Dial Tone from the Call Progress Tones package) or generic (e.g., cg/* to report all the tones from the Call Progress Tones package) using wildcards. The <h248-code> element has a single attribute, "package." The attribute "package" provides the name of the CFW package, compliant with Section 8.13.1.1 (Section 13.1.1 of RFC 6230), in which the specified codes are requested.

11.5.2.5.1.3.5 <mixing-modes> The <mixing-modes> element requests information relating to support for audio and video mixing, more specifically a list of supported algorithms to mix audio and a list of supported video presentation layouts. The element MAY be present. The <mixing-modes> element has no attributes. The <mixing-modes> element has zero or more of the following child elements:

- <audio-mixing-modes>: Describes the requested algorithms for audio mixing. The <audio-mixing-modes> element has no attributes. The <audio-mixing-modes> element has one child element. The child element, <audio-mixing-mode>, contains a requested mixing algorithm. Valid values for the <audio-mixing-mode> element are algorithm names, e.g., "nbest" and "controller" as defined in RFC 6505 (see Chapter 9). The element has a single attribute, "package." The attribute "package" provides the name of the CFW package, compliant with Section 8.13.1.1 (Section 13.1.1 of RFC 6230), for which the algorithm support is requested.
- <video-mixing-modes>: Describes the requested video presentation layouts for video mixing. The <video-mixing-modes> element has two attributes: "vas" and "activespeakermix." The "vas" attribute is of type Boolean with a value of "true" indicating that the Consumer client requires automatic Voice Activated Switching. The "activespeakermix" attribute is of type Boolean with a value of "true" indicating that the Consumer client requires an additional video stream for the loudest speaker participant without its contribution. The <video-mixing-modes> element has one child element. The child element, <video-mixing-mode>, contains the name of a specific video presentation layout. The name may refer to one of the predefined video layouts defined in the XCON conference information data model or to non-XCON layouts as well as long as they are appropriately prefixed. The <video-mixing-mode> element has a single attribute, "package." The attribute "package" provides the name of the CFW package, compliant with Section 8.13.1.1 (Section 13.1.1 of RFC 6230), for which the algorithm support is requested.

11.5.2.5.1.3.6 <application-data> The <application-data> element provides an arbitrary string of characters as mixer application-level data. This data is meant to have meaning only at the application-level logic and as such, is not otherwise restricted by this specification. The set of allowed characters is the same as those in XML (tab, carriage return, line feed, and the legal characters of Unicode and ISO/IEC 10646 [1] (see also Section 2.2 of <www.w3.org/TR/xml/>). The element MAY be present.

The <application-data> element has no attributes.

The <application-data> element has no child elements.

11.5.2.5.1.3.7 <location> The <location> element requests a civic location for a mixer MS. The request makes use of the Civic Address Schema standardized in RFC 5139. The element MAY be present. More precisely, this section is entirely optional, and it is implementation specific to fill it with just the details each implementer deems necessary for any optimization that may be needed. The <location> element has no attributes. The <location> element has a single child element:

- <civicAddress>: Describes the civic address location of the requested MS, whose representation refers to Section 4 of RFC 5139.

11.5.2.5.1.3.8 <encryption> The <encryption> element allows a Consumer client to request support for encrypting mixed RTP media streams using RFC 3711. The element MAY be present. If the element is present, then the MS supports DTLS-SRTP (RFC 5763).

The <encryption> element has no attributes.

The <encryption> element has no child elements.

11.5.2.6 Media Service Resource Response

This section provides the element definitions for use in Consumer interface responses. The responses are carried in the <mediaResourceResponse> element.

11.5.2.6.1 <mediaResourceResponse>

The <mediaResourceResponse> element provides information for clients receiving response information from an external MRB entity. The <mediaResourceResponse> element has two mandatory attributes: "id" and "status." The "id" attribute must contain the same value that the client provided in the "id" attribute in the <mediaResourceRequest> to which the response is mapped. The "status" attribute indicates the status code of the operation. The status codes defined for "status" are shown in Table 11.2.

If a new media resource request made by a client application has been accepted, the MRB MUST reply with a <mediaResourceResponse> with status code 200. The same rule applies whenever a request to update (action="update") or remove (action="remove") an existing transaction can be fulfilled by the MRB. A media resource request, nevertheless, may fail for several reasons. In such a case, the status codes defined in Table 11.2 must be used instead. Specifically, if the MRB fails to handle a request due to a syntax error in the request itself (e.g., incorrect XML, violation of the schema constraints, or invalid values in any of the attributes/ elements), the MRB MUST reply with a <mediaResourceResponse> with status code 400. If a syntactically correct request fails because the request also includes any attribute/element the MRB does not understand, the MRB MUST reply with a <mediaResourceResponse> with status code 420.

If a syntactically correct request fails because it contains a wrong sequence number, that is, a "seq" value not consistent with the increment the MRB expects according to Section 11.5.2.3, the MRB MUST reply with a <mediaResourceResponse> with status code 405. If a syntactically correct request fails because the MRB could not find any MS able to fulfill the requirements presented by the AS in its request, the MRB MUST reply with a <mediaResourceResponse> with status code 408.

TABLE 11.2
<mediaResourceResponse> Status Codes

Code	Description
200	OK
400	Syntax error
405	Wrong sequence number
408	Unable to find Resource
410	Unable to remove Resource
420	Unsupported attribute or element

If a syntactically correct request fails because the MRB could not update an existing request according to the new requirements presented by the AS in its request, the MRB MUST reply with a <mediaResourceResponse> with status code 409. If a syntactically correct request fails because the MRB could not remove an existing request and release the related resources as requested by the AS, the MRB MUST reply with a <mediaResourceResponse> with status code 410.

Further details on status codes 409 and 410 are included in Section 11.5.2.3, where the leasing mechanism, along with its related scenarios, is described in more detail. The <mediaResourceResponse> element has <response-session-info> as a child element. This element is used to describe the response of a Consumer interface query and is covered in the following sub-section.

11.5.2.6.1.1 <response-session-info> The <response-session-info> element is included in Consumer responses. This applies to responses to both requests for new resources and requests to update an existing media resource session. The ability to change and remove an existing media resource session is described in more detail in Section 11.5.2.3. If the request was successful, the <mediaResourceResponse> MUST have one <response-session-info> child, which describes the media resource session addressed by the request. If the request was not successful, the <mediaResourceResponse> MUST NOT have a <response-session-info> child.

The <response-session-info> element has no attributes.

The <response-session-info> element has zero or more of the following child elements:

- <session-id>: A unique identifier that explicitly references an existing media resource session on the MRB. The identifier is included to update the existing session and is described in more detail in Section 11.5.2.3.
- <seq>: Used in association with the <session-id> element in a subsequent request to update an existing media resource session on an MRB. The <seq> number is incremented from its original value returned in response to the initial request for media resources. More information on its use is provided in Section 11.5.2.3.
- <expires>: Includes the number of seconds that the media resources are reserved as part of this interaction. If the lease is not refreshed before expiry, the MRB will reclaim the resources, and they will no longer be guaranteed.

It is RECOMMENDED that a minimum value of 300 seconds be used for the value of the "expires" attribute. It is also RECOMMENDED that a Consumer client refresh the lease at an interval that is not too close to the expiry time. A value of 80% of the timeout period could be used. For example, if the timeout period is 300 seconds, the Consumer client would refresh the transaction at 240 seconds. More information on its use is provided in Section 11.5.2.3. <media-server-address>: Provides information to reach the MS handling the requested media resource. One or more instances of these elements may appear.

The <media-server-address> element has a single attribute named "uri," which supplies an SIP URI that reaches the specified MS. It also has three optional elements: <connection-id>, <ivr-sessions>, and <mixers>. The <ivr-sessions> and <mixers> elements are defined in Sections 11.5.2.5.1.2.1 and 11.5.2.5.1.3.1, respectively, and have the same meaning but are applied to individual MS instances as a subset of the overall resources reported in the <connection-id> element. If multiple MSs are assigned in an IAMM operation, exactly one <media-server-address> element, more specifically the MS that provided the media dialog or CFW response, will have a <connection-id> element. Additional information relating to the use of the <connection-id> element for media dialogs is included in Section 11.6.

11.5.3 In-Line Unaware MRB Interface

An entity acting as an In-line MRB can act in one of two roles for a request, as introduced in Section 11.4.2: The IUMM of operation and the IAMM of operation. This section further describes IUMM.

It should be noted that the introduction of an MRB entity into the network, as specified in this specification, requires interfaces to be implemented by those requesting MS resources (for example, an AS). This applies when using the Consumer interface as discussed in Sections 11.5.2.1 (Query mode) and 11.5.2.2 (IAMM).

An MRB entity can also act in a client-unaware mode when deployed into the network. This allows any SIP-compliant client entity, as defined by RFC 3261 (also see Section 1.2) and its extensions, to send requests to an MRB, which in turn, will select an appropriate MS based on knowledge of MS resources it currently has available transparently to the client entity. Using an MRB in this mode allows easy migration of current applications and services that are unaware of the MRB concept and would simply require a configuration change resulting in the MRB being set as an SIP outbound proxy for clients requiring media services.

With IUMM, the MRB may conclude that an assigned media resource is no longer needed when it receives an SIP BYE from the AS or MS ending the SIP dialog that initiated the request. As with IAMM, in IUMM, the SIP INVITE from the AS could convey the application/sdp payload to set up either a media dialog or a Control Framework Control Channel. In either case, in order to permit the AS to associate a media dialog with a Control Channel to the same MS, using the procedures of RFC 6230 (see Chapter 8) Section 11.6, the MRB should be acting as an SIP proxy (and not a B2BUA). This allows the SIP URI of the targeted MS to be transparently passed back to the AS in the SIP response, resulting in a direct SIP dialog between the AS and the MS. While IUMM has the lowest impact on legacy AS, it also provides the lowest versatility. See Section 11.8.

11.6 MRB ACTING AS A B2BUA

An MRB entity can act as an SIP B2BUA or an SIP Proxy Server as defined in RFC 3261 (also see Section 1.2). When an MRB acts as a B2BUA, issues can arise when using Media Control Channel packages such as the IVR (RFC 6231 – see Section 12) and mixer (RFC 6505 – see Section 9) packages. Specifically, the framework attribute "connectionid" as provided in Section 8.14 (Appendix A "Common Package Components" of RFC 6230) uses a concatenation of the SIP dialog identifiers to be used for referencing SIP dialogs within the Media Control Channel.

When a request traverses an MRB acting as a B2BUA, the SIP dialog identifiers change, and so the "connectionid" cannot be used as intended due to this change. For this reason, when an MRB wishes to act as an SIP B2BUA when handling a request from an AS to set up a media dialog to an MS, it MUST include the optional <connection-id> element in a Consumer interface response with a value that provides the equivalent for the "connectionid" ("Local Dialog Tag" + "Remote Dialog Tag") for the far side of the B2BUA. If present, this value MUST be used as the value for the "connectionid" in packages where the Common Package Components are used. The <connection-id> element MUST NOT be included in an HTTP Consumer interface response.

It is important to point out that although more MS instances may be returned in a Consumer response (i.e., the MRB has assigned more than one MS to a Consumer request to fulfill the AS requirements), in IAMM, the MRB will only act as a B2BUA with a single MS. In this case, exactly one <media-server-address> element, describing the media dialog or CFW response, will have a <connection-id> element that will not be included in any additional <media-server-address> elements.

11.7 MULTIMODAL MRB IMPLEMENTATIONS

An MRB implementation may operate multimodally with a collection of AS clients all sharing the same pool of media resources. That is, an MRB may be simultaneously operating in Query mode, IAMM, and IUMM. It knows in which mode to act on any particular request from a client, depending on the context of the request:

- If the received request is an HTTP POST message with application/ mrb-consumer+xml content, then the MRB processes it in Query mode.
- If the received request is an SIP INVITE with application/ mrb-consumer+xml content and application/sdp content, then the MRB processes it in IAMM.
- If the received request is an SIP INVITE without application/ mrb-consumer+xml content but with application/sdp content, then the MRB processes it in IUMM.

11.8 RELATIVE MERITS OF QUERY MODE, IAMM, AND IUMM

At a high level, the possible AS MRB interactions can be distinguished by the following basic types:

a. Query mode – the client is requesting the assignment by the MRB of suitable MS resources
b. IAMM/media dialog – the client is requesting the assignment by the MRB of suitable MS resources and the establishment of a media dialog to one of the MSs
c. IAMM/Control Channel – the client is requesting the assignment by the MRB of suitable MS resources and the establishment of a CFW Control Channel to one of the MSs
d. IUMM/media dialog – the client is requesting the establishment of a media dialog to an MS resource
e. IUMM/Control Channel – the client is requesting the establishment of a CFW Control Channel to an MS resource

Each type of interaction has advantages and disadvantages, where such considerations relate to the versatility of what the MRB can provide, technical aspects such as efficiency in different application scenarios, complexity, delay, use with legacy ASs, or use with the CFW. Depending on the characteristics of a particular setting that an MRB is intended to support, some of these interaction types may be more appropriate than others. This section provides a few observations on relative merits but is not intended to be exhaustive. Some constraints of a given interaction type may be subtle.

- Operation with other types of media control: Any of the types of interactions work with the mechanisms described in RFC 4240 and RFC 5552 (see Chapter 13) where initial control instructions are conveyed in the SIP INVITE from the AS for the media dialog to the MS, and subsequent instructions may be fetched using HTTP. Query mode (a), IAMM/media dialog (b), and IUMM/media dialog (d) work with the Media Server Markup Language (MSML) as per RFC 5707 or the Media Server Control Markup Language (MSCML) as per RFC 5022.
- As stated previously, IUMM has no interface impacts on an AS. When using IUMM, the AS does not specify the characteristics of the type of media resource it requires, as the <mediaResourceRequest> element is not passed to the MRB. For IUMM/media dialog (d), the MRB can deduce an appropriate media resource on a best-effort basis using information gleaned from examining information in the SIP INVITE. This includes the SDP information for the media dialog or initial control information in the SIP Request-URI as per RFC 4240. With IUMM/Control Channel (e), there is even less information for the MRB to use.
- If using IUMM/Control Channel (e), the subsequent sending of the media dialog to the MS should not be done using IUMM/media dialog. That is, the SIP signaling to send the media dialog to the selected MS must be directly between the AS and that MS, and not through the MRB. Unless resources can be confidentially identified, the MRB could send the media dialog to a different MS. Likewise, if using IUMM/media dialog (d), the subsequent establishment of a Control Channel should not be done with IUMM/Control Channel (e) unless definitive information is available.

- Query mode (a) and IAMM/Control Channel (c) lend themselves to requesting a pool of media resources (e.g., a number of IVR or conferencing ports) in advance of use and retaining use over a period of time, independently of whether there are media dialogs to those resources at any given moment, whereas the other types of interactions do not. This also applies to making a subsequent request to increase or decrease the amount of resources previously awarded.
- While Query mode (a) and IAMM/Control Channel (c) are the most versatile interaction types, the former is completely decoupled from the use or non-use of a Control Channel, whereas the latter requires the use of a Control Channel.
- When CFW Control Channels are to be used in conjunction with the use of an MRB, Query mode (a) would typically result in fewer such channels being established over time as compared with IAMM/Control Channel (c). That is because the latter would involve setting up an additional Control Channel every time an AS has a new request for an MRB for media resources.

11.9 EXAMPLES

This section provides examples of both the Publish and Consumer interfaces. Both the Query mode and In-line mode are addressed. Note that due to RFC formatting conventions, this section often splits HTTP, SIP/SDP, and CFW across lines whose content would exceed 72 characters. A backslash character marks where this line folding has taken place. This backslash, and its trailing CRLF and whitespace, would not appear in the actual protocol contents. Also, note that the indentation of the XML content is only provided for readability: actual messages will follow strict XML syntax, which allows but does not require indentation.

11.9.1 Publish Example

The following example assumes that a Control Channel has been established and synced as described in the CFW (RFC 6230 – see Chapter 8). Figure 11.9 shows the subscription/notification mechanism the Publish interface is based on, as defined in Section 11.5.1. The MRB subscribes for information at the MS (message A1.), and the MS accepts the subscription (A2.). Notifications are triggered by the MS (B1.) and acknowledged by the MRB (B2.).

The rest of this section includes a full dump of the messages associated with the previous sequence diagram, specifically:

1. The subscription (A1.) in an <mrbrequest> (CFW CONTROL)
2. The MS accepting the subscription (A2.) in an <mrbresponse> (CFW 200)
3. A notification (B1.) in an <mrbnotification> (CFW CONTROL);
4. The ack to the notification (B2.) in a framework-level 200 message (CFW 200).

```
A1. MRB --> MS (CONTROL, publish request)

    CFW lidc30BZObiC CONTROL
    Control-Package: mrb-publish/1.0
    Content-Type: application/mrb-publish+xml
    Content-Length: 337

    <?xml version="1.0" encoding="UTF-8" standalone="yes"?>
    <mrbpublish version="1.0" xmlns="urn:ietf:params:xml:ns:mrb-publish">
        <mrbrequest>
            <subscription action="create" seqnumber="1" id="p0T65U">
```

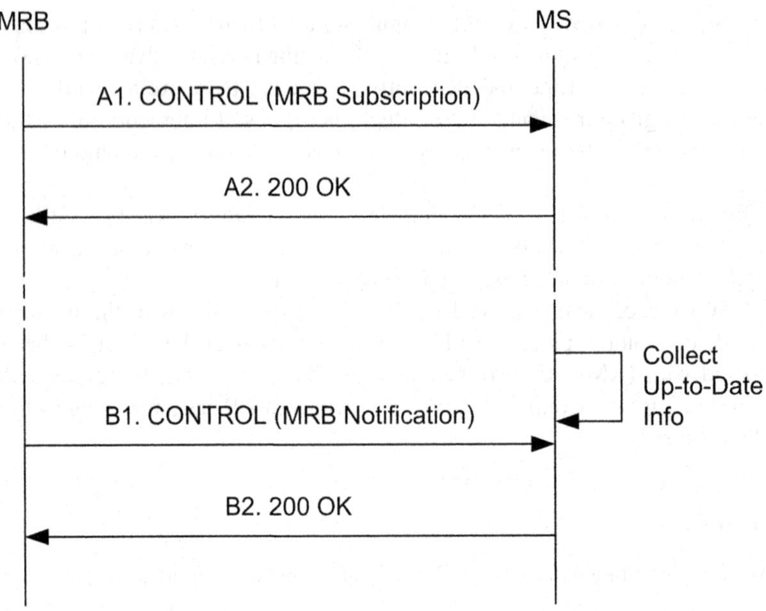

FIGURE 11.9 Publish Example: Sequence Diagram.

```
            <expires>600</expires>
            <minfrequency>20</minfrequency>
            <maxfrequency>20</maxfrequency>
        </subscription>
    </mrbrequest>
</mrbpublish>

A2. MRB <-- MS (200 to CONTROL, request accepted)

CFW lidc30BZObiC 200
Timeout: 10
Content-Type: application/mrb-publish+xml
Content-Length: 139

<mrbpublish version="1.0" xmlns="urn:ietf:params:xml:ns:mrb-publish">
    <mrbresponse status="200" reason="OK: Request accepted"/>
</mrbpublish>

B1. MRB <-- MS (CONTROL, event notification from MS)

CFW 03fff52e7b7a CONTROL
Control-Package: mrb-publish/1.0
Content-Type: application/mrb-publish+xml
Content-Length: 4226

<?xml version="1.0" encoding="UTF-8" standalone="yes"?>
    <mrbpublish version="1.0"
            xmlns="urn:ietf:params:xml:ns:mrb-publish">
        <mrbnotification seqnumber="1" id="QQ6J3c">
            <media-server-id>a1b2c3d4</media-server-id>
            <supported-packages>
                <package name="msc-ivr/1.0"/>
                <package name="msc-mixer/1.0"/>
```

```
        <package name="mrb-publish/1.0"/>
        <package name="msc-example-pkg/1.0"/>
    </supported-packages>
    <active-rtp-sessions>
        <rtp-codec name="audio/basic">
            <decoding>10</decoding>
            <encoding>20</encoding>
        </rtp-codec>
    </active-rtp-sessions>
    <active-mixer-sessions>
        <active-mix conferenceid="7cfgs43">
            <rtp-codec name="audio/basic">
                <decoding>3</decoding>
                <encoding>3</encoding>
            </rtp-codec>
        </active-mix>
    </active-mixer-sessions>
    <non-active-rtp-sessions>
        <rtp-codec name="audio/basic">
            <decoding>50</decoding>
            <encoding>40</encoding>
        </rtp-codec>
    </non-active-rtp-sessions>
    <non-active-mixer-sessions>
        <non-active-mix available="15">
            <rtp-codec name="audio/basic">
                <decoding>15</decoding>
                <encoding>15</encoding>
            </rtp-codec>
        </non-active-mix>
    </non-active-mixer-sessions>
    <media-server-status>active</media-server-status>
        <supported-codecs>
            <supported-codec name="audio/basic">
                <supported-codec-package name="msc-ivr/1.0">
                    <supported-action>encoding</supported-action>
                    <supported-action>decoding</supported-action>
                </supported-codec-package>
                <supported-codec-package name="msc-mixer/1.0">
                    <supported-action>encoding</supported-action>
                    <supported-action>decoding</supported-action>
                </supported-codec-package>
        </supported-codec>
    </supported-codecs>
    <application-data>TestbedPrototype</application-data>
    <file-formats>
        <supported-format name="audio/x-wav">
            <supported-file-package>
                msc-ivr/1.0
            </supported-file-package>
        </supported-format>
    </file-formats>
    <max-prepared-duration>
        <max-time max-time-seconds="3600">
            <max-time-package>msc-ivr/1.0</max-time-package>
        </max-time>
```

```
    </max-prepared-duration>
    <dtmf-support>
       <detect>
          <dtmf-type package="msc-ivr/1.0"
             name="RFC4733"/>
          <dtmf-type package="msc-mixer/1.0"
             name="RFC4733"/>
       </detect>
       <generate>
          <dtmf-type package="msc-ivr/1.0"
             name="RFC4733"/>
          <dtmf-type package="msc-mixer/1.0"
             name="RFC4733"/>
       </generate>
       <passthrough>
          <dtmf-type package="msc-ivr/1.0"
             name="RFC4733"/>
          <dtmf-type package="msc-mixer/1.0"
             name="RFC4733"/>
       </passthrough>
    </dtmf-support>
    <mixing-modes>
       <audio-mixing-modes>
          <audio-mixing-mode package="msc-ivr/1.0">
             nbest
          </audio-mixing-mode>
       </audio-mixing-modes>
       <video-mixing-modes activespeakermix="true" vas="true">
          <video-mixing-mode package="msc-mixer/1.0">
             single-view
          </video-mixing-mode>
       <video-mixing-mode package="msc-mixer/1.0">
             dual-view
       </video-mixing-mode>
       <video-mixing-mode package="msc-mixer/1.0">
             dual-view-crop
       </video-mixing-mode>
       <video-mixing-mode package="msc-mixer/1.0">
             dual-view-2x1
       </video-mixing-mode>
       <video-mixing-mode package="msc-mixer/1.0">
             dual-view-2x1-crop
       </video-mixing-mode>
       <video-mixing-mode package="msc-mixer/1.0">
             quad-view
       </video-mixing-mode>
       <video-mixing-mode package="msc-mixer/1.0">
          multiple-5x1
       </video-mixing-mode>
       <video-mixing-mode package="msc-mixer/1.0">
          multiple-3x3
       </video-mixing-mode>
       <video-mixing-mode package="msc-mixer/1.0">
          multiple-4x4
       </video-mixing-mode>
    </video-mixing-modes>
 </mixing-modes>
```

```
        <supported-tones>
            <supported-country-codes>
                <country-code package="msc-
                    ivr/1.0">GB</country-code>
                <country-code package="msc-
                    ivr/1.0">IT</country-code>
                <country-code package="msc-
                    ivr/1.0">US</country-code>
            </supported-country-codes>
            <supported-h248-codes>
                <h248-code package="msc-ivr/1.0">cg/*</h248-code>
                <h248-code package="msc-
                    ivr/1.0">biztn/ofque</h248-
                    code>
                <h248-code package="msc-
                    ivr/1.0">biztn/erwt</h248-
                    code>
                <h248-code package="msc-
                   mixer/1.0">conftn/*</h248-
                    code>
            </supported-h248-codes>
        </supported-tones>
        <file-transfer-modes>
            <file-transfer-mode package="msc-ivr/1.0"
                name="HTTP"/>
            </file-transfer-modes>
            <asr-tts-support>
                <asr-support>
                    <language xml:lang="en"/>
                </asr-support>
                <tts-support>
                    <language xml:lang="en"/>
                </tts-support>
            </asr-tts-support>
            <vxml-support>
                <vxml-mode package="msc-ivr/1.0"
                    support="RFC6231"/>
            </vxml-support>
            <media-server-location>
                <civicAddress xml:lang="it">
                    <country>IT</country>
                    <A1>Campania</A1>
                    <A3>Napoli</A3>
                        <A6>Via Claudio</A6>
                        <HNO>21</HNO>
                        <LMK>University of Napoli Federico II</LMK>
                        <NAM>Dipartimento di Informatica e
                           Sistemistica</NAM>
                        <PC>80210</PC>
                    </civicAddress>
                </media-server-location>
                <label>TestbedPrototype-01</label>
                <media-server-address>sip:MS1@ms.example.net
                        </media-server-address>
            <encryption/>
    </mrbnotification>
</mrbpublish>
```

```
B2. MRB -> MS (200 to CONTROL)

CFW 03fff52e7b7a 200
```

11.9.2 Consumer Examples

As specified in Section 11.5.2, the Consumer interface can be involved in two different modes: Query and In-line aware. When in Query mode, Consumer messages are transported in HTTP messages: An example of such an approach is presented in Section 11.9.2.1. When in In-line aware mode, messages are instead transported as part of SIP negotiations: Considering that SIP negotiations may be related to either the creation of a Control Channel or to a User Agent Client (UAC) media dialog, two separate examples of such an approach are presented in Section 11.9.2.2.

11.9.2.1 Query Example

The following example (Figure 11.10) assumes that the interested AS already knows the HTTP URL where an MRB is listening for Consumer messages. Figure 11.10 shows the HTTP-based transaction between the AS and the MRB. The AS sends a Consumer request as payload of an HTTP POST message (1.), and the MRB provides an answer in an HTTP 200 OK message (2.). Specifically, as will be shown in the examples, the AS is interested in 100 IVR ports: the MRB finds two MSs that can satisfy the request (one providing 60 ports and the other providing 40 ports) and reports them to the AS.

The rest of this section includes a full dump of the messages associated with the previous sequence diagram, specifically:

1. The Consumer request (1.) in a <mediaResourceRequest> (HTTP POST, Content-Type 'application/mrb-consumer+xml')
2. The Consumer response (2.) in a <mediaResourceResponse> (HTTP 200 OK, Content-Type 'application/mrb-consumer+xml')

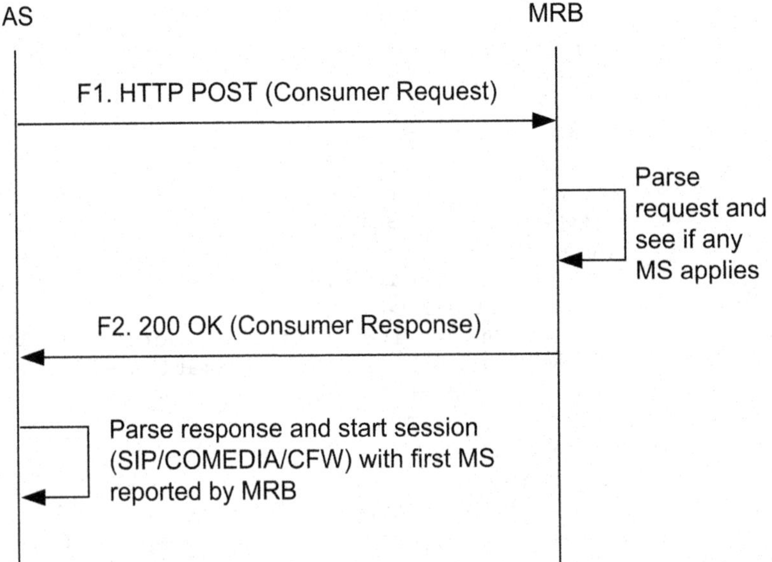

FIGURE 11.10 Consumer Example (Query): Sequence Diagram.

1. AS --> MRB (HTTP POST, Consumer request)

```
POST /Mrb/Consumer HTTP/1.1
Content-Length: 893
Content-Type: application/mrb-consumer+xml
Host: mrb.example.net:8080
Connection: Keep-Alive
User-Agent: Apache-HttpClient/4.0.1 (java 1.5)

<?xml version="1.0" encoding="UTF-8" standalone="yes"?>

<mrbconsumer version="1.0" xmlns="urn:ietf:params:xml:ns:mrb-consumer">
    <mediaResourceRequest id="gh11x23v">
    <generalInfo>
        <packages>
            <package>msc-ivr/1.0</package>
            <package>msc-mixer/1.0</package>
        </packages>
    </generalInfo>
    <ivrInfo>
        <ivr-sessions>
            <rtp-codec name="audio/basic">
                <decoding>100</decoding>
                <encoding>100</encoding>
            </rtp-codec>
        </ivr-sessions>
        <file-formats>
            <required-format name="audio/x-wav"/>
        </file-formats>
        <file-transfer-modes>
            <file-transfer-mode package="msc-ivr/1.0"
                name="HTTP"/>
        </file-transfer-modes>
    </ivrInfo>
    </mediaResourceRequest>
</mrbconsumer>
```

2. AS <- MRB (200 to POST, Consumer response)

```
HTTP/1.1 200 OK
X-Powered-By: Servlet/2.5
Server: Sun GlassFish Communications Server 1.5
Content-Type: application/mrb-consumer+xml;charset=ISO-8859-1
Content-Length: 1133
Date: Mon, 12 Apr 2011 14:59:26 GMT

<?xml version="1.0" encoding="UTF-8" standalone="yes"?>
<mrbconsumer version="1.0" xmlns="urn:ietf:params:xml:ns:mrb-consumer" >
    <mediaResourceResponse reason="Resource found" status="200"
            id="gh11x23v">
        <response-session-info>
            <session-id>5t3Y4IQ84gY1</session-id>
            <seq>9</seq>
            <expires>3600</expires>
            <media-server-address
```

```
                uri="sip:MediaServer@ms.
                    example.com:5080">
                <ivr-sessions>
                    <rtp-codec name="audio/basic">
                        <decoding>60</decoding>
                        <encoding>60</encoding>
                    </rtp-codec>
                </ivr-sessions>
            </media-server-address>
            <media-server-address
                    uri="sip:OtherMediaServer@pool.
                    example.net:5080">
                <ivr-sessions>
                    <rtp-codec name="audio/basic">
                        <decoding>40</decoding>
                        <encoding>40</encoding>
                    </rtp-codec>
                </ivr-sessions>
            </media-server-address>
        </response-session-info>
    </mediaResourceResponse>
</mrbconsumer>
```

As the example shows, the request and response are associated by means of the "id" attribute (id="gh11x23v"). The MRB has picked "9" as the random sequence number that needs to be incremented by the AS for the subsequent request associated with the same session. The rest of the scenario is omitted for brevity. After having received the "mediaResourceResponse," the AS has the URIs of two MSs able to fulfill its media requirements and can start a control dialog with one or both of them.

11.9.2.2 IAMM Examples

Two separate examples are presented for the IAMM case: In fact, IAMM can take advantage of two different approaches with respect to the SIP dialogs to be exploited to carry Consumer messages, i.e., i) a SIP control dialog to create a Control Channel and ii) a UAC media dialog to attach to a MS. To make things clearer for the reader, the same Consumer request as the one presented in the Query mode will be sent in order to clarify how the behavior of the involved parties may differ.

11.9.2.2.1 IAMM Example: CFW-Based Approach

The following example assumes that the interested AS already knows the SIP URI of an MRB. Figure 11.11 shows the first approach, i.e., SIP-based transactions between the AS, the MRB, and one MS that the MRB chooses from the two that are allocated to fulfill the request. The diagram is more complex than before. This is basically a scenario envisaging the MRB as a B2BUA. The AS sends a SIP INVITE (1.) containing both a CFW-related SDP and a Consumer request (multipart body).

The MRB sends a provisional response to the AS (2.) and starts working on the request. First of all, it makes use of the Consumer request from the AS to determine which MSs should be exploited. Once the right MSs have been chosen (MS1 and MS2 in the example), the MRB sends a new SIP INVITE (3.) to one of the MSs (MS1 in the example) by just including the SDP part of the original request. That MS negotiates this INVITE as specified in (RFC 6230 – see Chapter 8) (4., 5., 6.), providing the MRB with its own CFW-related SDP.

The MRB replies to the original AS INVITE, preparing an SIP 200 OK with another multipart body (7.): this multipart body includes the Consumer response used by the MRB to determine the right MSs and the SDP returned by the MS (MS1) in (5.). The AS finally acknowledges the 200 OK

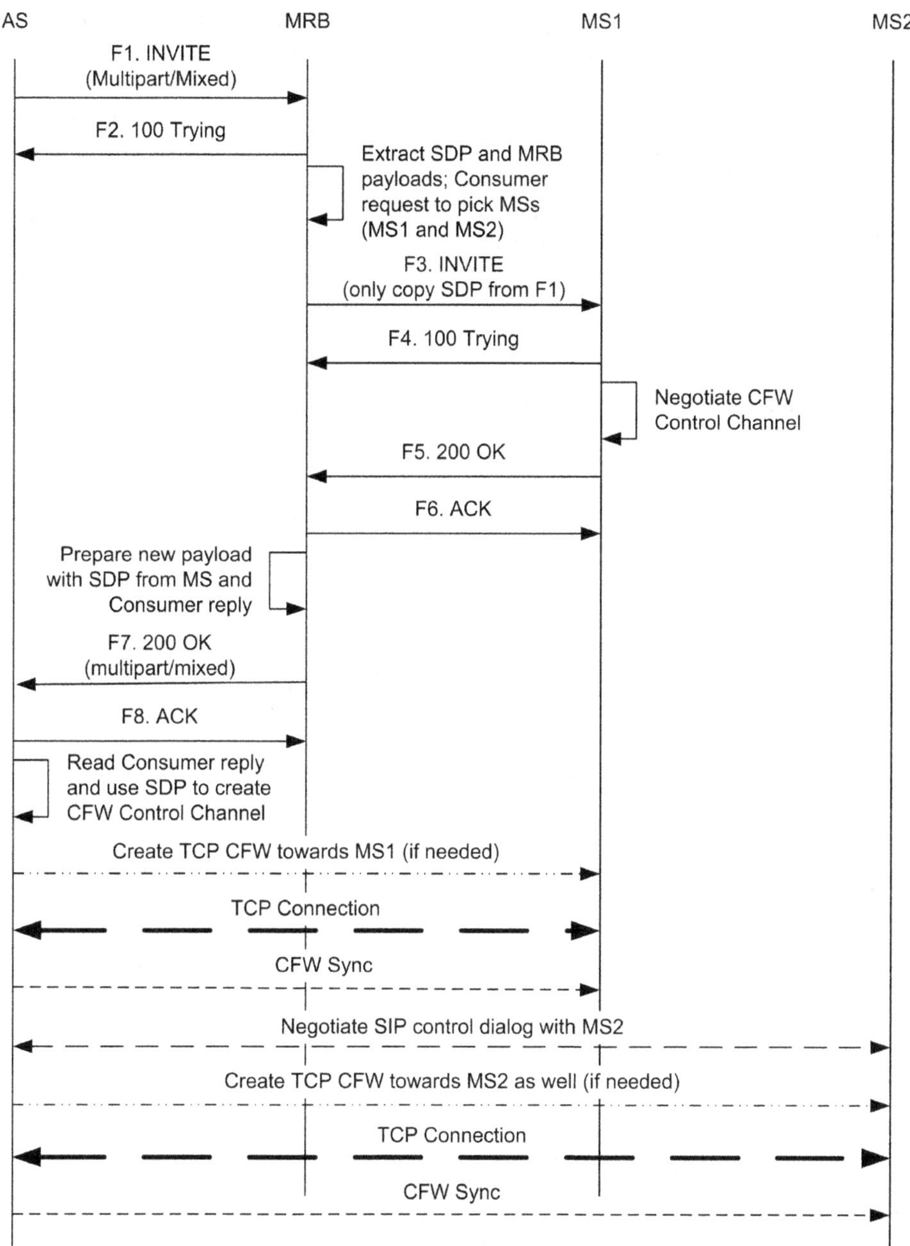

FIGURE 11.11 Consumer Example (IAMM/Control Channel): Sequence Diagram. (Copyright: IETF.)

(8.) and can start a CFW connection towards that MS (MS1). Since the MRB provided the AS with two MS instances to fulfill its requirements, the AS can use the URI in the <media-server-address> element in the <mediaResourceResponse> that describes the other MS to establish a CFW channel with that MS (MS2) as well.

Please note that to ease the reading of the protocol contents, a simple "=_Part" is used whenever a boundary for a "multipart/mixed" payload is provided instead of the actual boundary that would be inserted in the SIP messages.

The rest of this section includes an almost full trace of the messages associated with the previous sequence diagram. Only the relevant SIP messages are shown (both the INVITEs and the 200

OKs), and only the relevant headers are preserved for brevity (Content-Type and multipart-related information). Specifically:

1. The original INVITE (F1.) containing both a CFW-related SDP (Connection-Oriented Media (COMEDIA) information to negotiate a new Control Channel) and a Consumer <mediaResourceRequest>
2. The INVITE sent by the MRB (acting as a B2BUA) to the MS (F3) containing only the CFW-related SDP from the original INVITE
3. The 200 OK sent by the MS back to the MRB (F5) to complete the CFW-related negotiation (SDP only)
4. The 200 OK sent by the MRB back to the AS in response to the original INVITE (F) containing both the CFW-related information sent by the MS and a Consumer <mediaResourceRequest> documenting the MRB's decision to use that MS.

```
F1. AS -> MRB (INVITE multipart/mixed)
   [..]
   Content-Type: multipart/mixed;boundary="=_Part"

   =_Part
   Content-Type: application/sdp
   v=0
   o=- 2890844526 2890842807 IN IP4 as.example.com
   s=MediaCtrl
   c=IN IP4 as.example.com
   t=0 0
   m=application 48035 TCP cfw
   a=connection:new
   a=setup:active
   a=cfw-id:vF0zD4xzUAW9
   a=ctrl-package:msc-mixer/1.0
   a=ctrl-package:msc-ivr/1.0

   =_Part
   Content-Type: application/mrb-consumer+xml

   <?xml version="1.0" encoding="UTF-8" standalone="yes"?>

   <mrbconsumer version="1.0"
         xmlns="urn:ietf:params:xml:ns:mrb-consumer">
      <mediaResourceRequest id="pz78hnq1">
         <generalInfo>
            <packages>
               <package>msc-ivr/1.0</package>
               <package>msc-mixer/1.0</package>
            </packages>
         </generalInfo>
         <ivrInfo>
            <ivr-sessions>
               <rtp-codec name="audio/basic">
                  <decoding>100</decoding>
                  <encoding>100</encoding>
               </rtp-codec>
            </ivr-sessions>
            <file-formats>
               <required-format name="audio/x-wav"/>
```

```
            </file-formats>
            <file-transfer-modes>
               <file-transfer-mode package="msc-ivr/1.0"
                     name="HTTP"/>
            </file-transfer-modes>
         </ivrInfo>
      </mediaResourceRequest>
   </mrbconsumer>

   =_Part
```

```
F3. MRB -> MS (INVITE sdp only)

   [..]
   Content-Type: application/sdp

   v=0
   o=- 2890844526 2890842807 IN IP4 as.example.com
   s=MediaCtrl
   c=IN IP4 as.example.com
   t=0 0
   m=application 48035 TCP cfw
   a=connection:new
   a=setup:active
   a=cfw-id:vF0zD4xzUAW9
   a=ctrl-package:msc-mixer/1.0
   a=ctrl-package:msc-ivr/1.0

5. MRB <- MS (200 OK sdp)

    [..]
   Content-Type: application/sdp
   v=0
   o=lminiero 2890844526 2890842808 IN IP4 ms.example.net
   s=MediaCtrl
   c=IN IP4 ms.example.net
   t=0 0
   m=application 7575 TCP cfw
   a=connection:new
   a=setup:passive
   a=cfw-id:vF0zD4xzUAW9
   a=ctrl-package:msc-mixer/1.0
   a=ctrl-package:msc-ivr/1.0
   a=ctrl-package:mrb-publish/1.0
   a=ctrl-package:msc-example-pkg/1.0

F7. AS <- MRB (200 OK multipart/mixed)

   [..]
   Content-Type: multipart/mixed;boundary="=_Part"

   =_Part
   Content-Type: application/sdp

   v=0
   o=lminiero 2890844526 2890842808 IN IP4 ms.example.net
```

```
s=MediaCtrl
c=IN IP4 ms.example.net
t=0 0
m=application 7575 TCP cfw
a=connection:new
a=setup:passive
a=cfw-id:vF0zD4xzUAW9
a=ctrl-package:msc-mixer/1.0
a=ctrl-package:msc-ivr/1.0
a=ctrl-package:mrb-publish/1.0
a=ctrl-package:msc-example-pkg/1.0

=_Part
Content-Type: application/mrb-consumer+xml

<?xml version="1.0" encoding="UTF-8" standalone="yes"?>

<mrbconsumer version="1.0"
       xmlns="urn:ietf:params:xml:ns:mrb-consumer" >
   <mediaResourceResponse reason="Resource found" status="200"
          id="pz78hnq1">
       <response-session-info>
          <session-id>z1skKYZQ3eFu</session-id>
          <seq>9</seq>
          <expires>3600</expires>
          <media-server-address
                 uri="sip:MediaServer@
                 ms.example.com:5080">
              <connection-id>32pbdxZ8:KQw677BF
                 </connection-id>
              <ivr-sessions>
                 <rtp-codec name="audio/basic">
                     <decoding>60</decoding>
                     <encoding>60</encoding>
                 </rtp-codec>
              </ivr-sessions>
          </media-server-address>
          <media-server-address
                 uri="sip:OtherMediaServer@
                 pool.example.net:5080">
              <ivr-sessions>
                 <rtp-codec name="audio/basic">
                     <decoding>40</decoding>
                     <encoding>40</encoding>
                 </rtp-codec>
              </ivr-sessions>
          </media-server-address>
       </response-session-info>
   </mediaResourceResponse>
</mrbconsumer>

=_Part
```

As the previous example illustrates, the only difference in the response that the MRB provides to the AS is in the “connection-id” attribute that is added to the first allocated MS instance: this allows the AS to understand that the MRB has sent the CFW channel negotiation to that specific MS and that the connection-id to be used is the one provided. This will be described in more detail in the following section for the media dialog–based approach. The continuation of the scenario (the AS connecting to MS1 to start the Control Channel and the related SYNC message, the AS connecting to MS2 as well later on, all the media dialogs being attached to either MS) is omitted for brevity.

11.9.2.2.2 IAMM Example: Media Dialog–Based Approach

The following example assumes that the interested AS already knows the SIP URI of an MRB. Figure 11.12 shows the second approach, i.e., SIP-based transactions between a SIP client, the AS, the MRB, and the MS that the MRB chooses. The interaction is basically the same as previous examples (e.g., contents of the multipart body), but considering that a new party is involved in the communication, the diagram is slightly more complex than before. As before, the MRB acts as a B2BUA. A UAC sends an SIP INVITE to an SIP URI handled by the AS, since it is interested to its services (1.).

The AS sends a provisional response (2.) and since it does not have the resources yet, sends to the MRB a new SIP INVITE (3.) containing both the UAC media-related SDP and a Consumer request (multipart body). The MRB sends a provisional response to the AS (4.) and starts working on the request. First of all, it makes use of the Consumer request from the AS to determine which MSs should be chosen. Once the MS has been chosen, the MRB sends a new SIP INVITE to one of the MSs by including the SDP part of the original request (5.).

The MS negotiates this INVITE as specified in RFC 6230 (see Chapter 8) (6., 7., 8.) to allocate the needed media resources to handle the new media dialog, eventually providing the MRB with its own media-related SDP. The MRB replies to the original AS INVITE preparing a SIP 200 OK with a multipart body (9.): this multipart body includes the Consumer response from the MRB indicating the chosen MSs and the SDP returned by the MS in (7.). The AS finally acknowledges the 200 OK (10.) and ends the scenario by eventually providing the UAC with the SDP it needs to set up the RTP channels with the chosen MS: a separate direct SIP control dialog may be initiated by the AS to the same MS in order to set up a Control Channel to manipulate the media dialog.

As with the IAMM/Control Channel example in the prior section, this example has the MRB selecting MS resources across two MS instances. The convention could be that the MRB sent the SIP INVITE to the first MS in the list provided to the AS in the Consumer response information. For the sake of brevity, considerations related to connecting to the other MSs as well are omitted, since they have already been addressed in the previous section. Please note that to ease the reading of the protocol contents, a simple "=_Part" is used whenever a boundary for a "multipart/mixed" payload is provided instead of the actual boundary that would be inserted in the SIP messages.

The rest of this section includes a trace of the messages associated with the previous sequence diagram. Only the relevant SIP messages are shown (both the INVITEs and the 200 OKs), and only the relevant headers are preserved for brevity (Content-Type, From/To, and multipart-related information). Specifically:

1. The original INVITE (1.) containing the media-related SDP sent by a UAC
2. The INVITE sent by the AS to the MRB (3.) containing both the media-related SDP and a Consumer <mediaResourceRequest>
3. The INVITE sent by the MRB (acting as a B2BUA) to the MS (5.) containing only the media-related SDP from the original INVITE
4. The 200 OK sent by the MRB back to the AS in response to the original INVITE (9.) containing both the media-related information sent by the MS and a Consumer <mediaResourceRequest> specificationing the MRB's decision to use that MS;
5. The 200 OK sent by the AS back to the UAC to have it set up the RTP channel(s) with the MS (11.)

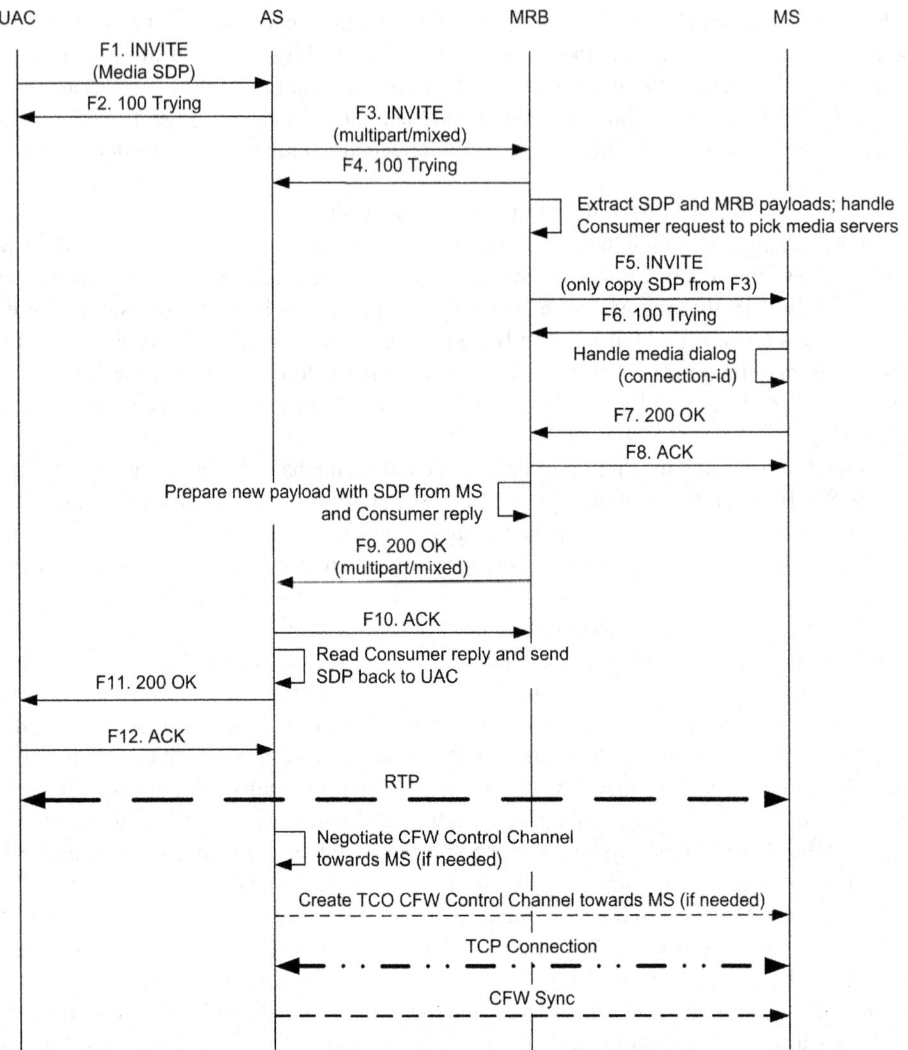

FIGURE 11.12 Consumer Example (IAMM/Media Dialog): Sequence Diagram. (Copyright: IETF.)

```
F1. UAC -> AS (INVITE with media SDP)

    [..]

    From: <sip:lminiero@users.example.com>;tag=1153573888
    To: <sip:mediactrlDemo@as.example.com>

    [..]

    Content-Type: application/sdp
    v=0
    o=lminiero 123456 654321 IN IP4 203.0.113.2
    s=A conversation
    c=IN IP4 203.0.113.2
    t=0 0
    m=audio 7078 RTP/AVP 0 3 8 101
    a=rtpmap:0 PCMU/8000/1
```

```
    a=rtpmap:3 GSM/8000/1
    a=rtpmap:8 PCMA/8000/1
    a=rtpmap:101 telephone-event/8000
    a=fmtp:101 0-11
    m=video 9078 RTP/AVP 98

F3. AS -> MRB (INVITE multipart/mixed)

     [..]

    From: <sip:ApplicationServer@as.example.com>;tag=fd4fush5
    To: <sip:Mrb@mrb.example.org>

    [..]

    Content-Type: multipart/mixed;boundary="=_Part"
    =_Part
    Content-Type: application/sdp
    v=0
    o=lminiero 123456 654321 IN IP4 203.0.113.2
    s=A conversation
    c=IN IP4 203.0.113.2
    t=0 0
    m=audio 7078 RTP/AVP 0 3 8 101
    a=rtpmap:0 PCMU/8000/1
    a=rtpmap:3 GSM/8000/1
    a=rtpmap:8 PCMA/8000/1
    a=rtpmap:101 telephone-event/8000
    a=fmtp:101 0-11
    m=video 9078 RTP/AVP 98
    =_Part
    Content-Type: application/mrb-consumer+xml

    <?xml version="1.0" encoding="UTF-8" standalone="yes"?>

    <mrbconsumer version="1.0"
          xmlns="urn:ietf:params:xml:ns:mrb-consumer">
       <mediaResourceRequest id="ns56g1x0">
          <generalInfo>
             <packages>
                <package>msc-ivr/1.0</package>
                <package>msc-mixer/1.0</package>
             </packages>
          </generalInfo>
          <ivrInfo>
             <ivr-sessions>
                <rtp-codec name="audio/basic">
                   <decoding>100</decoding>
                   <encoding>100</encoding>
                </rtp-codec>
             </ivr-sessions>
             <file-formats>
                <required-format name="audio/x-wav"/>
             </file-formats>
                <file-transfer-modes>
                   <file-transfer-mode package="msc-ivr/1.0"
```

```
                        name="HTTP"/>
                </file-transfer-modes>
            </ivrInfo>
        </mediaResourceRequest>
    </mrbconsumer>

    =_Part
```

```
F5. MRB -> MS (INVITE sdp only)

     [..]

    From: <sip:Mrb@mrb.example.org:5060>;tag=32pbdxZ8
    To: <sip:MediaServer@ms.example.com:5080>

    [..]

    Content-Type: application/sdp
    v=0
    o=lminiero 123456 654321 IN IP4 203.0.113.2
    s=A conversation
    c=IN IP4 203.0.113.2
    t=0 0
    m=audio 7078 RTP/AVP 0 3 8 101
    a=rtpmap:0 PCMU/8000/1
    a=rtpmap:3 GSM/8000/1
    a=rtpmap:8 PCMA/8000/1
    a=rtpmap:101 telephone-event/8000
    a=fmtp:101 0-11
    m=video 9078 RTP/AVP 98
```

```
F7. MRB <- MS (200 OK sdp)

    [..]

    From: <sip:Mrb@mrb.example.org:5060>;tag=32pbdxZ8
    To: <sip:MediaServer@ms.example.com:5080>;tag=KQw677BF

    [..]

    Content-Type: application/sdp
    v=0
    o=lminiero 123456 654322 IN IP4 203.0.113.1
    s=MediaCtrl
    c=IN IP4 203.0.113.1
    t=0 0
    m=audio 63442 RTP/AVP 0 3 8 101
    a=rtpmap:0 PCMU/8000
    a=rtpmap:3 GSM/8000
    a=rtpmap:8 PCMA/8000
    a=rtpmap:101 telephone-event/8000
    a=fmtp:101 0-15
    a=ptime:20
    a=label:7eda834
    m=video 33468 RTP/AVP 98
```

```
    a=rtpmap:98 H263-1998/90000
    a=fmtp:98 CIF=2
    a=label:0132ca2

F9. AS <- MRB (200 OK multipart/mixed)

    [..]

    From: <sip:ApplicationServer@as.example.com>;tag=fd4fush5
    To: <sip:Mrb@mrb.example.org>;tag=117652221

    [..]

    Content-Type: multipart/mixed;boundary="=_Part"

    =_Part
    Content-Type: application/sdp

    v=0
    o=lminiero 123456 654322 IN IP4 203.0.113.1
    s=MediaCtrl
    c=IN IP4 203.0.113.1
    t=0 0
    m=audio 63442 RTP/AVP 0 3 8 101
    a=rtpmap:0 PCMU/8000
    a=rtpmap:3 GSM/8000
    a=rtpmap:8 PCMA/8000
    a=rtpmap:101 telephone-event/8000
    a=fmtp:101 0-15
    a=ptime:20
    a=label:7eda834
    m=video 33468 RTP/AVP 98
    a=rtpmap:98 H263-1998/90000
    a=fmtp:98 CIF=2
    a=label:0132ca2

    =_Part
    Content-Type: application/mrb-consumer+xml

    <?xml version="1.0" encoding="UTF-8" standalone="yes"?>

    <mrbconsumer version="1.0"
          xmlns="urn:ietf:params:xml:ns:mrb-consumer" >
       <mediaResourceResponse reason="Resource found" status="200"
             id="ns56g1x0">
          <response-session-info>
             <session-id>z1skKYZQ3eFu</session-id>
             <seq>9</seq>
             <expires>3600</expires>
             <media-server-address
                   uri="sip:MediaServer@
                   ms.example.com:5080">
                <connection-id>32pbdxZ8:
                      KQw677BF</connection-id>
                <ivr-sessions>
                   <rtp-codec name="audio/basic">
```

```
                    <decoding>60</decoding>
                    <encoding>60</encoding>
                </rtp-codec>
            </ivr-sessions>
        </media-server-address>
        <media-server-address
                uri="sip:OtherMediaServer@
                        pool.example.net:5080">
            <ivr-sessions>
                <rtp-codec name="audio/basic">
                    <decoding>40</decoding>
                    <encoding>40</encoding>
                </rtp-codec>
            </ivr-sessions>
        </media-server-address>
    </response-session-info>
  </mediaResourceResponse>
</mrbconsumer>

=_Part
```

```
F11. UAC <- AS (200 OK sdp)

    [..]

    From: <sip:lminiero@users.example.com>;tag=1153573888
    To: <sip:mediactrlDemo@as.example.com>;tag=bcd47c32

    [..]
    Content-Type: application/sdp
    v=0
    o=lminiero 123456 654322 IN IP4 203.0.113.1
    s=MediaCtrl
    c=IN IP4 203.0.113.1
    t=0 0
    m=audio 63442 RTP/AVP 0 3 8 101
    a=rtpmap:0 PCMU/8000
    a=rtpmap:3 GSM/8000
    a=rtpmap:8 PCMA/8000
    a=rtpmap:101 telephone-event/8000
    a=fmtp:101 0-15
    a=ptime:20
    a=label:7eda834
    m=video 33468 RTP/AVP 98
    a=rtpmap:98 H263-1998/90000
    a=fmtp:98 CIF=2
    a=label:0132ca2
```

As the examples illustrate, as in the IAMM/Control Channel example, the MRB provides the AS with a <media-server-address> element in the Consumer response: the "uri" attribute identifies the specific MS to which the MRB has sent the SDP media negotiation, and the "connection-id" enables the AS to identify to the MS the dialog between the MRB and the MS. This attribute is needed, since according to the framework specification (RFC 6230 – see Chapter 8), the connection-id is built out of the From/To tags of the dialog between the MRB and the MS; since the MRB acts as a B2BUA in this scenario, without that attribute, the AS does not know the relevant tags, thus preventing the CFW protocol from working as expected. The continuation of the scenario (the AS connecting to the MS to start the Control Channel, the SYNC message, etc.) is omitted for brevity.

11.10 MEDIA SERVICE RESOURCE PUBLISHER INTERFACE XML SCHEMA

This section gives the XML schema definition [5,6] of the "application/mrb-publish+xml" format.

```
<?xml version="1.0" encoding="UTF-8"?>

<xsd:schema targetNamespace="urn:ietf:params:xml:ns:mrb-publish"
        elementFormDefault="qualified" blockDefault="#all"
        xmlns="urn:ietf:params:xml:ns:mrb-publish"
        xmlns:fw="urn:ietf:params:xml:ns:control:
              framework-attributes"
        xmlns:ca="urn:ietf:params:xml:ns:pidf:geopriv10:
              civicAddr"
        xmlns:xsd="http://www.w3.org/2001/XMLSchema">
   <xsd:annotation>
      <xsd:specificationation>
         IETF MediaCtrl MRB 1.0
         This is the schema of the IETF MediaCtrl MRB package.
         The schema namespace is urn:ietf:params:xml:ns:mrb-publish
      </xsd:specificationation>
   </xsd:annotation>

   <!--
   ############################################################
   SCHEMA IMPORTS
   ############################################################
   -->

   <xsd:import namespace="http://www.w3.org/XML/1998/namespace"
         schemaLocation="http://www.w3.org/2001/xml.xsd">
      <xsd:annotation>
         <xsd:specificationation>
            This import brings in the XML attributes for
            xml:base, xml:lang, etc.
         </xsd:specificationation>
      </xsd:annotation>
   </xsd:import>
   <xsd:import
         namespace="urn:ietf:params:xml:ns:control:
         framework-attributes"
         schemaLocation="framework.xsd">
      <xsd:annotation>
         <xsd:specificationation>
            This import brings in the framework attributes for
            conferenceid and connectionid.
         </xsd:specificationation>
      </xsd:annotation>
   </xsd:import>
   <xsd:import
         namespace="urn:ietf:params:xml:ns:pidf:geopriv10:civicAddr"
         schemaLocation="civicAddress.xsd">
      <xsd:annotation>
         <xsd:specificationation>
            This import brings in the civicAddress specification
            from RFC 5139.
         </xsd:specificationation>
      </xsd:annotation>
   </xsd:import>
```

```
<!--
#####################################################
Extensible core type
#####################################################
-->

<xsd:complexType name="Tcore">
   <xsd:annotation>
      <xsd:specificationation>
         This type is extended by other (non-mixed)
         component types to
         allow attributes from other namespaces.
      </xsd:specificationation>
   </xsd:annotation>
   <xsd:sequence/>
   <xsd:anyAttribute namespace="##other" processContents="lax" />
</xsd:complexType>

<!--
#####################################################
TOP-LEVEL ELEMENT: mrbpublish
#####################################################
-->

<xsd:complexType name="mrbpublishType">
   <xsd:complexContent>
      <xsd:extension base="Tcore">
         <xsd:sequence>
            <xsd:choice>
               <xsd:element ref="mrbrequest" />
               <xsd:element ref="mrbresponse" />
               <xsd:element ref=
                     "mrbnotification" />
               <xsd:any namespace="##other"
                     minOccurs="0"
                     maxOccurs="unbounded" processContents="lax" />
            </xsd:choice>
         </xsd:sequence>
         <xsd:attribute name="version" type="version.datatype"
               use="required" />
         <xsd:anyAttribute namespace="##other"
               processContents="lax" />
      </xsd:extension>
   </xsd:complexContent>
</xsd:complexType>
<xsd:element name="mrbpublish" type="mrbpublishType" />

<!--
#####################################################
mrbrequest TYPE
#####################################################
-->

<!-- mrbrequest -->

<xsd:complexType name="mrbrequestType">
   <xsd:complexContent>
```

```
        <xsd:extension base="Tcore">
            <xsd:sequence>
                <xsd:element ref="subscription" />
                    <xsd:any namespace="##other"
                        minOccurs="0"
                        maxOccurs="unbounded" processContents="lax" />
            </xsd:sequence>
            <xsd:anyAttribute namespace="##other"
                processContents="lax" />
        </xsd:extension>
    </xsd:complexContent>
</xsd:complexType>
<xsd:element name="mrbrequest" type="mrbrequestType" />

<!-- subscription -->

<xsd:complexType name="subscriptionType">
    <xsd:complexContent>
        <xsd:extension base="Tcore">
            <xsd:sequence>
                <xsd:element name="expires"
                        type="xsd:nonNegativeInteger"
                        minOccurs="0" maxOccurs="1" />
                <xsd:element name="minfrequency"
                        type="xsd:nonNegativeInteger"
                        minOccurs="0" maxOccurs="1" />
                <xsd:element name="maxfrequency"
                        type="xsd:nonNegativeInteger"
                        minOccurs="0" maxOccurs="1" />
                <xsd:any namespace="##other"
                        minOccurs="0"
                        maxOccurs="unbounded" processContents="lax" />
            </xsd:sequence>
            <xsd:attribute name="id" type="id.datatype"
                    use="required" />
            <xsd:attribute name="seqnumber"
                    type="xsd:nonNegativeInteger"
                    use="required" />
            <xsd:attribute name="action"
                    type="action.datatype"
                    use="required" />
            <xsd:anyAttribute namespace="##other"
                    processContents="lax" />
        </xsd:extension>
    </xsd:complexContent>
</xsd:complexType>
<xsd:element name="subscription" type="subscriptionType" />

<!--
####################################################
mrbresponse TYPE
####################################################
-->

<!-- mrbresponse -->

<xsd:complexType name="mrbresponseType">
```

```
    <xsd:complexContent>
        <xsd:extension base="Tcore">
            <xsd:sequence>
                <xsd:element ref="subscription"
                        minOccurs="0"
                        maxOccurs="1" />
                <xsd:any namespace="##other"
                        minOccurs="0"
                        maxOccurs="unbounded" processContents="lax" />
            </xsd:sequence>
            <xsd:attribute name="status" type="status.datatype"
                    use="required" />
            <xsd:attribute name="reason" type="xsd:string" />
            <xsd:anyAttribute namespace="##other"
                    processContents="lax" />
        </xsd:extension>
    </xsd:complexContent>
</xsd:complexType>
<xsd:element name="mrbresponse" type="mrbresponseType" />

<!--
####################################################
mrbnotification TYPE
####################################################
-->

<!-- mrbnotification -->

<xsd:complexType name="mrbnotificationType">
    <xsd:complexContent>
        <xsd:extension base="Tcore">
            <xsd:sequence>
                <xsd:element name="media-server-id"
                        type="subscriptionid.datatype"/>
                <xsd:element ref="supported-packages"
                        minOccurs="0" />
                <xsd:element ref="active-rtp-sessions"
                        minOccurs="0" />
                <xsd:element ref="active-mixer-sessions"
                        minOccurs="0" />
                <xsd:element ref="non-active-rtp-sessions"
                        minOccurs="0" />
                <xsd:element ref="non-active-mixer-
                        sessions" minOccurs="0" />
                <xsd:element ref="media-server-status"
                        minOccurs="0" />
                <xsd:element ref="supported-codecs"
                        minOccurs="0" />
                <xsd:element ref="application-data"
                        minOccurs="0"
                        maxOccurs="unbounded" />
                <xsd:element ref="file-formats"
                        minOccurs="0" />
                <xsd:element ref="max-prepared-duration"
                        minOccurs="0" />
                <xsd:element ref="dtmf-support"
                        minOccurs="0" />
```

```
                <xsd:element ref="mixing-modes"
                      minOccurs="0" />
                <xsd:element ref="supported-tones"
                      minOccurs="0" />
                <xsd:element ref="file-transfer-modes"
                      minOccurs="0" />
                <xsd:element ref="asr-tts-support"
                      minOccurs="0" />
                <xsd:element ref="vxml-support"
                      minOccurs="0" />
                <xsd:element ref="media-server-location"
                      minOccurs="0" />
                <xsd:element ref="label" minOccurs="0" />
                <xsd:element ref="media-server-address"
                      minOccurs="0" />
                <xsd:element ref="encryption"
                      minOccurs="0" />
                <xsd:any namespace="##other"
                      minOccurs="0"
                      maxOccurs="unbounded" processContents="lax" />
            </xsd:sequence>
            <xsd:attribute name="id"
                  type="subscriptionid.datatype"
                        use="required" />
            <xsd:attribute name="seqnumber"
                  type="xsd:nonNegativeInteger"
                  use="required" />
            <xsd:anyAttribute namespace="##other"
                  processContents="lax" />
        </xsd:extension>
    </xsd:complexContent>
</xsd:complexType>
<xsd:element name="mrbnotification" type="mrbnotificationType" />

<!-- supported-packages -->

<xsd:complexType name="supported-packagesType">
    <xsd:complexContent>
        <xsd:extension base="Tcore">
            <xsd:sequence>
                <xsd:element ref="package" minOccurs="0"
                      maxOccurs="unbounded" />
                <xsd:any namespace="##other"
                      minOccurs="0"
                      maxOccurs="unbounded" processContents="lax" />
                </xsd:sequence>
                <xsd:anyAttribute namespace="##other"
                      processContents="lax" />
        </xsd:extension>
    </xsd:complexContent>
</xsd:complexType>
<xsd:element name="supported-packages" type="supported-
        packagesType"/>
<xsd:complexType name="packageType">
    <xsd:complexContent>
        <xsd:extension base="Tcore">
            <xsd:sequence>
```

```
            <xsd:any namespace="##other"
                   minOccurs="0"
                   maxOccurs="unbounded" processContents="lax" />
         </xsd:sequence>
         <xsd:attribute name="name" type="xsd:string"
               use="required" />
         <xsd:anyAttribute namespace="##other"
               processContents="lax" />
      </xsd:extension>
   </xsd:complexContent>
</xsd:complexType>
<xsd:element name="package" type="packageType" />

<!-- active-rtp-sessions -->

<xsd:complexType name="active-rtp-sessionsType">
   <xsd:complexContent>
      <xsd:extension base="Tcore">
         <xsd:sequence>
            <xsd:element ref="rtp-codec"
                   minOccurs="0"
                   maxOccurs="unbounded" />
            <xsd:any namespace="##other"
                   minOccurs="0"
                   maxOccurs="unbounded" processContents="lax" />
         </xsd:sequence>
         <xsd:anyAttribute namespace="##other"
               processContents="lax" />
      </xsd:extension>
   </xsd:complexContent>
</xsd:complexType>
<xsd:element name="active-rtp-sessions" type="active-rtp-
      sessionsType"/>
<xsd:complexType name="rtp-codecType">
   <xsd:complexContent>
      <xsd:extension base="Tcore">
         <xsd:sequence>
            <xsd:element name="decoding"
                   type="xsd:nonNegativeInteger" />
            <xsd:element name="encoding"
                   type="xsd:nonNegativeInteger" />
            <xsd:any namespace="##other"
                   minOccurs="0"
                   maxOccurs="unbounded" processContents="lax" />
         </xsd:sequence>
         <xsd:attribute name="name" type="xsd:string"
               use="required" />
         <xsd:anyAttribute namespace="##other"
               processContents="lax" />
      </xsd:extension>
   </xsd:complexContent>
</xsd:complexType>
<xsd:element name="rtp-codec" type="rtp-codecType" />

<!-- active-mixer-sessions -->
```

```
<xsd:complexType name="active-mixer-sessionsType">
   <xsd:complexContent>
      <xsd:extension base="Tcore">
         <xsd:sequence>
            <xsd:element ref="active-mix"
                  minOccurs="0"
                  maxOccurs="unbounded" />
            <xsd:any namespace="##other"
                  minOccurs="0"
                  maxOccurs="unbounded" processContents="lax" />
         </xsd:sequence>
         <xsd:anyAttribute namespace="##other"
               processContents="lax" />
      </xsd:extension>
   </xsd:complexContent>
</xsd:complexType>
<xsd:element name="active-mixer-sessions"
      type="active-mixer-sessionsType" />
<xsd:complexType name="active-mixType">
   <xsd:complexContent>
      <xsd:extension base="Tcore">
         <xsd:sequence>
            <xsd:element ref="rtp-codec"
                  minOccurs="0"
                  maxOccurs="unbounded" />
            <xsd:any namespace="##other"
                  minOccurs="0"
                  maxOccurs="unbounded" processContents="lax" />
         </xsd:sequence>
         <xsd:attributeGroup ref="fw:
               framework-attributes" />
         <xsd:anyAttribute namespace="##other"
               processContents="lax" />
      </xsd:extension>
   </xsd:complexContent>
</xsd:complexType>
<xsd:element name="active-mix" type="active-mixType" />

<!-- non-active-rtp-sessions -->

<xsd:complexType name="non-active-rtp-sessionsType">
   <xsd:complexContent>
      <xsd:extension base="Tcore">
         <xsd:sequence>
            <xsd:element ref="rtp-codec"
                  minOccurs="0"
                  maxOccurs="unbounded" />
            <xsd:any namespace="##other"
                  minOccurs="0"
                  maxOccurs="unbounded" processContents="lax" />
         </xsd:sequence>
         <xsd:anyAttribute namespace="##other"
               processContents="lax" />
      </xsd:extension>
   </xsd:complexContent>
```

```
</xsd:complexType>
<xsd:element name="non-active-rtp-sessions"
      type="non-active-rtp-sessionsType" />

<!-- non-active-mixer-sessions -->

<xsd:complexType name="non-active-mixer-sessionsType">
   <xsd:complexContent>
      <xsd:extension base="Tcore">
         <xsd:sequence>
            <xsd:element ref="non-active-mix"
                  minOccurs="0"
                  maxOccurs="unbounded" />
            <xsd:any namespace="##other"
                  minOccurs="0"
                  maxOccurs="unbounded" processContents="lax" />
         </xsd:sequence>
         <xsd:anyAttribute namespace="##other"
               processContents="lax" />
      </xsd:extension>
   </xsd:complexContent>
</xsd:complexType>
<xsd:element name="non-active-mixer-sessions"
      type="non-active-mixer-sessionsType" />
<xsd:complexType name="non-active-mixType">
   <xsd:complexContent>
      <xsd:extension base="Tcore">
         <xsd:sequence>
            <xsd:element ref="rtp-codec" />
               <xsd:any namespace="##other"
                     minOccurs="0"
                     maxOccurs="unbounded" processContents="lax" />
         </xsd:sequence>
         <xsd:attribute name="available"
               type="xsd:nonNegativeInteger"
               use="required" />
         <xsd:anyAttribute namespace="##other"
               processContents="lax" />
      </xsd:extension>
   </xsd:complexContent>
</xsd:complexType>
<xsd:element name="non-active-mix" type="non-active-mixType" />

<!-- media-server-status -->


<xsd:element name="media-server-status" type="msstatus.datatype" />

<!-- supported-codecs -->

<xsd:complexType name="supported-codecsType">
   <xsd:complexContent>
      <xsd:extension base="Tcore">
         <xsd:sequence>
            <xsd:element ref="supported-codec"
                  minOccurs="0" maxOccurs="unbounded" />
            <xsd:any namespace="##other"
```

```
                    minOccurs="0"
                    maxOccurs="unbounded" processContents="lax" />
            </xsd:sequence>
            <xsd:anyAttribute namespace="##other"
                    processContents="lax" />
        </xsd:extension>
    </xsd:complexContent>
</xsd:complexType>
<xsd:element name="supported-codecs" type="supported-codecsType" />
<xsd:complexType name="supported-codecType">
    <xsd:complexContent>
        <xsd:extension base="Tcore">
            <xsd:sequence>
                <xsd:element ref="supported-codec-
                    package"
                    minOccurs="0" maxOccurs="unbounded" />
                <xsd:any namespace="##other"
                    minOccurs="0"
                    maxOccurs="unbounded" processContents="lax" />
            </xsd:sequence>
            <xsd:attribute name="name" type="xsd:string"
                    use="required" />
            <xsd:anyAttribute namespace="##other"
                    processContents="lax" />
        </xsd:extension>
    </xsd:complexContent>
</xsd:complexType>
<xsd:element name="supported-codec" type="supported-codecType" />
<xsd:complexType name="supported-codec-packageType">
    <xsd:complexContent>
        <xsd:extension base="Tcore">
            <xsd:sequence>
                <xsd:element name="supported-action"
                    type="actions.datatype"
                    minOccurs="0" maxOccurs="unbounded" />
                <xsd:any namespace="##other"
                    minOccurs="0"
                    maxOccurs="unbounded" processContents="lax" />
            </xsd:sequence>
            <xsd:attribute name="name" type="xsd:string"
                    use="required" />
            <xsd:anyAttribute namespace="##other"
                    processContents="lax" />
        </xsd:extension>
    </xsd:complexContent>
</xsd:complexType>
<xsd:element name="supported-codec-package"
        type="supported-codec-packageType" />

<!-- application-data -->

<xsd:element name="application-data" type="appdata.datatype" />

<!-- file-formats -->

<xsd:complexType name="file-formatsType">
    <xsd:complexContent>
```

```
        <xsd:extension base="Tcore">
            <xsd:sequence>
                <xsd:element ref="supported-format"
                        minOccurs="0" maxOccurs="unbounded" />
                <xsd:any namespace="##other"
                        minOccurs="0"
                        maxOccurs="unbounded" processContents="lax" />
            </xsd:sequence>
            <xsd:anyAttribute namespace="##other"
                    processContents="lax" />
        </xsd:extension>
    </xsd:complexContent>
</xsd:complexType>
<xsd:element name="file-formats" type="file-formatsType" />
<xsd:complexType name="supported-formatType">
    <xsd:complexContent>
        <xsd:extension base="Tcore">
            <xsd:sequence>
                <xsd:element ref="supported-file-package"
                        minOccurs="0" maxOccurs="unbounded" />
                <xsd:any namespace="##other"
                        minOccurs="0"
                        maxOccurs="unbounded" processContents="lax" />
            </xsd:sequence>
            <xsd:attribute name="name" type="xsd:string"
                    use="required" />
            <xsd:anyAttribute namespace="##other"
                    processContents="lax" />
        </xsd:extension>
    </xsd:complexContent>
</xsd:complexType>
<xsd:element name="supported-format" type="supported-formatType" />
<xsd:element name="supported-file-package"
       type="xsd:string" />

<!-- max-prepared-duration -->

<xsd:complexType name="max-prepared-durationType">
    <xsd:complexContent>
        <xsd:extension base="Tcore">
            <xsd:sequence>
                <xsd:element ref="max-time" />
                <xsd:any namespace="##other"
                        minOccurs="0"
                        maxOccurs="unbounded" processContents="lax" />
            </xsd:sequence>
            <xsd:anyAttribute namespace="##other"
                    processContents="lax" />
        </xsd:extension>
    </xsd:complexContent>
</xsd:complexType>
<xsd:element name="max-prepared-duration"
       type="max-prepared-durationType" />
<xsd:complexType name="max-timeType">
    <xsd:complexContent>
        <xsd:extension base="Tcore">
            <xsd:sequence>
```

```
                <xsd:element name="max-time-package"
                      type="xsd:string" />
                <xsd:any namespace="##other"
                      minOccurs="0"
                      maxOccurs="unbounded" processContents="lax" />
            </xsd:sequence>
            <xsd:attribute name="max-time-seconds"
                  type="xsd:nonNegativeInteger"
                  use="required" />
            <xsd:anyAttribute namespace="##other"
                  processContents="lax" />
        </xsd:extension>
    </xsd:complexContent>
</xsd:complexType>
<xsd:element name="max-time" type="max-timeType" />

<!-- dtmf-support -->

<xsd:complexType name="dtmf-supportType">
    <xsd:complexContent>
        <xsd:extension base="Tcore">
            <xsd:sequence>
                <xsd:element ref="detect" />
                <xsd:element ref="generate" />
                <xsd:element ref="passthrough" />
                <xsd:any namespace="##other"
                      minOccurs="0"
                      maxOccurs="unbounded" processContents="lax" />
            </xsd:sequence>
            <xsd:anyAttribute namespace="##other"
                  processContents="lax" />
        </xsd:extension>
    </xsd:complexContent>
</xsd:complexType>
<xsd:element name="dtmf-support" type="dtmf-supportType" />
<xsd:complexType name="detectType">
    <xsd:complexContent>
        <xsd:extension base="Tcore">
            <xsd:sequence>
            <xsd:element ref="dtmf-type"
                  minOccurs="0" maxOccurs="unbounded" />
            <xsd:any namespace="##other"
                  minOccurs="0"
                  maxOccurs="unbounded" processContents="lax" />
            </xsd:sequence>
            <xsd:anyAttribute namespace="##other"
                  processContents="lax" />
        </xsd:extension>
    </xsd:complexContent>
</xsd:complexType>
<xsd:element name="detect" type="detectType" />
<xsd:complexType name="generateType">
    <xsd:complexContent>
        <xsd:extension base="Tcore">
            <xsd:sequence>
                <xsd:element ref="dtmf-type"
                      minOccurs="0" maxOccurs="unbounded" />
```

```
            <xsd:any namespace="##other"
                  minOccurs="0"
                  maxOccurs="unbounded" processContents="lax" />
         </xsd:sequence>
         <xsd:anyAttribute namespace="##other"
               processContents="lax" />
      </xsd:extension>
   </xsd:complexContent>
</xsd:complexType>
<xsd:element name="generate" type="generateType" />
<xsd:complexType name="passthroughType">
   <xsd:complexContent>
      <xsd:extension base="Tcore">
         <xsd:sequence>
         <xsd:element ref="dtmf-type"
               minOccurs="0" maxOccurs="unbounded" />
         <xsd:any namespace="##other"
               minOccurs="0"
               maxOccurs="unbounded" processContents="lax" />
         </xsd:sequence>
         <xsd:anyAttribute namespace="##other"
               processContents="lax" />
      </xsd:extension>
   </xsd:complexContent>
</xsd:complexType>
<xsd:element name="passthrough" type="passthroughType" />
<xsd:complexType name="dtmf-typeType">
   <xsd:complexContent>
      <xsd:extension base="Tcore">
         <xsd:sequence>
            <xsd:any namespace="##other" minOccurs="0"
                  maxOccurs="unbounded" processContents="lax" />
         </xsd:sequence>
         <xsd:attribute name="name" type="dtmf.datatype"
               use="required" />
         <xsd:attribute name="package" type="xsd:string"
               use="required" />
         <xsd:anyAttribute namespace="##other"
               processContents="lax" />
      </xsd:extension>
   </xsd:complexContent>
</xsd:complexType>
<xsd:element name="dtmf-type" type="dtmf-typeType" />

<!-- mixing-modes -->

<xsd:complexType name="mixing-modesType">
   <xsd:complexContent>
      <xsd:extension base="Tcore">
         <xsd:sequence>
         <xsd:element ref="audio-mixing-modes"
               minOccurs="0" maxOccurs="1" />
         <xsd:element ref="video-mixing-modes"
               minOccurs="0" maxOccurs="1" />
         <xsd:any namespace="##other"
               minOccurs="0"
```

```
                maxOccurs="unbounded" processContents="lax" />
            </xsd:sequence>
            <xsd:anyAttribute namespace="##other"
                processContents="lax" />
        </xsd:extension>
    </xsd:complexContent>
</xsd:complexType>
<xsd:element name="mixing-modes" type="mixing-modesType" />
<xsd:complexType name="audio-mixing-modesType">
    <xsd:complexContent>
        <xsd:extension base="Tcore">
            <xsd:sequence>
            <xsd:element ref="audio-mixing-mode"
                minOccurs="0" maxOccurs="unbounded" />
            <xsd:any namespace="##other"
                minOccurs="0"
                maxOccurs="unbounded" processContents="lax" />
            </xsd:sequence>
            <xsd:anyAttribute namespace="##other"
                processContents="lax" />
        </xsd:extension>
    </xsd:complexContent>
</xsd:complexType>
<xsd:element name="audio-mixing-modes" type="audio-mixing-
    modesType" />
<xsd:complexType name="audio-mixing-modeType" mixed="true">
    <xsd:sequence>
        <xsd:any namespace="##other" minOccurs="0"
            maxOccurs="unbounded" processContents="lax" />
    </xsd:sequence>
    <xsd:attribute name="package" type="xsd:string" use="required" />
    <xsd:anyAttribute namespace="##other" processContents="lax" />
</xsd:complexType>
<xsd:element name="audio-mixing-mode" type="audio-mixing-
    modeType" />
<xsd:complexType name="video-mixing-modesType">
    <xsd:complexContent>
        <xsd:extension base="Tcore">
            <xsd:sequence>
                <xsd:element ref="video-mixing-mode"
                    minOccurs="0" maxOccurs="unbounded" />
                <xsd:any namespace="##other"
                    minOccurs="0"
                    maxOccurs="unbounded" processContents="lax" />
            </xsd:sequence>
            <xsd:attribute name="vas" type="boolean.datatype"
                default="false" />
            <xsd:attribute name="activespeakermix"
                type="boolean.datatype"
                default="false" />
            <xsd:anyAttribute namespace="##other" processContents="lax" />
        </xsd:extension>
    </xsd:complexContent>
</xsd:complexType>
<xsd:element name="video-mixing-modes" type="video-mixing-modesType" />
<xsd:complexType name="video-mixing-modeType" mixed="true">
```

```
    <xsd:sequence>
        <xsd:any namespace="##other" minOccurs="0"
                maxOccurs="unbounded" processContents="lax" />
    </xsd:sequence>
    <xsd:attribute name="package" type="xsd:string"
            use="required" />
    <xsd:anyAttribute namespace="##other" processContents="lax" />
</xsd:complexType>
<xsd:element name="video-mixing-mode" type="video-mixing-
        modeType" />

<!-- supported-tones -->

<xsd:complexType name="supported-tonesType">
    <xsd:complexContent>
        <xsd:extension base="Tcore">
            <xsd:sequence>
                <xsd:element ref="supported-country-
                        codes"
                        minOccurs="0" maxOccurs="1" />
                <xsd:element ref="supported-h248-codes"
                        minOccurs="0" maxOccurs="1" />
                <xsd:any namespace="##other"
                        minOccurs="0"
                        maxOccurs="unbounded" processContents="lax" />
            </xsd:sequence>
            <xsd:anyAttribute namespace="##other"
                    processContents="lax" />
        </xsd:extension>
    </xsd:complexContent>
</xsd:complexType>
<xsd:element name="supported-tones" type="supported-tonesType" />
<xsd:complexType name="supported-country-codesType">
    <xsd:complexContent>
        <xsd:extension base="Tcore">
            <xsd:sequence>
                <xsd:element ref="country-code"
                        minOccurs="0" maxOccurs="unbounded" />
                <xsd:any namespace="##other"
                        minOccurs="0"
                        maxOccurs="unbounded" processContents="lax" />
            </xsd:sequence>
            <xsd:anyAttribute namespace="##other"
                    processContents="lax" />
        </xsd:extension>
    </xsd:complexContent>
</xsd:complexType>
<xsd:element name="supported-country-codes"
        type="supported-country-codesType" />
<xsd:complexType name="country-codeType" mixed="true">
    <xsd:sequence>
        <xsd:any namespace="##other" minOccurs="0"
                maxOccurs="unbounded" processContents="lax" />
    </xsd:sequence>
    <xsd:attribute name="package" type="xsd:string"
            use="required" />
```

```
        <xsd:anyAttribute namespace="##other" processContents="lax" />
    </xsd:complexType>
    <xsd:element name="country-code" type="country-codeType" />
    <xsd:complexType name="supported-h248-codesType">
        <xsd:complexContent>
            <xsd:extension base="Tcore">
                <xsd:sequence>
                    <xsd:element ref="h248-code"
                            minOccurs="0" maxOccurs="unbounded" />
                    <xsd:any namespace="##other"
                            minOccurs="0"
                            maxOccurs="unbounded" processContents="lax" />
                </xsd:sequence>
                <xsd:anyAttribute namespace="##other"
                        processContents="lax" />
            </xsd:extension>
        </xsd:complexContent>
    </xsd:complexType>
    <xsd:element name="supported-h248-codes"
            type="supported-h248-codesType" />
    <xsd:complexType name="h248-codeType" mixed="true">
        <xsd:sequence>
            <xsd:any namespace="##other" minOccurs="0"
                    maxOccurs="unbounded" processContents="lax" />
        </xsd:sequence>
        <xsd:attribute name="package" type="xsd:string"
                use="required" />
        <xsd:anyAttribute namespace="##other" processContents="lax" />
    </xsd:complexType>
    <xsd:element name="h248-code" type="h248-codeType" />

    <!-- file-transfer-modes -->

    <xsd:complexType name="file-transfer-modesType">
        <xsd:complexContent>
            <xsd:extension base="Tcore">
                <xsd:sequence>
                    <xsd:element ref="file-transfer-mode"
                            minOccurs="0" maxOccurs="unbounded" />
                    <xsd:any namespace="##other"
                            minOccurs="0"
                            maxOccurs="unbounded" processContents="lax" />
                </xsd:sequence>
                <xsd:anyAttribute namespace="##other"
                        processContents="lax" />
            </xsd:extension>
        </xsd:complexContent>
    </xsd:complexType>
    <xsd:element name="file-transfer-modes"
type="file-transfer-modesType" />
    <xsd:complexType name="file-transfer-modeType">
        <xsd:complexContent>
            <xsd:extension base="Tcore">
                <xsd:sequence>
                    <xsd:any namespace="##other"
                            minOccurs="0"
                            maxOccurs="unbounded" processContents="lax" />
```

```
            </xsd:sequence>
            <xsd:attribute name="name"
                  type="transfermode.datatype"
                  use="required" />
            <xsd:attribute name="package" type="xsd:string"
                  use="required" />
            <xsd:anyAttribute namespace="##other"
                  processContents="lax" />
        </xsd:extension>
    </xsd:complexContent>
</xsd:complexType>
<xsd:element name="file-transfer-mode"
        type="file-transfer-modeType" />

<!-- asr-tts-support -->

<xsd:complexType name="asr-tts-supportType">
    <xsd:complexContent>
        <xsd:extension base="Tcore">
            <xsd:sequence>
                <xsd:element ref="asr-support"
                      minOccurs="0" maxOccurs="1" />
                <xsd:element ref="tts-support"
                      minOccurs="0" maxOccurs="1" />
                <xsd:any namespace="##other"
                      minOccurs="0"
                      maxOccurs="unbounded" processContents="lax" />
            </xsd:sequence>
            <xsd:anyAttribute namespace="##other"
                  processContents="lax" />
        </xsd:extension>
    </xsd:complexContent>
</xsd:complexType>
<xsd:element name="asr-tts-support" type="asr-tts-supportType" />
<xsd:complexType name="asr-supportType">
    <xsd:complexContent>
        <xsd:extension base="Tcore">
            <xsd:sequence>
                <xsd:element ref="language"
                      minOccurs="0" maxOccurs="unbounded" />
                <xsd:any namespace="##other"
                      minOccurs="0"
                      maxOccurs="unbounded" processContents="lax" />
            </xsd:sequence>
            <xsd:anyAttribute namespace="##other"
                  processContents="lax" />
        </xsd:extension>
    </xsd:complexContent>
</xsd:complexType>
<xsd:element name="asr-support" type="asr-supportType" />
<xsd:complexType name="tts-supportType">
    <xsd:complexContent>
        <xsd:extension base="Tcore">
            <xsd:sequence>
                <xsd:element ref="language"
                      minOccurs="0" maxOccurs="unbounded" />
                <xsd:any namespace="##other"
```

```
                        minOccurs="0"
                        maxOccurs="unbounded" processContents="lax" />
            </xsd:sequence>
            <xsd:anyAttribute namespace="##other"
                    processContents="lax" />
        </xsd:extension>
    </xsd:complexContent>
</xsd:complexType>
<xsd:element name="tts-support" type="tts-supportType" />
<xsd:complexType name="languageType">
    <xsd:complexContent>
        <xsd:extension base="Tcore">
            <xsd:sequence>
                <xsd:any namespace="##other"
                        minOccurs="0"
                        maxOccurs="unbounded" processContents="lax" />
            </xsd:sequence>
            <xsd:attribute ref="xml:lang" />
            <xsd:anyAttribute namespace="##other"
                    processContents="lax" />
        </xsd:extension>
    </xsd:complexContent>
</xsd:complexType>
<xsd:element name="language" type="languageType" />

<!-- media-server-location -->

<xsd:complexType name="media-server-locationType">
    <xsd:complexContent>
        <xsd:extension base="Tcore">
            <xsd:sequence>
                <xsd:element name="civicAddress"
                        type="ca:civicAddress"
                        minOccurs="1" maxOccurs="1" />
                <xsd:any namespace="##other"
                        minOccurs="0"
                        maxOccurs="unbounded" processContents="lax" />
            </xsd:sequence>
            <xsd:anyAttribute namespace="##other"
                    processContents="lax" />
        </xsd:extension>
    </xsd:complexContent>
</xsd:complexType>
<xsd:element name="media-server-location"
        type="media-server-locationType" />

<!-- vxml-support -->

<xsd:complexType name="vxml-supportType">
    <xsd:complexContent>
        <xsd:extension base="Tcore">
            <xsd:sequence>
                <xsd:element ref="vxml-mode"
                        minOccurs="0" maxOccurs="unbounded" />
                <xsd:any namespace="##other"
                        minOccurs="0"
                        maxOccurs="unbounded" processContents="lax" />
```

```
            </xsd:sequence>
            <xsd:anyAttribute namespace="##other"
                  processContents="lax" />
        </xsd:extension>
    </xsd:complexContent>
</xsd:complexType>
<xsd:element name="vxml-support" type="vxml-supportType" />
<xsd:complexType name="vxml-modeType">
    <xsd:complexContent>
        <xsd:extension base="Tcore">
            <xsd:sequence>
                <xsd:any namespace="##other"
                      minOccurs="0"
                      maxOccurs="unbounded" processContents="lax" />
            </xsd:sequence>
            <xsd:attribute name="package" type="xsd:string"
                  use="required" />
            <xsd:attribute name="support"
                  type="vxml.datatype" use="required" />
            <xsd:anyAttribute namespace="##other"
                  processContents="lax" />
        </xsd:extension>
    </xsd:complexContent>
</xsd:complexType>
<xsd:element name="vxml-mode" type="vxml-modeType" />

<!-- label -->

<xsd:element name="label" type="label.datatype" />

<!-- media-server-address -->

<xsd:element name="media-server-address" type="xsd:anyURI" />

<!-- encryption -->

<xsd:complexType name="encryptionType">
    <xsd:complexContent>
        <xsd:extension base="Tcore">
            <xsd:sequence>
                <xsd:any namespace="##other"
                      minOccurs="0"
                      maxOccurs="unbounded" processContents="lax" />
            </xsd:sequence>
            <xsd:anyAttribute namespace="##other"
                  processContents="lax" />
        </xsd:extension>
    </xsd:complexContent>
</xsd:complexType>
<xsd:element name="encryption" type="encryptionType" />

<!--
####################################################
DATATYPES
####################################################
-->
```

```
<xsd:simpleType name="version.datatype">
   <xsd:restriction base="xsd:NMTOKEN">
      <xsd:enumeration value="1.0" />
   </xsd:restriction>
</xsd:simpleType>
<xsd:simpleType name="id.datatype">
   <xsd:restriction base="xsd:NMTOKEN" />
</xsd:simpleType>
<xsd:simpleType name="status.datatype">
   <xsd:restriction base="xsd:positiveInteger">
      <xsd:pattern value="[0-9][0-9][0-9]" />
   </xsd:restriction>
</xsd:simpleType>
<xsd:simpleType name="msstatus.datatype">
   <xsd:restriction base="xsd:NMTOKEN">
      <xsd:enumeration value="active" />
      <xsd:enumeration value="deactivated" />
      <xsd:enumeration value="unavailable" />
   </xsd:restriction>
</xsd:simpleType>
<xsd:simpleType name="action.datatype">
   <xsd:restriction base="xsd:NMTOKEN">
      <xsd:enumeration value="create" />
      <xsd:enumeration value="update" />
      <xsd:enumeration value="remove" />
   </xsd:restriction>
</xsd:simpleType>
<xsd:simpleType name="actions.datatype">
   <xsd:restriction base="xsd:NMTOKEN">
      <xsd:enumeration value="encoding" />
      <xsd:enumeration value="decoding" />
      <xsd:enumeration value="passthrough" />
   </xsd:restriction>
</xsd:simpleType>
<xsd:simpleType name="appdata.datatype">
   <xsd:restriction base="xsd:string" />
</xsd:simpleType>
<xsd:simpleType name="dtmf.datatype">
   <xsd:restriction base="xsd:NMTOKEN"/>
</xsd:simpleType>
<xsd:simpleType name="transfermode.datatype">
   <xsd:restriction base="xsd:NMTOKEN" />
</xsd:simpleType>
<xsd:simpleType name="boolean.datatype">
   <xsd:restriction base="xsd:NMTOKEN">
      <xsd:enumeration value="true" />
      <xsd:enumeration value="false" />
   </xsd:restriction>
</xsd:simpleType>
<xsd:simpleType name="vxml.datatype">
   <xsd:restriction base="xsd:NMTOKEN"/>
</xsd:simpleType>
<xsd:simpleType name="label.datatype">
   <xsd:restriction base="xsd:NMTOKEN" />
</xsd:simpleType>
<xsd:simpleType name="subscriptionid.datatype">
   <xsd:restriction base="xsd:NMTOKEN" />
```

```
    </xsd:simpleType>
</xsd:schema>
```

11.11 MEDIA SERVICE RESOURCE CONSUMER INTERFACE XML SCHEMA

This section gives the XML schema definition [5,6] of the "application/mrb-consumer+xml" format.

```
<?xml version="1.0" encoding="UTF-8"?>

<xsd:schema targetNamespace="urn:ietf:params:xml:ns:mrb-consumer"
      elementFormDefault="qualified" blockDefault="#all"
      xmlns="urn:ietf:params:xml:ns:mrb-consumer"
      xmlns:ca="urn:ietf:params:xml:ns:pidf:geopriv10:
      civicAddr"
      xmlns:xsd="http://www.w3.org/2001/XMLSchema">
   <xsd:annotation>
      <xsd:specificationation>
            IETF MediaCtrl MRB 1.0
            This is the schema of the IETF MediaCtrl MRB Consumer
            interface. The schema namespace is
            urn:ietf:params:xml:ns:mrb-consumer.
      </xsd:specificationation>
   </xsd:annotation>

   <!--
   ############################################################
   SCHEMA IMPORTS
   ############################################################
   -->

   <xsd:import namespace="http://www.w3.org/XML/1998/namespace"
         schemaLocation="http://www.w3.org/2001/xml.xsd">
      <xsd:annotation>
         <xsd:specificationation>
            This import brings in the XML attributes for
            xml:base, xml:lang, etc.
         </xsd:specificationation>
      </xsd:annotation>
   </xsd:import>
   <xsd:import
         namespace="urn:ietf:params:xml:ns:pidf:geopriv10:civicAddr"
         schemaLocation="civicAddress.xsd">
      <xsd:annotation>
         <xsd:specificationation>
            This import brings in the civicAddress specification
      from RFC 5139.
         </xsd:specificationation>
      </xsd:annotation>
   </xsd:import>

   <!--
   ####################################################
   Extensible core type
   ####################################################
   -->
```

```
<xsd:complexType name="Tcore">
   <xsd:annotation>
      <xsd:specificationation>
         This type is extended by other (non-mixed) component types to
         allow attributes from other namespaces.
      </xsd:specificationation>
   </xsd:annotation>
   <xsd:sequence/>
   <xsd:anyAttribute namespace="##other" processContents="lax" />
</xsd:complexType>

<!--
#####################################################
TOP-LEVEL ELEMENT: mrbconsumer
#####################################################
-->

<xsd:complexType name="mrbconsumerType">
   <xsd:complexContent>
      <xsd:extension base="Tcore">
         <xsd:sequence>
            <xsd:choice>
               <xsd:element ref=
                     "mediaResourceRequest" />
               <xsd:element ref="mediaResourceResponse" />
               <xsd:any namespace="##other"
                     minOccurs="0"
                     maxOccurs="unbounded" processContents="lax" />
            </xsd:choice>
         </xsd:sequence>
         <xsd:attribute name="version"
               type="version.datatype"
               use="required" />
         <xsd:anyAttribute namespace="##other"
               processContents="lax" />
      </xsd:extension>
   </xsd:complexContent>
</xsd:complexType>
<xsd:element name="mrbconsumer" type="mrbconsumerType" />

<!--
#####################################################
mediaResourceRequest TYPE
#####################################################
-->

<!-- mediaResourceRequest -->

<xsd:complexType name="mediaResourceRequestType">
   <xsd:complexContent>
      <xsd:extension base="Tcore">
         <xsd:sequence>
            <xsd:element ref="generalInfo"
                  minOccurs="0" />
            <xsd:element ref="ivrInfo"
                  minOccurs="0" />
```

```
                <xsd:element ref="mixerInfo"
                        minOccurs="0" />
                <xsd:any namespace="##other"
                        minOccurs="0"
                        maxOccurs="unbounded" processContents="lax" />
            </xsd:sequence>
            <xsd:attribute name="id" type="xsd:string"
                    use="required" />
            <xsd:anyAttribute namespace="##other"
                    processContents="lax" />
        </xsd:extension>
    </xsd:complexContent>
</xsd:complexType>
<xsd:element name="mediaResourceRequest"
        type="mediaResourceRequestType" />

<!--
####################################################
generalInfo TYPE
####################################################
-->

<!-- generalInfo -->
<xsd:complexType name="generalInfoType">
    <xsd:complexContent>
        <xsd:extension base="Tcore">
            <xsd:sequence>
                <xsd:element ref="session-info"
                        minOccurs="0" />
                <xsd:element ref="packages"
                        minOccurs="0" />
                <xsd:any namespace="##other"
                        minOccurs="0"
                        maxOccurs="unbounded" processContents="lax" />
            </xsd:sequence>
            <xsd:anyAttribute namespace="##other"
                    processContents="lax" />
        </xsd:extension>
    </xsd:complexContent>
</xsd:complexType>
<xsd:element name="generalInfo" type="generalInfoType" />

<!-- session-info -->

<xsd:complexType name="session-infoType">
    <xsd:complexContent>
        <xsd:extension base="Tcore">
            <xsd:sequence>
                <xsd:element name="session-id"
                        type="id.datatype"/>
                <xsd:element name="seq"
                        type="xsd:nonNegativeInteger"/>
                <xsd:element name="action"
                        type="action.datatype"/>
                <xsd:any namespace="##other"
                        minOccurs="0"
```

```
                    maxOccurs="unbounded" processContents="lax" />
            </xsd:sequence>
            <xsd:anyAttribute namespace="##other"
                    processContents="lax" />
        </xsd:extension>
    </xsd:complexContent>
</xsd:complexType>
<xsd:element name="session-info" type="session-infoType" />

<!-- packages -->

<xsd:complexType name="packagesType">
    <xsd:complexContent>
        <xsd:extension base="Tcore">
            <xsd:sequence>
                <xsd:element name="package" type="xsd:string"
                        minOccurs="0"
                        maxOccurs="unbounded" />
                <xsd:any namespace="##other" minOccurs="0"
                        maxOccurs="unbounded" processContents="lax" />
            </xsd:sequence>
            <xsd:anyAttribute namespace="##other" processContents="lax" />
        </xsd:extension>
    </xsd:complexContent>
</xsd:complexType>
<xsd:element name="packages" type="packagesType"/>

<!--
#####################################################
ivrInfo TYPE
#####################################################
-->

<!-- ivrInfo -->

<xsd:complexType name="ivrInfoType">
    <xsd:complexContent>
        <xsd:extension base="Tcore">
            <xsd:sequence>
                <xsd:element ref="ivr-sessions"
                        minOccurs="0" />
                <xsd:element ref="file-formats"
                        minOccurs="0" />
                <xsd:element ref="dtmf-type"
                        minOccurs="0" />
                <xsd:element ref="tones" minOccurs="0" />
                <xsd:element ref="asr-tts"
                        minOccurs="0" />
                <xsd:element ref="vxml" minOccurs="0" />
                <xsd:element ref="location"
                        minOccurs="0" />
                <xsd:element ref="encryption"
                        minOccurs="0" />
                <xsd:element ref="application-data"
                        minOccurs="0" />
                <xsd:element ref="max-prepared-duration"
```

```
                    minOccurs="0" />
                <xsd:element ref="file-transfer-modes" minOccurs="0" />
                <xsd:any namespace="##other"
                    minOccurs="0"
                    maxOccurs="unbounded" processContents="lax" />
            </xsd:sequence>
            <xsd:anyAttribute namespace="##other"
                processContents="lax" />
        </xsd:extension>
    </xsd:complexContent>
</xsd:complexType>
<xsd:element name="ivrInfo" type="ivrInfoType" />

<!--
#####################################################
mixerInfo TYPE
#####################################################
-->

<!-- mixerInfo -->

<xsd:complexType name="mixerInfoType">
    <xsd:complexContent>
        <xsd:extension base="Tcore">
            <xsd:sequence>
                <xsd:element ref="mixers"
                    minOccurs="0"/>
                <xsd:element ref="file-formats"
                    minOccurs="0"/>
                <xsd:element ref="dtmf-type"
                    minOccurs="0"/>
                <xsd:element ref="tones" minOccurs="0"/>
                <xsd:element ref="mixing-modes"
                    minOccurs="0"/>
                <xsd:element ref="application-data"
                    minOccurs="0"/>
                <xsd:element ref="location"
                    minOccurs="0"/>
                <xsd:element ref="encryption"
                    minOccurs="0"/>
                <xsd:any namespace="##other"
                    minOccurs="0"
                    maxOccurs="unbounded" processContents="lax" />
            </xsd:sequence>
            <xsd:anyAttribute namespace="##other"
                processContents="lax" />
        </xsd:extension>
    </xsd:complexContent>
</xsd:complexType>
<xsd:element name="mixerInfo" type="mixerInfoType" />

<!--
#####################################################
mediaResourceResponse TYPE
#####################################################
-->
```

```
<!-- mediaResourceResponse -->

<xsd:complexType name="mediaResourceResponseType">
    <xsd:complexContent>
        <xsd:extension base="Tcore">
            <xsd:sequence>
                <xsd:element ref="response-session-info"
                        minOccurs="0" />
                <xsd:any namespace="##other"
                        minOccurs="0"
                        maxOccurs="unbounded" processContents="lax" />
            </xsd:sequence>
            <xsd:attribute name="id" type="xsd:string"
                    use="required" />
            <xsd:attribute name="status" type="status.datatype"
                    use="required" />
            <xsd:attribute name="reason" type="xsd:string" />
            <xsd:anyAttribute namespace="##other"
                    processContents="lax" />
        </xsd:extension>
    </xsd:complexContent>
</xsd:complexType>
<xsd:element name="mediaResourceResponse"
        type="mediaResourceResponseType" />

<!--
####################################################
ELEMENTS
####################################################
-->

<!-- response-session-info -->

<xsd:complexType name="response-session-infoType">
<xsd:complexContent>
<xsd:extension base="Tcore">
<xsd:sequence>
<xsd:element name="session-id" type="id.datatype"/>
<xsd:element name="seq" type="xsd:nonNegativeInteger"/>
<xsd:element name="expires" type="xsd:nonNegativeInteger"/>
<xsd:element ref="media-server-address"
        minOccurs="0" maxOccurs="unbounded" />
<xsd:any namespace="##other" minOccurs="0"
        maxOccurs="unbounded" processContents="lax" />
</xsd:sequence>
<xsd:anyAttribute namespace="##other"
        processContents="lax" />
</xsd:extension>
</xsd:complexContent>
</xsd:complexType>
<xsd:element name="response-session-info"
        type="response-session-infoType" />

<!-- media-server-address -->

<xsd:complexType name="media-server-addressTYPE">
```

```
      <xsd:complexContent>
         <xsd:extension base="Tcore">
            <xsd:sequence>
               <xsd:element name="connection-id"
                     type="xsd:string"
                     minOccurs="0" maxOccurs="unbounded" />
               <xsd:element ref="ivr-sessions"
                     minOccurs="0"/>
               <xsd:element ref="mixers"
                     minOccurs="0"/>
               <xsd:any namespace="##other"
                     minOccurs="0"
                     maxOccurs="unbounded" processContents="lax" />
            </xsd:sequence>
            <xsd:attribute name="uri" type="xsd:anyURI"
                  use="required" />
            <xsd:anyAttribute namespace="##other"
                  processContents="lax" />
         </xsd:extension>
      </xsd:complexContent>
   </xsd:complexType>
   <xsd:element name="media-server-address"
         type="media-server-addressTYPE" />

   <!-- ivr-sessions -->

   <xsd:complexType name="ivr-sessionsType">
      <xsd:complexContent>
         <xsd:extension base="Tcore">
            <xsd:sequence>
               <xsd:element ref="rtp-codec"
                     minOccurs="0"
                     maxOccurs="unbounded" />
               <xsd:any namespace="##other"
                     minOccurs="0"
                     maxOccurs="unbounded" processContents="lax" />
            </xsd:sequence>
            <xsd:anyAttribute namespace="##other"
                  processContents="lax" />
         </xsd:extension>
      </xsd:complexContent>
   </xsd:complexType>
   <xsd:element name="ivr-sessions" type="ivr-sessionsType" />
   <xsd:complexType name="rtp-codecType">
      <xsd:complexContent>
         <xsd:extension base="Tcore">
            <xsd:sequence>
               <xsd:element name="decoding"
                     type="xsd:nonNegativeInteger" />
               <xsd:element name="encoding"
                     type="xsd:nonNegativeInteger" />
               <xsd:any namespace="##other"
                     minOccurs="0"
                     maxOccurs="unbounded" processContents="lax" />
            </xsd:sequence>
            <xsd:attribute name="name" type="xsd:string"
```

```
                use="required" />
            <xsd:anyAttribute namespace="##other"
                processContents="lax" />
        </xsd:extension>
    </xsd:complexContent>
</xsd:complexType>
<xsd:element name="rtp-codec" type="rtp-codecType" />

<!-- file-formats -->

<xsd:complexType name="file-formatsType">
<xsd:complexContent>
<xsd:extension base="Tcore">
<xsd:sequence>
<xsd:element ref="required-format"
        minOccurs="0" maxOccurs="unbounded" />
<xsd:any namespace="##other" minOccurs="0"
        maxOccurs="unbounded" processContents="lax" />
</xsd:sequence>
<xsd:anyAttribute namespace="##other" processContents="lax" />
</xsd:extension>
</xsd:complexContent>
</xsd:complexType>
<xsd:element name="file-formats" type="file-formatsType" />
<xsd:complexType name="required-formatType">
    <xsd:complexContent>
        <xsd:extension base="Tcore">
            <xsd:sequence>
                <xsd:element ref="required-file-package"
                        minOccurs="0" maxOccurs="unbounded" />
                <xsd:any namespace="##other"
                        minOccurs="0"
                        maxOccurs="unbounded" processContents="lax" />
            </xsd:sequence>
            <xsd:attribute name="name" type="xsd:string"
                use="required" />
            <xsd:anyAttribute namespace="##other"
                processContents="lax" />
        </xsd:extension>
    </xsd:complexContent>
</xsd:complexType>
<xsd:element name="required-format" type="required-formatType" />
<xsd:complexType name="required-file-packageType">
    <xsd:complexContent>
        <xsd:extension base="Tcore">
            <xsd:sequence>
                <xsd:element name="required-file-package-
                        name" type="xsd:string"
                        minOccurs="0" maxOccurs="unbounded" />
                <xsd:any namespace="##other"
                        minOccurs="0"
                        maxOccurs="unbounded" processContents="lax" />
            </xsd:sequence>
            <xsd:anyAttribute namespace="##other"
                processContents="lax" />
        </xsd:extension>
```

```
    </xsd:complexContent>
</xsd:complexType>
<xsd:element name="required-file-package"
       type="required-file-packageType" />

<!-- dtmf-type -->

<xsd:complexType name="dtmfType">
    <xsd:complexContent>
        <xsd:extension base="Tcore">
            <xsd:sequence>
                <xsd:element ref="detect" />
                <xsd:element ref="generate" />
                <xsd:element ref="passthrough" />
                <xsd:any namespace="##other"
                        minOccurs="0"
                        maxOccurs="unbounded" processContents="lax" />
            </xsd:sequence>
            <xsd:anyAttribute namespace="##other"
                    processContents="lax" />
        </xsd:extension>
    </xsd:complexContent>
</xsd:complexType>
<xsd:element name="dtmf" type="dtmfType" />
<xsd:complexType name="detectType">
    <xsd:complexContent>
        <xsd:extension base="Tcore">
            <xsd:sequence>
                <xsd:element ref="dtmf-type"
                        minOccurs="0" maxOccurs="unbounded" />
                <xsd:any namespace="##other"
                        minOccurs="0"
                        maxOccurs="unbounded" processContents="lax" />
            </xsd:sequence>
            <xsd:anyAttribute namespace="##other"
                    processContents="lax" />
        </xsd:extension>
    </xsd:complexContent>
</xsd:complexType>
<xsd:element name="detect" type="detectType" />
<xsd:complexType name="generateType">
    <xsd:complexContent>
        <xsd:extension base="Tcore">
            <xsd:sequence>
                <xsd:element ref="dtmf-type"
                        minOccurs="0" maxOccurs="unbounded" />
                <xsd:any namespace="##other"
                        minOccurs="0"
                        maxOccurs="unbounded" processContents="lax" />
            </xsd:sequence>
            <xsd:anyAttribute namespace="##other"
                    processContents="lax" />
        </xsd:extension>
    </xsd:complexContent>
</xsd:complexType>
<xsd:element name="generate" type="generateType" />
```

```
<xsd:complexType name="passthroughType">
   <xsd:complexContent>
      <xsd:extension base="Tcore">
         <xsd:sequence>
            <xsd:element ref="dtmf-type"
                  minOccurs="0" maxOccurs="unbounded" />
            <xsd:any namespace="##other"
                  minOccurs="0"
                  maxOccurs="unbounded" processContents="lax" />
         </xsd:sequence>
         <xsd:anyAttribute namespace="##other"
               processContents="lax" />
      </xsd:extension>
   </xsd:complexContent>
</xsd:complexType>
<xsd:element name="passthrough" type="passthroughType" />
<xsd:complexType name="dtmf-typeType">
   <xsd:complexContent>
      <xsd:extension base="Tcore">
         <xsd:sequence>
            <xsd:any namespace="##other"
                  minOccurs="0"
                  maxOccurs="unbounded" processContents="lax" />
         </xsd:sequence>
         <xsd:attribute name="name" type="dtmf.datatype"
               use="required" />
         <xsd:attribute name="package" type="xsd:string"
               use="required" />
         <xsd:anyAttribute namespace="##other"
               processContents="lax" />
      </xsd:extension>
   </xsd:complexContent>
</xsd:complexType>
<xsd:element name="dtmf-type" type="dtmf-typeType" />

<!-- tones -->

<xsd:complexType name="required-tonesType">
   <xsd:complexContent>
      <xsd:extension base="Tcore">
         <xsd:sequence>
            <xsd:element ref="country-codes"
                  minOccurs="0" maxOccurs="1" />
            <xsd:element ref="h248-codes"
                  minOccurs="0" maxOccurs="1" />
            <xsd:any namespace="##other"
                  minOccurs="0"
                  maxOccurs="unbounded" processContents="lax" />
         </xsd:sequence>
         <xsd:anyAttribute namespace="##other"
               processContents="lax" />
      </xsd:extension>
   </xsd:complexContent>
</xsd:complexType>
<xsd:element name="tones" type="required-tonesType" />
<xsd:complexType name="required-country-codesType">
```

```
    <xsd:complexContent>
        <xsd:extension base="Tcore">
            <xsd:sequence>
                <xsd:element ref="country-code"
                        minOccurs="0" maxOccurs="unbounded" />
                <xsd:any namespace="##other"
                        minOccurs="0"
                        maxOccurs="unbounded" processContents="lax" />
            </xsd:sequence>
            <xsd:anyAttribute namespace="##other"
                    processContents="lax" />
        </xsd:extension>
    </xsd:complexContent>
</xsd:complexType>
<xsd:element name="country-codes"
        type="required-country-codesType" />
<xsd:complexType name="country-codeType" mixed="true">
    <xsd:sequence>
        <xsd:any namespace="##other" minOccurs="0"
            maxOccurs="unbounded" processContents="lax" />
    </xsd:sequence>
    <xsd:attribute name="package" type="xsd:string"
            use="required" />
    <xsd:anyAttribute namespace="##other" processContents="lax" />
</xsd:complexType>
<xsd:element name="country-code" type="country-codeType" />
<xsd:complexType name="required-h248-codesType">
    <xsd:complexContent>
        <xsd:extension base="Tcore">
            <xsd:sequence>
                <xsd:element ref="h248-code"
                        minOccurs="0" maxOccurs="unbounded" />
                <xsd:any namespace="##other"
                        minOccurs="0"
                        maxOccurs="unbounded" processContents="lax" />
            </xsd:sequence>
            <xsd:anyAttribute namespace="##other"
                    processContents="lax" />
        </xsd:extension>
    </xsd:complexContent>
</xsd:complexType>
<xsd:element name="h248-codes"
        type="required-h248-codesType" />
<xsd:complexType name="h248-codeType" mixed="true">
    <xsd:sequence>
        <xsd:any namespace="##other" minOccurs="0"
                maxOccurs="unbounded" processContents="lax" />
    </xsd:sequence>
    <xsd:attribute name="package" type="xsd:string"
            use="required" />
    <xsd:anyAttribute namespace="##other" processContents="lax" />
</xsd:complexType>
<xsd:element name="h248-code" type="h248-codeType" />

<!-- asr-tts -->

<xsd:complexType name="asr-ttsType">
```

```
        <xsd:complexContent>
            <xsd:extension base="Tcore">
                <xsd:sequence>
                    <xsd:element ref="asr-support"
                            minOccurs="0" maxOccurs="1" />
                    <xsd:element ref="tts-support"
                            minOccurs="0" maxOccurs="1" />
                    <xsd:any namespace="##other"
                            minOccurs="0"
                            maxOccurs="unbounded" processContents="lax" />
                </xsd:sequence>
                <xsd:anyAttribute namespace="##other"
                        processContents="lax" />
            </xsd:extension>
        </xsd:complexContent>
    </xsd:complexType>
    <xsd:element name="asr-tts" type="asr-ttsType" />
    <xsd:complexType name="asr-supportType">
        <xsd:complexContent>
            <xsd:extension base="Tcore">
                <xsd:sequence>
                    <xsd:element ref="language"
                            minOccurs="0" maxOccurs="unbounded" />
                    <xsd:any namespace="##other"
                            minOccurs="0"
                            maxOccurs="unbounded" processContents="lax" />
                </xsd:sequence>
                <xsd:anyAttribute namespace="##other"
                        processContents="lax" />
            </xsd:extension>
        </xsd:complexContent>
    </xsd:complexType>
    <xsd:element name="asr-support" type="asr-supportType" />
    <xsd:complexType name="tts-supportType">
        <xsd:complexContent>
            <xsd:extension base="Tcore">
                <xsd:sequence>
                    <xsd:element ref="language"
                            minOccurs="0" maxOccurs="unbounded" />
                    <xsd:any namespace="##other"
                            minOccurs="0"
                            maxOccurs="unbounded" processContents="lax" />
                </xsd:sequence>
                <xsd:anyAttribute namespace="##other"
                        processContents="lax" />
            </xsd:extension>
        </xsd:complexContent>
    </xsd:complexType>
    <xsd:element name="tts-support" type="tts-supportType" />
    <xsd:complexType name="languageType">
        <xsd:complexContent>
            <xsd:extension base="Tcore">
                <xsd:sequence>
                    <xsd:any namespace="##other"
                            minOccurs="0"
                            maxOccurs="unbounded" processContents="lax" />
                </xsd:sequence>
```

```
            <xsd:attribute ref="xml:lang" />
            <xsd:anyAttribute namespace="##other"
                  processContents="lax" />
        </xsd:extension>
    </xsd:complexContent>
</xsd:complexType>
<xsd:element name="language" type="languageType" />

<!-- vxml -->

<xsd:complexType name="vxmlType">
    <xsd:complexContent>
        <xsd:extension base="Tcore">
            <xsd:sequence>
                <xsd:element ref="vxml-mode"
                      minOccurs="0" maxOccurs="unbounded" />
                <xsd:any namespace="##other"
                      minOccurs="0"
                      maxOccurs="unbounded" processContents="lax" />
            </xsd:sequence>
            <xsd:anyAttribute namespace="##other"
                  processContents="lax" />
        </xsd:extension>
    </xsd:complexContent>
</xsd:complexType>
<xsd:element name="vxml" type="vxmlType" />
<xsd:complexType name="vxml-modeType">
    <xsd:complexContent>
        <xsd:extension base="Tcore">
            <xsd:sequence>
                <xsd:any namespace="##other"
                      minOccurs="0"
                      maxOccurs="unbounded" processContents="lax" />
            </xsd:sequence>
            <xsd:attribute name="package" type="xsd:string"
                  use="required" />
            <xsd:attribute name="require"
                  type="vxml.datatype" use="required" />
            <xsd:anyAttribute namespace="##other"
                  processContents="lax" />
        </xsd:extension>
    </xsd:complexContent>
</xsd:complexType>
<xsd:element name="vxml-mode" type="vxml-modeType" />

<!-- location -->

<xsd:complexType name="locationType">
    <xsd:complexContent>
        <xsd:extension base="Tcore">
            <xsd:sequence>
                <xsd:element ref="ca:civicAddress"
                      minOccurs="1" maxOccurs="1" />
                <xsd:any namespace="##other"
                      minOccurs="0"
                      maxOccurs="unbounded" processContents="lax" />
```

```
            </xsd:sequence>
            <xsd:anyAttribute namespace="##other"
                  processContents="lax" />
        </xsd:extension>
    </xsd:complexContent>
</xsd:complexType>
<xsd:element name="location" type="locationType" />

<!-- encryption -->

<xsd:complexType name="encryptionType">
    <xsd:complexContent>
        <xsd:extension base="Tcore">
            <xsd:sequence>
                <xsd:any namespace="##other"
                        minOccurs="0"
                        maxOccurs="unbounded" processContents="lax" />
            </xsd:sequence>
            <xsd:anyAttribute namespace="##other"
                  processContents="lax" />
        </xsd:extension>
    </xsd:complexContent>
</xsd:complexType>
<xsd:element name="encryption" type="encryptionType" />

<!-- application-data -->

<xsd:element name="application-data" type="appdata.datatype" />

<!-- max-prepared-duration -->

<xsd:complexType name="max-prepared-durationType">
    <xsd:complexContent>
        <xsd:extension base="Tcore">
            <xsd:sequence>
                <xsd:element ref="max-time" />
                <xsd:any namespace="##other"
                        minOccurs="0"
                        maxOccurs="unbounded" processContents="lax" />
            </xsd:sequence>
            <xsd:anyAttribute namespace="##other"
                  processContents="lax" />
        </xsd:extension>
    </xsd:complexContent>
</xsd:complexType>
<xsd:element name="max-prepared-duration"
        type="max-prepared-durationType" />
<xsd:complexType name="max-timeType">
    <xsd:complexContent>
        <xsd:extension base="Tcore">
            <xsd:sequence>
                <xsd:element name="max-time-package"
                        type="xsd:string" />
                <xsd:any namespace="##other"
                        minOccurs="0"
                        maxOccurs="unbounded" processContents="lax" />
```

```
          </xsd:sequence>
          <xsd:attribute name="max-time-seconds"
                type="xsd:nonNegativeInteger"
                use="required" />
          <xsd:anyAttribute namespace="##other"
                processContents="lax" />
       </xsd:extension>
    </xsd:complexContent>
</xsd:complexType>
<xsd:element name="max-time" type="max-timeType" />

<!-- file-transfer-modes -->

<xsd:complexType name="file-transfer-modesType">
    <xsd:complexContent>
       <xsd:extension base="Tcore">
          <xsd:sequence>
             <xsd:element ref="file-transfer-mode"
                   minOccurs="0" maxOccurs="unbounded" />
             <xsd:any namespace="##other"
                   minOccurs="0"
                   maxOccurs="unbounded" processContents="lax" />
          </xsd:sequence>
          <xsd:anyAttribute namespace="##other"
                processContents="lax" />
       </xsd:extension>
    </xsd:complexContent>
</xsd:complexType>
<xsd:element name="file-transfer-modes"
       type="file-transfer-modesType" />
<xsd:complexType name="file-transfer-modeType">
    <xsd:complexContent>
       <xsd:extension base="Tcore">
          <xsd:sequence>
             <xsd:any namespace="##other"
                   minOccurs="0"
                   maxOccurs="unbounded" processContents="lax" />
          </xsd:sequence>
          <xsd:attribute name="name"
                type="transfermode.datatype"
                use="required" />
          <xsd:attribute name="package" type="xsd:string"
                use="required" />
          <xsd:anyAttribute namespace="##other"
                processContents="lax" />
       </xsd:extension>
    </xsd:complexContent>
</xsd:complexType>
<xsd:element name="file-transfer-mode"
       type="file-transfer-modeType" />

<!-- mixers -->

<xsd:complexType name="mixerssessionsType">
    <xsd:complexContent>
       <xsd:extension base="Tcore">
          <xsd:sequence>
```

```
                <xsd:element ref="mix" minOccurs="0"
                        maxOccurs="unbounded" />
                <xsd:any namespace="##other" minOccurs="0"
                        maxOccurs="unbounded" processContents="lax" />
            </xsd:sequence>
            <xsd:anyAttribute namespace="##other"
                    processContents="lax" />
        </xsd:extension>
    </xsd:complexContent>
</xsd:complexType>
<xsd:element name="mixers" type="mixerssessionsType" />
<xsd:complexType name="mixType">
    <xsd:complexContent>
        <xsd:extension base="Tcore">
            <xsd:sequence>
                <xsd:element ref="rtp-codec"
                        minOccurs="0"
                        maxOccurs="unbounded" />
                <xsd:any namespace="##other"
                        minOccurs="0"
                        maxOccurs="unbounded" processContents="lax" />
            </xsd:sequence>
            <xsd:attribute name="users"
                    type="xsd:nonNegativeInteger"
                    use="required" />
            <xsd:anyAttribute namespace="##other"
                    processContents="lax" />
        </xsd:extension>
    </xsd:complexContent>
</xsd:complexType>
<xsd:element name="mix" type="mixType" />

<!-- mixing-modes -->

<xsd:complexType name="mixing-modesType">
    <xsd:complexContent>
        <xsd:extension base="Tcore">
            <xsd:sequence>
                <xsd:element ref="audio-mixing-modes"
                        minOccurs="0" maxOccurs="1" />
                <xsd:element ref="video-mixing-modes"
                        minOccurs="0" maxOccurs="1" />
                <xsd:any namespace="##other"
                        minOccurs="0"
                        maxOccurs="unbounded" processContents="lax" />
            </xsd:sequence>
            <xsd:anyAttribute namespace="##other"
                    processContents="lax" />
        </xsd:extension>
    </xsd:complexContent>
</xsd:complexType>
<xsd:element name="mixing-modes" type="mixing-modesType" />
<xsd:complexType name="audio-mixing-modesType">
    <xsd:complexContent>
        <xsd:extension base="Tcore">
            <xsd:sequence>
                <xsd:element ref="audio-mixing-mode"
```

```
                    minOccurs="0" maxOccurs="unbounded" />
                <xsd:any namespace="##other"
                    minOccurs="0"
                    maxOccurs="unbounded" processContents="lax" />
            </xsd:sequence>
            <xsd:anyAttribute namespace="##other"
                processContents="lax" />
        </xsd:extension>
    </xsd:complexContent>
</xsd:complexType>
<xsd:element name="audio-mixing-modes"
        type="audio-mixing-modesType" />
<xsd:complexType name="audio-mixing-modeType" mixed="true">
    <xsd:sequence>
        <xsd:any namespace="##other" minOccurs="0"
            maxOccurs="unbounded" processContents="lax" />
    </xsd:sequence>
    <xsd:attribute name="package" type="xsd:string"
        use="required" />
    <xsd:anyAttribute namespace="##other" processContents="lax" />
</xsd:complexType>
<xsd:element name="audio-mixing-mode"
        type="audio-mixing-modeType" />
<xsd:complexType name="video-mixing-modesType">
    <xsd:complexContent>
        <xsd:extension base="Tcore">
            <xsd:sequence>
                <xsd:element ref="video-mixing-mode"
                    minOccurs="0" maxOccurs="unbounded" />
                <xsd:any namespace="##other"
                    minOccurs="0"
                    maxOccurs="unbounded" processContents="lax" />
            </xsd:sequence>
            <xsd:attribute name="vas" type="boolean.datatype"
                default="false" />
            <xsd:attribute name="activespeakermix"
                type="boolean.datatype"
                default="false" />
            <xsd:anyAttribute namespace="##other"
                processContents="lax" />
        </xsd:extension>
    </xsd:complexContent>
</xsd:complexType>
<xsd:element name="video-mixing-modes"
        type="video-mixing-modesType" />
<xsd:complexType name="video-mixing-modeType" mixed="true">
    <xsd:sequence>
        <xsd:any namespace="##other" minOccurs="0"
            maxOccurs="unbounded" processContents="lax" />
    </xsd:sequence>
    <xsd:attribute name="package" type="xsd:string"
        use="required" />
    <xsd:anyAttribute namespace="##other" processContents="lax" />
</xsd:complexType>
<xsd:element name="video-mixing-mode"
        type="video-mixing-modeType" />
```

```
<!--
##################################################
DATATYPES
##################################################
-->

<xsd:simpleType name="version.datatype">
   <xsd:restriction base="xsd:NMTOKEN">
      <xsd:enumeration value="1.0" />
   </xsd:restriction>
</xsd:simpleType>
<xsd:simpleType name="id.datatype">
   <xsd:restriction base="xsd:NMTOKEN" />
</xsd:simpleType>
<xsd:simpleType name="status.datatype">
   <xsd:restriction base="xsd:positiveInteger">
      <xsd:pattern value="[0-9][0-9][0-9]" />
   </xsd:restriction>
</xsd:simpleType>
<xsd:simpleType name="transfermode.datatype">
   <xsd:restriction base="xsd:NMTOKEN"/>
</xsd:simpleType>
<xsd:simpleType name="action.datatype">
   <xsd:restriction base="xsd:NMTOKEN">
      <xsd:enumeration value="remove" />
      <xsd:enumeration value="update" />
   </xsd:restriction>
</xsd:simpleType>
<xsd:simpleType name="dtmf.datatype">
   <xsd:restriction base="xsd:NMTOKEN"/>
</xsd:simpleType>
<xsd:simpleType name="boolean.datatype">
   <xsd:restriction base="xsd:NMTOKEN">
      <xsd:enumeration value="true" />
      <xsd:enumeration value="false" />
   </xsd:restriction>
</xsd:simpleType>
<xsd:simpleType name="vxml.datatype">
   <xsd:restriction base="xsd:NMTOKEN"/>
</xsd:simpleType>
<xsd:simpleType name="appdata.datatype">
   <xsd:restriction base="xsd:string" />
</xsd:simpleType>
</xsd:schema>
```

11.12 SECURITY CONSIDERATIONS

The MRB network entity has two primary interfaces – Publish and Consumer – that carry sensitive information and must therefore be appropriately protected and secured.

The Publish interface, as defined in and described in Section 11.5.1, uses the CFW (RFC 6230 – see Chapter 8) as a mechanism to connect an MRB to a MS. It is very important that the communication between the MRB and the MS is secured: a malicious entity may change or even delete subscriptions to a MS, thus affecting the view the MRB has of the resources actually available on a MS and leading it to incorrect selection when media resources are being requested by an AS. A malicious entity may even manipulate available resources on a MS, for example, to make the MRB think no resources are available at all. Considering that the Publish interface is a CFW Control Package, the same security considerations included in the CFW specification apply here to protect interactions between an MRB and an MS.

The Publish interface also allows a MS, as explained in Section 11.5.1.5.18, to provide more or less accurate information about its geographic location, should ASs be interested in such details when looking for services at an MRB. While the usage of this information is entirely optional, and the level of detail to be provided is implementation specific, it is important to draw attention to the potential security issues that the disclosure of such addresses may introduce. As such, it is important to make sure that MRB implementations do not disclose this information as is to interested ASs but only exploit those addresses as part of computation algorithms to pick the most adequate resources ASs may be looking for.

The Consumer interface, as defined in and described in Section 11.5.2, conceives transactions based on a session ID. These transactions may be transported by means of either HTTP messages or SIP dialogs. This means that malicious users could be able to disrupt or manipulate an MRB session should they have access to the above-mentioned session ID or replicate it somehow: for instance, a malicious entity could modify an existing session between an AS and the MRB, e.g., requesting fewer resources than originally requested to cause media dialogs to be rejected by the AS or requesting many more resources instead to try to lock as many as possible of (if not all) the resources an MRB can provide, thus making them unavailable to other legitimate ASs in subsequent requests.

In order to prevent this, it is strongly advised that MRB implementations generate session identifiers that are very hard to replicate in order to minimize the chances that malicious users could gain access to valid identifiers by just guessing or by means of brute-force attacks. It is very important, of course, to also secure the way that these identifiers are transported by the involved parties, in both requests and responses, in order to prevent network attackers from intercepting Consumer messages and having access to session IDs. The Consumer interface uses either the HTTP or the SIP as the mechanism for clients to connect to an MRB to request media resources. In the case where HTTP is used, any binding using the Consumer interface MUST be capable of being transacted over transport layer security (TLS), as described in RFC 2818. In the case where SIP is used, the same security considerations included in the CFW specification apply here to protect interactions between a client requesting media resources and an MRB.

Should a valid session ID be compromised somehow (that is, intercepted or just guessed by a malicious user), as a further means to prevent disruption, the Consumer interface also prescribes the use of a sequence number in its transactions. This sequence number is to be increased after each successful transaction, starting from a first value randomly generated by the MRB when the session is first created, and it must match in every request/response. While this adds complexity to the protocol (implementations must pay attention to those sequence numbers, since wrong values will cause "Wrong sequence number" errors and the failure of the related requests), it is an important added value for security. In fact, considering that different transactions related to the same session could be transported in different, unrelated HTTP messages (or SIP INVITEs in cases where the In-line mode is being used), this sequence number protection prevents the chances of session replication or disruption, especially in cases where the session ID has been compromised: that is, it should make it harder for malicious users to manipulate or remove a session for which they have obtained the session ID. It is strongly advised that the MRB does not choose 1 as the first sequence number for a new session but rather, picks a random value to start from. The reaction to transactions that are out of sequence is left to MRB implementations: a related error code is available, but implementations may decide to enforce further limitations or actions upon the receipt of too many failed attempts in a row or of what look like blatant attempts to guess what the current, valid sequence number is.

It is also worth noting that in In-line mode (both IAMM and IUMM), the MRB may act as a B2BUA. This means that when acting as a B2BUA, the MRB may modify SIP bodies: It is the case, for instance, for the IAMM handling multipart/mixed payloads. This impacts the ability to use any SIP security feature that protects the body (e.g., RFC 4474, S/MIME, etc.), unless the MRB acts as a mediator for the security association. This should be taken into account when implementing an MRB compliant with this specification. Both the Publishing interface and the Consumer interface may address the location of an MS: the Publishing interface may be used to inform the MRB where an MS is located (approximately or precisely), and the Consumer interface may be used to ask for a MS located somewhere in a particular region (e.g., a conference bridge close to San Francisco). Both MS and MRB implementers need to take this into account when deciding whether or not to make this location information available, and if so, how many bits of information really need to be made available for brokering purposes.

It is worthwhile to cover authorization issues related to this specification. Neither the Publishing interface nor the Consumer interface provides an explicit means for implementing authentication, i.e., they do not contain specific protocol interactions to ensure that authorized ASs can make use of the services provided by an MRB instance. Considering that both interfaces are transported using well-established protocols (HTTP, SIP, and CFW), support for such functionality can be expressed by means of the authentication mechanisms provided by the protocols themselves.

Therefore, any MRB-aware entity ASs, MSs, and MRBs themselves) MUST support HTTP and SIP Digest access authentication. The usage of such Digest access authentications is recommended and not mandatory, which means that MRB-aware entities MAY exploit it in deployment. An MRB may want to enforce further constraints on the interactions between an AS/MS and an MRB. For example, it may choose to only accept requests associated with a specific session ID from the IP address that originated the first request or may just make use of pre-shared certificates to assess the identity of legitimate ASs and/or MSs.

11.13 IANA CONSIDERATIONS

There are several IANA considerations associated with this specification. The detail can be seen in RFC 6917.

11.14 SUMMARY

We have articulated in detail that the centralized MS architecture of the XCON system, where media servers are located in a given single geographical location, cannot provide scalability for the large-scale multipoint multimedia conference that involves many parties who are located in different gegraphical locations across the whole nation, multiple nations, or world. We have explained why a "logically centralized" MS architecture of the XCON system is highly scalable where media server are physically distributed in different geographical locations in relation to the distribution of the conferencing parties located across the large geographical locations including the entire globe. "Logicaally Centralized" means that all the end-users (e.g., conference participnats) will be dialing to a single number centering around the foci/focus while media services will be offered by many different media servers located in many different geographical locations, transaparently giving the appearance as multimedia conference services are being provided a single enity. It is the job of the foci/focus to coordinate among all the media servers making sure that services are offered by each kind of media servers to its designated group of end-users/participants who are located within the shortest geographical distance having the lowest coast while maitainig load balancing and quality-of-services constraints.

In this context, we have explained how the CFW (RFC 6230 – see Chapter 8) for decentralization of the MSs and their resources. However, we have specified the actual interfaces that enable implementations of 1:M and M:N relationships between the ASs and the MSs using the MRB. The SIP (RFC 3261 – also see Sections 1.3 and 1.5) signaling protocol is used for inter-server communications at the application layer, with dynamic signaling schemes facilitating deployment of distributed scalable MSs architecture for XCOM conference systems offering interoperability in multivendor environments. By the way, we have explained why MRB needs to act as a B2BUA in SIP (RFC 3261 – also see Sections 1.3 and 1.5) signaling terminology.

We have discussed different MRB deployment scenarios for one-to-many and many-many AS and MS relationships: Query MRB, which allows an MRB client to request a specific MS, and In-line MRB, which allows an MRB client to use the media resource control and media dialog signaling of MRB in choosing a specific MS remaining in the signaling path between the AS and MS resources, eliminating the concept of "Query." Another deployment model, known as hybrid Query MRB, whereby an MRB might be co-hosted as a hybrid logical entity with an AS is also discussed.

Interestingly, the In-line MRB can be split into two distinct logical roles, which are explained: IUMM, which allows an MRB to act on behalf of clients requiring media services who are not aware of an MRB or its operation, and IAMM, which allows an MRB to act on behalf of clients requiring media services who are aware of an MRB and its operation.

In this section, the MRB interface specifications that provide a toolkit for a variety of deployment architectures where media resource brokering can take place are defined only for three cases: MS Resource Publish Interface, which is responsible for providing an MRB with appropriate MS resource information; Media Service Resource Consumer Interface, which provides the ability for clients of an MRB, such as ASs, to request an appropriate MS to satisfy specific criteria; and In-Line Unaware MRB Interface, which allows an entity acting as an In-line MRB to act in one of two roles for a request – either IUMM of operation or IAMM of operation. Note that it is beyond the scope of RFC 6917 to define exactly how to construct an MRB using the interfaces described in this section.

Another kind of implementations, known as Multimodal MRB Implementations, is discussed, whereby an MRB might be implemented to operate multimodally with a collection of AS clients all sharing the same pool of MSs and media resources; that is, an MRB may be simultaneously operating in Query mode, IAMM, and IUMM. It knows in which mode to act on any particular request from a client, depending on the context of the request. The relative merits of Query Mode, IAMM, and IUMM are explained in great detail.

Some examples of MRB Publish and Consumer interfaces are provided to give in-depth understanding of implementations for providing interoperability. In addition, XML schemas for both Media Service Resource Publisher and Consumer interfaces are defined. Finally, the security of the MRB deployments using different networking configurations is discussed. Note that IANA registrations are needed for the following for RFC 6917: Media Control Channel Framework Package, Different Media Types, URN Sub-Namespace Registration for mrb-consumer, and XML Schema Registration for mrb-publish and mrb-consumer. Interested readers need to see RFC 6917.

11.15 PROBLEMS

1. What are the problems of the centralized MS architecture in XCOM conference systems? Draw XCON conference systems and justify all the technical issues for a large-scale network.
2. How does the CFW framework become relevant in solving the problems of the centralized MS of XCON conference systems? How do you use the CFW framework in finding a solution like MRB in many-many AS and MS relationships? Justify your answer in detailed discussion using architectural schematic diagrams.
3. What is the role of SIP signaling as an inter-server communications protocol for XCON conference systems in view of the 3pcc? Describe in detail.
4. What is meant by 1:M one-many, M:1 many-to-one, and M:N many-many relationships between ASs and MSs/resources? Draw an architectural relationship and explain in detail.
5. What is meant by different MRB deployments: Query MRB, In-line MRB, and Hybrid Query MRB? Explain in detail using deployment configurations in relationship to XCON conference systems architecture, shown in Section 1.5.

6. Explain in detail the differences between IUMM and IAMM MRB architecture with detail of different MRB deployments in the context of XCON conference systems architecture, shown in Section 1.5.
7. What is meant by an MRB interface? What are these three MRB interfaces: MS Resource Publish Interface, Media Service Resource Consumer Interface, and In-Line Unaware MRB Interface? Explain in detail, drawing the logical architectural diagrams for each of these interfaces.
8. What is meant by multimodal MRB implementation? Explain by drawing the logical architecture in relationship to XCON conference systems architecture, shown in Section 1.5.
9. Explain in detail all the features that have been articulated in XML schemas for both Media Service Resource Publisher and Consumer Interfaces.
10. What do we do to prevent malicious attacks against Public and Consumer MRB Interfaces? Explain in detail.
11. How do you prevent attacks changing the contents of SIP signaling messages in view of the fact that both malicious users and MRB interfaces, being B2BUA, are able to do so? Explain in detail.
12. How does the use of pre-shared certificates to assess the identity of legitimate ASs and/or MSs help to improve security of the MRB interfaces? Explain the relative merits of this security scheme in the context of the XCON conference systems architecture shown in Section 1.5.

REFERENCES

1. International Organization for Standardization. (2012). Information Technology – Universal Coded Character Set (UCS). *ISO Standard 10646.*
2. International Organization for Standardization. (2006). Codes for the Representation of Names of Countries and Their Subdivisions - Part 1: Country Codes. *ISO Standard 3166-1:2006.*
3. International Telecommunication Union. (2002, December). Bearer Independent Call Bearer Control Protocol. *ITU-T Recommendation Q.1950.*
4. [ISO.639.2002] International Organization for Standardization. (2002). Codes for the Representation of Names of Languages – Part 1: Alpha-2 Code. *ISO Standard 639.*
5. Thompson, H. et al. (2004, October). XML Schema Part 1: Structures Second Edition. *World Wide Web Consortium Recommendation REC-xmlschema-1-20041028.* <http://www.w3.org/TR/2004/REC-xmlschema-1-20041028>.
6. Biron, P. and A. Malhotra. (2004, October). XML Schema Part 2: Datatypes Second Edition. *World Wide Web Consortium Recommendation REC-xmlschema-2-20041028.* <http://www.w3.org/TR/2004/REC-xmlschema-2-20041028>.

12 Media Control Channel Framework for Interactive Voice Response

12.1 INTRODUCTION

The Media Control Channel Framework (RFC 6230 – see Chapter 8) provides a generic approach for establishment and reporting capabilities of remotely initiated commands. The Channel Framework – an equivalent term for the Media Control Channel Framework – utilizes many functions provided by the session initiation protocol (SIP) (RFC 3261 – also see Section 1.2) for the rendezvous and establishment of a reliable channel for control interactions. The Control Framework also introduces the concept of a Control Package. A Control Package is an explicit usage of the Control Framework for a particular interaction set. This specification defines a Control Package for interactive voice response (IVR) dialogs on media connections and conferences. The term "dialog" in this specification refers to an IVR dialog and is completely unrelated to the notion of an SIP dialog. The term "IVR" is used in its inclusive sense, allowing media other than voice for dialog interaction.

The package (RFC 6231) that is described in this Chapter defines dialog management request elements for preparing, starting, and terminating dialog interactions as well as associated responses and notifications. Dialog interactions are specified using a dialog language, where the language specifies a well-defined syntax and semantics for permitted operations (play a prompt, record input from the user, etc.). This package defines a lightweight IVR dialog language (supporting prompt playback, runtime controls, dual-tone multi-frequency (DTMF) collection, and media recording) and allows other dialog languages to be used. These dialog languages are specified inside dialog management elements for preparing and starting dialog interactions. The package also defines elements for auditing package capabilities and IVR dialogs. This package has been designed to satisfy IVR requirements specified in "Media Server Control Protocol Requirements" (RFC 5167 – see Section 8.2) – more specifically, REQ-MCP-28, REQ-MCP-29, and REQ-MCP-30. It achieves this by building upon two major approaches to IVR dialog design.

These approaches address a wide range of IVR use cases and are used in many applications that are extensively deployed today. First, the package is designed to provide the major IVR functionality of SIP media server languages such as netann (RFC 4240), Media Server Control Markup Language (MSCML) (RFC 5022), and Media Server Markup Language (MSML) (RFC5707), which themselves build upon more traditional non-SIP languages ([1], RFC 2897). A key differentiator is that this package provides IVR functionality using the Channel Framework. Second, its design is aligned with key concepts of the web model as defined in W3C Voice Browser languages. The key dialog management mechanism is closely aligned with Call Control XML (CCXML) [2].

The dialog functionality defined in this package can be largely seen as a subset of Voice Extensible Markup Language (VoiceXML) [3, 4]: where possible, basic prompting, DTMF collection, and media recording features are incorporated, but not any advanced VoiceXML constructs (such as <form>, its interpretation algorithm, or a dynamic data model). As W3C develops VoiceXML 3.0 [5], we expect to see further alignment, especially in providing a set of basic independent primitive elements (such as prompt, collect, record, and runtime controls) that can be reused in different dialog languages.

By reusing and building upon design patterns from these approaches to IVR languages, this package is intended to provide a foundation that is familiar to current IVR developers and sufficient for most IVR applications, as well as a path to other languages that address more advanced applications. This Control Package defines a lightweight IVR dialog language. The scope of this dialog language is the following IVR functionality:

- playing one or more media resources as a prompt to the user
- runtime controls (including videocassette recorder (VCR) controls like speed and volume)
- collecting DTMF input from the user according to a grammar
- recording user media input

Out of scope for this dialog language are more advanced functions including ASR (automatic speech recognition), TTS (text-to-speech), fax, automatic prompt recovery ("media fallback"), and media transformation. Such functionality can be addressed by other dialog languages (such as VoiceXML) used with this package, extensions to this package (addition of foreign elements or attributes from another namespace), or other Control Packages. The functionality of this package is defined by messages, containing XML [6] elements, transported using the Media Control Channel Framework. The XML elements can be divided into three types: dialog management elements; a dialog element that defines a lightweight IVR dialog language used with dialog management elements; and finally, elements for auditing package capabilities as well as dialogs managed by the package.

Dialog management elements are designed to manage the general lifecycle of a dialog. Elements are provided for preparing a dialog, starting the dialog on a conference or connection, and terminating the execution of a dialog. Each of these elements is contained in a Media Control Channel Framework CONTROL message sent to the Media Server. When the appropriate action has been executed, the Media Server sends a REPORT message (or a 200 response to the CONTROL message if it can execute in time) with a response element indicating whether or not the operation was successful (e.g., if the dialog cannot be started, then the error is reported in this response).

Once a dialog has been successfully started, the Media Server can send further event notifications in a framework CONTROL message. This package defines two event notifications: a DTMF event indicating the DTMF activity, and a dialogexit event indicating that the dialog has exited. If the dialog has executed successfully, the dialogexit event includes information collected during the dialog. If an error occurs during execution (e.g., a media resource failed to play, no recording resource available, etc.), then error information is reported in the dialogexit event. Once a dialogexit event is sent, the dialog lifecycle is terminated.

The dialog management elements for preparing and starting a dialog specify the dialog using a dialog language. A dialog language has well-defined syntax and semantics for defined dialog operations. Typically, dialog languages are written in XML, where the root element has a designated XML namespace, and when used as standalone specifications, have an associated MIME media type. For example, VoiceXML is an XML dialog language with the root element <vxml> with the designated namespace "http://www.w3.org/2001/vxml," and standalone specifications are associated with the MIME media type "application/ voicexml+xml" (RFC 4267).

This Control Package defines its own lightweight IVR dialog language. The language has a root element (<dialog>) with the same designated namespace as used for other elements defined in this package (see Section 12.8.2). The root element contains child elements for playing prompts to the user, specifying runtime controls, collecting DTMF input from the user, and recording media input from the user. The child elements can co-occur so as to provide "play announcement," "prompt and collect," and "prompt and record" functionality. The dialog management elements for preparing and starting a dialog can specify the dialog language either by including inline a fragment with the root element or by referencing an external dialog specification. The dialog language defined in this package is specified inline. Other dialog languages, such as VoiceXML, can be used by referencing an external dialog specification.

The specification is organized as follows. Section 12.3 describes how this Control Package fulfills the requirements for a Media Control Channel Framework Control Package. Section 12.4 describes the syntax and semantics of defined elements, including dialog management (Section 12.4.2), the IVR dialog element (Section 12.4.3), and audit elements (Section 12.4.4). Section 12.5 describes an XML schema for these elements and provides extensibility by allowing attributes and elements from other namespaces. Section 12.6 provides examples of package usage. Section 12.7 describes important security considerations for use of this Control Package. Section 12.8 provides information on IANA registration of this Control Package, including its name, XML namespace, and MIME media type. It also establishes a registry for prompt variables. Finally, Section 12.9 provides additional information on using VoiceXML when supported as an external dialog language.

12.2 CONVENTIONS AND TERMINOLOGY

The following additional terms are defined for use in this specification (also see Section 1.6, Table 12.2):

- Dialog: A dialog performs media interaction with a user following the concept of an IVR dialog (this sense of "dialog" is completely unrelated to an SIP dialog). A dialog is specified as inline XML or via a uniform resource identifier (URI) reference to an external dialog specification. Traditional IVR dialogs typically feature capabilities such as playing audio prompts, collecting DTMF input, and recording audio input from the user. More inclusive definitions include support for other media types, runtime controls, synthesized speech, recording and playback of video, recognition of spoken input, and mixed initiative conversations.
- Application Server: An SIP (RFC 3261 – also see Section 1.2) Application Server (AS) hosts and executes services such as interactive media and conferencing in an operator's network. An AS influences and impacts the SIP session, in particular by terminating SIP sessions on a Media Server that is under its control.
- Media Server: A Media Server (MS) processes media streams on behalf of an AS by offering functionality such as interactive media, conferencing, and transcoding to the end user. Interactive media functionality is realized by way of dialogs that are initiated by the AS.

12.3 CONTROL PACKAGE DEFINITION

This section fulfills the mandatory requirement for information that MUST be specified during the definition of a Control Framework Package, as detailed in Section 8.7 (Section 7 of RFC 6230).

12.3.1 Control Package Name

The Control Framework requires a Control Package to specify and register a unique name. The name of this Control Package is "msc-ivr/1.0" (Media Server Control – Interactive Voice Response – version 1.0). IANA registration is specified in Section 12.8. Since this is the initial ("1.0") version of the Control Package, there are no backwards-compatibility issues to address.

12.3.2 Framework Message Usage

The Control Framework requires a Control Package to explicitly detail the CONTROL messages that can be used as well as to provide an indication of directionality between entities. This will include which role type is allowed to initiate a request type. This package specifies Control and response messages in terms of XML elements defined in Section 12.4, where the message bodies have the MIME media type defined in Section 8.4. These elements describe requests, responses, and notifications, and all are contained within a root <mscivr> element (Section 12.4.1).

In this package, the MS operates as a Control Server in receiving requests from, and sending responses to, the AS (operating as Control Client). Dialog management requests and responses are defined in Section 12.4.2. Audit requests and responses are defined in Section 12.4.4. Dialog management and audit responses are carried in a framework 200 response or REPORT message bodies. This package's response codes are defined in Section 12.4.5. Note that package responses are different from framework response codes. Framework error response codes (see Section 8.7 [Section 7 of RFC 6230]) are used when the request or event notification is invalid; for example, a request is invalid XML (400) or not understood (500). The MS also operates as a Control Client in sending event notification to the AS (Control Server). Event notifications (Section 12.4.2.5) are carried in CONTROL message bodies. The AS MUST respond with a Control Framework 200 response.

12.3.3 Common XML Support

The Control Framework requires a Control Package definition to specify whether the attributes for media dialog or conference references are required. This package requires that the XML schema in Section 8.14.1 (Appendix A.1 of RFC 6230) MUST be supported for media dialogs and conferences. The package uses "connectionid" and "conferenceid" attributes for various element definitions (Section 12.4). The XML schema (Section 12.5) imports the definitions of these attributes from the framework schema.

12.3.4 CONTROL Message Body

The Control Framework requires a Control Package to define the control body that can be contained within a CONTROL command request and to indicate the location of detailed syntax definitions and semantics for the appropriate body types. When operating as Control Server, the MS receives Control message bodies with the MIME media type defined in Section 12.8.4 and containing an <mscivr> element (Section 14.4.1) with either a dialog management or an audit request child element. The following dialog management request elements are carried in CONTROL message bodies to the MS: <dialogprepare> (Section 12.4.2.1), <dialogstart> (Section 12.4.2.2), and <dialogterminate> (Section 12.4.2.3) elements. The <audit> request element (Section 12.4.4.1) is also carried in CONTROL message bodies. When operating as Control Client, the MS sends CONTROL messages with the MIME media type defined in Section 8.4 and a body containing an <mscivr> element (Section 12.4.1) with a notification <event> child element (Section 12.4.2.5).

12.3.5 REPORT Message Body

The Control Framework requires a Control Package definition to define the REPORT body that can be contained within a REPORT command request, or that no report package body is required. This section indicates the location of detailed syntax definitions and semantics for the appropriate body types. When operating as Control Server, the MS sends REPORT bodies with the MIME media type defined in Section 12.8.4 and containing a <mscivr> element (Section 12.4.1) with a response child element. The response element for dialog management requests is a <response> element (Section 12.4.2.4). The response element for an audit request is an <auditresponse> element (Section 12.4.4.2).

12.3.6 Audit

The Control Framework encourages Control Packages to specify whether auditing is available and how it is triggered as well as the query/response formats. This Control Package supports auditing of package capabilities and dialogs on the MS. An audit request is carried in a CONTROL message (see Section 3.4) and an audit response in a REPORT message (or a 200 response to the CONTROL if it can execute the audit in time) (see Section 12.3.5). The syntax and semantics of audit request and response elements are defined in Section 12.4.4.

12.3.7 Examples

The Control Framework recommends Control Packages to provide a range of message flows that represent common flows using the package and this framework specification. This Control Package provides examples of such message flows in Section 12.6.

12.4 ELEMENT DEFINITIONS

This section defines the XML elements for this package. The elements are defined in the XML namespace specified in Section 8.2. The root element is <mscivr> (Section 12.4.1). All other XML elements (requests, responses, and notification elements) are contained within it. Child elements describe dialog management (Section 12.4.2) and audit (Section 12.4.4) functionality. The IVR dialog element (contained within dialog management elements) is defined in Section 12.4.3. Response status codes are defined in Section 12.4.5 and type definitions in Section 12.4.6. Implementation of this Control Package MUST address the security considerations described in Section 12.7.

Implementation of this Control Package MUST adhere to the syntax and semantics of XML elements defined in this section and the schema (Section 12.5). Since XML schema is unable to support some types of syntactic constraints (such as attribute and element co-occurrence), some elements in this package specify additional syntactic constraints in their textual definition. If there is a difference in constraints between the XML schema and the textual description of elements in this section, the textual definition takes priority. The XML schema supports extensibility by allowing attributes and elements from other namespaces. Implementations MAY support additional capabilities by means of attributes and elements from other (foreign) namespaces. Attributes and elements from foreign namespaces are not described in this section.

Some elements in this Control Package contain attributes whose value is a URI. These elements include: <dialogprepare> (Section 12.4.2.1), <dialogstart> (Section 12.4.2.2), <media> (Section 12.4.3.1.5), <grammar> (Section 12.4.3.1.3.1), and <record> (Section 12.4.3.1.4). The MS MUST support both HTTP (RFC 2616) and HTTPS (RFC 2818) protocol schemes for fetching and uploading resources, and the MS MAY support other schemes. The implementation SHOULD support storage of authentication information as part of its configuration, including security certificates for use with HTTPS. If the implementation wants to support user authentication, user certifications and passwords can also be stored as part of its configuration, or the implementation can extend the schema (adding, for example, an http-password attribute in its own namespace) and then map user authentication information onto the appropriate headers following the HTTP authentication model (RFC 2616).

Some elements in this Control Package contain attributes whose value is descriptive text primarily for diagnostic use. The implementation can indicate the language used in the descriptive text by means of a "desclang" attribute (RFCs 2277 and 5646). The desclang attribute can appear on the root element as well as selected subordinate elements (see Section 12.4.1). The desclang attribute value on the root element applies to all desclang attributes in subordinate elements unless the subordinate element has an explicit desclang attribute that overrides it. Usage examples are provided in Section 12.6.

12.4.1 <mscivr>

The <mscivr> element has the following attributes (in addition to standard XML namespace attributes such as xmlns): version: a string specifying the mscivr package version. The value is fixed as "1.0" for this version of the package. The attribute is mandatory. desclang: specifies the language used in descriptive text attributes of subordinate elements (unless the subordinate element provides a desclang attribute that overrides the value for its descriptive text attributes). The descriptive text attributes on subordinate elements include: the reason attribute on <response> (Section 14.4.2.4), <dialogexit> (Section 14.4.2.5.1), and <auditresponse> (Section 12.4.4.2); desc attribute on <variabletype> and <format> (Section 12.4.4.2.2.5.1). A valid value is a language identifier (Section 12.4.6.11). The attribute is optional. The default value is i-default (BCP 47 [RFC 5646]). The <mscivr> element has the following defined child elements, only one of which can occur:

1. dialog management elements defined in Section 12.4.2: <dialogprepare> prepare a dialog. See Section 12.4.2.1. <dialogstart> start a dialog. See Section 12.4.2.2. <dialogterminate> terminate a dialog. See Section 12.4.2.3. <response> response to a dialog request. See Section 12.4.2.4. <event> dialog or subscription notification. See Section 12.4.2.5.
2. audit elements defined in Section 12.4.4: <audit> audit package capabilities and managed dialogs. See Section 12.4.4.1. <auditresponse> response to an audit request. See Section 12.4.4.2.

For example, a request to the MS to start an IVR dialog playing a prompt:

```
<mscivr version="1.0" xmlns="urn:ietf:params:xml:ns:msc-ivr">
   <dialogstart connectionid="ssd3r3:sds345b">
      <dialog>
         <prompt>
            <media loc="http://www.example.com/
                  welcome.wav"/>
         </prompt>
      </dialog>
   </dialogstart>
</mscivr>
```

and a response from the MS that the dialog started successfully:

```
<mscivr version="1.0" xmlns="urn:ietf:params:xml:ns:msc-ivr">
   <response status="200" dialogid="d1"/>
</mscivr>
```

and finally, a notification from the MS indicating that the dialog exited upon completion of playing the prompt:

```
<mscivr version="1.0" xmlns="urn:ietf:params:xml:ns:msc-ivr"
      desclang="en">
   <event dialogid="d1">
      <dialogexit status="1" reason="successful
            completion of the dialog">
         <promptinfo termmode="completed"/>
      </dialogexit>
   </event>
</mscivr>
```

The language of the descriptive text in the reason attribute of <dialogexit> is explicitly indicated by the desclang attribute of the <mscivr> root element.

12.4.2 Dialog Management Elements

This section defines the dialog management XML elements for this Control Package. These elements are divided into requests, responses, and notifications. Request elements are sent to the MS to request a specific dialog operation to be executed. The following request elements are defined:

- <dialogprepare>: prepare a dialog for later execution
- <dialogstart>: start a (prepared) dialog on a connection or conference
- <dialogterminate>: terminate a dialog

Responses from the MS describe the status of the requested operation. Responses are specified in a <response> element (Section 12.4.2.4) that includes a mandatory attribute describing the status in terms of a numeric code. Response status codes are defined in Section 12.4.5. The MS MUST respond to a request message with a response message. If the MS is not able to process the request and carry out the dialog operation, the request has failed, and the MS MUST indicate the class of failure using an appropriate 4xx response code. Unless an error response code is specified for a class of error within this section, implementations follow Section 12.4.5 in determining the appropriate status code for the response. Notifications are sent from the MS to provide updates on the status of a dialog or operations defined within the dialog. Notifications are specified in an <event> element (Section 12.4.2.5).

The MS implementation MUST adhere to the dialog lifecycle shown in Figure 12.1, where each dialog has the following states:

- IDLE: The dialog is uninitialized.
- PREPARING: The dialog is being prepared. The dialog is assigned a valid dialog identifier (see later). If an error occurs, the dialog transitions to the TERMINATED state, and the MS MUST send a response indicating the error. If the dialog is terminated before preparation is complete, the dialog transitions to the TERMINATED state, and the MS MUST send a 410 response (Section 12.4.5) for the prepare request.
- PREPARED: The dialog has been successfully prepared, and the MS MUST send a 200 response indicating that the prepare operation was successful. If the dialog is terminated, then the MS MUST send a 200 response, the dialog transitions to the TERMINATED state, and the MS MUST send a dialogexit notification event (see Section 12.4.2.5.1). If the duration for which the dialog remains in the PREPARED state exceeds the maximum preparation duration, the dialog transitions to the TERMINATED state, and the MS MUST send a dialogexit notification with the appropriate error status code (see Section 12.4.2.5.1). A maximum preparation duration of 300 s is RECOMMENDED.
- STARTING: The dialog is being started. If the dialog has not already been prepared, it is first prepared and assigned a valid dialog identifier (see later). If an error occurs, the dialog transitions to the TERMINATED state, and the MS MUST send a response indicating the error. If the dialog is terminated, the dialog transitions to the TERMINATED state, and the MS MUST send a 410 response (Section 12.4.5) for the start request.
- STARTED: The dialog has been successfully started and is now active. The MS MUST send a 200 response indicating that the start operation was successful. If any dialog events occur that were subscribed to, the MS MUST send a notification when the dialog event occurs. When the dialog exits (due to normal termination, an error, or a terminate request), the MS MUST send a dialogexit notification event (see Section 12.4.2.5.1), and the dialog transitions to the TERMINATED state.
- TERMINATED: the dialog is terminated, and its dialog identifier is no longer valid. Dialog notifications MUST NOT be sent for this dialog. Each dialog has a valid identifier until it

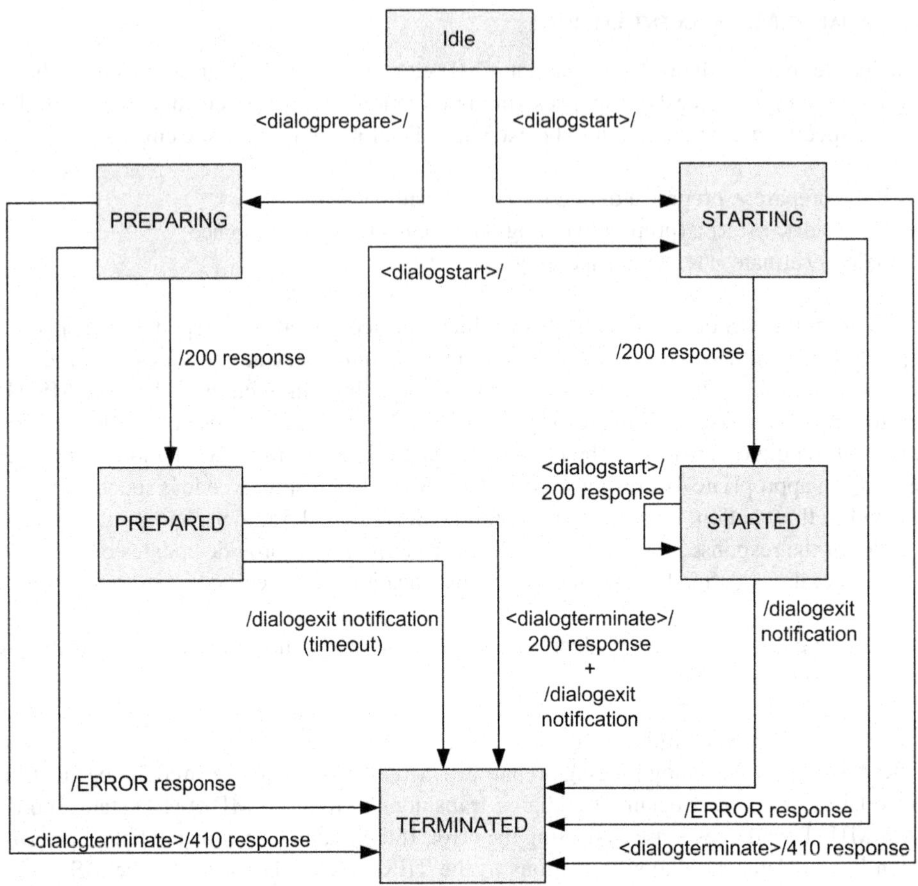

FIGURE 12.1 Dialog Lifecycle (Copyright: IETF)

transitions to a TERMINATED state. The dialog identifier is assigned by the MS unless the <dialogprepare> or <dialogstart> request already specifies a identifier (dialogid) that is not associated with any other dialog on the MS. Once a dialog is in a TERMINATED state, its dialog identifier is no longer valid and can be reused for another dialog.

The identifier is used to reference the dialog in subsequent requests, responses, and notifications. In a <dialogstart> request, the dialog identifier can be specified in the prepareddialogid attribute indicating the prepared dialog to start. In <dialogterminate> and <audit> requests, the dialog identifier is specified in the dialogid attribute, indicating which dialog is to be terminated or audited, respectively. If these requests specify a dialog identifier already associated with another dialog on the MS, the MS sends a response with a 405 status code (see Section 12.4.5) and the same dialogid as in the request. The MS MUST specify a dialog identifier in notifications associated with the dialog. The MS MUST specify a dialog identifier in responses unless it is a response to a syntactically invalid request.

For a given dialog, the <dialogprepare> or <dialogstart> request elements specify the dialog content to execute either by including inline a <dialog> element (the dialog language defined in this package; see Section 12.4.3) or by referencing an external dialog specification (a dialog language defined outside this package). When referencing an external dialog specification, the request element contains a URI reference to the remote specification (specifying the dialog definition) and optionally, a type attribute indicating the MIME media type associated with the dialog specification.

Consequently, the dialog language associated with a dialog on the MS is identified either inline by a <dialog> child element or by an src attribute referencing a specification containing the dialog language. The MS MUST support inline the IVR dialog language defined in Section 12.4.3. The MS MAY support other dialog languages by reference.

12.4.2.1 <dialogprepare>

The <dialogprepare> request is sent to the MS to request preparation of a dialog. Dialog preparation consists of (a) retrieving an external dialog specification and/or external resources referenced within an inline <dialog> element and (b) validating the dialog specification syntactically and semantically. A prepared dialog is executed when the MS receives a <dialogstart> request referencing the prepared dialog identifier (see Section 12.4.2.2). The <dialogprepare> element has the following attributes:

- src: Specifies the location of an external dialog specification to prepare. A valid value is a URI (see Section 12.4.6.9). The MS MUST support both HTTP (RFC 2616) and HTTPS (RFC 2818) schemes, and the MS MAY support other schemes. If the URI scheme is unsupported, the MS sends a <response> with a 420 status code (Section 12.4.5). If the specification cannot be retrieved within the timeout interval, the MS sends a <response> with a 409 status code. If the specification contains a type of dialog language that the MS does not support, the MS sends a <response> with a 421 status code. The attribute is optional. There is no default value.
- type: Specifies the type of the external dialog specification indicated in the "src" attribute. A valid value is a MIME media type (see Section 12.4.6.10). If the URI scheme used in the src attribute defines a mechanism for establishing the authoritative MIME media type of the media resource, the value returned by that mechanism takes precedence over this attribute. The attribute is optional. There is no default value.
- maxage: Used to set the max-age value of the "Cache-Control" header in conjunction with an external dialog specification fetched using HTTP, as per RFC 2616. A valid value is a non-negative integer (see Section 12.4.6.4). The attribute is optional. There is no default value.
- maxstale: Used to set the max-stale value of the "Cache-Control" header in conjunction with an external dialog specification fetched using HTTP, as per RFC 2616. A valid value is a non-negative integer (see Section 12.4.6.4). The attribute is optional. There is no default value.
- fetchtimeout: The maximum timeout interval to wait when fetching an external dialog specification. A valid value is a Time Designation (see Section 12.4.6.7). The attribute is optional. The default value is 30 s.
- dialogid: String indicating a unique name for the dialog. If a dialog with the same name already exists on the MS, the MS sends a <response> with a 405 status code (Section 12.4.5). If this attribute is not specified, the MS MUST create a unique name for the dialog (see Section 12.4.2 for dialog identifier assignment). The attribute is optional. There is no default value. The <dialogprepare> element has the following sequence of child elements: <dialog> an IVR dialog (Section 12.4.3) to prepare. The element is optional.
- <params>: Specifies input parameters (Section 12.4.2.6) for dialog languages defined outside this specification. The element is optional. If a parameter is not supported by the MS for the external dialog language, the MS sends a <response> with a 427 status code (Section 12.4.5).

The dialog to prepare can be specified either inline with a <dialog> child element or externally (for dialog languages defined outside this specification) using the src attribute. It is a syntax error if both an inline <dialog> element and a src attribute are specified, and the MS sends a <response> with a 400 status code (see Section 12.4.5). The type, maxage, maxstale, and fetchtimeout attributes are only relevant when a dialog is specified as an external specification. For example, a <dialogprepare> request to prepare an inline IVR dialog with a single prompt:

```
<mscivr version="1.0" xmlns="urn:ietf:params:xml:ns:msc-ivr">
   <dialogprepare>
      <dialog>
         <prompt>
            <media loc="http://www.example.com/welcome.wav"/>
         </prompt>
      </dialog>
   </dialogprepare>
</mscivr>
```

In this example, a request with a specified dialogid to prepare a VoiceXML dialog specification located externally:

```
<mscivr version="1.0" xmlns="urn:ietf:params:xml:ns:msc-ivr">
   <dialogprepare dialogid="d2" type="application/voicexml+xml"
         src="http://www.example.com/mydialog.vxml"
         fetchtimeout="15s"/>
</mscivr>
```

Since MS support for dialog languages other than the IVR dialog language defined in this package is optional, if the MS does not support the dialog language, it will send a response with the status code 421 (Section 12.4.5). Further information on using VoiceXML can be found in Section 12.9.

12.4.2.2 <dialogstart>

The <dialogstart> element is sent to the MS to start a dialog. If the dialog has not been prepared, the dialog is prepared (retrieving external specification and/or external resources referenced within <dialog> element and the dialog specification validated syntactically and semantically). Media processors (e.g., DTMF and prompt queue) are activated and associated with the specified connection or conference. The <dialogstart> element has the following attributes:

- src: Specifies the location of an external dialog specification to start. A valid value is a URI (see Section 12.4.6.9). The MS MUST support both HTTP (RFC 2616) and HTTPS (RFC 2818) schemes, and the MS MAY support other schemes. If the URI scheme is unsupported, the MS sends a <response> with a 420 status code (Section 12.4.5). If the specification cannot be retrieved with the timeout interval, the MS sends a <response> with a 409 status code. If the specification contains a type of dialog language that the MS does not support, the MS sends a <response> with a 421 status code. The attribute is optional. There is no default value.
- type: Specifies the type of the external dialog specification indicated in the "src" attribute. A valid value is a MIME media type (see Section 12.4.6.10). If the URI scheme used in the src attribute defines a mechanism for establishing the authoritative MIME media type of the media resource, the value returned by that mechanism takes precedence over this attribute. The attribute is optional. There is no default value.
- maxage: Used to set the max-age value of the "Cache-Control" header in conjunction with an external dialog specification fetched using HTTP, as per RFC 2616. A valid value is a non-negative integer (see Section 12.4.6.4). The attribute is optional. There is no default value.
- maxstale: Used to set the max-stale value of the "Cache-Control" header in conjunction with an external dialog specification fetched using HTTP, as per RFC 2616. A valid value

is a non-negative integer (see Section 12.4.6.4). The attribute is optional. There is no default value.

- fetchtimeout: The maximum timeout interval to wait when fetching an external dialog specification. A valid value is a Time Designation (see Section 12.4.6.7). The attribute is optional. The default value is 30 s.
- dialogid: String indicating a unique name for the dialog. If a dialog with the same name already exists on the MS, the MS sends a <response> with a 405 status code (Section 12.4.5). If neither the dialogid attribute nor the "prepareddialogid" attribute is specified, the MS MUST create a unique name for the dialog (see Section 12.4.2 for dialog identifier assignment). The attribute is optional. There is no default value.
- prepareddialogid: String identifying a dialog previously prepared using a dialogprepare (Section 12.4.2.1) request. If neither the dialogid attribute nor the prepareddialogid attribute is specified, the MS MUST create a unique name for the dialog (see Section 12.4.2 for dialog identifier assignment). The attribute is optional. There is no default value.
- connectionid: String identifying the SIP dialog connection on which this dialog is to be started (see Section 8.14.1 [Appendix A.1 of RFC 6230]). The attribute is optional. There is no default value.
- conferenceid: String identifying the conference on which this dialog is to be started (see Section 8.14.1 [Appendix A.1 of RFC 6230]). The attribute is optional. There is no default value. Exactly one of the connectionid or conferenceid attributes MUST be specified. If both the connectionid and conferenceid attributes are specified or neither is specified, it is a syntax error, and the MS sends a <response> with a 400 status code (Section 12.4.5). It is an error if the connection or conference referenced by a specific connectionid or conferenceid attribute is not available on the MS at the time the <dialogstart> request is executed. If an invalid connectionid is specified, the MS sends a <response> with a 407 status code (Section 12.4.5). If an invalid conferenceid is specified, the MS sends a <response> with a 408 status code. The <dialogstart> element has the following sequence of child elements:
  - <dialog>: Specifies an IVR dialog (Section 12.4.3) to execute. The element is optional.
  - <subscribe>: Specifies subscriptions to dialog events (Section 12.4.2.2.1). The element is optional.
  - <params>: Specifies input parameters (Section 12.4.2.6) for dialog languages defined outside this specification. The element is optional. If a parameter is not supported by the MS for the external dialog language, the MS sends a <response> with a 427 status code (Section 12.4.5).
  - <stream>: Determines the media stream(s) associated with the connection or conference on which the dialog is executed (Section 12.4.2.2.2). The <stream> element is optional. Multiple <stream> elements can be specified.

The dialog to start can be specified either (a) inline with a <dialog> child element, (b) externally using the src attribute (for dialog languages defined outside this specification), or (c) by referencing a previously prepared dialog using the prepareddialogid attribute. If exactly one of the src attribute, the prepareddialogid, or a <dialog> child element is not specified, it is a syntax error, and the MS sends a <response> with a 400 status code (Section 12.4.5). If the prepareddialogid and dialogid attributes are specified, it is also a syntax error, and the MS sends a <response> with a 400 status code. The type, maxage, maxstale, and fetchtimeout attributes are only relevant when a dialog is specified as an external specification.

The <stream> element provides explicit control over which media streams on the connection or conference are used during dialog execution. For example, if a connection supports both audio and video streams, a <stream> element could be used to indicate that only the audio stream is used in receive mode. In cases where there are multiple media streams of the same type for a dialog, the AS MUST use <stream> elements to explicitly specify the configuration. If no <stream> elements are specified, then the default media configuration is that defined for the connection or conference. If a <stream> element is in conflict (a) with another <stream> element, (b) with specified connection or conference media capabilities, or (c) with a session description protocol (SDP) label value as part of the connectionid (see Section 8.14.1 [Appendix A.1 of RFC 6230]), then the MS sends a <response> with a 411 status code (Section 12.4.5).

If the media stream configuration is not supported by the MS, then the MS sends a <response> with a 428 status code (Section 12.4.5). The MS MAY support multiple simultaneous dialogs being started on the same connection or conference. For example, the same connection can receive different media streams (e.g., audio and video) from different dialogs or receive (and implicitly mix where appropriate) the same type of media streams from different dialogs. If the MS does not support starting another dialog on the same connection or conference, it sends a <response> with a 432 status code (Section 12.4.5) when it receives the second (or subsequent) dialog request. For example, a request to start an ivr dialog on a connection subscribing to DTMF notifications:

```
<mscivr version="1.0" xmlns="urn:ietf:params:xml:ns:msc-ivr">
   <dialogstart connectionid="connection1">
      <dialog>
         <prompt>
            <media loc="http://www.example.com/
                  getpin.wav"/>
         </prompt>
         <collect maxdigits="2"/>
      </dialog>
      <subscribe>
         <dtmfsub matchmode="all"/>
      </subscribe>
   </dialogstart>
</mscivr>
```

In this example, the dialog is started on a conference where the conference only receives an audio media stream from the dialog:

```
<mscivr version="1.0" xmlns="urn:ietf:params:xml:ns:msc-ivr">
   <dialogstart conferenceid="conference1">
      <dialog>
         <record maxtime="384000s"/>
      </dialog>
      <stream media="audio" direction="recvonly"/>
   </dialogstart>
</mscivr>
```

12.4.2.2.1 <subscribe>

The <subscribe> element allows the AS to subscribe to, and be notified of, specific events that occur during execution of the dialog. Notifications of dialog events are delivered using the <event> element (see Section 12.4.2.5). The <subscribe> element has no attributes. The <subscribe> element has the following sequence of child elements (0 or more occurrences):

- <dtmfsub>: Subscription to DTMF input during the dialog (Section 12.4.2.2.1.1). The element is optional.

If a request has a <subscribe> with no child elements, the MS treats the request as if no <subscribe> element were specified. The MS MUST support <dtmfsub> subscription for the IVR dialog language defined in this specification (Section 12.4.3). It MAY support other dialog subscriptions (specified using attributes and child elements from a foreign namespace). If the MS does not support a subscription specified in a foreign namespace, the MS sends a response with a 431 status code (see Section 12.4.5).

12.4.2.2.1.1 <dtmfsub> The <dtmfsub> element has the following attributes: matchmode: controls which DTMF input is subscribed to. Valid values are "all" – notify all DTMF key presses received during the dialog; "collect" – notify only DTMF input matched by the collect operation (Section 12.4.3.1.3); and "control" – notify only DTMF input matched by the runtime control operation (Section 12.4.3.1.2). The attribute is optional. The default value is "all." The <dtmfsub> element has no child elements. DTMF notifications are delivered in the <dtmfnotify> element (Section 12.4.2.5.2). For example, the AS wishes to subscribe to DTMF key press matching a runtime control:

```
<mscivr version="1.0" xmlns="urn:ietf:params:xml:ns:msc-ivr">
   <dialogstart dialogid="d3" connectionid="connection1">
      <dialog>
         <prompt>
            <media loc="http://www.example.com/
                  getpin.wav"/>
         </prompt>
         <control ffkey="2" rwkey="3"/>
      </dialog>
      <subscribe>
         <dtmfsub matchmode="control"/>
      </subscribe>
   </dialogstart>
</mscivr>
```

Each time a "2" or "3" DTMF input is received, the MS sends a notification event:

```
<mscivr version="1.0" xmlns="urn:ietf:params:xml:ns:msc-ivr">
   <event dialogid="d3">
      <dtmfnotify matchmode="collect" dtmf="2"
            timestamp="2008-05-12T12:13:14Z"/>
   </event>
</mscivr>
```

12.4.2.2.2 <stream>

The <stream> element has the following attributes:

- media: A string indicating the type of media associated with the stream. A valid value is a MIME type-name as defined in Section 4.2 of RFC 4288. The following values MUST be used for common types of media: "audio" for audio media and "video" for video media. See [7] for registered MIME type names. The attribute is mandatory.
- label: A string indicating the SDP label associated with a media stream RFC 4574. The attribute is optional. direction: A string indicating the direction of the media flow relative to the endpoint conference or connection. Defined values are "sendrecv" (the endpoint can send media to, and receive media from, the dialog), "sendonly" (the endpoint can only send media to the dialog), "recvonly" (the endpoint can only receive media from the dialog), and "inactive" (stream is not to be used). The default value is "sendrecv." The attribute is optional.

The <stream> element has the following sequence of child elements:

- <region>: An element to specify the area within a mixer video layout where a media stream is displayed (Section 12.4.2.2.2.1). The element is optional.
- <priority>: An element to configure priority associated with the stream in the conference mix (Section 12.4.2.2.2.2). The element is optional.

If conferenceid is not specified, or if the "media" attribute does not have the value of "video," then the MS ignores the <region> and <priority> elements. For example, assume a user agent connection with multiple audio and video streams associated with the user and a separate web camera. In this case, the dialog could be started to record only the audio and video streams associated with the user:

```
<mscivr version="1.0" xmlns="urn:ietf:params:xml:ns:msc-ivr">
   <dialogstart connectionid="connection1">
      <dialog>
         <record maxtime="384000s"/>
      </dialog>
      <stream media="audio" label="camaudio" direction="inactive"/>
      <stream media="video" label="camvideo" direction="inactive"/>
      <stream media="audio" label="useraudio" direction="sendonly"/>
      <stream media="video" label="uservideo" direction="sendonly"/>
   </dialogstart>
</mscivr>
```

Using the <region> element, the dialog can be started on a conference mixer so that the video output from the dialog is directed to a specific area within a video layout. For example:

```
<mscivr version="1.0" xmlns="urn:ietf:params:xml:ns:msc-ivr">
   <dialogstart conferenceid="conference1">
      <dialog>
         <prompt>
            <media loc="http://www.example.com/
                  presentation.3gp"/>
         </prompt>
      </dialog>
      <stream media="video" direction="recvonly">
         <region>1</region>
      </stream>
   </dialogstart>
</mscivr>
```

12.4.2.2.2.1 <region> The <region> element is used to specify a named area within a presentation layout where a video media stream is displayed. The MS could, for example, play video media into an area of a video layout where the layout and its named regions are specified using the Mixer Control Package (RFC 6505 – see Chapter 9). The <region> element has no attributes, and its content model specifies the name of the region. If the region name is invalid, then the MS reports a 416 status code (Section 12.4.5) in the response to the request element containing the <region> element.

12.4.2.2.2.2 <priority> The <priority> element is used to explicitly specify the priority of the dialog for presentation in a conference mix. The <priority> element has no attributes, and its content model specifies a positive integer (see Section 12.4.6.5). The lower the value, the higher the priority.

12.4.2.3 <dialogterminate>

A dialog can be terminated by sending a <dialogterminate> request element to the MS. The <dialogterminate> element has the following attributes: dialogid: String identifying the dialog to terminate. If the specified dialog identifier is invalid, the MS sends a response with a 405 status code (Section 12.4.5). The attribute is mandatory. immediate: Indicates whether or not a dialog in the STARTED state is to be terminated immediately (in other states, termination is always immediate). A valid value is a Boolean (see Section 12.4.6.1). A value of true indicates that the dialog is terminated immediately, and the MS MUST send a dialogexit notification (Section 12.4.2.5.1) without report information. A value of false indicates that the dialog terminates after the current iteration, and the MS MUST send a dialogexit notification with report information. The attribute is optional. The default value is false.

The MS MUST reply to the <dialogterminate> request with a <response> element (Section 12.4.2.4) reporting whether or not the dialog was terminated successfully. For example, immediately terminating a STARTED dialog with dialogid "d4:"

```
<mscivr version="1.0" xmlns="urn:ietf:params:xml:ns:msc-ivr">
   <dialogterminate dialogid="d4" immediate="true"/>
</mscivr>
```

If the dialog is terminated successfully, then the response to the dialogterminate request would be:

```
<mscivr version="1.0" xmlns="urn:ietf:params:xml:ns:msc-ivr">
   <response status="200" dialogid="d4"/>
</mscivr>
```

12.4.2.4 <response>

Responses to dialog management requests are specified with a <response> element. The <response> element has the following attributes: status: Numeric code indicating the response status. Valid values are defined in Section 12.4.5. The attribute is mandatory.

- reason: String specifying a reason for the response status. The attribute is optional. There is no default value.
- desclang: Specifies the language used in the value of the reason attribute. A valid value is a language identifier (Section 12.4.6.11). The attribute is optional. If not specified, the value of the desclang attribute on <mscivr> (Section 12.4.1) applies.
- dialogid: String identifying the dialog. If the request specifies a dialogid, then that value is used. Otherwise, with <dialogprepare> and <dialogstart> requests, the dialogid generated by the MS is used. If there is no available dialogid because the request is syntactically invalid (e.g., a <dialogterminate> request with no dialogid attribute specified), then the value is the empty string. The attribute is mandatory.
- connectionid: String identifying the SIP dialog connection associated with the dialog (see Section 8.14.1 [Appendix A.1 of RFC 6230]). The attribute is optional. There is no default value.
- conferenceid: String identifying the conference associated with the dialog (see Section 8.14 [Appendix A.1 of RFC 6230]). The attribute is optional. There is no default value.

For example, a response when a dialog was prepared successfully:

```
<mscivr version="1.0" xmlns="urn:ietf:params:xml:ns:msc-ivr">
   <response status="200" dialogid="d5"/>
</mscivr>
```

The response if dialog preparation failed due to an unsupported dialog language:

```
<mscivr version="1.0" xmlns="urn:ietf:params:xml:ns:msc-ivr">
   <response status="421" dialogid="d5"
      reason="Unsupported dialog language:
         application/voicexml+xml"/>
</mscivr>
```

In this example, a <dialogterminate> request does not specify a dialogid:

```
<mscivr version="1.0" xmlns="urn:ietf:params:xml:ns:msc-ivr">
   <dialogterminate/>
</mscivr>
```

The response status indicates a 400 (Syntax error) status code, and the dialogid attribute has an empty string value:

```
<mscivr version="1.0" xmlns="urn:ietf:params:xml:ns:msc-ivr">
   <response status="400" dialogid=" "
         reason="Attribute required: dialogid"/>
</mscivr>
```

12.4.2.5 <event>

When a dialog generates a notification event, the MS sends the event using an <event> element. The <event> element has the following attribute:

- dialogid: String identifying the dialog that generated the event. The attribute is mandatory.

The <event> element has the following child elements, only one of which can occur:

- <dialogexit>: Indicates that the dialog has exited (Section 12.4.2.5.1).
- <dtmfnotify>: Indicates that a DTMF key press occurred (Section 12.4.2.5.2).

12.4.2.5.1 <dialogexit>

The <dialogexit> event indicates that a prepared or active dialog has exited because it is complete, it has been terminated, or an error occurred during execution (for example, a media resource cannot be played). This event MUST be sent by the MS when the dialog exits. The <dialogexit> element has the following attributes:

status: A status code indicating the status of the dialog when it exits. A valid value is a non-negative integer (see Section 12.4.6.4). The MS MUST support the following values:

- 0 indicates the dialog has been terminated by a <dialogterminate> request.
- 1 indicates successful completion of the dialog.
- 2 indicates that the dialog terminated because the connection or conference associated with the dialog has terminated.
- 3 indicates that the dialog terminated due to exceeding its maximum duration.
- 4 indicates that the dialog terminated due to an execution error.

All other valid but undefined values are reserved for future use, where new status codes are assigned using the Standards Action process defined in RFC 5226. The AS MUST treat any status code it does not recognize as being equivalent to 4 (dialog execution error). The attribute is mandatory.

- reason: A textual description that the MS SHOULD use to provide a reason for the status code, e.g., details about an error. A valid value is a string (see Section 12.4.6.6). The attribute is optional. There is no default value.

- desclang: Specifies the language used in the value of the reason attribute. A valid value is a language identifier (Section 12.4.6.11). The attribute is optional. If it is not specified, the value of the desclang attribute on <mscivr> (Section 12.4.1) applies.

The <dialogexit> element has the following sequence of child elements:

- <promptinfo>: Reports information (Section 12.4.3.2.1) about the prompt execution in an IVR <dialog>. The element is optional.
- <controlinfo>: Reports information (Section 12.4.3.2.2) about the control execution in an IVR <dialog>. The element is optional.
- <collectinfo>: Reports information (Section 12.4.3.2.3) about the collect execution in an IVR <dialog>. The element is optional.
- <recordinfo>: Reports information (Section 12.4.3.2.4) about the record execution in an IVR <dialog>. The element is optional.
- <params>: Reports exit parameters (Section 12.4.2.6) for a dialog language defined outside this specification. The element is optional.

For example, when an active <dialog> exits normally, the MS sends a dialogexit <event> reporting information:

```
<mscivr version="1.0" xmlns="urn:ietf:params:xml:ns:msc-ivr">
   <event dialogid="d6">
      <dialogexit status="1">
         <collectinfo dtmf="1234" termmode="match"/>
      </dialogexit>
   </event>
</mscivr>
```

12.4.2.5.2 <dtmfnotify>

The <dtmfnotify> element provides a notification of DTMF input received during the active dialog as requested by a <dtmfsub> subscription (Section 12.4.2.2.1). The <dtmfnotify> element has the following attributes:

- matchmode: Indicates the matching mode specified in the subscription request. Valid values are as follows:
  - "all" – All DTMF key presses notified individually.
  - "collect" – Only DTMF input matched by the collect operation notified.
  - "control" – Only DTMF input matched by the control operation notified. The attribute is optional. The default value is "all."
- dtmf: DTMF key presses received according to the matchmode. A valid value is a DTMF string (see Section 12.4.6.3) with no space between characters. The attribute is mandatory.
- timestamp: Indicates the time (on the MS) at which the last key press occurred according to the matchmode. A valid value is a dateTime expression (Section 12.4.6.12). The attribute is mandatory.

For example, a notification of DTMF input matched during the collect operation:

```
<mscivr version="1.0" xmlns="urn:ietf:params:xml:ns:msc-ivr">
   <event dialogid="d3">
      <dtmfnotify matchmode="collect" dtmf="3123"
            timestamp="2008-05-12T12:13:14Z"/>
   </event>
</mscivr>
```

12.4.2.6 <params>

The <params> element is a container for <param> elements (Section 12.4.2.6.1). The <params> element has no attributes, but the following child elements are defined (0 or more):

- <param>: Specifies a parameter name and value (Section 12.4.2.6.1).

For example, usage with a dialog language defined outside this specification to send additional parameters into the dialog:

```
<mscivr version="1.0" xmlns="urn:ietf:params:xml:ns:msc-ivr">
   <dialogstart type="application/x-dialog"
         src="nfs://nas01/dialog4" connectionid="c1">
      <params>
         <param name="mode">playannouncement</param>
         <param name="prompt1">nfs://nas01/media1.3gp</param>
         <param name="prompt2">nfs://nas01/media2.3gp</param>
      </params>
   </dialogstart>
</mscivr>
```

12.4.2.6.1 <param>

The <param> element describes a parameter name and value. The <param> element has the following attributes:

- **name:** A string indicating the name of the parameter. The attribute is mandatory.
- **type:** Specifies a type indicating how the inline value of the parameter is to be interpreted. A valid value is a MIME media type (see Section 12.4.6.10). The attribute is optional. The default value is "text/plain."
- **encoding:** specifies a content-transfer-encoding schema applied to the inline value of the parameter on top of the MIME media type specified with the type attribute. A valid value is a content-transfer-encoding schema as defined by the "mechanism" token in Section 6.1 of RFC 2045. The attribute is optional. There is no default value.

The <param> element content model is the value of the parameter. Note that a value that contains XML characters (e.g., "<") needs to be escaped following standard XML conventions. For example, usage with a dialog language defined outside this specification to receive parameters from the dialog when it exits:

```
<mscivr version="1.0" xmlns="urn:ietf:params:xml:ns:msc-ivr">
   <event dialogid="d6">
      <dialogexit status="1">
         <params>
            <param name="mode">recording</param>
            <param name="recording1" type="audio/x-wav"
                  encoding="base64">
               <![CDATA[R0lGODlhZABqALMAA
                  FrMYr/BvlKOVJKO
                  g2xZUKmenMfDw8tgWJpV]]>
            </param>
         </params>
      </dialogexit>
   </event>
</mscivr>
```

12.4.3 IVR Dialog Elements

This section describes the IVR dialog language defined as part of this specification. The MS MUST support this dialog language. The <dialog> element is an execution container for operations of playing prompts (Section 12.4.3.1.1), runtime controls (Section 12.4.3.1.2), collecting DTMF (Section 12.4.3.1.3), and recording user input (Section 12.4.3.1.4). Results of the dialog execution (Section 12.4.3.2) are reported in a dialogexit notification event. Using these elements, three common dialog models are supported:

- **playannouncements:** Only a <prompt> element is specified in the container. The prompt media resources are played in sequence.
- **promptandcollect:** A <collect> element is specified and optionally, a <prompt> element. If a <prompt> element is specified and bargein is enabled, playing of the prompt is terminated when bargein occurs, and DTMF collection is initiated; otherwise, the prompt is played to completion before DTMF collection is initiated. If no prompt element is specified, DTMF collection is initiated immediately.
- **promptandrecord:** A <record> element is specified and optionally, a <prompt> element. If a <prompt> element is specified and bargein is enabled, playing of the prompt is terminated when bargein occurs, and recording is initiated; otherwise, the prompt is played to completion before recording is initiated. If no prompt element is specified, recording is initiated immediately.

In addition, this dialog language supports runtime ("VCR") controls enabling a user to control prompt playback using DTMF. Each of the core elements – <prompt>, <control>, <collect>, and <record> – is specified so that their execution and reporting are largely self-contained. This facilitates their reuse in other dialog container elements. Note that DTMF and bargein behavior affects multiple elements and is addressed in the relevant element definitions. Execution results are reported in the <dialogexit> notification event with child elements defined in Section 12.4.3.2. If the dialog terminated normally (i.e., not due to an error or to a <dialogterminate> request), then the MS MUST report the results for the operations specified in the dialog:

- <prompt>: <promptinfo> (see Section 12.4.3.2.1) with at least the termmode attribute specified
- <control>: <controlinfo> (see Section 12.4.3.2.2) if any runtime controls are matched
- <collect>: <collectinfo> (see Section 12.4.3.2.3) with the dtmf and termmode attributes specified
- <record>: <recordinfo> (see Section 12.4.3.2.4) with at least the termmode attribute and one <mediainfo> element specified

The media format requirements for IVR dialogs are undefined. This package is agnostic to the media types and codecs for media resources and recording that need to be supported by an implementation. For example, an MS implementation might only support audio and in particular, the "audio/basic" codec for media playback and recording. However, when executing a dialog, if an MS encounters a media type or codec that it cannot process, the MS MUST stop further processing and report the error using the dialogexit notification.

12.4.3.1 <dialog>

An IVR dialog to play prompts to the user, allow runtime controls, collect DTMF, or record input. The dialog is specified using a <dialog> element. A <dialog> element has the following attributes:

repeatCount: Number of times the dialog is to be executed. A valid value is a non-negative integer (see Section 12.4.6.4). A value of 0 indicates that the dialog is repeated until halted by other means. The attribute is optional. The default value is 1. repeatDur: Maximum duration for dialog execution. A valid value is a time designation (see Section 12.4.6.7). If no value is specified, then there is no limit on the duration of the dialog. The attribute is optional. There is no default value. repeatUntilComplete: Indicates whether the MS terminates dialog execution when an input operation is completed successfully. A valid value is a Boolean (see Section 12.4.6.1). A value of true indicates that dialog execution is terminated when an input operation associated with its child elements is completed successfully (see execution model later for precise conditions). A value of false indicates that dialog execution is terminated by other means. The attribute is optional. The default value is false. The repeatDur attribute takes priority over the repeatCount attribute in determining maximum duration of the dialog. See "repeatCount" and "repeatDur" in the Synchronized Multimedia Integration Language (SMIL) [8] for further information. In the situation where a dialog is repeated more than once, only the results of operations in the last dialog iteration are reported. The <dialog> element has the following sequence of child elements (at least one, any order):

- <prompt>: Defines media resources to play in sequence (see Section 12.4.3.1.1). The element is optional.
- <control>: Defines how DTMF is used for runtime controls (see Section 12.4.3.1.2). The element is optional.
- <collect>: Defines how DTMF is collected (see Section 12.4.3.1.3). The element is optional.
- <record>: Defines how recording takes place (see Section 12.4.3.1.4). The element is optional.

Although the behavior when both <collect> and <record> elements are specified in a request is not defined in this Control Package, the MS MAY support this configuration. If the MS does not support this configuration, the MS sends a <response> with a 433 status code. The MS has the following execution model for the IVR dialog after initialization (initialization errors are reported by the MS in the response):

1. If an error occurs during execution, then the MS terminates the dialog and reports the error in the <dialogexit> event by setting the status attribute (see Section 12.4.3.2). Details about the error are specified in the reason attribute.
2. The MS initializes a counter to 0.
3. The MS starts a duration timer for the value of the repeatDur attribute. If the timer expires before the dialog is complete, then the MS terminates the dialog and sends a dialogexit whose status attribute is set to 3 (see Section 12.4.2.5.1). The MS MAY report information in the dialogexit gathered in the last execution cycle (if any).
4. The MS initiates a dialog execution cycle. Each cycle executes the operations associated with the child elements of the dialog. If a <prompt> element is specified, then execute the element's prompt playing operation and activate any controls (if the <control> element is specified). If no <prompt> is specified, or when a specified <prompt> terminates, then start the collect operation or the record operation if the <collect> or <record> elements, respectively, are specified. If subscriptions are specified for the dialog, then the MS sends a notification event when the specified event occurs. If execution of a child element results in an error, the MS terminates dialog execution (and stops other child element operations), and the MS sends a dialogexit status event, reporting any information gathered.
5. If the dialog execution cycle completes successfully, then the MS increments the counter by one. The MS terminates dialog execution if either of the following conditions is true:
   - The value of the repeatCount attribute is greater than zero, and the counter is equal to the value of the repeatCount attribute.
   - The value of the repeatUntilComplete attribute is true, and one of the following conditions is true:

- + <collect> reports termination status of "match" or "stopped."
- + <record> reports termination status of "stopped," "dtmf," "maxtime," or "finalsilence."

When the MS terminates dialog execution, it sends a dialogexit (with a status of 1) reporting operation information collected in the last dialog execution cycle only. Otherwise, another dialog execution cycle is initiated.

12.4.3.1.1 <prompt>

The <prompt> element specifies a sequence of media resources to play back in specification order. A <prompt> element has the following attributes:

- xml:base: A string declaring the base URI from which relative URIs in child elements are resolved prior to fetching. A valid value is a URI (see Section 12.4.6.9). The attribute is optional. There is no default value.
- bargein: Indicates whether user input stops prompt playback unless the input is associated with a specified runtime <control> operation (input matching control operations never interrupts prompt playback). A valid value is a Boolean (see Section 12.4.6.1). A value of true indicates that bargein is permitted, and prompt playback is stopped. A value of false indicates that bargein is not permitted: user input does not terminate prompt playback. The attribute is optional. The default value is true. The <prompt> element has the following child elements (at least one, any order, multiple occurrences of elements permitted):
  - <media>: Specifies a media resource (see Section 12.4.3.1.5) to play. The element is optional.
  - <variable>: Specifies a variable media announcement (see Section 12.4.3.1.1.1) to play. The element is optional.
  - <dtmf>: Generates one or more DTMF tones (see Section 12.4.3.1.1.2) to play. The element is optional.
  - <par>: specifies media resources to play in parallel (see Section 12.4.3.1.1.3). The element is optional.

If the MS does not support the configuration required for prompt playback to the output media streams, and a more specific error code is not defined for its child elements, the MS sends a <response> with a 429 status code (Section 12.4.5). The MS MAY support transcoding between the media resource format and the output stream format. The MS has the following execution model for prompt playing after initialization:

1. The MS initiates prompt playback playing its child elements (<media>, <variable>, <dtmf>, and <par>) one after another in specification order.
2. If any error (including fetching and rendering errors) occurs during prompt execution, then the MS terminates playback and reports its error status to the dialog container (see Section 12.4.3) with a <promptinfo> (see Section 12.4.3.2.1) where the termmode attribute is set to stopped and any additional information is set.
3. If DTMF input is received, and the value of the bargein attribute is true, then the MS terminates prompt playback and reports its execution status to the dialog container (see Section 4.3) with a <promptinfo> (see Section 12.4.3.2.1) where the termmode attribute is set to bargein and any additional information is set.
4. If prompt playback is stopped by the dialog container, then the MS reports its execution status to the dialog container (see Section 12.4.3) with a <promptinfo> (see Section 12.4.3.2.1) where the termmode attribute is set to stopped and any additional information is set.

5. If prompt playback completes successfully, then the MS reports its execution status to the dialog container (see Section 12.4.3) with a <promptinfo> (see Section 12.4.3.2.1) where the termmode attribute is set to completed and any additional information is set.

12.4.3.1.1.1 <variable> The <variable> element specifies variable announcements using predefined media resources. Each variable has at least a type (e.g., date) and a value (e.g., 2008-02-25). The value is rendered according to the prompt variable type (e.g., 2008-02-25 is rendered as the date February 25, 2008). The precise mechanism for generating variable announcements (including the location of associated media resources) is implementation specific. A <variable> element has the following attributes:

- type: Specifies the type of prompt variable to render. This specification defines three values – date (Section 12.4.3.1.1.1.1), time (Section 12.4.3.1.1.1.2), and digits (Section 12.4.3.1.1.1.3). All other valid but undefined values are reserved for future use, where new values are assigned as described in Section 12.8.5. A valid value is a string (see Section 12.4.6.6). The attribute is mandatory.
- value: Specifies a string to be rendered according to the prompt variable type. A valid value is a string (see Section 12.4.6.6). The attribute is mandatory.
- format: Specifies format information that the prompt variable type uses to render the value attribute. A valid value is a string (see Section 12.4.6.6). The attribute is optional. There is no default value.
- gender: Specifies the gender that the prompt variable type uses to render the value attribute. Valid values are "male" or "female." The attribute is optional. There is no default value.
- xml:lang: Specifies the language that the prompt variable type uses to render the value attribute. A valid value is a language identifier (see Section 12.4.6.11). The attribute is optional. There is no default value. The <variable> element has no children.

This specification is agnostic to the type and codec of media resources into which variables are rendered as well as the rendering mechanism itself. For example, an MS implementation supporting audio rendering could map the <variable> into one or more audio media resources. This package is agnostic to which <variable> types are supported by an implementation. If a <variable> element configuration specified in a request is not supported by the MS, the MS sends a <response> with a 425 status code (Section 12.4.5).

12.4.3.1.1.1.1 Date Type The date variable type provides a mechanism for dynamically rendering a date prompt.

The <variable> type attribute MUST have the value "date." The <variable> format attribute MUST be one of the following values and comply with its rendering of the value attribute:

mdy, indicating that the <variable> value attribute is to be rendered as sequence composed of month, then day, then year

- ymd, indicating that the <variable> value attribute is to be rendered as sequence composed of year, then month, then day
- dym, indicating that the <variable> value attribute is to be rendered as sequence composed of day, then year, then month
- dm, indicating that the <variable> value attribute is to be rendered as sequence composed of day then month
- The <variable> value attribute MUST comply with a lexical representation of date where

yyyy '-' mm '-' dd as defined in Section 12.3.2.9 of Biron and Malhotra [9]. For example, <variable type="date" format="dmy" value="2010-11-25" xml:lang="en" gender="male"/> describes a variable date prompt where the date can be rendered in audio as “twenty-fifth of November two thousand and ten” using a list of <media> resources:

```
<media loc="nfs://voicebase/en/male/25th.wav"/>
<media loc="nfs://voicebase/en/male/of.wav"/>
<media loc="nfs://voicebase/en/male/november.wav"/>
<media loc="nfs://voicebase/en/male/2000.wav"/>
<media loc="nfs://voicebase/en/male/and.wav"/>
<media loc="nfs://voicebase/en/male/10.wav"/>
```

12.4.3.1.1.1.2 Time Type The time variable type provides a mechanism for dynamically rendering a time prompt. The <variable> type attribute MUST have the value “time.” The <variable> format attribute MUST be one of the following values and comply with its rendering of the value attribute:

- t12, indicating that the <variable> value attribute is to be rendered as a time in traditional 12-hour format using am or pm (for example, “twenty-five minutes past 2 pm” for “14:25”)
- t24, indicating that the <variable> value attribute is to be rendered as a time in 24-hour format (for example, “fourteen twenty-five” for “14:25”)

The <variable> value attribute MUST comply with a lexical representation of time where hh ‘:’ mm (‘:’ ss)? as defined in Section 3.2.8 of Biron and Malhotra [9].

12.4.3.1.1.1.3 Digits Type The digits variable type provides a mechanism for dynamically rendering a digit sequence. The <variable> type attribute MUST have the value “digits.” The <variable> format attribute MUST be one of the following values and comply with its rendering of the value attribute:

- gen, indicating that the <variable> value attribute is to be rendered as a general digit string (for example, “one two three” for “123”)
- crn, indicating that the <variable> value attribute is to be rendered as a cardinal number (for example, “one hundred and twentythree” for “123”)
- ord, indicating that the <variable> value attribute is to be rendered as an ordinal number (for example, “one hundred and twentythird” for “123”)

The <variable> value attribute MUST comply with the lexical representation d+ i.e., one or more digits.

12.4.3.1.1.2 <dtmf> The <dtmf> element specifies a sequence of DTMF tones for output. DTMF tones could be generated using <media> resources where the output is transported as real-time transport protocol (RTP) audio packets. However, <media> resources are not sufficient for cases where DTMF tones are to be transported as DTMF RTP (RFC 4733) or in event packages. A <dtmf> element has the following attributes:

- digits: Specifies the DTMF sequence to output. A valid value is a DTMF string (see Section 12.4.6.3). The attribute is mandatory.
- level: Used to define the power level for which the DTMF tones will be generated. Values are expressed in dBm0. A valid value is an integer in the range of 0 to −96 dBm0. Larger negative values express lower power levels. Note that values lower than −55 dBm0 will be rejected by most receivers (TR-TSY-000181, ITU-T Q.24A). The attribute is optional. The default value is −6 dBm0.

- duration: Specifies the duration for which each DTMF tone is generated. A valid value is a time designation (see Section 12.4.6.7). The MS MAY round the value if it only supports discrete durations. The attribute is optional. The default value is 100 ms.
- interval: Specifies the duration of a silence interval following each generated DTMF tone. A valid value is a time designation (see Section 12.4.6.7). The MS MAY round the value if it only supports discrete durations. The attribute is optional. The default value is 100 ms. The <dtmf> element has no children. If a <dtmf> element configuration is not supported, the MS sends a <response> with a 426 status code (Section 12.4.5).

12.4.3.1.1.3 <par> The <par> element allows media resources to be played in parallel. Each of its child elements specifies a media resource (or a sequence of media resources using the <seq> element). When playback of the <par> element is initiated, the MS begins playback of all its child elements at the same time. This element is modeled after the <par> element in SMIL [8]. The <par> element has the following attributes:

- endsync: Indicates when playback of the element is complete. Valid values are "first" (indicates that the element is complete when any child element reports that it is complete) and "last" (indicates that it is complete when every child element is complete). The attribute is optional. The default value is "last." If the value is "first," then playback of other child elements is stopped when one child element reports it is complete. The <par> element has the following child elements (at least one, any order, multiple occurrences of each element permitted):
  - <seq>: Specifies a sequence of media resources to play in parallel with other <par> child elements (see Section 12.4.3.1.1.3.1). The element is optional.
  - <media>: Specifies a media resource (see Section 12.4.3.1.5) to play. The MS is responsible for assigning the appropriate media stream(s) when more than one is available. The element is optional.
  - <variable>: Specifies a variable media announcement (see Section 12.4.3.1.1.1) to play. The element is optional.
  - <dtmf>: Generates one or more DTMF tones (see Section 12.4.3.1.1.2) to play. The element is optional.

It is RECOMMENDED that a <par> element contains only one <media> element of the same media type (i.e., same type-name as defined in Section 12.4.6.10). If a <par> element configuration is not supported, the MS sends a <response> with a 435 status code (Section 4.5). Runtime <control>s (Section 12.4.3.1.2) apply to each child element playing in parallel. For example, pause and resume controls cause all child elements to be paused and resumed, respectively.

If the <par> element is stopped by the prompt container (e.g., bargein or dialog termination), then playback of all child elements is stopped. The playback duration (Section 12.4.3.2.1) reported for the <par> element is the duration of parallel playback, not the cumulative duration of each child element played in parallel. For example, a request to play back audio and video media in parallel:

```
<mscivr version="1.0" xmlns="urn:ietf:params:xml:ns:msc-ivr">
   <dialogstart connectionid="c1">
      <dialog>
         <prompt>
            <par>
               <media type="audio/x-wav"
                     loc="http://www.example.
                     com/media/
                     comments.wav"/>
               <media type="video/3gpp;codecs='s263'"
```

```
                                  loc="http://www.example.com/
                                  media/camera.3gp"/>
                    </par>
                </prompt>
            </dialog>
        </dialogstart>
    </mscivr>
```

When the <prompt> element is executed, it begins playback of its child element in specification-order sequence. In this case, there is only one child element, a <par> element itself containing audio and video <media> child elements. Consequently, playback of both audio and video media resources is initiated at the same time. Since the endsync attribute is not specified, the default value "last" applies. The <par> element playback is complete when the media resource with the longest duration is complete.

12.4.3.1.1.3.1 <seq> The <seq> element specifies media resources to be played back in sequence. This allows a sequence of media resources to be played at the same time as other children of a <par> element are played in parallel, for example, a sequence of audio resources while a video resource is played in parallel. This element is modeled after the <seq> element in SMIL [8]. The <seq> element has no attributes. The <seq> element has the following child elements (at least one, any order, multiple occurrences of each element permitted):

- <media>: Specifies a media resource (see Section 12.4.3.1.5) to play. The element is optional.
- <variable>: Specifies a variable media announcement (see Section 12.4.3.1.1.1) to play. The element is optional.
- <dtmf>: Generates one or more DTMF tones (see Section 12.4.3.1.1.2) to play. The element is optional.

Playback of a <seq> element is complete when all child elements in the sequence are complete. If the <seq> element is stopped by the <par> container, then playback of the current child element is stopped (remaining child elements in the sequence are not played). For example, a request to play a sequence of audio resources in parallel with a video media:

```
<mscivr version="1.0" xmlns="urn:ietf:params:xml:ns:msc-ivr">
    <dialogstart connectionid="c1">
        <dialog>
            <prompt>
                <par endsync="first">
                    <seq>
                        <media type="audio/x-wav"
                               loc="http://www.example.
                                      com/media/date.wav"/>
                        <media type="audio/x-wav"
                               loc="http://www.example.
                               com/media/intro.wav"/>
                        <media type="audio/x-wav"
                               loc="http://www.example.
                               com/media/main.wav"/>
                        <media type="audio/x-wav"
                               loc="http://www.example.
                               com/media/end.wav"/>
                    </seq>
```

```
                <media type="video/3gpp;codecs='s263'"
                       loc="rtsp://www.example.com/
                       media/camera.3gp"/>
            </par>
        </prompt>
    </dialog>
  </dialogstart>
</mscivr>
```

When the <prompt> element is executed, it begins playback of the <par> element containing a <seq> element and a video <media> element. The <seq> element itself contains a sequence of audio <media> elements. Consequently, playback of the video media resource is initiated at the same time as playback of the sequence of the audio media resources is initiated. Each audio resource is played back after the previous one completes. Since the endsync attribute is set to "first," the <par> element playback is complete when either all the audio resources in <seq> have been played to completion or the video <media> is complete, whichever occurs first.

12.4.3.1.2 <control>

The <control> element defines how DTMF input is mapped to runtime controls, including prompt playback controls. DTMF input matching these controls MUST NOT cause prompt playback to be interrupted (i.e., no prompt bargein) but causes the appropriate operation to be applied, for example, speeding up prompt playback. DTMF input matching these controls has priority over <collect> input for the duration of prompt playback. If an incoming DTMF character matches a specified runtime control, then the DTMF character is consumed: it is not added to the digit buffer and so is not available to the <collect> operation. Once prompt playback is complete, runtime controls are no longer active. The <control> element has the following attributes:

- **gotostartkey:** Maps a DTMF key to skip directly to the start of the prompt. A valid value is a DTMF character (see Section 12.4.6.2). The attribute is optional. There is no default value.
- **gotoendkey:** Maps a DTMF key to skip directly to the end of the prompt. A valid value is a DTMF character (see Section 12.4.6.2). The attribute is optional. There is no default value.
- **skipinterval:** Indicates how far an MS skips backwards or forwards through prompt playback when the rewind (rwkey) of fast forward key (ffkey) is pressed. A valid value is a Time Designation (see Section 12.4.6.7). The attribute is optional. The default value is 6 s.
- **ffkey:** Maps a DTMF key to a fast forward operation equal to the value of "skipinterval." A valid value is a DTMF character (see Section 12.4.6.2). The attribute is optional. There is no default value.
- **rwkey:** Maps a DTMF key to a rewind operation equal to the value of "skipinterval." A valid value is a DTMF character (see Section 12.4.6.2). The attribute is optional. There is no default value.
- **pauseinterval:** Indicates how long an MS pauses prompt playback when the pausekey is pressed. A valid value is a Time Designation (see Section 12.4.6.7). The attribute is optional. The default value is 10 s.
- **pausekey:** Maps a DTMF key to a pause operation equal to the value of "pauseinterval." A valid value is a DTMF character (see Section 12.4.6.2). The attribute is optional. There is no default value.
- **resumekey:** Maps a DTMF key to a resume operation. A valid value is a DTMF character (see Section 12.4.6.2). The attribute is optional. There is no default value.
- **volumeinterval:** Indicates the increase or decrease in playback volume (relative to the current volume) when the volupkey or voldnkey is pressed. A valid value is a percentage (see Section 12.4.6.8). The attribute is optional. The default value is 10%.

- **volupkey:** Maps a DTMF key to a volume increase operation equal to the value of "volumeinterval." A valid value is a DTMF character (see Section 12.4.6.2). The attribute is optional. There is no default value.
- **voldnkey:** Maps a DTMF key to a volume decrease operation equal to the value of "volumeinterval." A valid value is a DTMF character (see Section 12.4.6.2). The attribute is optional. There is no default value.
- **speedinterval:** Indicates the increase or decrease in playback speed (relative to the current speed) when the speedupkey or speeddnkey is pressed. A valid value is a percentage (see Section 12.4.6.8). The attribute is optional. The default value is 10%.
- **speedupkey:** Maps a DTMF key to a speed increase operation equal to the value of the speedinterval attribute. A valid value is a DTMF character (see Section 12.4.6.2). The attribute is optional. There is no default value.
- **speeddnkey:** Maps a DTMF key to a speed decrease operation equal to the value of the speedinterval attribute. A valid value is a DTMF character (see Section 12.4.6.2). The attribute is optional. There is no default value.
- **external:** Allows one or more DTMF keys to be declared as external controls (for example, video camera controls); the MS can send notifications when a matching key is activated using <dtmfnotify> (Section 12.4.2.5.2). A valid value is a DTMF string (see Section 12.4.6.3). The attribute is optional. There is no default value. If the same DTMF is specified in more than one DTMF key control attribute – except the pausekey and resumekey attributes – the MS sends a <response> with a 413 status code (Section 12.4.5). The MS has the following execution model for runtime control after initialization:
    1. If an error occurs during execution, then the MS terminates runtime control, and the error is reported to the dialog container. The MS MAY report controls executed successfully before the error in <controlinfo> (see Section 12.4.3.2.2).
    2. Runtime controls are active only during prompt playback (if no <prompt> element is specified, then runtime controls are ignored). If DTMF input matches any specified keys (for example, the ffkey), then the MS applies the appropriate operation immediately. If a seek operation (ffkey, rwkey) attempts to go beyond the beginning or end of the prompt queue, then the MS automatically truncates it to the prompt queue beginning or end, respectively. If a volume operation (voldnkey, volupkey) attempts to go beyond the minimum or maximum volume supported by the platform, then the MS automatically limits the operation to minimum or maximum supported volume, respectively. If a speed operation (speeddnkey, speedupkey) attempts to go beyond the minimum or maximum playback speed supported by the platform, then the MS automatically limits the operation to minimum or maximum supported speed, respectively. If the pause operation attempts to pause output when it is already paused, then the operation is ignored. If the resume operation attempts to resume when the prompts are not paused, then the operation is ignored. If a seek, volume, or speed operation is applied when output is paused, then the MS also resumes output automatically.
    3. If DTMF control subscription has been specified for the dialog, then each DTMF match of a control operation is reported in a <dtmfnotify> notification event (Section 12.4.2.5.2).
    4. When the dialog exits, all control matches are reported in a <controlinfo> element (Section 12.4.3.2.2).

12.4.3.1.3 <collect>

The <collect> element defines how DTMF input is collected. The <collect> element has the following attributes:

- **cleardigitbuffer:** indicates whether the digit buffer is to be cleared. A valid value is a Boolean (see Section 12.4.6.1). A value of true indicates that the digit buffer is to be cleared. A value of false indicates that the digit buffer is not to be cleared. The attribute is optional. The default value is true.
- **timeout:** indicates the maximum time to wait for user input to begin. A valid value is a Time Designation (see Section 12.4.6.7). The attribute is optional. The default value is 5 s.
- **interdigittimeout:** indicates the maximum time to wait for another DTMF when the collected input is incomplete with respect to the grammar. A valid value is a Time Designation (see Section 12.4.6.7). The attribute is optional. The default value is 2 s.
- **termtimeout:** indicates the maximum time to wait for the termchar character when the collected input is complete with respect to the grammar. A valid value is a Time Designation (see Section 12.4.6.7). The attribute is optional. The default value is 0 s (no delay).
- **escapekey:** specifies a DTMF key that indicates collected grammar matches are discarded and the DTMF collection is to be reinitiated. A valid value is a DTMF character (see Section 12.4.6.2). The attribute is optional. There is no default value.
- **termchar:** specifies a DTMF character for terminating DTMF input collection using the internal grammar. It is ignored when a custom grammar is specified. A valid value is a DTMF character (see Section 12.4.6.2). To disable termination by a conventional DTMF character, set the parameter to an unconventional character like "A." The attribute is optional. The default value is "#."
- **maxdigits:** The maximum number of digits to collect using an internal digits (0–9 only) grammar. It is ignored when a custom grammar is specified. A valid value is a positive integer (see Section 12.4.6.5). The attribute is optional. The default value is 5.

The following matching priority is defined for incoming DTMF:

- **termchar attribute**, **escapekey attribute**, and then as part of a grammar. For example, if "1" is defined as the escapekey attribute and as part of a grammar, then its interpretation as an escapekey takes priority.

The <collect> element has the following child element:

- **<grammar>:** indicates a custom grammar format (see Section 12.4.3.1.3.1). The element is optional. The custom grammar takes priority over the internal grammar. If a <grammar> element is specified, the MS MUST use it for DTMF collection.

The MS has the following execution model for DTMF collection after initialization:

1. The DTMF collection buffer MUST NOT receive DTMF input matching <control> operations (see Section 12.4.3.1.2).
2. If an error occurs during execution, then the MS terminates collection and reports the error to the dialog container (see Section 12.4.3). The MS MAY report DTMF collected before the error in <collectinfo> (see Section 12.4.3.2.3).
3. The MS clears the digit buffer if the value of the cleardigitbuffer attribute is true.
4. The MS activates an initial timer with the duration of the value of the timeout attribute. If the initial timer expires before any DTMF input is received, then collection execution terminates, the <collectinfo> (see Section 12.4.3.2.3) has the termmode attribute set to noinput, and the execution status is reported to the dialog container.
5. When the first DTMF collect input is received, the initial timer is canceled, and DTMF collection begins. Each DTMF input is collected unless it matches the value of the escapekey attribute or the termchar attribute when the internal grammar is used. Collected input is matched against the grammar to determine whether it is valid and if valid, whether

collection is complete. Valid DTMF patterns are either a simple digit string, where the maximum length is determined by the maxdigits attribute and that can be optionally terminated by the character in the termchar attribute, or a custom DTMF grammar specified with the <grammar> element.

6. After escapekey input, or a valid input that does not complete the grammar, the MS activates a timer for the value of the interdigittimeout attribute or the termtimeout attribute. The MS only uses the termtimeout value when the grammar does not allow any additional input; otherwise, the MS uses the interdigittimeout.
7. If DTMF collect input matches the value of the escapekey attribute, then the MS re-initializes DTMF collection: i.e., the MS discards collected DTMFs already matched against the grammar, and the MS attempts to match incoming DTMF (including any pending in the digit buffer) as described in Step 5.
8. If the collect input is not valid with respect to the grammar, or an interdigittimeout timer expires, the MS terminates collection execution and reports execution status to the dialog container with a <collectinfo> (see Section 12.4.3.2.3) where the termmode attribute is set to nomatch.
9. If the collect input completes the grammar, or if a termtimeout timer expires, then the MS terminates collection execution and reports execution status to the dialog container with <collectinfo> (see Section 12.4.3.2.3) where the termmode attribute is set to match.

12.4.3.1.3.1 <grammar> The <grammar> element allows a custom grammar, inline or external, to be specified. Custom grammars permit the full range of DTMF characters including "*" and "#" to be specified for DTMF pattern matching. The <grammar> element has the following attributes:

1. **src:** Specifies the location of an external grammar specification. A valid value is a URI (see Section 12.4.6.9). The MS MUST support both HTTP (RFC 2616) and HTTPS (RFC 2818) schemes, and the MS MAY support other schemes. If the URI scheme is unsupported, the MS sends a <response> with a 420 status code (Section 12.4.5). If the resource cannot be retrieved within the timeout interval, the MS sends a <response> with a 409 status code. If the grammar format is not supported, the MS sends a <response> with a 424 status code. The attribute is optional. There is no default value.
2. **type:** Identifies the preferred type of the grammar specification identified by the src attribute. A valid value is a MIME media type (see Section 12.4.6.10). If the URI scheme used in the src attribute defines a mechanism for establishing the authoritative MIME media type of the media resource, the value returned by that mechanism takes precedence over this attribute. The attribute is optional. There is no default value.
3. **fetchtimeout:** The maximum interval to wait when fetching a grammar resource. A valid value is a Time Designation (see Section 12.4.6.7). The attribute is optional. The default value is 30 s. The <grammar> element allows inline grammars to be specified. XML grammar formats MUST use a namespace other than the one used in this specification. Non-XML grammar formats MAY use a CDATA section. The MS MUST support the Speech Recognition Grammar Specification [10] XML grammar format ("application/srgs+xml") and MS MAY support the Key Press Markup Language (KPML) (RFC 4730) or other grammar formats. If the grammar format is not supported by the MS, then the MS sends a <response> with a 424 status code (Section 12.4.5).

For example, the following fragment shows DTMF collection with an inline SRGS grammar:

```
<collect cleardigitbuffer="false" timeout="20s" interdigittimeout="1s">
   <grammar>
      <grammar xmlns=http://www.w3.org/2001/06/grammar
            version="1.0" mode="dtmf">
```

```
            <rule id="digit">
                <one-of>
                    <item>0</item>
                    <item>1</item>
                    <item>2</item>
                    <item>3</item>
                    <item>4</item>
                    <item>5</item>
                    <item>6</item>
                    <item>7</item>
                    <item>8</item>
                    <item>9</item>
                </one-of>
            </rule>
            <rule id="pin" scope="public">
                <one-of>
                    <item>
                    <item repeat="4">
                    <ruleref uri="#digit"/>
                    </item>#</item>
                    <item>* 9</item>
                </one-of>
            </rule>
        </grammar>
    </grammar>
</collect>
```

The same grammar could also be referenced externally (and take advantage of HTTP caching):

```
<collect cleardigitbuffer="false" timeout="20s">
    <grammar type="application/srgs+xml"
            src="http://example.org/pin.grxml"/>
</collect>
```

12.4.3.1.4 <record>

The <record> element specifies how media input is recorded. The <record> element has the following attributes:

- timeout: Indicates the time to wait for user input to begin. A valid value is a Time Designation (see Section 12.4.6.7). The attribute is optional. The default value is 5 s.
- vadinitial: Controls whether voice activity detection (VAD) is used to initiate the recording operation. A valid value is a Boolean (see Section 4.6.1). A value of true indicates that the MS MUST initiate recording if the VAD detects voice on the configured inbound audio streams. A value of false indicates that the MS MUST NOT initiate recording using VAD. The attribute is optional. The default value is false.
- vadfinal: Controls whether VAD is used to terminate the recording operation. A valid value is a Boolean (see Section 12.4.6.1). A value of true indicates that the MS MUST terminate recording if the VAD detects a period of silence (whose duration is specified by the finalsilence attribute) on configured inbound audio streams. A value of false indicates that the MS MUST NOT terminate recording using VAD. The attribute is optional. The default value is false.
- dtmfterm: Indicates whether the recording operation is terminated by DTMF input. A valid value is a Boolean (see Section 12.4.6.1). A value of true indicates that recording is terminated by DTMF input. A value of false indicates that recording is not terminated by DTMF input. The attribute is optional. The default value is true.

- maxtime: indicates the maximum duration of the recording. A valid value is a Time Designation (see Section 12.4.6.7). The attribute is optional. The default value is 15 s.
- beep: Indicates whether a "beep" is to be played immediately prior to initiation of the recording operation. A valid value is a Boolean (see Section 12.4.6.1). The attribute is optional. The default value is false.
- finalsilence: Indicates the interval of silence that indicates the end of voice input. This interval is not part of the recording itself. This parameter is ignored if the vadfinal attribute has the value false. A valid value is a Time Designation (see Section 12.4.6.7). The attribute is optional. The default value is 5 s.
- append: Indicates whether recorded data is appended or not to a recording location if a resource already exists. A valid value is a Boolean (see Section 12.4.6.1). A value of true indicates that recorded data is appended to the existing resource at a recording location. A value of false indicates that recorded data is to overwrite the existing resource. The attribute is optional. The default value is false. When a recording location is specified using the HTTP or HTTPS protocol, the recording operation SHOULD be performed using the HTTP GET and PUT methods, unless the HTTP server provides a special interface for recording uploads and appends (e.g., using POST). When the append attribute has the value false, the recording data is uploaded to the specified location using HTTP PUT and replaces any data at that location on the HTTP origin server. When append has the value true, the existing data (if any) is first downloaded from the specified location using HTTP GET; then, the recording data is appended to the existing recording (note that this might require codec conversion and modification to the existing data); and then, the combined recording is uploaded to the specified location using HTTP PUT. HTTP errors are handled as described in RFC 2616. When the recording location is specified using protocols other than HTTP or HTTPS, the mapping of the append operation onto the upload protocol scheme is implementation specific. If either the vadinitial or vadfinal attribute is set to true, and the MS does not support VAD, the MS sends a <response> with a 434 status code (Section 12.4.5).

The <record> element has the following child element (0 or more occurrences):

- <media>: specifies the location and type of the media resource for uploading recorded data (see Section 12.4.3.1.5).

The MS MUST support both HTTP (RFC 2616) and HTTPS (RFC 2818) schemes for uploading recorded data, and the MS MAY support other schemes. The MS uploads recorded data to this resource as soon as possible after recording is complete. The element is optional. If multiple <media> elements are specified, then media input is to be recorded in parallel to multiple resource locations. If no <media> child element is specified, the MS MUST record media input, but the recording location and the recording format are implementation specific (e.g., the MS records audio in the WAV format to a local disk accessible by HTTP). The recording location and format are reported in <recordinfo> (Section 12.4.3.2.4) when the dialog terminates. The recording MUST be available from this location until the connection or conference associated with the dialog on the MS terminates.

If the MS does not support the configuration required for recording from the input media streams to one or more <media> elements, and a more specific error code is not defined for its child elements, the MS sends a <response> with a 423 status code (Section 12.4.5). Note that an MS MAY support uploading recorded data to recording locations at the same time as the recording operation takes place. Such implementations need to be aware of the requirements of certain recording formats (e.g., WAV) for metadata at the beginning of the uploaded file, that the finalsilence interval is not part of the recording, and how these requirements interact with the URI scheme. The MS has the following execution model for recording after initialization:

1. If an error occurs during execution (e.g., authentication or communication error when trying to upload to a recording location), then the MS terminates record execution and reports the error to the dialog container (see Section 12.4.3). The MS MAY report data recorded before the error in <recordinfo> (see Section 12.4.3.2.4).
2. If DTMF input (not matching a <control> operation) is received during prompt playback, and the prompt bargein attribute is set to true, then the MS activates the record execution. Otherwise, the MS activates it after the completion of prompt playback.
3. If a beep attribute with the value of true is specified, then the MS plays a beep tone.
4. The MS activates a timer with the duration of the value of the timeout attribute. If the timer expires before the recording operation begins, then the MS terminates the recording execution and reports the status to dialog container with <recordinfo> (see Section 12.4.3.2.4) where the termmode attribute is set to noinput.
5. Initiation of the recording operation depends on the value of the vadinitial attribute. If vadinitial has the value false, then the recording operation is initiated immediately. Otherwise, the recording operation is initiated when voice activity is detected.
6. When the recording operation is initiated, a timer is started for the value of the maxtime attribute (maximum duration of the recording). If the timer expires before the recording operation is complete, then the MS terminates recording execution and reports the execution status to the dialog container with <recordinfo> (see Section 12.4.3.2.4) where the termmode attribute is set to maxtime.
7. During the record operation input, media streams are recording to a location and format specified in one or more <media> child elements. If no <media> child element is specified, the MS records input to an implementation-specific location and format.
8. If the dtmfterm attribute has the value true, and DTMF input is detected during the record operation, then the MS terminates recording, and its status is reported to the dialog container with a <recordinfo> (see Section 12.4.3.2.4) where the termmode attribute is set to dtmf.
9. If the vadfinal attribute has the value true, then the MS terminates the recording operation when a period of silence, with the duration specified by the value of the finalsilence attribute, is detected. This period of silence is not part of the final recording. The status is reported to the dialog container with a <recordinfo> (see Section 12.4.3.2.4) where the termmode attribute is set to finalsilence.

For example, a request to record audio and video input to separate locations:

```
<mscivr version="1.0" xmlns="urn:ietf:params:xml:ns:msc-ivr">
   <dialogstart connectionid="c1">
      <dialog>
         <record maxtime="30s" vadinitial="false"
               vadfinal="false">
            <media type="audio/x-wav"
                  loc="http://www.example.com/
                  upload/audio.wav"/>
            <media type="video/3gpp;codecs='s263'"
                  loc="http://www.example.com/
                  upload/video.3gp"/>
         </record>
      </dialog>
   </dialogstart>
</mscivr>
```

When the <record> element is executed, it immediately begins recording of the audio and video (since vadinitial is false), where the destination locations are specified in the <media> child elements. Recording is completed when the duration reaches 30 s or the connection is terminated.

12.4.3.1.5 <media>

The <media> element specifies a media resource to play back from (see Section 12.4.3.1.1) or record to (see Section 12.4.3.1.4). In the playback case, the resource is retrieved, and in the recording case, recording data is uploaded to the resource location. A <media> element has the following attributes:

- loc: Specifies the location of the media resource. A valid value is a URI (see Section 12.4.6.9). The MS MUST support both HTTP (RFC 2616) and HTTPS (RFC 2818) schemes, and the MS MAY support other schemes. If the URI scheme is not supported by the MS, the MS sends a <response> with a 420 status code (Section 12.4.5). If the resource is to be retrieved, but the MS cannot retrieve it within the timeout interval, the MS sends a <response> with a 409 status code. If the format of the media resource is not supported, the MS sends a <response> with a 429 status code. The attribute is mandatory.
- type: Specifies the type of the media resource indicated in the loc attribute. A valid value is a MIME media type (see Section 12.4.6.10) that depending on its definition, can include additional parameters (e.g., RFC 4281). If the URI scheme used in the loc attribute defines a mechanism for establishing the authoritative MIME media type of the media resource, the value returned by that mechanism takes precedence over this attribute. If additional media parameters are specified, the MS MUST use them to determine media processing. For example, RFC 4281 defines a "codec" parameter for media types like video/3gpp that would determine which media streams are played or recorded. The attribute is optional. There is no default value.
- fetchtimeout: The maximum interval to wait when fetching a media resource. A valid value is a Time Designation (see Section 12.4.6.7). The attribute is optional. The default value is 30 s.
- soundLevel: Playback soundLevel (volume) for the media resource. A valid value is a percentage (see Section 12.4.6.8). The value indicates increase or decrease relative to the original recorded volume of the media. A value of 100% (the default) plays the media at its recorded volume, a value of 200% will play the media at twice its recorded volume, 50% at half its recorded volume, a value of 0% will play the media silently, and so on. See "soundLevel" in SMIL [8] for further information. The attribute is optional. The default value is 100%.
- clipBegin: Offset from start of media resource to begin playback. A valid value is a Time Designation (see Section 12.4.6.7). The offset is measured in normal media playback time from the beginning of the media resource. If the clipBegin offset is after the end of media (or the clipEnd offset), no media is played. See "clipBegin" in SMIL [8] for further information. The attribute is optional. The default value is 0 s.
- clipEnd: offset from start of media resource to end playback. A valid value is a Time Designation (see Section 12.4.6.7). The offset is measured in normal media playback time from the beginning of the media resource. If the clipEnd offset is after the end of the media, then the media is played to the end. If clipBegin is after clipEnd, then no media is played. See "clipEnd" in SMIL [8] for further information. The attribute is optional. There is no default value. The fetchtimeout, soundLevel, clipBegin, and clipEnd attributes are only relevant in the playback use case. The MS ignores these attributes when using the <media> for recording. The <media> element has no children.

12.4.3.2 Exit Information

When the dialog exits, information about the specified operations is reported in a <dialogexit> notification event (Section 12.4.2.5.1).

12.4.3.2.1 <promptinfo>

The <promptinfo> element reports the information about prompt execution. It has the following attributes:

- duration: Indicates the duration of prompt playback in milliseconds. A valid value is a non-negative integer (see Section 12.4.6.4). The attribute is optional. There is no default value.
- termmode: indicates how playback was terminated. Valid values are "stopped," "completed," or "bargein." The attribute is mandatory. The <promptinfo> element has no child elements.

12.4.3.2.2 <controlinfo>

The <controlinfo> element reports information about control execution. The <controlinfo> element has no attributes and has 0 or more <controlmatch> child elements, each describing an individual runtime control match.

12.4.3.2.2.1 <controlmatch> The <controlmatch> element has the following attributes:

- dtmf: DTMF input triggering the runtime control. A valid value is a DTMF string (see Section 12.4.6.3) with no space between characters. The attribute is mandatory.
- timestamp: indicates the time (on the MS) at which the control was triggered. A valid value is a dateTime expression (Section 12.4.6.12). The attribute is mandatory. The <controlmatch> element has no child elements.

12.4.3.2.3 <collectinfo>

The <collectinfo> element reports the information about collect execution. The <collectinfo> element has the following attributes:
dtmf: DTMF input collected from the user. A valid value is a DTMF string (see Section 12.4.6.3) with no space between characters. The attribute is optional. There is no default value.
termmode: Indicates how collection was terminated. Valid values are "stopped," "match," "noinput," or "nomatch." The attribute is mandatory. The <collectinfo> element has no child elements.

12.4.3.2.4 <recordinfo>

The <recordinfo> element reports information about record execution (Section 12.4.3.1.4).
The <recordinfo> element has the following attributes:

- termmode: Indicates how recording was terminated. Valid values are "stopped," "noinput," "dtmf," "maxtime," and "finalsilence." The attribute is mandatory.
- duration: Indicates the duration of the recording in milliseconds. A valid value is a non-negative integer (see Section 12.4.6.4). The attribute is optional. There is no default value.

The <recordinfo> element has the following child element (0 or more occurrences):

- <mediainfo>: Indicates information about a recorded media resource (see Section 12.4.3.2.4.1). The element is optional.

When the record operation is successful, the MS MUST specify a <mediainfo> element for each recording location. For example, if the <record> element contained three <media> child elements, then the <recordinfo> would contain three <mediainfo> child elements.

12.4.3.2.4.1 <mediainfo> The <mediainfo> element reports information about a recorded media resource. The <mediainfo> element has the following attributes:

- loc: Indicates the location of the media resource. A valid value is a URI (see Section 12.4.6.9). The attribute is mandatory.

- type: Indicates the format of the media resource. A valid value is a MIME media type (see Section 12.4.6.10). The attribute is mandatory.
- size: indicates the size of the media resource in bytes. A valid value is a non-negative integer (see Section 12.4.6.4). The attribute is optional. There is no default value.

12.4.4 Audit Elements

The audit elements defined in this section allow the MS to be audited for package capabilities as well as dialogs managed by the package. Auditing is particularly important for two use cases. First, it enables discovery of package capabilities supported on an MS before an AS starts a dialog on connection or conference. The AS can then use this information to create request elements using supported capabilities and in the case of codecs, to negotiate an appropriate SDP for a user agent's connection. Second, auditing enables the discovery of the existence and status of dialogs currently managed by the package on the MS. This could be used when one AS takes over management of the dialogs if the AS that initiated the dialogs fails or is no longer available (see security considerations described in Chapter 7).

12.4.4.1 <audit>

The <audit> request element is sent to the MS to request information about the capabilities of, and dialogs currently managed with, this Control Package. Capabilities include supported dialog languages, grammar formats, record and media types, and codecs. Dialog information includes the status of managed dialogs as well as codecs. The <audit> element has the following attributes:

- capabilities: Indicates whether package capabilities are to be audited. A valid value is a Boolean (see Section 12.4.6.1). A value of true indicates that capability information is to be reported. A value of false indicates that capability information is not to be reported. The attribute is optional. The default value is true.
- dialogs: Indicates whether dialogs currently managed by the package are to be audited. A valid value is a Boolean (see Section 12.4.6.1). A value of true indicates that dialog information is to be reported. A value of false indicates that dialog information is not to be reported. The attribute is optional. The default value is true.
- dialogid: String identifying a specific dialog to audit. The MS sends a response with a 406 status code (Section 12.4.5) if the specified dialog identifier is invalid. The attribute is optional. There is no default value. If the dialogs attribute has the value true, and dialogid attribute is specified, then only audit information about the specified dialog is reported. If the dialogs attribute has the value false, then no dialog audit information is reported even if a dialogid attribute is specified.

The <audit> element has no child elements. When the MS receives an <audit> request, it MUST reply with an <auditresponse> element (Section 12.4.4.2), which includes a mandatory attribute describing the status in terms of a numeric code. Response status codes are defined in Section 4.5. If the request is successful, the <auditresponse> contains (depending on attribute values) a <capabilities> element (Section 12.4.4.2.2) reporting package capabilities and a <dialogs> element (Section 12.4.4.2.3) reporting managed dialog information. If the MS is not able to process the request and carry out the audit operation, the audit request has failed, and the MS MUST indicate the class of failure using an appropriate 4xx response code. Unless an error response code is specified for a class of error within this section, implementations follow Section 12.4.5 in determining the appropriate status code for the response. For example, a request to audit capabilities and dialogs managed by the package:

```
<mscivr version="1.0" xmlns="urn:ietf:params:xml:ns:msc-ivr">
   <audit/>
</mscivr>
```

In this example, only capabilities are to be audited:

```
<mscivr version="1.0" xmlns="urn:ietf:params:xml:ns:msc-ivr">
   <audit dialogs="false"/>
</mscivr>
```

With this example, only a specific dialog is to be audited:

```
<mscivr version="1.0" xmlns="urn:ietf:params:xml:ns:msc-ivr">
   <audit capabilities="false" dialogid="d4"/>
</mscivr>
```

12.4.4.2 <auditresponse>

The <auditresponse> element describes a response to an <audit> request. The <auditresponse> element has the following attributes:

- status: Numeric code indicating the audit response status. The attribute is mandatory. Valid values are defined in Section 12.4.5.
- reason: String specifying a reason for the status. The attribute is optional.
- desclang: Specifies the language used in the value of the reason attribute. A valid value is a language identifier (Section 12.4.6.11). The attribute is optional. If it is not specified, the value of the desclang attribute on <mscivr> (Section 12.4.1) applies.

The <auditresponse> element has the following sequence of child elements:

- <capabilities> element (Section12. 4.4.2.2) describing capabilities of the package. The element is optional.
- <dialogs> element (Section 12.4.4.2.3) describing information about managed dialogs. The element is optional.

For example, a successful response to an <audit> request requesting capabilities and dialogs information:

```
<mscivr version="1.0" xmlns="urn:ietf:params:xml:ns:msc-ivr">
   <auditresponse status="200">
      <capabilities>
         <dialoglanguages>
            <mimetype>application/voicexml+xml
               </mimetype>
         </dialoglanguages>
         <grammartypes/>
         <recordtypes>
            <mimetype>audio/x-wav</mimetype>
            <mimetype>video/3gpp</mimetype>
         </recordtypes>
         <prompttypes>
            <mimetype>audio/x-wav</mimetype>
            <mimetype>video/3gpp</mimetype>
         </prompttypes>
         <variables>
```

```
          <variabletype type="date" desc="value
                formatted as YYYYMMDD">
             <format desc="month year
                   day">mdy</format>
             <format desc="year month
                   day">ymd</format>
             <format desc="day month
                   year">dmy</format>
             <format desc="day month">dm</format>
          </variabletype>
       </variables>
       <maxprepareduration>600s</maxprepareduration>
       <maxrecordduration>1800s</maxrecordduration>
       <codecs>
          <codec name="video">
                <subtype>H263</subtype>
          </codec>
          <codec name="video">
             <subtype>H264</subtype>
          </codec>
          <codec name="audio">
             <subtype>PCMU</subtype>
          </codec>
          <codec name="audio">
             <subtype>PCMA</subtype>
          </codec>
          <codec name="audio">
             <subtype>telephone-event</subtype>
          </codec>
       </codecs>
    </capabilities>
    <dialogs>
       <dialogaudit dialogid="4532" state="preparing"/>
       <dialogaudit dialogid="4599" state="prepared"/>
       <dialogaudit dialogid="1234" state="started"
             conferenceid="conf1">
          <codecs>
             <codec name="audio">
                <subtype>PCMA</subtype>
             </codec>
             <codec name="audio">
                <subtype>telephone-event</subtype>
             </codec>
          </codecs>
       </dialogaudit>
    </dialogs>
 </auditresponse>
</mscivr>
```

12.4.4.2.1 <codecs>

The <codecs> provides audit information about codecs. The <codecs> element has no attributes. The <codecs> element has the following sequence of child elements (0 or more occurrences):

- <codec>: Audit information for a codec (Section 12.4.4.2.1.1). The element is optional.

For example, a fragment describing two codecs:

```
<codecs>
   <codec name="audio">
      <subtype>PCMA</subtype>
   </codec>
   <codec name="audio">
      <subtype>telephone-event</subtype>
   </codec>
</codecs>
```

12.4.4.2.1.1 <codec> The <codec> element describes a codec on the MS. The element is modeled on the <codec> element in the XCON conference information data model (RFC 6501 – see Chapter 4) but allows additional information (e.g., rate, speed, etc.) to be specified. The <codec> element has the following attributes:

- name: Indicates the type name of the codec's media format as defined in Internet Assigned Numbers Authority (IANA) [7]. A valid value is a "type-name" as defined in Section 4.2 of RFC 4288. The attribute is mandatory.

The <codec> element has the following sequence of child elements:

- <subtype>: Element whose content model describes the subtype of the codec's media format as defined in IANA [7]. A valid value is a "subtype-name" as defined in Section 4.2 of RFC 4288. The element is mandatory.
- <params>: element (Section 12.4.2.6) describing additional information about the codec.

This package is agnostic to the names and values of the codec parameters supported by an implementation. The element is optional.

For example, a fragment with a <codec> element describing the H263 video codec:

```
<codec name="video">
   <subtype>H263</subtype>
</codec>
```

12.4.4.2.2 <capabilities>

The <capabilities> element provides audit information about package capabilities. The <capabilities> element has no attributes. The <capabilities> element has the following sequence of child elements:

- <dialoglanguages>: Element (Section 12.4.4.2.2.1) describing additional dialog languages supported by the MS. The element is mandatory.
- <grammartypes>: Element (Section 12.4.4.2.2.2) describing supported <grammar> (Section 12.4.3.1.3.1) format types. The element is mandatory.
- <recordtypes>: Element (Section 12.4.4.2.2.3) describing <media> (Section 12.4.3.1.5) format types supported for <record> (Section 12.4.3.1.4). The element is mandatory.
- <prompttypes>: Element (Section 12.4.4.2.2.4) describing supported <media> (Section 12.4.3.1.5) format types for playback within a <prompt> (Section 12.4.3.1.1). The element is mandatory.
- <variables>: Element (Section 12.4.4.2.2.5) describing supported types and formats for the <variable> element (Section 12.4.3.1.1.1). The element is mandatory.
- <maxpreparedduration>: Element (Section 12.4.4.2.2.6) describing the supported maximum duration for a prepared dialog following a <dialogprepare> (Section 12.4.2.1) request. The element is mandatory.

- <maxrecordduration>: Element (Section 12.4.4.2.2.7) describing the supported maximum duration for a recording <record> (Section 12.4.3.1.4) request. The element is mandatory.
- <codecs>: Element (Section 12.4.4.2.1) describing codecs available to the package. The element is mandatory.

For example, a fragment describing capabilities:

```
<capabilities>
   <dialoglanguages>
      <mimetype>application/voicexml+xml</mimetype>
   </dialoglanguages>
   <grammartypes/>
   <recordtypes>
      <mimetype>audio/x-wav</mimetype>
      <mimetype>video/3gpp</mimetype>
   </recordtypes>
   <prompttypes>
      <mimetype>audio/x-wav</mimetype>
      <mimetype>video/3gpp</mimetype>
   </prompttypes>
   <variables/>
   <maxpreparedduration>30s</maxpreparedduration>
   <maxrecordduration>60s</maxrecordduration>
   <codecs>
      <codec name="video">
         <subtype>H263</subtype>
      </codec>
      <codec name="video">
         <subtype>H264</subtype>
      </codec>
      <codec name="audio">
         <subtype>PCMU</subtype>
      </codec>
      <codec name="audio">
         <subtype>PCMA</subtype>
      </codec>
      <codec name="audio">
         <subtype>telephone-event</subtype>
      </codec>
   </codecs>
</capabilities>
```

12.4.4.2.2.1 <dialoglanguages> The <dialoglanguages> element provides information about additional dialog languages supported by the package. Dialog languages are identified by their associated MIME media types. The MS MUST NOT include the mandatory dialog language for this package (Section 12.4.3). The <dialoglanguages> element has no attributes. The <dialoglanguages> element has the following sequence of child elements (0 or more occurrences):

- <mimetype>: Element whose content model describes a MIME media type (Section 4.6.10) associated with a supported dialog language. The element is optional.

12.4.4.2.2.2 <grammartypes> The <grammartypes> element provides information about <grammar> format types supported by the package. The MS MUST NOT include the mandatory

SRGS format type, "application/srgs+xml" (Section 12.4.3.1.3.1). The <grammartypes> element has no attributes. The <grammartypes> element has the following sequence of child elements (0 or more occurrences):

- <mimetype>: Element whose content model describes a mime type (Section 12.4.6.10). The element is optional.

12.4.4.2.2.3 <recordtypes> The <recordtypes> element provides information about media resource format types of <record> supported by the package (Section 12.4.3.1.4). The <recordtypes> element has no attributes. The <recordtypes> element has the following sequence of child elements (0 or more occurrences):

- <mimetype>: Element whose content model describes a mime type (Section 12.4.6.10). The element is optional.

12.4.4.2.2.4 <prompttypes> The <prompttypes> element provides information about media resource format types of <prompt> supported by the package (Section 12.4.3.1.1). The <prompttypes> element has no attributes. The <prompttypes> element has the following sequence of child elements (0 or more occurrences):

- <mimetype>: Element whose content model describes a mime type (Section 12.4.6.10). The element is optional.

12.4.4.2.2.5 <variables> The <variables> element provides information about types and formats for the <variable> element (Section 12.4.3.1.1.1) supported by the package. The <variables> element has no attributes. The <variables> element has the following sequence of child elements (0 or more occurrences):

<variabletype>: Element describing the formats support for a given type (Section 12.4.4.2.2.5.1). The element is optional. For example, a fragment describing support for <variable> with a "date" type according to the formats specified in Section 12.4.3.1.1.1.1:

```
<variables>
   <variabletype type="date" desc="value formatted as YYYYMMDD">
      <format desc="month year day">mdy</format>
      <format desc="year month day">ymd</format>
      <format desc="day month year">dmy</format>
      <format desc="day month">dm</format>
   </variabletype>
</variables>
```

12.4.4.2.2.5.1 <variabletype> The <variabletype> element describes the formats supported for <variable> supported type. The <variabletype> element has the following attributes:

- type: Indicates a supported value associated with the type attribute of the <variable> element. The attribute is mandatory.
- desc: A string providing some textual description of the type and format. The attribute is optional.
- desclang: specifies the language used in the value of the desc attribute. A valid value is a language identifier (Section 12.4.6.11). The attribute is optional. If it is not specified, the value of the desclang attribute on <mscivr> (Section 12.4.1) applies.

The <variabletype> element has the following sequence of child elements (0 or more occurrences):

- <format>: Element with a desc attribute (optional description), desclang (optional language identifier for the description), and a content model describing a supported format in the <variable> format attribute. The element is optional.

12.4.4.2.2.6 <maxprepareddduration> The <maxprepareddduration> element describes the maximum duration for a dialog to remain in the prepared state (Section 12.4.2) following a <dialogprepare> (Section 12.4.2.1) request. The <maxprepareddduration> element has no attributes. The <maxprepareddduration> element has a content model describing the maximum prepared dialog duration as a time designation (Section 12.4.6.7).

12.4.4.2.2.7 <maxrecordduration> The <maxrecordduration> element describes the maximum recording duration for a <record> (Section 12.4.3.1.4) request supported by the MS. The <maxrecordduration> element has no attributes. The <maxrecordduration> element has a content model describing the maximum duration of recording as a time designation (Section 12.4.6.7).

12.4.4.2.3 <dialogs>

The <dialogs> element provides audit information about dialogs. The <dialogs> element has no attributes. The <dialogs> element has the following sequence of child elements (0 or more occurrences):

- <dialogaudit>: Audit information for a dialog (Section 12.4.4.2.3.1). The element is optional.

12.4.4.2.3.1 <dialogaudit> The <dialogaudit> element has the following attributes:

- dialogid: String identifying the dialog. The attribute is mandatory.
- state: String indicating the state of the dialog. Valid values are preparing, prepared, starting, and started. The attribute is mandatory.
- connectionid: String identifying the SIP dialog connection associated with the dialog (see Section 8.14.1 [Appendix A.1 of RFC 6230]). The attribute is optional. There is no default value.
- conferenceid: String identifying the conference associated with the dialog (see Section 8.14.1 [Appendix A.1 of RFC 6230]). The attribute is optional. There is no default value.

The <dialogaudit> element has the following child element: <codecs> element describing codecs used in the dialog. See Section 12.4.4.2.1. The element is optional. For example, a fragment describing a started dialog that is using PCMU and telephony-event audio codecs:

```
<dialogaudit dialogid="1234" state="started" conferenceid="conf1">
   <codecs>
      <codec name="audio">
         <subtype>PCMU</subtype>
      </codec>
      <codec name="audio">
         <subtype>telephone-event</subtype>
      </codec>
   </codecs>
</dialogaudit>
```

12.4.5 Response Status Codes

This section describes the response codes in Table 12.1 for the status attribute of dialog management <response> (Section 12.4.2.4) and audit <auditresponse> (Section 12.4.4.2) responses. The

TABLE 12.1
Status Codes

Code	Summary	Description	Informational: AS Possible Recovery Action
200	OK	Request has succeeded.	
400	Syntax error	Request is syntactically invalid: It is not valid with respect to the XML schema specified in Section 12.5, or it violates a co-occurrence constraint for a request element defined in Section 12.4.	Change the request so that it is syntactically valid.
405	dialogid already exists	Request uses a dialogid identifier for a new dialog that is already used by another dialog on the MS (see Section 12.4.2).	Send a request for a new dialog without specifying the dialogid and let MS generate a unique dialogid in the response.
406	dialogid does not exist	Request uses a dialogid identifier for a dialog that does not exist on the MS (see Section 12.4.2).	Send an <audit> request (Section 12.4.4.1) requesting the list of dialogid identifiers already used by the MS and then use one of the listed dialog identifiers.
407	connectionid does not exist	Request uses a connectionid identifier for a connection that does not exist on the MS.	Use another method to determine which connections are available on the MS.
408	conferenceid does not exist	Request uses a conferenceid identifier for a connection that does not exist on the MS.	Use another method to determine which conferences are available on the MS.
409	Resource cannot be retrieved	Request uses a URI to reference an external resource (e.g. dialog, media, or grammar) that cannot be retrieved within the timeout interval.	Check that the resource URI is valid and can be reached from the MS, and that the appropriate authentication is used.
410	Dialog execution canceled	Request to prepare or start a dialog that has been terminated by a <dialogterminate/> request (see Section 12.4.2).	
411	Incompatible stream configuration	Request specifies a media stream configuration that is incomplete with itself or the connection or conference capabilities (see Section 12.4.2.2).	Change the media stream configuration to match the capabilities of the connection or conference.
412	Media stream not available	Request specifies an operation for which a media stream is not available. For example, playing a video media resource on an connection or conference without video streams.	Check the media stream capability of the connection or conference and use an operation that uses these capabilities.
413	Control keys with same value	Request contains a <control> element (Section 12.4.3.1.2) where some keys have the same value.	Use different keys for the different control operations.
419	Other execution error	Requested operation cannot be executed by the MS.	Use a URI scheme that is supported.
421	Unsupported dialog language	Request references an external dialog language not supported by the MS.	Send an <audit> request (Section 12.4.4.1) requesting the MS capabilities and then use one of the listed dialog languages.

(Continued)

TABLE 12.1 (CONTINUED)
Status Codes

Code	Summary	Description	Informational: AS Possible Recovery Action
422	Unsupported playback format	Request references a media resource for playback whose format is not supported by the MS.	Send an <audit> request (Section 12.4.4.1) requesting the MS capabilities and then use one of the listed playback media formats.
423	Unsupported record format	Request references a media resource for recording whose format is not supported by the MS.	Send an <audit> request (Section 12.4.4.1) requesting the MS capabilities and then use one of the listed record media formats.
424	Unsupported grammar format	Request references a grammar whose format is not supported by the MS.	Send an <audit> request (Section 12.4.4.1) requesting the MS capabilities and then use one of the listed grammar types.
425	Unsupported variable configuration	Request contains a prompt <variable> element (Section 12.4.3.1.1.1) not supported by the MS.	Send an <audit> request (Section 4.4.1) requesting the MS capabilities and then use one of the listed variable types.
426	Unsupported DTMF configuration	Request contains a prompt <dtmf> element (Section 12.4.3.1.1.2) not supported by the MS.	
427	Unsupported parameter	Request contains a prompt <param> element (Section 12.4.2.6.1) not supported by the MS.	
428	Unsupported media stream configuration	Request contains a prompt <stream> element (Section 12.4.2.2.2) not supported by the MS.	
429	Unsupported playback configuration	Request contains a prompt <prompt> element (Section 12.4.3.1.1) that the MS is unable to play on the available output media streams.	
430	Unsupported record configuration	Request contains a prompt <record> element (Section 12.4.3.1.4) that the MS is unable to record with on the available input media streams.	
431	Unsupported foreign namespace attribute or element	Request contains attributes or elements from another namespace that the MS does not support.	
432	Unsupported multiple dialog capability	Request tries to start another dialog on the same conference or connection where a dialog is already running.	
433	Unsupported collect or record capability	Request contains <collect> and <record> elements, and the MS does not support these operations simultaneously.	
434	Unsupported VAD capability	Request contains a <record> element where VAD is required, but the MS does not support VAD.	
435	Unsupported parallel playback	Request contains a prompt <par> element whose configuration is not supported by the MS.	
439	Other unsupported capability	Request requires another capability not supported by the MS.	

MS MUST support the status response codes defined here. All other valid but undefined values are reserved for future use, where new status codes are assigned using the Standards Action process defined in RFC 5226. The AS MUST treat any responses it does not recognize as being equivalent to the x00 response code for all classes. For example, if an AS receives an unrecognized response code of 499, it can safely assume that there was something wrong with its request and treat the response as if it had received a 400 (Syntax error) response code. 4xx responses are definite failure responses from a particular MS.

The reason attribute in the response SHOULD identify the failure in more detail; for example, "Mandatory attribute missing: src in media element" for a 400 (Syntax error) response code. The AS SHOULD NOT retry the same request without modification (for example, correcting a syntax error or changing the connectionid to use one available on the MS). However, the same request to a different MS might be successful, for example, if another MS supports a capability required in the request. 4xx failure responses can be grouped into three classes: Failure due to a syntax error in the request (400); failure due to an error executing the request on the MS (405–419); and failure due to the request requiring a capability not supported by the MS (420–439). In cases where more than one request code could be reported for a failure, the MS SHOULD use the most specific error code of the failure class for the detected error. For example, if the MS detects that the dialogid in the request is invalid, then it uses a 406 status code. However, if the MS merely detects that an execution error occurred, then 419 is used.

12.4.6 Type Definitions

This section defines types referenced in attribute and element definitions.

12.4.6.1 Boolean

The value space of Boolean is the set {true, false, 1, 0} as defined in Section 3.2.2 of Biron and Malhotra [9]. In accordance with this definition, the concept of false can be lexically represented by the strings "0" and "false" and the concept of true by the strings "1" and "true;" implementations MUST support both styles of lexical representation.

12.4.6.2 DTMFChar

A DTMF character. The value space is the set {0, 1, 2, 3, 4, 5, 6, 7, 8, 9, #, *, A, B, C, D}.

12.4.6.3 DTMFString

A string composed of one or more DTMFChars.

12.4.6.4 Non-Negative Integer

The value space of a non-negative integer is the infinite set {0,1,2,...} as defined in Section 3.3.20 of Biron and Malhotra [9]. Implementation Note: It is RECOMMENDED that implementations at least support a maximum value of a 32-bit integer (2,147,483,647).

12.4.6.5 Positive Integer

The value space of positive integer is the infinite set {1,2,...} as defined in Section 3.3.25 of Biron and Malhotra [9].

Implementation Note: It is RECOMMENDED that implementations at least support a maximum value of a 32-bit integer (2,147,483,647).

12.4.6.6 String

A string in the character encoding associated with the XML element as defined in Section 3.2.1 of Biron and Malhotra [9].

12.4.6.7 Time Designation

A time designation consists of a non-negative real number followed by a time unit identifier. The time unit identifiers are "ms" (milliseconds) and "s" (seconds). Examples include: "3 s," "850 ms," "0.7 s," ".5 s," and "+1.5 s."

12.4.6.8 Percentage

A percentage consists of a positive integer followed by "%." Examples include: "100%," "500%," and "10%."

12.4.6.9 URI

Uniform Resource Indicator as defined in RFC 3986.

12.4.6.10 MIME Media Type

A string formatted as an IANA MIME media type [11]. The ABNF (RFC 5234) production for the string is: type = type-name "/" subtype-name *(";" parameter) parameter = parameter-name "=" value, where "type-name" and "subtype-name" are defined in Section 4.2 of RFC 4288, "parameter-name" is defined in Section 4.3 of RFC 4288, and "value" is defined in Section 5.1 of RFC 2045.

12.4.6.11 Language Identifier

A language identifier labels information content as being of a particular human language variant. Following the XML specification for language identification [6], a legal language identifier is identified by a RFC 5646 code and matched according to RFC 4647.

12.4.6.12 DateTime

A string formatted according to the XML schema definition of a dateTime type [9].

12.5 FORMAL SYNTAX

This section defines the XML schema for the IVR Control Package. The schema is normative. The schema defines datatypes, attributes, dialog management, and IVR dialog elements in the urn:ietf:params:xml:ns:msc-ivr namespace. In most elements, the order of child elements is significant. The schema is extensible: Elements allow attributes and child elements from other namespaces. Elements from outside this package's namespace can occur after elements defined in this package. The schema is dependent upon the schema (framework.xsd) defined in Section 8.14.1 (Appendix A.1 of the Control Framework RFC 6230). It is also dependent upon the W3C (xml.xsd) schema for definitions of XML attributes (e.g., xml:base).

```
<?xml version="1.0" encoding="UTF-8"?>
<xsd:schema targetNamespace="urn:ietf:params:xml:ns:msc-ivr"
      elementFormDefault="qualified" blockDefault="#all"
      xmlns="urn:ietf:params:xml:ns:msc-ivr"
      xmlns:fw="urn:ietf:params:xml:ns:control:framework-attributes"
      xmlns:xsd="http://www.w3.org/2001/XMLSchema">
   <xsd:annotation>
      <xsd:specificationation>
         IETF MediaCtrl IVR 1.0 (20110104)
         This is the schema of the IETF MediaCtrl IVR Control
         Package.
         The schema namespace is urn:ietf:params:xml:ns:msc-ivr
      </xsd:specificationation>
   </xsd:annotation>
```

```
<!--
###############################################################
SCHEMA IMPORTS
###############################################################
-->

<xsd:import namespace="http://www.w3.org/XML/1998/namespace"
      schemaLocation="http://www.w3.org/2001/xml.xsd">
   <xsd:annotation>
      <xsd:specificationation>
         This import brings in the XML attributes for
         xml:base, xml:lang, etc
         See http://www.w3.org/2001/xml.xsd for latest version
      </xsd:specificationation>
   </xsd:annotation>
</xsd:import>
<xsd:import
      namespace="urn:ietf:params:xml:ns:control:
      framework-attributes"
      schemaLocation="framework.xsd">
   <xsd:annotation>
      <xsd:specificationation>
         This import brings in the framework attributes for
         conferenceid and connectionid.
      </xsd:specificationation>
   </xsd:annotation>
</xsd:import>

<!--
######################################################
Extensible core type
######################################################
-->

<xsd:complexType name="Tcore">
   <xsd:annotation>
      <xsd:specificationation>
         This type is extended by other (non-mixed) component
         types to allow attributes from other namespaces.
      </xsd:specificationation>
   </xsd:annotation>
   <xsd:sequence/>
   <xsd:anyAttribute namespace="##other" processContents="lax" />
</xsd:complexType>

<!--
######################################################
TOP LEVEL ELEMENT: mscivr
######################################################
-->

<xsd:complexType name="mscivrType">
   <xsd:complexContent>
      <xsd:extension base="Tcore">
         <xsd:sequence>
            <xsd:choice>
```

```
                    <xsd:element ref="dialogprepare" />
                    <xsd:element ref="dialogstart" />
                    <xsd:element ref=
                          "dialogterminate" />
                    <xsd:element ref="response" />
                    <xsd:element ref="event" />
                    <xsd:element ref="audit" />
                    <xsd:element ref="auditresponse" />
                    <xsd:any namespace="##other" minOccurs="0"
                          maxOccurs="unbounded" processContents="lax" />
                </xsd:choice>
            </xsd:sequence>
            <xsd:attribute name="version"
                  type="version.datatype"
                  use="required" />
            <xsd:attribute name="desclang"
                  type="xsd:language"
                  default="i-default" />
        </xsd:extension>
    </xsd:complexContent>
</xsd:complexType>
<xsd:element name="mscivr" type="mscivrType" />

<!--
################################################################
DIALOG MANAGEMENT TYPES
################################################################
-->

<!-- dialogprepare -->

<xsd:complexType name="dialogprepareType">
    <xsd:complexContent>
        <xsd:extension base="Tcore">
            <xsd:sequence>
                <xsd:element ref="dialog" minOccurs="0"
                      maxOccurs="1" />
                <xsd:element ref="params" minOccurs="0"
                      maxOccurs="1" />
                <xsd:any namespace="##other"
                      minOccurs="0"
                      maxOccurs="unbounded" processContents="lax" />
            </xsd:sequence>
            <xsd:attribute name="src" type="xsd:anyURI" />
            <xsd:attribute name="type"
                  type="mime.datatype"/>
            <xsd:attribute name="maxage"
                  type="xsd:nonNegativeInteger"/>
            <xsd:attribute name="maxstale"
                  type="xsd:nonNegativeInteger"/>
            <xsd:attribute name="fetchtimeout"
                  type="timedesignation.datatype" default="30s" />
            <xsd:attribute name="dialogid"
                  type="dialogid.datatype" />
        </xsd:extension>
    </xsd:complexContent>
```

```
</xsd:complexType>
<xsd:element name="dialogprepare" type="dialogprepareType" />

<!-- dialogstart -->

<xsd:complexType name="dialogstartType">
   <xsd:complexContent>
      <xsd:extension base="Tcore">
         <xsd:sequence>
            <xsd:element ref="dialog" minOccurs="0"
                  maxOccurs="1" />
            <xsd:element ref="subscribe" minOccurs="0"
                  maxOccurs="1" />
            <xsd:element ref="params" minOccurs="0"
                  maxOccurs="1" />
            <xsd:element ref="stream" minOccurs="0"
                  maxOccurs="unbounded" />
            <xsd:any namespace="##other" minOccurs="0"
                  maxOccurs="unbounded" processContents="lax" />
         </xsd:sequence>
         <xsd:attribute name="src" type="xsd:anyURI" />
         <xsd:attribute name="type" type="mime.datatype"/>
         <xsd:attribute name="maxage"
               type="xsd:nonNegativeInteger"/>
         <xsd:attribute name="maxstale"
               type="xsd:nonNegativeInteger"/>
         <xsd:attribute name="fetchtimeout"
               type="timedesignation.datatype" default="30s" />
         <xsd:attribute name="dialogid"
               type="dialogid.datatype" />
         <xsd:attribute name="prepareddialogid"
               type="dialogid.datatype" />
         <xsd:attributeGroup ref="fw:framework-attributes" />
      </xsd:extension>
   </xsd:complexContent>
</xsd:complexType>
<xsd:element name="dialogstart" type="dialogstartType" />

<!-- dialogterminate -->

<xsd:complexType name="dialogterminateType">
   <xsd:complexContent>
      <xsd:extension base="Tcore">
         <xsd:sequence>
            <xsd:any namespace="##other"
                  minOccurs="0"
                  maxOccurs="unbounded" processContents="lax" />
         </xsd:sequence>
         <xsd:attribute name="dialogid"
               type="dialogid.datatype" use="required" />
         <xsd:attribute name="immediate"
               type="xsd:boolean" default="false" />
      </xsd:extension>
   </xsd:complexContent>
</xsd:complexType>
<xsd:element name="dialogterminate" type="dialogterminateType" />
```

```
<!-- response -->

<xsd:complexType name="responseType">
    <xsd:complexContent>
        <xsd:extension base="Tcore">
            <xsd:sequence>
                <xsd:any namespace="##other"
                        minOccurs="0"
                        maxOccurs="unbounded" processContents="lax" />
            </xsd:sequence>
            <xsd:attribute name="status" type="status.datatype"
                    use="required" />
            <xsd:attribute name="reason" type="xsd:string" />
            <xsd:attribute name="desclang"
                    type="xsd:language"/>
            <xsd:attribute name="dialogid"
                    type="dialogid.datatype" use="required" />
            <xsd:attributeGroup ref=
                    "fw:framework-attributes" />
        </xsd:extension>
    </xsd:complexContent>
</xsd:complexType>
<xsd:element name="response" type="responseType" />

<!-- event -->

<xsd:complexType name="eventType">
    <xsd:complexContent>
        <xsd:extension base="Tcore">
            <xsd:sequence>
                <xsd:choice>
                    <xsd:element ref="dialogexit"
                            minOccurs="0"
                            maxOccurs="1" />
                    <xsd:element ref="dtmfnotify"
                            minOccurs="0"
                            maxOccurs="1" />
                    <xsd:any namespace="##other"
                            minOccurs="0"
                            maxOccurs="unbounded" processContents="lax" />
                </xsd:choice>
            </xsd:sequence>
            <xsd:attribute name="dialogid"
                    type="dialogid.datatype" use="required" />
        </xsd:extension>
    </xsd:complexContent>
</xsd:complexType>
<xsd:element name="event" type="eventType" />

<!-- dialogexit-->

<xsd:complexType name="dialogexitType">
    <xsd:complexContent>
        <xsd:extension base="Tcore">
            <xsd:sequence>
                <xsd:element ref="promptinfo"
```

```
                    minOccurs="0"
                    maxOccurs="1" />
                <xsd:element ref="controlinfo"
                    minOccurs="0"
                    maxOccurs="1" />
                <xsd:element ref="collectinfo"
                    minOccurs="0"
                    maxOccurs="1" />
                <xsd:element ref="recordinfo"
                    minOccurs="0"
                    maxOccurs="1" />
                <xsd:element ref="params" minOccurs="0"
                    maxOccurs="1" />
                <xsd:any namespace="##other"
                    minOccurs="0"
                    maxOccurs="unbounded" processContents="lax" />
            </xsd:sequence>
            <xsd:attribute name="status"
                type="xsd:nonNegativeInteger" use="required" />
            <xsd:attribute name="reason" type="xsd:string" />
            <xsd:attribute name="desclang"
                type="xsd:language"/>
        </xsd:extension>
    </xsd:complexContent>
</xsd:complexType>
<xsd:element name="dialogexit" type="dialogexitType" />

<!-- dtmfnotify-->

<xsd:complexType name="dtmfnotifyType">
    <xsd:complexContent>
        <xsd:extension base="Tcore">
            <xsd:sequence>
                <xsd:any namespace="##other"
                    minOccurs="0"
                    maxOccurs="unbounded" processContents="lax" />
            </xsd:sequence>
            <xsd:attribute name="matchmode"
                type="matchmode.datatype" default="all" />
            <xsd:attribute name="dtmf"
                type="dtmfstring.datatype"
                use="required" />
            <xsd:attribute name="timestamp"
                type="xsd:dateTime"
                use="required" />
        </xsd:extension>
    </xsd:complexContent>
</xsd:complexType>
<xsd:element name="dtmfnotify" type="dtmfnotifyType" />

<!-- promptinfo -->

<xsd:complexType name="promptinfoType">
    <xsd:complexContent>
        <xsd:extension base="Tcore">
            <xsd:sequence>
```

```
                <xsd:any namespace="##other"
                      minOccurs="0"
                      maxOccurs="unbounded" processContents="lax" />
            </xsd:sequence>
            <xsd:attribute name="duration"
                  type="xsd:nonNegativeInteger" />
            <xsd:attribute name="termmode"
                  type="prompt_termmode.datatype" use="required" />
        </xsd:extension>
    </xsd:complexContent>
</xsd:complexType>
<xsd:element name="promptinfo" type="promptinfoType" />

<!-- controlinfo -->

<xsd:complexType name="controlinfoType">
    <xsd:complexContent>
        <xsd:extension base="Tcore">
            <xsd:sequence>
                <xsd:element ref="controlmatch"
                      minOccurs="0"
                      maxOccurs="unbounded" />
                <xsd:any namespace="##other"
                      minOccurs="0"
                      maxOccurs="unbounded" processContents="lax" />
            </xsd:sequence>
        </xsd:extension>
    </xsd:complexContent>
</xsd:complexType>
<xsd:element name="controlinfo" type="controlinfoType" />

<!-- controlmatch -->

<xsd:complexType name="controlmatchType">
    <xsd:complexContent>
        <xsd:extension base="Tcore">
            <xsd:sequence>
                <xsd:any namespace="##other"
                      minOccurs="0"
                      maxOccurs="unbounded" processContents="lax" />
            </xsd:sequence>
            <xsd:attribute name="dtmf"
                  type="dtmfstring.datatype" />
            <xsd:attribute name="timestamp"
                  type="xsd:dateTime" />
        </xsd:extension>
    </xsd:complexContent>
</xsd:complexType>
<xsd:element name="controlmatch" type="controlmatchType" />

<!-- collectinfo -->

<xsd:complexType name="collectinfoType">
    <xsd:complexContent>
        <xsd:extension base="Tcore">
            <xsd:sequence>
```

```
                <xsd:any namespace="##other"
                        minOccurs="0"
                        maxOccurs="unbounded" processContents="lax" />
            </xsd:sequence>
            <xsd:attribute name="dtmf"
                    type="dtmfstring.datatype" />
            <xsd:attribute name="termmode"
                    type="collect_termmode.datatype" use="required" />
        </xsd:extension>
    </xsd:complexContent>
</xsd:complexType>
<xsd:element name="collectinfo" type="collectinfoType" />

<!-- recordinfo -->

<xsd:complexType name="recordinfoType">
    <xsd:complexContent>
        <xsd:extension base="Tcore">
            <xsd:sequence>
                <xsd:element ref="mediainfo"
                        minOccurs="0"
                        maxOccurs="unbounded" />
                    <xsd:any namespace="##other"
                        minOccurs="0"
                        maxOccurs="unbounded" processContents="lax" />
            </xsd:sequence>
            <xsd:attribute name="duration"
                    type="xsd:nonNegativeInteger" />
            <xsd:attribute name="termmode"
                    type="record_termmode.datatype" use="required" />
        </xsd:extension>
    </xsd:complexContent>
</xsd:complexType>
<xsd:element name="recordinfo" type="recordinfoType" />

<!-- mediainfo -->

<xsd:complexType name="mediainfoType">
    <xsd:complexContent>
        <xsd:extension base="Tcore">
            <xsd:sequence>
                <xsd:any namespace="##other"
                        minOccurs="0"
                        maxOccurs="unbounded" processContents="lax" />
            </xsd:sequence>
            <xsd:attribute name="loc" type="xsd:anyURI"
                    use="required" />
            <xsd:attribute name="type" type="mime.datatype"
                    use="required"/>
            <xsd:attribute name="size"
                    type="xsd:nonNegativeInteger" />
        </xsd:extension>
    </xsd:complexContent>
</xsd:complexType>
<xsd:element name="mediainfo" type="mediainfoType" />
```

```
    <!-- subscribe -->

    <xsd:complexType name="subscribeType">
       <xsd:complexContent>
          <xsd:extension base="Tcore">
             <xsd:sequence>
                <xsd:element ref="dtmfsub" minOccurs="0"
                      maxOccurs="unbounded" />
                <xsd:any namespace="##other"
                      minOccurs="0"
                      maxOccurs="unbounded" processContents="lax" />
             </xsd:sequence>
          </xsd:extension>
       </xsd:complexContent>
    </xsd:complexType>
    <xsd:element name="subscribe" type="subscribeType" />

 <!-- dtmfsub -->

    <xsd:complexType name="dtmfsubType">
       <xsd:complexContent>
          <xsd:extension base="Tcore">
             <xsd:sequence>
                <xsd:any namespace="##other"
                      minOccurs="0"
                      maxOccurs="unbounded" processContents="lax" />
             </xsd:sequence>
             <xsd:attribute name="matchmode"
                   type="matchmode.datatype" default="all" />
          </xsd:extension>
       </xsd:complexContent>
    </xsd:complexType>
    <xsd:element name="dtmfsub" type="dtmfsubType" />

    <!-- params -->

    <xsd:complexType name="paramsType">
       <xsd:complexContent>
          <xsd:extension base="Tcore">
             <xsd:sequence>
                <xsd:element ref="param" minOccurs="0"
                      maxOccurs="unbounded" />
                <xsd:any namespace="##other"
                      minOccurs="0"
                      maxOccurs="unbounded" processContents="lax" />
             </xsd:sequence>
          </xsd:extension>
       </xsd:complexContent>
    </xsd:complexType>
    <xsd:element name="params" type="paramsType" />

    <!-- param -->
    <!-- doesn't extend tCore since its content model is mixed -->

    <xsd:complexType name="paramType" mixed="true">
       <xsd:sequence/>
```

```
    <xsd:attribute name="name" type="xsd:string" use="required" />
    <xsd:attribute name="type" type="mime.datatype"
          default="text/plain"/>
    <xsd:attribute name="encoding" type="xsd:string"/>
    <xsd:anyAttribute namespace="##other" processContents="lax" />
</xsd:complexType>
<xsd:element name="param" type="paramType" />

<!-- stream -->

<xsd:complexType name="streamType">
    <xsd:complexContent>
        <xsd:extension base="Tcore">
            <xsd:sequence>
                <xsd:element ref="region" minOccurs="0"
                      maxOccurs="1" />
                <xsd:element ref="priority" minOccurs="0"
                      maxOccurs="1" />
                <xsd:any namespace="##other"
                      minOccurs="0"
                      maxOccurs="unbounded" processContents="lax" />
            </xsd:sequence>
            <xsd:attribute name="media"
                  type="media.datatype"
                  use="required" />
            <xsd:attribute name="label" type=
                  "label.datatype" />
            <xsd:attribute name="direction"
                  type="direction.datatype" default="sendrecv" />
        </xsd:extension>
    </xsd:complexContent>
</xsd:complexType>
<xsd:element name="stream" type="streamType" />

<!-- region -->

<xsd:simpleType name="regionType">
    <xsd:restriction base="xsd:NMTOKEN"/>
</xsd:simpleType>
<xsd:element name="region" type="regionType" />

<!-- priority -->

<xsd:simpleType name="priorityType">
    <xsd:restriction base="xsd:positiveInteger" />
</xsd:simpleType>
<xsd:element name="priority" type="priorityType" />

<!-- dialog -->

<xsd:complexType name="dialogType">
    <xsd:complexContent>
        <xsd:extension base="Tcore">
            <xsd:sequence>
                <xsd:element ref="prompt" minOccurs="0"
                      maxOccurs="1" />
```

```
                <xsd:element ref="control" minOccurs="0"
                      maxOccurs="1" />
                <xsd:element ref="collect" minOccurs="0"
                      maxOccurs="1" />
                <xsd:element ref="record" minOccurs="0"
                      maxOccurs="1" />
                <xsd:any namespace="##other"
                      minOccurs="0"
                      maxOccurs="unbounded" processContents="lax" />
            </xsd:sequence>
            <xsd:attribute name="repeatCount"
                  type="xsd:nonNegativeInteger"
                  default="1" />
            <xsd:attribute name="repeatDur"
                  type="timedesignation.datatype" />
            <xsd:attribute name="repeatUntilComplete"
                  type="xsd:boolean" default="false"/>
        </xsd:extension>
    </xsd:complexContent>
</xsd:complexType>
<xsd:element name="dialog" type="dialogType" />

<!-- prompt -->

<xsd:complexType name="promptType">
    <xsd:complexContent>
        <xsd:extension base="Tcore">
            <xsd:choice minOccurs="1"
                  maxOccurs="unbounded">
                <xsd:element ref="media" />
                <xsd:element ref="variable" />
                <xsd:element ref="dtmf" />
                <xsd:element ref="par" />
                <xsd:any namespace="##other"
                      processContents="lax" />
            </xsd:choice>
            <xsd:attribute ref="xml:base" />
            <xsd:attribute name="bargein" type="xsd:boolean"
                  default="true" />
        </xsd:extension>
    </xsd:complexContent>
</xsd:complexType>
<xsd:element name="prompt" type="promptType" />

<!-- media -->

<xsd:complexType name="mediaType">
    <xsd:complexContent>
        <xsd:extension base="Tcore">
            <xsd:sequence>
                <xsd:any namespace="##other"
                      minOccurs="0"
                      maxOccurs="unbounded" processContents="lax" />
            </xsd:sequence>
            <xsd:attribute name="loc" type="xsd:anyURI"
                  use="required" />
```

```
            <xsd:attribute name="type" type=
                  "mime.datatype" />
            <xsd:attribute name="fetchtimeout"
                  type="timedesignation.datatype" default="30s" />
            <xsd:attribute name="soundLevel"
                  type="percentage.datatype"
                  default="100%" />
            <xsd:attribute name="clipBegin"
                  type="timedesignation.datatype" default="0s" />
            <xsd:attribute name="clipEnd"
                  type="timedesignation.datatype"/>
         </xsd:extension>
      </xsd:complexContent>
   </xsd:complexType>
   <xsd:element name="media" type="mediaType" />

   <!-- variable -->

   <xsd:complexType name="variableT">
      <xsd:complexContent>
         <xsd:extension base="Tcore">
            <xsd:sequence>
               <xsd:any namespace="##other"
                     minOccurs="0"
                     maxOccurs="unbounded" processContents="lax" />
            </xsd:sequence>
            <xsd:attribute name="value" type="xsd:string"
                  use="required" />
            <xsd:attribute name="type" type="xsd:string"
                  use="required" />
            <xsd:attribute name="format" type="xsd:string" />
            <xsd:attribute name="gender"
                  type="gender.datatype" />
            <xsd:attribute ref="xml:lang" />
         </xsd:extension>
      </xsd:complexContent>
   </xsd:complexType>
   <xsd:element name="variable" type="variableT" />

   <!-- dtmf -->

   <xsd:complexType name="dtmfType">
      <xsd:complexContent>
         <xsd:extension base="Tcore">
            <xsd:sequence>
               <xsd:any namespace="##other"
                     minOccurs="0"
                     maxOccurs="unbounded" processContents="lax" />
            </xsd:sequence>
            <xsd:attribute name="digits"
                  type="dtmfstring.datatype"
                  use="required" />
            <xsd:attribute name="level" type="xsd:integer"
                  default="-6" />
            <xsd:attribute name="duration"
                  type="timedesignation.datatype" default="100ms" />
```

```
            <xsd:attribute name="interval"
                    type="timedesignation.datatype" default="100ms" />
        </xsd:extension>
    </xsd:complexContent>
</xsd:complexType>
<xsd:element name="dtmf" type="dtmfType" />

<!-- par -->

<xsd:complexType name="parType">
    <xsd:complexContent>
        <xsd:extension base="Tcore">
            <xsd:choice minOccurs="1"
                    maxOccurs="unbounded">
                <xsd:element ref="media" />
                <xsd:element ref="variable" />
                <xsd:element ref="dtmf" />
                <xsd:element ref="seq" />
                <xsd:any namespace="##other"
                        processContents="lax" />
            </xsd:choice>
            <xsd:attribute name="endsync"
                    type="endsync.datatype"
                    default="last"/>
        </xsd:extension>
    </xsd:complexContent>
</xsd:complexType>
<xsd:element name="par" type="parType" />

<!-- seq -->

<xsd:complexType name="seqType">
    <xsd:complexContent>
        <xsd:extension base="Tcore">
            <xsd:choice minOccurs="1"
                    maxOccurs="unbounded">
                    <xsd:element ref="media" />
                    <xsd:element ref="variable" />
                    <xsd:element ref="dtmf" />
                    <xsd:any namespace="##other"
                            processContents="lax" />
            </xsd:choice>
        </xsd:extension>
    </xsd:complexContent>
</xsd:complexType>
<xsd:element name="seq" type="seqType" />

<!-- control -->

<xsd:complexType name="controlType">
    <xsd:complexContent>
        <xsd:extension base="Tcore">
            <xsd:sequence>
                <xsd:any namespace="##other"
                        minOccurs="0"
                        maxOccurs="unbounded" processContents="lax" />
```

```
            </xsd:sequence>
            <xsd:attribute name="skipinterval"
                  type="timedesignation.datatype" default="6s" />
            <xsd:attribute name="ffkey"
                  type="dtmfchar.datatype" />
            <xsd:attribute name="rwkey"
                  type="dtmfchar.datatype" />
            <xsd:attribute name="pauseinterval"
                  type="timedesignation.datatype" default="10s" />
            <xsd:attribute name="pausekey"
                  type="dtmfchar.datatype" />
            <xsd:attribute name="resumekey"
                  type="dtmfchar.datatype" />
            <xsd:attribute name="volumeinterval"
                  type="percentage.datatype"
                  default="10%" />
            <xsd:attribute name="volupkey"
                  type="dtmfchar.datatype" />
            <xsd:attribute name="voldnkey"
                  type="dtmfchar.datatype" />
            <xsd:attribute name="speedinterval"
                  type="percentage.datatype"
                  default="10%" />
            <xsd:attribute name="speedupkey"
                  type="dtmfchar.datatype" />
            <xsd:attribute name="speeddnkey"
                  type="dtmfchar.datatype" />
            <xsd:attribute name="gotostartkey"
                  type="dtmfchar.datatype" />
            <xsd:attribute name="gotoendkey"
                  type="dtmfchar.datatype" />
            <xsd:attribute name="external"
                  type="dtmfstring.datatype" />
        </xsd:extension>
    </xsd:complexContent>
</xsd:complexType>
<xsd:element name="control" type="controlType" />

<!-- collect -->

<xsd:complexType name="collectType">
    <xsd:complexContent>
        <xsd:extension base="Tcore">
            <xsd:sequence>
                <xsd:element ref="grammar"
                      minOccurs="0"
                      maxOccurs="1" />
                <xsd:any namespace="##other"
                      minOccurs="0"
                      maxOccurs="unbounded" processContents="lax" />
            </xsd:sequence>
            <xsd:attribute name="cleardigitbuffer"
                  type="xsd:boolean" default="true" />
            <xsd:attribute name="timeout"
                  type="timedesignation.datatype" default="5s" />
            <xsd:attribute name="interdigittimeout"
                  type="timedesignation.datatype" default="2s" />
```

```
            <xsd:attribute name="termtimeout"
                  type="timedesignation.datatype" default="0s" />
            <xsd:attribute name="escapekey"
                  type="dtmfchar.datatype" />
            <xsd:attribute name="termchar"
                  type="dtmfchar.datatype" default="#" />
            <xsd:attribute name="maxdigits"
                  type="xsd:positiveInteger" default="5" />
         </xsd:extension>
      </xsd:complexContent>
   </xsd:complexType>
   <xsd:element name="collect" type="collectType" />

   <!-- grammar -->
   <!-- doesn't extend tCore since its content model is mixed -->

   <xsd:complexType name="grammarType" mixed="true">
      <xsd:sequence>
         <xsd:any namespace="##other" minOccurs="0"
               maxOccurs="unbounded" processContents="lax" />
      </xsd:sequence>
      <xsd:attribute name="src" type="xsd:anyURI" />
      <xsd:attribute name="type" type="mime.datatype" />
      <xsd:attribute name="fetchtimeout"
            type="timedesignation.datatype" default="30s" />
      <xsd:anyAttribute namespace="##other" processContents="lax" />
   </xsd:complexType>
   <xsd:element name="grammar" type="grammarType" />

   <!-- record -->

   <xsd:complexType name="recordType">
      <xsd:complexContent>
         <xsd:extension base="Tcore">
            <xsd:sequence>
               <xsd:element ref="media" minOccurs="0"
                     maxOccurs="unbounded" />
               <xsd:any namespace="##other"
                     minOccurs="0"
                     maxOccurs="unbounded" processContents="lax" />
            </xsd:sequence>
            <xsd:attribute name="timeout"
                  type="timedesignation.datatype" default="5s" />
            <xsd:attribute name="beep" type="xsd:boolean"
                  default="false" />
            <xsd:attribute name="vadinitial"
                  type="xsd:boolean" default="false" />
            <xsd:attribute name="vadfinal"
                  type="xsd:boolean" default="false" />
            <xsd:attribute name="dtmfterm"
                  type="xsd:boolean" default="true" />
            <xsd:attribute name="maxtime"
                  type="timedesignation.datatype" default="15s" />
            <xsd:attribute name="finalsilence"
                  type="timedesignation.datatype" default="5s" />
            <xsd:attribute name="append" type="xsd:boolean"
                  default="false" />
```

```
        </xsd:extension>
    </xsd:complexContent>
</xsd:complexType>
<xsd:element name="record" type="recordType" />

<!--
#####################################################
AUDIT TYPES
#####################################################
-->

<!-- audit -->

<xsd:complexType name="auditType">
    <xsd:complexContent>
        <xsd:extension base="Tcore">
            <xsd:sequence>
                <xsd:any namespace="##other"
                        minOccurs="0"
                        maxOccurs="unbounded" processContents="lax" />
            </xsd:sequence>
            <xsd:attribute name="capabilities"
                    type="xsd:boolean" default="true" />
            <xsd:attribute name="dialogs"
                    type="xsd:boolean" default="true" />
            <xsd:attribute name="dialogid"
                    type="dialogid.datatype"/>
        </xsd:extension>
    </xsd:complexContent>
</xsd:complexType>
<xsd:element name="audit" type="auditType" />

<!-- auditresponse -->

<xsd:complexType name="auditresponseType">
    <xsd:complexContent>
        <xsd:extension base="Tcore">
            <xsd:sequence>
                <xsd:element ref="capabilities"
                        minOccurs="0"
                        maxOccurs="1" />
                <xsd:element ref="dialogs" minOccurs="0"
                        maxOccurs="1" />
                <xsd:any namespace="##other"
                        minOccurs="0"
                        maxOccurs="unbounded" processContents="lax" />
            </xsd:sequence>
            <xsd:attribute name="status" type="status.datatype"
                    use="required" />
            <xsd:attribute name="reason" type="xsd:string" />
            <xsd:attribute name="desclang"
                    type="xsd:language"/>
        </xsd:extension>
    </xsd:complexContent>
</xsd:complexType>
```

```
<xsd:element name="auditresponse" type="auditresponseType" />

<!-- codec -->

<xsd:complexType name="codecType">
   <xsd:complexContent>
      <xsd:extension base="Tcore">
         <xsd:sequence>
            <xsd:element ref="subtype" minOccurs="1"
                  maxOccurs="1" />
            <xsd:element ref="params" minOccurs="0"
                  maxOccurs="1" />
            <xsd:any namespace="##other"
                  minOccurs="0"
                  maxOccurs="unbounded" processContents="lax" />
         </xsd:sequence>
         <xsd:attribute name="name" type="xsd:string"
               use="required" />
      </xsd:extension>
   </xsd:complexContent>
</xsd:complexType>
<xsd:element name="codec" type="codecType" />

<!-- subtype -->

<xsd:simpleType name="subtypeType">
   <xsd:restriction base="xsd:string" />
</xsd:simpleType>
<xsd:element name="subtype" type="subtypeType" />

<!-- codecs -->

<xsd:complexType name="codecsType">
   <xsd:complexContent>
      <xsd:extension base="Tcore">
         <xsd:sequence>
            <xsd:element ref="codec" minOccurs="0"
                  maxOccurs="unbounded" />
            <xsd:any namespace="##other"
                  minOccurs="0"
                  maxOccurs="unbounded" processContents="lax" />
         </xsd:sequence>
      </xsd:extension>
   </xsd:complexContent>
</xsd:complexType>
<xsd:element name="codecs" type="codecsType" />

<!-- capabilities -->

<xsd:complexType name="capabilitiesType">
   <xsd:complexContent>
      <xsd:extension base="Tcore">
         <xsd:sequence>
            <xsd:element ref="dialoglanguages"
                  minOccurs="1"
```

```
                maxOccurs="1" />
            <xsd:element ref="grammartypes"
                minOccurs="1"
                maxOccurs="1" />
            <xsd:element ref="recordtypes"
                minOccurs="1"
                maxOccurs="1" />
            <xsd:element ref="prompttypes"
                minOccurs="1"
                maxOccurs="1" />
            <xsd:element ref="variables"
                minOccurs="1"
                maxOccurs="1" />
            <xsd:element ref="maxprepareddduration"
                minOccurs="1"
                maxOccurs="1" />
            <xsd:element ref="maxrecordduration"
                minOccurs="1"
                maxOccurs="1" />
            <xsd:element ref="codecs" minOccurs="1"
                maxOccurs="1" />
            <xsd:any namespace="##other"
                minOccurs="0"
                maxOccurs="unbounded" processContents="lax" />
        </xsd:sequence>
      </xsd:extension>
    </xsd:complexContent>
</xsd:complexType>
<xsd:element name="capabilities" type="capabilitiesType" />

<!-- mimetype -->

<xsd:element name="mimetype" type="mime.datatype" />

<!-- dialoglanguages -->

<xsd:complexType name="dialoglanguagesType">
    <xsd:complexContent>
      <xsd:extension base="Tcore">
        <xsd:sequence>
            <xsd:element ref="mimetype"
                minOccurs="0"
                maxOccurs="unbounded" />
            <xsd:any namespace="##other"
                minOccurs="0"
                maxOccurs="unbounded" processContents="lax" />
        </xsd:sequence>
      </xsd:extension>
    </xsd:complexContent>
</xsd:complexType>
<xsd:element name="dialoglanguages" type="dialoglanguagesType" />

<!-- grammartypes -->

<xsd:complexType name="grammartypesType">
    <xsd:complexContent>
```

```
            <xsd:extension base="Tcore">
                <xsd:sequence>
                    <xsd:element ref="mimetype"
                            minOccurs="0"
                            maxOccurs="unbounded" />
                    <xsd:any namespace="##other"
                            minOccurs="0"
                            maxOccurs="unbounded" processContents="lax" />
                </xsd:sequence>
            </xsd:extension>
        </xsd:complexContent>
    </xsd:complexType>
    <xsd:element name="grammartypes" type="grammartypesType" />

    <!-- recordtypes -->

    <xsd:complexType name="recordtypesType">
        <xsd:complexContent>
            <xsd:extension base="Tcore">
                <xsd:sequence>
                    <xsd:element ref="mimetype"
                            minOccurs="0"
                            maxOccurs="unbounded" />
                    <xsd:any namespace="##other"
                            minOccurs="0"
                            maxOccurs="unbounded" processContents="lax" />
                </xsd:sequence>
            </xsd:extension>
        </xsd:complexContent>
    </xsd:complexType>
    <xsd:element name="recordtypes" type="recordtypesType" />

    <!-- prompttypes -->

    <xsd:complexType name="prompttypesType">
        <xsd:complexContent>
            <xsd:extension base="Tcore">
                <xsd:sequence>
                    <xsd:element ref="mimetype"
                            minOccurs="0"
                            maxOccurs="unbounded" />
                        <xsd:any namespace="##other"
                            minOccurs="0"
                            maxOccurs="unbounded" processContents="lax" />
                </xsd:sequence>
            </xsd:extension>
        </xsd:complexContent>
    </xsd:complexType>
    <xsd:element name="prompttypes" type="prompttypesType" />

    <!-- variables -->

    <xsd:complexType name="variablesType">
        <xsd:complexContent>
            <xsd:extension base="Tcore">
                <xsd:sequence>
```

```
                <xsd:element ref="variabletype"
                        minOccurs="0"
                        maxOccurs="unbounded" />
                <xsd:any namespace="##other"
                        minOccurs="0"
                        maxOccurs="unbounded" processContents="lax" />
            </xsd:sequence>
        </xsd:extension>
    </xsd:complexContent>
</xsd:complexType>
<xsd:element name="variables" type="variablesType" />
<xsd:complexType name="variabletypeType">
    <xsd:complexContent>
        <xsd:extension base="Tcore">
            <xsd:sequence>
                <xsd:element ref="format" minOccurs="0"
                        maxOccurs="unbounded" />
                <xsd:any namespace="##other"
                        minOccurs="0"
                        maxOccurs="unbounded" processContents="lax" />
            </xsd:sequence>
            <xsd:attribute name="type" type="xsd:string"
                    use="required" />
            <xsd:attribute name="desc" type="xsd:string"/>
            <xsd:attribute name="desclang"
                    type="xsd:language"/>
        </xsd:extension>
    </xsd:complexContent>
</xsd:complexType>
<xsd:element name="variabletype" type="variabletypeType" />

<!-- format -->
<!-- doesn't extend tCore since its content model is mixed -->

<xsd:complexType name="formatType" mixed="true">
    <xsd:sequence>
        <xsd:any namespace="##other" minOccurs="0"
                maxOccurs="unbounded" processContents="lax" />
        </xsd:sequence>
        <xsd:attribute name="desc" type="xsd:string" />
        <xsd:attribute name="desclang" type="xsd:language"/>
        <xsd:anyAttribute namespace="##other" processContents="lax" />
</xsd:complexType>
<xsd:element name="format" type="formatType" />

<!-- maxpreparedduration -->

<xsd:element name="maxpreparedduration"
        type="timedesignation.datatype"/>

<!-- maxrecordduration -->

<xsd:element name="maxrecordduration"
        type="timedesignation.datatype"/>

<!-- dialogs -->
```

```
<xsd:complexType name="dialogsType">
   <xsd:complexContent>
      <xsd:extension base="Tcore">
         <xsd:sequence>
            <xsd:element ref="dialogaudit"
                  minOccurs="0"
                  maxOccurs="unbounded" />
               <xsd:any namespace="##other"
                  minOccurs="0"
                  maxOccurs="unbounded" processContents="lax" />
         </xsd:sequence>
      </xsd:extension>
   </xsd:complexContent>
</xsd:complexType>
<xsd:element name="dialogs" type="dialogsType" />

<!-- dialogaudit -->

<xsd:complexType name="dialogauditType">
   <xsd:complexContent>
      <xsd:extension base="Tcore">
         <xsd:sequence>
            <xsd:element ref="codecs" minOccurs="0"
                  maxOccurs="1" />
            <xsd:any namespace="##other"
                  minOccurs="0"
                  maxOccurs="unbounded" processContents="lax" />
         </xsd:sequence>
         <xsd:attribute name="dialogid"
               type="dialogid.datatype" use="required" />
         <xsd:attribute name="state" type="state.datatype"
               use="required" />
         <xsd:attributeGroup ref="fw:
               framework-attributes" />
      </xsd:extension>
   </xsd:complexContent>
</xsd:complexType>
<xsd:element name="dialogaudit" type="dialogauditType" />

<!--
#####################################################
DATATYPES
#####################################################
-->

<xsd:simpleType name="version.datatype">
   <xsd:restriction base="xsd:NMTOKEN">
      <xsd:enumeration value="1.0" />
   </xsd:restriction>
</xsd:simpleType>
<xsd:simpleType name="mime.datatype">
   <xsd:restriction base="xsd:string" />
</xsd:simpleType>
<xsd:simpleType name="dialogid.datatype">
   <xsd:restriction base="xsd:string" />
</xsd:simpleType>
```

```
<xsd:simpleType name="gender.datatype">
   <xsd:restriction base="xsd:NMTOKEN">
      <xsd:enumeration value="female" />
      <xsd:enumeration value="male" />
   </xsd:restriction>
</xsd:simpleType>
<xsd:simpleType name="state.datatype">
   <xsd:restriction base="xsd:NMTOKEN">
      <xsd:enumeration value="preparing" />
      <xsd:enumeration value="prepared" />
      <xsd:enumeration value="starting" />
      <xsd:enumeration value="started" />
   </xsd:restriction>
</xsd:simpleType>
<xsd:simpleType name="status.datatype">
   <xsd:restriction base="xsd:positiveInteger">
      <xsd:pattern value="[0-9][0-9][0-9]" />
   </xsd:restriction>
</xsd:simpleType>
<xsd:simpleType name="media.datatype">
   <xsd:restriction base="xsd:string" />
</xsd:simpleType>
<xsd:simpleType name="label.datatype">
   <xsd:restriction base="xsd:string" />
</xsd:simpleType>
<xsd:simpleType name="direction.datatype">
   <xsd:restriction base="xsd:NMTOKEN">
      <xsd:enumeration value="sendrecv" />
      <xsd:enumeration value="sendonly" />
      <xsd:enumeration value="recvonly" />
      <xsd:enumeration value="inactive" />
   </xsd:restriction>
</xsd:simpleType>
<xsd:simpleType name="timedesignation.datatype">
   <xsd:annotation>
      <xsd:specificationation>
         Time designation following Time in CSS2
     </xsd:specificationation>
   </xsd:annotation>
   <xsd:restriction base="xsd:string">
      <xsd:pattern value="(\+)?([0-9]*\.)?[0-9]+(ms|s)" />
   </xsd:restriction>
</xsd:simpleType>
<xsd:simpleType name="dtmfchar.datatype">
   <xsd:annotation>
      <xsd:specificationation>
         DTMF character [0-9#*A-D]
      </xsd:specificationation>
   </xsd:annotation>
   <xsd:restriction base="xsd:string">
      <xsd:pattern value="[0-9#*A-D]" />
   </xsd:restriction>
</xsd:simpleType>
<xsd:simpleType name="dtmfstring.datatype">
   <xsd:annotation>
      <xsd:specificationation>
```

```
            DTMF sequence [0-9#*A-D]
         </xsd:specificationation>
      </xsd:annotation>
      <xsd:restriction base="xsd:string">
         <xsd:pattern value="([0-9#*A-D])+" />
      </xsd:restriction>
   </xsd:simpleType>
   <xsd:simpleType name="percentage.datatype">
      <xsd:annotation>
         <xsd:specificationation>
            whole integer followed by '%'
         </xsd:specificationation>
      </xsd:annotation>
      <xsd:restriction base="xsd:string">
         <xsd:pattern value="([0-9])+%" />
      </xsd:restriction>
   </xsd:simpleType>
   <xsd:simpleType name="prompt_termmode.datatype">
      <xsd:restriction base="xsd:NMTOKEN">
         <xsd:enumeration value="completed" />
         <xsd:enumeration value="bargein" />
         <xsd:enumeration value="stopped" />
      </xsd:restriction>
   </xsd:simpleType>
   <xsd:simpleType name="collect_termmode.datatype">
      <xsd:restriction base="xsd:NMTOKEN">
         <xsd:enumeration value="match" />
         <xsd:enumeration value="noinput" />
         <xsd:enumeration value="nomatch" />
         <xsd:enumeration value="stopped" />
      </xsd:restriction>
   </xsd:simpleType>
   <xsd:simpleType name="record_termmode.datatype">
      <xsd:restriction base="xsd:NMTOKEN">
         <xsd:enumeration value="noinput" />
         <xsd:enumeration value="dtmf" />
         <xsd:enumeration value="maxtime" />
         <xsd:enumeration value="finalsilence" />
         <xsd:enumeration value="stopped" />
      </xsd:restriction>
   </xsd:simpleType>
   <xsd:simpleType name="matchmode.datatype">
      <xsd:restriction base="xsd:NMTOKEN">
         <xsd:enumeration value="all" />
         <xsd:enumeration value="collect" />
         <xsd:enumeration value="control" />
      </xsd:restriction>
   </xsd:simpleType>
   <xsd:simpleType name="endsync.datatype">
      <xsd:restriction base="xsd:NMTOKEN">
         <xsd:enumeration value="first" />
         <xsd:enumeration value="last" />
      </xsd:restriction>
   </xsd:simpleType>
</xsd:schema>
```

12.6 EXAMPLES

This section provides examples of the IVR control package.

12.6.1 AS-MS Dialog Interaction Examples

The following example assumes that a Control Channel has been established and synced as described in the Media Control Channel Framework (RFC 6230 – see Chapter 8). The XML messages are in angled brackets (with the root <mscivr> omitted); the REPORT status is in round brackets. Other aspects of the protocol are omitted for readability.

12.6.1.1 Starting an IVR Dialog

An IVR dialog is started successfully, and dialogexit notification <event> is sent from the MS to the AS when the dialog exits normally.

12.6.1.2 IVR Dialog Fails to Start

An IVR dialog fails to start due to an unknown dialog language. The <response> is reported in a framework 200 message.

12.6.1.3 Preparing and Starting an IVR Dialog

An IVR dialog is prepared and started successfully, and then the dialog exits normally.

12.6.1.4 Terminating a Dialog

An IVR dialog is started successfully and then terminated by the AS. The dialogexit event is sent to the AS when the dialog exits.

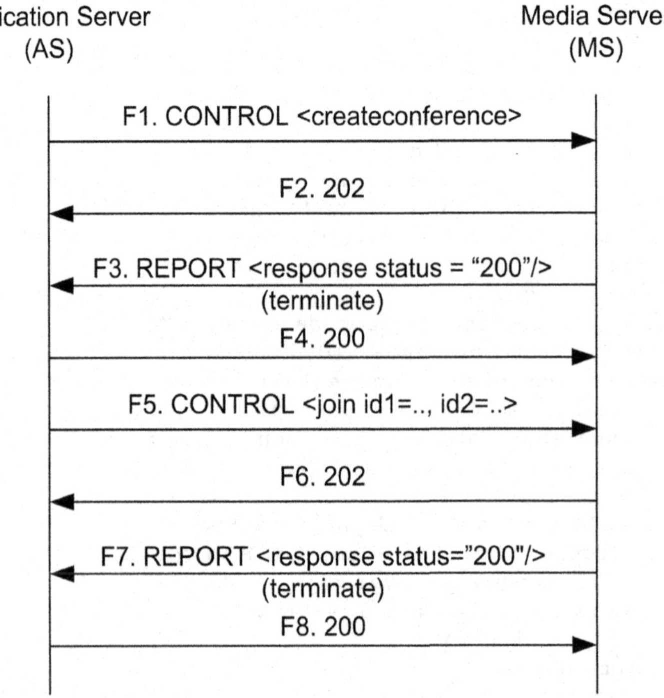

FIGURE 12.2 Terminating a Dialog.

Note that in (6), the <response> payload to the <dialogterminate/> request is carried on a framework 200 response, since it could complete the requested operation before the transaction timeout.

12.6.2 IVR Dialog Examples

The following examples show how <dialog> is used with <dialogprepare>, <dialogstart>, and <event> elements to play prompts, set runtime controls, collect DTMF input, and record user input. The examples do not specify all messages between the AS and the MS.

12.6.2.1 Playing Announcements

This example prepares an announcement composed of two prompts where the dialog repeatCount is set to 2.

```
<mscivr version="1.0" xmlns="urn:ietf:params:xml:ns:msc-ivr">
   <dialogprepare>
      <dialog repeatCount="2">
         <prompt>
            <media loc="http://www.example.com/media/
                  Number_09.wav"/>
            <media loc="http://www.example.com/media
                  /Number_11.wav"/>
         </prompt>
      </dialog>
   </dialogprepare>
</mscivr>
```

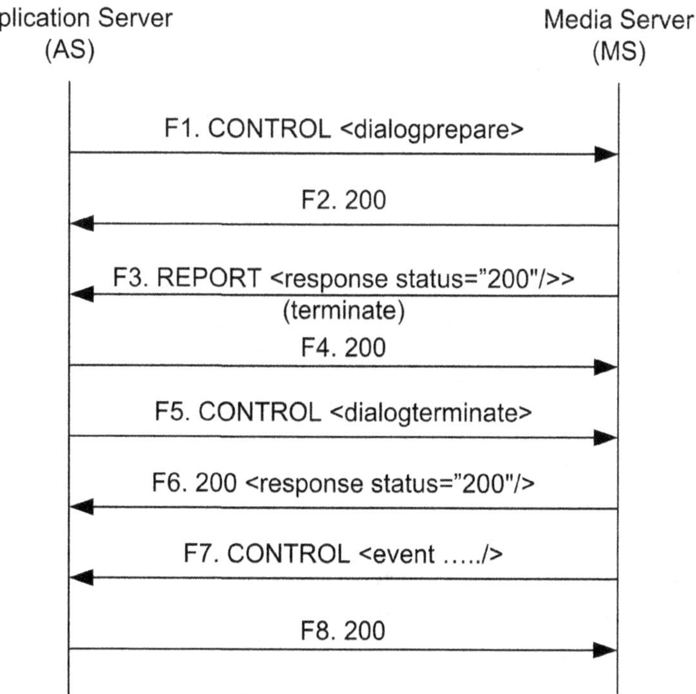

FIGURE 12.3 Starting of an IVR Dialog between a Application Server and Media Server.

If the dialog is prepared successfully, a <response> is returned with status 200 and a dialog identifier assigned by the MS:

```
<mscivr version="1.0" xmlns="urn:ietf:params:xml:ns:msc-ivr">
   <response status="200" dialogid="vxi78"/>
</mscivr>
```

The prepared dialog is then started on a conference playing the prompts twice:

```
<mscivr version="1.0" xmlns="urn:ietf:params:xml:ns:msc-ivr">
   <dialogstart prepareddialogid="vxi78"
conferenceid="conference11"/>
</mscivr>
```

In the case of a successful dialog, the output is provided in <event>; for example:

```
<mscivr version="1.0" xmlns="urn:ietf:params:xml:ns:msc-ivr">
   <event dialogid="vxi78">
      <dialogexit status="1">
         <promptinfo termmode="completed" duration="24000"/>
      </dialogexit>
   </event>
</mscivr>
```

12.6.2.2 Prompt and Collect

In this example, a prompt is played, and then the MS waits for 30 s for a two digit sequence:

```
<mscivr version="1.0" xmlns="urn:ietf:params:xml:ns:msc-ivr">
   <dialogstart connectionid="7HDY839:HJKSkyHS~HUwkuh7ns">
      <dialog>
         <prompt>
            <media loc="http://www.example.com/prompt1.wav"/>
         </prompt>
         <collect timeout="30s" maxdigits="2"/>
      </dialog>
   </dialogstart>
</mscivr>
```

If no user input is collected within 30 s, then the following notification event will be returned:

```
<mscivr version="1.0" xmlns="urn:ietf:params:xml:ns:msc-ivr">
   <event dialogid="vxi81">
      <dialogexit status="1" >
         <promptinfo termmode="completed" duration="4000"/>
            <collectinfo termmode="noinput"/>
      </dialogexit>
   </event>
</mscivr>
```

The collect operation can be specified without a prompt. Here, the MS just waits for DTMF input from the user (the maxdigits attribute of <collect> defaults to 5):

```
<mscivr version="1.0" xmlns="urn:ietf:params:xml:ns:msc-ivr">
   <dialogstart connectionid="7HDY839:HJKSkyHS~HUwkuh7ns">
      <dialog>
```

```
            <collect/>
        </dialog>
    </dialogstart>
</mscivr>
```

If the dialog is successful, then dialogexit <event> contains the dtmf collected in its result parameter:

```
<mscivr version="1.0" xmlns="urn:ietf:params:xml:ns:msc-ivr">
    <event dialogid="vxi80">
        <dialogexit status="1">
            <collectinfo dtmf="12345" termmode="match"/>
        </dialogexit>
    </event>
</mscivr>
```

And finally, in this example, one of the input parameters is invalid:

```
<mscivr version="1.0" xmlns="urn:ietf:params:xml:ns:msc-ivr">
    <dialogstart connectionid="7HDY839:HJKSkyHS~HUwkuh7ns">
        <dialog repeatCount="two">
            <prompt>
                <media loc="http://www.example.com/
                        prompt1.wav"/>
            </prompt>
            <collect cleardigitbuffer="true"
                    timeout="4s" interdigittimeout="2s"
                    termtimeout="0s" maxdigits="2"/>
        </dialog>
    </dialogstart>
</mscivr>
```

The error is reported in the response:

```
<mscivr version="1.0" xmlns="urn:ietf:params:xml:ns:msc-ivr">
    <response status="400" dialogid="vxi82"
            reason="repeatCount attribute value invalid: two"/>
</mscivr>
```

12.6.2.3 Prompt and Record

In this example, the user is prompted, and then, their input is recorded for a maximum of 30 seconds.

```
<mscivr version="1.0" xmlns="urn:ietf:params:xml:ns:msc-ivr">
    <dialogstart connectionid="7HDY839:HJKSkyHS~HUwkuh7ns">
        <dialog>
            <prompt>
                <media loc="http://www.example.com/media/
                        sayname.wav"/>
            </prompt>
            <record dtmfterm="false" maxtime="30s" beep="true"/>
        </dialog>
    </dialogstart>
</mscivr>
```

If this is successful, and the recording is terminated by DTMF, the following is returned in a dialogexit <event>:

```
<mscivr version="1.0" xmlns="urn:ietf:params:xml:ns:msc-ivr">
    <event dialogid="vxi83">
        <dialogexit status="1">
            <recordinfo termmode="dtmf">
                <mediainfo type="audio/x-wav"
                        loc="http://www.example.com/
                        recording1.wav"/>
            </recordinfo>
        </dialogexit>
    </event>
</mscivr>
```

12.6.2.4 Runtime Controls

In this example, a prompt is played with the collect operation and runtime controls activated.

```
<mscivr version="1.0" xmlns="urn:ietf:params:xml:ns:msc-ivr">
    <dialogstart connectionid="7HDY839:HJKSkyHS~HUwkuh7ns">
        <dialog>
            <prompt bargein="true">
                <media loc="http://www.example.com/
                        prompt1.wav"/>
            </prompt>
            <control ffkey="5" rwkey="6" speedupkey="3"
                    speeddnkey="4"/>
            <collect maxdigits="2"/>
        </dialog>
    </dialogstart>
</mscivr>
```

Once the dialog is active, the user can press keys 3, 4, 5, and 6 to execute runtime controls on the prompt queue. The keys do not cause bargein to occur. If the user presses any other key, then the prompt is interrupted, and DTMF collect begins. Note that runtime controls are not active during the collect operation. When the dialog is completed successfully, then both control and collect information is reported.

```
<mscivr version="1.0" xmlns="urn:ietf:params:xml:ns:msc-ivr">
    <event dialogid="vxi81">
        <dialogexit status="1">
            <promptinfo termmode="bargein"/>
                <controlinfo>
                    <controlmatch dtmf="4" timestamp="2008-05-
                            12T12:13:14Z"/>
                    <controlmatch dtmf="3" timestamp="2008-05-
                            12T12:13:15Z"/>
                    <controlmatch dtmf="5" timestamp="2008-05-
                            12T12:13:16Z"/>
                </controlinfo>
                <collectinfo termmode="match" dtmf="14"/>
        </dialogexit>
    </event>
</mscivr>
```

12.6.2.5 Subscriptions and Notifications

In this example, a looped dialog is started with subscription for notifications each time the user input matches the collect grammar:

```
<mscivr version="1.0" xmlns="urn:ietf:params:xml:ns:msc-ivr">
   <dialogstart connectionid="7HDY839:HJKSkyHS">
      <dialog repeatCount="0">
         <collect maxdigits="2"/>
      </dialog>
      <subscribe>
         <dtmfsub matchmode="collect"/>
      </subscribe>
   </dialogstart>
</mscivr>
```

Each time the user inputs the DTMF matching the grammar, the following notification event will be sent:

```
<mscivr version="1.0" xmlns="urn:ietf:params:xml:ns:msc-ivr">
   <event dialogid="vxi81">
      <dtmfnotify matchmode="collect" dtmf="12"
            timestamp="2008-05-12T12:13:14Z"/>
   </event>
</mscivr>
```

If no user input was provided, or the input did not match the grammar, the dialog would continue to loop until terminated (or an error occurred).

12.6.2.6 Dialog Repetition until DTMF Collection Complete

This example is a prompt and collect dialog to collect the PIN from the user. The repeatUntilComplete attribute in the <dialog> is set to true in this case, so that when the grammar collection is complete, the MS automatically terminates the dialog repeat cycle and reports the results in a <dialogexit> event.

```
<mscivr version="1.0" xmlns="urn:ietf:params:xml:ns:msc-ivr">
   <dialogstart connectionid="7HDY839:HJKSkyHS">
      <dialog repeatCount="3" repeatUntilComplete="true">
         <prompt bargein="true">
            <media loc="http://example.com/
                  please_enter_your_pin.vox"/>
         </prompt>
         <collect maxdigits="4"/>
      </dialog>
   </dialogstart>
</mscivr>
```

If the user barges in on the prompt and <collect> receives DTMF input matching the grammar, the dialog cycle is considered complete, and the MS returns the following:

```
<mscivr version="1.0" xmlns="urn:ietf:params:xml:ns:msc-ivr">
   <event dialogid="vxi81">
      <dialogexit status="1">
         <promptinfo duration="3654" termmode="bargein"/>
         <collectinfo dtmf="1234" termmode="match"/>
```

```
        </dialogexit>
    </event>
</mscivr>
```

If no user input was provided, or the input did not match the grammar, the dialog would loop for a maximum of three times.

12.6.3 Other Dialog Languages

The following example requests that a VoiceXML dialog is started:

```
<mscivr version="1.0" xmlns="urn:ietf:params:xml:ns:msc-ivr">
    <dialogstart dialogid="d2"
            connectionid="7HDY839:HJKSkyHS"
            type="application/voicexml+xml"
            src="http://www.example.com/mydialog.vxml"
            fetchtimeout="15s">
        <params>
            <param name="prompt1">nfs://
                    nas01/media1.3gp</param>
            <param name="prompt2">nfs://
                    nas01/media2.3gp</param>
        </params>
    </dialogstart>
</mscivr>
```

If the MS does not support this dialog language, then the response would have the status code 421 (Section 12.4.5). However, if it does support the VoiceXML dialog language, it would respond with a 200 status, activate the VoiceXML dialog, and make the <params> available to the VoiceXML script as described in Section 12.9. When the VoiceXML dialog exits, exit namelist parameters are specified using <params> in the dialogexit event:

```
<mscivr version="1.0" xmlns="urn:ietf:params:xml:ns:msc-ivr">
    <event dialogid="d2">
        <dialogexit status="1">
            <params>
                <param name="username">peter</param>
                <param name="pin">1234</param>
            </params>
        </dialogexit>
    </event>
</mscivr>
```

12.6.4 Foreign Namespace Attributes and Elements

An MS can support attributes and elements from foreign namespaces within the <mscivr> element. For example, the MS could support a <listen> element (in a foreign namespace) for speech recognition by analogy to how <collect> supports DTMF collection. In the following example, a prompt and collect request is extended with a <listen> element:

```
<mscivr version="1.0" xmlns="urn:ietf:params:xml:ns:msc-ivr"
        xmlns:ex="http://www.example.com/mediactrl/extensions/1">
    <dialogstart connectionid="7HDY839:HJKSkyHS~HUwkuh7ns">
        <dialog>
```

```
            <prompt>
               <media loc="http://www.example.com/
                     prompt1.wav"/>
            </prompt>
            <collect timeout="30s" maxdigits="4"/>
            <ex:listen maxtimeout="30s" >
               <ex:grammar src="http://example.org/pin.grxml"/>
            </ex:listen>
         </dialog>
      </dialogstart>
   </mscivr>
```

In the <mscivr> root element, the xmlns:ex attribute declares that "ex" is associated with the foreign namespace URI "http://www.example.com/mediactrl/extensions/1." The <ex:listen>, its attributes, and child elements are associated with this namespace. This <listen> could be defined so that it activates a Speech Recognition Grammar Specification (SRGS) grammar and listens for user input matching the grammar in a similar manner to DTMF collection. If an MS receives this request but does not support the <listen> element, then it will send a 431 response:

```
<mscivr version="1.0" xmlns="urn:ietf:params:xml:ns:msc-ivr">
   <response status="431" dialogid="d560"
         reason="unsupported foreign listen element"/>
</mscivr>
```

If the MS does support this foreign element, it will send a 200 response and start the dialog with speech recognition. When the dialog exits, it provides information about the <listen> execution within <dialogexit>, again using elements in a foreign namespace such as <listeninfo>:

```
<mscivr version="1.0" xmlns="urn:ietf:params:xml:ns:msc-ivr"
      xmlns:ex="http://www.example.com/mediactrl/extensions/1">
   <event dialogid="d560">
      <dialogexit status="1">
         <ex:listeninfo speech="1 2 3 4" termmode="match"/>
      </dialogexit>
   </event>
</mscivr>
```

Note that in reply the AS sends a Control Framework 200 response, even though the notification event contains an element in a foreign namespace that it might not understand.

12.7 SECURITY CONSIDERATIONS

As this Control Package processes XML markup, implementations MUST address the security considerations of RFC 3023. Implementations of this Control Package MUST address security, confidentiality, and integrity of messages transported over the Control Channel as described in Section 8.12 (Section 12 of "Media Control Channel Framework" – RFC 6230), including Transport Level Protection, Control Channel Policy Management, and Session Establishment. In addition, implementations MUST address security, confidentiality, and integrity of user agent sessions with the MS in terms of both SIP signaling and associated RTP media flow; see RFC 6230, described in Chapter 8 of this book, for further details on this topic. Finally, implementations MUST address security, confidentiality, and integrity of sessions where, following a URI scheme, an MS uploads recordings or retrieves specifications and resources (e.g., fetching a grammar specification from a web server using HTTPS). Adequate transport protection and authentication are critical, especially when the implementation is deployed in open networks. If the implementation fails to correctly address these issues, it risks exposure to malicious attacks, including (but not limited to):

- **Denial of Service:** An attacker could insert a request message into the transport stream causing specific dialogs on the MS to be terminated immediately: For example, <dialogterminate dialogid="XXXX" immediate="true">, where the value of "XXXX" could be guessed or discovered by auditing active dialogs on the MS using an <audit> request. Likewise, an attacker could impersonate the MS and insert error responses into the transport stream, so denying the AS access to package capabilities.
- **Resource Exhaustion:** An attacker could insert into the Control Channel new request messages (or modify existing ones) with, for instance, <dialogprepare> elements with a very long fetchtimeout attribute and a bogus source uniform resource locator (URL). At some point, this will exhaust the number of connections that the MS is able to make.
- **Phishing:** An attacker with access to the Control Channel could modify the "loc" attribute of the <media> element in a dialog to point to some other audio file that had different information from the original. This modified file could include a different phone number for people to call if they want more information or need to provide additional information (such as governmental, corporate, or financial information).
- **Data Theft:** An attacker could modify a <record> element in the Control Channel so as to add a new recording location:

```
<mscivr version="1.0" xmlns="urn:ietf:params:xml:ns:msc-ivr">
   <dialogstart>
      <dialog>
         <record>
            <media type="audio/x-wav" loc="(Good URI)"/>
            <media type="audio/x-wav" loc=
                  "(Attacker's URI)"/>
         </record>
      </dialog>
   </dialogstart>
</mscivr>
```

  The recorded data would be uploaded to two locations indicated by the "{Good URI}" and the "{Attacker's URI}." This allows the attacker to steal the recorded audio (which could include sensitive or confidential information) without the originator of the request necessarily being aware of the theft. The Media Control Channel Framework permits additional security policy management, including resource access and Control Channel usage, to be specified at the Control Package level beyond that specified for the Media Control Channel Framework described in Section 8.12.3 of this book (Section 12.3 of RFC 6230). Since the creation of IVR dialogs is associated with media processing resources (e.g., DTMF detectors, media playback and recording, etc.) on the MS, the security policy for this Control Package needs to address how such dialogs are securely managed across more than one Control Channel. Such a security policy is only useful for secure, confidential, and integrity-protected channels. The identity of Control Channels is determined by the channel identifier, i.e., the value of the cfw-id attribute in the SDP and "Dialog-ID" header in the channel protocol specified in RFC 6230 (see Chapter 8). Channels are the same if they have the same identifier; otherwise, they are different.

This Control Package imposes the following additional security policies:

- **Responses:** The MS MUST only send a response to a dialog management or audit request using the same Control Channel as the one used to send the request.

- **Notifications:** The MS MUST only send notification events for a dialog using the same Control Channel as the one by which it received the request creating the dialog.
- **Auditing:** The MS MUST only provide audit information about dialogs that have been created on the same Control Channel as the one upon which the <audit> request is sent.
- **Rejection:** The MS SHOULD reject requests to audit or manipulate an existing dialog on the MS if the channel is not the same as the one used when the dialog was created. The MS rejects a request by sending a Control Framework 403 response that is described in Section 8.7.4 and Section 8.12.3 of this book (Section 7.4 and Section 12.3 of RFC 6230). For example, if a channel with identifier "cfw1234" has been used to send a request to create a particular dialog and the MS receives on channel "cfw98969" a request to audit or terminate the dialog, then the MS sends a 403 framework response.

There can be valid reasons why an implementation does not reject an audit or dialog manipulation request on a different channel from the one that created the dialog. For example, a system administrator might require a separate channel to audit dialog resources created by system users and to terminate dialogs consuming excessive system resources. Alternatively, a system monitor or resource broker might require a separate channel to audit dialogs managed by this package on an MS. However, the full implications need to be understood by the implementation and carefully weighed before these reasons are accepted as valid. If the reasons are not valid in their particular circumstances, the MS rejects such requests.

There can also be valid reasons for "channel handover," including high availability support or where one AS needs to take over management of dialogs after the AS that created them has failed. This could be achieved by the Control Channels using the same channel identifier, one after another. For example, assume a channel is created with the identifier "cfw1234," and the channel is used to create dialogs on the MS. This channel (and associated SIP dialog) then terminates due to a failure on the AS. As permitted by the Control Framework, the channel identifier "cfw1234" could then be reused, so that another channel is created with the same identifier "cfw1234," allowing it to "take over" management of the dialogs on the MS. Again, the implementation needs to understand the full implications and carefully weigh them before accepting these reasons as valid. If the reasons are not valid for their particular circumstances, the MS uses the appropriate SIP mechanisms to prevent session establishment when the same channel identifier is used in setting up another Control Channel described in Section 8.4 of this book (Section 4 of RFC 6230).

12.8 IANA CONSIDERATIONS

IANA has registered a new Media Control Channel Framework Package, a new XML namespace, a new XML schema, and a new MIME type. IANA has further created a new registry for IVR prompt variable types. Interested readers need to see RFC 6231 (also see Section 1.2).

12.9 USING VOICEXML AS A DIALOG LANGUAGE

The IVR control package allows, but does not require, the MS to support other dialog languages by referencing an external dialog specification. This section provides MS implementations that support the VoiceXML dialog language [3–5] with additional details about using these dialogs in this package. This section is normative for an MS that supports the VoiceXML dialog language. Note that VoiceXML media services with SIP are described in RFC 5552 (see Chapter 13). This section covers preparing (Section 12.9.1), starting (Section 12.9.2), terminating (Section 12.9.3), and exiting (Section 12.9.4) VoiceXML dialogs as well as handling VoiceXML call transfer (Section 12.9.5). Note that RFC 5552 (see Chapter 13) also supports SIP interface-based VoiceXML media services.

12.9.1 Preparing a VoiceXML Dialog

A VoiceXML dialog is prepared by sending the MS a request containing a <dialogprepare> element (Section 12.4.2.1). The type attribute is set to "application/voicexml+xml" and the src attribute to the URI of the VoiceXML specification that is to be prepared by the MS. For example:

```
<mscivr version="1.0" xmlns="urn:ietf:params:xml:ns:msc-ivr">
    <dialogprepare type="application/voicexml+xml"
            src="http://www.example.com/mydialog.vxml"
            fetchtimeout="15s"/>
</mscivr>
```

The VoiceXML dialog environment uses the <dialogprepare> request as an opportunity to fetch and validate the initial specification indicated by the src attribute along with any resources referenced in the VoiceXML specification marked as prefetchable. The maxage and maxstale attributes, if specified, control how the initial VoiceXML specification is fetched using HTTP (see RFC 2616). Note that the fetchtimeout attribute is not defined in VoiceXML for an initial specification, but the MS MUST support this attribute in its VoiceXML environment.

If a <params> child element of <dialogprepare> is specified, then the MS MUST map the parameter information into a VoiceXML session variable object as described in Section 12.9.2.3. The success or failure of the VoiceXML specification preparation is reported in the MS response. For example, if the VoiceXML specification cannot be retrieved, then a 409 error response is returned. If the specification is syntactically invalid according to VoiceXML, then a 400 response is returned. If successful, the response includes a dialogid attribute whose value the AS can use in <dialogstart> element to start the prepared dialog.

12.9.2 Starting a VoiceXML Dialog

A VoiceXML dialog is started by sending the MS a request containing a <dialogstart> element (Section 12.4.2.2). If a VoiceXML dialog has already been prepared using <dialogprepare>, then the MS starts the dialog indicated by the prepareddialogid attribute. Otherwise, a new VoiceXML dialog can be started by setting the type attribute to "application/voicexml+xml" and the src attribute to the URI of the VoiceXML specification. For example:

```
<mscivr version="1.0" xmlns="urn:ietf:params:xml:ns:msc-ivr">
    <dialogstart connectionid="ssd3r3:sds345b"
            type="application/voicexml+xml"
            src="http://www.example.com/mydialog.vxml"
            fetchtimeout="15s"/>
</mscivr>
```

The maxage and maxstale attributes, if specified, control how the initial VoiceXML specification is fetched using HTTP (see RFC 2616). Note that the fetchtimeout attribute is not defined in VoiceXML for an initial specification, but the MS MUST support this attribute in its VoiceXML environment. Note also that support for <dtmfsub> subscriptions (Section 12.4.2.2.1.1) and their associated dialog notification events is not defined in VoiceXML. If such a subscription is specified in a <dialogstart> request, then the MS sends a 439 error response (see Section 12.4.5).

The success or failure of starting a VoiceXML dialog is reported in the MS response as described in Section 12.4.2.2. When the MS starts a VoiceXML dialog, the MS MUST map session information into a VoiceXML session variable object. There are three types of session information: Protocol information (Section 12.9.2.1), media stream information (Section 12.9.2.2), and parameter information (Section 12.9.2.3).

12.9.2.1 Session Protocol Information

If the connectionid attribute is specified, the MS assigns protocol information from the SIP dialog associated with the connection to the following session variables in VoiceXML:

- session.connection.local.uri: Evaluates to the SIP URI specified in the "To:" header of the initial INVITE.
- session.connection.remote.uri: Evaluates to the SIP URI specified in the "From:" header of the initial INVITE.
- session.connection.originator: Evaluates to the value of session.connection.remote (MS receives inbound connections but does not create outbound connections).
- session.connection.protocol.name: Evaluates to "sip." Note that this is intended to reflect the use of SIP in general and does not distinguish between whether the connection accesses the MS via SIP or SIP Secure (SIPS) procedures.
- session.connection.protocol.version: Evaluates to "2.0."
- session.connection.redirect: This array is populated by information contained in the "History-Info" header (RFC 4244) in the initial INVITE or is otherwise undefined. Each entry (hi-entry) in the "History-Info" header is mapped, in the order it appeared in the "History-Info" header, into an element of the session.connection.redirect array. Properties of each element of the array are determined as follows:
  - uri: Set to the hi-targeted-to-uri value of the History-Info entry.
  - pi: Set to "true" if hi-targeted-to-uri contains a "Privacy=history" parameter, or if the INVITE "Privacy" header includes "history;" "false" otherwise.
  - si: Set to the value of the "si" parameter if it exists; undefined otherwise.
  - reason: Set verbatim to the value of the "Reason" parameter of hitargeted-to-uri.
  - session.connection.aai: Evaluates to the value of a SIP header with the name "aai" if present; undefined otherwise.
  - session.connection.protocol.sip.requesturi: This is an associative array where the array keys and values are formed from the URI parameters on the SIP Request-URI of the initial INVITE. The array key is the URI parameter name. The corresponding array value is obtained by evaluating the URI parameter value as a string. In addition, the array's toString() function returns the full SIP Request-URI.
  - session.connection.protocol.sip.headers: This is an associative array where each key in the array is the non-compact name of a SIP header in the initial INVITE converted to lowercase (note that the case conversion does not apply to the header value). If multiple header fields of the same field name are present, the values are combined into a single comma-separated value. Implementations MUST at a minimum include the "Call-ID" header and MAY include other headers. For example, session.connection.protocol.si p.headers["call-id"] evaluates to the Call-ID of the SIP dialog.

If a conferenceid attribute is specified, then the MS populates the following session variables in VoiceXML:

- session.conference.name: Evaluates to the value of the conferenceid attribute.

12.9.2.2 Session Media Stream Information

The media streams of the connection or conference to use for the dialog are described in Section 12.4.2.2, including use of <stream> elements (Section 12.4.2.2.2) if specified. The MS maps media stream information into the VoiceXML session variable session.connection.protocol.sip.media for a connection and session.conference.media for a conference. In both variables, the value of the variable is an array where each array element is an object with the following properties:

- type: This required property indicates the type of the media associated with the stream (see Section 12.4.2.2.2 <stream> type attribute definition).
- direction: This required property indicates the directionality of the media relative to the endpoint of the dialog (see Section 12.4.2.2.2 <stream> direction attribute definition).
- format: This property is optional. If defined, the value of the property is an array. Each array element is an object that specifies information about one format of the media stream. The object contains at least one property called name, whose value is the subtype name of the media format (RFC 4855). Other properties may be defined with string values; these correspond to required and, if defined, optional parameters of the format.

As a consequence of this definition, when a connectionid is specified, there is an array entry in session.connection.protocol.sip.media for each media stream used by the VoiceXML dialog. For an example, consider a connection with bidirectional G.711 mu-law audio sampled at 8 kHz, where the dialog is started with:

```
<mscivr version="1.0" xmlns="urn:ietf:params:xml:ns:msc-ivr">
   <dialogstart connectionid="ssd3r3:sds345b"
         type="application/voicexml+xml"
         src="http://www.example.com/mydialog.vxml"
         fetchtimeout="15s">
      <stream media="audio" direction="recvonly"/>
   </dialogstart>
</mscivr>
```

In this case, session.connection.protocol.sip.media[0].type evaluates to "audio," session.connection.protocol.sip.media[0].direction evaluates to "recvonly" (i.e., the endpoint only receives media from the dialog – the endpoint does not send media to the dialog), session.connection.protocol.sip.media[0].format[0].name evaluates to "PCMU," and session.connection.protocol.sip.media[0].format[0].rate evaluates to "8000." Note that the session variable is updated if the connection or conference media session characteristics for the VoiceXML dialog change (e.g., due to a SIP re-INVITE).

12.9.2.3 Session Parameter Information

Parameter information is specified in the <params> child element of <dialogprepare> and <dialogstart> elements, where each parameter is specified using a <param> element. The MS maps parameter information into VoiceXML session variables as follows:

session.values: This is an associative array mapped to the <params> element. It is undefined if no <params> element is specified. If a <params> element is specified in both <dialogprepare> and <dialogstart> elements for the same dialog, then the array is first initialized with the <params> specified in the <dialogprepare> element and then updated with the <params> specified in the <dialogstart> element; in cases of conflict, the <dialogstart> parameter value take priority. Array keys and values are formed from <param> children of the <params> element. Each array key is the value of the name attribute of a <param> element. If the same name is used in more than one <param> element, then the array key is associated with the last <param> in specification order. The corresponding value for each key is an object with two required properties: a "type" property evaluating to the value of the type attribute and a "content" property evaluating to the content of the <param>. In addition, this object's toString() function returns the value of the "content" property as a string.

For example, a VoiceXML dialog started with one parameter:

```
<mscivr version="1.0" xmlns="urn:ietf:params:xml:ns:msc-ivr">
   <dialogstart connectionid="ssd3r3:sds345b"
```

```
            type="application/voicexml+xml"
            src="http://www.example.com/mydialog.vxml"
            fetchtimeout="15s">
        <params>
            <param name="mode">playannouncement</param>
        </params>
    </dialogstart>
</mscivr>
```

In this case, session.values would be defined with one item in the array where session.values['mode'].type evaluates to "text/plain" (the default value), session.values['mode'].content evaluates to "playannouncement," and session.values['mode'].toString() also evaluates to "playannouncement." The MS sends an error response (see Section 12.4.2.2) if a <param> is not supported by the MS (e.g., the parameter type is not supported).

12.9.3 Terminating a VoiceXML Dialog

When the MS receives a request with a <dialogterminate> element (Section 12.4.2.3), the MS throws a "connection.disconnect.hangup" event into the specified VoiceXML dialog. Note that if the immediate attribute has the value true, then the MS MUST NOT return <params> information when the VoiceXML dialog exits (even if the VoiceXML dialog provides such information) – see Section 12.9.4. If the connection or conference associated with the VoiceXML dialog terminates, then the MS throws a "connection.disconnect.hangup" event into the specified VoiceXML dialog.

12.9.4 Exiting a VoiceXML Dialog

The MS sends a <dialogexit> notification event (Section 12.4.2.5.1) when the VoiceXML dialog is complete, has been terminated, or exits due to an error. The <dialogexit> status attribute specifies the status of the VoiceXML dialog when it exits, and its <params> child element specifies information, if any, returned from the VoiceXML dialog. A VoiceXML dialog exits when it processes a <disconnect> element, an <exit> element, or an implicit exit according to the VoiceXML form interpretation algorithm (FIA). If the VoiceXML dialog executes a <disconnect> and then subsequently executes an <exit> with namelist information, the namelist information from the <exit> element is discarded.

The MS reports namelist variables in the <params> element of the <dialogexit>. Each <param> reports on a namelist variable. The MS set the <param> name attribute to the name of the VoiceXML variable. The MS sets the <param> type attribute according to the type of the VoiceXML variable. The MS sets the <param> type to "text/plain" when the VoiceXML variable is a simple ECMAScript value. If the VoiceXML variable is a recording, the MS sets the <param> type to the MIME media type of the recording and encodes the recorded content as CDATA in the <param> (see Section 12.4.2.6.1 for an example). If the VoiceXML variable is a complex ECMAScript value (e.g., object, array, etc.), the MS sets the <param> type to "application/json" and converts the variable value to its JSON value equivalent (RFC 4627). The behavior resulting from specifying an ECMAScript object with circular references is not defined. Note: RFC 5551 makes a "downref" normative reference to RFC 4627 – an informational specification describing a proprietary (but extremely popular) format.

If the expr attribute is specified on the VoiceXML <exit> element instead of the namelist attribute, the MS creates a <param> element with the reserved name "__exit." If the value is an ECMAScript literal, the <param> type is "text/plain," and the content is the literal value. If the value is a variable, the <param> type and content are set in the same way as a namelist variable; for example, an expr attribute referencing a variable with a simple ECMAScript value has the type "text/plain," and the content is set to the ECMAScript value. To allow the AS to differentiate between a <dialogexit>

TABLE 12.2
VoiceXML <exit> Mapping Examples

<exit> Usage	<params> Result
<exit>	<params> <param name="__reason">exit</param> </params>
<exit expr="5">	<params> <param name="__reason">5</param> </params>
<exit expr=" 'done' ">	<params> <param name="__reason">exit</param> <param name="__exit">'done'</param> </params>
<exit expr="userAuthorized">	<params> <param name="__reason">exit<param name="__exit">'true'</param> </params>
<exit namelist="pin errors">	<params> <param name="__reason">exit</param> <param name="pin">1234</param> <param name="errors">0</param </params>

notification event resulting from a VoiceXML <disconnect> from one resulting from an <exit>, the MS creates a <param> with the reserved name "__reason," the type "text/plain," a value of "disconnect" (without brackets) to reflect the use of VoiceXML's <disconnect> element, and the value of "exit" (without brackets) to an explicit <exit> in the VoiceXML dialog. If the VoiceXML session terminates for other reasons (such as encountering an error), this parameter MAY be omitted or take on platform-specific values prefixed with an underscore. Table 12.2 provides some examples of VoiceXML <exit> usage and the corresponding <params> element in the <dialogexit> notification event. It assumes the following VoiceXML variable names and values: userAuthorized=true, pin=1234, and errors=0. The <param> type attributes ('text/plain') are omitted for clarity.

12.9.5 Call Transfer

While VoiceXML is at its core a dialog language, it also provides optional call transfer capability. It is NOT RECOMMENDED to use VoiceXML's call transfer capability in networks involving application servers. Rather, the AS itself can provide call routing functionality by taking signaling actions based on the data returned to it, either through VoiceXML's own data submission mechanisms or through the mechanism described in Section 12.9.4. If the MS encounters a VoiceXML dialog requesting call transfer capability, the MS SHOULD raise an error event in the VoiceXML dialog execution context: an error.unsupported.transfer.blind event if blind transfer is requested, error.unsupported.transfer.bridge if bridge transfer is requested, or error.unsupported.transfer.consultation if consultation transfer is requested.

12.10 SUMMARY

We have described the IVR control package based on CFW package (RFC 6230 – see Chapter 8) using SIP (RFC 3261 – also see Sections 1.2 and 1.5) protocol for the rendezvous and establishment of a reliable channel for control interactions. The term "IVR" is used in its inclusive sense, allowing media other than voice for dialog interaction. The package is designed meeting the specific MS control protocol requirements (RFC 5167 – see Section 8.1): REQ-28, REQ-29, and REQ-30. However, the package has equivalent capabilities to those of MSCML (RFC 5022) and MSML (RFC 5707), but using SIP as the control signaling protocol. Moreover, the package also defines elements for auditing package capabilities and IVR dialogs, and its design is aligned with key concepts of the web model as defined in W3C Voice Browser languages along with the key dialog management mechanism being closely aligned with Call Control XML [2]. Moreover, the functionality of this package is defined by messages, containing XML [6] elements, transported using the CFW (RFC 6230 – see Chapter 8) dividing the XML elements into types: a. Dialog management elements, b. Dialog element that defines a lightweight IVR dialog language used with dialog management elements, and c. Elements for auditing package capabilities as well as dialogs managed by the package.

More specifically, the package contains the IVR control package definition and name, usage of CFW messages, common XML support, CONTROL and REPORT message body, and audit along with some examples. It also describes the syntax and semantics of defined elements, including dialog management, the IVR dialog element, audit elements, response status codes, and type definitions. It describes an XML schema for these elements and provides extensibility by allowing attributes and elements from other namespaces. Some examples for package usage are provided, as follows: IVR Dialog, Other Dialog Languages, and Foreign Namespace Attributes and Elements. The important security schemes that need to be implemented for the IVR package to protect against cyberattacks are described. Finally, additional information on using VoiceXML when supported as an external dialog language is described at great length. By the way, RFC 5552 (see Section 13) also supports SIP interface-based VoiceXML media services. Note that RFC 6231 has IANA registration for this IVR control package, including its name, XML namespace, and MIME media type. Interested readers are requested to see RFC 6231.

12.11 PROBLEMS

1. Explain in detail how the IVR control package is designed to meet the requirements (REQ-28, REQ-29, and REQ-30) of the MS control protocol (RFC 5167).
2. How is the CFW (RFC 6230) used in designing the IVR control package?
3. How can this IVR control package (RFC 6231) using SIP as the inter-server signaling protocol claim that it has equivalent capabilities to those used in MSCML (RFC 5022) and MSML (RFC 5707)? Justify your answer.
4. How is this IVR control package (RFC 6231) aligned with the W3C web model defined in W3C Voice Browser languages? Explain in detail, justifying your answer.
5. How are the specific capabilities of the IVR control package defined? Explain in detail.
6. How does the IVR control package align with the web model of W3C Voice Browser languages? Explain in detail.
7. How is the key dialog management mechanism of the IVR control package closely aligned with Call Control XML (CCXML)? Explain in detail.
8. Explain all functionalities of XML elements along with the transport using the CFW framework: a. Dialog management elements, b. Dialog element that defines a lightweight IVR dialog language used with dialog management elements, and c. Elements for auditing package capabilities and dialogs managed by the package.
9. Describe the following features of the IVR control package: IVR Dialog, Other Dialog Languages, and Foreign Namespace Attributes and Elements.
10. What are the important security schemes that need to be implemented for the IVR package to protect against cyberattacks?

REFERENCES

1. Gateway Control Protocol: Advanced Media Server Packages. *ITU-T Recommendation H.248.9.*
2. Auburn, R J. (2011, July). Voice Browser Call Control: CCXML Version 1.0. *W3C Recommendation.*
3. McGlashan, S. et al. (2004, March). Voice Extensible Markup Language (VoiceXML) Version 2.0. *W3C Recommendation.*
4. Oshry, M. et al. (2007, June). Voice Extensible Markup Language (VoiceXML) Version 2.1. *W3C Recommendation.*

5. McGlashan, S. et al. (2010, August). Voice Extensible Markup Language (VoiceXML) Version 3.0. *W3C Working Draft.*
6. Bray, T. et al. (2004, February). Extensible Markup Language (XML) 1.0 (Third Edition). *W3C Recommendation.*
7. IANA. RTP Payload Types. <http://www.iana.org>.
8. Jansen, J. et al. (2005, December). Synchronized Multimedia Integration Language (SMIL 2.1). *World Wide Web Consortium Recommendation REC-SMIL2-20051213.* <http://www.w3.org/TR/2005/REC-SMIL2-20051213>.
9. Biron, P. and A. Malhotra. (2004, October). XML Schema Part 2: Datatypes Second Edition. *W3C Recommendation.*
10. Hunt, A. and S. McGlashan. (2004, March). Speech Recognition Grammar Specification Version 1.0. *W3C Recommendation.*
11. IANA. MIME Media Types. <http://www.iana.org>.

13 SIP Interface to VoiceXML Media Services

13.1 INTRODUCTION

Voice Extensible Markup Language (VoiceXML) [1, 2] is a World Wide Web Consortium (W3C) standard for creating audio and video dialogs that feature synthesized speech, digitized audio, recognition of spoken and dual tone multi-frequency (DTMF) key input, recording of audio and video, telephony, and mixed-initiative conversations. VoiceXML allows Web-based development and content delivery paradigms to be used with interactive video and voice response applications. This specification describes a SIP (RFC 3261 – also see Section 1.2) interface to VoiceXML media services. Commonly, Application Servers controlling Media Servers use this protocol for pure VoiceXML processing capabilities. Session initiation protocol (SIP) is responsible for initiating a media session to the VoiceXML Media Server and simultaneously triggering the execution of a specified VoiceXML application. This protocol is an adjunct to the full MEDIACTRL protocol and packages mechanism.

The interface described here (RFC 5552) leverages a mechanism for identifying dialog media services first described in RFC 4240. The interface has been updated and extended to support the W3C Recommendation for VoiceXML 2.0 [1] and VoiceXML 2.1 [2]. A set of commonly implemented functions and extensions have been specified, including VoiceXML dialog preparation, outbound calling, video media support, and transfers. VoiceXML session variable mappings have been defined for SIP with an extensible mechanism for passing application-specific values into the VoiceXML application. Mechanisms for returning data to the Application Server have also been added.

13.1.1 USE CASES

The VoiceXML media service user in this specification is generically referred to as an Application Server. In practice, it is intended that the interface defined by this specification be applicable across a wide range of use cases. Several intended use cases are described in the following.

13.1.1.1 Interactive Voice Response (IVR) Services with Application Servers

SIP Application Servers provide services to users of the network. Typically, there may be several Application Servers in the same network, each specialized in providing a particular service. Throughout this specification and without loss of generality, we posit the presence of an Application Server specialized in providing IVR services. A typical configuration for this use case is illustrated in Figure 13.1).

Assuming the Application Server also supports HTTP, the VoiceXML application may be hosted on it and served up via HTTP (RFC 2616). Note, however, that the Web model allows the VoiceXML application to be hosted on a separate (HTTP) Application Server from the (SIP) Application Server that interacts with the VoiceXML Media Server via this specification. It is also possible for a static VoiceXML application to be stored locally on the VoiceXML Media Server, leveraging the VoiceXML 2.1 [2] <data> mechanism to interact with a Web/Application Server when dynamic behavior is required. The viability of static VoiceXML applications is further enhanced by the mechanisms defined in Section 2.4, through which the Application Server can make session-specific

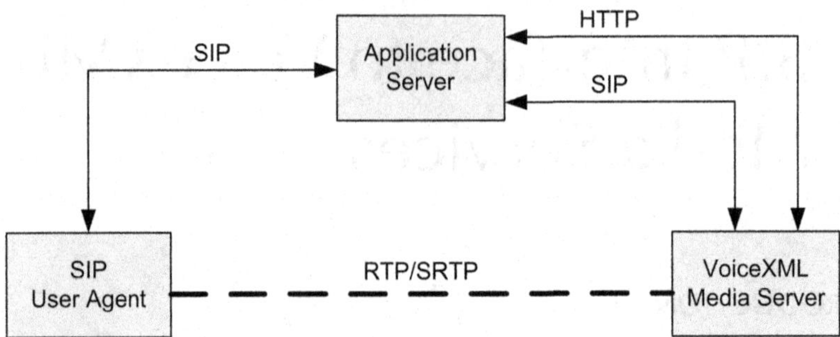

FIGURE 13.1 IVR Services with Application Servers. (Copyright: IETF.)

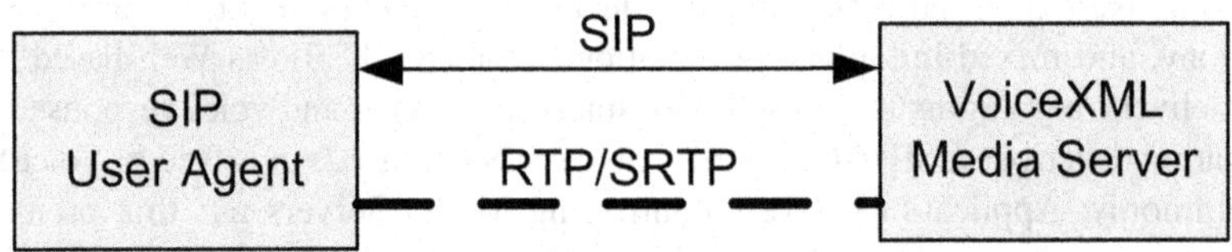

FIGURE 13.2 PSTN IVR Service Node.

information available within the VoiceXML session context. The approach described in this specification is sometimes termed the "delegation model" – the Application Server is essentially delegating programmatic control of the human–machine interactions to one or more VoiceXML specifications running on the VoiceXML Media Server. During the human–machine interactions, the Application Server remains in the signaling path and can respond to results returned from the VoiceXML Media Server or other external network events.

13.1.1.2 PSTN IVR Service Node

While this specification is intended to enable enhanced use of VoiceXML as a component of larger systems and services, it is intended that devices that are completely unaware of this specification remain capable of invoking VoiceXML services offered by a VoiceXML Media Server compliant with this specification. A typical configuration for this use case is shown in Figure 13.2).

Note also that beyond the invocation and termination of a VoiceXML dialog, the semantics defined for call transfers using REFER are intended to be compatible with standard, existing Internet protocol (IP)/PSTN (public switched telephone network) gateways.

13.1.1.3 3GPP IP Multimedia Subsystem (IMS) Media Resource Function (MRF)

The 3rd Generation Partnership Project (3GPP) IMS [3] defines an MRF used to offer media processing services such as conferencing, transcoding, and prompt/collect. The capabilities offered by VoiceXML are ideal for offering richer media processing services in the context of the MRF (Figure 13.3). In this architecture, the interface defined here corresponds to the "Mr" interface to the MRFC (MRF Controller); the implementation of this interface might use separated MRFC and MRFP (MRF Processor) elements (as per the IMS architecture) or might be an integrated MRF (as is common practice).

This diagram is highly simplified and shows a subset of nodes typically involved in MRF interactions. It should be noted that while the MRF will primarily be used by the Application Server via the Serving Call Session Control Function (S-CSCF), it is also possible for calls to be routed directly to the MRF without the involvement of an Application Server. Although the above is described in terms of the 3GPP IMS architecture, it is intended that it is also applicable to 3GPP2, Next Generation Network (NGN), and PacketCable architectures that are converging with 3GPP IMS standards.

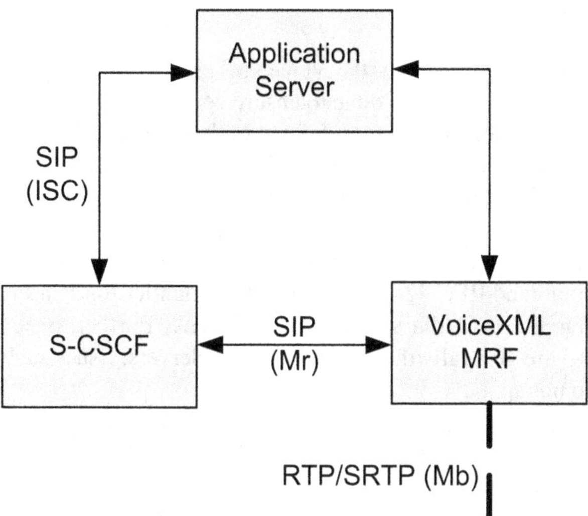

FIGURE 13.3 3GGP IMS Media Secure Server Function. (Copyright: IETF.)

13.1.1.4 Call Control eXtensible Markup Language (CCXML) ↔ VoiceXML Interaction

CCXML 1.0 [4] applications provide services mainly through controlling the interaction between Connections, Conferences, and Dialogs. Although CCXML is capable of supporting arbitrary dialog environments, VoiceXML is commonly used as a dialog environment in conjunction with CCXML applications; CCXML is specifically designed to effectively support the use of VoiceXML. CCXML 1.0 [4] defines language elements that allow Dialogs to be prepared, started, and terminated; it further allows data to be returned by the dialog environment, call transfers to be requested (by the dialog) and responded to by the CCXML application, and arbitrary eventing between the CCXML application and running dialog application.

The interface described in this specification can be used by CCXML 1.0 implementations to control VoiceXML Media Servers. Note, however, that some CCXML language features require eventing facilities between CCXML and VoiceXML sessions that go beyond what is defined in this specification. For example, VoiceXML-controlled call transfers and mid-dialog, application-defined events cannot be fully realized using this specification alone. An SIP event package (RFC 3265) MAY be used in addition to this specification to provide extended eventing.

13.1.1.5 Other Use Cases

In addition to the use cases described in some detail here, there are a number of other intended use cases that are not described in detail, such as:

1. Use of a VoiceXML Media Server as an adjunct to an IP-based Private Branch Exchange/Automatic Call Distributor (PBX/ACD), possibly to provide voicemail/messaging, automated attendant, or other capabilities
2. Invocation and control of a VoiceXML session that provides the voice modality component in a multimodal system

13.2 VOICEXML SESSION ESTABLISHMENT AND TERMINATION

This section describes how to establish a VoiceXML session, with or without preparation, and how to terminate a session. This section also addresses how session information is made available to VoiceXML applications.

13.2.1 Service Identification

The SIP Request-URI is used to identify the VoiceXML media service. The user part of the SIP Request-URI is fixed to "dialog." This is done to ensure compatibility with that of RFC 4240, since this specification extends the dialog interface defined in that specification and because this convention from RFC 4240 is widely adopted by existing Media Servers. Standardizing the SIP Request-URI, including the user part, also improves interoperability between Application Servers and Media Servers and reduces the provisioning overhead that would be required if use of a Media Server by an Application Server required an individually provisioned uniform resource identifier (URI). In this respect, this specification (and RFC 4240) does not add semantics to the user part, but rather, standardizes the way that targets on Media Servers are provisioned. Further, since Application Servers – and not human beings – are generally the clients of Media Servers, issues such as interpretation and internationalization do not apply.

Exposing a VoiceXML media service with a well-known address may enhance the possibility of exploitation: The VoiceXML Media Server is RECOMMENDED to use standard SIP mechanisms to authenticate endpoints, as discussed in Section 13.7. The initial VoiceXML specification is specified with the "voicexml" parameter. In addition, parameters are defined that control how the VoiceXML Media Server fetches the specified VoiceXML specification. The list of parameters (also see Section 1.5, Table 1.2) defined by this specification is as follows (note that the parameter names are case-insensitive):

- voicexml: URI of the initial VoiceXML specification to fetch. This will typically contain an HTTP URI but may use other URI schemes, for example, to refer to local, static VoiceXML specifications. If the "voicexml" parameter is omitted, the VoiceXML Media Server may select the initial VoiceXML specification by other means, such as by applying a default, or may reject the request.
- maxage: Used to set the max-age value of the Cache-Control header in conjunction with VoiceXML specifications fetched using HTTP, as per RFC 2616. If omitted, the VoiceXML Media Server will use a default value.
- maxstale: Used to set the max-stale value of the Cache-Control header in conjunction with VoiceXML specifications fetched using HTTP, as per RFC 2616. If omitted, the VoiceXML Media Server will use a default value.
- method: Used to set the HTTP method applied in the fetch of the initial VoiceXML specification. Allowed values are "get" or "post" (case-insensitive). Default is "get."
- postbody: Used to set the application/x-www-form-urlencoded HTML 4.01 encoded [5] HTTP body for "post" requests (or is otherwise ignored).
- ccxml: Used to specify a "JSON value" (RFC 4627) that is mapped to the session.connection.ccxml VoiceXML session variable – see Section 13.2.4.
- aai: Used to specify a "JSON value" (RFC 4627) that is mapped to the session.connection.aai VoiceXML session variable – see Section 13.2.4.

Other application-specific parameters may be added to the Request-URI and are exposed in VoiceXML session variables (see Section 13.2.4). Formally, the Request-URI for the VoiceXML media service has a fixed user part "dialog." Seven URI parameters are defined (see the definition of uri-parameter in Section 25.1 of RFC 3261).

```
dialog-param      =    "voicexml=" vxml-url ; vxml-url follows the URI
                       ; syntax defined in RFC 3986
maxage-param      =    "maxage=" 1*DIGIT
maxstale-param    =    "maxstale=" 1*DIGIT
method-param      =    "method=" ("get" / "post")
postbody-param    =    "postbody=" token
```

```
ccxml-param        =    "ccxml=" json-value
aai-param          =    "aai=" json-value
json-value         =    false /
                        null /
                        true /
                        object /
                        array /
                        number /
                        string ; defined in RFC 4627
```

Parameters of the Request-URI in subsequent re-INVITEs are ignored. One consequence of this is that the VoiceXML Media Server cannot be instructed by the Application Server to change the executing VoiceXML Application after a VoiceXML Session has been started. Special characters contained in the dialog-param, postbody-param, ccxml-param, and aai-param values must be URL-encoded ("escaped") as required by the SIP URI syntax, for example, "?" (%3f), "=" (%3d), and ";" (%3b). The VoiceXML Media Server MUST therefore unescape these parameter values before making use of them or exposing them to running VoiceXML applications. It is important that the VoiceXML Media Server only unescape the parameter values once, since the desired VoiceXML URI value could itself be URL-encoded, for example. Since some applications may choose to transfer confidential information, the VoiceXML Media Server MUST support the sips: scheme as discussed in Section 13.7.

Informative note: With respect to the postbody-param value, since the application/x-www-form-urlencoded content itself escapes nonalphanumeric characters by inserting %HH replacements, the escaping rules in the previous paragraph will result in the "%2 characters being further escaped in addition to the "&" and "=" name/value separators.

As an example, the following SIP Request-URI identifies the use of VoiceXML media services, with "http://appserver.example.com/promptcollect.vxml" as the initial VoiceXML specification, to be fetched with max-age/max-stale values of 3600 s/0 s, respectively:

```
sip:dialog@mediaserver.example.com; \
voicexml=http://appserver.example.com/promptcollect.vxml; \
maxage=3600;maxstale=0
```

13.2.2 Initiating a VoiceXML Session

A VoiceXML Session is initiated via the Application Server using a SIP INVITE. Typically, the Application Server will be specialized in providing VoiceXML services. At a minimum, the Application Server may behave as a simple proxy by rewriting the Request-URI received from the user agent to a Request-URI suitable for consumption by the VoiceXML Media Server (as specified in Section 13.2.1). For example, a user agent might present a dialed number: tel:+1-201-555-0123 that the Application Server maps to a directory assistance application on the VoiceXML Media Server with a Request-URI of:

```
sip:dialog@ms1.example.com; \
voicexml=http://as1.example.com/da.vxml
```

Certain header values in the INVITE message to the VoiceXML Media Server are mapped into VoiceXML session variables and are specified in Section 13.2.4. On receipt of the INVITE, the VoiceXML Media Server issues a provisional response, 100 Trying, and commences the fetch of the initial VoiceXML specification. The 200 OK response indicates that the VoiceXML specification has been fetched and parsed correctly and is ready for execution. Application execution commences on receipt of the ACK (except if the dialog is being prepared as specified in Section 13.2.3).

Note that the 100 Trying response will usually be sent on receipt of the INVITE in accordance with RFC 3261 (also see Section 1.2), since the VoiceXML Media Server cannot in general guarantee that the initial fetch will complete in less than 200 ms. However, certain implementations may be able to guarantee response times to the initial INVITE, and thus may not need to send a 100 Trying response. As an optimization, prior to sending the 200 OK response, the VoiceXML Media Server MAY execute the application up to the point of the first VoiceXML waiting state or prompt flush.

A VoiceXML Media Server, like any SIP user agent, may be unable to accept the INVITE request for a variety of reasons. For instance, a session description protocol (SDP) offer contained in the INVITE might require the use of codecs that are not supported by the Media Server. In such cases, the Media Server should respond as defined by RFC 3261 (also see Section 1.2). However, there are error conditions specific to VoiceXML, as follows:

1. If the Request-URI does not conform to this specification, a 400 Bad Request MUST be returned (unless it is used to select other services not defined by this specification).
2. If a URI parameter in the Request-URI is repeated, then the request MUST be rejected with a 400 Bad Request response.
3. If the Request-URI does not include a "voicexml" parameter, and the VoiceXML Media Server does not elect to use a default page, the VoiceXML Media Server MUST return a final response of 400 Bad Request, and it SHOULD include a Warning header with a 3-digit code of 399 and a human-readable error message.
4. If the VoiceXML specification cannot be fetched or parsed, the VoiceXML Media Server MUST return a final response of 500 Server Internal Error and SHOULD include a Warning header with a 3-digit code of 399 and a human-readable error message.

Informative note: Certain applications may pass a significant amount of data to the VoiceXML dialog in the form of Request-URI parameters. This may cause the total size of the INVITE request to exceed the maximum transmission unit (MTU) of the underlying network. In such cases, applications/implementations must take care either to use a transport appropriate to these larger messages (such as transmission control protocol [TCP]) or to use alternative means of passing the required information to the VoiceXML dialog (such as supplying a unique session identifier in the initial VoiceXML URI and later using that identifier as a key to retrieve data from the HTTP server).

13.2.3 Preparing a VoiceXML Session

In certain scenarios, it is beneficial to prepare a VoiceXML Session for execution prior to running it. A previously prepared VoiceXML Session is expected to execute with minimal delay when instructed to do so. If a media-less SIP dialog is established with the initial INVITE to the VoiceXML Media Server, the VoiceXML application will not execute after receipt of the ACK. To run the VoiceXML application, the Application Server must issue a re-INVITE to establish a media session.

A media-less SIP dialog can be established by sending an SDP containing no media lines in the initial INVITE. Alternatively, if no SDP is sent in the initial INVITE, the VoiceXML Media Server will include an offer in the 200 OK message, which can be responded to with an answer in the ACK with the media port(s) set to 0. Once a VoiceXML application is running, a re-INVITE that disables the media streams (i.e., sets the ports to 0) will not otherwise affect the executing application (except that recognition actions initiated while the media streams are disabled will result in noinput timeouts).

13.2.4 Session Variable Mappings

The standard VoiceXML session variables are assigned values according to:

- session.connection.local.uri: Evaluates to the SIP URI specified in the To: header of the initial INVITE.

- session.connection.remote.uri: Evaluates to the SIP URI specified in the From: header of the initial INVITE.
- session.connection.redirect: This array is populated by information contained in the History-Info (RFC 4244) header in the initial INVITE or is otherwise undefined. Each entry (hi-entry) in the History-Info header is mapped, in reverse order, into an element of the session.connection.redirect array.
  Properties of each element of the array are determined as follows:
  - uri – Set to the hi-targeted-to-uri value of the History-Info entry
  - pi – Set to "true" if hi-targeted-to-uri contains a "Privacy=history" parameter or if the INVITE Privacy header includes "history;" "false" otherwise
  - si – Set to the value of the "si" parameter if it exists, undefined otherwise
  - reason – Set verbatim to the value of the "Reason" parameter of hi-targeted-to-uri
- session.connection.protocol.name: Evaluates to "sip." Note that this is intended to reflect the use of SIP in general and does not distinguish between whether the Media Server was accessed via SIP or SIPS procedures.
- session.connection.protocol.version: Evaluates to "2.0."
- session.connection.protocol.sip.headers: This is an associative array where each key in the array is the non-compact name of a SIP header in the initial INVITE converted to lowercase (note that the case conversion does not apply to the header value). If multiple header fields of the same field name are present, the values are combined into a single comma-separated value. Implementations MUST at a minimum include the Call-ID header and MAY include other headers. For example,
- session.connection.protocol.sip.headers["call-id"] evaluates to the Call-ID of the SIP dialog.
- session.connection.protocol.sip.requesturi: This is an associative array where the array keys and values are formed from the URI parameters on the SIP Request-URI of the initial INVITE. The array key is the URI parameter name converted to lowercase (note that the case conversion does not apply to the parameter value). The corresponding array value is obtained by evaluating the URI parameter value as a "JSON value" (RFC 4627) in the case of the ccxml-param and aai-param values and otherwise as a string. In addition, the array's toString() function returns the full SIP Request-URI. For example, assuming a Request-URI of:
  sip:dialog@example.com;voicexml=http://example.com;aai=
  %7b"x":1%2c"y":true%7d
- then session.connection.protocol.sip.requesturi["voicexml"] evaluates to http://example.com.
- session.connection.protocol.sip.requesturi["aai"].x evaluates to 1 (type Number).
- session.connection.protocol.sip.requesturi["aai"].y evaluates to true (type Boolean).
- session.connection.protocol.sip.requesturi evaluates to the complete Request-URI (type String) "sip:dialog@example.com;voicexml=http://example.com; aai={"x":1,"y":true}."
- session.connection.aai: Evaluates to session.connection.protocol.sip.requesturi["aai"].
- session.connection.ccxml: Evaluates to session.connection.protocol.sip.requesturi["ccxml"].
- session.connection.protocol.sip.media: This is an array where each array element is an object with the following properties:
  - type: – This required property indicates the type of the media associated with the stream. The value is a string. It is strongly recommended that the following values are used for common types of media: "audio" for audio media and "video" for video media.
  - direction: – This required property indicates the directionality of the media relative to session.connection.originator. Defined values are sendrecv, sendonly, recvonly, and inactive.

- format: – This property is optional. If defined, the value of the property is an array. Each array element is an object that specifies information about one format of the media (there is an array element for each payload type on the m-line). The object contains at least one property called "name" whose value is the MIME subtype of the media format (MIME subtypes are registered in RFC 4855).
- Other properties may be defined with string values; these correspond to required and if defined, optional parameters of the format.

As a consequence of this definition, there is an array entry in session.connection.protocol.sip.media for each non-disabled m-line for the negotiated media session. Note that this session variable is updated if the media session characteristics for the VoiceXML Session change (i.e., due to a re-INVITE). For an example, consider a connection with bidirectional G.711 mu-law "audio" sampled at 8 kHz. In this case,

- session.initiation.protocol.sip.media[0].type evaluates to "audio."
- session.initiation.protocol.sip.media[0].direction to "sendrecv."
- session.initiation.protocol.sip.media[0].format[0].name evaluates to "audio/PCMU."
- session.initiation.protocol.sip.media[0].format[0].rate evaluates to "8000."

Note that when accessing SIP headers and Request-URI parameters via the session.connection.protocol.sip.headers and session.connection.protocol.sip.requesturi associative arrays defined here, applications can choose between two semantically equivalent ways of referring to the array. For example, either of the following can be used to access a Request-URI parameter named "foo:"

- session.initiation.protocol.sip.requesturi["foo"]
- session.initiation.protocol.sip.requesturi.foo

However, it is important to note that not all SIP header names or Request-URI parameter names are valid ECMAScript identifiers, and as such, can only be accessed using the first form (array notation). For example, the Call-ID header can only be accessed as session.connection.protocol.sip.headers["call-id"]; attempting to access the same value as session.initiation.protocol.sip.headers.call-id would result in an error.

13.2.5 Terminating a VoiceXML Session

The Application Server can terminate a VoiceXML Session by issuing a BYE to the VoiceXML Media Server. Upon receipt of a BYE in the context of an existing VoiceXML Session, the VoiceXML Media Server MUST send a 200 OK response and MUST throw a "connection.disconnect.hangup" event to the VoiceXML application. If the Reason header (RFC 3326) is present on the BYE request, then the value of the Reason header is provided verbatim via the "_message" variable within the catch element's anonymous variable scope.

The VoiceXML Media Server may also initiate termination of the session by issuing a BYE request. This will typically occur as a result of encountering a <disconnect> or <exit> in the VoiceXML application due to the VoiceXML application running to completion or due to unhandled errors within the VoiceXML application. See Chapter 4 for mechanisms to return data to the Application Server.

13.2.6 Examples

13.2.6.1 Basic Session Establishment

This example illustrates an Application Server setting up a VoiceXML Session (Figure 13.4) on behalf of a user agent.

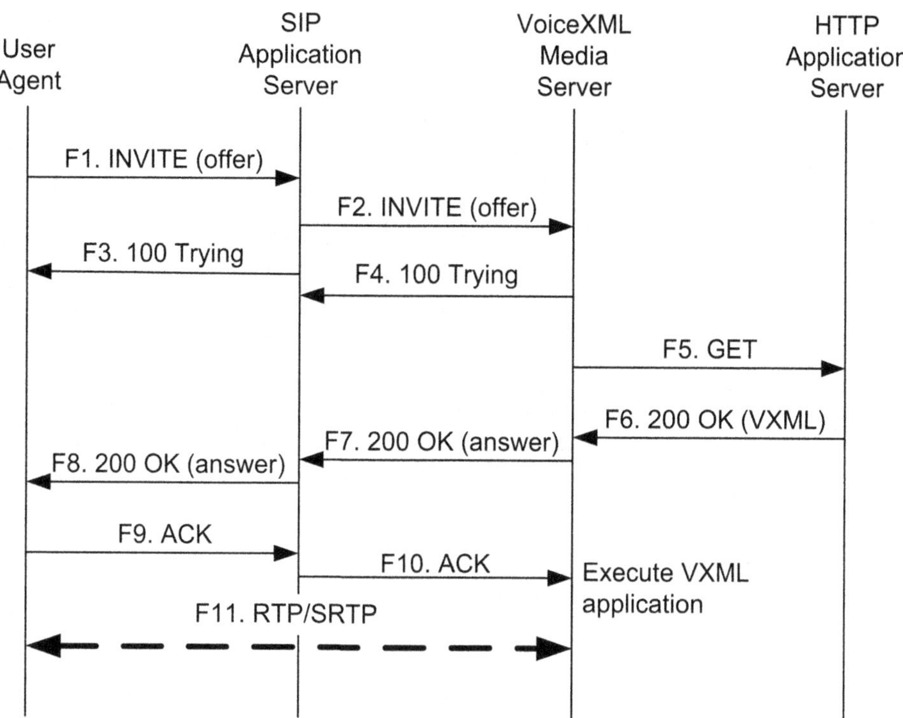

FIGURE 13.4 Basic VoiceXML Session Establishment. (Copyright: IETF.)

13.2.6.2 VoiceXML Session Preparation

This example (Figure 13.5) demonstrates the preparation of a VoiceXML Session. In this example, the VoiceXML session is prepared prior to placing an outbound call to a user agent and is started as soon as the user agent answers. The [answerl:0] notation is used to indicate an SDP answer with the media ports set to 0.

Implementation detail: offer2′ is derived from offer2 – it duplicates the m-lines and a-lines from offer2. However, offer2′ differs from offer2, since it must contain the same o-line as used in answerl:0 but with the version number incremented. Also, if offerl has more m-lines than offer2, then offer2′ must be padded with extra (rejected) m-lines.

13.2.6.3 Media Resource Control Protocol (MRCP) Connection Establishment

MRCP (RFC 6787 – see Chapter 14) is a protocol that enables clients such as a VoiceXML Media Server to control media service resources such as speech synthesizers, recognizers, verifiers, and identifiers residing in servers on the network. The example in Figure 13.6 illustrates how a VoiceXML Media Server may establish an MRCP session in response to an initial INVITE.

In this example, the VoiceXML Media Server is responsible for establishing a session with the MRCPv2 (RFC 6787 – see Chapter 14) Media Resource Server prior to sending the 200 OK response to the initial INVITE. The VoiceXML Media Server will perform the appropriate offer/answer with the MRCPv2 Media Resource Server based on the SDP capabilities of the Application Server and the MRCPv2 Media Resource Server. The VoiceXML Media Server will change the offer received from step (1) to establish an MRCPv2 session in step (5) and will rewrite the SDP to include an m-line for each MRCPv2 resource to be used and other required SDP modifications as specified by MRCPv2. Once the VoiceXML Media Server performs the offer/answer with the MRCPv2 Media Resource Server, it will establish an MRCPv2 control channel in step (8). The MRCPv2 resource is deallocated when the VoiceXML Media Server receives or sends a BYE (not shown).

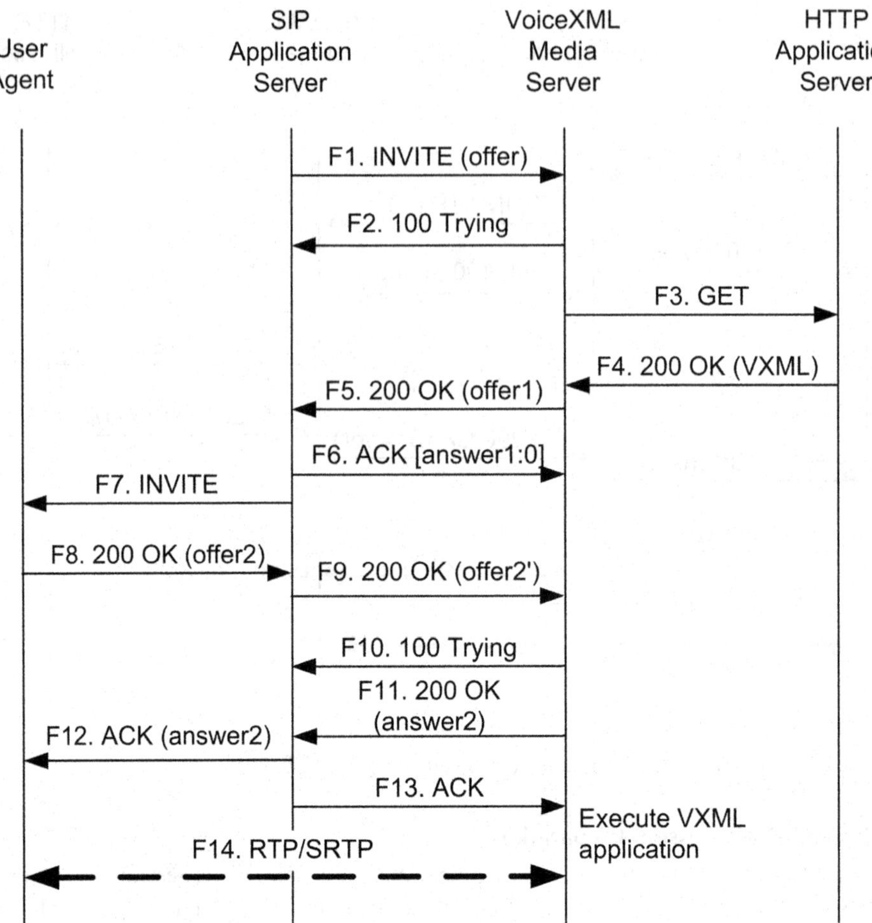

FIGURE 13.5 VoiceXML Session Preparation.

13.3 MEDIA SUPPORT

This section describes the mandatory and optional media support required by this interface.

13.3.1 Offer/Answer

The VoiceXML Media Server MUST support the standard offer/answer mechanism of RFC 3264. In particular, if an SDP offer is not present in the INVITE, the VoiceXML Media Server will make an offer in the 200 OK response listing its supported codecs.

13.3.2 Early Media

The VoiceXML Media Server MAY support early establishment of media streams as described in RFC 3960. This allows the Application Server to establish media streams between a user agent and the VoiceXML Media Server in parallel with the initial VoiceXML specification being processed (which may involve dynamic VoiceXML page generation and interaction with databases or other systems). This is useful primarily for minimizing the delay in starting a VoiceXML session, particularly in cases where a session with the user agent already exists, but the media stream associated

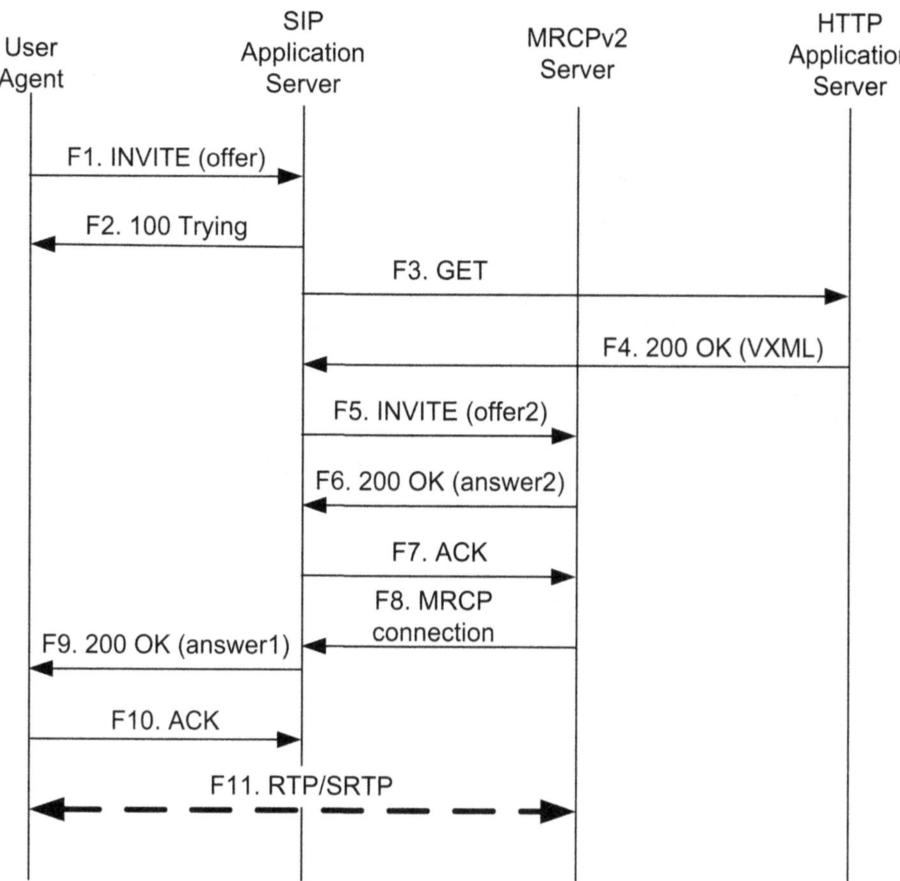

FIGURE 13.6 MRCP Connection Establishment for VoiceXML Services. (Copyright: IETF.)

with that session needs to be redirected to a VoiceXML Media Server. The flow in Figure 13.7 demonstrates the use of early media (using the Gateway model defined in RFC 3960).

Although RFC 3960 prefers the use of the Application Server model for early media over the Gateway model, the primary issue with the Gateway model – forking – is significantly less common when issuing requests to VoiceXML Media Servers. This is because VoiceXML Media Servers respond to all requests with 200 OK responses in the absence of unusual errors, and they typically do so within several hundreds of milliseconds. This makes them unlikely targets in forking scenarios, since alternative targets of the forking process would virtually never be able to respond more quickly than an automated system unless they are themselves automated systems – in which case there is little point in setting up a response time race between two automated systems. Issues with ringing tone generation in the Gateway model are also mitigated, both by the typically quick 200 OK response time and because this specification mandates that no media packets are generated until the receipt of an ACK (thus eliminating the need for the user agent to perform media packet analysis). Note that the offer of early media by a VoiceXML Media Server does not imply that the referenced VoiceXML application can always be fetched and executed successfully. For instance, if the HTTP Application Server were to return a 4xx response in step (10) earlier, or if the provided VoiceXML content was not valid, the VoiceXML Media Server would still return a 500 response (as per Section 13.2.2). At this point, it would be the responsibility of the Application Server to tear down any media streams established with the Media Server.

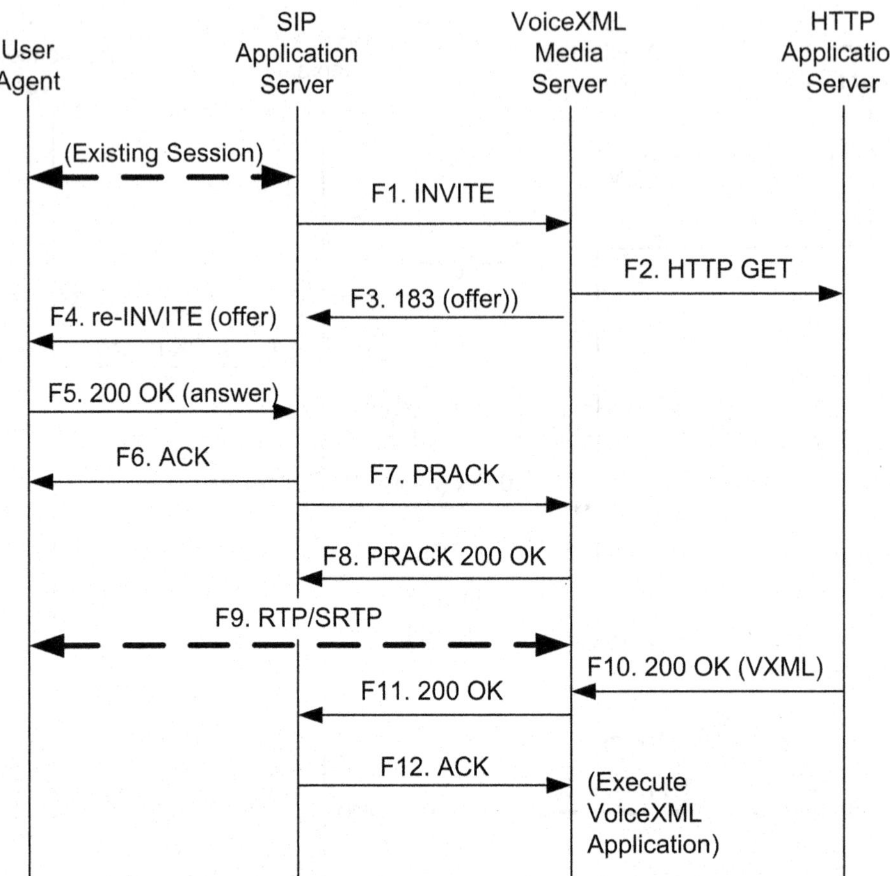

FIGURE 13.7 VoiceXML Early Media Session Using the Gateway Model. (Copyright: IETF.)

13.3.3 Modifying the Media Session

The VoiceXML Media Server MUST allow the media session to be modified via a re-INVITE and SHOULD support the UPDATE method (RFC 3311) for the same purpose. In particular, it MUST be possible to change streams between sendrecv, sendonly, and recvonly as specified in RFC 3264. Unidirectional streams are useful for announcement- or listening-only (hotword). The preferred mechanism for putting the media session on hold is specified in RFC 3264, i.e., the user agent modifies the stream to be sendonly and mutes its own stream. Modification of the media session does not affect VoiceXML application execution (except that recognition actions initiated while on hold will result in noinput timeouts).

13.3.4 Audio and Video Codecs

For the purposes of achieving a basic level of interoperability, this section specifies a minimal subset of codecs and RTP (RFC 3550) payload formats that MUST be supported by the VoiceXML Media Server. For audio-only applications, G.711 mu-law and A-law MUST be supported using the RTP payload type 0 and 8 (RFC 3551). Other codecs and payload formats MAY be supported. Video telephony applications, which employ a video stream in addition to the audio stream, are possible in VoiceXML 2.0/2.1 through the use of multimedia file container formats such as the .3gp [6] and .mp4 formats [7]. Video support is optional for this specification. If video is supported, then:

1. H.263 Baseline (RFC 4629) MUST be supported. For legacy reasons, the 1996 version of H.263 MAY be supported using the RTP payload format defined in [RFC2190] (payload type 34 (RFC 3551).
2. Adaptive multi-rate (AMR) narrow band audio (RFC 4867) SHOULD be supported.
3. MPEG-4 video (RFC 3016) SHOULD be supported.
4. MPEG-4 Advanced Audio Coding (AAC) audio (RFC 3016) SHOULD be supported.
5. Other codecs and payload formats MAY be supported.

Video record operations carried out by the VoiceXML Media Server typically require receipt of an intra-frame before the recording can commence. The VoiceXML Media Server SHOULD use the mechanism described in (RFC 4585) to request that a new intra-frame be sent. Since some applications may choose to transfer confidential information, the VoiceXML Media Server MUST support Secure RTP (SRTP) (RFC 3711) as discussed in Section 13.7.

13.3.5 DTMF

DTMF events (RFC 4733) MUST be supported. When the user agent does not indicate support for RFC 4733, the VoiceXML Media Server MAY perform DTMF detection using other means, such as detecting DTMF tones in the audio stream. Implementation note: The reason only (RFC 4733) telephone-events must be used when the user agent indicates support of it is to avoid the risk of double detection of DTMF if detection on the audio stream is simultaneously applied.

13.4 RETURNING DATA TO THE APPLICATION SERVER

This section discusses the mechanisms for returning data (e.g., collected utterance or digit information) from the VoiceXML Media Server to the Application Server.

13.4.1 HTTP Mechanism

At any time during the execution of the VoiceXML application, data can be returned to the Application Server via HTTP using standard VoiceXML elements such as <submit> or <subdialog>. Notably, the <data> element in VoiceXML 2.1 [2] allows data to be sent to the Application Server efficiently without requiring a VoiceXML page transition and is ideal for short VoiceXML applications such as "prompt and collect." For most applications, it is necessary to correlate the information being passed over HTTP with a particular VoiceXML Session. One way this can be achieved is to include the SIP Call-ID (accessible in VoiceXML via the session.connection.protocol.sip.headers array) within the HTTP POST fields. Alternatively, a unique "POST-back URI" can be specified as an application-specific URI parameter in the Request-URI of the initial INVITE (accessible in VoiceXML via the session.connection.protocol.sip.requesturi array). Since some applications may choose to transfer confidential information, the VoiceXML Media Server MUST support the https: scheme as discussed in Section 13.7.

13.4.2 SIP Mechanism

Data can be returned to the Application Server via the expr or namelist attribute on <exit> or the namelist attribute on <disconnect>. A VoiceXML Media Server MUST support encoding of the expr/namelist data in the message body of a BYE request sent from the VoiceXML Media Server as a result of encountering the <exit> or <disconnect> element. A VoiceXML Media Server MAY support inclusion of the expr/namelist data in the message body of the 200 OK message in response to a received BYE request (i.e., when the VoiceXML application responds to the connection.

disconnect.hangup event and subsequently executes an <exit> element with the expr or namelist attribute specified). Note that sending expr/namelist data in the 200 OK response requires that the VoiceXML Media Server delay the final response to the received BYE request until the VoiceXML application's post-disconnect final processing state terminates. This mechanism is subject to the constraint that the VoiceXML Media Server must respond before the User Agent Client's (UAC's) timer F expires (defaults to 32 seconds). Moreover, for unreliable transports, the UAC will retransmit the BYE request according to the rules of RFC 3261. The VoiceXML Media Server SHOULD implement the recommendations of RFC 4320 regarding when to send the 100 Trying provisional response to the BYE request.

If a VoiceXML application executes a <disconnect> [2] and then subsequently executes an <exit> with namelist information, the namelist information from the <exit> element is discarded. Namelist variables are first converted to their "JSON value" equivalent RFC 4627 and encoded in the message body using the application/x-www-form-urlencoded format content type [5]. The behavior resulting from specifying a recording variable in the namelist or an ECMAScript object with circular references is not defined. If the expr attribute is specified on the <exit> element instead of the namelist attribute, the reserved name __exit is used.

To allow the Application Server to differentiate between a BYE resulting from a <disconnect> from one resulting from an <exit>, the reserved name __reason is used, with a value of "disconnect" (without brackets) to reflect the use of VoiceXML's <disconnect> element, and a value of "exit" (without brackets) to an explicit <exit> in the VoiceXML specification. If the session terminates for other reasons (such as the Media Server encountering an error), this parameter may be omitted or may take on platform-specific values prefixed with an underscore.

This specification extends the application/x-www-form-urlencoded by replacing non-ASCII characters with one or more octets of the UTF-8 representation of the character, with each octet in turn replaced by %HH, where HH represents the uppercase hexadecimal notation for the octet value and % is a literal character. As a consequence, the Content-Type header field in a BYE message containing expr/namelist data MUST be set to application/x-www-form-urlencoded;charset=utf-8. Table 13.1 provides some examples of <exit> usage and the corresponding result content, assuming the following VoiceXML variables and values:

```
userAuthorized = true
pin = 1234
errors = 0
```

For example, consider the VoiceXML snippet:

```
...
<exit namelist="id pin"/>
...
```

TABLE 13.1
<exit> Usage Examples

<exit> Usage	**Result Content**
<exit/>	__reason=exit
<exit expr=" 'done' "/>	__exit="done"&__reason=exit
<exit expr="userAuthorized"/>	__exit=true&__reason=exit
<exit namelist="pin erros"/>	

If id equals 1234 and pin equals 9999, say, the BYE message would look similar to:

```
BYE sip:user@pc33.example.com SIP/2.0
Via: SIP/2.0/UDP 192.0.2.4;branch=z9hG4bKnashds10
Max-Forwards: 70
From: sip:dialog@example.com;tag=a6c85cf
To: sip:user@example.com;tag=1928301774
Call-ID: a84b4c76e66710
CSeq: 231 BYE
Content-Type: application/x-www-form-urlencoded;charset=utf-8
Content-Length: 30
id=1234&pin=9999&__reason=exit
```

Since some applications may choose to transfer confidential information, the VoiceXML Media Server MUST support the S/MIME encoding of SIP message bodies as discussed in Section 13.7.

13.5 OUTBOUND CALLING

Outbound calls can be triggered via the Application Server using third-party call control (RFC 3725). Flow IV from RFC 3725 is recommended in conjunction with the VoiceXML Session preparation mechanism. This flow has several advantages over others, namely:

1. Selection of a VoiceXML Media Server and preparation of the VoiceXML application can occur before the call is placed to avoid the callee experiencing delays.
2. Avoidance of timing difficulties that could occur with other flows due to the time taken to fetch and parse the initial VoiceXML specification.
3. The flow is IPv6 compatible. An example flow for an Application Server–initiated outbound call is provided in Section 13.2.6.2.

13.6 CALL TRANSFER

While VoiceXML is at its core a dialog language, it also provides optional call transfer capability. VoiceXML's transfer capability is particularly suited to the PSTN IVR Service Node use case described in Section 13.1.1.2. It is NOT RECOMMENDED to use VoiceXML's call transfer capability in networks involving Application Servers. Rather, the Application Server itself can provide call routing functionality by taking signaling actions based on the data returned to it from the VoiceXML Media Server via HTTP or in the SIP BYE message.

If VoiceXML transfer is supported, the mechanism described in this section MUST be employed. The transfer flows specified here are selected on the basis that they provide the best interworking across a wide range of SIP devices. CCXML↔VoiceXML implementations, which require tight-coupling in the form of bidirectional eventing to support all transfer types defined in VoiceXML, may benefit from other approaches, such as the use of SIP event packages (RFC 3265). In what follows, the provisional responses have been omitted for clarity.

13.6.1 BLIND

The blind-transfer sequence is initiated by the VoiceXML Media Server via a REFER message (RFC 3265) on the original SIP dialog. The Refer-To header contains the URI for the called party, as specified via the dest or destexpr attributes on the VoiceXML <transfer> tag. If the REFER request is accepted, in which case the VoiceXML Media Server will receive a 2xx response, the VoiceXML Media Server throws the connection.disconnect.transfer event and will terminate the VoiceXML Session with a BYE message. For blind transfers, implementations MAY use RFC 4488 to suppress

TABLE 13.2
Possible SIP Responses for Blind Transfer Errors

SIP Response	<transfer> variable / event
404 Not Found	error.connection.baddestination
405 Method Not Allowed	error.unsupported.transfer.blind
503 Service Unavailable	error.connection.noresource
(no response)	network_busy
(Other 3xx/4xx/5xx/6xx)	unknown

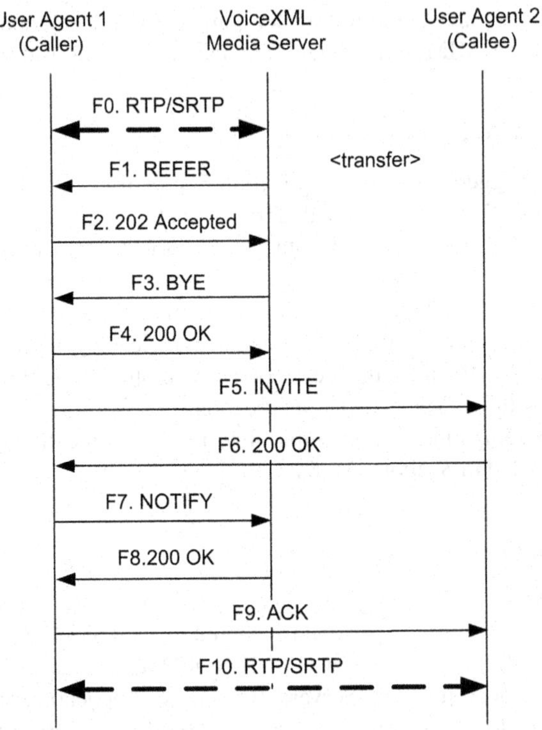

FIGURE 13.8 Blind Transfer Using SIP REFER Method. (Copyright: IETF.)

the implicit subscription associated with the REFER message. If the REFER request results in a non-2xx response, the <transfer>'s form item variable (or event raised) depends on the SIP response and is specified in Table 13.2. Note that this indicates that the transfer request was rejected.

An example (Figure 13.8) is illustrated (provisional responses and NOTIFY messages corresponding to provisional responses have been omitted for clarity).

If the aai or aaiexpr attribute is present on <transfer>, it is appended to the Refer-To URI as a parameter named "aai" in the REFER method. Reserved characters are URL-encoded as required for SIP/SIPS URIs (RFC 3261; also see Section 1.2). The mapping of values outside the ASCII range is platform specific.

13.6.2 Bridge

The bridge transfer function results in the creation of a small multi-party session involving the Caller, the VoiceXML Media Server, and the Callee. The VoiceXML Media Server invites the

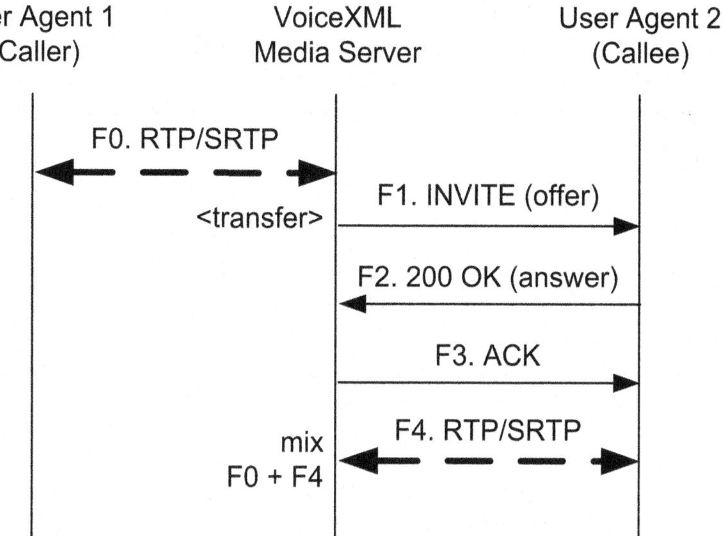

FIGURE 13.9 SIP Call Bridge Transfer.

Callee to the session and will eject the Callee if the transfer is terminated. If the aai or aaiexpr attribute is present on <transfer>, it is appended to the Request-URI in the INVITE as a URI parameter named "aai." Reserved characters are URL-encoded as required for SIP/SIPS URIs (RFC 3261 also see Section 1.2). The mapping of values outside the ASCII range is platform specific.

During the transfer attempt, audio specified in the transferaudio attribute of <transfer> is streamed to User Agent 1. A VoiceXML Media Server MAY play early media received from the Callee to the Caller if the transferaudio attribute is omitted. The bridge transfer sequence is illustrated in the following. The VoiceXML Media Server (acting as a UAC) makes a call to User Agent 2 with the same codecs used by User Agent 1. When the call setup is complete, RTP flows between User Agent 2 and the VoiceXML Media Server. This stream is mixed with User Agent 1's (Figure 13.9).

If a final response is not received from User Agent 2 from the INVITE, and the connecttimeout expires (specified as an attribute of <transfer>), the VoiceXML Media Server will issue a CANCEL to terminate the transaction, and the <transfer>'s form item variable is set to noanswer. If INVITE results in a non-2xx response, the <transfer>'s form item variable (or event raised) depends on the SIP response and is specified in Table 13.3.

Once the transfer is established, the VoiceXML Media Server can "listen" to the media stream from User Agent 1 to perform speech or DTMF hotword, which when matched, results in a near-end disconnect, i.e., the VoiceXML Media Server issues a BYE to User Agent 2 and the VoiceXML

TABLE 13.3
Possible SIP Error Messages for Call Bridge Transfer

SIP Response	<transfer> variable / event
404 Not Found	error.connection.baddestination
405 Method Not Allowed	error.unsupported.transfer.bridge
408 Request Timeout	noanswe
486 Busy Here	busy
503 Service Unavailable	error.connection.noresource
(no response)	network_busy
(Other 3xx/4xx/5xx/6xx)	unknown

application continues with User Agent 1. A BYE will also be issued to User Agent 2 if the call duration exceeds the maximum duration specified in the maxtime attribute on <transfer>. If User Agent 2 issues a BYE during the transfer, the transfer terminates, and the VoiceXML <transfer>'s form item variable receives the value far_end_disconnect. If User Agent 1 issues a BYE during the transfer, the transfer terminates, and the VoiceXML event connection.disconnect.transfer is thrown.

13.6.3 Consultation

The consultation transfer (also called attended transfer [RFC 5359]) is similar to a blind transfer except that the outcome of the transfer call setup is known and the Caller is not dropped as a result of an unsuccessful transfer attempt. Consultation transfer commences with the same flow as for bridge transfer except that the RTP streams are not mixed at step (4), and error.unsupported.transfer.consultation supplants error.unsupported.transfer.bridge. Assuming that a new SIP dialog with User Agent 2 is created, the remainder of the sequence follows as illustrated (provisional responses and NOTIFY messages corresponding to provisional responses have been omitted for clarity). Consultation transfer makes use of the Replaces: header (RFC 3891) such that User Agent 1 calls User Agent 2 and replaces the latter's SIP dialog with the VoiceXML Media Server with a new SIP dialog between the Caller and Callee (Figure 13.10).

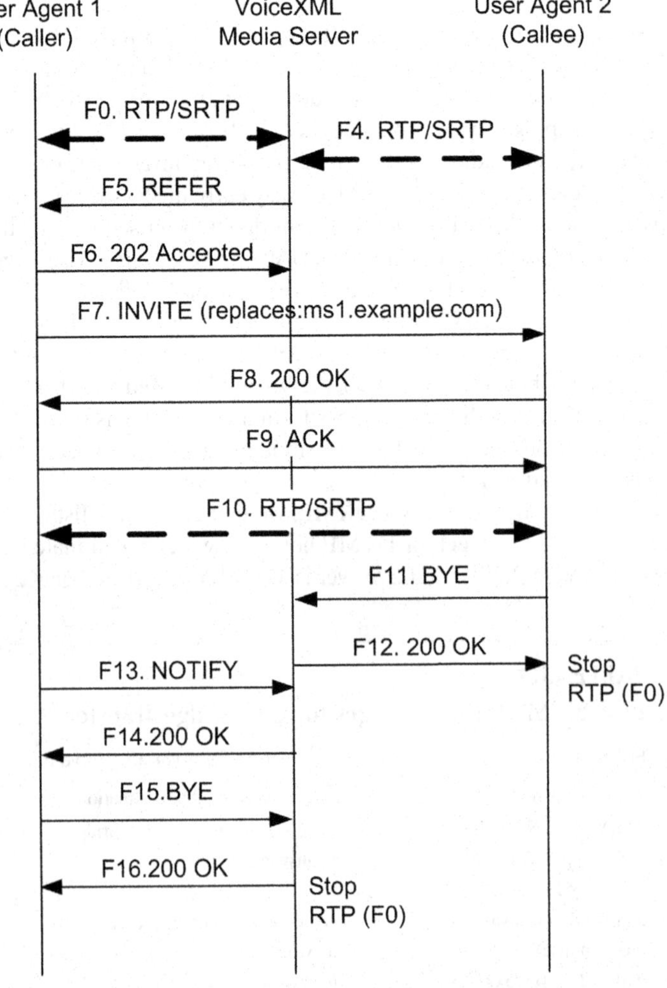

FIGURE 13.10 Call Transfer for Consultation/Called Attended Transfer. (Copyright: IETF.)

If a response other than 202 Accepted is received in response to the REFER request sent to User Agent 1, the transfer terminates, and an error.unsupported.transfer.consultation event is raised. In addition, a BYE is sent to User Agent 2 to terminate the established outbound leg. The VoiceXML Media Server uses receipt of a NOTIFY message with a sipfrag message of 200 OK to determine that the consultation transfer has succeeded. When this occurs, the connection.disconnect.transfer event will be thrown to the VoiceXML application, and a BYE is sent to User Agent 1 to terminate the session. A NOTIFY message with a non-2xx final response sipfrag message body will result in the transfer terminating and the associated VoiceXML input item variable being set to "unknown." Note that as a consequence of this mechanism, implementations MUST NOT use RFC 4488 to suppress the implicit subscription associated with the REFER message for consultation transfers.

13.7 SECURITY CONSIDERATIONS

Exposing a VoiceXML media service with a well-known address may enhance the possibility of exploitation (for example, an invoked network service may trigger a billing event). The VoiceXML Media Server is RECOMMENDED to use standard SIP mechanisms (RFC 3261; also see Section 1.2) to authenticate requesting endpoints and authorize per local policy. Some applications may choose to transfer confidential information to or from the VoiceXML Media Server. To provide data confidentiality, the VoiceXML Media Server MUST implement the sips: and https: schemes in addition to S/MIME message body encoding as described in RFC 3261 (also see Section 1.2).

The VoiceXML Media Server MUST support Secure RTP (SRTP) (RFC 3711) to provide confidentiality, authentication, and replay protection for RTP media streams (including RTCP control traffic). To mitigate the possibility of denial-of-service attacks, the VoiceXML Media Server is RECOMMENDED (in addition to authenticating and authorizing endpoints described earlier) to provide mechanisms for implementing local policies, such as the time-limiting of VoiceXML application execution.

13.8 IANA CONSIDERATIONS

The Internet Assigned Numbers Authority (IANA) has registered some parameters in the SIP/SIPS URI Parameters registry following the Specification Required policy of RFC 3969. See detail in RFC 5552.

13.9 SUMMARY

We have described briefly the VoiceXML [1, 2] that has been standardized by W3C for creating audio and video dialogs that feature synthesized speech, digitized audio, recognition of spoken and DTMF key input, recording of audio and video, telephony, and mixed-initiative conversations. VoiceXML, including Web-based VoiceXML development and content delivery paradigms, needs to be used with interactive video and voice response applications. However, this section (RFC 5552) specifies a SIP (RFC 3261 – also see Sections 1.2 and 1.5) interface to VoiceXML media services. The channel control framework (CFW) termed "MediaCtrl," which is based on SIP protocol described in RFC 6230 (see Chapter 8), is used for this VoiceXML media services package.

We have explained that SIP is responsible for initiating a media session to the VoiceXML Media Server and simultaneously triggering the execution of a specified VoiceXML application, while the Application Servers controlling media servers use SIP protocol for pure VoiceXML processing capabilities. In fact, users invoke VoiceXML services by invoking the Application Servers, and in turn, the Application Servers use SIP protocol for invoking the VoiceXML media services to users. We have also articulated several use cases for the VoiceXML media services: IVR Services with Application Servers, PSTN IVR Service Node, 3GPP IMS MRF, CCXML and VoiceXML Interaction, IP-based PBX/ACD, voicemail/messaging, automated attendant, voice modality component in a multimodal system, and other capabilities. The main features of the SIP-based VoiceXML media services interface are: VoiceXML Session Establishment and Termination, Media Support, Offer/Answer, Returning Data to the Application Servers, Outbound Calling, and Call Transfer.

In Session Establishment and Termination, we have described how to establish a VoiceXML Session, with or without preparation, and how to terminate a session, including addressing how session information is made available to VoiceXML applications. The SIP Request-URI, which is standardized to identify the VoiceXML media service, enables interoperability between Application Servers and Media Servers and reduces the provisioning overhead. More specifically, we have described the Initiation of VoiceXML Session, Preparation of VoiceXML Session, Mapping of Session Variable, and Termination of VoiceXML Session. In addition, we have provided some examples for explaining these capabilities: Basic Session Establishment, VoiceXML Session Preparation, and MRCP Establishment. Note that MRCP (Media Resource Control), specified in RFC 6787 (see Section 14), like SIPREC (SIP-based Media Recording), described in RFC 6341 (see Section 10.1), can be used for media recording.

In Media Support, we have explained the mandatory and optional media support required by this interface. The SDP Offer/Answer model is used by SIP for negotiations of different types of media for a given session: Early Media, Different Types of Audio and Video Codecs, and DTMF. In addition, we have articulated how the VoiceXML Media Server MUST allow the media session to be modified via a re-INVITE and SHOULD support the UPDATE method. In Returning Data to the Application Server, we have discussed the mechanisms for returning data (for example, collected utterance or digit information) from the VoiceXML Media Server to the Application Server using either the HTTP mechanism or the SIP mechanism. We have described that Outbound calls can be triggered via the Application Server using third-party call control, described in Flow IV of RFC 3725 (see Section 1.3.1.4), which is recommended in conjunction with the VoiceXML Session preparation mechanism because of its several advantages.

We have explained that the Call Transfer is the optional capability for the VoiceXML services. The PSTN IVR Service use case described here is particularly suitable for VoiceXML Call Transfer services. However, it is NOT RECOMMENDED to use VoiceXML's call transfer capability in networks involving Application Servers. Rather, it is RECOMMENDED that the Application Server itself can provide call routing functionality by taking signaling actions based on the data returned to it from the VoiceXML Media Server via HTTP or in the SIP BYE message. Note that implementations of CCXML and VoiceXML interworking, which require tight-coupling in the form of bidirectional eventing to support all transfer types defined in VoiceXML, may benefit from other approaches, such as the use of SIP event packages (RFC 3265). In addition, we have specified that the VoiceXML Call Transfer mechanisms, if supported using blind transfer, bridge transfer, or consultation transfer, MUST follow the procedures described in this section (RFC 5552). Finally, we have discussed some specific security concerns but have RECOMMENDED the use of SIP security mechanisms for VoiceXML media services. Note that IANA has registered some parameters in the SIP/SIPS URI, and interested readers are requested to see RFC 5552.

13.10 PROBLEMS

1. Briefly explain the capabilities of W3C VoiceXML standards. How do the SIP-based VoiceXML media services use the CFW (RFC 6230 – see Chapter 8) in invoking all the capabilities of W3C VoiceXML?
2. Draw the architectural diagram for the XCON conference systems, showing how SIP is used for initiating a media session to the VoiceXML Media Server and simultaneously triggering the execution of a specified VoiceXML application, while the Application Servers controlling Media Servers use SIP protocol for pure VoiceXML processing capabilities. Explain the functional features for this scenario.
3. Explain the following SIP-based VoiceXML media services, including call flows: a. IVR Services with Application Servers, b. PSTN IVR Service Node, c. 3GPP IMS MRF, d.

CCXML and VoiceXML Interaction, e. IP-based PBX/ACD, f. voicemail/messaging, g. automated attendant, and e. voice modality component in a multimodal system.

4. Describe the call flows of the SIP-based VoiceXML, media services for the following features: a. Basic Session Establishment, b. VoiceXML Session Preparation, and c. MRCP Establishment.
5. Specify the following with call flows for the SIP-based VoiceXML media services: SDP Offer/Answer model used by SIP for negotiations of different types of media for a given session: a. Early Media, b. Different Types of Audio and Video Codecs, and c. DTMF.
6. How do you modify a SIP-based VoiceXML on-going media session in XCON conference systems? Explain in detail using call flows.
7. In Returning Data to the Application Server, how do you return data (for example, collected utterance or digit information) from the VoiceXML Media Server to the Application Server: a. using HTTP mechanism and b. using SIP mechanism? Explain in detail using call flows for the XCON conference systems.
8. How do you trigger outbound calls via the Application Server for the SIP-based VoiceXML Media Server for the XCON conference systems? Explain in detail using call flows. What are the advantages of your call configurations compared with other configurations, as there can be several mechanisms for the setup of the outbound calls?
9. Explain the PSTN IVR Service use case showing call flows for the SIP-based VoiceXML media Call Transfer services in the context of the XCON conference systems.
10. Why is it NOT RECOMMENDED to use VoiceXML's call transfer capability in networks involving Application Servers? Explain in detail using detail call flow for the XCON conference systems.
11. Why is it RECOMMENDED that the Application Server itself can provide call routing functionality for the Call Transfer function of the SIP-based VoiceXML media services? Explain in detail showing call flows using the XCOM conference systems.
12. For implementations of CCXML and VoiceXML interworking, why is it beneficial to use SIP event packages described in RFC 3265 for Call Transfer of SIP-based VoiceXML media services? Justify your answer with call flows using the XCOM conference systems.
13. Explain in detail, with call flow procedures that MUST be followed per RFC 5552 for the following Call Transfer mechanisms of the SIP-based VoiceXML media services in the context of the XCON conference systems: a. Blind Transfer, b. Bridge Transfer, and d. Consultation Transfer.
14. Describe in detail the specific SIP-based security mechanisms that need to be used for SIP-based VoiceXML services.

REFERENCES

1. McGlashan, S. et al. (2004, March) Voice Extensible Markup Language (VoiceXML) Version 2.0. *W3C Recommendation.*
2. Oshry, M. et al. (2005 June) Voice Extensible Markup Language (VoiceXML) Version.1. *W3C Candidate Recommendation.*
3. 3rd Generation Partnership Project: Network Architecture (Release 6). *3GPP TS 23.002 v6.6.0.* (2004, December).
4. Auburn, R J. (2005, June). Voice Browser Call Control: CCXML Version 1.0. *W3C Working Draft.*
5. Raggett, D., Le Hors, A., and Jacobs, I. (1999, December) HTML 4.01 Specification. *W3C Recommendation.*
6. Transparent End-To-End Packet Switched Streaming Service (PSS); 3GPP File Format (3GP). *3GPP TS 26.244v6.4.0.* (2004, December).
7. Information Technology. Coding of Audio-Visual Objects. MP4 File Format. *ISO/IEC ISO/IEC 14496–14:2003.* (2003, October).

14 Media Resource Control Protocol Version 2

14.1 INTRODUCTION

Media Resource Control Protocol Version 2 (MRCPv2) (RFC 6787) that is described in this Chapter is designed to allow a client device to control media processing resources on the network. Some of these media processing resources include speech recognition engines, speech synthesis engines, speaker verification, and speaker identification engines. MRCPv2 enables the implementation of distributed interactive voice response platforms using VoiceXML [1] browsers or other client applications while maintaining separate back-end speech processing capabilities on specialized speech processing servers. MRCPv2 is based on the earlier Media Resource Control Protocol version 1 (MRCPvl) – RFC 4463, but note that there is no interworking between RFC 6787-MRCPv2 described in this chapter and RFC 4463-MRCPvl. MRCPvl (RFC 4463) was developed jointly by Cisco Systems, Inc., Nuance Communications, and Speechworks, Inc. Although some of the method names are similar, the way in which these methods are communicated is different. There are also more resources and more methods for each resource. The first version of MRCP was essentially taken only as input to the development of this protocol. There is no expectation that an MRCPv2 client will work with an MRCPvl server or vice versa. There is no migration plan or gateway definition between the two protocols.

The protocol requirements of Speech Services Control (SPEECHSC) (RFC 4313) include that the solution must be capable of reaching a media processing server, setting up communication channels to the media resources, and sending and receiving control messages and media streams to/from the server. The session initiation protocol (SIP) (RFC 3261 – also see Section 1.2) meets these requirements. The proprietary version of MRCP ran over the real-time streaming protocol (RTSP) (RFC 2326). At the time work on MRCPv2 was begun, the consensus was that this use of RTSP would break the RTSP protocol or cause backwards-compatibility problems, something forbidden by Section 3.2 of RFC 4313. This is the reason why MRCPv2 does not run over RTSP.

MRCPv2 leverages these capabilities by building upon SIP and the session description protocol (SDP) (RFC 4566). MRCPv2 uses SIP to set up and tear down media and control sessions with the server. In addition, the client can use an SIP re-INVITE method (an INVITE dialog sent within an existing SIP session) to change the characteristics of these media and control session while maintaining the SIP dialog between the client and server. SDP is used to describe the parameters of the media sessions associated with that dialog. It is mandatory to support SIP as the session establishment protocol to ensure interoperability. Other protocols can be used for session establishment by prior agreement. This specification only describes the use of SIP and SDP.

MRCPv2 uses SIP and SDP to create the speech client/server dialog and set up the media channels to the server. It also uses SIP and SDP to establish MRCPv2 control sessions between the client and the server for each media processing resource required for that dialog. The MRCPv2 protocol exchange between the client and the media resource is carried on that control session. MRCPv2 exchanges do not change the state of the SIP dialog, the media sessions, or other parameters of the dialog initiated via SIP. MRCPv2 controls and affects the state of the media processing resource associated with the MRCPv2 session(s).

MRCPv2 defines the messages to control the different media processing resources and the state machines required to guide their operation. It also describes how these messages are carried over a transport layer protocol such as the transmission control protocol (TCP) (RFC 0793) or the transport layer security (TLS) protocol (RFC 5246). (Note: the stream control transmission protocol

(SCTP) (RFC 4960) is a viable transport for MRCPv2 as well, but the mapping onto SCTP is not described in this specification.)

14.2 CONVENTIONS

Since many of the definitions and syntax are identical to those for the hypertext transfer protocol – HTTP/1.1 (RFC 2616), this specification refers to the section where they are defined rather than copying it. For brevity, notation "HX.Y" is to be taken to refer to Section X.Y of RFC 2616. All the mechanisms specified in this specification are described in both prose and an augmented Backus–Naur form (ABNF – RFC 5234). The complete message format in ABNF form is provided in Section 14.15 and is the normative format definition. Note that productions may be duplicated within the main body of the specification for reading convenience. If a production in the body of the text conflicts with one in the normative definition, the latter rules.

14.2.1 Definitions

(Also see Section 1.6, Table 1.2)

- Media Resource: An entity on the speech processing server that can be controlled through MRCPv2.
- MRCP Server: Aggregate of one or more "Media Resource" entities on a server, exposed through MRCPv2. Often, "server" in this specification refers to an MRCP server.
- MRCP Client: An entity controlling one or more Media Resources through MRCPv2 ("Client" for short).
- DTMF: Dual-tone multi-frequency; a method of transmitting key presses in-band, either as actual tones (Q.23 [2]) or as named tone events (RFC 4733).
- Endpointing: The process of automatically detecting the beginning and end of speech in an audio stream. This is critical both for speech recognition and for automated recording, such as one would find in voice mail systems.
- Hotword Mode: A mode of speech recognition where a stream of utterances is evaluated for match against a small set of command words. This is generally employed either to trigger some action or to control the subsequent grammar to be used for further recognition.

14.2.2 State-Machine Diagrams

The state-machine diagrams in this specification do not show every possible method call. Rather, they reflect the state of the resource based on the methods that have moved to IN-PROGRESS or COMPLETE states (see Section 5.3). Note that since PENDING requests essentially have not affected the resource yet and are in the queue to be processed, they are not reflected in the state-machine diagrams.

14.2.3 URI Schemes

This specification defines many protocol headers that contain URIs (uniform resource identifiers [RFC 3986]) or lists of URIs for referencing media. The entire specification, including the Security Considerations section (Section 14.12), assumes that HTTP or HTTP over TLS (HTTPS) (RFC 2818) will be used as the URI addressing scheme unless otherwise stated. However, implementations MAY support other schemes (such as "file"), provided they have addressed any security considerations described in this specification and any others particular to the specific scheme. For example, implementations where the client and server both reside on the same physical hardware and the file system is secured by traditional user-level file access controls could be reasonable candidates for supporting the "file" scheme.

14.3 ARCHITECTURE

A system using MRCPv2 consists of a client that requires the generation and/or consumption of media streams and a media resource server that has the resources or "engines" to process these streams as input or generate these streams as output. The client uses SIP and SDP to establish an MRCPv2 control channel with the server to use its media processing resources. MRCPv2 servers are addressed using SIP URIs. SIP uses SDP with the offer/answer model described in RFC 3264 to set up the MRCPv2 control channels and describe their characteristics. A separate MRCPv2 session is needed to control each of the media processing resources associated with the SIP dialog between the client and the server. Within an SIP dialog, the individual resource control channels for the different resources are added or removed through SDP offer/answer carried in a SIP re-INVITE transaction.

The server, through the SDP exchange, provides the client with a difficult-to-guess, unambiguous channel identifier and a TCP port number (see Section 14.4.2). The client MAY then open a new TCP connection with the server on this port number. Multiple MRCPv2 channels can share a TCP connection between the client and the server. All MRCPv2 messages exchanged between the client and the server carry the specified channel identifier, which the server MUST ensure is unambiguous among all MRCPv2 control channels that are active on that server. The client uses this channel identifier to indicate the media processing resource associated with that channel. For information on message framing, see Section 14.5. SIP also establishes the media sessions between the client (or other source/sink of media) and the MRCPv2 server using SDP "m=" lines. One or more media processing resources may share a media session under an SIP session, or each media processing resource may have its own media session. Figure 14.1 shows the general architecture of a system that uses MRCPv2. To simplify, only a few resources are shown.

14.3.1 MRCPv2 Media Resource Types

An MRCPv2 server may offer one or more of the following media processing resources to its clients:

- Basic Synthesizer: A speech synthesizer resource that has very limited capabilities and can generate its media stream exclusively from concatenated audio clips. The speech data is described using a limited subset of the Speech Synthesis Markup Language (SSML) [3]

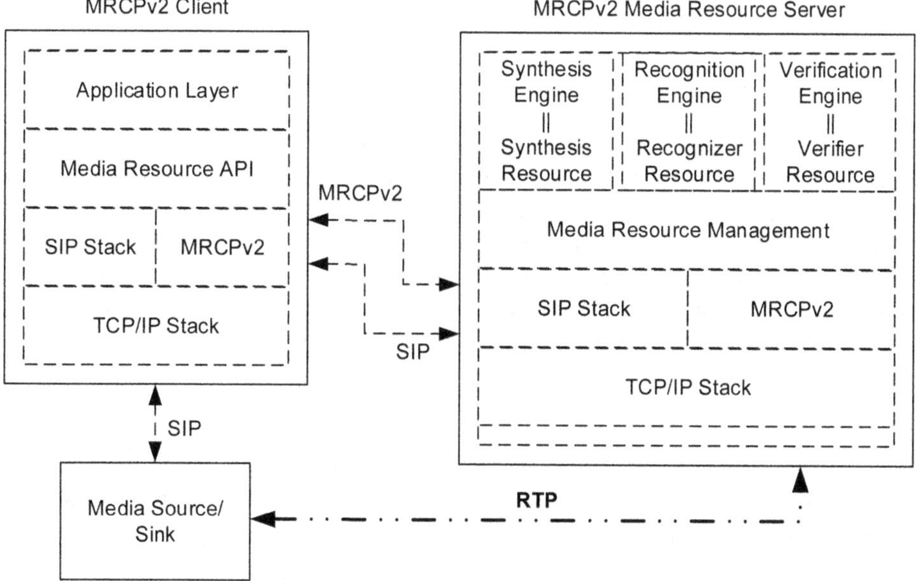

FIGURE 14.1 Architectural Diagram. (Copyright: IETF.)

elements. A basic synthesizer MUST support the SSML tags <speak>, <audio>, <say-as>, and <mark>.

- Speech Synthesizer: A full-capability speech synthesis resource that can render speech from text. Such a synthesizer MUST have full SSML (W3C.REC-speech-synthesis-20040907) support.
- Recorder: A resource capable of recording audio and providing a URI pointer to the recording. A recorder MUST provide endpointing capabilities for suppressing silence at the beginning and end of a recording and MAY also suppress silence in the middle of a recording. If such suppression is done, the recorder MUST maintain timing metadata to indicate the actual timestamps of the recorded media.
- DTMF Recognizer: A recognizer resource capable of extracting and interpreting DTMF [2] digits in a media stream and matching them against a supplied digit grammar. It could also do a semantic interpretation based on semantic tags in the grammar.
- Speech Recognizer: A full speech recognition resource that is capable of receiving a media stream containing audio and interpreting the recognition results. It also has a natural language semantic interpreter to post-process the recognized data according to the semantic data in the grammar and provide semantic results along with the recognized input. The recognizer MAY also support enrolled grammars, where the client can enroll and create new personal grammars for use in future recognition operations.
- Speaker Verifier: A resource capable of verifying the authenticity of a claimed identity by matching a media stream containing spoken input to a pre-existing voiceprint. This may also involve matching the caller's voice against more than one voiceprint, also called multi-verification or speaker identification.

14.3.2 Server and Resource Addressing

The MRCPv2 server is a generic SIP server and is thus addressed by an SIP URI (RFC 3261 – also see Chapter 2). For example:

sip:mrcpv2@example.net or
sips:mrcpv2@example.net

14.4 MRCPV2 BASICS

MRCPv2 requires a connection-oriented transport layer protocol such as TCP to guarantee reliable sequencing and delivery of MRCPv2 control messages between the client and the server. In order to meet the requirements for security enumerated in SPEECHSC requirements (RFC 4313), clients and servers MUST implement TLS as well. One or more connections between the client and the server can be shared among different MRCPv2 channels to the server. The individual messages carry the channel identifier to differentiate messages on different channels. MRCPv2 encoding is text based with mechanisms to carry embedded binary data. This allows arbitrary data like recognition grammars, recognition results, synthesizer speech markup, etc., to be carried in MRCPv2 messages. For information on message framing, see Section 14.5.

14.4.1 Connecting to the Server

MRCPv2 employs SIP, in conjunction with SDP, as the session establishment and management protocol. The client reaches an MRCPv2 server using conventional INVITE and other SIP requests for establishing, maintaining, and terminating SIP dialogs. The SDP offer/answer exchange model over SIP is used to establish a resource control channel for each resource. The SDP offer/answer exchange is also used to establish media sessions between the server and the source or sink of audio.

14.4.2 Managing Resource Control Channels

The client needs a separate MRCPv2 resource control channel to control each media processing resource under the SIP dialog. A unique channel identifier string identifies these resource control channels. The channel identifier is a difficult-to-guess, unambiguous string followed by an "@" and then by a string token specifying the type of resource. The server generates the channel identifier and MUST make sure it does not clash with the identifier of any other MRCP channel currently allocated by that server. MRCPv2 defines the following Internet Assigned Numbers Authority (IANA)-registered types of media processing resources. Additional resource types and their associated methods/events and state machines (Table 14.1) may be added as described in Section 14.13 in IANA.

The SIP INVITE or re-INVITE transaction and the SDP offer/answer exchange it carries contain "m=" lines describing the resource control channel to be allocated. There MUST be one SDP "m=" line for each MRCPv2 resource to be used in the session. This "m=" line MUST have a media type field of "application" and a transport type field of either "TCP/MRCPv2" or "TCP/TLS/MRCPv2." The port number field of the "m=" line MUST contain the "discard" port of the transport protocol (port 9 for TCP) in the SDP offer from the client and MUST contain the TCP listen port on the server in the SDP answer. The client may then either set up a TCP or TLS connection to that server port or share an already established connection to that port. Since MRCPv2 allows multiple sessions to share the same TCP connection, multiple "m=" lines in a single SDP specification MAY share the same port field value; MRCPv2 servers MUST NOT assume any relationship between resources using the same port other than the sharing of the communication channel.

MRCPv2 resources do not use the port or format field of the "m=" line to distinguish themselves from other resources using the same channel. The client MUST specify the resource type identifier in the resource attribute associated with the control "m=" line of the SDP offer. The server MUST respond with the full Channel-Identifier (which includes the resource type identifier and a difficult-to-guess, unambiguous string) in the "channel" attribute associated with the control "m=" line of the SDP answer. To remain backwards compatible with conventional SDP usage, the format field of the "m=" line MUST have the arbitrarily selected value of "1." When the client wants to add a media processing resource to the session, it issues a new SDP offer, according to the procedures of RFC 3264, in an SIP re-INVITE request. The SDP offer/answer exchange carried by this SIP transaction contains one or more additional control "m=" lines for the new resources to be allocated to the session. The server, on seeing the new "m=" line, allocates the resources (if they are available) and responds with a corresponding control "m=" line in the SDP answer carried in the SIP response. If the new resources are not available, the re-INVITE receives an error message, and existing media processing going on before the re-INVITE will continue as it was before. It is not possible to allocate more than one resource of each type. If a client requests more than one resource of any type, the server MUST behave as if the resources of that type (beyond the first one) were not available.

MRCPv2 clients and servers using TCP as a transport protocol MUST use the procedures specified in RFC 4145 for setting up the TCP connection, with the considerations described hereby.

TABLE 14.1
Resource Types

Resource Type	Resource Description	Described in
speechrecog	Speech Recognizer	Section 14.9
dtmfrecog	DTMF Recognizer	Section 14.9
speechsynth	Speech Synthesizer	Section 14.8
basicsynth	Basic Synthesizer	Section 14.8
speakverify	Speaker Verification	Section 14.11
recorder	Speech Recorder	Section 14.10

Similarly, MRCPv2 clients and servers using TCP/TLS as a transport protocol MUST use the procedures specified in RFC 4572 for setting up the TLS connection, with the considerations described hereby. The a=setup attribute, as described in RFC 4145, MUST be "active" for the offer from the client and MUST be "passive" for the answer from the MRCPv2 server. The a=connection attribute MUST have a value of "new" on the very first control "m=" line offer from the client to an MRCPv2 server. Subsequent control "m=" line offers from the client to the MRCP server MAY contain "new" or "existing," depending on whether the client wants to set up a new connection or share an existing connection, respectively. If the client specifies a value of "new," the server MUST respond with a value of "new." If the client specifies a value of "existing," the server MUST respond. The legal values in the response are "existing" if the server prefers to share an existing connection or "new" if not. In the latter case, the client MUST initiate a new transport connection.

When the client wants to deallocate the resource from this session, it issues a new SDP offer, according to RFC 3264, where the control "m=" line port MUST be set to 0. This SDP offer is sent in a SIP re-INVITE request. This deallocates the associated MRCPv2 identifier and resource. The server MUST NOT close the TCP or TLS connection if it is currently being shared among multiple MRCP channels. When all MRCP channels that may be sharing the connection are released, and/or the associated SIP dialog is terminated, the client or server terminates the connection. When the client wants to tear down the whole session and all its resources, it MUST issue an SIP BYE request to close the SIP session. This will deallocate all the control channels and resources allocated under the session.

All servers MUST support TLS. Servers MAY use TCP without TLS in controlled environments (e.g., not in the public Internet) where both nodes are inside a protected perimeter, for example, preventing access to the MRCP server from remote nodes outside the controlled perimeter. It is up to the client, through the SDP offer, to choose which transport it wants to use for an MRCPv2 session. Aside from the exceptions given earlier, when using TCP, the "m=" lines MUST conform to RFC 4145, which describes the usage of SDP for connection-oriented transport. When using TLS, the SDP "m=" line for the control stream MUST conform to Connection-Oriented Media (COMEDIA) over TLS (RFC 4572), which specifies the usage of SDP for establishing a secure connection-oriented transport over TLS.

14.4.3 SIP Session Example

This first example shows the power of using SIP to route to the appropriate resource. In the example, note the use of a request to a domain's speech server service in the INVITE to mresources@example.com. The SIP routing machinery in the domain locates the actual server, mresources@server.example.com, which is returned in the 200 OK. Note that "cmid" is defined in Section 14.4.4. This example (Figure 14.2) exchange adds a resource control channel for a synthesizer. Since a synthesizer also generates an audio stream, this interaction also creates a receive-only real-time protocol (RTP) (RFC 3550) media session for the server to send audio to. The SIP dialog with the media source/sink is independent of MRCP and is not shown.

This example (Figure 14.3) exchange continues from the previous figure and allocates an additional resource control channel for a recognizer. Since a recognizer would need to receive an audio stream for recognition, this interaction also updates the audio stream to sendrecv, making it a two-way RTP media session.

This example (Figure 14.4) exchange continues from the previous figure and deallocates the recognizer channel. Since a recognizer no longer needs to receive an audio stream, this interaction also updates the RTP media session to recvonly.

14.4.4 Media Streams and RTP Ports

Since MRCPv2 resources either generate or consume media streams, the client or the server needs to associate media sessions with their corresponding resource or resources. More than one resource could be associated with a single media session, or each resource could be assigned a separate

```
C->S: INVITE sip:mresources@example.com SIP/2.0
      Via:SIP/2.0/TCP client.atlanta.example.com:5060;
      branch=z9hG4bK74bf1
      Max-Forwards:6
      To:MediaServer <sip:mresources@example.com>
      From:sarvi <sip:sarvi@example.com>;tag=1928301774
      Call-ID:a84b4c76e66710
      CSeq:314161 INVITE
      Contact:<sip:sarvi@client.example.com>
      Content-Type:application/sdp
      Content-Length:...

      v=0
      o=sarvi 2890844526 2890844526 IN IP4 192.0.2.12
      s=-
      c=IN IP4 192.0.2.12
      t=0 0
      m=application 9 TCP/MRCPv2 1
      a=setup:active
      a=connection:new
      a=resource:speechsynth
      a=cmid:1
      m=audio 49170 RTP/AVP 0
      a=rtpmap:0 pcmu/8000
      a=recvonly
      a=mid:1

S->C: SIP/2.0 200 OK
      Via:SIP/2.0/TCP client.atlanta.example.com:5060;
      branch=z9hG4bK74bf1;received=192.0.32.10
      To:MediaServer <sip:mresources@example.com>;tag=62784
      From:sarvi <sip:sarvi@example.com>;tag=1928301774
      Call-ID:a84b4c76e66710
      CSeq:314161 INVITE
      Contact:<sip:mresources@server.example.com>
      Content-Type:application/sdp
      Content-Length:...

      v=0
      o=- 2890842808 2890842808 IN IP4 192.0.2.11
      s=-
      c=IN IP4 192.0.2.11
      t=0 0
      m=application 32416 TCP/MRCPv2 1
      a=setup:passive
      a=connection:new
      a=channel:32AECB234338@speechsynth
      a=cmid:1
      m=audio 48260 RTP/AVP 0
      a=rtpmap:0 pcmu/8000
      a=sendonly
      a=mid:1

C->S: ACK sip:mresources@server.example.com SIP/2.0
      Via:SIP/2.0/TCP client.atlanta.example.com:5060;
      branch=z9hG4bK74bf2
      Max-Forwards:6
      To:MediaServer <sip:mresources@example.com>;tag=62784
      From:Sarvi <sip:sarvi@example.com>;tag=1928301774
      Call-ID:a84b4c76e66710
      CSeq:314161 ACK
      Content-Length:0
```

FIGURE 14.2 Example: Add Synthesizer Control Channel.

media session. Also, note that more than one media session can be associated with a single resource if need be, but this scenario is not useful for the current set of resources. For example, a synthesizer and a recognizer could be associated to the same media session (m=audio line) if it is opened in "sendrecv" mode. Alternatively, the recognizer could have its own "sendonly" audio session, and the synthesizer could have its own "recvonly" audio session.

```
C->S: INVITE sip:mresources@server.example.com SIP/2.0
      Via:SIP/2.0/TCP client.atlanta.example.com:5060;
      branch=z9hG4bK74bf3
      Max-Forwards:6
      To:MediaServer <sip:mresources@example.com>;tag=62784
      From:sarvi <sip:sarvi@example.com>;tag=1928301774
      Call-ID:a84b4c76e66710
      CSeq:314162 INVITE
      Contact:<sip:sarvi@client.example.com>
      Content-Type:application/sdp
      Content-Length:...

      v=0
      o=sarvi 2890844526 2890844527 IN IP4 192.0.2.12
      s=-
      c=IN IP4 192.0.2.12
      t=0 0
      m=application 9 TCP/MRCPv2 1
      a=setup:active
      a=connection:existing
      a=resource:speechsynth
      a=cmid:1
      m=audio 49170 RTP/AVP 0 96
      a=rtpmap:0 pcmu/8000
      a=rtpmap:96 telephone-event/8000
      a=fmtp:96 0-15
      a=sendrecv
      a=mid:1
      m=application 9 TCP/MRCPv2 1
      a=setup:active
      a=connection:existing
      a=resource:speechrecog
      a=cmid:1

S->C: SIP/2.0 200 OK
      Via:SIP/2.0/TCP client.atlanta.example.com:5060;
      branch=z9hG4bK74bf3;received=192.0.32.10
      To:MediaServer <sip:mresources@example.com>;tag=62784
      From:sarvi <sip:sarvi@example.com>;tag=1928301774
      Call-ID:a84b4c76e66710
      CSeq:314162 INVITE
      Contact:<sip:mresources@server.example.com>
      Content-Type:application/sdp
      Content-Length:...

      v=0
      o=- 2890842808 2890842809 IN IP4 192.0.2.11
      s=-
      c=IN IP4 192.0.2.11
      t=0 0
      m=application 32416 TCP/MRCPv2 1
      a=setup:passive
      a=connection:existing
      a=channel:32AECB234338@speechsynth
      a=cmid:1
      m=audio 48260 RTP/AVP 0 96
      a=rtpmap:0 pcmu/8000
      a=rtpmap:96 telephone-event/8000
      a=fmtp:96 0-15
      a=sendrecv
      a=mid:1
      m=application 32416 TCP/MRCPv2 1
      a=setup:passive
      a=connection:existing
      a=channel:32AECB234338@speechrecog
      a=cmid:1

C->S: ACK sip:mresources@server.example.com SIP/2.0
      Via:SIP/2.0/TCP client.atlanta.example.com:5060;
      branch=z9hG4bK74bf4
      Max-Forwards:6
      To:MediaServer <sip:mresources@example.com>;tag=62784
      From:Sarvi <sip:sarvi@example.com>;tag=1928301774
      Call-ID:a84b4c76e66710
      CSeq:314162 ACK
      Content-Length:0
```

FIGURE 14.3 Example: Add Recognizer.

The association between control channels and their corresponding media sessions is established using a new "resource channel media identifier" media-level attribute ("cmid"). Valid values of this attribute are the values of the "mid" attribute defined in RFC 5888. If there is more than one audio "m=" line, then each audio "m=" line MUST have a "mid" attribute. Each control "m=" line MAY have one or more "cmid" attributes that match the resource control channel to the "mid" attributes

```
C->S: INVITE sip:mresources@server.example.com SIP/2.0
      Via:SIP/2.0/TCP client.atlanta.example.com:5060;
      branch=z9hG4bK74bf5
      Max-Forwards:6
      To:MediaServer <sip:mresources@example.com>;tag=62784
      From:sarvi <sip:sarvi@example.com>;tag=1928301774
      Call-ID:a84b4c76e66710
      CSeq:314163 INVITE
      Contact:<sip:sarvi@client.example.com>
      Content-Type:application/sdp
      Content-Length:...

      v=0
      o=sarvi 2890844526 2890844528 IN IP4 192.0.2.12
      s=-
      c=IN IP4 192.0.2.12
      t=0 0
      m=application 9 TCP/MRCPv2 1
      a=resource:speechsynth
      a=cmid:1
      m=audio 49170 RTP/AVP 0
      a=rtpmap:0 pcmu/8000
      a=recvonly
      a=mid:1
      m=application 0 TCP/MRCPv2 1
      a=resource:speechrecog
      a=cmid:1

S->C: SIP/2.0 200 OK
      Via:SIP/2.0/TCP client.atlanta.example.com:5060;
      branch=z9hG4bK74bf5;received=192.0.32.10
      To:MediaServer <sip:mresources@example.com>;tag=62784
      From:sarvi <sip:sarvi@example.com>;tag=1928301774
      Call-ID:a84b4c76e66710
      CSeq:314163 INVITE
      Contact:<sip:mresources@server.example.com>
      Content-Type:application/sdp
      Content-Length:...

      v=0
      o=- 2890842808 2890842810 IN IP4 192.0.2.11
      s=-
      c=IN IP4 192.0.2.11
      t=0 0
      m=application 32416 TCP/MRCPv2 1
      a=channel:32AECB234338@speechsynth
      a=cmid:1
      m=audio 48260 RTP/AVP 0
      a=rtpmap:0 pcmu/8000
      a=sendonly
      a=mid:1
      m=application 0 TCP/MRCPv2 1
      a=channel:32AECB234338@speechrecog
      a=cmid:1

C->S: ACK sip:mresources@server.example.com SIP/2.0
      Via:SIP/2.0/TCP client.atlanta.example.com:5060;
      branch=z9hG4bK74bf6
      Max-Forwards:6
      To:MediaServer <sip:mresources@example.com>;tag=62784
      From:Sarvi <sip:sarvi@example.com>;tag=1928301774
      Call-ID:a84b4c76e66710
      CSeq:314163 ACK
      Content-Length:0
```

FIGURE 14.4 Example: Deallocate Recognizer.

of the audio "m=" lines it is associated with. Note that if a control "m=" line does not have a "cmid" attribute, it will not be associated with any media. The operations on such a resource will hence be limited. For example, if it is a recognizer resource, the RECOGNIZE method requires an associated media to process, while the INTERPRET method does not. The formatting of the "cmid" attribute is described by the following ABNF:

```
cmid-attribute         =     "a=cmid:" identification-tag
identification-tag     =     token
```

To allow this flexible mapping of media sessions to MRCPv2 control channels, a single audio "m=" line can be associated with multiple resources, or each resource can have its own audio "m=" line. For example, if the client wants to allocate a recognizer and a synthesizer and associate them with a single two-way audio stream, the SDP offer would contain two control "m=" lines and a single audio "m=" line with an attribute of "sendrecv." Each of the control "m=" lines would have a "cmid" attribute whose value matches the "mid" of the audio "m=" line. If, on the other hand, the client wants to allocate a recognizer and a synthesizer each with its own separate audio stream, the SDP offer would carry two control "m=" lines (one for the recognizer and another for the synthesizer) and two audio "m=" lines (one with the attribute "sendonly" and another with attribute "recvonly"). The "cmid" attribute of the recognizer control "m=" line would match the "mid" value of the "sendonly" audio "m=" line, and the "cmid" attribute of the synthesizer control "m=" line would match the "mid" attribute of the "recvonly" "m=" line.

When a server receives media (e.g., audio) on a media session that is associated with more than one media processing resource, it is the responsibility of the server to receive and fork the media to the resources that need to consume it. If multiple resources in an MRCPv2 session are generating audio (or other media) to be sent on a single associated media session, it is the responsibility of the server either to multiplex the multiple streams onto the single RTP session or to contain an embedded RTP mixer (see RFC 3550) to combine the multiple streams into one. In the former case, the media stream will contain RTP packets generated by different sources, and hence, the packets will have different synchronization source identifiers (SSRCs). In the latter case, the RTP packets will contain multiple contributing source identifiers (CSRCs) corresponding to the original streams before being combined by the mixer. If an MRCPv2 server implementation neither multiplexes nor mixes, it MUST disallow the client from associating multiple such resources to a single audio stream by rejecting the SDP offer with a SIP 488 "Not Acceptable" error. Note that there is a large installed base that will return a SIP 501 "Not Implemented" error in this case. To facilitate interoperability with this installed base, new implementations SHOULD treat a 501 in this context as a 488 when it is received from an element known to be a legacy implementation.

14.4.5 MRCPv2 Message Transport

The MRCPv2 messages defined in this specification are transported over a TCP or TLS connection between the client and the server. The method for setting up this transport connection and the resource control channel is discussed in Sections 4.1 and 4.2. Multiple resource control channels between a client and a server that belong to different SIP dialogs can share one or more TLS or TCP connections between them; the server and client MUST support this mode of operation. Clients and servers MUST use the MRCPv2 channel identifier, carried in the Channel-Identifier header field in individual MRCPv2 messages, to differentiate MRCPv2 messages from different resource channels (see Section 6.2.1 for details). All MRCPv2 servers MUST support TLS. Servers MAY use TCP without TLS in controlled environments (e.g., not in the public Internet) where both nodes are inside a protected perimeter, for example, preventing access to the MRCP server from remote nodes outside the controlled perimeter. It is up to the client to choose which mode of transport it wants to use for an MRCPv2 session. Most examples from here on show only the MRCPv2 messages and do not show the SIP messages that may have been used to establish the MRCPv2 control channel.

14.4.6 MRCPv2 Session Termination

If an MRCP client notices that the underlying connection has been closed for one of its MRCP channels, and it has not previously initiated a re-INVITE to close that channel, it MUST send a BYE to close down the SIP dialog and all other MRCP channels. If an MRCP server notices that the underlying connection has been closed for one of its MRCP channels, and it has not previously received and accepted a re-INVITE closing that channel, then it MUST send a BYE to close down the SIP dialog and all other MRCP channels.

14.5 MRCPV2 SPECIFICATION

Except as otherwise indicated, MRCPv2 messages are Unicode encoded in UTF-8 (RFC 3629) to allow many different languages to be represented. DEFINE-GRAMMAR (Section 9.8), for example, is one such exception, since its body can contain arbitrary XML in arbitrary (but specified via XML) encodings. MRCPv2 also allows message bodies to be represented in other character sets (for example, ISO 8859-1 [4]) because in some locales, other character sets are already in widespread use. The MRCPv2 headers (the first line of an MRCP message) and header field names use only the US-ASCII subset of UTF-8. Lines are terminated by CRLF (carriage return, then line feed). Also, some parameters in the message may contain binary data or a record spanning multiple lines. Such fields have a length value associated with the parameter, which indicates the number of octets immediately following the parameter.

14.5.1 Common Protocol Elements

The MRCPv2 message set consists of requests from the client to the server, responses from the server to the client, and asynchronous events from the server to the client. All these messages consist of a start-line, one or more header fields, an empty line (i.e., a line with nothing preceding the CRLF) indicating the end of the header fields, and an optional message body.

```
generic-message     =     start-line
                          message-header
                          CRLF
                          [ message-body ]
message-body        =     *OCTET
start-line          =     request-line / response-line / event-line
message-header      =     1*(generic-header /
                          resource-header / generic-field)
                          resource-header = synthesizer-header
                          / recognizer-header
                          / recorder-header
                          / verifier-header
```

The message-body contains resource-specific and message-specific data. The actual media types used to carry the data are specified in the sections defining the individual messages. Generic header fields are described in Section 14.6.2. If a message contains a message body, the message MUST contain content-headers indicating the media type and encoding of the data in the message body. Request, response, and event messages (described in the following sections) include the version of MRCP that the message conforms to. Version compatibility rules follow H3.1 regarding version ordering, compliance requirements, and upgrading of version numbers. The version information is indicated by "MRCP" (as opposed to "HTTP" in H3.1) or "MRCP/2.0" (as opposed to "HTTP/1.1" in H3.1). To be compliant with this specification, clients and servers sending MRCPv2 messages MUST indicate an mrcp-version of "MRCP/2.0." ABNF productions using mrcp-version can be found in Sections 14.5.2, 14.5.3, and 14.5.5.

```
mrcp-version  =  "MRCP" "/" 1*2DIGIT "." 1*2DIGIT
```

The message-length field specifies the length of the message in octets, including the start-line, and MUST be the second token from the beginning of the message. This is to make the framing and parsing of the message simpler to do. This field specifies the length of the message, including data that may be encoded into the body of the message. Note that this value MAY be given as a fixed-length integer that is zero-padded (with leading zeros) in order to eliminate or reduce inefficiency in cases where the message-length value would change as a result of the length of the message-length token itself. This value, as with all lengths in MRCP, is to be interpreted as a base-10 number. In particular, leading zeros do not indicate that the value is to be interpreted as a base-8 number.

```
message-length  =  1*19DIGIT
```

The following sample MRCP exchange demonstrates proper message-length values. The values for message-length have been removed from all other examples in the specification and replaced by "..." to reduce confusion in the case of minor message-length computation errors in those examples.

```
C->S: MRCP/2.0 877 INTERPRET 543266
      Channel-Identifier:32AECB23433801@speechrecog
      Interpret-Text:may I speak to Andre Roy
      Content-Type:application/srgs+xml
      Content-ID:<request1@form-level.store>
      Content-Length:661

      <?xml version="1.0"?>

      <!-- the default grammar language is US English -->

      <grammar xmlns="http://www.w3.org/2001/06/grammar"
            xml:lang="en-US" version="1.0" root="request">

         <!-- single language attachment to tokens -->

         <rule id="yes">
            <one-of>
               <item xml:lang="fr-CA">oui</item>
               <item xml:lang="en-US">yes</item>
            </one-of>
         </rule>

         <!-- single language attachment to a rule expansion -->

         <rule id="request">
               may I speak to
            <one-of xml:lang="fr-CA">
               <item>Michel Tremblay</item>
               <item>Andre Roy</item>
            </one-of>
         </rule>
      </grammar>

S->C: MRCP/2.0 82 543266 200 IN-PROGRESS
      Channel-Identifier:32AECB23433801@speechrecog

S->C: MRCP/2.0 634 INTERPRETATION-COMPLETE 543266 200
      COMPLETE
      Channel-Identifier:32AECB23433801@speechrecog
```

```
   Completion-Cause:000 success
   Content-Type:application/nlsml+xml
   Content-Length:441

   <?xml version="1.0"?>
   <result xmlns="urn:ietf:params:xml:ns:mrcpv2"
         xmlns:ex="http://www.example.com/example"
         grammar="session:request1@form-level.store">
      <interpretation>
         <instance name="Person">
            <ex:Person>
               <ex:Name> Andre Roy </ex:Name>
            </ex:Person>
         </instance>
      <input> may I speak to Andre Roy </input>
   </interpretation>
</result>
```

All MRCPv2 messages, responses, and events MUST carry the Channel-Identifier header field so that the server or client can differentiate messages from different control channels that may share the same transport connection. In the resource-specific header field descriptions in Sections 14.8.11, a header field is disallowed on a method (request, response, or event) for that resource unless specifically listed as being allowed. Also, the phrasing "This header field MAY occur on method X" indicates that the header field is allowed on that method but is not required to be used in every instance of that method.

14.5.2 Request

An MRCPv2 request consists of a Request line followed by the message header section and an optional message body containing data specific to the request message. The Request message from a client to the server includes within the first line the method to be applied, a method tag for that request, and the version of the protocol in use.

```
request-line    =    mrcp-version SP message-length SP method-name
                     SP request-id CRLF
```

The mrcp-version field is the MRCP protocol version that is being used by the client. The message-length field specifies the length of the message, including the start-line. Details about the mrcp-version and message-length fields are given in Section 14.5.1. The method-name field identifies the specific request that the client is making to the server. Each resource supports a subset of the MRCPv2 methods. The subset for each resource is defined in the section of the specification for the corresponding resource.

```
method-name     =    generic-method
                     / synthesizer-method
                     / recognizer-method
                     / recorder-method
                     / verifier-method
```

The request-id field is a unique identifier representable as an unsigned 32-bit integer created by the client and sent to the server. Clients MUST utilize monotonically increasing request-ids for consecutive requests within an MRCP session. The request-id space is linear (i.e., not mod-32), so the space does not wrap, and validity can be checked with a simple unsigned comparison operation. The client may choose any initial value for its first request, but a small integer is RECOMMENDED to avoid exhausting the space in long sessions. If the server receives duplicate or out-of-order requests, the server MUST reject the request with a response code of 410. Since request-ids are scoped to the MRCP session, they are unique across all TCP connections and all resource channels in the session. The server resource MUST use the client-assigned identifier in its response to the request. If the request does not complete synchronously, future asynchronous events associated with this request MUST carry the client-assigned request-id.

```
request-id    =    1*10DIGIT
```

14.5.3 Response

After receiving and interpreting the request message for a method, the server resource responds with an MRCPv2 response message. The response consists of a response line followed by the message header section and an optional message body containing data specific to the method.

```
response-line    =    mrcp-version SP message-length SP request-id
                      SP status-code SP request-state CRLF
```

The mrcp-version field MUST contain the version of the request if supported; otherwise, it MUST contain the highest version of MRCP supported by the server. The message-length field specifies the length of the message, including the start-line. Details about the mrcp-version and message-length fields are given in Section 14.5.1. The request-id used in the response MUST match the one sent in the corresponding request message. The status-code field is a 3-digit code representing the success or failure or other status of the request.

```
status-code    =    3DIGIT
```

The request-state field indicates whether the action initiated by the Request is PENDING, IN-PROGRESS, or COMPLETE. The COMPLETE status means that the request was processed to completion and that there will be no more events or other messages from that resource to the client with that request-id. The PENDING status means that the request has been placed in a queue and will be processed in first-in-first-out (FIFO) order. The IN-PROGRESS status means that the request is being processed and is not yet complete. A PENDING or IN-PROGRESS status indicates that further Event messages may be delivered with that request-id.

```
request-state    =    "COMPLETE"
                      / "IN-PROGRESS"
                      / "PENDING"
```

14.5.4 Status Codes

The status codes are classified under the Success (2xx) (Table 14.2), Client Failure (4xx) (Table 14.3), and Server Failure (5xx) codes (Table 14.4).

TABLE 14.2
Success

Code	Meaning
200	Success
201	Success with some optional header fields ignored

TABLE 14.3
Client Failure (4xx)

Code	Meaning
401	Method not allowed
402	Method not valid in this state
403	Unsupported header field
404	Illegal value for header field. This is the error for a syntax violation
405	Resource not allocated for this session or does not exist
406	Mandatory Header Field Missing
407	Method or Operation Failed (e.g., Grammar compilation failed in the recognizer). Detailed cause codes might be available through a resource-specific header
408	Unrecognized or unsupported message entity
409	Unsupported Header Field Value. This is a value that is syntactically legal but exceeds the implementation's capabilities or expectations
410	Non-Monotonic or Out-of-order sequence number in request
411–420	Reserved for future assignment

TABLE 14.4
Server Failure (5xx)

Code	Meaning
501	Server Internal Error
502	Protocol Version not supported
503	Reserved for future assignment
504	Message too large

14.5.5 Events

The server resource may need to communicate a change in state or the occurrence of a certain event to the client. These messages are used when a request does not complete immediately and the response returns a status of PENDING or IN-PROGRESS. The intermediate results and events of the request are indicated to the client through the event message from the server. The event message consists of an event header line followed by the message header section and an optional message body containing data specific to the event message. The header line has the request-id of the corresponding request and status value. The request-state value is COMPLETE if the request is done and this was the last event; else, it is IN-PROGRESS.

```
event-line  =  mrcp-version SP message-length SP event-name
SP request-id SP request-state CRLF
```

The mrcp-version used here is identical to the one used in the Request/Response line and indicates the highest version of MRCP running on the server. The message-length field specifies the length of the message, including the start-line. Details about the mrcp-version and message-length fields are given in Section 5.1. The event-name identifies the nature of the event generated by the media resource. The set of valid event names depends on the resource generating it. See the corresponding resource-specific section of the specification.

```
event-name      =      synthesizer-event
                       / recognizer-event
                       / recorder-event
                       / verifier-event
```

The request-id used in the event MUST match the one sent in the request that caused this event. The request-state indicates whether the Request/Command causing this event is complete or still in progress and whether it is the same as the one mentioned in Section 5.3. The final event for a request has a COMPLETE status, indicating the completion of the request.

14.6 MRCPV2 GENERIC METHODS, HEADERS, AND RESULT STRUCTURE

MRCPv2 supports a set of methods and header fields that are common to all resources. These are discussed here; resource-specific methods and header fields are discussed in the corresponding resource-specific section of the specification.

14.6.1 Generic Methods

MRCPv2 supports two generic methods for reading and writing the state associated with a resource.

```
generic-method     =     "SET-PARAMS"
                         / "GET-PARAMS"
```

These are described in the following subsections.

14.6.1.1 SET-PARAMS

The SET-PARAMS method, from the client to the server, tells the MRCPv2 resource to define parameters for the session, such as voice characteristics and prosody on synthesizers, recognition timers on recognizers, etc. If the server accepts and sets all parameters, it MUST return a response status-code of 200. If it chooses to ignore some optional header fields that can be safely ignored without affecting operation of the server, it MUST return 201. If one or more of the header fields being sent is incorrect, error 403, 404, or 409 MUST be returned as follows:

- If one or more of the header fields being set has an illegal value, the server MUST reject the request with a 404 Illegal Value for Header Field.
- If one or more of the header fields being set is unsupported for the resource, the server MUST reject the request with a 403 Unsupported Header Field, except as described in the next paragraph.
- If one or more of the header fields being set has an unsupported value, the server MUST reject the request with a 409 Unsupported Header Field Value, except as described in the next paragraph.

If both error 404 and another error have occurred, only error 404 MUST be returned. If both errors 403 and 409 have occurred, but not error 404, only error 403 MUST be returned. If error 403, 404, or 409 is returned, the response MUST include the bad or unsupported header fields and their values exactly as they were sent from the client. Session parameters modified using SET-PARAMS do not override parameters explicitly specified on individual requests or requests that are IN-PROGRESS.

```
C->S:  MRCP/2.0 ... SET-PARAMS 543256
       Channel-Identifier:32AECB23433802@speechsynth
       Voice-gender:female
       Voice-variant:3
```

```
S->C:  MRCP/2.0 ... 543256 200 COMPLETE
       Channel-Identifier:32AECB23433802@speechsynth
```

14.6.1.2 GET-PARAMS

The GET-PARAMS method, from the client to the server, asks the MRCPv2 resource for its current session parameters, such as voice characteristics and prosody on synthesizers, recognition timers on recognizers, etc. For every header field the client sends in the request without a value, the server MUST include the header field and its corresponding value in the response. If no parameter header fields are specified by the client, then the server MUST return all the settable parameters and their values in the corresponding header section of the response, including vendor-specific parameters. Such wildcard parameter requests can be very processing-intensive, since the number of settable parameters can be large, depending on the implementation. Hence, it is RECOMMENDED that the client does not use the wildcard GET-PARAMS operation very often. Note that GET-PARAMS returns header field values that apply to the whole session and not values that have a request-level scope. For example, Input-Waveform-URI is a request-level header field and thus, would not be returned by GET-PARAMS.

If all the header fields requested are supported, the server MUST return a response status-code of 200. If some of the header fields being retrieved are unsupported for the resource, the server MUST reject the request with a 403 Unsupported Header Field. Such a response MUST include the unsupported header fields exactly as they were sent from the client, without values.

```
C->S:  MRCP/2.0 ... GET-PARAMS 543256
       Channel-Identifier:32AECB23433802@speechsynth
       Voice-gender:
       Voice-variant:
       Vendor-Specific-Parameters:com.example.param1;
       com.example.param2

S->C:  MRCP/2.0 ... 543256 200 COMPLETE
       Channel-Identifier:32AECB23433802@speechsynth
       Voice-gender:female
       Voice-variant:3
       Vendor-Specific-Parameters:com.example.param1="Company Name";
       com.example.param2="124324234@example.com"
```

14.6.2 Generic Message Headers

All MRCPv2 header fields, which include both the generic-headers defined in the following subsections and the resource-specific header fields defined later, follow the same generic format as that given in Section 3.1 of RFC 5322. Each header field consists of a name followed by a colon (":") and the value. Header field names are case-insensitive. The value MAY be preceded by any amount of LWS (linear white space), though a single SP (space) is preferred. Header fields may extend over multiple lines by preceding each extra line with at least one SP or HT (horizontal tab).

```
generic-field   =   field-name ":" [ field-value ]
field-name      =   token
field-value     =   *LWS field-content *( CRLF 1*LWS field-content)
field-content   =   <the OCTETs making up the field-value
                    and consisting of either *TEXT or combinations
                    of token, separators, and quoted-string>
```

The field-content does not include any leading or trailing LWS (i.e., linear white space occurring before the first non-whitespace character of the field-value or after the last non-whitespace character of the field-value). Such leading or trailing LWS MAY be removed without changing the semantics of the field value. Any LWS that occurs between field-content MAY be replaced with a single SP before interpreting the field value or forwarding the message downstream.

MRCPv2 servers and clients MUST NOT depend on header field order. It is RECOMMENDED to send general-header fields first, followed by request-header or response-header fields, and ending with the entity-header fields. However, MRCPv2 servers and clients MUST be prepared to process the header fields in any order. The only exception to this rule is when there are multiple header fields with the same name in a message. Multiple header fields with the same name MAY be present in a message if and only if the entire value for that header field is defined as a comma-separated list [i.e., #(values)].

Since vendor-specific parameters may be order-dependent, it MUST be possible to combine multiple header fields of the same name into one "name:value" pair without changing the semantics of the message by appending each subsequent value to the first, each separated by a comma. The order in which header fields with the same name are received is therefore significant to the interpretation of the combined header field value, and thus, an intermediary MUST NOT change the order of these values when a message is forwarded.

```
generic-header     =     channel-identifier
                         / accept
                         / active-request-id-list
                         / proxy-sync-id
                         / accept-charset
                         / content-type
                         / content-id
                         / content-base
                         / content-encoding
                         / content-location
                         / content-length
                         / fetch-timeout
                         / cache-control
                         / logging-tag
                         / set-cookie
                         / vendor-specific
```

14.6.2.1 Channel-Identifier

All MRCPv2 requests, responses, and events MUST contain the Channel-Identifier header field. The value is allocated by the server when a control channel is added to the session and communicated to the client by the "a=channel" attribute in the SDP answer from the server. The header field value consists of two parts separated by the "@" symbol. The first part is an unambiguous string identifying the MRCPv2 session. The second part is a string token that specifies one of the media processing resource types listed in Section 13.3.1. The unambiguous string (first part) MUST be difficult to guess, unique among the resource instances managed by the server, and common to all resource channels with that server established through a single SIP dialog.

```
channel-identifier    =    "Channel-Identifier" ":" channel-id CRLF
channel-id            =    1*alphanum "@" 1*alphanum
```

14.6.2.2 Accept

The Accept header field follows the syntax defined in H14.1. The semantics are also identical, with the exception that if no Accept header field is present, the server MUST assume a default value that is specific to the resource type that is being controlled. This default value can be changed for a resource on a session by sending this header field in a SET-PARAMS method. The current default value of this header field for a resource in a session can be found through a GET-PARAMS method. This header field MAY occur on any request.

14.6.2.3 Active-Request-Id-List

In a request, this header field indicates the list of request-ids to which the request applies. This is useful when there are multiple requests that are PENDING or IN-PROGRESS, and the client wants this request to apply to one or more of these specifically. In a response, this header field returns the list of request-ids that the method modified or affected. There could be one or more requests in a request-state of PENDING or IN-PROGRESS. When a method affecting one or more PENDING or IN-PROGRESS requests is sent from the client to the server, the response MUST contain the list of request-ids that were affected or modified by this command in its header section.

The Active-Request-Id-List is only used in requests and responses, not in events. For example, if a STOP request with no Active-Request-Id-List is sent to a synthesizer resource that has one or more SPEAK requests in the PENDING or IN-PROGRESS state, all SPEAK requests MUST be cancelled, including the one IN-PROGRESS. The response to the STOP request contains in the Active-Request-Id-List value the request-ids of all the SPEAK requests that were terminated. After sending the STOP response, the server MUST NOT send any SPEAK-COMPLETE or RECOGNITION-COMPLETE events for the terminated requests.

```
active-request-id-list   =    "Active-Request-Id-List" ":"
                              request-id *("," request-id) CRLF
```

14.6.2.4 Proxy-Sync-Id

When any server resource generates a "barge-in-able" event, it also generates a unique tag. The tag is sent as this header field's value in an event to the client. The client then acts as an intermediary among the server resources and sends a BARGE-IN-OCCURRED method to the synthesizer server resource with the Proxy-Sync-Id it received from the server resource. When the recognizer and synthesizer resources are part of the same session, they may choose to work together to achieve quicker interaction and response. Here, the Proxy-Sync-Id helps the resource receiving the event, intermediated by the client, to decide whether this event has been processed through a direct interaction of the resources. This header field MAY occur only on events and the BARGE-IN-OCCURRED method. The name of this header field contains the word "proxy" only for historical reasons and does not imply that a proxy server is involved.

```
proxy-sync-id   =   "Proxy-Sync-Id" ":" 1*VCHAR CRLF
```

14.6.2.5 Accept-Charset

See H14.2. This specifies the acceptable character sets for entities returned in the response or events associated with this request. This is useful in specifying the character set to use in the Natural Language Semantic Markup Language (NLSML) results of a RECOGNITION-COMPLETE event. This header field is only used on requests.

14.6.2.6 Content-Type

See H14.17. MRCPv2 supports a restricted set of registered media types for content, including speech markup, grammar, and recognition results. The content types applicable to each MRCPv2 resource-type are specified in the corresponding section of the specification and are registered in the MIME Media Types registry maintained by IANA. The multipart content type "multipart/mixed" is supported to communicate multiple of the above-mentioned contents, in which case the body parts MUST NOT contain any MRCPv2-specific header fields. This header field MAY occur on all messages.

```
content-type        =    "Content-Type" ":" media-type-value CRLF
media-type-value    =    type "/" subtype *( ";" parameter )
type                =    token
subtype             =    token
parameter           =    attribute "=" value
attribute           =    token
value               =    token / quoted-string
```

14.6.2.7 Content-ID

This header field contains an ID or name for the content by which it can be referenced. This header field operates according to the specification in RFC 2392 and is required for content disambiguation in multipart messages. In MRCPv2, whenever the associated content is stored by either the client or the server, it MUST be retrievable using this ID. Such content can be referenced later in a session by addressing it with the "session" URI scheme described in Section 14.13.6. This header field MAY occur on all messages.

14.6.2.8 Content-Base

The Content-Base entity-header MAY be used to specify the base URI for resolving relative URIs within the entity.

```
content-base    =    "Content-Base" ":" absoluteURI CRLF
```

Note, however, that the base URI of the contents within the entity-body may be redefined within that entity-body. An example of this would be multipart media, which in turn, can have multiple entities within it. This header field MAY occur on all messages.

14.6.2.9 Content-Encoding

The Content-Encoding entity-header is used as a modifier to the Content-Type. When present, its value indicates what additional content encoding has been applied to the entity-body and thus, what decoding mechanisms must be applied in order to obtain the Media Type referenced by the Content-Type header field. Content-Encoding is primarily used to allow a specification to be compressed without losing the identity of its underlying media type. Note that the SIP session can be used to determine accepted encodings (see Section 14.7). This header field MAY occur on all messages.

```
content-encoding    =    "Content-Encoding" ":"
                         *WSP content-coding
                         *(*WSP "," *WSP content-coding *WSP )
                         CRLF
```

Content codings are defined in H3.5. An example of its use is Content-Encoding:gzip. If multiple encodings have been applied to an entity, the content encodings MUST be listed in the order in which they were applied.

14.6.2.10 Content-Location

The Content-Location entity-header MAY be used to supply the resource location for the entity enclosed in the message when that entity is accessible from a location separate from the requested resource's URI. Refer to H14.14.

```
content-location    =    "Content-Location" ":"
                         ( absoluteURI / relativeURI ) CRLF
```

The Content-Location value is a statement of the location of the resource corresponding to this particular entity at the time of the request. This header field is provided for optimization purposes only. The receiver of this header field MAY assume that the entity being sent is identical to what would have been retrieved or might already have been retrieved from the Content-Location URI. For example, if the client provided a grammar markup inline, and it had previously retrieved it from a certain URI, that URI can be provided as part of the entity, using the Content-Location header field. This allows a resource like the recognizer to look into its cache to see whether this grammar was previously retrieved, compiled, and cached. In this case, it might optimize by using the previously compiled grammar object. If the Content-Location is a relative URI, the relative URI is interpreted relative to the Content-Base URI. This header field MAY occur on all messages.

14.6.2.11 Content-Length

This header field contains the length of the content of the message body (i.e., after the double CRLF following the last header field). Unlike in HTTP, it MUST be included in all messages that carry content beyond the header section. If it is missing, a default value of zero is assumed. Otherwise, it is interpreted according to H14.13. When a message having no use for a message body contains one, i.e., the Content-Length is non-zero, the receiver MUST ignore the content of the message body. This header field MAY occur on all messages.

```
content-length    =    "Content-Length" ":" 1*19DIGIT CRLF
```

14.6.2.12 Fetch Timeout

When the recognizer or synthesizer needs to fetch specifications or other resources, this header field controls the corresponding URI access properties. This defines the timeout for content that the server may need to fetch over the network. The value is interpreted to be in milliseconds and ranges from 0 to an implementation-specific maximum value. It is RECOMMENDED that servers be cautious about accepting long timeout values. The default value for this header field is implementation specific. This header field MAY occur in DEFINEGRAMMAR, RECOGNIZE, SPEAK, SET-PARAMS, or GET-PARAMS.

```
fetch-timeout    =    "Fetch-Timeout" ":" 1*19DIGIT CRLF
```

14.6.2.13 Cache-Control

If the server implements content caching, it MUST adhere to the cache correctness rules of HTTP 1.1 (RFC 2616) when accessing and caching stored content. In particular, the "expires" and "cache-control" header fields of the cached URI or specification MUST be honored and take precedence over the Cache-Control defaults set by this header field. The Cache-Control directives are used to define the default caching algorithms on the server for the session or request. The scope of the directive is based on the method it is sent on. If the directive is sent on a SET-PARAMS method, it applies for all requests for external specifications the server makes during that session, unless it is overridden by a Cache-Control header field on an individual request.

If the directives are sent on any other requests, they apply only to external specification requests the server makes for that request. An empty Cache-Control header field on the GET-PARAMS method is a request for the server to return the current Cache-Control directives setting on the server. This header field MAY occur only on requests.

```
cache-control     =    "Cache-Control" ":"
                       [*WSP cache-directive
                       *( *WSP "," *WSP cache-directive *WSP )]
                       CRLF
                       cache-directive = "max-age" "=" delta-seconds
                       / "max-stale" [ "=" delta-seconds ]
                       / "min-fresh" "=" delta-seconds
delta-seconds     =    1*19DIGIT
```

Here, delta-seconds is a decimal time value specifying the number of seconds since the instant when the message response or data was received by the server. The different cache-directive options allow the client to ask the server to override the default cache expiration mechanisms:

- **max-age:** Indicates that the client can tolerate the server using content whose age is no greater than the specified time in seconds. Unless a "max-stale" directive is also included, the client is not willing to accept a response based on stale data.
- **min-fresh:** Indicates that the client is willing to accept a server response with cached data whose expiration is no less than its current age plus the specified time in seconds. If the server's cache time-to-live exceeds the client-supplied min-fresh value, the server MUST NOT utilize cached content.
- **max-stale:** Indicates that the client is willing to allow a server to utilize cached data that has exceeded its expiration time. If "max-stale" is assigned a value, then the client is willing to allow the server to use cached data that has exceeded its expiration time by no more than the specified number of seconds. If no value is assigned to "max-stale," then the client is willing to allow the server to use stale data of any age.

If the server cache is requested to use stale response/data without validation, it MAY do so only if this does not conflict with any "MUST"-level requirements concerning cache validation (e.g., a "mustrevalidate" Cache-Control directive in the HTTP 1.1 specification pertaining to the corresponding URI). If both the MRCPv2 Cache-Control directive and the cached entry on the server include "max-age" directives, then the lesser of the two values is used for determining the freshness of the cached entry for that request.

14.6.2.14 Logging-Tag

This header field MAY be sent as part of a SET-PARAMS/GET-PARAMS method to set or retrieve the logging tag for logs generated by the server. Once set, the value persists until a new value is set or the session ends. The MRCPv2 server MAY provide a mechanism to create subsets of its output logs so that system administrators can examine or extract only the log file portion during which the logging tag was set to a certain value. It is RECOMMENDED that clients include in the logging tag information to identify the MRCPv2 client user agent, so that one can determine which MRCPv2 client request generated a given log message at the server. It is also RECOMMENDED that MRCPv2 clients do not log personally identifiable information such as credit card numbers and national identification numbers.

```
logging-tag  =  "Logging-Tag" ":" 1*UTFCHAR CRLF
```

14.6.2.15 Set-Cookie

Since the associated HTTP client on an MRCPv2 server fetches specifications for processing on behalf of the MRCPv2 client, the cookie store in the HTTP client of the MRCPv2 server is treated as an extension of the cookie store in the HTTP client of the MRCPv2 client. This requires that the MRCPv2 client and server be able to synchronize their common cookie store as needed. To enable the MRCPv2 client to push its stored cookies to the MRCPv2 server and get new cookies from the MRCPv2 server stored back to the MRCPv2 client, the Set-Cookie entity-header field MAY be included in MRCPv2 requests to update the cookie store on a server and be returned in final MRCPv2 responses or events to subsequently update the client's own cookie store. The stored cookies on the server persist for the duration of the MRCPv2 session and MUST be destroyed at the end of the session. To ensure support for cookies, MRCPv2 clients and servers MUST support the Set-Cookie entity-header field. Note that it is the MRCPv2 client that determines which, if any, cookies are sent to the server. There is no requirement that all cookies be shared. Rather, it is RECOMMENDED that MRCPv2 clients communicate only cookies needed by the MRCPv2 server to process its requests.

```
set-cookie          =    "Set-Cookie:" cookies CRLF
cookies             =    cookie *("," *LWS cookie)
cookie              =    attribute "=" value *(";" cookie-av)
cookie-av           =    "Comment" "=" value
                         / "Domain" "=" value
                         / "Max-Age" "=" value
                         / "Path" "=" value
                         / "Secure"
                         / "Version" "=" 1*19DIGIT
                         / "Age" "=" delta-seconds
set-cookie          =    "Set-Cookie:" SP set-cookie-string
set-cookie-string   =    cookie-pair *( ";" SP cookie-av )
cookie-pair         =    cookie-name "=" cookie-value
cookie-name         =    token
cookie-value        =    *cookie-octet /
                         ( DQUOTE *cookie-octet DQUOTE )
cookie-octet        =    %x21 / %x23-2B / %x2D-3A / %x3C-5B
                         / %x5D-7E
token               =    <token, defined in RFC 2616, Section 2.2>
cookie-av           =    expires-av / max-age-av / domain-av /
                         path-av / secure-av / httponly-av /
                         extension-av / age-av
expires-av          =    "Expires=" sane-cookie-date
sane-cookie-date    =    <rfc1123-date, defined in RFC 2616,
                         Section 3.3.1>
max-age-av          =    "Max-Age=" non-zero-digit *DIGIT
non-zero-digit      =    %x31-39
domain-av           =    "Domain=" domain-value
domain-value        =    <subdomain>
path-av             =    "Path=" path-value
path-value          =    <any CHAR except CTLs or ";">
secure-av           =    "Secure"
httponly-av         =    "HttpOnly"
extension-av        =    <any CHAR except CTLs or ";">
age-av              =    "Age=" delta-seconds
```

The Set-Cookie header field is specified in RFC 6265. The “Age” attribute is introduced in this specification to indicate the age of the cookie and is OPTIONAL. An MRCPv2 client or server MUST calculate the age of the cookie according to the age calculation rules in the HTTP/1.1 specification (RFC 2616) and append the “Age” attribute accordingly. This attribute is provided because time may have passed since the client received the cookie from an HTTP server. Rather than having the client reduce Max-Age by the actual age, it passes Max-Age verbatim and appends the “Age” attribute, thus maintaining the cookie as received while still accounting for the fact that time has passed.

The MRCPv2 client or server MUST supply defaults for the “Domain” and “Path” attributes, as specified in RFC 6265, if they are omitted by the HTTP origin server. Note that there is no leading dot present in the “Domain” attribute value in this case. Although an explicitly specified “Domain” value received via the HTTP protocol may be modified to include a leading dot, an MRCPv2 client or server MUST NOT modify the “Domain” value when received via the MRCPv2 protocol. An MRCPv2 client or server MAY combine multiple cookie header fields of the same type into a single “field-name:field-value” pair as described in Section 14.6.2.

The Set-Cookie header field MAY be specified in any request that subsequently results in the server performing an HTTP access. When a server receives new cookie information from an HTTP origin server, and assuming the cookie store is modified according to RFC 6265, the server MUST return the new cookie information in the MRCPv2 COMPLETE response or event, as appropriate, to allow the client to update its own cookie store. The SET-PARAMS request MAY specify the Set-Cookie header field to update the cookie store on a server. The GET-PARAMS request MAY be used to return the entire cookie store of "Set-Cookie" type cookies to the client.

14.6.2.16 Vendor-Specific Parameters

This set of header fields allows the client to set or retrieve vendor-specific parameters.

```
vendor-specific          =   "Vendor-Specific-Parameters" ":"
                             [vendor-specific-av-pair
                             *(";" vendor-specific-av-pair)] CRLF
vendor-specific-av-pair  =   vendor-av-pair-name "=" value
vendor-av-pair-name      =   1*UTFCHAR
```

Header fields of this form MAY be sent in any method (request) and are used to manage implementation-specific parameters on the server side. The vendor-av-pair-name follows the reverse Internet Domain Name convention (see Section 14.13.1.6 for syntax and registration information). The value of the vendor attribute is specified after the "=" symbol and MAY be quoted. For example:

```
com.example.companyA.paramxyz=256
com.example.companyA.paramabc=High
com.example.companyB.paramxyz=Low
```

When used in GET-PARAMS to get the current value of these parameters from the server, this header field value MAY contain a semicolon-separated list of implementation-specific attribute names.

14.6.3 Generic Result Structure

Result data from the server for the Recognizer and Verifier resources is carried as a typed media entity in the MRCPv2 message body of various events. NLSML, an XML markup based on an early draft from the W3C, is the default standard for returning results to the client. Hence, all servers implementing these resource types MUST support the media type "application/nlsml+xml." The Extensible MultiModal Annotation (EMMA) [5] format can be used to return results as well. This can be done by negotiating the format at session establishment time with SDP (a=resultformat:application/emma+xml) or with SIP (Allow/Accept). With SIP, for example, if a client wants results in EMMA, an MRCPv2 server can route the request to another server that supports EMMA by inspecting the SIP header fields rather than having to inspect the SDP.

MRCPv2 uses this representation to convey content among the clients and servers that generate and make use of the markup. MRCPv2 uses NSLML specifically to convey recognition, enrollment, and verification results (Figure 14.5) between the corresponding resource on the MRCPv2 server and the MRCPv2 client. Details of this result format are fully described in Section 14.6.3.1.

14.6.3.1 Natural Language Semantics Markup Language

NLSML is an XML data structure with elements and attributes designed to carry result information from recognizer (including enrollment) and verifier resources. The normative definition of NLSML is the RelaxNG schema in Section 14.16.1. Note that the elements and attributes of this format are defined in the MRCPv2 namespace. In the result structure, they must either be prefixed by a namespace prefix declared within the result or be children of an element identified as

```
Content-Type:application/nlsml+xml
Content-Length:...

<?xml version="1.0"?>

<result xmlns="urn:ietf:params:xml:ns:mrcpv2"
            xmlns:ex="http://www.example.com/example"
            grammar="http://theYesNoGrammar">
      <interpretation>
            <instance>
                  <ex:response>yes</ex:response>
            </instance>
            <input>OK</input>
      </interpretation>
</result>
```

FIGURE 14.5 Result Example.

belonging to the respective namespace. For details on how to use XML Namespaces, see Layman et al. [6]. Section 2 of Layman et al. [6] provides details on how to declare namespaces and namespace prefixes.

The root element of NLSML is <result>. Optional child elements are <interpretation>, <enrollment-result>, and <verification-result>, at least one of which must be present. A single <result> MAY contain any or all of the optional child elements. Details of the <result> and <interpretation> elements and their subelements and attributes can be found in Section 14.9.6. Details of the <enrollment-result> element and its subelements can be found in Section 14.9.7. Details of the <verification-result> element and its subelements can be found in Section 14.11.5.2.

14.7 RESOURCE DISCOVERY

Server resources may be discovered and their capabilities learned by clients through standard SIP machinery. The client MAY issue a SIP OPTIONS transaction to a server, which has the effect of requesting the capabilities of the server. The server MUST respond to such a request with an SDP-encoded description of its capabilities according to RFC 3264. The MRCPv2 capabilities are described by a single "m=" line containing the media type "application" and transport type "TCP/TLS/MRCPv2" or "TCP/MRCPv2." There MUST be one "resource" attribute for each media resource that the server supports, and it has the resource type identifier as its value. The SDP description MUST also contain "m=" lines describing the audio capabilities and the coders the server supports. In this example, the client uses the SIP OPTIONS method (Figure 14.6) to query the capabilities of the MRCPv2 server.

14.8 SPEECH SYNTHESIZER RESOURCE

This resource processes text markup provided by the client and generates a stream of synthesized speech in real time. Depending upon the server implementation and capability of this resource, the

```
C->S:
OPTIONS sip:mrcp@server.example.com SIP/2.0
Via:SIP/2.0/TCP client.atlanta.example.com:5060;
branch=z9hG4bK74bf7
Max-Forwards:6
To:<sip:mrcp@example.com>
From:Sarvi <sip:sarvi@example.com>;tag=1928301774
Call-ID:a84b4c76e66710
CSeq:63104 OPTIONS
Contact:<sip:sarvi@client.example.com>
Accept:application/sdp
Content-Length:0

S->C:
SIP/2.0 200 OK
Via:SIP/2.0/TCP client.atlanta.example.com:5060;
branch=z9hG4bK74bf7;received=192.0.32.10
To:<sip:mrcp@example.com>;tag=62784
From:Sarvi <sip:sarvi@example.com>;tag=1928301774
Call-ID:a84b4c76e66710
CSeq:63104 OPTIONS
Contact:<sip:mrcp@server.example.com>
Allow:INVITE, ACK, CANCEL, OPTIONS, BYE

Accept:application/sdp
Accept-Encoding:gzip
Accept-Language:en
Supported:foo
Content-Type:application/sdp
Content-Length:...

v=0
o=sarvi 2890844536 2890842811 IN IP4 192.0.2.12
s=-
i=MRCPv2 server capabilities
c=IN IP4 192.0.2.12/127
t=0 0
m=application 0 TCP/TLS/MRCPv2 1
a=resource:speechsynth
a=resource:speechrecog
a=resource:speakverify
m=audio 0 RTP/AVP 0 3
a=rtpmap:0 PCMU/8000
a=rtpmap:3 GSM/8000
```

FIGURE 14.6 Using SIP OPTIONS for MRCPv2 Server Capability Discovery.

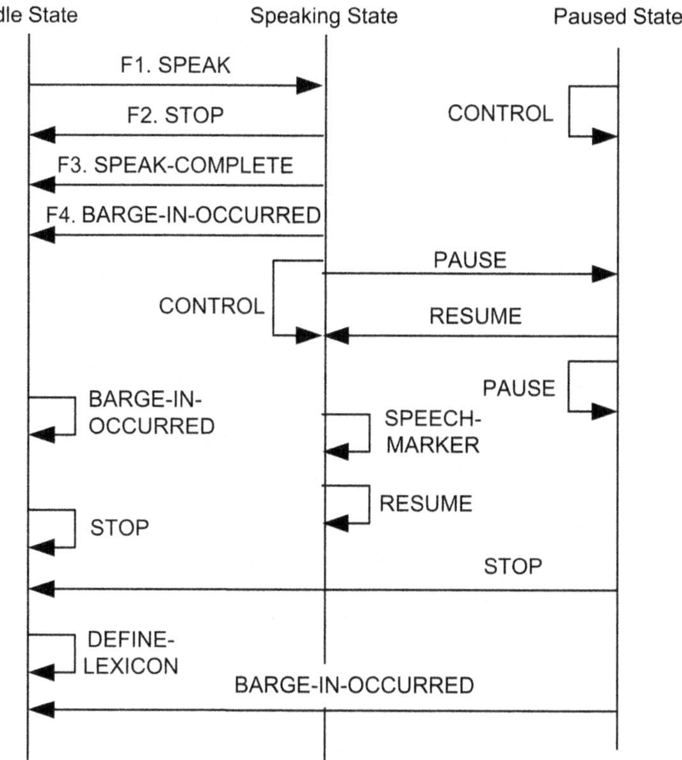

FIGURE 14.7 Synthesizer State Machine. (Copyright: IETF.)

client can also dictate parameters of the synthesized speech, such as voice characteristics, speaker speed, etc. The synthesizer resource is controlled by MRCPv2 requests from the client. Similarly, the resource can respond to these requests or generate asynchronous events to the client to indicate conditions of interest to the client during the generation of the synthesized speech stream. This section applies for the following resource types:

- speechsynth
- basicsynth

The capabilities of these resources are defined in Section 14.3.1.

14.8.1 Synthesizer State Machine

The synthesizer maintains a state machine to process MRCPv2 requests from the client. The state transitions shown in Figure 14.7 describe the states of the synthesizer and reflect the state of the request at the head of the synthesizer resource queue. A SPEAK request in the PENDING state can be deleted or stopped by a STOP request without affecting the state of the resource.

14.8.2 Synthesizer Methods

The synthesizer supports the following methods:

```
synthesizer-method      =       "SPEAK"
                                / "STOP"
                                / "PAUSE"
```

```
                              / "RESUME"
                              / "BARGE-IN-OCCURRED"
                              / "CONTROL"
                              / "DEFINE-LEXICON"
```

14.8.3 Synthesizer Events

The synthesizer can generate the following events:

```
synthesizer-event     =     "SPEECH-MARKER"
                              / "SPEAK-COMPLETE"
```

14.8.4 Synthesizer Header Fields

A synthesizer method can contain header fields containing request options and information to augment the Request, Response, or Event it is associated with.

```
synthesizer-header    =    jump-size
                           / kill-on-barge-in
                           / speaker-profile
                           / completion-cause
                           / completion-reason
                           / voice-parameter
                           / prosody-parameter
                           / speech-marker
                           / speech-language
                           / fetch-hint
                           / audio-fetch-hint
                           / failed-uri
                           / failed-uri-cause
                           / speak-restart
                           / speak-length
                           / load-lexicon
                           / lexicon-search-order
```

14.8.4.1 Jump-Size

This header field MAY be specified in a CONTROL method and controls the amount to jump forwards or backwards in an active SPEAK request. A "+" or "-" indicates a relative value to what is being currently played. This header field MAY also be specified in a SPEAK request as a desired offset into the synthesized speech. In this case, the synthesizer MUST begin speaking from this amount of time into the speech markup. Note that an offset that extends beyond the end of the produced speech will result in audio of length zero. The different speech length units supported are dependent on the synthesizer implementation. If the synthesizer resource does not support a unit for the operation, the resource MUST respond with a status-code of 409 "Unsupported Header Field Value."

```
jump-size                =   "Jump-Size" ":" speech-length-value CRLF
speech-length-value      =   numeric-speech-length
                             / text-speech-length
text-speech-length       =   1*UTFCHAR SP "Tag"
numeric-speech-length    =   ("+" / "-") positive-speech-length
positive-speech-length   =   1*19DIGIT SP numeric-speech-unit
numeric-speech-unit      =   "Second"
                             / "Word"
                             / "Sentence"
                             / "Paragraph"
```

14.8.4.2 Kill-On-Barge-In

This header field MAY be sent as part of the SPEAK method to enable "kill-on-barge-in" support. If it is enabled, the SPEAK method is interrupted by DTMF input detected by a signal detector resource or by the start of speech sensed or recognized by the speech recognizer resource.

```
kill-on-barge-in    =    "Kill-On-Barge-In" ":" BOOLEAN CRLF
```

The client MUST send a BARGE-IN-OCCURRED method to the synthesizer resource when it receives a barge-in-able event from any source. This source could be a synthesizer resource or signal detector resource and MAY be either local or distributed. If this header field is not specified in a SPEAK request or explicitly set by a SET-PARAMS, the default value for this header field is "true." If the recognizer or signal detector resource is on the same server as the synthesizer, and both are part of the same session, the server MAY work with both to provide internal notification to the synthesizer, so that audio may be stopped without having to wait for the client's BARGE-IN-OCCURRED event. It is generally RECOMMENDED, when playing a prompt to the user with Kill-On-Barge-In and asking for input, that the client issue the RECOGNIZE request ahead of the SPEAK request for optimum performance and user experience. In this way, it is guaranteed that the recognizer is online before the prompt starts playing, and the user's speech will not be truncated at the beginning (especially for power users).

14.8.4.3 Speaker-Profile

This header field MAY be part of the SET-PARAMS/GET-PARAMS or SPEAK request from the client to the server and specifies a URI that references the profile of the speaker. Speaker profiles are collections of voice parameters like gender, accent, etc.

```
speaker-profile    =    "Speaker-Profile" ":" uri CRLF
```

14.8.4.4 Completion-Cause

This header field MUST be specified in a SPEAK-COMPLETE event coming from the synthesizer resource to the client (Table 14.5). This indicates the reason why the SPEAK request completed.

```
completion-cause    =    "Completion-Cause" ":" 3DIGIT SP
                         1*VCHAR CRLF
```

TABLE 14.5
Synthesizer Resource Completion Cause-Codes

Cause-Code	Cause-Name	Description
000	normal	SPEAK completed normally
001	barge-in	SPEAK request was terminated because of barge-in
002	parse-failure	SPEAK request terminated because of a failure to parse the speech markup text
003	uri-failure	SPEAK request terminated because access to one of the URIs failed
004	error	SPEAK request terminated prematurely due to synthesizer error
005	language-unsupported	Language not supported
006	lexicon-load-failure	Lexicon loading failed
007	cancelled	A prior SPEAK request failed while this one was still in the queue

14.8.4.5 Completion-Reason

This header field MAY be specified in a SPEAK-COMPLETE event coming from the synthesizer resource to the client. This contains the reason text behind the SPEAK request completion. This header field communicates text describing the reason for the failure, such as an error in parsing the speech markup text.

```
completion-reason    =    "Completion-Reason" ":"
                          quoted-string CRLF
```

The completion reason text is provided for client use in logs and for debugging and instrumentation purposes. Clients MUST NOT interpret the completion reason text.

14.8.4.6 Voice-Parameter

This set of header fields defines the voice of the speaker.

```
voice-parameter      =  voice-gender
                        / voice-age
                        / voice-variant
                        / voice-name
voice-gender         =  "Voice-Gender:" voice-gender-value CRLF
voice-gender-value   =  "male"
                        / "female"
                        / "neutral"
voice-age            =  "Voice-Age:" 1*3DIGIT CRLF
voice-variant        =  "Voice-Variant:" 1*19DIGIT CRLF
voice-name           =  "Voice-Name:"
                        1*UTFCHAR *(1*WSP 1*UTFCHAR) CRLF
```

The "Voice-" parameters are derived from the similarly named attributes of the voice element specified in W3C's SSML [3]. Legal values for these parameters are as defined in that specification. These header fields MAY be sent in SET-PARAMS or GET-PARAMS requests to define or get default values for the entire session or MAY be sent in the SPEAK request to define default values for that SPEAK request. Note that SSML content can itself set these values internal to the SSML specification, of course.

Voice parameter header fields MAY also be sent in a CONTROL method to affect a SPEAK request in progress and change its behavior on the fly. If the synthesizer resource does not support this operation, it MUST reject the request with a status-code of 403 "Unsupported Header Field."

14.8.4.7 Prosody-Parameters

This set of header fields defines the prosody of the speech.

```
prosody-parameter     =    "Prosody-" prosody-param-name ":"
                           prosody-param-value CRLF
prosody-param-name    =    1*VCHAR
                           prosody-param-value = 1*VCHAR
```

prosody-param-name is any one of the attribute names under the prosody element specified in W3C's SSML Specification [3]. The prosodyparam-value is any one of the value choices of the corresponding prosody element attribute from that specification. These header fields MAY be sent in SET-PARAMS or GET-PARAMS requests to define or get default values for the entire session or MAY be sent in the SPEAK request to define default values for that SPEAK request. Furthermore, these attributes can be part of the speech text marked up in SSML.

The prosody parameter header fields in the SET-PARAMS or SPEAK request only apply if the speech data is of type "text/plain" and does not use a speech markup format. These prosody parameter header fields MAY also be sent in a CONTROL method to affect a SPEAK request in progress and change its behavior on the fly. If the synthesizer resource does not support this operation, it MUST respond back to the client with a status-code of 403 "Unsupported Header Field."

14.8.4.8 Speech-Marker

This header field contains timestamp information in a "timestamp" field. This is a network time protocol (NTP) (RFC 5905) timestamp, a 64-bit number in decimal form. It MUST be synced with the RTP (RFC 3550) timestamp of the media stream through the RTCP (RFC 3550). Markers are bookmarks that are defined within the markup. Most speech markup formats provide mechanisms to embed marker fields within speech texts. The synthesizer generates SPEECH-MARKER events when it reaches these marker fields. This header field MUST be part of the SPEECH-MARKER event and contain the marker tag value after the timestamp, separated by a semicolon. In these events, the timestamp marks the time the text corresponding to the marker was emitted as speech by the synthesizer.

This header field MUST also be returned in responses to STOP, CONTROL, and BARGE-IN-OCCURRED methods, in the SPEAK-COMPLETE event, and in an IN-PROGRESS SPEAK response. In these messages, if any markers have been encountered for the current SPEAK, the marker tag value MUST be the last embedded marker encountered. If no markers have yet been encountered for the current SPEAK, only the timestamp is REQUIRED. Note that in these events, the purpose of this header field is to provide timestamp information associated with important events within the lifecycle of a request (start of SPEAK processing, end of SPEAK processing, or receipt of CONTROL/STOP/BARGE-IN-OCCURRED).

```
timestamp           =    "timestamp" "=" time-stamp-value
time-stamp-value    =    1*20DIGIT
speech-marker       =    "Speech-Marker" ":"
timestamp
                         [";" 1*(UTFCHAR / %x20)] CRLF
```

14.8.4.9 Speech-Language

This header field specifies the default language of the speech data if the language is not specified in the markup. The value of this header field MUST follow RFC 5646 for its values. The header field MAY occur in SPEAK, SET-PARAMS, or GET-PARAMS requests.

```
speech-language  =  "Speech-Language" ":" 1*VCHAR CRLF
```

14.8.4.10 Fetch-Hint

When the synthesizer needs to fetch specifications or other resources like speech markup or audio files, this header field controls the corresponding URI access properties. This provides client policy on when the synthesizer should retrieve content from the server. A value of "prefetch" indicates that the content MAY be downloaded when the request is received, whereas "safe" indicates that content MUST NOT be downloaded until actually referenced. The default value is "prefetch." This header field MAY occur in SPEAK, SET-PARAMS, or GET-PARAMS requests.

```
fetch-hint  =  "Fetch-Hint" ":" ("prefetch" / "safe") CRLF
```

14.8.4.11 Audio-Fetch-Hint

When the synthesizer needs to fetch specifications or other resources like speech audio files, this header field controls the corresponding URI access properties. This provides client policy on whether or not the synthesizer is permitted to attempt to optimize speech by prefetching audio. The value is either "safe," to say that audio is only fetched when it is referenced, never before; "prefetch," to permit, but not require, the implementation to pre-fetch the audio; or "stream," to allow it to stream the audio fetches. The default value is "prefetch." This header field MAY occur in SPEAK, SET-PARAMS, or GET-PARAMS requests.

```
audio-fetch-hint    =    "Audio-Fetch-Hint" ":"
                         ("prefetch" / "safe" / "stream") CRLF
```

14.8.4.12 Failed-URI

When a synthesizer method needs a synthesizer to fetch or access a URI, and the access fails, the server SHOULD provide the failed URI in this header field in the method response, unless there are multiple URI failures, in which case the server MUST provide one of the failed URIs in this header field in the method response.

```
failed-uri    =    "Failed-URI" ":" absoluteURI CRLF
```

14.8.4.13 Failed-URI-Cause

When a synthesizer method needs a synthesizer to fetch or access a URI, and the access fails, the server MUST provide the URI-specific or protocol-specific response code for the URI in the Failed-URI header field in the method response through this header field. The value encoding is UTF-8 (RFC 3629) to accommodate any access protocol – some access protocols might have a response string instead of a numeric response code.

```
failed-uri-cause    =  "Failed-URI-Cause" ":" 1*UTFCHAR CRLF
```

14.8.4.14 Speak-Restart

When a client issues a CONTROL request to a currently speaking synthesizer resource to jump backwards, and the target jump point is before the start of the current SPEAK request, the current SPEAK request MUST restart from the beginning of its speech data, and the server's response to the CONTROL request MUST contain this header field with a value of "true," indicating a restart.

```
speak-restart    =    "Speak-Restart" ":" BOOLEAN CRLF
```

14.8.4.15 Speak-Length

This header field MAY be specified in a CONTROL method to control the maximum length of speech to speak relative to the current speaking point in the currently active SPEAK request. If it is numeric, the value MUST be a positive integer. If a header field with a Tag unit is specified, then the speech output continues until the tag is reached or the SPEAK request is completed, whichever comes first. This header field MAY be specified in a SPEAK request to indicate the length to speak from the speech data and is relative to the point in speech where the SPEAK request starts. The different speech length units supported are synthesizer implementation dependent. If a server does not support the specified unit, the server MUST respond with a status-code of 409 "Unsupported Header Field Value."

```
speak-length               =  "Speak-Length" ":" positive-length-value
                              CRLF
                              positive-length-value = positive-speech-length
                              / text-speech-length
text-speech-length         =  1*UTFCHAR SP "Tag"
positive-speech-length     =  1*19DIGIT SP numeric-speech-unit
numeric-speech-unit        =  "Second"
                              / "Word"
                              / "Sentence"
                              / "Paragraph"
```

14.8.4.16 Load-Lexicon

This header field is used to indicate whether a lexicon has to be loaded or unloaded. The value "true" means to load the lexicon if not already loaded, and the value "false" means to unload the lexicon if it is loaded. The default value for this header field is "true." This header field MAY be specified in a DEFINE-LEXICON method.

```
load-lexicon    =    "Load-Lexicon" ":" BOOLEAN CRLF
```

14.8.4.17 Lexicon-Search-Order

This header field is used to specify a list of active pronunciation lexicon URIs and the search order among the active lexicons. Lexicons specified within the SSML specification take precedence over the lexicons specified in this header field. This header field MAY be specified in the SPEAK, SET-PARAMS, and GET-PARAMS methods.

```
lexicon-search-order = "Lexicon-Search-Order" ":"
                       "<" absoluteURI ">" *(" " "<" absoluteURI ">")
                       CRLF
```

14.8.5 Synthesizer Message Body

A synthesizer message can contain additional information associated with the Request, Response, or Event in its message body.

14.8.5.1 Synthesizer Speech Data

Marked-up text for the synthesizer to speak is specified as a typed media entity in the message body. The speech data to be spoken by the synthesizer can be specified inline by embedding the data in the message body or by reference by providing a URI for accessing the data. In either case, the data and the format used to mark up the speech need to be of a content type supported by the server. All MRCPv2 servers containing synthesizer resources MUST support both plain text speech data and W3C's SSML (Figure 14.8) [3] and hence, MUST support the media types "text/plain" and "application/ssml+xml." Other formats MAY be supported.

If the speech data is to be fetched by URI reference, the media type "text/uri-list" (see RFC 2483) is used to indicate one or more URIs that when dereferenced, will contain the content to be spoken. If a list of speech URIs is specified, the resource MUST speak the speech data provided by each URI in the order in which the URIs are specified in the content.

MRCPv2 clients and servers MUST support the "multipart/mixed" media type (Figure 14.9). This is the appropriate media type to use when providing a mix of URI and inline speech data. Embedded within the multipart content block, there MAY be content for the "text/uri-list," "application/ssml+xml," and/or "text/plain" media types. The character set and encoding used in the speech data are specified according to standard media type definitions. The multipart content MAY also contain actual audio data. Clients may have recorded audio clips stored in memory or on a local device and wish to play them as part of the SPEAK request. The audio portions MAY be sent by the client as part of the multipart content block. This audio is referenced in the speech markup data, which is another part in the multipart content block according to the "multipart/mixed" media type specification.

14.8.5.2 Lexicon Data

Synthesizer lexicon data from the client to the server can be provided inline or by reference. Either way, it is carried as typed media in the message body of the MRCPv2 request message (see Section 14.8.14). When a lexicon is specified inline in the message, the client MUST provide a

```
Content-Type:text/uri-list
Content-Length:...
http://www.example.com/ASR-Introduction.ssml
http://www.example.com/ASR-Specification-Part1.ssml
http://www.example.com/ASR-Specification-Part2.ssml
http://www.example.com/ASR-Conclusion.ssml
URI List Example
Content-Type:application/ssml+xml
Content-Length:...

<?xml version="1.0"?>

<speak version="1.0"
            xmlns="http://www.w3.org/2001/10/synthesis"
            xmlns:xsi="http://www.w3.org/2001/XMLSchema-instance"
            xsi:schemaLocation="http://www.w3.org/2001/10/synthesis
            http://www.w3.org/TR/speech-synthesis/synthesis.xsd"
            xml:lang="en-US">
      <p>
            <s>You have 4 new messages.</s>
            <s>The first is from Aldine Turnbet
                  and arrived at <break/>
                  <say-as interpret-as="vxml:time">0345p</say-as>.</s>
            <s>The subject is <prosody
                  rate="-20%">ski trip</prosody></s>
      </p>
</speak>
```

FIGURE 14.8 SSML Example.

Content-ID for that lexicon as part of the content header fields. The server MUST store the lexicon associated with that Content-ID for the duration of the session. A stored lexicon can be overwritten by defining a new lexicon with the same Content-ID.

Lexicons that have been associated with a Content-ID can be referenced through the "session" URI scheme (see Section 14.13.6). If lexicon data is specified by external URI reference, the media type "text/uri-list" (see RFC 2483) is used to list the one or more URIs that may be dereferenced to obtain the lexicon data. All MRCPv2 servers MUST support the "http" and "https" URI access mechanisms and MAY support other mechanisms. If the data in the message body consists of a mix of URI and inline lexicon data, the "multipart/mixed" media type is used. The character set and encoding used in the lexicon data may be specified according to standard media type definitions.

14.8.6 SPEAK Method

The SPEAK request (Figure 14.10) provides the synthesizer resource with the speech text and initiates speech synthesis and streaming. The SPEAK method MAY carry voice and prosody header fields that alter the behavior of the voice being synthesized as well as a typed media message body containing the actual marked-up text to be spoken. The SPEAK method implementation MUST do a fetch of all external URIs that are part of that operation. If caching is implemented, this URI

```
Content-Type:multipart/mixed; boundary="break"

--break

Content-Type:text/uri-list
Content-Length:...
http://www.example.com/ASR-Introduction.ssml
http://www.example.com/ASR-Specification-Part1.ssml
http://www.example.com/ASR-Specification-Part2.ssml
http://www.example.com/ASR-Conclusion.ssml

--break

Content-Type:application/ssml+xml
Content-Length:...

<?xml version="1.0"?>

<speak version="1.0"
          xmlns="http://www.w3.org/2001/10/synthesis"
          xmlns:xsi="http://www.w3.org/2001/XMLSchema-instance"
          xsi:schemaLocation="http://www.w3.org/2001/10/synthesis
          http://www.w3.org/TR/speech-synthesis/synthesis.xsd"
          xml:lang="en-US">
     <p>
          <s>You have 4 new messages.</s>
          <s>The first is from Stephanie Williams
               and arrived at <break/>
               <say-as interpret-as="vxml:time">0342p</say-as>.</s>
          <s>The subject is <prosody
               rate="-20%">ski trip</prosody></s>
     </p>
</speak>

--break--
```

FIGURE 14.9 Multipart Example.

fetching MUST conform to the cache-control hints and parameter header fields associated with the method in deciding whether it is to be fetched from cache or from the external server. If these hints/parameters are not specified in the method, the values set for the session using SET-PARAMS/GET-PARAMS apply. If it was not set for the session, their default values apply.

When applying voice parameters, there are three levels of precedence. The highest precedence is for those specified within the speech markup text, followed by those specified in the header fields of the SPEAK request and hence, that apply for that SPEAK request only, followed by the session default values that can be set using the SET-PARAMS request and apply for subsequent methods

```
C->S:  MRCP/2.0 ... SPEAK 543257

              Channel-Identifier:32AECB23433802@speechsynth
              Voice-gender:neutral
              Voice-Age:25
              Prosody-volume:medium
              Content-Type:application/ssml+xml
              Content-Length:...
              <?xml version="1.0"?>

       <speak version="1.0"
       xmlns="http://www.w3.org/2001/10/synthesis"
              xmlns:xsi="http://www.w3.org/2001/XMLSchema-instance"
              xsi:schemaLocation="http://www.w3.org/2001/10/synthesis
              http://www.w3.org/TR/speech-synthesis/synthesis.xsd"
              xml:lang="en-US">
       <p>
              <s>You have 4 new messages.</s>
              <s>The first is from Stephanie Williams and arrived at
                     <break/>
                     <say-as interpret-as="vxml:time">0342p</say-as>.
              </s>
              <s>The subject is
                     <prosody rate="-20%">ski trip</prosody>
              </s>
       </p>
       </speak>

S->C:  MRCP/2.0 ... 543257 200 IN-PROGRESS
       Channel-Identifier:32AECB23433802@speechsynth
       Speech-Marker:timestamp=857206027059

S->C:  MRCP/2.0 ... SPEAK-COMPLETE 543257 COMPLETE
       Channel-Identifier:32AECB23433802@speechsynth
       Completion-Cause:000 normal
       Speech-Marker:timestamp=857206027059
```

FIGURE 14.10 SPEAK Example.

invoked during the session. If the resource was idle at the time the SPEAK request arrived at the server, and the SPEAK method is being actively processed, the resource responds immediately with a success status code and a request-state of IN-PROGRESS. If the resource is in the speaking or paused state when the SPEAK method arrives at the server, i.e., it is in the middle of processing a previous SPEAK request, the status returns success with a request-state of PENDING. The server places the SPEAK request in the synthesizer resource request queue. The request queue operates strictly FIFO: requests are processed serially in order of receipt. If the current SPEAK fails, all SPEAK methods in the pending queue are cancelled, and each generates a SPEAK-COMPLETE event with a Completion-Cause of "cancelled." For the synthesizer resource, SPEAK is the only method that can return a request-state of IN-PROGRESS or PENDING. When the text has been synthesized and played into the media stream, the resource issues a SPEAK-COMPLETE event with the request-id of the SPEAK request and a request-state of COMPLETE.

14.8.7 STOP

The STOP method (Figure 10.11) from the client to the server tells the synthesizer resource to stop speaking if it is speaking something. The STOP request can be sent with an Active-Request-Id-List

```
C->S: MRCP/2.0 ... SPEAK 543258
Channel-Identifier:32AECB23433802@speechsynth
Content-Type:application/ssml+xml
Content-Length:...

<?xml version="1.0"?>

<speak version="1.0"
                xmlns="http://www.w3.org/2001/10/synthesis"
                xmlns:xsi="http://www.w3.org/2001/XMLSchema-instance"
                xsi:schemaLocation="http://www.w3.org/2001/10/
                synthesis
                http://www.w3.org/TR/speech-synthesis/synthesis.xsd"
                xml:lang="en-US">
        <p>
                <s>You have 4 new messages.</s>
                <s>The first is from Stephanie Williams and arrived at
                        <break/>
                        <say-as interpret-as="vxml:time">0342p</say-as>.</s>
                <s>The subject is
                        <prosody rate="-20%">ski trip</prosody></s>
        </p>
</speak>

S->C: MRCP/2.0 ... 543258 200 IN-PROGRESS
      Channel-Identifier:32AECB23433802@speechsynth
      Speech-Marker:timestamp=857206027059

C->S: MRCP/2.0 ... STOP 543259
      Channel-Identifier:32AECB23433802@speechsynth

S->C: MRCP/2.0 ... 543259 200 COMPLETE
      Channel-Identifier:32AECB23433802@speechsynth
      Active-Request-Id-List:543258
      Speech-Marker:timestamp=857206039059
```

FIGURE 14.11 STOP Example.

header field to stop the zero or more specific SPEAK requests that may be in queue and return a response status-code of 200 "Success." If no Active-Request-Id-List header field is sent in the STOP request, the server terminates all outstanding SPEAK requests. If a STOP request successfully terminated one or more PENDING or IN-PROGRESS SPEAK requests, then the response MUST contain an Active-Request-Id-List header field enumerating the SPEAK request-ids that were terminated. Otherwise, there is no Active-Request-Id-List header field in the response. No SPEAK-COMPLETE events are sent for such terminated requests.

If a SPEAK request that was IN-PROGRESS and speaking was stopped, the next pending SPEAK request, if any, becomes IN-PROGRESS at the resource and enters the speaking state. If a SPEAK request that was IN-PROGRESS and paused was stopped, the next pending SPEAK request, if any, becomes IN-PROGRESS and enters the paused state.

14.8.8 BARGE-IN-OCCURRED

The BARGE-IN-OCCURRED method, when used with the synthesizer resource, provides a client that has detected a barge-in-able event with a means to communicate the occurrence of the event to the synthesizer resource. This method is useful in two scenarios:

1. The client has detected DTMF digits in the input media or some other barge-in-able event and wants to communicate that to the synthesizer resource.
2. The recognizer resource and the synthesizer resource are in different servers. In this case, the client acts as an intermediary for the two servers. It receives an event from the recognition resource and sends a BARGE-IN-OCCURRED request to the synthesizer. In such cases, the BARGE-IN-OCCURRED method would also have a Proxy-Sync-Id header field received from the resource generating the original event.

If a SPEAK request is active with kill-on-barge-in enabled (see Section 14.8.4.2), and the BARGE-IN-OCCURRED event is received, the synthesizer MUST immediately stop streaming out audio. It MUST also terminate any speech requests queued behind the current active one, irrespective of whether or not they have barge-in enabled. If a barge-in-able SPEAK request was playing and it was terminated, the response MUST contain an Active-Request-Id-List header field listing the request-ids of all SPEAK requests that were terminated. The server generates no SPEAK-COMPLETE events for these requests.

If there were no SPEAK requests terminated by the synthesizer resource as a result of the BARGE-IN-OCCURRED method, the server MUST respond to the BARGE-IN-OCCURRED with a status-code of 200 "Success," and the response MUST NOT contain an Active-Request-Id-List header field. If the synthesizer and recognizer resources are part of the same MRCPv2 session, they can be optimized for a quicker kill-on-barge-in response if the recognizer and synthesizer interact directly. In these cases, the client MUST still react to a START-OF-INPUT event from the recognizer by invoking the BARGE-IN-OCCURRED method to the synthesizer. The client MUST invoke the BARGE-IN-OCCURRED if it has any outstanding requests to the synthesizer resource in either the PENDING or the IN-PROGRESS state.

```
C->S: MRCP/2.0 ... SPEAK 543258
      Channel-Identifier:32AECB23433802@speechsynth
      Voice-gender:neutral
      Voice-Age:25
      Prosody-volume:medium
      Content-Type:application/ssml+xml
      Content-Length:...

      <?xml version="1.0"?>
```

```
<speak version="1.0"
      xmlns="http://www.w3.org/2001/10/synthesis"
      xmlns:xsi="http://www.w3.org/2001/XMLSchema-instance"
      xsi:schemaLocation="http://www.w3.org/2001/10
      /synthesis
      http://www.w3.org/TR/speech-synthesis/synthesis.xsd"
      xml:lang="en-US">
   <p>
      <s>You have 4 new messages.</s>
      <s>The first is from Stephanie Williams and arrived at
         <break/>
         <say-as interpret-as="vxml:time">0342p</say-as>.</s>
      <s>The subject is
         <prosody rate="-20%">ski trip</prosody></s>
   </p>
</speak>

S->C: MRCP/2.0 ... 543258 200 IN-PROGRESS
      Channel-Identifier:32AECB23433802@speechsynth
      Speech-Marker:timestamp=857206027059

C->S: MRCP/2.0 ... BARGE-IN-OCCURRED 543259
      Channel-Identifier:32AECB23433802@speechsynth
      Proxy-Sync-Id:987654321

S->C: MRCP/2.0 ... 543259 200 COMPLETE
      Channel-Identifier:32AECB23433802@speechsynth
      Active-Request-Id-List:543258
      Speech-Marker:timestamp=857206039059
      BARGE-IN-OCCURRED Example
```

14.8.9 PAUSE

The PAUSE method (Figure 14.12) from the client to the server tells the synthesizer resource to pause speech output if it is speaking something. If a PAUSE method is issued on a session when a SPEAK is not active, the server MUST respond with a status-code of 402 "Method not valid in this state." If a PAUSE method is issued on a session when a SPEAK is active and paused, the server MUST respond with a status-code of 200 "Success." If a SPEAK request was active, the server MUST return an Active-Request-Id-List header field whose value contains the request-id of the SPEAK request that was paused.

14.8.10 RESUME

The RESUME method (Figure 14.13) from the client to the server tells a paused synthesizer resource to resume speaking. If a RESUME request is issued on a session with no active SPEAK request, the server MUST respond with a status-code of 402 "Method not valid in this state." If a RESUME request is issued on a session with an active SPEAK request that is speaking (i.e., not paused), the server MUST respond with a status-code of 200 "Success." If a SPEAK request was paused, the server MUST return an Active-Request – value contains the request-id of the SPEAK request that was resumed.

14.8.11 CONTROL

The CONTROL method (Figure 14.14) from the client to the server tells a synthesizer that is speaking to modify what it is speaking on the fly. This method is used to request the synthesizer to jump

```
C->S: MRCP/2.0 ... SPEAK 543258
Channel-Identifier:32AECB23433802@speechsynth
Voice-gender:neutral
Voice-Age:25
Prosody-volume:medium
Content-Type:application/ssml+xml
Content-Length:...

<?xml version="1.0"?>

<speak version="1.0"
            xmlns="http://www.w3.org/2001/10/synthesis"
            xmlns:xsi="http://www.w3.org/2001/XMLSchema-instance"
            xsi:schemaLocation="http://www.w3.org/2001/10/synthesis
            http://www.w3.org/TR/speech-synthesis/synthesis.xsd"
            xml:lang="en-US">
      <p>
            <s>You have 4 new messages.</s>
            <s>The first is from Stephanie Williams and arrived at
                  <break/>
                  <say-as interpret-as="vxml:time">0342p</say-as>.</s>
            <s>The subject is
                  <prosody rate="-20%">ski trip</prosody></s>
      </p>
</speak>

S->C: MRCP/2.0 ... 543258 200 IN-PROGRESS
      Channel-Identifier:32AECB23433802@speechsynth
      Speech-Marker:timestamp=857206027059

C->S: MRCP/2.0 ... PAUSE 543259
      Channel-Identifier:32AECB23433802@speechsynth

S->C: MRCP/2.0 ... 543259 200 COMPLETE
      Channel-Identifier:32AECB23433802@speechsynth
      Active-Request-Id-List:543258
```

FIGURE 14.12 PAUSE Example.

forwards or backwards in what it is speaking or change speaker rate, speaker parameters, etc. It affects only the currently IN-PROGRESS SPEAK request. Depending on the implementation and capability of the synthesizer resource, it may or may not support the various modifications indicated by header fields in the CONTROL request. When a client invokes a CONTROL method to jump forwards, and the operation goes beyond the end of the active SPEAK method's text, the CONTROL request still succeeds. The active SPEAK request completes and returns a SPEAK-COMPLETE event following the response to the CONTROL method. If there are more SPEAK requests in the queue, the synthesizer resource starts at the beginning of the next SPEAK request in the queue.

```
C->S: MRCP/2.0 ... SPEAK 543258
      Channel-Identifier:32AECB23433802@speechsynth
      Voice-gender:neutral
      Voice-age:25
      Prosody-volume:medium
      Content-Type:application/ssml+xml
      Content-Length:...

      <?xml version="1.0"?>

      <speak version="1.0"
                  xmlns="http://www.w3.org/2001/10/synthesis"
                  xmlns:xsi="http://www.w3.org/2001/
                  XMLSchema-instance"
                  xsi:schemaLocation="http://www.w3.org/2001/10
                  /synthesis
                  http://www.w3.org/TR/speech-synthesis/synthesis.xsd"
                  xml:lang="en-US">
            <p>
                  <s>You have 4 new messages.</s>
                  <s>The first is from Stephanie Williams and arrived at
                        <break/>
                        <say-as interpret-as="vxml:time">
                        0342p</say-as>.</s>
                  <s>The subject is
                        <prosody rate="-20%">ski trip</prosody></s>
            </p>
      </speak>

S->C: MRCP/2.0 ... 543258 200 IN-PROGRESS@speechsynth
      Channel-Identifier:32AECB23433802
      Speech-Marker:timestamp=857206027059

C->S: MRCP/2.0 ... PAUSE 543259
      Channel-Identifier:32AECB23433802@speechsynth

S->C: MRCP/2.0 ... 543259 200 COMPLETE
      Channel-Identifier:32AECB23433802@speechsynth
      Active-Request-Id-List:543258

C->S: MRCP/2.0 ... RESUME 543260
      Channel-Identifier:32AECB23433802@speechsynth

S->C: MRCP/2.0 ... 543260 200 COMPLETE
      Channel-Identifier:32AECB23433802@speechsynth
      Active-Request-Id-List:543258
```

FIGURE 14.13 RESUME Example.

```
C->S: MRCP/2.0 ... SPEAK 543258
Channel-Identifier:32AECB23433802@speechsynth
Voice-gender:neutral
Voice-age:25
Prosody-volume:medium
Content-Type:application/ssml+xml
Content-Length:...

<?xml version="1.0"?>

<speak version="1.0"
         xmlns="http://www.w3.org/2001/10/synthesis"
         xmlns:xsi="http://www.w3.org/2001/XMLSchema-instance"
         xsi:schemaLocation="http://www.w3.org/2001/10/synthesis
         http://www.w3.org/TR/speech-synthesis/synthesis.xsd"
         xml:lang="en-US">
    <p>
        <s>You have 4 new messages.</s>
        <s>The first is from Stephanie Williams
            and arrived at <break/>
            <say-as interpret-as="vxml:time">0342p</say-as>.</s>
        <s>The subject is <prosody
            rate="-20%">ski trip</prosody></s>
    </p>
</speak>

S->C: MRCP/2.0 ... 543258 200 IN-PROGRESS
      Channel-Identifier:32AECB23433802@speechsynth
      Speech-Marker:timestamp=857205016059

C->S: MRCP/2.0 ... CONTROL 543259
      Channel-Identifier:32AECB23433802@speechsynth
      Prosody-rate:fast

S->C: MRCP/2.0 ... 543259 200 COMPLETE
      Channel-Identifier:32AECB23433802@speechsynth
      Active-Request-Id-List:543258
      Speech-Marker:timestamp=857206027059

C->S: MRCP/2.0 ... CONTROL 543260
      Channel-Identifier:32AECB23433802@speechsynth
      Jump-Size:-15 Words

S->C: MRCP/2.0 ... 543260 200 COMPLETE
      Channel-Identifier:32AECB23433802@speechsynth
      Active-Request-Id-List:543258
      Speech-Marker:timestamp=857206039059
```

FIGURE 14.14 CONTROL Example.

When a client invokes a CONTROL method to jump backwards, and the operation jumps to the beginning or beyond the beginning of the speech data of the active SPEAK method, the CONTROL request still succeeds. The response to the CONTROL request contains the speak-restart header field, and the active SPEAK request restarts from the beginning of its speech data. These two behaviors can be used to rewind or fast-forward across multiple speech requests if the client wants to break up a speech markup text into multiple SPEAK requests. If a SPEAK request was active when the CONTROL method was received, the server MUST return an Active-Request-Id-List header field containing the request-id of the SPEAK request that was active.

14.8.12 SPEAK-COMPLETE

This is an Event message from the synthesizer resource to the client that indicates the corresponding SPEAK request was completed. The request-id field matches the request-id of the SPEAK request that initiated the speech that just completed. The request-state field is set to COMPLETE by the server (Figure 14.15), indicating that this is the last event with the corresponding request-id. The Completion-Cause header field specifies the cause code pertaining to the status and reason of request completion, such as the SPEAK completed normally or because of an error, kill-on-barge-in, etc.

14.8.13 SPEECH-MARKER

This is an event generated by the synthesizer resource to the client when the synthesizer encounters a marker tag in the speech markup it is currently processing. The value of the request-id field MUST match that of the corresponding SPEAK request. The request-statefield MUST have the value "IN-PROGRESS," as the speech is still not complete. The value of the speech marker tag hit, describing where the synthesizer is in the speech markup, MUST be returned in the Speech-Marker header field along with an NTP timestamp indicating the instant in the output speech stream when the marker was encountered. The SPEECH-MARKER event (Figure 14.16) MUST also be generated with a null marker value and output NTP timestamp when a SPEAK request in Pending-State (i.e., in the queue) changes state to IN-PROGRESS and starts speaking. The NTP timestamp MUST be synchronized with the RTP timestamp used to generate the speech stream through standard RTCP machinery.

14.8.14 DEFINE-LEXICON

The DEFINE-LEXICON method, from the client to the server, provides a lexicon and tells the server to load or unload the lexicon (see Section 14.8.4.16). The media type of the lexicon is provided in the Content-Type header (see Section 14.8.5.2). One such media type is "application/pls+xml" for the Pronunciation Lexicon Specification (PLS) [7] and RFC 4267. If the server resource is in the speaking or paused state, the server MUST respond with a failure status-code of 402 "Method not valid in this state." If the resource is in the idle state and is able to successfully load/unload the lexicon, the status MUST return a 200 "Success" status-code, and the request-state MUST be COMPLETE.

If the synthesizer could not define the lexicon for some reason, for example, because the download failed or the lexicon was in an unsupported form, the server MUST respond with a failure status-code of 407 and a Completion-Cause header field describing the failure reason.

14.9 SPEECH RECOGNIZER RESOURCE

The speech recognizer resource receives an incoming voice stream and provides the client with an interpretation of what was spoken in textual form. The recognizer resource is controlled by MRCPv2 requests from the client. The recognizer resource can both respond to these requests and

```
C->S: MRCP/2.0 ... SPEAK 543260
      Channel-Identifier:32AECB23433802@speechsynth
      Voice-gender:neutral
      Voice-age:25
      Prosody-volume:medium
      Content-Type:application/ssml+xml
      Content-Length:...

      <?xml version="1.0"?>

      <speak version="1.0"
                  xmlns="http://www.w3.org/2001/10/synthesis"
                  xmlns:xsi="http://www.w3.org/2001/
                  XMLSchema-instance"
                  xsi:schemaLocation="http://www.w3.org/2001/10/
                  synthesis
                  http://www.w3.org/TR/speech-synthesis/synthesis.xsd"
                  xml:lang="en-US">
            <p>
                  <s>You have 4 new messages.</s>
                  <s>The first is from Stephanie Williams
                        and arrived at <break/>
                  <say-as interpret-as="vxml:time">0342p</say-as>.</s>
                  <s>The subject is
                        <prosody rate="-20%">ski trip</prosody></s>
            </p>
      </speak>

S->C: MRCP/2.0 ... 543260 200 IN-PROGRESS
      Channel-Identifier:32AECB23433802@speechsynth
      Speech-Marker:timestamp=857206027059

S->C: MRCP/2.0 ... SPEAK-COMPLETE 543260 COMPLETE
      Channel-Identifier:32AECB23433802@speechsynth
      Completion-Cause:000 normal
      Speech-Marker:timestamp=857206039059
```

FIGURE 14.15 SPEAK-COMPLETE Example.

generate asynchronous events to the client to indicate conditions of interest during the processing of the method. This section applies to the following resource types:

1. speechrecog
2. dtmfrecog

The difference between these two resources is in their level of support for recognition grammars. The "dtmfrecog" resource type is capable of recognizing only DTMF digits and hence, accepts

```
C->S: MRCP/2.0 ... SPEAK 543261
      Channel-Identifier:32AECB23433802@speechsynth
      Voice-gender:neutral
      Voice-age:25
      Prosody-volume:medium
      Content-Type:application/ssml+xml
      Content-Length:...

      <?xml version="1.0"?>
      <speak version="1.0"
                  xmlns="http://www.w3.org/2001/10/synthesis"
                  xmlns:xsi="http://www.w3.org/2001/
                  XMLSchema-instance"
                  xsi:schemaLocation="http://www.w3.org/2001/
                  10/synthesis
                  http://www.w3.org/TR/speech-synthesis/synthesis.xsd"
                  xml:lang="en-US">
        <p>
            <s>You have 4 new messages.</s>
            <s>The first is from Stephanie Williams
                  and arrived at <break/>
                  <say-as interpret-as="vxml:time">0342p</say-as>.
            </s>
            <mark name="here"/>
            <s>The subject is
                  <prosody rate="-20%">ski trip</prosody>
            </s>
            <mark name="ANSWER"/>
        </p>
      </speak>

S->C: MRCP/2.0 ... 543261 200 IN-PROGRESS
      Channel-Identifier:32AECB23433802@speechsynth
      Speech-Marker:timestamp=857205015059

S->C: MRCP/2.0 ... SPEECH-MARKER 543261 IN-PROGRESS
      Channel-Identifier:32AECB23433802@speechsynth
      Speech-Marker:timestamp=857206027059;here

S->C: MRCP/2.0 ... SPEECH-MARKER 543261 IN-PROGRESS
      Channel-Identifier:32AECB23433802@speechsynth
      Speech-Marker:timestamp=857206039059;ANSWER

S->C: MRCP/2.0 ... SPEAK-COMPLETE 543261 COMPLETE
      Channel-Identifier:32AECB23433802@speechsynth
      Completion-Cause:000 normal
      Speech-Marker:timestamp=857207689259;ANSWER
```

FIGURE 14.16 SPEECH-MARKER Example.

only DTMF grammars. It only generates barge-in for DTMF inputs and ignores speech. The "speechrecog" resource type can recognize regular speech as well as DTMF digits and hence, MUST support grammars describing either speech or DTMF. This resource generates barge-in events for speech and/or DTMF. By analyzing the grammars that are activated by the RECOGNIZE method, it determines whether a barge-in should occur for speech and/or DTMF. When the recognizer decides it needs to generate a barge-in, it also generates a START-OF-INPUT event to the client. The recognizer resource MAY support recognition in the normal or hotword modes or both (although note that a single "speechrecog" resource does not perform normal and hotword mode recognition simultaneously). For implementations where a single recognizer resource does not support both modes, or simultaneous normal and hotword recognition is desired, the two modes can be invoked through separate resources allocated to the same SIP dialog (with different MRCP session identifiers) and share the RTP audio feed. The capabilities of the recognizer resource are enumerated as follows:

Normal Mode Recognition: Normal mode recognition tries to match all the speech or DTMF against the grammar and returns a no-match status if the input fails to match or the method times out.

Hotword Mode Recognition: Hotword mode is where the recognizer looks for a match against specific speech grammar or DTMF sequence and ignores speech or DTMF that does not match. The recognition completes only if there is a successful match of grammar, if the client cancels the request, or if there is a non-input or recognition timeout. Voice Enrolled Grammars: A recognizer resource MAY optionally support Voice Enrolled Grammars. With this functionality, enrollment is performed using a person's voice. For example, a list of contacts can be created and maintained by recording the person's names using the caller's voice. This technique is sometimes also called speaker-dependent recognition.

Interpretation: A recognizer resource MAY be employed strictly for its natural language interpretation capabilities by supplying it with a text string as input instead of speech. In this mode, the resource takes text as input and produces an "interpretation" of the input according to the supplied grammar. Voice enrollment has the concept of an enrollment session. A session to add a new phrase to a personal grammar involves the initial enrollment followed by a repeat of enough utterances before committing the new phrase to the personal grammar. Each time an utterance is recorded, it is compared for similarity with the other samples, and a clash test is performed against other entries in the personal grammar to ensure there are no similar and confusable entries. Enrollment is done using a recognizer resource. Controlling which utterances are to be considered for enrollment of a new phrase is done by setting a header field (see Section 14.9.4.39) in the Recognize request. Interpretation is accomplished through the INTERPRET method (Section 14.9.20) and the Interpret-Text header field (Section 14.9.4.30).

14.9.1 Recognizer State Machine

The recognizer resource maintains a state machine to process MRCPv2 requests from the client (Figure 14.17).

If a recognizer resource supports voice enrolled grammars, starting an enrollment session does not change the state of the recognizer resource. Once an enrollment session is started, then utterances are enrolled by calling the RECOGNIZE method repeatedly. The state of the speech recognizer resource goes from IDLE to RECOGNIZING state each time RECOGNIZE is called.

14.9.2 Recognizer Methods

The recognizer supports the following methods.

```
recognizer-method   =   recog-only-method
                        / enrollment-method
recog-only-method   =   "DEFINE-GRAMMAR"
                        / "RECOGNIZE"
```

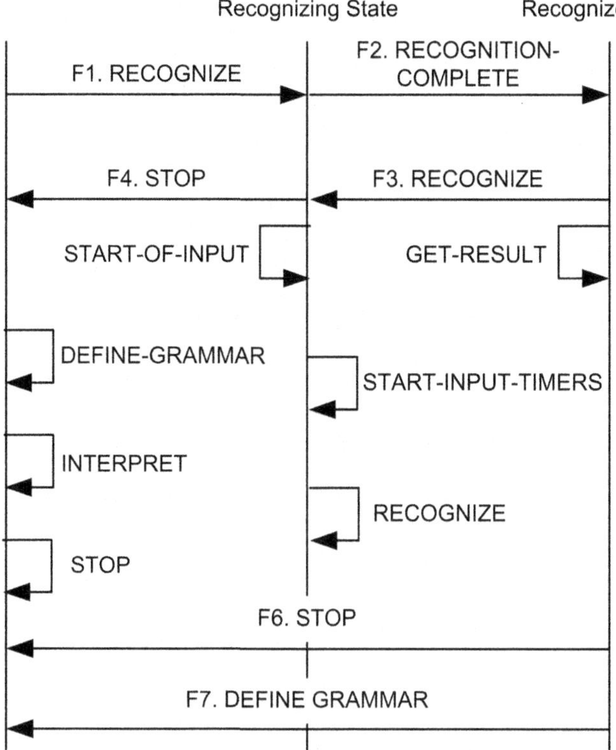

FIGURE 14.17 Recognizer State Machine.

```
                          / "INTERPRET"
                          / "GET-RESULT"
                          / "START-INPUT-TIMERS"
                          / "STOP"
```

It is OPTIONAL for a recognizer resource to support voice enrolled grammars. If the recognizer resource does support voice enrolled grammars, it MUST support the following methods.

```
enrollment-method      =      "START-PHRASE-ENROLLMENT"
                              / "ENROLLMENT-ROLLBACK"
                              / "END-PHRASE-ENROLLMENT"
                              / "MODIFY-PHRASE"
                              / "DELETE-PHRASE"
```

14.9.3 Recognizer Events

The recognizer can generate the following events.

```
recognizer-event    =    "START-OF-INPUT"
                         / "RECOGNITION-COMPLETE"
                         / "INTERPRETATION-COMPLETE"
```

14.9.4 Recognizer Header Fields

A recognizer message can contain header fields containing request options and information to augment the Method, Response, or Event message it is associated with.

```
recognizer-header      =      recog-only-header
                              / enrollment-header
recog-only-header      =      confidence-threshold
                              / sensitivity-level
                              / speed-vs-accuracy
                              / n-best-list-length
                              / no-input-timeout
                              / input-type
                              / recognition-timeout
                              / waveform-uri
                              / input-waveform-uri
                              / completion-cause
                              / completion-reason
                              / recognizer-context-block
                              / start-input-timers
                              / speech-complete-timeout
                              / speech-incomplete-timeout
                              / dtmf-interdigit-timeout
                              / dtmf-term-timeout
                              / dtmf-term-char
                              / failed-uri
                              / failed-uri-cause
                              / save-waveform
                              / media-type
                              / new-audio-channel
                              / speech-language
                              / ver-buffer-utterance
                              / recognition-mode
                              / cancel-if-queue
                              / hotword-max-duration
                              / hotword-min-duration
                              / interpret-text
                              / dtmf-buffer-time
                              / clear-dtmf-buffer
                              / early-no-match
```

If a recognizer resource supports voice enrolled grammars, the following header fields are also used.

```
enrollment-header   =   num-min-consistent-pronunciations
                        / consistency-threshold
                        / clash-threshold
                        / personal-grammar-uri
                        / enroll-utterance
                        / phrase-id
                        / phrase-nl
                        / weight
                        / save-best-waveform
                        / new-phrase-id
                        / confusable-phrases-uri
                        / abort-phrase-enrollment
```

For enrollment-specific header fields that can appear as part of SET-PARAMS or GET-PARAMS methods, the following general rule applies: The START-PHRASE-ENROLLMENT method MUST be invoked before these header fields may be set through the SET-PARAMS method or retrieved through the GET-PARAMS method. Note that the Waveform-URI header field of the Recognizer resource can also appear in the response to the END-PHRASE-ENROLLMENT method.

14.9.4.1 Confidence-Threshold

When a recognizer resource recognizes or matches a spoken phrase with some portion of the grammar, it associates a confidence level with that match. The Confidence-Threshold header field tells the recognizer resource what confidence level the client considers a successful match. This is a float value between 0.0 and 1.0 indicating the recognizer's confidence in the recognition. If the recognizer determines that there is no candidate match with a confidence that is greater than the confidence threshold, then it MUST return no-match as the recognition result. This header field MAY occur in RECOGNIZE, SET-PARAMS, or GET-PARAMS. The default value for this header field is implementation specific, as is the interpretation of any specific value for this header field. Although values for servers from different vendors are not comparable, it is expected that clients will tune this value over time for a given server.

```
confidence-threshold    =    "Confidence-Threshold" ":" FLOAT CRLF
```

14.9.4.2 Sensitivity-Level

To filter out background noise and not mistake it for speech, the recognizer resource supports a variable level of sound sensitivity. The Sensitivity-Level header field is a float value between 0.0 and 1.0 and allows the client to set the sensitivity level for the recognizer. This header field MAY occur in RECOGNIZE, SET-PARAMS, or GET-PARAMS. A higher value for this header field means higher sensitivity. The default value for this header field is implementation specific, as is the interpretation of any specific value for this header field. Although values for servers from different vendors are not comparable, it is expected that clients will tune this value over time for a given server.

```
sensitivity-level    =    "Sensitivity-Level" ":" FLOAT CRLF
```

14.9.4.3 Speed-Vs-Accuracy

Depending on the implementation and capability of the recognizer resource, it may be tunable towards Performance or Accuracy. Higher accuracy may mean more processing and higher CPU utilization, meaning fewer active sessions per server, and vice versa. The value is a float between 0.0 and 1.0. A value of 0.0 means fastest recognition. A value of 1.0 means best accuracy. This header field MAY occur in RECOGNIZE, SET-PARAMS, or GET-PARAMS. The default value for this header field is implementation specific. Although values for servers from different vendors are not comparable, it is expected that clients will tune this value over time for a given server.

```
speed-vs-accuracy    =    "Speed-Vs-Accuracy" ":" FLOAT CRLF
```

14.9.4.4 N-Best-List-Length

When the recognizer matches an incoming stream with the grammar, it may come up with more than one alternative match because of confidence levels in certain words or conversation paths. If this header field is not specified, by default, the recognizer resource returns only the best match above the confidence threshold. The client, by setting this header field, can ask the recognition resource to send it more than one alternative. All alternatives must still be above the Confidence-Threshold. A value greater than one does not guarantee that the recognizer will provide the requested number of alternatives. This header field MAY occur in RECOGNIZE, SET-PARAMS, or GET-PARAMS. The minimum value for this header field is 1. The default value for this header field is 1.

```
n-best-list-length    =    "N-Best-List-Length" ":" 1*19DIGIT CRLF
```

14.9.4.5 Input-Type

When the recognizer detects barge-in-able input and generates a START-OF-INPUT event, that event MUST carry this header field to specify whether the input that caused the barge-in was DTMF or speech.

```
input-type   =   "Input-Type" ":" inputs CRLF
inputs       =   "speech" / "dtmf"
```

14.9.4.6 No-Input-Timeout

When recognition is started and there is no speech detected for a certain period of time, the recognizer can send a RECOGNITION-COMPLETE event to the client with a Completion-Cause of "no-inputtimeout" and terminate the recognition operation. The client can use the No-Input-Timeout header field to set this timeout. The value is in milliseconds and can range from 0 to an implementation-specific maximum value. This header field MAY occur in RECOGNIZE, SET-PARAMS, or GET-PARAMS. The default value is implementation specific.

```
no-input-timeout    =    "No-Input-Timeout" ":" 1*19DIGIT CRLF
```

14.9.4.7 Recognition-Timeout

When recognition is started and there is no match for a certain period of time, the recognizer can send a RECOGNITION-COMPLETE event to the client and terminate the recognition operation. The Recognition-Timeout header field allows the client to set this timeout value. The value is in milliseconds. The value for this header field ranges from 0 to an implementation-specific maximum value. The default value is 10 seconds. This header field MAY occur in RECOGNIZE, SET-PARAMS, or GET-PARAMS.

```
recognition-timeout    =    "Recognition-Timeout" ":" 1*19DIGIT CRLF
```

14.9.4.8 Waveform-URI

If the Save-Waveform header field is set to "true," the recognizer MUST record the incoming audio stream of the recognition into a stored form and provide a URI for the client to access it. This header field MUST be present in the RECOGNITION-COMPLETE event if the Save-Waveform header field was set to "true." The value of the header field MUST be empty if there was some error condition preventing the server from recording. Otherwise, the URI generated by the server MUST be unambiguous across the server and all its recognition sessions. The content associated with the URI MUST be available to the client until the MRCPv2 session terminates.

Similarly, if the Save-Best-Waveform header field is set to "true," the recognizer MUST save the audio stream for the best repetition of the phrase that was used during the enrollment session. The recognizer MUST then record the recognized audio and make it available to the client by returning a URI in the Waveform-URI header field in the response to the END-PHRASE-ENROLLMENT method. The value of the header field MUST be empty if there was some error condition preventing the server from recording. Otherwise, the URI generated by the server MUST be unambiguous across the server and all its recognition sessions. The content associated with the URI MUST be available to the client until the MRCPv2 session terminates. See the discussion on the sensitivity of saved waveforms in Chapter 12. The server MUST also return the size in octets and the duration in milliseconds of the recorded audio waveform as parameters associated with the header field.

```
waveform-uri    =    "Waveform-URI" ":" ["<" uri ">"
                     ";" "size" "=" 1*19DIGIT
                     ";" "duration" "=" 1*19DIGIT] CRLF
```

14.9.4.9 Media-Type

This header field MAY be specified in the SET-PARAMS, GET-PARAMS, or the RECOGNIZE methods and tells the server resource the media type in which to store captured audio or video, such as the one captured and returned by the Waveform-URI header field.

```
media-type   =   "Media-Type" ":" media-type-value CRLF
```

14.9.4.10 Input-Waveform-URI

This optional header field specifies a URI pointing to audio content to be processed by the RECOGNIZE operation. This enables the client to request recognition from a specified buffer or audio file.

```
input-waveform-uri    =    "Input-Waveform-URI" ":" uri CRLF
```

14.9.4.11 Completion-Cause

This header field MUST be part of a RECOGNITION-COMPLETE event coming from the recognizer resource to the client. It indicates the reason behind the RECOGNIZE method completion. This header field MUST be sent in the DEFINE-GRAMMAR and RECOGNIZE responses if they return with a failure status and a COMPLETE state. In the following ABNF, the cause-code contains a numerical value selected from the Cause-Code column of the following table. The cause-name contains the corresponding token selected from the Cause-Name column.

```
completion-cause    =    "Completion-Cause" ":" cause-code SP
                         cause-name CRLF
cause-code          =    3DIGIT
cause-name          =    *VCHAR
```

Cause-Code	Cause-Name	Description
000	success	RECOGNIZE completed with a match or DEFINE-GRAMMAR succeeded in downloading and compiling the grammar.
001	no-match	RECOGNIZE completed, but no match was found.
002	no-input-timeout	RECOGNIZE completed without a match due to a no-input-timeout.
003	hotword-maxtime	RECOGNIZE in hotword mode completed without a match due to a recognition-timeout.
004	grammar-load-failure	RECOGNIZE failed due to grammar load failure.
005	grammar-compilation-failure	RECOGNIZE failed due to grammar compilation failure.
006	recognizer-error	RECOGNIZE request terminated prematurely due to a recognizer error.
007	speech-too-early	RECOGNIZE request terminated because speech was too early. This happens when the audio stream is already "in-speech" when the RECOGNIZE request was received.
008	success-maxtime	RECOGNIZE request terminated because speech was too long but whatever was spoken till that point was a full match.
009	uri-failure	Failure accessing a URI.
010	language-unsupported	Language not supported.
011	cancelled	A new RECOGNIZE cancelled this one, or a prior RECOGNIZE failed while this one was still in the queue.
012	semantics-failure	Recognition succeeded, but semantic interpretation of the recognized input failed. The RECOGNITION-COMPLETE event MUST contain the Recognition result with only input text and no interpretation.
013	partial-match	Speech Incomplete Timeout expired before there was a full match. But whatever was spoken till that point was a partial match to one or more grammars.
014	partial-match-maxtime	The Recognition-Timeout expired before full match was achieved. But whatever was spoken till that point was a partial match to one or more grammars.
015	no-match-maxtime	The Recognition-Timeout expired. Whatever was spoken till that point did not match any of the grammars. This cause could also be returned if the recognizer does not support detecting partial grammar matches.
016	grammar-definition-failure	Any DEFINE-GRAMMAR error other than grammar-load-failure and grammar-compilation-failure.

14.9.4.12 Completion-Reason

This header field MAY be specified in a RECOGNITION-COMPLETE event coming from the recognizer resource to the client. This contains the reason text behind the RECOGNIZE request completion. The server uses this header field to communicate text describing the reason for the failure, such as the specific error encountered in parsing a grammar markup. The completion reason text is provided for client use in logs and for debugging and instrumentation purposes. Clients MUST NOT interpret the completion reason text.

```
completion-reason    =    "Completion-Reason" ":"
                          quoted-string CRLF
```

14.9.4.13 Recognizer-Context-Block

This header field MAY be sent as part of the SET-PARAMS or GET-PARAMS request. If the GET-PARAMS method contains this header field with no value, then it is a request to the recognizer to return the recognizer context block. The response to such a message MAY contain a recognizer context block as a typed media message body. If the server returns a recognizer context block, the response MUST contain this header field, and its value MUST match the Content-ID of the corresponding media block.

If the SET-PARAMS method contains this header field, it MUST also contain a message body containing the recognizer context data and a Content-ID matching this header field value. This Content-ID MUST match the Content-ID that came with the context data during the GET-PARAMS operation. An implementation choosing to use this mechanism to hand off recognizer context data between servers MUST distinguish its implementation-specific block of data by using an IANA-registered content type in the IANA Media Type vendor tree.

```
recognizer-context-block    =    "Recognizer-Context-Block" ":"
                                 [1*VCHAR] CRLF
```

14.9.4.14 Start-Input-Timers

This header field MAY be sent as part of the RECOGNIZE request. A value of false tells the recognizer to start recognition but not to start the no-input timer yet. The recognizer MUST NOT start the timers until the client sends a START-INPUT-TIMERS request to the recognizer. This is useful in the scenario when the recognizer and synthesizer engines are not part of the same session. In such configurations, when a kill-on-barge-in prompt is being played (see Section 14.8.4.2), the client wants the RECOGNIZE request to be simultaneously active so that it can detect and implement kill-on-barge-in. However, the recognizer SHOULD NOT start the no-input timers until the prompt is finished. The default value is "true."

```
start-input-timers    =  "Start-Input-Timers" ":" BOOLEAN CRLF
```

14.9.4.15 Speech-Complete-Timeout

This header field specifies the length of silence required following user speech before the speech recognizer finalizes a result (either accepting it or generating a no-match result). The Speech-Complete-Timeout value applies when the recognizer currently has a complete match against an active grammar and specifies how long the recognizer MUST wait for more input before declaring a match. By contrast, the Speech-Incomplete-Timeout is used when the speech is an incomplete match to an active grammar. The value is in milliseconds.

```
speech-complete-timeout    =    "Speech-Complete-Timeout" ":" 1*
                                19DIGIT CRLF
```

A long Speech-Complete-Timeout value delays the result to the client and therefore, makes the application's response to a user slow. A short Speech-Complete-Timeout may lead to an utterance being broken up inappropriately. Reasonable speech complete timeout values are typically in the range of 0.3 to 1.0 seconds. The value for this header field ranges from 0 to an implementation-specific maximum value. The default value for this header field is implementation specific. This header field MAY occur in RECOGNIZE, SET-PARAMS, or GET-PARAMS.

14.9.4.16 Speech-Incomplete-Timeout

This header field specifies the required length of silence following user speech after which a recognizer finalizes a result. The incomplete timeout applies when the speech prior to the silence is an incomplete match of all active grammars. In this case, once the timeout is triggered, the partial result is rejected (with a Completion-Cause of "partial-match"). The value is in milliseconds. The value for this header field ranges from 0 to an implementation-specific maximum value. The default value for this header field is implementation specific.

```
speech-incomplete-timeout     =     "Speech-Incomplete-Timeout" ":"
                                    1*19DIGIT CRLF
```

The Speech-Incomplete-Timeout also applies when the speech prior to the silence is a complete match of an active grammar but where it is possible to speak further and still match the grammar. By contrast, the Speech-Complete-Timeout is used when the speech is a complete match to an active grammar and no further spoken words can continue to represent a match.

A long Speech-Incomplete-Timeout value delays the result to the client and therefore, makes the application's response to a user slow. A short Speech-Incomplete-Timeout may lead to an utterance being broken up inappropriately. The Speech-Incomplete-Timeout is usually longer than the Speech-Complete-Timeout to allow users to pause mid-utterance (for example, to breathe). This header field MAY occur in RECOGNIZE, SET-PARAMS, or GET-PARAMS.

14.9.4.17 DTMF-Interdigit-Timeout

This header field specifies the interdigit timeout value to use when recognizing DTMF input. The value is in milliseconds. The value for this header field ranges from 0 to an implementation-specific maximum value. The default value is 5 seconds. This header field MAY occur in RECOGNIZE, SET-PARAMS, or GET-PARAMS.

```
dtmf-interdigit-timeout   =   "DTMF-Interdigit-Timeout" ":"
                              1*19DIGIT CRLF
```

14.9.4.18 DTMF-Term-Timeout

This header field specifies the terminating timeout to use when recognizing DTMF input. The DTMF-Term-Timeout applies only when no additional input is allowed by the grammar; otherwise, the DTMF-Interdigit-Timeout applies. The value is in milliseconds. The value for this header field ranges from 0 to an implementation-specific maximum value. The default value is 10 seconds. This header field MAY occur in RECOGNIZE, SET-PARAMS, or GET-PARAMS.

```
dtmf-term-timeout    =    "DTMF-Term-Timeout" ":"
                          1*19DIGIT CRLF
```

14.9.4.19 DTMF-Term-Char

This header field specifies the terminating DTMF character for DTMF input recognition. The default value is NULL, which is indicated by an empty header field value. This header field MAY occur in RECOGNIZE, SET-PARAMS, or GET-PARAMS.

```
dtmf-term-char    =    "DTMF-Term-Char" ":" VCHAR CRLF
```

14.9.4.20 Failed-URI

When a recognizer needs to fetch or access a URI and the access fails, the server SHOULD provide the failed URI in this header field in the method response, unless there are multiple URI failures, in which case one of the failed URIs MUST be provided in this header field in the method response.

```
failed-uri    =    "Failed-URI" ":" absoluteURI CRLF
```

14.9.4.21 Failed-URI-Cause

When a recognizer method needs a recognizer to fetch or access a URI and the access fails, the server MUST provide the URI-specific or protocol-specific response code for the URI in the Failed-URI header field through this header field in the method response. The value encoding is UTF-8 (RFC 3629) to accommodate any access protocol, some of which might have a response string instead of a numeric response code.

```
failed-uri-cause    =    "Failed-URI-Cause" ":" 1*UTFCHAR CRLF
```

14.9.4.22 Save-Waveform

This header field allows the client to request the recognizer resource to save the audio input to the recognizer. The recognizer resource MUST then attempt to record the recognized audio, without endpointing, and make it available to the client in the form of a URI returned in the Waveform-URI header field in the RECOGNITION-COMPLETE event. If there was an error in recording the stream, or the audio content is otherwise not available, the recognizer MUST return an empty Waveform-URI header field. The default value for this field is "false." This header field MAY occur in RECOGNIZE, SET-PARAMS, or GET-PARAMS. See the discussion on the sensitivity of saved waveforms in Section 14.12.

```
save-waveform    =    "Save-Waveform" ":" BOOLEAN CRLF
```

14.9.4.23 New-Audio-Channel

This header field MAY be specified in a RECOGNIZE request and allows the client to tell the server that from this point on, further input audio comes from a different audio source, channel, or speaker. If the recognizer resource had collected any input statistics or adaptation state, the recognizer resource MUST do what is appropriate for the specific recognition technology, which includes, but is not limited to, discarding any collected input statistics or adaptation state before starting the RECOGNIZE request. Note that if there are multiple resources that are sharing a media stream and are collecting or using this data, and the client issues this header field to one of the resources, the reset operation applies to all resources that use the shared media stream. This helps in a number of use cases, including where the client wishes to reuse an open recognition session with an existing media session for multiple telephone calls.

```
new-audio-channel    =    "New-Audio-Channel" ":" BOOLEAN
                          CRLF
```

14.9.4.24 Speech-Language

This header field specifies the language of recognition grammar data within a session or request, if it is not specified within the data. The value of this header field MUST follow RFC 5646 for its values. This MAY occur in DEFINE-GRAMMAR, RECOGNIZE, SET-PARAMS, or GET-PARAMS requests.

```
speech-language    =    "Speech-Language" ":" 1*VCHAR CRLF
```

14.9.4.25 Ver-Buffer-Utterance

This header field lets the client request the server to buffer the utterance associated with this recognition request into a buffer available to a co-resident verifier resource. The buffer is shared across resources within a session and is allocated when a verifier resource is added to this session. The client MUST NOT send this header field unless a verifier resource is instantiated for the session. The buffer is released when the verifier resource is released from the session.

14.9.4.26 Recognition-Mode

This header field specifies what mode the RECOGNIZE method will operate in. The value choices are "normal" or "hotword." If the value is "normal," the RECOGNIZE starts matching speech and DTMF to the grammars specified in the RECOGNIZE request. If any portion of the speech does not match the grammar, the RECOGNIZE command completes with a no-match status. Timers may be active to detect speech in the audio (see Section 14.9.4.14), so the RECOGNIZE method may complete because of a timeout waiting for speech. If the value of this header field is "hotword," the RECOGNIZE method operates in hotword mode, where it only looks for the particular keywords or DTMF sequences specified in the grammar and ignores silence or other speech in the audio stream. The default value for this header field is "normal." This header field MAY occur on the RECOGNIZE method.

```
recognition-mode    =    "Recognition-Mode" ":"
                         "normal" / "hotword" CRLF
```

14.9.4.27 Cancel-If-Queue

This header field specifies what will happen if the client attempts to invoke another RECOGNIZE method when this RECOGNIZE request is already in progress for the resource. The value for this header field is a Boolean. A value of "true" means the server MUST terminate this RECOGNIZE request, with a Completion-Cause of "cancelled," if the client issues another RECOGNIZE request for the same resource. A value of "false" for this header field indicates to the server that this RECOGNIZE request will continue to completion, and if the client issues more RECOGNIZE requests to the same resource, they are queued. When the currently active RECOGNIZE request is stopped or completes with a successful match, the first RECOGNIZE method in the queue becomes active. If the current RECOGNIZE fails, all RECOGNIZE methods in the pending queue are cancelled, and each generates a RECOGNITION-COMPLETE event with a Completion-Cause of "cancelled." This header field MUST be present in every RECOGNIZE request. There is no default value.

```
cancel-if-queue    =    "Cancel-If-Queue" ":" BOOLEAN CRLF
```

14.9.4.28 Hotword-Max-Duration

This header field MAY be sent in a hotword mode RECOGNIZE request. It specifies the maximum length of an utterance (in seconds) that will be considered for hotword recognition. This header field, along with Hotword-Min-Duration, can be used to tune performance by preventing the recognizer from evaluating utterances that are too short or too long to be one of the hotwords in the grammar(s). The value is in milliseconds. The default is implementation dependent. If present in a RECOGNIZE request specifying a mode other than "hotword," the header field is ignored.

```
hotword-max-duration    =    "Hotword-Max-Duration" ":" 1*19DIGIT
                             CRLF
```

14.9.4.29 Hotword-Min-Duration

This header field MAY be sent in a hotword mode RECOGNIZE request. It specifies the minimum length of an utterance (in seconds) that will be considered for hotword recognition. This header field, along with Hotword-Max-Duration, can be used to tune performance by preventing the recognizer from evaluating utterances that are too short or too long to be one of the hotwords in the grammar(s). The value is in milliseconds. The default value is implementation dependent. If present in a RECOGNIZE request specifying a mode other than "hotword", the header field is ignored.

```
hotword-min-duration     =     "Hotword-Min-Duration" ":"
                               1*19DIGIT CRLF
```

14.9.4.30 Interpret-Text

The value of this header field is used to provide a pointer to the text for which a natural language interpretation is desired. The value is either a URI or text. If the value is a URI, it MUST be a Content-ID that refers to an entity of type "text/plain" in the body of the message. Otherwise, the server MUST treat the value as the text to be interpreted. This header field MUST be used when invoking the INTERPRET method.

```
interpret-text    =    "Interpret-Text" ":" 1*VCHAR CRLF
```

14.9.4.31 DTMF-Buffer-Time

This header field MAY be specified in a GET-PARAMS or SET-PARAMS method and is used to specify the amount of time, in milliseconds, of the type-ahead buffer for the recognizer. This is the buffer that collects DTMF digits as they are pressed even when there is no RECOGNIZE command active. When a subsequent RECOGNIZE method is received, it MUST look to this buffer to match the RECOGNIZE request. If the digits in the buffer are not sufficient, then it can continue to listen to more digits to match the grammar. The default size of this DTMF buffer is platform specific.

```
dtmf-buffer-time    =    "DTMF-Buffer-Time" ":" 1*19DIGIT CRLF
```

14.9.4.32 Clear-DTMF-Buffer

This header field MAY be specified in a RECOGNIZE method and is used to tell the recognizer to clear the DTMF type-ahead buffer before starting the RECOGNIZE. The default value of this header field is "false," which does not clear the type-ahead buffer before starting the RECOGNIZE method. If this header field is specified to be "true," then the RECOGNIZE will clear the DTMF buffer before starting recognition. This means that digits pressed by the caller before the RECOGNIZE command was issued are discarded.

```
clear-dtmf-buffer    =    "Clear-DTMF-Buffer" ":" BOOLEAN CRLF
```

14.9.4.33 Early-No-Match

This header field MAY be specified in a RECOGNIZE method and is used to tell the recognizer that it MUST NOT wait for the end of speech before processing the collected speech to match active grammars. A value of "true" indicates that the recognizer MUST do early matching. The default value for this header field if not specified is "false." If the recognizer does not support the processing of the collected audio before the end of speech, this header field can be safely ignored.

```
early-no-match    =    "Early-No-Match" ":" BOOLEAN CRLF
```

14.9.4.34 Num-Min-Consistent-Pronunciations

This header field MAY be specified in a START-PHRASE-ENROLLMENT, SET-PARAMS, or GET-PARAMS method and is used to specify the minimum number of consistent pronunciations that must be obtained to voice enroll a new phrase. The minimum value is 1. The default value is implementation specific and MAY be greater than 1.

```
num-min-consistent-pronunciations  =
        "Num-Min-Consistent-Pronunciations" ":" 1*19DIGIT CRLF
```

14.9.4.35 Consistency-Threshold

This header field MAY be sent as part of the START-PHRASE-ENROLLMENT, SET-PARAMS, or GET-PARAMS method. Used during voice enrollment, this header field specifies how similar to a previously enrolled pronunciation of the same phrase an utterance needs to be in order to be considered "consistent." The higher the threshold, the closer the match between an utterance and previous pronunciations must be for the pronunciation to be considered consistent. The range for this threshold is a float value between 0.0 and 1.0. The default value for this header field is implementation specific.

```
consistency-threshold   =   "Consistency-Threshold" ":" FLOAT CRLF
```

14.9.4.36 Clash-Threshold

This header field MAY be sent as part of the START-PHRASE-ENROLLMENT, SET-PARAMS, or GET-PARAMS method. Used during voice enrollment, this header field specifies how similar the pronunciations of two different phrases can be before they are considered to be clashing. For example, pronunciations of phrases such as "John Smith" and "Jon Smits" may be so similar that they are difficult to distinguish correctly. A smaller threshold reduces the number of clashes detected. The range for this threshold is a float value between 0.0 and 1.0. The default value for this header field is implementation specific. Clash testing can be turned off completely by setting the Clash-Threshold header field value to 0.

```
clash-threshold   =   "Clash-Threshold" ":" FLOAT CRLF
```

14.9.4.37 Personal-Grammar-URI

This header field specifies the speaker-trained grammar to be used or referenced during enrollment operations. Phrases are added to this grammar during enrollment. For example, a contact list for user "Jeff" could be stored at the Personal-Grammar-URI "http://myserver.example.com/myenro llmentdb/jeff-list." The generated grammar syntax MAY be implementation specific. There is no default value for this header field. This header field MAY be sent as part of the START-PHRASE-ENROLLMENT, SET-PARAMS, or GET-PARAMS method.

```
personal-grammar-uri   =   "Personal-Grammar-URI" ":" uri CRLF
```

14.9.4.38 Enroll-Utterance

This header field MAY be specified in the RECOGNIZE method. If this header field is set to "true" and an Enrollment is active, the RECOGNIZE command MUST add the collected utterance to the personal grammar that is being enrolled. The way in which this occurs is engine specific and may be an area of future standardization. The default value for this header field is "false."

```
enroll-utterance   =   "Enroll-Utterance" ":" BOOLEAN CRLF
```

14.9.4.39 Phrase-Id

This header field in a request identifies a phrase in an existing personal grammar for which enrollment is desired. It is also returned to the client in the RECOGNIZE complete event. This header field MAY occur in START-PHRASE-ENROLLMENT, MODIFY-PHRASE, or DELETE-PHRASE requests. There is no default value for this header field.

```
phrase-id   =   "Phrase-ID" ":" 1*VCHAR CRLF
```

14.9.4.40 Phrase-NL

This string specifies the interpreted text to be returned when the phrase is recognized. This header field MAY occur in START-PHRASE-ENROLLMENT and MODIFY-PHRASE requests. There is no default value for this header field.

```
phrase-nl    =    "Phrase-NL" ":" 1*UTFCHAR CRLF
```

14.9.4.41 Weight

The value of this header field represents the occurrence likelihood of a phrase in an enrolled grammar. When using grammar enrollment, the system is essentially constructing a grammar segment consisting of a list of possible match phrases. This can be thought of to be similar to the dynamic construction of a <one-of> tag in the W3C grammar specification. Each enrolled phrase becomes an item in the list that can be matched against spoken input similar to the <item> within a <one-of> list. This header field allows you to assign a weight to the phrase (i.e., <item> entry) in the <one-of> list that is enrolled. Grammar weights are normalized to a sum of one at grammar compilation time, so a weight value of 1 for each phrase in an enrolled grammar list indicates all items in that list have the same weight. This header field MAY occur in START-PHRASE-ENROLLMENT and MODIFY-PHRASE requests. The default value for this header field is implementation specific.

```
weight    =    "Weight" ":" FLOAT CRLF
```

14.9.4.42 Save-Best-Waveform

This header field allows the client to request the recognizer resource to save the audio stream for the best repetition of the phrase that was used during the enrollment session. The recognizer MUST attempt to record the recognized audio and make it available to the client in the form of a URI returned in the Waveform-URI header field in the response to the END-PHRASE-ENROLLMENT method. If there was an error in recording the stream, or the audio data is otherwise not available, the recognizer MUST return an empty Waveform-URI header field. This header field MAY occur in the START-PHRASE-ENROLLMENT, SET-PARAMS, and GET-PARAMS methods.

```
save-best-waveform    =    "Save-Best-Waveform" ":" BOOLEAN CRLF
```

14.9.4.43 New-Phrase-Id

This header field replaces the ID used to identify the phrase in a personal grammar. The recognizer returns the new ID when using an enrollment grammar. This header field MAY occur in MODIFY-PHRASE requests.

```
new-phrase-id    =    "New-Phrase-ID" ":" 1*VCHAR CRLF
```

14.9.4.44 Confusable-Phrases-URI

This header field specifies a grammar that defines invalid phrases for enrollment. For example, typical applications do not allow an enrolled phrase that is also a command word. This header field MAY occur in RECOGNIZE requests that are part of an enrollment session.

```
confusable-phrases-uri    =    "Confusable-Phrases-URI" ":" uri CRLF
```

14.9.4.45 Abort-Phrase-Enrollment

This header field MAY be specified in the END-PHRASE-ENROLLMENT method to abort the phrase enrollment rather than committing the phrase to the personal grammar.

```
abort-phrase-enrollment    =    "Abort-Phrase-Enrollment" ":"
                                BOOLEAN CRLF
```

14.9.5 Recognizer Message Body

A recognizer message can carry additional data associated with the request, response, or event. The client MAY provide the grammar to be recognized in DEFINE-GRAMMAR or RECOGNIZE requests. When one or more grammars are specified using the DEFINE-GRAMMAR method, the server MUST attempt to fetch, compile, and optimize the grammar before returning a response to the DEFINE-GRAMMAR method. A RECOGNIZE request MUST completely specify the grammars to be active during the recognition operation except when the RECOGNIZE method is being used to enroll a grammar. During grammar enrollment, such grammars are OPTIONAL. The server resource sends the recognition results in the RECOGNITION-COMPLETE event and the GET-RESULT response. Grammars and recognition results are carried in the message body of the corresponding MRCPv2 messages.

14.9.5.1 Recognizer Grammar Data

Recognizer grammar data from the client to the server can be provided inline or by reference. Either way, grammar data is carried as typed media entities in the message body of the RECOGNIZE or DEFINE-GRAMMAR request. All MRCPv2 servers MUST accept grammars in the XML form (media type "application/srgs+xml") of the W3C's XML-based Speech Recognition Grammar Specification (SRGS) Markup Format [8] and MAY accept grammars in other formats. Examples include but are not limited to:

- the ABNF form (media type "application/srgs") of SRGS
- Sun's Java Speech Grammar Format (JSGF) [9]

Additionally, MRCPv2 servers MAY support the Semantic Interpretation for Speech Recognition (SISR) [10] specification. When a grammar is specified inline in the request, the client MUST provide a Content-ID for that grammar as part of the content header fields. If there is no space on the server to store the inline grammar, the request MUST return with a Completion-Cause code of 016 "grammar-definition-failure." Otherwise, the server MUST associate the inline grammar block with that Content-ID and MUST store it on the server for the duration of the session. However, if the Content-ID is redefined later in the session through a subsequent DEFINE-GRAMMAR, the inline grammar previously associated with the Content-ID MUST be freed. If the Content-ID is redefined through a subsequent DEFINE-GRAMMAR with an empty message body (i.e., no grammar definition), then in addition to freeing any grammar previously associated with the Content-ID, the server MUST clear all bindings and associations to the Content-ID. Unless and until subsequently redefined, this URI MUST be interpreted by the server as one that has never been set. Grammars that have been associated with a Content-ID can be referenced through the "session" URI scheme (see Section 14.13.6). For example:

```
session:help@root-level.store
```

Grammar data MAY be specified using external URI references. To do so, the client uses a body of media type "text/uri-list" (see RFC 2483) to list the one or more URIs that point to the grammar data. The client can use a body of media type "text/ grammar-ref-list" (see Section 14.13.5.1) if it wants to assign weights to the list of grammar URIs. All MRCPv2 servers MUST support grammar access using the "http" and "https" URI schemes. If the grammar data the client wishes to be used on a request consists of a mix of URI and inline grammar data, the client uses the "multipart/mixed" media type to enclose the "text/uri-list," "application/srgs," or "application/srgs+xml" content entities. The character set and encoding used in the grammar data are specified using standard media type definitions. When more than one grammar URI or inline grammar block is specified in a message body of the RECOGNIZE request, the server interprets this as a list of grammar alternatives to match against (Figures 14.18 and 14.19).

```
Content-Type:application/srgs+xml
Content-ID:<request1@form-level.store>
Content-Length:...

<?xml version="1.0"?>

<!-- the default grammar language is US English -->

<grammar xmlns="http://www.w3.org/2001/06/grammar"
        xml:lang="en-US" version="1.0" root="request">

    <!-- single language attachment to tokens -->

    <rule id="yes">
        <one-of>
            <item xml:lang="fr-CA">oui</item>
            <item xml:lang="en-US">yes</item>
        </one-of>
    </rule>

    <!-- single language attachment to a rule expansion -->

    <rule id="request">
            may I speak to
        <one-of xml:lang="fr-CA">
            <item>Michel Tremblay</item>
            <item>Andre Roy</item>
        </one-of>
    </rule>

    <!-- multiple language attachment to a token -->

    <rule id="people1">
        <token lexicon="en-US,fr-CA"> Robert </token>
    </rule>

    <!-- the equivalent single-language attachment expansion -->

    <rule id="people2">
        <one-of>
            <item xml:lang="en-US">Robert</item>
            <item xml:lang="fr-CA">Robert</item>
        </one-of>
    </rule>
</grammar>
```

FIGURE 14.18 SRGS Grammar Example.

```
Content-Type:text/uri-list
Content-Length:...

session:help@root-level.store
http://www.example.com/Directory-Name-List.grxml
http://www.example.com/Department-List.grxml
http://www.example.com/TAC-Contact-List.grxml
session:menu1@menu-level.store
Grammar Reference Example
Content-Type:multipart/mixed; boundary="break"

--break

Content-Type:text/uri-list
Content-Length:...
http://www.example.com/Directory-Name-List.grxml
http://www.example.com/Department-List.grxml
http://www.example.com/TAC-Contact-List.grxml

--break

Content-Type:application/srgs+xml
Content-ID:<request1@form-level.store>
Content-Length:...

<?xml version="1.0"?>

<!-- the default grammar language is US English -->

<grammar xmlns="http://www.w3.org/2001/06/grammar"
         xml:lang="en-US" version="1.0">

     <!-- single language attachment to tokens -->

     <rule id="yes">
          <one-of>
               <item xml:lang="fr-CA">oui</item>
               <item xml:lang="en-US">yes</item>
          </one-of>
     </rule>

     <!-- single language attachment to a rule expansion -->

     <rule id="request">
               may I speak to
          <one-of xml:lang="fr-CA">
               <item>Michel Tremblay</item>
               <item>Andre Roy</item>
          </one-of>
     </rule>

   <!-- multiple language attachment to a token -->

   <rule id="people1">
        <token lexicon="en-US,fr-CA"> Robert </token>
   </rule>

   <!-- the equivalent single-language attachment expansion -->

   <rule id="people2">
        <one-of>
             <item xml:lang="en-US">Robert</item>
             <item xml:lang="fr-CA">Robert</item>
        </one-of>
   </rule>
</grammar>

--break--
```

FIGURE 14.19 Mixed Grammar Reference Example.

```
Content-Type:application/nlsml+xml
Content-Length:...

<?xml version="1.0"?>

<result xmlns="urn:ietf:params:xml:ns:mrcpv2"
            xmlns:ex="http://www.example.com/example"
            grammar="http://www.example.com/theYesNoGrammar">
      <interpretation>
            <instance>
                  <ex:response>yes</ex:response>
            </instance>
            <input>OK</input>
      </interpretation>
</result>
```

FIGURE 14.20 Result Example.

14.9.5.2 Recognizer Result Data

Recognition (Figure 14.20) results are returned to the client in the message body of the RECOGNITION-COMPLETE event or the GET-RESULT response message as described in Section 14.6.3. Element and attribute descriptions for the recognition portion of the NLSML format are provided in Section 14.9.6 with a normative definition of the schema in Section 14.16.1.

14.9.5.3 Enrollment Result Data

Enrollment results are returned to the client in the message body of the RECOGNITION-COMPLETE event as described in Section 14.6.3. Element and attribute descriptions for the enrollment portion of the NLSML format are provided in Section 14.9.7 with a normative definition of the schema in Section 14.16.2.

14.9.5.4 Recognizer Context Block

When a client changes servers while operating on the behalf of the same incoming communication session, this header field allows the client to collect a block of opaque data from one server and provide it to another server. This capability is desirable if the client needs different language support or because the server issued a redirect. Here, the first recognizer resource may have collected acoustic and other data during its execution of recognition methods. After a server switch, communicating this data may allow the recognizer resource on the new server to provide better recognition. This block of data is implementation specific and MUST be carried as media type "application/octets" in the body of the message.

This block of data is communicated in the SET-PARAMS and GET-PARAMS method/response messages. In the GET-PARAMS method, if an empty Recognizer-Context-Block header field is present, then the recognizer SHOULD return its vendor-specific context block, if any, in the message body as an entity of media type "application/octets" with a specific Content-ID. The Content-ID value MUST also be specified in the Recognizer-Context-Block header field in the GET-PARAMS response. The SET-PARAMS request wishing to provide this vendor-specific data MUST send it in the message body as a typed entity with the same Content-ID that it received from the GET-PARAMS. The Content-ID MUST also be sent in the Recognizer-Context-Block header field of the SET-PARAMS message. Each speech recognition implementation choosing to use this mechanism to hand off recognizer context data among servers MUST distinguish its implementation-specific block of data from other implementations by choosing a Content-ID that is recognizable among the participating servers and unlikely to collide with values chosen by another implementation.

14.9.6 Recognizer Results

The recognizer portion of NLSML (see Section 14.6.3.1) represents information automatically extracted from a user's utterances by a semantic interpretation component, where "utterance" is to be taken in the general sense of a meaningful user input in any modality supported by the MRCPv2 implementation.

14.9.6.1 Markup Functions

MRCPv2 recognizer resources employ NLSML to interpret natural language speech input and to format the interpretation for consumption by an MRCPv2 client. The elements of the markup fall into the following general functional categories: interpretation, side information, and multimodal integration.

14.9.6.1.1 Interpretation

Elements and attributes represent the semantics of a user's utterance, including the <result>, <interpretation>, and <instance> elements. The <result> element contains the full result of processing one utterance. It MAY contain multiple <interpretation> elements if the interpretation of the utterance results in multiple alternative meanings due to uncertainty in speech recognition or natural language understanding. There are at least two reasons for providing multiple interpretations:

1. The client application might have additional information, for example, information from a database, that would allow it to select a preferred interpretation from among the possible interpretations returned from the semantic interpreter.
2. A client-based dialog manager (e.g., VoiceXML [8]) that was unable to select between several competing interpretations could use this information to go back to the user and find out what was intended. For example, it could issue a SPEAK request to a synthesizer resource to emit "Did you say 'Boston' or 'Austin'?"

14.9.6.1.2 Side Information

These are elements and attributes representing additional information about the interpretation, over and above the interpretation itself. Side information includes:

1. Whether an interpretation was achieved (the <nomatch> element) and the system's confidence in an interpretation (the "confidence" attribute of <interpretation>)
2. Alternative interpretations (<interpretation>)
3. Input formats and automatic speech recognition (ASR) information: The <input> element, representing the input to the semantic interpreter.

14.9.6.1.3 Multimodal Integration

When more than one modality is available for input, the interpretation of the inputs needs to be coordinated. The "mode" attribute of <input> supports this by indicating whether the utterance was input by speech, DTMF, pointing, etc. The "timestamp-start" and "timestamp-end" attributes of <input> also provide for temporal coordination by indicating when inputs occurred.

14.9.6.2 Overview of Recognizer Result Elements and Their Relationships

The recognizer elements in NLSML fall into two categories:

1. Description of the input that was processed
2. Description of the meaning that was extracted from the input

Next to each element are its attributes. In addition, some elements can contain multiple instances of other elements. For example, a <result> can contain multiple <interpretation> elements, each of which is taken to be an alternative. Similarly, <input> can contain multiple child <input> elements, which are taken to be cumulative. To illustrate the basic usage of these elements, as a simple example, consider the utterance "OK" (interpreted as "yes"). The example illustrates how that utterance and its interpretation would be represented in the NLSML markup.

```
<?xml version="1.0"?>

<result xmlns="urn:ietf:params:xml:ns:mrcpv2"
      xmlns:ex="http://www.example.com/example"
      grammar="http://www.example.com/theYesNoGrammar">
<interpretation>
   <instance>
      <ex:response>yes</ex:response>
   </instance>
   <input>OK</input>
</interpretation>
</result>
```

This example includes only the minimum required information. There is an overall <result> element, which includes one interpretation and an input element. The interpretation contains the application-specific element "<response>," which is the semantically interpreted result.

14.9.6.3 Elements and Attributes

14.9.6.3.1 <result> Root Element

The root element of the markup is <result>. The <result> element includes one or more <interpretation> elements. Multiple interpretations can result from ambiguities in the input or in the semantic interpretation. If the "grammar" attribute does not apply to all the interpretations in the result, it can be overridden for individual interpretations at the <interpretation> level.

Attributes:

1. **grammar:** The grammar or recognition rule matched by this result. The format of the grammar attribute will match the rule reference semantics defined in the grammar specification. Specifically, the rule reference is in the external XML form for grammar rule references. The markup interpreter needs to know the grammar rule that is matched by the utterance, because multiple rules may be simultaneously active. The value is the grammar URI used by the markup interpreter to specify the grammar. The grammar can be overridden by a grammar attribute in the <interpretation> element if the input was ambiguous as to which grammar it matched. If all interpretation elements within the result element contain their own grammar attributes, the attribute can be dropped from the result element.

```
<?xml version="1.0"?>

<result xmlns="urn:ietf:params:xml:ns:mrcpv2"
      grammar="http://www.example.com/grammar">
      <interpretation>
```

```
        ....
        </interpretation>
</result>
```

14.9.6.3.2 <interpretation> Element

An <interpretation> element contains a single semantic interpretation.

Attributes:

1. **confidence:** A float value from 0.0 to 1.0 indicating the semantic analyzer's confidence in this interpretation. A value of 1.0 indicates maximum confidence. The values are implementation dependent but are intended to align with the value interpretation for the confidence MRCPv2 header field defined in Section 14.9.4.1. This attribute is OPTIONAL.
2. **grammar:** The grammar or recognition rule matched by this interpretation (if needed to override the grammar specification at the <interpretation> level). This attribute is only needed under <interpretation> if it is necessary to override a grammar that was defined at the <result> level. Note that the grammar attribute for the interpretation element is optional if and only if the grammar attribute is specified in the <result> element.

Interpretations MUST be sorted best-first by some measure of "goodness." The goodness measure is "confidence" if present; otherwise, it is some implementation-specific indication of quality. The grammar is expected to be specified most frequently at the <result> level. However, it can be overridden at the <interpretation> level, because it is possible that different interpretations may match different grammar rules. The <interpretation> element includes an optional <input> element containing the input being analyzed and at least one <instance> element containing the interpretation of the utterance.

```
<interpretation confidence="0.75"
        grammar="http://www.example.com/grammar">
        ...
</interpretation>
```

14.9.6.3.3 <instance> Element

The <instance> element contains the interpretation of the utterance. When the SISR format is used, the <instance> element contains the XML serialization of the result using the approach defined in that specification. When there is semantic markup in the grammar that does not create semantic objects but instead, only does a semantic translation of a portion of the input, such as translating "coke" to "coca-cola," the instance contains the whole input but with the translation applied. The NLSML looks like the markup in Figure 14.21. If there are no semantic objects created, nor any semantic translation, the instance value is the same as the input value.

Attributes:

a. confidence: Each element of the instance MAY have a confidence attribute, defined in the NLSML namespace. The confidence attribute contains a float value in the range from 0.0 to 1.0 reflecting the system's confidence in the analysis of that slot. A value of 1.0 indicates maximum confidence. The values are implementation dependent but are intended to align with the value interpretation for the MRCPv2 header field Confidence-Threshold defined in Section 14.9.4.1. This attribute is OPTIONAL.

14.9.6.3.4 <input> Element

The <input> element is the text representation of a user's input. It includes an optional "confidence" attribute, which indicates the recognizer's confidence in the recognition result (as opposed to the

```
<instance>
        <nameAddress>
                <street confidence="0.75">123 Maple Street</street>
                <city>Mill Valley</city>
                <state>CA</state>
                <zip>90952</zip>
        </nameAddress>
</instance>
<input>
        My address is 123 Maple Street,
        Mill Valley, California, 90952
</input>
<instance>
        I would like to buy a coca-cola
</instance>
<input>
        I would like to buy a coke
</input>
```

FIGURE 14.21 NSLML Example.

confidence in the interpretation, which is indicated by the "confidence" attribute of <interpretation>). Optional "timestamp-start" and "timestamp-end" attributes indicate the start and end times of a spoken utterance in ISO 8601 format [11].

Attributes:

1. timestamp-start: The time at which the input began (optional)
2. timestamp-end: The time at which the input ended (optional)
3. mode: The modality of the input, for example, speech, DTMF, etc. (optional)
4. confidence: The confidence of the recognizer in the correctness of the input in the range 0.0 to 1.0 (optional)

Note that it may not make sense for temporally overlapping inputs to have the same mode; however, this constraint is not expected to be enforced by implementations. When there is no time zone designator, ISO 8601 time representations default to local time. There are three possible formats for the <input> element.

1. The <input> element can contain simple text:

```
<input>onions</input>
```

A future possibility is for <input> to contain not only text but additional markup representing prosodic information that was contained in the original utterance and extracted by the speech recognizer. This depends on the availability of ASRs that are capable of producing prosodic information. MRCPv2 clients MUST be prepared to receive such markup and MAY make use of it.

2. An <input> tag can also contain additional <input> tags. Having additional input elements allows the representation to support future multimodal inputs as well as finer-grained speech information, such as timestamps for individual words and word-level confidences.

```
<input>
   <input mode="speech" confidence="0.5"
         timestamp-start="2000-04-03T0:00:00"
         timestamp-end="2000-04-03T0:00:00.2">fried</input>
   <input mode="speech" confidence="1.0"
         timestamp-start="2000-04-03T0:00:00.25"
         timestamp-end="2000-04-03T0:00:00.6">onions</input>
</input>
```

3. Finally, the <input> element can contain <nomatch> and <noinput> elements, which describe situations in which the speech recognizer received input that it was unable to process or did not receive any input at all, respectively.

14.9.6.3.5 <nomatch> Element

The <nomatch> element under <input> is used to indicate that the semantic interpreter was unable to successfully match any input with confidence above the threshold. It can optionally contain the text of the best of the (rejected) matches.

```
<interpretation>
   <instance/>
   <input confidence="0.1">
      <nomatch/>
   </input>
   </interpretation>
   <interpretation>
      <instance/>
      <input mode="speech" confidence="0.1">
         <nomatch>I want to go to New York</nomatch>
      </input>
</interpretation>
```

14.9.6.3.6 <noinput> Element

<noinput> indicates that there was no input – a timeout occurred in the speech recognizer due to silence.

```
<interpretation>
   <instance/>
      <input>
         <noinput/>
      </input>
</interpretation>
```

If there are multiple levels of inputs, the most natural place for <nomatch> and <noinput> elements to appear is under the highest level of <input> for <noinput> and under the appropriate level of <interpretation> for <nomatch>. So, <noinput> means "no input at all," and <nomatch> means "no match in speech modality" or "no match in DTMF modality." For example, to represent garbled speech combined with DTMF "1 2 3 4," the markup would be:

```
<input>
    <input mode="speech"><nomatch/></input>
    <input mode="dtmf">1 2 3 4</input>
</input>
```

Note: while <noinput> could be represented as an attribute of input, <nomatch> cannot, since it could potentially include PCDATA content with the best match. For parallelism, <noinput> is also an element.

14.9.7 Enrollment Results

All enrollment elements are contained within a single <enrollment-result> element under <result>. The elements are described in the following and have the schema defined in Section 16.2. The following elements are defined:

1. num-clashes
2. num-good-repetitions
3. num-repetitions-still-needed
4. consistency-status
5. clash-phrase-ids
6. transcriptions
7. confusable-phrases

14.9.7.1 <num-clashes> Element

The <num-clashes> element contains the number of clashes that this pronunciation has with other pronunciations in an active enrollment session. The associated Clash-Threshold header field determines the sensitivity of the clash measurement. Note that clash testing can be turned off completely by setting the Clash-Threshold header field value to 0.

14.9.7.2 <num-good-repetitions> Element

The <num-good-repetitions> element contains the number of consistent pronunciations obtained so far in an active enrollment session.\

14.9.7.3 <num-repetitions-still-needed> Element

The <num-repetitions-still-needed> element contains the number of consistent pronunciations that must still be obtained before the new phrase can be added to the enrollment grammar. The number of consistent pronunciations required is specified by the client in the request header field Num-Min-Consistent-Pronunciations. The returned value must be 0 before the client can successfully commit a phrase to the grammar by ending the enrollment session.

14.9.7.4 <consistency-status> Element

The <consistency-status> element is used to indicate how consistent the repetitions are when learning a new phrase. It can have the values of consistent, inconsistent, and undecided.

14.9.7.5 <clash-phrase-ids> Element

The <clash-phrase-ids> element contains the phrase IDs of clashing pronunciation(s), if any. This element is absent if there are no clashes.

14.9.7.6 <transcriptions> Element

The <transcriptions> element contains the transcriptions returned in the last repetition of the phrase being enrolled.

14.9.7.7 <confusable-phrases> Element

The <confusable-phrases> element contains a list of phrases from a command grammar that are confusable with the phrase being added to the personal grammar. This element MAY be absent if there are no confusable phrases.

14.9.8 DEFINE-GRAMMAR

The DEFINE-GRAMMAR method, from the client to the server, provides one or more grammars and requests the server to access, fetch, and compile the grammars as needed. The DEFINE-GRAMMAR method implementation MUST do a fetch of all external URIs that are part of that operation. If caching is implemented, this URI fetching MUST conform to the cache control hints and parameter header fields associated with the method in deciding whether the URIs should be fetched from cache or from the external server. If these hints/parameters are not specified in the method, the values set for the session using SET-PARAMS/GET-PARAMS apply. If it was not set for the session, their default values apply.

If the server resource is in the recognition state, the DEFINEGRAMMAR request MUST respond with a failure status. If the resource is in the idle state and is able to successfully process the supplied grammars, the server MUST return a success code status, and the request-state MUST be COMPLETE. If the recognizer resource could not define the grammar for some reason (for example, if the download failed, the grammar failed to compile, or the grammar was in an unsupported form), the MRCPv2 response for the DEFINE-GRAMMAR method MUST contain a failure status-code of 407 and contain a Completion-Cause header field describing the failure reason.

```
C->S: MRCP/2.0 ... DEFINE-GRAMMAR 543257
      Channel-Identifier:32AECB23433801@speechrecog
      Content-Type:application/srgs+xml
      Content-ID:<request1@form-level.store>
      Content-Length:...

      <?xml version="1.0"?>

      <!-- the default grammar language is US English -->

      <grammar xmlns="http://www.w3.org/2001/06/grammar"
            xml:lang="en-US" version="1.0">

         <!-- single language attachment to tokens -->

         <rule id="yes">
            <one-of>
               <item xml:lang="fr-CA">oui</item>
               <item xml:lang="en-US">yes</item>
            </one-of>
         </rule>

         <!-- single language attachment to a rule expansion -->

         <rule id="request">
               may I speak to
            <one-of xml:lang="fr-CA">
               <item>Michel Tremblay</item>
               <item>Andre Roy</item>
            </one-of>
         </rule>
      </grammar>
```

```
S->C: MRCP/2.0 ... 543257 200 COMPLETE
      Channel-Identifier:32AECB23433801@speechrecog
      Completion-Cause:000 success

C->S: MRCP/2.0 ... DEFINE-GRAMMAR 543258
      Channel-Identifier:32AECB23433801@speechrecog
      Content-Type:application/srgs+xml
      Content-ID:<helpgrammar@root-level.store>
      Content-Length:...

      <?xml version="1.0"?>

      <!-- the default grammar language is US English -->

      <grammar xmlns="http://www.w3.org/2001/06/grammar"
            xml:lang="en-US" version="1.0">

         <rule id="request">
            I need help
         </rule>
      </grammar>

S->C: MRCP/2.0 ... 543258 200 COMPLETE
      Channel-Identifier:32AECB23433801@speechrecog
      Completion-Cause:000 success

C->S: MRCP/2.0 ... DEFINE-GRAMMAR 543259
      Channel-Identifier:32AECB23433801@speechrecog
      Content-Type:application/srgs+xml
      Content-ID:<request2@field-level.store>
      Content-Length:...

      <?xml version="1.0" encoding="UTF-8"?>

      <!DOCTYPE grammar PUBLIC "-//W3C//DTD GRAMMAR 1.0//EN"
            "http://www.w3.org/TR/speech-grammar/grammar.dtd">
      <grammar xmlns="http://www.w3.org/2001/06/grammar" xml:lang="en"
            xmlns:xsi="http://www.w3.org/2001/XMLSchema-instance"
            xsi:schemaLocation="http://www.w3.org/2001/06/grammar
            http://www.w3.org/TR/speech-grammar/grammar.xsd"
            version="1.0" mode="voice" root="basicCmd">
         <meta name="author" content="Stephanie Williams"/>
         <rule id="basicCmd" scope="public">
            <example> please move the window </example>
            <example> open a file </example>
               <ruleref
                     uri="http://grammar.example.com/politeness.
                     grxml#startPolite"/>

               <ruleref uri="#command"/>
               <ruleref
                     uri="http://grammar.example.com/
                     politeness.grxml#endPolite"/>
         </rule>
         <rule id="command">
            <ruleref uri="#action"/> <ruleref uri="#object"/>
         </rule>
```

```
        <rule id="action">
            <one-of>
                <item weight="10"> open <tag>open</tag> </item>
                <item weight="2"> close <tag>close</tag> </item>
                <item weight="1"> delete <tag>delete</tag> </item>
                <item weight="1"> move <tag>move</tag> </item>
            </one-of>
        </rule>
        <rule id="object">
            <item repeat="0-1">
                <one-of>
                    <item> the </item>
                    <item> a </item>
                </one-of>
            </item>
            <one-of>
                <item> window </item>
                <item> file </item>
                <item> menu </item>
            </one-of>
        </rule>
    </grammar>

S->C: MRCP/2.0 ... 543259 200 COMPLETE
      Channel-Identifier:32AECB23433801@speechrecog
      Completion-Cause:000 success

C->S: MRCP/2.0 ... RECOGNIZE 543260
      Channel-Identifier:32AECB23433801@speechrecog
      N-Best-List-Length:2
      Content-Type:text/uri-list
      Content-Length:...

      session:request1@form-level.store
      session:request2@field-level.store
      session:helpgramar@root-level.store

S->C: MRCP/2.0 ... 543260 200 IN-PROGRESS
      Channel-Identifier:32AECB23433801@speechrecog

S->C: MRCP/2.0 ... START-OF-INPUT 543260 IN-PROGRESS
      Channel-Identifier:32AECB23433801@speechrecog

S->C: MRCP/2.0 ... RECOGNITION-COMPLETE 543260 COMPLETE
      Channel-Identifier:32AECB23433801@speechrecog
      Completion-Cause:000 success
      Waveform-URI:<http://web.media.com/session123/audio.wav>;
      size=124535;duration=2340
      Content-Type:application/x-nlsml
      Content-Length:...

      <?xml version="1.0"?>

      <result xmlns="urn:ietf:params:xml:ns:mrcpv2"
            xmlns:ex="http://www.example.com/example"
            grammar="session:request1@form-level.store">
        <interpretation>
```

```
        <instance name="Person">
            <ex:Person>
                <ex:Name> Andre Roy </ex:Name>
            </ex:Person>
        </instance>
        <input> may I speak to Andre Roy </input>
    </interpretation>
</result>
```

DEFINE GRAMMAR EXAMPLE

14.9.9 RECOGNIZE

The RECOGNIZE method from the client to the server requests the recognizer to start recognition and provides it with one or more grammar references for grammars to match against the input media. The RECOGNIZE method can carry header fields to control the sensitivity, confidence level, and level of detail in results provided by the recognizer. These header field values override the current values set by a previous SET-PARAMS method. The RECOGNIZE method can request the recognizer resource to operate in normal or hotword mode as specified by the Recognition-Mode header field. The default value is "normal." If the resource could not start a recognition, the server MUST respond with a failure status-code of 407 and a Completion-Cause header field in the response describing the cause of failure.

The RECOGNIZE request uses the message body to specify the grammars applicable to the request. The active grammar(s) for the request can be specified in one of three ways. If the client needs to explicitly control grammar weights for the recognition operation, it MUST employ method 3 in the following. The order of these grammars specifies the precedence of the grammars that is used when more than one grammar in the list matches the speech; in this case, the grammar with the higher precedence is returned as a match. This precedence capability is useful in applications like VoiceXML browsers to order grammars specified at the dialog, specification, and root level of a VoiceXML application.

1. The grammar MAY be placed directly in the message body as typed content. If more than one grammar is included in the body, the order of inclusion controls the corresponding precedence for the grammars during recognition, with earlier grammars in the body having a higher precedence than later ones.
2. The body MAY contain a list of grammar URIs specified in content of media type "text/uri-list" (RFC 2483). The order of the URIs determines the corresponding precedence for the grammars during recognition, with highest precedence first and decreasing for each URI thereafter.
3. The body MAY contain a list of grammar URIs specified in content of media type "text/grammar-ref-list." This type defines a list of grammar URIs and allows each grammar URI to be assigned a weight in the list. This weight has the same meaning as the weights described in Section 14.2.4.1 of the SRGS [8].

In addition to performing recognition on the input, the recognizer MUST also enroll the collected utterance in a personal grammar if the Enroll-Utterance header field is set to true and an Enrollment is active (via an earlier execution of the START-PHRASE-ENROLLMENT method). If so, and if the RECOGNIZE request contains a Content-ID header field, then the resulting grammar (which includes the personal grammar as a sub-grammar) can be referenced through the "session" URI scheme (see Section 14.13.6). If the resource was able to successfully start the recognition, the server MUST return a success status-code and a request-state of IN-PROGRESS. This means that the recognizer is active and that the client MUST be prepared to receive further events with this request-id.

If the resource was able to queue the request, the server MUST return a success code and request-state of PENDING. This means that the recognizer is currently active with another request and that this request has been queued for processing. If the resource could not start a recognition, the server MUST respond with a failure status-code of 407 and a Completion-Cause header field in the response describing the cause of failure. For the recognizer resource, RECOGNIZE and INTERPRET are the only requests that return a request-state of IN-PROGRESS, meaning that recognition is in progress. When the recognition completes by matching one of the grammar alternatives or by a timeout without a match or for some other reason, the recognizer resource MUST send the client a RECOGNITION-COMPLETE event (or INTERPRETATION-COMPLETE, if INTERPRET was the request) with the result of the recognition and a request-state of COMPLETE.

Large grammars can take a long time for the server to compile. For grammars that are used repeatedly, the client can improve server performance by issuing a DEFINE-GRAMMAR request with the grammar ahead of time. In such a case, the client can issue the RECOGNIZE request and reference the grammar through the "session" URI scheme (see Section 13.6). This also applies in general if the client wants to repeat recognition with a previous inline grammar. The RECOGNIZE method implementation MUST do a fetch of all external URIs that are part of that operation. If caching is implemented, this URI fetching MUST conform to the cache control hints and parameter header fields associated with the method in deciding whether it should be fetched from cache or from the external server. If these hints/parameters are not specified in the method, the values set for the session using SET-PARAMS/GET-PARAMS apply. If it was not set for the session, their default values apply.

Note that since the audio and the messages are carried over separate communication paths, there may be a race condition between the start of the flow of audio and the receipt of the RECOGNIZE method. For example, if an audio flow is started by the client at the same time as the RECOGNIZE method is sent, either the audio or the RECOGNIZE can arrive at the recognizer first. As another example, the client may choose to continuously send audio to the server and signal the server to recognize using the RECOGNIZE method. Mechanisms to resolve this condition are outside the scope of this specification. The recognizer can expect the media to start flowing when it receives the RECOGNIZE request, but it MUST NOT buffer anything it receives beforehand in order to preserve the semantics that application authors expect with respect to the input timers.

When a RECOGNIZE method has been received, the recognition is initiated on the stream. The No-Input-Timer MUST be started at this time if the Start-Input-Timers header field is specified as "true." If this header field is set to "false," the No-Input-Timer MUST be started when it receives the START-INPUT-TIMERS method from the client. The Recognition-Timeout MUST be started when the recognition resource detects speech or a DTMF digit in the media stream.

For recognition when not in hotword mode:

When the recognizer resource detects speech or a DTMF digit in the media stream, it MUST send the START-OF-INPUT event. When enough speech has been collected for the server to process, the recognizer can try to match the collected speech with the active grammars. If the speech collected at this point fully matches with any of the active grammars, the Speech-Complete-Timer is started. If it matches partially with one or more of the active grammars, with more speech needed before a full match is achieved, then the Speech-Incomplete-Timer is started.

1. When the No-Input-Timer expires, the recognizer MUST complete with a Completion-Cause code of "no-input-timeout."
2. The recognizer MUST support detecting a no-match condition upon detecting end of speech. The recognizer MAY support detecting a no-match condition before waiting for end-of-speech. If this is supported, this capability is enabled by setting the Early-No-Match header field to "true." Upon detecting a no-match condition, the RECOGNIZE MUST return with "no-match."
3. When the Speech-Incomplete-Timer expires, the recognizer SHOULD complete with a Completion-Cause code of "partial-match" unless the recognizer cannot differentiate a

partial-match, in which case it MUST return a Completion-Cause code of "no-match." The recognizer MAY return results for the partially matched grammar.
4. When the Speech-Complete-Timer expires, the recognizer MUST complete with a Completion-Cause code of "success."
5. When the Recognition-Timeout expires, one of the following MUST happen:
   a. If there was a partial-match, the recognizer SHOULD complete with a Completion-Cause code of "partial-matchmaxtime" unless the recognizer cannot differentiate a partial-match, in which case it MUST complete with a Completion-Cause code of "no-match-maxtime." The recognizer MAY return results for the partially matched grammar.
   b. If there was a full-match, the recognizer MUST complete with a Completion-Cause code of "success-maxtime."
   c. If there was a no match, the recognizer MUST complete with a Completion-Cause code of "no-match-maxtime."

For recognition in hotword mode:

Note that for recognition in hotword mode, the START-OF-INPUT event is not generated when speech or a DTMF digit is detected.

1. When the No-Input-Timer expires, the recognizer MUST complete with a Completion-Cause code of "no-input-timeout."
2. If at any point a match occurs, the RECOGNIZE MUST complete with a Completion-Cause code of "success."
3. When the Recognition-Timeout expires and there is not a match, the RECOGNIZE MUST complete with a Completion-Cause code of "hotword-maxtime."
4. When the Recognition-Timeout expires and there is a match, the RECOGNIZE MUST complete with a Completion-Cause code of "success-maxtime."
5. When the Recognition-Timeout is running but the detected speech/ DTMF has not resulted in a match, the Recognition-Timeout MUST be stopped and reset. It MUST then be restarted when speech/DTMF is again detected.

The following is a complete example of using RECOGNIZE. It shows the call to RECOGNIZE, the IN-PROGRESS and START-OF-INPUT status messages, and the final RECOGNITION-COMPLETE message containing the result.

```
C->S:  MRCP/2.0 ... RECOGNIZE 543257
       Channel-Identifier:32AECB23433801@speechrecog
       Confidence-Threshold:0.9
       Content-Type:application/srgs+xml
       Content-ID:<request1@form-level.store>
       Content-Length:...

       <?xml version="1.0"?>

       <!-- the default grammar language is US English -->

       <grammar xmlns="http://www.w3.org/2001/06/grammar"
             xml:lang="en-US" version="1.0" root="request">
          <!-- single language attachment to tokens -->

          <rule id="yes">
             <one-of>
                <item xml:lang="fr-CA">oui</item>
                <item xml:lang="en-US">yes</item>
             </one-of>
```

```
        </rule>

        <!-- single language attachment to a rule expansion -->

        <rule id="request">
           may I speak to
              <one-of xml:lang="fr-CA">
                 <item>Michel Tremblay</item>
                 <item>Andre Roy</item>
              </one-of>
        </rule>
     </grammar>

S->C: MRCP/2.0 ... 543257 200 IN-PROGRESS
      Channel-Identifier:32AECB23433801@speechrecog

S->C: MRCP/2.0 ... START-OF-INPUT 543257 IN-PROGRESS
      Channel-Identifier:32AECB23433801@speechrecog

S->C: MRCP/2.0 ... RECOGNITION-COMPLETE 543257 COMPLETE
      Channel-Identifier:32AECB23433801@speechrecog
      Completion-Cause:000 success
      Waveform-URI:<http://web.media.com/session123/audio.wav>;
      size=424252;duration=2543
      Content-Type:application/nlsml+xml
      Content-Length:...

      <?xml version="1.0"?>

      <result xmlns="urn:ietf:params:xml:ns:mrcpv2"
            xmlns:ex="http://www.example.com/example"
            grammar="session:request1@form-level.store">
         <interpretation>
            <instance name="Person">
               <ex:Person>
                  <ex:Name> Andre Roy </ex:Name>
               </ex:Person>
            </instance>
            <input> may I speak to Andre Roy </input>
         </interpretation>
      </result>
```

The following is an example of calling RECOGNIZE with a different grammar.

No status or completion messages are shown in this example, although they would of course occur in normal usage.

```
C->S: MRCP/2.0 ... RECOGNIZE 543257
      Channel-Identifier:32AECB23433801@speechrecog
      Confidence-Threshold:0.9
      Fetch-Timeout:20
      Content-Type:application/srgs+xml
      Content-Length:...

      <?xml version="1.0"? Version="1.0" mode="voice"
            root="Basic md">
         <rule id="rule_list" scope="public">
            <one-of>
```

```
            <item weight=10>
                <ruleref uri=
                        "http://grammar.example.com/world-cities.grxml
                        #canada"/>
            </item>
            <item weight=1.5>
                <ruleref uri=
                        "http://grammar.example.com/world-cities.grxml
                        #america"/>
            </item>
            <item weight=0.5>
                <ruleref uri=
                        "http://grammar.example.com/world-cities.grxml
                        #india"/>
            </item>
        </one-of>
    </rule>
```

14.9.10 STOP

The STOP method from the client to the server tells the resource to stop recognition if a request is active. If a RECOGNIZE request is active and the STOP request successfully terminated it, then the response header section contains an Active-Request-Id-List header field containing the request-id of the RECOGNIZE request that was terminated. In this case, no RECOGNITION-COMPLETE event is sent for the terminated request. If there was no recognition active, then the response MUST NOT contain an Active-Request-Id-List header field. Either way, the response MUST contain a status-code of 200 "Success."

```
C->S: MRCP/2.0 ... RECOGNIZE 543257
      Channel-Identifier:32AECB23433801@speechrecog
      Confidence-Threshold:0.9
      Content-Type:application/srgs+xml
      Content-ID:<request1@form-level.store>
      Content-Length:...

      <?xml version="1.0"?>

      <!-- the default grammar language is US English -->

      <grammar xmlns="http://www.w3.org/2001/06/grammar"
            xml:lang="en-US" version="1.0" root="request">
         <!-- single language attachment to tokens -->

            <rule id="yes">
               <one-of>
                  <item xml:lang="fr-CA">oui</item>
                  <item xml:lang="en-US">yes</item>
               </one-of>
            </rule>

         <!-- single language attachment to a rule expansion -->

         <rule id="request">
               may I speak to
            <one-of xml:lang="fr-CA">
               <item>Michel Tremblay</item>
```

```
            <item>Andre Roy</item>
         </one-of>
      </rule>
   </grammar>

S->C: MRCP/2.0 ... 543257 200 IN-PROGRESS
      Channel-Identifier:32AECB23433801@speechrecog

C->S: MRCP/2.0 ... STOP 543258 200
      Channel-Identifier:32AECB23433801@speechrecog

S->C: MRCP/2.0 ... 543258 200 COMPLETE
      Channel-Identifier:32AECB23433801@speechrecog
      Active-Request-Id-List:543257
```

14.9.11 GET-RESULT

The GET-RESULT method from the client to the server MAY be issued when the recognizer resource is in the recognized state. This request allows the client to retrieve results for a completed recognition. This is useful if the client decides it wants more alternatives or more information. When the server receives this request, it re-computes and returns the results according to the recognition constraints provided in the GET-RESULT request. The GET-RESULT request can specify constraints such as a different confidence-threshold or n-best-list-length. This capability is OPTIONAL for MRCPv2 servers, and the automatic speech recognition engine in the server MUST return a status of unsupported feature if it is not supported.

```
C->S: MRCP/2.0 ... GET-RESULT 543257
      Channel-Identifier:32AECB23433801@speechrecog
      Confidence-Threshold:0.9

S->C: MRCP/2.0 ... 543257 200 COMPLETE
      Channel-Identifier:32AECB23433801@speechrecog
      Content-Type:application/nlsml+xml
      Content-Length:...

      <?xml version="1.0"?>

      <result xmlns="urn:ietf:params:xml:ns:mrcpv2"
            xmlns:ex="http://www.example.com/example"
            grammar="session:request1@form-level.store">
         <interpretation>
            <instance name="Person">
               <ex:Person>
                  <ex:Name> Andre Roy </ex:Name>
               </ex:Person>
            </instance>
            <input> may I speak to Andre Roy </input>
         </interpretation>
      </result>
```

14.9.12 START-OF-INPUT

This is an event from the server to the client indicating that the recognizer resource has detected speech or a DTMF digit in the media stream. This event is useful in implementing kill-on-barge-in scenarios when a synthesizer resource is in a different session from the recognizer resource and hence, is not aware of an incoming audio source (see Section 14.8.4.2). In these cases, it is up to the client to act as an intermediary and respond to this event by issuing a BARGE-IN-OCCURRED event to the synthesizer resource. The recognizer resource also MUST send a Proxy-Sync-Id header field with a unique value for this event. This event MUST be generated by the server, irrespective of whether or not the synthesizer and recognizer are on the same server.

14.9.13 START-INPUT-TIMERS

This request is sent from the client to the recognizer resource when it knows that a kill-on-barge-in prompt has finished playing (see Section 14.8.4.2). This is useful in the scenario when the recognition and synthesizer engines are not in the same session. When a kill-on-barge-in prompt is being played, the client may want a RECOGNIZE request to be simultaneously active so that it can detect and implement kill-on-barge-in. But at the same time, the client does not want the recognizer to start the no-input timers until the prompt is finished. The Start-Input-Timers header field in the RECOGNIZE request allows the client to say whether or not the timers should be started immediately. If not, the recognizer resource MUST NOT start the timers until the client sends a START-INPUT-TIMERS method to the recognizer.

14.9.14 RECOGNITION-COMPLETE

This is an event from the recognizer resource to the client indicating that the recognition completed. The recognition result is sent in the body of the MRCPv2 message. The request-state field MUST be COMPLETE, indicating that this is the last event with that request-id and that the request with that request-id is now complete. The server MUST maintain the recognizer context containing the results and the audio waveform input of that recognition until the next RECOGNIZE request is issued for that resource or the session terminates. If the server returns a URI to the audio waveform, it MUST do so in a Waveform-URI header field in the RECOGNITION-COMPLETE event. The client can use this URI to retrieve or play back the audio. Note: If an enrollment session was active, the RECOGNITION-COMPLETE event can contain either recognition or enrollment results depending on what was spoken. The following example shows a complete exchange with a recognition result.

```
C->S: MRCP/2.0 ... RECOGNIZE 543257
       Channel-Identifier:32AECB23433801@speechrecog
       Confidence-Threshold:0.9
       Content-Type:application/srgs+xml
       Content-ID:<request1@form-level.store>
       Content-Length:...

       <?xml version="1.0"?>

       <!-- the default grammar language is US English -->

       <grammar xmlns="http://www.w3.org/2001/06/grammar"
             xml:lang="en-US" version="1.0" root="request">

       <!-- single language attachment to tokens -->

          <rule id="yes">
             <one-of>
                <item xml:lang="fr-CA">oui</item>
                <item xml:lang="en-US">yes</item>
             </one-of>
          </rule>

          <!-- single language attachment to a rule expansion -->
```

```
        <rule id="request">
              may I speak to
           <one-of xml:lang="fr-CA">
              <item>Michel Tremblay</item>
              <item>Andre Roy</item>
           </one-of>
        </rule>
     </grammar>

S->C: MRCP/2.0 ... 543257 200 IN-PROGRESS
      Channel-Identifier:32AECB23433801@speechrecog

S->C: MRCP/2.0 ... START-OF-INPUT 543257 IN-PROGRESS
      Channel-Identifier:32AECB23433801@speechrecog

S->C: MRCP/2.0 ... RECOGNITION-COMPLETE 543257 COMPLETE
      Channel-Identifier:32AECB23433801@speechrecog
      Completion-Cause:000 success
      Waveform-URI:<http://web.media.com/session123/audio.wav>;
      size=342456;duration=25435
      Content-Type:application/nlsml+xml
      Content-Length:...

      <?xml version="1.0"?>
      <result xmlns="urn:ietf:params:xml:ns:mrcpv2"
            xmlns:ex="http://www.example.com/example"
            grammar="session:request1@form-level.store">
         <interpretation>
            <instance name="Person">
               <ex:Person>
                  <ex:Name> Andre Roy </ex:Name>
               </ex:Person>
            </instance>
            <input> may I speak to Andre Roy </input>
         </interpretation>
      </result>
```

If the result had instead been an enrollment result, the final message from the server could have been:

```
S->C: MRCP/2.0 ... RECOGNITION-COMPLETE 543257 COMPLETE
      Channel-Identifier:32AECB23433801@speechrecog
      Completion-Cause:000 success
      Content-Type:application/nlsml+xml
      Content-Length:...

      <?xml version= "1.0"?>

      <result xmlns="urn:ietf:params:xml:ns:mrcpv2"
            grammar="Personal-Grammar-URI">
         <enrollment-result>
            <num-clashes> 2 </num-clashes>
            <num-good-repetitions> 1 </num-good-repetitions>
            <num-repetitions-still-needed>
               1
            </num-repetitions-still-needed>
            <consistency-status> consistent </consistency-status>
```

```
        <clash-phrase-ids>
            <item> Jeff </item> <item> Andre </item>
        </clash-phrase-ids>
        <transcriptions>
        <item> m ay b r ow k er </item>
        <item> m ax r aa k ah </item>
        </transcriptions>
        <confusable-phrases>
            <item>
                <phrase> call </phrase>
                <confusion-level> 10 </confusion-level>
            </item>
        </confusable-phrases>
    </enrollment-result>
</result>
```

14.9.15 START-PHRASE-ENROLLMENT

The START-PHRASE-ENROLLMENT method from the client to the server starts a new phrase enrollment session during which the client can call RECOGNIZE multiple times to enroll a new utterance in a grammar. An enrollment session consists of a set of calls to RECOGNIZE in which the caller speaks a phrase several times so that the system can "learn" it. The phrase is then added to a personal grammar (speaker-trained grammar) so that the system can recognize it later. Only one phrase enrollment session can be active at a time for a resource. The Personal-Grammar-URI identifies the grammar that is used during enrollment to store the personal list of phrases. Once RECOGNIZE is called, the result is returned in a RECOGNITION-COMPLETE event and will contain either an enrollment result OR a recognition result for a regular recognition.

Calling END-PHRASE-ENROLLMENT ends the ongoing phrase enrollment session, which is typically done after a sequence of successful calls to RECOGNIZE. This method can be called to commit the new phrase to the personal grammar or to abort the phrase enrollment session. The grammar to contain the new enrolled phrase, specified by Personal-Grammar-URI, is created if it does not exist. Also, the personal grammar MUST ONLY contain phrases added via a phrase enrollment session.

The Phrase-ID passed to this method is used to identify this phrase in the grammar and will be returned as the speech input when doing a RECOGNIZE on the grammar. The Phrase-NL similarly is returned in a RECOGNITION-COMPLETE event in the same manner as other Natural Language (NL) in a grammar. The tag-format of this NL is implementation specific. If the client has specified Save-Best-Waveform as true, then the response after ending the phrase enrollment session MUST contain the location/URI of a recording of the best repetition of the learned phrase.

```
C->S: MRCP/2.0 ... START-PHRASE-ENROLLMENT 543258
       Channel-Identifier:32AECB23433801@speechrecog
       Num-Min-Consistent-Pronunciations:2
       Consistency-Threshold:30
       Clash-Threshold:12
       Personal-Grammar-URI:<personal grammar uri>
       Phrase-Id:<phrase id>
       Phrase-NL:<NL phrase>
       Weight:1
       Save-Best-Waveform:true

S->C: MRCP/2.0 ... 543258 200 COMPLETE
       Channel-Identifier:32AECB23433801@speechrecog
```

14.9.16 ENROLLMENT-ROLLBACK

The ENROLLMENT-ROLLBACK method discards the last live utterance from the RECOGNIZE operation. The client can invoke this method when the caller provides undesirable input such as non-speech noises, sidespeech, commands, utterance from the RECOGNIZE grammar, etc. Note that this method does not provide a stack of rollback states. Executing ENROLLMENT-ROLLBACK twice in succession without an intervening recognition operation has no effect the second time.

```
C->S: MRCP/2.0 ... ENROLLMENT-ROLLBACK 543261
       Channel-Identifier:32AECB23433801@speechrecog

S->C:  MRCP/2.0 ... 543261 200 COMPLETE
       Channel-Identifier:32AECB23433801@speechrecog
```

14.9.17 END-PHRASE-ENROLLMENT

The client MAY call the END-PHRASE-ENROLLMENT method ONLY during an active phrase enrollment session. It MUST NOT be called during an ongoing RECOGNIZE operation. To commit the new phrase in the grammar, the client MAY call this method once successive calls to RECOGNIZE have succeeded and Num-Repetitions-Still-Needed has been returned as 0 in the RECOGNITION-COMPLETE event. Alternatively, the client MAY abort the phrase enrollment session by calling this method with the Abort-Phrase-Enrollment header field. If the client has specified Save-Best-Waveform as "true" in the START-PHRASE-ENROLLMENT request, then the response MUST contain a Waveform-URI header whose value is the location/URI of a recording of the best repetition of the learned phrase.

```
C->S: MRCP/2.0 ... END-PHRASE-ENROLLMENT 543262
       Channel-Identifier:32AECB23433801@speechrecog

S->C:  MRCP/2.0 ... 543262 200 COMPLETE
       Channel-Identifier:32AECB23433801@speechrecog
       Waveform-URI:<http://mediaserver.com/recordings/file1324.wav>;
       size=242453;duration=25432
```

14.9.18 MODIFY-PHRASE

The MODIFY-PHRASE method sent from the client to the server is used to change the phrase ID, NL phrase, and/or weight for a given phrase in a personal grammar. If no fields are supplied, then calling this method has no effect.

```
C->S: MRCP/2.0 ... MODIFY-PHRASE 543265
       Channel-Identifier:32AECB23433801@speechrecog
       Personal-Grammar-URI:<personal grammar uri>
       Phrase-Id:<phrase id>
       New-Phrase-Id:<new phrase id>
       Phrase-NL:<NL phrase>
       Weight:1

S->C:  MRCP/2.0 ... 543265 200 COMPLETE
       Channel-Identifier:32AECB23433801@speechrecog
```

14.9.19 DELETE-PHRASE

The DELETE-PHRASE method sent from the client to the server is used to delete a phase that is in a personal grammar and was added through voice enrollment or text enrollment. If the specified phrase does not exist, this method has no effect.

```
C->S: MRCP/2.0 ... DELETE-PHRASE 543266
       Channel-Identifier:32AECB23433801@speechrecog
       Personal-Grammar-URI:<personal grammar uri>
       Phrase-Id:<phrase id>

S->C:  MRCP/2.0 ... 543266 200 COMPLETE
       Channel-Identifier:32AECB23433801@speechrecog
```

14.9.20 INTERPRET

The INTERPRET method from the client to the server takes as input an Interpret-Text header field containing the text for which the semantic interpretation is desired, and returns, via the INTERPRETATION-COMPLETE event, an interpretation result that is very similar to the one returned from a RECOGNIZE method invocation. Only portions of the result relevant to acoustic matching are excluded from the result. The Interpret-Text header field MUST be included in the INTERPRET request. Recognizer grammar data is treated in the same way as it is when issuing a RECOGNIZE method call. If a RECOGNIZE, RECORD, or another INTERPRET operation is already in progress for the resource, the server MUST reject the request with a response having a status-code of 402 "Method not valid in this state" and a COMPLETE request state.

```
C->S: MRCP/2.0 ... INTERPRET 543266
       Channel-Identifier:32AECB23433801@speechrecog
       Interpret-Text:may I speak to Andre Roy
       Content-Type:application/srgs+xml
       Content-ID:<request1@form-level.store>
       Content-Length:...

       <?xml version="1.0"?>

       <!-- the default grammar language is US English -->

       <grammar xmlns="http://www.w3.org/2001/06/grammar"
             xml:lang="en-US" version="1.0" root="request">
          <!-- single language attachment to tokens -->

          <rule id="yes">
             <one-of>
                <item xml:lang="fr-CA">oui</item>
                <item xml:lang="en-US">yes</item>
             </one-of>
          </rule>

          <!-- single language attachment to a rule expansion -->
             <rule id="request">
                   may I speak to
                <one-of xml:lang="fr-CA">
                   <item>Michel Tremblay</item>
                   <item>Andre Roy</item>
                </one-of>
```

```
        </rule>
      </grammar>

S->C: MRCP/2.0 ... 543266 200 IN-PROGRESS
      Channel-Identifier:32AECB23433801@speechrecog

S->C: MRCP/2.0 ... INTERPRETATION-COMPLETE 543266 200 COMPLETE
      Channel-Identifier:32AECB23433801@speechrecog
      Completion-Cause:000 success
      Content-Type:application/nlsml+xml
      Content-Length:...

      <?xml version="1.0"?>

      <result xmlns="urn:ietf:params:xml:ns:mrcpv2"
            xmlns:ex="http://www.example.com/example"
            grammar="session:request1@form-level.store">
         <interpretation>
            <instance name="Person">
               <ex:Person>
                  <ex:Name> Andre Roy </ex:Name>
               </ex:Person>
            </instance>
            <input> may I speak to Andre Roy </input>
         </interpretation>
      </result>
```

14.9.21 INTERPRETATION-COMPLETE

This event from the recognizer resource to the client indicates that the INTERPRET operation is complete. The interpretation result is sent in the body of the MRCP message. The request state MUST be set to COMPLETE. The Completion-Cause header field MUST be included in this event and MUST be set to an appropriate value from the list of cause codes.

```
C->S: MRCP/2.0 ... INTERPRET 543266
      Channel-Identifier:32AECB23433801@speechrecog
      Interpret-Text:may I speak to Andre Roy
      Content-Type:application/srgs+xml
      Content-ID:<request1@form-level.store>
      Content-Length:...

      <?xml version="1.0"?>

      <!-- the default grammar language is US English -->

      <grammar xmlns="http://www.w3.org/2001/06/grammar"
            xml:lang="en-US" version="1.0" root="request">

         <!-- single language attachment to tokens -->
            <rule id="yes">
               <one-of>
                  <item xml:lang="fr-CA">oui</item>
                  <item xml:lang="en-US">yes</item>
               </one-of>
            </rule>
```

```
        <!-- single language attachment to a rule expansion -->
            <rule id="request">
                    may I speak to
                <one-of xml:lang="fr-CA">
                    <item>Michel Tremblay</item>
                    <item>Andre Roy</item>
                </one-of>
            </rule>
        </grammar>

S->C:   MRCP/2.0 ... 543266 200 IN-PROGRESS
        Channel-Identifier:32AECB23433801@speechrecog

S->C:   MRCP/2.0 ... INTERPRETATION-COMPLETE 543266 200 COMPLETE
        Channel-Identifier:32AECB23433801@speechrecog
        Completion-Cause:000 success
        Content-Type:application/nlsml+xml
        Content-Length:...

        <?xml version="1.0"?>

        <result xmlns="urn:ietf:params:xml:ns:mrcpv2"
                xmlns:ex="http://www.example.com/example"
                grammar="session:request1@form-level.store">
            <interpretation>
                <instance name="Person">
                    <ex:Person>
                        <ex:Name> Andre Roy </ex:Name>
                    </ex:Person>
                </instance>
                <input> may I speak to Andre Roy </input>
            </interpretation>
        </result>
```

14.9.22 DTMF Detection

Digits received as DTMF tones are delivered to the recognition resource in the MRCPv2 server in the RTP stream according to RFC 4733. The ASR MUST support RFC 4733 to recognize digits, and it MAY support recognizing DTMF tones [2] in the audio.

14.10 RECORDER RESOURCE

This resource captures received audio and video and stores it as content pointed to by a URI. The main usages of recorders are:

1. To capture speech audio that may be submitted for recognition at a later time
2. To record voice or video mails

Both these applications require functionality above and beyond those specified by protocols such as RTSP (RFC 2326). This includes audio endpointing (i.e., detecting speech or silence). The support for video is OPTIONAL and is mainly capturing video mails that may require the speech or audio processing mentioned earlier. A recorder MUST provide endpointing capabilities for suppressing silence at the beginning and end of a recording, and it MAY also suppress silence in the middle of a recording. If such suppression is done, the recorder MUST maintain timing metadata to indicate the

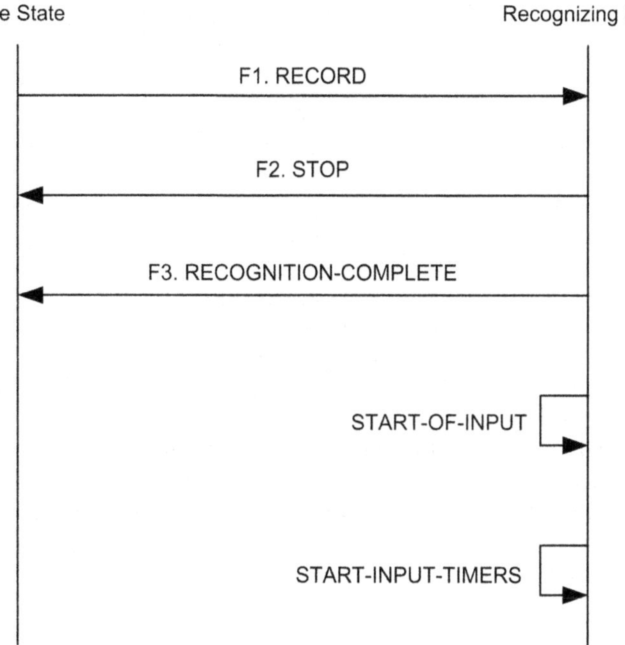

FIGURE 14.22 Recorder State Machine.

actual time stamps of the recorded media. See the discussion on the sensitivity of saved waveforms in Section 14.12.

14.10.1 Recorder State Machine

The recorder state machine is shown in Figure 14.22.

14.10.2 Recorder Methods

The recorder resource supports the following methods.

```
recorder-method    =    "RECORD"
                        / "STOP"
                        / "START-INPUT-TIMERS"
```

14.10.3 Recorder Events

The recorder resource can generate the following events.

```
recorder-event    =    "START-OF-INPUT"
                       / "RECORD-COMPLETE"
```

14.10.4 Recorder Header Fields

Method invocations for the recorder resource can contain resource-specific header fields containing request options and information to augment the Method, Response, or Event message it is associated with.

```
recorder-header    =    sensitivity-level
                        / no-input-timeout
```

```
/ completion-cause
/ completion-reason
/ failed-uri
/ failed-uri-cause
/ record-uri
/ media-type
/ max-time
/ trim-length
/ final-silence
/ capture-on-speech
/ ver-buffer-utterance
/ start-input-timers
/ new-audio-channel
```

14.10.4.1 Sensitivity-Level

To filter out background noise and not mistake it for speech, the recorder can support a variable level of sound sensitivity. The Sensitivity-Level header field is a float value between 0.0 and 1.0 and allows the client to set the sensitivity level for the recorder. This header field MAY occur in RECORD, SET-PARAMS, or GET-PARAMS. A higher value for this header field means higher sensitivity. The default value for this header field is implementation specific.

```
sensitivity-level    =    "Sensitivity-Level" ":" FLOAT CRLF
```

14.10.4.2 No-Input-Timeout

When recording is started and there is no speech detected for a certain period of time, the recorder can send a RECORD-COMPLETE event to the client and terminate the record operation. The No-Input-Timeout header field can set this timeout value. The value is in milliseconds. This header field MAY occur in RECORD, SET-PARAMS, or GET-PARAMS. The value for this header field ranges from 0 to an implementation-specific maximum value. The default value for this header field is implementation specific.
no-input-timeout = "No-Input-Timeout" ":" 1*19DIGIT CRLF

14.10.4.3 Completion-Cause

This header field MUST be part of a RECORD-COMPLETE event from the recorder resource to the client. This indicates the reason behind the RECORD method completion. This header field MUST be sent in the RECORD responses if they return with a failure status and a COMPLETE state. In the following ABNF, the "cause-code" contains a numerical value selected from the Cause-Code column of Table 14.6. The "cause-name" contains the corresponding token selected from the Cause-Name column.

TABLE 14.6
Completion Cause for Speech Recognizer Resource

Cause-Code	Cause-Name	Description
000	success-silence	RECORD completed with a silence at the end
001	success-maxtime	RECORD completed after reaching maximum recording time specified in record method
002	no-input-timeout	RECORD failed due to no input
003	uri-failure	Failure accessing the record URI
004	error	RECORD request terminated prematurely due to a recorder error

```
completion-cause     =    "Completion-Cause" ":" cause-code SP
                          cause-name CRLF
cause-code           =    3DIGIT
cause-name           =    *VCHAR
```

14.10.4.4 Completion-Reason

This header field MAY be present in a RECORD-COMPLETE event coming from the recorder resource to the client. It contains the reason text behind the RECORD request completion. This header field communicates text describing the reason for the failure. The completion reason text is provided for client use in logs and for debugging and instrumentation purposes. Clients MUST NOT interpret the completion reason text.

```
completion-reason     =     "Completion-Reason" ":"
                            quoted-string CRLF
```

14.10.4.5 Failed-URI

When a recorder method needs to post the audio to a URI, and access to the URI fails, the server MUST provide the failed URI in this header field in the method response.

```
failed-uri          =       "Failed-URI" ":" absoluteURI CRLF
```

14.10.4.6 Failed-URI-Cause

When a recorder method needs to post the audio to a URI, and access to the URI fails, the server MAY provide the URI-specific or protocol-specific response code through this header field in the method response. The value encoding is UTF-8 (RFC 3629) to accommodate any access protocol – some access protocols might have a response string instead of a numeric response code.

```
failed-uri-cause     =    "Failed-URI-Cause" ":" 1*UTFCHAR
                          CRLF
```

14.10.4.7 Record-URI

When a recorder method contains this header field, the server MUST capture the audio and store it. If the header field is present but specified with no value, the server MUST store the content locally and generate a URI that points to it. This URI is then returned in either the STOP response or the RECORD-COMPLETE event. If the header field in the RECORD method specifies a URI, the server MUST attempt to capture and store the audio at that location. If this header field is not specified in the RECORD request, the server MUST capture the audio, MUST encode it, and MUST send it in the STOP response or the RECORD-COMPLETE event as a message body. In this case, the response carrying the audio content MUST include a Content ID (cid) (RFC 2392) value in this header pointing to the Content-ID in the message body. The server MUST also return the size in octets and the duration in milliseconds of the recorded audio waveform as parameters associated with the header field. Implementations MUST support "http" (RFC 2616), "https" (RFC 2818), "file" (RFC 3986), and "cid" (RFC 2392) schemes in the URI. Note that implementations already exist that support other schemes.

```
record-uri          =       "Record-URI" ":" ["<" uri ">"
                            ";" "size" "=" 1*19DIGIT
                            ";" "duration" "=" 1*19DIGIT] CRLF
```

14.10.4.8 Media-Type

A RECORD method MUST contain this header field, which specifies to the server the media type of the captured audio or video.

```
media-type     =     "Media-Type" ":" media-type-value CRLF
```

14.10.4.9 Max-Time

When recording is started, this specifies the maximum length of the recording in milliseconds, calculated from the time the actual capture and store begins, and is not necessarily the time when the RECORD method is received. It specifies the duration before silence suppression, if any, has been applied by the recorder resource. After this time, the recording stops, and the server MUST return a RECORD-COMPLETE event to the client having a request-state of COMPLETE. This header field MAY occur in RECORD, SET-PARAMS, or GETPARAMS. The value for this header field ranges from 0 to an implementation-specific maximum value. A value of 0 means infinity, and hence, the recording continues until one or more of the other stop conditions are met. The default value for this header field is 0.

```
max-time    =    "Max-Time" ":" 1*19DIGIT CRLF
```

14.10.4.10 Trim-Length

This header field MAY be sent on a STOP method and specifies the length of audio to be trimmed from the end of the recording after the stop. The length is interpreted to be in milliseconds. The default value for this header field is 0.

```
trim-length    =    "Trim-Length" ":" 1*19DIGIT CRLF
```

14.10.4.11 Final-Silence

When the recorder is started and the actual capture begins, this header field specifies the length of silence in the audio that is to be interpreted as the end of the recording. This header field MAY occur in RECORD, SET-PARAMS, or GET-PARAMS. The value for this header field ranges from 0 to an implementation-specific maximum value and is interpreted to be in milliseconds. A value of 0 means infinity, and hence, the recording will continue until one of the other stop conditions are met. The default value for this header field is implementation specific.

```
final-silence    =    "Final-Silence" ":" 1*19DIGIT CRLF
```

14.10.4.12 Capture-On-Speech

If "false," the recorder MUST start capturing immediately when started. If "true," the recorder MUST wait for the endpointing functionality to detect speech before it starts capturing. This header field MAY occur in the RECORD, SET-PARAMS, or GET-PARAMS. The value for this header field is a Boolean. The default value for this header field is "false."

```
capture-on-speech    =    "Capture-On-Speech " ":" BOOLEAN CRLF
```

14.10.4.13 Ver-Buffer-Utterance

This header field is the same as the one described for the verifier resource (see Section 14.11.4.14). This tells the server to buffer the utterance associated with this recording request into the verification buffer. Sending this header field is permitted only if the verification buffer is for the session. This buffer is shared across resources within a session. It is instantiated when a verifier resource is added to this session and is released when the verifier resource is released from the session.

14.10.4.14 Start-Input-Timers

This header field MAY be sent as part of the RECORD request. A value of "false" tells the recorder resource to start the operation but not to start the no-input timer until the client sends a START-INPUT-TIMERS request to the recorder resource. This is useful in the scenario when the recorder and synthesizer resources are not part of the same session. When a kill-on-barge-in prompt is being played, the client may want the RECORD request to be simultaneously active so that it can detect and implement kill-on-barge-in (see Section 8.4.2). But at the same time, the client does not want the recorder resource to start the no-input timers until the prompt is finished. The default value is "true."

```
   start-input-timers      =      "Start-Input-Timers" ":"
                                  BOOLEAN CRLF
```

14.10.4.15 New-Audio-Channel

This header field is the same as the one described for the recognizer resource (see Section 14.9.4.23).

14.10.5 Recorder Message Body

If the RECORD request did not have a Record-URI header field, the STOP response or the RECORD-COMPLETE event MUST contain a message body carrying the captured audio. In this case, the message carrying the audio content has a Record-URI header field with a Content ID value pointing to the message body entity that contains the recorded audio. See Section 14.10.4.7 for details.

14.10.6 RECORD

The RECORD request places the recorder resource in the recording state. Depending on the header fields specified in the RECORD method, the resource may start recording the audio immediately or wait for the endpointing functionality to detect speech in the audio. The audio is then made available to the client either in the message body or as specified by Record-URI. The server MUST support the "https" URI scheme and MAY support other schemes. Note that due to the sensitive nature of voice recordings, any protocols used for dereferencing SHOULD employ integrity and confidentiality unless other means, such as use of a controlled environment (see Section 14.4.2), are employed.

If a RECORD operation is already in progress, invoking this method causes the server to issue a response having a status-code of 402 "Method not valid in this state" and a request-state of COMPLETE. If the Record-URI is not valid, a status-code of 404 "Illegal Value for Header Field" is returned in the response. If it is impossible for the server to create the requested stored content, a status-code of 407 "Method or Operation Failed" is returned. If the type specified in the Media-Type header field is not supported, the server MUST respond with a status-code of 409 "Unsupported Header Field Value" with the Media-Type header field in its response.

When the recording operation is initiated, the response indicates an IN-PROGRESS request state. The server MAY generate a subsequent START-OF-INPUT event when speech is detected. Upon completion of the recording operation, the server generates a RECORD-COMPLETE event.

```
C->S: MRCP/2.0 ... RECORD 543257
       Channel-Identifier:32AECB23433802@recorder
       Record-URI:<file://mediaserver/recordings/myfile.wav>
       Media-Type:audio/wav
       Capture-On-Speech:true
       Final-Silence:300
       Max-Time:6000

S->C:  MRCP/2.0 ... 543257 200 IN-PROGRESS
       Channel-Identifier:32AECB23433802@recorder

S->C:  MRCP/2.0 ... START-OF-INPUT 543257 IN-PROGRESS
       Channel-Identifier:32AECB23433802@recorder

S->C:  MRCP/2.0 ... RECORD-COMPLETE 543257 COMPLETE
       Channel-Identifier:32AECB23433802@recorder
       Completion-Cause:000 success-silence
       Record-URI:<file://mediaserver/recordings/myfile.wav>;
       size=242552;duration=25645
```

RECORD Example

```
C->S: MRCP/2.0 ... RECORD 543257
      Channel-Identifier:32AECB23433802@recorder
      Record-URI:<file://mediaserver/recordings/myfile.wav>
      Capture-On-Speech:true
      Final-Silence:300
      Max-Time:6000

S->C: MRCP/2.0 ... 543257 200 IN-PROGRESS
      Channel-Identifier:32AECB23433802@recorder

S->C: MRCP/2.0 ... START-OF-INPUT 543257 IN-PROGRESS
      Channel-Identifier:32AECB23433802@recorder

C->S: MRCP/2.0 ... STOP 543257
      Channel-Identifier:32AECB23433802@recorder
      Trim-Length:200
      S->C: MRCP/2.0 ... 543257 200 COMPLETE
      Channel-Identifier:32AECB23433802@recorder
      Record-URI:<file://mediaserver/recordings/myfile.wav>;
      size=324253;duration=24561
      Active-Request-Id-List:543257
```

FIGURE 14.23 STOP Example.

14.10.7 STOP

The STOP method (Figure 14.23) moves the recorder from the recording state back to the idle state. If a RECORD request is active, and the STOP request successfully terminates it, then the STOP response MUST contain an Active-Request-Id-List header field containing the RECORD request-id that was terminated. In this case, no RECORD-COMPLETE event is sent for the terminated request. If there was no recording active, then the response MUST NOT contain an Active-Request-Id-List header field. If the recording was a success, the STOP response MUST contain a Record-URI header field pointing to the recorded audio content or to a typed entity in the body of the STOP response containing the recorded audio. The STOP method MAY have a Trim-Length header field, in which case the specified length of audio is trimmed from the end of the recording after the stop. In any case, the response MUST contain a status-code of 200 "Success."

14.10.8 RECORD-COMPLETE

If the recording completes due to no input, silence after speech, or reaching the max-time, the server MUST generate the RECORD-COMPLETE event (Figure 14.24) to the client with a request-state of COMPLETE. If the recording was a success, the RECORD-COMPLETE event contains a Record-URI header field pointing to the recorded audio file on the server or to a typed entity in the message body containing the recorded audio.

```
C->S: MRCP/2.0 ... RECORD 543257
      Channel-Identifier:32AECB23433802@recorder
      Record-URI:<file://mediaserver/recordings/myfile.wav>
      Capture-On-Speech:true
      Final-Silence:300
      Max-Time:6000

S->C: MRCP/2.0 ... 543257 200 IN-PROGRESS
      Channel-Identifier:32AECB23433802@recorder

S->C: MRCP/2.0 ... START-OF-INPUT 543257 IN-PROGRESS
      Channel-Identifier:32AECB23433802@recorder

S->C: MRCP/2.0 ... RECORD-COMPLETE 543257 COMPLETE
      Channel-Identifier:32AECB23433802@recorder
      Completion-Cause:000 success
      Record-URI:<file://mediaserver/recordings/myfile.wav>;
      size=325325;duration=24652
```

FIGURE 14.24 RECORD-COMPLETE Example.

14.10.9 START-INPUT-TIMERS

This request is sent from the client to the recorder resource when it discovers that a kill-on-barge-in prompt has finished playing (see Section 14.8.4.2). This is useful in the scenario when the recorder and synthesizer resources are not in the same MRCPv2 session. When a kill-on-barge-in prompt is being played, the client wants the RECORD request to be simultaneously active so that it can detect and implement kill-on-barge-in. But at the same time, the client does not want the recorder resource to start the no-input timers until the prompt is finished. The Start-Input-Timers header field in the RECORD request allows the client to say whether the timers should be started or not. In the preceding case, the recorder resource does not start the timers until the client sends a START-INPUT-TIMERS method to the recorder.

14.10.10 START-OF-INPUT

The START-OF-INPUT event is returned from the server to the client once the server has detected speech. This event is always returned by the recorder resource when speech has been detected. The recorder resource also MUST send a Proxy-Sync-Id header field with a unique value for this event.

```
S->C: MRCP/2.0 ... START-OF-INPUT 543259 IN-PROGRESS
      Channel-Identifier:32AECB23433801@recorder
      Proxy-Sync-Id:987654321
```

14.11 SPEAKER VERIFICATION AND IDENTIFICATION

This section describes the methods, responses, and events employed by MRCPv2 for doing speaker verification/identification. Speaker verification is a voice authentication methodology that can be used to identify the speaker in order to grant the user access to sensitive information and transactions. Because speech is a biometric, a number of essential security considerations related to biometric authentication technologies apply to its implementation and usage. Implementers should carefully read Section 14.12 in this specification and the corresponding section of the SPEECHSC requirements (RFC 4313). Implementers and deployers of this technology are strongly encouraged to check the state of the art for any new risks and solutions that might have been developed.

In speaker verification, a recorded utterance is compared with a previously stored voiceprint, which is in turn associated with a claimed identity for that user. Verification typically consists of two phases: A designation phase to establish the claimed identity of the caller and an execution phase in which a voiceprint is either created (training) or used to authenticate the claimed identity (verification).

Speaker identification is the process of associating an unknown speaker with a member in a population. It does not employ a claim of identity. When an individual claims to belong to a group (e.g., one of the owners of a joint bank account), a group authentication is performed. This is generally implemented as a kind of verification involving comparison with more than one voice model. It is sometimes called "multi-verification." If the individual speaker can be identified from the group, this may be useful for applications where multiple users share the same access privileges to some data or application. Speaker identification and group authentication are also done in two phases, a designation phase and an execution phase.

Note that from a functionality standpoint, identification can be thought of as a special case of group authentication (if the individual is identified), where the group is the entire population, although the implementation of speaker identification may be different from the way group authentication is performed. To accommodate single-voiceprint verification, verification against multiple voiceprints, group authentication, and identification, this specification provides a single set of methods that can take a list of identifiers, called "voiceprint identifiers," and return a list of identifiers, with a score for each that represents how well the input speech matched each identifier. The input and output lists of identifiers do not have to match, allowing a vendor-specific group identifier to be used as input to indicate that identification is to be performed. In this specification, the terms "identification" and "multi-verification" are used to indicate that the input represents a group (potentially the entire population) and that results for multiple voiceprints may be returned.

It is possible for a verifier resource to share the same session with a recognizer resource or to operate independently. In order to share the same session, the verifier and recognizer resources MUST be allocated from within the same SIP dialog. Otherwise, an independent verifier resource, running on the same physical server or a separate one, will be set up. Note that in addition to allowing both resources to be allocated in the same INVITE, it is possible to allocate one initially and the other later via a re-INVITE. Some of the speaker verification methods, described later, apply only to a specific mode of operation.

The verifier resource has a verification buffer associated with it (see Section 14.11.4.14). This allows the storage of speech utterances for the purposes of verification, identification, or training from the buffered speech. This buffer is owned by the verifier resource, but other input resources (such as the recognizer resource or recorder resource) may write to it. This allows the speech received as part of a recognition or recording operation to be later used for verification, identification, or training. Access to the buffer is limited to one operation at time. Hence, when the resource is doing read, write, or delete operations, such as a RECOGNIZE with ver-buffer-utterance turned on, another operation involving the buffer fails with a status-code of 402. The verification buffer can be cleared by a CLEAR-BUFFER request from the client and is freed when the verifier resource is deallocated or the session with the server terminates. The verification buffer is different from collecting waveforms and processing them using either the real-time audio stream or stored audio, because this buffering mechanism does not simply accumulate speech to a buffer. The verification buffer MAY contain additional information gathered by the recognizer resource that serves to improve verification performance.

14.11.1 Speaker Verification State Machine

Speaker verification (Figure 14.25) may operate in a training or a verification session. Starting one of these sessions does not change the state of the verifier resource; i.e., it remains idle. Once a verification or training session is started, then utterances are trained or verified by calling the VERIFY or VERIFY-FROM-BUFFER method. The state of the verifier resources goes from IDLE to VERIFYING state each time VERIFY or VERIFY-FROM-BUFFER is called.

14.11.2 Speaker Verification Methods

The verifier resource supports the following methods.

```
verifier-method    =    "START-SESSION"
                        / "END-SESSION"
                        / "QUERY-VOICEPRINT"
                        / "DELETE-VOICEPRINT"
                        / "VERIFY"
                        / "VERIFY-FROM-BUFFER"
                        / "VERIFY-ROLLBACK"
                        / "STOP"
                        / "CLEAR-BUFFER"
                        / "START-INPUT-TIMERS"
                        / "GET-INTERMEDIATE-RESULT"
```

These methods allow the client to control the mode and target of verification or identification operations within the context of a session. All the verification input operations that occur within a session can be used to create, update, or validate against the voiceprint specified during the session. At the beginning of each session, the verifier resource is reset to the state it had prior to any previous verification session. Verification/identification operations can be executed against live or buffered audio. The verifier resource provides methods for collecting and evaluating live audio data and methods for controlling the verifier resource and adjusting its configured behavior. There are no dedicated methods for collecting buffered audio data. This is accomplished by calling VERIFY, RECOGNIZE, or RECORD, as appropriate for the resource, with the header field Ver-Buffer-Utterance. Then, when the following method is called, verification is performed using the set of buffered audio.

1. VERIFY-FROM-BUFFER

The following methods are used for verification of live audio utterances:

1. VERIFY
2. START-INPUT-TIMERS

The following methods are used for configuring the verifier resource and for establishing resource states:

1. START-SESSION
2. END-SESSION
3. QUERY-VOICEPRINT
4. DELETE-VOICEPRINT
5. VERIFY-ROLLBACK
6. STOP
7. CLEAR-BUFFER

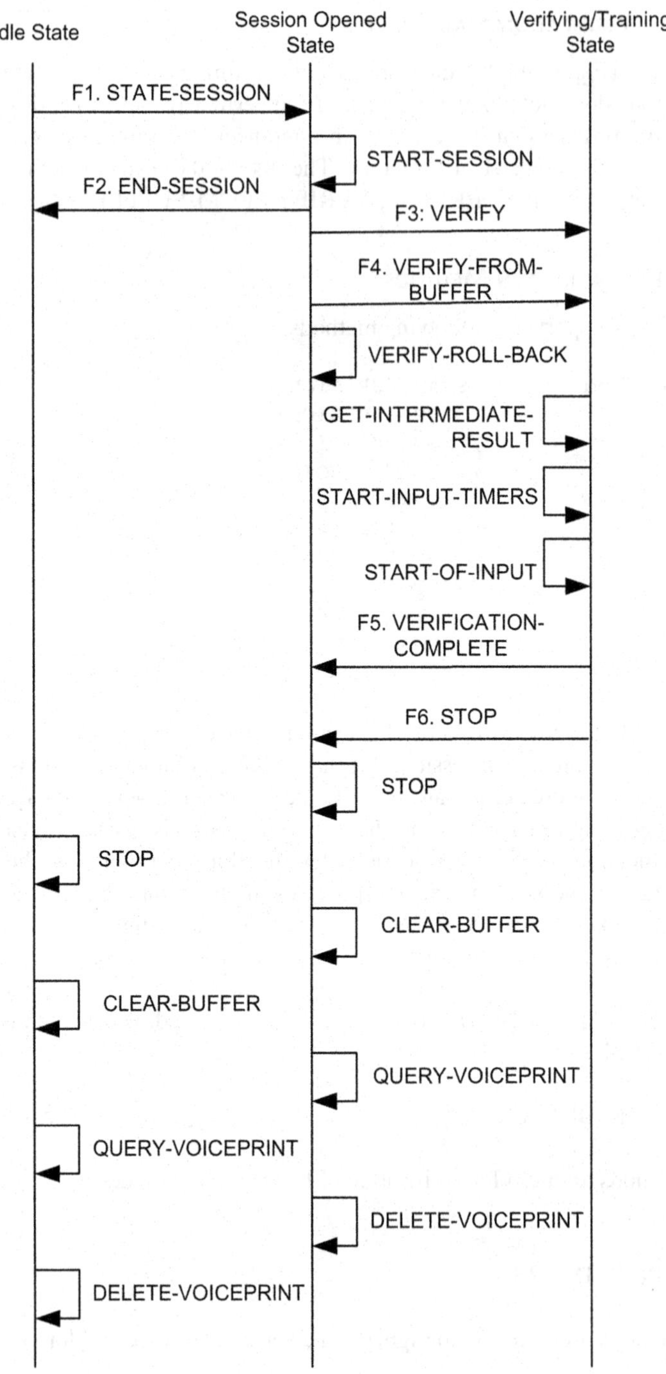

FIGURE 14.25 Verifier Resource State Machine. (Copyright: IETF.)

The following method allows the polling of a verification in progress for intermediate results.

1. GET-INTERMEDIATE-RESULT

14.11.3 Verification Events

The verifier resource generates the following events.

```
verifier-event    =      "VERIFICATION-COMPLETE"
                         / "START-OF-INPUT"
```

14.11.4 Verification Header Fields

A verifier resource message can contain header fields containing request options and information to augment the Request, Response, or Event message it is associated with.

```
verification-header    =    repository-uri
                            / voiceprint-identifier
                            / verification-mode
                            / adapt-model
                            / abort-model
                            / min-verification-score
                            / num-min-verification-phrases
                            / num-max-verification-phrases
                            / no-input-timeout
                            / save-waveform
                            / media-type
                            / waveform-uri
                            / voiceprint-exists
                            / ver-buffer-utterance
                            / input-waveform-uri
                            / completion-cause
                            / completion-reason
                            / speech-complete-timeout
                            / new-audio-channel
                            / abort-verification
                            / start-input-timers
```

14.11.4.1 Repository-URI

This header field specifies the voiceprint repository to be used or referenced during speaker verification or identification operations. This header field is required in the START-SESSION, QUERY-VOICEPRINT, and DELETE-VOICEPRINT methods.

```
repository-uri   =  "Repository-URI" ":" uri CRLF
```

14.11.4.2 Voiceprint-Identifier

This header field specifies the claimed identity for verification applications. The claimed identity MAY be used to specify an existing voiceprint or to establish a new voiceprint. This header field MUST be present in the QUERY-VOICEPRINT and DELETE-VOICEPRINT methods. The Voiceprint-Identifier MUST be present in the START-SESSION method for verification operations. For identification or multi-verification operations, this header field MAY contain a list of voiceprint identifiers separated by semicolons. For identification operations, the client MAY also specify a voiceprint group identifier instead of a list of voiceprint identifiers.

```
voiceprint-identifier     =     "Voiceprint-Identifier" ":"
                                vid *[";" vid] CRLF
                                vid = 1*VCHAR ["." 1*VCHAR]
```

14.11.4.3 Verification-Mode

This header field specifies the mode of the verifier resource and is set by the START-SESSION method. Acceptable values indicate whether the verification session will train a voiceprint ("train") or verify/identify using an existing voiceprint ("verify"). Training and verification sessions both require the voiceprint Repository-URI to be specified in the START-SESSION. In many usage scenarios, however, the system does not know the speaker's claimed identity until a recognition operation has, for example, recognized an account number to which the user desires access. In order to allow the first few utterances of a dialog to be both recognized and verified, the verifier resource on the MRCPv2 server retains a buffer. In this buffer, the MRCPv2 server accumulates recognized utterances. The client can later execute a verification method and apply the buffered utterances to the current verification session.

Some voice user interfaces may require additional user input that should not be subject to verification. For example, the user's input may have been recognized with low confidence and thus require a confirmation cycle. In such cases, the client SHOULD NOT execute the VERIFY or VERIFY-FROM-BUFFER methods to collect and analyze the caller's input. A separate recognizer resource can analyze the caller's response without any participation by the verifier resource. Once the following conditions have been met:

1. the voiceprint identity has been successfully established through the Voiceprint-Identifier header fields of the START-SESSION method
2. the verification mode has been set to one of "train" or "verify"

   the verifier resource can begin providing verification information during verification operations. If the verifier resource does not reach one of the two major states ("train" or "verify"), it MUST report an error condition in the MRCPv2 status code to indicate why the verifier resource is not ready for the corresponding usage. The value of verification-mode is persistent within a verification session. If the client attempts to change the mode during a verification session, the verifier resource reports an error, and the mode retains its current value.

```
verification-mode          =    "Verification-Mode" ":"
                                verification-mode-string
verification-mode-string   =    "train" / "verify"
```

14.11.4.4 Adapt-Model

This header field indicates the desired behavior of the verifier resource after a successful verification operation. If the value of this header field is "true," the server SHOULD use audio collected during the verification session to update the voiceprint to account for ongoing changes in a speaker's incoming speech characteristics unless local policy prohibits updating the voiceprint. If the value is "false" (the default), the server MUST NOT update the voiceprint. This header field MAY occur in the START-SESSION method.

```
adapt-mode    =    "Adapt-Model" ":" BOOLEAN CRLF
```

14.11.4.5 Abort-Model

The Abort-Model header field indicates the desired behavior of the verifier resource upon session termination. If the value of this header field is "true," the server MUST discard any pending changes to a voiceprint due to verification training or verification adaptation. If the value is "false" (the default), the server MUST commit any pending changes for a training session or a successful verification session to the voiceprint repository. A value of "true" for Abort-Model overrides a value of "true" for the Adapt-Model header field. This header field MAY occur in the END-SESSION method.

```
abort-model     =     "Abort-Model" ":" BOOLEAN CRLF
```

14.11.4.6 Min-Verification-Score

The Min-Verification-Score header field, when used with a verifier resource through a SET-PARAMS, GET-PARAMS, or START-SESSION method, determines the minimum verification score for which a verification decision of "accepted" may be declared by the server. This is a float value between –1.0 and 1.0. The default value for this header field is implementation specific.

```
min-verification-score    =     "Min-Verification-Score" ":"
                                [ %x2D ] FLOAT CRLF
```

14.11.4.7 Num-Min-Verification-Phrases

The Num-Min-Verification-Phrases header field is used to specify the minimum number of valid utterances before a positive decision is given for verification. The value for this header field is an integer, and the default value is 1. The verifier resource MUST NOT declare a verification "accepted" unless Num-Min-Verification-Phrases valid utterances have been received. The minimum value is 1. This header field MAY occur in START-SESSION, SET-PARAMS, or GET-PARAMS.

```
num-min-verification-phrases  =  "Num-Min-Verification-Phrases" ":"
                                 1*19DIGIT CRLF
```

14.11.4.8 Num-Max-Verification-Phrases

The Num-Max-Verification-Phrases header field is used to specify the number of valid utterances required before a decision is forced for verification. The verifier resource MUST NOT return a decision of "undecided" once Num-Max-Verification-Phrases have been collected and used to determine a verification score. The value for this header field is an integer, and the minimum value is 1. The default value is implementation specific. This header field MAY occur in START-SESSION, SET-PARAMS, or GET-PARAMS.

```
num-max-verification-phrases   =   "Num-Max-Verification-Phrases" ":"
                                   1*19DIGIT CRLF
```

14.11.4.9 No-Input-Timeout

The No-Input-Timeout header field sets the length of time from the start of the verification timers (see START-INPUT-TIMERS) until the VERIFICATION-COMPLETE server event message declares that no input has been received (i.e., has a Completion-Cause of no-input-timeout). The value is in milliseconds. This header field MAY occur in VERIFY, SET-PARAMS, or GET-PARAMS. The value for this header field ranges from 0 to an implementation-specific maximum value. The default value for this header field is implementation specific.

```
no-input-timeout    =    "No-Input-Timeout" ":" 1*19DIGIT CRLF
```

14.11.4.10 Save-Waveform

This header field allows the client to request that the verifier resource save the audio stream that was used for verification/ identification. The verifier resource MUST attempt to record the audio and make it available to the client in the form of a URI returned in the Waveform-URI header field in the VERIFICATION-COMPLETE event. If there was an error in recording the stream, or the audio content is otherwise not available, the verifier resource MUST return an empty Waveform-URI header field. The default value for this header field is "false." This header field MAY appear in the VERIFY method. Note that this header field does not appear in the VERIFY-FROM-BUFFER method, since it only controls whether or not to save the waveform for live verification/identification operations.

```
save-waveform  =  "Save-Waveform" ":" BOOLEAN CRLF
```

14.11.4.11 Media-Type

This header field MAY be specified in the SET-PARAMS, GET-PARAMS, or VERIFY methods and tells the server resource the media type of the captured audio or video, such as the one captured and returned by the Waveform-URI header field.

```
media-type   =  "Media-Type" ":" media-type-value
CRLF
```

14.11.4.12 Waveform-URI

If the Save-Waveform header field is set to "true," the verifier resource MUST attempt to record the incoming audio stream of the verification into a file and provide a URI for the client to access it. This header field MUST be present in the VERIFICATION-COMPLETE event if the Save-Waveform header field was set to true by the client. The value of the header field MUST be empty if there was some error condition preventing the server from recording. Otherwise, the URI generated by the server MUST be globally unique across the server and all its verification sessions. The content MUST be available via the URI until the verification session ends. Since the Save-Waveform header field applies only to live verification/identification operations, the server can return the Waveform-URI only in the VERIFICATION-COMPLETE event for live verification/identification operations. The server MUST also return the size in octets and the duration in milliseconds of the recorded audio waveform as parameters associated with the header field.

```
waveform-uri   =  "Waveform-URI" ":" ["<" uri ">"
";" "size" "=" 1*19DIGIT
";" "duration" "=" 1*19DIGIT] CRLF
```

14.11.4.13 Voiceprint-Exists

This header field MUST be returned in QUERY-VOICEPRINT and DELETE-VOICEPRINT responses. This is the status of the voiceprint specified in the QUERY-VOICEPRINT method. For the DELETE-VOICEPRINT method, this header field indicates the status of the voiceprint at the moment the method execution started.

```
voiceprint-exists  =  "Voiceprint-Exists" ":" BOOLEAN CRLF
```

14.11.4.14 Ver-Buffer-Utterance

This header field is used to indicate that this utterance could be later considered for speaker verification. In this way, a client can request the server to buffer utterances while doing regular recognition or verification activities, and speaker verification can later be requested on the buffered utterances. This header field is optional in the RECOGNIZE, VERIFY, and RECORD methods. The default value for this header field is "false."

```
ver-buffer-utterance   =   "Ver-Buffer-Utterance" ":" BOOLEAN
CRLF
```

14.11.4.15 Input-Waveform-URI

This header field specifies stored audio content that the client requests the server to fetch and process according to the current verification mode, either to train the voiceprint or to verify a claimed

TABLE 14.7
Completion Cause for Speaker Verification and Identification

Cause-Code	Cause-Name	Description
000	success	VERIFY or VERIFY-FROM-BUFFER request completed successfully. The verify decision can be "accepted," "rejected," or "undecided."
001	error	VERIFY or VERIFY-FROM-BUFFER request terminated prematurely due to a verifier resource or system error.
002	no-input-timeout	VERIFY request completed with no result due to a no-input-timeout.
003	too-much-speech-timeout	VERIFY request completed with no result due to too much speech.
004	speech-too-early	VERIFY request completed with no result due to speech too soon.
005	buffer-empty	VERIFY-FROM-BUFFER request completed with no result due to empty buffer.
006	out-of-sequence	Verification operation failed due to out-of-sequence method invocations, for example, calling VERIFY before QUERY-VOICEPRINT.
007	repository-uri-failure	Failure accessing Repository-URI.
008	repository-uri-missing	Repository-URI is not specified
009	voiceprint-id-missing	Voiceprint-Identifier is not specified.
010	voiceprint-id-not-exist	Voiceprint-Identifier does not exist in the voiceprint repository.
011	speech-not-usable	VERIFY request completed with no result because the speech was not usable (too noisy, too short, etc.)

identity. This header field enables the client to implement the buffering use case where the recognizer and verifier resources are in different sessions, and the verification buffer technique cannot be used. It MAY be specified on the VERIFY request.

```
input-waveform-uri    =    "Input-Waveform-URI" ":" uri CRLF
```

14.11.4.16 Completion-Cause

This header field MUST be part of a VERIFICATION-COMPLETE event from the verifier resource to the client. This indicates the cause of VERIFY or VERIFY-FROM-BUFFER method completion. This header field MUST be sent in the VERIFY, VERIFY-FROM-BUFFER, and QUERY-VOICEPRINT responses if they return with a failure status and a COMPLETE state. In the following ABNF, the "cause-code" contains a numerical value selected from the Cause-Code column of Table 14.7. The "cause-name" contains the corresponding token selected from the Cause-Name column.

```
completion-cause    =    "Completion-Cause" ":" cause-code SP
                         cause-name CRLF
cause-code          =    3DIGIT
cause-name          =    *VCHAR
```

14.11.4.17 Completion-Reason

This header field MAY be specified in a VERIFICATION-COMPLETE event coming from the verifier resource to the client. It contains the reason text behind the VERIFY request completion. This header field communicates text describing the reason for the failure. The completion reason text is provided for client use in logs and for debugging and instrumentation purposes. Clients MUST NOT interpret the completion reason text.

```
completion-reason     =     "Completion-Reason" ":"
                            quoted-string CRLF
```

14.11.4.18 Speech-Complete-Timeout

This header field is the same as the one described for the Recognizer resource. See Section 9.4.15. This header field MAY occur in VERIFY, SET-PARAMS, or GET-PARAMS.

14.11.4.19 New-Audio-Channel

This header field is the same as the one described for the Recognizer resource. See Section 9.4.23. This header field MAY be specified in a VERIFY request.

14.11.4.20 Abort-Verification

This header field MUST be sent in a STOP request to indicate whether or not to abort a VERIFY method in progress. A value of "true" requests the server to discard the results. A value of "false" requests the server to return in the STOP response the verification results obtained up to the point it received the STOP request.

```
abort-verification    =    "Abort-Verification " ":" BOOLEAN CRLF
```

14.11.4.21 Start-Input-Timers

This header field MAY be sent as part of a VERIFY request. A value of "false" tells the verifier resource to start the VERIFY operation but not to start the no-input timer yet. The verifier resource MUST NOT start the timers until the client sends a START-INPUT-TIMERS request to the resource. This is useful in the scenario when the verifier and synthesizer resources are not part of the same session. In this scenario, when a kill-on-barge-in prompt is being played, the client may want the VERIFY request to be simultaneously active so that it can detect and implement kill-on-barge-in (see Section 8.4.2). But at the same time, the client does not want the verifier resource to start the no-input timers until the prompt is finished. The default value is "true."

```
start-input-timers    =    "Start-Input-Timers" ":"
                           BOOLEAN CRLF
```

14.11.5 Verification Message Body

A verification response or event message can carry additional data, as described in the following sub-section.

14.11.5.1 Verification Result Data

Verification results are returned to the client in the message body of the VERIFICATION-COMPLETE event or the GET-INTERMEDIATE-RESULT response message as described in Section 14.14.6.3. Element and attribute descriptions for the verification portion of the NLSML format are provided in Section 14.11.5.2 with a normative definition of the schema in Section 14.16.3.

14.11.5.2 Verification Result Elements

All verification elements are contained within a single <verification-result> element under <result>. The elements are described in the following and have the schema defined in Section 16.2. The following elements are defined:

1. <voiceprint>
2. <incremental>
3. <cumulative>
4. <decision>
5. <utterance-length>
6. <device>
7. <gender>

8. <adapted>
9. <verification-score>
10. <vendor-specific-results>

14.11.5.2.1 <voiceprint> Element

This element in the verification results provides information on how the speech data matched a single voiceprint. The result data returned MAY have more than one such entity in the case of identification or multi-verification. Each <voiceprint> element and the XML data within the element describe verification result information for how well the speech data matched that particular voiceprint. The list of <voiceprint> element data are ordered according to their cumulative verification match scores, with the highest score first.

14.11.5.2.2 <cumulative> Element

Within each <voiceprint> element, there MUST be a <cumulative> element with the cumulative scores of how well multiple utterances matched the voiceprint.

14.11.5.2.3 <incremental> Element

The first <voiceprint> element MAY contain an <incremental> element with the incremental scores of how well the last utterance matched the voiceprint.

14.11.5.2.4 <Decision> Element

This element is found within the <incremental> or <cumulative> element within the verification results. Its value indicates the verification decision. It can have the values of "accepted," "rejected," or "undecided."

14.11.5.2.5 <utterance-length> Element

This element MAY occur within either the <incremental> or <cumulative> elements within the first <voiceprint> element. Its value indicates the size in milliseconds of the last utterance or the cumulated set of utterances, respectively.

14.11.5.2.6 <device> Element

This element is found within the <incremental> or <cumulative> element within the verification results. Its value indicates the apparent type of device used by the caller as determined by the verifier resource. It can have the values of "cellular-phone," "electret-phone," "carbon-button-phone," or "unknown."

14.11.5.2.7 <gender> Element

This element is found within the <incremental> or <cumulative> element within the verification results. Its value indicates the apparent gender of the speaker as determined by the verifier resource. It can have the values of "male," "female," or "unknown."

14.11.5.2.8 <adapted> Element

This element is found within the first <voiceprint> element within the verification results. When verification is trying to confirm the voiceprint, this indicates whether the voiceprint has been adapted as a consequence of analyzing the source utterances. It is not returned during verification training. The value can be "true" or "false."

14.11.5.2.9 <verification-score> Element

This element is found within the <incremental> or <cumulative> element within the verification results. Its value indicates the score of the last utterance as determined by verification. During verification, the higher the score, the more likely it is that the speaker is the same one as the one who spoke the voiceprint utterances. During training, the higher the score, the more likely the

```
<?xml version="1.0"?>

<result xmlns="urn:ietf:params:xml:ns:mrcpv2"
            grammar="What-Grammar-URI">
      <verification-result>
            <voiceprint id="johnsmith">
                  <adapted> true </adapted>
                  <incremental>
                        <utterance-length> 500 </utterance-length>
                        <device> cellular-phone </device>
                        <gender> male </gender>
                        <decision> accepted </decision>
                        <verification-score> 0.98514 </verification-score>
                  </incremental>
                  <cumulative>
                        <utterance-length> 10000 </utterance-length>
                        <device> cellular-phone </device>
                        <gender> male </gender>
                        <decision> accepted </decision>
                        <verification-score> 0.96725</verification-score>
                  </cumulative>
            </voiceprint>
            <voiceprint id="marysmith">
                  <cumulative>
                        <verification-score> 0.93410 </verification-score>
                  </cumulative>
            </voiceprint>
            <voiceprint uri="juniorsmith">
                  <cumulative>
                        <verification-score> 0.74209 </verification-score>
                  </cumulative>
            </voiceprint>
      </verification-result>
</result>
```

FIGURE 14.26 Verification Results Example 1.

speaker is to have spoken all the analyzed utterances. The value is a floating point between –1.0 and 1.0. If there are no such utterances, the score is –1. Note that the verification score is not a probability value.

14.11.5.2.10 <vendor-specific-results> Element

MRCPv2 servers MAY send verification results (Figure 14.26) containing implementation-specific data that augment the information provided by the MRCPv2-defined elements. Such data might be useful to clients who have private knowledge of how to interpret these schema extensions. Implementation-specific additions to the verification results schema MUST belong to the vendor's own namespace. In the result structure, either they MUST be indicated by a namespace prefix declared within the result or they MUST be children of an element identified as belonging to the respective namespace. The following example shows the results of three voiceprints. Note that the

```
<?xml version="1.0"?>

<result xmlns="urn:ietf:params:xml:ns:mrcpv2"
        xmlns:xmpl="http://www.example.org/2003/12/mrcpv2"
        grammar="What-Grammar-URI">
<verification-result>
    <voiceprint id="johnsmith">
        <incremental>
            <utterance-length> 500 </utterance-length>
            <device> cellular-phone </device>
            <gender> male </gender>
            <verification-score> 0.88514 </verification-score>
            <xmpl:raspiness> high </xmpl:raspiness>
            <xmpl:emotion> sadness </xmpl:emotion>
        </incremental>
        <cumulative>
            <utterance-length> 10000 </utterance-length>
            <device> cellular-phone </device>
            <gender> male </gender>
            <decision> rejected </decision>
            <verification-score> 0.9345 </verification-score>
        </cumulative>
    </voiceprint>
</verification-result>
</result>
```

FIGURE 14.27 Verification Results Example 2.

first one has crossed the verification score threshold, and the speaker has been accepted. The voiceprint was also adapted with the most recent utterance.

In this next example (Figure 14.27), the verifier has enough information to decide to reject the speaker.

14.11.6 START-SESSION

The START-SESSION method starts a speaker verification or speaker identification session. Execution of this method places the verifier resource into its initial state. If this method is called during an ongoing verification session, the previous session is implicitly aborted. If this method is invoked when VERIFY or VERIFY-FROM-BUFFER is active, the method fails, and the server returns a status-code of 402. Upon completion of the START-SESSION method, the verifier resource MUST have terminated any ongoing verification session and cleared any voiceprint designation.

A verification session is associated with the voiceprint repository to be used during the session. This is specified through the Repository-URI header field (see Section 14.11.4.1). The START-SESSION method also establishes, through the Voiceprint-Identifier header field, which voiceprints are to be matched or trained during the verification session. If this is an Identification session, or if the client wants to do Multi-Verification, the Voiceprint-Identifier header field contains a list of semicolon-separated voiceprint identifiers.

The Adapt-Model header field MAY also be present in the START-SESSION request to indicate whether or not to adapt a voiceprint based on data collected during the session (if the voiceprint verification phase succeeds). By default, the voiceprint model MUST NOT be adapted with data from a verification session. The START-SESSION also determines whether the session is for training or verification of a voiceprint. Hence, the Verification-Mode header field MUST be sent in every START-SESSION request. The value of the Verification-Mode header field MUST be either "train" or "verify."

Before a verification/identification session is started, the client may only request that VERIFY-ROLLBACK and generic SET-PARAMS and GET-PARAMS operations be performed on the verifier resource. The server MUST return status-code 402 "Method not valid in this state" for all other verification operations. A verifier resource MUST NOT have more than a single session active at one time.

```
C->S: MRCP/2.0 ... START-SESSION 314161
      Channel-Identifier:32AECB23433801@speakverify
      Repository-URI:http://www.example.com/voiceprintdbase/
      Voiceprint-Mode:verify
      Voiceprint-Identifier:johnsmith.voiceprint
      Adapt-Model:true

S->C: MRCP/2.0 ... 314161 200 COMPLETE
      Channel-Identifier:32AECB23433801@speakverify
```

14.11.7 END-SESSION

The END-SESSION method terminates an ongoing verification session and releases the verification voiceprint resources. The session may terminate in one of three ways:

1. abort – The voiceprint adaptation or creation may be aborted so that the voiceprint remains unchanged (or is not created).
2. commit – When terminating a voiceprint training session, the new voiceprint is committed to the repository.
3. adapt – An existing voiceprint is modified using a successful verification.

The Abort-Model header field MAY be included in the END-SESSION to control whether or not to abort any pending changes to the voiceprint. The default behavior is to commit (not abort) any pending changes to the designated voiceprint. The END-SESSION method may be safely executed multiple times without first executing the START-SESSION method. Any additional executions of this method without an intervening use of the START-SESSION method have no effect on the verifier resource. The following example assumes there is either a training session or a verification session in progress.

```
C->S: MRCP/2.0 ... END-SESSION 314174
      Channel-Identifier:32AECB23433801@speakverify
      Abort-Model:true

S->C: MRCP/2.0 ... 314174 200 COMPLETE
      Channel-Identifier:32AECB23433801@speakverify
```

14.11.8 QUERY-VOICEPRINT

The QUERY-VOICEPRINT method is used to get status information on a particular voiceprint and can be used by the client to ascertain whether a voiceprint or repository exists and whether it contains trained voiceprints. The response to the QUERY-VOICEPRINT request contains an indication of the status of the designated voiceprint in the Voiceprint-Exists header field, allowing the client to determine whether to use the current voiceprint for verification, train a new voiceprint, or choose a different voiceprint. A voiceprint is completely specified by providing a repository location and a voiceprint identifier. The particular voiceprint or identity within the repository is specified by a string identifier that is unique within the repository. The Voiceprint-Identifier header field carries this unique voiceprint identifier within a given repository. The following example assumes that a verification session is in progress and the voiceprint exists in the voiceprint repository.

```
C->S: MRCP/2.0 ... QUERY-VOICEPRINT 314168
      Channel-Identifier:32AECB23433801@speakverify
      Repository-URI:http://www.example.com/voiceprints/
      Voiceprint-Identifier:johnsmith.voiceprint

S->C: MRCP/2.0 ... 314168 200 COMPLETE
      Channel-Identifier:32AECB23433801@speakverify
      Repository-URI:http://www.example.com/voiceprints/
      Voiceprint-Identifier:johnsmith.voiceprint
      Voiceprint-Exists:true
```

The following example assumes that the URI provided in the Repository-URI header field is a bad URI.

```
C->S: MRCP/2.0 ... QUERY-VOICEPRINT 314168
      Channel-Identifier:32AECB23433801@speakverify
      Repository-URI:http://www.example.com/bad-uri/
      Voiceprint-Identifier:johnsmith.voiceprint

S->C: MRCP/2.0 ... 314168 405 COMPLETE
      Channel-Identifier:32AECB23433801@speakverify
      Repository-URI:http://www.example.com/bad-uri/
      Voiceprint-Identifier:johnsmith.voiceprint
      Completion-Cause:007 repository-uri-failure
```

14.11.9 DELETE-VOICEPRINT

The DELETE-VOICEPRINT method removes a voiceprint from a repository. This method MUST carry the Repository-URI and Voiceprint-Identifier header fields. An MRCPv2 server MUST reject a DELETE-VOICEPRINT request with a 401 status code unless the MRCPv2 client has been authenticated and authorized. Note that MRCPv2 does not have a standard mechanism for this. See Section 12.8. If the corresponding voiceprint does not exist, the DELETE-VOICEPRINT method MUST return a 200 status code. The following example demonstrates a DELETE-VOICEPRINT operation to remove a specific voiceprint.

```
C->S: MRCP/2.0 ... DELETE-VOICEPRINT 314168
      Channel-Identifier:32AECB23433801@speakverify
      Repository-URI:http://www.example.com/bad-uri/
      Voiceprint-Identifier:johnsmith.voiceprint

S->C: MRCP/2.0 ... 314168 200 COMPLETE
      Channel-Identifier:32AECB23433801@speakverify
```

14.11.10 VERIFY

The VERIFY method is used to request that the verifier resource either train/adapt the voiceprint or verify/identify a claimed identity. If the voiceprint is new or was deleted by a previous DELETE-VOICEPRINT method, the VERIFY method trains the voiceprint. If the voiceprint already exists, it is adapted and not retrained by the VERIFY command.

```
C->S: MRCP/2.0 ... VERIFY 543260
       Channel-Identifier:32AECB23433801@speakverify

S->C:  MRCP/2.0 ... 543260 200 IN-PROGRESS
       Channel-Identifier:32AECB23433801@speakverify
```

When the VERIFY request completes, the MRCPv2 server MUST send a VERIFICATION-COMPLETE event to the client.

14.11.11 VERIFY-FROM-BUFFER

The VERIFY-FROM-BUFFER method directs the verifier resource to verify buffered audio against a voiceprint. Only one VERIFY or VERIFY-FROM-BUFFER method may be active for a verifier resource at a time. The buffered audio is not consumed by this method, and thus, VERIFY-FROM-BUFFER may be invoked multiple times by the client to attempt verification against different voiceprints. For the VERIFY-FROM-BUFFER method, the server MAY optionally return an IN-PROGRESS response before the VERIFICATION-COMPLETE event.

When the VERIFY-FROM-BUFFER method is invoked, and the verification buffer is in use by another resource sharing it, the server MUST return an IN-PROGRESS response and wait until the buffer is available to it. The verification buffer is owned by the verifier resource but is shared with write access from other input resources on the same session. Hence, it is considered to be in use if there is a read or write operation such as a RECORD or RECOGNIZE with the Ver-Buffer-Utterance header field set to "true" on a resource that shares this buffer. Note that if a RECORD or RECOGNIZE method returns with a failure cause code, the VERIFY-FROM-BUFFER request waiting to process that buffer MUST also fail with a Completion-Cause of 005 (buffer-empty).

The following example (Figure 14.28) illustrates the usage of some buffering methods. In this scenario, the client had first performed a live verification, but the utterance had been rejected. In the meantime, the utterance is also saved to the audio buffer. Then, another voiceprint is used to do verification against the audio buffer and the utterance is accepted. For the example, we assume that both Num-Min-Verification-Phrases and Num-Max-Verification-Phrases are 1.

14.11.12 VERIFY-ROLLBACK

The VERIFY-ROLLBACK method (Figure 14.29) discards the last buffered utterance or discards the last live utterances (when the mode is "train" or "verify"). The client will probably want to invoke this method when the user provides undesirable input such as non-speech noises, sidespeech, out-of-grammar utterances, commands, etc. Note that this method does not provide a stack of rollback states. Executing VERIFY-ROLLBACK twice in succession without an intervening recognition operation has no effect on the second attempt.

14.11.13 STOP

The STOP method (Figure 14.30) from the client to the server tells the verifier resource to stop the VERIFY or VERIFY-FROM-BUFFER request if one is active. If such a request is active, and the STOP request successfully terminates it, then the response header section contains an

```
C->S: MRCP/2.0 ... START-SESSION 314161
      Channel-Identifier:32AECB23433801@speakverify
      Verification-Mode:verify
      Adapt-Model:true
      Repository-URI:http://www.example.com/voiceprints
      Voiceprint-Identifier:johnsmith.voiceprint

S->C: MRCP/2.0 ... 314161 200 COMPLETE
      Channel-Identifier:32AECB23433801@speakverify

C->S: MRCP/2.0 ... VERIFY 314162
      Channel-Identifier:32AECB23433801@speakverify
      Ver-buffer-utterance:true

S->C: MRCP/2.0 ... 314162 200 IN-PROGRESS

S->C: MRCP/2.0 ... VERIFICATION-COMPLETE 314162 COMPLETE
      Channel-Identifier:32AECB23433801@speakverify
      Completion-Cause:000 success
      Content-Type:application/nlsml+xml
      Content-Length:...

      <?xml version="1.0"?>

      <result xmlns="urn:ietf:params:xml:ns:mrcpv2"
                  grammar="What-Grammar-URI">
            <verification-result>
                  <voiceprint id="johnsmith">
                        <incremental>
                              <utterance-length> 500 </utterance-length>
                              <device> cellular-phone </device>
                              <gender> female </gender>
                              <decision> rejected </decision>
                              <verification-score> 0.05465 </verification-score>
                        </incremental>
                        <cumulative>
                              <utterance-length> 500 </utterance-length>
                              <device> cellular-phone </device>
                              <gender> female </gender>
                              <decision> rejected </decision>
                              <verification-score> 0.05465 </verification-score>
                        </cumulative>
                  </voiceprint>
            </verification-result>
      </result>

C->S: MRCP/2.0 ... QUERY-VOICEPRINT 314163
      Channel-Identifier:32AECB23433801@speakverify
      Repository-URI:http://www.example.com/voiceprints/
```

FIGURE 14.28 VERIFY-FROM-BUFFER Example.

```
        Voiceprint-Identifier:johnsmith

S->C:   MRCP/2.0 ... 314163 200 COMPLETE
        Channel-Identifier:32AECB23433801@speakverify
        Repository-URI:http://www.example.com/voiceprints/
        Voiceprint-Identifier:johnsmith.voiceprint
        Voiceprint-Exists:true

C->S:   MRCP/2.0 ... START-SESSION 314164
        Channel-Identifier:32AECB23433801@speakverify
        Verification-Mode:verify
        Adapt-Model:true
        Repository-URI:http://www.example.com/voiceprints
        Voiceprint-Identifier:marysmith.voiceprint

S->C:   MRCP/2.0 ... 314164 200 COMPLETE
        Channel-Identifier:32AECB23433801@speakverify

C->S:   MRCP/2.0 ... VERIFY-FROM-BUFFER 314165
        Channel-Identifier:32AECB23433801@speakverify

S->C:   MRCP/2.0 ... 314165 200 IN-PROGRESS
        Channel-Identifier:32AECB23433801@speakverify

S->C:   MRCP/2.0 ... VERIFICATION-COMPLETE 314165 COMPLETE
        Channel-Identifier:32AECB23433801@speakverify
        Completion-Cause:000 success
        Content-Type:application/nlsml+xml
        Content-Length:...

        <?xml version="1.0"?>

        <result xmlns="urn:ietf:params:xml:ns:mrcpv2"
                        grammar="What-Grammar-URI">
                <verification-result>
                        <voiceprint id="marysmith">
                                <incremental>
                                        <utterance-length> 1000 </utterance-length>
                                        <device> cellular-phone </device>
                                        <gender> female </gender>
                                        <decision> accepted </decision>
                                        <verification-score> 0.98 </verification-score>
                                </incremental>
                                <cumulative>
                                        <utterance-length> 1000 </utterance-length>
                                        <device> cellular-phone </device>
                                        <gender> female </gender>
                                        <decision> accepted </decision>
                                        <verification-score> 0.98 </verification-score>
                                </cumulative>

                        </voiceprint>
                </verification-result>
        </result>

C->S:   MRCP/2.0 ... END-SESSION 314166
        Channel-Identifier:32AECB23433801@speakverify

S->C:   MRCP/2.0 ... 314166 200 COMPLETE
        Channel-Identifier:32AECB23433801@speakverify
```

FIGURE 14.28 Continued.

```
C->S: MRCP/2.0 ... VERIFY-ROLLBACK 314165
      Channel-Identifier:32AECB23433801@speakverify

S->C: MRCP/2.0 ... 314165 200 COMPLETE
      Channel-Identifier:32AECB23433801@speakverify
```

FIGURE 14.29 VERIFY-ROLLBACK Example.

```
C->S: MRCP/2.0 ... VERIFY 314177
      Channel-Identifier:32AECB23433801@speakverify

S->C: MRCP/2.0 ... 314177 200 IN-PROGRESS
      Channel-Identifier:32AECB23433801@speakverify

C->S: MRCP/2.0 ... STOP 314178
      Channel-Identifier:32AECB23433801@speakverify

S->C: MRCP/2.0 ... 314178 200 COMPLETE
      Channel-Identifier:32AECB23433801@speakverify
      Active-Request-Id-List:314177
```

FIGURE 14.30 STOP Verification Example.

Active-Request-Id-List header field containing the request-id of the VERIFY or VERIFY-FROM-BUFFER request that was terminated. In this case, no VERIFICATION-COMPLETE event is sent for the terminated request. If there was no verify request active, then the response MUST NOT contain an Active-Request-Id-List header field. Either way, the response MUST contain a status-code of 200 "Success." The STOP method can carry an Abort-Verification header field, which specifies whether the verification result until that point should be discarded or returned. If this header field is not present, or if the value is "true," the verification result is discarded, and the STOP response does not contain any result data. If the header field is present, and its value is "false," the STOP response MUST contain a Completion-Cause header field and carry the Verification result data in its body. An aborted VERIFY request does an automatic rollback and hence, does not affect the cumulative score. A VERIFY request that was stopped with no Abort-Verification header field or with the Abort-Verification header field set to "false" does affect cumulative scores and would need to be explicitly rolled back if the client does not want the verification result considered in the cumulative scores. The following example assumes that a voiceprint identity has already been established.

14.11.14 START-INPUT-TIMERS

This request is sent from the client to the verifier resource to start the no-input timer, usually once the client has ascertained that any audio prompts to the user have played to completion.

```
C->S: MRCP/2.0 ... START-INPUT-TIMERS 543260
      Channel-Identifier:32AECB23433801@speakverify
```

```
S->C: MRCP/2.0 ... 543260 200 COMPLETE
      Channel-Identifier:32AECB23433801@speakverify
```

14.11.15 VERIFICATION-COMPLETE

The VERIFICATION-COMPLETE event follows a call to VERIFY or VERIFY-FROM-BUFFER and is used to communicate the verification results to the client. The event message body contains only verification results.

```
S->C: MRCP/2.0 ... VERIFICATION-COMPLETE 543259 COMPLETE
      Completion-Cause:000 success
      Content-Type:application/nlsml+xml
      Content-Length:...

      <?xml version="1.0"?>

      <result xmlns="urn:ietf:params:xml:ns:mrcpv2"
            grammar="What-Grammar-URI">
         <verification-result>
            <voiceprint id="johnsmith">
               <incremental>
                  <utterance-length> 500 </utterance-length>
                  <device> cellular-phone </device>
                  <gender> male </gender>
                  <decision> accepted </decision>
                  <verification-score> 0.85 </verification-score>
               </incremental>
               <cumulative>
                  <utterance-length> 1500 </utterance-length>
                  <device> cellular-phone </device>
                  <gender> male </gender>
                  <decision> accepted </decision>
                  <verification-score> 0.75 </verification-score>
               </cumulative>
            </voiceprint>
         </verification-result>
      </result>
```

14.11.16 START-OF-INPUT

The START-OF-INPUT event is returned from the server to the client once the server has detected speech. This event is always returned by the verifier resource when speech has been detected, irrespective of whether or not the recognizer and verifier resources share the same session.

```
S->C: MRCP/2.0 ... START-OF-INPUT 543259 IN-PROGRESS
      Channel-Identifier:32AECB23433801@speakverify
```

14.11.17 CLEAR-BUFFER

The CLEAR-BUFFER method can be used to clear the verification buffer. This buffer is used to buffer speech during recognition, record, or verification operations that may later be used by VERIFY-FROM-BUFFER. As noted before, the buffer associated with the verifier resource is shared by other input resources like recognizers and recorders. Hence, a CLEAR-BUFFER request fails if the verification buffer is in use. This can happen when any one of the input resources that share this buffer has an active read or write operation such as RECORD, RECOGNIZE, or VERIFY with the Ver-Buffer-Utterance header field set to "true."

```
C->S: MRCP/2.0 ... CLEAR-BUFFER 543260
       Channel-Identifier:32AECB23433801@speakverify

S->C:  MRCP/2.0 ... 543260 200 COMPLETE
       Channel-Identifier:32AECB23433801@speakverify
```

14.11.18 GET-INTERMEDIATE-RESULT

A client can use the GET-INTERMEDIATE-RESULT method to poll for intermediate results of a verification request that is in progress. Invoking this method does not change the state of the resource. The verifier resource collects the accumulated verification results and returns the information in the method response. The message body in the response to a GET-INTERMEDIATE-RESULT REQUEST contains only verification results. The method response MUST NOT contain a Completion-Cause header field, as the request is not yet complete. If the resource does not have a verification in progress, the response has a 402 failure status-code and no result in the body.

```
C->S: MRCP/2.0 ... GET-INTERMEDIATE-RESULT 543260
       Channel-Identifier:32AECB23433801@speakverify

S->C:  MRCP/2.0 ... 543260 200 COMPLETE
       Channel-Identifier:32AECB23433801@speakverify
       Content-Type:application/nlsml+xml
       Content-Length:...

       <?xml version="1.0"?>

       <result xmlns="urn:ietf:params:xml:ns:mrcpv2"
             grammar="What-Grammar-URI">
          <verification-result>
             <voiceprint id="marysmith">
                <incremental>
                   <utterance-length> 50 </utterance-length>
                   <device> cellular-phone </device>
                   <gender> female </gender>
                   <decision> undecided </decision>
                   <verification-score> 0.85 </verification-score>
                </incremental>
                <cumulative>
                   <utterance-length> 150 </utterance-length>
                   <device> cellular-phone </device>
                   <gender> female </gender>
                   <decision> undecided </decision>
                   <verification-score> 0.65 </verification-score>
                </cumulative>
             </voiceprint>
          </verification-result>
       </result>
```

14.12 SECURITY CONSIDERATIONS

MRCPv2 is designed to comply with the security-related requirements specificationed in the SPEECHSC requirements (RFC4313). Implementers and users of MRCPv2 are strongly encouraged to read the Security Considerations section of RFC 4313, because that specification contains discussion of a number of important security issues associated with the utilization of speech, such as biometric authentication technology, and on the threats against systems that store recorded speech, contain large corpora of voiceprints, and send and receive sensitive information based on voice input to a recognizer or speech output from a synthesizer. Specific security measures employed by MRCPv2 are summarized in the following sub-sections. See the corresponding sections of this specification for how the security-related machinery is invoked by individual protocol operations.

14.12.1 Rendezvous and Session Establishment

MRCPv2 control sessions are established as media sessions described by SDP within the context of a SIP dialog. In order to ensure secure rendezvous between MRCPv2 clients and servers, the following are required:

1. The SIP implementation in MRCPv2 clients and servers MUST support SIP digest authentication (RFC 3261 – also see Section 1.2) and SHOULD employ it.
2. The SIP implementation in MRCPv2 clients and servers MUST support "sips" URIs and SHOULD employ "sips" URIs; this includes that clients and servers SHOULD set up TLS (RFC 5246) connections.
3. If media stream cryptographic keying is done through SDP (e.g., using RFC 4568), the MRCPv2 clients and servers MUST employ the "sips" URI.
4. When TLS is used for SIP, the client MUST verify the identity of the server to which it connects, following the rules and guidelines defined in RFC 5922.

14.12.2 Control Channel Protection

Sensitive data is carried over the MRCPv2 control channel. This includes things like the output of speech recognition operations, speaker verification results, input to text-to-speech conversion, personally identifying grammars, etc. For this reason, MRCPv2 servers must be properly authenticated, and the control channel must permit the use of both confidentiality and integrity for the data. To ensure control channel protection, MRCPv2 clients and servers MUST support TLS and SHOULD utilize it by default unless alternative control channel protection is used. When TLS is used, the client MUST verify the identity of the server to which it connects, following the rules and guidelines defined in RFC 4572. If there are multiple TLS-protected channels between the client and the server, the server MUST NOT send a response to the client over a channel for which the TLS identities of the server or client differ from the channel over which the server received the corresponding request. Alternative control-channel protection MAY be used if desired (e.g., Security Architecture for the Internet Protocol (IPsec) (RFC 4301).

14.12.3 Media Session Protection

Sensitive data is also carried on media sessions terminating on MRCPv2 servers (the other end of a media channel may or may not be on the MRCPv2 client). This data includes the user's spoken utterances and the output of text-to-speech operations. MRCPv2 servers MUST support a security mechanism for protection of audio media sessions. MRCPv2 clients that originate or consume audio similarly MUST support a security mechanism for protection of the audio. One such mechanism is the secure real-time transport protocol (SRTP) (RFC 3711).

14.12.4 Indirect Content Access

MCRPv2 employs content indirection extensively. Content may be fetched and/or stored based on URI addressing on systems other than the MRCPv2 client or server. Not all the stored content is necessarily sensitive (e.g., XML schemas), but the majority generally needs protection, and some indirect content, such as voice recordings and voiceprints, is extremely sensitive and must always be protected. MRCPv2 clients and servers MUST implement HTTPS for indirect content access and SHOULD employ secure access for all sensitive indirect content. Other secure URI schemes such as Secure FTP (FTPS) (RFC 4217) MAY also be used. See Section 14.6.2.15 for the header fields used to transfer cookie information between the MRCPv2 client and server if needed for authentication.

Access to URIs provided by servers introduces risks that need to be considered. Although RFC 6454 discusses and focuses on a same-origin policy, to which MRCPv2 does not restrict URIs, it still provides an excellent description of the pitfalls of blindly following server-provided URIs in Section 3 of the RFC. Servers also need to be aware that clients could provide URIs to sites designed to tie up the server in long or otherwise problematic specification fetches. MRCPv2 servers, and the services they access, MUST always be prepared for the possibility of such a denial-of-service attack.

MRCPv2 makes no inherent assumptions about the lifetime and access controls associated with a URI. For example, if neither authentication nor scheme-specific access controls are used, a leak of the URI is equivalent to a leak of the content. Moreover, MRCPv2 makes no specific demands on the lifetime of a URI. If a server offers a URI, and the client takes a long, long time to access that URI, the server may have removed the resource in the interim time period. MRCPv2 deals with this case by using the URI access scheme's "resource not found" error, such as 404 for HTTPS. How long a server should keep a dynamic resource available is highly application and context dependent. However, the server SHOULD keep the resource available for a reasonable amount of time to make it likely that the client will have the resource available when the client needs the resource. Conversely, to mitigate state exhaustion attacks, MRCPv2 servers are not obligated to keep resources and resource state in perpetuity. The server SHOULD delete dynamically generated resources associated with an MRCPv2 session when the session ends.

One method to avoid resource leakage is for the server to use difficult-to-guess, one-time resource URIs. In this instance, there can be only a single access to the underlying resource using the given URI. A downside to this approach is if an attacker uses the URI before the client uses the URI; then, the client is denied the resource. Other methods would be to adopt a mechanism similar to the URLAUTH IMAP extension (RFC 4467), where the server sets cryptographic checks on URI usage, as well as capabilities for expiration, revocation, and so on. Specifying such a mechanism is beyond the scope of this specification.

14.12.5 Protection of Stored Media

MRCPv2 applications often require the use of stored media. Voice recordings are both stored (e.g., for diagnosis and system tuning) and fetched (for replaying utterances into multiple MRCPv2 resources). Voiceprints are fundamental to the speaker identification and verification functions. This data can be extremely sensitive and can present substantial privacy and impersonation risks if stolen. Systems employing MRCPv2 SHOULD be deployed in ways that minimize these risks. The SPEECHSC requirements RFC (RFC 4313) contains a more extensive discussion of these risks and ways they may be mitigated.

14.12.6 DTMF and Recognition Buffers

DTMF buffers and recognition buffers may grow large enough to exceed the capabilities of a server, and the server MUST be prepared to gracefully handle resource consumption. A server

MAY respond with the appropriate recognition incomplete if the server is in danger of running out of resources.

14.12.7 Client-Set Server Parameters

In MRCPv2, there are some tasks, such as URI resource fetches, that the server does on behalf of the client. To control this behavior, MRCPv2 has a number of server parameters that a client can configure. With one such parameter, Fetch-Timeout (Section 14.6.2.12), a malicious client could set a very large value and then request the server to fetch a non-existent specification. It is RECOMMENDED that servers be cautious about accepting long timeout values or abnormally large values for other client-set parameters.

14.12.8 DELETE-VOICEPRINT and Authorization

Since this specification does not mandate a specific mechanism for authentication and authorization when requesting DELETE-VOICEPRINT (Section 14.11.9), there is a risk that an MRCPv2 server may not do such a check for authentication and authorization. In practice, each provider of voice biometric solutions does insist on its own authentication and authorization mechanism, outside this specification, so this is not likely to be a major problem. If in the future, voice biometric providers standardize on such a mechanism, then a future version of MRCP can mandate it.

14.13 IANA CONSIDERATIONS

MRCPv2 has created IANA registers for registration of many parameters. IANA assignment/registration policies are described in RFC 5226. Interested readers are requested to see RFC 6787 for registrations of MRCPv2 parameters.

14.14 EXAMPLES

14.14.1 Message Flow

The following is an example of a typical MRCPv2 session of speech synthesis and recognition between a client and a server. Although the SDP "s=" attribute in these examples has a text description value to assist in understanding the examples, please keep in mind that RFC 3264 recommends that messages actually put on the wire use a space or a dash. The following figure illustrates opening a session to the MRCPv2 server. This exchange does not allocate a resource or setup media. It simply establishes an SIP session with the MRCPv2 server.

```
C->S:
       INVITE sip:mresources@example.com SIP/2.0
       Via:SIP/2.0/TCP client.atlanta.example.com:5060;
       branch=z9hG4bK74bg1
       Max-Forwards:6
       To:MediaServer <sip:mresources@example.com>
       From:sarvi <sip:sarvi@example.com>;tag=1928301774
       Call-ID:a84b4c76e66710
       CSeq:323123 INVITE
       Contact:<sip:sarvi@client.example.com>
       Content-Type:application/sdp
       Content-Length:...

       v=0
       o=sarvi 2614933546 2614933546 IN IP4 192.0.2.12
```

```
        s=Set up MRCPv2 control and audio
        i=Initial contact
        c=IN IP4 192.0.2.12

S->C:
        SIP/2.0 200 OK
        Via:SIP/2.0/TCP client.atlanta.example.com:5060;
        branch=z9hG4bK74bg1;received=192.0.32.10
        To:MediaServer <sip:mresources@example.com>;tag=62784
        From:sarvi <sip:sarvi@example.com>;tag=1928301774
        Call-ID:a84b4c76e66710
        CSeq:323123 INVITE
        Contact:<sip:mresources@server.example.com>
        Content-Type:application/sdp
        Content-Length:...

        v=0
        o=- 3000000001 3000000001 IN IP4 192.0.2.11
        s=Set up MRCPv2 control and audio
        i=Initial contact
        c=IN IP4 192.0.2.11

C->S:
        ACK sip:mresources@server.example.com SIP/2.0
        Via:SIP/2.0/TCP client.atlanta.example.com:5060;
        branch=z9hG4bK74bg2
        Max-Forwards:6
        To:MediaServer <sip:mresources@example.com>;tag=62784
        From:Sarvi <sip:sarvi@example.com>;tag=1928301774
        Call-ID:a84b4c76e66710
        CSeq:323123 ACK
        Content-Length:0
```

The client requests the server to create a synthesizer resource control channel to do speech synthesis. This also adds a media stream to send the generated speech. Note that in this example, the client requests a new MRCPv2 TCP stream between the client and the server. In the following requests, the client will ask to use the existing connection.

```
C->S:
        INVITE sip:mresources@server.example.com SIP/2.0
        Via:SIP/2.0/TCP client.atlanta.example.com:5060;
        branch=z9hG4bK74bg3
        Max-Forwards:6
        To:MediaServer <sip:mresources@example.com>;tag=62784
        From:sarvi <sip:sarvi@example.com>;tag=1928301774
        Call-ID:a84b4c76e66710
        CSeq:323124 INVITE
        Contact:<sip:sarvi@client.example.com>
        Content-Type:application/sdp
        Content-Length:...

        v=0
        o=sarvi 2614933546 2614933547 IN IP4 192.0.2.12
        s=Set up MRCPv2 control and audio
        i=Add TCP channel, synthesizer and one-way audio
        c=IN IP4 192.0.2.12
```

```
        t=0 0
        m=application 9 TCP/MRCPv2 1
        a=setup:active
        a=connection:new
        a=resource:speechsynth
        a=cmid:1
        m=audio 49170 RTP/AVP 0 96
        a=rtpmap:0 pcmu/8000
        a=rtpmap:96 telephone-event/8000
        a=fmtp:96 0-15
        a=recvonly
        a=mid:1

S->C:
        SIP/2.0 200 OK
        Via:SIP/2.0/TCP client.atlanta.example.com:5060;
        branch=z9hG4bK74bg3;received=192.0.32.10
        To:MediaServer <sip:mresources@example.com>;tag=62784
        From:sarvi <sip:sarvi@example.com>;tag=1928301774
        Call-ID:a84b4c76e66710
        CSeq:323124 INVITE
        Contact:<sip:mresources@server.example.com>
        Content-Type:application/sdp
        Content-Length:...

        v=0
        o=- 3000000001 3000000002 IN IP4 192.0.2.11
        s=Set up MRCPv2 control and audio
        i=Add TCP channel, synthesizer and one-way audio
        c=IN IP4 192.0.2.11
        t=0 0
        m=application 32416 TCP/MRCPv2 1
        a=setup:passive
        a=connection:new
        a=channel:32AECB23433801@speechsynth
        a=cmid:1
        m=audio 48260 RTP/AVP 0
        a=rtpmap:0 pcmu/8000
        a=sendonly
        a=mid:1

C->S:
        ACK sip:mresources@server.example.com SIP/2.0
        Via:SIP/2.0/TCP client.atlanta.example.com:5060;
        branch=z9hG4bK74bg4
        Max-Forwards:6
        To:MediaServer <sip:mresources@example.com>;tag=62784
        From:Sarvi <sip:sarvi@example.com>;tag=1928301774
        Call-ID:a84b4c76e66710
        CSeq:323124 ACK
        Content-Length:0
```

This exchange allocates an additional resource control channel for a recognizer. Since a recognizer would need to receive an audio stream for recognition, this interaction also updates the audio stream to sendrecv, making it a two-way audio stream.

```
C->S:
       INVITE sip:mresources@server.example.com SIP/2.0
       Via:SIP/2.0/TCP client.atlanta.example.com:5060;
       branch=z9hG4bK74bg5
       Max-Forwards:6
       To:MediaServer <sip:mresources@example.com>;tag=62784
       From:sarvi <sip:sarvi@example.com>;tag=1928301774
       Call-ID:a84b4c76e66710
       CSeq:323125 INVITE
       Contact:<sip:sarvi@client.example.com>
       Content-Type:application/sdp
       Content-Length:...

       v=0
       o=sarvi 2614933546 2614933548 IN IP4 192.0.2.12
       s=Set up MRCPv2 control and audio
       i=Add recognizer and duplex the audio
       c=IN IP4 192.0.2.12
       t=0 0
       m=application 9 TCP/MRCPv2 1
       a=setup:active
       a=connection:existing
       a=resource:speechsynth
       a=cmid:1
       m=audio 49170 RTP/AVP 0 96
       a=rtpmap:0 pcmu/8000
       a=rtpmap:96 telephone-event/8000
       a=fmtp:96 0-15
       a=recvonly
       a=mid:1
       m=application 9 TCP/MRCPv2 1
       a=setup:active
       a=connection:existing
       a=resource:speechrecog
       a=cmid:2
       m=audio 49180 RTP/AVP 0 96
       a=rtpmap:0 pcmu/8000
       a=rtpmap:96 telephone-event/8000
       a=fmtp:96 0-15
       a=sendonly
       a=mid:2

S->C:
       SIP/2.0 200 OK
       Via:SIP/2.0/TCP client.atlanta.example.com:5060;
       branch=z9hG4bK74bg5;received=192.0.32.10
       To:MediaServer <sip:mresources@example.com>;tag=62784
       From:sarvi <sip:sarvi@example.com>;tag=1928301774
       Call-ID:a84b4c76e66710
       CSeq:323125 INVITE
       Contact:<sip:mresources@server.example.com>
       Content-Type:application/sdp
       Content-Length:...

       v=0
       o=- 3000000001 3000000003 IN IP4 192.0.2.11
```

```
    s=Set up MRCPv2 control and audio
    i=Add recognizer and duplex the audio
    c=IN IP4 192.0.2.11
    t=0 0
    m=application 32416 TCP/MRCPv2 1
    a=channel:32AECB23433801@speechsynth
    a=cmid:1
    m=audio 48260 RTP/AVP 0
    a=rtpmap:0 pcmu/8000
    a=sendonly
    a=mid:1
    m=application 32416 TCP/MRCPv2 1
    a=channel:32AECB23433801@speechrecog
    a=cmid:2
    m=audio 48260 RTP/AVP 0
    a=rtpmap:0 pcmu/8000
    a=rtpmap:96 telephone-event/8000
    a=fmtp:96 0-15
    a=recvonly
    a=mid:2

C->S:
    ACK sip:mresources@server.example.com SIP/2.0
    Via:SIP/2.0/TCP client.atlanta.example.com:5060;
    branch=z9hG4bK74bg6
    Max-Forwards:6
    To:MediaServer <sip:mresources@example.com>;tag=62784
    From:Sarvi <sip:sarvi@example.com>;tag=1928301774
    Call-ID:a84b4c76e66710
    CSeq:323125 ACK
    Content-Length:0
```

A MRCPv2 SPEAK request initiates speech.

```
C->S:
    MRCP/2.0 ... SPEAK 543257
    Channel-Identifier:32AECB23433801@speechsynth
    Kill-On-Barge-In:false
    Voice-gender:neutral
    Voice-age:25
    Prosody-volume:medium
    Content-Type:application/ssml+xml
    Content-Length:...

    <?xml version="1.0"?>

    <speak version="1.0"
          xmlns="http://www.w3.org/2001/10/synthesis"
          xmlns:xsi="http://www.w3.org/2001/XMLSchema-instance"
          xsi:schemaLocation="http://www.w3.org/2001/10/synthesis
          http://www.w3.org/TR/speech-synthesis/synthesis.xsd"
          xml:lang="en-US">
       <p>
          <s>You have 4 new messages.</s>
          <s>The first is from Stephanie Williams
             <mark name="Stephanie"/>
```

```
              and arrived at <break/>
              <say-as interpret-as="vxml:time">0345p</say-as>.
          </s>
          <s>The subject is <prosody
                  rate="-20%">ski trip</prosody></s>
      </p>
  </speak>
```

```
S->C:
      MRCP/2.0 ... 543257 200 IN-PROGRESS
      Channel-Identifier:32AECB23433801@speechsynth
      Speech-Marker:timestamp=857205015059
```

The synthesizer hits the special marker in the message to be spoken and faithfully informs the client of the event.

```
S->C: MRCP/2.0 ... SPEECH-MARKER 543257 IN-PROGRESS
      Channel-Identifier:32AECB23433801@speechsynth
      Speech-Marker:timestamp=857206027059;Stephanie
      The synthesizer finishes with the SPEAK request.
      S->C: MRCP/2.0 ... SPEAK-COMPLETE 543257 COMPLETE
      Channel-Identifier:32AECB23433801@speechsynth
      Speech-Marker:timestamp=857207685213;Stephanie
```

The recognizer is issued a request to listen for the customer choices.

```
      C->S: MRCP/2.0 ... RECOGNIZE 543258
      Channel-Identifier:32AECB23433801@speechrecog
      Content-Type:application/srgs+xml
      Content-Length:...

      <?xml version="1.0"?>

      <!-- the default grammar language is US English -->

      <grammar xmlns="http://www.w3.org/2001/06/grammar"
              xml:lang="en-US" version="1.0" root="request">
         <!-- single language attachment to a rule expansion -->

         <rule id="request">
            Can I speak to
            <one-of xml:lang="fr-CA">
               <item>Michel Tremblay</item>
               <item>Andre Roy</item>
            </one-of>
         </rule>
      </grammar>

S->C: MRCP/2.0 ... 543258 200 IN-PROGRESS
      Channel-Identifier:32AECB23433801@speechrecog
```

The client issues the next MRCPv2 SPEAK method.

```
C->S: MRCP/2.0 ... SPEAK 543259
      Channel-Identifier:32AECB23433801@speechsynth
      Kill-On-Barge-In:true
```

```
        Content-Type:application/ssml+xml
        Content-Length:...

        <?xml version="1.0"?>

        <speak version="1.0"
              xmlns="http://www.w3.org/2001/10/synthesis"
              xmlns:xsi="http://www.w3.org/2001/XMLSchema-instance"
              xsi:schemaLocation="http://www.w3.org/2001/10/synthesis
              http://www.w3.org/TR/speech-synthesis/synthesis.xsd"
              xml:lang="en-US">

              Welcome to ABC corporation.</s>
              Who would you like to talk to?</s>
           </p>
        </speak>

S->C:   MRCP/2.0 ... 543259 200 IN-PROGRESS
        Channel-Identifier:32AECB23433801@speechsynth
        Speech-Marker:timestamp=857207696314
```

This next section of this ongoing example demonstrates how kill-on-barge-in support works. Since this last SPEAK request had Kill-On-Barge-In set to "true," when the recognizer (the server) generated the START-OF-INPUT event while a SPEAK was active, the client immediately issued a BARGE-IN-OCCURRED method to the synthesizer resource. The speech synthesizer then terminated playback and notified the client. The completion-cause code provided the indication that this was a kill-on-barge-in interruption rather than a normal completion.

Note that since the recognition and synthesizer resources are in the same session on the same server, to obtain a faster response, the server might have internally relayed the start-of-input condition to the synthesizer directly before receiving the expected BARGE-INOCCURRED event. However, any such communication is outside the scope of MRCPv2.

```
S->C:   MRCP/2.0 ... START-OF-INPUT 543258 IN-PROGRESS
        Channel-Identifier:32AECB23433801@speechrecog
        Proxy-Sync-Id:987654321

C->S:   MRCP/2.0 ... BARGE-IN-OCCURRED 543259
        Channel-Identifier:32AECB23433801@speechsynth
        Proxy-Sync-Id:987654321

S->C:   MRCP/2.0 ... 543259 200 COMPLETE
        Channel-Identifier:32AECB23433801@speechsynth
        Active-Request-Id-List:543258
        Speech-Marker:timestamp=857206096314

S->C:   MRCP/2.0 ... SPEAK-COMPLETE 543259 COMPLETE
        Channel-Identifier:32AECB23433801@speechsynth
        Completion-Cause:001 barge-in
        Speech-Marker:timestamp=857207685213
```

The recognizer resource matched the spoken stream to a grammar and generated results. The result of the recognition is returned by the server as part of the RECOGNITION-COMPLETE event.

```
S->C:   MRCP/2.0 ... RECOGNITION-COMPLETE 543258 COMPLETE
```

```
Channel-Identifier:32AECB23433801@speechrecog
Completion-Cause:000 success
Waveform-URI:<http://web.media.com/session123/audio.wav>;
size=423523;duration=25432
Content-Type:application/nlsml+xml
Content-Length:...

<?xml version="1.0"?>

<result xmlns="urn:ietf:params:xml:ns:mrcpv2"
      xmlns:ex="http://www.example.com/example"
      grammar="session:request1@form-level.store">
   <interpretation>
      <instance name="Person">
         <ex:Person>
            <ex:Name> Andre Roy </ex:Name>
         </ex:Person>
      </instance>
      <input> may I speak to Andre Roy </input>
   </interpretation>
</result>
```

Since the client was now finished with the session, including all resources, it issued a SIP BYE request to close the SIP session. This caused all control channels and resources allocated under the session to be deallocated.

```
C->S: BYE sip:mresources@server.example.com SIP/2.0
      Via:SIP/2.0/TCP client.atlanta.example.com:5060;
      branch=z9hG4bK74bg7
      Max-Forwards:6
      From:Sarvi <sip:sarvi@example.com>;tag=1928301774
      To:MediaServer <sip:mresources@example.com>;tag=62784
      Call-ID:a84b4c76e66710
      CSeq:323126 BYE
      Content-Length:0
```

14.14.2 Recognition Result Examples

14.14.2.1 Simple ASR Ambiguity

System: To which city will you be traveling?
User: I want to go to Pittsburgh.

```
<?xml version="1.0"?>
<result xmlns="urn:ietf:params:xml:ns:mrcpv2"
      xmlns:ex="http://www.example.com/example"
      grammar="http://www.example.com/flight">
   <interpretation confidence="0.6">
      <instance>
         <ex:airline>
            <ex:to_city>Pittsburgh</ex:to_city>
         </ex:airline>
      </instance>
      <input mode="speech">
         I want to go to Pittsburgh
      </input>
```

```
</interpretation>
<interpretation confidence="0.4"
   <instance>
      <ex:airline>
         <ex:to_city>Stockholm</ex:to_city>
      </ex:airline>
   </instance>
   <input>I want to go to Stockholm</input>
</interpretation>
</result>
```

14.14.2.2 Mixed Initiative

System: What would you like?

User: I would like 2 pizzas, one with pepperoni and cheese, one with sausage and a bottle of coke, to go.

This example includes an order object, which in turn contains objects named "food_item," "drink_item," and "delivery_method." The representation assumes that there are no ambiguities in the speech or natural language processing. Note that this representation also assumes some level of intra-sentential anaphora resolution, i.e., to resolve the two "one"s as "pizza."

```
<?xml version="1.0"?>

<nl:result xmlns:nl="urn:ietf:params:xml:ns:mrcpv2"
      xmlns="http://www.example.com/example"
      grammar="http://www.example.com/foodorder">
   <nl:interpretation confidence="1.0" >
      <nl:instance>
         <order>
            <food_item confidence="1.0">
               <pizza>
                  <ingredients confidence="1.0">
                     pepperoni
                  </ingredients>
                  <ingredients confidence="1.0">
                     cheese
                  </ingredients>
               </pizza>
               <pizza>
                  <ingredients>sausage</ingredients>
               </pizza>
            </food_item>
            <drink_item confidence="1.0">
               <size>2-liter</size>
            </drink_item>
            <delivery_method>to go</delivery_method>
         </order>
      </nl:instance>
      <nl:input mode="speech">I would like 2 pizzas,
            one with pepperoni and cheese, one with sausage
            and a bottle of coke, to go.
      </nl:input>
   </nl:interpretation>
</nl:result>
```

14.14.2.3 DTMF Input

A combination of DTMF input and speech is represented using nested input elements. For example:

User: My pin is (dtmf 1 2 3 4)

```
<input>
   <input mode="speech" confidence ="1.0"
         timestamp-start="2000-04-03T0:00:00"
         timestamp-end="2000-04-03T0:00:01.5">My pin is
   </input>
   <input mode="dtmf" confidence ="1.0"
         timestamp-start="2000-04-03T0:00:01.5"
         timestamp-end="2000-04-03T0:00:02.0">1 2 3 4
   </input>
</input>
```

Note that grammars that recognize mixtures of speech and DTMF are not currently possible in SRGS; however, this representation might be needed for other applications of NLSML, and this mixture capability might be introduced in future versions of SRGS.

14.14.2.4 Interpreting Meta-Dialog and Meta-Task Utterances

Natural language communication makes use of meta-dialog and meta-task utterances. This specification is flexible enough that meta-utterances can be represented on an application-specific basis without requiring other standard markup. Here are two examples of how meta-task and meta-dialog utterances might be represented.

System: What toppings do you want on your pizza?
User: What toppings do you have?

```
<interpretation grammar="http://www.example.com/toppings">
   <instance>
      <question>
         <questioned_item>toppings</questioned_item>
         <questioned_property>
               availability
         </questioned_property>
      </question>
   </instance>
   <input mode="speech">
         what toppings do you have?
   </input>
</interpretation>

User: slow down.

<interpretation
      grammar="http://www.example.com/
      generalCommandsGrammar">
   <instance>
      <command>
         <action>reduce speech rate</action>
         <doer>system</doer>
      </command>
   </instance>
```

```
    <input mode="speech">slow down</input>
</interpretation>
```

14.14.2.5 Anaphora and Deixis

This specification can be used on an application-specific basis to represent utterances that contain unresolved anaphoric and deictic references. Anaphoric references, which include pronouns and definite noun phrases that refer to something that was mentioned in the preceding linguistic context, and deictic references, which refer to something that is present in the non-linguistic context, present similar problems in that there may not be sufficient unambiguous linguistic context to determine what their exact role in the interpretation should be. In order to represent unresolved anaphora and deixis using this specification, one strategy would be for the developer to define a more surface-oriented representation that leaves the specific details of the interpretation of the reference open. (This assumes that a later component is responsible for actually resolving the reference.) Example (ignoring the issue of representing the input from the pointing gesture):

System: What do you want to drink?
User: I want this. (Clicks on picture of large root beer.)

```
<?xml version="1.0"?>

<nl:result xmlns:nl="urn:ietf:params:xml:ns:mrcpv2"
        xmlns="http://www.example.com/example"
        grammar="http://www.example.com/beverages.grxml">
    <nl:interpretation>
        <nl:instance>
            <doer>I</doer>
            <action>want</action>
            <object>this</object>
        </nl:instance>
        <nl:input mode="speech">I want this</nl:input>
    </nl:interpretation>
</nl:result>
```

14.14.2.6 Distinguishing Individual Items from Sets with One Member

For programming convenience, it is useful to be able to distinguish between individual items and sets containing one item in the XML representation of semantic results. For example, a pizza order might consist of exactly one pizza, but a pizza might contain zero or more toppings. Since there is no standard way of marking this distinction directly in XML, in the current framework, the developer is free to adopt any conventions that would convey this information in the XML markup. One strategy would be for the developer to wrap the set of items in a grouping element, as in the following example.

```
<order>
    <pizza>
        <topping-group>
            <topping>mushrooms</topping>
        </topping-group>
    </pizza>
    <drink>coke</drink>
</order>
```

In this example, the programmer can assume that there is supposed to be exactly one pizza and one drink in the order, but the fact that there is only one topping is an accident of this particular pizza order.

Note that the client controls both the grammar and the semantics to be returned upon grammar matches, so the user of MRCPv2 is fully empowered to cause results to be returned in NLSML in such a way that the interpretation is clear to that user.

14.14.2.7 Extensibility

Extensibility in NLSML is provided via result content flexibility, as described in the discussions of meta-utterances and anaphora. NLSML can easily be used in sophisticated systems to convey application-specific information that more basic systems would not make use of, for example, defining speech acts.

14.15 ABNF NORMATIVE DEFINITION

The following productions make use of the core rules defined in Section B.1 of RFC 5234 [RFC5234].

```
LWS             =    [*WSP CRLF] 1*WSP ; linear whitespace
SWS             =    [LWS] ; sep whitespace
UTF8-NONASCII   =    %xC0-DF 1UTF8-CONT
                     / %xE0-EF 2UTF8-CONT
                     / %xF0-F7 3UTF8-CONT
                     / %xF8-FB 4UTF8-CONT
                     / %xFC-FD 5UTF8-CONT
UTF8-CONT       =    %x80-BF
UTFCHAR         =    %x21-7E
                     / UTF8-NONASCII

param           =    *pchar
quoted-string   =    SWS DQUOTE *(qdtext / quoted-pair )
                     DQUOTE
qdtext          =    LWS / %x21 / %x23-5B / %x5D-7E
                     / UTF8-NONASCII
quoted-pair     =    "\" (%x00-09 / %x0B-0C / %x0E-7F)
token           =    1*(alphanum / "-" / "." / "!" / "%" / "*"
                     / "_" / "+" / "`" / "'" / "~" )
reserved        =    ";" / "/" / "?" / ":" / "@" / "&" / "="
                     / "+" / "$" / ","
mark            =    "-" / "_" / "." / "!" / "~" / "*" / "'"
                     / "(" / ")"
unreserved      =    alphanum / mark
pchar           =    unreserved / escaped
                     / ":" / "@" / "&" / "=" / "+" / "$" / ","
alphanum        =    ALPHA / DIGIT
BOOLEAN         =    "true" / "false"
FLOAT           =    *DIGIT ["." *DIGIT]
escaped         =    "%" HEXDIG HEXDIG
fragment        =    *uric
uri             =    [ absoluteURI / relativeURI ]
                       [ "#" fragment ]
absoluteURI     =    scheme ":" ( hier-part / opaque-part )
relativeURI     =    ( net-path / abs-path / rel-path )
                       [ "?" query ]
hier-part       =    ( net-path / abs-path ) [ "?" query ]
net-path        =    "//" authority [ abs-path ]
abs-path        =    "/" path-segments
rel-path        =    rel-segment [ abs-path ]
rel-segment     =    1*( unreserved / escaped / ";" / "@"
```

```
                          / "&" / "=" / "+" / "$" / "," )
opaque-part         =     uric-no-slash *uric
uric                =     reserved / unreserved / escaped
uric-no-slash       =     unreserved / escaped / ";" / "?" / ":"
                          / "@" / "&" / "=" / "+" / "$" / ","
path-segments       =     segment *( "/" segment )
segment             =     *pchar *( ";" param )
scheme              =     ALPHA *( ALPHA / DIGIT / "+" / "-" / "." )
authority           =     srvr / reg-name
srvr                =     [ [ userinfo "@" ] hostport ]
reg-name            =     1*( unreserved / escaped / "$" / ","
                          / ";" / ":" / "@" / "&" / "=" / "+" )
query               =     *uric
userinfo            =     ( user ) [ ":" password ] "@"
user                =     1*( unreserved / escaped
                          / user-unreserved )
user-unreserved     =     "&" / "=" / "+" / "$" / "," / ";"
                          / "?" / "/"
password            =     *( unreserved / escaped
                          / "&" / "=" / "+" / "$" / "," )
hostport            =     host [ ":" port ]
host                =     hostname / IPv4address / IPv6reference
hostname            =     *( domainlabel "." ) toplabel [ "." ]
domainlabel         =     alphanum / alphanum *( alphanum / "-" )
                          alphanum
toplabel            =     ALPHA / ALPHA *( alphanum / "-" )
                          alphanum

IPv4address         =     1*3DIGIT "." 1*3DIGIT "." 1*3DIGIT "."
                          1*3DIGIT
IPv6reference       =     "[" IPv6address "]"
IPv6address         =     hexpart [ ":" IPv4address ]
hexpart             =     hexseq / hexseq "::" [ hexseq ] / "::"
                          [ hexseq ]
hexseq              =     hex4 *( ":" hex4)
hex4                =     1*4HEXDIG
port                =     1*19DIGIT
                          ; generic-message is the top-level rule
generic-message     =     start-line message-header CRLF
                          [ message-body ]
message-body        =     *OCTET
start-line          =     request-line / response-line / event-line
request-line        =     mrcp-version SP message-length SP method-name
                          SP request-id CRLF
response-line       =     mrcp-version SP message-length SP request-id
                          SP status-code SP request-state CRLF
event-line          =     mrcp-version SP message-length SP event-name
                          SP request-id SP request-state CRLF
method-name         =     generic-method
                          / synthesizer-method
                          / recognizer-method
                          / recorder-method
                          / verifier-method
generic-method      =     "SET-PARAMS"
                          / "GET-PARAMS"
request-state       =     "COMPLETE"
                          / "IN-PROGRESS"
                          / "PENDING"
```

```
event-name               =    synthesizer-event
                              / recognizer-event
                              / recorder-event
                              / verifier-event
message-header           =    1*(generic-header / resource-header /
generic-field)
generic-field            =    field-name ":" [ field-value ]
field-name               =    token
field-value              =    *LWS field-content *( CRLF 1*LWS
                              field-content)
field-content            =    <the OCTETs making up the field-value
                              and consisting of either *TEXT or
                              combinations
                              of token, separators, and quoted-string>
resource-header          =    synthesizer-header
                              / recognizer-header
                              / recorder-header
                              / verifier-header
generic-header           =    channel-identifier
                              / accept
                              / active-request-id-list
                              / proxy-sync-id
                              / accept-charset
                              / content-type
                              / content-id
                              / content-base
                              / content-encoding
                              / content-location
                              / content-length
                              / fetch-timeout
                              / cache-control
                              / logging-tag
                              / set-cookie
                              / vendor-specific
                              ; -- content-id is as defined in RFC 2392,
                              RFC 2046 and RFC 5322
                              ; -- accept and accept-charset are as
                              defined in RFC 2616
mrcp-version             =    "MRCP" "/" 1*2DIGIT "." 1*2DIGIT
message-length           =    1*19DIGIT
request-id               =    1*10DIGIT
status-code              =    3DIGIT
channel-identifier       =    "Channel-Identifier" ":"
                              channel-id CRLF
channel-id               =    1*alphanum "@" 1*alphanum
active-request-id-list   =    "Active-Request-Id-List" ":"
                              request-id *("," request-id) CRLF
proxy-sync-id            =    "Proxy-Sync-Id" ":" 1*VCHAR CRLF
content-base             =    "Content-Base" ":" absoluteURI CRLF
content-length           =    "Content-Length" ":" 1*19DIGIT CRLF
content-type             =    "Content-Type" ":" media-type-value CRLF
media-type-value         =    type "/" subtype *( ";" parameter )
type                     =    token
subtype                  =    token
parameter                =    attribute "=" value
attribute                =    token
value                    =    token / quoted-string
content-encoding         =    "Content-Encoding" ":"
```

```
                                 *WSP content-coding
                                 *(*WSP "," *WSP content-coding *WSP )
                                 CRLF
content-coding              =    token
content-location            =    "Content-Location" ":"
                                 ( absoluteURI / relativeURI ) CRLF
cache-control               =    "Cache-Control" ":"
                                 [*WSP cache-directive
                                 *( *WSP "," *WSP cache-directive *WSP )]
                                 CRLF
fetch-timeout               =    "Fetch-Timeout" ":" 1*19DIGIT CRLF
cache-directive             =    "max-age" "=" delta-seconds
                                 / "max-stale" ["=" delta-seconds ]
                                 / "min-fresh" "=" delta-seconds
delta-seconds               =    1*19DIGIT
logging-tag                 =    "Logging-Tag" ":" 1*UTFCHAR CRLF
vendor-specific             =    "Vendor-Specific-Parameters" ":"
                                 [vendor-specific-av-pair
                                 *(";" vendor-specific-av-pair)] CRLF
vendor-specific-av-pair     =    vendor-av-pair-name "="
                                 value
vendor-av-pair-name         =    1*UTFCHAR
set-cookie                  =    "Set-Cookie:" SP set-cookie-string
set-cookie-string           =    cookie-pair *( ";" SP cookie-av )
cookie-pair                 =    cookie-name "=" cookie-value
cookie-name                 =    token
cookie-value                =    *cookie-octet / ( DQUOTE *cookie-octet
                                 DQUOTE )
cookie-octet                =    %x21 / %x23-2B / %x2D-3A / %x3C-5B /
                                 %x5D-7E
token                       =    <token, defined in [RFC2616], Section 2.2>
cookie-av                   =    expires-av / max-age-av / domain-av /
                                 path-av / secure-av / httponly-av /
                                 extension-av / age-av
expires-av                  =    "Expires=" sane-cookie-date
sane-cookie-date            =    <rfc1123-date, defined in RFC 2616, Section
                                 3.3.1>
max-age-av                  =    "Max-Age=" non-zero-digit *DIGIT
non-zero-digit              =    %x31-39
domain-av                   =    "Domain=" domain-value
domain-value                =    <subdomain>
path-av                     =    "Path=" path-value
path-value                  =    <any CHAR except CTLs or ";">
secure-av                   =    "Secure"
httponly-av                 =    "HttpOnly"
extension-av                =    <any CHAR except CTLs or ";">
age-av                      =    "Age=" delta-seconds
                                     ; Synthesizer ABNF
synthesizer-method          =    "SPEAK"
                                 / "STOP"
                                 / "PAUSE"
                                 / "RESUME"
                                 / "BARGE-IN-OCCURRED"
                                 / "CONTROL"
                                 / "DEFINE-LEXICON"
synthesizer-event           =    "SPEECH-MARKER"
                                 / "SPEAK-COMPLETE"
```

```
synthesizer-header          =   jump-size
                                / kill-on-barge-in
                                / speaker-profile
                                / completion-cause
                                / completion-reason
                                / voice-parameter
                                / prosody-parameter
                                / speech-marker
                                / speech-language
                                / fetch-hint
                                / audio-fetch-hint
                                / failed-uri
                                / failed-uri-cause
                                / speak-restart
                                / speak-length
                                / load-lexicon
                                / lexicon-search-order
jump-size                   =   "Jump-Size" ":" speech-length-value CRLF
speech-length-value         =   numeric-speech-length
                                / text-speech-length
text-speech-length          =   1*UTFCHAR SP "Tag"
numeric-speech-length       =   ("+" / "-") positive-speech-length
positive-speech-length      =   1*19DIGIT SP numeric-speech-unit
numeric-speech-unit         =   "Second"
                                / "Word"
                                / "Sentence"
                                / "Paragraph"
kill-on-barge-in            =   "Kill-On-Barge-In" ":" BOOLEAN
                                CRLF
speaker-profile             =   "Speaker-Profile" ":" uri CRLF
completion-cause            =   "Completion-Cause" ":" cause-code SP
                                cause-name CRLF
cause-code                  =   3DIGIT
cause-name                  =   *VCHAR
completion-reason           =   "Completion-Reason" ":"
                                quoted-string CRLF
voice-parameter             =   voice-gender
                                / voice-age
                                / voice-variant
                                / voice-name
voice-gender                =   "Voice-Gender:" voice-gender-value CRLF
voice-gender-value          =   "male"
                                / "female"
                                / "neutral"
voice-age                   =   "Voice-Age:" 1*3DIGIT CRLF
voice-variant               =   "Voice-Variant:" 1*19DIGIT CRLF
voice-name                  =   "Voice-Name:"
                                1*UTFCHAR *(1*WSP 1*UTFCHAR) CRLF
prosody-parameter           =   "Prosody-" prosody-param-name ":"
prosody-param-value CRLF
prosody-param-name          =   1*VCHAR
prosody-param-value         =   1*VCHAR
timestamp                   =   "timestamp" "=" time-stamp-value
time-stamp-value            =   1*20DIGIT
speech-marker               =   "Speech-Marker" ":"
                                timestamp
                                [";" 1*(UTFCHAR / %x20)] CRLF
```

```
speech-language          =     "Speech-Language" ":" 1*VCHAR CRLF
fetch-hint               =     "Fetch-Hint" ":" ("prefetch" / "safe")
CRLF
audio-fetch-hint         =     "Audio-Fetch-Hint" ":"
                               ("prefetch" / "safe" / "stream") CRLF
failed-uri               =     "Failed-URI" ":" absoluteURI CRLF
failed-uri-cause         =     "Failed-URI-Cause" ":" 1*UTFCHAR CRLF
speak-restart            =     "Speak-Restart" ":" BOOLEAN CRLF
speak-length             =     "Speak-Length" ":" positive-length-value
                               CRLF
positive-length-value    =     positive-speech-length
                               / text-speech-length
load-lexicon             =     "Load-Lexicon" ":" BOOLEAN CRLF
lexicon-search-order     =     "Lexicon-Search-Order" ":"
                               "<" absoluteURI ">" *(" " "<" absoluteURI
                               ">") CRLF
                               ; Recognizer ABNF
recognizer-method        =     recog-only-method
                               / enrollment-method
recog-only-method        =     "DEFINE-GRAMMAR"
                               / "RECOGNIZE"
                               / "INTERPRET"
                               / "GET-RESULT"
                               / "START-INPUT-TIMERS"
                               / "STOP"
enrollment-method        =     "START-PHRASE-ENROLLMENT"
                               / "ENROLLMENT-ROLLBACK"
                               / "END-PHRASE-ENROLLMENT"
                               / "MODIFY-PHRASE"
                               / "DELETE-PHRASE"
recognizer-event         =     "START-OF-INPUT"
                               / "RECOGNITION-COMPLETE"
                               / "INTERPRETATION-COMPLETE"
recognizer-header        =     recog-only-header
                               / enrollment-header
recog-only-header        =     confidence-threshold
                               / sensitivity-level
                               / speed-vs-accuracy
                               / n-best-list-length
                               / input-type
                               / no-input-timeout
                               / recognition-timeout
                               / waveform-uri
                               / input-waveform-uri
                               / completion-cause
                               / completion-reason
                               / recognizer-context-block
                               / start-input-timers
                               / speech-complete-timeout
                               / speech-incomplete-timeout
                               / dtmf-interdigit-timeout
                               / dtmf-term-timeout
                               / dtmf-term-char
                               / failed-uri
                               / failed-uri-cause
                               / save-waveform
                               / media-type
```

```
                                 / new-audio-channel
                                 / speech-language
                                 / ver-buffer-utterance
                                 / recognition-mode
                                 / cancel-if-queue
                                 / hotword-max-duration
                                 / hotword-min-duration
                                 / interpret-text
                                 / dtmf-buffer-time
                                 / clear-dtmf-buffer
                                 / early-no-match
enrollment-header             =  num-min-consistent-pronunciations
                                 / consistency-threshold
                                 / clash-threshold
                                 / personal-grammar-uri
                                 / enroll-utterance
                                 / phrase-id
                                 / phrase-nl
                                 / weight
                                 / save-best-waveform
                                 / new-phrase-id
                                 / confusable-phrases-uri
                                 / abort-phrase-enrollment
confidence-threshold          =  "Confidence-Threshold" ":"
                                 FLOAT CRLF
sensitivity-level             =  "Sensitivity-Level" ":" FLOAT
                                 CRLF
speed-vs-accuracy             =  "Speed-Vs-Accuracy" ":" FLOAT
                                 CRLF
n-best-list-length            =  "N-Best-List-Length" ":" 1*19DIGIT
                                 CRLF
input-type                    =  "Input-Type" ":" inputs CRLF
inputs                        =  "speech" / "dtmf"
no-input-timeout              =  "No-Input-Timeout" ":" 1*19DIGIT
                                 CRLF
recognition-timeout           =  "Recognition-Timeout" ":" 1*19DIGIT
                                 CRLF
waveform-uri                  =  "Waveform-URI" ":" ["<" uri ">"
                                 ";" "size" "=" 1*19DIGIT
                                 ";" "duration" "=" 1*19DIGIT] CRLF
recognizer-context-block      =  "Recognizer-Context-Block" ":"
                                 [1*VCHAR] CRLF
start-input-timers            =  "Start-Input-Timers" ":"
                                 BOOLEAN CRLF
speech-complete-timeout       =  "Speech-Complete-Timeout" ":"
                                 1*19DIGIT CRLF
speech-incomplete-timeout     =  "Speech-Incomplete-Timeout" ":"
                                 1*19DIGIT CRLF
dtmf-interdigit-timeout       =  "DTMF-Interdigit-Timeout" ":"
                                 1*19DIGIT CRLF
dtmf-term-timeout             =  "DTMF-Term-Timeout" ":" 1*19DIGIT
                                 CRLF
dtmf-term-char                =  "DTMF-Term-Char" ":" VCHAR CRLF
save-waveform                 =  "Save-Waveform" ":" BOOLEAN CRLF
new-audio-channel             =  "New-Audio-Channel" ":"
                                 BOOLEAN CRLF
recognition-mode              =  "Recognition-Mode" ":"
```

```
                                   "normal" / "hotword" CRLF
cancel-if-queue              =     "Cancel-If-Queue" ":" BOOLEAN CRLF
hotword-max-duration         =     "Hotword-Max-Duration" ":"
                                   1*19DIGIT CRLF
hotword-min-duration         =     "Hotword-Min-Duration" ":"
                                   1*19DIGIT CRLF
interpret-text               =     "Interpret-Text" ":" 1*VCHAR CRLF
dtmf-buffer-time             =     "DTMF-Buffer-Time" ":" 1*19DIGIT CRLF
clear-dtmf-buffer            =     "Clear-DTMF-Buffer" ":" BOOLEAN CRLF
early-no-match               =     "Early-No-Match" ":" BOOLEAN CRLF
num-min-consistent-pronunciations    =
"Num-Min-Consistent-Pronunciations" ":"
                                   1*19DIGIT CRLF
consistency-threshold        =     "Consistency-Threshold" ":" FLOAT
                                   CRLF
clash-threshold              =     "Clash-Threshold" ":" FLOAT CRLF
personal-grammar-uri         =     "Personal-Grammar-URI" ":" uri CRLF
enroll-utterance             =     "Enroll-Utterance" ":" BOOLEAN CRLF
phrase-id                    =     "Phrase-ID" ":" 1*VCHAR CRLF
phrase-nl                    =     "Phrase-NL" ":" 1*UTFCHAR CRLF
weight                       =     "Weight" ":" FLOAT CRLF
save-best-waveform           =     "Save-Best-Waveform" ":"
                                   BOOLEAN CRLF
new-phrase-id                =     "New-Phrase-ID" ":" 1*VCHAR CRLF
confusable-phrases-uri       =     "Confusable-Phrases-URI" ":"
                                   uri CRLF
abort-phrase-enrollment      =     "Abort-Phrase-Enrollment" ":"
                                   BOOLEAN CRLF
                                   ; Recorder ABNF
recorder-method              =     "RECORD"
                                   / "STOP"
                                   / "START-INPUT-TIMERS"
recorder-event               =     "START-OF-INPUT"
                                   / "RECORD-COMPLETE"
recorder-header              =     sensitivity-level
                                   / no-input-timeout
                                   / completion-cause
                                   / completion-reason
                                   / failed-uri
                                   / failed-uri-cause
                                   / record-uri
                                   / media-type
                                   / max-time
                                   / trim-length
                                   / final-silence
                                   / capture-on-speech
                                   / ver-buffer-utterance
                                   / start-input-timers
                                   / new-audio-channel
record-uri                   =     "Record-URI" ":" [ "<" uri ">"
                                   ";" "size" "=" 1*19DIGIT
                                   ";" "duration" "=" 1*19DIGIT] CRLF
media-type                   =     "Media-Type" ":" media-type-value CRLF
max-time                     =     "Max-Time" ":" 1*19DIGIT CRLF
trim-length                  =     "Trim-Length" ":" 1*19DIGIT CRLF
final-silence                =     "Final-Silence" ":" 1*19DIGIT CRLF
capture-on-speech            =     "Capture-On-Speech " ":"
```

```
                                           BOOLEAN CRLF
                                           ; Verifier ABNF
verifier-method                    =       "START-SESSION"
                                           / "END-SESSION"
                                           / "QUERY-VOICEPRINT"
                                           / "DELETE-VOICEPRINT"
                                           / "VERIFY"
                                           / "VERIFY-FROM-BUFFER"
                                           / "VERIFY-ROLLBACK"
                                           / "STOP"
                                           / "CLEAR-BUFFER"
                                           / "START-INPUT-TIMERS"
                                           / "GET-INTERMEDIATE-RESULT"
verifier-event                     =       "VERIFICATION-COMPLETE"
                                           / "START-OF-INPUT"
verifier-header                    =       repository-uri
                                           / voiceprint-identifier
                                           / verification-mode
                                           / adapt-model
                                           / abort-model
                                           / min-verification-score
                                           / num-min-verification-phrases
                                           / num-max-verification-phrases
                                           / no-input-timeout
                                           / save-waveform
                                           / media-type
                                           / waveform-uri
                                           / voiceprint-exists
                                           / ver-buffer-utterance
                                           / input-waveform-uri
                                           / completion-cause
                                           / completion-reason
                                           / speech-complete-timeout
                                           / new-audio-channel
                                           / abort-verification
                                           / start-input-timers
                                           / input-type
repository-uri                     =       "Repository-URI" ":" uri CRLF
voiceprint-identifier              =       "Voiceprint-Identifier" ":"
                                           vid *[";" vid] CRLF
vid                                =       1*VCHAR ["." 1*VCHAR]
verification-mode                  =       "Verification-Mode" ":"
                                           verification-mode-string
verification-mode-string           =       "train" / "verify"
adapt-model                        =       "Adapt-Model" ":" BOOLEAN CRLF
abort-model                        =       "Abort-Model" ":" BOOLEAN CRLF
min-verification-score             =       "Min-Verification-Score" ":"
                                           [ %x2D ] FLOAT CRLF
num-min-verification-phrases       =       "Num-Min-Verification-Phrases"
                                                ":" 1*19DIGIT CRLF
num-max-verification-phrases       =       "Num-Max-Verification-Phrases"
                                               ":" 1*19DIGIT CRLF
voiceprint-exists                  =       "Voiceprint-Exists" ":"
                                           BOOLEAN CRLF
ver-buffer-utterance               =       "Ver-Buffer-Utterance" ":"
                                           BOOLEAN CRLF
input-waveform-uri                 =       "Input-Waveform-URI" ":" uri CRLF
```

```
abort-verification                =      "Abort-Verification " ":"
                                         BOOLEAN CRLF
```

The following productions add a new SDP session-level attribute. See Paragraph 5.

```
cmid-attribute        =     "a=cmid:" identification-tag
identification-tag    =     token
```

14.16 XML SCHEMAS

14.16.1 NLSML Schema Definition

```
<?xml version="1.0" encoding="UTF-8"?>

<xs:schema xmlns:xs="http://www.w3.org/2001/XMLSchema"
      targetNamespace="urn:ietf:params:xml:ns:mrcpv2"
      xmlns="urn:ietf:params:xml:ns:mrcpv2"
      elementFormDefault="qualified"
      attributeFormDefault="unqualified" >
   <xs:annotation>
      <xs:specificationation>
         Natural Language Semantic Markup Schema
      </xs:specificationation>
   </xs:annotation>
   <xs:include schemaLocation="enrollment-schema.rng"/>
   <xs:include schemaLocation="verification-schema.rng"/>
   <xs:element name="result">
      <xs:complexType>
         <xs:sequence>
            <xs:element name="interpretation" maxOccurs="unbounded">
               <xs:complexType>
                  <xs:sequence>
                     <xs:element name="instance">
                        <xs:complexType
                              mixed="true">
                        <xs:sequence
                              minOccurs="0">
                           <xs:any namespace="##other" processContents
                              ="lax"/>
                        </xs:sequence>
                        </xs:complexType>
                     </xs:element>
                     <xs:element name="input" minOccurs="0">
                        <xs:complexType mixed="true">
                           <xs:choice>
                              <xs:element
                                    name="noinput"
                                    minOccurs="0"/>
                              <xs:element name="nomatch"
                                    minOccurs="0"/>
                              <xs:element name="input"
                                    minOccurs="0"/>
                           </xs:choice>
                           <xs:attribute name="mode"
                                 type="xs:string" default="speech"/>
```

```
                                <xs:attribute
                                      name="confidence"
                                      type="confidenceinfo" default="1.0"/>
                                <xs:attribute
                                      name="timestamp-
                                      start"
                                      type="xs:string"/>
                                <xs:attribute
                                      name="timestamp-
                                      end"
                                      type="xs:string"/>
                             </xs:complexType>
                          </xs:element>
                       </xs:sequence>
                       <xs:attribute name="confidence" type="confidenceinfo"
                             default="1.0"/>
                       <xs:attribute name="grammar" type="xs:anyURI"
                             use="optional"/>
                    </xs:complexType>
                 </xs:element>
                 <xs:element name="enrollment-result"
                       type="enrollment-contents"/>
                 <xs:element name="verification-result"
                       type="verification-contents"/>
              </xs:sequence>
              <xs:attribute name="grammar" type="xs:anyURI"
                    use="optional"/>
           </xs:complexType>
        </xs:element>
        <xs:simpleType name="confidenceinfo">
           <xs:restriction base="xs:float">
              <xs:minInclusive value="0.0"/>
              <xs:maxInclusive value="1.0"/>
           </xs:restriction>
        </xs:simpleType>
     </xs:schema>
```

14.16.2 Enrollment Results Schema Definition

```
   <?xml version="1.0" encoding="UTF-8"?>

   <!-- MRCP Enrollment Schema

   (See http://www.oasis-open.org/committees/relax-ng/spec.html)
   -->

   <grammar datatypeLibrary="http://www.w3.org/2001/XMLSchema-datatypes"
         ns="urn:ietf:params:xml:ns:mrcpv2"
         xmlns="http://relaxng.org/ns/structure/1.0">
      <start>
         <element name="enrollment-result">
            <ref name="enrollment-content"/>
         </element>
      </start>
      <define name="enrollment-content">
```

```
    <interleave>
       <element name="num-clashes">
          <data type="nonNegativeInteger"/>
       </element>
       <element name="num-good-repetitions">
          <data type="nonNegativeInteger"/>
       </element>
       <element name="num-repetitions-still-needed">
          <data type="nonNegativeInteger"/>
       </element>
       <element name="consistency-status">
          <choice>
             <value>consistent</value>
             <value>inconsistent</value>
             <value>undecided</value>
          </choice>
       </element>
       <optional>
          <element name="clash-phrase-ids">
             <oneOrMore>
                <element name="item">
                   <data type="token"/>
                </element>
             </oneOrMore>
          </element>
       </optional>
       <optional>
          <element name="transcriptions">
             <oneOrMore>
                <element name="item">
                   <text/>
                </element>
             </oneOrMore>
          </element>
       </optional>
       <optional>
          <element name="confusable-phrases">
             <oneOrMore>
                <element name="item">
                   <text/>
                </element>
             </oneOrMore>
          </element>
       </optional>
    </interleave>
  </define>
</grammar>
```

14.16.3 Verification Results Schema Definition

```
<?xml version="1.0" encoding="UTF-8"?>

<!-- MRCP Verification Results Schema

(See http://www.oasis-open.org/committees/relax-ng/spec.html)
   -->
```

```
<grammar datatypeLibrary="http://www.w3.org/2001/XMLSchema-datatypes"
      ns="urn:ietf:params:xml:ns:mrcpv2"
      xmlns="http://relaxng.org/ns/structure/1.0">
   <start>
      <element name="verification-result">
         <ref name="verification-contents"/>
      </element>
   </start>
   <define name="verification-contents">
      <element name="voiceprint">
         <ref name="firstVoiceprintContent"/>
      </element>
      <zeroOrMore>
         <element name="voiceprint">
            <ref name="restVoiceprintContent"/>
         </element>
      </zeroOrMore>
   </define>
   <define name="firstVoiceprintContent">
      <attribute name="id">
         <data type="string"/>
      </attribute>
      <interleave>
         <optional>
            <element name="adapted">
               <data type="boolean"/>
            </element>
         </optional>
         <optional>
            <element name="needmoredata">
               <ref name="needmoredataContent"/>
            </element>
         </optional>
         <optional>
            <element name="incremental">
               <ref name="firstCommonContent"/>
            </element>
         </optional>
         <element name="cumulative">
            <ref name="firstCommonContent"/>
         </element>
      </interleave>
   </define>
   <define name="restVoiceprintContent">
      <attribute name="id">
         <data type="string"/>
      </attribute>
      <element name="cumulative">
         <ref name="restCommonContent"/>
      </element>
   </define>
   <define name="firstCommonContent">
      <interleave>
         <element name="decision">
            <ref name="decisionContent"/>
         </element>
```

```
            <optional>
                <element name="utterance-length">
                    <ref name="utterance-lengthContent"/>
                </element>
            </optional>
            <optional>
                <element name="device">
                    <ref name="deviceContent"/>
                </element>
            </optional>
            <optional>
                <element name="gender">
                    <ref name="genderContent"/>
                </element>
            </optional>
            <zeroOrMore>
                <element name="verification-score">
                    <ref name="verification-scoreContent"/>
                </element>
            </zeroOrMore>
        </interleave>
    </define>
    <define name="restCommonContent">
        <interleave>
            <optional>
                <element name="decision">
                    <ref name="decisionContent"/>
                </element>
            </optional>
            <optional>
                <element name="device">
                    <ref name="deviceContent"/>
                </element>
            </optional>
            <optional>
                <element name="gender">
                    <ref name="genderContent"/>
                </element>
            </optional>
            <zeroOrMore>
                <element name="verification-score">
                    <ref name="verification-scoreContent"/>
                </element>
            </zeroOrMore>
        </interleave>
    </define>
    <define name="decisionContent">
        <choice>
            <value>accepted</value>
            <value>rejected</value>
            <value>undecided</value>
        </choice>
    </define>
    <define name="needmoredataContent">
        <data type="boolean"/>
    </define>
```

```
        <define name="utterance-lengthContent">
            <data type="nonNegativeInteger"/>
        </define>
        <define name="deviceContent">
            <choice>
                <value>cellular-phone</value>
                <value>electret-phone</value>
                <value>carbon-button-phone</value>
                <value>unknown</value>
            </choice>
        </define>
        <define name="genderContent">
            <choice>
                <value>male</value>
                <value>female</value>
                <value>unknown</value>
            </choice>
        </define>
        <define name="verification-scoreContent">
            <data type="float">
                <param name="minInclusive">-1</param>
                <param name="maxInclusive">1</param>
            </data>
        </define>
    </grammar>
```

14.17 SUMMARY

We have described the Media Resource Control Protocol Version 2 (MRCPv2) that allows client hosts to control media service resources, such as speech synthesizer, recorder, speech recognizers, DTMF recognizer, speaker verifier, and speaker identifier, residing in servers on the network. MRCPv2 messages are Unicode encoded in UTF-8 (RFC 3629) to allow many different languages to be represented for speech. MRCPv2 requires a connection-oriented transport layer protocol such as TCP to guarantee reliable sequencing and delivery of MRCPv2 control messages between the client and the server and uses TLS for transport layer security. However, MRCPv2 needs to use SIP (RFC 3261 – also see Sections 1.2 and 1.5) to set up and tear down the media session between the server and the client, thereby establishing control channels termed "SIP Dialogs" for transferring media resources, while SDP is used to negotiate the MRCP media resources and associated parameters. Once this is done, the MRCPv2 exchange operates over the control session established earlier, allowing the client to control the media processing resources on the speech resource server.

We have described the basic MRCPv2 server and client architecture, explaining how binary encoded MRCP media payload and control messages embedded in RTP are transported over the resource control channel managed by SIP via the secured TLS/TCP/IP network. The MRCPv2 resource types are named "speechrecog," "dtmfrecog," "speechsynth," "basicsynth," "speakverify," and "recorder" for speech recognizer, DTMF recognizer, speech synthesizer, speaker verification, and speech recorder, respectively. MRCPv2 specifications are defined with Common Protocol Elements, Request, Response, Status Codes, and Events messages and their grammars. Once this is done, the MRCPv2 exchange operates over the control session established earlier, allowing the client to control the media processing resources on the speech resource server. MRCPv2 also allows message bodies to be represented in other character sets (for example, ISO 8859-1.1987 [4]).

In this chapter, the MRCPv2 Generic Methods, Headers, and Result Structure are specified. Two important generic methods – SET-PARAMS and GET-PARAMS – are defined. The SET-PARAMS method, from the client to the server, tells the MRCPv2 resource to define parameters for the session, such as voice characteristics and prosody on synthesizers, recognition timers on recognizers, and others. The GET-PARAMS method, from the client to the server, asks the MRCPv2 resource for its current session parameters, such as voice characteristics and prosody on synthesizers, recognition timers on recognizers, and others. Two types of headers are defined in detail: Generic Message Headers and Resource-Specific Headers. All header field names are case-insensitive. Generic Result Structure data from the server for the Recognizer and Verifier resources is carried as a typed media entity in the MRCPv2 message body of various events.

The resource discovery in MRCPv2 is done via using SIP OPTIONS method. The client MAY issue a SIP OPTIONS transaction to a server, which has the effect of requesting the capabilities of the server. The server MUST respond to such a request with an SDP-encoded description of its capabilities according to RFC 3264.

The speech synthesizer resource is controlled by MRCPv2 requests from the client. Similarly, the resource can respond to these requests or generate asynchronous events to the client to indicate conditions of interest to the client during the generation of the synthesized speech stream. However, the speech synthesizer resource processes text markup provided by the client and generates a stream of synthesized speech in real time. Depending upon the server implementation and the capability of this resource, the client can also dictate parameters of the synthesized speech, such as voice characteristics, speaker speed, and others. Methods such as SPEAK, STOP, BARGE-IN-OCCURRED, PAUSE, RESUME, CONTROL, SPEAK-COMPLETE, SPEECH-MARKER, and DEFINE-LEXICON that are used by the synthesizer resource are explained in detail, using examples as appropriate. In addition, the synthesizer state-machine, events, header fields, and message body are specified.

Like the speech synthesizer resource, the speech recognizer resource is also controlled by MRCPv2 requests from the client. The recognizer resource can both respond to these requests and generate asynchronous events to the client to indicate conditions of interest during the processing of the method. Actually, the recognizer resource receives an incoming voice stream and provides the client with an interpretation of what was spoken in textual form. Detailed examples for use of the methods such as DEFINE-GRAMMAR, RECOGNIZE, STOP, GET-RESULT, START-OF-INPUT, START-INPUT-TIMERS, RECOGNITION-COMPLETE, START-PHRASE-ENROLLMENT, ENROLLMENT-ROLLBACK, END-PHRASE-ENROLLMENT, MODIFY-PHRASE, DELETE-PHRASE, INTERPRET, and INTERPRETATION-COMPLETE by the speech recognizer resource are provided. The Recognizer State Machine, Recognizer Events, Recognizer Header Fields, Recognizer Message Body, and Recognizer Results of the Speech Recognizer are also specified.

The MRCPv2 server collects the digits of DTMF tones from the RTP media stream according to RFC 4733. The ASR MUST support RFC 4733 to recognize digits, and it MAY use this in the audio. MRCPv2 has the audio and video media recording capability. In fact, a URI points to the content of the stored audio and video resources that are captured. The main usages of media recorders are to capture speech audio that may be submitted for recognition at a later time and to record voice or video mails. By the way, both these applications require functionality above and beyond that specified by protocols such as RTSP (RFC 2326). This includes audio endpointing (i.e., detecting speech or silence). The support for video is OPTIONAL and is mainly to capture video mails that may require the speech or audio processing mentioned earlier. A recorder MUST provide endpointing capabilities for suppressing silence at the beginning and end of a recording, and it MAY also suppress silence in the middle of a recording. If such suppression is done, the recorder MUST maintain timing metadata to indicate the actual time stamps of the recorded media. The media recording processes are explained with detailed call flows.

We have explained that the speaker verification is a voice authentication methodology that can be used to identify the speaker in order to grant the user access to sensitive information and transactions. Because speech is a biometric, a number of essential security considerations related to biometric authentication technologies apply to its implementation and usage. The methods, responses, and events employed by MRCPv2 for doing speaker verification and identification are explained in full detail, including call flows. The usages of START-SESSION, END-SESSION, QUERY-VOICEPRINT, DELETE-VOICEPRINT, VERIFY, VERIFY-FROM-BUFFER, VERIFY-ROLLBACK, STOP, START-INPUT-TIMERS, VERIFICATION-COMPLETE, START-OF-INPUT, CLEAR-BUFFER, and GET-INTERMEDIATE-RESULT methods by the Speaker Verification and Identification are provided in detail. In addition, Speaker Verification State Machine, Verification Methods, Verification Events, Verification Header Fields, and Verification Message Body are specified.

We have provided examples of end-to-end message flows and recognition results. The examples in the recognition result include Simple ASR Ambiguity, Mixed Initiative, DTMF Input, Interpreting Meta-Dialog and Meta-Task Utterances, Anaphora and Deixis, Distinguishing Individual Items from Sets with One Member, and Extensibility. The ABNF Normative Definition for MRCPv2 is provided. The XML Schemas for MRCPv2 including NLSML Schema, Enrollment Results Schema, and Verification Results Schema are specified.

Finally, we have discussed how the security of handling media resources is paramount, because media resources like voice and video can be used for authentication. The real-time presentation of the media stream using MRCPv2 must be handled with the utmost care in addition to access to and retrieval of the stored media. We have explained all the security details with clarity distinguishing between the mandatory and the optional features for implementation.

14.18 PROBLEMS

1. What are the general capabilities of MRCPv2 (RFC 6787)? Are Audio and Video media recording also a part of this? Explain briefly for each of these capabilities. What is the reason for using Binary Unicode for MRCPv2 messages?
2. What are the differences between MRCPvl (RFC 4463) and MRCPv2 (RFC 6787)? Explain in detail.
3. Draw the MRCPv2 basic architecture and explain the characteristics with communications between the MRCPv2 server and client for media resource services such as speech synthesizer, recorder, speech recognizers, DTMF recognizer, speaker verifier, and speaker identifier residing in servers on the network.
4. Why does MRCPv2 need SIP and SDP for media services? What is the role of RTP, TLS, TCP, and IP for these media services? Explain in detail.
5. If the MRCPv2 media services are offered integrated XCON conference systems, what will the combined architecture look like? Draw the integrated conference architecture and describe the functional call flows for each of these media services (without showing MRCPv2, RTP, TLS, TCP, and IP protocol messages): speech synthesizer, recorder, speech recognizers, DTMF recognizer, speaker verifier, and speaker identifier.
6. Describe the MRCPv2 SET-PARAMS and GET-PARAMS methods, showing a use case with call flows. Explain the features of Generic Message Headers and Resource-Specific Headers of MRCPv2.
7. How does the MRCPv2 discover resources? Explain in detail after drawing the detailed call flows for a given use case.
8. What are the functions of the MRCPv2 speech synthesizer resource? Explain each of the methods that the speech synthesizer resource uses.
9. Explain with call flows the speech synthesizer resource state-machine, events, header fields, and message body, taking example use cases.
10. What are the functions of the MRCPv2 speech recognizer resource? Explain with call flows each of the methods that the speech synthesizer resource uses.

11. Explain with call flows the speech recognizer resource Recognizer State Machine, Recognizer Events, Recognizer Header Fields, Recognizer Message Body, and Recognizer Results by taking example use cases.
12. How does the MRCPv2 server collect the digits of DTMF tones? What is the ASR? How does the ASR support recognizing DTMF tones? Explain, showing call flows.
13. Explain with detail call flows how the MRCPv2 supports audio and video recording, including suppressing silence in audio. Explain why does RSTP (RFC 2326) not support by MRCPv2?
14. How does MRCPv2 speaker verification authenticate a speaker to grant access to the resources? Explain with call flows each of the methods that are used by the MRCPv2 Speaker Verification and Identification.
15. Explain with call flows the Speaker Verification State Machine, Verification Methods, Verification Events, Verification Header Fields, and Verification Message Body.
16. Explain with call flows using MRCPv2 messages flows for speech recognition as follows: Simple ASR Ambiguity, Mixed Initiative, DTMF Input, Interpreting Meta-Dialog and Meta-Task Utterances, Anaphora and Deixis, Distinguishing Individual Items from Sets with One Member, and Extensibility.
17. Explain the salient features of XML Schemas for MRCPv2, including NLSML Schema, Enrollment Results Schema, and Verification Results Schema.
18. Describe in detail how the security will be implemented in MRCPv2 for the following cases: Rendezvous and Session Establishment, Control Channel Protection, Media Session Protection, Indirect Content Access, Protection of Stored Media, DTMF and Recognition Buffers, Client-Set Server Parameters, DELETE-VOICEPRINT, and Authorization.

REFERENCES

1. Danielsen, P. et al. (2004, March) Voice Extensible Markup Language (VoiceXML) Version 2.0. *World Wide Web Consortium Recommendation REC-voicexml20-20040316.* <http://www.w3.org/TR/2004/REC-voicexml20-20040316>.
2. International Telecommunications Union. (1993) Technical Features of Push-Button Telephone Sets. *ITU-T Q.23.*
3. Walker, M., Burnett, D., and Hunt, A. (2004, September) Speech Synthesis Markup Language (SSML) Version 1.0. *World Wide Web Consortium Recommendation REC-speech-synthesis-20040907.* <http://www.w3.org/TR/2004/REC-speech-synthesis-20040907>.
4. International Organization for Standardization. (1987) Information Technology –8-Bit Single Byte Coded Graphic - Character Sets - Part 1: Latin Alphabet No. 1, JTC1/SC2. *ISO Standard 8859-1.*
5. Johnston, M. et al. (2009, February) EMMA: Extensible MultiModal Annotation Markup Language. *World Wide Web Consortium Recommendation REC-emma-20090210.* <http://www.w3.org/TR/2009/REC-emma-20090210>.
6. Layman, A. et al. (2004, February) Namespaces in XML 1.1. *World Wide Web Consortium First Edition REC-xml-names11-20040204.* <http://www.w3.org/TR/2004/REC-xml-names11-20040204>.
7. Baggia, P. et al. (2008, October) Pronunciation Lexicon Specification (PLS). *World Wide Web Consortium Recommendation REC-pronunciation-lexicon-20081014.* <http://www.w3.org/TR/2008/REC-pronunciation-lexicon-20081014>.
8. McGlashan, S. and Hunt, A. (2004, March) Speech Recognition Grammar Specification Version 1.0. *World Wide Web Consortium Recommendation REC-speech-grammar-20040316.* <http://www.w3.org/TR/2004/REC-speech-grammar-20040316>.
9. Sun Microsystems. (1998, October) Java Speech Grammar Format Version 1.0.
10. Tichelen, L. and Burke, D. (2007, April) Semantic Interpretation for Speech Recognition (SISR) Version 1.0. *World Wide Web Consortium Recommendation REC-semanticinterpretation- 20070405.* <http://www.w3.org/TR/2007/REC-semantic-interpretation-20070405>.
11. International Organization for Standardization. (1988, June) Data Elements and Interchange Formats - Information Interchange - Representation of Dates and Times. *ISO Standard 8601.*

15 Media Control Channel Framework (CFW) Call Flow Examples

15.1 INTRODUCTION

This specification (RFC 7085) that is described in this Chapter provides a list of typical MEDIACTRL Media Control Channel Framework (RFC6230 – see Section 8) call flows. The motivation for this comes from our implementation experience with the framework and its protocol. This drove us to write a simple guide to the use of the several interfaces between Application Servers (ASs) and MEDIACTRL-based Media Servers (MSs) and a base reference specification for other implementors and protocol researchers. Following this spirit, this specification covers several aspects of the interaction between ASs and MSs. However, in the context of this specification, the call flows almost always depict the interaction between a single AS and a single MS. In Section 15.7, some flows involving more entities by means of a Media Resource Broker (MRB) compliant with RFC 6917 are presented. To help readers understand all the flows (as related to both session initiation protocol [SIP] dialogs and Media Control Channel Framework [CFW] transactions), the domains hosting the AS and the MS in all the scenarios are called "as.example.com" and 'ms.example.net," respectively, per RFC 2606. The flows will often focus more on the CFW (RFC6230 – see Chapter 8) interaction rather than on the other involved protocols, e.g., SIP (RFC3261 – also see Section 1.2), the session description protocol (SDP) (RFC 3264), or real-time transport protocol (RTP) (RFC 3550).

In the next paragraphs, a brief overview of our implementation approach is described, with a particular focus on protocol-related aspects. This involves state diagrams that depict both the client side (the AS) and the server side (the MS). Of course, this section is not at all to be considered a mandatory approach to the implementation of the framework. It is only meant to help readers understand how the framework works from a practical point of view.

Once we are done with these preliminary considerations, in the subsequent sections, real-life scenarios are addressed. In this context, first of all, the establishment of the Control Channel is dealt with.

After that, some use-case scenarios involving the most typical multimedia applications are depicted and described. It is worth pointing out that this specification is not meant in any way to be a self-contained guide to implementing a MEDIACTRL-compliant framework. The specifications are a mandatory read for all implementors, especially because this specification follows their guidelines but does not delve into the details of every aspect of the protocol.

15.2 CONVENTIONS

Note that due to RFC formatting conventions, SIP/SDP and CFW lines whose content exceeds 72 characters are split across lines. This line folding is marked by a backslash at the end of the first line. This backslash, the preceding whitespace, the following CRLF, and the whitespace beginning the next line would not appear in the actual protocol contents. Note also that the indentation of the XML content is only provided for readability. Actual messages will follow strict XML syntax, which allows, but does not require, indentation. Due to the same limit of 72 characters per line, this specification also sometimes splits the content of XML elements across lines. Please be aware that when this happens, no whitespace is actually meant to be at either the beginning or the end of the element content. Note also that a few diagrams show arrows that go from a network entity to itself. It is worth pointing out that such arrows do not represent any transaction message but rather, are meant as an indication to the reader that the involved network entity made a decision, within its application logic, according to the input it previously received.

15.3 TERMINOLOGY

This specification uses the same terminology as RFC6230 (see Chapter 8), RFC6231 (see Chapter 4), RFC6505 (see Chapter 9), and RFC6917 (see Chapter 11). The following terms are only a summarization of the terms most commonly used in this context and are mostly derived from the terminology (also see Section 1.6, Table 1.2) used in the related specifications:

- COMEDIA: Connection-oriented media (i.e., transmission control protocol [TCP] and transport layer security [TLS]). Also used to signify the support in SDP for connection-oriented media and the RFCs that define that support (RFC 4145 and RFC 4572).
- Application Server: An entity that requests media processing and manipulation from an MS; typical examples are back-to-back user agents (B2BUAs) and endpoints requesting manipulation of a third party's media stream.
- Media Server: An entity that performs a service, such as media processing, on behalf of an AS; typical provided functions are mixing, announcement, tone detection and generation, and play and record services.
- Control Channel: A reliable connection between an AS and an MS that is used to exchange framework messages.
- VCR controls: Runtime control of aspects of an audio playback like speed and volume, via dual-tone multi-frequency [DTMF] signals sent by the user, in a manner that resembles the functions of a VCR (video cassette recorder) controller.

15.4 A PRACTICAL APPROACH

In this specification, we embrace an engineering approach to the description of a number of interesting scenarios that can be realized through the careful orchestration of the CFW entities, namely, the AS and the MS. We will demonstrate, through detailed call flows, how a variegated bouquet of services (ranging from very simple scenarios to much more complicated examples) can be implemented with the functionality currently offered, within the main MEDIACTRL framework, by the Control Packages that have been made available to date. The specification aims to be a useful guide for those interested in investigating the inter-operation among MEDIACTRL components as well as being a base reference specification for application developers willing to build advanced services on top of the base infrastructure made available by the framework.

15.4.1 State Diagrams

In this section, we present an "informal" view of the main MEDIACTRL protocol interactions in the form of state diagrams. Each diagram is indeed a classical representation of a Mealy automaton, comprising a number of possible protocol states, indicated with rectangular boxes. Transitions between states are indicated through edges, with each edge labeled with a slash-separated pair representing a specific input together with the associated output (a dash in the output position means that for that particular input, no output is generated from the automaton). Some of the inputs are associated with MEDIACTRL protocol messages arriving at a MEDIACTRL component while it is in a certain state. This is the case for "CONTROL," "REPORT" (in its various "flavors" – pending, terminate, etc.), "200," "202," and "Error" (error messages correspond to specific numeric codes).

Further inputs represent triggers arriving at the MEDIACTRL automaton from the upper layer, namely, the Application Programming Interface (API) used by programmers while implementing MEDIACTRL-enabled services. Such inputs have been indicated with the term "API" followed by the message that the API itself is triggering (as an example, "API terminate" is a request to send a "REPORT" message with a status of "terminate" to the peering component).

Four diagrams are provided. Two of them (Figures 15.1 and 15.2) describe normal operation of the framework. Figure 15.3 contains two diagrams describing asynchronous event notifications. Figure 15.1 embraces the MS perspective, whereas Figure 15.2 shows the AS side. The upper part of Figure 15.3 shows how events are generated, on the MS side, by issuing a CONTROL message addressed to the AS; events are acknowledged by the AS through standard 200 responses. Hence, the behavior of the AS, which mirrors that of the MS, is depicted in the lower part of the figure.

Coming back to Figure 15.1, the diagram shows that the MS activates upon reception of CONTROL messages coming from the AS. The CONTROL messages instruct the MS regarding the execution of a specific command that belongs to one of the available Control Packages. The execution of the received command can either be quick or require some time. In the former case, right after completing its operation, the MS sends back to the AS a 200 message, which basically acknowledges correct termination of the invoked task. In the latter case, the MS first sends back an interlocutory 202 message containing a "Timeout" value, which lets it enter a different state ("202" sent), while in the new state, the MS keeps on performing the invoked task. If the task does not complete in the provided timeout, the server will update the AS on the other side of the Control Channel by periodically issuing "REPORT update" messages; each such message has to be acknowledged by the AS (through a "200" response). Eventually, when the MS is done with the required service, it sends to the AS a "REPORT terminate" message. The transaction is concluded when the AS acknowledges receipt of the message. It is worth pointing out that the MS may send a 202 response after it determines that the request does not contain any errors that cannot be reported in a later REPORT terminate request instead. After the MS sends a 202 response, any error that it (or the API) finds in the request is reported in the final REPORT terminate request. Again, the behavior

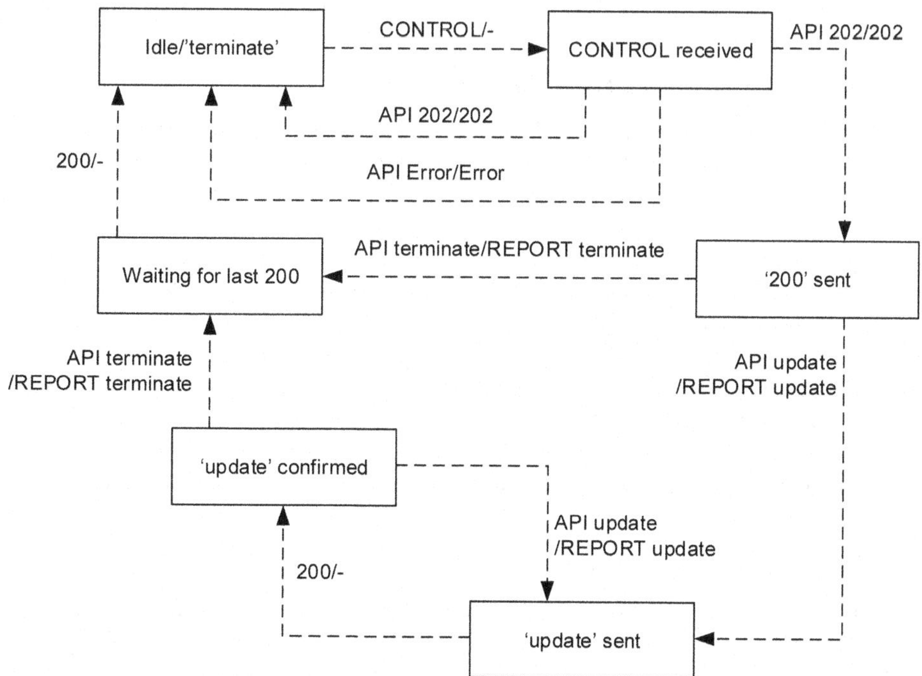

FIGURE 15.1 Media Server CFW State Diagram (200 = Successful and 202 = Timeout). (Copyright: IETF.)

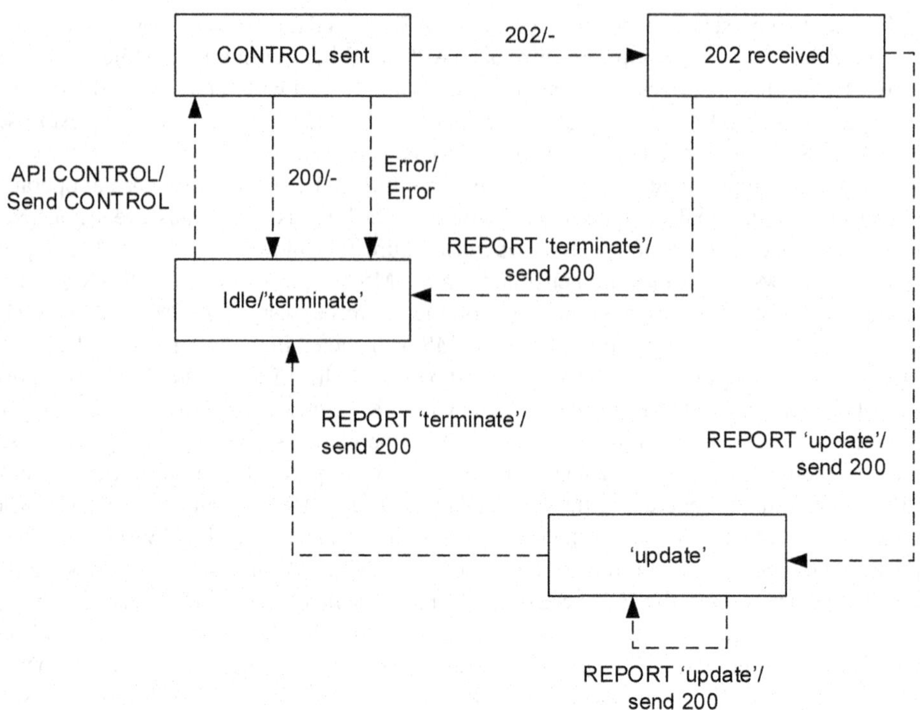

FIGURE 15.2 Application Server CFW State Diagram. (Copyright: IETF.)

of the AS, as depicted in Figure 15.2, mirrors the above-described actions undertaken at the MS side. The figures also show the cases in which transactions cannot be successfully completed due to abnormal conditions; such conditions always trigger the creation and transmission of a specific "Error" message that as mentioned previously, is reported as a numeric error code.

15.5 CONTROL CHANNEL ESTABLISHMENT

As specified in RFC 6230 (see Chapter 8), the preliminary step to any interaction between an AS and an MS is the establishment of a Control Channel between the two (Figure 15.4). As explained in the next sub-section, this is accomplished by means of a connection-oriented media (COMEDIA) (RFCs 4145 and 4572) negotiation. This negotiation allows a reliable connection to be created between the AS and the MS. It is here that the AS and the MS agree on the transport-level protocol to use TCP/stream control transmission protocol (SCTP) and whether any application-level security is needed or not (e.g., TLS).

For the sake of simplicity, we assume that an unencrypted TCP connection is negotiated between the two involved entities. Once they have connected, a SYNC message sent by the AS to the MS consolidates the Control Channel. An example of how a keep-alive message is triggered is also presented in the following paragraphs. For the sake of completeness, this section also includes a couple of common mistakes that can occur when dealing with the Control Channel establishment.

15.5.1 COMEDIA Negotiation

As a first step, the AS and the MS establish a Control SIP dialog. This is usually originated by the AS itself. The AS generates a SIP INVITE message containing in its SDP body information about the TCP connection it wants to establish with the MS. In the provided example (see Figure 15.5 and the following call flow), the AS wants to actively open a new TCP connection, which on its side

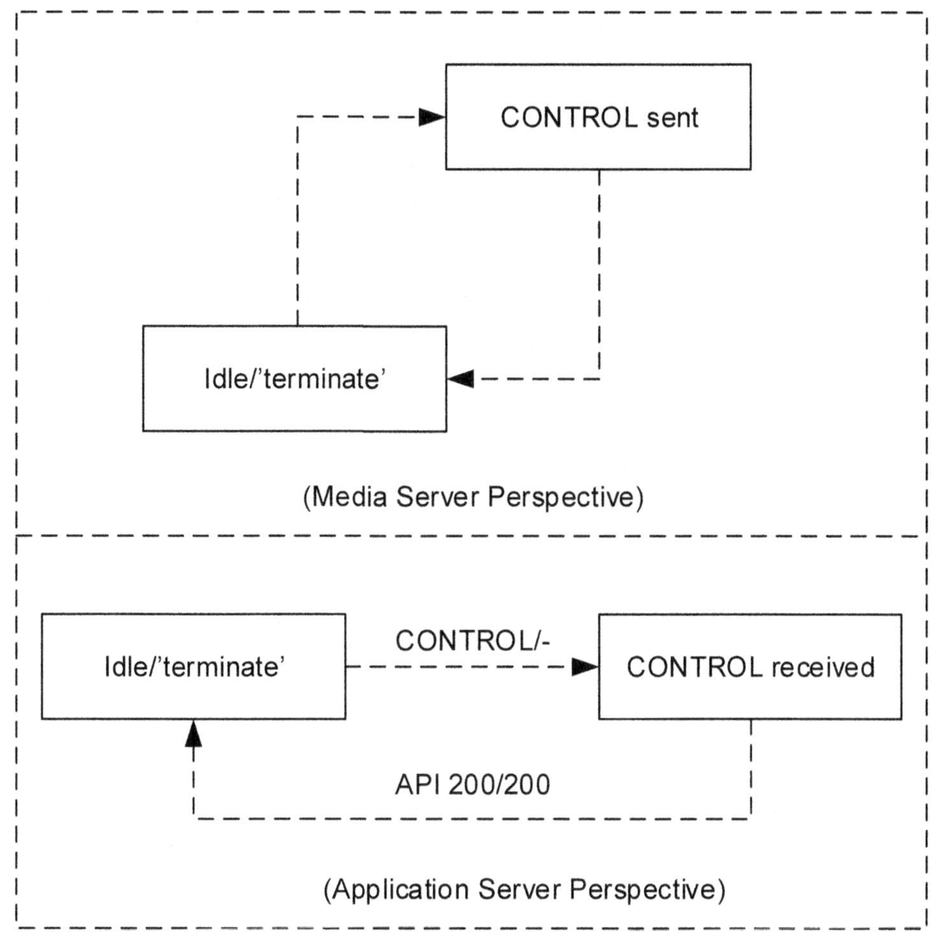

FIGURE 15.3 Event Notifications. (Copyright: IETF.)

will be bound to port 5757. If the request is fine, the MS answers by communicating to the AS the transport address to connect to in order to establish the TCP connection. In the provided example, the MS will listen on port 7575. Once this negotiation is over, the AS can effectively connect to the MS. The negotiation includes additional attributes. The "cfw-id" attribute is the most important, since it specifies the Dialog-ID, which in turn will be subsequently referred to by both the AS and the MS as specified in RFC 6230 (see Chapter 8).

```
F1. AS -> MS (SIP INVITE)
------------------------
INVITE sip:MediaServer@ms.example.net:5060 SIP/2.0
Via: SIP/2.0/UDP 203.0.113.1:5060;\
branch=z9hG4bK-d8754z-9b07c8201c3aa510-1---d8754z-;rport=5060
Max-Forwards: 70
Contact: <sip:ApplicationServer@203.0.113.1:5060>
To: <sip:MediaServer@ms.example.net:5060>
From: <sip:ApplicationServer@as.example.com:5060>;tag=4354ec63
Call-ID: MDk2YTk1MDU3YmVkZjgzYTQwYmJlNjE5NTA4ZDQ1OGY.
CSeq: 1 INVITE
Allow: INVITE, ACK, CANCEL, OPTIONS, BYE, UPDATE, REGISTER
Content-Type: application/sdp
Content-Length: 203
```

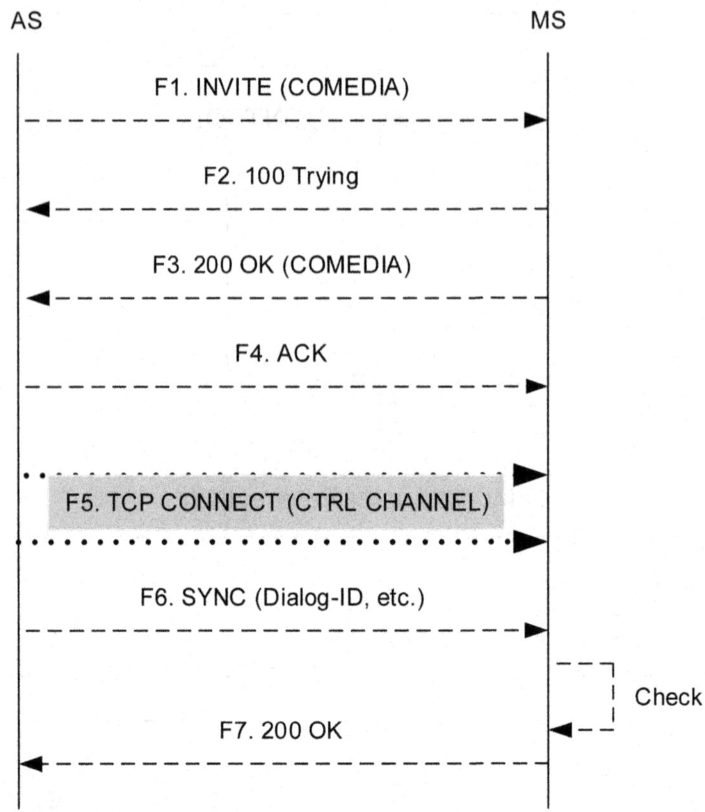

FIGURE 15.4 Control Channel Establishment. (Copyright: IETF.)

```
v=0
o=lminiero 2890844526 2890842807 IN IP4 as.example.com
s=MediaCtrl
c=IN IP4 as.example.com
t=0 0
m=application 5757 TCP cfw
a=connection:new
a=setup:active
a=cfw-id:5feb6486792a

F2. AS <- MS (SIP 100 Trying)
----------------------------
SIP/2.0 100 Trying
Via: SIP/2.0/UDP 203.0.113.1:5060; \
branch=z9hG4bK-d8754z-9b07c8201c3aa510-1---d8754z-;rport=5060
To: <sip:MediaServer@ms.example.net:5060>;tag=499a5b74
From: <sip:ApplicationServer@as.example.com:5060>;tag=4354ec63
Call-ID: MDk2YTk1MDU3YmVkZjgzYTQwYmJlNjE5NTA4ZDQ1OGY.
CSeq: 1 INVITE
Content-Length: 0

F3. AS <- MS (SIP 200 OK)
------------------------
SIP/2.0 200 OK
Via: SIP/2.0/UDP 203.0.113.1:5060; \
```

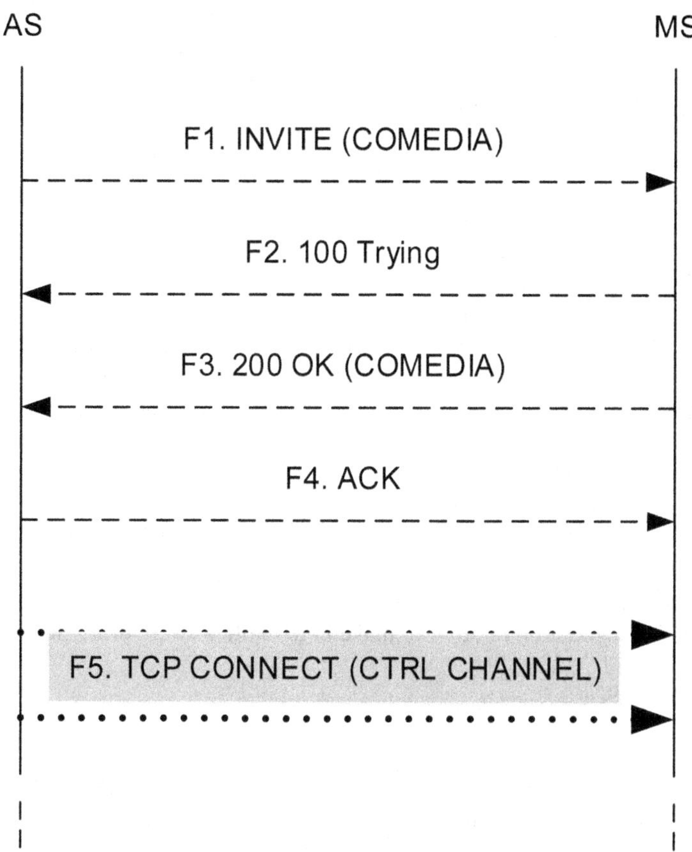

FIGURE 15.5 COMEDIA Negotiation: Sequence Diagram. (Copyright: IETF.)

```
branch=z9hG4bK-d8754z-9b07c8201c3aa510-1---d8754z-;rport=5060
Contact: <sip:MediaServer@ms.example.net:5060>
To: <sip:MediaServer@ms.example.net:5060>;tag=499a5b74
From: <sip:ApplicationServer@as.example.com:5060>;tag=4354ec63
Call-ID: MDk2YTk1MDU3YmVkZjgzYTQwYmJlNjE5NTA4ZDQ1OGY.
CSeq: 1 INVITE
Allow: INVITE, ACK, CANCEL, OPTIONS, BYE, UPDATE, REGISTER
Content-Type: application/sdp
Content-Length: 199
v=0
o=lminiero 2890844526 2890842808 IN IP4 ms.example.net
s=MediaCtrl
c=IN IP4 ms.example.net
t=0 0
m=application 7575 TCP cfw
a=connection:new
a=setup:passive
a=cfw-id:5feb6486792a

F4. AS -> MS (SIP ACK)
---------------------
ACK sip:MediaServer@ms.example.net:5060 SIP/2.0
Via: SIP/2.0/UDP 203.0.113.1:5060; \
branch=z9hG4bK-d8754z-22940f5f4589701b-1---d8754z-;rport
```

```
Max-Forwards: 70
Contact: <sip:ApplicationServer@203.0.113.1:5060>
To: <sip:MediaServer@ms.example.net:5060>;tag=499a5b74
From: <sip:ApplicationServer@as.example.com:5060>;tag=4354ec63
Call-ID: MDk2YTk1MDU3YmVkZjgzYTQwYmJlNjE5NTA4ZDQ1OGY.
CSeq: 1 ACK
Content-Length: 0
```

15.5.2 SYNC

Once the AS and the MS have successfully established a TCP connection, an additional step is needed before the Control Channel can be used. In fact, as seen in the previous sub-section, the first interaction between the AS and the MS happens by means of a SIP dialog, which in turn, allows the creation of the TCP connection. This introduces the need for a proper correlation between the above-mentioned entities (SIP dialog and TCP connection), so that the MS can be sure that the connection came from the AS that requested it.

This is accomplished by means of a dedicated framework message called a SYNC message. This SYNC message (Figure 15.6) uses a unique identifier called the Dialog-ID to validate the Control Channel. This identifier, as introduced previously, is meant to be globally unique and as such, is properly generated by the caller (the AS in the call flow) and added as an SDP media attribute (cfw-id) to the COMEDIA negotiation in order to make both entities aware of its value:

```
a=cfw-id:5feb6486792a
^^^^^^^^^^^^
```

It also offers an additional negotiation mechanism. In fact, the AS uses the SYNC not only to properly correlate, as explained before, but also to negotiate with the MS the Control Packages in which it is interested, as well as to agree on a "Keep-Alive" timer needed by both the AS and the MS so that they will know if problems on the connection occur. In the provided example (see Figure 15.6 and the related call flow), the AS sends a SYNC with a Dialog-ID constructed as needed (using the "cfw-id" attribute from the SIP dialog) and requests access to two Control Packages: specifically, the interactive voice response (IVR) package and the Mixer package. The AS also instructs the MS that a 100-second timeout is to be used for keep-alive messages. The MS validates the request by matching the received Dialog-ID with the SIP dialog values, and, assuming that it supports the

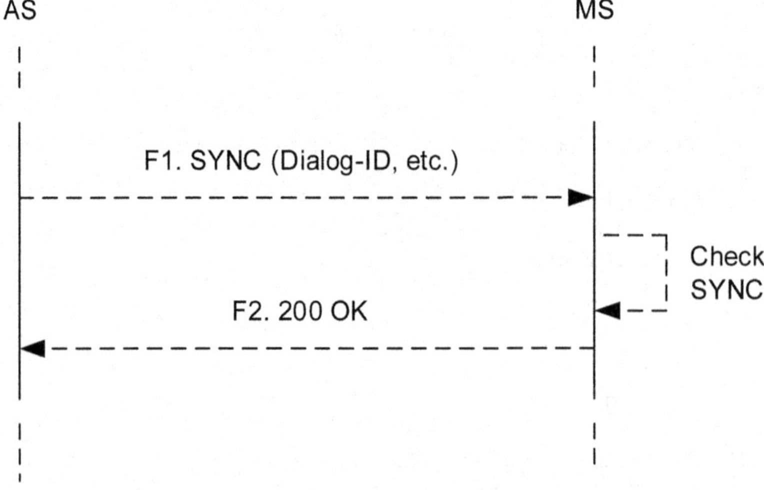

FIGURE 15.6 SYNC: Sequence Diagram

Control Packages the AS requested access to (and for the sake of this specification we assume that it does), it answers with a 200 message. Additionally, the MS provides the AS with a list of other unrequested packages it supports (in this case, just a dummy package providing testing functionality).

```
F1. AS -> MS (CFW SYNC)
----------------------
CFW 6e5e86f95609 SYNC
Dialog-ID: 5feb6486792a
Keep-Alive: 100
Packages: msc-ivr/1.0,msc-mixer/1.0

F2. AS <- MS (CFW 200)
---------------------
CFW 6e5e86f95609 200
Keep-Alive: 100
Packages: msc-ivr/1.0,msc-mixer/1.0
Supported: msc-example-pkg/1.0
```

The framework-level transaction identifier is obviously the same in both the request and the response (6e5e86f95609), since the AS needs to be able to match the response to the original request. At this point, the Control Channel is finally established, and it can be used by the AS to request services from the MS.

15.5.3 K-ALIVE

RFC 6230 (see Chapter 8) provides a mechanism for implementing a keep-alive functionality. Such a mechanism is especially useful whenever any network address translator (NAT) or firewall sits in the path between an AS and an MS. In fact, NATs and firewalls may have timeout values for the TCP connections they handle, which means that if no traffic is detected on these connections within a specific time, they could be shut down. This could be the case for a Control Channel established between an AS and an MS but not used for some time. For this reason, RFC 6230 (see Chapter 8) specifies a dedicated framework message (K-ALIVE) that the AS and MS can use in order to generate traffic on the TCP connection and keep it alive.

As discussed in Section 15.5.2, the timeout value for the keep-alive mechanism is set by the SYNC request. Specifically, in the example, the AS specified a value of 100 seconds. In fact, the timeout value is not actually negotiated between the AS and MS, as it is simply specified by whichever endpoint takes the active role. The 100-second value is compliant with how NATs and firewalls are usually implemented, since in most cases, the timeout value they use before shutting TCP connections down is around 2 minutes. Such a value has a strong meaning within the context of this mechanism. In fact, it means that the active role (the AS, in this case) has to send a K-ALIVE message before those 100 seconds pass; otherwise, the passive role (the MS) will tear down the connection, treating it like a timeout. RFC6230 (see Chapter 8) suggests a more conservative approach towards handling this timeout value, suggesting that the K-ALIVE message be triggered before 80% of the negotiated time passes (80 seconds, in this case). This is exactly the case presented in Figure 15.7.

After the Control Channel has been established (COMEDIA+SYNC), both the AS and the MS start local "Keep-Alive" timers mapped to the negotiated keep-alive timeout value (100 seconds). When about 80 seconds have passed since the start of the timer (80% of 100 seconds), the AS sends a framework-level K-ALIVE message to the MS. The message as seen in the protocol message dump is very lightweight, since it only includes a single line with no additional header. When the MS receives the K-ALIVE message, it resets its local "Keep-Alive" timer and sends a 200 message back as confirmation. As soon as the AS receives the 200 message, it resets its local "Keep-Alive"

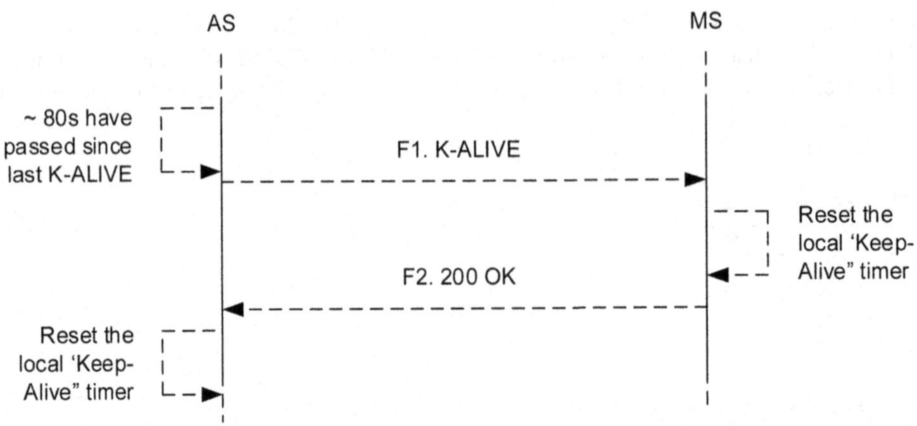

FIGURE 15.7 K-ALIVE: Sequence Diagram.

timer as well, and the mechanism starts over again. The actual transaction steps are presented in the following.

```
F1. AS -> MS (K-ALIVE)
---------------------
CFW 518ba6047880 K-ALIVE

F2. AS <- MS (CFW 200)
---------------------
CFW 518ba6047880 200
```

If the timer expired in either the AS or the MS (i.e., the K-ALIVE or the 200 arrived after the 100 seconds), the connection and the associated SIP control dialog would be torn down by the entity detecting the timeout, thus ending the interaction between the AS and the MS.

15.5.4 Wrong Behavior

This section will briefly address some types of behavior that could represent the most common mistakes when dealing with the establishment of a Control Channel between an AS and an MS. These scenarios are obviously of interest, since they result in the AS and the MS being unable to interact with each other. Specifically, these simple scenarios will be described:

F1. an AS providing the MS with a wrong Dialog-ID in the initial SYNC

F2. an AS sending a generic CONTROL message instead of SYNC as a first transaction

The first scenario is depicted in Figure 15.8.

This scenario is similar to the scenario presented in Section 5.2, but with a difference: instead of using the correct, expected Dialog-ID in the SYNC message (5feb6486792a, the one negotiated via COMEDIA), the AS uses a wrong value (4hrn7490012c). This causes the SYNC transaction to fail. First of all, the MS sends a framework-level 481 message. This response, when given in reply to a SYNC message, means that the SIP dialog associated with the provided Dialog-ID (the wrong identifier) does not exist. The Control Channel must be torn down as a consequence, and so the MS also closes the TCP connection from which it received the SYNC message. The AS at this point is supposed to tear down its SIP control dialog as well, and so it sends a SIP BYE to the MS.

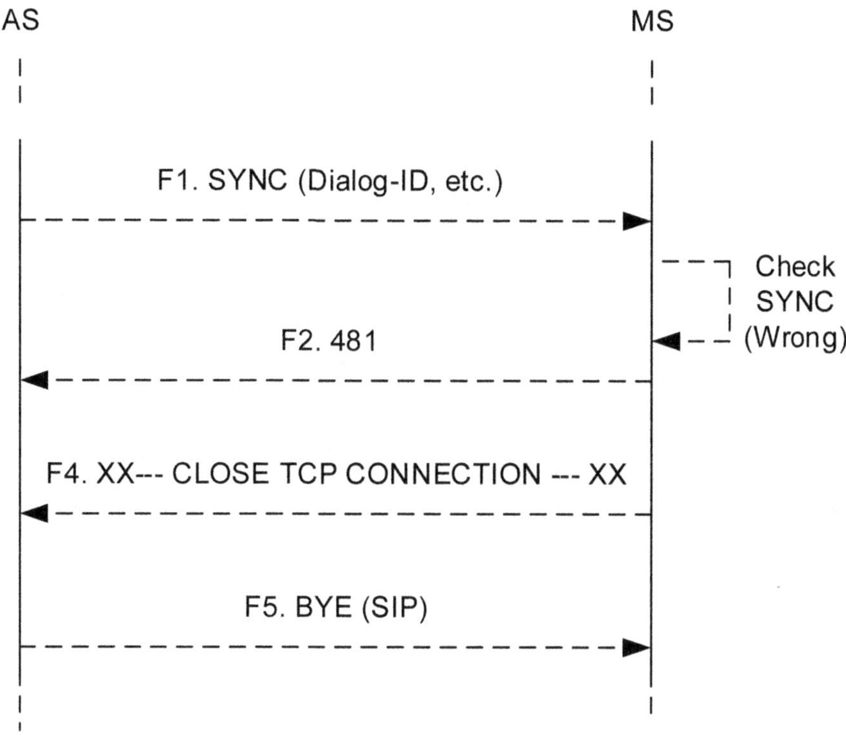

FIGURE 15.8 SYNC with Wrong Dialog-ID: Sequence Diagram. (Copyright: IETF.)

The actual transaction is presented in the following.

```
F1. AS -> MS (CFW SYNC with wrong Dialog-ID)
--------------------------------------------
CFW 2b4dd8724f27 SYNC
Dialog-ID: 4hrn7490012c
Keep-Alive: 100
Packages: msc-ivr/1.0,msc-mixer/1.0

F2. AS <- MS (CFW 481)
---------------------
CFW 2b4dd8724f27 481
```

The second scenario is depicted in Figure 15.9.

This scenario demonstrates another common mistake that could occur when trying to set up a Control Channel. In fact, RFC6230 (see Chapter 8) mandates that the first transaction after the COMEDIA negotiation must be a SYNC to conclude the setup. If the AS, instead of triggering a SYNC message as expected, sends a different message to the MS (in the following example, it tries to send an <audit> message addressed to the IVR Control Package), the MS treats it like an error. As a consequence, the MS replies with a framework-level 403 message (Forbidden) and, just as before, closes the TCP connection and waits for the related SIP control dialog to be torn down. The actual transaction is presented in the following.

```
F1. AS -> MS (CFW CONTROL instead of SYNC)
------------------------------------------
CFW 101fbbd62c35 CONTROL
```

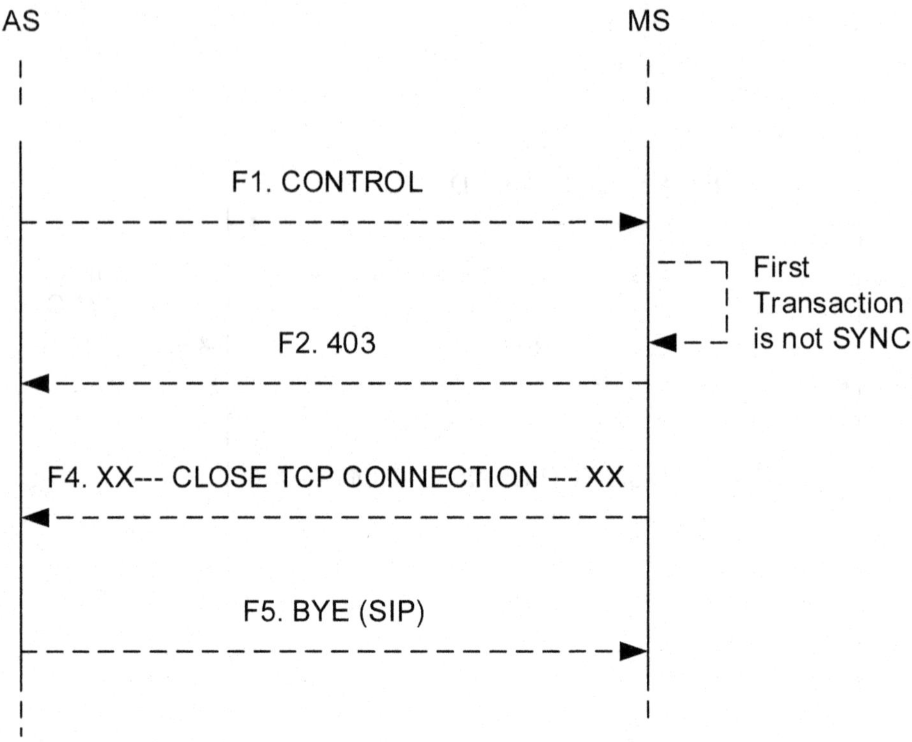

FIGURE 15.9 Incorrect First Transaction: Sequence Diagram. (Copyright: IETF.)

```
Control-Package: msc-ivr/1.0
Content-Type: application/msc-ivr+xml
Content-Length: 78
<mscivr version="1.0" xmlns="urn:ietf:params:xml:ns:msc-ivr">
<audit/>
</mscivr>

F2. AS <- MS (CFW 403 Forbidden)
--------------------------------
CFW 101fbbd62c35 403
```

15.6 USE-CASE SCENARIOS AND EXAMPLES

The following scenarios have been chosen for their common presence in many rich real-time multimedia applications. Each scenario is depicted as a set of call flows involving both the SIP/SDP signaling (UACs↔AS↔MS) and the Control Channel communication (AS↔MS). All the examples assume that a Control Channel has already been correctly established and SYNCed between the reference AS and MS. Also, unless stated otherwise, the same User Agent Client (UAC) session is referenced in all the examples that will be presented in this specification. The UAC session is assumed to have been created as described in Figure 15.10.

Note well: This is only an example of a possible approach involving a third-party call control (3pcc) negotiation among the UAC, the AS, and the MS, and as such, is not at all to be considered the mandatory way, or the best common practice, in the presented scenario. RFC 3725 (see Section 1.3) provides several different solutions and many details about how 3pcc can be realized, with pros and cons. It is also worth pointing out that the two INVITEs displayed in the figure are different SIP dialogs.

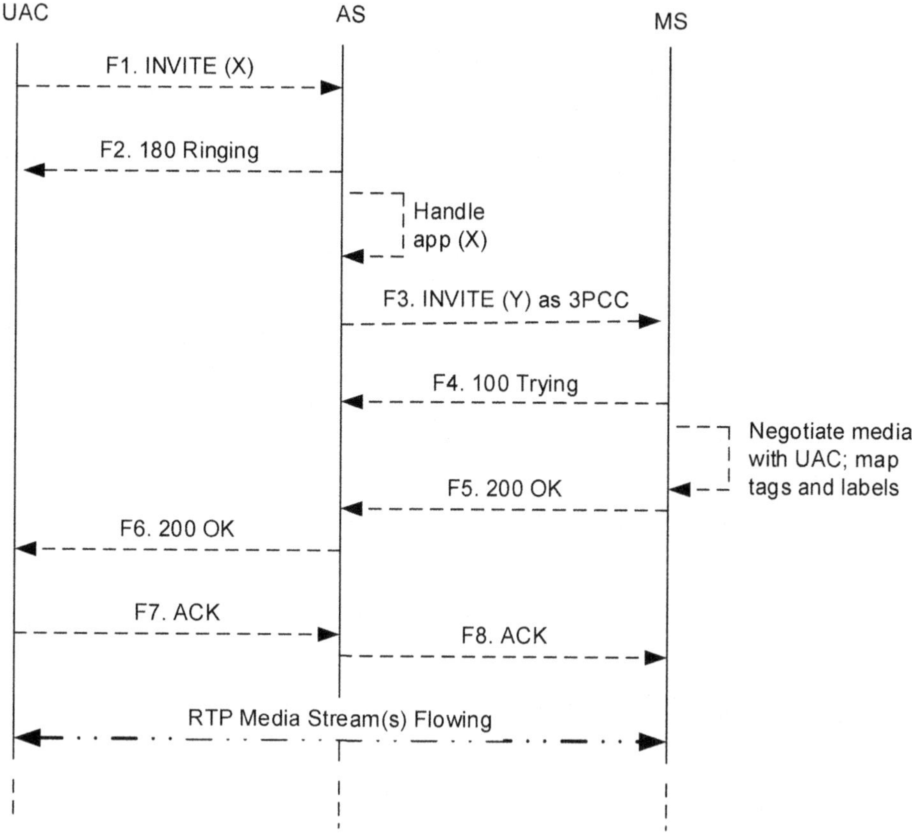

FIGURE 15.10 3pcc Sequence Diagram. (Copyright: IETF.)

The UAC first places a call to a SIP URI for which the AS is responsible. The specific URI is not relevant to the examples, since the application logic behind the mapping between a URI and the service it provides is a matter that is important only to the AS. So, a generic "sip:mediactrlDemo@as.example.com" is used in all the examples, whereas the service this URI is associated with in the AS logic is mapped scenario by scenario to the case under examination.

The UAC INVITE is treated as envisaged in RFC 5567 (see Chapter 3). The INVITE is forwarded by the AS to the MS via a third party (e.g., the 3pcc approach), without the SDP provided by the UAC being touched, in order to have the session fully negotiated by the MS according to its description. The MS matches the UAC's offer with its own capabilities and provides its answer in a 200 OK. This answer is then forwarded, again without the SDP contents being touched, by the AS to the target UAC. In this way, while the SIP signaling from the UAC is terminated in the AS, all the media will start flowing directly between the UAC and the MS.

As a consequence of this negotiation, one or more media connections are created between the MS and the UAC. They are then addressed, when needed, by the AS and the MS by means of the concatenation of tags, as specified in RFC 6230 (see Chapter 8). How the identifiers are created and addressed is explained by using the sample signaling provided in the following lines.

```
F1. UAC -> AS (SIP INVITE)
-------------------------

INVITE sip:mediactrlDemo@as.example.com SIP/2.0
Via: SIP/2.0/UDP 203.0.113.2:5063;rport;branch=z9hG4bK1396873708
From: <sip:lminiero@users.example.com>;tag=1153573888
To: <sip:mediactrlDemo@as.example.com>
```

```
Call-ID: 1355333098
CSeq: 20 INVITE
Contact: <sip:lminiero@203.0.113.2:5063>
Content-Type: application/sdp
Max-Forwards: 70
User-Agent: Linphone/2.1.1 (eXosip2/3.0.3)
Subject: Phone call
Expires: 120
Content-Length: 330
v=0
o=lminiero 123456 654321 IN IP4 203.0.113.2
s=A conversation
c=IN IP4 203.0.113.2
t=0 0
m=audio 7078 RTP/AVP 0 3 8 101
a=rtpmap:0 PCMU/8000/1
a=rtpmap:3 GSM/8000/1
a=rtpmap:8 PCMA/8000/1
a=rtpmap:101 telephone-event/8000
a=fmtp:101 0-11
m=video 9078 RTP/AVP 98
a=rtpmap:98 H263-1998/90000
a=fmtp:98 CIF=1;QCIF=1

F2. UAC <- AS (SIP 180 Ringing)
-------------------------------
SIP/2.0 180 Ringing
Via: SIP/2.0/UDP 203.0.113.2:5063;rport=5063; \
branch=z9hG4bK1396873708
Contact: <sip:mediactrlDemo@as.example.com>
To: <sip:mediactrlDemo@as.example.com>;tag=bcd47c32
From: <sip:lminiero@users.example.com>;tag=1153573888
Call-ID: 1355333098
CSeq: 20 INVITE
Content-Length: 0

F3. AS -> MS (SIP INVITE)
------------------------
INVITE sip:MediaServer@ms.example.net:5060;transport=UDP SIP/2.0
Via: SIP/2.0/UDP 203.0.113.1:5060; \
branch=z9hG4bK-d8754z-8723e421ebc45f6b-1---d8754z-;rport
Max-Forwards: 70
Contact: <sip:ApplicationServer@203.0.113.1:5060>
To: <sip:MediaServer@ms.example.net:5060>
From: <sip:ApplicationServer@as.example.com:5060>;tag=10514b7f
Call-ID: NzI0ZjQ0ZTBlMTEzMGU1ZjVhMjk5NTliMmJmZjE0NDQ.
CSeq: 1 INVITE
Allow: INVITE, ACK, CANCEL, OPTIONS, BYE, UPDATE, REGISTER
Content-Type: application/sdp
Content-Length: 330
v=0
o=lminiero 123456 654321 IN IP4 203.0.113.2
s=A conversation
c=IN IP4 203.0.113.2
t=0 0
m=audio 7078 RTP/AVP 0 3 8 101
a=rtpmap:0 PCMU/8000/1
```

```
a=rtpmap:3 GSM/8000/1
a=rtpmap:8 PCMA/8000/1
a=rtpmap:101 telephone-event/8000
a=fmtp:101 0-11
m=video 9078 RTP/AVP 98
a=rtpmap:98 H263-1998/90000
a=fmtp:98 CIF=1;QCIF=1

F4. AS <- MS (SIP 100 Trying)
----------------------------
SIP/2.0 100 Trying
Via: SIP/2.0/UDP 203.0.113.1:5060; \
branch=z9hG4bK-d8754z-8723e421ebc45f6b-1---d8754z-;rport=5060
To: <sip:MediaServer@ms.example.net:5060>;tag=6a900179
From: <sip:ApplicationServer@as.example.com:5060>;tag=10514b7f
Call-ID: NzI0ZjQ0ZTBlMTEzMGU1ZjVhMjk5NTliMmJmZjE0NDQ.
CSeq: 1 INVITE
Content-Length: 0

F5. AS <- MS (SIP 200 OK)
------------------------
SIP/2.0 200 OK
Via: SIP/2.0/UDP 203.0.113.1:5060; \
branch=z9hG4bK-d8754z-8723e421ebc45f6b-1---d8754z-;rport=5060
Contact: <sip:MediaServer@ms.example.net:5060>
To: <sip:MediaServer@ms.example.net:5060>;tag=6a900179
From: <sip:ApplicationServer@as.example.com:5060>;tag=10514b7f
Call-ID: NzI0ZjQ0ZTBlMTEzMGU1ZjVhMjk5NTliMmJmZjE0NDQ.
CSeq: 1 INVITE
Allow: INVITE, ACK, CANCEL, OPTIONS, BYE, UPDATE, REGISTER
Content-Type: application/sdp
Content-Length: 374
v=0
o=lminiero 123456 654322 IN IP4 ms.example.net
s=MediaCtrl
c=IN IP4 ms.example.net
t=0 0
m=audio 63442 RTP/AVP 0 3 8 101

a=rtpmap:0 PCMU/8000
a=rtpmap:3 GSM/8000
a=rtpmap:8 PCMA/8000
a=rtpmap:101 telephone-event/8000
a=fmtp:101 0-15
a=ptime:20
a=label:7eda834
m=video 33468 RTP/AVP 98
a=rtpmap:98 H263-1998/90000
a=fmtp:98 CIF=2
a=label:0132ca2

F6. UAC <- AS (SIP 200 OK)
-------------------------
SIP/2.0 200 OK
Via: SIP/2.0/UDP 203.0.113.2:5063;rport=5063; \
branch=z9hG4bK1396873708
Contact: <sip:mediactrlDemo@as.example.com>
```

```
To: <sip:mediactrlDemo@as.example.com>;tag=bcd47c32
From: <sip:lminiero@users.example.com>;tag=1153573888
Call-ID: 1355333098
CSeq: 20 INVITE
Allow: INVITE, ACK, CANCEL, OPTIONS, BYE, UPDATE, REGISTER
Content-Type: application/sdp
Content-Length: 374
v=0
o=lminiero 123456 654322 IN IP4 ms.example.net
s=MediaCtrl
c=IN IP4 ms.example.net
t=0 0
m=audio 63442 RTP/AVP 0 3 8 101
a=rtpmap:0 PCMU/8000
a=rtpmap:3 GSM/8000
a=rtpmap:8 PCMA/8000
a=rtpmap:101 telephone-event/8000
a=fmtp:101 0-15
a=ptime:20
a=label:7eda834
m=video 33468 RTP/AVP 98
a=rtpmap:98 H263-1998/90000
a=fmtp:98 CIF=2
a=label:0132ca2

F7. UAC -> AS (SIP ACK)
----------------------
ACK sip:mediactrlDemo@as.example.com SIP/2.0
Via: SIP/2.0/UDP 203.0.113.2:5063;rport;branch=z9hG4bK1113338059
From: <sip:lminiero@users.example.com>;tag=1153573888
To: <sip:mediactrlDemo@as.example.com>;tag=bcd47c32
Call-ID: 1355333098
CSeq: 20 ACK
Contact: <sip:lminiero@203.0.113.2:5063>
Max-Forwards: 70
User-Agent: Linphone/2.1.1 (eXosip2/3.0.3)
Content-Length: 0

F8. AS -> MS (SIP ACK)
---------------------
ACK sip:MediaServer@ms.example.net:5060;transport=UDP SIP/2.0
Via: SIP/2.0/UDP 203.0.113.1:5060; \
branch=z9hG4bK-d8754z-5246003419ccd662-1---d8754z-;rport
Max-Forwards: 70
Contact: <sip:ApplicationServer@203.0.113.1:5060>
To: <sip:MediaServer@ms.example.net:5060;tag=6a900179
From: <sip:ApplicationServer@as.example.com:5060>;tag=10514b7f
Call-ID: NzI0ZjQ0ZTBlMTEzMGU1ZjVhMjk5NTliMmJmZjE0NDQ.
CSeq: 1 ACK
Content-Length: 0
```

As a result of the 3pcc negotiation just presented, the following relevant information is retrieved:

1. The "From" and "To" tags (10514b7f and 6a900179, respectively) of the AS↔MS session:

   From: <sip:ApplicationServer@as.example.com:5060>;tag=10514b7f
   ^^^^^^^^

To: <sip:MediaServer@ms.example.net:5060>;tag=6a900179
^^^^^^^^

2. The labels (RFC 4574) associated with the negotiated media connections, in this case an audio stream (7eda834) and a video stream (0132ca2):

   m=audio 63442 RTP/AVP 0 3 8 101
   [..]
   a=label:7eda834
   ^^^^^^^

   m=video 33468 RTP/AVP 98
   [..]
   a=label:0132ca2
   ^^^^^^^

These four identifiers allow the AS and MS to univocally and unambiguously address to each other the connections associated with the related UAC. Specifically:

1. 10514b7f:6a900179, the concatenation of the "From" and "To" tags through a colon (":") token, addresses all the media connections between the MS and the UAC.
2. 10514b7f:6a900179 ↔ 7eda834, the association of the previous value with the label attribute, addresses only one of the media connections of the UAC session (in this case, the audio stream). Since, as will be made clearer in the example scenarios, the explicit identifiers in requests can only address "from:tag" connections, an additional mechanism will be required to have a finer control of individual media streams (i.e., by means of the <stream> element in package-level requests). The mapping that the AS makes between the UACs↔AS and the AS↔MS SIP dialogs is out of scope for this specification. We just assume that the AS knows how to address the right connection according to the related session it has with a UAC (e.g., to play an announcement to a specific UAC). This is obviously very important, since the AS is responsible for all the business logic of the multimedia application it provides.

15.6.1 Echo Test

The echo test is the simplest example scenario that can be achieved by means of an MS. It basically consists of a UAC directly or indirectly "talking" to itself. A media perspective of such a scenario is depicted in Figure 15.11.

From the framework point of view, when the UAC's leg is not attached to anything yet, what appears is shown in Figure 15.12: since there is no connection involving the UAC yet, the frames it might be sending are discarded, and nothing is sent to it (except for silence, if its transmission is requested).

Starting from these considerations, two different approaches to the echo test scenario are explored in this specification:

1. A Direct Echo Test approach, where the UAC directly talks to itself
2. A Recording-based Echo Test approach, where the UAC indirectly talks to itself

15.6.1.1 Direct Echo Test

In the Direct Echo Test approach, the UAC is directly connected to itself. This means that as depicted in Figure 15.13, each frame the MS receives from the UAC is sent back to it in real time.

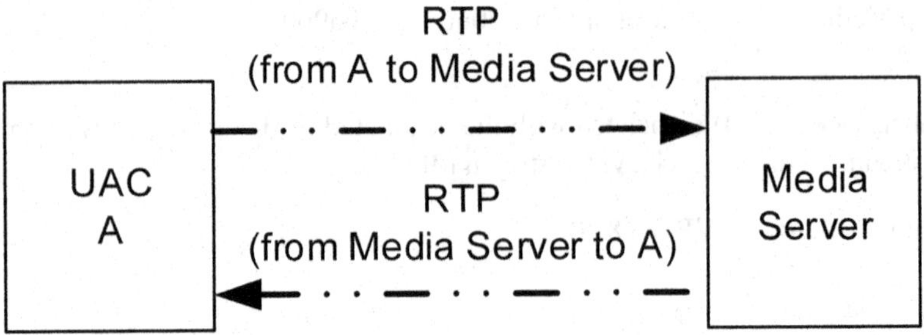

FIGURE 15.11 Echo Test: Media Perspective.

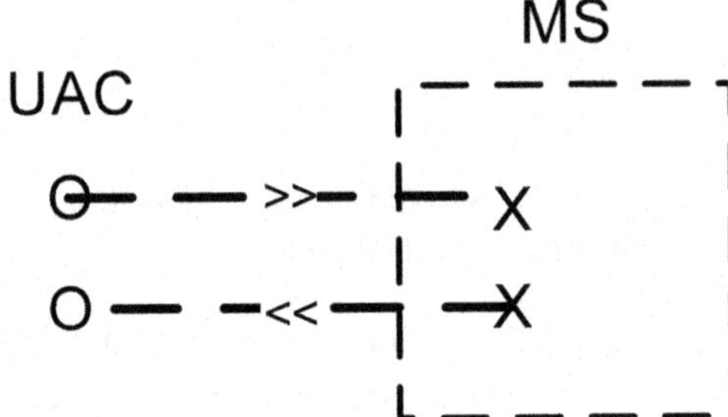

FIGURE 15.12 Echo Test: UAC Media Leg Not Attached.

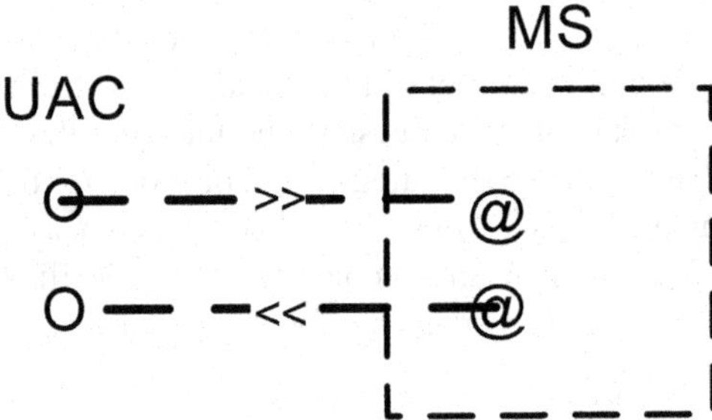

FIGURE 15.13 Echo Test: Direct Echo (Self-Connection).

In the framework, this can be achieved by means of the Mixer Control Package (RFC 6505 – see Chapter 9), which is in charge of joining connections and conferences. A sequence diagram of a potential transaction is depicted in Figure 15.14:

The transaction steps have been numbered and are explained below:

- The AS requests the joining of the connection to itself by sending to the MS a CONTROL request (1) that is specifically meant for the conferencing Control Package (msc-mixer/1.0). A <join> request is used for this purpose, and since the connection must be attached to itself, both id1 and id2 attributes are set to the same value, i.e., the connectionid.
- The MS, having checked the validity of the request, enforces the joining of the connection to itself. This means that all the frames sent by the UAC are sent back to it. To report the result of the operation, the MS sends a 200 OK (2) in reply to the AS, thus ending the transaction. The transaction ended successfully, as indicated by the body of the message (the 200 status code in the <response> tag).

The complete transaction – that is, the full bodies of the exchanged messages – is provided in the following lines:

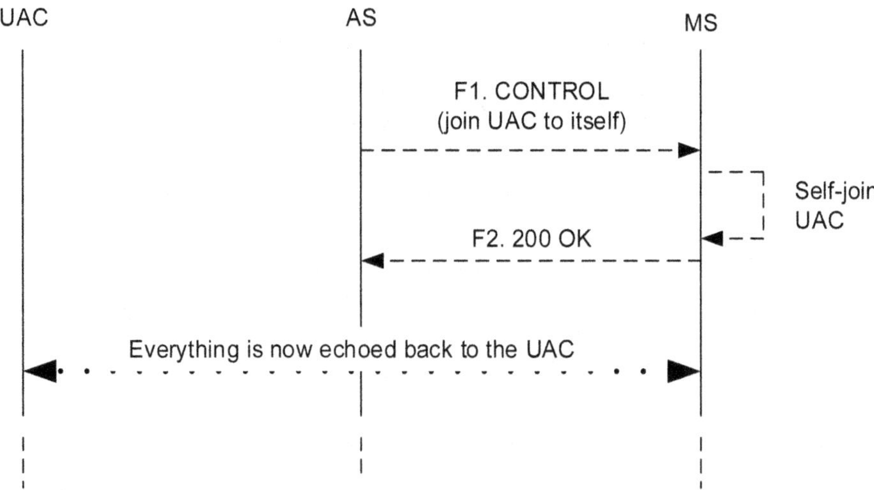

FIGURE 15.14 Self-Connection: Framework Transaction.

```
F1. AS -> MS (CFW CONTROL)
-------------------------
CFW 4fed9bf147e2 CONTROL
Control-Package: msc-mixer/1.0
Content-Type: application/msc-mixer+xml
Content-Length: 130
<mscmixer version="1.0" xmlns="urn:ietf:params:xml:ns:msc-mixer">
      <join id1="10514b7f:6a900179" id2="10514b7f:6a900179"/>
</mscmixer>

F2. AS <- MS (CFW 200 OK)
------------------------
CFW 4fed9bf147e2 200
Timeout: 10
Content-Type: application/msc-mixer+xml
Content-Length: 125
<mscmixer version="1.0" xmlns="urn:ietf:params:xml:ns:msc-mixer">
      <response status="200" reason="Join successful"/>
</mscmixer>
```

15.6.1.2 Echo Test Based on Recording

In the Recording-based Echo Test approach, the UAC is NOT directly connected to itself but rather, indirectly. This means that, as depicted in Figure 15.15, each frame the MS receives from the UAC is first recorded; then, when the recording process is ended, the whole recorded frames are played back to the UAC as an announcement.

In the framework, this can be achieved by means of the IVR Control Package (RFC 6231 – also see Section 1.2), which is in charge of both the recording and the playout phases. However, the whole scenario cannot be accomplished in a single transaction; at least two steps, in fact, need to be performed:

1. First, a recording (preceded by an announcement, if requested) must take place.
2. Then, a playout of the previously recorded media must occur.

This means that two separate transactions need to be invoked. A sequence diagram of a potential multiple transaction is depicted in Figure 15.16.

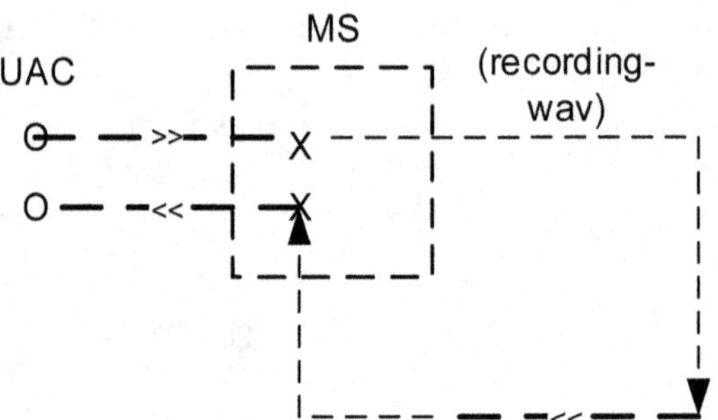

FIGURE 15.15 Echo Test: Recording Involved.

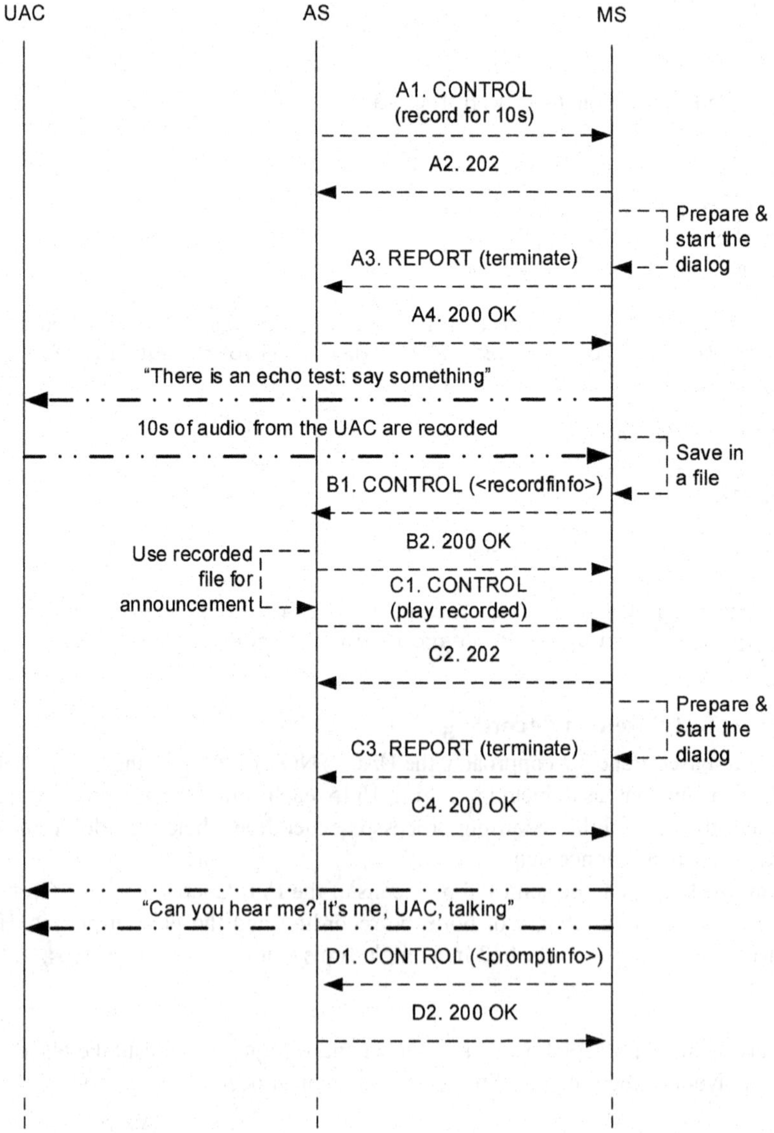

UAC
AS
MS
A1. CONTROL
(record for 10s)
A2. 202
Prepare &
start the
dialog
A3. REPORT (terminate)
A4. 200 OK
"There is an echo test: say something"
10s of audio from the UAC are recorded
Save in
a file
B1. CONTROL (<recordfinfo>)
B2. 200 OK
Use recorded
file for
announcement
C1. CONTROL
(play recorded)
C2. 202
Prepare &
start the
dialog
C3. REPORT (terminate)
C4. 200 OK
"Can you hear me? It's me, UAC, talking"
D1. CONTROL (<promptinfo>)
D2. 200 OK

FIGURE 15.16 Recording-Based Echo: Two Framework Transactions. (Copyright: IETF.)

The first obvious difference that stands out when looking at the diagram is that unlike the Direct Echo scenario, the MS does not reply with a 200 message to the CONTROL request originated by the AS. Instead, a 202 provisional message is sent first, followed by a REPORT message. The 202+REPORT(s) mechanism is used whenever the MS wants to tell the AS that the requested operation might take more time than the limit specified in the definition of the Control Package. So while the <join> operation in the Direct Echo scenario was expected to be fulfilled in a very short time, the IVR request was assumed to last longer. A 202 message provides a timeout value and tells the AS to wait a bit, since the preparation of the dialog might not happen immediately. In this example, the preparation ends before the timeout, and so the transaction is concluded with a "REPORT terminate," which reports the end of the transaction (as did the 200 message in the previous example). If the preparation took longer than the timeout, an additional "REPORT update" would have been sent with a new timeout value, and so on, until completion by means of a "REPORT terminate."

Note that the REPORT mechanism depicted is only shown to clarify its behavior. In fact, the 202+REPORT mechanism is assumed to be involved only when the requested transaction is expected to take a long time (e.g., retrieving a large media file for a prompt from an external server). In this scenario, the transaction would be prepared in much less time and as a consequence, would very likely be completed within the context of a simple CONTROL+200 request/response. The following scenarios will only involve 202+REPORTs when they are strictly necessary.

Regarding the dialog itself, note how the AS-originated CONTROL transactions are terminated as soon as the requested dialogs start. As specified in RFC 6231 (see Chapter 12), the MS uses a framework CONTROL message to report the result of the dialog and how it has proceeded. The two transactions (the AS-generated CONTROL request and the MS-generated CONTROL event) are correlated by means of the associated dialog identifier, as explained in the following. As before, the transaction steps have been numbered. The two transactions are distinguished by the preceding letter (A,B = recording, C,D = playout).

- The AS, as a first transaction, invokes a recording on the UAC connection by means of a CONTROL request (A1). The body is for the IVR package (msc-ivr/1.0) and requests the start (<dialogstart>) of a new recording context (<record>). The recording must be preceded by an announcement (<prompt>), must not last longer than 10 s (maxtime), and cannot be interrupted by a DTMF tone (dtmfterm=false). This is only done once (the missing repeatCount attribute is 1 by default for a <dialog>), which means that if the recording does not succeed the first time, the transaction must fail. A video recording is requested (considering that the associated connection includes both audio and video, and no restriction is enforced on streams to record), which is to be fed by both of the negotiated media streams. A beep has to be played (beep=true) right before the recording starts, to notify the UAC.
- As seen before, the MS sends a provisional 202 response to let the AS know that the operation might need some time.
- In the meantime, the MS prepares the dialog (e.g., by retrieving the announcement file, for which an HTTP URL is provided, and by checking that the request is well formed), and if all is fine, it starts it, notifying the AS with a new REPORT (A3) with a terminated status. As explained previously, interlocutory REPORT messages with an update status would have been sent if the preparation took longer than the timeout provided in the 202 message (e.g., if retrieving the resource via HTTP took longer than expected). Once the dialog has been prepared and started, the UAC connection is then passed to the IVR package, which first plays the announcement on the connection, followed by a beep, and then records all the incoming frames to a buffer. The MS also provides the AS with a unique dialog identifier (dialogid) that will be used in all subsequent event notifications concerning the dialog it refers to.

- The AS acks the latest REPORT (A4), thus terminating this transaction, and waits for the results.
- Once the recording is over, the MS prepares a notification CONTROL (B1). The <event> body is prepared with an explicit reference to the previously provided dialog identifier in order to make the AS aware of the fact that the notification is related to that specific dialog. The event body is then completed with the recording-related information (<recordinfo>), in this case the path to the recorded file (here, an HTTP URL) that can be used by the AS for anything it needs. The payload also contains information about the prompt (<prompt-info>), which is, however, not relevant to the scenario.
- The AS concludes this first recording transaction by acking the CONTROL event (B2).

Now that the first transaction has ended, the AS has the 10-s recording of the UAC talking and can let the UAC hear it by having the MS play it for the UAC as an announcement:

- In the second transaction, the AS invokes a playout on the UAC connection by means of a new CONTROL request (C1). The body is once again for the IVR package (msc-ivr/1.0), but this time, it requests the start (<dialogstart>) of a new announcement context (<prompt>). The file to be played is the file that was recorded before (<media>).
- Again, the usual provisional 202 (C2) takes place.
- In the meantime, the MS prepares and starts the new dialog and notifies the AS with a new REPORT (C3) with a terminated status. The connection is then passed to the IVR package, which plays the file on it.
- The AS acks the terminating REPORT (C4), now waiting for the announcement to end.
- Once the playout is over, the MS sends a CONTROL event (D1) that contains in its body (<promptinfo>) information about the justconcluded announcement. As before, the proper dialogid is used as a reference to the correct dialog.
- The AS concludes this second and last transaction by acking the CONTROL event (D2).

As in the previous paragraph, the whole CFW interaction is provided for a more in-depth evaluation of the protocol interaction.

```
A1. AS -> MS (CFW CONTROL, record)
----------------------------------
CFW 796d83aa1ce4 CONTROL
Control-Package: msc-ivr/1.0
Content-Type: application/msc-ivr+xml
Content-Length: 265
<mscivr version="1.0" xmlns="urn:ietf:params:xml:ns:msc-ivr">
    <dialogstart connectionid="10514b7f:6a900179">
        <dialog>
            <prompt>
                <media loc="http://www.example.com/
                        demo/echorecord.mpg"/>
            </prompt>
            <record beep="true" maxtime="10s"/>
        </dialog>
    </dialogstart>
</mscivr>

A2. AS <- MS (CFW 202)
----------------------
CFW 796d83aa1ce4 202
Timeout: 5
```

```
A3. AS <- MS (CFW REPORT terminate)
-----------------------------------
CFW 796d83aa1ce4 REPORT
Seq: 1
Status: terminate
Timeout: 25
Content-Type: application/msc-ivr+xml
Content-Length: 137
<mscivr version="1.0" xmlns="urn:ietf:params:xml:ns:msc-ivr">
   <response status="200" reason="Dialog started"
      dialogid="68d6569"/>
</mscivr>

A4. AS -> MS (CFW 200, ACK to 'REPORT terminate')
-------------------------------------------------
CFW 796d83aa1ce4 200
Seq: 1

B1. AS <- MS (CFW CONTROL event)
--------------------------------
CFW 0eb1678c0bfc CONTROL
Control-Package: msc-ivr/1.0
Content-Type: application/msc-ivr+xml
Content-Length: 403
<mscivr version="1.0" xmlns="urn:ietf:params:xml:ns:msc-ivr">
   <event dialogid="68d6569">
      <dialogexit status="1" reason="Dialog successfully completed">
         <promptinfo duration="9987" termmode="completed"/>
            <recordinfo duration="10017" termmode="maxtime">
               <mediainfo loc="http://www.example.net/recordings/
                     recording-68d6569.mpg" type="video/mpeg"
                     size="591872"/>
            </recordinfo>
      </dialogexit>
   </event>
</mscivr>

B2. AS -> MS (CFW 200, ACK to 'CONTROL event')
----------------------------------------------
CFW 0eb1678c0bfc 200

C1. AS -> MS (CFW CONTROL, play)
--------------------------------
CFW 1632eead7e3b CONTROL
Control-Package: msc-ivr/1.0
Content-Type: application/msc-ivr+xml
Content-Length: 241
<mscivr version="1.0" xmlns="urn:ietf:params:xml:ns:msc-ivr">
   <dialogstart connectionid="10514b7f:6a900179">
      <dialog>
         <prompt>
            <media loc="http://www.example.net/recordings/recording
                  68d6569.mpg"/>
         </prompt>
      </dialog>
   </dialogstart>
</mscivr>
```

```
C2. AS <- MS (CFW 202)
----------------------
CFW 1632eead7e3b 202
Timeout: 5

C3. AS <- MS (CFW REPORT terminate)
-----------------------------------
CFW 1632eead7e3b REPORT
Seq: 1
Status: terminate
Timeout: 25
Content-Type: application/msc-ivr+xml
Content-Length: 137
<mscivr version="1.0" xmlns="urn:ietf:params:xml:ns:msc-ivr">
   <response status="200" reason="Dialog started"
         dialogid="5f5cb45"/>
</mscivr>

C4. AS -> MS (CFW 200, ACK to 'REPORT terminate')
-------------------------------------------------
CFW 1632eead7e3b 200
Seq: 1

D1. AS <- MS (CFW CONTROL event)
--------------------------------
CFW 502a5fd83db8 CONTROL
Control-Package: msc-ivr/1.0
Content-Type: application/msc-ivr+xml
Content-Length: 230
<mscivr version="1.0" xmlns="urn:ietf:params:xml:ns:msc-ivr">
   <event dialogid="5f5cb45">
      <dialogexit status="1" reason="Dialog successfully completed">
         <promptinfo duration="10366" termmode="completed"/>
      </dialogexit>
   </event>
</mscivr>

D2. AS -> MS (CFW 200, ACK to 'CONTROL event')
----------------------------------------------
CFW 502a5fd83db8 200
```

15.6.2 Phone Call

Another scenario that might involve the interaction between an AS and an MS is the classic phone call between two UACs. In fact, even though the most straightforward way to achieve this would be to let the UACs negotiate the session and the media to be used between them, there are cases when the services provided by an MS might also prove useful for such phone calls. One of these cases is when the two UACs have no common supported codecs: having the two UACs directly negotiate the session would result in a session with no available media. Involving the MS as a transcoder would in this case still allow the two UACs to communicate. Another interesting case is when the AS (or any other entity on whose behalf the AS is working) is interested in manipulating or monitoring the media session between the UACs, e.g., to record the conversation. A similar scenario will be dealt with in Section 15.6.2.2.

Before looking at how such a scenario might be accomplished by means of the CFW, it is worth mentioning what the SIP signaling involving all the interested parties might look like. In fact, in

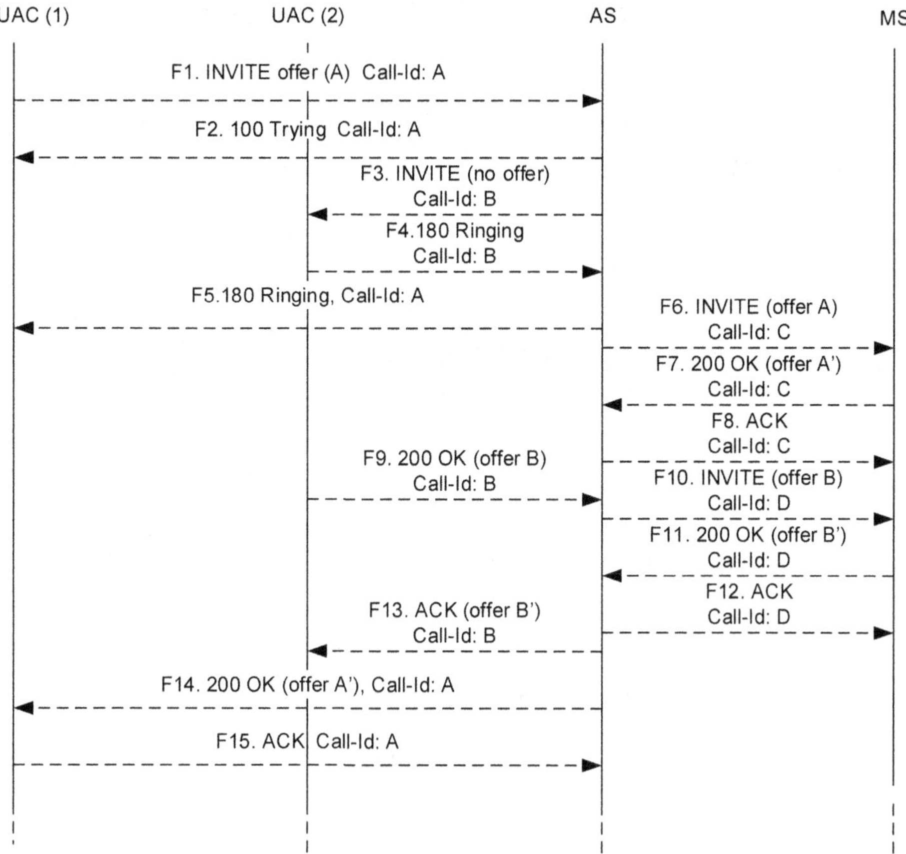

FIGURE 15.17 Phone Call: Example of 3pcc.

such a scenario, a 3pcc approach is absolutely needed. An example is provided in Figure 15.17. Again, the presented example is not at all to be considered best common practice when 3pcc is needed in a MEDIACTRL-based framework. It is only described in order to help the reader more easily understand what the requirements are on the MS side and as a consequence, what information might be required. RFC 3725 (see Section 1.3) provides a much more detailed overview on 3pcc patterns in several use cases. Only an explanatory sequence diagram is provided without delving into the details of the exchanged SIP messages.

In this example, UAC1 wants to place a phone call to UAC2. To do so, it sends an INVITE to the AS with its offer A. The AS sends an offerless INVITE to UAC2. When UAC2 responds with a 180, the same message is forwarded by the AS to UAC1 to notify it that the callee is ringing. In the meantime, the AS also adds a leg to the MS for UAC1, as explained at the beginning of Section 15.6. To do so, it of course uses the offer A that UAC1 made. Once UAC2 accepts the call by providing its own offer B in the 200, the AS also adds a leg for offer B to the MS. At this point, the negotiation can be completed by providing the two UACs with the SDP answer negotiated by the MS with them (A′ and B′, respectively). Of course, this is only one way to deal with the signaling and shall not be considered an absolutely mandatory approach.

Once the negotiation is over, the two UACs are not in communication yet. In fact, it is up to the AS now to actively trigger the MS to somehow attach their media streams to each other by referring to the connection identifiers associated with the UACs, as explained previously. This specification presents two different approaches that might be followed, according to what needs to be accomplished. A generic media perspective of the phone call scenario is depicted in Figure 15.18. The MS is basically in the media path between the two UACs.

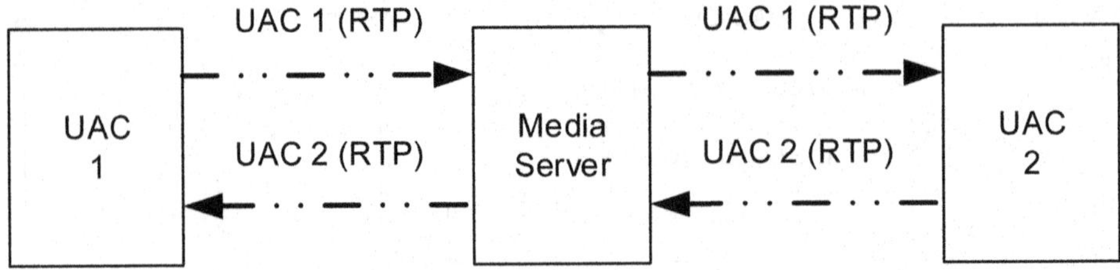

FIGURE 15.18 Phone Call: Media Perspective.

FIGURE 15.19 Phone Call: UAC Media Leg Not Attached.

FIGURE 15.20 Phone Call: Direct Connection.

From the framework point of view, when the UACs' legs are not attached to anything yet, what appears is shown in Figure 15.19: since there are no connections involving the UACs yet, the frames they might be sending are discarded, and nothing is sent to them (except for silence, if its transmission is requested).

15.6.2.1 Direct Connection

The Direct Connection approach is the easiest, and a more straightforward, approach to get the phone call between the two UACs to work. The idea is basically the same as that of the Direct Echo approach. A <join> directive is used to directly attach one UAC to the other by exploiting the MS to only deal with the transcoding/ adaption of the flowing frames, if needed. This approach is depicted in Figure 15.20.

The framework transactions needed to accomplish this scenario are very trivial and easy to understand. They are basically the same as those presented in the Direct Echo Test (Figure 15.21) scenario; the only difference is in the provided identifiers. In fact, this time, the MS is not supposed to attach the UACs' media connections to themselves but has to join the media connections of two different UACs, i.e., UAC1 and UAC2. This means that in this transaction, id1 and i2 will have to address the media connections of UAC1 and UAC2. In the case of a successful transaction, the MS takes care of forwarding all media coming from UAC1 to UAC2 and vice versa, transparently taking care of any required transcoding steps, if necessary.

```
1. AS -> MS (CFW CONTROL)
-------------------------

CFW 0600855d24c8 CONTROL
Control-Package: msc-mixer/1.0
```

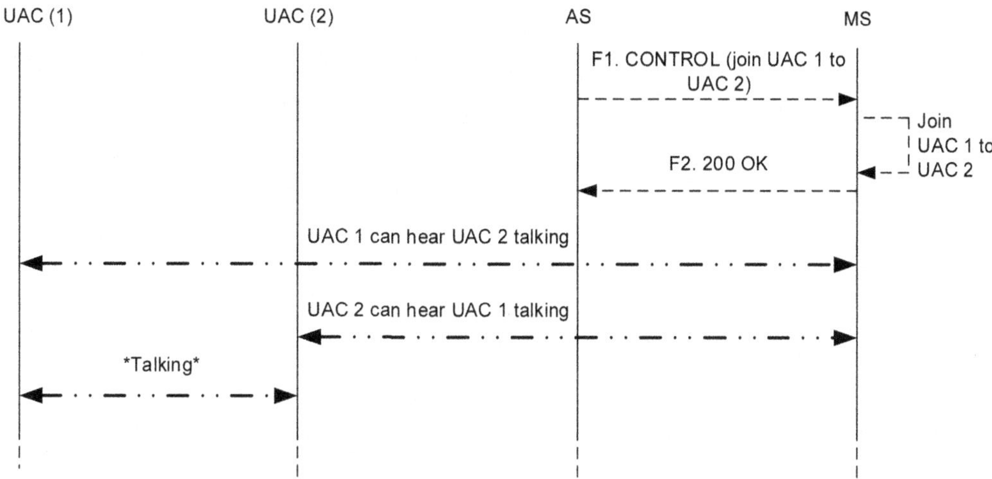

FIGURE 15.21 Direct Connection: Framework Transactions (Copyright: IETF)

```
Content-Type: application/msc-mixer+xml
Content-Length: 130
<mscmixer version="1.0" xmlns="urn:ietf:params:xml:ns:msc-mixer">
   <join id1="10514b7f:6a900179" id2="e1e1427c:1c998d22"/>
</mscmixer>

2. AS <- MS (CFW 200 OK)
------------------------
CFW 0600855d24c8 200
Timeout: 10
Content-Type: application/msc-mixer+xml
Content-Length: 125
<mscmixer version="1.0" xmlns="urn:ietf:params:xml:ns:msc-mixer">
   <response status="200" reason="Join successful"/>
</mscmixer>
```

Such a simple approach has its drawbacks. For instance, with such an approach, recording a conversation between two users might be tricky to accomplish. In fact, since no mixing would be involved, only the single connections (UAC1↔MS and UAC2↔MS) could be recorded. If the AS wants a conversation-recording service to be provided anyway, it needs additional business logic on its side. An example of such a use case is provided in Section 15.6.2.3.

15.6.2.2 Conference-Based Approach

The approach described in Section 15.6.2.1 surely works for a basic phone call but as explained previously, might have some drawbacks whenever more advanced features are needed. For instance, one cannot record the whole conversation – only the single connections – since no mixing is involved. Additionally, even the single task of playing an announcement over the conversation could be complex, especially if the MS does not support implicit mixing over media connections. For this reason, in more advanced cases, a different approach might be taken, like the conference-based approach described in this section.

The idea is to use a mixing entity in the MS that acts as a bridge between the two UACs. The presence of this entity allows more customization of what needs to be done with the conversation, like the recording of the conversation that has been provided as an example. The approach is depicted in Figure 15.22. The mixing functionality in the MS will be described in more detail in the following section (which deals with many conference-related scenarios), so only some hints will be provided here for basic comprehension of the approach.

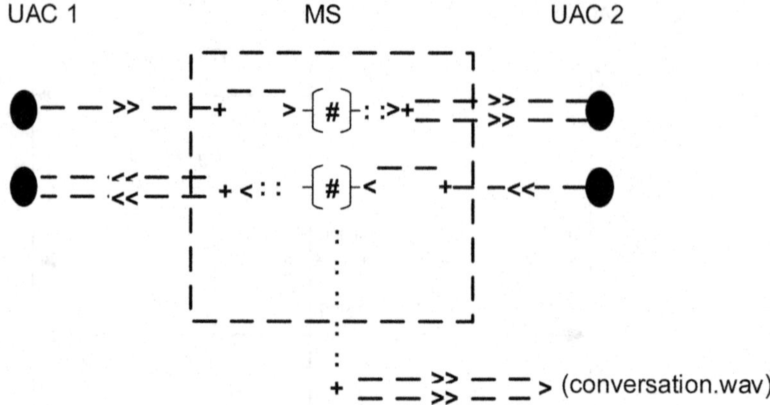

FIGURE 15.22 Phone Call: Conference-Based Approach. (Copyright: IETF.)

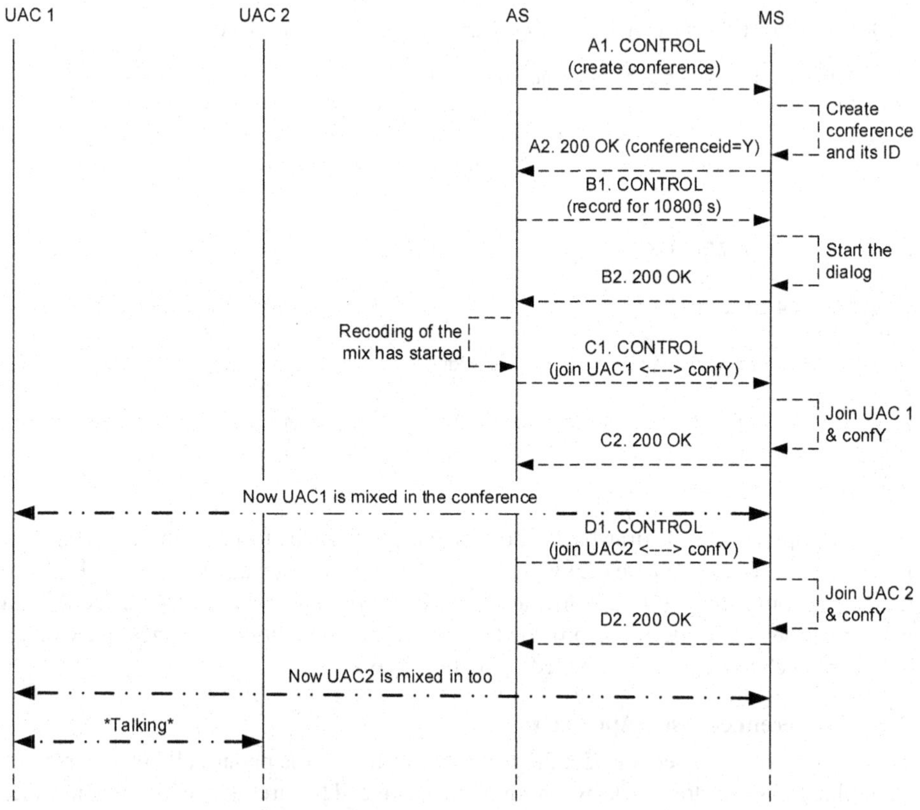

FIGURE 15.23 Conference-Based Approach: Framework Transactions. (Copyright: IETF.)

To identify a single sample scenario, let us consider a phone call that the AS wants to record.

Figure 15.23 shows how this could be accomplished in the Media Control Channel Framework. This example, as usual, hides the previous interaction between the UACs and the AS and instead focuses on the Control Channel operations and what follows.

The AS uses two different packages to accomplish this scenario: the Mixer package (to create the mixing entity and join the UACs) and the IVR package (to record what happens in the conference). The framework transaction steps can be described as follows:

- First of all, the AS creates a new hidden conference by means of a <createconference> request (A1). This conference is properly configured according to the use it is assigned to. In fact, since only two participants will be joined to it, both "reserved-talkers" and "reserved-listeners" are set to 2, just as the "n" value for the N-best audio mixing algorithm. The video layout is also set accordingly (<single-view>/<dual-view>).
- The MS sends notification of the successful creation of the new conference in a 200 framework message (A2). The identifier assigned to the conference, which will be used in subsequent requests addressed to it, is 6013f1e.
- The AS requests a new recording for the newly created conference. To do so, it places a proper request to the IVR package (B1). The AS is interested in a video recording (type=video/mpeg), which must not last longer than 3 hours (maxtime=10800s), after which the recording must end. Additionally, no beep must be played on the conference (beep=false), and the recording must start immediately whether or not any audio activity has been reported (vadinitial=false is the default value for <record>).
- The transaction is handled by the MS, and when the dialog has been successfully started, a 200 OK is issued to the AS (B2). The message contains the dialogid associated with the dialog (00b29fb), which the AS must refer to for later notifications.
- At this point, the AS attaches both UACs to the conference with two separate <join> directives (C1/D1). When the MS confirms the success of both operations (C2/D2), the two UACs are actually in contact with each other (even though indirectly, since a hidden conference they are unaware of is on their path), and their media contribution is recorded.

```
A1. AS -> MS (CFW CONTROL, createconference)
-------------------------------------------
CFW 238e1f2946e8 CONTROL
Control-Package: msc-mixer
Content-Type: application/msc-mixer+xml
Content-Length: 395
<mscmixer version="1.0" xmlns="urn:ietf:params:xml:ns:msc-mixer">
   <createconference reserved-talkers="2" reserved-listeners="2">
      <audio-mixing type="nbest" n="2"/>
         <video-layouts>
            <video-layout min-participants='1'>
               <single-view/>
            </video-layout>
            <video-layout min-participants='2'>
               <dual-view/>
            </video-layout>
         </video-layouts>
         <video-switch>
            <controller/>
         </video-switch>
   </createconference>
</mscmixer>

A2. AS <- MS (CFW 200 OK)
-------------------------
CFW 238e1f2946e8 200
Timeout: 10
Content-Type: application/msc-mixer+xml
Content-Length: 151
<mscmixer version="1.0" xmlns="urn:ietf:params:xml:ns:msc-mixer">
```

```
   <response status="200" reason="Conference created"
      conferenceid="6013f1e"/>
</mscmixer>

B1. AS -> MS (CFW CONTROL, record)
----------------------------------
CFW 515f007c5bd0 CONTROL
Control-Package: msc-ivr
Content-Type: application/msc-ivr+xml
Content-Length: 226
<mscivr version="1.0" xmlns="urn:ietf:params:xml:ns:msc-ivr">
   <dialogstart conferenceid="6013f1e">
      <dialog>
         <record beep="false" type="video/mpeg" maxtime="10800s"/>
      </dialog>
   </dialogstart>
</mscivr>
B2. AS <- MS (CFW 200 OK)
-------------------------
CFW 515f007c5bd0 200
Timeout: 10
Content-Type: application/msc-ivr+xml
Content-Length: 137
<mscivr version="1.0" xmlns="urn:ietf:params:xml:ns:msc-ivr">
   <response status="200" reason="Dialog started" dialogid="00b29fb"/>
</mscivr>
C1. AS -> MS (CFW CONTROL, join)
--------------------------------
CFW 0216231b1f16 CONTROL
Control-Package: msc-mixer
Content-Type: application/msc-mixer+xml
Content-Length: 123
<mscmixer version="1.0" xmlns="urn:ietf:params:xml:ns:msc-mixer">
   <join id1="10514b7f:6a900179" id2="6013f1e"/>
</mscmixer>
C2. AS <- MS (CFW 200 OK)
-------------------------
CFW 0216231b1f16 200
Timeout: 10
Content-Type: application/msc-mixer+xml
Content-Length: 125
<mscmixer version="1.0" xmlns="urn:ietf:params:xml:ns:msc-mixer">
   <response status="200" reason="Join successful"/>
</mscmixer>
D1. AS -> MS (CFW CONTROL, join)
--------------------------------
CFW 140e0f763352 CONTROL
Control-Package: msc-mixer
Content-Type: application/msc-mixer+xml
Content-Length: 124
<mscmixer version="1.0" xmlns="urn:ietf:params:xml:ns:msc-mixer">
   <join id1="219782951:0b9d3347" id2="6013f1e"/>
</mscmixer>
D2. AS <- MS (CFW 200 OK)
-------------------------
CFW 140e0f763352 200
Timeout: 10
```

```
Content-Type: application/msc-mixer+xml
Content-Length: 125
<mscmixer version="1.0" xmlns="urn:ietf:params:xml:ns:msc-mixer">
   <response status="200" reason="Join successful"/>
</mscmixer>
```

The recording of the conversation can subsequently be accessed by the AS by waiting for an event notification from the MS. This event, which will be associated with the previously started recording dialog, will contain the URI of the recorded file. Such an event may be triggered by either a natural completion of the dialog (e.g., the dialog has reached its programmed 3 hours) or any interruption of the dialog itself (e.g., the AS actively requests that the recording be interrupted, since the call between the UACs ended).

15.6.2.3 Recording a Conversation

The previous section described how to take advantage of the conferencing functionality of the Mixer package in order to allow the recording of phone calls in a simple way. However, using a dedicated mixer just for a phone call might be considered overkill. This section shows how recording a conversation and subsequently playing it out can be accomplished without a mixing entity involved in the call, i.e., by using the Direct Connection approach as described in Section 15.6.2.1.

As explained previously, if the AS wants to record a phone call between two UACs, the use of just the <join> directive without a mixer forces the AS to just rely on separate recording commands. That is, the AS can only instruct the MS to separately record the media flowing on each media leg: a recording for all the data coming from UAC1 and a different recording for all the data coming from UAC2. If someone subsequently wants to access the whole conversation, the AS may take at least two different approaches:

1. It may mix the two recordings itself (e.g., by delegating it to an offline mixing entity) in order to obtain a single file containing the combination of the two recordings. In this way, a simple playout as described in Section 15.6.1.2 would suffice.
2. Alternatively, it may take advantage of the mixing functionality provided by the MS itself. One way to do this is to create a hidden conference on the MS, attach the UAC as a passive participant to it, and play the separate recordings on the conference as announcements. In this way, the UAC accessing the recording would experience both the recordings at the same time.

The second approach is considered in this section. The framework transaction as described in Figure 15.24 assumes that a recording has already been requested for both UAC1 and UAC2, that the phone call has ended, and that the AS has successfully received the URIs to both the recordings from the MS. Such steps are not described again, since they would be quite similar to the steps described in Section 15.6.1.2. As mentioned previously, the idea is to use a properly constructed hidden conference to mix the two separate recordings on the fly and present them to the UAC. It is, of course, up to the AS to subsequently unjoin the user from the conference and destroy the conference itself once the playout of the recordings ends for any reason.

Figure 15.24 assumes that a recording of both the channels (UAC1 and UAC2) has already taken place. Later, when we desire to play the whole conversation to a new user, UAC3, the AS may take care of the presented transactions. The framework transaction steps are only apparently more complicated than those presented so far. The only difference, in fact, is that transactions C and D are concurrent, since the recordings must be played together.

- First of all, the AS creates a new conference to act as a mixing entity (A1). The settings for the conference are chosen according to the use case, e.g., the video layout, which is fixed to <dual-view>, and the switching type to <controller>. When the conference has been successfully created (A2), the AS takes note of the conference identifier.

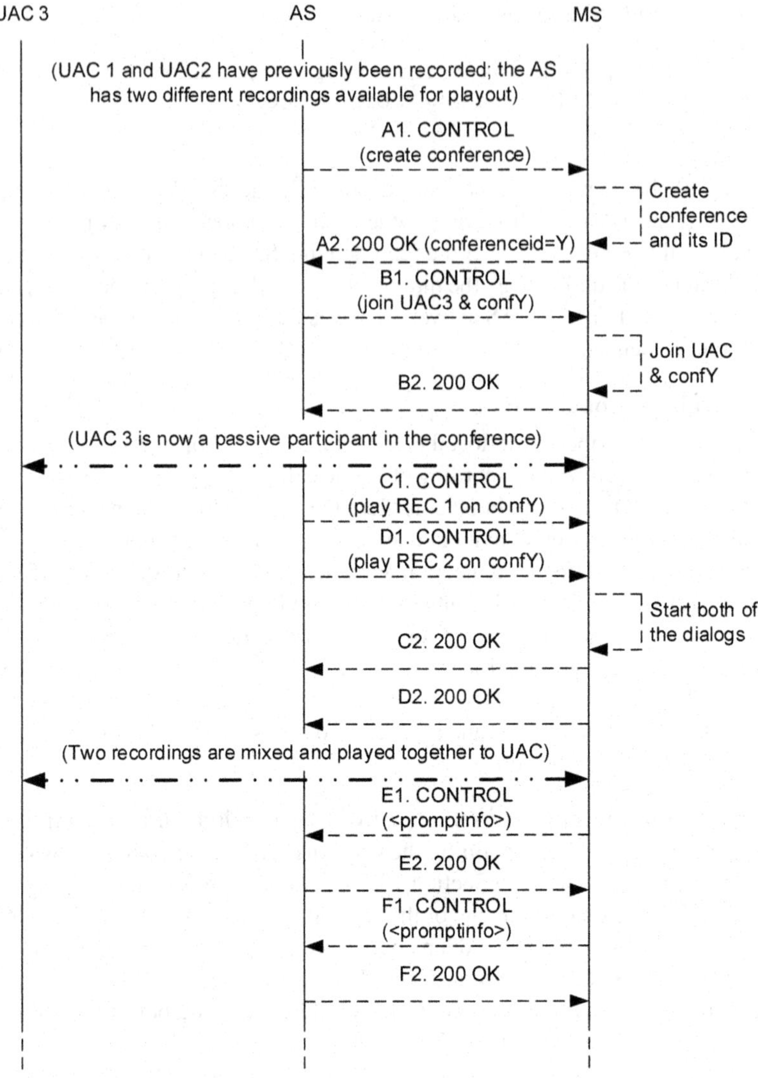

FIGURE 15.24 Phone Call: Playout of a Recorded Conversation. (Copyright: IETF.)

- At this point, UAC3 is attached to the conference as a passive user (B1). There would be no point in letting the user contribute to the conference mix, since he will only need to watch a recording. In order to specify his passive status, both the audio and video streams for the user are set to "recvonly." If the transaction succeeds, the MS notifies the AS (B2).
- Once the conference has been created, and UAC3 has been attached to it, the AS can request the playout of the recordings; in order to do so, it requests two concurrent <prompt> directives (C1 and D1), addressing the recording of UAC1 (REC1) and UAC2 (REC2), respectively. Both the prompts must be played on the previously created conference and not to UAC3 directly, as can be deduced from the "conferenceid" attribute of the <dialog> element.
- The transactions "live their lives" exactly as explained for previous <prompt> examples. The originating transactions are first prepared and started (C2, D2), and then, as soon as the playout ends, a related CONTROL message is triggered by the MS (E1, F1). This notification may contain a <promptinfo> element with information about how the playout proceeded (e.g., whether the playout completed normally or was interrupted by a DTMF tone, etc.).

```
A1. AS -> MS (CFW CONTROL, createconference)
--------------------------------------------
CFW 506e039f65bd CONTROL
Control-Package: msc-mixer/1.0
Content-Type: application/msc-mixer+xml
Content-Length: 312
<mscmixer version="1.0" xmlns="urn:ietf:params:xml:ns:msc-mixer">
   <createconference reserved-talkers="0" reserved-listeners="1">
      <audio-mixing type="controller"/>
         <video-layouts>
            <video-layout min-participants='1'>
               <dual-view/>
            </video-layout>
         </video-layouts>
         <video-switch>
            <controller/>
         </video-switch>
   </createconference>
</mscmixer>

A2. AS <- MS (CFW 200 OK)
-------------------------
CFW 506e039f65bd 200
Timeout: 10
Content-Type: application/msc-mixer+xml
Content-Length: 151
<mscmixer version="1.0" xmlns="urn:ietf:params:xml:ns:msc-mixer">
   <response status="200" reason="Conference created"
      conferenceid="2625069"/>
</mscmixer>

B1. AS -> MS (CFW CONTROL, join)
--------------------------------
CFW 09202baf0c81 CONTROL
Control-Package: msc-mixer/1.0
Content-Type: application/msc-mixer+xml
Content-Length: 214
<mscmixer version="1.0" xmlns="urn:ietf:params:xml:ns:msc-mixer">
<join id1="aafaf62d:0eac5236" id2="2625069">
<stream media="audio" direction="recvonly"/>
<stream media="video" direction="recvonly"/>
</join>
</mscmixer>

B2. AS <- MS (CFW 200 OK)
-------------------------
CFW 09202baf0c81 200
Timeout: 10
Content-Type: application/msc-mixer+xml
Content-Length: 125
<mscmixer version="1.0" xmlns="urn:ietf:params:xml:ns:msc-mixer">
   <response status="200" reason="Join successful"/>
</mscmixer>

C1. AS -> MS (CFW CONTROL, play recording from UAC1)
----------------------------------------------------
CFW 3c2a08be4562 CONTROL
```

```
Control-Package: msc-ivr/1.0
Content-Type: application/msc-ivr+xml
Content-Length: 229
<mscivr version="1.0" xmlns="urn:ietf:params:xml:ns:msc-ivr">
   <dialogstart conferenceid="2625069">
      <dialog>
         <prompt>
            <media loc="http://www.example.net/recordings/
                  recording-4ca9fc2.mpg"/>
         </prompt>
      </dialog>
   </dialogstart>
</mscivr>

D1. AS -> MS (CFW CONTROL, play recording from UAC2)
-----------------------------------------------------
CFW 1c268d810baa CONTROL
Control-Package: msc-ivr/1.0
Content-Type: application/msc-ivr+xml
Content-Length: 229
<mscivr version="1.0" xmlns="urn:ietf:params:xml:ns:msc-ivr">
   <dialogstart conferenceid="2625069">
      <dialog>
         <prompt>
            <media loc="http://www.example.net/recordings/
                  recording-39dfef4.mpg"/>
         </prompt>
      </dialog>
   </dialogstart>
</mscivr>

C2. AS <- MS (CFW 200 OK)
-------------------------
CFW 1c268d810baa 200
Timeout: 10
Content-Type: application/msc-ivr+xml
Content-Length: 137
<mscivr version="1.0" xmlns="urn:ietf:params:xml:ns:msc-ivr">
   <response status="200" reason="Dialog started"
         dialogid="7a457cc"/>
</mscivr>

D2. AS <- MS (CFW 200 OK)
-------------------------
CFW 3c2a08be4562 200
Timeout: 10
Content-Type: application/msc-ivr+xml
Content-Length: 137
<mscivr version="1.0" xmlns="urn:ietf:params:xml:ns:msc-ivr">
   <response status="200" reason="Dialog started"
         dialogid="1a0c7cf"/>
</mscivr>

E1. AS <- MS (CFW CONTROL event, playout of recorded UAC1 ended)
----------------------------------------------------------------
CFW 77aec0735922 CONTROL
Control-Package: msc-ivr/1.0
```

```
Content-Type: application/msc-ivr+xml
Content-Length: 230
<mscivr version="1.0" xmlns="urn:ietf:params:xml:ns:msc-ivr">
   <event dialogid="7a457cc">
      <dialogexit status="1" reason="Dialog successfully completed">
         <promptinfo duration="10339" termmode="completed"/>
      </dialogexit>
   </event>
</mscivr>

E2. AS -> MS (CFW 200, ACK to 'CONTROL event')
----------------------------------------------
CFW 77aec0735922 200

F1. AS <- MS (CFW CONTROL event, playout of recorded UAC2 ended)
----------------------------------------------------------------
CFW 62726ace1660 CONTROL
Control-Package: msc-ivr/1.0
Content-Type: application/msc-ivr+xml
Content-Length: 230
<mscivr version="1.0" xmlns="urn:ietf:params:xml:ns:msc-ivr">
   <event dialogid="1a0c7cf">
      <dialogexit status="1" reason="Dialog successfully completed">
         <promptinfo duration="10342" termmode="completed"/>
      </dialogexit>
   </event>
</mscivr>

F2. AS -> MS (CFW 200, ACK to 'CONTROL event')
----------------------------------------------
CFW 62726ace1660 200
```

15.6.3 Conferencing

One of the most important services the MS must be able to provide is mixing. This involves mixing media streams from different sources and delivering the resulting mix(es) to each interested party, often according to per-user policies, settings, and encoding. A typical scenario involving mixing is, of course, media conferencing. In such a scenario, the media sent by each participant is mixed, and each participant typically receives the overall mix, excluding its own contribution and encoded in the format it negotiated. This example points out in a quite clear way how mixing must take care of the profile of each involved entity. A media perspective of such a scenario is depicted in Figure 15.25.

From the framework point of view, when the UACs' legs are not attached to anything yet, what appears is shown in Figure 15.26: since there are no connections involving the UACs yet, the frames they might be sending are discarded, and nothing is sent back to them (except for silence, if its transmission is requested).

The next sub-sections will cover several typical scenarios involving mixing and conferencing as a whole, specifically:

1. Simple Bridging scenario, which is a very basic (i.e., no "special effects;" just mixing involved) conference involving one or more participants
2. Rich Conference scenario, which enriches the Simple Bridging scenario by adding additional features typically found in conferencing systems (e.g., DTMF collection for PIN-based conference access, private and global announcements, recordings, and so on)

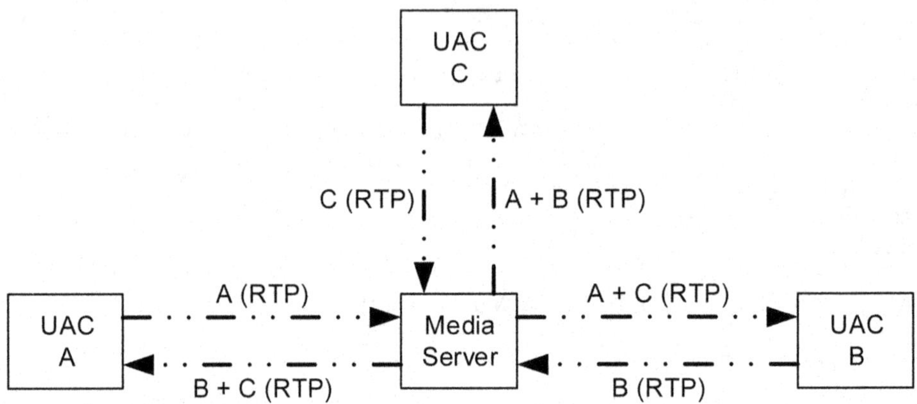

FIGURE 15.25 Conference: Media Perspective. (Copyright: IETF.)

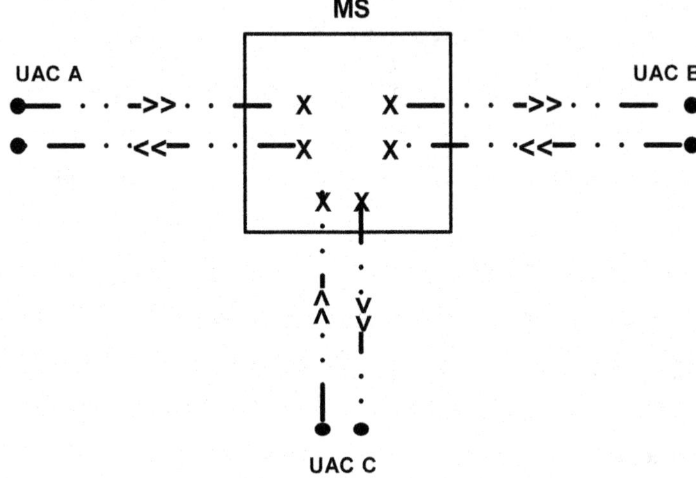

FIGURE 15.26 Conference: UAC Legs Not Attached. (Copyright: IETF.)

3. Coaching scenario, which is a more complex scenario that involves per-user mixing (customers, agents, and coaches don't all get the same mixes)
4. Sidebars scenario, which adds more complexity to the previous conferencing scenarios by involving sidebars (i.e., separate conference instances that only exist within the context of a parent conference instance) and the custom media delivery that follows
5. Floor Control scenario, which provides some guidance on how floor control could be involved in a MEDIACTRL-based media conference

All the above-mentioned scenarios depend on the availability of a mixing entity. Such an entity is provided in the CFW by the conferencing package. Besides allowing for the interconnection of media sources as seen in the Direct Echo Test section, this package enables the creation of abstract connections that can be joined to multiple connections. These abstract connections, called conferences, mix the contribution of each attached connection and feed them accordingly (e.g., a connection with a "sendrecv" property would be able to contribute to the mix and listen to it, while a connection with a "recvonly" property would only be able to listen to the overall mix but not actively contribute to it).

That said, each of the above-mentioned scenarios will start more or less in the same way: by the creation of a conference connection (or more than one, as needed in some cases) to be subsequently

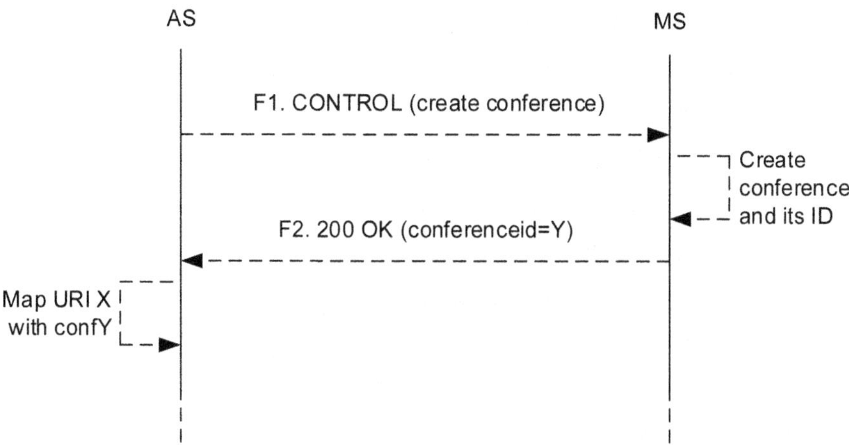

FIGURE 15.27 Conference: Framework Transactions. (Copyright: IETF.)

referred to when it comes to mixing. A typical framework transaction to create a new conference instance in the CFW is depicted in Figure 15.27:

The call flow is quite straightforward and can typically be summarized in the following steps:

- The AS invokes the creation of a new conference instance by means of a CONTROL request (1); this request is addressed to the conferencing package (msc-mixer/1.0) and contains in the body the directive (<createconference>) with all the desired settings for the new conference instance. In the following example, the mixing policy is to mix the five ("reserved-talkers") loudest speakers (nbest), while 10 listeners at maximum are allowed. Video settings are configured, including the mechanism used to select active video sources (<controller>, meaning that the AS will explicitly instruct the MS about it) and details about the video layouts to make available. In this example, the AS is instructing the MS to use a <single-view> layout when only one video source is active, to pass to a <quad-view> layout when at least two video sources are active, and to use a <multiple-5x1> layout whenever the number of sources is at least five. Finally, the AS also subscribes to the "active-talkers" event, which means it wants to be informed (at a rate of 4 seconds) whenever an active participant is speaking.
- The MS creates the conference instance, assigns a unique identifier to it (6146dd5), and completes the transaction with a 200 response (2).
- At this point, the requested conference instance is active and ready to be used by the AS. It is then up to the AS to integrate the use of this identifier in its application logic.

```
F1. AS -> MS (CFW CONTROL)
-------------------------
CFW 3032e5fb79a1 CONTROL
Control-Package: msc-mixer/1.0
Content-Type: application/msc-mixer+xml
Content-Length: 489
<mscmixer version="1.0" xmlns="urn:ietf:params:xml:ns:msc-mixer">
    <createconference reserved-talkers="5" reserved-listeners="10">
        <audio-mixing type="nbest"/>
            <video-layouts>
                <video-layout min-participants='1'>
                    <single-view/>
                </video-layout>
                <video-layout min-participants='2'>
```

```
            <quad-view/>
         </video-layout>
         <video-layout min-participants='5'>
            <multiple-5x1/>
         </video-layout>
      </video-layouts>
      <video-switch>
         <controller/>
      </video-switch>
      <subscribe>
         <active-talkers-sub interval="4"/>
      </subscribe>
   </createconference>
</mscmixer>
```

```
F2. AS <- MS (CFW 200)
---------------------
CFW 3032e5fb79a1 200
Timeout: 10
Content-Type: application/msc-mixer+xml
Content-Length: 151
<mscmixer version="1.0" xmlns="urn:ietf:params:xml:ns:msc-mixer">
   <response status="200" reason="Conference created"
         conferenceid="6146dd5"/>
</mscmixer>
```

15.6.3.1 Simple Bridging

As mentioned previously, the simplest way that an AS can use a conference instance is simple bridging. In this scenario, the conference instance just acts as a bridge for all the participants that are attached to it. The bridge takes care of transcoding, if needed (in general, different participants may use different codecs for their streams), echo cancellation (each participant will receive the overall mix, excluding its own contribution), and per-participant mixing (each participant may receive different mixed streams according to what it needs/is allowed to send/receive). This assumes, of course, that each interested participant must be somehow joined to the bridge in order to indirectly communicate with the other participants. From the media perspective, the scenario can be seen as depicted in Figure 15.28.

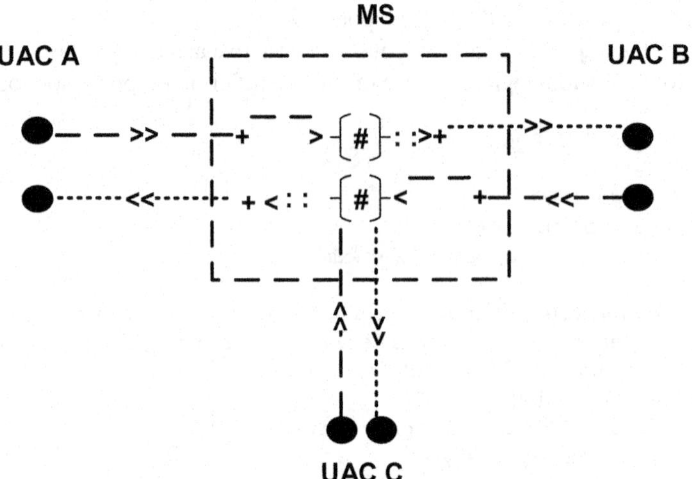

FIGURE 15.28 Conference: Simple Bridging. (Copyright: IETF.)

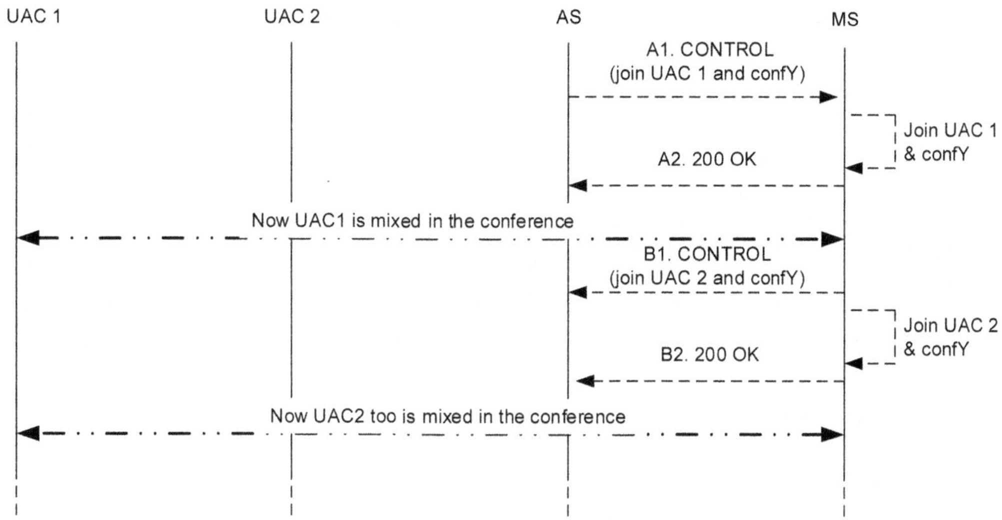

FIGURE 15.29 Simple Bridging: Framework Transactions (1). (Copyright: IETF.)

In the framework, the first step is obviously to create a new conference instance, as seen in the introductory section (Figure 15.27).

Assuming that a conference instance has already been created, bridging participants to it is quite straightforward and can be accomplished as seen in the Direct Echo Test scenario. The only difference here is that each participant is not directly connected to itself (Direct Echo) or to another UAC (Direct Connection) but to the bridge instead. Figure 15.29 shows the example of two different UACs joining the same conference. The example, as usual, hides the previous interaction between each of the two UACs and the AS and instead, focuses on what the AS does in order to actually join the participants to the bridge so that they can interact in a conference. Please note also that to make the diagram more readable, two different identifiers (UAC1 and UAC2) are used in place of the identifiers previously employed to introduce the scenario (UAC A, B, C).

The framework transaction steps are actually quite trivial and easy to understand, since they are very similar to some previously described scenarios. The AS joins both UAC1 (id1 in A1) and UAC2 (id1 in B1) to the conference (id2 in both transactions). As a result of these two operations, both UACs are mixed in the conference. Since no <stream> is explicitly provided in any of the transactions, all the media from the UACs (audio/video) are attached to the conference (as long as the conference has been properly configured to support both, of course).

```
A1. AS -> MS (CFW CONTROL)
--------------------------
CFW 434a95786df8 CONTROL
Control-Package: msc-mixer/1.0
Content-Type: application/msc-mixer+xml
Content-Length: 120
<mscmixer version="1.0" xmlns="urn:ietf:params:xml:ns:msc-mixer">
   <join id1="e1e1427c:1c998d22" id2="6146dd5"/>
</mscmixer>

A2. AS <- MS (CFW 200 OK)
-------------------------
CFW 434a95786df8 200
Timeout: 10
Content-Type: application/msc-mixer+xml
Content-Length: 125
```

```
<mscmixer version="1.0" xmlns="urn:ietf:params:xml:ns:msc-mixer">
   <response status="200" reason="Join successful"/>
</mscmixer>

B1. AS -> MS (CFW CONTROL)
--------------------------
CFW 5c0cbd372046 CONTROL
Control-Package: msc-mixer/1.0
Content-Type: application/msc-mixer+xml
Content-Length: 120
<mscmixer version="1.0" xmlns="urn:ietf:params:xml:ns:msc-mixer">
   <join id1="10514b7f:6a900179" id2="6146dd5"/>
</mscmixer>

B2. AS <- MS (CFW 200 OK)
-------------------------
CFW 5c0cbd372046 200
Timeout: 10
Content-Type: application/msc-mixer+xml
Content-Length: 125
<mscmixer version="1.0" xmlns="urn:ietf:params:xml:ns:msc-mixer">
   <response status="200" reason="Join successful"/>
</mscmixer>
```

Once one or more participants have been attached to the bridge, their connections and how their media are handled by the bridge can be dynamically manipulated by means of another directive called <modifyjoin>. A typical use case for this directive is the change of direction of an existing media (e.g., a previously speaking participant is muted, which means that its media direction changes from "sendrecv" to "recvonly"). Figure 15.30 shows how a framework transaction requesting such a directive might appear.

The directive used to modify an existing join configuration is <modifyjoin>, and its syntax is exactly the same as the syntax required in <join> instructions. In fact, the same syntax is used for identifiers (id1/id2). Whenever a <modifyjoin> is requested, and id1 and id2 address one or more joined connections, the AS is requesting a change of the join configuration. In this case, the AS instructs the MS to mute (<stream> media=audio, direction=recvonly) UAC1 (id1=UAC1) in the conference (id2) it has been attached to previously. Any other connection existing between them is left untouched.

It is worth noting that the <stream> settings are enforced according to both the provided direction AND the id1 and id2 identifiers. For instance, in this example, id1 refers to UAC1, while id2 refers to the conference in the MS. This means that the required modifications have to be applied to the stream specified in the <stream> element of the message along the direction that goes from

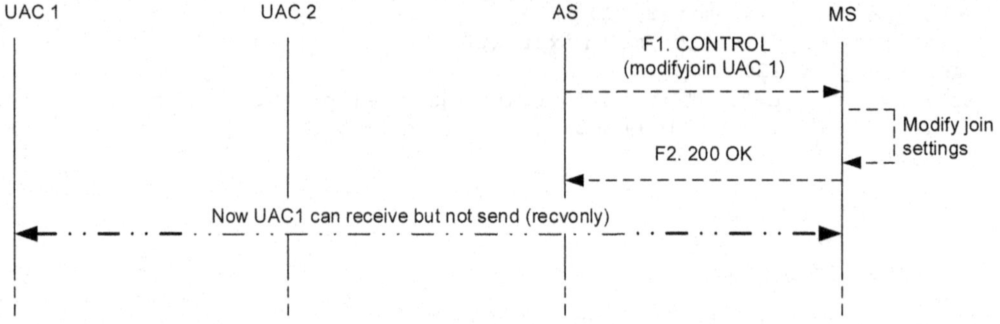

FIGURE 15.30 Simple Bridging: Framework Transactions (2). (Copyright: IETF.)

"id1" to "id2" (as specified in the <modifyjoin> element of the message). In the provided example, the AS wants to mute UAC1 with respect to the conference. To do so, the direction is set to "recvonly," meaning that for what affects id1, the media stream is only to be received.

If id1 referred to the conference and id2 to UAC1, to achieve the same result, the direction would have to be set to "sendonly," meaning "id1 (the conference) can only send to id2 (UAC1), and no media stream must be received." Additional settings for a <stream> audio volume, region assignments, and so on) follow the same approach, as discussed in subsequent sections. (e.g., Sections 15.6.3.3–15.6.4.3).

```
F1. AS -> MS (CFW CONTROL)
-------------------------
CFW 57f2195875c9 CONTROL
Control-Package: msc-mixer/1.0
Content-Type: application/msc-mixer+xml
Content-Length: 182
<mscmixer version="1.0" xmlns="urn:ietf:params:xml:ns:msc-mixer">
   <modifyjoin id1="e1e1427c:1c998d22" id2="6146dd5">
      <stream media="audio" direction="recvonly"/>
   </modifyjoin>
</mscmixer>

   F2. AS <- MS (CFW 200 OK)
------------------------
CFW 57f2195875c9 200
Timeout: 10
Content-Type: application/msc-mixer+xml
Content-Length: 123
<mscmixer version="1.0" xmlns="urn:ietf:params:xml:ns:msc-mixer">
   <response status="200" reason="Join modified"/>
</mscmixer>
```

15.6.3.2 Rich Conference Scenario

The previous scenario can be enriched with additional features often found in existing conferencing systems. Typical examples include IVR-based menus (e.g., the DTMF collection for PIN-based conference access), partial and complete recordings in the conference (e.g., for the "state your name" functionality and recording of the whole conference), private and global announcements, and so on. All of this can be achieved by means of the functionality provided by the MS. In fact, even if the conferencing and IVR features come from different packages, the AS can interact with both of them and achieve complex results by correlating the effects of different transactions in its application logic.

From the media and framework perspective, a typical Rich Conference scenario can be seen as depicted in Figure 15.31.

To identify a single sample scenario, let us consider this sequence for a participant joining a conference (which again, we assume has already been created):

1. The UAC as usual INVITEs a URI associated with a conference, and the AS follows the previously explained procedure to have the UAC negotiate a new media session with the MS.
2. The UAC is presented with an IVR menu, in which it is requested to input a PIN code to access the conference.
3. If the PIN is correct, the UAC is asked to state its name so that it can be recorded.
4. The UAC is attached to the conference, and the previously recorded name is announced globally to the conference to advertise its arrival.

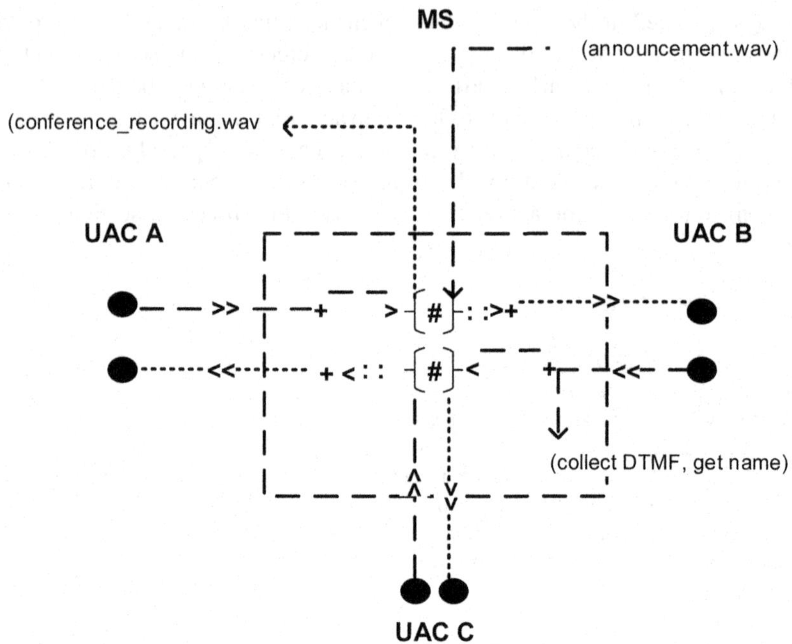

FIGURE 15.31 Conference: Rich Conference Scenario. (Copyright: IETF.)

Figure 15.32 shows a single UAC joining a conference. The example, as usual, hides the previous interaction between the UAC and the AS and instead, focuses on what the AS does to actually interact with the participant and join it to the conference bridge.

As can be deduced from Figure 15.32, the AS, in its business logic, correlates the results of different transactions, addressed to different packages, to implement a conferencing scenario more complex than the Simple Bridging scenario previously described. The framework transaction steps are as follows:

- Since this is a private conference, the UAC is to be presented with a request for a password, in this case a PIN number. To do so, the AS instructs the MS (A1) to collect a series of DTMF digits from the specified UAC (connectionid=UAC). The request includes both a voice message (<prompt>) and the described digit collection context (<collect>). The PIN is assumed to be a 4-digit number, and so the MS has to collect 4 digits maximum (maxdigits=4).
- The DTMF digit buffer must be cleared before collecting (cleardigitbuffer=true), and the UAC can use the star key to restart the collection (escapekey=*), e.g., if the UAC is aware that he mistyped any of the digits and wants to start again.
- The transaction goes on as usual (A2), with the transaction being handled and notification of the dialog start being sent in a 200 OK. After that, the UAC is actually presented with the voice message and is subsequently requested to input the required PIN number.
- We assume that the UAC typed the correct PIN number (1234), which is reported by the MS to the AS by means of the usual MS-generated CONTROL event (B1). The AS correlates this event to the previously started dialog by checking the referenced dialogid (06d1bac) and acks the event (B2). It then extracts the information it needs from the event (in this case, the digits provided by the MS) from the <controlinfo> container (dtmf=1234) and verifies that it is correct.
- Since the PIN is correct, the AS can proceed to the next step, i.e., asking the UAC to state his name, in order to subsequently play the recording on the conference to report the new participant. Again, this is done with a request to the IVR package (C1). The AS

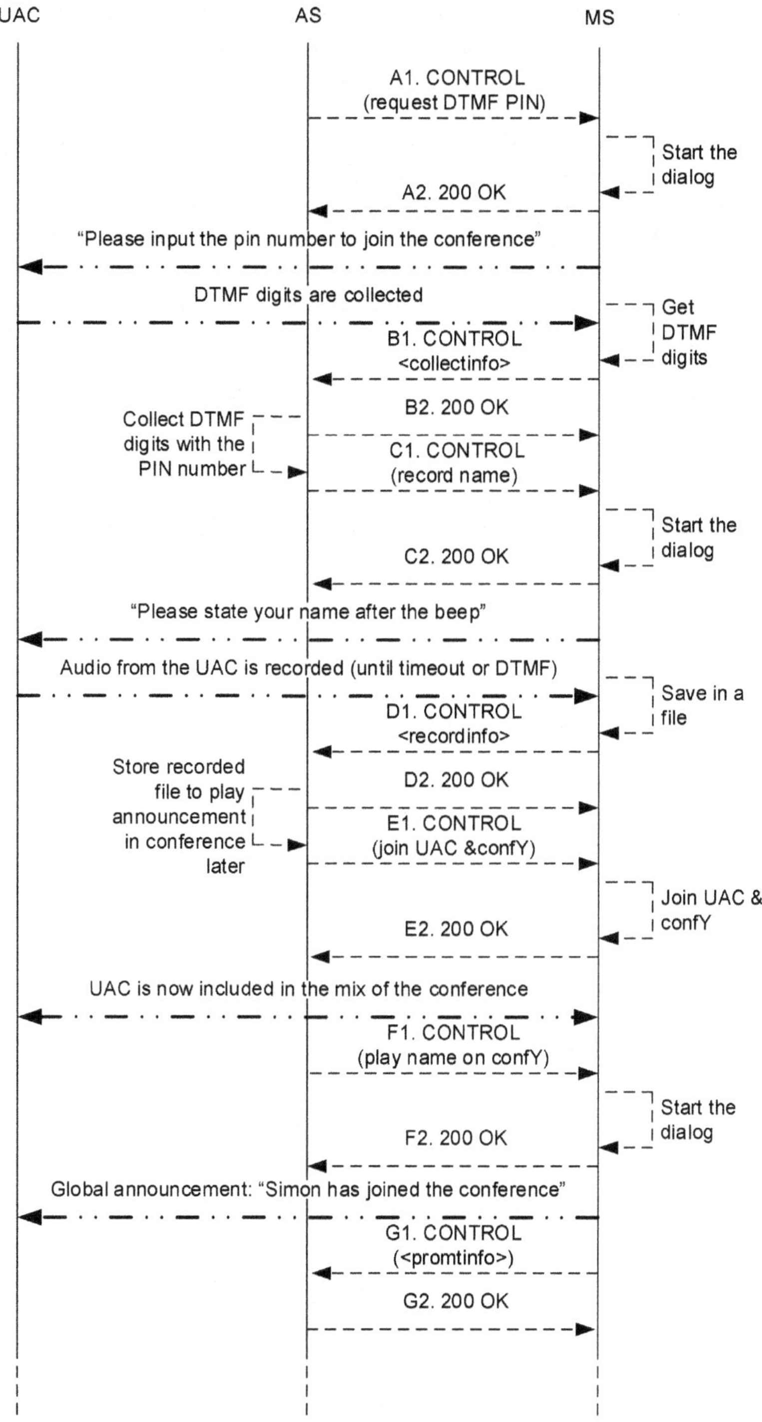

FIGURE 15.32 Rich Conference Scenario: Framework Transactions. (Copyright: IETF.)

instructs the MS to play a voice message ("state your name after the beep"), to be followed by a recording of only the audio from the UAC (in stream, media=audio/sendonly, while media=video/inactive). A beep must be played right before the recording starts (beep=true), and the recording must only last 3 seconds (maxtime=3s), since it is only needed as a brief announcement.

- Without delving again into the details of a recording-related transaction (C2), the AS finally gets the URI of the requested recording (D1, acked in D2).
- At this point, the AS attaches the UAC (id1) to the conference (id2), just as explained for the Simple Bridging scenario (E1/E2).
- Finally, to notify the other participants that a new participant has arrived, the AS requests a global announcement on the conference. This is a simple <prompt> request to the IVR package (F1), as explained in previous sections (e.g., Section 15.6.1.2, among others), but with a slight difference: the target of the prompt is not a connectionid (a media connection) but the conference itself (conferenceid=6146dd5). As a result of this transaction, the announcement will be played on all the media connections attached to the conference that are allowed to receive media from it. The AS specifically requests that two media files be played:
    1. The media file containing the recorded name of the new user as retrieved in D1 ("Simon...")
    2. A pre-recorded media file explaining what happened ("... has joined the conference")

The transaction then follows its usual flow (F2), and the event that sends notification regarding the end of the announcement (G1, acked in G2) concludes the scenario.

```
A1. AS -> MS (CFW CONTROL, collect)
-----------------------------------
CFW 50e56b8d65f9 CONTROL
Control-Package: msc-ivr/1.0
Content-Type: application/msc-ivr+xml
Content-Length: 311
<mscivr version="1.0" xmlns="urn:ietf:params:xml:ns:msc-ivr">
   <dialogstart connectionid="10514b7f:6a900179">
      <dialog>
         <prompt>
            <media loc="http://www.example.net
               /prompts/conf-getpin.wav" type="audio/x-wav"/>
         </prompt>
         <collect maxdigits="4" escapekey="*" cleardigitbuffer="true"/>
      </dialog>
   </dialogstart>
</mscivr>

A2. AS <- MS (CFW 200 OK)
-------------------------
CFW 50e56b8d65f9 200
Timeout: 10
Content-Type: application/msc-ivr+xml
Content-Length: 137
<mscivr version="1.0" xmlns="urn:ietf:params:xml:ns:msc-ivr">
   <response status="200" reason="Dialog started" dialogid="06d1bac"/>
</mscivr>

B1. AS <- MS (CFW CONTROL event)
--------------------------------
CFW 166d68a76659 CONTROL
Control-Package: msc-ivr/1.0
Content-Type: application/msc-ivr+xml
Content-Length: 272
<mscivr version="1.0" xmlns="urn:ietf:params:xml:ns:msc-ivr">
```

```
    <event dialogid="06d1bac">
        <dialogexit status="1" reason="Dialog successfully completed">
            <promptinfo duration="2312" termmode="completed"/>
            <collectinfo dtmf="1234" termmode="match"/>
        </dialogexit>
    </event>
</mscivr>

B2. AS -> MS (CFW 200, ACK to 'CONTROL event')
----------------------------------------------
CFW 166d68a76659 200
C1. AS -> MS (CFW CONTROL, record)
----------------------------------
CFW 61fd484f196e CONTROL
Control-Package: msc-ivr/1.0
Content-Type: application/msc-ivr+xml
Content-Length: 373
<mscivr version="1.0" xmlns="urn:ietf:params:xml:ns:msc-ivr">
    <dialogstart connectionid="10514b7f:6a900179">
        <dialog>
            <prompt>
                <media loc="http://www.example.net/prompts/conf-rec-
                        name.wav" type="audio/x-wav"/>
            </prompt>
            <record beep="true" maxtime="3s"/>
        </dialog>
        <stream media="audio" direction="sendonly"/>
        <stream media="video" direction="inactive"/>
    </dialogstart>
</mscivr>

C2. AS <- MS (CFW 200 OK)
-------------------------
CFW 61fd484f196e 200
Timeout: 10
Content-Type: application/msc-ivr+xml
Content-Length: 137
<mscivr version="1.0" xmlns="urn:ietf:params:xml:ns:msc-ivr">
    <response status="200" reason="Dialog started" dialogid="1cf0549"/>
</mscivr>D1. AS <- MS (CFW CONTROL event)
--------------------------------
CFW 3ec13ab96224 CONTROL
Control-Package: msc-ivr/1.0
Content-Type: application/msc-ivr+xml
Content-Length: 402
<mscivr version="1.0" xmlns="urn:ietf:params:xml:ns:msc-ivr">
    <event dialogid="1cf0549">
        <dialogexit status="1" reason="Dialog successfully completed">
            <promptinfo duration="4988" termmode="completed"/>
            <recordinfo duration="3000" termmode="maxtime">
                <mediainfo loc="http://www.example.net/
                        recordings/recording-1cf0549.wav"
                        type="audio/x-wav" size="48044"/>
            </recordinfo>
        </dialogexit>
```

```
   </event>
</mscivr>

D2. AS -> MS (CFW 200, ACK to 'CONTROL event')
----------------------------------------------
CFW 3ec13ab96224 200
E1. AS -> MS (CFW CONTROL, join)
--------------------------------
CFW 261d188b63b7 CONTROL
Control-Package: msc-mixer/1.0
Content-Type: application/msc-mixer+xml
Content-Length: 120
<mscmixer version="1.0" xmlns="urn:ietf:params:xml:ns:msc-mixer">
   <join id1="10514b7f:6a900179" id2="6146dd5"/>
</mscmixer>

E2. AS <- MS (CFW 200 OK)
-------------------------
CFW 261d188b63b7 200
Timeout: 10
Content-Type: application/msc-mixer+xml
Content-Length: 125
<mscmixer version="1.0" xmlns="urn:ietf:params:xml:ns:msc-mixer">
   <response status="200" reason="Join successful"/>
</mscmixer>

F1. AS -> MS (CFW CONTROL, play)
--------------------------------
CFW 718c30836f38 CONTROL
Control-Package: msc-ivr/1.0
Content-Type: application/msc-ivr+xml
Content-Length: 334
<mscivr version="1.0" xmlns="urn:ietf:params:xml:ns:msc-ivr">
   <dialogstart conferenceid="6146dd5">
      <dialog>
         <prompt>
            <media loc="http://www.example.net/recordings/
                  recording-1cf0549.wav" type="audio/x-
                  wav"/>
            <media loc="http://www.example.net/prompts/
                  conf-hasjoin.wav" type="audio/x-wav"/>
         </prompt>
      </dialog>
   </dialogstart>
</mscivr>

F2. AS <- MS (CFW 200 OK)
-------------------------
CFW 718c30836f38 200
Timeout: 10
Content-Type: application/msc-ivr+xml
Content-Length: 137
<mscivr version="1.0" xmlns="urn:ietf:params:xml:ns:msc-ivr">
   <response status="200" reason="Dialog started" dialogid="5f4bc7e"/>
</mscivr>
```

```
G1. AS <- MS (CFW CONTROL event)
--------------------------------
CFW 6485194f622f CONTROL
Control-Package: msc-ivr/1.0
Content-Type: application/msc-ivr+xml
Content-Length: 229
<mscivr version="1.0" xmlns="urn:ietf:params:xml:ns:msc-ivr">
   <event dialogid="5f4bc7e">
      <dialogexit status="1" reason="Dialog successfully completed">
         <promptinfo duration="1838" termmode="completed"/>
      </dialogexit>
   </event>
</mscivr>

G2. AS -> MS (CFW 200, ACK to 'CONTROL event')
----------------------------------------------
CFW 6485194f622f 200
```

15.6.3.3 Coaching Scenario

Another typical conference-based use case is the so-called Coaching scenario. In such a scenario, a customer (called "A" in the following example) places a call to a business call center. An agent (B) is assigned to the customer. A coach (C), who cannot be heard by the customer, provides the agent with whispered suggestions about what to say. This scenario is also described in RFC 4597.

As can be deduced from the scenario description, per-user policies for media mixing and delivery, i.e., who can hear what, are very important. The MS must make sure that only the agent can hear the coach's suggestions. Since this is basically a multiparty call (despite what the customer might be thinking), a mixing entity is needed in order to accomplish the scenario requirements. To summarize:

- The customer (A) must only hear what the agent (B) says.
- The agent (B) must be able to hear both A and the coach (C).
- C must be able to hear both A and B in order to give B the right suggestions and also be aware of the whole conversation.

From the media and framework perspective, such a scenario can be seen as depicted in Figure 15.33.

From the framework point of view, when the previously mentioned legs are not attached to anything yet, what appears is shown in Figure 15.34.

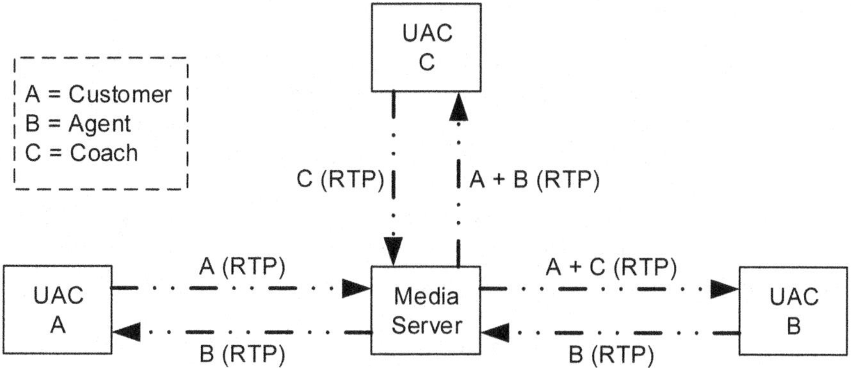

FIGURE 15.33 Coaching Scenario: Media Perspective. (Copyright: IETF.)

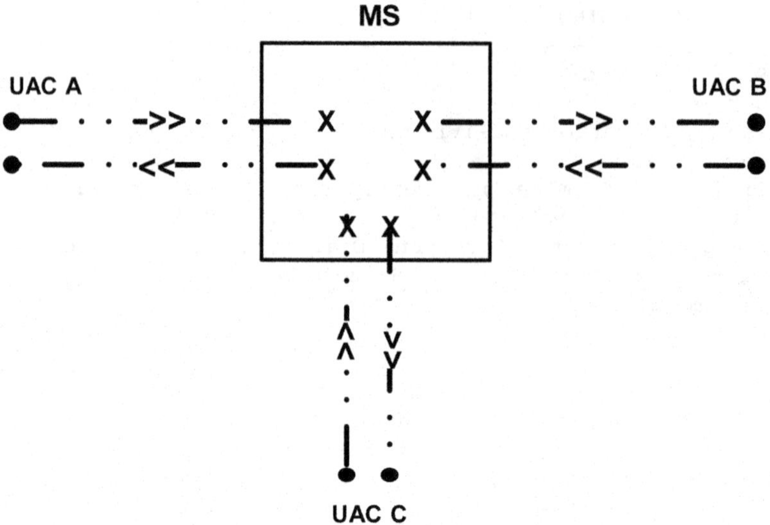

FIGURE 15.34 Coaching Scenario: UAC Legs Not Attached.

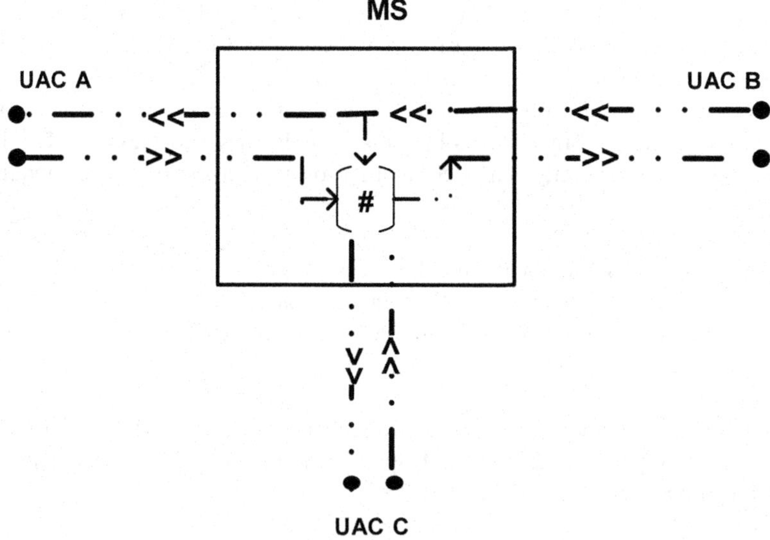

FIGURE 15.35 Coaching Scenario: UAC Legs Mixed and Attached.

By contrast, what the scenario should look like is depicted in Figure 15.35. The customer receives media directly from the agent ("recvonly"), while all the three involved participants contribute to a hidden conference. Of course, the customer is not allowed to receive the mixed flows from the conference ("sendonly"), unlike the agent and the coach, who must both be aware of the whole conversation ("sendrecv").

In the framework, this can be achieved by means of the Mixer Control Package, which as demonstrated in the previous conferencing examples, can be exploited whenever mixing and joining entities are needed. The needed steps can be summarized in the following list:

1. First of all, a hidden conference is created.
2. Then, the three participants are all attached to it, each with a custom mixing policy, specifically:

- The customer (A) as "sendonly"
- The agent (B) as "sendrecv"
- The coach (C) as "sendrecv" and with a gain of −3 dB to halve the volume of its own contribution (so that the agent actually hears the customer at a louder volume and hears the coach whispering)

3. Finally, the customer is joined to the agent as a passive receiver ("recvonly").

A diagram of such a sequence of transactions is depicted in Figure 15.36.

1. First of all, the AS creates a new hidden conference by means of a <createconference> request (A1). This conference is properly configured according to the use it is assigned to, i.e., to mix all the involved parties accordingly. Since only three participants will be joined to it, "reserved-talkers" is set to 3. "reserved-listeners," on the other hand, is set to 2, since

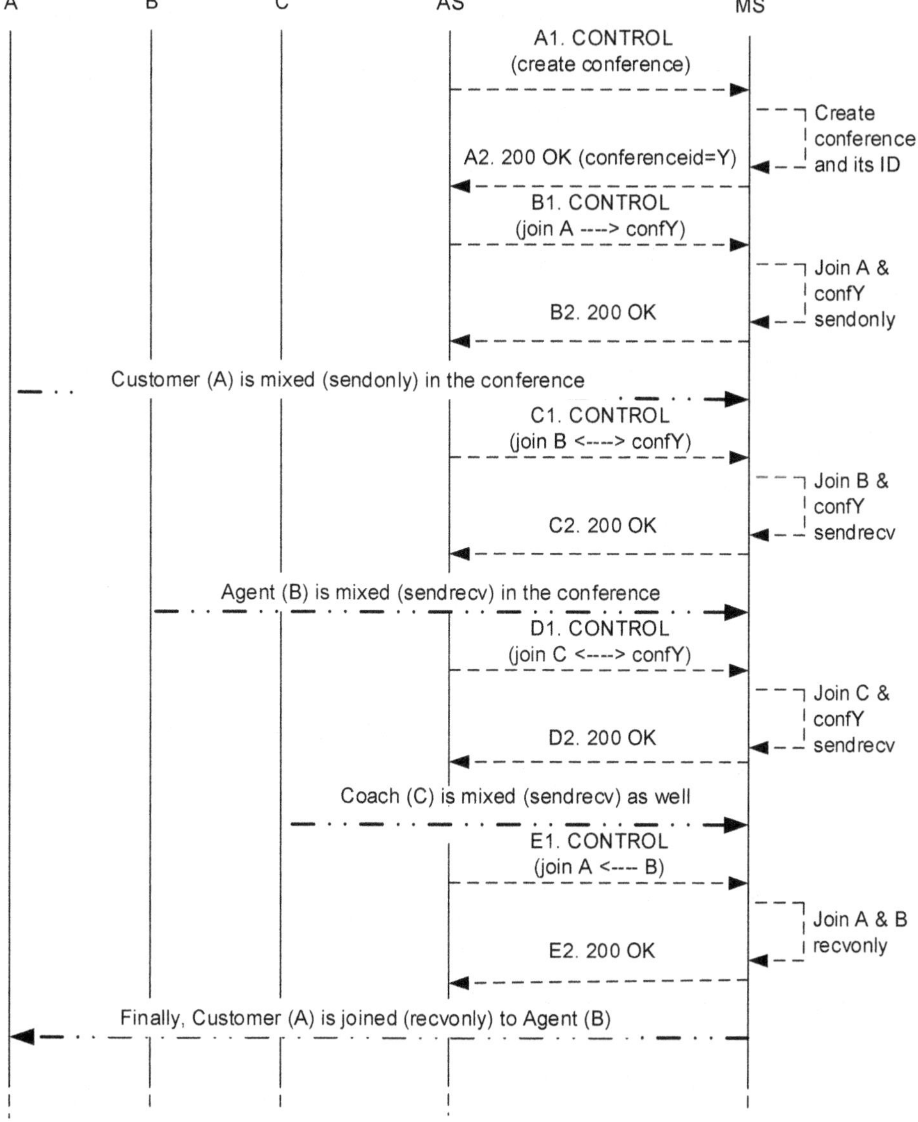

FIGURE 15.36 Coaching Scenario: Framework Transactions. (Copyright: IETF.)

only the agent and the coach must receive a mix, while the customer must be unaware of the coach. Finally, the video layout is set to <dual-view> for the same reason, since only the customer and the agent must appear in the mix.

2. The MS sends notification of the successful creation of the new conference in a 200 framework message (A2). The identifier assigned to the conference, which will be used in subsequent requests addressed to it, is 1df080e.
3. Now that the conference has been created, the AS joins the three actors to it with different policies; namely, (i) the customer (A) is joined as "sendonly" to the conference (B1), (ii) the agent (B) is joined as "sendrecv" to the conference (C1), and (iii) the coach (C) is joined as "sendrecv" (but audio only) to the conference and with a lower volume (D1). The custom policies are enforced by means of properly constructed <stream> elements.
4. The MS takes care of the requests and acks them (B2, C2, D2). At this point, the conference will receive media from all the actors but will only provide the agent and the coach with the resulting mix.
5. To complete the scenario, the AS joins A with B directly as "recvonly" (E1). The aim of this request is to provide A with media too, namely, the media contributed by B. In this way, A is unaware of the fact that its media are accessed by C by means of the hidden mixer.
   - The MS takes care of this request too and acks it (E2), concluding the scenario.

```
A1. AS -> MS (CFW CONTROL, createconference)
--------------------------------------------
CFW 238e1f2946e8 CONTROL
Control-Package: msc-mixer
Content-Type: application/msc-mixer+xml
Content-Length: 329
<mscmixer version="1.0" xmlns="urn:ietf:params:xml:ns:msc-mixer">
   <createconference reserved-talkers="3" reserved-listeners="2">
      <audio-mixing type="nbest"/>
         <video-layouts>
            <video-layout min-participants='1'>
               <dual-view/>
            </video-layout>
         </video-layouts>
         <video-switch>
            <controller/>
         </video-switch>
   </createconference>
</mscmixer>

A2. AS <- MS (CFW 200 OK)
-------------------------
CFW 238e1f2946e8 200
Timeout: 10
Content-Type: application/msc-mixer+xml
Content-Length: 151
<mscmixer version="1.0" xmlns="urn:ietf:params:xml:ns:msc-mixer">
   <response status="200" reason="Conference created"
         conferenceid="1df080e"/>
</mscmixer>

B1. AS -> MS (CFW CONTROL, join)
--------------------------------
CFW 2eb141f241b7 CONTROL
Control-Package: msc-mixer
```

```
Content-Type: application/msc-mixer+xml
Content-Length: 226
<mscmixer version="1.0" xmlns="urn:ietf:params:xml:ns:msc-mixer">
   <join id1="10514b7f:6a900179" id2="1df080e">
      <stream media="audio" direction="sendonly"/>
      <stream media="video" direction="sendonly"/>
   </join>
</mscmixer>

B2. AS <- MS (CFW 200 OK)
-------------------------
CFW 2eb141f241b7 200
Timeout: 10
Content-Type: application/msc-mixer+xml
Content-Length: 125
<mscmixer version="1.0" xmlns="urn:ietf:params:xml:ns:msc-mixer">
   <response status="200" reason="Join successful"/>
</mscmixer>

C1. AS -> MS (CFW CONTROL, join)
--------------------------------
CFW 515f007c5bd0 CONTROL
Control-Package: msc-mixer
Content-Type: application/msc-mixer+xml
Content-Length: 122
<mscmixer version="1.0" xmlns="urn:ietf:params:xml:ns:msc-mixer">
   <join id1="756471213:c52ebf1b" id2="1df080e"/>
</mscmixer>

C2. AS <- MS (CFW 200 OK)
-------------------------
CFW 515f007c5bd0 200
Timeout: 10
Content-Type: application/msc-mixer+xml
Content-Length: 125
<mscmixer version="1.0" xmlns="urn:ietf:params:xml:ns:msc-mixer">
   <response status="200" reason="Join successful"/>
</mscmixer>

D1. AS -> MS (CFW CONTROL, join)
--------------------------------
CFW 0216231b1f16 CONTROL
Control-Package: msc-mixer
Content-Type: application/msc-mixer+xml
Content-Length: 221
<mscmixer version="1.0" xmlns="urn:ietf:params:xml:ns:msc-mixer">
   <join id1="z9hG4bK19461552:1353807a" id2="1df080e">
      <stream media="audio">
         <volume controltype="setgain" value="-3"/>
      </stream>
   </join>
</mscmixer>

D2. AS <- MS (CFW 200 OK)
-------------------------
CFW 0216231b1f16 200
Timeout: 10
```

```
Content-Type: application/msc-mixer+xml
Content-Length: 125
<mscmixer version="1.0" xmlns="urn:ietf:params:xml:ns:msc-mixer">
   <response status="200" reason="Join successful"/>
</mscmixer>

E1. AS -> MS (CFW CONTROL, join)
--------------------------------
CFW 140e0f763352 CONTROL
Control-Package: msc-mixer
Content-Type: application/msc-mixer+xml
Content-Length: 236
<mscmixer version="1.0" xmlns="urn:ietf:params:xml:ns:msc-mixer">
   <join id1="10514b7f:6a900179" id2="756471213:c52ebf1b">
      <stream media="audio" direction="recvonly"/>
      <stream media="video" direction="recvonly"/>
   </join>
</mscmixer>

E2. AS <- MS (CFW 200 OK)
-------------------------
CFW 140e0f763352 200
Timeout: 10
Content-Type: application/msc-mixer+xml
Content-Length: 125
<mscmixer version="1.0" xmlns="urn:ietf:params:xml:ns:msc-mixer">
   <response status="200" reason="Join successful"/>
</mscmixer>
```

15.6.3.4 Sidebars

Within the context of conferencing, there could be a need for so-called sidebars, or side conferences. This would be the case, for instance, if two or more participants of a conference were willing to create a side conference among each other while still receiving part of the original conference mix in the background. Motivations for such a use case can be found in both RFC 4597 and RFC 5239 (see Chapter 2). It is clear that in such a case, the side conference is actually a separate conference but must also somehow be related to the original conference. Although the application-level relationship is out of scope for this specification (this "belongs" to Centralized Conferencing [XCON]), the media stream relationship is more relevant here, because there is a stronger relationship at the media level from the MEDIACTRL point of view. Consequently, it is interesting to analyze how sidebars could be used to construct the conference mixes according to the MEDIACTRL specification.

The scenario presented in this section is a conference hosting four different participants: A, B, C, and D. All these participants are attached to the conference as active senders and receivers of the existing media streams. At a certain point in time, two participants (B and D) decide to create a sidebar just for them. The sidebar they want to create is constructed so that only audio is involved. The audio mix of the sidebar must not be made available to the main conference. The mix of the conference must be attached to the sidebar, but with a lower volume (30%), because it is just background to the actual conversation. This would allow both B and D to talk to each other without A and C listening to them, while B and D could still have an overview of what is happening in the main conference. From the media and framework perspective, such a scenario can be seen as depicted in Figure 15.37.

From the framework point of view, when all the participants are attached to the main conference, what appears is shown in Figure 15.38.

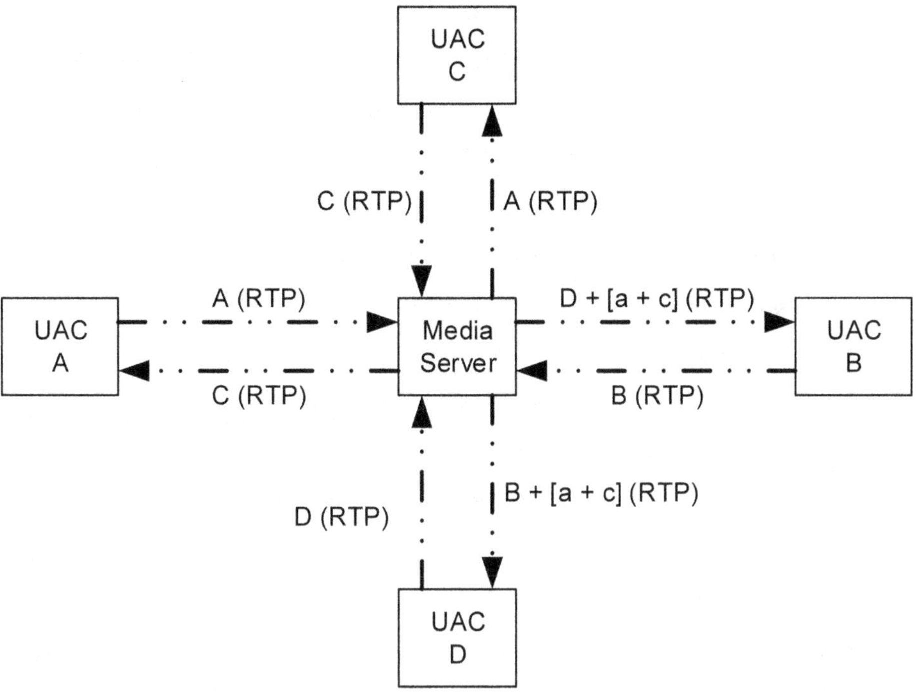

FIGURE 15.37 Sidebars: Media Perspective.

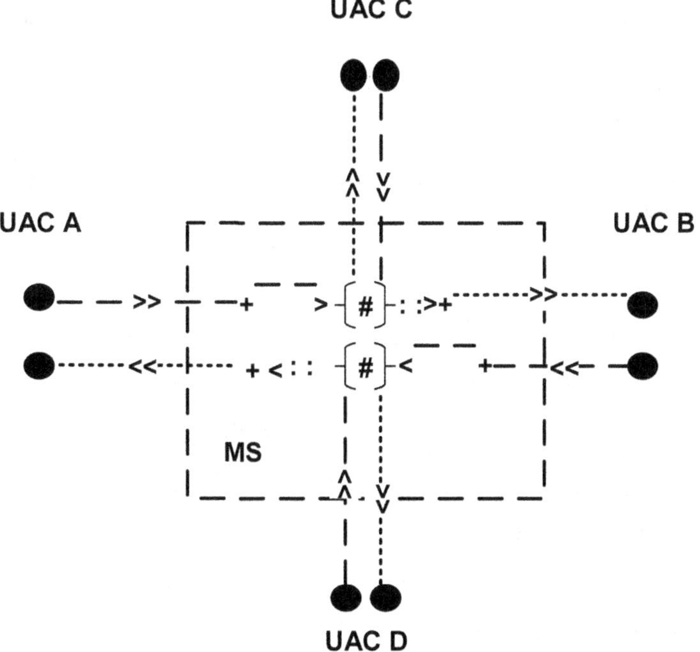

FIGURE 15.38 Sidebars: UAC Legs in Main Conference.

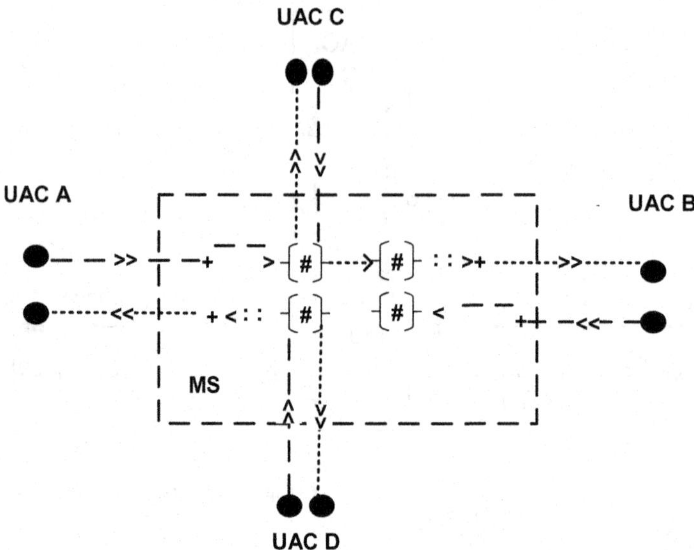

FIGURE 15.39 Sidebars: UAC Legs Mixed and Attached.

By contrast, what the scenario should look like is depicted in Figure 15.39. A new mixer is created to host the sidebar. The main mix is then attached as "sendonly" to the sidebar mix at a lower volume (in order to provide the sidebar users with a background of the main conference). The two interested participants (B and D) have their audio leg detached from the main conference and attached to the sidebar. This detachment can be achieved by either actually detaching the leg or just modifying the status of the leg to "inactive." Note that this only affects the audio stream: the video of the two users is still attached to the main conference, and what happens at the application level may or may not have been changed accordingly (e.g., XCON protocol interactions). Please note that the main conference is assumed to be in place and the involved participants (A, B, C, and D) attached ("sendrecv") to it.

The situation may subsequently be reverted (e.g., destroying the sidebar conference and reattaching B and D to the main conference mix) in the same way. The AS would just need to unjoin B and D from the hidden conference and change their connection with the main conference back to "sendrecv." After unjoining the main mix and the sidebar mix, the sidebar conference could then be destroyed. For brevity, and because similar transactions have already been presented, these steps are not described here. In the framework, just as in the previous section, the presented scenario can again be achieved by means of the Mixer Control Package. The needed steps can be summarized in the following list:

1. First of all, a hidden conference is created (the sidebar mix).
2. Then, the main conference mix is attached to it as "sendonly" and with a gain of −5 dB to limit the volume of its own contribution to 30% (so that B and D can hear each other at a louder volume while still listening to what is happening in the main conference in the background).
3. B and D are detached from the main mix for audio (<modifyjoin> with "inactive" status).
4. B and D are attached to the hidden sidebar mix for audio.

Note that for detaching B and D from the main mix, a <modifyjoin> with an "inactive" status is used instead of an <unjoin>. The motivation for this is related to how a subsequent rejoining of B and D to the main mix could take place. In fact, by using <modifyjoin>, the resources created when first joining B and D to the main mix remain in place, even if marked as unused at the moment.

An <unjoin>, on the other hand, would actually free those resources, which in turn, could be granted to other participants joining the conference in the meantime. This means that when needing to reattach B and D to the main mix, the MS may not have the resources to do so, resulting in an undesired failure. A diagram of such a sequence of transactions (where confX is the identifier of the pre-existing main conference mix) is depicted in Figure 15.40.

- First of all, the hidden conference mix is created (A1). The request is basically the same as that presented in previous sections, i.e., a <createconference> request addressed to the Mixer package. The request is very lightweight and asks the MS to only reserve two

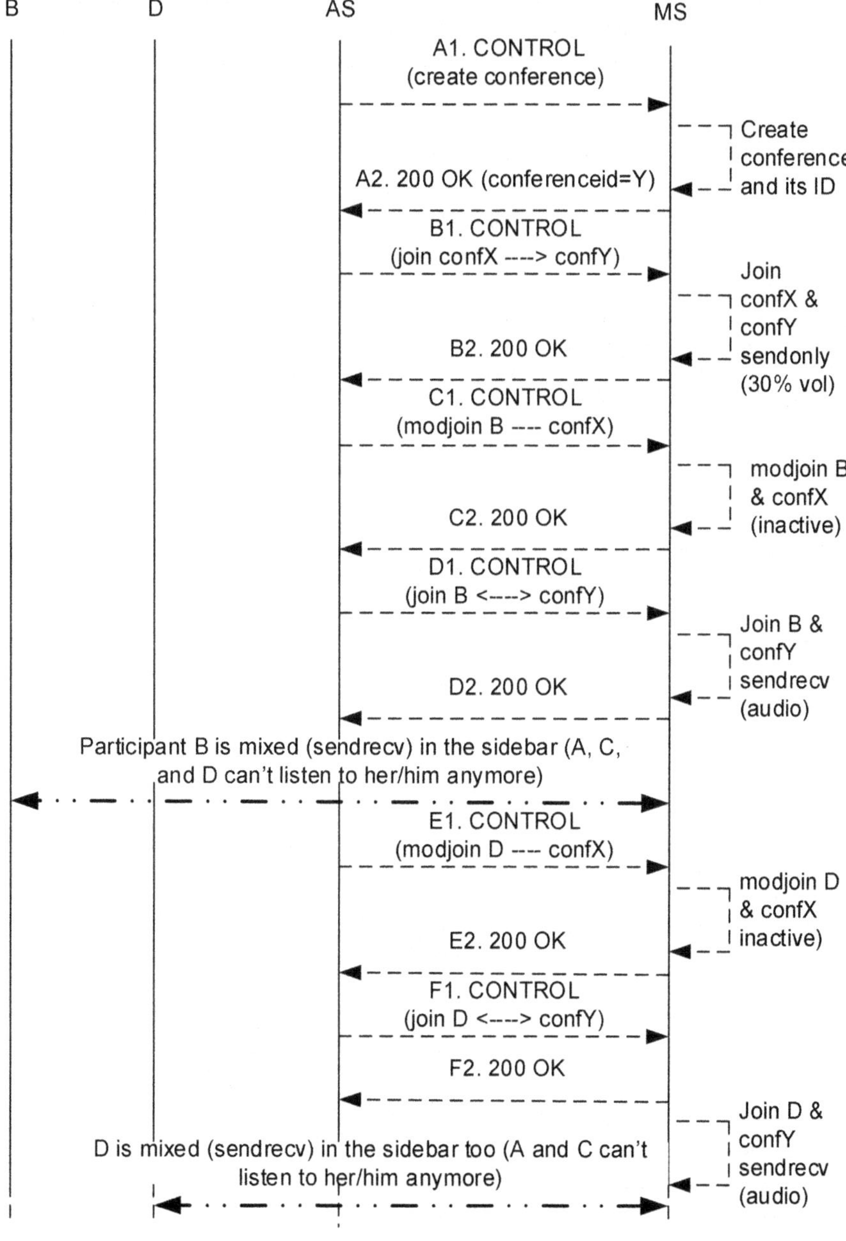

FIGURE 15.40 Sidebars: Framework Transactions. (Copyright: IETF.)

listener seats ("reserved-listeners," since only B and D have to hear something) and three talker seats ("reserved-listeners," because in addition to B and D, the main mix is also an active contributor to the sidebar). The mixing will be driven by directives from the AS (mix-type=controller). When the mix is successfully created, the MS provides the AS with its identifier (519c1b9).

- As a first transaction after that, the AS joins the audio from the main conference and the newly created sidebar conference mix (B1). Only audio needs to be joined (media=audio), with a "sendonly" direction (i.e., only flowing from the main conference to the sidebar and not vice versa) and at 30% volume with respect to the original stream (setgain=–5 dB). A successful completion of the transaction is reported to the AS (B2).
- At this point, the AS makes the connection of B (2133178233:18294826) and the main conference (2f5ad43) inactive by means of a <modifyjoin> directive (C1). The request is taken care of by the MS (C2), and B is actually excluded from the mix for sending as well as receiving.
- After that, the AS (D1) joins B (2133178233:18294826) to the sidebar mix created previously (519c1b9). The MS attaches the requested connections and sends confirmation to the AS (D2).
- The same transactions already done for B are done for D as well (id1=1264755310:2beeae5b), i.e., the <modifyjoin> to make the connection in the main conference inactive (E1-2) and the <join> to attach D to the sidebar mix (F1-2). At the end of these transactions, A and C will only listen to each other, while B and D will listen to each other and to the conference mix as a comfortable background.

```
A1. AS -> MS (CFW CONTROL, createconference)
--------------------------------------------
CFW 7fdcc2331bef CONTROL
Control-Package: msc-mixer/1.0
Content-Type: application/msc-mixer+xml
Content-Length: 198
<mscmixer version="1.0" xmlns="urn:ietf:params:xml:ns:msc-mixer">
   <createconference reserved-talkers="3" reserved-listeners="2">
      <audio-mixing type="controller"/>
   </createconference>
</mscmixer>

A2. AS <- MS (CFW 200 OK)
-------------------------
CFW 7fdcc2331bef 200
Timeout: 10
Content-Type: application/msc-mixer+xml
Content-Length: 151
<mscmixer version="1.0" xmlns="urn:ietf:params:xml:ns:msc-mixer">
   <response status="200" reason="Conference created"
         conferenceid="519c1b9"/>
</mscmixer>

B1. AS -> MS (CFW CONTROL, join with setgain)
--------------------------------------------
CFW 4e6afb6625e4 CONTROL
Control-Package: msc-mixer/1.0
Content-Type: application/msc-mixer+xml
Content-Length: 226
<mscmixer version="1.0" xmlns="urn:ietf:params:xml:ns:msc-mixer">
   <join id1="2f5ad43" id2="519c1b9">
```

```
        <stream media="audio" direction="sendonly">
            <volume controltype="setgain" value="-5"/>
        </stream>
    </join>
</mscmixer>

B2. AS <- MS (CFW 200 OK)
-------------------------
CFW 4e6afb6625e4 200
Timeout: 10
Content-Type: application/msc-mixer+xml
Content-Length: 125
<mscmixer version="1.0" xmlns="urn:ietf:params:xml:ns:msc-mixer">
    <response status="200" reason="Join successful"/>
</mscmixer>

C1. AS -> MS (CFW CONTROL, modifyjoin with 'inactive' status)
-------------------------------------------------------------
CFW 3f2dba317c83 CONTROL
Control-Package: msc-mixer/1.0
Content-Type: application/msc-mixer+xml
Content-Length: 193
<mscmixer version="1.0" xmlns="urn:ietf:params:xml:ns:msc-mixer">
    <modifyjoin id1="2133178233:18294826" id2="2f5ad43">
        <stream media="audio" direction="inactive"/>
    </modifyjoin>
</mscmixer>

C2. AS <- MS (CFW 200 OK)
-------------------------
CFW 3f2dba317c83 200
Timeout: 10
Content-Type: application/msc-mixer+xml
Content-Length: 123
<mscmixer version="1.0" xmlns="urn:ietf:params:xml:ns:msc-mixer">
    <response status="200" reason="Join modified"/>
</mscmixer>

D1. AS -> MS (CFW CONTROL, join to sidebar)
-------------------------------------------
CFW 2443a8582d1d CONTROL
Control-Package: msc-mixer/1.0
Content-Type: application/msc-mixer+xml
Content-Length: 181
<mscmixer version="1.0" xmlns="urn:ietf:params:xml:ns:msc-mixer">
    <join id1="2133178233:18294826" id2="519c1b9">
        <stream media="audio" direction="sendrecv"/>
    </join>
</mscmixer>

D2. AS <- MS (CFW 200 OK)
-------------------------
CFW 2443a8582d1d 200
Timeout: 10
Content-Type: application/msc-mixer+xml
Content-Length: 125
```

```
<mscmixer version="1.0" xmlns="urn:ietf:params:xml:ns:msc-mixer">
   <response status="200" reason="Join successful"/>
</mscmixer>

E1. AS -> MS (CFW CONTROL, modifyjoin with 'inactive' status)
-----------------------------------------------------------------
CFW 436c6125628c CONTROL
Control-Package: msc-mixer/1.0
Content-Type: application/msc-mixer+xml
Content-Length: 193
<mscmixer version="1.0" xmlns="urn:ietf:params:xml:ns:msc-mixer">
   <modifyjoin id1="1264755310:2beeae5b" id2="2f5ad43">
      <stream media="audio" direction="inactive"/>
   </modifyjoin>
</mscmixer>

E2. AS <- MS (CFW 200 OK)
-------------------------
CFW 436c6125628c 200
Timeout: 10
Content-Type: application/msc-mixer+xml
Content-Length: 123
<mscmixer version="1.0" xmlns="urn:ietf:params:xml:ns:msc-mixer">
   <response status="200" reason="Join modified"/>
</mscmixer>

F1. AS -> MS (CFW CONTROL, join to sidebar)
-------------------------------------------
CFW 7b7ed00665dd CONTROL
Control-Package: msc-mixer/1.0
Content-Type: application/msc-mixer+xml
Content-Length: 181
<mscmixer version="1.0" xmlns="urn:ietf:params:xml:ns:msc-mixer">
   <join id1="1264755310:2beeae5b" id2="519c1b9">
      <stream media="audio" direction="sendrecv"/>
   </join>
</mscmixer>

F2. AS <- MS (CFW 200 OK)
-------------------------
CFW 7b7ed00665dd 200
Timeout: 10
Content-Type: application/msc-mixer+xml
Content-Length: 125
<mscmixer version="1.0" xmlns="urn:ietf:params:xml:ns:msc-mixer">
   <response status="200" reason="Join successful"/>
</mscmixer>
```

15.6.3.5 Floor Control

As described in RFC 4597, floor control is a feature typically needed and employed in most conference scenarios. In fact, while not a mandatory feature to implement when realizing a conferencing application, it provides additional control of the media streams contributed by participants, thus allowing moderation of the available resources. The Centralized Conferencing (XCON) framework RFC 5239 (see Chapter 2) suggests the use of the binary floor control protocol (BFCP) (RFC 4582 – see Chapter 6) to achieve the aforementioned functionality.

That said, a combined use of floor control functionality and the tools made available by the MEDIACTRL specification for conferencing would definitely be interesting to investigate.

RFC 5567 (see Section 8.2) introduces two different approaches to integrating floor control with the MEDIACTRL architecture:

(i) A topology where the Floor Control Server is co-located with the AS
(ii) A topology where the Floor Control Server is co-located with the MS

The two approaches are obviously different with respect to the amount of information the AS and the MS have to deal with, especially when thinking about the logic behind the floor queues and automated decisions. Nevertheless, considering how the CFW Framework is conceived, approach (ii) would need a dedicated package (be it an extension or a totally new package) in order to make the MS aware of floor control and allow the MS to interact with the interested UAC accordingly. At the time of writing, such a package does not exist yet, and as a consequence, only approach (i) will be dealt with in the presented scenario.

The scenario will then assume that the Floor Control Server (FCS) is co-located with the AS. This means that all the BFCP requests will be sent by Floor Control Participants (FCPs) to the FCS, which will make the AS directly aware of the floor statuses. For the sake of simplicity, the scenario assumes that the involved participants are already aware of all the identifiers needed in order to make BFCP requests for a specific conference. Such information may have been carried according to the COMEDIA negotiation as specified in RFC 4583.

It is important to note that such information must not reach the MS. This means that within the context of the 3pcc mechanism that may have been used in order to attach a UAC to the MS, all the BFCP-related information negotiated by the AS and the UAC must be removed before making the negotiation available to the MS, which may be unable to understand the specification. A simplified example of how this could be achieved is presented in Figure 15.41.

Please note that within the context of this example scenario, different identifiers may be used to address the same entity. Specifically, in this case, the UAC (the endpoint sending and receiving media) is also a FCP, as it negotiates a BFCP channel too.

From the media and framework perspective, such a scenario does not differ much from the conferencing scenarios presented earlier. It is more interesting to focus on the chosen topology for the scenario, as depicted in Figure 15.42.

The AS, besides maintaining the already-known SIP signaling among the involved parties, also acts as the FCS for the participants in the conferences for which it is responsible. In the scenario, two Floor Control Participants are involved: a basic Participant (FCP) and a Chair (FCC).

As in all the previously described conferencing examples, in the framework, this can be achieved by means of the Mixer Control Package. Assuming that the conference has been created, the participant has been attached ("recvonly") to it, and the participant is aware of the involved BFCP identifiers, the needed steps can be summarized in the following list:

1. The assigned chair, FCC, sends a subscription for events related to the floor for which it is responsible (FloorQuery).
2. The FCP sends a BFCP request (FloorRequest) to access the audio resource ("I want to speak").
3. The FCS (AS) sends a provisional response to the FCP (FloorRequestStatus PENDING) and handles the request in its queue. Since a chair is assigned to this floor, the request is forwarded to the FCC for a decision (FloorStatus).
4. The FCC makes a decision and sends it to the FCS (ChairAction ACCEPTED).
5. The FCS takes note of the decision and updates the queue accordingly. The decision is sent to the FCP (FloorRequestStatus ACCEPTED). The floor has not been granted yet.
6. As soon as the queue allows it, the floor is actually granted to the FCP. The AS, which is co-located with the FCS, understands in its business logic that such an event is associated with the audio resource being granted to the FCP. As a consequence, a <modifyjoin>

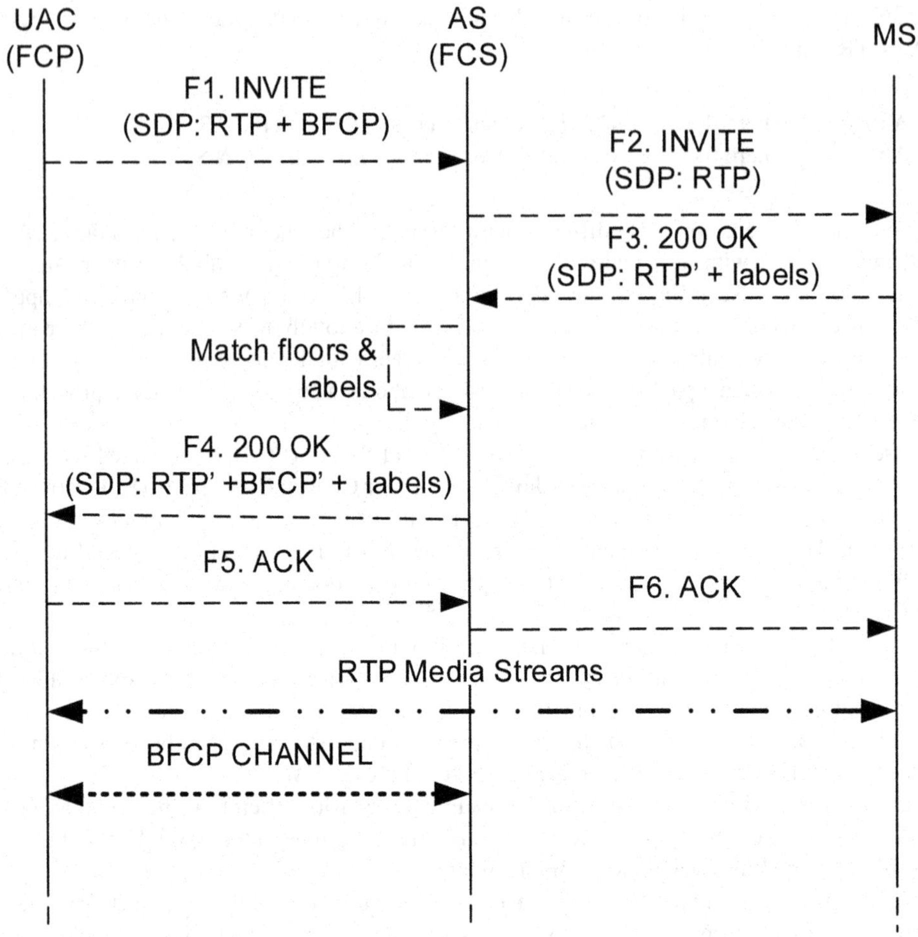

UAC
(FCP)
AS
(FCS)
MS
F1. INVITE
(SDP: RTP + BFCP)
F2. INVITE
(SDP: RTP)
F3. 200 OK
(SDP: RTP' + labels)
Match floors &
labels
F4. 200 OK
(SDP: RTP' +BFCP' + labels)
F5. ACK
F6. ACK
RTP Media Streams
BFCP CHANNEL

FIGURE 15.41 Floor Control: Example of Negotiation. (Copyright: IETF.)

UAC
(FCP)
BFCP
AS
(FCS)
BFCP
Floor Control
Chair
(FCC)
CFW
RTP
MS

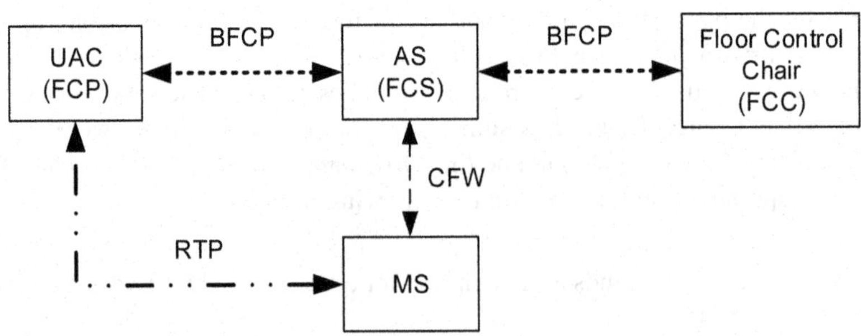

FIGURE 15.42 Floor Control: Overall Perspective.

('sendrecv') is sent through the Control Channel to the MS in order to unmute the FCP UAC in the conference.

7. The FCP is notified of this event (FloorRequestStatus GRANTED), thus ending the scenario.

A diagram of such a sequence of transactions (also involving the BFCP message flow at a higher level) is depicted in Figure 15.43.

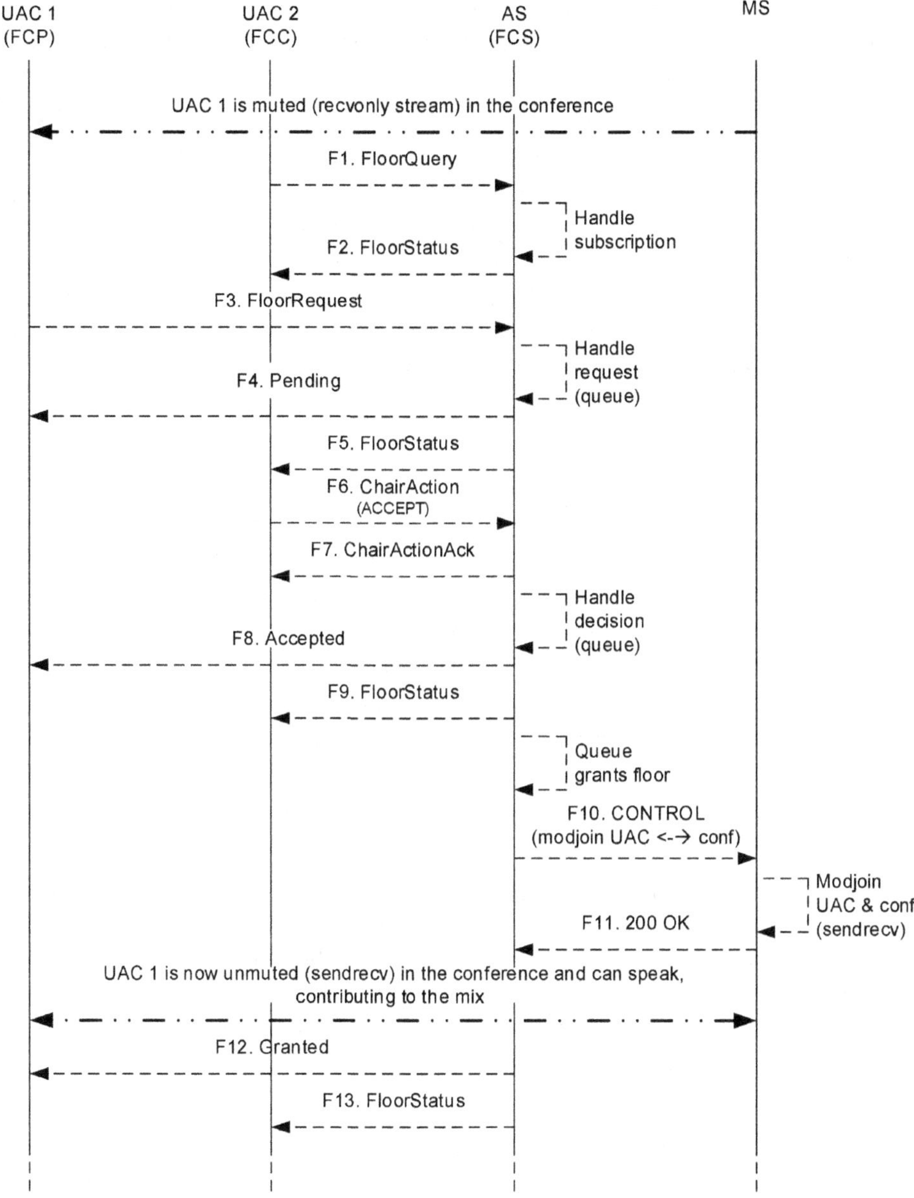

FIGURE 15.43 Floor Control: Framework Transactions. (Copyright: IETF.)

As can easily be deduced from this diagram, the complex interaction at the BFCP level only results in a single transaction at the MEDIACTRL level. In fact, the purpose of the BFCP transactions is to moderate access to the audio resource, which means providing the event trigger to MEDIACTRL-based conference manipulation transactions. Before being granted the floor, the FCP UAC is excluded from the conference mix at the MEDIACTRL level ("recvonly"). As soon as the floor has been granted, the FCP UAC is included in the mix. In MEDIACTRL words:

- Since the FCP UAC must be included in the audio mix, a <modifyjoin> is sent to the MS in a CONTROL directive. The <modifyjoin> has as identifiers the connectionid associated with the FCP UAC (e1e1427c:1c998d22) and the conferenceid of the mix (cf45ee2).

The <stream> element tells the MS that the audio media stream between the two must become bidirectional ("sendrecv"), changing the previous status ("recvonly"). Please note that in this case, only audio was involved in the conference; if video were involved as well, and video had to be unchanged, a <stream> directive for video would have to be placed in the request as well in order to maintain its current status.

```
1. AS -> MS (CFW CONTROL)
-------------------------
CFW gh67ffg56w21 CONTROL
Control-Package: msc-mixer/1.0
Content-Type: application/msc-mixer+xml
Content-Length: 182
<mscmixer version="1.0" xmlns="urn:ietf:params:xml:ns:msc-mixer">
   <modifyjoin id1="e1e1427c:1c998d22" id2="cf45ee2">
      <stream media="audio" direction="sendrecv"/>
   </modifyjoin>
</mscmixer>

2. AS <- MS (CFW 200 OK)
------------------------
CFW gh67ffg56w21 200
Timeout: 10
Content-Type: application/msc-mixer+xml
Content-Length: 123
<mscmixer version="1.0" xmlns="urn:ietf:params:xml:ns:msc-mixer">
   <response status="200" reason="Join modified"/>
</mscmixer>
```

15.6.4 Additional Scenarios

This section includes additional scenarios that can be of interest when dealing with AS↔MS flows. The aim of the following sub-sections is to present the use of features peculiar to the IVR package: Specifically, variable announcements, VCR prompts, parallel playback, recurring dialogs, and custom grammars. To describe how call flows involving such features might happen, three sample scenarios have been chosen:

1. Voice Mail (variable announcements for digits, VCR controls)
2. Current Time (variable announcements for date and time, parallel playback)
3. DTMF-driven Conference Manipulation (recurring dialogs, custom grammars)

15.6.4.1 Voice Mail

An application that typically uses the services an MS can provide is Voice Mail. In fact, while it is clear that many of its features are part of the application logic (e.g., the mapping of a uniform resource identifier (URI) with a specific user's voice mailbox, the list of messages and their properties, and so on), the actual media work is accomplished through the MS. Features needed by a Voice Mail application include the ability to record a stream and play it back at a later time, give verbose announcements regarding the status of the application, control the playout of recorded messages by means of VCR controls, and so on. These features are all supported by the MS through the IVR package.

Without delving into the details of a full Voice Mail application and all its possible use cases, this section will cover a specific scenario and try to deal with as many interactions as possible that may happen between the AS and the MS in such a context. This scenario, depicted as a sequence diagram in Figure 15.44, will be as follows:

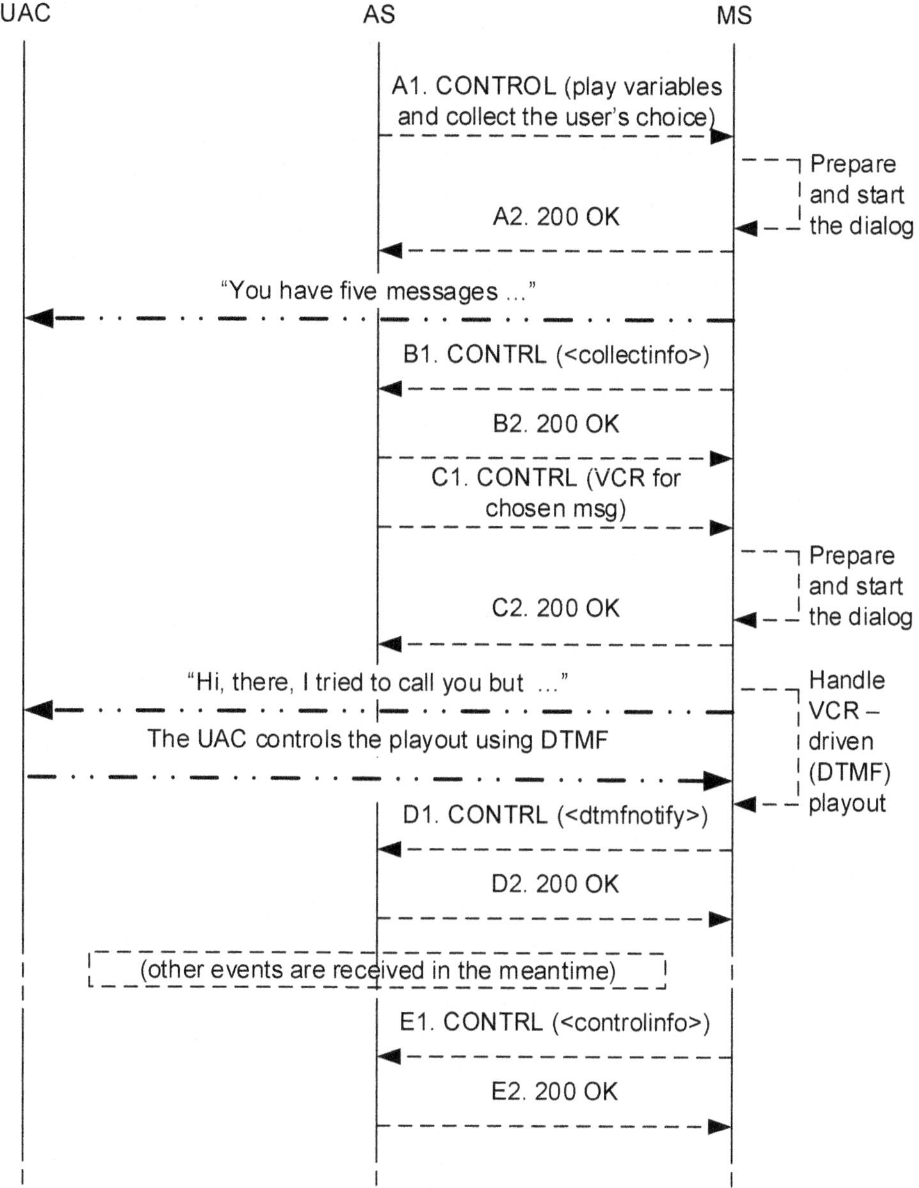

FIGURE 15.44 Voice Mail: Framework Transactions. (Copyright: IETF.)

1. The UAC INVITEs a URI associated with his mailbox, and the AS follows the previously explained procedure to have the UAC negotiate a new media session with the MS.
2. The UAC is first prompted with an announcement giving him the number of available new messages in the mailbox. After that, the UAC can choose which message to access by sending a DTMF tone.
3. The UAC is then presented with a VCR-controlled announcement, in which the chosen received mail is played back to him. VCR controls allow him to navigate through the prompt.

This is quite an oversimplified scenario, considering that it does not even allow the UAC to delete old messages or organize them but just to choose which received message to play. Nevertheless, it gives us the chance to deal with variable announcements and VCR controls – two typical features that a Voice Mail application would almost always take advantage of. Other features that a Voice Mail application would rely upon (e.g., recording streams, event-driven IVR menus, and so on) have been introduced in previous sections, and so representing them would be redundant. This means that the presented call flows assume that some messages have already been recorded and are available at reachable locations. The example also assumes that the AS has placed the recordings in its own storage facilities, since it is not safe to rely upon the internal MS storage, which is likely to be temporary.

The framework transaction steps are as follows:

- The first transaction (A1) is addressed to the IVR package (mscivr). It is basically an RFC 6231 (also see Section 1.2) "promptandcollect" dialog but with a slight difference: Some of the prompts to play are actual audio files, for which a URI is provided (media loc="xxx"), while others are so-called <variable> prompts; these <variable> prompts are actually constructed by the MS itself according to the directives provided by the AS. In this example, the sequence of prompts requested by the AS is as follows:
  1. Play a wav file ("you have...")
  2. Play a digit ("five...") by building it (variable: digit=5)
  3. Play a wav file ("messages...")
- A DTMF collection is requested as well (<collect>) to be taken after the prompts have been played. The AS is only interested in a single digit (maxdigits=1).
- The transaction is handled by the MS, and if everything works fine (i.e., the MS retrieved all the audio files and successfully built the variable announcements), the dialog is started; its start is reported, together with the associated identifier (5db01f4), to the AS in a terminating 200 OK message (A2).
- The AS then waits for the dialog to end in order to retrieve the results in which it is interested (in this case, the DTMF tone the UAC chooses, since it will affect which message will have to be played subsequently).
- The UAC hears the prompts and chooses a message to play. In this example, he wants to listen to the first message and so inputs "1." The MS intercepts this tone and notifies the AS of it in a newly created CONTROL event message (B1); this CONTROL includes information about how each single requested operation ended (<promptinfo> and <collectinfo>). Specifically, the event states that the prompt ended normally (termmode=completed) and that the subsequently collected tone is 1 (dtmf=1). The AS acks the event (B2), since the dialogid provided in the message is the same as that of the previously started dialog.
- At this point, the AS uses the value retrieved from the event to proceed with its business logic. It decides to present the UAC with a VCR-controllable playout of the requested message. This is done with a new request to the IVR package (C1), which contains two operations: <prompt> to address the media file to play (an old recording) and <control> to instruct the MS about how the playout of this media file shall be controlled via DTMF tones provided by the UAC (in this example, different DTMF digits are associated with different actions, e.g., pause/resume, fast forward, rewind, and so on). The AS also subscribes to DTMF events related to this control operation (matchmode=control), which means that the MS is to trigger an event any time that a DTMF associated with a control operation (e.g., 7=pause) is intercepted.
- The MS prepares the dialog and, when the playout starts, sends notification in a terminating 200 OK message (C2). At this point, the UAC is presented with the prompt and can use DTMF digits to control the playback.
- As explained previously, any DTMF associated with a VCR operation is then reported to the AS, together with a timestamp stating when the event happened. An example is provided (D1), in which the UAC pressed the fast-forward key (6) at a specific time. Of course, as for any other MS-generated event, the AS acks it (D2).

- When the playback ends (whether because the media reached its termination or because any other interruption occurred), the MS triggers a concluding event with information about the whole dialog (E1). This event, besides including information about the prompt itself (<promptinfo>), also includes information related to the VCR operations (<controlinfo>), that is, all the VCR controls the UAC used (fast forward/rewind/pause/resume in this example) and when it happened. The final ack by the AS (E2) concludes the scenario.

```
A1. AS -> MS (CFW CONTROL, play and collect)
--------------------------------------------
CFW 2f931de22820 CONTROL
Control-Package: msc-ivr/1.0
Content-Type: application/msc-ivr+xml
Content-Length: 429
<mscivr version="1.0" xmlns="urn:ietf:params:xml:ns:msc-ivr">
   <dialogstart connectionid="10514b7f:6a900179">
      <dialog>
         <prompt>
            <media loc="http://www.example.net/prompts/
                 vm-youhave.wav" type="audio/x-wav"/>
            <variable value="5" type="digits"/>
            <media loc="http://www.example.net/prompts/
                 vm-messages.wav" type="audio/x-wav"/>
         </prompt>
         <collect maxdigits="1" escapekey="*"
               cleardigitbuffer="true"/>
      </dialog>
   </dialogstart>
</mscivr>

A2. AS <- MS (CFW 200 OK)
-------------------------
CFW 2f931de22820 200
Timeout: 10
Content-Type: application/msc-ivr+xml
Content-Length: 137
<mscivr version="1.0" xmlns="urn:ietf:params:xml:ns:msc-ivr">
   <response status="200" reason="Dialog started" dialogid="5db01f4"/>
</mscivr>

B1. AS <- MS (CFW CONTROL event)
--------------------------------
CFW 7c97adc41b3e CONTROL
Control-Package: msc-ivr/1.0
Content-Type: application/msc-ivr+xml
Content-Length: 270
<mscivr version="1.0" xmlns="urn:ietf:params:xml:ns:msc-ivr">
   <event dialogid="5db01f4">
      <dialogexit status="1" reason="Dialog successfully completed">
         <promptinfo duration="11713" termmode="completed"/>
         <collectinfo dtmf="1" termmode="match"/>
      </dialogexit>
   </event>
</mscivr>

B2. AS -> MS (CFW 200, ACK to 'CONTROL event')
----------------------------------------------
CFW 7c97adc41b3e 200
```

```
C1. AS -> MS (CFW CONTROL, VCR)
-------------------------------
CFW 3140c24614bb CONTROL
Control-Package: msc-ivr/1.0
Content-Type: application/msc-ivr+xml
Content-Length: 423
<mscivr version="1.0" xmlns="urn:ietf:params:xml:ns:msc-ivr">
   <dialogstart connectionid="10514b7f:6a900179">
      <dialog>
         <prompt bargein="false">
            <media loc="http://www.example.com/messages/
                  recording-4ca9fc2.mpg"/>
         </prompt>
         <control gotostartkey="1" gotoendkey="3"
               ffkey="6" rwkey="4" pausekey="7" resumekey="9"
               volupkey="#" voldnkey="*"/>
      </dialog>
      <subscribe>
         <dtmfsub matchmode="control"/>
      </subscribe>
   </dialogstart>
</mscivr>

C2. AS <- MS (CFW 200 OK)
-------------------------
CFW 3140c24614bb 200
Timeout: 10
Content-Type: application/msc-ivr+xml
Content-Length: 137
<mscivr version="1.0" xmlns="urn:ietf:params:xml:ns:msc-ivr">
   <response status="200" reason="Dialog started" dialogid="3e936e0"/>
</mscivr>

D1. AS <- MS (CFW CONTROL event, dtmfnotify)
--------------------------------------------
CFW 361840da0581 CONTROL
Control-Package: msc-ivr/1.0
Content-Type: application/msc-ivr+xml
Content-Length: 179
<mscivr version="1.0" xmlns="urn:ietf:params:xml:ns:msc-ivr">
   <event dialogid="3e936e0">
      <dtmfnotify matchmode="control" dtmf="6"
            timestamp="2008-12-16T12:58:36Z"/>
   </event>
</mscivr>

D2. AS -> MS (CFW 200, ACK to 'CONTROL event')
----------------------------------------------
CFW 361840da0581 200
[..] The other VCR DTMF notifications are skipped for brevity [..]

E1. AS <- MS (CFW CONTROL event, dialogexit)
--------------------------------------------
CFW 3ffab81c21e9 CONTROL
Control-Package: msc-ivr/1.0
Content-Type: application/msc-ivr+xml
Content-Length: 485
```

```
<mscivr version="1.0" xmlns="urn:ietf:params:xml:ns:msc-ivr">
   <event dialogid="3e936e0">
      <dialogexit status="1" reason="Dialog successfully completed">
         <promptinfo duration="10270" termmode="completed"/>
         <controlinfo>
            <controlmatch dtmf="6" timestamp="2008-12-
                  16T12:58:36Z"/>
            <controlmatch dtmf="4" timestamp="2008-12-
                  16T12:58:37Z"/>
            <controlmatch dtmf="7" timestamp="2008-12-
                  16T12:58:38Z"/>
            <controlmatch dtmf="9" timestamp="2008-12-
                  16T12:58:40Z"/>
         </controlinfo>
      </dialogexit>
   </event>
</mscivr>

E2. AS -> MS (CFW 200, ACK to 'CONTROL event')
----------------------------------------------
CFW 3ffab81c21e9 200
```

15.6.4.2 Current Time

An interesting scenario to create with the help of features provided by the MS is what is typically called "Current Time." A UAC calls a URI, which presents the caller with the current date and time. As can easily be deduced by the very nature of the application, variable announcements play an important role in this scenario. In fact, rather than having the AS build an announcement according to the current time using different framework messages, it is much easier to rely upon the "variable announcements" mechanism provided by the IVR package, as variable announcements provide several ways to deal with dates and times.

To make the scenario more interesting and have it cover more functionality, the application is also assumed to have background music played during the announcement. Because most of the announcements will be variable, a means is needed to have more streams played in parallel on the same connection. This can be achieved in two different ways:

1. Two separate and different dialogs, playing the variable announcements and the background track, respectively
2. A single dialog implementing a parallel playback

The first approach assumes that the available MS implements implicit mixing, which may or may not be supported, since it is a recommended feature but not mandatory. The second approach assumes that the MS implements support for more streams of the same media type (in this case, audio) in the same dialog, which, exactly as for the case of implicit mixing, is not to be taken for granted. Because the first approach is quite straightforward and easy to understand, the following scenario uses the second approach and assumes that the available MS supports parallel playback of more audio tracks within the context of the same dialog. That said, the covered scenario, depicted as a sequence diagram in Figure 15.45, will be as follows:

1. The UAC INVITEs a URI associated with the Current Time application, and the AS follows the previously explained procedure to have the UAC negotiate a new media session with the MS.
2. The UAC is presented with an announcement including (i) a voice stating the current date and time; (ii) a background music track; and (iii) a mute background video track.

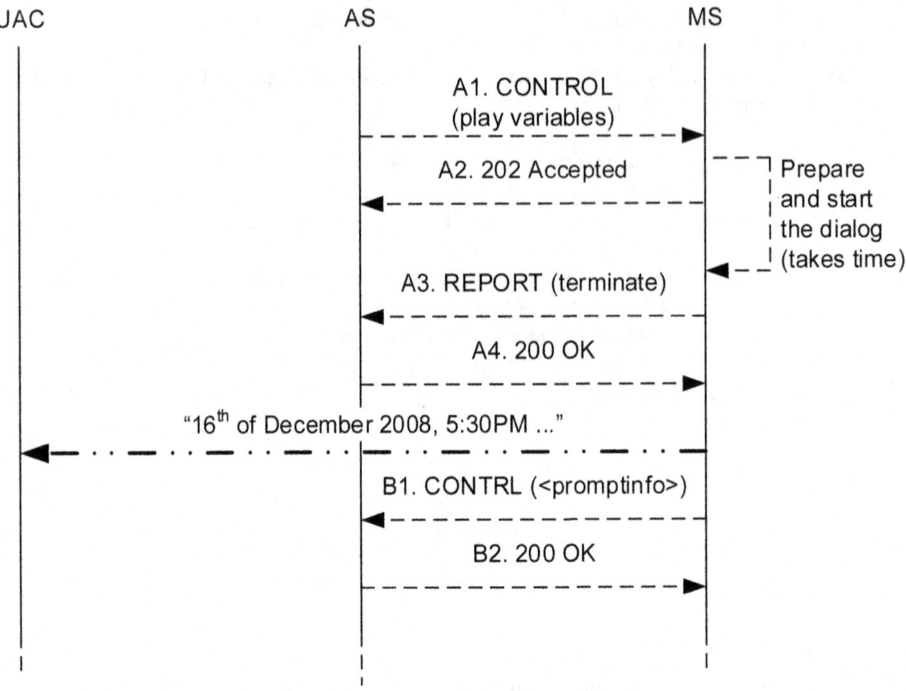

FIGURE 15.45 Current Time: Framework Transactions. (Copyright: IETF.)

The framework transaction steps are as follows:

- The first transaction (A1) is addressed to the IVR package (mscivr); it is basically an (RFC 6231 – see Chapter 12) "playannouncements" dialog, but unlike all the scenarios presented so far, it includes directives for a parallel playback, as indicated by the <par> element. There are three flows to play in parallel:
  - A sequence (<seq>) of variable and static announcements (the actual time and date)
  - A music track ("media=music.wav") to be played in the background at a lower volume (soundLevel=50%)
  - A mute background video track (media=clock.mpg)
- The global announcement ends when the longest of the three parallel steps ends (endsync=last). This means that if one of the steps ends before the others, the step is muted for the rest of the playback. In this example, the series of static and <variable> announcements is requested by the AS:
  - Play a wav file ("Tuesday...")
  - Play a date ("16th of December 2008...") by building it (variable: date with a ymd=year/month/day format)
  - Play a time ("5:31 PM...") by building it (variable: time with a t12=12 hour day format, am/pm)
- The transaction is extended by the MS (A2) with a new timeout, as in this case, the MS needs some more time to retrieve all the needed media files. Should the new timeout expire as well, the MS would send a further message to extend it again (a REPORT update), but for the sake of simplicity, we assume that in this scenario, it is not needed. If everything went fine (i.e., the MS retrieved all the audio files and successfully built the variable announcements, and it supports parallel playback as requested), the dialog is started. Its start is reported, together with the associated identifier (415719e), to the AS in a terminating REPORT message (A3).

- The AS acks the REPORT (A4) and waits for the dialog to end in order to either conclude the application or proceed to further steps if required by the application itself.
- When the last of the three parallel announcements ends, the dialog terminates, and an event (B1) is triggered to the AS with the relevant information (promptinfo). The AS acks (B2) and terminates the scenario.

```
A1. AS -> MS (CFW CONTROL, play)
--------------------------------
CFW 0c7680191bd2 CONTROL
Control-Package: msc-ivr/1.0
Content-Type: application/msc-ivr+xml
Content-Length: 506
<mscivr version="1.0" xmlns="urn:ietf:params:xml:ns:msc-ivr">
   <dialogstart connectionid="21c8e07b:055a893f">
      <dialog>
         <prompt bargein="true">
            <par endsync="last">
               <seq>
                  <media loc="http://www.example.com/day-
                        2.wav"/>
                  <variable value="2008-12-16" type="date"
                        format="ymd"/>
                  <variable value="17:31" type="time"
                        format="t12"/>
               </seq>
               <media loc="http://www.example.com/music.wav"
                     soundLevel="50%"/>
               <media loc="http://www.example.com/
                     clock.mpg"/>
            </par>
         </prompt>
      </dialog>
   </dialogstart>
</mscivr>

A2. AS <- MS (CFW 202)
----------------------
CFW 0c7680191bd2 202
Timeout: 5

A3. AS <- MS (CFW REPORT terminate)
-----------------------------------
CFW 0c7680191bd2 REPORT
Seq: 1
Status: terminate
Timeout: 10
Content-Type: application/msc-ivr+xml
Content-Length: 137
<mscivr version="1.0" xmlns="urn:ietf:params:xml:ns:msc-ivr">
   <response status="200" reason="Dialog started" dialogid="415719e"/>
</mscivr>

A4. AS -> MS (CFW 200, ACK to 'REPORT terminate')
-------------------------------------------------
CFW 0c7680191bd2 200
Seq: 1
```

```
B1. AS <- MS (CFW CONTROL event)
--------------------------------
CFW 4481ca0c4fca CONTROL
Control-Package: msc-ivr/1.0
Content-Type: application/msc-ivr+xml
Content-Length: 229
<mscivr version="1.0" xmlns="urn:ietf:params:xml:ns:msc-ivr">
   <event dialogid="415719e">
      <dialogexit status="1" reason="Dialog successfully completed">
         <promptinfo duration="8046" termmode="completed"/>
      </dialogexit>
   </event>
</mscivr>

B2. AS -> MS (CFW 200, ACK to 'CONTROL event')
----------------------------------------------
CFW 4481ca0c4fca 200
```

15.6.4.3 DTMF-Driven Conference Manipulation

To complete the scenarios presented in Section 15.6.3, this section deals with how the AS can use the MS to detect DTMF tones from conference participants and take actions on the conference accordingly. A typical example is when participants in a conference are provided with specific codes to:

- Mute/unmute themselves in the conference.
- Change their volume in the conference or the volume of the conference itself.
- Change the video layout in the conference, if allowed.
- Kick abusive users out of the conference; and so on. To achieve all this, the simplest thing an AS can do is to prepare a recurring DTMF collection for each participant with specific grammars to match. If the collected tones match the grammar, the MS will notify the AS of the tones and start the collection again. Upon receipt of <collectinfo> events, the AS will in turn originate the proper related request, e.g., a <modifyjoin> on the participant's stream with the conference. This is made possible by three features provided by the IVR package:
  1. The "repeatCount" attribute
  2. The subscription mechanism
  3. The Speech Recognition Grammar Specification (SRGS) [1]

The first feature allows recurring instances of the same dialog without the need for additional requests upon completion of the dialog itself. In fact, the "repeatCount" attribute indicates how many times the dialog has to be repeated. When the attribute has the value 0, it means that the dialog has to be repeated indefinitely, meaning that it is up to the AS to destroy it by means of a <dialogterminate> request when the dialog is no longer needed.

The second feature allows the AS to subscribe to events related to the IVR package without waiting for the dialog to end, e.g., matching DTMF collections in this case. Finally, the last feature allows custom matching grammars to be specified. In this way, only a subset of the possible DTMF strings can be specified, so that only those matches in which the AS is interested are reported. Grammars other than SRGS may be supported by the MS and will achieve the same result. This specification will only describe the use of an SRGS grammar, since support for SRGS is mandated in RFC 6231 (also see Section 1.2).

To identify a single sample scenario, we assume that a participant has successfully joined a conference, e.g., as detailed in Figure 15.32. We also assume that the following codes are to be provided within the conference to participants in order to let them take advantage of advanced features:

1. *6 to mute/unmute themselves (on/off trigger)
2. *1 to lower their own volume in the conference and *3 to raise it
3. *7 to lower the volume of the conference stream they are receiving and *9 to raise it
4. *0 to leave the conference

This means that six different codes are supported and are to be matched in the requested DTMF collection. All other codes are collected by the MS but discarded, and no event is triggered to the AS. Because all the codes have the "*" (star) DTMF code in common, the following is an example of an SRGS grammar that may be used in the request by the AS:

```
<grammar mode="dtmf" version="1.0"
      xmlns="http://www.w3.org/2001/06/grammar">
   <rule id="digit">
      <one-of>
         <item>0</item>
         <item>1</item>
         <item>3</item>
         <item>6</item>
         <item>7</item>
         <item>9</item>
      </one-of>
   </rule>
   <rule id="action" scope="public">
      <item>
         *
            <item><ruleref uri="#digit"/></item>
      </item>
   </rule>
</grammar>
```

As can be deduced by looking at the grammar, the presented SRGS XML code specifies exactly the requirements for the collections to match. The rule is to match any string that has a star ("*") followed by a single supported digit (0, 1, 3, 6, 7, or 9). Such grammar, as stated in RFC 6231 (also see Section 1.2), may be either provided inline in the request itself or referenced externally by means of the "src" attribute. In the example scenario, we will put it inline, but an external reference to the same specification would achieve exactly the same result. Figure 15.46 shows how the AS might request the recurring collection for a UAC. As before, the example assumes that the UAC is already a participant in the conference.

As can be deduced from the sequence diagram in Figure 15.46), the AS, in its business logic, correlates the results of different transactions, addressed to different packages, to implement a more complex conferencing scenario. In fact, <dtmfnotify> events are used to take actions according to the functions of the DTMF codes. The framework transaction steps are as follows:

- The UAC is already in the conference, and so the AS starts a recurring collect with a grammar to match. This is done by placing a CONTROL request addressed to the IVR package (A1). The operation to implement is a <collect>, and we are only interested in two-digit DTMF strings (maxdigits). The AS is not interested in a DTMF terminator (termchar is set to a non-conventional DTMF character), and the DTMF escape key is set to "#" (the default is "*," which would conflict with the code syntax for the conference and so needs to be changed). A custom SRGS grammar is provided inline (<grammar> with mode=dtmf). The whole dialog is to be repeated indefinitely (dialog has repeat-Count=0), and the AS wants to be notified when matching collections occur (dtmfsub with matchmode=collect).

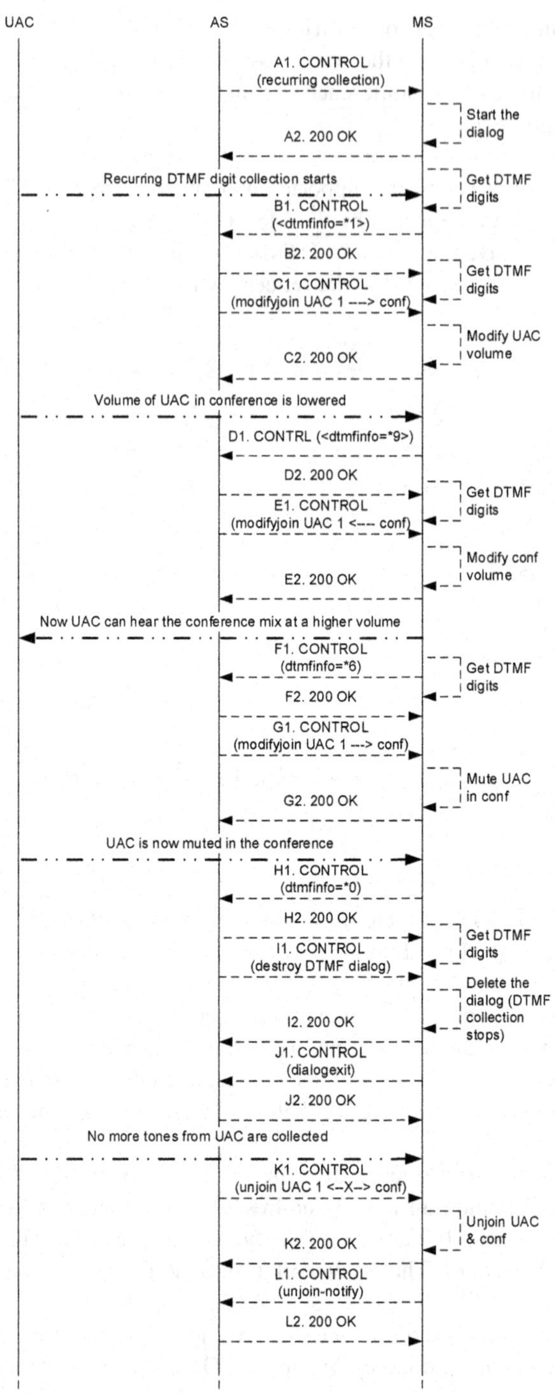

FIGURE 15.46 DTMF-Driven Conference Manipulation – Framework Transactions. (Copyright: IETF.)

- The request is handled by the MS (A2) and then successfully started (dialogid=01d1b38). This means that the MS has started collecting DTMF tones from the UAC.
- The MS collects a matching DTMF string from the UAC (*1). Since the AS subscribed to this kind of event, a CONTROL event notification (dtmfnotify) is triggered by the MS (B1), including the collected tones. Since the dialog is recurring, the MS immediately restarts the collection.

- The AS acks the event (B2) and in its business logic, understands that the code "*1" means that the UAC wants its own volume to be lowered in the conference mix. The AS is able to associate the event with the right UAC by referring to the attached dialogid (01d1b38). It then acts accordingly by sending a <modifyjoin> (C1) that does exactly this: the provided <stream> child element instructs the MS to modify the volume of the UAC→conference audio flow (setgain=–5 dB 'sendonly'). Note that the "setgain" value is the absolute volume level. If the user's request is to lower the volume level, the AS must remember the previously set volume level and from it calculate the new volume level. Note how the request also includes directives for the inverse direction. This verbose approach is needed; otherwise, the MS would not only change the volume in the requested direction but would also disable the media flow in the other direction. Having a proper <stream> addressing the UAC←conf media flow as well ensures that this does not happen.
- The MS successfully enforces the requested operation (C2), changing the volume.
- A new matching DTMF string from the UAC is collected (*9). As before, an event is triggered to the AS (D1).
- The AS acks the event (D2) and matches the new code ("*9") with its related operation (raise the volume of the conference mix for the UAC), taking the proper action. A different <modifyjoin> is sent (E1) with the new instructions (setgain=+3 dB "recvonly"). The same considerations regarding how the incremental operation should be mapped to the command apply here as well. Note also how a <stream> for the inverse direction ('sendonly') is again provided just as a placeholder, which basically instructs the MS that the settings for that direction are not to be changed, maintaining the previous directives of (C1).
- The MS successfully enforces this requested operation as well (E2), changing the volume in the specified direction.
- At this point, a further matching DTMF string from the UAC is collected (*6) and sent to the AS (F1).
- After the required ack (F2), the AS reacts by implementing the action associated with the new code ("*6"), by which the UAC requested that it be muted within the conference. A new <modifyjoin> is sent (G1) with a properly constructed payload (setstate=mute "sendonly"), and the MS enforces it (G2).
- A last (in this scenario) matching DTMF string is collected by the MS (*0). As with all the previous codes, notification of this string is sent to the AS (H1).
- The AS acks the event (H2) and understands that the UAC wants to leave the conference now (since the code is *0). This means that a series of actions must be taken:
  - The recurring collection is stopped, since it is no longer needed.
  - The UAC is unjoined from the conference it is in.
  - Additional operations might be considered, e.g., a global announcement stating that the UAC is leaving, but for the sake of conciseness, such operations are not listed here.

  The former is accomplished by means of a <dialogterminate> request (I1) to the IVR package (dialogid=01d1b38) and the latter by means of an <unjoin> request (K1) to the Mixer package.
- The <dialogterminate> request is handled by the MS (I2), and the dialog is terminated successfully. As soon as the dialog has actually been terminated, a <dialogexit> event is triggered as well to the AS (J1). This event has no report of the result of the last iteration (since the dialog was terminated abruptly with an immediate=true) and is acked by the AS (J2) to finally complete the dialog lifetime.
- The <unjoin> request is immediately enforced (K2). As a consequence of the unjoin operation, an <unjoin-notify> event notification is triggered by the MS (L1) to confirm to the AS that the requested entities are no longer attached to each other. The status in the event is set to 0, which as stated in the specification, means that the join has been terminated by an <unjoin> request. The ack from the AS (L2) concludes this scenario.

```
A1. AS -> MS (CFW CONTROL, recurring collect with grammar)
----------------------------------------------------------
CFW 238e1f2946e8 CONTROL
Control-Package: msc-ivr/1.0
Content-Type: application/msc-ivr+xml
Content-Length: 809
<mscivr version="1.0" xmlns="urn:ietf:params:xml:ns:msc-ivr">
   <dialogstart connectionid="14849028:37fc2523">
      <dialog repeatCount="0">
         <collect maxdigits="2" termchar="A" escapekey="#">
            <grammar>
               <grammar version="1.0" mode="dtmf"
                     xmlns="http://www.w3.org/2001/06/gramm
                     ar">
                  <rule id="digit">
                     <one-of>
                        <item>0</item>
                        <item>1</item>
                        <item>3</item>
                        <item>6</item>
                        <item>7</item>
                        <item>9</item>
                     </one-of>
                  </rule>
                  <rule id="action" scope="public">
                     <example>*3</example>
                        <one-of>
                           <item>
                              *
                                 <ruleref uri="#digit"/>
                           </item>
                     </one-of>
                  </rule>
               </grammar>
            </grammar>
         </collect>
      </dialog>
      <subscribe>
         <dtmfsub matchmode="collect"/>
      </subscribe>
   </dialogstart>
</mscivr>

A2. AS <- MS (CFW 200 OK)
-------------------------
CFW 238e1f2946e8 200
Timeout: 10
Content-Type: application/msc-ivr+xml
Content-Length: 137
<mscivr version="1.0" xmlns="urn:ietf:params:xml:ns:msc-ivr">
   <response status="200" reason="Dialog started" dialogid="01d1b38"/>
</mscivr>

B1. AS <- MS (CFW CONTROL dtmfnotify event)
-------------------------------------------
CFW 1dd62e043c00 CONTROL
Control-Package: msc-ivr/1.0
```

```
Content-Type: application/msc-ivr+xml
Content-Length: 180
<mscivr version="1.0" xmlns="urn:ietf:params:xml:ns:msc-ivr">
   <event dialogid="01d1b38">
      <dtmfnotify matchmode="collect" dtmf="*1"
            timestamp="2008-12-17T17:20:53Z"/>
      </event>
</mscivr>

B2. AS -> MS (CFW 200, ACK to 'CONTROL event')
----------------------------------------------
CFW 1dd62e043c00 200

C1. AS -> MS (CFW CONTROL, modifyjoin with setgain)
---------------------------------------------------
CFW 0216231b1f16 CONTROL
Control-Package: msc-mixer/1.0
Content-Type: application/msc-mixer+xml
Content-Length: 290
<mscmixer version="1.0" xmlns="urn:ietf:params:xml:ns:msc-mixer">
   <modifyjoin id1="873975758:a5105056" id2="54b4ab3">
      <stream media="audio" direction="sendonly">
         <volume controltype="setgain" value="-5"/>
      </stream>
      <stream media="audio" direction="recvonly"/>
   </modifyjoin>
</mscmixer>

C2. AS <- MS (CFW 200 OK)
-------------------------
CFW 0216231b1f16 200
Timeout: 10
Content-Type: application/msc-mixer+xml
Content-Length: 123
<mscmixer version="1.0" xmlns="urn:ietf:params:xml:ns:msc-mixer">
   <response status="200" reason="Join modified"/>
</mscmixer>

D1. AS <- MS (CFW CONTROL dtmfnotify event)
-------------------------------------------
CFW 4d674b3e0862 CONTROL
Control-Package: msc-ivr/1.0
Content-Type: application/msc-ivr+xml
Content-Length: 180
<mscivr version="1.0" xmlns="urn:ietf:params:xml:ns:msc-ivr">
   <event dialogid="01d1b38">
      <dtmfnotify matchmode="collect" dtmf="*9"
            timestamp="2008-12-17T17:20:57Z"/>
   </event>
</mscivr>

D2. AS -> MS (CFW 200, ACK to 'CONTROL event')
----------------------------------------------
CFW 4d674b3e0862 200

E1. AS -> MS (CFW CONTROL, modifyjoin with setgain)
---------------------------------------------------
```

```
CFW 140e0f763352 CONTROL
Control-Package: msc-mixer/1.0
Content-Type: application/msc-mixer+xml
Content-Length: 292
<mscmixer version="1.0" xmlns="urn:ietf:params:xml:ns:msc-mixer">
   <modifyjoin id1="873975758:a5105056" id2="54b4ab3">
      <stream media="audio" direction="recvonly">
         <volume controltype="setgain" value="+3"/>
      </stream>
      <stream media="audio" direction="sendonly"/>
   </modifyjoin>
</mscmixer>

E2. AS <- MS (CFW 200 OK)
------------------------
CFW 140e0f763352 200
Timeout: 10
Content-Type: application/msc-mixer+xml
Content-Length: 123
<mscmixer version="1.0" xmlns="urn:ietf:params:xml:ns:msc-mixer">
   <response status="200" reason="Join modified"/>
</mscmixer>

F1. AS <- MS (CFW CONTROL dtmfnotify event)
------------------------------------------
CFW 478ed6f1775b CONTROL
Control-Package: msc-ivr/1.0
Content-Type: application/msc-ivr+xml
Content-Length: 180
<mscivr version="1.0" xmlns="urn:ietf:params:xml:ns:msc-ivr">
   <event dialogid="01d1b38">
      <dtmfnotify matchmode="collect" dtmf="*6"
            timestamp="2008-12-17T17:21:02Z"/>
   </event>
</mscivr>

F2. AS -> MS (CFW 200, ACK to 'CONTROL event')
---------------------------------------------
CFW 478ed6f1775b 200
G1. AS -> MS (CFW CONTROL, modifyjoin with setstate)
---------------------------------------------------
CFW 7fdcc2331bef CONTROL
Control-Package: msc-mixer/1.0
Content-Type: application/msc-mixer+xml
Content-Length: 295
<mscmixer version="1.0" xmlns="urn:ietf:params:xml:ns:msc-mixer">
   <modifyjoin id1="873975758:a5105056" id2="54b4ab3">
      <stream media="audio" direction="sendonly">
         <volume controltype="setstate" value="mute"/>
      </stream>
      <stream media="audio" direction="recvonly"/>
   </modifyjoin>
</mscmixer>

G2. AS <- MS (CFW 200 OK)
------------------------
CFW 7fdcc2331bef 200
```

```
Timeout: 10
Content-Type: application/msc-mixer+xml
Content-Length: 123
<mscmixer version="1.0" xmlns="urn:ietf:params:xml:ns:msc-mixer">
   <response status="200" reason="Join modified"/>
</mscmixer>

H1. AS <- MS (CFW CONTROL dtmfnotify event)
-------------------------------------------
CFW 750b917a5d4a CONTROL
Control-Package: msc-ivr/1.0
Content-Type: application/msc-ivr+xml
Content-Length: 180
<mscivr version="1.0" xmlns="urn:ietf:params:xml:ns:msc-ivr">
   <event dialogid="01d1b38">
      <dtmfnotify matchmode="collect" dtmf="*0"
            timestamp="2008-12-17T17:21:05Z"/>
   </event>
</mscivr>

H2. AS -> MS (CFW 200, ACK to 'CONTROL event')
----------------------------------------------
CFW 750b917a5d4a 200

I1. AS -> MS (CFW CONTROL, dialogterminate)
-------------------------------------------
CFW 515f007c5bd0 CONTROL
Control-Package: msc-ivr/1.0
Content-Type: application/msc-ivr+xml
Content-Length: 128
<mscivr version="1.0" xmlns="urn:ietf:params:xml:ns:msc-ivr">
   <dialogterminate dialogid="01d1b38" immediate="true"/>
</mscivr>

I2. AS <- MS (CFW 200 OK)
-------------------------
CFW 515f007c5bd0 200
Timeout: 10
Content-Type: application/msc-ivr+xml
Content-Length: 140
<mscivr version="1.0" xmlns="urn:ietf:params:xml:ns:msc-ivr">
   <response status="200" reason="Dialog terminated"
         dialogid="01d1b38"/>
</mscivr>

J1. AS <- MS (CFW CONTROL dialogexit event)
-------------------------------------------
CFW 76adc41122c1 CONTROL
Control-Package: msc-ivr/1.0
Content-Type: application/msc-ivr+xml
Content-Length: 155
<mscivr version="1.0" xmlns="urn:ietf:params:xml:ns:msc-ivr">
   <event dialogid="01d1b38">
      <dialogexit status="0" reason="Dialog terminated"/>
   </event>
</mscivr>
```

```
J2. AS -> MS (CFW 200, ACK to 'CONTROL event')
----------------------------------------------
CFW 76adc41122c1 200

K1. AS -> MS (CFW CONTROL, unjoin)
----------------------------------
CFW 4e6afb6625e4 CONTROL
Control-Package: msc-mixer/1.0
Content-Type: application/msc-mixer+xml
Content-Length: 127
<mscmixer version="1.0" xmlns="urn:ietf:params:xml:ns:msc-mixer">
    <unjoin id1="873975758:a5105056" id2="54b4ab3"/>
</mscmixer>

K2. AS <- MS (CFW 200 OK)
-------------------------
CFW 4e6afb6625e4 200
Timeout: 10
Content-Type: application/msc-mixer+xml
Content-Length: 122
<mscmixer version="1.0" xmlns="urn:ietf:params:xml:ns:msc-mixer">
    <response status="200" reason="Join removed"/>
</mscmixer>

L1. AS <- MS (CFW CONTROL unjoin-notify event)
----------------------------------------------
CFW 577696293504 CONTROL
Control-Package: msc-mixer/1.0
Content-Type: application/msc-mixer+xml
Content-Length: 157
<mscmixer version="1.0" xmlns="urn:ietf:params:xml:ns:msc-mixer">
    <event>
        <unjoin-notify status="0"
                id1="873975758:a5105056" id2="54b4ab3"/>
    </event>
</mscmixer>

L2. AS -> MS (CFW 200, ACK to 'CONTROL event')
----------------------------------------------
CFW 577696293504 200
```

15.7 MEDIA RESOURCE BROKERING

All the flows presented so far describe the interaction between a single AS and a single MS. This is the simplest topology that can be envisaged in a MEDIACTRL-compliant architecture, but it is not the only topology that is allowed. RFC 5567 (see Chapter 3) presents several possible topologies that potentially involve several AS and several MS as well. To properly allow for such topologies, an additional element called the Media Resource Broker (MRB) has been introduced in the MEDIACTRL architecture. Such an entity, and the protocols needed to interact with it, has been standardized in RFC 6917 (see Chapter 11).

An MRB is basically a locator that is aware of a pool of MSs and makes them available to interested ASs according to their requirements. For this reason, two different interfaces have been introduced:

- The Publishing interface (Section 15.7.1), which allows an MRB to subscribe for notifications at the MS it is handling (e.g., available and occupied resources, current state, etc.).
- The Consumer interface (Section 15.7.2), which allows an interested AS to query an MRB for an MS capable of fulfilling its requirements.

The flows in the following sections will present some typical use-case scenarios involving an MRB and the different topologies in which it has been conceived to work. Additionally, a few considerations on the handling of media dialogs whenever an MRB is involved are presented in Section 15.7.3.

15.7.1 Publishing Interface

An MRB uses the MS's Publishing interface to acquire relevant information. This Publishing interface, as specified in RFC 6917 (see Chapter 11), is made available as a Control Package for the CFW. This means that in order to receive information from an MS, an MRB must negotiate a Control Channel as explained in Chapter 5. This package allows an MRB to either request information in the form of a direct request/answer from an MS or subscribe for events.

Of course, since the MRB is interested in the Publishing interface, the previously mentioned negotiation must be changed in order to take into account the need for the MRB Control Package. The name of this package is "mrb-publish/1.0," which means that the SYNC might look like the following:

```
1. MRB -> MS (CFW SYNC)
   ----------------------
   CFW 6b8b4567327b SYNC
   Dialog-ID: z9hG4bK-4542-1-0
   Keep-Alive: 100
   Packages: msc-ivr/1.0,msc-mixer/1.0,mrb-publish/1.0

2. MRB <- MS (CFW 200)
   ---------------------
   CFW 6b8b4567327b 200
   Keep-Alive: 100
   Packages: msc-ivr/1.0,msc-mixer/1.0,mrb-publish/1.0
   Supported: msc-example-pkg/1.0
```

The meaning of this negotiation was presented previously. It is enough to point out that the MRB in this case adds a new item to the "Packages" it needs support for (mrb-publish/1.0). In this case, the MS supports it, and in fact, it is added to the negotiated packages in the reply:

Packages: msc-ivr/1.0,msc-mixer/1.0,mrb-publish/1.0

The MS as described in Section 15.5, on the other hand, did not have support for that package, since only "msc-example-pkg/1.0" was part of the 'Supported' list.

Figure 15.47 presents a ladder diagram of a typical interaction based on the MRB Control Package.

In this example, the MRB subscribes for information at the specified MS, and events are triggered on a regular, negotiated basis. All these messages flow through the Control Channel, as do all of the messages discussed in this specification. The framework transaction steps are as follows:

- The MRB sends a new CONTROL message (A1) addressed to the MRB package (mrb-publish/1.0); it is a subscription for information (<subscription>), and the MRB is asking to be notified at least every 10 minutes (<minfrequency>) or if required, every 30 seconds

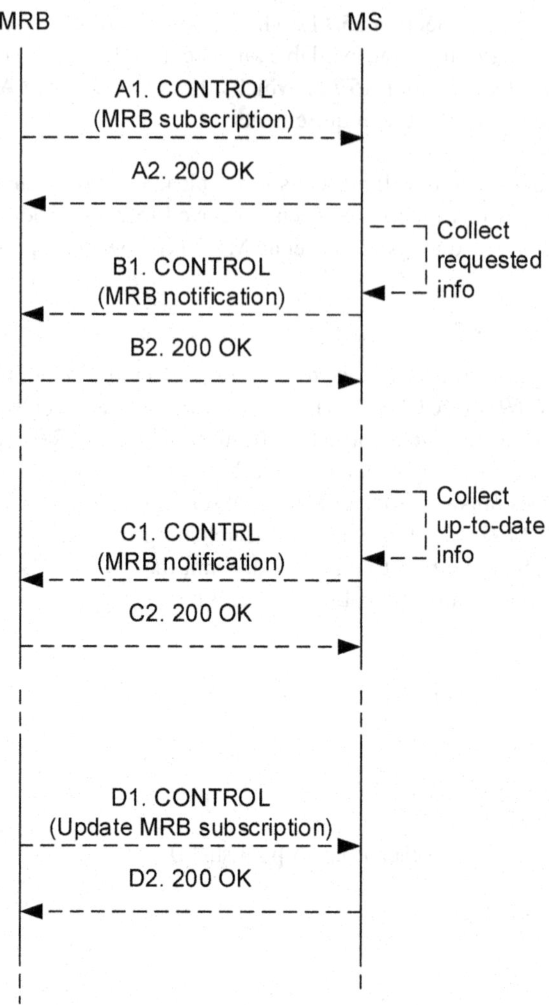

FIGURE 15.47 Media Resource Brokering: Subscription and Notification. (Copyright: IETF.)

at maximum. The subscription must last 30 minutes (<expires>), after which no additional notifications must be sent.

- The MS acknowledges the request (A2) and sends notification of the success of the request in a 200 OK message (<mrbresponse>).
- The MS prepares and sends the first notification to the MRB (B1). As has been done with other packages, the notification has been sent as an MS-generated CONTROL message; it is a notification related to the request in the first message, because the "id" (p0T65U) matches that request. All the information that the MRB subscribed for is provided in the payload.
- The MRB acknowledges the notification (B2) and uses the retrieved information to update its own information as part of its business logic.
- The previous step (the MRB acknowledges notifications and uses the retrieved information) repeats at the required frequency with up-to-date information.
- After a while, the MRB updates its subscription (D1) to get more frequent updates (minfrequency=1, an update every second at least). The MS accepts the update (D2), although it may adjust the frequency in the reply according to its policies (e.g., a lower rate, such as minfrequency=30). The notifications continue, but at the newer rate; the expiration is also updated accordingly (600 seconds again, since the update refreshes it).

```
A1. MRB -> MS (CONTROL, publish request)
----------------------------------------
CFW lidc30BZObiC CONTROL
Control-Package: mrb-publish/1.0
Content-Type: application/mrb-publish+xml
Content-Length: 337
<mrbpublish version="1.0" xmlns="urn:ietf:params:xml:ns:mrb-publish">
   <mrbrequest>
      <subscription action="create" seqnumber="1" id="p0T65U">
         <expires>60</expires>
         <minfrequency>600</minfrequency>
         <maxfrequency>30</maxfrequency>
      </subscription>
   </mrbrequest>
</mrbpublish>

A2. MRB <- MS (200 to CONTROL, request accepted)
------------------------------------------------
CFW lidc30BZObiC 200
Timeout: 10
Content-Type: application/mrb-publish+xml
Content-Length: 139
<mrbpublish version="1.0" xmlns="urn:ietf:params:xml:ns:mrb-publish">
   <mrbresponse status="200" reason="OK: Request accepted"/>
</mrbpublish>

B1. MRB <- MS (CONTROL, event notification from MS)
---------------------------------------------------
CFW 03fff52e7b7a CONTROL
Control-Package: mrb-publish/1.0
Content-Type: application/mrb-publish+xml
Content-Length: 4157
<mrbpublish version="1.0"
      xmlns="urn:ietf:params:xml:ns:mrb-publish">
   <mrbnotification seqnumber="1" id="p0T65U">
      <media-server-id>a1b2c3d4</media-server-id>
         <supported-packages>
            <package name="msc-ivr/1.0"/>
            <package name="msc-mixer/1.0"/>
            <package name="mrb-publish/1.0"/>
            <package name="msc-example-pkg/1.0"/>
         </supported-packages>
         <active-rtp-sessions>
            <rtp-codec name="audio/basic">
               <decoding>10</decoding>
               <encoding>20</encoding>
            </rtp-codec>
         </active-rtp-sessions>
         <active-mixer-sessions>
            <active-mix conferenceid="7cfgs43">
               <rtp-codec name="audio/basic">
                  <decoding>3</decoding>
                  <encoding>3</encoding>
               </rtp-codec>
            </active-mix>
         </active-mixer-sessions>
```

```
<non-active-rtp-sessions>
   <rtp-codec name="audio/basic">
      <decoding>50</decoding>
      <encoding>40</encoding>
   </rtp-codec>
</non-active-rtp-sessions>
<non-active-mixer-sessions>
   <non-active-mix available="15">
      <rtp-codec name="audio/basic">
         <decoding>15</decoding>
         <encoding>15</encoding>
      </rtp-codec>
   </non-active-mix>
</non-active-mixer-sessions>
<media-server-status>active</media-server-status>
   <supported-codecs>
      <supported-codec name="audio/basic">
         <supported-codec-package name="msc-ivr/1.0">
            <supported-action>encoding</supported-action>
            <supported-action>decoding</supported-action>
         </supported-codec-package>
         <supported-codec-package name="msc-mixer/1.0">
            <supported-action>encoding</supported-action>
            <supported-action>decoding</supported-action>
         </supported-codec-package>
      </supported-codec>
   </supported-codecs>
   <application-data>TestbedPrototype</application-data>
   <file-formats>
      <supported-format name="audio/x-wav">
         <supported-file-package>
               msc-ivr/1.0
         </supported-file-package>
      </supported-format>
   </file-formats>
   <max-prepared-duration>
      <max-time max-time-seconds="3600">
         <max-time-package>msc-ivr/1.0</max-time-package>
      </max-time>
   </max-prepared-duration>
   <dtmf-support>
      <detect>
         <dtmf-type package="msc-ivr/1.0" name="RFC4733"/>
         <dtmf-type package="msc-mixer/1.0" name="RFC4733"/>
      </detect>
      <generate>
         <dtmf-type package="msc-ivr/1.0" name="RFC4733"/>
         <dtmf-type package="msc-mixer/1.0" name="RFC4733"/>
      </generate>
      <passthrough>
         <dtmf-type package="msc-ivr/1.0" name="RFC4733"/>
         <dtmf-type package="msc-mixer/1.0" name="RFC4733"/>
      </passthrough>
   </dtmf-support>
   <mixing-modes>
      <audio-mixing-modes>
```

```
            <audio-mixing-mode package="msc-ivr/1.0"> \
                  nbest \
            </audio-mixing-mode>
         </audio-mixing-modes>
         <video-mixing-modes activespeakermix="true" vas="true">
            <video-mixing-mode package="msc-mixer/1.0"> \
               single-view \
            </video-mixing-mode>
            <video-mixing-mode package="msc-mixer/1.0"> \
               dual-view \
            </video-mixing-mode>
            <video-mixing-mode package="msc-mixer/1.0"> \
               dual-view-crop \
            </video-mixing-mode>
            <video-mixing-mode package="msc-mixer/1.0"> \
               dual-view-2x1 \
            </video-mixing-mode>
            <video-mixing-mode package="msc-mixer/1.0"> \
               dual-view-2x1-crop \
            </video-mixing-mode>
            <video-mixing-mode package="msc-mixer/1.0"> \
               quad-view \
            </video-mixing-mode>
            <video-mixing-mode package="msc-mixer/1.0"> \
               multiple-5x1 \
            </video-mixing-mode>
            <video-mixing-mode package="msc-mixer/1.0"> \
               multiple-3x3 \
            </video-mixing-mode>
            <video-mixing-mode package="msc-mixer/1.0"> \
               multiple-4x4 \
            </video-mixing-mode>
         </video-mixing-modes>
      </mixing-modes>
      <supported-tones>
         <supported-country-codes>
            <country-code package="msc-ivr/1.0">GB</country-code>
            <country-code package="msc-ivr/1.0">IT</country-code>
            <country-code package="msc-ivr/1.0">US</country-code>
         </supported-country-codes>
         <supported-h248-codes>
            <h248-code package="msc-ivr/1.0">cg/*</h248-code>
            <h248-code package="msc-ivr/1.0">biztn/ofque</h248-
                  code>
            <h248-code package="msc-ivr/1.0">biztn/erwt</h248-
                  code>
            <h248-code package="msc-mixer/1.0">conftn/*</h248-
                  code>
         </supported-h248-codes>
      </supported-tones>
      <file-transfer-modes>
         <file-transfer-mode package="msc-ivr/1.0" name="HTTP"/>
      </file-transfer-modes>
      <asr-tts-support>
         <asr-support>
            <language xml:lang="en"/>
```

```
                </asr-support>
                <tts-support>
                    <language xml:lang="en"/>
                </tts-support>
            </asr-tts-support>
            <vxml-support>
                <vxml-mode package="msc-ivr/1.0" support="rfc6231"/>
            </vxml-support>
            <media-server-location>
                <civicAddress xml:lang="it">
                    <country>IT</country>
                        <A1>Campania</A1>
                        <A3>Napoli</A3>
                        <A6>Via Claudio</A6>
                        <HNO>21</HNO>
                        <LMK>University of Napoli Federico II</LMK>
                        <NAM>Dipartimento di Informatica e
                                Sistemistica</NAM>
                        <PC>80210</PC>
                </civicAddress>
            </media-server-location>
            <label>TestbedPrototype-01</label>
                <media-server-address>
                        sip:MediaServer@ms.example.net
                </media-server-address>
                <encryption/>
    </mrbnotification>
</mrbpublish>

B2. MRB -> MS (200 to CONTROL)
------------------------------
CFW 03fff52e7b7a 200
(C1 and C2 omitted for brevity)

D1. MRB -> MS (CONTROL, publish request)
----------------------------------------
CFW pyu788fc32wa CONTROL
Control-Package: mrb-publish/1.0
Content-Type: application/mrb-publish+xml
Content-Length: 342
<?xml version="1.0" encoding="UTF-8" standalone="yes"?>
<mrbpublish version="1.0" xmlns="urn:ietf:params:xml:ns:mrb-publish">
    <mrbrequest>
        <subscription action="update" seqnumber="2" id="p0T65U">
            <expires>600</expires>
            <minfrequency>1</minfrequency>
        </subscription>
    </mrbrequest>
</mrbpublish>

D2. MRB <- MS (200 to CONTROL, request accepted)
------------------------------------------------
CFW pyu788fc32wa 200
Timeout: 10
Content-Type: application/mrb-publish+xml
Content-Length: 332
```

```
<mrbpublish version="1.0" xmlns="urn:ietf:params:xml:ns:mrb-publish">
   <mrbresponse status="200" reason="OK: Request accepted">
      <subscription action="create" seqnumber="2" id="p0T65U">
         <expires>600</expires>
         <minfrequency>30</minfrequency>
      </subscription>
   </mrbresponse>
</mrbpublish>
```


15.7.2 Consumer Interface

Whereas the Publishing interface is used by an MS to publish its functionality and up-to-date information to an MRB, the Consumer interface is used by an interested AS to access a resource. An AS can use the Consumer interface to contact an MRB and describe the resources it needs. The MRB then replies with the needed information: Specifically, the address of an MS that is capable of meeting the requirements.

However, unlike the Publishing interface, the Consumer interface is not specified as a Control Package. Rather, it is conceived as an XML-based protocol that can be transported by means of either HTTP or SIP, as will be shown in the following sections. As specified in RFC 6917 (see Chapter 11), the Consumer interface can be involved in two topologies: Query mode and Inline mode. In the Query mode (Section 15.7.2.1), the Consumer requests and responses are conveyed by means of HTTP.

Once the AS gets the answer, the usual MEDIACTRL interactions occur between the AS and the MS chosen by the MRB. By contrast, in the Inline mode, the MRB is in the path between the AS and the pool of MS it is handling. In this case, an AS can place Consumer requests using SIP as a transport by means of a multipart payload (Section 15.7.2.2) containing the Consumer request itself and an SDP related either to the creation of a Control Channel or to a UAC media dialog. This is called Inline-aware mode, since it assumes that the interested AS knows that an MRB is in place and knows how to talk to it.

The MRB is also conceived to work with ASs that are unaware of its functionality, i.e., unaware of the Consumer interface. In this kind of scenario, the Inline mode is still used, but with the AS thinking the MRB it is talking to is actually an MS. This approach is called Inline-unaware mode (Section 15.7.2.3).

15.7.2.1 Query Mode

As discussed in the previous section, in the Query mode, the AS sends Consumer requests by means of HTTP. Specifically, an HTTP POST is used to convey the request. The MRB is assumed to send its response by means of an HTTP 200 OK reply. Since a successful Consumer response contains information to contact a specific MS (the MS the MRB has deemed most capable of fulfilling the AS's requirements), an AS can subsequently directly contact the MS, as described in Section 15.5. This means that in the Query mode, the MRB acts purely as a locator, and then the AS and the MS can talk 1:1. Figure 15.48 presents a ladder diagram of a typical Consumer request in the Query topology.

In this example, the AS is interested in an MS meeting a defined set of requirements. The MS must:

1. Support both the IVR and Mixer packages
2. Provide at least 10 G.711 encoding/decoding RTP sessions for IVR purposes
3. Support HTTP-based streaming and support for the audio/x-wav file format in the IVR package

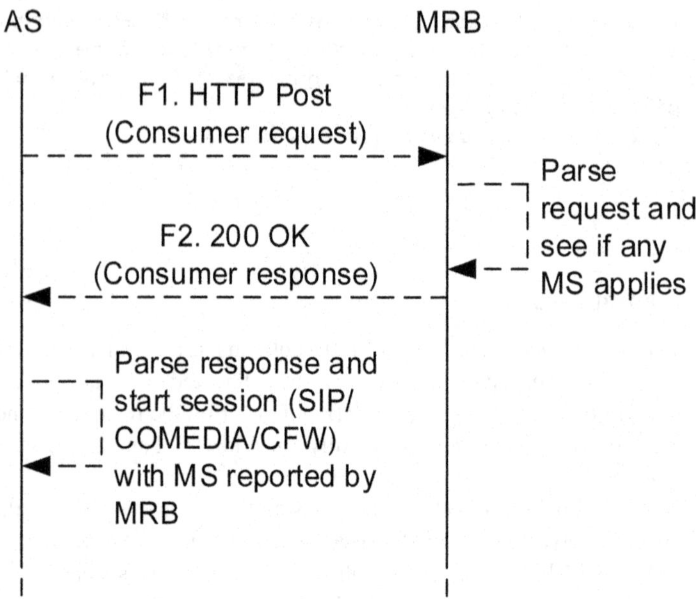

FIGURE 15.48 Media Resource Brokering: Query Mode. (Copyright: IETF.)

These requirements are properly formatted according to the MRB Consumer syntax. The framework transaction steps are as follows:

- The AS sends an HTTP POST message to the MRB (1). The payload is, of course, the Consumer request, which is reflected by the Content-Type header (application/mrb-consumer+xml). The Consumer request (<mediaResourceRequest>, uniquely identified by its "id" attribute set to the random value "n3un93wd"), includes some general requirements (<generalInfo>) and some IVR-specific requirements (<ivrInfo>). The general part of the requests contains the set of required packages (<packages>). The IVR-specific section contains requirements concerning the number of required IVR sessions (<ivr-sessions>), the file formats that are to be supported (<file-formats>), and the required file transfer capabilities (<file-transfer-modes>).
- The MRB gets the request and parses it. Then, according to its business logic, it realizes it cannot find a single MS capable of targeting the request and as a consequence, picks two MS instances that can handle 60 and 40 of the requested sessions, respectively. It prepares a Consumer response (2) to provide the AS with the requested information. The response (<mediaResourceResponse>, which includes the same "id" attribute as the request) indicates success (status=200) and includes the relevant information (<response-session-info>). Specifically, the response includes transaction-related information (the same session-id and seq provided by the AS in its request to allow proper request/response matching) together with information on the duration of the reservation (expires=3600, i.e., after an hour, the request will expire) and the SIP addresses of the chosen MSs.

Note how the sequence number the MRB returned is not 1. According to the MRB specification, this is the starting value to increment for the sequence number to be used in subsequent requests. This means that should the AS want to update or remove the session, it should use 10 as a value for the sequence number in the related request. According to Section 11.12 (Section 12 of RFC 6917), this random value for the first sequence number is also a way to help prevent a malicious entity from messing with or disrupting another AS session with the MRB. In fact, sequence numbers in requests and responses have to match, and failure to provide the correct sequence number would result in session failure and a 405 error message.

```
F1. AS -> MRB (HTTP POST, Consumer request)
-------------------------------------------
POST /Mrb/Consumer HTTP/1.1
Content-Length: 893
Content-Type: application/mrb-consumer+xml
Host: mrb.example.com:8080
Connection: Keep-Alive
User-Agent: Apache-HttpClient/4.0.1 (java 1.5)
<?xml version="1.0" encoding="UTF-8" standalone="yes"?>
<mrbconsumer version="1.0" xmlns="urn:ietf:params:xml:ns:mrb-consumer">
   <mediaResourceRequest id="n3un93wd">
      <generalInfo>
         <packages>
            <package>msc-ivr/1.0</package>
            <package>msc-mixer/1.0</package>
         </packages>
      </generalInfo>
      <ivrInfo>
         <ivr-sessions>
            <rtp-codec name="audio/basic">
               <decoding>100</decoding>
               <encoding>100</encoding>
            </rtp-codec>
         </ivr-sessions>
         <file-formats>
            <required-format name="audio/x-wav"/>
         </file-formats>
         <file-transfer-modes>
            <file-transfer-mode package="msc-ivr/1.0"
                  name="HTTP"/>
         </file-transfer-modes>
      </ivrInfo>
   </mediaResourceRequest>
</mrbconsumer>

F2. AS <- MRB (200 to POST, Consumer response)
----------------------------------------------
HTTP/1.1 200 OK
X-Powered-By: Servlet/2.5
Server: Sun GlassFish Communications Server 1.5
Content-Type: application/mrb-consumer+xml;charset=ISO-8859-1
Content-Length: 1146
Date: Thu, 28 Jul 2011 10:34:45 GMT
<?xml version="1.0" encoding="UTF-8" standalone="yes"?>
<mrbconsumer version="1.0" xmlns="urn:ietf:params:xml:ns:mrb-consumer">
   <mediaResourceResponse reason="Resource found" status="200"
         id="n3un93wd">
      <response-session-info>
         <session-id>z603G3yaUzM8</session-id>
            <seq>9</seq>
               <expires>3600</expires>
               <media-server-address
                     uri="sip:MediaServer@ms.example.com:5080">
```

```
                <ivr-sessions>
                    <rtp-codec name="audio/basic">
                        <decoding>60</decoding>
                        <encoding>60</encoding>
                    </rtp-codec>
                </ivr-sessions>
            </media-server-address>
            <media-server-address
                    uri="sip:OtherMediaServer@pool.example.net:5080">
        <ivr-sessions>
            <rtp-codec name="audio/basic">
                <decoding>40</decoding>
                <encoding>40</encoding>
            </rtp-codec>
        </ivr-sessions>
    </media-server-address>
  </response-session-info>
 </mediaResourceResponse>
</mrbconsumer>
```

For the sake of conciseness, the subsequent steps are not presented. They are very trivial, since they basically consist of the AS issuing a COMEDIA negotiation with either of the obtained MSs, as already presented in Chapter 5. The same can be said with respect to attaching UAC media dialogs. In fact, since after the Query, the AS↔MS interaction becomes 1:1, UAC media dialogs can be redirected directly to the proper MS using the 3pcc (see Section 1.3) approach, e.g., as shown in Figure 15.10.

15.7.2.2 Inline-Aware Mode

Unlike the Query mode, in the Inline-Aware MRB Mode (IAMM), the AS sends Consumer requests by means of SIP. Of course, saying that the transport changes from HTTP to SIP is not as trivial as it seems. In fact, HTTP and SIP behave in very different ways, and this is reflected in the way the Inline-aware mode is conceived.

An AS willing to issue a Consumer request by means of SIP has to do so by means of an INVITE. As specified in RFC 6917 (see Chapter 11), the payload of the INVITE cannot contain only the Consumer request itself. In fact, the Consumer request is assumed to be carried within an SIP transaction. A Consumer session is not strictly associated with the lifetime of any SIP transaction, meaning that Consumer requests belonging to the same session may be transported over different SIP messages; therefore, a hangup on any of these SIP dialogs would not affect a Consumer session.

That said, as specificationed in RFC 6230 (see Chapter 8), RFC 6917 (see Chapter 11) envisages two kinds of SIP dialogs over which a Consumer request may be sent: An SIP control dialog (an SIP dialog sent by the AS in order to set up a Control Channel) and a UAC media dialog (an SIP dialog sent by the AS in order to attach a UAC to an MS). In both cases, the AS would prepare a multipart/mixed payload to achieve both ends, i.e., receiving a reply to its Consumer request and effectively carrying on the negotiation described in the SDP payload.

The behaviors in the two cases, which are called the CFW-based approach and the media dialog–based approach, respectively, are only slightly different, but both will be presented to clarify how they could be exploited. To make things clearer for the reader, the same Consumer request as the Consumer request presented in the Query mode will be sent in order to clarify how the behavior of the involved parties may differ.

15.7.2.2.1 Inline-Aware Mode: CFW-Based Approach

Figure 15.49 presents a ladder diagram of a typical Consumer request in the CFW-based Inline-aware topology.

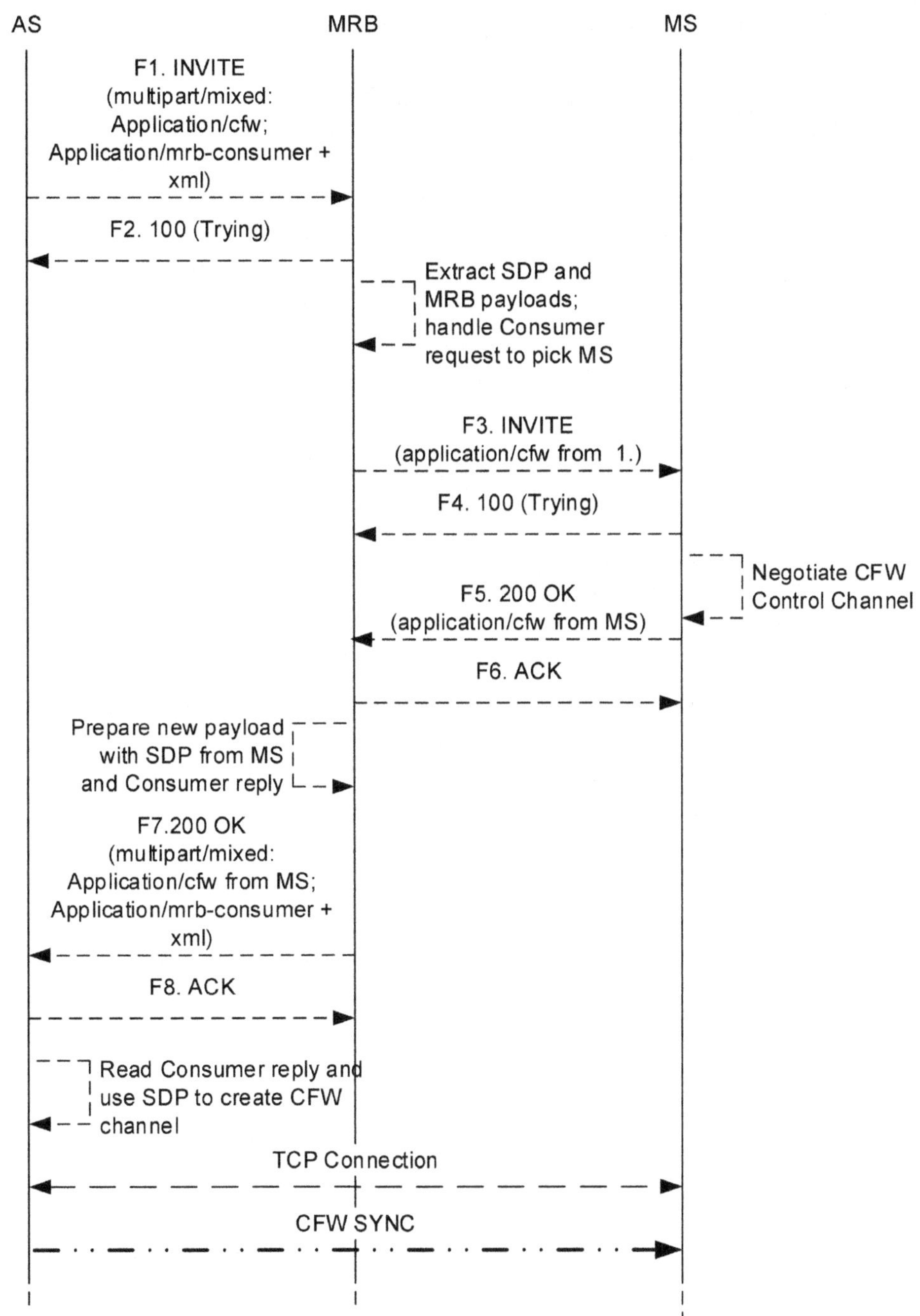

FIGURE 15.49 Media Resource Brokering: CFW-Based Inline-Aware Mode. (Copyright: IETF.)

To make the scenario easier to understand, we assume that the AS is interested in exactly the same set of requirements as those presented in Section 15.7.2.1. This means that the Consumer request originated by the AS will be the same as before, with only the transport/topology changing.

Please note that to make the protocol contents easier to read, a simple "Part" is used whenever a boundary for a multipart/mixed payload is provided instead of the actual boundary that would be inserted in the SIP messages. The framework transaction steps (for simplicity's sake, only the payloads, and not the complete SIP transactions, are reported) are as follows:

```
F1. AS -> MRB (INVITE multipart/mixed)
-------------------------------------
[..]
Content-Type: multipart/mixed;boundary="Part"
--Part
Content-Type: application/sdp
v=0
o=- 2890844526 2890842807 IN IP4 as.example.com
s=MediaCtrl
c=IN IP4 as.example.com
t=0 0
m=application 48035 TCP cfw
a=connection:new
a=setup:active
a=cfw-id:vF0zD4xzUAW9
--Part
Content-Type: application/mrb-consumer+xml
<?xml version="1.0" encoding="UTF-8" standalone="yes"?>
<mrbconsumer version="1.0"
       xmlns="urn:ietf:params:xml:ns:mrb-consumer">
   <mediaResourceRequest id="fr34asx1">
       <generalInfo>
           <packages>
               <package>msc-ivr/1.0</package>
               <package>msc-mixer/1.0</package>
           </packages>
       </generalInfo>
       <ivrInfo>
           <ivr-sessions>
               <rtp-codec name="audio/basic">
                   <decoding>100</decoding>
                   <encoding>100</encoding>
               </rtp-codec>
           </ivr-sessions>
           <file-formats>
               <required-format name="audio/x-wav"/>
           </file-formats>
           <file-transfer-modes>
               <file-transfer-mode package="msc-ivr/1.0"
                       name="HTTP"/>
               </file-transfer-modes>
       </ivrInfo>
   </mediaResourceRequest>
</mrbconsumer>
--Part

F3. MRB -> MS (INVITE SDP only)
------------------------------
[..]
Content-Type: application/sdp
v=0
o=- 2890844526 2890842807 IN IP4 as.example.com
s=MediaCtrl
c=IN IP4 as.example.com
t=0 0
m=application 48035 TCP cfw
a=connection:new
```

```
a=setup:active
a=cfw-id:vF0zD4xzUAW9

F5. MRB <- MS (200 OK SDP)
-------------------------
[..]
Content-Type: application/sdp
v=0
o=lminiero 2890844526 2890842808 IN IP4 ms.example.net
s=MediaCtrl
c=IN IP4 ms.example.net
t=0 0
m=application 7575 TCP cfw
a=connection:new
a=setup:passive
a=cfw-id:vF0zD4xzUAW9

F7. AS <- MRB (200 OK multipart/mixed)
-------------------------------------
[..]
Content-Type: multipart/mixed;boundary="Part"
--Part
Content-Type: application/sdp
v=0
o=lminiero 2890844526 2890842808 IN IP4 ms.example.net
s=MediaCtrl
c=IN IP4 ms.example.net
t=0 0
m=application 7575 TCP cfw
a=connection:new
a=setup:passive
a=cfw-id:vF0zD4xzUAW9
--Part
Content-Type: application/mrb-consumer+xml

<?xml version="1.0" encoding="UTF-8" standalone="yes"?>
<mrbconsumer version="1.0"
      xmlns="urn:ietf:params:xml:ns:mrb-consumer">
   <mediaResourceResponse reason="Resource found" status="200"
         id="fr34asx1">
      <response-session-info>
         <session-id>z603G3yaUzM8</session-id>
         <seq>9</seq>
         <expires>3600</expires>
         <media-server-address
               uri="sip:MediaServer@ms.example.com:5080">
            <connection-id>32pbdxZ8:KQw677BF</connection-id>
            <ivr-sessions>
               <rtp-codec name="audio/basic">
                  <decoding>60</decoding>
                  <encoding>60</encoding>
               </rtp-codec>
            </ivr-sessions>
         </media-server-address>
         <media-server-address
               uri="sip:OtherMediaServer@pool.example.net:5080">
            <ivr-sessions>
```

```
                <rtp-codec name="audio/basic">
                    <decoding>40</decoding>
                    <encoding>40</encoding>
                </rtp-codec>
            </ivr-sessions>
        </media-server-address>
    </response-session-info>
  </mediaResourceResponse>
</mrbconsumer>
--Part
```

The sequence diagram and the dumps effectively show the different approach with respect to the Query mode. The SIP INVITE sent by the AS (1.) includes both a Consumer request (the same as before) and an SDP to negotiate a CFW channel with an MS. The MRB takes care of the request exactly as before (provisioning two MS instances) but with a remarkable difference: First of all, it picks one of the two MS instances on behalf of the AS (negotiating the Control Channel in steps 3 to 6) and only then replies to the AS with both the MS side of the SDP negotiation (with information on how to set up the Control Channel) and the Consumer response itself.

The Consumer response is also slightly different in the first place. In fact, as can be seen in 7., there's an additional element (<connection-id>) that the MRB has added to the message. This element contains the "connection-id" that the AS and MS would have built out of the "From" and "To"' tags, as explained in Chapter 6, had the AS contacted the MS directly. Since the MRB has actually done the negotiation on behalf of the AS, without this information, the AS and MS would refer to different connectionid attributes to target the same dialog, thus causing the CFW protocol not to behave as expected. This aspect will be more carefully described in the next section (for the media dialog–based approach), since the "connection-id" attribute is strictly related to media sessions.

As before, for the sake of conciseness, the subsequent steps of the previous transaction are quite trivial and therefore, are not presented. In fact, as shown in the flow, the SIP negotiation has resulted in both the AS and the chosen MS negotiating a Control Channel. This means that the AS is only left to instantiate the Control Channel and send CFW requests according to its application logic.

It is worthwhile to highlight the fact that as in the Query example, the AS gets the addresses of both of the chosen MS in this example as well, since a Consumer transaction has taken place. This means that just as in the Query case, any UAC media dialog can be redirected directly to the proper MS using the 3pcc (see Section 1.3) approach, e.g., as shown in Figure 15.10, rather than again using the MRB as a Proxy/B2BUA. Of course, a separate SIP control dialog would be needed before attempting to use the second MS instance.

15.7.2.2.2 Inline-Aware Mode: Media Dialog-Based Approach

There is a second way to take advantage of the IAMM mode, i.e., exploiting SIP dialogs related to UAC media dialogs as "vessels" for Consumer messages. As will be made clearer in the following sequence diagram and protocol dumps, this scenario does not differ much from the scenario presented in Section 15.7.2.2.1 with respect to the Consumer request/response, but it may be useful to compare these two scenarios and show how they may differ with respect to the management of the media dialog itself and any CFW Control Channel that may be involved.

Figure 15.50 presents a ladder diagram of a typical Consumer request in the media dialog-based Inline-aware topology.

To make the scenario easier to understand, we assume that the AS is interested in exactly the same set of requirements as those presented in Section 15.7.2.1. This means that the Consumer request originated by the AS will be the same as before, with only the transport/topology changing.

Again, please note that to make the protocol contents easier to read, a simple "Part" is used whenever a boundary for a multipart/mixed payload is provided instead of the actual boundary that would be inserted in the SIP messages. The framework transaction steps (for simplicity's sake,

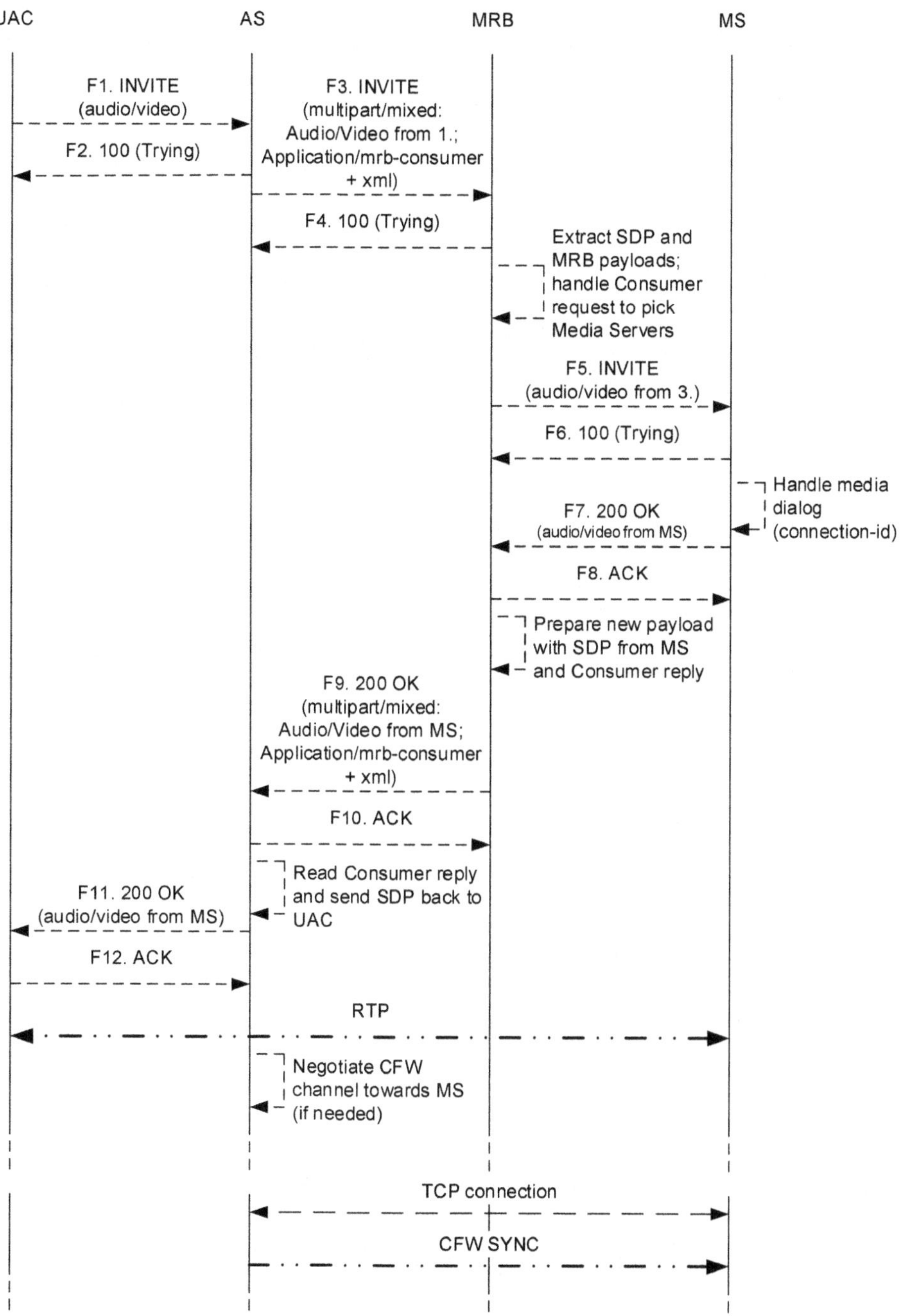

FIGURE 15.50 Media Resource Brokering: Media Dialog–Based Inline-Aware Mode. (Copyright: IETF.)

only the relevant headers and payloads, and not the complete SIP transactions, are reported) are as follows:

```
F1. UAC -> AS (INVITE with media SDP)
-------------------------------------
[..]
From: <sip:lminiero@users.example.com>;tag=1153573888
To: <sip:mediactrlDemo@as.example.com>
```

```
[..]
Content-Type: application/sdp
v=0
o=lminiero 123456 654321 IN IP4 203.0.113.2
s=A conversation
c=IN IP4 203.0.113.2
t=0 0
m=audio 7078 RTP/AVP 0 3 8 101
a=rtpmap:0 PCMU/8000/1
a=rtpmap:3 GSM/8000/1
a=rtpmap:8 PCMA/8000/1
a=rtpmap:101 telephone-event/8000
a=fmtp:101 0-11
m=video 9078 RTP/AVP 98

3. AS -> MRB (INVITE multipart/mixed)
-------------------------------------
[..]
From: <sip:ApplicationServer@as.example.com>;tag=fd4fush5
To: <sip:Mrb@mrb.example.org>
[..]
Content-Type: multipart/mixed;boundary="Part"
--Part
Content-Type: application/sdp
v=0
o=lminiero 123456 654321 IN IP4 203.0.113.2
s=A conversation
c=IN IP4 203.0.113.2
t=0 0
m=audio 7078 RTP/AVP 0 3 8 101
a=rtpmap:0 PCMU/8000/1
a=rtpmap:3 GSM/8000/1
a=rtpmap:8 PCMA/8000/1
a=rtpmap:101 telephone-event/8000
a=fmtp:101 0-11
m=video 9078 RTP/AVP 98
--Part
Content-Type: application/mrb-consumer+xml
<?xml version="1.0" encoding="UTF-8" standalone="yes"?>
<mrbconsumer version="1.0"
      xmlns="urn:ietf:params:xml:ns:mrb-consumer">
   <mediaResourceRequest id="bnv3xc45">
      <generalInfo>
         <packages>
            <package>msc-ivr/1.0</package>
            <package>msc-mixer/1.0</package>
         </packages>
      </generalInfo>
      <ivrInfo>
         <ivr-sessions>
            <rtp-codec name="audio/basic">
               <decoding>100</decoding>
               <encoding>100</encoding>
            </rtp-codec>
         </ivr-sessions>
         <file-formats>
```

```
            <required-format name="audio/x-wav"/>
         </file-formats>
         <file-transfer-modes>
            <file-transfer-mode package="msc-ivr/1.0"
                  name="HTTP"/>
         </file-transfer-modes>
      </ivrInfo>
   </mediaResourceRequest>
</mrbconsumer>
--Part

F5. MRB -> MS (INVITE SDP only)
------------------------------
[..]
From: <sip:Mrb@mrb.example.org:5060>;tag=32pbdxZ8
To: <sip:MediaServer@ms.example.com:5080>
[..]
Content-Type: application/sdp
v=0
o=lminiero 123456 654321 IN IP4 203.0.113.2
s=A conversation
c=IN IP4 203.0.113.2
t=0 0
m=audio 7078 RTP/AVP 0 3 8 101
a=rtpmap:0 PCMU/8000/1
a=rtpmap:3 GSM/8000/1
a=rtpmap:8 PCMA/8000/1
a=rtpmap:101 telephone-event/8000
a=fmtp:101 0-11
m=video 9078 RTP/AVP 98

F7. MRB <- MS (200 OK SDP)
-------------------------
[..]
From: <sip:Mrb@mrb.example.org:5060>;tag=32pbdxZ8
To: <sip:MediaServer@ms.example.com:5080>;tag=KQw677BF
[..]
Content-Type: application/sdp
v=0
o=lminiero 123456 654322 IN IP4 203.0.113.1
s=MediaCtrl
c=IN IP4 203.0.113.1
t=0 0
m=audio 63442 RTP/AVP 0 3 8 101
a=rtpmap:0 PCMU/8000
a=rtpmap:3 GSM/8000
a=rtpmap:8 PCMA/8000
a=rtpmap:101 telephone-event/8000
a=fmtp:101 0-15
a=ptime:20
a=label:7eda834
m=video 33468 RTP/AVP 98
a=rtpmap:98 H263-1998/90000
a=fmtp:98 CIF=2
a=label:0132ca2
```

```
F9. AS <- MRB (200 OK multipart/mixed)
------------------------------------
[..]
From: <sip:ApplicationServer@as.example.com>;tag=fd4fush5
To: <sip:Mrb@mrb.example.org>;tag=117652221
[..]
Content-Type: multipart/mixed;boundary="Part"
--Part
Content-Type: application/sdp
v=0
o=lminiero 123456 654322 IN IP4 203.0.113.1
s=MediaCtrl
c=IN IP4 203.0.113.1
t=0 0
m=audio 63442 RTP/AVP 0 3 8 101
a=rtpmap:0 PCMU/8000
a=rtpmap:3 GSM/8000
a=rtpmap:8 PCMA/8000
a=rtpmap:101 telephone-event/8000
a=fmtp:101 0-15
a=ptime:20
a=label:7eda834
m=video 33468 RTP/AVP 98
a=rtpmap:98 H263-1998/90000
a=fmtp:98 CIF=2
a=label:0132ca2
--Part
Content-Type: application/mrb-consumer+xml

<?xml version="1.0" encoding="UTF-8" standalone="yes"?>
<mrbconsumer version="1.0"
      xmlns="urn:ietf:params:xml:ns:mrb-consumer" >
   <mediaResourceResponse reason="Resource found" status="200"
         id="bnv3xc45">
      <response-session-info>
         <session-id>z1skKYZQ3eFu</session-id>
         <seq>9</seq>
         <expires>3600</expires>
         <media-server-address
               uri="sip:MediaServer@ms.example.com:5080">
            <connection-id>32pbdxZ8:KQw677BF</connection-id>
            <ivr-sessions>
               <rtp-codec name="audio/basic">
                  <decoding>60</decoding>
                  <encoding>60</encoding>
               </rtp-codec>
            </ivr-sessions>
         </media-server-address>
         <media-server-address
               uri="sip:OtherMediaServer@pool.example.net:5080">
            <ivr-sessions>
               <rtp-codec name="audio/basic">
                  <decoding>40</decoding>
                  <encoding>40</encoding>
               </rtp-codec>
            </ivr-sessions>
         </media-server-address>
```

```
        </response-session-info>
    </mediaResourceResponse>
</mrbconsumer>
--Part

F11. UAC <- AS (200 OK SDP)
--------------------------
[..]
From: <sip:lminiero@users.example.com>;tag=1153573888
To: <sip:mediactrlDemo@as.example.com>;tag=bcd47c32
[..]
Content-Type: application/sdp
v=0
o=lminiero 123456 654322 IN IP4 203.0.113.1
s=MediaCtrl
c=IN IP4 203.0.113.1
t=0 0
m=audio 63442 RTP/AVP 0 3 8 101
a=rtpmap:0 PCMU/8000
a=rtpmap:3 GSM/8000
a=rtpmap:8 PCMA/8000
a=rtpmap:101 telephone-event/8000
a=fmtp:101 0-15
a=ptime:20
a=label:7eda834
m=video 33468 RTP/AVP 98
a=rtpmap:98 H263-1998/90000
a=fmtp:98 CIF=2
a=label:0132ca2
```

The first obvious difference is that the first INVITE (1.) is not originated by the AS itself (the AS was willing to set up a Control Channel in the previous example) but by an authorized UAC (e.g., to take advantage of a media service provided by the AS). As such, the first INVITE only contains an SDP to negotiate an audio and video channel. The AS in its business logic needs to attach this UAC to an MS according to some specific requirements (e.g., the called URI is associated to a specific service) and as such, prepares a Consumer request to be sent to the MRB in order to obtain a valid MS for that purpose. As before, the Consumer request is sent together with the SDP to the MRB (3.).

The MRB extracts the Consumer payload and takes care of it as usual; it picks two MS instances and attaches the UAC to the first MS instance (5.). Once the MS has successfully negotiated the audio and video streams (7.), the MRB takes note of the "connection-id" associated with this call (which will be needed afterwards in order to manipulate the audio and video streams for this user) and sends back to the AS both the SDP returned by the MS and the Consumer response (9.). The AS extracts the Consumer response and takes note of both the MS instances it has been given and the connection-id information. It then completes the scenario by sending back to the UAC the SDP returned by the MS (11.).

At this point, the UAC has successfully been attached to an MS. The AS only needs to set up a Control Channel to that MS, if needed. This step may not be required, especially if the Consumer request is an update to an existing session rather than the preparation of a new session. Assuming that a Control Channel towards that MS does not exist yet, the AS creates it as usual by sending an INVITE directly to the MS for which it has an address. Once done with that, it can start manipulating the audio and video streams of the UAC.

To do so, it refers to the <connection-id> element as reported by the MRB rather than relying on the <connection-id> element that it is aware of. In fact, the AS is aware of a connection-id value (fd4fush5: 117652221, built out of the messages exchanged with the MRB), while the MS is aware of another (32pbdxZ8:KQw677BF, built out of the MRB-MS interaction). The right connection-id is, of course, the one the MS is aware of, and as such, the AS refers to that connection-id, which the MRB added to the Consumer response just for that purpose.

15.7.2.3 Inline-Unaware Mode

Whereas in the Inline-aware mode, the AS knows it is sending an INVITE to an MRB and not to an MS and acts accordingly (using the multipart/mixed payload to query for an MS able to fulfill its requirements), in the Inline-Unaware MRB Mode (IUMM), the AS does not distinguish an MRB from an MS. This means that an MRB-unaware AS having access to an MRB talks to it as if it were a generic MEDIACTRL MS: i.e., the AS negotiates a Control Channel directly with the MRB and attaches its media dialogs there as well. Of course, since the MRB does not provide any MS functionality by itself, it must act as a Proxy/B2BUA between the AS and an MS for both the Control Channel dialog and the media dialogs. According to implementation or deployment choices, simple redirects could also be exploited for that purpose.

The problem is that without any Consumer request being placed by the MRB-unaware AS, the MRB cannot rely on AS-originated directives to pick one MS rather than another. In fact, the MRB cannot know what the AS is looking for. The MRB is then assumed to pick one according to its logic, which is implementation specific.

Figure 15.51 presents a ladder diagram of a typical Consumer request in the Inline-unaware topology.

As can be seen in the diagram, in this topology, the MRB basically acts as a 3pcc between the AS and the chosen MS. The same can be said with respect to attaching UAC media dialogs. The MRB remembers the MS it has chosen for the AS, and for every UAC media dialog the AS tries to attach to the MRB, it makes sure that it is somehow actually redirected to the MS.

No content for the presented messages is provided in this section, as in the IUMM mode, no Consumer transaction is involved. In this example, a simple RFC 6230 (see Chapter 8) Control Channel negotiation occurs where the MRB acts as an intermediary, that is, picking an MS for the AS according to some logic. In this case, in fact, the AS does not support the MRB specification and so just tries to set up a Control Channel according to its own logic.

It is worth pointing out that the MRB may actually enforce its decision about the MS to grant to the AS in different ways. Specifically, the sentence "redirect the INVITE" that is used in Figure 15.51 does not necessarily mean that a SIP 302 message should be used for that purpose. A simple way to achieve this may be provisioning the unaware AS with different URIs, all actually transparently handled by the MRB itself; this would allow the MRB to simply map those URIs to different MS instances. The SIP "Contact" header may also be used by the MRB in a reply to an INVITE coming from an AS to provide the actual URI on which the chosen MS might be reached. A motivation for such a discussion, and more details on this topic, is provided in Section 15.7.3.2.

15.7.3 Handling Media Dialogs

It is worthwhile to briefly address how media dialogs would be managed whenever an MRB is involved in the following scenarios. In fact, the presence of an MRB may introduce an additional complexity compared with the quite straightforward 1:1 AS-MS topology.

15.7.3.1 Query and Inline-Aware Mode

Normally, especially in the Query and IAMM case, the MRB would only handle Consumer requests by an AS, and after that, the AS and the MS picked by the MRB for a specific request would talk directly to each other by means of SIP. This is made possible by the fact that the AS gets the MS SIP URI in reply to its request. In this case, an AS can simply relay media dialogs associated with that session to the right MS to have them handled accordingly. Of course, in order for this to work, it is assumed that the AS creates a Control Channel to a chosen MS before it has any requests to

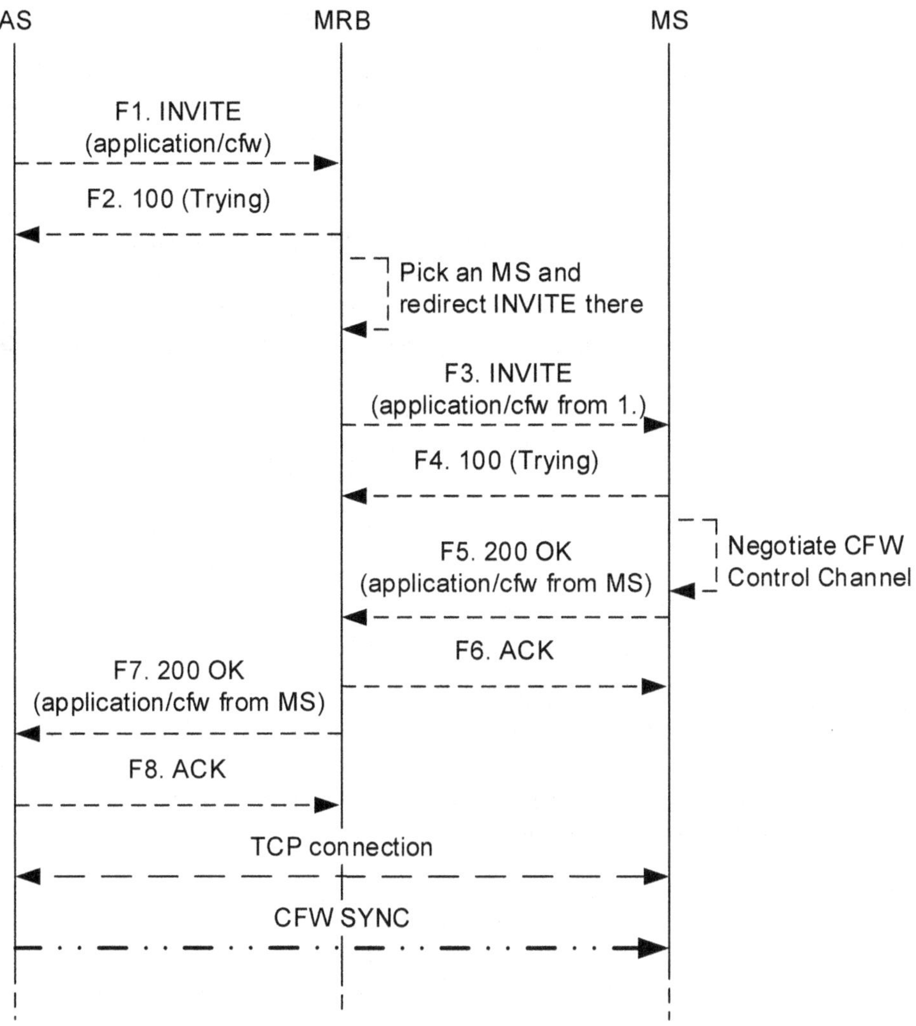

FIGURE 15.51 Media Resource Brokering: Inline-Unaware Mode. (Copyright: IETF.)

service. An example of such a scenario is presented in Figure 15.52. Please note that this diagram and subsequent diagrams in this section are simplified with respect to the actual protocol interactions. For instance, the whole SIP transactions are not presented, and only the originating messages are presented in order to clarify the scenario in a simple way.

As can be deduced from the diagram, the interactions among the components are quite straightforward. The AS knows which MS it has been assigned to (as a consequence of the MRB Consumer request, whether it has been achieved by means of HTTP or SIP), and so it can easily attach any UAC accessing its functionality to the MS itself and manipulate its media connections by using the CFW Control Channel as usual.

In such a scenario, the MRB is only involved as a locator. Once the MRB provides the AS with the URI of the required resource, it does not interfere with subsequent interactions unless it wants to perform monitoring (e.g., by exploiting the Publishing information reported by the MS). As a consequence, the scenario basically becomes 1:1 between the AS and the MS again.

Nevertheless, there are cases when having an MRB in the SIP signaling path as well might be a desired feature, e.g., for more control over the use of the resources. Considering how the Consumer interface has been envisaged, this feature is easily achievable, with no change to the

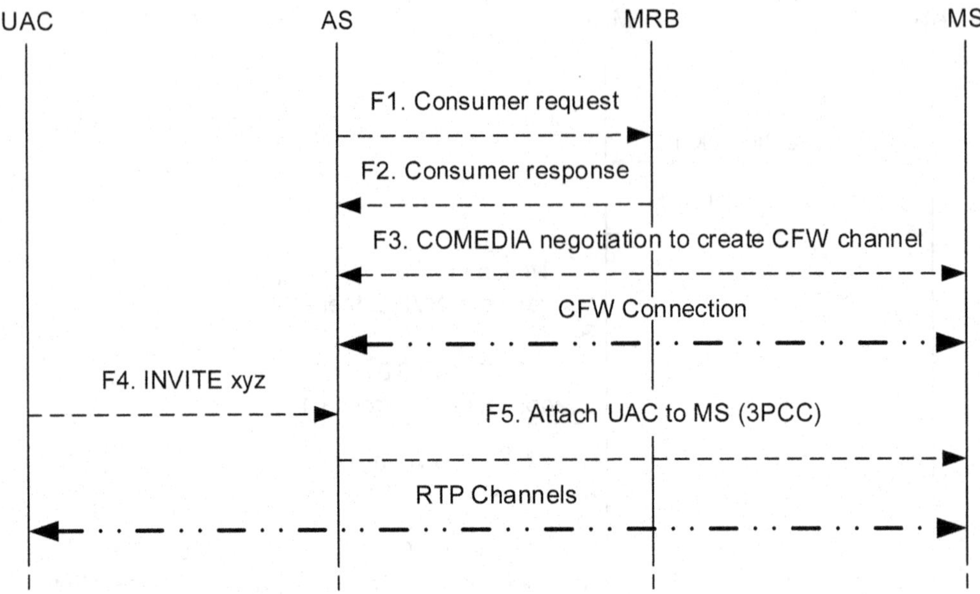

FIGURE 15.52 Handling Media Dialogs in Query/IAMM. (Copyright: IETF.)

protocol required at all. Specifically, in order to achieve such functionality, the MRB may reply to a Consumer request with a URI for which the MRB is responsible (rather than the MS SIP URI as discussed previously) and map this URI to the actual MS URI in its business logic; this would be transparent to the AS. In this way, the AS would interact with the MRB as if it were the MS itself. Figure 15.53 shows how the scenario would change in this case.

This time, even though the MRB has picked a specific MS after a request from an AS, it replies with another SIP URI, a URI it would reply to itself. The AS would contact that URI in order to negotiate the Control Channel, and the MRB would proxy/forward the request to the actual MS transparently. Eventually, the Control Channel would be instantiated between the AS and the MS. The same happens for UACs handled by the AS; the AS would forward the calls to the URI provided to it, the one handled by the MRB, which would in turn relay the call to the MS in order to have the proper RTP channels created between the UAC and the MS.

This scenario is not very different from the previous scenario except that the MRB is now on the signaling path for both the SIP control dialog and the SIP media dialogs, allowing it to have more control of the resources (e.g., triggering a BYE if a resource has expired). There are several possible approaches an MRB might take to allocate URIs to map to a requested MS. For example, an MRB might use SIP URI parameters to generate multiple SIP URIs that are unique but that all route to the same host and port, e.g., sip:MrbToMs@mrb.example.com:5080;p=1234567890. Alternatively, the MRB might simply allocate a pool of URIs for which it would be responsible and manage the associations with the requested MS services accordingly.

15.7.3.2 Inline-Unaware Mode

As mentioned previously, in the IUMM case, the AS would interact with the MRB as if it were the MS itself. One might argue that this would make the AS act as it would in the IAMM case. This is not the case, however, since the AS actually provided the MRB with information about the resources it required, leading to the selection of a proper MS, while in the IUMM case, the MRB would have to pick an MS with no help from the AS at all.

That said, the IUMM case is also very interesting with respect to media dialog management. In fact, in the MRB-unaware mode, there would be no Consumer request, and an AS would actually

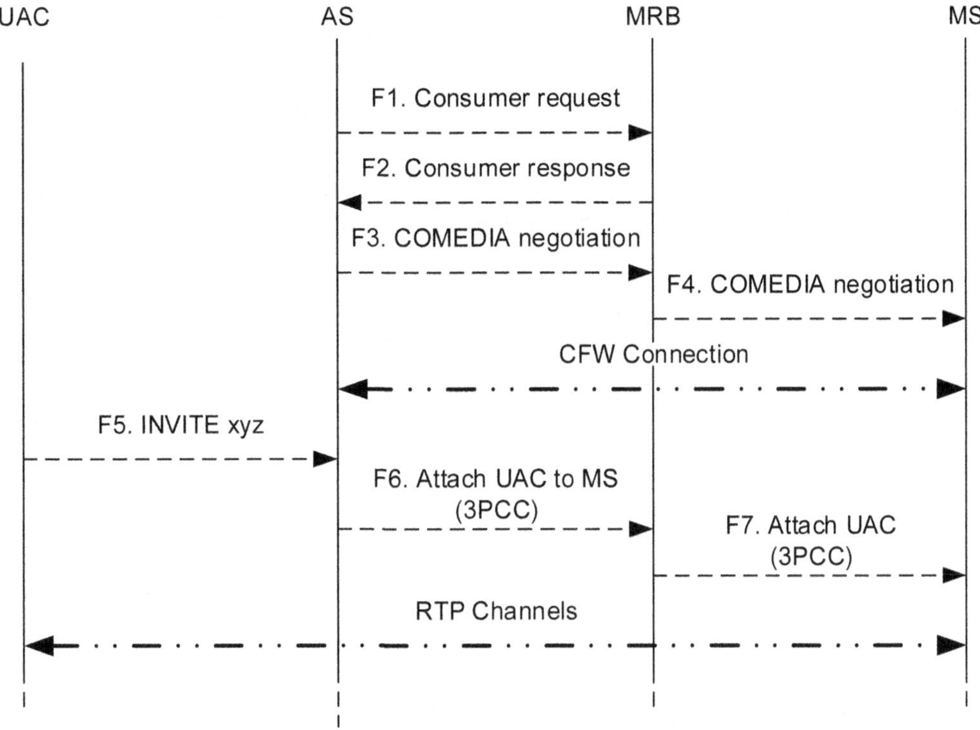

FIGURE 15.53 Handling Media Dialogs in Query/IAMM: MRB in the Signaling Path. (Copyright: IETF.)

see the MRB as an MS. Unlike the previous scenarios, because there is no AS↔MRB interaction and as such, no MS selection process, the MRB would likely be in the signaling path anyway, at least when the AS first shows up.

The MRB could either redirect the AS to an MS directly or transparently act as a Proxy/B2BUA and contact an MS (according to implementation-specific policies) on behalf of the unaware AS. While apparently not a problem, this raises an issue when the same unaware AS has several sessions with different MSs. The AS would only see one "virtual" MS (the MRB), and so it would relay all calls there, making it hard for the MRB to understand where these media dialogs should belong: Specifically, whether the UAC calling belongs to the AS application logic leading to MS1 or MS2, or somewhere else.

One possible, and very simple, approach to take care of this issue is to always relay the SIP dialogs from the same unaware AS to the same MS, as depicted in Figure 15.54.

In this example, the AS creates two different Control Channel sessions (A and B) to address two different business logic implementations; e.g., the AS SIP URI "xyz" (associated with CFW session A) may be an IVR pizza-ordering application, while the AS SIP URI "jkl" (associated with CFW session B) may be associated with a conference room. It is quite clear, then, that if the MRB forwarded the two CFW sessions to two different MSs, the handling of UAC media dialogs would prove troublesome, because the MRB would have difficulty figuring out whether UAC1 should be attached to the MS managing CFW session A or the MS managing CFW session B. In this example, forwarding all CFW sessions and UAC media dialogs coming from the same MRB-unaware AS to the same MS would work as expected.

The MRB would, in fact, leave the mapping of media dialogs and CFW sessions up to the AS. This approach, while very simple and indeed, not very scalable, would actually help take care of the issue. In fact, no matter how many separate Control Channels the AS might have with the MRB/MS (in this example, Control Channel A would be mapped to application xyz and Control Channel

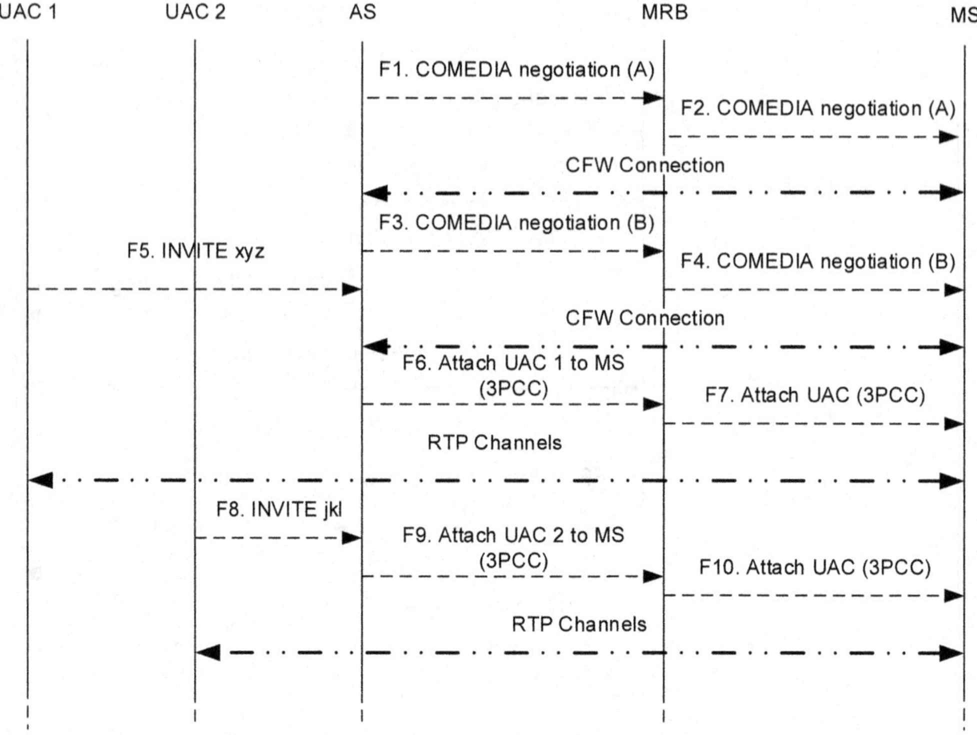

FIGURE 15.54 Handling Media Dialogs in IUMM: Always the Same MS. (Copyright: IETF.)

B to application jkl), the termination point would still always be the same MS, which would consequently be the destination for all media dialogs as well.

To overcome the scalability limitations of such an approach, at least in regard to the MRB being in the SIP signaling path for all calls, a different approach needs to be exploited. In fact, especially in the case of different applications handled by the same unaware AS, it makes sense to try to exploit different MSs for that purpose and to correctly track media dialogs being forwarded accordingly. This means that the MRB must find a way to somehow redirect the unaware AS to different MSs when it predicts or realizes that a different application logic is involved. To do so, the MRB might use different approaches. One approach would use redirection, e.g., by means of a SIP 302 message in reply to a Control Channel negotiation originated by an unaware AS. Such an approach is depicted in Figure 15.55.

With this approach, the MRB might redirect the AS to a specific MS whenever a new Control Channel is to be created, and as a consequence, the AS would redirect the related calls there. This is similar to the first approach of the Query/IAMM case, with the difference that no Consumer request would be involved. The scenario would again fall back to a 1:1 topology between the AS and the MS, making the interactions quite simple.

Just as before, the MRB might be interested in being in the signaling path for the SIP dialogs instead of just acting as a locator. A third potential approach could be implementing the "virtual" URIs handled by the MRB, as described in the previous section. Rather than resorting to explicit redirection or always using the same MS, the MRB may redirect new SIP control dialogs to one of its own URIs, using the same approach previously presented in Figure 15.53. Such an approach, as applied to the IUMM case, is depicted in Figure 15.56.

It is worth pointing out, though, that in both cases, there are scenarios where there could be no assurance that the 302 sent by the MRB would be seen by the AS. In fact, should a proxy be between the AS and the MRB, such a proxy could itself act on the 302. To properly cope with such an issue,

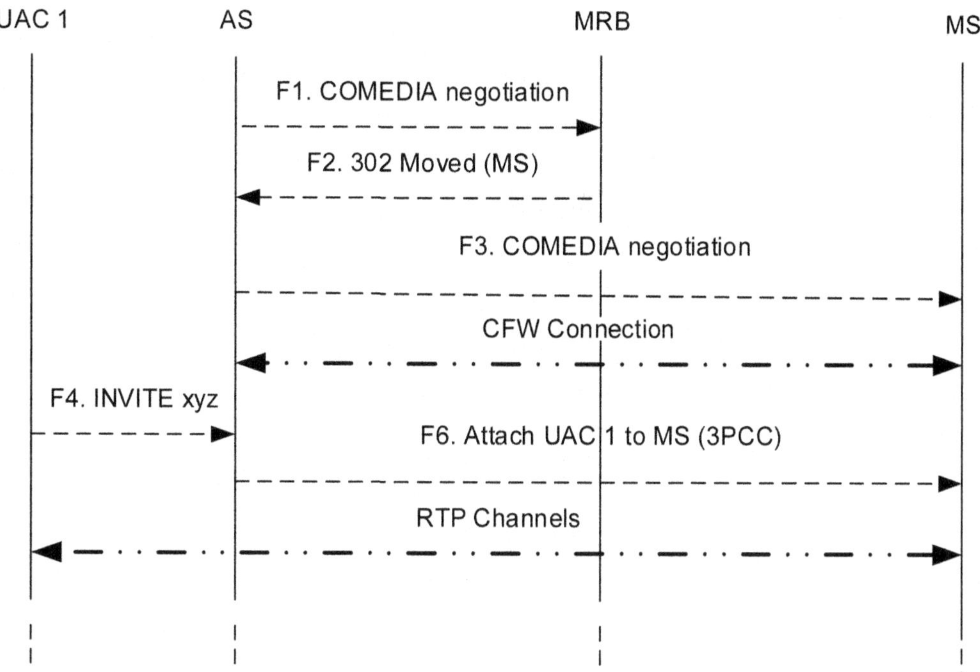

FIGURE 15.55 Handling Media Dialogs in IUMM: Redirection. (Copyright: IETF.)

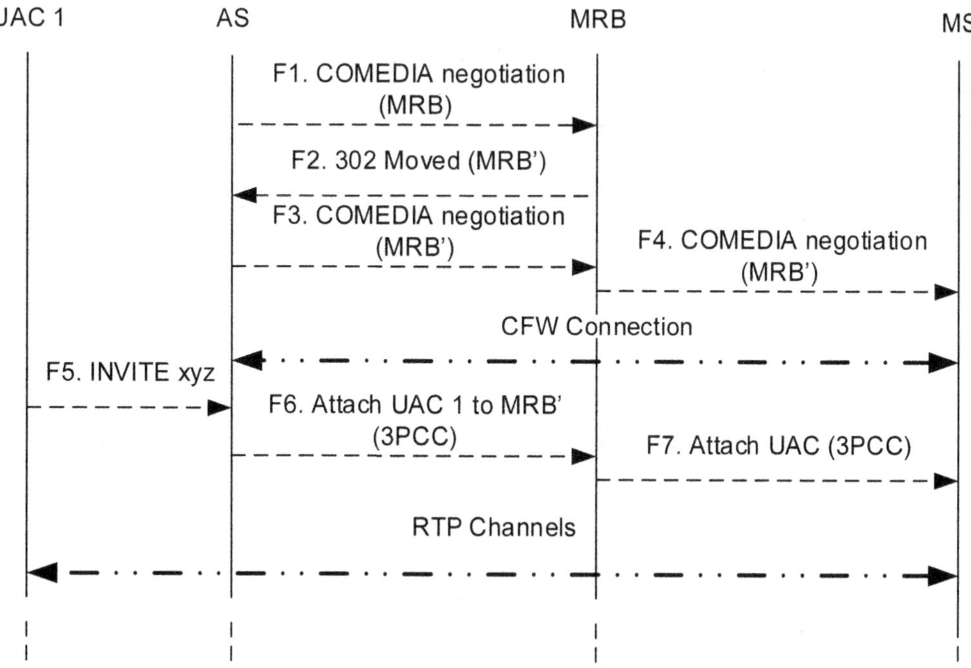

FIGURE 15.56 Handling Media Dialogs in IUMM: MRB in the Signaling Path. (Copyright: IETF.)

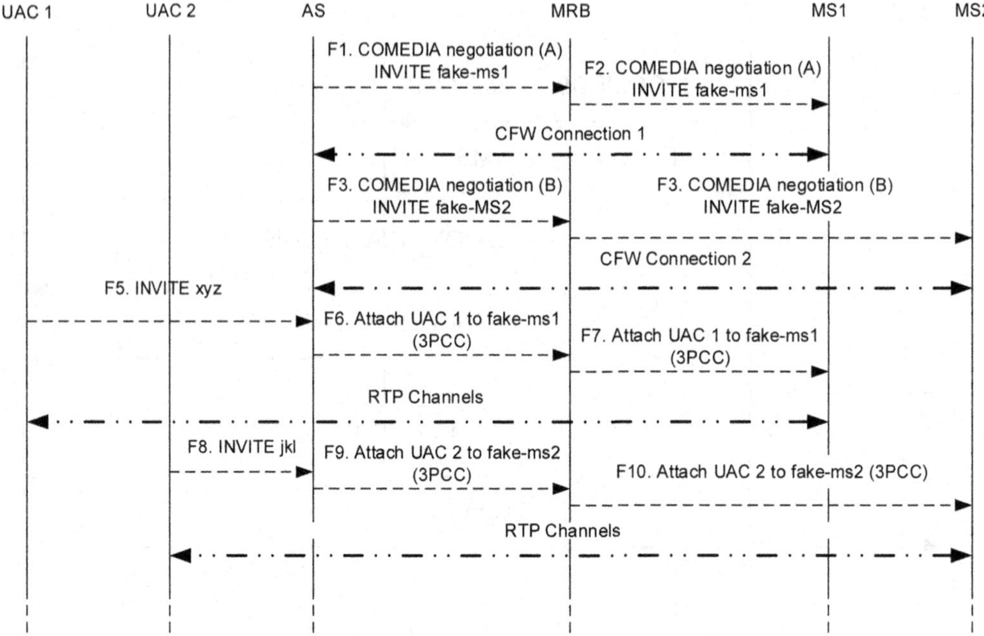

FIGURE 15.57 Handling Media Dialogs in IUMM: Provisioned URIs. (Copyright: IETF.)

the MRB might also use the "Contact" header in the SIP responses to the INVITE to address the right MS. Although the AS is not required to use the information in such a header to reach the MS, it could be reasonable to exploit it for that purpose, as it would take care of the proxy scenario mentioned.

To conclude, there is a further approach an MRB might try to exploit to take care of the IUMM case. Since, as explained before, the issues related to the IUMM case mostly relate to the fact that the MRB is seen as a single MS instance by the AS, a simple way to overcome this might be to make the MRB look like a set of different MSs right away; this can be done by simply provisioning the unaware AS with a series of different URIs, all handled by the MRB itself acting as a pool of "virtual" MSs. In this way, the AS may be designed to use different MSs for different classes of calls, e.g., for different applications it is managing (two in the example presented in this section), and as such, would contact two different provisioned URIs to create two distinct Control Channels towards two different MSs. Since both the URIs would be handled by the MRB, the MRB can use them to determine to which MS each call should be directed. Expanding on Figure 15.54 by removing the constraint to always use the same MS, this new scenario might look like that depicted in Figure 15.57.

In this new example, we still assume that the same unaware AS is handling two different applications, still associated with the same URIs as before. This time, though, we also assume that the AS has been designed to try to use different MS instances to handle the two very different applications for which it is responsible. We also assume that it has been configured to be able to use two different MS instances, reachable at SIP URI "fake-ms1" and "fake-ms2," respectively, and both actually handled by the MRB transparently. This results, just as before, in two different Control Channels (A and B) being created but this time, towards two different MSs.

Specifically, the MRB makes sure that for this AS, the Control Channel negotiation towards "fake-ms1" is actually redirected to MS1. At the same time, "fake-ms2" is associated with MS2. Once the AS has set up the Control Channels with both of the MSs, it is ready to handle media dialogs. UAC1 calls the SIP URI "xyz" on the AS to order a pizza. The AS attaches the media dialog to the MS it knows is responsible for that branch of application logic, i.e., "fake-ms1." The MRB, in turn, makes sure that it reaches the right MS instance, MS1. Later on, a different user, UAC2, calls

SIP URI "jkl" to join a conference room. This time, the AS attaches this new media dialog to the MS instance handling the conference application, i.e., "fake-ms2."

Again, the MRB makes sure that it is actually MS2 that receives the dialog. Again, this diagram is only meant to describe how the MRB might enforce its decisions. Just as described in the previous examples, the MRB may choose to either act as a Proxy/B2BUA between the AS and the MS instances or redirect the AS to the right MS instances when they are first contacted (e.g., by means of the Contact header and/or a SIP redirect, as explained before) and let the AS attach the media dialogs by itself.

15.7.3.3 CFW Protocol Behavior

As shown in the previous diagrams, no matter what the topology, the AS and MS usually end up with a direct connection with respect to the CFW Control Channel. As such, it can be expected that the CFW protocol will continue to work as it should, and as a consequence, all the call flows presented in this specification can easily be reproduced in those circumstances as well. Nevertheless, one aspect needs to be considered very carefully. It is worthwhile to remind readers that both the AS and the MS use some SIP-related information to address the entities they manipulate.

This is the case, for instance, for the <connectionid> element to which both the AS and the MS refer when addressing a specific UAC. As explained in Section 15.6, this "connectionid" identifier is constructed by concatenating the "From" and "To" tags extracted from a SIP header: specifically, from the headers of the AS↔MS leg that allows a UAC to be attached to the MS. The presence of an additional component in the path between the AS and the MS, the MRB, might alter these tags, thus causing the AS to use tags (AS↔MRB) different from those used by the MS (MRB↔MS). This would result in the AS and MS using different "connectionid" identifiers to address the same UAC, thus preventing the protocol from working as expected. As a consequence, it is very important that any MRB implementation takes very good care to preserve the integrity of the involved SIP headers when proxying/forwarding SIP dialogs between the AS and MS in order not to "break" the behavior of the protocol.

Let us take, for instance, the scenario depicted in Figure 15.53, especially steps 6 and 7, which specifically address a UAC being attached by an AS to an MS via the MRB. Let us assume that Figure 15.58 shows what happens to the "From" and "To" headers in that scenario when dealing with the 3pcc approach to attach a specific UAC to the MS.

In this example, once done with the 3pcc, and now that the UAC is attached to the MS, the AS and the MS end up with different interpretations of what the "connectionid" for the UAC should be. In fact, the AS builds a "connectionid" using the tags it is aware of (a1b2c3:d4e5f6), while the MS builds a different identifier after receiving different information from the MRB (aaabbb:cccddd).

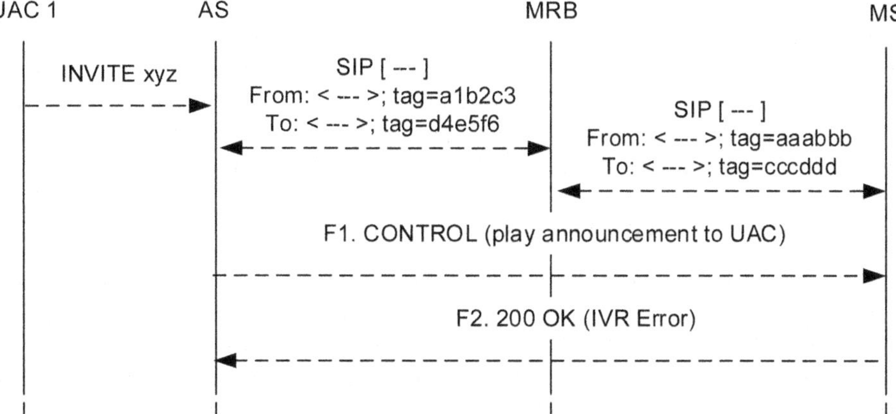

FIGURE 15.58 CFW Protocol Behavior in the Case of Manipulated Tags. (Copyright: IETF.)

As a consequence, when the AS tries to play an announcement to the UAC using the connectionid it correctly constructed, the MS just as correctly replies with an error, since it does not know that identifier. This is correct protocol behavior, because in this case, it was caused by misuse of the information needed for it to work as expected.

```
F1. AS -> MS (CFW CONTROL, play)
--------------------------------
CFW ffhg45dzf123 CONTROL
Control-Package: msc-ivr/1.0
Content-Type: application/msc-ivr+xml
Content-Length: 284
<mscivr version="1.0" xmlns="urn:ietf:params:xml:ns:msc-ivr">
   <dialogstart connectionid="a1b2c3:d4e5f6">
      <dialog>
         <prompt>
            <media loc="http://www.example.net/hello.wav"/>
         </prompt>
      </dialog>
   </dialogstart>
</mscivr>

F2. AS <- MS (CFW 200 OK)
-------------------------
CFW ffhg45dzf123 200
Timeout: 10
Content-Type: application/msc-ivr+xml
Content-Length: 148
<mscivr version="1.0" xmlns="urn:ietf:params:xml:ns:msc-ivr">
   <response status="407" reason="connectionid does not exist"
         dialogid=""/>
</mscivr>
```

In an even worse scenario, the connectionid might actually exist but might be mapped to a different UAC. In such a case, the transaction would succeed, but a completely different UAC would be involved, thus causing a silent failure that neither the AS nor the MS would be aware of. That said, proper management of these sensitive pieces of information by the MRB would prevent such failure scenarios from happening. How this issue is taken care of in the IAMM case (both CFW-based and media dialog–based) has already been described. Addressing this issue for the IUMM case is not specificationed in RFC 6917 (see Chapter 11) as being explicitly out of scope and as such, may be implementation specific.

The same applies to SDP fields as well. In fact, the AS and MS use ad hoc SDP attributes to instantiate a Control Channel, as they use SDP labels to address specific media connections of a UAC media dialog when a fine-grained approach is needed. As a consequence, any MRB implementation should limit any SDP manipulation as much as possible or at least take very good care not to cause changes that could "break" the expected behavior of the CFW protocol.

15.8 SECURITY CONSIDERATIONS

All the MEDIACTRL specifications have strong statements regarding security considerations within the context of the interactions occurring at all levels among the involved parties. Considering the sensitive nature of the interaction between AS and MS, particular efforts have been devoted to providing guidance on how to secure what flows through a Control Channel. In fact, transactions concerning dialogs, connections, and mixes are quite strongly related to resources actually being deployed and used in the MS. This means that it is in the interest of both the AS and the MS that resources created and handled by an entity are not manipulated by a potentially malicious third party if permission was not granted.

Because strong statements are provided in the aforementioned specifications, and these specifications provide good guidance to implementors with respect to these issues, this section will only provide the reader with some MEDIACTRL call flows that show how a single secured MS is assumed to reply to different ASs when receiving requests that may cross the bounds within which each AS is constrained. This would be the case, for instance, for generic auditing requests or explicit conference manipulation requests where the involved identifiers are "To" address a very specific scenario, let us assume that two different ASs, AS1 and AS2, have established a Control Channel with the same MS. Considering the SYNC transaction that an AS and an MS use to set up a Control Channel, the MS is able to discern the requests coming from AS1 from the requests coming from AS2. In fact, as explained in Sections 15.5.1 and 15.5.2, an AS and an MS negotiate a cfw-id attribute in the SDP, and the same value is subsequently used in the SYNC message on the Control Channel that is created after the negotiation, thus reassuring both the AS and the MS that the Control Channel they share is, in fact, the channel they negotiated in the first place.

Let us also assume that AS1 has created a conference mix (confid=74b6d62) to which it has attached some participants within the context of its business logic, while AS2 has created a currently active IVR dialog (dialogid=dfg3252) with a user agent it is handling (237430727:a338e95f). AS2 has also joined two connections to each other (1:75d4dd0d and 1:b9e6a659). Clearly, it is highly desirable that AS1 not be aware of what AS2 is doing with the MS, and vice versa, and that they not be allowed to manipulate each other's resources. The following transactions will occur:

1. AS1 places a generic audit request to both the Mixer and IVR packages.
2. AS2 places a generic audit request to both the Mixer and IVR packages.
3. AS1 tries to terminate the dialog created by AS2 (6791fee).
4. AS2 tries to join a user agent it handles (1:272e9c05) to the conference mix created by AS1 (74b6d62).

A sequence diagram of the above-mentioned transactions is depicted in Figure 15.59, which shows how the MS is assumed to reply in all cases in order to avoid security issues:

The expected outcome of the transaction is that the MS partially "lies" to both AS1 and AS2 when replying to the audit requests (not all of the identifiers are reported, but only those identifiers with which each AS is directly involved), and the MS denies the requests for the unauthorized operations (403). Looking at each transaction separately:

- In the first transaction (A1), AS1 places a generic <audit> request to the IVR package. The request is generic, since no attributes are passed as part of the request, meaning that AS1 is interested in the MS capabilities as well as all the dialogs that the MS is currently handling. As can be seen in the reply (A2), the MS only reports in the <auditresponse> the package capabilities, while the <dialogs> element is empty; this is because the only dialog the MS is handling has actually been created by AS2, which causes the MS not to report the related identifier (6791fee) to AS1. In fact, AS1 could use that identifier to manipulate the dialog, e.g., by tearing it down and thus causing the service to be interrupted without AS2's intervention.
- In the second transaction (B1), AS1 places an identical <audit> request to the Mixer package. The request is again generic, meaning that AS1 is interested in the package capabilities as well as all the mixers and connections that the package is handling at the moment. This time, the MS reports not only capabilities (B2) but information about mixers and connections as well. However, this information is not complete; in fact, only information about mixers and connections originated by AS1 is reported (mixer 74b6d62 and its participants), while the information originated by AS2 is omitted in the report. The motivation is the same as before.

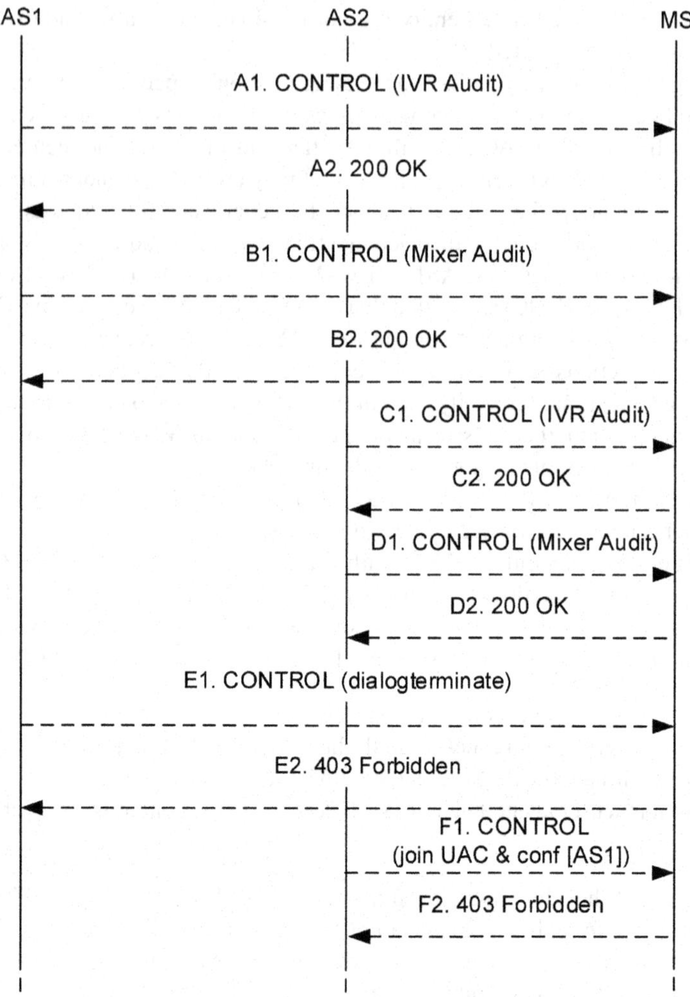

FIGURE 15.59 Security Considerations: Framework Transaction. (Copyright: IETF.)

- In the third and fourth transactions (C1 and D1), it is AS2 that places an <audit> request to both the IVR and Mixer packages. As with the previous transactions, the audit requests are generic. Looking at the replies (C2 and D2), it is obvious that the capabilities section is identical to the replies given to AS1. In fact, the MS has no reason to "lie" about what it can do. The <dialogs> and <mixers> sections are totally different. AS2, in fact, receives information about its own IVR dialog (6791fee), which was omitted in the reply to AS1, while it only receives information about the only connection it created (1:75d4dd0d and 1:b9e6a659) without any details related to the mixers and connections originated by AS1.
- In the fifth transaction (E1), AS1, instead of just auditing the packages, tries to terminate (<dialogterminate>) the dialog created by AS2 (6791fee). Since the identifier has not been reported by the MS in the reply to the previous audit request, we assume that AS1 accessed it via a different out-of-band mechanism. This is assumed to be an unauthorized operation because the above-mentioned dialog is outside the bounds of AS1; therefore, the MS, instead of handling the syntactically correct request, replies (E2) with a framework-level 403 message (Forbidden), leaving the dialog untouched.
- Similarly, in the sixth and last transaction (F1), AS2 tries to attach (<join>) one of the UACs it is handling to the conference mix created by AS1 (74b6d62). Just as in the previous transaction, the identifier is assumed to have been accessed by AS2 via some out-of-band

mechanism, since the MS did not report it in the reply to the previous audit request. While one of the identifiers (the UAC) is actually handled by AS2, the other (the conference mix) is not; therefore, as with the fifth transaction, this last transaction is regarded by the MS as outside the bounds of AS2. For the same reason as before, the MS replies (F2) with a framework-level 403 message (Forbidden), leaving the mix and the UAC unjoined.

```
A1. AS1 -> MS (CFW CONTROL, audit IVR)
-------------------------------------------------------
CFW 140e0f763352 CONTROL
Control-Package: msc-ivr/1.0
Content-Type: application/msc-ivr+xml
Content-Length: 81
<mscivr version="1.0" xmlns="urn:ietf:params:xml:ns:msc-ivr">
   <audit/>
</mscivr>

A2. AS1 <- MS (CFW 200, auditresponse)
-------------------------------------
CFW 140e0f763352 200
Timeout: 10
Content-Type: application/msc-ivr+xml
Content-Length: 1419
<mscivr version="1.0" xmlns="urn:ietf:params:xml:ns:msc-ivr">
   <auditresponse status="200">
      <capabilities>
         <dialoglanguages/>
         <grammartypes/>
         <recordtypes>
            <mimetype>audio/x-wav</mimetype>
            <mimetype>video/mpeg</mimetype>
         </recordtypes>
         <prompttypes>
            <mimetype>audio/x-wav</mimetype>
            <mimetype>video/mpeg</mimetype>
         </prompttypes>
         <variables>
            <variabletype type="date"
                  desc="value formatted as YYYY-MM-DD">
               <format desc="month day year">mdy</format>
               <format desc="year month day">ymd</format>
               <format desc="day month year">dmy</format>
               <format desc="day month">dm</format>
            </variabletype>
            <variabletype type="time" desc="value formatted as HH:MM">
               <format desc="24 hour format">t24</format>
               <format desc="12 hour format with am/pm">t12</format>
            </variabletype>
            <variabletype type="digits" desc="value formatted as D+">
               <format desc="general digit string">gen</format>
               <format desc="cardinal">crn</format>
               <format desc="ordinal">ord</format>
            </variabletype>
         </variables>
         <maxprepareduration>60s</maxprepareduration>
         <maxrecordduration>1800s</maxrecordduration>
         <codecs>
            <codec name="audio"><subtype>basic</subtype></codec>
```

```
            <codec name="audio"><subtype>gsm</subtype></codec>
            <codec name="video"><subtype>h261</subtype></codec>
            <codec name="video"><subtype>h263</subtype></codec>
            <codec name="video"><subtype>h263-
                  1998</subtype></codec>
            <codec name="video"><subtype>h264</subtype></codec>
         </codecs>
      </capabilities>
      <dialogs>
      </dialogs>
   </auditresponse>
</mscivr>

B1. AS1 -> MS (CFW CONTROL, audit mixer)
----------------------------------------
CFW 0216231b1f16 CONTROL
Control-Package: msc-mixer/1.0
Content-Type: application/msc-mixer+xml
Content-Length: 87
<mscmixer version="1.0" xmlns="urn:ietf:params:xml:ns:msc-mixer">
   <audit/>
</mscmixer>

B2. AS1 <- MS (CFW 200, auditresponse)
--------------------------------------
CFW 0216231b1f16 200
Timeout: 10
Content-Type: application/msc-mixer+xml
Content-Length: 903
<mscmixer version="1.0" xmlns="urn:ietf:params:xml:ns:msc-mixer">
   <auditresponse status="200">
      <capabilities>
         <codecs>
            <codec name="audio"><subtype>basic</subtype></codec>
            <codec name="audio"><subtype>gsm</subtype></codec>
            <codec name="video"><subtype>h261</subtype></codec>
            <codec name="video"><subtype>h263</subtype></codec>
            <codec name="video"><subtype>h263-1998</subtype></codec>
            <codec name="video"><subtype>h264</subtype></codec>
         </codecs>
      </capabilities>
      <mixers>
         <conferenceaudit conferenceid="74b6d62">
         <participants>
            <participant id="1864574426:e2192766"/>
            <participant id="1:5a97fd79"/>
         </participants>
         <video-layout min-participants="1">
            <quad-view/>
         </video-layout>
         </conferenceaudit>
         <joinaudit id1="1864574426:e2192766" id2="74b6d62"/>
         <joinaudit id1="1:5a97fd79" id2="74b6d62"/>
      </mixers>
   </auditresponse>
</mscmixer>
```

```
C1. AS2 -> MS (CFW CONTROL, audit IVR)
-------------------------------------
CFW 0216231b1f16 CONTROL
Control-Package: msc-ivr/1.0
Content-Type: application/msc-ivr+xml
Content-Length: 81
<mscivr version="1.0" xmlns="urn:ietf:params:xml:ns:msc-ivr">
   <audit/>
</mscivr>

C2. AS2 <- MS (CFW 200, auditresponse)
-------------------------------------
CFW 0216231b1f16 200
Timeout: 10
Content-Type: application/msc-ivr+xml
Content-Length: 1502
<mscivr version="1.0" xmlns="urn:ietf:params:xml:ns:msc-ivr">
   <auditresponse status="200">
      <capabilities>
      <dialoglanguages/>
      <grammartypes/>
      <recordtypes>
         <mimetype>audio/wav</mimetype>
         <mimetype>video/mpeg</mimetype>
      </recordtypes>
      <prompttypes>
         <mimetype>audio/wav</mimetype>
         <mimetype>video/mpeg</mimetype>
      </prompttypes>
      <variables>
         <variabletype type="date"
               desc="value formatted as YYYY-MM-DD">
            <format desc="month day year">mdy</format>
            <format desc="year month day">ymd</format>
            <format desc="day month year">dmy</format>
            <format desc="day month">dm</format>
         </variabletype>
         <variabletype type="time" desc="value formatted as HH:MM">
            <format desc="24 hour format">t24</format>
            <format desc="12 hour format with am/pm">t12</format>
         </variabletype>
         <variabletype type="digits" desc="value formatted as D+">
            <format desc="general digit string">gen</format>
            <format desc="cardinal">crn</format>
            <format desc="ordinal">ord</format>
         </variabletype>
      </variables>
      <maxprepareduration>60s</maxprepareduration>
      <maxrecordduration>1800s</maxrecordduration>
      <codecs>
         <codec name="audio"><subtype>basic</subtype></codec>
         <codec name="audio"><subtype>gsm</subtype></codec>
         <codec name="video"><subtype>h261</subtype></codec>
         <codec name="video"><subtype>h263</subtype></codec>
         <codec name="video"><subtype>h263-1998</subtype></codec>
         <codec name="video"><subtype>h264</subtype></codec>
```

```
        </codecs>
        </capabilities>
        <dialogs>
        <dialogaudit dialogid="6791fee" state="started"
              connectionid="237430727:a338e95f"/>
        </dialogs>
    </auditresponse>
</mscivr>

D1. AS2 -> MS (CFW CONTROL, audit mixer)
----------------------------------------
CFW 515f007c5bd0 CONTROL
Control-Package: msc-mixer/1.0
Content-Type: application/msc-mixer+xml
Content-Length: 87
<mscmixer version="1.0" xmlns="urn:ietf:params:xml:ns:msc-mixer">
    <audit/>
</mscmixer>

D2. AS2 <- MS (CFW 200, auditresponse)
--------------------------------------
CFW 515f007c5bd0 200
Timeout: 10
Content-Type: application/msc-mixer+xml
Content-Length: 548
<mscmixer version="1.0" xmlns="urn:ietf:params:xml:ns:msc-mixer">
    <auditresponse status="200">
        <capabilities>
           <codecs>
              <codec name="audio"><subtype>basic</subtype></codec>
              <codec name="audio"><subtype>gsm</subtype></codec>
              <codec name="video"><subtype>h261</subtype></codec>
              <codec name="video"><subtype>h263</subtype></codec>
              <codec name="video"><subtype>h263-1998</subtype></codec>
              <codec name="video"><subtype>h264</subtype></codec>
           </codecs>
        </capabilities>
        <mixers>
           <joinaudit id1="1:75d4dd0d" id2="1:b9e6a659"/>
        </mixers>
    </auditresponse>
</mscmixer>

E1. AS1 -> MS (CFW CONTROL, dialogterminate)
--------------------------------------------
CFW 7fdcc2331bef CONTROL
Control-Package: msc-ivr/1.0
Content-Type: application/msc-ivr+xml
Content-Length: 127
<mscivr version="1.0" xmlns="urn:ietf:params:xml:ns:msc-ivr">
    <dialogterminate dialogid="6791fee" immediate="true"/>
</mscivr>

E2. AS1 <- MS (CFW 403 Forbidden)
---------------------------------
CFW 7fdcc2331bef 403
```

```
F1. AS2 -> MS (CFW CONTROL, join to conference)
---------------------------------------------
CFW 140e0f763352 CONTROL
Control-Package: msc-mixer/1.0
Content-Type: application/msc-mixer+xml
Content-Length: 117
<mscmixer version="1.0" xmlns="urn:ietf:params:xml:ns:msc-mixer">
   <join id1="1:272e9c05" id2="74b6d62"/>
</mscmixer>

F2. AS2 <- MS (CFW 403 Forbidden)
--------------------------------
CFW 140e0f763352 403
```

15.9 SUMMARY

We have described the call flows for the XCON conference systems using the CFW-based MSs and their interactions with ASs through the interfaces specified throughout all the sections. The call flows described here cover only certain aspects of interactions between the AS and the MS based on implementation experiences. A number of interesting scenarios that can be realized through the careful orchestration of the CFW entities like AS and MS demonstrate how a variegated bouquet of services (ranging from very simple scenarios to much more complicated examples) can be implemented with the functionality currently offered within this framework by the Control Packages available to date, offering scalability using distributed MSs and media resources to build the large-scale multimedia conferencing services network. Many more powerful real-time multimedia conferencing services can be provided using XCON conference systems when more advanced intelligent Control Packages are developed using the CFW in the future.

We have described the state diagrams (Section 15.4) of the MS, AS, and notification service how MEDIACTRL (CFW) protocol messages interact between AS, MS, and notification server via APIs. The state diagrams are constructed using a Mealy automation machine comprising a number of possible protocol states. The output values of the Mealy finite-state machine are determined by both its current state and its current input. Inputs represent triggers arriving at the MEDIACTRL automaton from the upper layer, namely, the API used by programmers while implementing MEDIACTRL-enabled services. Such inputs have been indicated with the term "API" followed by the message that the API itself is triggering.

We have described the call flows (Section 15.5) showing the interaction between an AS and an MS for the Control Channel establishment. The channel is set up by means of a connection-oriented media (COMEDIA) (RFCs 4145 and 4572) negotiation using TLS/TCP/SCTP with the assumption that an unencrypted TCP connection is negotiated between the two involved entities. COMEDIA negotiations have also included port numbers of both AS and MS and additional attributes "cfw-id." The "cfw-id" attribute is the most important, since it specifies the Dialog-ID, which in turn is used by both the AS and the MS as specified in RFC 6230 (see Chapter 8). Once they have connected, in another call flows diagram, a SYNC message sent by the AS to the MS consolidates the Control Channel created over the TCP connection.

The media channel CFW connection needs to be kept alive, even when no traffic flows for some time, for the duration of the multimedia conference call. However, NATs and/or firewalls that may reside between the AS and MS path may disconnect TCP connects because of their timeout schemes. A separate call flows diagram shows an example of how a keep-alive (K-ALIVE) message is triggered, implementing RFC 6230 mechanisms that allow sending some dummy traffic for NATs/firewalls, preventing timeouts and thereby, keeping the TCP connection alive.

In implementations, some types of behavior could represent the most common mistakes when dealing with the establishment of a Control Channel between an AS and an MS. For completeness, we have also described a couple of common mistakes using call flows, such as the AS providing the MS with a wrong Dialog-ID in the initial SYNC and the AS sending a generic CONTROL message instead of SYNC as a first transaction.

We have described some chosen XCON conference scenarios (Section 15.6). At first, we have taken a 3pcc (RFC 3725) scenario where an UAC calls the controller to set up the multipoint (three or more parties) conferences. A controller being a black box sets up the calls among the multiple parties and facilitates negotiations among the parties. Once negotiations are completed, the controller offers media services among the multiple parties as well per RFC 3725 that defines no interfaces between AS and MS. Consequently, the large-scale multipoint conferences are not scalable using RFC 3725. However, RFC 6230 (see Chapter 8) defines standardized interfaces between AS and MS. If the same 3pcc multipoint (three or more parties) is offered using the media control CFW protocol per RFC 3260, a service provider can deploy AS and MS as two separate distinct entities manufactured by two different vendors. In this 3pcc scenario, the highly powerful negotiations capabilities allow one or media connections between the MS and the individual UACs triggered by the AS using the media control CFW (RFC 6230 – see Section 8), and hence making the XCON conference systems scalable through physical separation between the AS and MS using SIP-based CFW protocol.

We have then provided call flows for XCON conference scenarios because of their common presence in many rich real-time multimedia applications, as follows: Echo Test, Phone Call, Conferencing, and Additional Scenarios. In those call flows, we have assumed that the Control Channel has been established between the entities (e.g., AS and MS) using COMEDIA negotiations and TCP connection confirmation using SYNC messages, and K-ALIVE messages are sent, keeping the Control Channel connection up as described earlier. This part of the call flows has not been repeated in these conference scenarios.

The echo test (Section 16.5.1) is required because whether or not a UAC's leg is connected to the MS needs to be known from the media control CFW point of view. The simplest example of the echo test can be achieved by means of an MS. In Direct Echo Test, it basically consists of a UAC directly or indirectly "talking" to itself with direct connection between the UAC and MS. Another alternative is the Recording-based Echo Test, where the UAC indirectly talks to itself. The details of call flows for each scenario are described along with differentiations between them.

In the phone call scenario (Section 15.6.2), we have discussed the call flows for 3pcc (RFC 3725 – see Section 1.3) with CFW having two UACs, AS, and MS. It is only described in order to help the reader more easily understand what the requirements are on the MS side and as a consequence, what information might be required. RFC 3725 (see Section 1.3) provides a much more detailed overview of 3pcc patterns in several use cases. Only an explanatory sequence diagram is provided without delving into the details of the exchanged SIP messages. In addition, we have discussed three more scenarios using media control CFW: Direct Connection, Conference-Based Approach, and Recording a Conversation. We have discussed pros and cons for each of these scenarios. The Conference-Based Approach can address more advanced features like media mixing easily, while the direct connection cannot. The example of Recording a Conversation and subsequently playing it out in real time during the conference without a mixing entity involved in the call is straightforward for the Conference-Based Approach but cannot be done for the Direct Connection configuration.

We have described more resourceful XCON multimedia conference scenarios termed "Conferencing" (Section 15.6.3) for simplicity in distinguishing from all other scenarios. In conferencing scenarios, we have considered that one of the most important services the MS must be able to provide is media mixing or media bridging. The mixing of each medium (e.g., audio or video) is fundamentally different. Moreover, mixing involves each kind of media stream from different sources needing to be bridged, and then the mixed media stream needing to be delivered to each interested conference participant, often according to per-user policies, settings, and encoding. A typical scenario involving mixing is, of course, media conferencing. In such a scenario, the media sent by each participant is mixed, and each participant typically receives the overall mix, excluding its own contribution and encoded in the format it negotiated. This example points out in a quite clear way how mixing must take care of the profile of each involved entity. We have considered five categories of Conference according to variations of their capabilities/features: Simple Bridging, Rich Conference Scenario, Coaching Scenario, Sidebars, and Floor Control.

We defined these five example Conference categories involving mixing and conferencing as a whole as stated here:

- Simple Bridging: This is a very basic (that is, no "special effects" – just mixing involved) conference involving one or more participants.
- Rich Conference: This enriches the Simple Bridging scenario by adding additional features typically found in conferencing systems (e.g., DTMF collection for PIN-based conference access, private and global announcements, recordings, and so on).
- Coaching: This is a more complex scenario that involves per-user mixing (customers, agents, and coaches do not all get the same mixes).
- Sidebars: This adds more complexity to the previous conferencing scenarios by involving sidebars (that is, separate conference instances that only exist within the context of a parent conference instance) and the custom media delivery that follows.
- Floor Control: This provides some guidance on how floor control could be involved in a media control CFW-based media conference.

It can be seen that all the above-mentioned scenarios depend on the availability of a mixing entity that is provided by the media control CFW by the conferencing package. In addition, allowing for the interconnection of media sources as seen in the Direct Echo Test section, this package enables the creation of abstract connections that can be joined to multiple connections. These abstract connections, called conferences, mix the contribution of each attached connection and feed them accordingly. Interestingly, each of the above-mentioned scenarios starts more or less in the same way: by the creation of conference connections that need media bridging. We have described each of these XCON conference systems with detailed call flows.

We have described some "Additional" conferencing scenarios (Section 15.5.4) that also involve AS and MS communications using the media control CFW: Voice Mail, Current Time, and DTMF-driven Conference Manipulation. Each of these features can be added on top of many earlier scenarios of XCON conference systems as value-added services. In Voice Mail, the service may include variable announcements for digits and VCR controls. All these features of Voice Mail, although they can be implemented a wide variety of ways, are supported by the MS through the IVR package using CFW. In DTMF-Driven Conference Manipulation, the AS can use the MS to detect DTMF tones from conference participants and take actions on the conference accordingly. A richer version of this feature may include mute/unmute themselves in the conference, change the video layout in the conference, if allowed, kick abusive users out of the conference, and so on. The MS can support these capabilities through the IVR package using CFW. We have also provided detailed call flows for all these additional scenarios.

We have described the MRB interfaces (Section 15.7) for dealing with multiple ASs and multiple MSs with an *M*:*N* relationship that creates distributed MSs and media resources architecture for building the large-scale XCON conference systems network. The *M*:*N* relationship between *M* number of application servers and *N* numbers of MSs introduced in the media control CFW architecture (RFC 5567 – see Chapter 3) requires MRB interfaces for handling this topology. In fact, we have specified the MRB entity and the protocols needed to interact with it in Section 15.7.

An MRB is basically a locator that is aware of a pool of MSs and makes them available to the interested ASs according to their requirements, and two different MRB interfaces are defined: Publishing interface, which allows an MRB to subscribe for notifications at the MS it is handling (e.g., available and occupied resources, current state, and others) and Consumer interface, which allows an interested AS to query an MRB for an MS capable of fulfilling its requirements.

The MRB Consumer interface can operate primarily in three modes: Query Mode, Inline-Aware Mode, and Inline-Unaware Mode. In Query Mode, MRB operations can further be divided into two modes: (a) Inline-Aware Mode: CFW-Based Approach and (b) Inline-Aware Mode: Media Dialog–Based Approach. In addition, we have described how the MRB handles media dialogs for three different modes: Query and Inline-Aware Mode, Inline-Unaware Mode, and CFW Protocol Behavior Mode. The key problems are solved by the MRB in handling these media dialogs to ensure a 1:1 relationship with certainty. Dialogs are established by the MRB modes of operations in *M*:*N* communications environments of multiple application and MSs resolving all ambiguities. We have provided detailed call flows for each of those MRB modes of operations along with the pros and cons of different kinds of implementations.

Finally, we have considered the detailed security (Section 15.8) specifically for the media control CFW where media channels are created in *M*:*N* AS and MS communications environments for XCON conference systems. In ensuring security for RTP media control channels opened over TLS/TCP/IP (or SCTP in lieu of TCP) in many-many AS and MS communications environments, the cfw-id attribute that is negotiated through SDP plays an important role. The same cfw-id is also used in the SYNC message on the Control Channel that is created after the negotiation, thus reassuring both the AS and the MS that the Control Channel they share is, in fact, the channel they negotiated in the first place. The authorization of using MSs and media resources ensuring *M*:*N* relationship with AS and MS is critical because each AS's authorization is mapped into specific MS resources on a one-to-one basis eliminating any confusions for malicious attacks.

15.10 PROBLEMS

1. What are the scenarios for XCON conference systems considered in this section (Chapter 15)? Do you consider all those conference scenarios practical? Justify your answer.
2. What is the Mealy state machine? Describe the MS, AS, and event notification server CFW state diagram. How does the notification server help the AS MS in offering XCON conference services? Explain using detail call flows.
3. What is COMEDIA? Why do we need to use COMEDIA transport for the AS and MS communications? Describe the COMEDIA negotiations for RTP media streams over TLS/TCP/SCTP. How does the SYNC message help the AS and the MS? How is the cfw-id attribute used for authorization of MS resources by the SYNC message?
4. Why do NATs and firewalls (FWs) cause problems for MS media streams for XCON conference systems? How does the K-ALIVE message solve the NAT/FW problem? Describe in detail with technical justifications using call flows.
5. Illustrate some of the wrong behaviors when dealing with the establishment of a Control Channel between an AS and an MS. Describe in detail using call flows.
6. Why is the echo test needed? What are the differences between the Direct and the Recording-based Echo Test? Describe both kinds of echo test with detailed call flows and with pros and cons.
7. What are the differences in the architecture and call falls between the use case chosen in the 3pcc as a telephone scenario explained using the media channel CFW in RFC 7058 (Section 15.6) and the "corresponding" 3pcc scenario described in RFC 3725 (see Section 1.3)? Describe in detail.
8. What are the differences between different telephone call scenarios: Direct Connection, Conference-Based Approach, and Recording a Conversation? Describe the call flows of these scenarios in the context of the XCON conference systems including the focus and the media mixture. How does each of these call scenarios differ from those of the 3pcc described in RFC 3725 (see Section 1.3)?
9. Why is media mixing needed for XCON conference systems? What are the basic functions of media bridging? Describe in detail both audio and video mixing. How do the MS

control architecture and the media control CFW described in RFCs 5567 (see Chapter 3) and 6230 (see Chapter 8), respectively, complement each other for media mixing? Describe audio and video mixing for a three-party XCON conference scenario using functional call flows using media control CFW.

10. How do the Conferencing scenarios (Section 15.6.3) differ from the Phone Call scenarios (Section 15.6.2) and Echo Test Scenarios (Section 15.6.1)? Describe in detail with functional call flows.
11. What are the differences between the following Conference scenarios (Section 15.6.3): Simple Bridging, Rich Conference, Coaching, Sidebars, and Floor Control? What are the common capabilities that are needed in all of those five scenarios? Draw the XCON conference systems architecture for each case and describe the detailed call flows. Differentiate the functional features for each of those scenarios.
12. Draw the XCON conference systems architecture for each case and describe the detailed call flows: Voice Mail, Current Time, and DTMF-driven Conference Manipulation. Describe in detail how the IVR can be used with the MS controlled by the AS.
13. Why does the XCON conference system need multiple application servers and multiple media servers? How does the media control CFW provide solutions for communications in *M*:*N* many-to-many AS and MS communication environments? What are the functional features of the MRB? Draw the XCON conference systems architecture for each case and describe the detail with functional call flows.
14. Why does the MRB need these interfaces: Publishing Interface and Consumer Interface? Describe the detailed functionalities for each of those MRB interfaces. How do MS control architecture and the media control CFW described in RFCs 5567 (see Chapter 3) and 6230 (see Chapter 8), respectively, play a role in defining both MRB interfaces? Describe in detail.
15. Draw the MRB Publishing Interface for an XCON conference system architecture and describe in detail, including the call flows.
16. Draw each of the following MRB Consumer Interface Operation Modes for an XCON conference system architecture and describe in detail, including the call flows: Query Mode, Inline-Aware Mode, and Inline-Unaware Mode.
17. What are the differences in operations of the CFW- and Media Dialog–based Approach of the MRB Inline-Aware Mode of operations? Draw the functional architecture and the detailed call flows for each of the approaches for Inline-Aware Mode operations of the MRB Consumer Interface: Inline-Aware Mode – CFW-Based Approach and Inline-Aware Mode – Media Dialog–based Approach.
18. How does the MRB solve the problems of 1:1 mapping between the AS and the MS with certainty in view of the *M*:*N* many-many AS-MS communications environments for the following modes of operations through handling of Media Dialogs: Query and Inline-Aware Mode, Inline-Unaware Mode, and CFW Protocol Behavior Mode?
19. How is the cfw-id attribute used for XCON conference systems to ensure the security of the RTP media control channels opened over TLS/TCP/IP (or SCTP in lieu of TCP) in many-many application and MSs' communications environments? Describe in detail using call flows.
20. How is the authorization ensured in accessing the XCON conference systems' resources such that each AS's authorization is mapped into a specific MS's resources on a one-to-one basis, eliminating any confusions for malicious attacks?

REFERENCE

1. Hunt, A. and McGlashan S. (2004, March) Speech Recognition Grammar Specification Version 1.0. *W3C Recommendation.*

16 Multistream Immersive Telepresence Conferencing Systems

16.1 REQUIREMENTS FOR TELEPRESENCE MULTISTREAMS

This specification (RFC 7262) discusses the requirements for specifications that enable telepresence interoperability by describing behaviors and protocols for Controlling Multiple Streams for Telepresence (CLUE). In addition, the problem statement and related definitions are also covered herein.

16.1.1 Introduction

Telepresence systems greatly improve collaboration. In a telepresence conference (as used herein), the goal is to create an environment that gives the users a feeling of (co-located) presence – the feeling that a local user is in the same room with other local users and remote parties. Currently, systems from different vendors often do not interoperate because they do the same tasks differently, as discussed in the Problem Statement section (Section 16.1.4).

The approach taken in this chapter is to set requirements for a future specification(s) that when fulfilled by an implementation of the specification(s), provide for interoperability between Internet Engineering Task Force (IETF) protocol–based telepresence systems. It is anticipated that a solution for the requirements set out in this chapter likely involves the exchange of adequate information about participating sites; this information is currently not standardized by the IETF.

The purpose of this specification is to describe the requirements for a specification that enables interworking between different SIP-based (RFC 3261 – also see Section 1.2) telepresence systems by exchanging and negotiating appropriate information. In the context of the requirements in this specification and related solution specifications, this includes both point-to-point session initiation protocol (SIP) sessions as well as SIP-based conferences as described in the SIP conferencing framework (RFC 4353) and the SIP-based conference control (RFC 4579) specifications. Non-IETF protocol–based systems, such as those based on ITU-T Rec. H.323 [1], are out of scope. These requirements are for the specification; they are not requirements on the telepresence systems implementing the solution/protocol that will be specified.

Today, telepresence systems of different vendors can follow radically different architectural approaches while offering a similar user experience. CLUE will not dictate telepresence architectural and implementation choices; however, it will describe a protocol architecture for CLUE and how it relates to other protocols. CLUE enables interoperability between telepresence systems by exchanging information about the systems' characteristics. Systems can use this information to control their behavior to allow interoperability between those systems.

A telepresence session requires at least one sending and one receiving endpoint. Multiparty telepresence sessions include more than two endpoints and centralized infrastructure such as multipoint control units (MCUs) or equivalent. CLUE specifies the syntax, semantics, and control flow of information to enable the best possible user experience at those endpoints. Sending endpoints, or MCUs, are not mandated to use any of the CLUE specifications that describe their capabilities, attributes, or behavior. Similarly, it is not envisioned that endpoints or MCUs will ever have to take information received into account.

However, by making available as much information as possible, and by taking into account as much information as has been received or exchanged, MCUs and endpoints are expected to select operation modes that enable the best possible user experience under their constraints. The specification structure is as follows: Definitions are set out, followed by a description of the problem of telepresence interoperability that led to this work. Then, the requirements for a specification addressing the current shortcomings are enumerated and discussed.

16.1.2 Terminology

See Section 1.6, Table 1.2.

16.1.3 Definitions

The following terms (also see Section 1.6, Table 1.2) are used throughout this specification and serve as a reference for other specifications.

- **Audio Mixing**: Refers to the accumulation of scaled audio signals to produce a single audio stream. See "RTP Topologies" (RFC 5117).
- **Conference:** Used as defined in "A Framework for Conferencing within the Session Initiation Protocol (SIP)" (RFC 4353).
- **Endpoint:** The logical point of final termination through receiving, decoding and rendering, and/or initiation through capturing, encoding, and sending of media streams. An endpoint consists of one or more physical devices that source and sink media streams and exactly one participant (RFC 4353) (which, in turn, includes exactly one SIP user agent). In contrast to an endpoint, an MCU may also send and receive media streams, but it is not the initiator or the final terminator in the sense that media is captured or rendered. Endpoints can be anything from multiscreen/multicamera rooms to handheld devices. Endpoint Characteristics: Include placement of capture and rendering devices, capture/render angle, resolution of cameras and screens, spatial location, and mixing parameters of microphones. Endpoint characteristics are not specific to individual media streams sent by the endpoint.
- **Layout:** How rendered media streams are spatially arranged with respect to each other on a telepresence endpoint with a single screen and a single loudspeaker, and how rendered media streams are arranged with respect to each other on a telepresence endpoint with multiple screens or loudspeakers. Note that audio as well as video are encompassed by the term layout – in other words, included is the placement of audio streams on loudspeakers as well as video streams on video screens.
- **Local:** Sender and/or receiver physically co-located ("local") in the context of the discussion.
- **MCU:** Multipoint control unit – a device that connects two or more endpoints together into one single multimedia conference (RFC 5117). An MCU may include a mixer (RFC 4353).
- **Media:** Any data that after suitable encoding, can be conveyed over real-time transfer protocol (RTP), including audio, video, or timed text.
- **Model:** A set of assumptions a telepresence system of a given vendor adheres to and expects the remote telepresence system(s) to also adhere to.
- **Remote:** Sender and/or receiver on the other side of the communication channel (depending on context); i.e., not local. A remote can be an endpoint or an MCU.
- **Render:** The process of generating a representation from a media, such as displayed motion video or sound emitted from loudspeakers.

- **Telepresence:** An environment that gives non-co-located users or user groups a feeling of (co-located) presence – the feeling that a local user is in the same room with other local users and the remote parties. The inclusion of remote parties is achieved through multimedia communication including at least audio and video signals of high fidelity.

16.1.4 Problem Statement

In order to create a "being there" experience characteristic of telepresence, media inputs need to be transported, received, and coordinated between participating systems. Different telepresence systems take diverse approaches in crafting a solution, or they implement similar solutions quite differently. They use disparate techniques, and they describe, control, and negotiate media in dissimilar fashions. Such diversity creates an interoperability problem. The same issues are solved in different ways by different systems, so that they are not directly interoperable. This makes interworking difficult at best and sometimes impossible.

Worse, even if those extensions are based on common standards such as SIP, many telepresence systems use proprietary protocol extensions to solve telepresence-related problems. Some degree of interworking between systems from different vendors is possible through transcoding and translation. This requires additional devices, which are expensive, are often not entirely automatic, and sometimes introduce unwelcome side effects, such additional delay or degraded performance. Specialized knowledge is currently required to operate a telepresence conference with endpoints from different vendors, for example, to configure transcoding and translating devices. Often, such conferences do not start as planned or are interrupted by difficulties that arise.as

The general problem that needs to be solved can be described as follows. Today, each endpoint renders the audio and video captures it receives according to an implicitly assumed model that stipulates how to produce a realistic depiction of the remote location. If all endpoints are manufactured by the same vendor, they all share the same implicit model and render the received captures correctly. However, if the devices are from different vendors, the models used for rendering presence can, and usually do, differ. The result can be that the telepresence systems actually connect, but the user experience will suffer; for example, one system assumes that the first video stream is captured from the right camera, whereas the other assumes that the first video stream is captured from the left camera.

If Alice and Bob are at different sites, Alice needs to tell Bob about the camera and sound equipment arrangement at her site so that Bob's receiver can create an accurate rendering of her site. Alice and Bob need to agree on what the salient characteristics are as well as how to represent and communicate them. Characteristics may include number, placement, capture/render angle, resolution of cameras and screens, spatial location, and audio mixing parameters of microphones. The telepresence multistream work seeks to describe the sender situation in a way that allows the receiver to render it realistically even though it may have a different rendering model than the sender.

16.1.5 Requirements

Although some aspects of these requirements can be met by existing technology, such as the session description protocol (SDP) (RFC 4566), they are stated here to have a complete record of the requirements for CLUE. Determining whether a requirement needs new work or not will be part of the solution development and is not discussed in this specification. Note that the term "solution" is used in these requirements to mean the protocol specifications, including extensions to existing protocols as well as any new protocols, developed to support the use cases. The solution might introduce additional functionality that is not mapped directly to these requirements; e.g., the detailed information carried in the signaling protocol(s). In cases where the requirements are directly relevant to specific use cases as described in RFC 7205 (see Section 16.2), a reference to the use case is provided.

- REQ-1: The solution MUST support a description of the spatial arrangement of source video images sent in video streams that enables a satisfactory reproduction at the receiver of the original scene. This applies to each site in a point-to point or a multipoint meeting and refers to the spatial ordering within a site, not to the ordering of images between sites. This requirement relates to all the use cases described in RFC 7205 (see Section 16.2).
  - REQ-1a: The solution MUST support a means of allowing the preservation of the order of images in the captured scene. For example, if John is to Susan's right in the image capture, John is also to Susan's right in the rendered image.
  - REQ-1b: The solution MUST support a means of allowing the preservation of order of images in the scene in two dimensions – horizontal and vertical.
  - REQ-1c: The solution MUST support a means to identify the relative location, within a scene, of the point of capture of individual video captures in three dimensions.
  - REQ-1d: The solution MUST support a means to identify the area of coverage, within a scene, of individual video captures in three dimensions.
- REQ-2: The solution MUST support a description of the spatial arrangement of captured source audio sent in audio streams that enables a satisfactory reproduction at the receiver in a spatially correct manner. This applies to each site in a point-to-point or a multipoint meeting and refers to the spatial ordering within a site, not the ordering of channels between sites. This requirement relates to all the use cases described in RFC 7205 (see Section 16.2) but is particularly important in the Heterogeneous Systems use case.
  - REQ-2a: The solution MUST support a means of preserving the spatial order of audio in the captured scene. For example, if John sounds as if he is on Susan's right in the captured audio, John's voice is also placed on Susan's right in the rendered image.
  - REQ-2b: The solution MUST support a means to identify the number and spatial arrangement of audio channels, including monaural, stereophonic (2.0), and 3.0 (left, center, right) audio channels.
  - REQ-2c: The solution MUST support a means to identify the point of capture of individual audio captures in three dimensions.
  - REQ-2d: The solution MUST support a means to identify the area of coverage of individual audio captures in three dimensions.
- REQ-3: The solution MUST enable individual audio streams to be associated with one or more video image captures, and individual video image captures to be associated with one or more audio captures, for the purpose of rendering proper position. This requirement relates to all the use cases described in RFC 7205 (see Section 16.2).
- REQ-4: The solution MUST enable interoperability between endpoints that have a different number of similar devices. For example, an endpoint may have one screen, one loudspeaker, one camera, and one microphone, and another endpoint may have three screens, two loudspeakers, three cameras, and two microphones. Or in a multipoint conference, an endpoint may have one screen, another may have two screens, and a third may have three screens. This includes endpoints where the number of devices of a given type is zero. This requirement relates to the Point-to-Point Meeting: Symmetric and Multipoint Meeting use cases described in RFC 7205 (see Section 16.2).
- REQ-5: The solution MUST support means of enabling interoperability between telepresence endpoints where cameras are of different picture aspect ratios.
- REQ-6: The solution MUST provide scaling information that enables rendering of a video image at the actual size of the captured scene.
- REQ-7: The solution MUST support means of enabling interoperability between telepresence endpoints where displays are of different resolutions.
- REQ-8: The solution MUST support methods for handling different bit rates in the same conference.

- REQ-9: The solution MUST support means of enabling interoperability between endpoints that send and receive different numbers of media streams. This requirement relates to the Heterogeneous Systems and Multipoint Meeting use cases.
- REQ-10: The solution MUST ensure that endpoints that support telepresence extensions can establish a session with an SIP endpoint that does not support the telepresence extensions. For example, in the case of an SIP endpoint that supports a single audio and a single video stream, an endpoint that supports the telepresence extensions would set up a session with a single audio and a single video stream using existing SIP and SDP mechanisms.
- REQ-11: The solution MUST support a mechanism for determining whether or not an endpoint or MCU is capable of telepresence extensions.
- REQ-12: The solution MUST support a means to enable more than two endpoints to participate in a teleconference. This requirement relates to the Multipoint Meeting use case.
- REQ-13: The solution MUST support both transcoding and switching approaches for providing multipoint conferences.
- REQ-14: The solution MUST support mechanisms to allow media from one source endpoint or/and multiple source endpoints to be sent to a remote endpoint at a particular point in time. Which media is sent at a point in time may be based on local policy.
- REQ-15: The solution MUST provide mechanisms to support the following:
  - Presentations with different media sources
  - Presentations for which the media streams are visible to all endpoints
  - Multiple, simultaneous presentation media streams, including presentation media streams that are spatially related to each other. The requirement relates to the Presentation use case.
- REQ-16: The specification of any new protocols for the solution MUST provide extensibility mechanisms.
- REQ-17: The solution MUST support a mechanism for allowing information about media captures to change during a conference.
- REQ-18: The solution MUST provide a mechanism for the secure exchange of information about the media captures.

16.1.6 Security Considerations

REQ-18 identifies the need to securely transport the information about media captures. It is important to note that session setup for a telepresence session will use SIP for basic session setup and either SIP or the centralized conferencing manipulation protocol (CCMP) (RFC 6503 – see Chapter 5) for a multiparty telepresence session. Information carried in the SIP signaling can be secured by the SIP security mechanisms as defined in RFC 3261 (also see Section 1.2). In the case of conference control using CCMP, the security model and mechanisms as defined in the Centralized Conferencing (XCON) Framework (RFC 5239 – see Chapter 2) and CCMP (RFC 6503 – see Chapter 5) specifications would meet the requirement. Any additional signaling mechanism used to transport the information about media captures needs to define the mechanisms by which the information is secure. The detail of the mechanisms needs to be defined and described in the CLUE framework specification and related solution specification(s).

16.2 USE CASES FOR TELEPRESENCE MULTISTREAMS

Telepresence conferencing (RFC 7205) systems seek to create an environment that gives users (or user groups) that are not co-located a feeling of co-located presence through multimedia communication that includes at least audio and video signals of high fidelity. A number of techniques for handling audio and video streams are used to create this experience. When these techniques are not similar, interoperability between different systems is difficult at best and often not possible. Conveying information about the relationships between multiple streams of media would enable senders and receivers to make choices to allow telepresence systems to interwork. This chapter describes the most typical and important use cases for sending multiple streams in a telepresence conference.

16.2.1 Introduction

Telepresence applications (RFC 7205) try to provide a "being there" experience for conversational video conferencing. Often, this telepresence application is described as "immersive telepresence" in order to distinguish it from traditional video conferencing and from other forms of remote presence not related to conversational video conferencing, such as avatars and robots. The salient characteristics of telepresence are often described as: being actual sized, providing immersive video, preserving interpersonal interaction, and allowing non-verbal communication.

Although telepresence systems are based on open standards such as the RTP (RFC 3550), SIP (RFC 3261 – also see Section 1.2), H.264 [3], and H.323 [1] suite of protocols, they cannot easily interoperate with each other without operator assistance and expensive additional equipment that translates from one vendor's protocol to another. The basic features that give telepresence its distinctive characteristics are implemented in disparate ways in different systems. Currently, telepresence systems from diverse vendors interoperate to some extent, but this is not supported in a standards-based fashion. Interworking requires that translation and transcoding devices be included in the architecture. Such devices increase latency, reducing the quality of interpersonal interaction.

The use of these devices is often not automatic; it frequently requires substantial manual configuration and a detailed understanding of the nature of underlying audio and video streams. This state of affairs is not acceptable for the continued growth of telepresence – these systems should have the same ease of interoperability as do telephones. Thus, a standard way of describing the multiple streams constituting the media flows and the fundamental aspects of their behavior would allow telepresence systems to interwork.

This specification (RFC 7205) presents a set of use cases describing typical scenarios. Requirements will be derived from these use cases in a separate specification. The use cases are described from the viewpoint of the users. They are illustrative of the user experience that needs to be supported. It is possible to implement these use cases in a variety of different ways. Many different scenarios need to be supported. This specification describes in detail the most common and basic use cases. These will cover most of the requirements. There may be additional scenarios that bring new features and requirements that can be used to extend the initial work.

Point-to-point and multipoint telepresence conferences are considered. In some use cases, the number of screens is the same at all sites; in others, the number of screens differs at different sites. Both use cases are considered. Also included is a use case describing display of presentation material or content. The multipoint use cases may include a variety of systems from conference room systems to handheld devices, and such a use case is described in the specification. This specification's structure is as follows: Section 16.2.2 gives an overview of scenarios, and Section 16.2.3 describes use cases.

16.2.2 Overview of Telepresence Scenarios

This section describes the general characteristics of the use cases and what the scenarios are intended to show. The typical setting is a business conference, which was the initial focus of telepresence. Recently, consumer products have begun to be developed. We specifically do not include in our scenarios the physical infrastructure aspects of telepresence, such as room construction, layout, and decoration. Furthermore, these use cases do not describe all the aspects needed to create the best user experience (for example, the human factors).

We also specifically do not attempt to precisely define the boundaries between telepresence systems and other systems, nor do we attempt to identify the "best" solution for each presented scenario. Telepresence systems are typically composed of one or more video cameras and encoders and one or more display screens of large size (diagonal around 60 inches). Microphones pick up sound, and audio codec(s) produce one or more audio streams. The cameras used to capture the telepresence users are referred to as "participant cameras" (and likewise for screens). There may also be other cameras, such as for specification display. These will be referred to as "presentation cameras" or "content cameras," which generally have different formats, aspect ratios, and frame rates from the participant cameras. The presentation streams may be shown on participant screens or on auxiliary display screens. A user's computer may also serve as a virtual content camera, generating an animation or playing a video for display to the remote participants. We describe such a telepresence system as sending one or more video streams, audio streams, and presentation streams to the remote system(s).

The fundamental parameters describing today's typical telepresence scenarios include:

1. The number of participating sites
2. The number of visible seats at a site
3. The number of cameras
4. The number and type of microphones
5. The number of audio channels
6. The screen size
7. The screen capabilities – such as resolution, frame rate, and aspect ratio
8. The arrangement of the screens in relation to each other
9. The number of primary screens at each site
10. The type and number of presentation screens
11. Multipoint conference display strategies – for example, the camera-to-screen mappings may be static or dynamic
12. The camera point of capture
13. The cameras' fields of view and how they spatially relate to each other

As discussed in the Introduction, the basic features that give telepresence its distinctive characteristics are implemented in disparate ways in different systems. There is no agreed-upon way to adequately describe the semantics of how streams of various media types relate to each other. Without a standard for stream semantics to describe the particular roles and activities of each stream in the conference, interoperability is cumbersome at best. In a multiple-screen conference, the video and audio streams sent from remote participants must be understood by receivers so that they can be presented in a coherent and life-like manner. This includes the ability to present remote participants at their actual size for their apparent distance while maintaining correct eye contact and gesticular cues and simultaneously providing a spatial audio sound stage that is consistent with the displayed video.

The receiving device that decides how to render incoming information needs to understand a number of variables, such as the spatial position of the speaker, the field of view of the cameras, the camera zoom, which media stream is related to each of the screens, etc. It is not simply that individual streams must be adequately described – to a large extent this already exists – but rather, that the semantics of the relationships between the streams must be communicated. Note that all of this is still required even if the basic aspects of the streams, such as the bit rate, frame rate, and aspect ratio, are known. Thus, this problem has aspects considerably beyond those encountered in interoperation of video conferencing systems that have a single camera/screen.

16.2.3 Use Cases

The use cases focus on typical implementations. There are a number of possible variants for these use cases; for example, the audio supported may differ at the end points (such as mono or stereo versus surround sound), etc. Many of these systems offer a "full conference room" solution, where local participants sit at one side of a table and remote participants are displayed as if they are sitting on the other side of the table. The cameras and screens are typically arranged to provide a panoramic view of the remote room (left to right from the local user's viewpoint).

The sense of immersion and non-verbal communication is fostered by a number of technical features, such as:

1. Good eye contact, which is achieved by careful placement of participants, cameras, and screens.
2. Camera field of view and screen sizes are matched so that the images of the remote room appear to be full size.
3. The left side of each room is presented on the right screen at the far end; similarly, the right side of the room is presented on the left screen. The effect of this is that participants of each site appear to be sitting across the table from each other. If two participants on the same site glance at each other, all participants can observe it. Likewise, if a participant at one site gestures to a participant on the other site, all participants observe the gesture itself and the participants it includes.

16.2.3.1 Point-to-Point Meeting: Symmetric

In this case, each of the two sites has an identical number of screens, with cameras having fixed fields of view, and one camera for each screen. The sound type is the same at each end. As an example, there could be three cameras and three screens in each room, with stereo sound being sent and received at each end. Each screen is paired with a corresponding camera. Each camera/screen pair is typically connected to a separate codec, producing an encoded stream of video for transmission to the remote site and receiving a similarly encoded stream from the remote site.

Each system has one or multiple microphones for capturing audio. In some cases, stereophonic microphones are employed. In other systems, a microphone may be placed in front of each participant (or pair of participants). In typical systems, all the microphones are connected to a single codec that sends and receives the audio streams as either stereo or surround sound. The number of microphones and the number of audio channels are often not the same as the number of cameras. Also, the number of microphones is often not the same as the number of loudspeakers.

The audio may be transmitted as multi-channel (stereo/surround sound) or as distinct and separate monophonic streams. Audio levels should be matched so that the sound levels at both sites are identical. Loudspeaker and microphone placements are chosen so that the sound "stage" (orientation of apparent audio sources) is coordinated with the video. That is, if a participant at one site speaks, the participants at the remote site perceive her voice as originating from her visual image. In order to accomplish this, the audio needs to be mapped at the received site in the same fashion as the video. That is, audio received from the right side of the room needs to be output from loudspeaker(s) on the left side at the remote site, and vice versa.

16.2.3.2 Point-to-Point Meeting: Asymmetric

In this case, each site has a different number of screens and cameras than the other site. The important characteristic of this scenario is that the number of screens is different between the two sites. This creates challenges that are handled differently by different telepresence systems. This use case builds on the basic scenario of three screens to three screens. Here, we use the common case of three screens and three cameras at one site, and one screen and one camera at the other site, connected by a point-to-point call. The screen sizes and camera fields of view at both sites are basically similar, such that each camera view is designed to show two people sitting side by side. Thus, the one-screen room has up to two people seated at the table, while the three-screen room may have up to six people at the table.

The basic considerations of defining left and right and indicating relative placement of the multiple audio and video streams are the same as in the 3-3 use case. However, handling the mismatch between the two sites of the number of screens and cameras requires more complicated maneuvers. For the video sent from the one-camera room to the three-screen room, usually what is done is to simply use one of the three screens and keep the second and third screens inactive or for example, put up the current date. This would maintain the "full-size" image of the remote side. For the other direction, the three-camera room sending video to the one-screen room, there are more complicated variations to consider. Here are several possible ways in which the video streams can be handled.

1. The one-screen system might simply show only one of the three camera images, since the receiving side has only one screen. Two people are seen at full size, but four people are not seen at all. The choice of which one of the three streams to display could be fixed or could be selected by the users. It could also be made automatically, based on who is speaking in the three-screen room, such that the people in the one-screen room always see the person who is speaking. If the automatic selection is done at the sender, the transmission of streams that are not displayed could be suppressed, which would avoid wasting bandwidth.
2. The one-screen system might be capable of receiving and decoding all three streams from all three cameras. The one-screen system could then compose the three streams into one local image for display on the single screen. All six people would be seen but smaller than full size. This could be done in conjunction with reducing the image resolution of the streams, such that encode/decode resources and bandwidth are not wasted on streams that will be downsized for display anyway.
3. The three-screen system might be capable of including all six people in a single stream to send to the one-screen system. For example, it could use PTZ (Pan Tilt Zoom) cameras to physically adjust the cameras such that one camera captures the whole room of six people. Or, it could recompose the three camera images into one encoded stream to send to the remote site. These variations also show all six people but at a reduced size.
4. Or, there could be a combination of these approaches, such as simultaneously showing the speaker in full size with a composite of all six participants in a smaller size. The receiving telepresence system needs to have information about the content of the streams it receives to make any of these decisions. If the systems are capable of supporting more than one strategy, there needs to be some negotiation between the two sites to figure out which of the possible variations they will use in a specific point-to-point call.

16.2.3.3 Multipoint Meeting

In a multipoint telepresence conference, there are more than two sites participating. Additional complexity is required to enable media streams from each participant to show up on the screens of the other participants. Clearly, there are a great number of topologies that can be used to display the streams from multiple sites participating in a conference. One major objective for telepresence is to be able to preserve the "being there" user experience. However, in multi-site conferences, it is often (in fact, usually) not possible to simultaneously provide full-size video, eye contact, and common perception of gestures and gaze by all participants. Several policies can be used for stream distribution and display: All provide good results, but they all make different compromises.

One common policy is called site switching. Let us say the speaker is at site A, and the other participants are at various "remote" sites. When the room at site A shown, all the camera images from site A are forwarded to the remote sites. Therefore, at each receiving remote site, all the screens display camera images from site A. This can be used to preserve full-size image display and also provide full visual context of the displayed far end, site A. In site switching, there is a fixed relation between the cameras in each room and the screens in remote rooms. The room or participants being shown are switched from time to time based on who is speaking or by manual control, e.g., from site A to site B. Segment switching is another policy choice. In segment switching (assuming still that site A is where the speaker is, and "remote" refers to all the other sites), rather than sending all the images from site A, only the speaker at site A is shown. The camera images of the current speaker and previous speakers (if any) are forwarded to the other sites in the conference. Therefore, the screens in each site are usually displaying images from different remote sites – the current speaker at site A and the previous ones. This strategy can be used to preserve full-size image display and also capture the non-verbal communication between the speakers. In segment switching, the display depends on the activity in the remote rooms (generally, but not necessarily, based on audio/speech detection).

A third possibility is to reduce the image size so that multiple camera views can be composited onto one or more screens. This does not preserve full-size image display, but it provides the most visual context (since more sites or segments can be seen). Typically, in this case, the display mapping is static; i.e., each part of each room is shown in the same location on the display screens throughout the conference. Other policies and combinations are also possible. For example, there can be a static display of all screens from all remote rooms, with part or all of one screen being used to show the current speaker at full size.

16.2.3.4 Presentation

In addition to the video and audio streams showing the participants, additional streams are used for presentations. In systems available today, generally, only one additional video stream is available for presentations. Often, this presentation stream is half-duplex in nature with presenters taking turns. The presentation stream may be captured from a PC screen, or it may come from a multimedia source such as a specification camera, camcorder, or DVD. In a multipoint meeting, the presentation streams for the currently active presentation are always distributed to all sites in the meeting, so that the presentations are viewed by all.

Some systems display the presentation streams on a screen that is mounted either above or below the three participant screens. Other systems provide screens on the conference table for observing presentations. If multiple presentation screens are used, they generally display identical content. There is considerable variation in the placement, number, and size of presentation screens. In some systems, presentation audio is pre-mixed with the room audio. In others, a separate presentation audio stream is provided (if the presentation includes audio).

In H.323 [1] systems, H.239 [2] is typically used to control the video presentation stream. In SIP systems, similar control mechanisms can be provided using the binary floor control protocol (BFCP) (RFC 4582 – see Chapter 6) for the presentation token. These mechanisms are suitable for managing a single presentation stream. Although today's systems remain limited to a single video presentation stream, there are obvious uses for multiple presentation streams:

1. Frequently, the meeting convener is following a meeting agenda, and it is useful for her to be able to show that agenda to all participants during the meeting. Other participants at various remote sites are able to make presentations during the meeting with the presenters taking turns. The presentations and the agenda are both shown, either on separate screens or perhaps rescaled and shown on a single screen.
2. A single multimedia presentation can itself include multiple video streams that should be shown together. For instance, a presenter may be discussing the fairness of media coverage. In addition to slides that support the presenter's conclusions, she also has video excerpts from various news programs that she shows to illustrate her findings. She uses a DVD player for the video excerpts so that she can pause and reposition the video as needed.
3. An educator who is presenting a multiscreen slide show. This show requires that the placement of the images on the multiple screens at each site be consistent. There are many other examples where multiple presentation streams are useful.

16.2.3.5 Heterogeneous Systems

It is common in meeting scenarios for people to join the conference from a variety of environments, using different types of endpoint devices. A multiscreen immersive telepresence conference may include someone on a PC-based video conferencing system, a participant calling in by phone, and (soon) someone on a handheld device. What experience/view will each of these devices have? Some may be able to handle multiple streams, and others can handle only a single stream. (Here, we are not talking about legacy systems, but rather, systems built to participate in such a conference, although they are single stream only.) In a single video stream, the stream may contain one or more compositions depending on the available screen space on the device. In most cases, an intermediate transcoding device will be relied upon to produce a single stream, perhaps with some kind of continuous presence.

Bit rates will vary – the handheld device and phone having lower bit rates than PC and multiscreen systems. Layout is accomplished according to different policies. For example, a handheld device and a PC may receive the active speaker stream. The decision can either be made explicitly by the receiver or by the sender if it can receive some kind of rendering hint. The same is true for audio – i.e., that it receives a mixed stream or a number of the loudest speakers if mixing is not available in the network.

For the PC-based conferencing participant, the user's experience depends on the application. It could be single stream, similar to a handheld device but with a bigger screen. Or, it could be multiple streams, similar to an immersive telepresence system but with a smaller screen. Control for manipulation of streams can be local in the software application or in another location and sent to the application over the network.

The handheld device is the most extreme. How will that participant be viewed and heard? It should be an equal participant, though the bandwidth will be significantly less than in an immersive system. A receiver may choose to display output coming from a handheld device differently based on the resolution, but that would be the case with any low-resolution video stream, e.g., from a powerful PC on a bad network.

The handheld device will send and receive a single video stream, which could be a composite or a subset of the conference. The handheld device could say what it wants or could accept whatever the sender (conference server or sending endpoint) thinks is best. The handheld device will have to signal any actions it wants to take in the same way as an immersive system signals actions.

16.2.3.6 Multipoint Education Usage

The importance of this example is that the multiple video streams are not used to create an immersive conferencing experience with panoramic views at all the sites. Instead, the multiple streams are dynamically used to enable full participation of remote students in a university class. In some instances, the same video stream is displayed on multiple screens in the room; in other instances, an available stream is not displayed at all.

The main site is a university auditorium that is equipped with three cameras. One camera is focused on the professor at the podium. A second camera is mounted on the wall behind the professor and captures the class in its entirety. The third camera is co-located with the second and is designed to capture a close-up view of a questioner in the audience. It automatically zooms in on that student using sound localization.

Although the auditorium is equipped with three cameras, it is only equipped with two screens. One is a large screen located at the front so that the class can see it. The other is located at the rear so that the professor can see it. When someone asks a question, the front screen shows the questioner. Otherwise, it shows the professor (ensuring that everyone can easily see her). The remote sites are typical immersive telepresence rooms, each with three camera/screen pairs. All remote sites display the professor on the center screen at full size. A second screen shows the entire classroom view when the professor is speaking. However, when a student asks a question, the second screen shows the close-up view of the student at full size.

Sometimes, the student is in the auditorium; sometimes, the speaking student is at another remote site. The remote systems never display the students who are actually in that room. If someone at a remote site asks a question, then the screen in the auditorium will show the remote student at full size (as if they were present in the auditorium itself). The screen in the rear also shows this questioner, allowing the professor to see and respond to the student without needing to turn her back on the main class. When no one is asking a question, the screen in the rear briefly shows a full-room view of each remote site in turn, allowing the professor to monitor the entire class (remote and local students).

The professor can also use a control on the podium to see a particular site – she can choose either a full-room view or a single-camera view. Realization of this use case does not require any negotiation between the participating sites. Endpoint devices (and a MCU, if present) need to know who is speaking and which video stream includes the view of that speaker. The remote systems need some knowledge of which stream should be placed in the center. The ability of the professor to see specific sites (or of the system to show all the sites in turn) would also require the auditorium system to know what sites are available and to be able to request a particular view of any site. Bandwidth is optimized if video that is not being shown at a particular site is not distributed to that site.

16.2.3.7 Multipoint Multiview (Virtual Space)

This use case describes a virtual space multipoint meeting with good eye contact and spatial layout of participants. The use case was proposed very early in the development of video conferencing systems as described in 1983 by Allardyce and Randal [4]. The use case is illustrated in Figure 2-5 of their report [4]. The virtual space expands the point-to-point case by having all multipoint conference participants "seated" in a virtual room. In this case, each participant has a fixed "seat" in the virtual room, so each participant expects to see a different view having a different participant on his left and right side. Today, the use case is implemented in multiple telepresence-type video conferencing systems on the market. The term "virtual space" was used in their report.

The main difference between the results obtained with modern systems and those from 1983 is larger screen sizes. Virtual space multipoint as defined here assumes endpoints with multiple cameras and screens. Usually, there is the same number of cameras and screens at a given endpoint. A camera is positioned above each screen. A key aspect of virtual space multipoint is the details of how the cameras are aimed. The cameras are each aimed on the same area of view of the participants at the site. Thus, each camera takes a picture of the same set of people but from a different angle. Each endpoint sender in the virtual space multipoint meeting therefore offers a choice of video streams to remote receivers, each stream representing a different viewpoint.

For example, a camera positioned above a screen to a participant's left may take video pictures of the participant's left ear, while at the same time, a camera positioned above a screen to the participant's right may take video pictures of the participant's right ear. Since a sending endpoint has a camera associated with each screen, an association is made between the receiving stream output on a particular screen and the corresponding sending stream from the camera associated with that screen. These associations are repeated for each screen/camera pair in a meeting. The result of this system is a horizontal arrangement of video images from remote sites, one per screen. The image from each screen is paired with the camera output from the camera above that screen, resulting in excellent eye contact.

16.2.3.8 Multiple Presentation Streams – Telemedicine

This use case describes a scenario where multiple presentation streams are used. In this use case, the local site is a surgery room connected to one or more remote sites that may have different capabilities. At the local site, three main cameras capture the whole room (the typical three-camera telepresence case). Also, multiple presentation inputs are available: a surgery camera that is used to provide a zoomed view of the operation, an endoscopic monitor, a fluoroscope (X-ray imaging), an ultrasound diagnostic device, an electrocardiogram (ECG) monitor, etc. These devices are used to provide multiple local video presentation streams to help the surgeon monitor the status of the patient and assist in the surgical process.

The local site may have three main screens and one (or more) presentation screen(s). The main screens can be used to display the remote experts. The presentation screen(s) can be used to display multiple presentation streams from local and remote sites simultaneously. The three main cameras capture different parts of the surgery room. The surgeon can decide the number, the size, and the placement of the presentations displayed on the local presentation screen(s). He can also indicate which local presentation captures are provided for the remote sites. The local site can send multiple presentation captures to remote sites, and it can receive from them multiple presentations related to the patient or the procedure.

One type of remote site is a single- or dual-screen and one-camera system used by a consulting expert. In the general case, the remote sites can be part of a multipoint telepresence conference. The presentation screens at the remote sites allow the experts to see the details of the operation and related data. As at the main site, the experts can decide the number, the size, and the placement of the presentations displayed on the presentation screens. The presentation screens can display presentation streams from the surgery room, from other remote sites, or from local presentation streams. Thus, the experts can also start sending presentation streams that can carry medical records, pathology data, their references and analysis, etc.

Another type of remote site is a typical immersive telepresence room with three camera/screen pairs, allowing more experts to join the consultation. These sites can also be used for education. The teacher, who is not necessarily the surgeon, and the students are in different remote sites. Students can observe and learn the details of the whole procedure, while the teacher can explain and answer questions during the operation. All remote education sites can display the surgery room. Another option is to display the surgery room on the center screen, and the rest of the screens can show the teacher and the student who is asking a question. For all these sites, multiple presentation screens can be used to enhance visibility: one screen for the zoomed surgery stream and the others for medical image streams, such as magnetic resonance (MRI) images, cardiograms, ultrasonic images, and pathology data.

16.2.4 Security Considerations

While there are likely to be security considerations for any solution for telepresence interoperability, this specification has no security considerations.

16.3 FRAMEWORK FOR TELEPRESENCE MULTISTREAMS

This specification [13] defines a framework for a protocol to enable devices in a telepresence conference to interoperate. The protocol enables communication of information about multiple media streams so that a sending system and a receiving system can make reasonable decisions about transmitting, selecting, and rendering the media streams. This protocol is used in addition to SIP signaling and SDP negotiation for setting up a telepresence session.

16.3.1 Introduction

Current telepresence systems, though based on open standards such as RTP (RFC 3550) and SIP (RFC3261 – also see Section 1.2), cannot easily interoperate with each other. A major factor limiting the interoperability of telepresence systems is the lack of a standardized way to describe and negotiate the use of multiple audio and video streams comprising the media flows. This specification provides a framework for protocols to enable interoperability by handling multiple streams in a standardized way. The framework is intended to support the use cases described in Use Cases for Telepresence Multistreams (RFC 7205 – see Section 16.2) and to meet the requirements in Requirements for Telepresence Multistreams (RFC 7262 – see Section 16.1). This includes cases using multiple media streams that are not necessarily telepresence.

This specification occasionally refers to the term "CLUE" in capital letters. CLUE is an acronym for "ControLling mUltiple streams for tElepresence," which is the name of the IETF working group in which this specification and certain companion specifications have been developed. Often, CLUE-something refers to something that has been designed by the CLUE working group; for example, this specification may be called the CLUE framework.

The basic session setup for the use cases is based on SIP (RFC 3261 – also see Section 1.2) and SDP offer/answer (RFC 3264). In addition to basic SIP and SDP offer/answer, CLUE-specific signaling is required to exchange the information describing the multiple media streams. The motivation for this framework, an overview of the signaling, and information required to be exchanged is described in subsequent sections of this specification. Companion specifications describe the signaling details (see Section 16.8) and the data model (see Section 16.4) and protocol (see Section 16.5).

16.3.2 Terminology

See Section 1.6, Table 1.2.

16.3.3 Definitions

The terms (also see Section 1.6, Table 1.2) defined below are used throughout this specification and companion specifications. In order to easily identify the use of a defined term, those terms are capitalized.

- **Advertisement:** A CLUE message a Media Provider sends to a Media Consumer describing specific aspects of the content of the media, and any restrictions it has in terms of being able to provide certain Streams simultaneously.
- **Audio Capture:** Media Capture for audio. Denoted as ACn in the examples in this specification.
- **Capture:** Same as Media Capture.
- **Capture Device:** A device that converts physical input, such as audio, video, or text, into an electrical signal, in most cases to be fed into a media encoder.
- **Capture Encoding:** A specific encoding of a Media Capture, to be sent by a Media Provider to a Media Consumer via RTP.
- **Capture Scene:** A structure representing a spatial region captured by one or more Capture Devices, each capturing media representing a portion of the region. The spatial region represented by a Capture Scene may correspond to a real region in physical space, such as a room. A Capture Scene includes attributes and one or more Capture Scene Views, with each view including one or more Media Captures.
- **Capture Scene View (CSV):** A list of Media Captures of the same media type that together form one way to represent the entire Capture Scene.
- **CLUE-capable device:** A device that supports the CLUE data channel (see Section 16.6), the CLUE protocol (see Section 16.5) and the principles of CLUE negotiation and seeks CLUE-enabled calls.
- **CLUE-enabled call:** A call in which two CLUE-capable devices have successfully negotiated support for a CLUE data channel in SDP (RFC 4566). A CLUE-enabled call is not necessarily immediately able to send CLUE-controlled media; negotiation of the data channel and of the CLUE protocol must complete first. Calls between two CLUE-capable

devices that have not yet successfully completed negotiation of support for the CLUE data channel in SDP are not considered CLUE-enabled.

- **Conference:** Used as defined in RFC 4353, A Framework for Conferencing within the Session Initiation Protocol (SIP).
- **Configure Message:** A CLUE message a Media Consumer sends to a Media Provider specifying which content and Media Streams it wants to receive, based on the information in a corresponding Advertisement message.
- **Consumer:** Short for Media Consumer.
- **Encoding:** Short for Individual Encoding.
- **Encoding Group:** A set of encoding parameters representing a total media encoding capability to be sub-divided across potentially multiple Individual Encodings.
- **Endpoint:** A CLUE-capable device that is the logical point of final termination through receiving, decoding and rendering, and/or initiation through capturing, encoding, and sending of media streams. An endpoint consists of one or more physical devices that source and sink media streams and exactly one (RFC 4353) Participant (which, in turn, includes exactly one SIP user agent). Endpoints can be anything from multiscreen/multicamera rooms to handheld devices.
- **Global View:** A set of references to one or more Capture Scene Views of the same media type that are defined within Scenes of the same advertisement. A Global View is a suggestion from the Provider to the Consumer for one set of CSVs that provide a useful representation of all the scenes in the advertisement.
- **Global View List:** A list of Global Views included in an Advertisement. A Global View List may include Global Views of different media types.
- **Individual Encoding:** A set of parameters representing a way to encode a Media Capture to become a Capture Encoding.
- **Multipoint Control Unit (MCU):** A CLUE-capable device that connects two or more endpoints together into one single multimedia conference (RFC 5117). An MCU includes an RFC 4353-like Mixer without the RFC 4353 requirement to send media to each participant.
- **Media:** Any data that after suitable encoding, can be conveyed over RTP, including audio, video, or timed text.
- **Media Capture:** A source of Media, such as from one or more Capture Devices or constructed from other Media streams.
- **Media Consumer:** A CLUE-capable device that is intended to receive Capture Encodings.
- **Media Provider:** A CLUE-capable device that is intended to send Capture Encodings.
- **Multiple Content Capture (MCC):** A Capture that mixes and/or switches other Captures of a single type (e.g., all audio or all video). Particular Media Captures may or may not be present in the resultant Capture Encoding, depending on time or space. Denoted as MCCn in the example cases in this specification.
- **Plane of Interest:** The spatial plane within a scene containing the most relevant subject matter.
- **Provider:** Same as Media Provider.
- **Render:** The process of generating a representation from media, such as displayed motion video or sound emitted from loudspeakers.
- **Scene:** Same as Capture Scene.
- **Simultaneous Transmission Set:** A set of Media Captures that can be transmitted simultaneously from a Media Provider.
- **Single Media Capture:** A capture that contains media from a single source capture device, e.g., an audio capture from a single microphone or a video capture from a single camera.
- **Spatial Relation:** The arrangement in space of two objects, in contrast to relation in time or other relationships.

- **Stream:** A Capture Encoding sent from a Media Provider to a Media Consumer via RTP (RFC 3550).
- **Stream Characteristics:** The media stream attributes commonly used in non-CLUE SIP/SDP environments (such as media codec, bit rate, resolution, profile/level etc.) as well as CLUE-specific attributes, such as the Capture ID or a spatial location.
- **Video Capture:** Media Capture for video. Denoted as VCn in the example cases in this specification.
- **Video Composite:** A single image that is formed, normally by an RTP mixer inside an MCU, by combining visual elements from separate sources.

16.3.4 Overview and Motivation

This section provides an overview of the functional elements defined in this specification to represent a telepresence or multistream system. The motivations for the framework described in this specification are also provided. Two key concepts introduced in this specification are the terms "Media Provider" and "Media Consumer." A Media Provider represents the entity that sends the media, and a Media Consumer represents the entity that receives the media. A Media Provider provides Media in the form of RTP packets; a Media Consumer consumes those RTP packets. Media Providers and Media Consumers can reside in Endpoints or in MCUs. A Media Provider in an Endpoint is usually associated with the generation of media for Media Captures; these Media Captures are typically sourced from cameras, microphones, and the like.

Similarly, the Media Consumer in an Endpoint is usually associated with renderers, such as screens and loudspeakers. In MCUs, Media Providers and Consumers can have the form of outputs and inputs, respectively, of RTP mixers, RTP translators, and similar devices. Typically, telepresence devices such as Endpoints and MCUs would perform as both Media Providers and Media Consumers, the former being concerned with those devices' transmitted media and the latter with those devices' received media. In a few circumstances, a CLUE-capable device includes only Consumer or Provider functionality, such as recorder-type Consumers or webcam-type Providers.

The motivations for the framework outlined in this specification include the following:

1. Endpoints in telepresence systems typically have multiple Media Capture and Media Render devices, e.g., multiple cameras and screens. While previous system designs were able to set up calls that would capture media using all cameras and display media on all screens, for example, there was no mechanism that could associate these Media Captures with each other in space and time in a crossvendor interoperable way.
2. The mere fact that there are multiple capturing and rendering devices, each of which may be configurable in aspects such as zoom, leads to the difficulty that a variable number of such devices can be used to capture different aspects of a region. The Capture Scene concept allows the description of multiple setups for those multiple capture devices that could represent sensible operation points of the physical capture devices in a room chosen by the operator. A Consumer can pick and choose from those configurations based on its rendering abilities and inform the Provider about its choices. Details are provided in Section 16.3.7.
3. In some cases, physical limitations or other reasons disallow the concurrent use of a device in more than one setup. For example, the center camera in a typical three-camera conference room can set its zoom objective either to capture only the middle few seats or all seats of a room, but not both concurrently. The Simultaneous Transmission Set concept allows a Provider to signal such limitations. Simultaneous Transmission Sets are part of the Capture Scene description and are discussed in Section 16.3.8.
4. Often, the devices in a room do not have the computational complexity or connectivity to deal with multiple encoding options simultaneously, even if each of these options is

sensible in certain scenarios, and even if the simultaneous transmission is also sensible (i.e. in the case of multicast media distribution to multiple endpoints). Such constraints can be expressed by the Provider using the Encoding Group concept, described in Section 16.3.9.

5. Due to the potentially large number of RTP streams required for a Multimedia Conference involving potentially many Endpoints, each of which can have many Media Captures and media renderers, it has become common to multiplex multiple RTP streams onto the same multiplexing point and the associated shortcomings such as network address translator (NAT)/firewall traversal. The large number of possible permutations of sensible options a Media Provider can make available to a Media Consumer makes a mechanism desirable that allows it to narrow down the number of possible options that a SIP offer/answer exchange has to consider. Such information is made available using protocol mechanisms specified in this specification and companion specifications. The Media Provider and Media Consumer may use information in CLUE messages to reduce the complexity of SIP offer/answer messages. Also, there are aspects of the control of both Endpoints and MCUs that dynamically change during the progress of a call, such as audio level–based screen switching, layout changes, and so on, which need to be conveyed. Note that these control aspects are complementary to those specified in traditional SIP-based conference management such as BFCP. An exemplary call flow can be found in Chapter 5.

Finally, all this information needs to be conveyed, and the notion of support for it needs to be established. This is done by the negotiation of a "CLUE channel," a data channel negotiated early during the initiation of a call. An Endpoint or MCU that rejects the establishment of this data channel, by definition, does not support CLUE-based mechanisms, whereas an Endpoint or MCU that accepts it is indicating support for CLUE as specified in this specification and its companion specifications.

16.3.5 Description of the Framework/Model

The CLUE framework specifies how multiple media streams are to be handled in a telepresence conference. A Media Provider (transmitting Endpoint or MCU) describes specific aspects of the content of the media and the media stream encodings it can send in an Advertisement, and the Media Consumer responds to the Media Provider by specifying which content and media streams it wants to receive in a Configure message. The Provider then transmits the asked-for content in the specified streams. This Advertisement and Configure typically occur during call initiation, after CLUE has been enabled in a call, but MAY also happen at any time throughout the call, whenever there is a change in what the Consumer wants to receive or (perhaps less commonly) the Provider can send. An Endpoint or MCU typically acts as both Provider and Consumer at the same time, sending Advertisements and sending Configurations in response to receiving Advertisements. (It is possible to be just one or the other.)

The data model (Section 16.4) is based around two main concepts: a Capture and an Encoding. A Media Capture (MC), such as of audio or video type, has attributes to describe the content a Provider can send. Media Captures are described in terms of CLUE-defined attributes, such as spatial relationships and purpose of the capture. Providers tell Consumers which Media Captures they can provide, described in terms of the Media Capture attributes.

A Provider organizes its Media Captures into one or more Capture Scenes, each representing a spatial region, such as a room. A Consumer chooses which Media Captures it wants to receive from the Capture Scenes. In addition, the Provider can send the Consumer a description of the Individual Encodings it can send in terms of identifiers that relate to items in SDP (RFC 4566). The Provider can also specify constraints on its ability to provide Media, and a sensible design choice for a Consumer is to take these into account when choosing the content and Capture Encodings it requests in the later offer/answer exchange. Some constraints are due to the physical limitations of

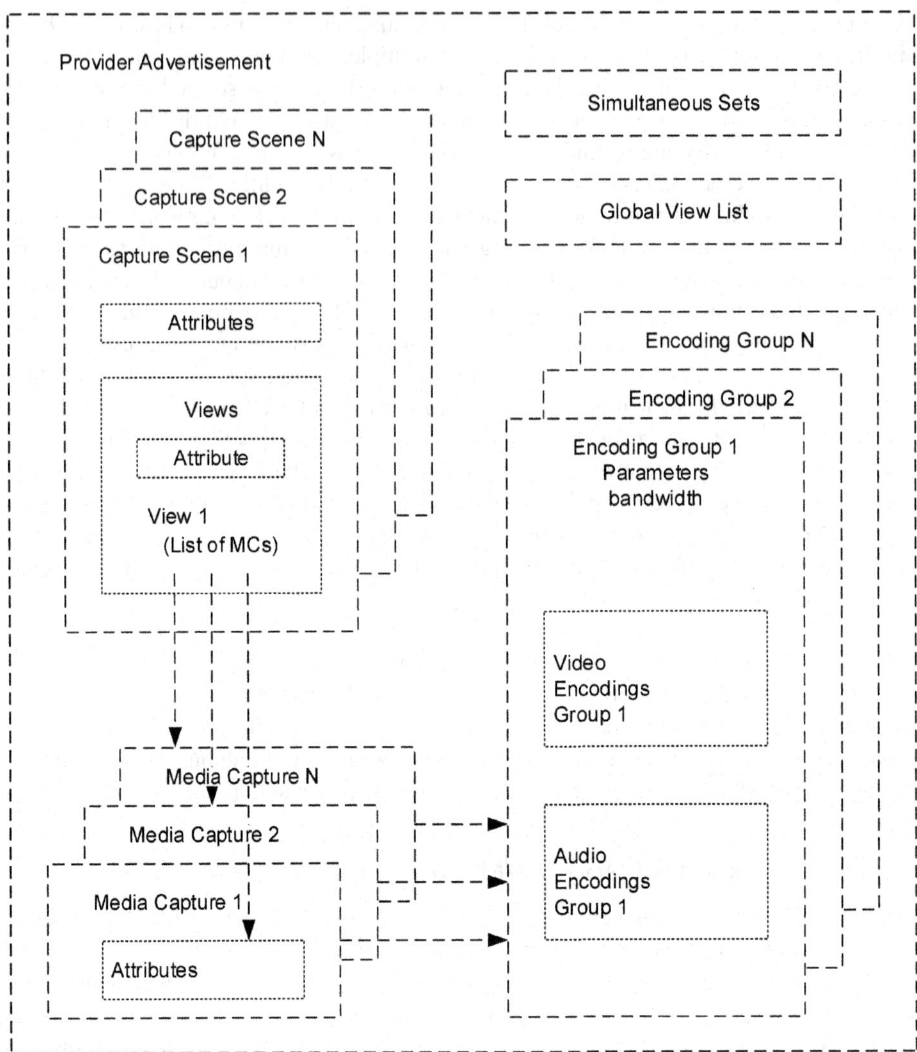

FIGURE 16.1 Advertisement Structure (Copyright: IETF)

devices – for example, a camera may not be able to provide zoom and non-zoom views simultaneously. Other constraints are system based, such as maximum bandwidth.

Figure 16.1 illustrates the information contained in an Advertisement.

A very brief outline of the call flow used by a simple system (two Endpoints) in compliance with this specification can be described as follows, and as shown in Figure 16.2.

An initial offer/answer exchange establishes a basic media session, for example audio-only, and a CLUE channel between two Endpoints. With the establishment of that channel, the endpoints have consented to use the CLUE protocol mechanisms and therefore, MUST adhere to the CLUE protocol suite as outlined herein. Over this CLUE channel, the Provider in each Endpoint conveys its characteristics and capabilities by sending an Advertisement as specified herein. The Advertisement is typically not sufficient to set up all media. The Consumer in the Endpoint receives the information provided by the Provider and can use it for several purposes. It uses it, along with information from an offer/answer exchange, to construct a CLUE Configure message to tell the Provider what the Consumer wishes to receive.

Also, the Consumer may use the information provided to tailor the SDP it is going to send during any following SIP offer/answer exchange and its reaction to SDP it receives in that step. It is

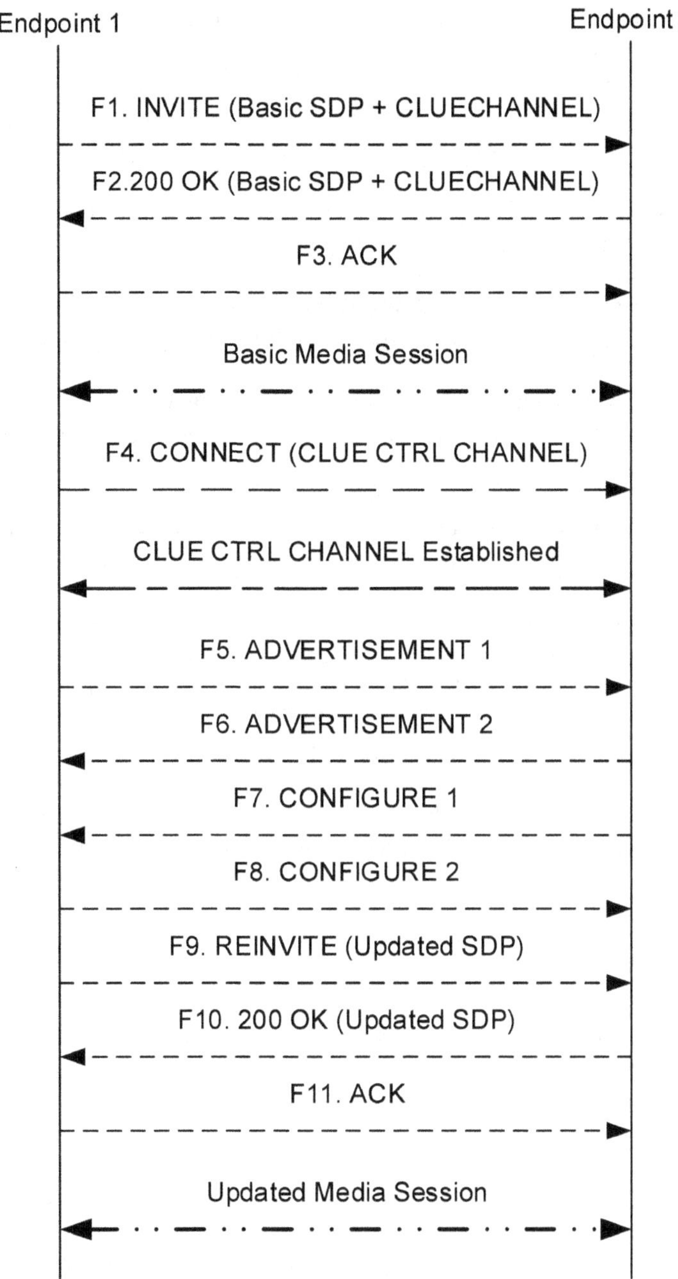

FIGURE 16.2 Basic Information Flow. (Copyright: IETF.)

often a sensible implementation choice to do so. Spatial relationships associated with the Media can be included in the Advertisement, and it is often sensible for the Media Consumer to take those spatial relationships into account when tailoring the SDP. The Consumer can also limit the number of encodings it must set up resources to receive and not waste resources on unwanted encodings, because it has the Provider's Advertisement information ahead of time to determine what it really wants to receive. The Consumer can also use the Advertisement information for local rendering decisions.

This initial CLUE exchange is followed by an SDP offer/answer exchange that not only establishes those aspects of the media that have not been "negotiated" over CLUE but also has the effect of setting up the media transmission itself, involving potentially security exchanges, Interactive Connectivity Establishment (ICE), and whatnot. This step is plain vanilla SIP.

During the lifetime of a call, further exchanges MAY occur over the CLUE channel. In some cases, those further exchanges lead to a modified system behavior of Provider or Consumer (or both) without any other protocol activity such as further offer/answer exchanges. For example, a Configure Message requesting the Provider to place a different Capture source into a Capture Encoding, signaled over the CLUE channel, ought not to lead to heavy-handed mechanisms like SIP re-invites. However, in other cases, after the CLUE negotiation, an additional offer/answer exchange becomes necessary. For example, if both sides decide to upgrade the call from a single screen to a multiscreen call, and more bandwidth is required for the additional video channels compared with what was previously negotiated using offer/answer, a new offer/answer exchange is required.

One aspect of the protocol outlined herein and specified in more detail in companion specifications is that it makes available to the Consumer information regarding the Provider's capabilities to deliver Media and attributes related to that Media, such as their spatial relationship. The operation of the renderer inside the Consumer is unspecified in that it can choose to ignore some information provided by the Provider and/or not render media streams available from the Provider (although the Consumer follows the CLUE protocol and therefore, gracefully receives and responds to the Provider's information using a Configure operation).

A CLUE-capable device interoperates with a device that does not support CLUE. The CLUE-capable device can determine, by the result of the initial offer/answer exchange, whether the other device supports and wishes to use CLUE. The specific mechanism for this is described in Section 16.8. If the other device does not use CLUE, then the CLUE-capable device falls back to behavior that does not require CLUE. As for the media, Provider and Consumer have an end-to-end communication relationship with respect to (RTP transported) media, and the mechanisms described herein and in companion specifications do not change the aspects of setting up those RTP flows and sessions. In other words, the RTP media sessions conform to the negotiated SDP whether or not CLUE is used.

16.3.6 Spatial Relationships

In order for a Consumer to perform a proper rendering, it is often necessary or at least helpful for the Consumer to have received spatial information about the streams it is receiving. CLUE defines a coordinate system that allows Media Providers to describe the spatial relationships of their Media Captures to enable proper scaling and spatially sensible rendering of their streams. The coordinate system is based on a few principles:

- Each Capture Scene has a distinct coordinate system, unrelated to the coordinate systems of other scenes.
- Simple systems that do not have multiple Media Captures to associate spatially need not use the coordinate model, although it can still be useful to provide an Area of Capture.
- Coordinates can be in real, physical units (millimeters), have an unknown scale, or have no physical scale. Systems that know their physical dimensions (for example, professionally installed Telepresence room systems) MUST provide those real-world measurements to enable the best user experience for advanced receiving systems that can utilize this information. Systems that do not know specific physical dimensions but still know relative distances MUST use "unknown scale." "No scale" is intended to be used only where Media Captures from different devices (with potentially different scales) will be forwarded alongside one another (e.g., in the case of an MCU).
    - "Millimeters" means that the scale is in millimeters.
    - "Unknown" means that the scale is not necessarily millimeters, but the scale is the same for every Capture in the Capture Scene.

    - "No Scale" means that the scale could be different for each capture – an MCU Provider that advertises two adjacent captures and picks sources (which can change quickly) from different endpoints might use this value; the scale could be different and changing for each capture. But the areas of capture still represent a spatial relation between captures.
- The coordinate system is right-handed Cartesian X, Y, Z with the origin at a spatial location of the Provider's choosing. The Provider MUST use the same coordinate system with the same scale and origin for all coordinates within the same Capture Scene.

The direction of increasing coordinate values is: X increases from left to right, from the point of view of an observer at the front of the room looking towards the back; Y increases from the front of the room to the back of the room; Z increases from low to high (i.e., floor to ceiling) Cameras in a scene typically point in the direction of increasing Y, from front to back. But there could be multiple cameras pointing in different directions. If the physical space does not have a well-defined front and back, the provider chooses any direction for X, Y, and Z that is consistent with right-handed coordinates.

16.3.7 Media Captures and Capture Scenes

This section describes how Providers can describe the content of media to Consumers.

16.3.7.1 Media Captures

Media Captures are the fundamental representations of streams that a device can transmit. What a Media Capture actually represents is flexible:

- It can represent the immediate output of a physical source (e.g., camera or microphone) or a "synthetic" source (e.g., laptop computer or DVD player).
- It can represent the output of an audio mixer or video composer.
- It can represent a concept such as "the loudest speaker."
- It can represent a conceptual position such as "the leftmost stream."

To identify and distinguish between multiple Capture instances, Captures have a unique identity: for instance, VC1, VC2 and AC1, AC2, where VC1 and VC2 refer to two different video captures, and AC1 and AC2 refer to two different audio captures.

Some key points about Media Captures:

- A Media Capture is of a single media type (e.g., audio or video).
- A Media Capture is defined in a Capture Scene and is given an Advertisement unique identity. The identity may be referenced outside the Capture Scene that defines it through a Multiple Content Capture (MCC).
- A Media Capture may be associated with one or more Capture Scene Views.
- A Media Capture has exactly one set of spatial information.
- A Media Capture can be the source of at most one Capture Encoding.

Each Media Capture can be associated with attributes to describe what it represents.

16.3.7.1.1 Media Capture Attributes

Media Capture Attributes describe information about the Captures. A Provider can use the Media Capture Attributes to describe the Captures for the benefit of the Consumer of the Advertisement message. All these attributes are optional. Media Capture Attributes include:

- Spatial information, such as point of capture, point on line of capture, and area of capture, all of which, in combination, define the capture field of, for example, a camera

- Other descriptive information to help the Consumer choose between captures (e.g., description, presentation, view, priority, language, person information, and type)

The following sub-sections define the Capture attributes.

16.3.7.1.1.1 Point of Capture The Point of Capture attribute is a field with a single Cartesian (X, Y, Z) point value that describes the spatial location of the capturing device (such as camera). For an Audio Capture with multiple microphones, the Point of Capture defines the nominal midpoint of the microphones.

16.3.7.1.1.2 Point on Line of Capture The Point on Line of Capture attribute is a field with a single Cartesian (X, Y, Z) point value that describes a position in space of a second point on the axis of the capturing device towards the direction it is pointing; the first point being the Point of Capture (see earlier). Together, the Point of Capture and Point on Line of Capture define the direction and axis of the capturing device, for example, the optical axis of a camera or the axis of a microphone. The Media Consumer can use this information to adjust how it renders the received media if it so chooses.

For an Audio Capture, the Media Consumer can use this information along with the Audio Capture Sensitivity Pattern to define a three-dimensional (3-D) volume of capture where sounds can be expected to be picked up by the microphone providing this specific audio capture. If the Consumer wants to associate an Audio Capture with a Video Capture, it can compare this volume with the area of capture for video media to provide a check on whether the audio capture is indeed spatially associated with the video capture. For example, a video area of capture that fails to intersect at all with the audio volume of capture, or is at such a long radial distance from the microphone point of capture that the audio level would be very low, would be inappropriate.

16.3.7.1.1.3 Area of Capture The area of capture is a field with a set of four (X, Y, Z) points as a value that describes the spatial location of what is being "captured." This attribute applies only to video captures, not other types of media. By comparing the area of capture for different Video Captures within the same Capture Scene, a Consumer can determine the spatial relationships between them and render them correctly.

The four points MUST be co-planar, forming a quadrilateral, which defines the Plane of Interest for the particular Media Capture. If the area of capture is not specified, it means that the Video Capture might be spatially related to other Captures in the same Scene, but there is no detailed information on the relationship. For a switched Capture that switches between different sections within a larger area, the area of capture MUST use coordinates for the larger potential area.

16.3.7.1.1.4 Mobility of Capture The Mobility of Capture attribute indicates whether or not the point of capture, line on point of capture, and area of capture values stay the same over time or are expected to change (potentially frequently). Possible values are static, dynamic, and highly dynamic. An example of "dynamic" is a camera mounted on a stand, which is occasionally hand-carried and placed at different positions in order to provide the best angle to capture a work task. A camera worn by a person who moves around the room is an example of "highly dynamic." In either case, the effect is that the capture point, capture axis, and area of capture change with time.

The capture point of a static Capture MUST NOT move for the life of the CLUE session. The capture point of dynamic Captures is categorized by a change in position followed by a reasonable period of stability – in the order of magnitude of minutes. Highly dynamic captures are categorized by a capture point that is constantly moving. If the "area of capture," "capture point," and "line of capture" attributes are included with dynamic or highly dynamic Captures, they indicate spatial information at the time of the Advertisement.

16.3.7.1.1.5 Audio Capture Sensitivity Pattern The Audio Capture Sensitivity Pattern attribute applies only to audio captures. This attribute gives information about the nominal sensitivity pattern of the microphone that is the source of the Capture. Possible values include patterns such as omni, shotgun, cardioid, and hyper-cardioid.

16.3.7.1.1.6 Description The Description attribute is a human-readable description (which could be in multiple languages) of the Capture.

16.3.7.1.1.7 Presentation The Presentation attribute indicates that the capture originates from a presentation device, that is, one that provides supplementary information to a conference through slides, video, still images, data, etc. Where more information is known about the capture, it MAY be expanded hierarchically to indicate the different types of presentation media, e.g., presentation. slides, presentation.image, etc.

Note: It is expected that a number of keywords will be defined that provide more detail on the type of presentation. Refer to Section 16.4 for how to extend the model.

16.3.7.1.1.8 View The View attribute is a field with enumerated values indicating what type of view the Capture relates to. The Consumer can use this information to help choose which Media Captures it wishes to receive. Possible values are:

- Room – Captures the entire scene
- Table – Captures the conference table with seated people
- Individual – Captures an individual person
- Lectern – Captures the region of the lectern including the presenter, for example, in a classroom-style conference room
- Audience – Captures a region showing the audience in a classroom-style conference room

16.3.7.1.1.9 Language The Language attribute indicates one or more languages used in the content of the Media Capture. Captures MAY be offered in different languages in the case of multilingual and/or accessible conferences. A Consumer can use this attribute to differentiate between them and pick the appropriate one.

Note that the Language attribute is defined and meaningful for both audio and video captures. In case of audio captures, the meaning is obvious. For a video capture, "Language" could, for example, be sign interpretation or text. The Language attribute is coded per RFC 5646.

16.3.7.1.1.10 Person Information The Person Information attribute allows a Provider to provide specific information regarding the people in a Capture (regardless of whether or not the capture has a Presentation attribute). The Provider may gather the information automatically or manually from a variety of sources; however, the xCard (RFC 6351) format is used to convey the information. This allows various information, such as Identification information (Section 6.2 of RFC 6350), Communication Information (Section 6.4 of RFC 6350), and Organizational information (Section 6.6 of RFC 6350), to be communicated. A Consumer may then automatically (i.e., via a policy) or manually select Captures based on information about who is in a Capture. It also allows a Consumer to render information regarding the people participating in the conference or to use it for further processing.

The Provider may supply a minimal set of information or a larger set of information. However it MUST be compliant to RFC 6350 and supply a "VERSION" and "FN" property. A Provider may supply multiple xCards per Capture of any KIND (Section 6.1.4 of RFC 6350). In order to keep CLUE messages compact, the Provider SHOULD use a uniform resource identifier (URI) to point to any LOGO, PHOTO, or SOUND contained in the xCARD rather than transmitting the LOGO, PHOTO, or SOUND data in a CLUE message.

16.3.7.1.1.11 Person Type The Person Type attribute indicates the type of people contained in the capture with respect to the meeting agenda (regardless of whether or not the capture has a Presentation attribute). As a capture may include multiple people, the attribute may contain multiple values. However, values MUST NOT be repeated within the attribute. An Advertiser associates the person type with an individual capture when it knows that a particular type is in the capture. If an Advertiser cannot link a particular type with some certainty to a capture, then it is not included. A Consumer, on reception of a capture with a person type attribute, knows with some certainly that the capture contains that person type. The capture may contain other person types, but the Advertiser has not been able to determine that this is the case.

The types of Captured people include:

- Chair – The person responsible for running the meeting according to the agenda.
- Vice-Chair – The person responsible for assisting the chair in running the meeting.
- Minute Taker – The person responsible for recording the minutes of the meeting.
- Attendee – The person has no particular responsibilities with respect to running the meeting.
- Observer – An Attendee without the right to influence the discussion.
- Presenter – The person is scheduled on the agenda to make a presentation in the meeting. Note: This is not related to any "active speaker" functionality.
- Translator – The person is providing some form of translation or commentary in the meeting.
- Timekeeper – The person is responsible for maintaining the meeting schedule.

Furthermore, the person type attribute may contain one or more strings allowing the Provider to indicate custom meeting-specific types.

16.3.7.1.1.12 Priority The Priority attribute indicates a relative priority between different Media Captures. The Provider sets this priority, and the Consumer MAY use the priority to help decide which Captures it wishes to receive. The "priority" attribute is an integer that indicates a relative priority between Captures. For example, it is possible to assign a priority between two presentation Captures that would allow a remote Endpoint to determine which presentation is more important.

Priority is assigned at the individual Capture level. It represents the Provider's view of the relative priority between Captures with a priority. The same priority number MAY be used across multiple Captures. It indicates that they are equally important. If no priority is assigned, no assumptions regarding relative importance of the Capture can be assumed.

16.3.7.1.1.13 Embedded Text The Embedded Text attribute indicates that a Capture provides embedded textual information. For example, the video Capture may contain speech to text information composed with the video image.

16.3.7.1.1.14 Related To The Related To attribute indicates that the Capture contains additional complementary information related to another Capture. The value indicates the identity of the other Capture to which this Capture is providing additional information. For example, a conference can utilize translators or facilitators that provide an additional audio stream (i.e., a translation or description or commentary of the conference). Where multiple captures are available, it may be advantageous for a Consumer to select a complementary Capture instead of or in addition to a Capture it relates to.

16.3.7.2 Multiple Content Capture

The MCC indicates that one or more Single Media Captures are multiplexed (temporally and/or spatially) or mixed in one Media Capture. Only one Capture type (i.e., audio, video, etc.) is allowed

TABLE 16.1
Multiple Content Capture Concept

Capture Scene #1	Attribute
VC1	{MC attributes}
VC2	{MC attributes}
VC3	{MC attributes}
MCC1 (VC1,VC2,VC3)	{MC and MCC attributes}
CSV (MCC1)	

in each MCC instance. The MCC may contain a reference to the Single Media Captures (which may have their own attributes) as well as attributes associated with the MCC itself. An MCC may also contain other MCCs. The MCC MAY reference Captures from within the Capture Scene that defines it or from other Capture Scenes. No ordering is implied by the order that Captures appear within a MCC.

A MCC MAY contain no references to other Captures to indicate that the MCC contains content from multiple sources, but no information regarding those sources is given. MCCs either contain the referenced Captures and no others or have no referenced captures and therefore, may contain any Capture. One or more MCCs may also be specified in a CSV. This allows an Advertiser to indicate that several MCC captures are used to represent a capture scene. Table 16.14 provides an example of this case. As outlined in Section 7.1, each instance of the MCC has its own Capture identity, i.e., MCC1. It allows all the individual captures contained in the MCC to be referenced by a single MCC identity.

Table 16.1 shows the use of an MCC.

This indicates that MCC1 is a single capture that contains the Captures VC1, VC2, and VC3 according to any MCC1 attributes.

16.3.7.2.1 Attributes

Media Capture Attributes may be associated with the MCC instance and the Single Media Captures that the MCC references. A Provider should avoid providing conflicting attribute values between the MCC and Single Media Captures. Where there is conflict, the attributes of the MCC override any that may be present in the individual Captures. A Provider MAY include as much or as little of the original source Capture information as it requires. There are MCC-specific attributes that MUST only be used with Multiple Content Captures. These are described in the following sections. The attributes described in Section 16.3.7.1.1 MAY also be used with MCCs.

The spatial-related attributes of an MCC indicate its area of capture and point of capture within the scene, just like any other media capture. The spatial information does not imply anything about how other captures are composed within an MCC. For example: A virtual scene could be constructed for the MCC capture with two Video Captures with a "MaxCaptures" attribute set to two and an "area of capture" attribute provided with an overall area. Each of the individual Captures could then also include an "area of capture" attribute with a subset of the overall area.

The Consumer would then know how each capture is related to others within the scene but not the relative position of the individual captures within the composed capture (Table 16.2).

The following sub-sections describe the MCC-only attributes.

16.3.7.2.1.1 Maximum Number of Captures within a MCC The Maximum Number of Captures MCC attribute indicates the maximum number of individual Captures that may appear in a Capture Encoding at a time. The actual number at any given time can be less than or equal to this

TABLE 16.2
Example of MCC and Single Media Capture Attributes

Capture Scene #1	Attribute
VC1	AreaofCapture = (0,0,0)(9,0,0) (0,0,9)(9,0,9)
VC2	AreaofCapture = (10,0,0)(19,0,0) (10,0,9)(19,0,9)
MCC1 (VC1,VC2)	MaxCaptures = 2
CSV (MCC1)	AreaofCapture = (0,0,0)(19,0,0) (0,0,9)(19,0,9)

maximum. It may be used to derive how the Single Media Captures within the MCC are composed/switched with regard to space and time.

A Provider can indicate that the number of Captures in a MCC Capture Encoding is equal "=" to the MaxCaptures value or that there may be any number of Captures up to and including "<=" the MaxCaptures value. This allows a Provider to distinguish between an MCC that purely represents a composition of sources and an MCC that represents switched or switched and composed sources. MaxCaptures may be set to one, so that only content related to one of the sources is shown in the MCC Capture Encoding at a time, or it may be set to any value up to the total number of Source Media Captures in the MCC.

The following bullets describe how the setting of MaxCapture versus the number of Captures in the MCC affects how sources appear in a Capture Encoding:

- When MaxCaptures is set to <= 1, and the number of Captures in the MCC is greater than 1 (or not specified) in the MCC, this is a switched case. Zero or one Captures may be switched into the Capture Encoding. Note: zero is allowed because of the "<=."
- hen MaxCaptures is set to = 1, and the number of Captures in the MCC is greater than 1 (or not specified) in the MCC, this is a switched case. Only one Capture source is contained in a Capture Encoding at a time.
- When MaxCaptures is set to <= N (with N > 1), and the number of Captures in the MCC is greater than N (or not specified), this is a switched and composed case. The Capture Encoding may contain purely switched sources (i.e., <=2 allows for one source on its own) or may contain composed and switched sources (i.e., a composition of two sources switched between the sources).
- When MaxCaptures is set to = N (with N > 1), and the number of Captures in the MCC is greater than N (or not specified), this is a switched and composed case. The Capture Encoding contains composed and switched sources (i.e., a composition of N sources switched between the sources). It is not possible to have a single source.
- When MaxCaptures is set to <= to the number of Captures in the MCC, this is a switched and composed case. The Capture Encoding may contain media switched between any number (up to the MaxCaptures) of composed sources.
- When MaxCaptures is set to = to the number of Captures in the MCC, this is a composed case. All the sources are composed into a single Capture Encoding.

If this attribute is not set, then as default, it is assumed that all source media capture content can appear concurrently in the Capture Encoding associated with the MCC.

For example: The use of MaxCaptures equal to 1 on a MCC with three Video Captures VC1, VC2, and VC3 would indicate that the Advertiser in the Capture Encoding would switch between VC1, VC2, and VC3, as there may be only a maximum of one Capture at a time.

TABLE 16.3
Example Policy MCC Attribute Usage

Capture Scene #1	Attribute
VC1	
VC2	
MCC1 (VC1,VC2)	Policy = SoundLevel:0
MCC2 (VC1,VC2)	MaxCaptures=1
CSV (MCC1, MCC2)	Policy = SoundLevel:0
	MaxCaptures=1

16.3.7.2.1.2 Policy The Policy MCC Attribute indicates the criteria that the Provider uses to determine when and/or where media content appears in the Capture Encoding related to the MCC. The attribute is in the form of a token that indicates the policy and an index representing an instance of the policy. The same index value can be used for multiple MCCs.

The tokens are:

- **SoundLevel**: This indicates that the content of the MCC is determined by a sound level detection algorithm. The loudest (active) speaker (or a previous speaker, depending on the index value) is contained in the MCC.
- **RoundRobin**: This indicates that the content of the MCC is determined by a time-based algorithm. For example: the Provider provides content from a particular source for a period of time and then provides content from another source, and so on.

An index is used to represent an instance in the policy setting. An index of 0 represents the most current instance of the policy, i.e., the active speaker; 1 represents the previous instance, i.e., the previous active speaker, and so on.

The following example (Table 16.3) shows a case where the Provider provides two media streams, one showing the active speaker and a second stream showing the previous speaker.

16.3.7.2.1.3 Synchronization Identity The Synchronization Identity MCC attribute indicates how the individual Captures in multiple MCC Captures are synchronized. To indicate that the Capture Encodings associated with MCCs contain Captures from the same source at the same time, a Provider should set the same Synchronization Identity on each of the concerned MCCs. It is the Provider that determines what the source for the Captures is, so a Provider can choose how to group together Single Media Captures into a combined "source" for the purpose of switching them together to keep them synchronized according to the SynchronizationID attribute. For example, when the Provider is in an MCU, it may determine that each separate CLUE Endpoint is a remote source of media. The Synchronization Identity may be used across media types, i.e., to synchronize audio- and video-related MCCs. Without this attribute, it is assumed that multiple MCCs may provide content from different sources at any particular point in time.

For example, see Table 16.4. This Advertisement would indicate that MCC1, MCC2, MCC3, and MCC4 make up a Capture Scene. There would be four Capture Encodings (one for each MCC). Because MCC1 and MCC2 have the same SynchronizationID, each Encoding from MCC1 and MCC2, respectively, would together have content from only Capture Scene 1 or only Capture Scene 2 or the combination of VC7 and VC8 at a particular point in time. In this case, the Provider has decided that the sources to be synchronized are Scene #1, Scene #2, and Scene #3 and #4 together. The Encoding from MCC3 will not be synchronized with MCC1 or MCC2. As MCC4 also has the same Synchronization Identity as MCC1 and MCC2, the content of the audio Encoding will be synchronized with the video content.

TABLE 16.4
Example Synchronization Identity MCC Attribute Usage

Capture Scene #1	Attribute
VC1	Description = Left
VC2	Description = Centre
VC3	Description = Right
AC1	Description = Room
CSV (VC1,VC2,VC3)	
CSV (AC1)	
Capture Scene #2	**Attribute**
VC4	Description = Left
VC5	Description = Centre
VC6	Description = Right
AC2	Description = Room
CSV (VC4,VC5,VC6)	
CSV (AC2)	
Capture Scene #3	**Attribute**
VC7	
AC3	
Capture Scene #4	**Attribute**
VC8	
AC4	
Capture Scene #5	**Attribute**
MCC1 (VC1,VC4,VC7)	SynchronizationID = 1
MCC2 (VC2,VC5,VC8)	MaxCaptures = 1
MCC3 (VC3,VC6)	SynchronizationID = 1
MCC4 (AC1, AC2, VC3, AC4)	MaxCaptures = 1
CSV (MCC1, MCC2, MCC3)	MaxCaptures = 1
CSV (MCC4)	SynchronizationID = 1
	MaxCaptures = 1

16.3.7.2.1.4 Allow Subset Choice The Allow Subset Choice MCC attribute is a Boolean value, indicating whether or not the Provider allows the Consumer to choose a specific subset of the Captures referenced by the MCC. If this attribute is true, and the MCC references other Captures, then the Consumer MAY select (in a Configure message) a specific subset of those Captures to be included in the MCC, and the Provider MUST then include only that subset. If this attribute is false, or the MCC does not reference other Captures, then the Consumer MUST NOT select a subset.

16.3.7.3 Capture Scene

In order for a Provider's individual Captures to be used effectively by a Consumer, the Provider organizes the Captures into one or more Capture Scenes, with the structure and contents of these Capture Scenes being sent from the Provider to the Consumer in the Advertisement. A Capture Scene is a structure representing a spatial region containing one or more Capture Devices, each capturing media representing a portion of the region.

A Capture Scene includes one or more CSV, with each CSV including one or more Media Captures of the same media type. There can also be Media Captures that are not included in a Capture Scene View. A Capture Scene represents, for example, the video image of a group of people seated next to each other, along with the sound of their voices, which could be represented by some number of VCs and ACs in the Capture Scene Views. An MCU can also describe in Capture Scenes what it constructs from media Streams it receives.

A Provider MAY advertise one or more Capture Scenes. What constitutes an entire Capture Scene is up to the Provider. A simple Provider might typically use one Capture Scene for participant media (live video from the room cameras) and another Capture Scene for a computer-generated presentation. In more complex systems, the use of additional Capture Scenes is also sensible. For example, a classroom may advertise two Capture Scenes involving live video, one including only the camera capturing the instructor (and associated audio) and the other including camera(s) capturing students (and associated audio).

A Capture Scene MAY (and typically will) include more than one type of media. For example, a Capture Scene can include several Capture Scene Views for Video Captures, and several Capture Scene Views for Audio Captures. A particular Capture MAY be included in more than one Capture Scene View.

A Provider MAY express spatial relationships between Captures that are included in the same Capture Scene. However, there is no spatial relationship between Media Captures from different Capture Scenes. In other words, Capture Scenes each use their own spatial measurement system, as outlined in Section 16.3.6.

A Provider arranges Captures in a Capture Scene to help the Consumer choose which captures it wants to render. The Capture Scene Views in a Capture Scene are different alternatives the Provider is suggesting for representing the Capture Scene. Each Capture Scene View is given an advertisement-unique identity. The order of Capture Scene Views within a Capture Scene has no significance. The Media Consumer can choose to receive all Media Captures from one Capture Scene View for each media type (e.g., audio and video), or it can pick and choose Media Captures regardless of how the Provider arranges them in Capture Scene Views. Different Capture Scene Views of the same media type are not necessarily mutually exclusive alternatives. Also note that the presence of multiple Capture Scene Views (with potentially multiple encoding options in each view) in a given Capture Scene does not necessarily imply that a Provider is able to serve all the associated media simultaneously (although the construction of such an over-rich Capture Scene is probably not sensible in many cases).

What a Provider can send simultaneously is determined through the Simultaneous Transmission Set mechanism, described in Chapter 8. Captures within the same Capture Scene View MUST be of the same media type – it is not possible to mix audio and video captures in the same Capture Scene View, for instance. The Provider MUST be capable of encoding and sending all Captures (that have an encoding group) in a single Capture Scene View simultaneously. The order of Captures within a Capture Scene View has no significance. A Consumer can decide to receive all the Captures in a single Capture Scene View, but a Consumer could also decide to receive just a subset of those captures. A Consumer can also decide to receive Captures from different Capture Scene Views, all subject to the constraints set by Simultaneous Transmission Sets, as discussed in Chapter 8. When a Provider advertises a Capture Scene with multiple CSVs, it is essentially signaling that there are multiple representations of the same Capture Scene available. In some cases, these multiple views would be used simultaneously (for instance, a "video view" and an "audio view"). In some cases, the views would conceptually be alternatives (for instance, a view consisting of three Video Captures covering the whole room versus a view consisting of just a single Video Capture covering only the center of a room). In this latter example, one sensible choice for a Consumer would be to indicate (through its Configure and possibly through an additional offer/answer exchange) the Captures of that Capture Scene View that most closely match the Consumer's number of display devices or screen layout.

The following is an example of four potential Capture Scene Views for an endpoint-style Provider:

1. (VC0, VC1, VC2) – Left, center and right camera Video Captures
2. (MCC3) – Video Capture associated with loudest room segment
3. (VC4) – Video Capture zoomed-out view of all people in the room
4. (AC0) – Main audio

The first view in this Capture Scene example is a list of Video Captures that have a spatial relationship to each other. Determination of the order of these captures (VC0, VC1, and VC2) for rendering purposes is accomplished through use of their area of capture attributes. The second view (MCC3) and the third view (VC4) are alternative representations of the same room's video, which might be better suited to some Consumers' rendering capabilities. The inclusion of the Audio Capture in the same Capture Scene indicates that AC0 is associated with all those Video Captures, meaning that it comes from the same spatial region. Therefore, if audio were to be rendered at all, this audio would be the correct choice irrespective of which Video Captures were chosen.

16.3.7.3.1 Capture Scene attributes

Capture Scene Attributes can be applied to Capture Scenes as well as to individual media captures. Attributes specified at this level apply to all constituent Captures. Capture Scene attributes include:

- Human-readable description of the Capture Scene, which could be in multiple languages
- xCard scene information
- Scale information (millimeters, unknown, no scale), as described in Section 16.4.16.6.

16.3.7.3.1.1 Scene Information The Scene information attribute provides information regarding the Capture Scene rather than individual participants. The Provider may gather the information automatically or manually from a variety of sources. The scene information attribute allows a Provider to indicate information, such as organizational or geographic information, allowing a Consumer to determine which Capture Scenes are of interest in order to then perform Capture selection. It also allows a Consumer to render information regarding the Scene or to use it for further processing.

As per Section 16.3.7.1.1.10, the xCard format is used to convey this information, and the Provider may supply a minimal set of information or a larger set of information. In order to keep CLUE messages compact, the Provider SHOULD use a URI to point to any LOGO, PHOTO, or SOUND contained in the xCARD rather than transmitting the LOGO, PHOTO, or SOUND data in a CLUE message.

16.3.7.3.2 Capture Scene View attributes

A Capture Scene can include one or more Capture Scene Views in addition to the Capture Scene–wide attributes described earlier. Capture Scene View attributes apply to the Capture Scene View as a whole, i.e., to all Captures that are part of the Capture Scene View. Capture Scene View attributes include:

- Human-readable description (which could be in multiple languages) of the Capture Scene View

16.3.7.4 Global View List

An Advertisement can include an optional Global View list. Each item in this list is a Global View. The Provider can include multiple Global Views to allow a Consumer to choose sets of captures appropriate to its capabilities or application. The choice of how to make these suggestions in the Global View list for what represents all the scenes for which the Provider can send media is up to the Provider. This is very similar to how each CSV represents a particular scene.

As an example, suppose an advertisement has three scenes, and each scene has three CSVs, ranging from one to three video captures in each CSV. The Provider is advertising a total of nine video Captures across three scenes. The Provider can use the Global View list to suggest alternatives for Consumers that cannot receive all nine video Captures as separate media streams. For accommodating a Consumer that wants to receive three video Captures, a Provider might suggest a Global View containing just a single CSV with three Captures and nothing from the other two scenes. Or, a Provider might suggest a Global View containing three different CSVs, one from each scene, with a single video Capture in each.

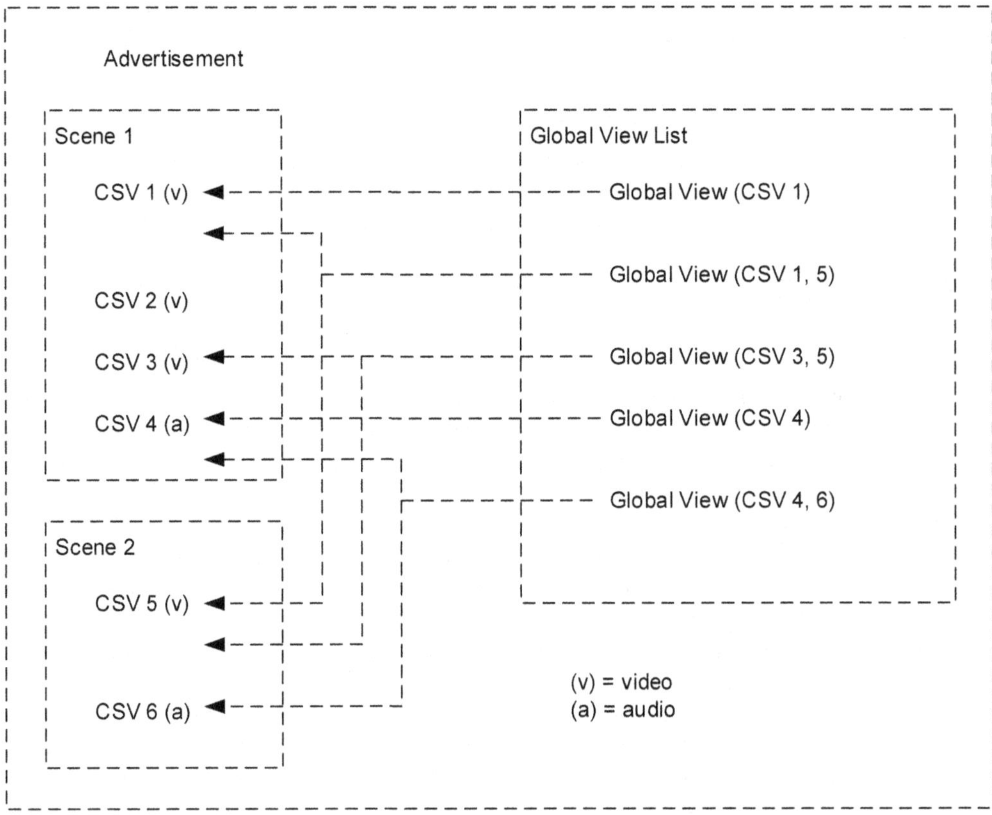

FIGURE 16.3 Global View List Structure. (Copyright: IETF.)

Some additional rules:

- The ordering of Global Views in the Global View list is insignificant.
- The ordering of CSVs within each Global View is insignificant.
- A particular CSV may be used in multiple Global Views.
- The Provider must be capable of encoding and sending all Captures within the CSVs of a given Global View simultaneously.

Figure 16.3 shows an example of the structure of Global Views in a Global View List.

16.3.8 Simultaneous Transmission Set Constraints

In many practical cases, a Provider has constraints or limitations on its ability to send Captures simultaneously. One type of limitation is caused by the physical limitations of capture mechanisms; these constraints are represented by a Simultaneous Transmission Set. The second type of limitation reflects the encoding resources available, such as bandwidth or video encoding throughput (macroblocks/second). This type of constraint is captured by Individual Encodings and Encoding Groups, discussed in the following.

Some Endpoints or MCUs can send multiple Captures simultaneously; however, sometimes there are constraints that limit which Captures can be sent simultaneously with other Captures. A device may not be able to be used in different ways at the same time. Provider Advertisements are made so that the Consumer can choose one of several possible mutually exclusive usages of the device. This type of constraint is expressed in a Simultaneous Transmission Set, which lists all the Captures of a particular media type (e.g. audio, video, text) that can be sent at the same time. There are different

TABLE 16.5
Two Simultaneous Transmission Sets

Simultaneous Sets
{VC0, VC1, VC2}
{VC0, VC3, VC2}

Simultaneous Transmission Sets for each media type in the Advertisement. This is easier to show in an example.

Consider the example of a room system where there are three cameras, each of which can send a separate Capture covering two persons each – VC0, VC1, and VC2. The middle camera can also zoom out (using an optical zoom lens) and show all six persons, VC3. But the middle camera cannot be used in both modes at the same time – it has to show either the space where two participants sit or the whole six seats, but not both at the same time. As a result, VC1 and VC3 cannot be sent simultaneously. Simultaneous Transmission Sets are expressed as sets of the Media Captures that the Provider could transmit at the same time (though in some cases, it is not intuitive to do so). If an MCC is included in a Simultaneous Transmission Set, it indicates that the Capture Encoding associated with it could be transmitted as the same time as the other Captures within the Simultaneous Transmission Set. It does not imply that the Single Media Captures contained in the MCC could all be transmitted at the same time.

In this example, the two Simultaneous Transmission Sets are shown in Table 16.5. If a Provider advertises one or more mutually exclusive Simultaneous Transmission Sets, then for each media type, the Consumer MUST ensure that it chooses Media Captures that lie wholly within one of those Simultaneous Transmission Sets.

A Provider OPTIONALLY can include the Simultaneous Transmission Sets in its Advertisement. These constraints apply across all the Capture Scenes in the Advertisement. It is a syntax conformance requirement that the Simultaneous Transmission Sets MUST allow all the media Captures in any particular Capture Scene View to be used simultaneously. Similarly, the Simultaneous Transmission Sets MUST reflect the simultaneity expressed by any Global View.

For shorthand convenience, a Provider MAY describe a Simultaneous Transmission Set in terms of Capture Scene Views and Capture Scenes. If a Capture Scene View is included in a Simultaneous Transmission Set, then all Media Captures in the Capture Scene View are included in the Simultaneous Transmission Set. If a Capture Scene is included in a Simultaneous Transmission Set, then all its Capture Scene Views (of the corresponding media type) are included in the Simultaneous Transmission Set. The end result reduces to a set of Media Captures, of a particular media type, in either case.

If an Advertisement does not include Simultaneous Transmission Sets, then the Provider MUST be able to simultaneously provide all the Captures from any one CSV of each media type from each Capture Scene. Likewise, if there are no Simultaneous Transmission Sets and there is a Global View list, then the Provider MUST be able to simultaneously provide all the Captures from any particular Global View (of each media type) from the Global View list.

If an Advertisement includes multiple Capture Scene Views in a Capture Scene, then the Consumer MAY choose one Capture Scene View for each media type or MAY choose individual Captures based on the Simultaneous Transmission Sets.

16.3.9 Encodings

Individual encodings and encoding groups are CLUE's mechanisms allowing a Provider to signal its limitations for sending Captures, or combinations of Captures, to a Consumer. Consumers can map the Captures they want to receive onto the Encodings with the encoding parameters they want.

For the relationship between the CLUE-specified mechanisms based on Encodings and the SIP offer/answer exchange, please refer to Section 16.3.5.

16.3.9.1 Individual Encodings

An Individual Encoding represents a way to encode a Media Capture as a Capture Encoding to be sent as an encoded media stream from the Provider to the Consumer. An Individual Encoding has a set of parameters characterizing how the media is encoded. Different media types have different parameters, and different encoding algorithms may have different parameters. An Individual Encoding can be assigned to at most one Capture Encoding at any given time.

Individual Encoding parameters are represented in SDP (RFC 4566), not in CLUE messages. For example, for a video encoding using H.26x compression technologies, this can include parameters such as:

- Maximum bandwidth
- Maximum picture size in pixels
- Maximum number of pixels to be processed per second

The bandwidth parameter is the only one that specifically relates to a CLUE Advertisement, as it can be further constrained by the maximum group bandwidth in an Encoding Group.

16.3.9.2 Encoding Group

An Encoding Group includes a set of one or more Individual Encodings and parameters that apply to the group as a whole. By grouping multiple individual Encodings together, an Encoding Group describes additional constraints on bandwidth for the group. A single Encoding Group MAY refer to Encodings for different media types.

The Encoding Group data structure contains:

- Maximum bitrate for all encodings in the group combined
- A list of identifiers for the Individual Encodings belonging to the group

When the Individual Encodings in a group are instantiated into Capture Encodings, each Capture Encoding has a bitrate that MUST be less than or equal to the max bitrate for the particular Individual Encoding. The "maximum bitrate for all encodings in the group" parameter gives the additional restriction that the sum of all the individual Capture Encoding bitrates MUST be less than or equal to this group value. Figure 16.4 illustrates one example of the structure of a media Provider's Encoding Groups and their contents.

A Provider advertises one or more Encoding Groups. Each Encoding Group includes one or more Individual Encodings. Each Individual Encoding can represent a different way of encoding media. For example, one Individual Encoding may be 1080p60 video, another could be 720p30, with a third being CIF, all in, for example, H.264 format. While a typical three codec/display system might have one Encoding Group per "codec box" (physical codec connected to one camera and one screen), there are many possibilities for the number of Encoding Groups a Provider may be able to offer and for the encoding values in each Encoding Group. There is no requirement for all Encodings within an Encoding Group to be instantiated at the same time.

16.3.9.3 Associating Captures with Encoding Groups

Each Media Capture, including MCCs, MAY be associated with one Encoding Group. To be eligible for configuration, a Media Capture MUST be associated with one Encoding Group, which is used to instantiate that Capture into a Capture Encoding. When an MCC is configured, all the Media Captures referenced by the MCC will appear in the Capture Encoding according to the attributes of the chosen encoding of the MCC. This allows an Advertiser to specify encoding attributes associated with the Media Captures without the need to provide an individual Capture Encoding for

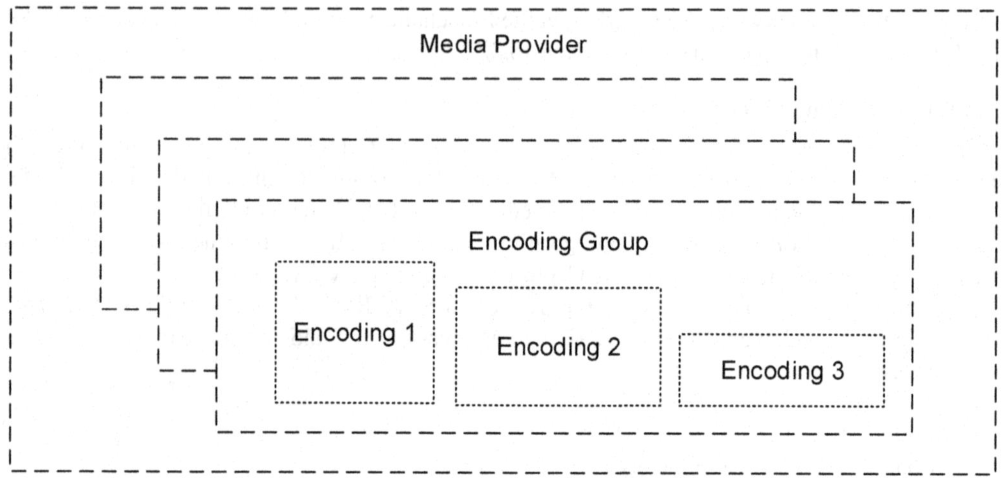

FIGURE 16.4 Encoding Group Structure. (Copyright: IETF.)

TABLE 16.6
Example Usage of Encoding with MCC and Source Captures

Capture Scene #1	Attribute
VC1	EncodeGroupID = 1
VC2	.
MCC1 (VC1,VC2)	EncodeGroupID = 2
CSV (VC1)	
CSV (MCC1)	

each of the inputs. If an Encoding Group is assigned to a Media Capture referenced by the MCC, it indicates that this Capture may also have an individual Capture Encoding. Table 16.6 shows some examples.

This would indicate that VC1 may be sent as its own Capture Encoding from EncodeGroupID=1 or that it may be sent as part of a Capture Encoding from EncodeGroupID=2 along with VC2. More than one Capture MAY use the same Encoding Group. The maximum number of Capture Encodings that can result from a particular Encoding Group constraint is equal to the number of individual Encodings in the group. The actual number of Capture Encodings used at any time MAY be lower than this maximum. Any of the Captures that use a particular Encoding Group can be encoded according to any of the Individual Encodings in the group. It is a protocol conformance requirement that the Encoding Groups MUST allow all the Captures in a particular Capture Scene View to be used simultaneously.

16.3.10 Consumer's Choice of Streams to Receive from the Provider

After receiving the Provider's Advertisement message (which includes media captures and associated constraints), the Consumer composes its reply to the Provider in the form of a Configure message. The Consumer is free to use the information in the Advertisement as it chooses, but there are a few obviously sensible design choices, which are outlined in the following. If multiple Providers connect to the same Consumer (i.e., in an MCU-less multiparty call), it is the responsibility of the Consumer to compose Configures for each Provider that fulfill each Provider's constraints as expressed in the Advertisement as well as its own capabilities.

In an MCU-based multiparty call, the MCU can logically terminate the Advertisement/Configure negotiation, in that it can hide the characteristics of the receiving endpoint and rely on its own capabilities (transcoding/transrating/...) to create Media Streams that can be decoded at the Endpoint Consumers. The timing of an MCU's sending of Advertisements (for its outgoing ports) and Configures (for its incoming ports in response to Advertisements received there) is up to the MCU and implementation dependent.

As a general outline, a Consumer can choose, based on the Advertisement it has received, which Captures it wishes to receive and which Individual Encodings it wants the Provider to use to encode the Captures. On receipt of an Advertisement with an MCC, the Consumer treats the MCC as per other non-MCC Captures, with the following differences:

- The Consumer would understand that the MCC is a Capture that includes the referenced individual Captures (or any Captures, if none are referenced) and that these individual Captures are delivered as part of the MCC's Capture Encoding.
- The Consumer may utilize any of the attributes associated with the referenced individual Captures and any Capture Scene attributes from where the individual Captures were defined to choose Captures and for rendering decisions.
- If the MCC attribute Allow Subset Choice is true, then the Consumer may or may not choose to receive all the indicated Captures. It can choose to receive a subset of Captures indicated by the MCC.

For example, if the Consumer receives:

MCC1(VC1,VC2,VC3){attributes}

A Consumer could choose all the Captures within a MCC; however, if the Consumer determines that it does not want VC3, it can return MCC1(VC1,VC2). If it wants all the individual Captures, then it returns only the MCC identity (i.e., MCC1). If the MCC in the advertisement does not reference any individual captures, or the Allow Subset Choice attribute is false, then the Consumer cannot choose what is included in the MCC; it is up to the Provider to decide.

A Configure Message includes a list of Capture Encodings. These are the Capture Encodings the Consumer wishes to receive from the Provider. Each Capture Encoding refers to one Media Capture and one Individual Encoding. For each Capture the Consumer wants to receive, it configures one of the Encodings in that Capture's Encoding Group. The Consumer does this by telling the Provider, in its Configure Message, which Encoding to use for each chosen Capture. Upon receipt of this Configure from the Consumer, common knowledge is established between Provider and Consumer regarding sensible choices for the media streams. The setup of the actual media channels, at least in the simplest case, is left to a following offer/answer exchange.

Optimized implementations may speed up the reaction to the offer/answer exchange by reserving the resources at the time of finalization of the CLUE handshake. CLUE advertisements and configure messages do not necessarily require a new SDP offer/answer for every CLUE message exchange, but the resulting encodings sent via RTP must conform to the most recent SDP offer/answer result. In order to meaningfully create and send an initial Configure, the Consumer needs to have received at least one Advertisement, and an SDP offer defining the Individual Encodings, from the Provider.

In addition, the Consumer can send a Configure at any time during the call. The Configure MUST be valid according to the most recently received Advertisement. The Consumer can send a Configure either in response to a new Advertisement from the Provider or on its own, for example, because of a local change in conditions (people leaving the room, connectivity changes, or multipoint-related considerations). When choosing which Media Streams to receive from the Provider, and the encoding characteristics of those Media Streams, the Consumer advantageously takes several things into account: Its local preference, simultaneity restrictions, and encoding limits.

16.3.10.1 Local preference

A variety of local factors influence the Consumer's choice of Media Streams to be received from the Provider:

- If the Consumer is an Endpoint, it is likely that it would choose, where possible, to receive video and audio Captures that match the number of display devices and the audio system it has.
- If the Consumer is an MCU, it may choose to receive loudest speaker streams (in order to perform its own media composition) and avoid pre-composed video Captures.
- User choice (for instance, selection of a new layout) may result in a different set of Captures, or different encoding characteristics, being required by the Consumer.

16.3.10.2 Physical simultaneity restrictions

Often, there are physical simultaneity constraints of the Provider that affect the Provider's ability to simultaneously send all the captures the Consumer would wish to receive. For instance, an MCU, when connected to a multicamera room system, might prefer to receive both individual video streams of the people present in the room and an overall view of the room from a single camera. Some Endpoint systems might be able to provide both of these sets of streams simultaneously, whereas others might not (if the overall room view were produced by changing the optical zoom level on the center camera, for instance).

16.3.10.3 Encoding and encoding group limits

Each of the Provider's encoding groups has limits on bandwidth, and the constituent potential encodings have limits on the bandwidth, computational complexity, video frame rate, and resolution that can be provided. When choosing the Captures to be received from a Provider, a Consumer device MUST ensure that the encoding characteristics requested for each individual Capture fit within the capability of the encoding it is being configured to use, as well as ensuring that the combined encoding characteristics for Captures fit within the capabilities of their associated encoding groups. In some cases, this could cause an otherwise "preferred" choice of capture encodings to be passed over in favor of different Capture Encodings – for instance, if a set of three Captures could only be provided at a low resolution, then a three-screen device could switch to favoring a single, higher-quality Capture Encoding.

16.3.11 Extensibility

One important characteristics of the Framework is its extensibility. The standard for interoperability and handling multiple streams must be future-proof. The framework itself is inherently extensible through expanding the data model types. For example:

- Adding more types of media, such as telemetry, can done by defining additional types of Captures in addition to audio and video.
- Adding new functionalities, such as 3-D video Captures, say, may require additional attributes describing the Captures.

The infrastructure is designed to be extended rather than requiring new infrastructure elements. Extension comes through adding to defined types.

16.3.12 Examples – Using the Framework (Informative)

This section gives some examples, first from the point of view of the Provider, then from that of the Consumer, then some multipoint scenarios.

16.3.12.1 Provider Behavior

This section shows some examples in more detail of how a Provider can use the framework to represent a typical case for telepresence rooms. First an endpoint is illustrated; then, an MCU case is shown.

16.3.12.1.1 Three screen Endpoint Provider

Consider an Endpoint with the following description:

Three cameras, three displays, a six-person table.

- Each camera can provide one Capture for each 1/3 section of the table.
- A single Capture representing the active speaker can be provided (voice activity–based camera selection to a given encoder input port implemented locally in the Endpoint).
- A single Capture representing the active speaker with the other two Captures shown picture in picture (PiP) within the stream can be provided (again, implemented inside the endpoint).
- A Capture showing a zoomed-out view of all six seats in the room can be provided.

The video and audio Captures for this Endpoint can be described as follows.

Video Captures:

- VC0 (the left camera stream), encoding group=EG0, view=table
- VC1 (the center camera stream), encoding group=EG1, view=table
- VC2 (the right camera stream), encoding group=EG2, view=table
- MCC3 (the loudest panel stream), encoding group=EG1, view=table, MaxCaptures=1, policy=SoundLevel
- MCC4 (the loudest panel stream with PiPs), encoding group=EG1, view=room, MaxCaptures=3, policy=SoundLevel
- VC5 (the zoomed-out view of all people in the room), encoding group=EG1, view=room
- VC6 (presentation stream), encoding group=EG1, presentation

Figure 16.5 is a top view of the room with three cameras, three displays, and six seats. Each camera captures two people. The six seats are not all in a straight line.

The two points labeled "b" and "c" are intended to be at the midpoint between the seating positions and where the fields of view of the cameras intersect.

The plane of interest for VC0 is a vertical plane that intersects points "a" and "b."

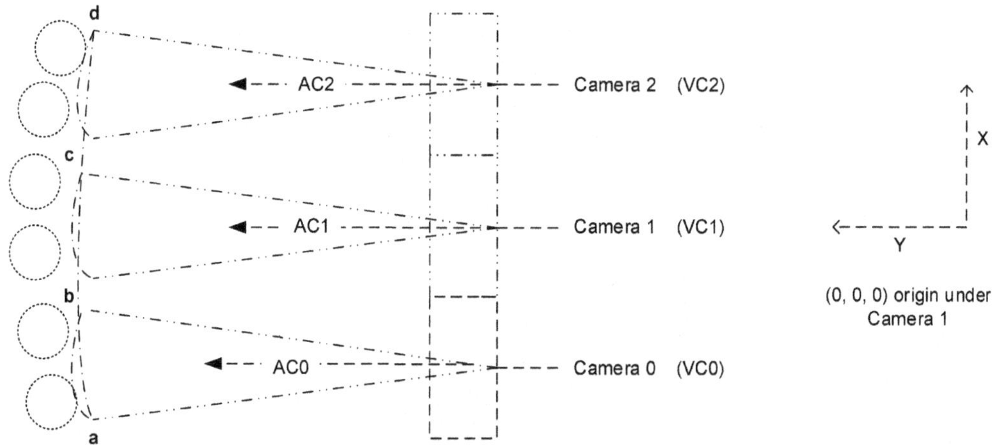

FIGURE 16.5 Room Layout Top View.

The plane of interest for VC1 intersects points "b" and "c." The plane of interest for VC2 intersects points "c" and "d."

This example uses an area scale of millimeters.

Areas of capture:

Parameter	Bottom left	Bottom right	Top left	Top right	Points of capture
VC0	(−2011,2850,0)	(−673,3000,0)	(−2011,2850,757)	(−673,3000,757)	(−1678,0,800)
VC1	(−673,3000,0)	(673,3000,0)	(−673,3000,757)	(673,3000,757)	(0,0,800)
VC2	(673,3000,0)	(2011,2850,0)	(673,3000,757)	(2011,3000,757)	(1678,0,800)
MCC3	(−2011,2850,0)	(2011,2850,0)	(−2011,2850,757)	(2011,3000,757)	none
MCC4	(−2011,2850,0)	(2011,2850,0)	(−2011,2850,757)	(2011,3000,757)	none
VC5	(−2011,2850,0)	(2011,2850,0)	(−2011,2850,757)	(2011,3000,757)	(0,0,800)
VC6	none				none

In this example, the right edge of the VC0 area lines up with the left edge of the VC1 area. It does not have to be this way. There could be a gap or an overlap. One additional thing to note for this example is that the distance from a to b is equal to the distance from b to c and the distance from c to d. All these distances are 1346 mm. This is the planar width of each area of capture for VC0, VC1, and VC2.

Note that the text in parentheses (e.g. "the left camera stream") is not explicitly part of the model; it is just explanatory text for this example and is not included in the model with the media captures and attributes. Also, MCC4 does not say anything about how a capture is composed, so the media consumer cannot tell based on this capture that MCC4 is composed of a "loudest panel with PiPs."

Audio Captures:

Three ceiling microphones are located between the cameras and the table at the same height as the cameras. The microphones point down at an angle towards the seating positions.

- AC0 (left), encoding group=EG3
- AC1 (right), encoding group=EG3
- AC2 (center) encoding group=EG3
- AC3 being a simple pre-mixed audio stream from the room (mono), encoding group=EG3
- AC4 audio stream associated with the presentation video (mono) encoding group=EG3, presentation

Parameter	Point of Capture	Point on Line of Capture
AC0	(−1342,2000,800)	(−1342,2925,379)
AC1	(1342,2000,800)	(1342,2925,379)
AC2	(0,2000,800)	(0,3000,379)
AC3	(0,2000,800)	(0,3000,379)
AC4	none	

The physical simultaneity information is:

Simultaneous transmission set #1 {VC0, VC1, VC2, MCC3, MCC4, VC6}
Simultaneous transmission set #2 {VC0, VC2, VC5, VC6}

This constraint indicates that it is not possible to use all the VCs at the same time. VC5 cannot be used at the same time as VC1 or MCC3 or MCC4. Also, using every member in the set simultaneously may not make sense – for example, MCC3(loudest) and MCC4 (loudest with PiP). In addition, there are encoding constraints that make choosing all the VCs in a set impossible. VC1, MCC3, MCC4, VC5, and VC6 all use EG1, and EG1 has only three encoding groups (ENCs).

This constraint shows up in the encoding groups, not in the simultaneous transmission sets.

In this example there are no restrictions on which Audio Captures can be sent simultaneously.

Encoding Groups:

This example (Figure 16.6) has three encoding groups associated with the video captures. Each group can have three encodings, but with each potential encoding having a progressively lower specification. In this example, 1080p60 transmission is possible (as ENC0 has a maxPps value compatible with this). Significantly, as up to three encodings are available per group, it is possible to transmit some video Captures simultaneously that are not in the same view in the Capture Scene: for example, VC1 and MCC3 at the same time. The following information about Encodings is a summary of what would be conveyed in SDP, not directly in the CLUE Advertisement.

For audio (Figure 16.7), there are five potential encodings available, so all five Audio Captures can be encoded at the same time.

encodeGroupID=EG0,	maxGroupBandwidth=6000000
encodeID=ENC0,	maxWidth=1920, maxHeight=1088, maxFrameRate=60, maxPps=124416000, maxBandwidth=4000000
encodeID=ENC1,	maxWidth=1280, maxHeight=720, maxFrameRate=30, maxPps=27648000, maxBandwidth=4000000
encodeID=ENC2,	maxWidth=960, maxHeight=544, maxFrameRate=30, maxPps=15552000, maxBandwidth=4000000
encodeGroupID=EG1,	maxGroupBandwidth=6000000
encodeID=ENC3,	maxWidth=1920, maxHeight=1088, maxFrameRate=60, maxPps=124416000, maxBandwidth=4000000
encodeID=ENC4,	maxWidth=1280, maxHeight=720, maxFrameRate=30, maxPps=27648000, maxBandwidth=4000000
encodeID=ENC5,	maxWidth=960, maxHeight=544, maxFrameRate=30, maxPps=15552000, maxBandwidth=4000000
encodeGroupID=EG2,	maxGroupBandwidth=6000000
encodeID=ENC6,	maxWidth=1920, maxHeight=1088, maxFrameRate=60, maxPps=124416000, maxBandwidth=4000000
encodeID=ENC7,	maxWidth=1280, maxHeight=720, maxFrameRate=30, maxPps=27648000, maxBandwidth=4000000
encodeID=ENC8,	maxWidth=960, maxHeight=544, maxFrameRate=30, maxPps=15552000, maxBandwidth=4000000

FIGURE 16.6 Example Encoding Groups for Video.

encodeGroupID=EG3,	maxGroupBandwidth=320000
encodeID=ENC9,	maxBandwidth=64000
encodeID=ENC10,	maxBandwidth=64000
encodeID=ENC11,	maxBandwidth=64000
encodeID=ENC12,	maxBandwidth=64000
encodeID=ENC13,	maxBandwidth=64000

FIGURE 16.7 Example Encoding Group for Audio.

TABLE 16.7
Example Capture Scene Views

Capture Scene #1
VC0, VC1, VC2
MCC3
MCC4
VC5
AC0, AC1, AC2
AC3
Capture Scene #2
VC6
AC4

Capture Scenes:

(Table 16.7) represents the Capture Scenes for this Provider. Recall that a Capture Scene is composed of alternative Capture Scene Views covering the same spatial region. Capture Scene #1 is for the main people captures, and Capture Scene #2 is for presentation.

Each row in the table is a separate Capture Scene View.

Different Capture Scenes are distinct from each other and are non-overlapping. A Consumer can choose a view from each Capture Scene. In this case, the three Captures VC0, VC1, and VC2 are one way of representing the video from the Endpoint. These three Captures should appear adjacent to each other. Alternatively, another way of representing the Capture Scene is with the capture MCC3, which automatically shows the person who is talking, and similarly for the MCC4 and VC5 alternatives.

As in the video case, the different views of audio in Capture Scene #1 represent the "same thing," in that one way to receive the audio is with the three Audio Captures (AC0, AC1, and AC2), and another way is with the mixed AC3. The Media Consumer can choose an audio CSV it is capable of receiving. The spatial ordering is understood by the Media Capture attributes Area of Capture, Point of Capture, and Point on Line of Capture.

A Media Consumer would likely want to choose a Capture Scene View to receive based in part on how many streams it can simultaneously receive. A consumer that can receive three video streams would probably prefer to receive the first view of Capture Scene #1 (VC0, VC1, and VC2) and not receive the other views. A consumer that can receive only one video stream would probably choose one of the other views. If the consumer can receive a presentation stream too, it would also choose to receive the only view from Capture Scene #2 (VC6).

16.3.12.1.2 Encoding Group Example

This is an example of an Encoding Group to illustrate how it can express dependencies between Encodings. The information below about Encodings is a summary of what would be conveyed in SDP, not directly in the CLUE Advertisement.

```
encodeGroupID=EG0,          maxGroupBandwidth=6000000
encodeID=VIDENC0,           maxWidth=1920, maxHeight=1088,
                            maxFrameRate=60, maxPps=62208000,
                            maxBandwidth=4000000
encodeID=VIDENC1,           maxWidth=1920, maxHeight=1088,
                            maxFrameRate=60,maxPps=62208000,
                            maxBandwidth=4000000
```

```
encodeID=AUDENC0,          maxBandwidth=96000
encodeID=AUDENC1,          maxBandwidth=96000
encodeID=AUDENC2,          maxBandwidth=96000
```

Here, the Encoding Group is EG0. Although the Encoding Group is capable of transmitting up to 6 Mbit/s, no individual video Encoding can exceed 4 Mbit/s. This encoding group also allows up to three audio encodings, AUDENC<0-2>. It is not required that audio and video encodings reside within the same encoding group, but if this is the case, then the group's overall maxBandwidth value is a limit on the sum of all audio and video encodings configured by the consumer. A system that does not wish or need to combine bandwidth limitations in this way should instead use separate encoding groups for audio and video in order for the bandwidth limitations on audio and video not to interact. Audio and video can be expressed in separate encoding groups, as in this illustration.

```
encodeGroupID=EG0,   maxGroupBandwidth=6000000
encodeID=VIDENC0,    maxWidth=1920, maxHeight=1088,
                     maxFrameRate=60, maxPps=62208000, maxBandwidth=4000000
encodeID=VIDENC1,    maxWidth=1920, maxHeight=1088,
                     maxFrameRate=60, maxPps=62208000,
                     maxBandwidth=4000000
encodeGroupID=EG1,   maxGroupBandwidth=500000
encodeID=AUDENC0,    maxBandwidth=96000
encodeID=AUDENC1,    maxBandwidth=96000
encodeID=AUDENC2,    maxBandwidth=96000
```

16.3.12.1.3 The MCU Case

This section shows (Table 16.8) how an MCU might express its Capture Scenes, intending to offer different choices for consumers that can handle different numbers of streams. Each MCC is for video. A single Audio Capture is provided for all single and multiscreen configurations that can be associated (e.g., lip-synced) with any combination of Video Captures (the MCCs) at the consumer.

If/when a presentation stream becomes active within the conference, the MCU might re-advertise the available media as in Table 16.9).

16.3.12.2 Media Consumer Behavior

This section gives an example of how a Media Consumer might behave when deciding how to request streams from the three screen endpoints described in the previous section. The receive side of a call needs to balance its requirements, based on number of screens and speakers, its decoding

TABLE 16.8
MCU Main Capture Scenes

Capture Scene #1	Remark
MCC	for a single-screen consumer
MC1, MCC2	for a two-screen consumer
MCC3, MCC4, MCC5	for a three-screen consumer
MCC6, MCC7, MCC8, MCC9	for a four-screen consumer
AC0	AC representing all participants
CVS (MCC0)	
CSV (MCC1, MCC2)	
CSV (MCC3, MCC4, MCC5)	
CSV (MCC6, MCC7, MCC8, MCC9)	
CSV (AC0)	

TABLE 16.9
MCU Presentation Capture Scene

Capture Scene #2	Note
VC10	video capture for presentation
AC1	presentation audio to accompany VC10
CSV (VC10)	
CSV (AC1)	

capabilities and available bandwidth, and the provider's capabilities in order to optimally configure the provider's streams. Typically, it would want to receive and decode media from each Capture Scene advertised by the Provider.

A sane, basic algorithm might be for the consumer to go through each Capture Scene View in turn and find the collection of Video Captures that best matches the number of screens it has (this might include consideration of screens dedicated to presentation video display rather than "people" video) and then decide between alternative views in the video Capture Scenes based on either hard-coded preferences or user choice. Once this choice has been made, the consumer would then decide how to configure the provider's encoding groups in order to make best use of the available network bandwidth and its own decoding capabilities.

16.3.12.2.1 One-screen Media Consumer

MCC3, MCC4, and VC5 are all different views by themselves, not grouped together in a single view, so the receiving device should choose one of these. The choice would come down to whether to see the greatest number of participants simultaneously at roughly equal precedence (VC5), a switched view of just the loudest region (MCC3), or a switched view with PiPs (MCC4). An end-point device with a small amount of knowledge of these differences could offer a dynamic choice of these options, incall, to the user.

16.3.12.2.2 Two-screen Media Consumer Configuring the Example

Mixing systems with an even number of screens, "2n," and those with "2n+1" cameras (and vice versa) is always likely to be the problematic case. In this instance, the behavior is likely to be determined by whether a "2 screen" system is really a "2 decoder" system, i.e., whether only one received stream can be displayed per screen or whether more than two streams can be received and spread across the available screen area. There are three possible behaviors here for the two-screen system when it learns that the far end is "ideally" expressed via three capture streams:

1. Fall back to receiving just a single stream (MCC3, MCC4, or VC5, as per the one screen consumer case earlier) and either leave one screen blank or use it for presentation if/when a presentation becomes active.
2. Receive three streams (VC0, VC1, and VC2) and display across two screens, either with each capture being scaled to 2/3 of a screen and the center capture being split across two screens or, as would be necessary if there were large bezels on the screens, with each stream being scaled to 1/2 the screen width and height and there being a fourth "blank" panel. This fourth panel could potentially be used for any presentation that became active during the call.
3. Receive three streams, decode all three, and use control information indicating which was the most active to switch between showing the left and center streams (one per screen) and the center and right streams.

For an endpoint capable of all three methods of working described, again, it might be appropriate to offer the user the choice of display mode.

16.3.12.2.3 Three-screen Media Consumer Configuring the Example

This is the most straightforward case – the Media Consumer would look to identify a set of streams to receive that best matched its available screens and so the VC0 plus VC1 plus VC2 should match optimally. The spatial ordering would give sufficient information for the correct Video Capture to be shown on the correct screen, and the consumer would either need to divide a single encoding group's capability by three to determine what resolution and frame rate to configure the provider with or to configure the individual Video Captures' Encoding Groups with what makes most sense (taking into account the receive side decode capabilities, overall call bandwidth, the resolution of the screens plus any user preferences such as motion vs. sharpness).

16.3.12.3 Multipoint Conference Utilizing MCCs

The use of MCCs allows the MCU to construct outgoing Advertisements describing complex media switching and composition scenarios. The following sections provide several examples. Note: In the examples, the identities of the CLUE elements (e.g., Captures and Capture Scene) in the incoming Advertisements overlap. This is because there is no co-ordination between the endpoints. The MCU is responsible for making these unique in the outgoing advertisement.

16.3.12.3.1 Single Media Captures and MCC in the Same Advertisement

Four endpoints are involved in a Conference where CLUE is used. An MCU acts as a middlebox between the endpoints with a CLUE channel between each endpoint and the MCU. The MCU receives the following Advertisements (Tables 16.10 through 16.12.

TABLE 16.10
Advertisement Received from Endpoint A

Capture Scene #1	**Description=AustralianConfRoom**
VC1	Description=Audience
CSV (VC1)	EncodeGroupID=1
Capture Scene #1	**Description=AustralianConfRoom**
VC1	Description=Audience
CSV (VC1)	EncodeGroupID=1

TABLE 16.11
Advertisement Received from Endpoint B

Capture Scene #1	**Description=AustralianConfRoom**
VC1	Description=Audience
VC2	EncodeGroupID=1
CSV (VC1, VC2)	Description=Audience
	EncodeGroupID=1

TABLE 16.12
Advertisement Received from Endpoint C

Capture Scene #1	**Description=USAConfRoom**
VC1	Description=Audience
CSV (VC1)	EncodeGroupID=1

TABLE 16.13
Advertisement Sent to Endpoint F – One Encoding

Capture Scene #1	**Description=AustralianConfRoom**
VC1	Description=Audience
CSV (VC1)	
Capture Scene #2	**Description=ChinaConfRoom**
VC2	Description=Speaker
VC3	Description=Audience
CSV (VC2, VC3)	
Capture Scene #3	**Description=USAConfRoom**
VC4	Description=Audience
CSV (VC4)	
Capture Scene #4	**Attribute**
MCC1 (VC1,VC2,VC3,VC4)	Policy=RoundRobin:1
CSV(MCC1)	MaxCaptures=1
	EncodingGroup=1

Note: Endpoint B indicates that it sends two streams.

If the MCU wanted to provide a MCC containing a round robin switched view of the audience from the three endpoints and the speaker, it could construct the Advertisement sent to Endpoint F in Table 16.13).

Alternatively, if the MCU wanted to provide the speaker as one media stream and the audiences as another, it could assign an encoding group to VC2 in Capture Scene 2 and provide a CSV in Capture Scene #4, as per Advertisement sent to Endpoint F (Table 16.14).

Therefore, a Consumer could choose whether or not to have a separate speaker-related stream and could choose which endpoints to see. If it wanted the second stream but not the Australian conference room, it could indicate the captures in the Configure message (Table 16.15).

TABLE 16.14
Advertisement Sent to Endpoint F – Two Encodings

Capture Scene #1	**Description=AustralianConfRoom**
VC1	Description=Audience
CSV (VC1)	
Capture Scene #2	**Description=ChinaConfRoom**
VC2	Description=Speaker
VC3	EncodingGroup=1
CSV (VC2, VC3)	Description=Audience
Capture Scene #3	**Description=USAConfRoom**
VC4	Description=Audience
CSV (VC4)	
Capture Scene #4	**Attribute**
MCC1 (VC1,VC3,VC4)	Policy=RoundRobin:1
CSV (MCC1)	MaxCaptures=1
CSV (MCC1, MCC2)	EncodingGroup=1
	AllowSubset=True
	MaxCaptures=1
	EncodingGroup=1

TABLE 16.15
MCU Case: Consumer Response

MCC1 (VC3,VC4)	Encoding
VC2	Encoding

TABLE 16.16
Advertisement Received from Endpoint D

Capture Scene #4	Description=AustralianConfRoom
MCC1 (VC1,VC2,VC3,VC4)	CaptureArea=Left
VC2	EncodingGroup=1
VC3	CaptureArea=Centre
CSV (VC1,VC2,VC3)	EncodingGroup=1
	CaptureArea=Right
	EncodingGroup=1

TABLE 16.17
Advertisement Received from Endpoint E

Capture Scene #4	Description=ChinaConfRoom
VC1	CaptureArea=Left
VC2	EncodingGroup=1
VC3	CaptureArea=Centre
CSV (VC1,VC2,VC3)	EncodingGroup=1
	CaptureArea=Right
	EncodingGroup=1

16.3.12.3.2 Several MCCs in the Same Advertisement

Multiple MCCs (Tables 16.16 and 16.17)can be used where multiple streams are used to carry media from multiple endpoints. For example: A conference has three endpoints, D, E, and F. Each endpoint has three video captures covering the left, middle, and right regions of each conference room. The MCU receives the following advertisements from D and E.

The MCU wants to offer Endpoint F three Capture Encodings. Each Capture Encoding would contain all the Captures from either Endpoint D or Endpoint E, depending on the active speaker.

The MCU sends the Advertisement in Table 16.18).

16.3.12.3.3 Heterogeneous Conference with Switching and Composition

Consider a conference between endpoints with the following characteristics:

Endpoint A – four screens, three cameras
Endpoint B – three screens, three cameras
Endpoint C – three screens, three cameras
Endpoint D – three screens, three cameras
Endpoint E – one screen, one camera
Endpoint F – two screens, one camera
Endpoint G – one screen, one camera

TABLE 16.18
Advertisement Sent to Endpoint F

Capture Scene #1	**Description=AustralianConfRoom**
VC1	
VC2	
VC3	
CSV (VC1, VC2, VC3)	
Capture Scene #2	**Description=ChinaConfRoom**
VC4	
VC5	
VC6	
CSV (VC4, VC5, VC6)	
Capture Scene #3	
MCC1 (VC1,VC4)	CaptureArea=Left
MCC2 (VC2, VC5)	MaxCaptures=1
MCC3 (VC3, VC6)	SynchronizationID=1
CSV (MCC1, MCC2, MCC3)	EncodingGroup=1
	CaptureArea=Centre
	MaxCaptures=1
	SynchronizationID=1
	EncodingGroup=1
	CaptureArea=Right
	MaxCaptures=1
	SynchronizationID=1
	EncodingGroup=1

This example focuses on what the user in one of the three-camera multiscreen endpoints sees. Call this person User A at Endpoint A. There are four large display screens at Endpoint A. Whenever somebody at another site is speaking, all the video captures from that endpoint are shown on the large screens. If the talker is at a three-camera site, then the video from those three cameras fills three of the screens. If the talker is at a single-camera site, then video from that camera fills one of the screens, while the other screens show video from other single-camera endpoints.

User A hears audio from the four loudest talkers.

User A can also see video from other endpoints, in addition to the current talker, although much smaller in size. Endpoint A has four screens (Figure 16.8), so one of those screens shows up to nine other Media Captures in a tiled fashion. When video from a three-camera endpoint appears in the tiled area, video from all three cameras appears together across the screen with correct spatial relationship among those three images.

User B at Endpoint B sees a similar arrangement, except that there are only three screens (Figure 16.9), so the nine other Media Captures are spread out across the bottom of the three displays in a PiP format. When video from a three-camera endpoint appears in the PiP area, video from all three cameras appears together across a single screen with the correct spatial relationship.

When somebody at a different endpoint becomes the current talker, then User A and User B both see the video from the new talker appear on their large screen area, while the previous talker takes one of the smaller tiled or PiP areas. The person who is the current talker does not see themselves; they see the previous talker in their large screen area.

One of the points of this example is that endpoints A and B each want to receive three capture encodings for their large display areas and nine encodings for their smaller areas. A and B are both able to send the same Configure message to the MCU, and each receives the same conceptual Media Captures from the MCU. The differences are in how they are rendered and are purely a local matter at A and B.

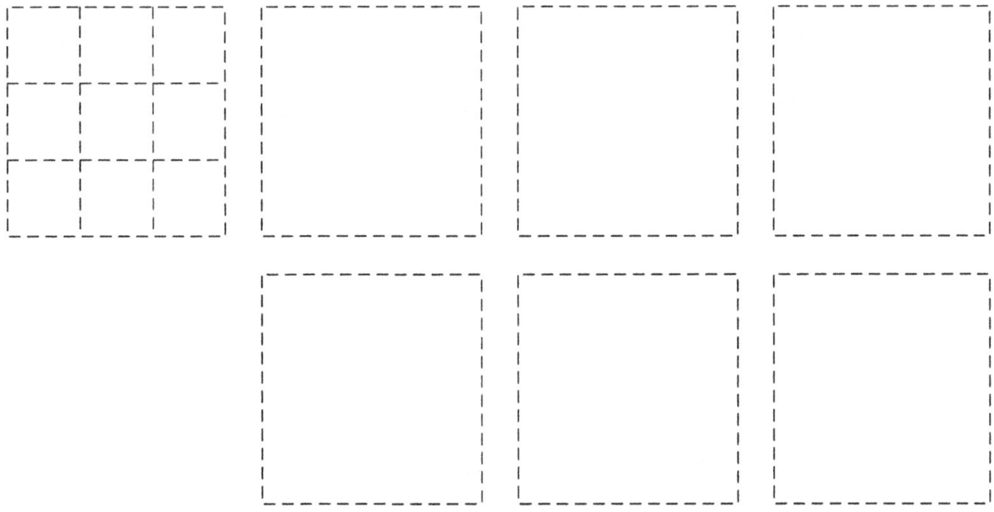

FIGURE 16.8 Endpoint A – Four-screen Display.

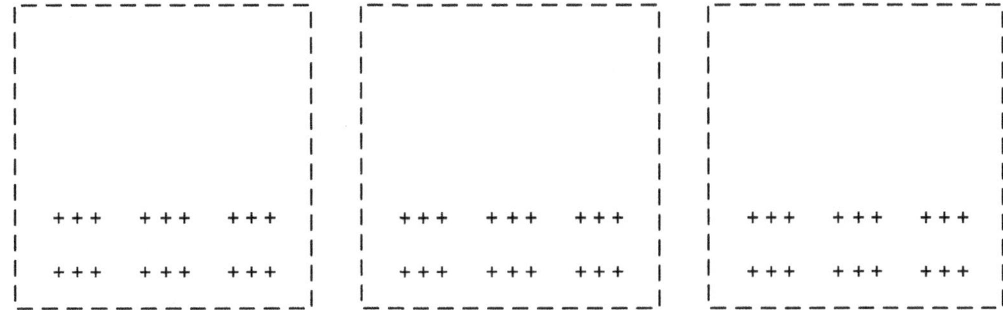

FIGURE 16.9 Endpoint B – Three-screen Display with PiPs.

The Advertisements for such a scenario are described in Tables 16.19 and 16.20).

Rather than considering what is displayed, CLUE concentrates more on what the MCU sends. The MCU does not know anything about the number of screens an endpoint has. As Endpoints A to D each advertise that three Captures make up a Capture Scene, the MCU offers these in a "site" switching mode. That is, there are three MCCs (and Capture Encodings) each switching between Endpoints. The MCU switches in the applicable media into the stream based on voice activity. Endpoint A will not see a capture from itself.

Using the MCC concept, the MCU would send the Advertisement in Table 16.21) to Endpoint A.

TABLE 16.19
Advertisement Received at the MCU from Endpoints A to D

Capture Scene #1	Description=Endpoint x
VC1	EncodingGroup=1
VC2	EncodingGroup=1
VC3	EncodingGroup=1
AC1	EncodingGroup=2
CSV1 (VC1, VC2, VC3)	
CSV2 (AC1)	

TABLE 16.20
Advertisement Received at the MCU from Endpoints E to G

Capture Scene #1	Description=Endpoint y
VC1	EncodingGroup=1
AC1	EncodingGroup=2
CSV1 (VC1)	
CSV2 (AC1)	

TABLE 16.21
Advertisement Sent to Endpoint A – Source Part

Capture Scene #1	Description=Endpoint B
VC4	CaptureArea=Left
VC5	CaptureArea=Center
VC6	CaptureArea=Right
AC1	
CSV (VC4,VC5,VC6)	
CSV (AC1)	
Capture Scene #2	**Description=Endpoint C**
VC4	CaptureArea=Left
VC5	CaptureArea=Center
VC6	CaptureArea=Right
AC1	
CSV (VC4,VC5,VC6)	
CSV (AC1)	
Capture Scene #3	**Description=Endpoint D**
VC10	CaptureArea=Left
VC11	CaptureArea=Center
VC12	CaptureArea=Right
AC3	
CSV (VC10,VC11,VC12)	
CSV (AC3)	
Capture Scene #4	**Description=Endpoint E**
VC13	
AC4	
CSV (VC13)	
CSV (AC4)	
Capture Scene #5	**Description=Endpoint F**
VC14	
AC5	
CSV (VC14)	
CSV (AC5)	
Capture Scene #6	**Description=Endpoint G**
VC15	
AC6	
CSV (VC15)	
CSV (AC6)	

TABLE 16.22
Advertisement Sent to Endpoint A – Switching Part

Capture Scene #7	Description=Output3streammix
MCC1 (VC4,VC7,VC10,VC13)	CaptureArea=Left MaxCaptures=1 SynchronizationID=1 Policy=SoundLevel:0 EncodingGroup=1
MCC2 (VC5,VC8,VC11,VC14)	CaptureArea=Center MaxCaptures=1 SynchronizationID=1 Policy=SoundLevel:0 EncodingGroup=1
MCC3 (VC6,VC9,VC12,VC15)	CaptureArea=Right MaxCaptures=1 SynchronizationID=1 Policy=SoundLevel:0 EncodingGroup=1
MCC4 ( ) (for audio)	CaptureArea=whole scene MaxCaptures=1 Policy=SoundLevel:0 EncodingGroup=2
MCC5 ( ) (for audio)	CaptureArea=whole scene MaxCaptures=1 Policy=SoundLevel:1 EncodingGroup=2
MCC6 ( ) (for audio)	CaptureArea=whole scene MaxCaptures=1 Policy=SoundLevel:2 EncodingGroup=2
MCC7 ( ) (for audio)	CaptureArea=whole scene MaxCaptures=1 Policy=SoundLevel:3 EncodingGroup=2
CSV (MCC1,MCC2,MCC3)	
CSV (MCC4,MCC5,MCC6, MCC7)	

This part of the Advertisement presents information about the sources to the MCC. The information is effectively the same as the received Advertisements, except that there are no Capture Encodings associated with them, and the identities have been re-numbered.

In addition to the source Capture information, the MCU advertises "site" switching of Endpoints B to G in three streams.

The previous part (Table 16.22) describes the switched three main streams that relate to site switching. MaxCaptures=1 indicates that only one Capture from the MCC is sent at a particular time. SynchronizationID=1 indicates that the source sending is synchronized. The provider can choose to group together VC13, VC14, and VC15 for the purpose of switching according to the SynchronizationID. Therefore, when the provider switches one of them into an MCC, it can also switch the others, even though they are not part of the same Capture Scene.

All the audio for the conference is included in this Scene #7. There is not necessarily a one-to-one relation between any audio capture and video capture in this scene. Typically, a change in loudest talker will cause the MCU to switch the audio streams more quickly than switching video streams.

TABLE 16.23
Advertisement Sent to Endpoint A – 9 Switched Part

Capture Scene #8	Description=Output9stream
MCC8 (VC4,VC5,VC6,VC7, VC8,VC9, VC10,VC11,VC12,VC13,VC14,VC15)	MaxCaptures=1 Policy=SoundLevel:0 EncodingGroup=1
MCC9(VC4,VC5,VC6,VC7,VC8,VC9,VC10,VC11, VC12,VC13,VC14,VC15)	MaxCaptures=1 Policy=SoundLevel:1 EncodingGroup=1
to	to
MCC16(VC4,VC5,VC6,VC7,VC8,VC9,VC10,VC11, VC12,VC13,VC14,VC15)	MaxCaptures=1 Policy=SoundLevel:8 EncodingGroup=1
CSV (MCC8,MCC9,MCC10,MCC11,MCC12, MCC13,MCC14,MCC15,MCC16)	

TABLE 16.24
Advertisement Sent to Endpoint A – 9 Composed Part

Capture Scene #9	Description=NineTiles
MCC13 (MCC8,MCC9,MCC10, MCC11, MCC12, MCC13, MCC14, MCC15, MCC16)	MaxCaptures=9 EncodingGroup=1
CSV (MCC13)	

The MCU can also supply nine media streams showing the active and previous eight speakers. See Table 16.23 for what it includes in the Advertisement.

The previous part indicates that there are nine capture encodings. Each of the Capture Encodings may contain any captures from any source site with a maximum of one Capture at a time. Which Capture is present is determined by the policy. The MCCs in this scene do not have any spatial attributes.

Note: The Provider alternatively could provide each of the preceding MCCs in its own Capture Scene. If the MCU wanted to provide a composed Capture Encoding containing all of the nine captures, it could advertise in addition (see Table 16.24).

As MaxCaptures is nine, this indicates that the capture encoding contains information from nine sources at a time. The Advertisement to Endpoint B is identical to that just described, other than that the captures from Endpoint A would be added, and the captures from Endpoint B would be removed. Whether the Captures are rendered on a four-screen display or a three-screen display is up to the Consumer to determine. The Consumer wants to place video captures from the same original source endpoint together, in the correct spatial order, but the MCCs do not have spatial attributes. So, the Consumer needs to associate incoming media packets with the original individual captures in the advertisement (such as VC4, VC5, and VC6) in order to know the spatial information it needs for correct placement on the screens. The Provider can use the RTCP CaptureId Source description (SDES) item and associated RTP header extension, as described in Section 16.7, to convey this information to the Consumer.

16.3.12.3.4 Heterogeneous Conference with Voice Activated Switching

This example illustrates how multipoint "voice activated switching" behavior can be realized, with an endpoint making its own decision about which of its outgoing video streams is considered the "active talker" from that endpoint. Then an MCU can decide which is the active talker among the whole conference.

TABLE 16.25
Advertisement Received at the MCU from Endpoints A and B

Capture Scene #1	Description=Endpoint x
VC1	CaptureArea=Left EncodingGroup=1
VC2	CaptureArea=Center EncodingGroup=1
VC3	CaptureArea=Right EncodingGroup=1
MCC1 (VC1, VC2, VC3)	MaxCaptures=1 CaptureArea=whole scene Policy=SoundLevel:0 EncodingGroup=1
AC1	CaptureArea=whole scene EncodingGroup=2
CSV1 (VC1, VC2, VC3)	
CSV2 (MCC1)	
CSV3(AC1)	

Consider a conference between endpoints with the following characteristics:

Endpoint A – three screens, three cameras
Endpoint B – three screens, three cameras
Endpoint C – one screen, one camera

This example (Table 16.25) focuses on what the user at Endpoint C sees. The user would like to see the video capture of the current talker without composing it with any other video capture. In this example, Endpoint C is capable of receiving only a single video stream. The following tables describe advertisements from A and B to the MCU, and from the MCU to C, that can be used to accomplish this.

Endpoints A and B are advertising each individual video capture and also a switched capture MCC1, which switches between the other three based on who is the active talker. These endpoints do not advertise distinct audio captures associated with each individual video capture, so it would be impossible for the MCU (as a media consumer) to make its own determination of which video capture is the active talker based just on information in the audio streams.

The MCU advertises one scene with four video MCCs (Table 16.26). Three of them in CSV1 give a left, center, and right view of the conference with "site switching." MCC4 provides a single video capture representing a view of the whole conference. The MCU intends MCC4 to be switched between all the other original source captures. In this example advertisement, the MCU is not giving all the information about all the other endpoints' scenes and which of those captures is included in the MCCs. The MCU could include all that information if it wanted to give the consumers more information, but it is not necessary for this example scenario.

The Provider advertises MCC5 and MCC6 for audio. Both are switched captures, with different SoundLevel policies indicating they are the top two dominant talkers. The Provider advertises CSV3 with both MCCs, suggesting that the Consumer should use both if it can.

Endpoint C, in its configure message to the MCU, requests to receive MCC4 for video and MCC5 and MCC6 for audio. In order for the MCU to get the information it needs to construct MCC4, it has to send configure messages to A and B asking to receive MCC1 from each of them along with their AC1 audio. Now, the MCU can use audio energy information from the two incoming audio streams from A and B to determine which of those alternatives is the current talker. Based on that, the MCU uses either MCC1 from A or MCC1 from B as the source of MCC4 to send to C.

TABLE 16.26
Advertisement Sent from the MCU to C

Capture Scene #7	Description=Output3streammix
MCC1 ( )	CaptureArea=Left MaxCaptures=1 SynchronizationID=1 Policy=SoundLevel:0 EncodingGroup=1
MCC2 ( )	CaptureArea=Center MaxCaptures=1 SynchronizationID=1 Policy=SoundLevel:0 EncodingGroup=1
MCC3 ( )	CaptureArea=Right MaxCaptures=1 SynchronizationID=1 Policy=SoundLevel:0 EncodingGroup=1
MCC4 ( )	CaptureArea=whole scene MaxCaptures=1 Policy=SoundLevel:0 EncodingGroup=1
MCC5 ( ) (for audio)	CaptureArea=whole scene MaxCaptures=1 Policy=SoundLevel:1 EncodingGroup=2
MCC6 ( ) (for audio)	CaptureArea=whole scene MaxCaptures=1 Policy=SoundLevel:2 EncodingGroup=2
CSV1 (MCC1, MCC2, MCC3)	
CSV2 (MCC4)	
CSV3 (MCC5, MCC6)	

16.3.13 IANA Considerations

None.

16.3.14 Security Considerations

There are several potential attacks related to telepresence, and specifically the protocols used by CLUE, in the case of conferencing sessions, due to the natural involvement of multiple endpoints and the many, often user-invoked, capabilities provided by the systems.

An MCU involved in a CLUE session can experience many of the same attacks as a conferencing system, such as that enabled by the XCON framework (RFC 5239 – see Chapter 2). Examples of attacks include the following: An endpoint attempting to listen to sessions in which it is not authorized to participate, an endpoint attempting to disconnect or mute other users, and theft of service by an endpoint in attempting to create telepresence sessions it is not allowed to create. Thus, it is RECOMMENDED that an MCU implementing the protocols necessary to support CLUE follow the security recommendations specified in the conference control protocol specifications. In the case of CLUE, SIP is the conferencing protocol; thus, the security considerations in RFC 4579

MUST be followed. Other security issues related to MCUs are discussed in the XCON framework (RFC 5239 – see Chapter 2). The use of xCard with potentially sensitive information provides another reason to implement the recommendations of Section 2.11 (Section 11 of RFC 5239).

One primary security concern, surrounding the CLUE framework introduced in this specification, involves securing the actual protocols and the associated authorization mechanisms. These concerns apply to endpoint-to-endpoint sessions as well as sessions involving multiple endpoints and MCUs. Figure 16.2 provides a basic flow of information exchange for CLUE and the protocols involved.

As described in Section 16.3.5, CLUE uses SIP/SDP to establish the session prior to exchanging any CLUE-specific information. Thus, the security mechanisms recommended for SIP (RFC 3261), including user authentication and authorization, MUST be supported. In addition, the media MUST be secured. DTLS/SRTP MUST be supported and SHOULD be used unless the media, which is based on RTP, is secured by other means (see RFC 7201 and RFC 7202). Media security is also discussed in Sections 16.8 and 16.7. Note that SIP call setup is done before any CLUE-specific information is available, so the authentication and authorization are based on the SIP mechanisms. The entity that will be authenticated may use the Endpoint identity or the endpoint user identity; this is an application issue and not a CLUE-specific issue.

A separate data channel is established to transport the CLUE protocol messages. The contents of the CLUE protocol messages are based on information introduced in this specification. The CLUE data model (Section 16.4) defines through an XML schema the syntax to be used. Some of the information that could possibly introduce privacy concerns is the xCard information as described in Section 16.3.7.1.1.10. The decision about which xCard information to send in the CLUE channel is an application policy for point-to-point and multipoint calls based on the authenticated identity, which can be the endpoint identity or that of the user of the endpoint. For example, the telepresence multipoint application can authenticate a user before starting a CLUE exchange with the telepresence system and have a policy per user.

In addition, the (text) description field in the Media Capture attribute (Section 16.3.7.1.1.6) could possibly reveal sensitive information or specific identities. The same would be true for the descriptions in the Capture Scene (Section 16.3.7.3.1) and Capture Scene View (Section 16.3.7.3.2) attributes. An implementation SHOULD give users control over what sensitive information is sent in an Advertisement. One other important consideration for the information in the xCard as well as the description field in the Media Capture and Capture Scene View attributes is that while the endpoints involved in the session have been authenticated, there is no assurance that the information in the xCard or description fields is authentic. Thus, this information MUST NOT be used to make any authorization decisions.

While other information in the CLUE protocol messages does not reveal specific identities, it can reveal characteristics and capabilities of the endpoints. That information could possibly uniquely identify specific endpoints. It might also be possible for an attacker to manipulate the information and disrupt the CLUE sessions. It would also be possible to mount a denial of service (DoS) attack on the CLUE endpoints if a malicious agent has access to the data channel. Thus, it MUST be possible for the endpoints to establish a channel that is secure against both message recovery and message modification. Further details on this are provided in the CLUE data channel solution specification (Section 16.4). There are also security issues associated with the authorization to perform actions at the CLUE endpoints to invoke specific capabilities (e.g., re-arranging screens, sharing content, etc.). However, the policies and security associated with these actions are outside the scope of this specification and the overall CLUE solution.

16.4 XML SCHEMA FOR THE CLUE DATA MODEL

This specification [14] provides an XML schema file for the definition of CLUE data model types. The term "CLUE" stands for "ControLling mUltiple streams for tElepresence" and is the name of the IETF working group in which this specification, as well as other companion specifications, has been developed. The specification defines a coherent structure for information associated with the description of a telepresence scenario.

16.4.1 Introduction

This specification provides an XML schema file for the definition of CLUE data model types. For the benefit of the reader, the term "CLUE" stands for "ControLling mUltiple streams for tElepresence" and is the name of the IETF working group in which this specification, as well as other companion specifications, has been developed. A thorough definition of the CLUE framework can be found in Section 16.3. The schema is also based on information contained in Section 16.3. It encodes information and constraints defined in the aforementioned specification in order to provide a formal representation of the concepts therein presented.

The specification aims at the definition of a coherent structure for information associated with the description of a telepresence scenario. Such information is used within the CLUE protocol messages (see Section 16.5) enabling the dialog between a Media Provider and a Media Consumer. CLUE protocol messages, indeed, are XML messages allowing (i) a Media Provider to advertise its telepresence capabilities in terms of media captures, capture scenes, and other features envisioned in the CLUE framework, according to the format herein defined, and (ii) a Media Consumer to request the desired telepresence options in the form of capture encodings, represented as described in this specification.

16.4.2 Terminology

See Section 1.6, Table 1.2.

16.4.3 Definitions

This specification refers to the same definitions (also see Section 1.6, Table 1.2) used in Section 16.3, except for the "CLUE Participant" definition. We briefly recall herein some of the main terms used in the specification.

- **Audio Capture:** Media Capture for audio. Denoted as ACn in the examples in this specification.
- **Capture:** Same as Media Capture.
- **Capture Device:** A device that converts physical input, such as audio, video, or text, into an electrical signal, in most cases to be fed into a media encoder.
- **Capture Encoding:** A specific encoding of a Media Capture, to be sent by a Media Provider to a Media Consumer via RTP.
- **Capture Scene:** A structure representing a spatial region captured by one or more Capture Devices, each capturing media representing a portion of the region. The spatial region represented by a Capture Scene MAY correspond to a real region in physical space, such as a room. A Capture Scene includes attributes and one or more Capture Scene Views, with each view including one or more Media Captures.
- **Capture Scene View:** A list of Media Captures of the same media type that together form one way to represent the entire Capture Scene.
- **CLUE Participant:** This term is imported from the CLUE protocol (see Section 16.5).
- **Consumer:** Short for Media Consumer.
- **Encoding or Individual Encoding**: A set of parameters representing a way to encode a Media Capture to become a Capture Encoding.
- **Encoding Group:** A set of encoding parameters representing a total media encoding capability to be sub-divided across potentially multiple Individual Encodings.

- **Endpoint:** A CLUE-capable device that is the logical point of final termination through receiving, decoding, and rendering and/or initiation through capturing, encoding, and sending of media streams. An endpoint consists of one or more physical devices that source and sink media streams and exactly one RFC 4553 Participant (which in turn, includes exactly one SIP user agent). Endpoints can be anything from multiscreen/multicamera rooms to handheld devices.
- **Media:** Any data that, after suitable encoding, can be conveyed over RTP, including audio, video, or timed text.
- **Media Capture:** A source of Media, such as from one or more Capture Devices or constructed from other Media streams.
- **Media Consumer:** A CLUE-capable device that is intended to receive Capture Encodings.
- **Media Provider:** A CLUE-capable device that is intended to send Capture Encodings.
- **MCC:** A Capture that mixes and/or switches other Captures of a single type (e.g., all audio or all video). Particular Media Captures may or may not be present in the resultant Capture Encoding, depending on time or space. Denoted as MCCn in the example cases in this specification.
- **Multipoint Control Unit** (**MCU**): A CLUE-capable device that connects two or more endpoints together into one single multimedia conference (RFC 7667). An MCU includes an RFC 4353-like Mixer without the RFC 4353 requirement to send media to each participant.
- **Plane of Interest:** The spatial plane containing the most relevant subject matter.
- **Provider:** Same as Media Provider.
- *Render:* The process of reproducing the received Streams, such as, for instance, displaying of the remote video on the Media Consumer's screens or playing of the remote audio through loudspeakers.
- **Scene:** Same as Capture Scene.
- **Simultaneous Transmission Set**: A set of Media Captures that can be transmitted simultaneously from a Media Provider.
- **Single Media Capture:** A capture that contains media from a single source capture device, e.g., an audio capture from a single microphone or a video capture from a single camera.
- **Spatial Relation:** The arrangement in space of two objects, in contrast to relation in time or other relationships.
- **Stream:** A Capture Encoding sent from a Media Provider to a Media Consumer via RTP (RFC 3550).
- **Stream Characteristics:** The union of the features used to describe a Stream in the CLUE environment and in the SIP-SDP environment.
- **Video Capture**: A Media Capture for video.

16.4.4 XML Schema

This section contains the CLUE data model schema definition. The element and attribute definitions are formal representations of the concepts needed to describe the capabilities of a Media Provider and the streams that are requested by a Media Consumer given the Media Provider's ADVERTISEMENT (see Section 16.5).

The main groups of information are:

<mediaCaptures>: The list of media captures available (Section 16.4.5)
<encodingGroups>: The list of encoding groups (Section 16.4.6)
<captureScenes>: The list of capture scenes (Section 16.4.7)
<simultaneousSets>: The list of simultaneous transmission sets (Section 16.4.8)
<globalViews>: The list of global views sets (Section 16.4.9)

<people>: Meta-data about the participants represented in the telepresence session (Section 16.4.21)

<captureEncodings>: The list of instantiated capture encodings (Section 16.4.10)

All of this refers to concepts that have been introduced in Section 16.3 and further detailed in this specification.

```
<?xml version="1.0" encoding="UTF-8" ?>
<xs:schema
      targetNamespace="urn:ietf:params:xml:ns:clue-info" xmlns:tns="urn:
      ietf:params:xml:ns:clue-info"
      xmlns:xs="http://www.w3.org/2001/XMLSchema"
      xmlns="urn:ietf:params:xml:ns:clue-info"
      xmlns:xcard="urn:ietf:params:xml:ns:vcard-4.0"
      elementFormDefault="qualified"
      attributeFormDefault="unqualified"
      version="1.0">
      <!-- Import xcard XML schema -->
      <xs:import namespace="urn:ietf:params:xml:ns:vcard-4.0"
      schemaLocation=
      "http://www.iana.org/assignments/xml-registry/schema/vcard-4.0.xsd"/>

      <!-- ELEMENT DEFINITIONS -->
      <xs:element name="mediaCaptures" type="mediaCapturesType"/>
      <xs:element name="encodingGroups" type="encodingGroupsType"/>
      <xs:element name="captureScenes" type="captureScenesType"/>
      <xs:element name="simultaneousSets" type="simultaneousSetsType"/>
      <xs:element name="globalViews" type="globalViewsType"/>
      <xs:element name="people" type="peopleType"/>
      <xs:element name="captureEncodings" type="captureEncodingsType"/>

      <!-- MEDIA CAPTURES TYPE -->
      <!-- envelope of media captures -->
         <xs:complexType name="mediaCapturesType">
            <xs:sequence>
               <xs:element name="mediaCapture" type="mediaCaptureType"
                     maxOccurs="unbounded"/>
            </xs:sequence>
         </xs:complexType>

      <!-- DESCRIPTION element -->
         <xs:element name="description">
            <xs:complexType>
               <xs:simpleContent>
                  <xs:extension base="xs:string">
                     <xs:attribute name="lang" type="xs:language"/>
                  </xs:extension>
               </xs:simpleContent>
            </xs:complexType>
         </xs:element>

      <!-- MEDIA CAPTURE TYPE -->
         <xs:complexType name="mediaCaptureType" abstract="true">
            <xs:sequence>
               <!-- mandatory fields -->
                  <xs:element name="captureSceneIDREF" type="xs:IDREF"/>
```

```
                <xs:choice>
                    <xs:sequence>
                        <xs:element name="spatialInformation"
                                type="tns:spatialInformationType"/>
                    </xs:sequence>
                    <xs:element name="nonSpatiallyDefinable"
                            type="xs:boolean" fixed="true"/>
                </xs:choice>
                <!-- for handling multi-content captures: -->
                <xs:choice>
                    <xs:sequence>
                        <xs:element name="synchronizationID"
                                type="xs:ID" minOccurs="0"/>
                        <xs:element name="content" type="contentType"
                                minOccurs="0"/>
                        <xs:element name="policy" type="policyType"
                                minOccurs="0"/>
                        <xs:element name="maxCaptures"
                                type="maxCapturesType" minOccurs="0"/>
                        <xs:element name="allowSubsetChoice"
                                type="xs:boolean" minOccurs="0"/>
                    </xs:sequence>
                    <xs:element name="individual" type="xs:boolean"
                            fixed="true"/>
                </xs:choice>
                <!-- optional fields -->
                <xs:element name="encGroupIDREF" type="xs:IDREF"
                        minOccurs="0"/>
                <xs:element ref="description" minOccurs="0"
                        maxOccurs="unbounded"/>
                <xs:element name="priority" type="xs:unsignedInt"
                        minOccurs="0"/>
                <xs:element name="lang" type="xs:language"
                        minOccurs="0"
                        maxOccurs="unbounded"/>
                <xs:element name="mobility" type="mobilityType"
                        minOccurs="0" />
                <xs:element ref="presentation" minOccurs="0" />
                <xs:element ref="embeddedText" minOccurs="0" />
                <xs:element ref="view" minOccurs="0" />
                <xs:element name="capturedPeople"
                        type="capturedPeopleType"
                        minOccurs="0"/>
                <xs:element name="relatedTo" type="xs:IDREF"
                        minOccurs="0"/>
        </xs:sequence>
        <xs:attribute name="captureID" type="xs:ID"
                use="required"/>
        <xs:attribute name="mediaType" type="xs:string"
                use="required"/>
    </xs:complexType>

<!-- POLICY TYPE -->
    <xs:simpleType name="policyType">
        <xs:restriction base="xs:string">
            <xs:pattern value="([a-zA-Z0-9])+[:]([0-9])+"/>
```

```
        </xs:restriction>
    </xs:simpleType>

<!-- CONTENT TYPE -->
<xs:complexType name="contentType">
    <xs:sequence>
        <xs:element name="mediaCaptureIDREF" type="xs:IDREF"
                minOccurs="0" maxOccurs="unbounded"/>
        <xs:element name="sceneViewIDREF" type="xs:IDREF"
                minOccurs="0" maxOccurs="unbounded"/>
        <xs:any namespace="##other" processContents="lax"
                minOccurs="0" maxOccurs="unbounded"/>
    </xs:sequence>
    <xs:anyAttribute namespace="##other" processContents="lax"/>
</xs:complexType>

<!-- MAX CAPTURES TYPE -->
<xs:simpleType name="positiveShort">
    <xs:restriction base="xs:unsignedShort">
        <xs:minInclusive value="1">
        </xs:minInclusive>
    </xs:restriction>
</xs:simpleType>
<xs:complexType name="maxCapturesType">
    <xs:simpleContent>
        <xs:extension base="positiveShort">
            <xs:attribute name="exactNumber"
                    type="xs:boolean"/>
        </xs:extension>
    </xs:simpleContent>
</xs:complexType>

<!-- CAPTURED PEOPLE TYPE -->
<xs:complexType name="capturedPeopleType">
    <xs:sequence>
        <xs:element name="personIDREF" type="xs:IDREF"
                maxOccurs="unbounded"/>
    </xs:sequence>
</xs:complexType>

<!-- PEOPLE TYPE -->
<xs:complexType name="peopleType">
    <xs:sequence>
        <xs:element name="person" type="personType"
                maxOccurs="unbounded"/>
    </xs:sequence>
</xs:complexType>

<!-- PERSON TYPE -->
<xs:complexType name="personType">
    <xs:sequence>
        <xs:element name="personInfo" type="xcard:vcardType"
                maxOccurs="1" minOccurs="0"/>
        <xs:element ref="personType" minOccurs="0"
                maxOccurs="unbounded" />
        <xs:any namespace="##other" processContents="lax"
                minOccurs="0" maxOccurs="unbounded"/>
```

```
        </xs:sequence>
        <xs:attribute name="personID" type="xs:ID" use="required"/>
        <xs:anyAttribute namespace="##other" processContents="lax"/>
    </xs:complexType>

    <!-- PERSON TYPE ELEMENT -->
    <xs:element name="personType" type="xs:string">
        <xs:annotation>
            <xs:specificationation>

Acceptable values (enumerations) for this type are managed by IANA in the
"CLUE Schema <personType> registry", accessible at TBD-IANA.

            </xs:specificationation>
        </xs:annotation>
    </xs:element>

    <!-- VIEW ELEMENT -->
    <xs:element name="view" type="xs:string">
        <xs:annotation>
            <xs:specificationation>

Acceptable values (enumerations) for this type are managed by IANA in the
"CLUE Schema <view> registry", accessible at TBD-IANA.

            </xs:specificationation>
        </xs:annotation>
    </xs:element>

    <!-- PRESENTATION ELEMENT -->
    <xs:element name="presentation" type="xs:string">
        <xs:annotation>
            <xs:specificationation>

Acceptable values (enumerations) for this type are managed by IANA in the
"CLUE Schema <presentation> registry," accessible at TBD-IANA.

            </xs:specificationation>
        </xs:annotation>
    </xs:element>

    <!-- SPATIAL INFORMATION TYPE -->
    <xs:complexType name="spatialInformationType">
        <xs:sequence>
            <xs:element name="captureOrigin" type="captureOriginType"
                    minOccurs="0"/>
            <xs:element name="captureArea" type="captureAreaType"
                    minOccurs="0"/>
            <xs:any namespace="##other" processContents="lax"
                    minOccurs="0" maxOccurs="unbounded"/>
        </xs:sequence>
        <xs:anyAttribute namespace="##other" processContents="lax"/>
    </xs:complexType>

    <!-- POINT TYPE -->
    <xs:complexType name="pointType">
        <xs:sequence>
```

```
            <xs:element name="x" type="xs:decimal"/>
            <xs:element name="y" type="xs:decimal"/>
            <xs:element name="z" type="xs:decimal"/>
        </xs:sequence>
    </xs:complexType>

    <!-- CAPTURE ORIGIN TYPE -->
    <xs:complexType name="captureOriginType">
        <xs:sequence>
            <xs:element name="capturePoint"
                    type="pointType"></xs:element>
            <xs:element name="lineOfCapturePoint" type="pointType"
                    minOccurs="0">
            </xs:element>
        </xs:sequence>
        <xs:anyAttribute namespace="##any" processContents="lax"/>
    </xs:complexType>

    <!-- CAPTURE AREA TYPE -->
    <xs:complexType name="captureAreaType">
        <xs:sequence>
            <xs:element name="bottomLeft" type="pointType"/>
            <xs:element name="bottomRight" type="pointType"/>
            <xs:element name="topLeft" type="pointType"/>
            <xs:element name="topRight" type="pointType"/>
        </xs:sequence>
    </xs:complexType>

    <!-- MOBILITY TYPE -->
    <xs:simpleType name="mobilityType">
        <xs:restriction base="xs:string">
            <xs:enumeration value="static" />
            <xs:enumeration value="dynamic" />
            <xs:enumeration value="highly-dynamic" />
        </xs:restriction>
    </xs:simpleType>

    <!-- TEXT CAPTURE TYPE -->
    <xs:complexType name="textCaptureType">
        <xs:complexContent>
            <xs:extension base="tns:mediaCaptureType">
                <xs:sequence>
                    <xs:any namespace="##other" processContents="lax"
                            minOccurs="0" maxOccurs="unbounded"/>
                </xs:sequence>
                <xs:anyAttribute namespace="##other"
                        processContents="lax"/>
            </xs:extension>
        </xs:complexContent>
    </xs:complexType>

    <!-- OTHER CAPTURE TYPE -->
    <xs:complexType name="otherCaptureType">
        <xs:complexContent>
            <xs:extension base="tns:mediaCaptureType">
                <xs:sequence>
```

```
                <xs:any namespace="##other"
                        processContents="lax" minOccurs="0"
                        maxOccurs="unbounded"/>
            </xs:sequence>
            <xs:anyAttribute namespace="##other"
                    processContents="lax"/>
        </xs:extension>
    </xs:complexContent>
</xs:complexType>

<!-- AUDIO CAPTURE TYPE -->
<xs:complexType name="audioCaptureType">
    <xs:complexContent>
        <xs:extension base="tns:mediaCaptureType">
            <xs:sequence>
                <xs:element ref="sensitivityPattern"
                        minOccurs="0" />
                <xs:any namespace="##other"
                        processContents="lax" minOccurs="0"
                        maxOccurs="unbounded"/>
            </xs:sequence>
            <xs:anyAttribute namespace="##other"
                    processContents="lax"/>
        </xs:extension>
    </xs:complexContent>
</xs:complexType>

<!-- SENSITIVITY PATTERN ELEMENT -->
<xs:element name="sensitivityPattern" type="xs:string">
    <xs:annotation>
        <xs:specificationation>

Acceptable values (enumerations) for this type are managed by IANA in the
"CLUE Schema <sensitivityPattern> registry", accessible at TBD-IANA.

        </xs:specificationation>
    </xs:annotation>
</xs:element>

<!-- VIDEO CAPTURE TYPE -->
<xs:complexType name="videoCaptureType">
    <xs:complexContent>
        <xs:extension base="tns:mediaCaptureType">
        <xs:sequence>
            <xs:any namespace="##other" processContents="lax"
                    minOccurs="0" maxOccurs="unbounded"/>
        </xs:sequence>
        <xs:anyAttribute namespace="##other" processContents="lax"/>
        </xs:extension>
    </xs:complexContent>
</xs:complexType>

<!-- EMBEDDED TEXT ELEMENT -->
<xs:element name="embeddedText">
    <xs:complexType>
        <xs:simpleContent>
```

```
            <xs:extension base="xs:boolean">
               <xs:attribute name="lang" type="xs:language"/>
            </xs:extension>
         </xs:simpleContent>
      </xs:complexType>
   </xs:element>

   <!-- CAPTURE SCENES TYPE -->
   <!-- envelope of capture scenes -->
   <xs:complexType name="captureScenesType">
      <xs:sequence>
         <xs:element name="captureScene"
               type="captureSceneType"maxOccurs="unbounded"/>
      </xs:sequence>
   </xs:complexType>

   <!-- CAPTURE SCENE TYPE -->
   <xs:complexType name="captureSceneType">
      <xs:sequence>
         <xs:element ref="description" minOccurs="0"
               maxOccurs="unbounded"/>
         <xs:element name="sceneInformation" type="xcard:vcardType"
               minOccurs="0"/>
         <xs:element name="sceneViews" type="sceneViewsType"
               minOccurs="0"/>
         <xs:any namespace="##other" processContents="lax"
               minOccurs="0" maxOccurs="unbounded"/>
      </xs:sequence>
      <xs:attribute name="sceneID" type="xs:ID" use="required"/>
      <xs:attribute name="scale" type="scaleType" use="required"/>
      <xs:anyAttribute namespace="##other" processContents="lax"/>
   </xs:complexType>

   <!-- SCALE TYPE -->
   <xs:simpleType name="scaleType">
      <xs:restriction base="xs:string">
         <xs:enumeration value="mm"/>
         <xs:enumeration value="unknown"/>
         <xs:enumeration value="noscale"/>
      </xs:restriction>
   </xs:simpleType>

   <!-- SCENE VIEWS TYPE -->
   <!-- envelope of scene views of a capture scene -->
   <xs:complexType name="sceneViewsType">
      <xs:sequence>
         <xs:element name="sceneView" type="sceneViewType"
               maxOccurs="unbounded"/>
      </xs:sequence>
   </xs:complexType>

   <!-- SCENE VIEW TYPE -->
   <xs:complexType name="sceneViewType">
      <xs:sequence>
         <xs:element ref="description" minOccurs="0"
         maxOccurs="unbounded"/>
         <xs:element name="mediaCaptureIDs"
```

```
                type="captureIDListType"/>
        </xs:sequence>
        <xs:attribute name="sceneViewID" type="xs:ID" use="required"/>
    </xs:complexType>

    <!-- CAPTURE ID LIST TYPE -->
    <xs:complexType name="captureIDListType">
        <xs:sequence>
            <xs:element name="mediaCaptureIDREF" type="xs:IDREF"
            maxOccurs="unbounded"/>
        </xs:sequence>
    </xs:complexType>

    <!-- ENCODING GROUPS TYPE -->
    <xs:complexType name="encodingGroupsType">
        <xs:sequence>
            <xs:element name="encodingGroup" type="tns:encodingGroupType"
                    maxOccurs="unbounded"/>
        </xs:sequence>
    </xs:complexType>

    <!-- ENCODING GROUP TYPE -->
    <xs:complexType name="encodingGroupType">
        <xs:sequence>
            <xs:element name="maxGroupBandwidth"
                    type="xs:unsignedLong"/>
            <xs:element name="encodingIDList"
                    type="encodingIDListType"/>
            <xs:any namespace="##other" processContents="lax"
                    minOccurs="0" maxOccurs="unbounded"/>
        </xs:sequence>
        <xs:attribute name="encodingGroupID" type="xs:ID" use="required"/>
        <xs:anyAttribute namespace="##any" processContents="lax"/>
    </xs:complexType>

    <!-- ENCODING ID LIST TYPE -->
    <xs:complexType name="encodingIDListType">
        <xs:sequence>
            <xs:element name="encodingID" type="xs:string"
                    maxOccurs="unbounded"/>
        </xs:sequence>
    </xs:complexType>

    <!-- SIMULTANEOUS SETS TYPE -->
    <xs:complexType name="simultaneousSetsType">
        <xs:sequence>
            <xs:element name="simultaneousSet"
                    type="simultaneousSetType" maxOccurs="unbounded"/>
        </xs:sequence>
    </xs:complexType>

    <!-- SIMULTANEOUS SET TYPE -->
    <xs:complexType name="simultaneousSetType">
        <xs:sequence>
            <xs:element name="mediaCaptureIDREF" type="xs:IDREF"
                    minOccurs="0" maxOccurs="unbounded"/>
```

```
        <xs:element name="sceneViewIDREF" type="xs:IDREF"
                minOccurs="0" maxOccurs="unbounded"/>
        <xs:element name="captureSceneIDREF" type="xs:IDREF"
                minOccurs="0" maxOccurs="unbounded"/>
        <xs:any namespace="##other" processContents="lax"
                minOccurs="0" maxOccurs="unbounded"/>
    </xs:sequence>
    <xs:attribute name="setID" type="xs:ID" use="required"/>
    <xs:attribute name="mediaType" type="xs:string"/>
    <xs:anyAttribute namespace="##any" processContents="lax"/>
</xs:complexType>

<!-- GLOBAL VIEWS TYPE -->
<xs:complexType name="globalViewsType">
    <xs:sequence>
        <xs:element name="globalView" type="globalViewType"
                maxOccurs="unbounded"/>
    </xs:sequence>
</xs:complexType>

<!-- GLOBAL VIEW TYPE -->
<xs:complexType name="globalViewType">
    <xs:sequence>
        <xs:element name="sceneViewIDREF" type="xs:IDREF"
                maxOccurs="unbounded"/>
        <xs:any namespace="##other" processContents="lax"
                minOccurs="0" maxOccurs="unbounded"/>
    </xs:sequence>
    <xs:attribute name="globalViewID" type="xs:ID"/>
    <xs:anyAttribute namespace="##any" processContents="lax"/>
</xs:complexType>

<!-- CAPTURE ENCODINGS TYPE -->
<xs:complexType name="captureEncodingsType">
    <xs:sequence>
        <xs:element name="captureEncoding"
                type="captureEncodingType"
                maxOccurs="unbounded"/>
    </xs:sequence>
</xs:complexType>

<!-- CAPTURE ENCODING TYPE -->
<xs:complexType name="captureEncodingType">
    <xs:sequence>
        <xs:element name="captureID" type="xs:string"/>
        <xs:element name="encodingID" type="xs:string"/>
        <xs:element name="configuredContent" type="contentType"
                minOccurs="0"/>
        <xs:any namespace="##other" processContents="lax"
                minOccurs="0" maxOccurs="unbounded"/>
    </xs:sequence>
    <xs:attribute name="ID" type="xs:ID" use="required"/>
    <xs:anyAttribute namespace="##any" processContents="lax"/>
</xs:complexType>

<!-- CLUE INFO ELEMENT -->
<xs:element name="clueInfo" type="clueInfoType"/>
```

```
        <!-- CLUE INFO TYPE -->
        <xs:complexType name="clueInfoType">
            <xs:sequence>
                <xs:element ref="mediaCaptures"/>
                <xs:element ref="encodingGroups"/>
                <xs:element ref="captureScenes"/>
                <xs:element ref="simultaneousSets" minOccurs="0"/>
                <xs:element ref="globalViews" minOccurs="0"/>
                <xs:element ref="people" minOccurs="0"/>
                <xs:any namespace="##other" processContents="lax"
                        minOccurs="0" maxOccurs="unbounded"/>
            </xs:sequence>
            <xs:attribute name="clueInfoID" type="xs:ID" use="required"/>
            <xs:anyAttribute namespace="##other" processContents="lax"/>
        </xs:complexType>
</xs:schema>
```

The following sections describe the XML schema in more detail. As a general remark, please notice that optional elements that do not define what their absence means are intended to be associated with undefined properties.

16.4.5 <mediaCaptures>

<mediaCaptures> represents the list of one or more media captures available at the Media Provider's side. Each media capture is represented by a <mediaCapture> element (Section 16.4.11).

16.4.6 <encodingGroups>

<encodingGroups> represents the list of the encoding groups organized on the Media Provider's side. Each encoding group is represented by an <encodingGroup> element (Section 16.4.18).

16.4.7 <captureScenes>

<captureScenes> represents the list of the capture scenes organized on the Media Provider's side. Each capture scene is represented by a <captureScene> element (Section 16.4.16).

16.4.8 <simultaneousSets>

<simultaneousSets> contains the simultaneous sets indicated by the Media Provider. Each simultaneous set is represented by a <simultaneousSet> element (Section 16.4.19).

16.4.9 <globalViews>

<globalViews> contains a set of alternative representations of all the scenes that are offered by a Media Provider to a Media Consumer. Each alternative is named "global view" and is represented by a <globalView> element (Section 16.4.20).

16.4.10 <captureEncodings>

<captureEncodings> is a list of capture encodings. It can represent the list of the desired capture encodings indicated by the Media Consumer or the list of instantiated captures on the provider's side. Each capture encoding is represented by a <captureEncoding> element (Section 16.4.22).

16.4.11 <MEDIACAPTURE>

A Media Capture is the fundamental representation of a media flow that is available on the provider's side. Media captures are characterized (i) by a set of features that are independent of the specific type of medium and (ii) by a set of features that are media specific. The features that are common to all media types appear within the media capture type, which has been designed as an abstract complex type. Media-specific captures, such as video captures, audio captures, and others, are specializations of that abstract media capture type, as in a typical generalization-specialization hierarchy.

The following is the XML Schema definition of the media capture type:

```
<!-- MEDIA CAPTURE TYPE -->
<xs:complexType name="mediaCaptureType" abstract="true">
   <xs:sequence>
      <!-- mandatory fields -->
      <xs:element name="captureSceneIDREF" type="xs:IDREF"/>
      <xs:choice>
         <xs:sequence>
            <xs:element name="spatialInformation"
                  type="tns:spatialInformationType"/>
         </xs:sequence>
         <xs:element name="nonSpatiallyDefinable" type="xs:boolean"
               fixed="true"/>
      </xs:choice>
         <!-- for handling multi-content captures: -->
      <xs:choice>
         <xs:sequence>
            <xs:element name="synchronizationID" type="xs:ID"
                  minOccurs="0"/>
            <xs:element name="content" type="contentType" minOccurs="0"/>
            <xs:element name="policy" type="policyType" minOccurs="0"/>
            <xs:element name="maxCaptures" type="maxCapturesType"
                  minOccurs="0"/>
            <xs:element name="allowSubsetChoice" type="xs:boolean"
                  minOccurs="0"/>
         </xs:sequence>
         <xs:element name="individual" type="xs:boolean" fixed="true"/>
      </xs:choice>
      <!-- optional fields -->
      <xs:element name="encGroupIDREF" type="xs:IDREF" minOccurs="0"/>
      <xs:element ref="description" minOccurs="0" maxOccurs="unbounded"/>
      <xs:element name="priority" type="xs:unsignedInt" minOccurs="0"/>
      <xs:element name="lang" type="xs:language" minOccurs="0"
            maxOccurs="unbounded"/>
      <xs:element name="mobility" type="mobilityType" minOccurs="0" />
      <xs:element ref="presentation" minOccurs="0" />
      <xs:element ref="embeddedText" minOccurs="0" />
      <xs:element ref="view" minOccurs="0" />
      <xs:element name="capturedPeople" type="capturedPeopleType"
            minOccurs="0"/>
      <xs:element name="relatedTo" type="xs:IDREF" minOccurs="0"/>
   </xs:sequence>
   <xs:attribute name="captureID" type="xs:ID" use="required"/>
   <xs:attribute name="mediaType" type="xs:string" use="required"/>
</xs:complexType>
```

16.4.11.1 captureID Attribute

The "captureID" attribute is a mandatory field containing the identifier of the media capture. Such an identifier serves as the way the capture is referenced from other data model elements (e.g., simultaneous sets, capture encodings, and others via <mediaCaptureIDREF>).

16.4.11.2 mediaType Attribute

The "mediaType" attribute is a mandatory attribute specifying the media type of the capture. Common standard values are "audio," "video," and "text," as defined in RFC 6838. Other values can be provided. It is assumed that implementations agree on the interpretation of those other values. The "mediaType" attribute is as generic as possible. Here is why: (i) The basic media capture type is an abstract one; (ii) "concrete" definitions for the standard (RFC 6838) audio, video, and text capture types have been specified; (iii) a generic "otherCaptureType" type has been defined; (iv) the "mediaType" attribute has been generically defined as a string, with no particular template. From these considerations, it is clear that if one chooses to rely on a brand new media type and wants to interoperate with others, an application-level agreement is needed on how to interpret such information.

16.4.11.3 <captureSceneIDREF>

<captureSceneIDREF> is a mandatory field containing the value of the identifier of the capture scene the media capture is defined in, i.e., the value of the sceneID (Section 16.4.16.3) attribute of that capture scene. Indeed, each media capture MUST be defined within one and only one capture scene. When a media capture is spatially definable, some spatial information is provided along with it in the form of point coordinates (see Section 16.4.11.5). Such coordinates refer to the space of coordinates defined for the capture scene containing the capture.

16.4.11.4 <encGroupIDREF>

<encGroupIDREF> is an optional field containing the identifier of the encoding group the media capture is associated with, i.e., the value of the encodingGroupID (Section 16.4.18.3) attribute of that encoding group. Media captures that are not associated with any encoding group cannot be instantiated as media streams.

16.4.11.5 <spatialInformation>

Media captures are divided into two categories: (i) Non-spatially definable captures and (ii) spatially definable captures. Captures are spatially definable when at least (i) it is possible to provide the coordinates of the device position within the telepresence room of origin (capture point) together with its capturing direction specified by a second point (point on line of capture) or (ii) it is possible to provide the represented area within the telepresence room by listing the coordinates of the four co-planar points identifying the plane of interest (area of capture).

The coordinates of the above-mentioned points MUST be expressed according to the coordinate space of the capture scene the media captures belong to. Non-spatially definable captures cannot be characterized within the physical space of the telepresence room of origin. Captures of this kind are, for example, those related to recordings, text captures, DVDs, registered presentations, or external streams that are played in the telepresence room and transmitted to remote sites. Spatially definable captures represent a part of the telepresence room. The captured part of the telepresence room is described by means of the <spatialInformation> element. By comparing the <spatialInformation> element of different media captures within the same capture scene, a consumer can better determine the spatial relationships between them and render them correctly. Non-spatially definable captures do not embed such an element in their XML description; they are instead characterized by having the <nonSpatiallyDefinable> tag set to "true" (see Section 16.4.11.6).

The definition of the spatial information type is the following:

```
<!-- SPATIAL INFORMATION TYPE -->
<xs:complexType name="spatialInformationType">
    <xs:sequence>
        <xs:element name="captureOrigin" type="captureOriginType"
                minOccurs="0"/>
        <xs:element name="captureArea" type="captureAreaType"
                minOccurs="0"/>
        <xs:any namespace="##other" processContents="lax"
                minOccurs="0" maxOccurs="unbounded"/>
    </xs:sequence>
    <xs:anyAttribute namespace="##other" processContents="lax"/>
</xs:complexType>
```

The <captureOrigin> contains the coordinates of the capture device that is taking the capture (i.e., the capture point), as well as, optionally, the pointing direction (i.e., the point on line of capture) (see Section 16.4.11.5.1). The <captureArea> is an optional field containing four points defining the captured area covered by the capture (see Section 16.4.11.5.2). The scale of the points coordinates is specified in the scale (Section 16.4.16.4) attribute of the capture scene the media capture belongs to. Indeed, all the spatially definable media captures referring to the same capture scene share the same coordinate system and express their spatial information according to the same scale.

16.4.11.5.1 <captureOrigin>

The <captureOrigin> element is used to represent the position and optionally, the line of capture of a capture device. <captureOrigin> MUST be included in spatially definable audio captures, while it is optional for spatially definable video captures.

The XML Schema definition of the <captureOrigin> element type is the following:

```
<!-- CAPTURE ORIGIN TYPE -->
<xs:complexType name="captureOriginType">
    <xs:sequence>
        <xs:element name="capturePoint" type="pointType"/>
        <xs:element name="lineOfCapturePoint" type="pointType"
                minOccurs="0"/>
    </xs:sequence>
    <xs:anyAttribute namespace="##any" processContents="lax"/>
</xs:complexType>
<!-- POINT TYPE -->
<xs:complexType name="pointType">
    <xs:sequence>
        <xs:element name="x" type="xs:decimal"/>
        <xs:element name="y" type="xs:decimal"/>
        <xs:element name="z" type="xs:decimal"/>
    </xs:sequence>
</xs:complexType>
```

The point type contains three spatial coordinates (x,y,z) representing a point in the space associated with a certain capture scene. The <captureOrigin> element includes a mandatory <capturePoint> element and an optional <lineOfCapturePoint> element, both of the type "pointType." <capturePoint> specifies the three coordinates identifying the position of the capture device. <lineOfCapturePoint> is another pointType element representing the "point on line of capture", which gives the pointing direction of the capture device. The coordinates of the point on line of capture MUST NOT be identical to the capture point coordinates. For a spatially definable video capture, if the point on line of capture is provided, it MUST belong to the region between the point of capture and the capture area. For a spatially definable audio capture, if the point on line of capture is not provided, the sensitivity pattern should be considered omnidirectional.

16.4.11.5.2 <captureArea>

<captureArea> is an optional element that can be contained within the spatial information associated with a media capture. It represents the spatial area captured by the media capture. <captureArea> MUST be included in the spatial information of spatially definable video captures, while it MUST NOT be associated with audio captures. The XML representation of that area is provided through a set of four point-type elements, <bottomLeft>, <bottomRight>, <topLeft>, and <topRight>, which MUST be co-planar. The four co-planar points are identified from the perspective of the capture device.

The XML schema definition is the following:

```
<!-- CAPTURE AREA TYPE -->
<xs:complexType name="captureAreaType">
   <xs:sequence>
      <xs:element name="bottomLeft" type="pointType"/>
      <xs:element name="bottomRight" type="pointType"/>
      <xs:element name="topLeft" type="pointType"/>
      <xs:element name="topRight" type="pointType"/>
   </xs:sequence>
</xs:complexType>
```

16.4.11.6 <nonSpatiallyDefinable>

When media captures are non-spatially definable, they MUST be marked with the Boolean <nonSpatiallyDefinable> element set to "true," and no <spatialInformation> MUST be provided. Indeed, <nonSpatiallyDefinable> and <spatialInformation> are mutually exclusive tags according to the <choice> section within the XML Schema definition of the media capture type.

16.4.11.7 <content>

A media capture can be (i) an individual media capture or (ii) an MCC. An MCC is made by different captures that can be arranged spatially (by a composition operation), or temporally (by a switching operation), or that can result from the orchestration of both the techniques. If a media capture is an MCC, then it MAY show in its XML data model representation the <content> element. It is composed by a list of media capture identifiers ("mediaCaptureIDREF") and capture scene view identifiers ("sceneViewIDREF"), where the latter are used as shortcuts to refer to multiple capture identifiers. The referenced captures are used to create the MCC according to a certain strategy. If the <content> element does not appear in a MCC, or it has no child elements, then the MCC is assumed to be made of multiple sources, but no information regarding those sources is provided.

```
<!-- CONTENT TYPE -->
<xs:complexType name="contentType">
   <xs:sequence>
      <xs:element name="mediaCaptureIDREF" type="xs:IDREF"
            minOccurs="0" maxOccurs="unbounded"/>
      <xs:element name="sceneViewIDREF" type="xs:IDREF"
            minOccurs="0" maxOccurs="unbounded"/>
      <xs:any namespace="##other" processContents="lax"
            minOccurs="0"
            maxOccurs="unbounded"/>
   </xs:sequence>
   <xs:anyAttribute namespace="##other" processContents="lax"/>
</xs:complexType>
```

16.4.11.8 <synchronizationID>

<synchronizationID> is an optional element for MCCs that contains a numeric identifier. MCCs marked with the same identifier in the <synchronizationID> contain at all times captures coming from the same sources. It is the Media Provider that determines what the source for the captures is. In this way, the Media Provider can choose how to group together single captures for the purpose of keeping them synchronized according to the <synchronizationID> element.

16.4.11.9 <allowSubsetChoice>

<allowSubsetChoice> is an optional Boolean element for MCCs. It indicates whether or not the Provider allows the Consumer to choose a specific subset of the captures referenced by the MCC. If this attribute is true, and the MCC references other captures, then the Consumer MAY specify in a CONFIGURE message a specific subset of those captures to be included in the MCC, and the Provider MUST then include only that subset. If this attribute is false, or the MCC does not reference other captures, then the Consumer MUST NOT select a subset. If <allowSubsetChoice> is not shown in the XML description of the MCC, its value is to be considered "false."

16.4.11.10 <policy>

<policy> is an optional element that can be used only for MCCs. It indicates the criteria applied to build the MCC using the media captures referenced in the <mediaCaptureIDREF> list. The <policy> value is in the form of a token that indicates the policy and an index representing an instance of the policy, separated by a ":" (e.g., SoundLevel:2, RoundRobin:0, etc.).

The XML schema defining the type of the <policy> element is the following:

```
<!-- POLICY TYPE -->
<xs:simpleType name="policyType">
   <xs:restriction base="xs:string">
      <xs:pattern value="([a-zA-Z0-9])+[:]([0-9])+"/>
   </xs:restriction>
</xs:simpleType>
```

At the time of writing, only two switching policies are defined in Section 16: SoundLevel: The content of the MCC is determined by a sound level detection algorithm. The loudest (active) speaker (or a previous speaker, depending on the index value) is contained in the MCC. Index 0 represents the most current instance of the policy, i.e., the currently active speaker, 1 represents the previous instance, i.e., the previous active speaker, and so on. RoundRobin: The content of the MCC is determined by a time-based algorithm. Other values for the <policy> element can be used. In this case, it is assumed that implementations agree on the meaning of those other values and/or those new switching policies are defined in later specifications.

16.4.11.11 <maxCaptures>

<maxCaptures> is an optional element that can be used only for MCC. It provides information about the number of media captures that can be represented in the MCC at a time. If <maxCaptures> is not provided, all the media captures listed in the <content> element can appear at a time in the capture encoding. The type definition is provided here.

```
<!-- MAX CAPTURES TYPE -->
<xs:simpleType name="positiveShort">
   <xs:restriction base="xs:unsignedShort">
      <xs:minInclusive value="1">
      </xs:minInclusive>
   </xs:restriction>
</xs:simpleType>
```

```
<xs:complexType name="maxCapturesType">
   <xs:simpleContent>
      <xs:extension base="positiveShort">
         <xs:attribute name="exactNumber"
               type="xs:boolean"/>
      </xs:extension>
   </xs:simpleContent>
</xs:complexType>
```

When the “exactNumber” attribute is set to “true,” it means that the <maxCaptures> element carries the exact number of the media captures appearing at a time. Otherwise, the number of the represented media captures MUST be considered “<=” the <maxCaptures> value. For instance, an audio MCC having the <maxCaptures> value set to 1 means that a media stream from the MCC will only contain audio from a single one of its constituent captures at a time. On the other hand, if the <maxCaptures> value is set to 4 and the exactNumber attribute is set to “true,” it means that the media stream received from the MCC will always contain a mix of audio from exactly four of its constituent captures.

16.4.11.12 <individual>

<individual> is a Boolean element that MUST be used for single content captures. Its value is fixed and set to “true.” This element indicates that the capture that is being described is not an MCC. Indeed, <individual> and the aforementioned tags related to MCC attributes (from Section 16.4.11.7 to Section 16.4.11.11) are mutually exclusive according to the <choice> section within the XML Schema definition of the media capture type.

16.4.11.13 <description>

<description> is used to provide human-readable textual information. This element is included in the XML definition of media captures, capture scenes, and capture scene views with the aim of providing human-readable description of, respectively, media captures, capture scenes, and capture scene views. According to the data model definition of a media capture (Section 16.4.11), zero or more <description> elements can be used, each providing information in a different language.

The <description> element definition is the following:

```
<!-- DESCRIPTION element -->
<xs:element name="description">
   <xs:complexType>
      <xs:simpleContent>
         <xs:extension base="xs:string">
            <xs:attribute name="lang" type="xs:language"/>
         </xs:extension>
      </xs:simpleContent>
   </xs:complexType>
</xs:element>
```

As can be seen, <description> is a string element with an attribute (“lang”) indicating the language used in the textual description. Such an attribute is compliant with the Language-Tag ABNF production from RFC 5646.

16.4.11.14 <priority>

<priority> is an optional unsigned integer field indicating the importance of a media capture according to the Media Provider’s perspective. It can be used on the receiver’s side to automatically identify the most relevant contribution from the Media Provider. The higher the importance, the

lower the contained value. If no priority is assigned, no assumptions regarding relative importance of the media capture can be assumed.

16.4.11.15 <lang>

<lang> is an optional element containing the language used in the capture. Zero or more <lang> elements can appear in the XML description of a media capture. Each such element has to be compliant with the Language-Tag ABNF production from RFC 5646.

16.4.11.16 <mobility>

<mobility> is an optional element indicating whether or not the capture device originating the capture may move during the telepresence session. That optional element can assume one of the three following values. Static SHOULD NOT change for the duration of the CLUE session across multiple ADVERTISEMENT messages. dynamic MAY change in each new ADVERTISEMENT message. It can be assumed to remain unchanged until there is a new ADVERTISEMENT message. highly-dynamic MAY change dynamically even between consecutive ADVERTISEMENT messages. The spatial information provided in an ADVERTISEMENT message is simply a snapshot of the current values at the time when the message is sent.

16.4.11.17 <relatedTo>

The optional <relatedTo> element contains the value of the captureID attribute (Section 16.4.11.1) of the media capture to which the considered media capture refers. The media capture marked with a <relatedTo> element can be, for example, the translation of the referred media capture in a different language.

16.4.11.18 <view>

The <view> element is an optional tag describing what is represented in the spatial area covered by a media capture. It has been specified as a simple string with an annotation pointing to an ad hoc defined Internet Assigned Numbers Authority (IANA) registry:

```
<!-- VIEW ELEMENT -->
<xs:element name="view" type="xs:string">
    <xs:annotation>
        <xs:specificationation>
            Acceptable values (enumerations) for this type are managed
            by IANA in the "CLUE Schema <view> registry",
            accessible at TBD-IANA.
        </xs:specificationation>
    </xs:annotation>
</xs:element>
```

The current possible values, as per the CLUE framework specification (see Section 16.3), are: "room," "table," "lectern," "individual," and "audience."

16.4.11.19 <presentation>

The <presentation> element is an optional tag used for media captures conveying information about presentations within the telepresence session. It has been specified as a simple string with an annotation pointing to an ad hoc defined IANA registry:

```
<!-- PRESENTATION ELEMENT -->
<xs:element name="presentation" type="xs:string">
    <xs:annotation>
        <xs:specificationation>
```

```
            Acceptable values (enumerations) for this type are
            managed by IANA in the "CLUE Schema <presentation>
            registry", accessible at TBD-IANA.
         </xs:specificationation>
      </xs:annotation>
   </xs:element>
```

The current possible values, as per the CLUE framework specification (Section 16.3), are "slides" and "images."

16.4.11.20 <embeddedText>

The <embeddedText> element is a Boolean element indicating that there is text embedded in the media capture (e.g., in a video capture). The language used in such embedded textual description is reported in <embeddedText> "lang" attribute.

The XML Schema definition of the <embeddedText> element is:

```
<!-- EMBEDDED TEXT ELEMENT -->
<xs:element name="embeddedText">
   <xs:complexType>
      <xs:simpleContent>
         <xs:extension base="xs:boolean">
            <xs:attribute name="lang" type="xs:language"/>
         </xs:extension>
      </xs:simpleContent>
   </xs:complexType>
</xs:element>
```

16.4.11.21 <capturedPeople>

This optional element is used to indicate which telepresence session participants are represented within the media captures. For each participant, a <personIDREF> element is provided.

16.4.11.21.1 <personIDREF>

<personIDREF> contains the identifier of the represented person, i.e., the value of the related personID attribute (Section 16.4.21.1.1). Meta-data about the represented participant can be retrieved by accessing the <people> list (Section 16.4.21).

16.4.12 Audio Captures

Audio captures inherit all the features of a generic media capture and present further audio-specific characteristics. The XML Schema definition of the audio capture type is reported here:

```
<!-- AUDIO CAPTURE TYPE -->
<xs:complexType name="audioCaptureType">
   <xs:complexContent>
      <xs:extension base="tns:mediaCaptureType">
         <xs:sequence>
            <xs:element ref="sensitivityPattern"
                  minOccurs="0" />
            <xs:any namespace="##other"
                  processContents="lax"
                  minOccurs="0" maxOccurs="unbounded"/>
         </xs:sequence>
         <xs:anyAttribute namespace="##other"
               processContents="lax"/>
```

```
        </xs:extension>
    </xs:complexContent>
</xs:complexType>
```

An example of audio-specific information that can be included is represented by the <sensitivityPattern> element. (Section 16.4.12.1).

16.4.12.1 <sensitivityPattern>

The <sensitivityPattern> element is an optional field describing the characteristics of the nominal sensitivity pattern of the microphone capturing the audio signal. It has been specified as a simple string with an annotation pointing to an ad hoc defined IANA registry:

```
<!-- SENSITIVITY PATTERN ELEMENT -->
<xs:element name="sensitivityPattern" type="xs:string">
    <xs:annotation>
        <xs:specificationation>
            Acceptable values (enumerations) for this type are
            managed by IANA in the "CLUE Schema <sensitivityPattern>
            registry", accessible at TBD-IANA.
        </xs:specificationation>
    </xs:annotation>
</xs:element>
```

The current possible values, as per the CLUE framework specification (see Section 16.3), are "uni," "shotgun," "omni," "figure8," "cardioid," and "hyper-cardioid."

16.4.13 Video Captures

Video captures, similarly to audio captures, extend the information of a generic media capture with video-specific features. The XML Schema representation of the video capture type is provided in the following:

```
<!-- VIDEO CAPTURE TYPE -->
<xs:complexType name="videoCaptureType">
    <xs:complexContent>
        <xs:extension base="tns:mediaCaptureType">
            <xs:sequence>
                <xs:any namespace="##other"
                        processContents="lax" minOccurs="0"
                        maxOccurs="unbounded"/>
            </xs:sequence>
            <xs:anyAttribute namespace="##other"
                    processContents="lax"/>
        </xs:extension>
    </xs:complexContent>
</xs:complexType>
```

16.4.14 Text Captures

Also, text captures can be described by extending the generic media capture information, similarly to audio captures and video captures. There are no known properties of a text-based media that are not already covered by the generic mediaCaptureType. Text captures are hence defined as follows:

```
<!-- TEXT CAPTURE TYPE -->
<xs:complexType name="textCaptureType">
   <xs:complexContent>
      <xs:extension base="tns:mediaCaptureType">
         <xs:sequence>
            <xs:any namespace="##other" processContents="lax"
                  minOccurs="0"
                  maxOccurs="unbounded"/>
         </xs:sequence>
         <xs:anyAttribute
               namespace="##other"processContents="lax"/>
      </xs:extension>
   </xs:complexContent>
</xs:complexType>
```

Text captures MUST be marked as non-spatially definable (i.e., they MUST present in their XML description the <nonSpatiallyDefinable> (Section 16.4.11.6) element set to "true").

16.4.15 Other Capture Types

Other media capture types can be described by using the CLUE data model. They can be represented by exploiting the "otherCaptureType" type. This media capture type is conceived to be filled in with elements defined within extensions of the current schema, i.e., with elements defined in other XML schemas (see Section 16.4.24 for an example). The otherCaptureType inherits all the features envisioned for the abstract mediaCaptureType.

The XML Schema representation of the otherCaptureType is the following:

```
<!-- OTHER CAPTURE TYPE -->
<xs:complexType name="otherCaptureType">
   <xs:complexContent>
      <xs:extension base="tns:mediaCaptureType">
         <xs:sequence>
            <xs:any namespace="##other"
                  processContents="lax" minOccurs="0"
                  maxOccurs="unbounded"/>
         </xs:sequence>
         <xs:anyAttribute namespace="##other"
               processContents="lax"/>
      </xs:extension>
   </xs:complexContent>
</xs:complexType>
```

When defining new media capture types that are going to be described by means of the <otherMediaCapture> element, spatial properties of such new media capture types SHOULD be defined (e.g., whether or not they are spatially definable, whether or not they should be associated with an area of capture, or other properties that may be defined).

16.4.16 <captureScene>

A Media Provider organizes the available captures in capture scenes in order to help the receiver both in the rendering and in the selection of the group of captures. Capture scenes are made of media captures and capture scene views, which are sets of media captures of the same media type. Each capture scene view is an alternative to represent completely a capture scene for a fixed media type.

The XML Schema representation of a <captureScene> element is the following:

```
<!-- CAPTURE SCENE TYPE -->
<xs:complexType name="captureSceneType">
    <xs:sequence>
        <xs:element ref="description" minOccurs="0"
                maxOccurs="unbounded"/>
        <xs:element name="sceneInformation" type="xcard:vcardType"
                minOccurs="0"/>
        <xs:element name="sceneViews" type="sceneViewsType"
                minOccurs="0"/>
        <xs:any namespace="##other" processContents="lax"
                minOccurs="0"
                maxOccurs="unbounded"/>
    </xs:sequence>
    <xs:attribute name="sceneID" type="xs:ID" use="required"/>
    <xs:attribute name="scale" type="scaleType" use="required"/>
    <xs:anyAttribute namespace="##other" processContents="lax"/>
</xs:complexType>
```

Each capture scene is identified by a “sceneID” attribute. The <captureScene> element can contain zero or more textual <description> elements, defined as in Section 16.4.11.13. Besides <description>, there is the optional <sceneInformation> element (Section 16.4.16.1), which contains structured information about the scene in the vcard format, and the optional <sceneViews> element (Section 16.4.16.2), which is the list of the capture scene views. When no <sceneViews> is provided, the capture scene is assumed to be made of all the media captures that contain the value of its sceneID attribute in their mandatory captureSceneIDREF attribute.

16.4.16.1 <sceneInformation>

The <sceneInformation> element contains optional information about the capture scene according to the vcard format, as specified in the Xcard (RFC 6351).

16.4.16.2 <sceneViews>

The <sceneViews> element is a mandatory field of a capture scene containing the list of scene views. Each scene view is represented by a <sceneView> element (Section 16.4.17).

```
<!-- SCENE VIEWS TYPE -->
<!-- envelope of scene views of a capture scene -->
<xs:complexType name="sceneViewsType">
    <xs:sequence>
        <xs:element name="sceneView" type="sceneViewType"
                maxOccurs="unbounded"/>
    </xs:sequence>
</xs:complexType>
```

16.4.16.3 sceneID attribute

The sceneID attribute is a mandatory attribute containing the identifier of the capture scene.

16.4.16.4 scale attribute

The scale attribute is a mandatory attribute that specifies the scale of the coordinates provided in the spatial information of the media capture belonging to the considered capture scene. The scale attribute can assume three different values. “mm” – The scale is in millimeters. Systems that know their physical dimensions (for example, professionally installed telepresence room systems) should always provide such real-world measurements. “unknown” – The scale is the same for every media capture in the capture scene, but the unit of measure is undefined. Systems that are not aware of specific physical dimensions yet still know relative distances should select “unknown” in the scale attribute of the capture scene to be described. “noscale” – There is no common physical scale among the media captures of the capture scene. This means that the scale could be different for each media capture.

```
<!-- SCALE TYPE -->
<xs:simpleType name="scaleType">
    <xs:restriction base="xs:string">
        <xs:enumeration value="mm"/>
        <xs:enumeration value="unknown"/>
        <xs:enumeration value="noscale"/>
    </xs:restriction>
</xs:simpleType>
```

16.4.17 <sceneView>

A <sceneView> element represents a capture scene view, which contains a set of media captures of the same media type describing a capture scene.

A <sceneView> element is characterized as follows.

```
<!-- SCENE VIEW TYPE -->
<xs:complexType name="sceneViewType">
    <xs:sequence>
        <xs:element ref="description" minOccurs="0"
                maxOccurs="unbounded"/>
        <xs:element name="mediaCaptureIDs"
                type="captureIDListType"/>
    </xs:sequence>
    <xs:attribute name="sceneViewID" type="xs:ID" use="required"/>
</xs:complexType>
```

One or more optional <description> elements provide human-readable information about what the scene view contains. <description> is defined as already seen in Section 16.4.11.13.

The remaining child elements are described in the following sub-sections.

16.4.17.1 <mediaCaptureIDs>

The <mediaCaptureIDs> is the list of the identifiers of the media captures included in the scene view. It is an element of the captureIDListType type, which is defined as a sequence of <mediaCaptureIDREF>, each containing the identifier of a media capture listed within the <mediaCaptures> element:

```
<!-- CAPTURE ID LIST TYPE -->
<xs:complexType name="captureIDListType">
    <xs:sequence>
        <xs:element name="mediaCaptureIDREF" type="xs:IDREF"
                maxOccurs="unbounded"/>
    </xs:sequence>
</xs:complexType>
```

16.4.17.2 sceneViewID attribute

The sceneViewID attribute is a mandatory attribute containing the identifier of the capture scene view represented by the <sceneView> element.

16.4.18 <encodingGroup>

The <encodingGroup> element represents an encoding group, which is made by a set of one or more individual encodings and some parameters that apply to the group as a whole. Encoding groups contain references to individual encodings that can be applied to media captures. The definition of the <encodingGroup> element is the following:

```
<!-- ENCODING GROUP TYPE -->
<xs:complexType name="encodingGroupType">
    <xs:sequence>
        <xs:element name="maxGroupBandwidth"
                type="xs:unsignedLong"/>
        <xs:element name="encodingIDList"
                type="encodingIDListType"/>
        <xs:any namespace="##other" processContents="lax"
                minOccurs="0" maxOccurs="unbounded"/>
    </xs:sequence>
    <xs:attribute name="encodingGroupID" type="xs:ID" use="required"/>
    <xs:anyAttribute namespace="##any" processContents="lax"/>
</xs:complexType>
```

In the following, the contained elements are further described.

16.4.18.1 <maxGroupBandwidth>

<maxGroupBandwidth> is an optional field containing the maximum bitrate expressed in bits per second that can be shared by the individual encodings included in the encoding group.

16.4.18.2 encodingIDList>

<encodingIDList> is the list of the individual encodings grouped together in the encoding group. Each individual encoding is represented through its identifier contained within an <encodingID> element.

```
<!-- ENCODING ID LIST TYPE -->
<xs:complexType name="encodingIDListType">
    <xs:sequence>
        <xs:element name="encodingID" type="xs:string"
                maxOccurs="unbounded"/>
    </xs:sequence>
</xs:complexType>
```

16.4.18.3 encodingGroupID attribute

The encodingGroupID attribute contains the identifier of the encoding group.

16.4.19 <simultaneousSet>

<simultaneousSet> represents a simultaneous transmission set, i.e., a list of captures of the same media type that can be transmitted at the same time by a Media Provider. There are different simultaneous transmission sets for each media type.

```
<!-- SIMULTANEOUS SET TYPE -->
<xs:complexType name="simultaneousSetType">
    <xs:sequence>
        <xs:element name="mediaCaptureIDREF" type="xs:IDREF"
                minOccurs="0" maxOccurs="unbounded"/>
```

```
            <xs:element name="sceneViewIDREF" type="xs:IDREF"
                  minOccurs="0" maxOccurs="unbounded"/>
            <xs:element name="captureSceneIDREF" type="xs:IDREF"
                  minOccurs="0" maxOccurs="unbounded"/>
            <xs:any namespace="##other" processContents="lax"
                  minOccurs="0" maxOccurs="unbounded"/>
        </xs:sequence>
        <xs:attribute name="setID" type="xs:ID" use="required"/>
        <xs:attribute name="mediaType" type="xs:string"/>
        <xs:anyAttribute namespace="##any" processContents="lax"/>
    </xs:complexType>
```

Besides the identifiers of the captures (<mediaCaptureIDREF> elements), the identifiers of capture scene views and of capture scene can also be exploited as shortcuts (<sceneViewIDREF> and <captureSceneIDREF> elements). As an example, let us consider the situation where there are two capture scene views (S1 and S7). S1 contains captures AC11, AC12, and AC13. S7 contains captures AC71 and AC72. Provided that AC11, AC12, AC13, AC71, and AC72 can be simultaneously sent to the media consumer, instead of having five <mediaCaptureIDREF> elements listed in the simultaneous set (i.e., one <mediaCaptureIDREF> for AC11, one for AC12, and so on), there can be just two <sceneViewIDREF> elements (one for S1 and one for S7).

16.4.19.1 setID attribute

The "setID" attribute is a mandatory field containing the identifier of the simultaneous set.

16.4.19.2 mediaType attribute

The "mediaType" attribute is an optional attribute containing the media type of the captures referenced by the simultaneous set. When only capture scene identifiers are listed within a simultaneous set, the media type attribute MUST appear in the XML description in order to determine which media captures can be simultaneously sent together.

16.4.19.3 <mediaCaptureIDREF>

<mediaCaptureIDREF> contains the identifier of the media capture that belongs to the simultaneous set.

16.4.19.4 <sceneViewIDREF>

<sceneViewIDREF> contains the identifier of the scene view containing a group of captures that are able to be sent simultaneously with the other captures of the simultaneous set.

16.4.19.5 <captureSceneIDREF>

<captureSceneIDREF> contains the identifier of the capture scene where all the included captures of a certain media type are able to be sent together with the other captures of the simultaneous set.

16.4.20 <GLOBALVIEW>

<globalView> is a set of captures of the same media type representing a summary of the complete Media Provider's offer. The content of a global view is expressed by leveraging only scene view identifiers, put within <sceneViewIDREF> elements. Each global view is identified by a unique identifier within the "globalViewID" attribute.

```
<!-- GLOBAL VIEW TYPE -->
<xs:complexType name="globalViewType">
    <xs:sequence>
```

```
        <xs:element name="sceneViewIDREF" type="xs:IDREF"
                maxOccurs="unbounded"/>
        <xs:any namespace="##other" processContents="lax"
                minOccurs="0" maxOccurs="unbounded"/>
    </xs:sequence>
    <xs:attribute name="globalViewID" type="xs:ID"/>
    <xs:anyAttribute namespace="##any" processContents="lax"/>
</xs:complexType>
```

16.4.21 <PEOPLE>

Information about the participants that are represented in the media captures is conveyed via the <people> element. As can be seen from the XML Schema depicted in the following, for each participant, a <person> element is provided.

```
<!-- PEOPLE TYPE -->
<xs:complexType name="peopleType">
    <xs:sequence>
        <xs:element name="person" type="personType"
                maxOccurs="unbounded"/>
    </xs:sequence>
</xs:complexType>
```

16.4.21.1 <person>

<person> includes all the metadata related to a person represented within one or more media captures. This element provides the vcard of the subject (via the <personInfo> element, see Section 16.4.21.1.2) and his conference role(s) (via one or more <personType> elements, see Section 16.421.1.3). Furthermore, it has a mandatory "personID" attribute (Section 16.4.21.1.1).

```
<!-- PERSON TYPE -->
<xs:complexType name="personType">
    <xs:sequence>
        <xs:element name="personInfo" type="xcard:vcardType"
                maxOccurs="1" minOccurs="0"/>
        <xs:element ref="personType" minOccurs="0"
                maxOccurs="unbounded" />
        <xs:any namespace="##other" processContents="lax"
                minOccurs="0" maxOccurs="unbounded"/>
    </xs:sequence>
    <xs:attribute name="personID" type="xs:ID" use="required"/>
    <xs:anyAttribute namespace="##other" processContents="lax"/>
</xs:complexType>
```

16.4.21.1.1 personID attribute

The "personID" attribute carries the identifier of a represented person. Such an identifier can be used to refer to the participant, as in the <capturedPeople> element in the media captures representation (Section 16.4.11.21).

16.4.21.1.2 <personInfo>

The <personInfo> element is the XML representation of all the fields composing a vcard as specified in the Xcard RFC 6351. The vcardType is imported by the Xcard XML Schema provided in Appendix A of RFC 7852. As this schema specifies, the <fn> element within <vcard> is mandatory.

16.4.21.1.3 <personType>

The value of the <personType> element determines the role of the represented participant within the telepresence session organization. It has been specified as a simple string with an annotation pointing to an ad hoc defined IANA registry:

```
<!-- PERSON TYPE ELEMENT -->
<xs:element name="personType" type="xs:string">
   <xs:annotation>
      <xs:specificationation>
         Acceptable values (enumerations) for this type are managed
         by IANA in the "CLUE Schema <personType> registry",
         accessible at TBD-IANA.
      </xs:specificationation>
   </xs:annotation>
</xs:element>
```

The current possible values, as per the CLUE framework specification (see Section 16.3), are "presenter," "timekeeper," "attendee," "minute taker," "translator," "chairman," "vicechairman," and "observer.". A participant can play more than one conference role. In that case, more than one <personType> element will appear in his description.

16.4.22 <CAPTUREENCODING>

A capture encoding is given from the association of a media capture with an individual encoding to form a capture stream as defined in Section 16.3. Capture encodings are used within CONFIGURE messages from a Media Consumer to a Media Provider for representing the streams desired by the Media Consumer. For each desired stream, the Media Consumer needs to be allowed to specify:

(i) The capture identifier of the desired capture that has been advertised by the Media Provider.
(ii) The encoding identifier of the encoding to use, among those advertised by the Media Provider.
(iii) Optionally, in the case of multi-content captures, the list of the capture identifiers of the desired captures. All the mentioned identifiers are intended to be included in the ADVERTISEMENT message that the CONFIGURE message refers to.

The XML model of <captureEncoding> is provided in the following.

```
<!-- CAPTURE ENCODING TYPE -->
<xs:complexType name="captureEncodingType">
   <xs:sequence>
      <xs:element name="captureID" type="xs:string"/>
      <xs:element name="encodingID" type="xs:string"/>
      <xs:element name="configuredContent" type="contentType"
            minOccurs="0"/>
      <xs:any namespace="##other" processContents="lax"
            minOccurs="0" maxOccurs="unbounded"/>
   </xs:sequence>
   <xs:attribute name="ID" type="xs:ID" use="required"/>
   <xs:anyAttribute namespace="##any" processContents="lax"/>
</xs:complexType>
```

16.4.22.1 <captureID>

<captureID> is the mandatory element containing the identifier of the media capture that has been encoded to form the capture encoding.

16.4.22.2 <encodingID>

<encodingID> is the mandatory element containing the identifier of the applied individual encoding.

16.4.22.3 <configuredContent>

<configuredContent> is an optional element to be used in the case of configuration of MCC. It contains the list of capture identifiers and capture scene view identifiers the Media Consumer wants within the MCC. That element is structured as the <content> element used to describe the content of an MCC. The total number of media captures listed in the <configuredContent> MUST be lower than or equal to the value carried within the <maxCaptures> attribute of the MCC.

16.4.23 <clueInfo>

The <clueInfo> element includes all the information needed to represent the Media Provider's description of its telepresence capabilities according to the CLUE framework. Indeed, it is made by:

- The list of the available media captures (<mediaCaptures> [Section 16.4.5])
- The list of encoding groups (<encodingGroups> [Section 16.4.6])
- The list of capture scenes (<captureScenes> [Section 16.4.7])
- The list of simultaneous transmission sets (<simultaneousSets> [Section 16.4.8])
- The list of global views sets (<globalViews> [Section 16.4.9])
- Meta-data about the participants represented in the telepresence session (<people> [Section 16.4.21])

It has been conceived only for data model testing purposes, and though it resembles the body of an ADVERTISEMENT message, it is not actually used in the CLUE protocol message definitions. The telepresence capabilities descriptions compliant to this data model specification that can be found in Sections 16.4.27 and 16.4.28 are provided by using the <clueInfo> element.

```
<!-- CLUE INFO TYPE -->
<xs:complexType name="clueInfoType">
    <xs:sequence>
        <xs:element ref="mediaCaptures"/>
        <xs:element ref="encodingGroups"/>
        <xs:element ref="captureScenes"/>
        <xs:element ref="simultaneousSets" minOccurs="0"/>
        <xs:element ref="globalViews" minOccurs="0"/>
        <xs:element ref="people" minOccurs="0"/>
        <xs:any namespace="##other" processContents="lax"
                minOccurs="0" maxOccurs="unbounded"/>
    </xs:sequence>
    <xs:attribute name="clueInfoID" type="xs:ID" use="required"/>
    <xs:anyAttribute namespace="##other" processContents="lax"/>
</xs:complexType>
```

16.4.24 XML Schema Extensibility

The telepresence data model defined in this specification is meant to be extensible. Extensions are accomplished by defining elements or attributes qualified by namespaces other than "urn:ietf:params:xml:ns:clue-info" and "urn:ietf:params:xml:ns:vcard-4.0" for use wherever the schema allows such extensions (i.e., where the XML Schema definition specifies "anyAttribute" or "anyElement"). Elements or attributes from unknown namespaces MUST be ignored. Extensibility was purposefully favored as much as possible based on expectations about custom implementations. Hence, the schema offers people enough flexibility to define custom extensions without losing compliance with the standard. This is achieved by leveraging <xs:any> elements and <xs: anyAttribute> attributes, which is a common approach with schemas, still matching the UPA (Unique Particle Attribution) constraint.

16.4.24.1 Example of Extension

When the CLUE data model is extended, a new schema with a new namespace associated with it needs to be specified. In the following, an example of extension is provided. The extension defines a new audio capture attribute ("newAudioFeature") and an attribute for characterizing the captures belonging to an "otherCaptureType" defined by the user. An XML specification compliant with the extension is also included.

The XML file results validated against the current CLUE data model schema.

```
<?xml version="1.0" encoding="UTF-8" ?>
<xs:schema
      targetNamespace="urn:ietf:params:xml:ns:clue-info-ext"
      xmlns:tns="urn:ietf:params:xml:ns:clue-info-ext"
      xmlns:clue-ext="urn:ietf:params:xml:ns:clue-info-ext"
      xmlns:xs="http://www.w3.org/2001/XMLSchema"
      xmlns="urn:ietf:params:xml:ns:clue-info-ext"
      xmlns:xcard="urn:ietf:params:xml:ns:vcard-4.0"
      xmlns:info="urn:ietf:params:xml:ns:clue-info"
      elementFormDefault="qualified"
      attributeFormDefault="unqualified">
      <!-- Import xcard XML schema -->
      <xs:import namespace="urn:ietf:params:xml:ns:vcard-4.0"
            schemaLocation=
            "http://www.iana.org/assignments/xml-registry/schema/
            vcard-4.0.xsd"/>
      <!-- Import CLUE XML schema -->
      <xs:import namespace="urn:ietf:params:xml:ns:clue-info"
            schemaLocation="clue-data-model-schema.xsd"/>
      <!-- ELEMENT DEFINITIONS -->
      <xs:element name="newAudioFeature" type="xs:string"/>
      <xs:element name="otherMediaCaptureTypeFeature"
            type="xs:string"/>
</xs:schema>

<?xml version="1.0" encoding="UTF-8" standalone="yes"?>
<clueInfo xmlns="urn:ietf:params:xml:ns:clue-info"
      xmlns:ns2="urn:ietf:params:xml:ns:vcard-4.0"
      xmlns:ns3="urn:ietf:params:xml:ns:clue-info-ext"
      clueInfoID="NapoliRoom">
   <mediaCaptures>
      <mediaCapture
            xmlns:xsi="http://www.w3.org/2001/XMLSchema-instance"
            xsi:type="audioCaptureType" captureID="AC0"
            mediaType="audio">
         <captureSceneIDREF>CS1</captureSceneIDREF>
         <nonSpatiallyDefinable>true</nonSpatiallyDefinable>
```

```
            <individual>true</individual>
            <encGroupIDREF>EG1</encGroupIDREF>
            <ns3:newAudioFeature>newAudioFeatureValue
            </ns3:newAudioFeature>
        </mediaCapture>
        <mediaCapture
                xmlns:xsi="http://www.w3.org/2001/XMLSchema-instance"
                xsi:type="otherCaptureType" captureID="OMC0"
                mediaType="other media type">
            <captureSceneIDREF>CS1</captureSceneIDREF>
            <nonSpatiallyDefinable>true</nonSpatiallyDefinable>
            <encGroupIDREF>EG1</encGroupIDREF>
            <ns3:otherMediaCaptureTypeFeature>OtherValue
            </ns3:otherMediaCaptureTypeFeature>
        </mediaCapture>
    </mediaCaptures>
    <encodingGroups>
        <encodingGroup encodingGroupID="EG1">
            <maxGroupBandwidth>300000</maxGroupBandwidth>
            <encodingIDList>
                <encodingID>ENC4</encodingID>
                <encodingID>ENC5</encodingID>
            </encodingIDList>
        </encodingGroup>
    </encodingGroups>
    <captureScenes>
        <captureScene scale="unknown" sceneID="CS1"/>
    </captureScenes>
</clueInfo>
```

16.4.25 Security Considerations

This specification defines an XML Schema data model for telepresence scenarios. The modeled information is identified in the CLUE framework as necessary in order to enable a full-fledged media stream negotiation and rendering. Indeed, the XML elements herein defined are used within CLUE protocol messages to describe both the media streams representing the Media Provider's telepresence offer and the desired selection requested by the Media Consumer. Security concerns described in Sections 16.3 and 16.4.15 apply to this specification.

Data model information carried within CLUE messages SHOULD be accessed only by authenticated endpoints. Indeed, authenticated access is strongly advisable, especially if you convey information about individuals (<personalInfo>) and/or scenes (<sceneInformation>). There might be more exceptions, depending on the level of criticality that is associated with the setup and configuration of a specific session. In principle, one might even decide that no protection at all is needed for a particular session; this is why authentication has not been identified as a mandatory requirement.

Going deeper into detail, some information published by the Media Provider might reveal sensitive data about who and what is represented in the transmitted streams. The vCard included in the <personInfo> elements (Section 16.4.21.1) mandatorily contains the identity of the represented person. Optionally, vCards can also carry the person's contact addresses together with his/her photo and other personal data. Similar privacy-critical information can be conveyed by means of <sceneInformation> elements (Section 16.4.1) describing the capture scenes. The <description> elements (Section 16.4.11.13) also can specify details about the content of media captures, capture scenes, and scene views that should be protected.

Integrity attacks to the data model information encapsulated in CLUE messages can invalidate the success of the telepresence session's setup by misleading the Media Consumer's and Media Provider's interpretation of the offered and desired media streams.

The assurance of the authenticated access and of the integrity of the data model information is up to the involved transport mechanisms, namely, the CLUE protocol (see Section 16.5) and the CLUE data channel (see Section 16.6). XML parsers need to be robust with respect to malformed specifications. Reading malformed specifications from unknown or untrusted sources could result in an attacker gaining privileges of the user running the XML parser. In an extreme situation, the entire machine could be compromised.

16.4.26 IANA Considerations

This specification registers a new XML namespace, a new XML schema, the MIME type for the schema, and four new registries associated, respectively, with acceptable <view>, <presentation>, <sensitivityPattern>, and <personType> values. Interested readers need to see the detail in Presta and Romano [14] or the latest version.

16.4.27 Sample XML File

The following XML specification represents a schema-compliant example of a CLUE telepresence scenario. Taking inspiration from the examples described in the framework specification (see Section 16.3), it provides the XML representation of an endpoint-style Media Provider's ADVERTISEMENT. There are three cameras, of which the central one is also capable of capturing a zoomed-out view of the overall telepresence room. Besides the three video captures coming from the cameras, the Media Provider makes available a further multi-content capture of the loudest segment of the room, obtained by switching the video source across the three cameras. For the sake of simplicity, only one audio capture is advertised for the audio of the whole room. The three cameras are placed in front of three participants (Alice, Bob, and Ciccio), whose vcard and conference role details are also provided. Media captures are arranged into four capture scene views:

1. (VC0, VC1, VC2) – Left, center, and right camera video captures
2. (VC3) – Video capture associated with loudest room segment
3. (VC4) – Video capture zoomed-out view of all people in the room
4. (AC0) – Main audio

There are two encoding groups: (i) EG0, for video encodings, and (ii) EG1, for audio encodings. As to the simultaneous sets, VC1 and VC4 cannot be transmitted simultaneously, since they are captured by the same device, i.e., the central camera (VC4 is a zoomed-out view, while VC1 is a focused view of the front participant). On the other hand, VC3 and VC4 cannot be simultaneous either, since VC3, the loudest segment of the room, might be at a certain point in time focusing on the central part of the room, i.e., the same as VC1. The simultaneous sets would then be the following:

SS1 made by VC3 and all the captures in the first capture scene view (VC0, VC1, and VC2); SS2 made by VC0, VC2, and VC4.

Note: Although the following XML schema was originally proposed (we have kept it here for reference only), actually, the XML schema defined in Section 16.5 supersedes this. Readers are advised to use the XML schema of Section 16.5.

```
<?xml version="1.0" encoding="UTF-8" standalone="yes"?>
<clueInfo xmlns="urn:ietf:params:xml:ns:clue-info"
      xmlns:ns2="urn:ietf:params:xml:ns:vcard-4.0"
      clueInfoID="NapoliRoom">
```

```
<mediaCaptures>
   <mediaCapture
         xmlns:xsi=“http://www.w3.org/2001/XMLSchema-instance”
         xsi:type=“audioCaptureType” captureID=“AC0” mediaType=“audio”>
      <captureSceneIDREF>CS1</captureSceneIDREF>
         <spatialInformation>
            <captureOrigin>
               <capturePoint>
                  0.0</x>
                  0.0</y>
                  10.0</z>
               </capturePoint>
               <lineOfCapturePoint>
                  0.0</x>
                  1.0</y>
                  10.0</z>
               </lineOfCapturePoint>
            </captureOrigin>
         </spatialInformation>
         <individual>true</individual>
         <encGroupIDREF>EG1</encGroupIDREF>
         <description lang=“en”>main audio from the room</description>
         <priority>1</priority>
         <lang>it</lang>
         <mobility>static</mobility>
         <view>room</view>
         <capturedPeople>
            <personIDREF>alice</personIDREF>
            <personIDREF>bob</personIDREF>
            <personIDREF>ciccio</personIDREF>
         </capturedPeople>
      </mediaCapture>
      <mediaCapture
            xmlns:xsi=“http://www.w3.org/2001/XMLSchema-instance”
            xsi:type=“videoCaptureType” captureID=“VC0”
            mediaType=“video”>
         <captureSceneIDREF>CS1</captureSceneIDREF>
         <spatialInformation>
            <captureOrigin>
               <capturePoint>
                  -2.0</x>
                  0.0</y>
                  10.0</z>
               </capturePoint>
            </captureOrigin>
            <captureArea>
               <bottomLeft>
                  -3.0</x>
                  20.0</y>
                  9.0</z>
               </bottomLeft>
               <bottomRight>
                  -1.0</x>
                  20.0</y>
                  9.0</z>
               </bottomRight>
```

```
                <topLeft>
                   -3.0</x>
                   20.0</y>
                   11.0</z>
                </topLeft>
                <topRight>
                   -1.0</x>
                   20.0</y>
                   11.0</z>
                </topRight>
             </captureArea>
       </spatialInformation>
       <individual>true</individual>
       <encGroupIDREF>EG0</encGroupIDREF>
       <description lang="en">left camera video capture
       </description>
       <priority>1</priority>
       <lang>it</lang>
       <mobility>static</mobility>
       <view>individual</view>
       <capturedPeople>
          <personIDREF>ciccio</personIDREF>
       </capturedPeople>
    </mediaCapture>
    <mediaCapture
          xmlns:xsi="http://www.w3.org/2001/XMLSchema-instance"
          xsi:type="videoCaptureType" captureID="VC1" mediaType="video">
       <captureSceneIDREF>CS1</captureSceneIDREF>
          <spatialInformation>
             <captureOrigin>
                <capturePoint>
                   0.0</x>
                   0.0</y>
                   10.0</z>
                </capturePoint>
             </captureOrigin>
             <captureArea>
                <bottomLeft>
                   -1.0</x>
                   20.0</y>
                   9.0</z>
                </bottomLeft>
                <bottomRight>
                   1.0</x>
                   20.0</y>
                   9.0</z>
                </bottomRight>
                <topLeft>
                   -1.0</x>
                   20.0</y>
                   11.0</z>
                </topLeft>
                <topRight>
                   1.0</x>
                   20.0</y>
                   11.0</z>
```

```
            </topRight>
         </captureArea>
      </spatialInformation>
      <individual>true</individual>
      <encGroupIDREF>EG0</encGroupIDREF>
      <description lang="en">central camera video capture</
      description>
      <priority>1</priority>
      <lang>it</lang>
      <mobility>static</mobility>
      <view>individual</view>
      <capturedPeople>
         <personIDREF>alice</personIDREF>
      </capturedPeople>
   </mediaCapture>
   <mediaCapture
         xmlns:xsi="http://www.w3.org/2001/XMLSchema-instance"
         xsi:type="videoCaptureType" captureID="VC2"
         mediaType="video">
      <captureSceneIDREF>CS1</captureSceneIDREF>
      <spatialInformation>
         <captureOrigin>
            <capturePoint>
               2.0</x>
               0.0</y>
               10.0</z>
            </capturePoint>
         </captureOrigin>
         <captureArea>
            <bottomLeft>
               1.0</x>
               20.0</y>
               9.0</z>
            </bottomLeft>
            <bottomRight>
               3.0</x>
               20.0</y>
               9.0</z>
            </bottomRight>
            <topLeft>
               1.0</x>
               20.0</y>
               11.0</z>
            </topLeft>
            <topRight>
               3.0</x>
               20.0</y>
               11.0</z>
            </topRight>
         </captureArea>
      </spatialInformation>
      <individual>true</individual>
      <encGroupIDREF>EG0</encGroupIDREF>
      <description lang="en">right camera video capture</
      description>
      <priority>1</priority>
      <lang>it</lang>
```

```
            <mobility>static</mobility>
            <view>individual</view>
            <capturedPeople>
                <personIDREF>bob</personIDREF>
            </capturedPeople>
        </mediaCapture>
        <mediaCapture
                xmlns:xsi=“http://www.w3.org/2001/XMLSchema-instance”
                xsi:type=“videoCaptureType” captureID=“VC3”
                mediaType=“video”>
            <captureSceneIDREF>CS1</captureSceneIDREF>
                <spatialInformation>
                    <captureArea>
                        <bottomLeft>
                            -3.0</x>
                            20.0</y>
                            9.0</z>
                        </bottomLeft>
                        <bottomRight>
                            3.0</x>
                            20.0</y>
                            9.0</z>
                        </bottomRight>
                        <topLeft>
                            -3.0</x>
                            20.0</y>
                            11.0</z>
                        </topLeft>
                        <topRight>
                            3.0</x>
                            20.0</y>
                            11.0</z>
                        </topRight>
                    </captureArea>
                </spatialInformation>
                <content>
                <sceneViewIDREF>SE1</sceneViewIDREF>
                </content>
                <policy>SoundLevel:0</policy>
                <encGroupIDREF>EG0</encGroupIDREF>
                <description lang=“en”>loudest room segment</
                description>
                <priority>2</priority>
                <lang>it</lang>
                <mobility>static</mobility>
                <view>individual</view>
            </mediaCapture>
            <mediaCapture
                    xmlns:xsi=“http://www.w3.org/2001/XMLSchema-instance”
                    xsi:type=“videoCaptureType” captureID=“VC4”
                    mediaType=“video”>
                <captureSceneIDREF>CS1</captureSceneIDREF>
                    <spatialInformation>
                        <captureOrigin>
                            <capturePoint>
                                0.0</x>
                                0.0</y>
```

```
                10.0</z>
            </capturePoint>
        </captureOrigin>
        <captureArea>
            <bottomLeft>
                -3.0</x>
                20.0</y>
                7.0</z>
            </bottomLeft>
            <bottomRight>
                3.0</x>
                20.0</y>
                7.0</z>
            </bottomRight>
            <topLeft>
                -3.0</x>
                20.0</y>
                13.0</z>
            </topLeft>
            <topRight>
                3.0</x>
                20.0</y>
                13.0</z>
            </topRight>
        </captureArea>
    </spatialInformation>
    <individual>true</individual>
    <encGroupIDREF>EG0</encGroupIDREF>
    <description lang="en">zoomed out view of all people
            in the room</description>
    <priority>2</priority>
    <lang>it</lang>
    <mobility>static</mobility>
    <view>room</view>
    <capturedPeople>
        <personIDREF>alice</personIDREF>
        <personIDREF>bob</personIDREF>
        <personIDREF>ciccio</personIDREF>
    </capturedPeople>
  </mediaCapture>
</mediaCaptures>
<encodingGroups>
  <encodingGroup encodingGroupID="EG0">
    <maxGroupBandwidth>600000</maxGroupBandwidth>
    <encodingIDList>
        <encodingID>ENC1</encodingID>
        <encodingID>ENC2</encodingID>
        <encodingID>ENC3</encodingID>
    </encodingIDList>
  </encodingGroup>
  <encodingGroup encodingGroupID="EG1">
    <maxGroupBandwidth>300000</maxGroupBandwidth>
    <encodingIDList>
        <encodingID>ENC4</encodingID>
        <encodingID>ENC5</encodingID>
    </encodingIDList>
```

```
        </encodingGroup>
      </encodingGroups>
      <captureScenes>
        <captureScene scale="unknown" sceneID="CS1">
          <sceneViews>
          <sceneView sceneViewID="SE1">
            <mediaCaptureIDs>
              <mediaCaptureIDREF>VC0
                </mediaCaptureIDREF>
              <mediaCaptureIDREF>VC1
                </mediaCaptureIDREF>
              <mediaCaptureIDREF>VC2
                </mediaCaptureIDREF>
            </mediaCaptureIDs>
          </sceneView>
          <sceneView sceneViewID="SE2">
            <mediaCaptureIDs>
              <mediaCaptureIDREF>VC3
                </mediaCaptureIDREF>
            </mediaCaptureIDs>
          </sceneView>
          <sceneView sceneViewID="SE3">
          <mediaCaptureIDs>
            <mediaCaptureIDREF>VC4
              </mediaCaptureIDREF>
                </mediaCaptureIDs>
          </sceneView>
          <sceneView sceneViewID="SE4">
            <mediaCaptureIDs>
              <mediaCaptureIDREF>AC0
                </mediaCaptureIDREF>
            </mediaCaptureIDs>
          </sceneView>
        </sceneViews>
      </captureScene>
    </captureScenes>
    <simultaneousSets>
      <simultaneousSet setID="SS1">
      <mediaCaptureIDREF>VC3</mediaCaptureIDREF>
      <sceneViewIDREF>SE1</sceneViewIDREF>
    </simultaneousSet>
      <simultaneousSet setID="SS2">
        <mediaCaptureIDREF>VC0</mediaCaptureIDREF>
        <mediaCaptureIDREF>VC2</mediaCaptureIDREF>
        <mediaCaptureIDREF>VC4</mediaCaptureIDREF>
      </simultaneousSet>
    </simultaneousSets>
    <people>
      <person personID="bob">
        <personInfo>
          <ns2:fn>
            <ns2:text>Bob</ns2:text>
          </ns2:fn>
        </personInfo>
      <personType>minute taker</personType>
    </person>
```

```
        <person personID="alice">
            <personInfo>
                <ns2:fn>
                    <ns2:text>Alice</ns2:text>
                </ns2:fn>
            </personInfo>
            <personType>presenter</personType>
        </person>
        <person personID="ciccio">
            <personInfo>
                <ns2:fn>
                    <ns2:text>Ciccio</ns2:text>
                </ns2:fn>
            </personInfo>
            <personType>chairman</personType>
            <personType>timekeeper</personType>
        </person>
    </people>
</clueInfo>
```

16.4.28 MCC Example

Enhancing the scenario presented in the previous example, the Media Provider is able to advertise a composed capture VC7 made by a big picture representing the current speaker (VC3) and two PiP boxes representing the previous speakers (the previous one – VC5 – and the oldest one – VC6). The provider does not want to instantiate and send VC5 and VC6, so it does not associate any encoding group with them. Their XML representations are provided for enabling the description of VC7.

Note: Although the following XML schema was originally proposed (we have kept it here for reference only), actually, the XML schema defined in Section 16.5 supersedes this. Readers are advised to use the XML schema of Section 16.5.

A possible description for that scenario could be the following:

```
<?xml version="1.0" encoding="UTF-8" standalone="yes"?>
<clueInfo xmlns="urn:ietf:params:xml:ns:clue-info"
    xmlns:ns2="urn:ietf:params:xml:ns:vcard-4.0"
    clueInfoID="NapoliRoom">
    <mediaCaptures>
        <mediaCapture
                xmlns:xsi="http://www.w3.org/2001/XMLSchema-instance"
                xsi:type="audioCaptureType" captureID="AC0"
                mediaType="audio">
                <captureSceneIDREF>CS1</captureSceneIDREF>
                <spatialInformation>
                    <captureOrigin>
                        <capturePoint>
                            0.0</x>
                            0.0</y>
                            10.0</z>
                        </capturePoint>
                        <lineOfCapturePoint>
                            0.0</x>
                            1.0</y>
                            10.0</z>
                        </lineOfCapturePoint>
                    </captureOrigin>
                </spatialInformation>
```

```
        <individual>true</individual>
        <encGroupIDREF>EG1</encGroupIDREF>
        <description lang="en">main audio from the room</
        description>
        <priority>1</priority>
        <lang>it</lang>
        <mobility>static</mobility>
        <view>room</view>
        <capturedPeople>
            <personIDREF>alice</personIDREF>
            <personIDREF>bob</personIDREF>
            <personIDREF>ciccio</personIDREF>
        </capturedPeople>
    </mediaCapture>
    <mediaCapture
            xmlns:xsi="http://www.w3.org/2001/XMLSchema-instance"
            xsi:type="videoCaptureType" captureID="VC0"
            mediaType="video">
        <captureSceneIDREF>CS1</captureSceneIDREF>
            <spatialInformation>
                <captureOrigin>
                    <capturePoint>
                        0.5</x>
                        1.0</y>
                        0.5</z>
                    </capturePoint>
                    <lineOfCapturePoint>
                        0.5</x>
                        0.0</y>
                        0.5</z>
                    </lineOfCapturePoint>
                </captureOrigin>
            </spatialInformation>
            <individual>true</individual>
            <encGroupIDREF>EG0</encGroupIDREF>
            <description lang="en">left camera video capture</
            description>
            <priority>1</priority>
            <lang>it</lang>
            <mobility>static</mobility>
            <view>individual</view>
            <capturedPeople>
                <personIDREF>ciccio</personIDREF>
            </capturedPeople>
    </mediaCapture>
    <mediaCapture
            xmlns:xsi="http://www.w3.org/2001/XMLSchema-instance"
            xsi:type="videoCaptureType" captureID="VC1"
                    mediaType="video">
            <captureSceneIDREF>CS1</captureSceneIDREF>
            <spatialInformation>
                <captureOrigin>
                    <capturePoint>
                        0.0</x>
                        0.0</y>
                        10.0</z>
                    </capturePoint>
```

```
            </captureOrigin>
            <captureArea>
            <bottomLeft>
                -1.0</x>
                20.0</y>
                9.0</z>
            </bottomLeft>
            <bottomRight>
                1.0</x>
                20.0</y>
                9.0</z>
            </bottomRight>
            <topLeft>
                -1.0</x>
                20.0</y>
                11.0</z>
            </topLeft>
            <topRight>
                1.0</x>
                20.0</y>
                11.0</z>
            </topRight>
        </captureArea>
    </spatialInformation>
    <individual>true</individual>
    <encGroupIDREF>EG0</encGroupIDREF>
    <description lang="en">central camera video capture</
    description>
    <priority>1</priority>
    <lang>it</lang>
    <mobility>static</mobility>
    <view>individual</view>
    <capturedPeople>
        <personIDREF>alice</personIDREF>
    </capturedPeople>
</mediaCapture>
<mediaCapture
        xmlns:xsi="http://www.w3.org/2001/XMLSchema-instance"
        xsi:type="videoCaptureType" captureID="VC2"
        mediaType="video">
    <captureSceneIDREF>CS1</captureSceneIDREF>
        <spatialInformation>
            <captureOrigin>
                <capturePoint>
                    2.0</x>
                    0.0</y>
                    10.0</z>
                </capturePoint>
            </captureOrigin>
            <captureArea>
            <bottomLeft>
                1.0</x>
                20.0</y>
                9.0</z>
            </bottomLeft>
            <bottomRight>
```

```
                3.0</x>
                20.0</y>
                9.0</z>
            </bottomRight>
            <topLeft>
                1.0</x>
                20.0</y>
                11.0</z>
            </topLeft>
            <topRight>
                3.0</x>
                20.0</y>
                11.0</z>
            </topRight>
        </captureArea>
    </spatialInformation>
    <individual>true</individual>
    <encGroupIDREF>EG0</encGroupIDREF>
    <description lang="en">right camera video capture</
    description>
    <priority>1</priority>
    <lang>it</lang>
    <mobility>static</mobility>
    <view>individual</view>
    <capturedPeople>
        <personIDREF>bob</personIDREF>
    </capturedPeople>
</mediaCapture>
<mediaCapture
        xmlns:xsi="http://www.w3.org/2001/XMLSchema-instance"
        xsi:type="videoCaptureType" captureID="VC3"
        mediaType="video">
    <captureSceneIDREF>CS1</captureSceneIDREF>
        <spatialInformation>
            <captureArea>
                <bottomLeft>
                    -3.0</x>
                    20.0</y>
                    9.0</z>
              </bottomLeft>
              <bottomRight>
                    3.0</x>
                    20.0</y>
                    9.0</z>
              </bottomRight>
              <topLeft>
                    -3.0</x>
                    20.0</y>
                    11.0</z>
              </topLeft>
              <topRight>
                    3.0</x>
                    20.0</y>
                    11.0</z>
              </topRight>
            </captureArea>
```

```
    </spatialInformation>
    <content>
        <sceneViewIDREF>SE1</sceneViewIDREF>
    </content>
    <policy>SoundLevel:0</policy>
    <encGroupIDREF>EG0</encGroupIDREF>
    <description lang="en">loudest room segment</
    description>
    <priority>2</priority>
    <lang>it</lang>
    <mobility>static</mobility>
    <view>individual</view>
</mediaCapture>
<mediaCapture
        xmlns:xsi="http://www.w3.org/2001/XMLSchema-instance"
        xsi:type="videoCaptureType" captureID="VC4"
        mediaType="video">
    <captureSceneIDREF>CS1</captureSceneIDREF>
        <spatialInformation>
        <captureOrigin>
            <capturePoint>
                0.0</x>
                0.0</y>
                10.0</z>
            </capturePoint>
        </captureOrigin>
        <captureArea>
            <bottomLeft>
                -3.0</x>
                20.0</y>
                7.0</z>
            </bottomLeft>
            <bottomRight>
                3.0</x>
                20.0</y>
                7.0</z>
            </bottomRight>
            <topLeft>
                -3.0</x>
                20.0</y>
                13.0</z>
            </topLeft>
            <topRight>
                3.0</x>
                20.0</y>
                13.0</z>
            </topRight>
        </captureArea>
    </spatialInformation>
<individual>true</individual>
<encGroupIDREF>EG0</encGroupIDREF>
<description lang="en">
    zoomed-out view of all people in the room
</description>
<priority>2</priority>
<lang>it</lang>
```

```
        <mobility>static</mobility>
        <view>room</view>
        <capturedPeople>
            <personIDREF>alice</personIDREF>
            <personIDREF>bob</personIDREF>
            <personIDREF>ciccio</personIDREF>
        </capturedPeople>
    </mediaCapture>
    <mediaCapture
            xmlns:xsi="http://www.w3.org/2001/XMLSchema-instance"
            xsi:type="videoCaptureType" captureID="VC5"
            mediaType="video">
        <captureSceneIDREF>CS1</captureSceneIDREF>
        <spatialInformation>
            <captureArea>
                <bottomLeft>
                    -3.0</x>
                    20.0</y>
                    9.0</z>
                </bottomLeft>
                <bottomRight>
                    3.0</x>
                    20.0</y>
                    9.0</z>
                </bottomRight>
                <topLeft>
                    -3.0</x>
                    20.0</y>
                    11.0</z>
                </topLeft>
                <topRight>
                        3.0</x>
                        20.0</y>
                        11.0</z>
                </topRight>
            </captureArea>
        </spatialInformation>
        <content>
            <sceneViewIDREF>SE1</sceneViewIDREF>
        </content>
        <policy>SoundLevel:1</policy>
        <description lang="en">penultimate loudest room segment
        </description>
        <lang>it</lang>
        <mobility>static</mobility>
        <view>individual</view>
    </mediaCapture>
    <mediaCapture
            xmlns:xsi="http://www.w3.org/2001/XMLSchema-instance"
            xsi:type="videoCaptureType" captureID="VC6"
            mediaType="video">
        <captureSceneIDREF>CS1</captureSceneIDREF>
        <spatialInformation>
            <captureArea>
                <bottomLeft>
                    -3.0</x>
```

```
                  20.0</y>
                  9.0</z>
               </bottomLeft>
               <bottomRight>
                  3.0</x>
                  20.0</y>
                  9.0</z>
               </bottomRight>
               <topLeft>
                  -3.0</x>
                  20.0</y>
                  11.0</z>
               </topLeft>
               <topRight>
                  3.0</x>
                  20.0</y>
                  11.0</z>
               </topRight>
            </captureArea>
         </spatialInformation>
         <content>
            <sceneViewIDREF>SE1</sceneViewIDREF>
         </content>
         <policy>SoundLevel:2</policy>
         <description lang="en">last but two loudest room segment
         </description>
         <lang>it</lang>
         <mobility>static</mobility>
         <view>individual</view>
      </mediaCapture>
<mediaCapture
      xmlns:xsi="http://www.w3.org/2001/XMLSchema-instance"
      xsi:type="videoCaptureType" captureID="VC7"
      mediaType="video">
   <captureSceneIDREF>CS1</captureSceneIDREF>
   <spatialInformation>
      <captureArea>
         <bottomLeft>
            -3.0</x>
            20.0</y>
            9.0</z>
         </bottomLeft>
         <bottomRight>
            3.0</x>
            20.0</y>
            9.0</z>
         </bottomRight>
         <topLeft>
            -3.0</x>
            20.0</y>
            11.0</z>
         </topLeft>
         <topRight>
            3.0</x>
            20.0</y>
            11.0</z>
```

```
                </topRight>
            </captureArea>
        </spatialInformation>
        <content>
            <mediaCaptureIDREF>VC3</mediaCaptureIDREF>
            <mediaCaptureIDREF>VC5</mediaCaptureIDREF>
            <mediaCaptureIDREF>VC6</mediaCaptureIDREF>
        </content>
        <maxCaptures exactNumber="true">3</maxCaptures>
        <encGroupIDREF>EG0</encGroupIDREF>
        <description lang="en">big picture of the current speaker +
            pips about previous speakers</description>
        <priority>3</priority>
        <lang>it</lang>
        <mobility>static</mobility>
        <view>individual</view>
    </mediaCapture>
</mediaCaptures>
<encodingGroups>
    <encodingGroup encodingGroupID="EG0">
        <maxGroupBandwidth>600000</maxGroupBandwidth>
        <encodingIDList>
            <encodingID>ENC1</encodingID>
            <encodingID>ENC2</encodingID>
            <encodingID>ENC3</encodingID>
        </encodingIDList>
    </encodingGroup>
    <encodingGroup encodingGroupID="EG1">
    <maxGroupBandwidth>300000</maxGroupBandwidth>
    <encodingIDList>
        <encodingID>ENC4</encodingID>
        <encodingID>ENC5</encodingID>
    </encodingIDList>
    </encodingGroup>
</encodingGroups>
<captureScenes>
    <captureScene scale="unknown" sceneID="CS1">
        <sceneViews>
            <sceneView sceneViewID="SE1">
                <description lang="en">participants' individual
                    videos</description>
                <mediaCaptureIDs>
                    <mediaCaptureIDREF>VC0</mediaCapture
                            IDREF>
                    <mediaCaptureIDREF>VC1</mediaCapture
                            IDREF>
                    <mediaCaptureIDREF>VC2</mediaCapture
                            IDREF>
                </mediaCaptureIDs>
            </sceneView>
            <sceneView sceneViewID="SE2">
                <description lang="en">loudest segment of the
                    room</description>
                <mediaCaptureIDs>
                    <mediaCaptureIDREF>VC3</mediaCaptureIDREF>
                </mediaCaptureIDs>
            </sceneView>
```

```
            <sceneView sceneViewID="SE5">
                <description lang="en">loudest segment of the room
                    + pips</description>
                <mediaCaptureIDs>
                    <mediaCaptureIDREF>VC7</mediaCapture
                            IDREF>
                </mediaCaptureIDs>
            </sceneView>
            <sceneView sceneViewID="SE4">
                <description lang="en">room audio</description>
                <mediaCaptureIDs>
                    <mediaCaptureIDREF>AC0</mediaCapture
                            IDREF>
                </mediaCaptureIDs>
            </sceneView>
            <sceneView sceneViewID="SE3">
                <description lang="en">room video</description>
                <mediaCaptureIDs>
                    <mediaCaptureIDREF>VC4</mediaCapture
                            IDREF>
                </mediaCaptureIDs>
            </sceneView>
        </sceneViews>
    </captureScene>
</captureScenes>
<simultaneousSets>
    <simultaneousSet setID="SS1">
        <mediaCaptureIDREF>VC3</mediaCaptureIDREF>
        <mediaCaptureIDREF>VC7</mediaCaptureIDREF>
        <sceneViewIDREF>SE1</sceneViewIDREF>
    </simultaneousSet>
    <simultaneousSet setID="SS2">
        <mediaCaptureIDREF>VC0</mediaCaptureIDREF>
        <mediaCaptureIDREF>VC2</mediaCaptureIDREF>
        <mediaCaptureIDREF>VC4</mediaCaptureIDREF>
    </simultaneousSet>
    </simultaneousSets>
    <people>
        <person personID="bob">
            <personInfo>
                <ns2:fn>
                    <ns2:text>Bob</ns2:text>
                </ns2:fn>
            </personInfo>
            <personType>minute taker</personType>
        </person>
        <person personID="alice">
        <personInfo>
            <ns2:fn>
                <ns2:text>Alice</ns2:text>
            </ns2:fn>
        </personInfo>
        <personType>presenter</personType>
    </person>
    <person personID="ciccio">
        <personInfo>
            <ns2:fn>
```

```
                <ns2:text>Ciccio</ns2:text>
            </ns2:fn>
        </personInfo>
        <personType>chairman</personType>
        <personType>timekeeper</personType>
    </person>
  </people>
</clueInfo>
```

16.5 CLUE PROTOCOL

The CLUE protocol [15] is an application protocol conceived for the description and negotiation of a telepresence session. The design of the CLUE protocol takes into account the requirements and the framework defined within the IETF CLUE working group. A companion specification delves into CLUE signaling details as well as the SIP/SDP session establishment phase. CLUE messages flow over the CLUE data channel, based on reliable and ordered stream control transmission protocol (SCTP) over DTLS transport. Message details, together with the behavior of CLUE Participants acting as Media Providers and/or Media Consumers, are herein discussed.

16.5.1 Introduction

The CLUE protocol is an application protocol used by two CLUE Participants to enhance the experience of a multimedia telepresence session. The main goals of the CLUE protocol are:

1. Enabling a Media Provider (MP) to properly announce its current telepresence capabilities to a Media Consumer in terms of available media captures, groups of encodings, simultaneity constraints, and other information defined in Section 16.3;
2. Enabling an MC to request the desired multimedia streams from the offering MP. CLUE-capable endpoints are connected by means of the CLUE data channel, an SCTP over DTLS channel, which is opened and established as described in Sections 16.8 and 16.6. CLUE protocol messages flowing over such a channel are detailed in this specification, both syntactically and semantically.

In Section 16.5.4, we provide a general overview of the CLUE protocol. CLUE protocol messages are detailed in Section 16.5.5. The CLUE Protocol state machines are introduced in Section 16.5.6. Versioning and extensions are discussed in Sections 16.5.7 and 16.5.8, respectively. The XML schema defining the CLUE messages is reported in Section 16.5.9.

16.5.2 Terminology

This specification refers to the same terminology used in Section 16.3 and in RFC 7262 (see Section 16.1). We briefly recall herein some of the main terms used in the specification. The definition (also see Section 1.6, Table 1.2) of "CLUE Participant" herein proposed is not imported from any of the preceding specifications.

- **Capture Encoding**: A specific encoding of a Media Capture to be sent via RTP (RFC 3550).
- **CLUE Participant (CP):** An entity able to use the CLUE protocol within a telepresence session. It can be an endpoint or an MCU able to use the CLUE protocol.
- **CLUE-capable device**: A device that supports the CLUE data channel (see Section 16.6), the CLUE protocol, and the principles of CLUE negotiation and seeks CLUE-enabled calls.

- **Endpoint**: A CLUE-capable device that is the logical point of final termination through receiving, decoding and rendering, and/or initiation through capturing, encoding, and sending of media streams. An endpoint consists of one or more physical devices that source and sink media streams and exactly one (RFC 4353) Participant (which in turn, includes exactly one SIP user agent). Endpoints can be anything from multiscreen/multicamera rooms to handheld devices.
- **MCU**: A CLUE-capable device that connects two or more endpoints together into one single multimedia conference (RFC 5117). An MCU includes an RFC 4353-like Mixer without the RFC 4353 requirement to send media to each participant.
- **Media**: Any data that after suitable encoding, can be conveyed over RTP, including audio, video, or timed text.
- **Media Capture**: A source of Media, such as from one or more Capture Devices or constructed from other Media streams.
- **Media Consumer**: A CP (i.e., an Endpoint or an MCU) able to receive Capture Encodings.
- **Media Provider (MP)**: A CP (i.e., an Endpoint or an MCU) able to send Capture Encodings.
- **Stream**: A Capture Encoding sent from a Media Provider to a Media Consumer via RTP (RFC 3550).

16.5.3 Conventions

Per RFC 2119.

16.5.4 Overview of the CLUE Protocol

The CLUE protocol is conceived to enable CLUE telepresence sessions. It is designed in order to address SDP limitations in terms of the description of some information about the multimedia streams that are involved in a real-time multimedia conference. Indeed, by simply using SDP, it is not possible to convey information about the features of the flowing multimedia streams that are needed to enable a "being there" rendering experience. Such information is contained in the CLUE framework specification (see Section 16.3) and formally defined and described in the CLUE data model specification (see Section 16.4). The CLUE protocol represents the mechanism for the exchange of telepresence information between CLUE Participants. It mainly provides the messages to enable a Media Provider to advertise its telepresence capabilities and to enable a Media Consumer to select the desired telepresence options. The CLUE protocol, as defined in the following, is a stateful, client-server, XML-based application protocol. CLUE protocol messages flow on a reliable and ordered SCTP over a DTLS transport channel connecting two CLUE Participants. Messages carry information taken from the XML-based CLUE data model (see Section 16.4). Three main communication phases can be identified:

1. Establishment of the CLUE data channel: In this phase, the CLUE data channel setup takes place. If it completes successfully, the CPs are able to communicate and start the initiation phase.
2. Negotiation of the CLUE protocol version and extensions (initiation phase): The CPs connected via the CLUE data channel agree on the version and on the extensions to be used during the telepresence session. Special CLUE messages are used for such a task ("options" and "optionsResponse"). The version and extensions negotiation can be performed once during the CLUE session and only at this stage. At the end of that basic negotiation, each CP starts its activity as a CLUE MP and/or CLUE MC.
3. CLUE telepresence capabilities description and negotiation: In this phase, the MP–MC dialogs take place on the data channel by means of the CLUE protocol messages.

As soon as the channel is ready, the CLUE Participants must agree on the protocol version and extensions to be used within the telepresence session. CLUE protocol version numbers are characterized by a major version number (single digit) and a minor version number (single digit), both unsigned integers, separated by a dot. While minor version numbers denote backwards compatible changes in the context of a given major version, different major version numbers generally indicate a lack of interoperability between the protocol implementations. In order to correctly establish a CLUE dialog, the involved CPs MUST have in common a major version number (see Section 16.5.7 for further details). The subset of the extensions that are allowed within the CLUE session is also determined in the initiation phase, this subset being the one including only the extensions that are supported by both parties. A mechanism for the negotiation of the CLUE protocol version and extensions is part of the initial phase. According to such a solution, the CP that is the CLUE Channel initiator (CI) issues a proper CLUE message ("options") to the CP that is the Channel Receiver (CR) specifying the supported version and extensions. The CR then answers by selecting the subset of the CI extensions that it is able to support and determines the protocol version to be used.

After the negotiation phase is completed, CLUE Participants describe and agree on the media flows to be exchanged. In many cases, CPs will seek to both transmit and receive media. Hence, in a call between two CPs, A and B, there would be two separate dialogs, as follows:

1. The one needed to describe and set up the media streams sent from A to B, i.e., the dialog between A's MP side and B's MC side
2. The one needed to describe and set up the media streams sent from B to A, i.e., the dialog between B's MP side and A's MC side

CLUE messages for the media session description and negotiation are designed by considering the MP side as the server side of the protocol, since it produces and provides media streams, and the MC side as the client side of the protocol, since it requests and receives media streams. The messages that are exchanged to set up the telepresence media session are described by focusing on a single MP–MC dialog.

The MP first advertises its available media captures and encoding capabilities to the MC, as well as its simultaneity constraints, according to the information model defined in Section 16.3. The CLUE message conveying the MP's multimedia offer is the "advertisement" message. This message leverages the XML data model definitions provided in Section 16.4.

The MC selects the desired streams of the MP by using the "configure" message, which makes reference to the information carried in the previously received "advertisement." Besides "advertisement" and "configure," other messages have been conceived in order to provide all the needed mechanisms and operations. Such messages are detailed in the following sections.

16.5.5 Protocol Messages

CLUE protocol messages are textual, XML-based messages that enable the configuration of the telepresence session. The formal definition of such messages is provided in the XML Schema provided at the end of this specification (Section 16.5.9). This section includes non-normative excerpts of the schema to aid in describing it. The XML definitions of the CLUE information provided in Section 16.4 are included within some CLUE protocol messages (namely, the "advertisement" and the "configure" messages) in order to use the concepts defined in Section 16.3.

The CLUE protocol messages are the following:

- options
- optionsResponse
- advertisement

```
<xs:complexType name="clueMessageType" abstract="true">
        <xs:sequence>
                <xs:element name="clueId" type="xs:string" minOccurs="0"/>
                <xs:element name="sequenceNr" type="xs:positiveInteger"/>
        </xs:sequence>
        <xs:attribute name="protocol" type="xs:string" fixed="CLUE"
                use="required"/>
        <xs:attribute name="v" type="versionType" use="required"/>
</xs:complexType>
<!-- VERSION TYPE -->
<xs:simpleType name="versionType">
        <xs:restriction base="xs:string">
                <xs:pattern value="[1-9][0-9]*\.[0-9]+" />
        </xs:restriction>
</xs:simpleType>
```

FIGURE 16.10 Structure of a CLUE Message.

- ack
- configure
- configureResponse

While the "options" and "optionsResponse" messages are exchanged in the initiation phase between the CPs, the other messages are involved in MP–MC dialogs. Each CLUE message inherits a basic structure depicted in the following excerpt (Figure 16.10):

- **clueId**: An optional XML element containing the identifier (in the form of a generic string) of the CP within the telepresence system.
- **sequenceNr**: An XML element containing the local message sequence number. The sender must increment the sequence numbers by one for each new message sent, and the receiver must remember the most recent sequence number received and send back a 402 error if it receives a message with an unexpected sequence number (e.g., sequence number gap, repeated sequence number, or sequence number too small). The initial sequence number can be chosen randomly by each party.
- **protocol**: A mandatory attribute set to "CLUE," identifying the protocol the messages refer to.
- **v**: A mandatory attribute carrying the version of the protocol. The content of the "v" attribute is composed by the major version number followed by a dot and then by the minor version number of the CLUE protocol in use. The major number cannot be "0," and if it is more than one digit, it cannot start with a "0." Allowed values are of this kind are, e.g., "1.3," "2.0," "20.44," etc.

This specification describes version 1.0. Each CP is responsible for creating and updating up to three independent streams of sequence numbers in messages it sends:

(i) One for the messages sent in the initiation phase
(ii) One for the messages sent as MP (if it is acting as an MP)
(iii) One for the messages sent as MC (if it is acting as an MC)

16.5.5.1 options

The "options" message is sent by the CP that is the CI to the CP that is the CR as soon as the CLUE data channel is ready. Besides the information envisioned in the basic structure, it specifies:

- <mediaProvider>: A mandatory Boolean field set to "true" if the CP is able to act as an MP
- <mediaConsumer>: A mandatory Boolean field set to "true" if the CP is able to act as an MC
- <supportedVersions>: The list of the supported versions
- <supportedExtensions>: The list of the supported extensions

The XML Schema of such a message is reported here (Figure 16.11):

<supportedVersions> contains the list of the versions that are supported by the CI, each one represented in a child <version> element. The content of each <version> element is a string made by the major version number followed by a dot and then by the minor version number (e.g., 1.3 or 2.4). Exactly one <version> element MUST be provided for each major version supported, containing the maximum minor version number of such a version, since all minor versions are backwards compatible. If no <supportedVersions> is carried within the "options" message, the CI supports only the version declared in the "v" attribute and all the versions having the same major version number and lower minor version number. For example, if the "v" attribute has a value of "3.4" and there is no <supportedVersions> tag in the "options" message, it means that the CI supports only major version 3 with all the minor versions comprised between 3.0 and 3.4, with version 3.4 included. If a <supportedVersion> is provided, at least one <version> tag MUST be included. In this case, the "v" attribute MUST be set to one of the versions that the CI supports, as per the <supportedVersions> list.

The <supportedExtensions> element specifies the list of extensions supported by the CI. If there is no <supportedExtensions> in the "options" message, the CI does not support anything other than what is envisioned in the versions it supports. For each extension, an <extension> element is provided. An extension is characterized by a name, an XML schema of reference where the extension is defined, and the version of the protocol to which the extension refers.

16.5.5.2 optionsResponse

CLUE response messages ("optionsResponse," "ack," "configureResponse") derive from a base type, which is defined as follows (Figure 16.12):

The elements <responseCode> and <reasonString> are populated as detailed in Section 16.5.5.7. The "optionsResponse" (Figure 16.13) is sent by a CR to a CI as a reply to the "options" message. The "optionsResponse" contains a mandatory response code and a reason string indicating the processing result of the "options" message. If the responseCode is between 200 and 299 inclusive, the response MUST also include <mediaProvider>, <mediaConsumer>, <version>, and <commonExtensions> elements; it MAY include them for any other response code. <mediaProvider> and <mediaConsumer> elements are associated with the supported roles (in terms of, respectively, MP and MC), similarly to what the CI does in the "options" message. The <version> field indicates the highest commonly supported version number. The content of the <version> element MUST be a string made of the major version number followed by a dot and then by the minor version number (e.g., 1.3 or 2.4). Finally, the commonly supported extensions are copied in the <commonExtensions> field.

Upon reception of the "optionsResponse," the version to be used is provided in the <version> tag of the message. The following CLUE messages MUST use such a version number in the "v" attribute. The allowed extensions in the CLUE dialog are those indicated in the <commonExtensions> of the "optionsResponse" message.

16.5.5.3 advertisement

The "advertisement" message is used by the MP to advertise the available media captures and related information to the MC. The MP sends an "advertisement" to the MC as soon as it is ready

```
CLUE OPTIONS -->
<xs:complexType name="optionsMessageType">
    <xs:complexContent>
        <xs:extension base="clueMessageType">
            <xs:sequence>
                <xs:element name="mediaProvider" type="xs:boolean" />
                <xs:element  name="mediaConsumer"  type="xs:boolean"
                />
                <xs:element name="supportedVersions"
                    type="versionsListType" minOccurs="0"/>
                <xs:element name="supportedExtensions"
                    type="extensionsListType" minOccurs="0"/>
                <xs:any namespace="##other" processContents="lax"
                    minOccurs="0"/>
            </xs:sequence>
            <xs:anyAttribute namespace="##other" processContents="lax"/>
        </xs:extension>
    </xs:complexContent>
</xs:complexType>
<!-- VERSIONS LIST TYPE -->
<xs:complexType name="versionsListType">
    <xs:sequence>
        <xs:element name="version" type="versionType" minOccurs="1"
            maxOccurs="unbounded"/>
        <xs:any namespace="##other" processContents="lax" minOccurs="0"/>

        </xs:sequence>
    <xs:anyAttribute namespace="##other" processContents="lax"/>
</xs:complexType>
<!-- EXTENSIONS LIST TYPE -->
<xs:complexType name="extensionsListType">
    <xs:sequence>
        <xs:element name="extension" type="extensionType" minOccurs="1"
            maxOccurs="unbounded"/>
        <xs:any namespace="##other" processContents="lax" minOccurs="0"/>
    </xs:sequence>
    <xs:anyAttribute namespace="##other" processContents="lax"/>
</xs:complexType>
<!-- EXTENSION TYPE -->
<xs:complexType name="extensionType">
    <xs:sequence>
        <xs:element name="name" type="xs:string" />
        <xs:element name="schemaRef" type="xs:anyURI" minOccurs="0"/>
        <xs:element name="version" type="versionType" minOccurs="0"/>
        <xs:any namespace="##other" processContents="lax" minOccurs="0"/>
    </xs:sequence>

    <xs:anyAttribute namespace="##other" processContents="lax"/>
</xs:complexType>
```

FIGURE 16.11 Structure of CLUE "options" Message.

after the successful completion of the initiation phase, i.e., as soon as the version and the extensions of the CLUE protocol are agreed between the CPs. During a single CLUE session, an MP may send new "advertisement" messages to replace the previous advertisement if, for instance, its CLUE telepresence media capabilities change mid-call.

```
<xs:complexType name="clueResponseType">
        <xs:complexContent>
                <xs:extension base="clueMessageType">
                        <xs:sequence>
                                <xs:element name="responseCode"
                                        type="responseCodeType"/>
                                <xs:element name="reasonString" type="xs:string"
                                        minOccurs="0"/>
                        </xs:sequence>
                </xs:extension>
        </xs:complexContent>
</xs:complexType>
```

FIGURE 16.12 Structure of CLUE Response Messages.

```
<!-- CLUE 'optionsResponse' -->
<xs:complexType name="optionsResponseMessageType">
        <xs:complexContent>
                <xs:extension base="clueResponseType">
                        <xs:sequence>
                                <xs:element name="mediaProvider"
                                        type="xs:boolean"
                                minOccurs="0"/>
                                <xs:element name="mediaConsumer"
                                        type="xs:boolean" minOccurs="0"/>
                                <xs:element name="version" type="versionType"
                                        minOccurs="0"/>
                                <xs:element name="commonExtensions"
                                        type="extensionsListType"
                                        minOccurs="0"/>
                                <xs:any namespace="##other"
                                        processContents="lax" minOccurs="0"/>
                        </xs:sequence>
                        <xs:anyAttribute namespace="##other"
                                processContents="lax"/>
                </xs:extension>
        </xs:complexContent>
</xs:complexType>
```

FIGURE 16.13 Structure of CLUE "optionsResponse" Message.

A new "advertisement" completely replaces the previous "advertisement." The "advertisement" structure is defined in the following schema excerpt (Figure 16.14). The "advertisement" contains elements compliant with the CLUE data model that characterize the MP's telepresence offer. Namely, such elements are: the list of the media captures (<mediaCaptures>), of the encoding groups (<encodingGroups>), of the capture scenes (<captureScenes>), of the simultaneous sets

```
<!-- CLUE ADVERTISEMENT MESSAGE TYPE -->
<xs:complexType name="advertisementMessageType">
    <xs:complexContent>
        <xs:extension base="clueMessageType">
            <xs:sequence>
                <!-- mandatory -->
                <xs:element name="mediaCaptures"
                    type="dm:mediaCapturesType"/>
                <xs:element name="encodingGroups"
                    type="dm:encodingGroupsType"/>
                <xs:element name="captureScenes"
                    type="dm:captureScenesType"/>
                <!-- optional -->
                <xs:element name="simultaneousSets"
                    type="dm:simultaneousSetsType"
                    minOccurs="0"/>
                <xs:element name="globalViews"
                    type="dm:globalViewsType"
                    minOccurs="0"/>
                <xs:element name="people" type="dm:peopleType"
                    minOccurs="0"/>
                <xs:any namespace="##other"
                    processContents="lax"
                    minOccurs="0"/>
            </xs:sequence>
            <xs:anyAttribute namespace="##other"
                processContents="lax"/>
        </xs:extension>
    </xs:complexContent>
</xs:complexType>
```

FIGURE 16.14 Structure of CLUE "advertisement" Message.

(<simultaneousSets>), of the global views (<globalViews>), and of the represented participants (<people>). Each of them is fully described in the CLUE framework specification and formally defined in the CLUE data model specification.

16.5.5.4 ack

The "ack" message is sent by a MC to a MP to acknowledge an 'advertisement' message. As can be seen from the message schema provided in the following excerpt (Figure 16.15), the "ack" contains a response code and a reason string for describing the processing result of the "advertisement." The <advSequenceNr> carries the sequence number of the "advertisement" message the "ack" refers to.

16.5.5.5 configure

The "configure" message is sent from an MC to an MP to list the advertised captures the MC wants to receive. The MC can send a "configure" after the reception of an "advertisement" or each time it wants to request other captures that have been previously advertised by the MP. The content of the "configure" message is shown in the following (Figure 16.16).

```
<!-- 'ack' MESSAGE TYPE -->
<xs:complexType name="advAcknowledgementMessageType">
        <xs:complexContent>
                <xs:extension base="clueResponseType">
                        <xs:sequence>
                        <xs:element name="advSequenceNr"
                                type="xs:positiveInteger"/>
                        <xs:any namespace="##other" processContents="lax"
                                minOccurs="0"/>
                        </xs:sequence>
                        <xs:anyAttribute namespace="##other"
                                processContents="lax"/>
                </xs:extension>
        </xs:complexContent>
</xs:complexType>
```

FIGURE 16.15 Structure of CLUE "ack" Message.

The <advSequenceNr> element contains the sequence number of the "advertisement" message the "configure" refers to. The optional <ack> element, when present, contains a success response code, as defined in Section 16.5.5.7. It indicates that the "configure" message also acknowledges with success the referred advertisement ("configure" + "ack" message) by applying in that way a piggybacking mechanism for simultaneously acknowledging and replying to the "advertisement" message. The <ack> element MUST NOT be present if an "ack" message has been already sent back to the MP. The most important content of the "configure" message is the list of the capture encodings provided in the <captureEncodings> element (see Section 16.4) for the definition of <captureEncodings>). Such an element contains a sequence of capture encodings, representing the streams to be instantiated.

16.5.5.6 configureResponse

The "configureResponse" message is sent from the MP to the MC to communicate the processing result of requests carried in the previously received "configure" message. It contains (Figure 16.17) a response code with a reason string indicating either the success or the failure (along with failure details) of a "configure" request processing. Following, the <confSequenceNr> field contains the sequence number of the "configure" message the response refers to. There is no partial execution of commands. As an example, if a MP is able to understand all the selected capture encodings except one, then the whole command fails, and nothing is instantiated.

16.5.5.7 Response Codes and Reason Strings

Response codes are defined as a sequence of three digits. A well-defined meaning is associated with the first digit. Response codes beginning with "2" are associated with successful responses. Response codes that do not begin with either "2" or "1" indicate an error response, i.e., that an error occurred while processing a CLUE request. In particular, response codes beginning with "3" indicate problems with the XML content of the message ("Bad syntax," "Invalid value," etc.), while response codes beginning with "4" refer to problems related to CLUE protocol semantics ("Invalid sequencing," "Version not supported," etc.). 200, 300, and 400 codes are considered catch-alls. Further response codes can be either defined in future versions of the protocol (by adding them to

```
<!-- CLUE 'configure' MESSAGE TYPE -->
<xs:complexType name="configureMessageType">
    <xs:complexContent>
        <xs:extension base="clueMessageType">
            <xs:sequence>
                <!-- mandatory fields -->
                <xs:element name="advSequenceNr"
                    type="xs:positiveInteger"/>
                <xs:element name="ack"
                    type="successResponseCodeType"
                        minOccurs="0"/>
                <xs:element name="captureEncodings"
                type="dm:captureEncodingsType"
                    minOccurs="0"/>
                <xs:any namespace="##other"
                    processContents="lax" minOccurs="0"/>
            </xs:sequence>
            <xs:anyAttribute namespace="##other"
                processContents="lax"/>
        </xs:extension>
    </xs:complexContent>
</xs:complexType>
```

FIGURE 16.16 Structure of CLUE "configure" Message.

the related IANA registry) or defined by leveraging the extension mechanism. In both cases, the new response codes MUST be registered with IANA.

Such new response codes MUST NOT overwrite the ones here defined, and they MUST respect the semantics of the first code digit. This specification does not define response codes starting with "1," and such response codes are not allowed to appear in major version 1 of the CLUE protocol. The range from 100 to 199 inclusive is reserved for future major versions of the protocol to define response codes for delayed or incomplete operations if necessary. Response codes starting with "5" through "9" are reserved for future major versions of the protocol to define new classes of response and are not allowed in major version 1 of the CLUE protocol. Response codes starting with "0" are not allowed. The response codes and strings defined for use with version 1 of the CLUE protocol are listed in Figure 16.18. The "Description" text contained in the table can be sent in the <reasonString> element of a response message. Implementations can (and are encouraged to) include more specific descriptions of the error condition, if possible.

16.5.6 Protocol State Machines

The CLUE protocol is an application protocol used between two CPs in order to properly configure a multimedia telepresence session. CLUE-protocol messages flow over the CLUE Data Channel, a DTLS/SCTP channel established as depicted in Section 16.6. We herein discuss the state machines associated, respectively, with the CP (Figure 16.19), with the MC process, and with the MP process. Endpoints often wish to both send and receive media, i.e., act as both MP and MC. As such, there will often be two sets of messages flowing in opposite directions; the state machines of these

```
<!-- 'configureResponse' MESSAGE TYPE -->
<xs:complexType name="configureResponseMessageType">
        <xs:complexContent>
                <xs:extension base="clueResponseType">
                        <xs:sequence>
                                <xs:element name="confSequenceNr"
                                        type="xs:positiveInteger" />
                                <xs:any namespace="##other"
                                        processContents="lax"
                                                minOccurs="0"/>
                        </xs:sequence>
                        <xs:anyAttribute namespace="##other"
                                processContents="lax" />
                </xs:extension>
        </xs:complexContent>
</xs:complexType>
```

FIGURE 16.17 Structure of CLUE "configureResponse" Message.

two flows do not interact with each other. Only the CLUE application logic is considered. The interaction of CLUE protocol and SDP negotiations for the media streams exchanged is treated in Section 16.8.

The main state machines focus on the behavior of the CP acting as a CLUE channel initiator/receiver (CI/CR). The initial state is the IDLE one. When in the IDLE state, the CLUE data channel is not established, and no CLUE-controlled media are exchanged between the two considered CLUE-capable devices (if there is an ongoing exchange of media streams, such media streams are not currently CLUE-controlled). When the CLUE data channel set up starts ("start channel"), the CP moves from the IDLE state to the CHANNEL SETUP state.

If the CLUE data channel is successfully set up ("channel established"), the CP moves from the CHANNEL SETUP state to the OPTIONS state. Otherwise, if "channel error," it moves back to the IDLE state. The same transition happens if the CLUE-enabled telepresence session ends ("session ends"), i.e., when an SDP negotiation for removing the CLUE data channel is performed. When in the OPTIONS state, the CP addresses the initiation phase where both parts agree on the version and on the extensions to be used in the subsequent CLUE messages exchange phase. If the CP is the CI, it sends an "options" message and waits for the "optionsResponse" message. If the CP is the CR, it waits for the "options" message and as soon as it arrives, replies with the "optionsResponse" message. If the negotiation is successfully completed ("OPTIONS phase success"), the CP moves from the OPTIONS state to the ACTIVE state. If the initiation phase fails ("OPTIONS phase failure"), the CP moves from the OPTIONS state to the IDLE state.

The initiation phase might fail because of one of the following reasons:

1. The CI receives an "optionsResponse" with an error response code.
2. The CI does not receive any "optionsResponse," and a timeout error is raised.
3. The CR does not receive any "options," and a timeout error is raised.

Response Code	Reason String	Description
200	Success	The request has been successfully processed
300	Low-level request error	A generic low-level request error has occurred
301	Bad syntax	The XML syntax of the message is not correct
302	Invalid value	The message contains an invalid parameter value
303	Conflicting values	The message contains values that cannot be used together
400	Semantic errors	Semantic errors in the received CLUE protocol message
401	Version not supported	The protocol version used in the message not supported
402	Invalid sequencing	Sequence number gap; repeated sequence number; sequence number outdated
403	Invalid identifier	The clueId used in the message is not valid or unknown
404	Advertisement Expired	The sequence number of the advertisement the configure refers to is out of date
405	Subset choice not allowed	The subset choice is not allowed for the specified Multiple Content Capture

FIGURE 16.18 CLUE Response Codes.

When in the ACTIVE state, the CP starts the envisioned sub-state machines (i.e., the MP state machine and the MC state machine) according to the roles it plays in the telepresence sessions. Such roles have been previously declared in the "options" and "optionsResponse" messages involved in the initiation phase (see OPTIONS Sections 16.5.5.1 and 16.5.5.2 for the details). When in the ACTIVE state, the CP delegates the sending and the processing of the CLUE messages to the appropriate MP/MC sub-state machines. If the CP receives a further "options"/"optionsResponse" message, it MUST ignore the message and stay in the ACTIVE state. The CP moves from the ACTIVE state to the IDLE one when the sub-state machines that had been activated are in the relative TERMINATED state (see Sections 16.5.6.1 and 16.5.6.2).

16.5.6.1 Media Provider's State Machine

As soon as the sub-state machine of the MP (Figure 16.20) is activated, it is in the ADV state. In the ADV state, the MP prepares the "advertisement" message reflecting its actual telepresence capabilities. After the "advertisement" has been sent ("advertisement sent"), the MP moves from the ADV state to the WAIT FOR ACK state. If an "ack" message with a successful response code arrives ("ack received"), the MP moves to the WAIT FOR CONF state. If a NACK arrives (i.e., an "ack"

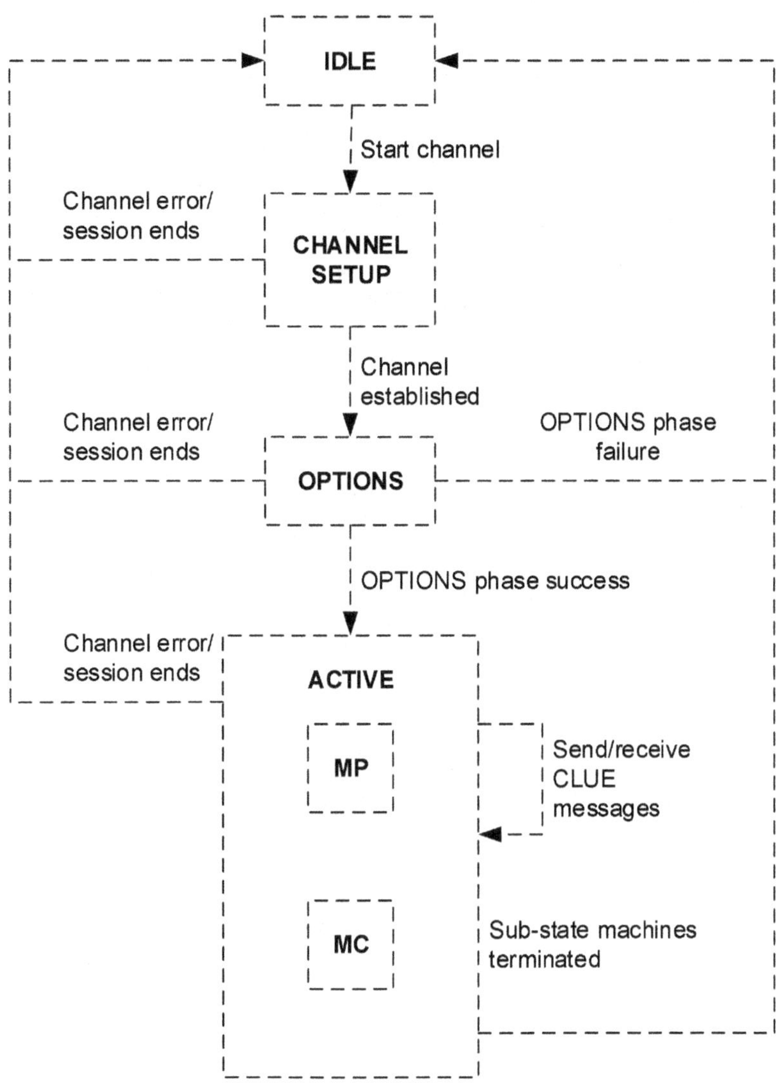

FIGURE 16.19 CLUE Participant State Machine. (Copyright: IETF.)

message with an error response code), and the number of NACKs for the issued "advertisement" is under the retry threshold ("NACKreceived && retry not exhausted"), the MP moves back to the ADV state for preparing a new "advertisement." If a NACK arrives, and the number of received NACKs for that "advertisement" overcomes the threshold ("NACK received && retry exhausted"), the MP goes to the MP-TERMINATED state. The same happens if the waiting time for the "ack" is fired a number of times under the retry threshold ("timeout && retry not exhausted"): also in this case, the MP goes back to the ADV state to send a new copy of the "advertisement." If the number of retries overcomes the threshold ("timeout && retry exhausted"), the MP moves from the WAIT FOR ACK state to the MP-TERMINATED state.

When in the WAIT FOR ACK state, if a "configure" message with the <ack> element set to TRUE arrives ("configure+ack received"), the MP goes directly to the CONF RESPONSE state. configure+ack messages referring to out-of-date (i.e., having a sequence number equal to or less than the highest seen so far) advertisements MUST be ignored, i.e., they do not trigger any state transition. If the telepresence settings of the MP change while in the WAIT FOR ACK state ("changed telepresence settings"), the MP switches from the WAIT FOR ACK state to the ADV state to create

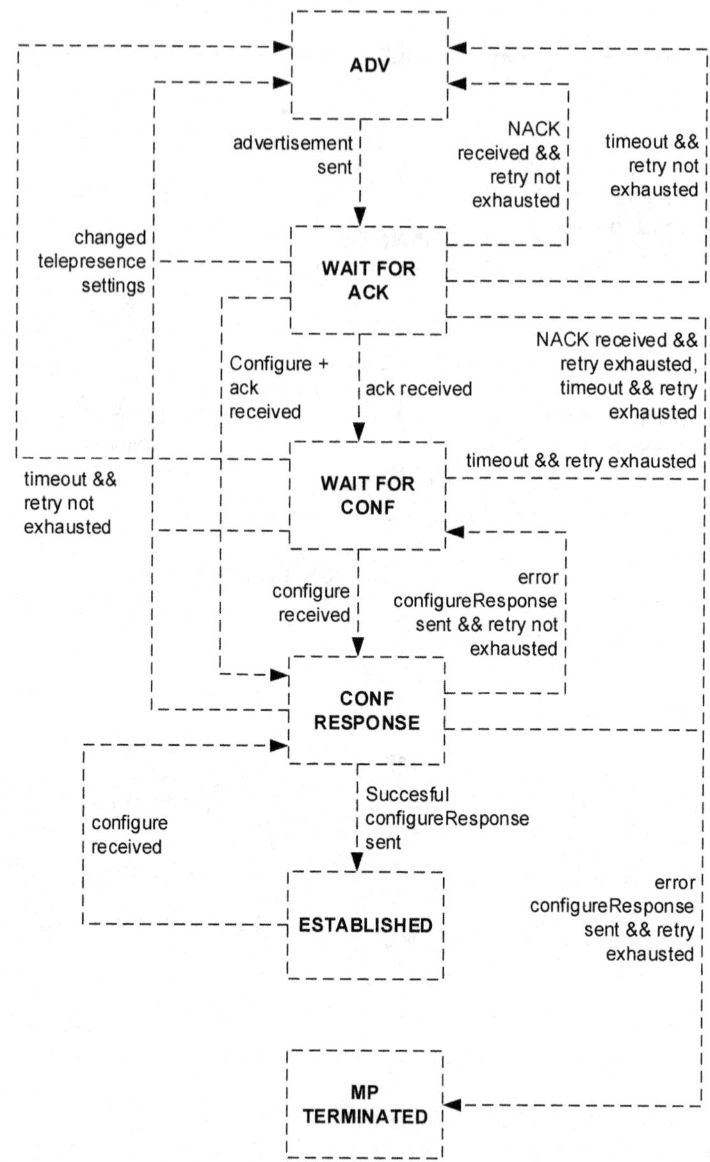

FIGURE 16.20 Media Provider's State Machine. (Copyright: IETF.)

a new "advertisement." When in the WAIT FOR CONF state, the MP listens to the channel for a "configure" request coming from the MC. If a "configure" arrives ("configure received"), the MP switches to the CONF RESPONSE state.

If the "configure" does not arrive within the timeout interval, and the retry threshold has not been overcome ("timeout && retry not exhausted"), the MP moves back to the ADV state. When the retry expires ("timeout && retry exhausted"), the MP moves to the MP TERMINATED state. If the telepresence settings change in the meanwhile ("changed telepresence settings"), the MP moves from the WAIT FOR CONF back to the ADV state to create the new "advertisement" to be sent to the MC.

The MP in the CONF RESPONSE state processes the received "configure" in order to produce a "configureResponse" message. If the MP successfully processes the MC's configuration, then it sends a 200 "configureResponse" ("success configureResponse sent") and moves to the

ESTABLISHED state. If there are errors in the "configure" processing, then the MP issues a "configureResponse" carrying an error response code, and if under the retry threshold ("error configureResponse sent && retry not exhausted"), it goes back to the WAIT FOR CONF state to wait for a new configuration request. If the number of trials exceeds the retry threshold ("error configureResponse sent && retry exhausted"), the state MP TERMINATED is reached. Finally, if there are changes in the MP's telepresence settings ("changed telepresence settings"), the MP switches to the ADV state.

The MP in the ESTABLISHED state has successfully negotiated the media streams with the MC by means of the CLUE messages. If there are changes in the MP's telepresence settings ("changed telepresence settings"), the MP moves back to the ADV state. In the ESTABLISHED state, the CLUE-controlled media streams of the session are those described in the last successfully processed "configure" message.

16.5.6.2 Media Consumer's State Machine

As soon as the sub-state machine of the MC (Figure 16.21) is activated, it is in the WAIT FOR ADV state. An MC in the WAIT FOR ADV state is waiting for an "advertisement" coming from the MP. If the "advertisement" arrives ("ADV received"), the MC reaches the ADV PROCESSING state. Otherwise, the MC stays in the WAIT FOR ADV state. In the ADV PROCESSING state, the "advertisement" is parsed by the MC. If the "advertisement" is successfully processed, there are two possibilities. In the former case, the MC issues a successful "ack" message to the MP ("ACK sent") and moves to the CONF state. This typically happens when the MC needs some more time to produce the "configure" message associated with the received "advertisement." In the latter case, the MC is able to immediately prepare and send back to the MP a "configure" message. Such a message will have the <ack> field set to "200" ("configure+ack sent") and will allow the MC to move directly to the WAIT FOR CONF RESPONSE state.

If the ADV processing is unsuccessful (bad syntax, missing XML elements, etc.), and the number of times this has happened is under the retry threshold, the MC sends a NACK message (i.e., an "ack" with an error response code) to the MP and optionally, further describes the problem via a proper reason phrase. In this way ("NACK sent && retry not exhausted"), the MC switches back to the WAIT FOR ADV state, waiting for a new "advertisement." If the NACK retry expires ("NACK sent && retry exhausted"), the MC moves to the MC TERMINATED state. When in the CONF state, the MC prepares the "configure" request to be issued to the MP on the basis of the previously ack-ed "advertisement." When the "configure" has been sent ("configure sent"), the MC moves to the WAIT FOR CONF RESPONSE state. If a new "advertisement" arrives in the meanwhile ("advertisement received"), the MC goes back to the ADV PROCESSING state. In the WAIT FOR CONF RESPONSE state, the MC waits for the MP's response to the issued "configure" or "configure+ack."

If a 200 "configureResponse" message is received ("successful configureResponse received"), it means that the MP and the MC have successfully agreed on the media streams to be shared. Then, the MC can move to the ESTABLISHED state. On the other hand, if an error response is received, and the associated retry counter does not overcome the threshold ("error configureResponse received && retry not exhausted"), the MC moves back to the CONF state to prepare a new "configure" request. In the case of "error configureResponse received && retry exhausted," the MC moves to the MC TERMINATED state. If no "configureResponse" arrives, and the number of timeouts is under the threshold ("timeout && retry not exhausted"), the MC moves to the CONF state and sends again the "configure" message. If no "configureResponse" arrives, and the number of timeouts is over the threshold ("timeout && retry exhausted"), the MC moves to the MC TERMINATED state. If a new "advertisement" is received in the WAIT FOR CONF RESPONSE state, the MC switches to the ADV PROCESSING state.

When the MC is in the ESTABLISHED state, the telepresence session configuration has been set up at the CLUE application level according to the MC's preferences. Both the MP and the MC have agreed on (and are aware of) the CLUE-controlled media streams to be exchanged within

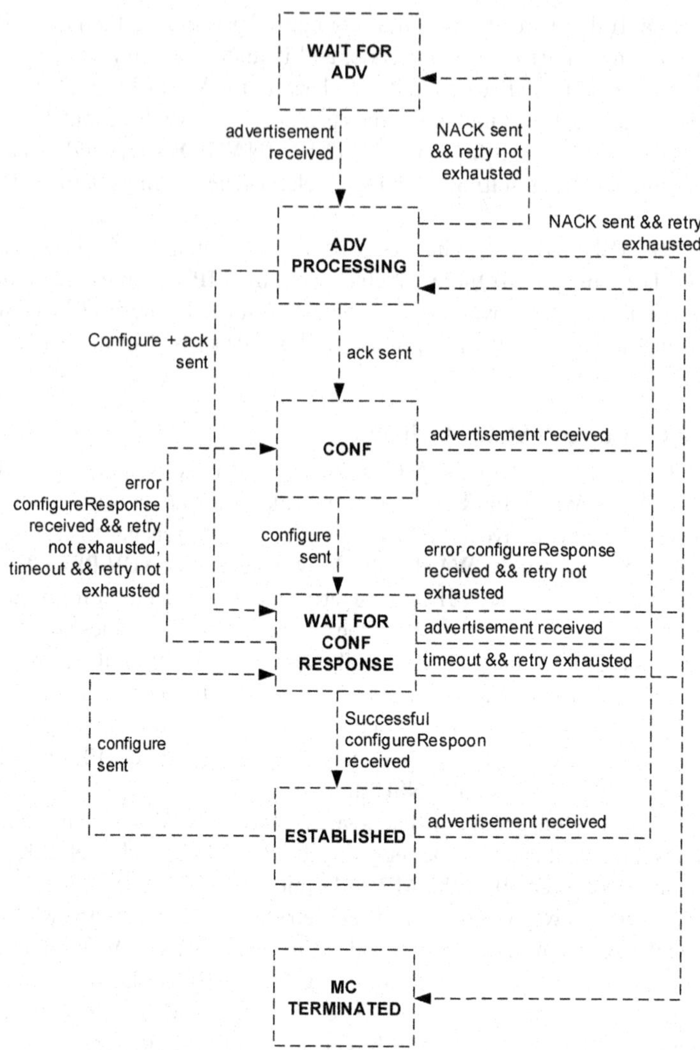

FIGURE 16.21 Media Consumer's State Machine. (Copyright: IETF.)

the call. While in the ESTABLISHED state, it might happen that the MC decides to change something in the call settings. The MC then issues a new "configure" ("configure sent") and goes to wait for the new "configureResponse" in the WAIT FOR CONF RESPONSE state. On the other hand, in the ESTABLISHED state, if a new "advertisement" arrives from the MP ("advertisement received"), it means that something has changed on the MP's side. The MC then moves to the ADV PROCESSING state.

16.5.7 Versioning

CLUE protocol messages are XML messages compliant with the CLUE protocol XML schema (see Section 16.4). The version of the protocol corresponds to the version of the schema. Both client and server have to test the compliance of the received messages with the XML schema of the CLUE protocol. If the compliance is not verified, the message cannot be processed further. Obviously, client and server cannot communicate if they do not share exactly the same XML schema. Such a schema is associated with the CLUE uniform resource name (URN) "urn:ietf:params:xml:ns:clue-protocol." If all CLUE-enabled devices use that schema, there will be no interoperability problems due to schema issues. This specification defines XML schema version 1.0. The version usage is similar in philosophy to XMPP (RFC 6120). A version number has major and minor components, each a non-negative integer. Major version changes denote non-interoperable changes. Minor version changes denote schema changes that are backwards compatible by ignoring unknown XML elements or other backwards compatible changes.

The minor versions of the XML schema MUST be backwards compatible, not only in terms of schema but also semantically and procedurally as well. This means that they should define further features and functionality besides those defined in the previous versions, in an incremental way, without impacting the basic rules defined in the previous version of the schema. In this way, if a MP is able to speak, e.g., version 1.5 of the protocol while the MC only understands version 1.4, the MP should have no problem in reverting the dialog to version 1.4 without exploiting 1.5 features and functionality. Version 1.4 is the one to be spoken and has to appear in the "v" attribute of the subsequent CLUE messages. In other words, in this example, the MP MUST use version 1.4 and downgrade to the lower version. This said, and coherently with the general IETF "protocol robustness principle" stating that "an implementation must be conservative in its sending behavior, and liberal in its receiving behavior" (RFC 1122), CLUE Participants MUST ignore unknown elements or attributes that are not envisioned in the negotiated protocol version and related extensions.

16.5.8 Extensions

Although the standard version of the CLUE protocol XML schema is designed to thoroughly cope with the requirements emerging from the application domain, new needs might arise, and extensions can be designed. Extensions specify information and behaviors that are not described in a certain version of the protocol and specified in the related RFC specification. Such information and behaviors can be optionally used in a CLUE dialog and MUST be negotiated in the CLUE initiation phase. They can relate to:

1. New information, to be carried in the existing messages. For example, more fields may be added within an existing message.
2. New messages. This is the case if there is no proper message for a certain task, so a brand new CLUE message needs to be defined.

As to the first type of extensions, it is possible to distinguish between protocol-specific and data model information. Indeed, CLUE messages are envelopes carrying both:

- XML elements defined within the CLUE protocol XML schema itself (protocol-specific information)
- Other XML elements compliant with the CLUE data model schema (data model information)

When new protocol-specific information is needed somewhere in the protocol messages, it can be added in place of the <any> elements and <anyAttribute> elements envisioned by the protocol schema. The policy currently defined in the protocol schema for handling <any> and <anyAttribute> elements is:

- elementFormDefault="qualified"
- attributeFormDefault="unqualified"

The new information must be qualified by namespaces other than "urn:ietf:params:xml:ns:clue-protocol" (the protocol URN) and "urn:ietf:params:xml:ns:clue-info" (the data model URN). Elements or attributes from unknown namespaces MUST be ignored. The other matter concerns data model information. Data model information is defined by the XML schema associated with the URN "urn:ietf:params:xml:ns:clue-info." Also, for the XML elements defined in such a schema, there are extensibility issues. Those issues are overcome by using <any> and <anyAttribute> placeholders. New information within data model elements can be added in place of <any> and <anyAttribute> schema elements, as long as they are properly namespace qualified.

On the other hand (second type of extensions), "extra" CLUE protocol messages, i.e., messages not envisioned in the latest standard version of the schema, may be needed. In that case, the messages and the associated behavior should be defined in external specifications that both communication parties must be aware of. Both types of extensions, i.e., new information and new messages, can be characterized by:

- A name.
- An external XML Schema defining the XML information and/or the XML messages representing the extension.
- The standard version of the protocol the extension refers to. Extensions are represented by means of the <extension> element as defined in Figure 16.22, which is carried within the "options" and "optionsResponse" messages to represent the extensions supported by both the CI and the CR.

Extensions MUST be defined in a separated XML schema file and SHOULD be provided with a companion specification describing their semantics and use.

16.5.8.1 Extension Example

An example of extension might be a "new" capture attribute (i.e., a capture attribute that is not envisioned in the current standard defining the CLUE data model in Section 16.4) needed to further describe a video capture. The CLUE data model specification (see Section 16.4) envisions the possibility of adding this kind of "extra" information in the description of a video capture by keeping the compatibility with the CLUE data model schema. This is made possible thanks to the presence of the <any> element in the XML definition of the video capture, allowing for introduction of a new XML field in the XML description. For the sake of convenience, the XML definition of a video capture taken from Section 16.4 is reported in Figure 16.23 that follows.

According to such a definition, a video capture might have, after the set of the generic media capture attributes, a set of new attributes defined elsewhere, i.e., in an XML schema defining an extension. The XML schema defining the extension might look like the following (Figure 16.24):

By using the preceding extension, a video capture can be further described in the advertisement using the <myVideoExtension> element containing two extra pieces of information (<newVideoAttribute1> and <newVideoAttribute2>) besides using the attributes envisioned for a generic media capture. As stated in this specification, both participants must be aware of the extension schema and

```
<xs:complexType name="extensionType">
        <xs:sequence>
            <xs:element name="name" type="xs:string" />
            <xs:element name="schemaRef" type="xs:anyURI" minOccurs="0"/>
            <xs:element name="version" type="versionType" minOccurs="0"/>
            <xs:any namespace="##other" processContents="lax" minOccurs="0"/>
        </xs:sequence>
        <xs:anyAttribute namespace="##other" processContents="lax"/>
</xs:complexType>
```

FIGURE 16.22 The <extension> Element.

```
<!-- VIDEO CAPTURE TYPE -->
<xs:complexType name="videoCaptureType">
    <xs:complexContent>
        <xs:extension base="tns:mediaCaptureType">
            <xs:sequence>
                <xs:any namespace="##other"
                    processContents="lax" minOccurs="0"
                    maxOccurs="unbounded"/>
            </xs:sequence>
            <xs:anyAttribute namespace="##other"
                processContents="lax"/>
        </xs:extension>
    </xs:complexContent>
</xs:complexType>
```

FIGURE 16.23 XML Definition of a CLUE Video Capture.

related semantics to use such an extension and must negotiate it via the "options" and "optionsResponse" mechanism.

16.5.9 XML Schema

In this section, the XML schema defining the CLUE messages is provided (Figure 16.25).

16.5.10 Examples

In this section we provide examples of "advertisement" messages representing the telepresence environment described in Section 16.4, Section 16.4.24 "Sample XML File," and Section 16.4.28 "MCC Example," respectively.

16.5.10.1 Simple "Advertisement"

Figure 16.26 presents a simple "advertisement" message. The associated MP's telepresence capabilities are described in Section 16.4, Section 16.4.24 "Sample XML File".

16.5.10.2 "Advertisement" with MCCs

Figure 16.27 presents a simple "advertisement" message containing a MCC. The associated MP's telepresence capabilities are described in Section 16.4, Section 16.4.28 "MCC Example."

16.5.11 Security Considerations

As a general consideration, we remark that the CLUE framework (and related protocol) has been conceived at the outset by embracing the security-by-design paradigm. This entails that a number of requirements have been identified and properly standardized as mandatory within the entire set of specifications associated with the CLUE architecture. Requirements include: (i) the use of cryptography and authentication; (ii) protection of all sensitive fields; (iii) mutual authentication between CLUE endpoints; (iv) the presence of authorization mechanisms; and (v) the presence of native defense mechanisms against malicious activities such as eavesdropping, selective modification, deletion, and replay (and related combinations thereof). Hence, the security of the single

```
<?xml version="1.0" encoding="UTF-8" ?>
<xs:schema version="1.0"
        targetNamespace="http://example.extensions.com/myVideoExtensions"
        xmlns:xs="http://www.w3.org/2001/XMLSchema"
        xmlns="http://example.extensions.com/myVideoExtensions"
        elementFormDefault="qualified"
        attributeFormDefault="unqualified">
        <!--
        This is the new element to be put in place of the <any>
        element in the video capture definition
        of the CLUE data model schema
        -->
        <xs:element name="myVideoExtension">
                <xs:complexType>
                        <xs:sequence>
                                <xs:element ref="newVideoAttribute1"/>
                                <xs:element ref="newVideoAttribute2"/>
                        </xs:sequence>
                </xs:complexType>
        </xs:element>
        <xs:element name="newVideoAttribute1" type="xs:string"/>
        <xs:element name = "newVideoAttribute2" type = "xs:boolean"/>
</xs:schema>
```

FIGURE 16.24 XML Schema Defining an Extension.

components of the CLUE solution cannot be evaluated independently of the integrated view of the final architecture.

The CLUE protocol is an application-level protocol allowing a Media Producer and a Media Consumer to negotiate a variegated set of parameters associated with the establishment of a telepresence session. This unavoidably exposes a CLUE-enabled telepresence system to a number of potential threats, most of which are extensively discussed in the framework specification (see Section 16.3). The security considerations section of the mentioned specification actually discusses issues associated with the setup and management of a telepresence session both in the basic case, involving two CLUE endpoints acting, respectively, as MP and MC, and in the more advanced scenario envisaging the presence of an MCU.

The framework specification also mentions that the information carried within CLUE protocol messages might contain sensitive data, which SHOULD hence be accessed only by authenticated endpoints. Security issues associated with the CLUE data model schema are discussed in Section 16.4. There is extra information carried by the CLUE protocol that is not associated with the CLUE data model schema and that exposes information that might be of concern. This information is primarily exchanged during the negotiation phase via the “options” and “optionsResponse” messages. In the CP state machine OPTIONS state, both parties agree on the version and on the extensions to be used in the subsequent CLUE messages exchange phase.

A malicious participant might either try to retrieve a detailed footprint of a specific CLUE protocol implementation during this initial setup phase or force the communicating party to use a non-up-to-date version of the protocol, which they know how to break. Indeed, exposing all the supported versions and extensions could conceivably leak information about the specific implementation of

```
<?xml version="1.0" encoding="UTF-8"?>
<xs:schema xmlns:xs="http://www.w3.org/2001/XMLSchema"
        xmlns="urn:ietf:params:xml:ns:clue-protocol"
        xmlns:dm="urn:ietf:params:xml:ns:clue-info"
        xmlns:tns="urn:ietf:params:xml:ns:clue-protocol"
        version="1.0"
        targetNamespace="urn:ietf:params:xml:ns:clue-protocol"
        elementFormDefault="qualified"
        attributeFormDefault="unqualified">
        <!-- Import data model schema -->
        <xs:import namespace="urn:ietf:params:xml:ns:clue-info"
                schemaLocation="clue-data-model-schema-17.xsd" />
        <!-- ELEMENT DEFINITIONS -->
        <xs:element name="options" type="optionsMessageType" />
        <xs:element name="optionsResponse" type="optionsResponseMessageType"/>
        <xs:element name="advertisement" type="advertisementMessageType"/>
        <xs:element name="ack" type="advAcknowledgementMessageType"/>
        <xs:element name="configure" type="configureMessageType"/>
        <xs:element name="configureResponse"
                type="configureResponseMessageType"/>
        <!-- CLUE MESSAGE TYPE -->
        <xs:complexType name="clueMessageType" abstract="true">
                <xs:sequence>
                        <xs:element name="clueId" type="xs:string" minOccurs="0" />
                        <xs:element name="sequenceNr" type="xs:positiveInteger" />
                </xs:sequence>
                <xs:attribute name="protocol" type="xs:string" fixed="CLUE"
                        use="required" />
                <xs:attribute name="v" type="versionType" use="required" />
        </xs:complexType>
        <!-- CLUE RESPONSE TYPE -->
        <xs:complexType name="clueResponseType">
                <xs:complexContent>
                        <xs:extension base="clueMessageType">
                                <xs:sequence>
                                        <xs:element name="responseCode"
                                                type="responseCodeType" />
                                        <xs:element name="reasonString" type="xs:string"
                                                minOccurs="0"/>
                                </xs:sequence>
                        </xs:extension>
                </xs:complexContent>
        </xs:complexType>
        <!-- VERSION TYPE -->
        <xs:simpleType name="versionType">
```

FIGURE 16.25 Schema defining CLUE messages.

the protocol. In theory an implementation could choose not to announce all the versions it supports if it wants to avoid such leakage, though at the expense of interoperability. With respect to these considerations, it is noted that the OPTIONS state is only reached after the CLUE data channel has been successfully set up.

```
        <xs:restriction base="xs:string">
                <xs:pattern value="[1-9][0-9]*\.[0-9]+" />
        </xs:restriction>
</xs:simpleType>
<!-- RESPONSE CODE TYPE -->
<xs:simpleType name="responseCodeType">
        <xs:restriction base="xs:integer">
                <xs:pattern value="[1-9][0-9][0-9]" />
        </xs:restriction>
</xs:simpleType>
<!-- SUCCESS RESPONSE CODE TYPE -->
<xs:simpleType name="successResponseCodeType">
        <xs:restriction base="xs:integer">
                <xs:pattern value="2[0-9][0-9]" />
        </xs:restriction>
</xs:simpleType>
<!-- CLUE OPTIONS -->
<xs:complexType name="optionsMessageType">
        <xs:complexContent>
                <xs:extension base="clueMessageType">
                        <xs:sequence>
                                <xs:element name="mediaProvider"
                                        type="xs:boolean"/>
                                <xs:element name="mediaConsumer"
                                        type="xs:boolean"/>
                                <xs:element name="supportedVersions"
                                        type="versionsListType"
                                        minOccurs="0" />
                                <xs:element name="supportedExtensions"
                                        type="extensionsListType"
                                        minOccurs="0"/>
                                <xs:any namespace="##other"
                                        processContents="lax"
                                        minOccurs="0"/>
                        </xs:sequence>
                        <xs:anyAttribute namespace="##other"
                                processContents="lax"/>
                </xs:extension>
        </xs:complexContent>
</xs:complexType>
<!-- VERSIONS LIST TYPE -->
<xs:complexType name="versionsListType">
        <xs:sequence>
                <xs:element name="version" type="versionType" minOccurs="1"
                        maxOccurs="unbounded"/>
                <xs:any namespace="##other" processContents="lax"
```

FIGURE 16.25 Continued.

This ensures that only authenticated parties can exchange "options" and related "optionsResponse" messages and hence, drastically reduces the attack surface that is exposed to malicious parties. The CLUE framework clearly states the requirement to protect CLUE protocol messages against threats deriving from the presence of a malicious agent capable of gaining access to the CLUE data channel. Such a requirement is met by the CLUE data channel

```
                    minOccurs="0"/>
        </xs:sequence>
                <xs:anyAttribute namespace="##other" processContents="lax" />
</xs:complexType>
<!-- EXTENSIONS LIST TYPE -->
<xs:complexType name="extensionsListType">
        <xs:sequence>
                <xs:element name="extension" type="extensionType"
                        minOccurs="1"
                        maxOccurs="unbounded"/>
                <xs:any namespace="##other" processContents="lax"
                        minOccurs="0"/>
        </xs:sequence>
        <xs:anyAttribute namespace="##other" processContents="lax" />
</xs:complexType>
<!-- EXTENSION TYPE -->
<xs:complexType name="extensionType">
        <xs:sequence>
                <xs:element name="name" type="xs:string" />
                <xs:element name="schemaRef" type="xs:anyURI"
                        minOccurs="0" />
                <xs:element name="version" type="versionType"
                        minOccurs="0" />
                <xs:any namespace="##other" processContents="lax"
                        minOccurs="0"/>
        </xs:sequence>
        <xs:anyAttribute namespace="##other" processContents="lax"/>
</xs:complexType>
<!-- CLUE 'optionsResponse' -->
<xs:complexType name="optionsResponseMessageType">
        <xs:complexContent>
                <xs:extension base="clueResponseType">
                        <xs:sequence>
                                <xs:element name="mediaProvider"
                                        type="xs:boolean"
                                        minOccurs="0"/>
                                <xs:element name="mediaConsumer"
                                        type="xs:boolean"
                                        minOccurs="0"/>
                                        <xs:element name="version"
                                                type="versionType"
                                                minOccurs="0"/>
                                <xs:element name="commonExtensions"
                                        type="extensionsListType"
                                        minOccurs="0"/>
                               <xs:any namespace="##other
```

FIGURE 16.25 Continued.

solution described in Section 16.6, which ensures protection from both message recovery and message tampering. With respect to this last point, any implementation of the CLUE protocol compliant with the CLUE specification MUST rely on the exchange of messages that flow on top of a reliable and ordered SCTP over DTLS transport channel connecting two CLUE Participants.

```
                                        processContents="lax"
                                        minOccurs="0"/>
                        </xs:sequence>
                        <xs:anyAttribute namespace="##other"
                                processContents="lax"/>
                </xs:extension>
        </xs:complexContent>
</xs:complexType>
<!-- CLUE ADVERTISEMENT MESSAGE TYPE -->
<xs:complexType name="advertisementMessageType">
        <xs:complexContent>
                <xs:extension base="clueMessageType">
                        <xs:sequence>
                                <!-- mandatory -->
                                <xs:element name="mediaCaptures"
                                        type="dm:mediaCapturesType"/>
                                <xs:element name="encodingGroups"
                                        type="dm:encodingGroupsType"/>
                                <xs:element name="captureScenes"
                                        type="dm:captureScenesType"/>
                                <!-- optional -->
                                <xs:element name="simultaneousSets"
                                        type="dm:simultaneousSetsType"
                                        minOccurs="0"/>
                                <xs:element name="globalViews"
                                        type="dm:globalViewsType"
                                        minOccurs="0"/>
                                <xs:element name="people" type="dm:peopleType"
                                        minOccurs="0"/>
                                <xs:any namespace="##other"
                                        processContents="lax"
                                        minOccurs="0"/>
                        </xs:sequence>
                        <xs:anyAttribute namespace="##other"
                                processContents="lax"/>
                </xs:extension>
        </xs:complexContent>
</xs:complexType>
<!-- 'ack' MESSAGE TYPE -->
<xs:complexType name="advAcknowledgementMessageType">
        <xs:complexContent>
                <xs:extension base="clueResponseType">
                        <xs:sequence>
                                <xs:element name="advSequenceNr"
                                        type="xs:positiveInteger"/>
                                <xs:any namespace="##other"
```

FIGURE 16.25 Continued.

```
                                        processContents="lax"
                                        minOccurs="0"/>
                              </xs:sequence>
                              <xs:anyAttribute namespace="##other"
                                        processContents="lax"/>
                    </xs:extension>
          </xs:complexContent>
</xs:complexType>
<!-- CLUE 'configure' MESSAGE TYPE -->
<xs:complexType name="configureMessageType">
          <xs:complexContent>
                    <xs:extension base="clueMessageType">
                              <xs:sequence>
                                        <!-- mandatory fields -->
                                        <xs:element name="advSequenceNr"
                                                  type="xs:positiveInteger"/>
                                        <xs:element name="ack"
                                                  type="successResponseCodeType"
                                                  minOccurs="0"/>
                                        <xs:element name="captureEncodings"
                                                  type="dm:captureEncodingsType"
                                                            minOccurs="0"/>
                                        <xs:any namespace="##other"
                                                  processContents="lax"
                                                  minOccurs="0"/>
                              </xs:sequence>
                              <xs:anyAttribute namespace="##other"
                                        processContents="lax"/>
                    </xs:extension>
          </xs:complexContent>
</xs:complexType>
<!-- 'configureResponse' MESSAGE TYPE -->
<xs:complexType name="configureResponseMessageType">
          <xs:complexContent>
                    <xs:extension base="clueResponseType">
                    <xs:sequence>
                              <xs:element name="confSequenceNr"
                                        type="xs:positiveInteger"/>
                              <xs:any namespace="##other" processContents="lax"
                                        minOccurs="0"/>
                    </xs:sequence>
                    <xs:anyAttribute namespace="##other" processContents="lax"/>
                    </xs:extension>
          </xs:complexContent>
</xs:complexType>
</xs:schema>
```

FIGURE 16.25 Continued.

```xml
<?xml version="1.0" encoding="UTF-8" standalone="yes"?>
<ns2:advertisement xmlns="urn:ietf:params:xml:ns:clue-info"
        xmlns:ns2="urn:ietf:params:xml:ns:clue-protocol"
        xmlns:ns3="urn:ietf:params:xml:ns:vcard-4.0" protocol="CLUE" v="1.0">
        <ns2:clueId>Napoli CLUE Endpoint</ns2:clueId>
        <ns2:sequenceNr>34</ns2:sequenceNr>
        <mediaCaptures>
                <mediaCapture
                        xmlns:xsi="http://www.w3.org/2001/XMLSchema-instance"
                        xsi:type="audioCaptureType" captureID="AC0"
                                mediaType="audio">
                        <captureSceneIDREF>CS1</captureSceneIDREF>
                        <spatialInformation>
                                <captureOrigin>
                                        <capturePoint>
                                                <x>0.0</x>
                                                <y>0.0</y>
                                                <z>10.0</z>
                                        </capturePoint>
                                        <lineOfCapturePoint>
                                                <x>0.0</x>
                                                <y>1.0</y>
                                                <z>10.0</z>
                                        </lineOfCapturePoint>
                                </captureOrigin>
                        </spatialInformation>
                        <individual>true</individual>
                        <encGroupIDREF>EG1</encGroupIDREF>
                        <description lang="en">main audio from the room</description>
                        <priority>1</priority>
                        <lang>it</lang>
                        <mobility>static</mobility>
                        <view>room</view>
                        <capturedPeople>
                                <personIDREF>alice</personIDREF>
                                <personIDREF>bob</personIDREF>
                                <personIDREF>ciccio</personIDREF>
                        </capturedPeople>
                </mediaCapture>
                <mediaCapture
                        xmlns:xsi="http://www.w3.org/2001/XMLSchema-instance"
                        xsi:type="videoCaptureType" captureID="VC0"
                                mediaType="video">
                        <captureSceneIDREF>CS1</captureSceneIDREF>
                        <spatialInformation>
                                <captureOrigin>
                                        <capturePoint>
                                                <x>-2.0</x>
                                                <y>0.0</y>
                                                <z>10.0</z>
                                        </capturePoint>
                                </captureOrigin>
                                <captureArea>
                                        <bottomLeft>
                                                <x>-3.0</x>
                                                <y>20.0</y>
                                                <z>9.0</z>
                                        </bottomLeft>
                                        <bottomRight>
                                                <x>-1.0</x>
                                                <y>20.0</y>
                                                <z>9.0</z>
                                        </bottomRight>
                                        <topLeft>
                                                <x>-3.0</x>
                                                <y>20.0</y>
                                                <z>11.0</z>
                                        </topLeft>
                                        <topRight>
                                                <x>-1.0</x>
                                                <y>20.0</y>
                                                <z>11.0</z>
                                        </topRight>
                                </captureArea>
                        </spatialInformation>
                        <individual>true</individual>
                        <encGroupIDREF>EG0</encGroupIDREF>
                        <description lang="en">left camera video capture</description>
                        <priority>1</priority>
                        <lang>it</lang>
```


FIGURE 16.26 'advertisement' message example.

```
            <personIDREF>alice</personIDREF>
            <personIDREF>bob</personIDREF>
            <personIDREF>ciccio</personIDREF>
        </capturedPeople>
</mediaCapture>
<mediaCapture
        xmlns:xsi="http://www.w3.org/2001/XMLSchema-instance"
        xsi:type="videoCaptureType" captureID="VC0"
            mediaType="video">
        <captureSceneIDREF>CS1</captureSceneIDREF>
        <spatialInformation>
            <captureOrigin>
                <capturePoint>
                    <x>-2.0</x>
                    <y>0.0</y>
                    <z>10.0</z>
                </capturePoint>
            </captureOrigin>
            <captureArea>
                <bottomLeft>
                    <x>-3.0</x>
                    <y>20.0</y>
                    <z>9.0</z>
                </bottomLeft>
                <bottomRight>
                    <x>-1.0</x>
                    <y>20.0</y>
                    <z>9.0</z>
                </bottomRight>
                <topLeft>
                    <x>-3.0</x>
                    <y>20.0</y>
                    <z>11.0</z>
                </topLeft>
                <topRight>
                    <x>-1.0</x>
                    <y>20.0</y>
                    <z>11.0</z>
                </topRight>
            </captureArea>
        </spatialInformation>
        <individual>true</individual>
        <encGroupIDREF>EG0</encGroupIDREF>
        <description lang="en">left camera video capture</description>
        <priority>1</priority>
        <lang>it</lang>
```

FIGURE 16.26 Continued.

```
        <mobility>static</mobility>
        <view>individual</view>
        <capturedPeople>
            <personIDREF>ciccio</personIDREF>
        </capturedPeople>
</mediaCapture>
<mediaCapture
        xmlns:xsi="http://www.w3.org/2001/XMLSchema-instance"
        xsi:type="videoCaptureType" captureID="VC1"
            mediaType="video">
<captureSceneIDREF>CS1</captureSceneIDREF>
<spatialInformation>
        <captureOrigin>
        <capturePoint>
            <x>0.0</x>
            <y>0.0</y>
            <z>10.0</z>
        </capturePoint>
        </captureOrigin>
        <captureArea>
            <bottomLeft>
                <x>-1.0</x>
                <y>20.0</y>
                <z>9.0</z>
            </bottomLeft>
            <bottomRight>
                <x>1.0</x>
                <y>20.0</y>
                <z>9.0</z>
            </bottomRight>
            <topLeft>
                <x>-1.0</x>
                <y>20.0</y>
                <z>11.0</z>
            </topLeft>
            <topRight>
                <x>1.0</x>
                <y>20.0</y>
                <z>11.0</z>
            </topRight>
        </captureArea>
</spatialInformation>
<individual>true</individual>
<encGroupIDREF>EG0</encGroupIDREF>
<description lang="en">central camera video capture</description>
<priority>1</priority>
```


FIGURE 16.26 Continued.

```
<lang>it</lang>
<mobility>static</mobility>
<view>individual</view>
<capturedPeople>
        <personIDREF>alice</personIDREF>
</capturedPeople>
</mediaCapture>
<mediaCapture
        xmlns:xsi="http://www.w3.org/2001/XMLSchema-instance"
        xsi:type="videoCaptureType" captureID="VC2"
                mediaType="video">
        <captureSceneIDREF>CS1</captureSceneIDREF>
        <spatialInformation>
                <captureOrigin>
                        <capturePoint>
                                <x>2.0</x>
                                <y>0.0</y>
                                <z>10.0</z>
                        </capturePoint>
                </captureOrigin>
                <captureArea>
                        <bottomLeft>
                                <x>1.0</x>
                                <y>20.0</y>
                                <z>9.0</z>
                        </bottomLeft>
                        <bottomRight>
                                <x>3.0</x>
                                <y>20.0</y>
                                <z>9.0</z>
                        </bottomRight>
                        <topLeft>
                                <x>1.0</x>
                                <y>20.0</y>
                                <z>11.0</z>
                        </topLeft>
                        <topRight>
                                <x>3.0</x>
                                <y>20.0</y>
                                <z>11.0</z>
                        </topRight>
                </captureArea>
        </spatialInformation>
        <individual>true</individual>
        <encGroupIDREF>EG0</encGroupIDREF>
        <description lang="en">right camera video capture</description>
```

FIGURE 16.26 Continued.

```
        <priority>1</priority>
        <lang>it</lang>
        <mobility>static</mobility>
        <view>individual</view>
        <capturedPeople>
                <personIDREF>bob</personIDREF>
        </capturedPeople>
        </mediaCapture>
<mediaCapture
        xmlns:xsi="http://www.w3.org/2001/XMLSchema-instance"
        xsi:type="videoCaptureType" captureID="VC3"
                mediaType="video">
        <captureSceneIDREF>CS1</captureSceneIDREF>
        <spatialInformation>
                <captureArea>
                        <bottomLeft>
                                <x>-3.0</x>
                                <y>20.0</y>
                                <z>9.0</z>
                        </bottomLeft>
                        <bottomRight>
                                <x>3.0</x>
                                <y>20.0</y>
                                <z>9.0</z>
                        </bottomRight>
                        <topLeft>
                                <x>-3.0</x>
                                <y>20.0</y>
                                <z>11.0</z>
                        </topLeft>
                        <topRight>
                                <x>3.0</x>
                                <y>20.0</y>
                                <z>11.0</z>
                        </topRight>
                </captureArea>
        </spatialInformation>
        <content>
                <sceneViewIDREF>SE1</sceneViewIDREF>
        </content>
        <policy>SoundLevel:0</policy>
        <encGroupIDREF>EG0</encGroupIDREF>
        <description lang="en">loudest room segment</description>
        <priority>2</priority>
        <lang>it</lang>
        <mobility>static</mobility>
```


FIGURE 16.26 Continued.

```
        <view>individual</view>
</mediaCapture>
<mediaCapture
        xmlns:xsi="http://www.w3.org/2001/XMLSchema-instance"
        xsi:type="videoCaptureType" captureID="VC4"
                mediaType="video">
        <captureSceneIDREF>CS1</captureSceneIDREF>
        <spatialInformation>
                <captureOrigin>
                        <capturePoint>
                                <x>0.0</x>
                                <y>0.0</y>
                                <z>10.0</z>
                        </capturePoint>
                </captureOrigin>
                <captureArea>
                        <bottomLeft>
                                <x>-3.0</x>
                                <y>20.0</y>
                                <z>7.0</z>
                        </bottomLeft>
                        <bottomRight>
                                <x>3.0</x>
                                <y>20.0</y>
                                <z>7.0</z>
                        </bottomRight>
                        <topLeft>
                                <x>-3.0</x>
                                <y>20.0</y>
                                <z>13.0</z>
                        </topLeft>
                        <topRight>
                                <x>3.0</x>
                                <y>20.0</y>
                                <z>13.0</z>
                        </topRight>
                </captureArea>
        </spatialInformation>
        <individual>true</individual>
        <encGroupIDREF>EG0</encGroupIDREF>
        <description lang="en">zoomed out view of all people in the
                room</description>
        <priority>2</priority>
        <lang>it</lang>
        <mobility>static</mobility>
        <view>room</view>
```

FIGURE 16.26 Continued.

```
            <capturedPeople>
                <personIDREF>alice</personIDREF>
                <personIDREF>bob</personIDREF>
                <personIDREF>ciccio</personIDREF>
            </capturedPeople>
        </mediaCapture>
</mediaCaptures>
<encodingGroups>
        <encodingGroup encodingGroupID="EG0">
            <maxGroupBandwidth>600000</maxGroupBandwidth>
            <encodingIDList>
                <encodingID>ENC1</encodingID>
                <encodingID>ENC2</encodingID>
                <encodingID>ENC3</encodingID>
            </encodingIDList>
        </encodingGroup>
        <encodingGroup encodingGroupID="EG1">
            <maxGroupBandwidth>300000</maxGroupBandwidth>
            <encodingIDList>
                <encodingID>ENC4</encodingID>
                <encodingID>ENC5</encodingID>
            </encodingIDList>
        </encodingGroup>
</encodingGroups>
<captureScenes>
        <captureScene scale="unknown" sceneID="CS1">
            <sceneViews>
                <sceneView sceneViewID="SE1">
                    <mediaCaptureIDs>
                        <mediaCaptureIDREF>VC0</mediaCapture
                            IDREF>
                        <mediaCaptureIDREF>VC1</mediaCapture
                            IDREF>
                        <mediaCaptureIDREF>VC2</mediaCapture
                            IDREF>
                    </mediaCaptureIDs>
                </sceneView>
                <sceneView sceneViewID="SE2">
                    <mediaCaptureIDs>
                        <mediaCaptureIDREF>VC3</mediaCapture
                            IDREF>
                    </mediaCaptureIDs>
                </sceneView>
                <sceneView sceneViewID="SE3">
                    <mediaCaptureIDs>
                        <mediaCaptureIDREF>VC4</mediaCapture
```

FIGURE 16.26 Continued.

```
                                        IDREF>
                                </mediaCaptureIDs>
                        </sceneView>
                        <sceneView sceneViewID="SE4">
                                <mediaCaptureIDs>
                                        <mediaCaptureIDREF>AC0</mediaCapture
                                        IDREF>
                                </mediaCaptureIDs>
                        </sceneView>
                </sceneViews>
        </captureScene>
    </captureScenes>
    <simultaneousSets>
        <simultaneousSet setID="SS1">
            <mediaCaptureIDREF>VC3</mediaCaptureIDREF>
            <sceneViewIDREF>SE1</sceneViewIDREF>
        </simultaneousSet>
        <simultaneousSet setID="SS2">
            <mediaCaptureIDREF>VC0</mediaCaptureIDREF>
            <mediaCaptureIDREF>VC2</mediaCaptureIDREF>
            <mediaCaptureIDREF>VC4</mediaCaptureIDREF>
        </simultaneousSet>
    </simultaneousSets>
    <people>
        <person personID="bob">
            <personInfo>
                <ns2:fn>
                    <ns2:text>Bob</ns2:text>
                </ns2:fn>
            </personInfo>
            <personType>minute taker</personType>
        </person>
        <person personID="alice">
            <personInfo>
                <ns3:fn>
                    <ns3:text>Alice</ns3:text>
                </ns3:fn>
            </personInfo>
            <personType>presenter</personType>
        </person>
        <person personID="ciccio">
            <personInfo>
                <ns3:fn>
                    <ns3:text>Ciccio</ns3:text>
                </ns3:fn>
            </personInfo>

            <personType>chairman</personType>
            <personType>timekeeper</personType>
        </person>
    </people>
</ns2:advertisement>
```

FIGURE 16.26 Continued.

```
<?xml version="1.0" encoding="UTF-8" standalone="yes"?>
<ns2:advertisement xmlns="urn:ietf:params:xml:ns:clue-info"
        xmlns:ns2="urn:ietf:params:xml:ns:clue-protocol"
        xmlns:ns3="urn:ietf:params:xml:ns:vcard-4.0" protocol="CLUE" v="1.0">
        <ns2:clueId>Napoli CLUE Endpoint</ns2:clueId>
        <ns2:sequenceNr>34</ns2:sequenceNr>
        <mediaCaptures>
                <mediaCapture
                        xmlns:xsi="http://www.w3.org/2001/XMLSchema-instance"
                        xsi:type="audioCaptureType" captureID="AC0"
                        mediaType="audio">
                        <captureSceneIDREF>CS1</captureSceneIDREF>
                        <spatialInformation>
                                <captureOrigin>
                                        <capturePoint>
                                                <x>0.0</x>
                                                <y>0.0</y>
                                                <z>10.0</z>
                                        </capturePoint>
                                        <lineOfCapturePoint>
                                                <x>0.0</x>
                                                <y>1.0</y>
                                                <z>10.0</z>
                                        </lineOfCapturePoint>
                                </captureOrigin>
                        </spatialInformation>
                        <individual>true</individual>
                        <encGroupIDREF>EG1</encGroupIDREF>
                        <description lang="en">main audio from the room</description>
                        <priority>1</priority>
                        <lang>it</lang>
                        <mobility>static</mobility>
                        <view>room</view>
```

FIGURE 16.27 Example of "advertisement" Message with MCC.

16.5.12 IANA Considerations

This specification registers a new XML namespace, a new XML schema, and the MIME type for the schema. This specification also registers the "CLUE" Application Service tag and the "CLUE" Application Protocol tag and defines registries for the CLUE messages and response codes. Interested readers need to see Presta and Romano [15] or its latest version.

16.6 CLUE PROTOCOL DATA CHANNEL

This specification [16] defines how to use the WebRTC data channel mechanism in order to realize a data channel, referred to as a CLUE data channel, for transporting CLUE protocol messages between two CLUE entities. The specification defines how to describe the SCTPoDTLS association used to realize the CLUE data channel using the SDP and defines usage of an

```
        <capturedPeople>
            <personIDREF>alice</personIDREF>
            <personIDREF>bob</personIDREF>
            <personIDREF>ciccio</personIDREF>
        </capturedPeople>
    </mediaCapture>
    <mediaCapture
        xmlns:xsi="http://www.w3.org/2001/XMLSchema-instance"
        xsi:type="videoCaptureType" captureID="VC0"
            mediaType="video">
        <captureSceneIDREF>CS1</captureSceneIDREF>
        <spatialInformation>
            <captureOrigin>
                <capturePoint>
                    <x>0.5</x>
                    <y>1.0</y>
                    <z>0.5</z>
                </capturePoint>
                <lineOfCapturePoint>
                    <x>0.5</x>
                    <y>0.0</y>
                    <z>0.5</z>
                </lineOfCapturePoint>
            </captureOrigin>
        </spatialInformation>
        <individual>true</individual>
        <encGroupIDREF>EG0</encGroupIDREF>
        <description lang="en">left camera video capture
        </description>
        <priority>1</priority>
        <lang>it</lang>
        <mobility>static</mobility>
        <view>individual</view>
        <capturedPeople>
            <personIDREF>ciccio</personIDREF>
        </capturedPeople>
    </mediaCapture>
    <mediaCapture
        xmlns:xsi="http://www.w3.org/2001/XMLSchema-instance"
        xsi:type="videoCaptureType" captureID="VC1"
            mediaType="video">
        <captureSceneIDREF>CS1</captureSceneIDREF>
        <spatialInformation>
            <captureOrigin>
                <capturePoint>
                    <x>0.0</x>
```


FIGURE 16.27 Continued.

SDP-based "SCTP over DTLS" data channel negotiation mechanism for establishing a CLUE data channel. Details and procedures associated with the CLUE protocol, and the SDP offer/answer procedures for negotiating usage of a CLUE data channel, are outside the scope of this specification.

```
                    <y>0.0</y>
                    <z>10.0</z>
                </capturePoint>
            </captureOrigin>
            <captureArea>
                <bottomLeft>
                    <x>-1.0</x>
                    <y>20.0</y>
                    <z>9.0</z>
                </bottomLeft>
                <bottomRight>
                    <x>1.0</x>
                    <y>20.0</y>
                    <z>9.0</z>
                </bottomRight>
                <topLeft>
                    <x>-1.0</x>
                    <y>20.0</y>
                    <z>11.0</z>
                </topLeft>
                <topRight>
                    <x>1.0</x>
                    <y>20.0</y>
                    <z>11.0</z>
                </topRight>
            </captureArea>
        </spatialInformation>
        <individual>true</individual>
        <encGroupIDREF>EG0</encGroupIDREF>
        <description lang="en">central camera video
        capture</description>
        <priority>1</priority>
        <lang>it</lang>
        <mobility>static</mobility>
        <view>individual</view>
        <capturedPeople>
            <personIDREF>alice</personIDREF>
        </capturedPeople>
    </mediaCapture>
    <mediaCapture
        xmlns:xsi="http://www.w3.org/2001/XMLSchema-instance"
        xsi:type="videoCaptureType" captureID="VC2"
            mediaType="video">
        <captureSceneIDREF>CS1</captureSceneIDREF>
        <spatialInformation>
            <captureOrigin>
```

FIGURE 16.27 Continued.

```
                    <x>2.0</x>
                    <y>0.0</y>
                    <z>10.0</z>
                </capturePoint>
            </captureOrigin>
            <captureArea>
                <bottomLeft>
                    <x>1.0</x>
                    <y>20.0</y>
                    <z>9.0</z>
                </bottomLeft>
                <bottomRight>
                    <x>3.0</x>
                    <y>20.0</y>
                    <z>9.0</z>
                </bottomRight>
                <topLeft>
                    <x>1.0</x>
                    <y>20.0</y>
                    <z>11.0</z>
                </topLeft>
                <topRight>
                    <x>3.0</x>
                    <y>20.0</y>
                    <z>11.0</z>
                </topRight>
            </captureArea>
        </spatialInformation>
        <individual>true</individual>
        <encGroupIDREF>EG0</encGroupIDREF>
        <description lang="en">right camera video capture</description>
        <priority>1</priority>
        <lang>it</lang>
        <mobility>static</mobility>
        <view>individual</view>
        <capturedPeople>
            <personIDREF>bob</personIDREF>
        </capturedPeople>
    </mediaCapture>
    <mediaCapture
        xmlns:xsi="http://www.w3.org/2001/XMLSchema-instance"
        xsi:type="videoCaptureType" captureID="VC3"
            mediaType="video">
        <captureSceneIDREF>CS1</captureSceneIDREF>
        <spatialInformation>
            <captureArea>
```


FIGURE 16.27 Continued.

```
                        <bottomLeft>
                                <x>-3.0</x>
                                <y>20.0</y>
                                <z>9.0</z>
                        </bottomLeft>
                        <bottomRight>
                                <x>3.0</x>
                                <y>20.0</y>
                                <z>9.0</z>
                        </bottomRight>
                        <topLeft>
                                <x>-3.0</x>
                                <y>20.0</y>
                                <z>11.0</z>
                        </topLeft>
                        <topRight>
                                <x>3.0</x>
                                <y>20.0</y>
                                <z>11.0</z>
                        </topRight>
                </captureArea>
        </spatialInformation>
        <content>
                <sceneViewIDREF>SE1</sceneViewIDREF>
        </content>
        <policy>SoundLevel:0</policy>
        <encGroupIDREF>EG0</encGroupIDREF>
        <description lang="en">loudest room segment</description>
        <priority>2</priority>
        <lang>it</lang>
        <mobility>static</mobility>
        <view>individual</view>
</mediaCapture>
<mediaCapture
        xmlns:xsi="http://www.w3.org/2001/XMLSchema-instance"
        xsi:type="videoCaptureType" captureID="VC4"
                mediaType="video">
        <captureSceneIDREF>CS1</captureSceneIDREF>
        <spatialInformation>
                <captureOrigin>
                        <capturePoint>
                                <x>0.0</x>
                                <y>0.0</y>
                                <z>10.0</z>
                        </capturePoint>
                </captureOrigin>
```

FIGURE 16.27 Continued.

```
            <captureArea>
                <bottomLeft>
                    <x>-3.0</x>
                    <y>20.0</y>
                    <z>7.0</z>
                </bottomLeft>
                <bottomRight>
                    <x>3.0</x>
                    <y>20.0</y>
                    <z>7.0</z>
                </bottomRight>
                <topLeft>
                    <x>-3.0</x>
                    <y>20.0</y>
                    <z>13.0</z>
                </topLeft>
                <topRight>
                    <x>3.0</x>
                    <y>20.0</y>
                    <z>13.0</z>
                </topRight>
            </captureArea>
        </spatialInformation>
        <individual>true</individual>
        <encGroupIDREF>EG0</encGroupIDREF>
        <description lang="en">
            zoomed out view of all people in the room
        </description>
        <priority>2</priority>
        <lang>it</lang>
        <mobility>static</mobility>
        <view>room</view>
        <capturedPeople>
            <personIDREF>alice</personIDREF>
            <personIDREF>bob</personIDREF>
            <personIDREF>ciccio</personIDREF>
        </capturedPeople>
    </mediaCapture>
    <mediaCapture
        xmlns:xsi="http://www.w3.org/2001/XMLSchema-instance"
        xsi:type="videoCaptureType" captureID="VC5"
            mediaType="video">
        <captureSceneIDREF>CS1</captureSceneIDREF>
        <spatialInformation>
            <captureArea>
                <bottomLeft>
```

FIGURE 16.27 Continued.

```
                    <x>-3.0</x>
                    <y>20.0</y>
                    <z>9.0</z>
                </bottomLeft>
                <bottomRight>
                    <x>3.0</x>
                    <y>20.0</y>
                    <z>9.0</z>
                </bottomRight>
                <topLeft>
                    <x>-3.0</x>
                    <y>20.0</y>
                    <z>11.0</z>
                </topLeft>
                <topRight>
                    <x>3.0</x>
                    <y>20.0</y>
                    <z>11.0</z>
                </topRight>
            </captureArea>
        </spatialInformation>
        <content>
            <sceneViewIDREF>SE1</sceneViewIDREF>
        </content>
        <policy>SoundLevel:1</policy>
        <description lang="en">penultimate loudest room
        segment</description>
        <lang>it</lang>
        <mobility>static</mobility>
        <view>individual</view>
    </mediaCapture>
    <mediaCapture
        xmlns:xsi="http://www.w3.org/2001/XMLSchema-instance"
        xsi:type="videoCaptureType" captureID="VC6"
            mediaType="video">
        <captureSceneIDREF>CS1</captureSceneIDREF>
        <spatialInformation>
            <captureArea>
                <bottomLeft>
                    <x>-3.0</x>
                    <y>20.0</y>
                    <z>9.0</z>
                </bottomLeft>
                <bottomRight>
                    <x>3.0</x>
                    <y>20.0</y>
```


FIGURE 16.27 Continued.

```
                        <z>9.0</z>
                </bottomRight>
                <topLeft>
                        <x>-3.0</x>
                        <y>20.0</y>
                        <z>11.0</z>
                </topLeft>
                <topRight>
                        <x>3.0</x>
                        <y>20.0</y>
                        <z>11.0</z>
                </topRight>
            </captureArea>
        </spatialInformation>
        <content>
            <sceneViewIDREF>SE1</sceneViewIDREF>
        </content>
        <policy>SoundLevel:2</policy>
        <description lang="en">last but two loudest room
            segment</description>
        <lang>it</lang>
        <mobility>static</mobility>
        <view>individual</view>
</mediaCapture>
<mediaCapture
        xmlns:xsi="http://www.w3.org/2001/XMLSchema-instance"
        xsi:type="videoCaptureType" captureID="VC7"
            mediaType="video">
        <captureSceneIDREF>CS1</captureSceneIDREF>
        <spatialInformation>
            <captureArea>
                <bottomLeft>
                        <x>-3.0</x>
                        <y>20.0</y>
                        <z>9.0</z>
                </bottomLeft>
                <bottomRight>
                        <x>3.0</x>
                        <y>20.0</y>
                        <z>9.0</z>
                </bottomRight>
                <topLeft>
                        <x>-3.0</x>
                        <y>20.0</y>
                        <z>11.0</z>
                </topLeft>
```

FIGURE 16.27 Continued.

```
                    <topRight>
                        <x>3.0</x>
                        <y>20.0</y>
                        <z>11.0</z>
                    </topRight>
                </captureArea>
            </spatialInformation>
            <content>
                <mediaCaptureIDREF>VC3</mediaCaptureIDREF>
                <mediaCaptureIDREF>VC5</mediaCaptureIDREF>
                <mediaCaptureIDREF>VC6</mediaCaptureIDREF>
            </content>
            <maxCaptures exactNumber="true">3</maxCaptures>
            <encGroupIDREF>EG0</encGroupIDREF>
            <description lang="en">big picture of the current speaker +
                pips about previous speakers</description>
            <priority>3</priority>
            <lang>it</lang>
            <mobility>static</mobility>
            <view>individual</view>
        </mediaCapture>
    </mediaCaptures>
    <encodingGroups>
        <encodingGroup encodingGroupID="EG0">
            <maxGroupBandwidth>600000</maxGroupBandwidth>
            <encodingIDList>
                <encodingID>ENC1</encodingID>
                <encodingID>ENC2</encodingID>
                <encodingID>ENC3</encodingID>
            </encodingIDList>
        </encodingGroup>
        <encodingGroup encodingGroupID="EG1">
            <maxGroupBandwidth>300000</maxGroupBandwidth>
            <encodingIDList>
                <encodingID>ENC4</encodingID>
                <encodingID>ENC5</encodingID>
            </encodingIDList>
        </encodingGroup>
    </encodingGroups>
    <captureScenes>
        <captureScene scale="unknown" sceneID="CS1">
            <sceneViews>
                <sceneView sceneViewID="SE1">
                    <description lang="en">participants' individual
                        videos</description>
                    <mediaCaptureIDs>
```


FIGURE 16.27 Continued.

```
                        <mediaCaptureIDREF>VC0</mediaCapture
                        IDREF>
                        <mediaCaptureIDREF>VC1</mediaCapture
                        IDREF>
                        <mediaCaptureIDREF>VC2</mediaCapture
                        IDREF>
                    </mediaCaptureIDs>
                </sceneView>
                <sceneView sceneViewID="SE2">
                    <description lang="en">loudest segment of the
                        room</description>
                    <mediaCaptureIDs>
                        <mediaCaptureIDREF>VC3</mediaCapture
                        IDREF>
                    </mediaCaptureIDs>
                </sceneView>
                <sceneView sceneViewID="SE5">
                    <description lang="en">loudest segment of the
                        room + pips</description>
                    <mediaCaptureIDs>
                        <mediaCaptureIDREF>VC7</mediaCapture
                        IDREF>
                    </mediaCaptureIDs>
                </sceneView>
                <sceneView sceneViewID="SE4">
                    <description lang="en">room audio</description>
                    <mediaCaptureIDs>
                        <mediaCaptureIDREF>AC0</mediaCapture
                        IDREF>
                    </mediaCaptureIDs>
                </sceneView>
                <sceneView sceneViewID="SE3">
                    <description lang="en">room video</description>
                    <mediaCaptureIDs>
                        <mediaCaptureIDREF>VC4</mediaCapture
                        IDREF>
                    </mediaCaptureIDs>
                </sceneView>
            </sceneViews>
        </captureScene>
</captureScenes>
<simultaneousSets>
        <simultaneousSet setID="SS1">
            <mediaCaptureIDREF>VC3</mediaCaptureIDREF>
            <mediaCaptureIDREF>VC7</mediaCaptureIDREF>
            <sceneViewIDREF>SE1</sceneViewIDREF>
```


FIGURE 16.27 Continued.

```
        </simultaneousSet>
        <simultaneousSet setID="SS2">
            <mediaCaptureIDREF>VC0</mediaCaptureIDREF>
            <mediaCaptureIDREF>VC2</mediaCaptureIDREF>
            <mediaCaptureIDREF>VC4</mediaCaptureIDREF>
        </simultaneousSet>
    </simultaneousSets>
    <people>
        <person personID="bob">
            <personInfo>
                <ns3:fn>
                    <ns3:text>Bob</ns2:text>
                </ns3:fn>
            </personInfo>
            <personType>minute taker</personType>
        </person>
        <person personID="alice">
            <personInfo>
                <ns3:fn>
                    <ns3:text>Alice</ns2:text>
                </ns3:fn>
            </personInfo>
            <personType>presenter</personType>
        </person>
        <person personID="ciccio">
            <personInfo>
                <ns3:fn>
                    <ns3:text>Ciccio</ns2:text>
                </ns3:fn>
            </personInfo>
            <personType>chairman</personType>
            <personType>timekeeper</personType>
        </person>
    </people>
</ns2:advertisement>
```


FIGURE 16.27 Continued.

16.6.1 Introduction

This specification defines how to use the WebRTC data channel mechanism [6] in order to realize a data channel, referred to as a CLUE data channel, for transporting CLUE protocol (see Section 16.5) messages between two CLUE entities.

The specification defines how to describe the SCTPoDTLS association [12] used to realize the CLUE data channel using the SDP (RFC 4566) and defines usage of the SDP-based "SCTP over DTLS" data channel negotiation mechanism [7]. This includes SCTP considerations specific to a

CLUE data channel, the SDP Media Description (m-line) values, and usage of SDP attributes specific to a CLUE data channel. Details and procedures associated with the CLUE protocol, and the SDP offer/answer (RFC 3264) procedures for negotiating usage of a CLUE data channel, are outside the scope of this specification.

Note: The usage of the data channel Establishment Protocol (DCEP) [8] for establishing a CLUE data channel is outside the scope of this specification.

16.6.2 Conventions

The conventions are used per RFC 2119.

SCTPoDTLS association refers to an SCTP association carried over an DTLS connection [12]. WebRTC data channel refers to a pair of SCTP streams over an SCTPoDTLS association that is used to transport non-media data between two entities, as defined in Jesup et al. [6]. CLUE data channel refers to a WebRTC data channel [6] realization, with a specific set of SCTP characteristics, with the purpose of transporting CLUE protocol (see Section 16.6) messages between two CLUE entities.

CLUE entity refers to an SIP user agent (UA) (RFC 3261 – also see Section 1.2) that supports the CLUE data channel and the CLUE protocol. CLUE session refers to an SIP session (RFC 3261 – also see Section 1.2) between two SIP UAs, where a CLUE data channel, associated with the SIP session, has been established between the SIP UAs. SCTP stream is defined in RFC 4960 as a unidirectional logical channel established from one to another associated SCTP endpoint, within which all user messages are delivered in sequence except for those submitted to the unordered delivery service. SCTP identifier is defined in RFC 4960 as an unsigned integer, which identifies an SCTP stream.

16.6.3 CLUE Data Channel

16.6.3.1 General

This section describes the realization of a CLUE data channel using the WebRTC data channel mechanism. This includes a set of SCTP characteristics specific to a CLUE data channel, the values of the m-line describing the SCTPoDTLS association associated with the WebRTC data channel, and the usage of the SDP-based "SCTP over DTLS" data channel negotiation mechanism for creating the CLUE data channel. As described in Jesup et al. [6], the SCTP streams realizing a WebRTC data channel must be associated with the same SCTP association. In addition, both SCTP streams realizing the WebRTC data channel must use the same SCTP stream identifier value. These rules also apply to a CLUE data channel. Within a given CLUE session, a CLUE entity MUST use a single CLUE data channel for transport of all CLUE messages towards its peer.

16.6.3.2 SCTP Considerations

16.6.3.2.1 General

As described in rtcweb data channel [6], different SCTP options (e.g. regarding ordered delivery) can be used for a data channel. This section describes the SCTP options used for a CLUE data channel. Section 16.6.3.3 describes how SCTP options are signaled using SDP.

Note: While SCTP allows SCTP options to be applied per SCTP message, rtcweb data channel [6] mandates that for a given data channel, the same SCTP options are applied to each SCTP message associated with that data channel.

16.6.3.2.2 SCTP Payload Protocol Identifier (PPID)

A CLUE entity MUST use the PPID value 51 when sending a CLUE message on a CLUE data channel.

Note: As described in rtcweb data channel [6], the PPID value 51 indicates that the SCTP message contains data encoded in a UTF-8 format. The PPID value 51 does not indicate what application protocol the SCTP message is associated with, only the format in which the data is encoded.

16.6.3.2.3 Reliability

The usage of SCTP for the CLUE data channel ensures reliable transport of CLUE protocol (see Section 16.5) messages. A CLUE entity MUST NOT use the partial reliability or limited retransmission SCTP extensions, described in RFC 3758, for the CLUE data channel.

Note: RTCweb data channel [6] requires the support of the partial reliability extension defined in RFC 3758. This is not needed for a CLUE data channel, as messages are required to always be sent reliably. RTCweb data channel [6] also mandates support of the limited retransmission policy defined in RFC 7498.

16.6.3.2.4 Order

A CLUE entity MUST use the ordered delivery SCTP service, as described in RFC 4960, for the CLUE data channel.

16.6.3.2.5 Stream Reset

A CLUE entity MUST support the stream reset extension defined in RFC 6525. As defined in RTCweb data channel [6], the dynamic address reconfiguration extension ("Supported Extensions Parameter" parameter) defined in RFC 5061 must be used to signal the support of the stream reset extension defined in RFC 6525. Other features of RFC 5061 MUST NOT be used for CLUE data channels.

16.6.3.2.6 SCTP Multihoming

SCTP multihoming is not supported for SCTPoDTLS associations and can therefore not be used for a CLUE data channel.

16.6.3.2.7 Close CLUE Data Channel

As described in RTCweb data protocol [8], in order to close a data channel, an entity sends an SCTP reset message (RFC 6525) on its outgoing SCTP stream associated with the data channel. When the remote peer receives the reset message, it also sends (unless already sent) a reset message on its outgoing SCTP stream associated with the data channel. The SCTPoDTLS association and other data channels established on the same association are not affected by the SCTP reset messages.

16.6.3.3 SDP Considerations

16.6.3.3.1 General

This section defines how to construct the SDP Media Description (m-line) for describing the SCTPoDTLS association used to realize a CLUE data channel. The section also defines how to use the SDP-based "SCTP over DTLS" data channel negotiation mechanism [7] for establishing a CLUE data channel on the SCTPoDTLS association.

Note: Other protocols than SDP for negotiating usage of an SCTPoDTLS association for realizing a CLUE data channel are outside the scope of this specification. Section 16.8 describes the SDP offer/answer procedures for negotiating a CLUE session, including the CLUE-controlled media streams and the CLUE data channel.

16.6.3.3.1.1 SDP Media Description Fields As defined in Holmberg et al. [5], the field values of an m-line describing an SCTPoDTLS association are set as in Table 16.27).

CLUE entities SHOULD NOT transport the SCTPoDTLS association used to realize the CLUE data channel over TCP (using the "TCP/DTLS/SCTP" proto value) unless it is known that UDP/DTLS/SCTP will not work (for instance, when the ICE mechanism (RFC 5245) is used, and the ICE procedures determine that TCP transport is required). As defined in Holmberg et al. [5], when

TABLE 16.27
SDP "proto" Field Values

Media	Port	Proto	fmt
"application"	UDP port value	"UDP/DTLS/SCTP"	application usage
"application"	TCP port value	"TCP/DTLS/SCTP"	application usage

TABLE 16.28
SDP dcmap Attribute Values

Stream-id	subprotocol	label	ordered	Max-retr	Max-time
Value of the SCTP stream used to realize the CLUE data channel	"CLUE"	Application specific	"true"	N/A	N/A

the SCTPoDTLS association is used to realize a WebRTC data channel, the value of the application usage part is "webrtc-datachannel."

16.6.3.3.1.2 SDP sctp-port Attribute As defined in Holmberg et al. [5], the SDP sctp-port attribute value is set to the SCTP port of the SCTPoDTLS association. A CLUE entity can choose any valid SCTP port value.

16.6.3.3.2 SDP dcmap Attribute

The values of the SDP dcmap attribute [7], associated with the m-line describing the SCTPoDTLS association used to realize the WebRTC data channel, are set as in Table 16.28).

Note: As CLUE entities are required to use ordered SCTP message delivery, with full reliability, according to the procedures in [7], the max-retr and max-time attribute parameters are not used when negotiating CLUE data channels.

16.6.3.3.3 SDP dcsa Attribute

The SDP dcsa attribute [7] is not used when establishing a CLUE data channel.

16.6.3.3.4 Example

Figure 16.28 shows an example of an SDP media description for a CLUE channel:

16.6.4 Security Considerations

This specification relies on the security properties of the WebRTC data channel described in RTCweb data channel [6], including reliance on DTLS. Since CLUE sessions are established using SIP/SDP, protecting the data channel against message modification and recovery requires the use of SIP authentication and authorization mechanisms described in RFC 3261 (also see Section 1.2) for session establishment prior to establishing the data channel.

```
m=application 54111 UDP/DTLS/SCTP webrtc-datachannel
a=sctp-port: 5000
a=dcmap:2 subprotocol="CLUE";ordered=true
```

FIGURE 16.28 SDP Media Description for a CLUE Data Channel.

16.6.5 IANA Considerations

New WebRTC data channel Protocol Value is registered in IANA. Interested readers need to see Holmberg [16] or its latest version.

16.7 MAPPING RTP STREAMS TO CLUE MEDIA CAPTURES

This specification [17] describes how the RTP is used in the context of the CLUE protocol (ControLling mUltiple streams for tElepresence). It also describes the mechanisms and recommended practice for mapping RTP media streams defined in SDP to CLUE Media Captures and defines a new RTP header extension (CaptureId).

16.7.1 Introduction

Telepresence systems can send and receive multiple media streams. The CLUE framework (see Section 16.3) defines Media Captures (MC) as a source of Media from one or more Capture Devices. A Media Capture may also be constructed from other Media streams. A middle box can express conceptual Media Captures that it constructs from Media streams it receives. An MCC is a special Media Capture composed of multiple Media Captures. SIP offer/answer (RFC 3264) uses SDP (RFC 4566) to describe the RTP (RFC 3550) media streams. Each RTP stream has a unique synchronization source (SSRC) within its RTP session. The content of the RTP stream is created by an encoder in the endpoint. This may be an original content from a camera or a content created by an intermediary device like an MCU.

This specification makes recommendations for the CLUE architecture about how RTP and RTCP streams should be encoded and transmitted and how their relation to CLUE Media Captures should be communicated. The proposed solution supports multiple RTP topologies (RFC 7667). With regard to the media (audio, video, and timed text), systems that support CLUE use RTP for the media, SDP for codec and media transport negotiation (CLUE individual encodings). and the CLUE protocol for Media Capture description and selection. In order to associate the media in the different protocols there are three mappings that need to be specified:

1. CLUE individual encodings to SDP
2. RTP streams to SDP (this is not a CLUE-specific mapping)
3. RTP streams to MC to map the received RTP stream to the current MC in the MCC

16.7.2 Terminology

The definitions (also see Section 1.6, Table 1.2) from the CLUE framework specification specified in Section 16.3.3 are used by this specification as well.

16.7.3 RTP Topologies for CLUE

The typical RTP topologies used by CLUE Telepresence systems specify different behaviors for RTP and RTCP distribution. A number of RTP topologies are described in RFC 7667. For CLUE telepresence, the relevant topologies include Point-to-Point as well as Media-Mixing mixers, Media-Switching mixers, and Selective Forwarding Middleboxes. In the Point-to-Point topology, one peer communicates directly with a single peer over unicast. There can be one or more RTP sessions, each sent on a separate 5-tuple and having a separate SSRC space, with each RTP session carrying multiple RTP streams identified by their SSRC. All SSRCs are recognized by the peers based on the information in the RTCP SDES report, which includes the CNAME and SSRC of the sent RTP streams. There are different Point-to-Point use cases, as specified in the CLUE use case in RFC 7205 (see Section 16.2). In some cases, a CLUE session that at a high-level, is point-to-point may nonetheless have an RTP stream that is best described by one of the mixer topologies. For example, a CLUE endpoint can produce composite or switched captures for use by a receiving system with fewer displays than the sender has cameras. The Media Capture may be described using an MCC.

For the Media Mixer topology (RFC 7667), the peers communicate only with the mixer. The mixer provides mixed or composited media streams, using its own SSRC for the sent streams. If needed by the CLUE endpoint, the conference roster information, including conference participants, endpoints, media, and media-id (SSRC), can be determined using the conference event package (RFC 4575) element. Media-switching mixers and Selective Forwarding Middleboxes behave as described in RFC 7667.

16.7.4 Mapping CLUE Capture Encodings to RTP Streams

The different topologies described in Section 16.7.3 create different SSRC distribution models and RTP stream multiplexing points. Most video conferencing systems today can separate multiple RTP sources by placing them into RTP sessions using the SDP description; the video conferencing application can also have some knowledge about the purpose of each RTP session. For example, video conferencing applications that have a primary video source and a slides video source can send each media source in a separate RTP session with a content attribute (RFC 4796), enabling different application behavior for each received RTP media source. Demultiplexing is straightforward, because each media capture is sent as a single RTP stream, with each RTP stream being sent in a separate RTP session on a distinct UDP 5-tuple. This will also be true for mapping the RTP streams to Media Captures Encodings if each Media Capture Encoding uses a separate RTP session and the consumer can identify it based on the receiving RTP port. In this case, SDP only needs to label the RTP session with an identifier that can be used to identify the Media Capture in the CLUE description. The SDP label attribute serves as this identifier.

Each Capture Encoding MUST be sent as a separate RTP stream. CLUE endpoints MUST support sending each such RTP stream in a separate RTP session signaled by an SDP m=line. They MAY also support sending some or all of the RTP streams in a single RTP session, using the mechanism described in SDP bundle negotiation [9] to relate RTP streams to SDP m=lines. MCCs bring another mapping issue, in that an MCC represents multiple Media Captures that can be sent as part of this MCC if configured by the consumer. When receiving an RTP stream that is mapped to the MCC, the consumer needs to know which original MC it is in order to get the MC parameters from the advertisement. If a consumer requested an MCC, the original MC does not have a capture encoding, so it cannot be associated with an m-line using a label as described in CLUE signaling (see Section 16.8). This is important, for example, to get correct scaling information for the original MC, which may be different for the various MCs that are contributing to the MCC.

16.7.5 MCC Constituent CaptureID definition

For an MCC, which can represent multiple switched MCs, there is a need to know which MC is represented in the current RTP stream at any given time. This requires a mapping from the SSRC of the RTP stream conveying a particular MCC to the constituent MC. In order to address this mapping, this specification defines an RTP header extension and SDES item that includes the captureID of the original MC, allowing the consumer to use the original source MC's attributes, such as spatial information. This mapping temporarily associates the SSRC of the RTP stream conveying a particular MCC with the captureID of the single original MC that is currently switched into the MCC. This mapping cannot be used for the composed case where more than one original MC is composed into the MCC simultaneously.

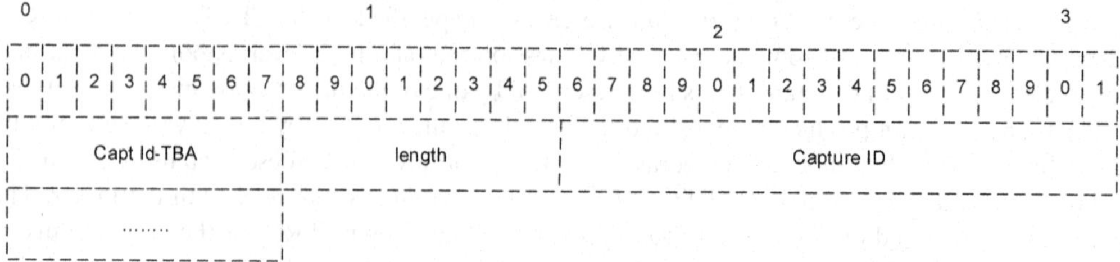

FIGURE 16.29 RTCP CaptureID SDES.

If there is only one MC in the MCC, then the MP MUST send the captureID of the current constituent MC in the RTP Header Extension and as an RTCP CaptureID SDES item. When the MP switches the MC it sends within an MCC, it MUST send the captureID value for the MC just switched into the MCC in an RTP Header Extension and as an RTCP CaptureID SDES item as specified in RFC 7941. If there is more than one MC composed into the MCC, then the MP MUST NOT send any of the MCs' captureIDs using this mechanism. However, if an MCC is sending contributing source (CSRC) information in the RTP header for a composed capture, it MAY send the captureID values in the RTCP SDES packets giving source information for the SSRC values sent as CSRCs.

If the MP sends the captureID of a single MC switched into an MCC and later, sends one composed stream of multiple MCs in the same MCC, it MUST send the special value "-," a single dash character, as the captureID RTP Header Extension and RTCP CaptureID SDES item. The single dash character indicates that there is no applicable value for the MCC constituent CaptureID. The media consumer interprets this as meaning that any previous CaptureID value associated with this SSRC no longer applies. As the CLUE data model schema (see Section 16.6) defines the captureID syntax as "xs:ID," the single dash character is not a legal captureID value, so there is no possibility of confusing it with an actual captureID.

16.7.5.1 RTCP CaptureID SDES Item

This specification specifies a new RTCP SDES item as depicted in Figure 16.29.

Note to the RFC Editor: Please replace TBA with the value assigned by IANA. This CaptureID is a variable-length UTF-8 string corresponding either to a CaptureID negotiated in the CLUE protocol or to the single character "-." This SDES item MUST be sent in an SDES packet within a compound RTCP packet unless support for reduced-size RTCP has been negotiated as specified in RFC 5506, in which case it can be sent as an SDES packet in a non-compound RTCP packet.

16.7.5.2 RTP Header Extension

The CaptureID is also carried in an RTP header extension (RFC 5285) using the mechanism defined in RFC 7941. Support is negotiated within SDP using the URN "urn:ietf:params:rtphdrext:sdes:CaptureID." The CaptureID is sent in an RTP Header Extension because for switched captures, receivers need to know which original MC corresponds to the media being sent for an MCC in order to correctly apply geometric adjustments to the received media. As discussed in RFC 7941, there is no need to send the CaptId Header Extension with all RTP packets. Senders MAY choose to send it only when a new MC is sent. If such a mode is being used, the header extension SHOULD be sent in the first few RTP packets to reduce the risk of losing it due to packet loss. See RFC 7941 for more discussion of this.

16.7.6 Examples

In this partial advertisement, the MP advertises a composed capture VC7 made of a big picture representing the current speaker (VC3) and two PiP boxes representing the previous speakers (the previous one – VC5 – and the oldest one – VC6).

```
<ns2:mediaCapture xmlns:xsi="http://www.w3.org/2001/XMLSchema-instance"
      xsi:type="ns2:videoCaptureType" captureID="VC7" mediaType="video">
   <ns2:captureSceneIDREF>CS1</ns2:captureSceneIDREF>
   <ns2:nonSpatiallyDefinable>true</ns2:nonSpatiallyDefinable>
   <ns2:content>
      <ns2:captureIDREF>VC3</ns2:captureIDREF>
      <ns2:captureIDREF>VC5</ns2:captureIDREF>
      <ns2:captureIDREF>VC6</ns2:captureIDREF>
   </ns2:content>
   <ns2:maxCaptures>3</ns2:maxCaptures>
   <ns2:allowSubsetChoice>false</ns2:allowSubsetChoice>
   <ns2:description lang="en">big picture of the current speaker
         pips about previous speakers</ns2:description>
   <ns2:priority>1</ns2:priority>
   <ns2:lang>it</ns2:lang>
   <ns2:mobility>static</ns2:mobility>
   <ns2:view>individual</ns2:view>
</ns2:mediaCapture>
```

In this case, the MP will send capture IDs VC3, VC5, or VC6 as an RTP header extension and RTCP SDES message for the RTP stream associated with the MC. Note that this is part of the full advertisement message example from the CLUE data model [I-D.ietf-clue-data-model-schema] example and is not a valid xml specification.

16.7.7 Communication Security

CLUE endpoints MUST support RTP/SAVPF profile and SRTP (RFC 3711). CLUE endpoints MUST support DTLS (RFC 6347) and DTLS-SRTP (RFC 5763 and RFC 5764) for SRTP keying. All media channels SHOULD be secure via SRTP and the RTP/SAVPF profile unless the RTP media and its associated RTCP are secure by other means (see RFC 7201 and RFC7202). All CLUE implementations MUST implement DTLS 1.0, with the ciphersuite TLS_ECDHE_ECDSA_WITH_AES_128_CBC_SHA with the P-256 curve [10]. The DTLS-SRTP protection profile SRTP_AES128_CM_HMAC_SHA1_80 MUST be supported for SRTP.Encrypted SRTP Header extensions (RFC 6904) MUST be supported.

Implementations SHOULD implement DTLS 1.2 with the TLS_ECDHE_ECDSA_WITH_AES_128_GCM_SHA256 ciphersuite. Implementations MUST favor ciphersuites that support PFS over non-PFS ciphersuites and SHOULD favor AEAD over non-AEAD ciphersuites. NULL Protection profiles MUST NOT be used for RTP or RTCP. CLUE endpoints MUST generate short-term persistent RTCP CNAMES, as specified in RFC 7022, and thus cannot be used for long-term tracking of the users.

16.8 SESSION SIGNALING FOR CONTROLLING MULTIPLE STREAMS FOR TELEPRESENCE (CLUE)

This specification [18] specifies how CLUE-specific signaling such as the CLUE protocol and the CLUE data channel are used in conjunction with each other and with existing signaling mechanisms such as SIP and SDP to produce a telepresence call.

16.8.1 Introduction

To enable devices to participate in a telepresence call, selecting the sources they wish to view, receiving those media sources, and displaying them in an optimal fashion, CLUE employs two principal and inter-related protocol negotiations. SDP (RFC 4566), conveyed via SIP (RFC 3261 – also see Section 1.2), is used to negotiate the specific media capabilities that can be delivered to specific addresses on a device. Meanwhile, CLUE protocol (see Section 16.5) messages, transported via a CLUE data channel (see Section 16.6), are used to negotiate the Capture Sources available, their attributes, and any constraints on their use. They also allow the far-end device to specify which Captures they wish to receive. Beyond negotiating the CLUE channel, SDP is also used to negotiate the details of supported media streams and the maximum capability of each of those streams. As the CLUE Framework (see Section 16.3) defines the manner in which the MP expresses their maximum encoding group capabilities, SDP is also used to express the encoding limits for each potential Encoding. Backwards-compatibility is an important consideration of the protocol: It is vital that a CLUE-capable device contacting a device that does not support CLUE is able to fall back to a fully functional non-CLUE call. The specification also defines how a non-CLUE call may be upgraded to CLUE in mid-call and similarly, how CLUE functionality can be removed mid-call to return to a standard non-CLUE call.

16.8.2 Terminology

This specification uses terminology defined in the CLUE Framework (see Section 16.3). A few additional terms (also see Section 1.6, Table 1.2) specific to this specification are defined as follows:

- **Non-CLUE device:** A device that supports standard SIP and SDP but either does not support CLUE or does, but does not currently wish to invoke CLUE capabilities.
- **CLUE-controlled media:** A media "m=" line that is under CLUE control; the Capture Source that provides the media on this "m=" line is negotiated in CLUE. See Section 16.8.4 for details of how this control is signaled in SDP. There is a corresponding "non-CLUE-controlled" media term.

16.8.3 Media Feature Tag Definition

The "sip.clue" media feature tag SIP (RFC 3840) indicates support for CLUE in SIP (3261 – also see Section 1.2) calls. A CLUE-capable device SHOULD include this media feature tag in its REGISTER requests and OPTION responses. It SHOULD also include the media feature tag in INVITE and UPDATE (RFC 3311) requests and responses. The presence of the media feature tag in the contact field of a request or response can be used to determine that the far end supports CLUE.

16.8.4 SDP Grouping Framework CLUE Extension Semantics

16.8.4.1 General

This section defines a new SDP Grouping Framework (RFC 5888) extension called "CLUE." The CLUE extension can be indicated using an SDP session-level "group" attribute. Each SDP media "m=" line that is included in this group, using SDP media-level mid attributes, is CLUE-controlled by a CLUE data channel also included in this CLUE group. Currently, only support for a single CLUE group is specified; support for multiple CLUE groups in a single session is outside the scope of this specification. A device MUST NOT include more than one CLUE group in its SDP message unless it is following a specification that defines how multiple CLUE channels are signaled and is either able to determine that the other side of the SDP exchange supports multiple CLUE channels or able to fail gracefully in the event that it does not.

16.8.4.2 The CLUE Data Channel and the CLUE Grouping Semantic

The CLUE data channel (see Section 16.6) is a bidirectional data channel [6] used for the transport of CLUE messages conveyed within an SCTP over DTLS connection. The CLUE signaling channel must be established before CLUE protocol messages can be exchanged and CLUE-controlled media can be sent. The data channel is negotiated over SDP as described in Drage et al. [7]. A CLUE-capable device wishing to negotiate CLUE MUST also include a CLUE group in their SDP offer or answer and include the "mid" of the "m=" line for the data channel in that group. The CLUE group MUST include the "mid" of the "m=" line for one (and only one) data channel. The presence of the data channel in the CLUE group in an SDP offer or answer also serves, along with the "sip.clue" media feature tag, as an indication that the device supports CLUE and wishes to upgrade the call to include CLUE-controlled media. A CLUE-capable device SHOULD include a data channel "m=" line in offers and when allowed by RFC 3264, answers.

16.8.4.3 CLUE-controlled Media and the CLUE Grouping Semantic

CLUE-controlled media lines in an SDP are "m=" lines in which the content of the media streams to be sent is negotiated via the CLUE protocol (see Section 16.5). For an "m=" line to be CLUE-controlled, its "mid" value MUST be included in the CLUE group. CLUE-controlled media is controlled by the CLUE protocol as negotiated on the CLUE data channel with a "mid" included in the CLUE group. "m=" lines not specified as under CLUE control follow normal rules for media streams negotiated in SDP as defined in specifications such as RFC 3264. The restrictions on CLUE-controlled media always apply to "m=" lines in an SDP offer or answer, even if negotiation of the data channel in SDP failed due to lack of CLUE support by the remote device or for any other reason, or in an offer if the recipient does not include the "mid" of the corresponding "m=" line in their CLUE group.

16.8.4.4 SDP Semantics for CLUE-controlled Media

16.8.4.4.1 Signaling CLUE Encodings

The CLUE Framework (see Section 16.3) defines the concept of "Encodings," which represent the sender's encode ability. Each Encoding the MP wishes to signal is signaled via an "m=" line of the appropriate media type, which MUST be marked as sendonly with the "a=sendonly" attribute or as inactive with the "a=inactive" attribute. The encoder limits of active (e.g., "a=sendonly") Encodings can then be expressed using existing SDP syntax. For instance, for H.264, see Table 6 in RFC 6184 for a list of valid parameters for representing encoder sender stream limits. These Encodings are CLUE-controlled and hence MUST include a "mid" in the CLUE group as defined previously.

As well as the normal restrictions defined in RFC 3264, the stream MUST be treated as if the "m=" line direction attribute had been set to "a=inactive" until the MP has received a valid CLUE CONFIGURE message specifying the Capture to be used for this stream. This means that RTP packets MUST NOT be sent until configuration is complete, while non-media packets such as STUN, RTCP, and DTLS MUST be sent as per their relevant specifications if negotiated. Every "m=" line representing a CLUE Encoding MUST contain a "label" attribute as defined in RFC 4574. This label is used to identify the Encoding by the sender in CLUE ADVERTISEMENT messages and by the receiver in CLUE CONFIGURE messages. Each label used for a CLUE-controlled "m=" line MUST be different from the label on all other "m=" lines in the CLUE group unless an "m=" line represents a dependent stream related to another "m=" line (such as an FEC stream), in which case it MUST have the same label value as the "m=" line on which it depends.

16.8.4.4.1.1 Referencing Encodings in the CLUE Protocol CLUE Encodings are defined in SDP but can be referenced from CLUE protocol messages – this is how the protocol defines which Encodings are part of an Encoding Group (in ADVERTISEMENT messages) and the Encoding with which to encode a specific Capture (in CONFIGURE messages). The labels on the CLUE-controlled "m=" lines are the references that are used in the CLUE protocol. Each <encID> (in encodingIDList) in a CLUE ADVERTISEMENT message SHOULD represent an Encoding defined in SDP; the specific Encoding referenced is a CLUE-controlled "m=" line in the most recent SDP sent by the sender of the ADVERTISEMENT message with a label value corresponding to the text content of the <encID>. Similarly, each <encodingID> (in captureEncodingType) in a CLUE CONFIGURE message SHOULD represent an Encoding defined in SDP; the specific Encoding referenced is a CLUE-controlled "m=" line in the most recent SDP received by the sender of the CONFIGURE message with a label value corresponding to the text content of the <encodingID>.

Note that the non-atomic nature of SDP/CLUE protocol interaction may mean that there are temporary periods where an <encID>/<encodingID> in a CLUE message does not reference an SDP "m=" line or where an Encoding represented in SDP is not referenced in a CLUE protocol message. See Section 16.8.5 for specifics.

16.8.4.4.2 Negotiating Receipt of CLUE Capture Encodings in SDP

A receiver who wishes to receive a CLUE stream via a specific Encoding requires an "a=recvonly" "m=" line that matches the "a=sendonly" Encoding. These "m=" lines are CLUE-controlled and hence MUST include their "mid" in the CLUE group. They MAY include a "label" attribute, but this is not required by CLUE, as only label values associated with "a=sendonly" Encodings are referenced by CLUE protocol messages.

16.8.4.5 SDP Offer/Answer Procedures

16.8.4.5.1 Generating the Initial Offer

A CLUE-capable device sending an initial SDP offer of a SIP session SHOULD include an "m=" line for the data channel to convey the CLUE protocol, along with a CLUE group containing the "mid" of the data channel "m=" line. For interoperability with non-CLUE devices, a CLUE-capable device sending an initial SDP offer SHOULD NOT include any "m=" line for CLUE-controlled media beyond the "m=" line for the CLUE data channel and SHOULD include at least one non-CLUE-controlled media "m=" line. If the device has evidence that the receiver is also CLUE-capable, for instance due to receiving an initial INVITE with no SDP but including a "sip.clue" media feature tag, the previous recommendation is waived, and the initial offer MAY contain "m=" lines for CLUE-controlled media. With the same interoperability recommendations as for Encodings, the sender of the initial SDP offer MAY also include "a=recvonly" media lines to pre-allocate "m=" lines to receive media. Alternatively, it MAY wait until CLUE protocol negotiation has completed before including these lines in a new offer/answer exchange – see Section 16.8.5 for recommendations.

16.8.4.5.2 Generating the Answer

16.8.4.5.2.1 Negotiating Use of CLUE and the CLUE Data Channel If the recipient of an initial offer is CLUE-capable, and the offer contains both an "m=" line for a data channel and a CLUE group containing the "mid" for that "m=" line, they SHOULD negotiate data channel support for an "m=" line and include the "mid" of that "m=" line in a corresponding CLUE group. A CLUE-capable recipient that receives an "m=" line for a data channel but no corresponding CLUE group containing the "mid" of that "m=" line MAY still include a corresponding data channel "m=" line if there are any other non-CLUE protocols it can convey over that channel but MUST NOT negotiate use of the CLUE protocol on this channel.

16.8.4.5.2.2 Negotiating CLUE-controlled Media If the initial offer contained "a=recvonly" CLUE-controlled media lines, the recipient SHOULD include corresponding "a=sendonly" CLUE-controlled media lines for accepted Encodings up to the maximum number of Encodings it wishes to advertise. As CLUE-controlled media, the "mid" of these "m=" lines must be included in the corresponding CLUE group. The recipient MUST set the direction of the corresponding "m=" lines of any remaining "a=recvonly" CLUE-controlled media lines received in the offer to "a=inactive."

If the initial offer contained "a=sendonly" CLUE-controlled media lines, the recipient MAY include corresponding "a=recvonly" CLUE-controlled media lines up to the maximum number of

Capture Encodings it wishes to receive. Alternatively, it MAY wait until CLUE protocol negotiation has completed before including these lines in a new offer/answer exchange – see Section 16.8.5 for recommendations. The recipient MUST set the direction of the corresponding "m=" lines of any remaining "a=recvonly" CLUE-controlled media lines received in the offer to "a=inactive."

16.8.4.5.2.3 Negotiating Non-CLUE-controlled Media A CLUE-controlled device implementation may prefer to render initial, single-stream audio and/or video for the user as rapidly as possible, transitioning to CLUE-controlled media once that has been negotiated. Alternatively, an implementation may wish to suppress initial media, only providing media once the final, CLUE-controlled streams have been negotiated. The receiver of the initial offer, if making the call CLUE-enabled with their SDP answer, can make their preference clear by their action in accepting or rejecting non-CLUE-controlled media lines. Rejecting these "m=" lines will ensure that no non-CLUE-controlled media flows before the CLUE-controlled media is negotiated. In contrast, accepting one or more non-CLUE-controlled "m=" lines in this initial answer will enable initial media to flow. If the answerer chooses to send initial non-CLUE-controlled media in a CLUE-enabled call, Section 16.8.4.5.4.1 addresses the need to disable it once CLUE-controlled media is fully negotiated.

16.8.4.5.3 Processing the Initial Offer/Answer Negotiation

In the event that both offer and answer include a data channel "m=" line with a mid value included in corresponding CLUE groups, CLUE has been successfully negotiated, and the call is now CLUE-enabled. If not, then the call is not CLUE-enabled.

16.8.4.5.3.1 Successful CLUE Negotiation In the event of successful CLUE-enablement of the call, devices MUST now begin negotiation of the CLUE channel; see Section 16.6 for negotiation details. If negotiation is successful, sending of CLUE protocol (see Section 16.5) messages can begin. A CLUE-capable device MAY choose not to send RTP on the non-CLUE-controlled channels during the period in which control of the CLUE-controlled media lines is being negotiated (though RTCP MUST still be sent and received as normal). However, a CLUE-capable device MUST still be prepared to receive media on non-CLUE-controlled media lines that have been successfully negotiated as defined in RFC 3264. If either side of the call wishes to add additional CLUE-controlled "m=" lines to send or receive CLUE-controlled media, they MAY now send a SIP request with a new SDP offer following the normal rules of SDP offer/answer and any negotiated extensions.

16.8.4.5.3.2 CLUE Negotiation Failure In the event that the negotiation of CLUE fails, and the call is not CLUE-enabled once the initial offer/answer negotiation completes, then CLUE is not in use in the call. The CLUE-capable devices MUST either revert to non-CLUE behavior or terminate the call.

16.8.4.5.4 Modifying the Session

16.8.4.5.4.1 Adding and Removing CLUE-controlled Media Subsequent offer/answer exchanges MAY add additional "m=" lines for CLUE-controlled media or activate or deactivate existing "m=" lines per the standard SDP mechanisms. In most cases, at least one additional exchange after the initial offer/answer exchange will be required before both sides have added all the Encodings and ability to receive Encodings that they desire. Devices MAY delay adding "a=recvonly" CLUE-controlled "m=" lines until after CLUE protocol negotiation completes – see Section 16.8.5 for recommendations. Once CLUE media has been successfully negotiated, devices SHOULD ensure that non-CLUE-controlled media is deactivated by setting their ports to 0 in cases where it corresponds to the media type of CLUE-controlled media that has been successfully negotiated. This deactivation may require an additional SDP exchange or may be incorporated into one that is part of the CLUE negotiation.

16.8.4.5.4.2 Enabling CLUE Mid-call A CLUE-capable device that receives an initial SDP offer from a non-CLUE device SHOULD include a new data channel "m=" line and corresponding CLUE group in any subsequent offers it sends to indicate that it is CLUE-capable. If in an ongoing non-CLUE call, an SDP offer/answer exchange completes with both sides having included a data channel "m=" line in their SDP and with the "mid" for that channel in a corresponding CLUE group, then the call is now CLUE-enabled; negotiation of the data channel and subsequently, the CLUE protocol begin.

16.8.4.5.4.3 Disabling CLUE Mid-call If during an ongoing CLUE-enabled call, a device wishes to disable CLUE, it can do so by following the procedures for closing a data channel defined in Drage et al. [7]: Sending a new SDP offer/answer exchange and subsequent Stream Control Transmission Protocol (SCTP) Stream Sequence Number (SSN) reset for the CLUE channel. It MUST also remove the CLUE group. Without the CLUE group, any "m=" lines that were previously CLUE-controlled no longer are; implementations MAY disable them by setting their ports to 0 or may continue to use them – in the latter case, how they are used is outside the scope of this specification.

If a device follows this procedure, or an SDP offer-answer negotiation completes in a fashion in which either the "m=" CLUE data channel line was not successfully negotiated and/or one side did not include the data channel in the CLUE group, then CLUE for this call is disabled. In the event that this occurs, CLUE is no longer enabled. Any active "m=" lines still included in the CLUE group are no longer CLUE-controlled, and the implementation MAY either disable them in a subsequent negotiation or continue to use them in some other fashion. If the data channel is still present but not included in the CLUE group, semantic CLUE protocol messages MUST no longer be sent.

16.8.4.5.4.4 CLUE Protocol Failure Mid-call In contrast to the specific disablement of the use of CLUE described previously, the CLUE channel may fail unexpectedly. Two circumstances where this can occur are:

- The CLUE data channel terminates, either gracefully or ungracefully, without any corresponding SDP renegotiation.
- The CLUE protocol enters an unrecoverable error state as defined in Section 16.6.6 of this book, either the "MP-TERMINATED" state for the MP or "MC-TERMINATED" for the Media Consumer. In this circumstance, implementations MUST continue to transmit and receive CLUE-controlled media on the basis of the last negotiated CLUE messages until the CLUE protocol is disabled mid-call by an SDP exchange as defined in Section 16.8.4.5.4.3. Implementations MAY choose to send such an SDP request to disable CLUE immediately or MAY continue in a call-preservation mode.

16.8.5 Interaction of CLUE Protocol and SDP Negotiations

Information about media streams in CLUE is split between two message types: SDP, which defines media addresses and limits, and the CLUE channel, which defines properties of Capture Devices available, scene information, and additional constraints. As a result, certain operations, such as advertising support for a new transmissible Capture with associated stream, cannot be performed atomically, as they require changes to both SDP and CLUE messaging. This section defines how the negotiation of the two protocols interact, provides some recommendations on dealing with intermediate stages in non-atomic operations, and mandates additional constraints on when CLUE-configured media can be sent.

16.8.5.1 Independence of SDP and CLUE Negotiation

To avoid the need to implement interlocking state machines with the potential to reach invalid states if messages were to be lost or be rewritten en-route by middle boxes, the state machines in SDP and CLUE operate independently. The state of the CLUE channel does not restrict when an implementation may send a new SDP offer or answer, and likewise, the implementation's ability to send a new CLUE ADVERTISEMENT or CONFIGURE message is not restricted by the results or the state of the most recent SDP negotiation (unless the SDP negotiation has removed the CLUE channel). The primary implication of this is that a device may receive an SDP with a CLUE Encoding for which it does not yet have Capture information or receive a CLUE CONFIGURE message specifying a Capture Encoding for which the far end has not negotiated a media stream in SDP.

CLUE messages contain an <encID> (in encodingIDList) or <encodingID> (in captureEncodingType), which is used to identify a specific encoding or captureEncoding in SDP; see Section 16.4 for specifics. The non-atomic nature of CLUE negotiation means that a sender may wish to send a new ADVERTISEMENT before the corresponding SDP message. As such, the sender of the CLUE message MAY include an <encID> that does not currently match a CLUE-controlled "m=" line label in SDP; A CLUE-capable implementation MUST NOT reject a CLUE protocol message solely because it contains <encID> elements that do not match a label in SDP.

The current state of the CP or Media Provider/Consumer state machines does not affect compliance with any of the normative language of RFC 3264. That is, they MUST NOT delay an ongoing SDP exchange as part of an SIP server or client transaction; an implementation MUST NOT delay an SDP exchange while waiting for CLUE negotiation to complete or for a CONFIGURE message to arrive. Similarly, a device in a CLUE-enabled call MUST NOT delay any mandatory state transitions in the CP or Media Provider/Consumer state machines due to the presence or absence of an ongoing SDP exchange.

A device with the CP state machine in the ACTIVE state MAY choose not to move from ESTABLISHED to ADV (MP state machine) or from ESTABLISHED to WAIT FOR CONF RESPONSE (Media Consumer state machine) based on the SDP state. See Section 16.5 for CLUE state machine specifics. Similarly, a device MAY choose to delay initiating a new SDP exchange based on the state of their CLUE state machines.

16.8.5.2 Constraints on Sending Media

While SDP and CLUE message states do not impose constraints on each other, both impose constraints on the sending of media – CLUE-controlled media MUST NOT be sent unless it has been negotiated in both CLUE and SDP: an implementation MUST NOT send a specific CLUE Capture Encoding unless its most recent SDP exchange contains an active media channel for that Encoding AND the far end has sent a CLUE CONFIGURE message specifying a valid Capture for that Encoding.

16.8.5.3 Recommendations for Operating with Non-atomic Operations

CLUE-capable devices MUST be able to handle states in which CLUE recently received SDP, irrespective of the order in which SDP and CLUE messages are received. While these mismatches will usually be transitory, a device's messages MUST make reference to EncodingIDs that do not match the most be able to cope with such mismatches remaining indefinitely. However, this specification makes some recommendations on message ordering for these non-atomic transitions. CLUE-capable devices SHOULD ensure that any inconsistencies between SDP and CLUE signaling are temporary by sending updated SDP or CLUE messages as soon as the relevant state machines and other constraints permit.

Generally, implementations that receive messages for which they have incomplete information SHOULD wait until they have the corresponding information they lack before sending messages to make changes related to that information. For example, an answerer that receives a new SDP offer with three new "a=sendonly" CLUE "m=" lines for which it has received no CLUE ADVERTISEMENT providing the corresponding capture information would typically include corresponding "a=inactive" lines in its answer and only make a new SDP offer with "a=recvonly" when and if a new ADVERTISEMENT arrives with Captures relevant to those Encodings.

Because of the constraints of SDP offer/answer, and because new SDP negotiations are generally more "costly" than sending a new CLUE message, implementations needing to make changes to both channels SHOULD prioritize sending the updated CLUE message over sending the new SDP message. The aim is for the recipient to receive the CLUE changes before the SDP changes, allowing the recipient to send their SDP answers without incomplete information, reducing the number of new SDP offers required.

16.8.6 Interaction of CLUE Protocol and RTP/RTCP CaptureID

The CLUE framework (see Section 16.3) allows for MCCs: Captures that contain multiple source Captures, whether composited into a single stream or switched based on some metric. The Captures that contribute to these MCCs may or may not be defined in the ADVERTISEMENT message. If they are defined, and the MCC is providing them in a switched format, the recipient may wish to determine which originating source Capture is currently being provided so that they can apply geometric corrections based on that Capture's geometry or take some other action based on the original Capture information. To do this, mapping RTP streams to CLUE media captures (see Section 16.7) allows the CaptureID of the originating Capture to be conveyed via RTP or RTCP. An MP sending switched media for an MCC with defined originating sources MUST send the CaptureID in both RTP and RTCP, as described in the mapping specification.

16.8.6.1 CaptureID Reception during MCC Redefinition

Because the RTP/RTCP CaptureID is delivered via a different channel than the ADVERTISEMENT in which the contents of the MCC are defined, there is an intrinsic race condition in cases in which the contents of an MCC are redefined. When an MP redefines an MCC that involves CaptureIDs, the reception of the relevant CaptureIDs by the recipient will either lead or lag the reception and processing of the new ADVERTISEMENT by the recipient. As such, a Media Consumer MUST NOT be disrupted by any of the following in any CLUE-controlled media stream it is receiving, whether that stream is for a static Capture or for an MCC (as any static Capture may be redefined to an MCC in a later ADVERTISEMENT):

- Receiving RTP or RTCP containing a CaptureID when the most recently processed ADVERTISEMENT means that none are expected
- Receiving RTP or RTCP without CaptureIDs when the most recently processed ADVERTISEMENT means that media CaptureIDs are expected
- Receiving a CaptureID in RTP or RTCP for a Capture defined in the most recently processed ADVERTISEMENT, but which the same ADVERTISEMENT does not include in the MCC
- Receiving a CaptureID in RTP or RTCP for a Capture not defined in the most recently processed ADVERTISEMENT

16.8.7 Multiplexing of CLUE-controlled Media using BUNDLE

16.8.7.1 Overview

A CLUE call may involve sending and/or receiving significant numbers of media streams. Conventionally, media streams are sent and received on unique ports. However, each separate port used for this purpose may impose costs that a device wishes to avoid, such as the need to open that port on firewalls and NATs, the need to collect ICE candidates, RFC 5245, etc. The BUNDLE extension for negotiations of media multiplexing using SDP [9] can be used to negotiate the multiplexing of multiple media lines onto a single 5-tuple for sending and receiving media, allowing devices in calls to another BUNDLE-supporting device to potentially avoid some of these costs. While CLUE-capable devices MAY support the BUNDLE extension for this purpose, supporting the extension is not mandatory for a device to be CLUE-compliant. A CLUE-capable device that supports BUNDLE SHOULD also support rtcpmux, specified in RFC 5761. However, a CLUE-capable device that supports rtcp-mux MAY or MAY NOT support BUNDLE.

16.8.7.2 Usage of BUNDLE with CLUE

This specification imposes no additional requirements or restrictions on the usage of BUNDLE when used with CLUE. There is no restriction on combining CLUE-controlled media lines and non-CLUE-controlled media lines in the same BUNDLE group or in multiple such groups. However, there are several steps an implementation may wish to take to ameliorate the cost and time requirements of extra SDP offer/answer exchanges between CLUE and BUNDLE.

16.8.7.2.1 Generating the Initial Offer

BUNDLE mandates that the initial SDP offer MUST use a unique address for each "m=" line with a non-zero port. Because CLUE implementations generally will not include CLUE-controlled media lines, with the exception of the data channel in the initial SDP offer, CLUE devices that support large numbers of streams can avoid ever having to open large numbers of ports if they successfully negotiate BUNDLE. An implementation that does include CLUE-controlled media lines in its initial SDP offer while also using BUNDLE must take care to avoid rendering its CLUE-controlled media lines unusable in the event that the far end does not negotiate BUNDLE. An implementation MUST NOT send any CLUE-controlled media lines in an initial offer with the "bundleonly" attribute unless it has established via some other channel that the recipient supports and is able to use BUNDLE.

16.8.7.2.2 Multiplexing of the Data Channel and RTP Media

BUNDLE-supporting CLUE-capable devices MAY include the data channel in the same BUNDLE group as RTP media. In this case, the device MUST be able to demultiplex the various transports – see media multiplexing using SDP in Holmberg et al. [9]. If the BUNDLE group includes other protocols than the data channel transported via DTLS, the device MUST also be able to differentiate the various protocols.

16.8.8 Example: A Call between Two CLUE-capable Endpoints

This example (Figure 16.30) illustrates a call between two CLUE-capable Endpoints. Alice, initiating the call, is a system with three cameras and three screens. Bob, receiving the call, is a system with two cameras and two screens. A call-flow diagram is presented, followed by a summary of each message. To manage the size of this section, the SDP snippets only illustrate video "m=" lines. SIP ACKs are not always discussed. Note that BUNDLE is not in use.

In SIP INVITE 1, Alice sends Bob a SIP INVITE including in the SDP body the basic audio and video capabilities and the data channel as per SCTP-based media transport in SDP [7]. Alice also includes the "sip.clue" media feature tag in the INVITE. A snippet of the SDP showing the grouping attribute and the video "m=" line is shown below. Alice has included a "CLUE" group and included the mid corresponding to a data channel in the group (3). Note that Alice has chosen not to include any CLUE-controlled media in the initial offer – the mid value of the video line is not included in the "CLUE" group.

```
...
a=group:CLUE 3
...
m=video 6002 RTP/AVP 96
a=rtpmap:96 H264/90000
a=fmtp:96 profile-level-id=42e016;max-mbps=108000;max-fs=3600
```

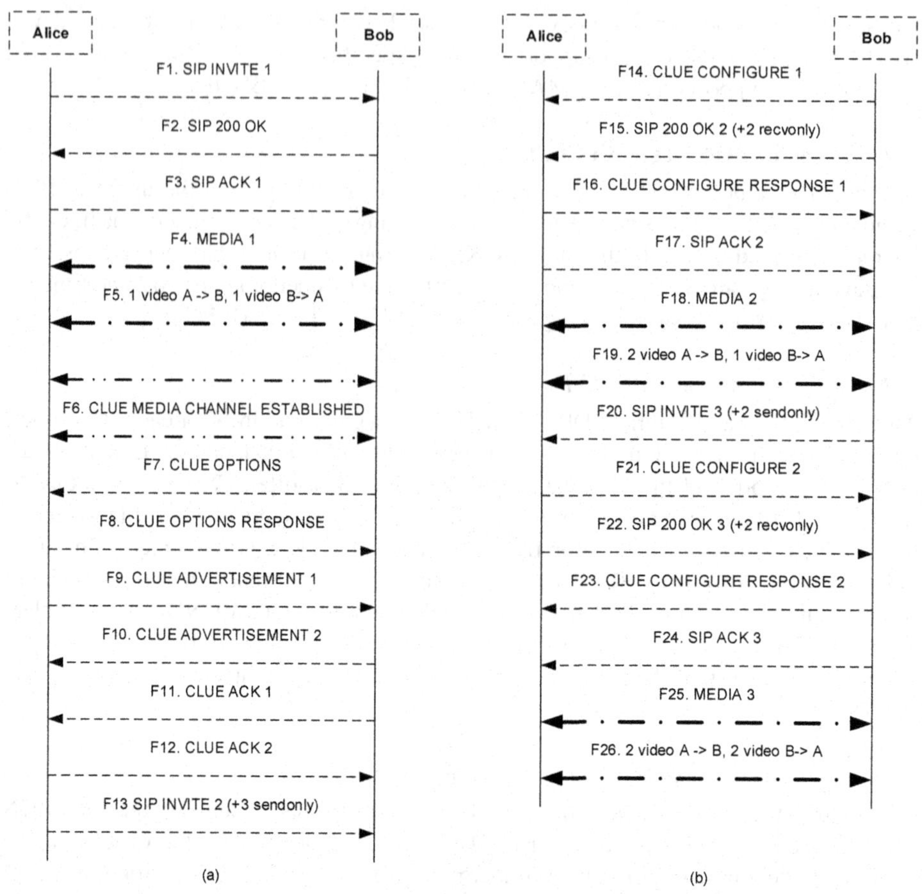

FIGURE 16.30 Call between Two CLUE-capable Endpoints. (Copyright: IETF.)

```
a=sendrecv
a=mid:2
...
m=application 6100 UDP/DTLS/SCTP webrtc-datachannel
a=setup:actpass
a=sctp-port: 5000
a=dcmap:2 subprotocol="CLUE";ordered=true
a=mid:3
```

Bob responds with a similar SDP in SIP 200 OK 1, which also has a "CLUE" group including the mid value of a data channel; due to their similarity, no SDP snippet is shown here. Bob wishes to receive initial media and so includes corresponding non-CLUE-controlled audio and video lines. Bob also includes the "sip.clue" media feature tag in the 200 OK. Alice and Bob are each now able to send a single audio and video stream. This is illustrated as MEDIA 1.

With the successful initial SDP offer/answer, Alice and Bob are also free to negotiate the CLUE data channel. This is illustrated as CLUE DATA CHANNEL ESTABLISHED. Once the data channel is established, CLUE protocol negotiation begins. In this case, Bob chose to be the DTLS client (sending a=active in his SDP answer) and hence is the CLUE CI and sends a CLUE OPTIONS message describing his version support. On receiving that message, Alice sends her corresponding CLUE OPTIONS RESPONSE.

With the OPTIONS phase complete, Alice now sends her CLUE ADVERTISEMENT (CLUE ADVERTISEMENT 1). She advertises three static Captures representing her three cameras. She also includes switched Captures suitable for two- and one-screen systems. All of these Captures are in a single Capture Scene, with suitable Capture Scene Views to tell Bob that he should subscribe to the three static Captures, the two switched Captures, or the one switched Capture. Alice has no simultaneity constraints, so she includes all six Captures in one simultaneous set. Finally, Alice includes an Encoding Group with three Encoding IDs: "encl," "enc2," and "enc3." These Encoding IDs are not currently valid but will match the next SDP offer she sends.

Bob received CLUE ADVERTISEMENT 1 but does not yet send a CONFIGURE message, because he has not yet received Alice's Encoding information, so as yet, he does not know whether she will have sufficient resources to send him the two streams he ideally wants at a quality he is happy with. Because Bob is not sending an immediate CONFIGURE with the "ack" element set, he must send an explicit ADVERTISEMENT ACKNOWLEDGEMENT message (CLUE ACK 1) to signal receipt of CLUE ADVERTISEMENT 1.

Bob also sends his CLUE ADVERTISEMENT (CLUE ADVERTISEMENT 2) – though the diagram shows that this occurs after Alice sends CLUE ADVERTISEMENT 1, Bob sends his ADVERTISEMENT independently and does not wait for CLUE ADVERTISEMENT 1 to arrive. He advertises two static Captures representing his cameras. He also includes a single composed Capture for single-screen systems, in which he will composite the two camera views into a single video stream. All three Captures are in a single Capture Scene, with suitable Capture Scene Views to tell Alice that she should subscribe to the two static Captures or the single composed Capture. Bob also has no simultaneity constraints, so he includes all three Captures in one simultaneous set. Bob also includes a single Encoding Group with two Encoding IDs: "foo" and "bar."

Similarly, Alice receives CLUE ADVERTISEMENT 2 but does not yet send a CONFIGURE message, because she has not yet received Bob's Encoding information, sending instead an ADVERTISEMENT ACKNOWLEDGEMENT (CLUE ACK 2). Both sides have now sent their CLUE ADVERTISEMENT messages, and an SDP exchange is required to negotiate Encodings. For simplicity, in this case, Alice is shown sending an INVITE with a new offer; in many implementations, both sides might send an INVITE, which would be resolved by use of the 491 Request Pending resolution mechanism from RFC 3261 (also see Section 1.2).

Alice now sends SIP INVITE 2. She maintains the sendrecv audio, video, and CLUE "m=" lines, and she adds three new sendonly "m=" lines to represent the three CLUE-controlled Encodings she can send. Each of these "m=" lines has a label corresponding to one of the Encoding IDs from CLUE ADVERTISEMENT 1. Each also has its mid added to the grouping attribute to show that it is controlled by the CLUE channel. A snippet of the SDP showing the grouping attribute, the data channel, and the video "m=" lines is shown in the following:

```
...
a=group:CLUE 3 4 5 6
...
m=video 6002 RTP/AVP 96
a=rtpmap:96 H264/90000
a=fmtp:96 profile-level-id=42e016;max-mbps=108000;max-fs=3600
a=sendrecv
a=mid:2
...
m=application 6100 UDP/DTLS/SCTP webrtc-datachannel
a=sctp-port: 5000
a=dcmap:2 subprotocol="CLUE";ordered=true
a=mid:3
...
m=video 6004 RTP/AVP 96
a=rtpmap:96 H264/90000
a=fmtp:96 profile-level-id=42e016
a=sendonly
```

```
a=mid:4
a=label:enc1
m=video 6006 RTP/AVP 96
a=rtpmap:96 H264/90000
a=fmtp:96 profile-level-id=42e016
a=sendonly
a=mid:5
a=label:enc2
m=video 6008 RTP/AVP 96
a=rtpmap:96 H264/90000
a=fmtp:96 profile-level-id=42e016
a=sendonly
a=mid:6
a=label:enc3
```

Bob now has all the information he needs to decide which streams to configure, allowing him to send both a CLUE CONFIGURE message and his SDP answer. As such, he now sends CLUE CONFIGURE 1. This requests the pair of switched Captures that represent Alice's scene, and he configures them with encoder ids "encl" and "enc2." Bob also sends his SDP answer as part of SIP 200 OK 2. Alongside his original audio, video, and CLUE "m=" lines, he includes three additional "m=" lines corresponding to the three added by Alice: two active recvonly "m=" lines and an inactive "m=" line for the third. He adds their mid values to the grouping attribute to show that they are controlled by the CLUE channel. A snippet of the SDP showing the grouping attribute and the video "m=" lines is shown in the following (mid 100 represents the CLUE channel, not shown):

```
...
a=group:CLUE 11 12 13 100
...
m=video 58722 RTP/AVP 96
a=rtpmap:96 H264/90000
a=fmtp:96 profile-level-id=42e016;max-mbps=108000;max-fs=3600
a=sendrecv
a=mid:10
...
m=video 58724 RTP/AVP 96
a=rtpmap:96 H264/90000
a=fmtp:96 profile-level-id=42e016;max-mbps=108000;max-fs=3600
a=recvonly
a=mid:11
m=video 58726 RTP/AVP 96
a=rtpmap:96 H264/90000
a=fmtp:96 profile-level-id=42e016;max-mbps=108000;max-fs=3600
a=recvonly
a=mid:12
m=video 58728 RTP/AVP 96
a=rtpmap:96 H264/90000
a=fmtp:96 profile-level-id=42e016;max-mbps=108000;max-fs=3600
a=inactive
a=mid:13
```

Alice receives Bob's message CLUE CONFIGURE 1 and sends CLUE CONFIGURE RESPONSE 1 to ack its reception. She does not yet send the Capture Encodings specified, because at this stage, she has not processed Bob's answer SDP and so has not negotiated the ability for Bob to receive these streams. On receiving SIP 200 OK 2 from Bob, Alice sends her SIP ACK (SIP ACK 2). She is now able to send the two streams of video Bob requested – this is illustrated as MEDIA 2.

The constraints of offer/answer meant that Bob could not include his encoding information as new "m=" lines in SIP 200 OK 2. As such, Bob now sends SIP INVITE 3 to generate a new offer. Along with all the streams from SIP 200 OK 2, Bob also includes two new sendonly streams. Each stream has a label corresponding to the Encoding IDs in his CLUE ADVERTISEMENT 2 message. He also adds their mid values to the grouping attribute to show that they are controlled by the CLUE channel. A snippet of the SDP showing the grouping attribute and the video "m=" lines is shown in the following (mid 100 represents the CLUE channel, not shown):

```
...
a=group:CLUE 11 12 14 15 100
...
m=video 58722 RTP/AVP 96
a=rtpmap:96 H264/90000
a=fmtp:96 profile-level-id=42e016;max-mbps=108000;max-fs=3600
a=sendrecv
a=mid:10
...
m=video 58724 RTP/AVP 96
a=rtpmap:96 H264/90000
a=fmtp:96 profile-level-id=42e016;max-mbps=108000;max-fs=3600
a=recvonly
a=mid:11
m=video 58726 RTP/AVP 96
a=rtpmap:96 H264/90000
a=fmtp:96 profile-level-id=42e016;max-mbps=108000;max-fs=3600
a=recvonly
a=mid:12
m=video 0 RTP/AVP 96
a=mid:13
m=video 58728 RTP/AVP 96
a=rtpmap:96 H264/90000
a=fmtp:96 profile-level-id=42e016
a=sendonly
a=label:foo
a=mid:14
m=video 58730 RTP/AVP 96
a=rtpmap:96 H264/90000
a=fmtp:96 profile-level-id=42e016
a=sendonly
a=label:bar
a=mid:15
```

Having received this, Alice now has all the information she needs to send her CLUE CONFIGURE message and her SDP answer. In CLUE CONFIGURE 2, she requests the two static Captures from Bob, to be sent on Encodings "foo" and "bar." Alice also sends SIP 200 OK 3, matching two recvonly "m=" lines to Bob's new sendonly lines. She includes their mid values in the grouping attribute to show that they are controlled by the CLUE channel. Alice also now deactivates the initial non-CLUE-controlled media, as bidirectional CLUE-controlled media is now available. A snippet of the SDP showing the grouping attribute and the video "m=" lines is shown in the following (mid 3 represents the data channel, not shown):

```
...
a=group:CLUE 3 4 5 7 8
...
m=video 0 RTP/AVP 96
```

```
a=mid:2
...
m=video 6004 RTP/AVP 96
a=rtpmap:96 H264/90000
a=fmtp:96 profile-level-id=42e016
a=sendonly
a=mid:4
a=label:enc1
m=video 6006 RTP/AVP 96
a=rtpmap:96 H264/90000
a=fmtp:96 profile-level-id=42e016
a=sendonly
a=mid:5
a=label:enc2
m=video 0 RTP/AVP 96
a=mid:6
m=video 6010 RTP/AVP 96
a=rtpmap:96 H264/90000
a=fmtp:96 profile-level-id=42e016;max-mbps=108000;max-fs=3600
a=recvonly
a=mid:7
m=video 6012 RTP/AVP 96
a=rtpmap:96 H264/90000
a=fmtp:96 profile-level-id=42e016;max-mbps=108000;max-fs=3600
a=recvonly
a=mid:8
```

Bob receives Alice's message CLUE CONFIGURE 2 and sends CLUE CONFIGURE RESPONSE 2 to ack its reception. Bob does not yet send the Capture Encodings specified, because he has not yet received and processed Alice's SDP answer and negotiated the ability to send these streams.

Finally, on receiving SIP 200 OK 3, Bob is now able to send the two streams of video Alice requested – this is illustrated as MEDIA 3. Both sides of the call are now sending multiple video streams with their sources defined via CLUE negotiation. As the call progresses, either side can send new ADVERTISEMENT or CONFIGURE message or new SDP offer/answers to add, remove, or change what they have available or want to receive.

16.8.9 Example: A Call between a CLUE-capable and non-CLUE Endpoint

In this brief example (Figure 16.31), Alice is a CLUE-capable Endpoint making a call to Bob, who is not CLUE-capable (i.e., is not able to use the CLUE protocol).

In SIP INVITE 1, Alice sends Bob an SIP INVITE, including in the SDP body the basic audio and video capabilities and the data channel as per SCTP-based media transport in SDP [7]. Alice also includes the "sip.clue" media feature tag in the INVITE. A snippet of the SDP showing the grouping attribute and the video "m=" line is shown in the following. Alice has included a "CLUE" group and included the mid corresponding to a data channel in the group (3). Note that Alice has chosen not to include any CLUE-controlled media in the initial offer – the mid value of the video line is not included in the "CLUE" group.

```
...
a=group:CLUE 3
...
m=video 6002 RTP/AVP 96
a=rtpmap:96 H264/90000
```

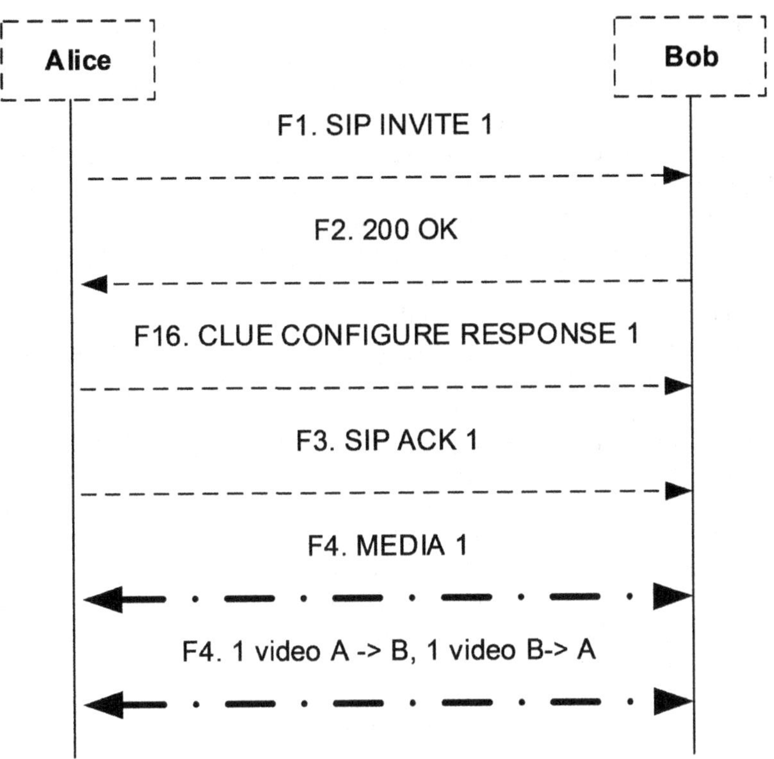

FIGURE 16.31 Call between a CLUE-capable and a non-CLUE Endpoint. (Copyright: IETF.)

```
a=fmtp:96 profile-level-id=42e016;max-mbps=108000;max-fs=3600
a=sendrecv
a=mid:2
...
m=application 6100 UDP/DTLS/SCTP webrtc-datachannel
a=sctp-port: 5000
a=dcmap:2 subprotocol="CLUE";ordered=true
a=mid:3
```

Bob is not CLUE-capable and hence, does not recognize the "CLUE" semantic for grouping attribute, nor does he support the data channel. IN SIP 200 OK 1, he responds with an answer with audio and video but with the data channel zeroed. From the lack of a CLUE group, Alice understands that Bob does not support CLUE or does not wish to use it. Both sides are now able to send a single audio and video stream to each other. Alice at this point begins to send her fallback video: in this case, probably a switched view from whichever camera shows the current loudest participant on her side.

16.9 SUMMARY

We have defined the telepresence as the multimedia conference systems that consist of high-definition, high-quality audio and video at a minimum video creating an environment among users or user groups who are not collocated having a feeling of collocated presence through multimedia communication, as if, an experience of "being there," creating a sense of co-location of all participants. In fact, we can describe telepresence as a special kind of high-quality immersive conference system, like that of the XCON conference system if the telepresence is made of centralized architecture. However, the formal definition is as follows: Telepresence is the real-time interactive audio, video, and data applications communications experience between all multipoint conference participants of being in the same space with a strong sense of realism and presence by optimizing a variety of attributes such as life-size images, high-fidelity and high-quality audio and video, spatial audio, preserving spatial relationships between streams, eye contact, gaze awareness, and body language.

In this section, we defined the requirements for telepresence multistreams (RFC 7262 – see Section 16.1). First, we addressed what problems we need to solve in telepresence that have not been offered in XCON conference systems. For example, in order to create a "being there" experience characteristic of telepresence, media inputs need to be transported, received, and coordinated between participating systems. However, there can be a great number of possible ways available for solving this telepresence problem, but none of them will be interoperable.

They use disparate techniques, and they describe, control, and negotiate media in dissimilar fashions. Such diversity creates an interoperability problem. It is worse when these telepresence problems are solved using extensions based on common standards such as SIP, as used in XCON conference systems with proprietary solutions. The interoperability through transcoding and translation that is needed for SIP-based XCON solutions makes the telepresence systems non-scalable and offers telepresence conferences with higher delays, degraded-performances, and/or degraded-qualities, defeating the very purpose of telepresence.

Second, we addressed the solution of telepresence multistream problems, from rendering of the audio and video captures, number, placement, capture/render angle, resolution of cameras and screens, and spatial location to audio mixing parameters of microphones seeking to describe the sender situation in a way that allows the receiver to render it realistically even though it may have a different rendering model than the sender. Finally, we defined the telepresence multistream requirements based on the approach described for developing common standards. The IETF CLUE working group has developed the telepresence requirements. The key has been that these requirements may lead to a solution that means the protocol specifications, including extensions to existing protocols as well as any new protocols, being developed to support the use cases (RFC 7205 – see Section 16.2). The solution might introduce additional functionality that is not mapped directly to these requirements.

With respect to security and for meeting some additional requirements of the telepresence systems, CLUE has adopted the following solutions: (a) In the case of conference control to use CCMP (RFC 6503 – see Section 5.10), (b) For security, to use the security model and mechanisms as defined in the XCON Framework (RFC 5239 – see Section 2.8), and (c) For other relevant features to use the reamaning CCMP specifications. Any additional signaling mechanism that is used to transport the information about media captures needs to be defined by which the information is secure.

We have described telepresence multistream use cases (RFC 7205 – see Section 16.2) for the following scenarios: a. Point-to-Point Meeting: Symmetric, b. Point-to-Point Meeting: Asymmetric, c. Multipoint Meeting, d. Presentation, e. Heterogeneous Systems, f. Multipoint Education Usage, g. Multipoint Multiview (Virtual Space), and h. Multiple Presentation Streams – Telemedicine. A key observation is that the fundamental parameters describing today's typical telepresence scenarios include: 1. Number of participating sites, 2. Number of visible seats at a site, 3. Number of cameras, 4. Number and type of microphones, 5. Number of audio channels, 6. Screen size, 7. Screen capabilities – such as resolution, frame rate, aspect ratio, 8. Arrangement of the screens in relation to each other, 9. Number of primary screens at each site, 10. Type and number of presentation screens, 11. Multipoint conference display strategies – for example, the camera-to-screen mappings may be static or dynamic, 12. Camera point of capture, and 13. Cameras fields of view and how they spatially relate to each other. In addition, the telepresence system does need to include the ability to present remote participants at their actual size for their apparent distance while maintaining correct eye contact and gesticular cues and simultaneously providing a spatial audio sound stage that is consistent with the displayed video.

In this chapter, we have described the CLUE framework for a protocol to enable devices in a telepresence conference to interoperate (Section 16.3) in multi-vendor environments. The protocol enables communication of information about multiple media streams so that a sending system and a receiving system can make reasonable decisions about transmitting, selecting, and rendering the media streams. Note that the protocol described in the framework is used in addition to SIP (RFC 3261– also see Sections 12 and 1.5) signaling and SDP negotiation for setting up a telepresence session. We have specified the Protocol Framework/Model Description, Spatial Relationships, Media Captures and Capture Scenes, Simultaneous Transmission Set Constraints, Encodings, Consumer's Choice of Streams to Receive from the Provider, and Extensibility in detail.

In addition, telepresence multistream examples are provided using the framework (informative) for the following scenarios: 1. Provider Behavior – 1.1 Three-screen Endpoint Provider, 1.2 Encoding Group Example, and 1.3 The MCU Case; 2. Media Consumer Behavior – 2.1 One-screen Media Consumer, 2.2 Two-screen Media Consumer configuring the example, and 2.3 Three-screen Media Consumer configuring the example; and 3. Multipoint Conference utilizing Multiple Content Captures – 3.1 Single Media Captures and MCC in the same Advertisement, 3.2 Several MCCs in the same Advertisement, 3.3 Heterogeneous conference with switching and composition, and 3.4 Heterogeneous conference with voice activated switching. The security for the CLUE framework is described, reusing the security schemes described for the XCON framework (RFC 5239 – see Section 2.8) and CCMP (RFC 6503 – see Section 5.10).

We have specified an XML schema (Section 16.4) file for the definition of CLUE data model types based on information contained in the CLUE framework (see Section 16.3). It encodes information and constraints defined in the CLUE framework in order to provide a formal representation of the concepts therein presented. The specification aims to define a coherent structure for information associated with the description of a telepresence scenario. This information is used within the CLUE protocol messages (see Section 16.5) enabling the dialog between an MP and a Media Consumer. CLUE protocol messages, indeed, are XML messages allowing (i) an MP to advertise its telepresence capabilities in terms of media captures, capture scenes, and other features envisioned in the CLUE framework, according to the format herein defined and (ii) a Media Consumer to request the desired telepresence options in the form of capture encodings, represented as described in this specification.

We have described the CLUE protocol (Section 16.5) for telepresence multistreams. It is an application protocol used by two CLUE Participants to enhance the experience of a multimedia telepresence session. It is designed in order to address SDP limitations in terms of the description of some information about the multimedia streams that are involved in a real-time multimedia conference. The CLUE protocol conveys information about the features of the flowing multimedia streams that are needed to enable a "being there" rendering experience that SDP cannot provide. The CLUE protocol is a stateful, client-server, XML-based application protocol. CLUE protocol messages flow on a reliable and ordered SCTP over DTLS transport channel connecting two CLUE Participants and Messages carrying information are taken from the XML-based CLUE data model (see Section 16.4). The protocol has three phases: 1. Establishment of the CLUE data channel, 2. Negotiation of the CLUE protocol version and extensions (initiation phase), and 3. CLUE telepresence capabilities description and negotiation.

The CLUE protocol has two main goals:

- Enabling an MP to properly announce its current telepresence capabilities to an MC in terms of available media captures, groups of encodings, simultaneity constraints, and other information defined in Section 16.3.
- Enabling an MC to request the desired multimedia streams from the offering MP. CLUE-capable endpoints are connected by means of the CLUE data channel, an SCTP over DTLS channel that is opened and established as described in Section 16.8 and Section 16.6. CLUE protocol messages flowing over such a channel are detailed in this specification, both syntactically and semantically.

The security of the CLUE framework and protocol has been conceived at the outset by embracing the security-by-design paradigm, identifying the mandatory implementations within the entire set of CLUE systems architecture as follows: (i) the use of cryptography and authentication; (ii) protection of all sensitive fields; (iii) mutual authentication between CLUE endpoints; (iv) the presence of authorization mechanisms; and (v) the presence of native defense mechanisms against malicious activities such as eavesdropping, selective modification, deletion, and replay (and related combinations thereof). Hence, the security of the single components of the CLUE solution must be evaluated with the integrated view of the final architecture.

We have described the CLUE protocol data channel, which is defined in the Section 16.6. The CLUE protocol data channel is used for transporting CLUE protocol messages between two CLUE entities. However, this data channel is based on the WebRTC data channel mechanisms in order to realize a data channel. As a result, this specification defines how to use the WebRTC data channel mechanism in order to realize a data channel, referred to as a CLUE data channel. More specifically, this specification defines how to describe the SCTP over DTLS association used to realize the CLUE data channel using the SDP and defines usage of the SDP-based "SCTP over DTLS" data channel negotiation mechanism for establishing a CLUE data channel. Note that details and procedures associated with the CLUE protocol and the SDP offer/answer procedures for negotiating usage of a CLUE data channel are not defined here. Interested readers need to see the usage of the WebRTC data channel establishment protocol (DCEP) [8] [for establishing a CLUE data channel.

We have specified the PPID, Reliability, Order, Stream Reset, Multihoming, and Close CLUE data channel of the STCP in using the WebRTC data channel mechanism in the context of CLUE protocol data channel. In addition, the general uses of the SDP are described, including SDP Media Description Fields, SDP sctp-port Attribute, SDP dcmap Attribute, and SDP dcsa Attribute. Finally, examples are provided of transferring CLUE protocol messages over the data channel.

We have described the mapping of RTP streams to CLUE media captures in this chapter (Section 16.7). The specification describes the mechanisms and recommended practice for mapping RTP media streams defined in SDP to CLUE media captures and defines a new RTP header extension (Captureld). We have described that a media capture can also be constructed from other media streams as described in the CLUE framework (see Section 16.3). A middle box can express conceptual media captures that it constructs from media streams it receives. An MCC is a special media capture composed of multiple media captures.

This specification recommends that the systems that support CLUE need to use the following: RTP for the media (audio, video, and timed text), SDP for codec and media transport negotiation (CLUE individual encodings), and the CLUE protocol for media capture description and selection. This specification defines three mappings in order to associate the media in the different protocols: 1. CLUE individual encodings to SDP, 2. RTP streams to SDP (this is not a CLUE-specific mapping), and 3. RTP streams to MC to map the received RTP stream to the current MC in the MCC. The communication security is specified for CLUE endpoints and media channels. In addition, the RTP security schemes for CLUE are described in great detail.

We have specified the CLUE session signaling protocol for controlling multiple streams for telepresence in this chapter (Section 16.8). This specification describes how CLUE-specific signaling such as the CLUE protocol and the CLUE data channel is used in conjunction with each other and with existing signaling mechanisms such as SIP and SDP to produce a telepresence call. The CLUE signaling protocol enables the CLUE endpoint devices to participate in a telepresence call through selecting the sources they wish to view, receiving those media sources, and displaying them in an optimal fashion. It is important to note that CLUE employs two principal and inter-related protocol negotiations to meet those objectives: SDP (RFC 4566), conveyed via SIP (RFC 3261 – also see Sections 1.2 and 1.5) is used to negotiate the specific media capabilities that can be delivered to specific addresses on a device.

Moreover, CLUE protocol (see Section 16.5) messages, which are transported via a CLUE data channel (see Section 16.6), are used to negotiate the Capture Sources' availability, their attributes, and any constraints on their use. They also allow the far-end device to specify which Captures it wishes to receive.

Beyond negotiating the CLUE channel, SDP is also used to negotiate the details of supported media streams and the maximum capability of each of those streams. As the CLUE Framework (see Section 16.3) defines the manner in which the MP expresses its maximum encoding group capabilities, SDP is also used to express the encoding limits for each potential Encoding. Provisions are also made for backwards compatibility between CLUE and non-CLUE devices.

More specifically, the CLUE signaling specification specifies Media Feature Tag Definition, SDP Grouping Framework CLUE Extension Semantics, Interaction of CLUE protocol and SDP negotiations, Interaction of CLUE protocol and RTP/RTCP CaptureID, and Multiplexing of CLUE-controlled media using BUNDLE. Two examples with detailed call flows are also described: Call between two CLUE-capable Endpoints and Call between a CLUE-capable and a non-CLUE Endpoint. The security mechanisms defined for the CLUE Framework (see Section 16.3) are good enough to meet the CLUE session signaling protocol.

16.10 PROBLEMS

1. What is telepresence? How does it differ from the usual XCON conference? Explain in detail using both quantitative and qualitative description for differentiation between the two.
2. What are the requirements of telepresence multistreams? How do the telepresence requirements differ from those of the XCON conference? Explain in detail.
3. How do the solutions of the telepresence multistreams will meet the requirements as defined as the objectives? What should be the common criteria for interoperability among different solutions, although each of those solutions for rendering may differ between the sender and the receiver? Explain in detail.
4. Describe each of these telepresence multistream use cases using both quantitative and qualitative description, including functional call flows: a. Point-to-Point Meeting: Symmetric, b. Point-to-Point Meeting: Asymmetric, c. Multipoint Meeting, d. Presentation, e. Heterogeneous Systems, f. Multipoint Education Usage, g. Multipoint Multiview (Virtual Space), and h. Multiple Presentation Streams – Telemedicine.
5. What are the fundamental parameters that need to be considered for all the telepresence use cases described in Problem 4? Explain in detail how each of these parameters influences the telepresence system realization for those use cases.
6. What are the technical criteria that telepresence needs to consider in presenting the remote participants? Explain how each of these technical criteria satisfies the need for creation of the "being there" experience characteristic of telepresence.
7. Why are the transcoding and translation of media streams not the solution for providing interoperability among disparate telepresence systems? Explain in detail the CLUE framework that enables devices in a telepresence conference to interoperate in multi-vendor environments.
8. Explain with functional call flows the following telepresence multistream scenarios using the CLUE framework: 1. Provider Behavior – 1.1 Three-screen Endpoint Provider, 1.2 Encoding Group Example, and 1.3 The MCU Case; 2. Media Consumer Behavior – 2.1 One-screen Media Consumer, 2.2 Two-screen Media Consumer configuring the example, and 2.3 Three-screen Media Consumer configuring the example; and 3. Multipoint Conference utilizing Multiple Content Captures – 3.1 Single Media Captures and MCC in the same Advertisement, 3.2 Several MCCs in the same Advertisement, 3.3 Heterogeneous conference with switching and composition, and 3.4 Heterogeneous conference with voice activated switching.
9. Describe in detail the security mechanisms for the CLUE framework.
10. Explain in detail the salient functional features of the XML schema for the CLUE data model. Describe the usages of XML schemas in the context of both CLUE framework and CLUE protocol.

11. Why is a separate CLUE protocol needed for telepresence in addition to SIP and SDP? How does the CLUE protocol enable a "being there" rendering experience for telepresence? What are the main goals of the CLUE protocol? How does the CLUE protocol carry messages taken from the XML-based CLUE data model on a reliable and ordered SCTP over DTLS transport channel connecting two CLUE participants? Explain in detail using functional call flows.
12. What is meant by the security-by-design paradigm for the CLUE protocol and framework? How does the integrated CLUE architecture that uses the CLUE framework, protocol, XML schemas of the CLUE data model, and CLUE data channel with SCTP over DTLS ensure security? Describe in detail.
13. Why does the CLUE data channel need to use the WebRTC data channel mechanisms in order to realize a data channel? Explain in detail using functional call flows how the CLUE protocol data channel is used for transporting between two CLUE entities.
14. Explain in detail, using functional call flows, how the CLUE data channel is established using SDP-based SCTP over DTLS data channel negotiations with the usage of the WebRTC DCEP for SDP offer/answer procedures.
15. Describe in detail, using functional call flows, the transferring of CLUE protocol messages over the data channel, using examples that specify the PPID, Reliability, Order, Stream Reset, Multihoming, and Close CLUE data channel of the STCP in using the WebRTC data channel mechanism.
16. Explain the following in detail: a. How is the RTP stream mapping to CLUE media captures performed? b. How can the media capture be constructed from other media streams, as described in the CLUE framework? c. How can the MCC be accomplished in CLUE?
17. Specify in detail the following mapping schemes in CLUE: a. CLUE individual encodings to SDP, b. RTP streams to SDP (this is not a CLUE-specific mapping), and c. RTP streams to MC to map the received RTP stream to the current MC in the MCC.
18. Describe in detail the communications and the RTP security for the CLUE data channel.
19. What does the CLUE session signaling protocol do? How does it produce a telepresence call using the CLUE protocol, CLUE data channel, SIP, and SDP? How does the telepresence call work in selecting different resources, including Capture Sources' availability, their attributes, and any constraints on their usages, negotiating the details of supported media streams and the maximum capability of each of those streams, and encoding limits, employing the CLUE signaling protocol?
20. Describe the following capabilities of the CLUE session signaling protocol: Media Feature Tag Definition, SDP Grouping Framework CLUE Extension Semantics, Interaction of CLUE protocol and SDP negotiations, Interaction of CLUE protocol and RTP/RTCP CaptureID, and Multiplexing of CLUE-controlled media using BUNDLE.
21. Explain the following telepresence scenarios with detailed call flows using the CLUE session signaling protocol and other associated CLUE capabilities: Call between two CLUE-capable Endpoints and Call between a CLUE-capable and a non-CLUE Endpoint.
22. Justify in detail why the security mechanisms described in the CLUE framework are sufficient in meeting the CLUE session signaling protocol.

REFERENCES

1. ITU-T. (2009, December) Packet-Based Multimedia Communications Systems. *ITU-T Recommendation H.323.*
2. ITU-T. (2005, September) Role Management and Additional Media Channels for H.300-Series Terminals. *ITU-T Recommendation H.239.*
3. ITU-T. (2013, April) Advanced Video Coding for Generic Audiovisual Services. *ITU-T Recommendation H.264.*

4. Allardyce, L. and Randall, L. (1983, April) Development of Teleconferencing Methodologies with Emphasis on Virtual Space Video and Interactive Graphics. <http://www.dtic.mil/docs/citations/ADA1 27738>.16.3.
5. Holmberg, C., Loreto, S., and Camarillo, G. (2017, April) Stream Control Transmission Protocol (SCTP)-Based Media Transport in the SDP. *draft-ietfmmusic- sctp-sdp-26.txt* (work in progress).
6. Jesup, R., Loreto, S., and Tuexen, M. (2015, January) WebRTC Data Channels. *draft-ietf-rtcweb-data-channel-13.txt* (work in progress).
7. Drage, K. et al. (2019, May) SDP-Based "SCTP Over DTLS" Data Channel Negotiation. *draft-ietf -mmusic-data-channel-sdpneg-28.txt* (work in progress).
8. Jesup, R., Loreto, S., and Tuexen, M. (2015, January) WebRTC Data Channel Establishment Protocol. *draft-ietf-rtcweb-data-protocol-13.txt* (work in progress).
9. Holmberg, C., Alvestrand, H., and Jennings, C. (2018, December) Negotiating Media Multiplexing Using the Session Description Protocol (SDP). *draft-ietf-mmusic-sdp-bundle-negotiation-54* (work in progress).
10. National Institute of Standards and Technology. (2013, July) Digital Signature Standard. *FIPS PUB 186-4*.
11. Lennox, J. et al. (2015, December) Sending Multiple Media Streams in a Single RTP Session. *draft-ietf-avtcore-rtp-multi-stream-11* (work in progress).
12. Stewart, R. and Loreto, R. S. (2015, January 24) DTLS Encapsulation of SCTP Packets. *draft-ietf-tsvwg-sctp-dtls-encaps-09.txt* (work in progress).
13. Duckworth, M. and Wenger, S. (2016, January 8) Framework for Telepresence Multi-Streams. *draft-ietf-clue-framework-25.txt* (work in progress).
14. Presta, R. and Romano, S. P. (2016, August 13) An XML Schema for the CLUE Data Model. *draft-ietf-clue-data-model-schema-17* (work in progress).
15. Presta, R. and Romano, S. P. (2017, October 6) CLUE Protocol. *draft-ietf-clue-protocol-14* (work in progress).
16. Holmberg, C. (2016, August 9) CLUE Protocol Data Channel. *draft-ietf-clue-datachannel-14* (work in progress).
17. Even, R. and Lennox, J. (2017, February 27) Mapping RTP Streams to CLUE Media Captures. *draft-ietf-clue-rtp-mapping-14.txt* (work in progress).
18. Hansen, R. et al. (2017, November 20) Session Signaling for Controlling Multiple Streams for Telepresence (CLUE). *draft-ietf-clue-signaling-13* (work in progress).

Appendix

RFC	Description
0793	Transmission Control Protocol. J. Postel. September 1981. (Obsoletes RFC 0761) (Updated by RFC 1122, RFC 3168, RFC 6093, RFC 6528) (Also STD 0007) (Status: Internet Standard)
1122	Requirements for Internet Hosts - Communication Layers. R. Braden. October 1989. (Updated by RFCs 1349, 4379, 5884, 6093, 6298, 6633, 6864, and 8029. Errata Exist) Status: Internet Standard)
2045	Multipurpose Internet Mail Extensions (MIME) Part One: Format of Internet Message Bodies. N. Freed, N. Borenstein. November 1996. (Obsoletes RFC 1521, RFC 1522, RFC 1590) (Updated by RFC 2184, RFC 2231, RFC 5335, RFC 6532) (Status: Draft Standard)
2046	Multipurpose Internet Mail Extensions (MIME) Part Two: Media Types. N. Freed, N. Borenstein. November 1996. (Obsoletes RFC 1521, RFC 1522, RFC 1590) (Updated by RFC 2646, RFC 3798, RFC 5147, RFC 6657) (Status: Proposed Standard)
2119	Key words for use in RFCs to Indicate Requirement Levels. S. Bradner. March 1997. (Updated by RFC 8174) (Also BCP0014) (Status: Best Current Practice)
2277	IETF Policy on Character Sets and Languages. H. Alvestrand. January 1998. (Status: Best Current Practice)
2326	Real Time Streaming Protocol (RTSP). H. Schulzrinne, A. Rao, R. Lanphier. April 1998. (Status: Proposed Standard)
2392	Content-ID and Message-ID Uniform Resource Locators. E. Levinson. August 1998. (Obsoletes RFC 2111) (Status: Proposed Standard)
2483	URI Resolution Services Necessary for URN Resolution M. Mealling, R. Daniel. January 1999. (Status: EXPERIMENTAL) (Stream: IETF, Area: app, WG: urn) (DOI: 10.17487/RFC2483)
2506	Media Feature Tag Registration Procedure K. Holtman, A. Mutz, T. Hardie. March 1999. (Also BCP0031) (Status: BEST CURRENT PRACTICE) (Stream: IETF, Area: app, WG: conneg) (DOI: 10.17487/RFC2506)
2616	Hypertext Transfer Protocol -- HTTP/1.1. R. Fielding, J. Gettys, J. Mogul, H. Frystyk, L. Masinter, P. Leach, T. Berners-Lee. June 1999. (Obsoletes RFC 2068) (Obsoleted by RFC 7230, RFC 7231, RFC 7232, RFC 7233, RFC 7234, RFC7235) (Updated by RFC 2817, RFC 5785, RFC 6266, RFC 0585) (Status: Draft Standard)
2617	HTTP Authentication: Basic and Digest Access Authentication. J. Franks, P. Hallam-Baker, J. Hostetler, S. Lawrence, P. Leach, A. Luotonen, L. Stewart. June 1999. (Format: TXT, HTML) (Obsoletes RFC2069) (Obsoleted by RFC7235, RFC7615, RFC7616, RFC7617) (Status: Draft Standard)
2804	IETF Policy on Wiretapping. IAB, IESG. May 2000. (Status: INFORMATIONAL)
2818	HTTP Over TLS. E. Rescorla. May 2000. (Format: TXT, HTML) (Updated by RFC5785, RFC7230) (Status: INFORMATIONAL)
2848	The PINT Service Protocol: Extensions to SIP and SDP for IP Access to Telephone Call Services. S. Petrack, L. Conroy. June 2000. (Status: Proposed Standard)
2898	KCS #5: Password-Based Cryptography Specification Version 2.0. B. Kaliski. September 2000. (Format: TXT, HTML) (Obsoleted by RFC8018) (Status: INFORMATIONAL)
2976	The SIP INFO Method. S. Donovan. October 2000. (Format: TXT, HTML) (Obsoleted by RFC6086) (Status: PROPOSED STANDARD)
3016	RTP Payload Format for MPEG-4 Audio/Visual Streams. Y. Kikuchi, T. Nomura, S. Fukunaga, Y. Matsui, H. Kimata. November 2000. Obsoleted by RFC 6416) (Status: Proposed Standard)
3023	XML Media Types. M. Murata, S. St. Laurent, D. Kohn. January 2001. (Obsoletes RFC2376) (Obsoleted by RFC7303) (Updates RFC2048) (Updated by RFC6839) (Status: PROPOSED STANDARD)
3261	SIP: Session Initiation Protocol. J. Rosenberg, H. Schulzrinne, G. Camarillo, A. Johnston, J. Peterson, R. Sparks, M. Handley, E. Schooler. June 2002. (Obsoletes RFC 2543) (Updated by RFC 3265, RFC 3853, RFC 4320, RFC 4916, RFC 5393, RFC 5621, RFC 5626, RFC 5630, RFC 5922, RFC 5954, RFC 6026, RFC 6141) (Status: Proposed Standard)

RFC	Description
3262	Reliability of Provisional Responses in Session Initiation Protocol (SIP). J. Rosenberg, H. Schulzrinne. June 2002. (Obsoletes RFC 2543) (Status: Proposed Standard)
3264	An Offer/Answer Model with Session Description Protocol (SDP). J. Rosenberg, H. Schulzrinne. June 2002. (Obsoletes RFC 2543) (Updated by RFC 6157) (Status: Proposed Standard)
3265	Session Initiation Protocol (SIP)-Specific Event Notification. A. B. Roach. June 2002. (Obsoletes RFC2543) (Obsoleted by RFC6665) (Updates RFC3261) (Updated by RFC5367, RFC5727, RFC6446 (Status: PROPOSED STANDARD)
3268	Advanced Encryption Standard (AES) Ciphersuites for Transport Layer Security (TLS). P. Chown. June 2002. (Obsoleted by RFC 5246) (Status: PROPOSED STANDARD)
3280	Internet X.509 Public Key Infrastructure Certificate and Certificate Revocation List (CRL) Profile. R. Housley, W. Polk, W. Ford, D. Solo. April 2002. (Obsoletes RFC 2459) (Obsoleted by RFC 5280) (Updated by RFC 4325, RFC 4630) (Status: Proposed Standard)
3303	Middlebox Communication Architecture and Framework. P. Srisuresh, J. Kuthan, J. Rosenberg, A. Molitor, A. Rayhan. August 2002. (Status: Informational)
3311	The Session Initiation Protocol (SIP) UPDATE Method. J. Rosenberg. October 2002.(Status: Proposed Standard)
3312	Integration of Resource Management and Session Initiation Protocol (SIP). G. Camarillo, Ed., W. Marshall, Ed., J. Rosenberg. October 2002. (Updated by RFC 4032, RFC 5027) (Status: Proposed Standard)
3323	A Privacy Mechanism for the Session Initiation Protocol (SIP). J. Peterson. November 2002. (Status: Proposed Standard)
3325	Private Extensions to the Session Initiation Protocol (SIP) for Asserted Identity within Trusted Networks. C. Jennings, J. Peterson, M. Watson. November 2002. (Updated by RFC5876, RFC8217) (Status: INFORMATIONAL)
3326	The Reason Header Field for the Session Initiation Protocol (SIP). H. Schulzrinne, D. Oran, G. Camarillo. December 2002. (Format: TXT, HTML) (Updated by RFC8606) (Status: PROPOSED STANDARD)
3339	Date and Time on the Internet: Timestamps. G. Klyne, C. Newman. July 2002. (Format: TXT, HTML) (Status: PROPOSED STANDARD)
3384	Lightweight Directory Access Protocol (version 3) Replication Requirements. E. Stokes, R. Weiser, R. Moats, R. Huber. October 2002. (Format: TXT, HTML) (Status: INFORMATIONAL)
3515	The Session Initiation Protocol (SIP) Refer Method. R. Sparks. April 2003. (Updated by RFC7647, RFC8217) (Status: PROPOSED STANDARD)
3550	RTP: A Transport Protocol for Real-Time Applications. H. Schulzrinne, S. Casner, R. Frederick, V. Jacobson. July 2003. (Obsoletes RFC 1889) (Updated by RFC 5506, RFC 5761, RFC 6051, RFC 6222, RFC 7022, RFC 7160, RFC 7164) (Also STD 0064) (Status: Internet Standard)
3629	UTF-8, a transformation format of ISO 10646. F. Yergeau. November 2003 (Obsoletes RFC 2279) (Also STD0063) (Status: Internet Standard)
3711	The Secure Real-time Transport Protocol (SRTP). M. Baugher, D. McGrew, M. Naslund, E. Carrara, K. Norrman. March 2004. (Updated by RFC 5506, RFC 6904) (Status: Proposed Standard)
3725	Best Current Practices for Third Party Call Control (3pcc) in the Session Initiation Protocol (SIP). J. Rosenberg, J. Peterson, H. Schulzrinne, G. Camarillo. April 2004. (Status: Best Current Practice)
3758	Stream Control Transmission Protocol (SCTP) Partial Reliability Extension. R. Stewart, M. Ramalho, Q. Xie, M. Tuexen, P. Conrad. May 2004 (Status: PROPOSED STANDARD)
3839	MIME Type Registrations for 3rd Generation Partnership Projec (3GPP) Multimedia files. R. Castagno, D. Singer. July 2004. (Format: TXT, HTML) (Updated by RFC6381) (Status: PROPOSED STANDARD)
3840	Indicating User Agent Capabilities in the Session Initiation Protocol (SIP). J. Rosenberg, H. Schulzrinne, P. Kyzivat. August 2004. (Status: PROPOSED STANDARD)
3841	3841 Caller Preferences for the Session Initiation Protocol (SIP). J. Rosenberg, H. Schulzrinne, P. Kyzivat. August 2004. (Status: Proposed Standard)
3891	The Session Initiation Protocol (SIP) "Replaces" Header. R. Mahy, B. Biggs, R. Dean. September 2004. (Status: PROPOSED STANDARD)
3960	Early Media and Ringing Tone Generation in the Session Initiation Protocol (SIP). G. Camarillo, H. Schulzrinne. December 2004 (Status: INFORMATIONAL)

RFC	Description
3966	The tel URI for Telephone Numbers. H. Schulzrinne. December 2004. (Status: Proposed Standard) (Obsoletes RFC 2806) (Updated by RFC 5341) (Status: Proposed Standard)
3986	Uniform Resource Identifier (URI): Generic Syntax. T. Berners-Lee, R. Fielding, L. Masinter.January 2005. (Obsoletes RFC 2732, RFC 2396, RFC 1808) (Updates RFC 1738) (Updated by RFC 6874, RFC 7320) (Also STD 0066) (Status: Internet Standard)
4086	RTP Payload for Text Conversation. G. Hellstrom, P. Jones. June 2005. (Obsoletes RFC2793) (Status: PROPOSED STANDARD)
4091	The Alternative Network Address Types (ANAT) Semantics for the Session Description Protocol (SDP) Grouping Framework. G. Camarillo, J. Rosenberg. June 2005. (Obsoleted by RFC 5245) (Status: Proposed Standard)
4092	Usage of the Session Description Protocol (SDP) Alternative Network Address Types (ANAT) Semantics in the Session Initiation Protocol (SIP). G. Camarillo, J. Rosenberg. June 2005. (Obsoleted by RFC 5245) (Status: Proposed Standard)
4103	RTP Payload for Text Conversation. G. Hellstrom, P. Jones. June 2005. (Obsoletes RFC2793) (Status: PROPOSED STANDARD)
4122	A Universally Unique IDentifier (UUID) URN Namespace. P. Leach, M. Mealling, R. Salz. July 2005. (Status: PROPOSED STANDARD)
4145	TCP-Based Media Transport in the Session Description Protocol (SDP). D. Yon, G. Camarillo. September 2005. (Updated by RFC 4572) (Status: Proposed Standard)
4217	Securing FTP with TLS. P. Ford-Hutchinson. October 2005. (Status: PROPOSED STANDARD)
4235	An INVITE-Initiated Dialog Event Package for the Session Initiation Protocol (SIP). J. Rosenberg, H. Schulzrinne, R. Mahy, Ed., November 2005. (Updated by RFC7463) (Status: PROPOSED STANDARD)
4240	Basic Network Media Services with SIP. E. Burger, Ed., J. Van Dyke, A. Spitzer. December 2005. (Status: INFORMATIONAL)
4244	An Extension to the Session Initiation Protocol (SIP) for Request History Information. M. Barnes, Ed.. November 2005. (Obsoleted by RFC7044) (Status: PROPOSED STANDARD)
4267	The W3C Speech Interface Framework Media Types: application/voicexml+xml, application/ssml+xml, application/srgs, application/srgs+xml, application/ccxml+xml,application/pls+xml. M. Froumentin. November 2005. (Status: INFORMATIONAL)
4279	Pre-Shared Key Ciphersuites for Transport Layer Security (TLS). P. Eronen, Ed., H. Tschofenig, Ed.. December 2005. (Status: PROPOSED STANDARD)
4281	The Codecs Parameter for "Bucket" Media Types. R. Gellens, D. Singer, P. Frojdh. November 2005. (Obsoleted by RFC6381) (Status: PROPOSED STANDARD)
4288	Media Type Specifications and Registration Procedures. N. Freed, J Klensin. December 2005 (Obsoletes RFC2048) (Obsoleted by RFC6838) (Status: BEST CURRENT PRACTICE)
4289	Multipurpose Internet Mail Extensions (MIME) Part Four: Registration Procedures. N. Freed, J. Klensin. (Format: TXT, HTML) (Obsoletes RFC2048) (Also BCP0013) (Status: BEST CURRENT PRACTICE)
4301	Security Architecture for the Internet Protocol. S. Kent, K. Seo. December 2005. (Obsoletes RFC 2401) (Updates RFC 3168) (Updated by RFC 6040) (Status: Proposed Standard)
4313	Requirements for Distributed Control of Automatic Speech Recognition (ASR), Speaker Identification/Speaker Verification (SI/SV), and Text-to-Speech (TTS) Resources. D. Oran. December 2005. (Status: INFORMATIONAL)
4317	Session Description Protocol (SDP) Offer/Answer Examples. A. Johnston, R. Sparks. December 2005. (Status: Informational)
4320	Internet Voice Messaging (IVM). S. McRae, G. Parsons. November 2005.(Status: PROPOSED STANDARD)
4346	The Transport Layer Security (TLS) Protocol Version 1.1. T. Dierks, E. Rescorla. April 2006. (Obsoletes RFC 2246) (Obsoleted by RFC 5246) (Updated by RFC 4366, RFC 4680, RFC 4681, RFC 5746, RFC 6176, RFC 7465) (Status: Proposed Standard)
4353	A Framework for Conferencing with the Session Initiation Protocol (SIP). J. Rosenberg. February 2006. (Status: INFORMATIONAL)

RFC	Description
4376	Requirements for Floor Control Protocols. P. Koskelainen, J. Ott, H. Schulzrinne, X. Wu. February 2006.) (Status: INFORMATIONAL)
4463	A Media Resource Control Protocol (MRCP) Developed by Cisco, Nuance, and Speechworks. S. Shanmugham, P. Monaco, B. Eberman. April 2006. (Status: INFORMATIONAL)
4467	Internet Message Access Protocol (IMAP) - URLAUTH Extension. M. Crispin. May 2006. (Updated by RFC5092, RFC5550) (Status: PROPOSED STANDARD)
4474	Enhancements for Authenticated Identity Management in the Session Initiation Protocol (SIP). J. Peterson, C. Jennings. August 2006. (Status: Proposed Standard)
4488	Suppression of Session Initiation Protocol (SIP) REFER Method Implicit Subscription. O. Levin. May 2006. (Status: PROPOSED STANDARD)
4566	Session Description Protocol. M. Handley, V. Jacobson, C. Perkins. July 2006. (Obsoletes RFC 2327, RFC 3266) (Status: Proposed Standard)
4579	Session Initiation Protocol (SIP) Call Control - Conferencing for User Agents. A. Johnston, O. Levin. August 2006. (Status: Best Current Practice)
4582	The Binary Floor Control Protocol (BFCP). G. Camarillo, J. Ott, K. Drage. November 2006. (Status: PROPOSED STANDARD)
4583	Session Description Protocol (SDP) Format for Binary Floor Control Protocol (BFCP) Streams. G. Camarillo. November 2006. (Status: Proposed Standard)
4585	Extended RTP Profile for Real-time Transport Control Protocol (RTCP)-Based Feedback (RTP/AVPF). J. Ott, S. Wenger, N. Sato, C. Burmeister, J. Rey. July 2006. (Updated by RFC 5506) (Status: Proposed Standard)
4597	Conferencing Scenarios. R. Even, N. Ismail. August 2006. (Status: Informational)
4627	The application/json Media Type for JavaScript Object Notation (JSON). D. Crockford. July 2006. (Format: TXT, HTML) (Obsoleted by RFC7159) (Status: INFORMATIONAL)
4629	RTP Payload Format for ITU-T Rec. H.263 Video. J. Ott, C. Bormann, G. Sullivan, S. Wenger, R. Even, Ed. January 2007. (Obsoletes RFC 2429) (Updates RFC 3555) (Status: Proposed Standard)
4647	Matching of Language Tags. A. Phillips, M. Davis. September 2006. (Obsoletes RFC3066) (Also BCP0047) (Status: BEST CURRENT PRACTICE)
4648	The Base16, Base32, and Base64 Data Encodings. S. Josefsson. October 2006. (Obsoletes RFC 3548) (Status: Proposed Standard)
4732	Virtual Private LAN Service (VPLS) Using Label Distribution Protocol (LDP) Signaling. M. Lasserre, Ed., V. Kompella, Ed.,January 2007.(Status: PROPOSED STANDARD)
4733	RTP Payload for DTMF Digits, Telephony Tones, and Telephony Signals. H. Schulzrinne, T. Taylor. December 2006 (Obsoletes RFC 2833) (Updated by RFC 4734, RFC 5244) (Status: Proposed Standard)
4796	The Session Description Protocol (SDP) Content Attribute. J. Hautakorpi, G. Camarillo. February 2007. (Status: Proposed Standard)
4855	Media Type Registration of RTP Payload Formats. S. Casner. February 2007. (Obsoletes RFC3555) (Status: PROPOSED STANDARD)
4867	RTP Payload Format and File Storage Format for the Adaptive Multi-Rate (AMR) and Adaptive Multi-Rate Wideband (AMR-WB) Audio Codecs. J. Sjoberg, M. Westerlund, A. Lakaniemi, Q. Xie. April 2007. (Obsoletes RFC 3267) (Status: Proposed Standard)
4916	Connected Identity in the Session Initiation Protocol (SIP). J. Elwell. June 2007. (Updates RFC 3261) (Status: Proposed Standard)
4943	Virtual Private LAN Service (VPLS) Using Label Distribution Protocol (LDP) Signaling. M. Lasserre, Ed., V. Kompella, Ed., January 2007.(Status: PROPOSED STANDARD)
4949	Internet Security Glossary, Version 2. R. Shirey. August 2007. (Obsoletes RFC2828) (Also FYI0036) (Status: INFORMATIONAL)
4960	Stream Control Transmission Protocol. R. Stewart, Ed. September 2007. (Obsoletes RFC 2960, RFC 3309) (Updated by RFC 6096, RFC 6335, RFC 7053) (Status: Proposed Standard)
4961	Neighbor Discovery for IP version 6 (IPv6). T. Narten, E. Nordmark, W. Simpson, H. Soliman. September 2007. (Obsoletes RFC 2461) (Updated by RFC 5942, RFC 6980, RFC7048, RFC7527, RFC 7559, RFC8028) (Status: Draft Standard)

RFC	Description
5018	Connection Establishment in the Binary Floor Control Protocol (BFCP). G. Camarillo. September 2007. (Status: PROPOSED STANDARD)
5022	Media Server Control Markup Language (MSCML) and Protocol. J. Van Dyke, E. Burger, Ed., A. Spitzer. September 2007. Obsoletes RFC4722) (Status: INFORMATIONAL)
5061	Stream Control Transmission Protocol (SCTP) Dynamic Address Reconfiguration. R. Stewart, Q. Xie, M. Tuexen, S. Maruyama, M. Kozuka. September 2007. (Status: PROPOSED STANDARD)
5104	Codec Control Messages in the RTP Audio-Visual Profile with Feedback (AVPF). S. Wenger, U. Chandra, M. Westerlund, B. Burman. February 2008. (Updated by RFC 7728, RFC8082) (Status: Proposed Standard)
5109	RTP Payload Format for Generic Forward Error Correction. A. Li, Ed. December 2007. (Obsoletes RFC 2733, RFC 3009) (Status: Proposed Standard)
5117	RTP Topologies. M. Westerlund, S. Wenger. January 2008. (Obsoleted by RFC 7667) (Status: Informational)
5124	Extended Secure RTP Profile for Real-time Transport Control Protocol (RTCP)-Based Feedback (RTP/SAVPF). J. Ott, E. Carrara. February 2008. (Status: Proposed Standard)
5139	Revised Civic Location Format for Presence Information Data Format Location Object (PIDF-LO). M. Thomson, J. Winterbottom. February 2008. (Updates RFC 4119) (Status: Proposed Standard)
5167	Media Server Control Protocol Requirements. M. Dolly, R. Even. March 2008.(Status: INFORMATIONAL)
5168	XML Schema for Media Control. O. Levin, R. Even, P. Hagendorf. March 2008. (Status: INFORMATIONAL)
5226	Guidelines for Writing an IANA Considerations Section in RFCs. T. Narten, H. Alvestrand. May 2008. (Obsoletes RFC 2434) (Also BCP 0026) (Status: Best Current Practice)
5234	Augmented BNF for Syntax Specifications: ABNF. D. Crocker, Ed., P. Overell. January 2008. (Obsoletes RFC 4234) (Updated by RFC 7405) (Also STD 0068) (Status: Internet Standard)
5239	A Framework for Centralized Conferencing. M. Barnes, C. Boulton, O. Levin. June 2008. (Status: Proposed Standard
5245	Interactive Connectivity Establishment (ICE): A Protocol for Network Address Translator (NAT) Traversal for Offer/Answer Protocols. J. Rosenberg. April 2010. (Obsoletes RFC 4091, RFC 4092) (Updated by RFC 6336) (Status: Proposed Standard)
5246	The Transport Layer Security (TLS) Protocol Version 1.2. T. Dierks, E. Rescorla. August 2008. (Obsoletes RFC 3268, RFC 4346, RFC 4366) (Updates RFC 4492) (Updated by RFC 5746, RFC 5878, RFC 6176, RFC 7465) (Status: Proposed Standard)
5256	Internet Message Access Protocol - SORT and THREAD Extensions. M.Crispin, K. Murchison. June 2008. (Updated by RFC5957) (Status: PROPOSED STANDARD)
5261	An Extensible Markup Language (XML) Patch Operations Framework Utilizing XML Path Language (XPath) Selectors. J. Urpalainen. September 2008. (Status: PROPOSED STANDARD)
5284	User-Defined Errors for RSVP. G. Swallow, A. Farrel. August 2008.(Status: PROPOSED STANDARD)
5285	A General Mechanism for RTP Header Extensions. D. Singer, H. Desineni. July 2008 (Status: Proposed Standard)
5322	Internet Message Format. P. Resnick, Ed.October 2008. (Obsoletes RFC 2822) (Updates RFC 4021) (Updated by RFC 6854) (Status: Draft Standard)
5359	Session Initiation Protocol Service Examples. A. Johnston, Ed., R.Sparks, C. Cunningham, S. Donovan, K. Summers. October 2008. HTML) (Also BCP0144) (Status: BEST CURRENT PRACTICE)
5369	Framework for Transcoding with the Session Initiation Protocol (SIP). G. Camarillo. October 2008. (Status: INFORMATIONAL)
5506	Support for Reduced-Size Real-Time Transport Control Protocol (RTCP): Opportunities and Consequences. I. Johansson, M. Westerlund. April 2009. (Updates RFC 3550, RFC 3711, RFC 4585) (Status: Proposed Standard)
5545	Internet Calendaring and Scheduling Core Object Specification (iCalendar). B. Desruisseaux, Ed.. September 2009 (Obsoletes RFC2445) (Updated by RFC5546, RFC6868, RFC7529, RFC7953, RFC7986) (Status: PROPOSED STANDARD)
5552	SIP Interface to VoiceXML Media Services. D. Burke, M. Scott. May 2009. (Status: PROPOSED STANDARD)
5567	An Architectural Framework for Media Server Control. T. Melanchuk, Ed., June 2009.(Status: INFORMATIONAL)
5626	Managing Client-Initiated Connections in the Session Initiation Protocol (SIP). C. Jennings, Ed., R. Mahy, Ed., F. Audet, Ed., October 2009. (Updates RFC3261, RFC3327) (Status: PROPOSED STANDARD)

RFC	Description
5630	The Use of the SIPS URI Scheme in the Session Initiation Protocol (SIP). F. Audet. October 2009. (Updates RFC3261, RFC3608) (Status: PROPOSED STANDARD)
5646	Tags for Identifying Languages. A. Phillips, Ed., M. Davis, Ed., September 2009. (Obsoletes RFC4646) (Also BCP0047) (Status: BEST CURRENT PRACTICE)
5707	Media Server Markup Language (MSML). A. Saleem, Y. Xin, G. Sharratt. February 2010. (Status: INFORMATIONAL)
5751	5751 Secure/Multipurpose Internet Mail Extensions (S/MIME) Version 3.2 Message Specification. B. Ramsdell, S. Turner. January 2010. (Obsoletes RFC 3851) (Status: Proposed Standard)
5761	Multiplexing RTP Data and Control Packets on a Single Port. C. Perkins, M. Westerlund. April 2010. (Updates RFC 3550, RFC 3551) (Status: Proposed Standard)
5763	Framework for Establishing a Secure Real-time Transport Protocol (SRTP) Security Context Using Datagram Transport Layer Security (DTLS). J. Fischl, H. Tschofenig, E. Rescorla. May 2010. (Status: Proposed Standard)
5764	5764 Datagram Transport Layer Security (DTLS) Extension to Establish Keys for the Secure Real-time Transport Protocol (SRTP). D. McGrew, E. Rescorla. May 2010. (Status: Proposed Standard)
5888	The Session Description Protocol (SDP) Grouping Framework. G. Camarillo, H. Schulzrinne. June 2010. (Obsoletes RFC 3388) (Status: Proposed Standard)
5905	Network Time Protocol Version 4: Protocol and Algorithm Specification. D. Mills, J. Martin, Ed., J. Burbank, W. Kasch. June 2010. (Obsoletes RFC1305, RFC4330) (Updated by RFC7822, RFC8573) (Status: PROPOSED STANDARD)
5922	Domain Certificates in the Session Initiation Protocol (SIP). V. Gurbani, S. Lawrence, A. Jeffrey. June 2010. (Updates RFC3261) (Status: PROPOSED STANDARD)
6120	Extensible Messaging and Presence Protocol (XMPP): Core. P. Saint-Andre. March 2011. (Obsoletes RFC3920) (Updated by RFC7590, RFC8553) (Status: PROPOSED STANDARD)
6157	IPv6 Transition in the Session Initiation Protocol (SIP). G. Camarillo, K. El Malki, V. Gurbani. April 2011. (Updates RFC 3264) (Status: Proposed Standard)
6184	RTP Payload Format for H.264 Video. Y.-K. Wang, R. Even, T. Kristensen, R. Jesup. May 2011. (Obsoletes RFC3984) (Status Proposed Standard)
6193	Media Description for the Internet Key Exchange Protocol (IKE) in the Session Description Protocol (SDP). M. Saito, D. Wing, M. Toyama. April 2011. (Status: Informational)
6230	Media Control Channel Framework. C. Boulton, T. Melanchuk, S. McGlashan. May 2011. (Status: PROPOSED STANDARD)
6231	An Interactive Voice Response (IVR) Control Package for the MediaControl Channel Framework. S. McGlashan, T. Melanchuk, C. Boulton. May 2011. (Updated by RFC6623) (Status: PROPOSED STANDARD)
6263	Application Mechanism for Keeping Alive the NAT Mappings Associated with RTP / RTP Control Protocol (RTCP) Flows. X. Marjou, A. Sollaud. June 2011. (Status: Proposed Standard)
6265	HTTP State Management Mechanism. A. Barth. April 2011. (Format: TXT, HTML) (Obsoletes RFC2965) (Status: PROPOSED STANDARD)
6341	Use Cases and Requirements for SIP-Based Media Recording (SIPREC). K. Rehor, Ed., L. Portman, Ed., A. Hutton, R. Jain. August 2011. (Status: INFORMATIONAL)
6347	The Address plus Port (A+P) Approach to the IPv4 Address Shortage. R. Bush, Ed.. August 2011. (Status: Experimental)
6350	vCard Format Specification. S. Perreault. August 2011. (Obsoletes RFC2425, RFC2426, RFC4770) (Updates RFC2739) (Updated by RFC6868) (Status: PROPOSED STANDARD)
6351	xCard: vCard XML Representation. S. Perreault. August 2011. (Updated by RFC6868) (Status: PROPOSED STANDARD)
6381	The 'Codecs' and 'Profiles' Parameters for "Bucket" Media Types. R. Gellens, D. Singer, P. Frojdh. August 2011. (Obsoletes RFC4281) (Updates RFC3839, RFC4393, RFC4337) (Status: PROPOSED STANDARD)
6454	The Web Origin Concept. A. Barth. December 2011. (Format: TXT, HTML) (Status: PROPOSED STANDARD)
6501	Conference Information Data Model for Centralized Conferencing (XCON). O. Novo, G. Camarillo, D. Morgan, J. Urpalainen. March 2012. (Status: PROPOSED STANDARD)
6502	Conference Event Package Data Format Extension for Centralized Conferencing (XCON). G. Camarillo, S. Srinivasan, R. Even, J. Urpalainen. March 2012. (Status: PROPOSED STANDARD)

RFC	Description
6503	Centralized Conferencing Manipulation Protocol. M. Barnes, C. Boulton, S. Romano, H. Schulzrinne. March 2012. (Status: Proposed Standard)
6505	A Mixer Control Package for the Media Control Channel Framework. S. McGlashan, T. Melanchuk, C. Boulton. March 2012. (Format: TXT, HTML) (Status: PROPOSED STANDARD)
6525	Stream Control Transmission Protocol (SCTP) Stream Reconfiguration. R. Stewart, M. Tuexen, P. Lei. February 2012. (Status: PROPOSED STANDARD)
6787	Media Resource Control Protocol Version 2 (MRCPv2). D. Burnett, S. Shanmugham. November 2012. (Status: PROPOSED STANDARD)
6838	Media Type Specifications and Registration Procedures. N. Freed, J. Klensin, T. Hansen. January 2013. (Obsoletes RFC4288) (Also BCP0013) (Status: BEST CURRENT PRACTICE)
6917	Media Resource Brokering. C. Boulton, L. Miniero, G. Munson. April 2013. (Format: TXT, HTML) (Status: PROPOSED STANDARD)
7022	Guidelines for Choosing RTP Control Protocol (RTCP) Canonical Names (CNAMEs). A. Begen, C. Perkins, D. Wing, E. Rescorla. September 2013. (Format: TXT, HTML) (Obsoletes RFC6222) (Updates RFC3550) (Status: PROPOSED STANDARD)
7058	Media Control Channel Framework (CFW) Call Flow Examples. A. Amirante, T. Castaldi, L. Miniero, S P. Romano. November 2013. (Status: INFORMATIONAL)
7201	Options for Securing RTP Sessions. M. Westerlund, C. Perkins. April 2014. (Format: TXT, HTML) (Status: INFORMATIONAL)
7202	Securing the RTP Framework: Why RTP Does Not Mandate a Single Media Security Solution. C. Perkins, M. Westerlund. April 2014 (Status: INFORMATIONAL)
7205	Use Cases for Telepresence Multistreams. A. Romanow, S. Botzko, M. Duckworth, R. Even, Ed.. April 2014. (Status: INFORMATIONAL)
7245	An Architecture for Media Recording Using the Session Initiation Protocol. A. Hutton, Ed., L. Portman, Ed., R. Jain, K. Rehor. May 2014. (Status: INFORMATIONAL)
7262	Requirements for Telepresence Multistreams. A. Romanow, S. Botzko, M. Barnes. June 2014. (Status: INFORMATIONAL)
7491	A PCE-Based Architecture for Application-Based Network Operations. D. King, A. Farrel. March 2015. (Status: INFORMATIONAL)
7498	Problem Statement for Service Function Chaining. P. Quinn, Ed., T. Nadeau, Ed.. April 2015. (Format: TXT, HTML) (Status: INFORMATIONAL)
7525	Recommendations for Secure Use of Transport Layer Security (TLS) and Datagram Transport Layer Security (DTLS). Y. Sheffer, R. Holz, P. Saint-Andre. May 2015. (Also BCP0195) (Status: BEST CURRENT PRACTICE)
7647	Clarifications for the Use of REFER with RFC 6665. R. Sparks, A.B. Roach. September 2015. (Updates RFC3515) (Status: PROPOSED STANDARD)
7667	RTP Topologies. M. Westerlund, S. Wenger. November 2015. (Obsoletes RFC5117) (Status: INFORMATIONAL)
7865	Session Initiation Protocol (SIP) Recording Metadata. R. Ravindranath, P. Ravindran, P. Kyzivat. May 2016. (Status: PROPOSED STANDARD)
7866	Session Recording Protocol. L. Portman, H. Lum, Ed., C. Eckel, A. Johnston, A. Hutton. May 2016. (Format: TXT, HTML) (Status: PROPOSED STANDARD)
7989	End-to-End Session Identification in IP-Based Multimedia Communication Networks. P. Jones, G. Salgueiro, C. Pearce, P. Giralt. October 2016. (Obsoletes RFC7329) (Status: PROPOSED STANDARD)
8217	Clarifications for When to Use the name-addr Production in SIP Messages. R. Sparks. August 2017. (Updates RFC3261, RFC3325, RFC3515, RFC3892, RFC4508, RFC5002, RFC5318, RFC5360, RFC5502) (Status: PROPOSED STANDARD)
8466	YANG Data Model for Layer 2 Virtual Private Network (L2VPN) Service Delivery. B. Wen, G. Fioccola, Ed., C. Xie, L. Jalil. October 2018. (Format: TXT, HTML) (Status: PROPOSED STANDARD)

Index

Printed in the United States
By Bookmasters